MW00844481

Acknowledgments

Grateful acknowledgment is given to the authors, artists, photographers, museums, publishers, and agents for permission to reprint copyrighted material. Every effort has been made to secure the appropriate permission. If any omissions have been made or if corrections are required, please contact the Publisher.

Photographic Credits

Front Cover: Krishan Lad/EyeEm/Getty Images
Back Cover: Yuhko Okada/EyeEm/Getty Images

Acknowledgments and credits continue on page 589.

For product information and technology assistance, contact us at Customer & Sales Support, **888-915-3276**

For permission to use material from this text or product, submit all requests online at **www.cengage.com/permissions**

Further permissions questions can be emailed to **permissionrequest@cengage.com**

National Geographic Learning |Cengage
200 Pier 4 Blvd., Suite 400
Boston, MA 02210

National Geographic Learning, a Cengage company, is a provider of quality core and supplemental educational materials for the PreK-12, adult education, and ELT markets. Cengage is a leading provider of customized learning solutions with employees residing in nearly 40 different countries and sales in more than 125 countries around the world. Find your local representative at **NGL.Cengage.com/RepFinder**.

Visit National Geographic Learning online at **NGL.Cengage.com**.

ISBN: 9780357859155

Printed in the United States of America

Print Number: 01
Print Year: 2022

FRONT PAGES OF THE TEACHER'S EDITION

PROGRAM CONSULTANTS

CATHERINE WORKMAN, PH.D.
Vice President of Science
National Geographic Society
Washington, D.C.

Dr. Catherine Workman develops and oversees the implementation and execution of the science components of the National Geographic Society's strategy and leads the science team. Trained as an evolutionary anthropologist, Catherine has extensive experience in the policy, strategy, management, metrics, and communication of biodiversity conservation, international grant-making, and combating wildlife trafficking. She collaborates with colleagues and partners to engage a range of audiences and stakeholders including nonscientific, underrepresented, and youth groups. Prior to joining National Geographic, Catherine worked in the Conservation Biology department at Denver Zoo, and was a 2014–2015 AAAS Science & Technology Policy Fellow at USAID, where she helped address issues from unsustainable fishing to illegal logging. Catherine has a Ph.D. in evolutionary anthropology from Duke University.

CATHERINE L. QUINLAN, ED.D.
Assistant Professor of Science Education
School of Education, Howard University
Washington, D.C.

Dr. Catherine L. Quinlan has over two decades of science-teaching experience in the K–12 and university settings, sixteen years of which were teaching high school biology. She holds a B.A. in English from Barnard College (premed), and an Ed.D. in Science Education from Teachers College, Columbia University. In her current position at Howard University School of Education, Dr. Quinlan prepares preservice teachers in science methods and education foundations for the K–12 classroom. Her research uses multidisciplinary and interdisciplinary approaches to look at the impact of cognitive, social, cultural, and historical factors on representation, and on Black students' persistence in STEM. Her research focuses on creating and evaluating a culturally representative science curriculum, with a pilot funded by NSF. Dr. Quinlan continues to bridge theory and practice with her representative chapter book series for children titled *Keystone Passage*, which grounds the realities of science, history, and culture in light fantasy. Her twitter handle is @ProfQuinlan.

PROGRAM REVIEWERS

CONTENT REVIEWERS

NADIA BAZIHIZINA, PH.D.
Research Fellow
University of Florence
Florence, Italy

SHANE CAMPBELL-STATON, PH.D.
Assistant Professor
Department of Ecology and Evolutionary Biology
Princeton University
Princeton, New Jersey

ELAINE Y. HSIAO, PH.D.
De Logi Associate Professor of Biological Sciences
Department of Integrative Biology and Physiology
University of California, Los Angeles
Los Angeles, California

ERICA SMITH, PH.D.
Biology Education Consultant
Chicago, Illinois

LAB REVIEWERS

TRICIA KERSHNER
Hilliard Davidson High School
Hilliard, Ohio

SARA E. LAHMAN, PH.D.
Associate Professor of Biology
University of Mount Olive
Mount Olive, North Carolina

ELIZABETH LUCAS
Hilliard Davidson High School
Hilliard, Ohio

TEACHER CONSULTANTS

JEFF BAIER
Webster Thomas High School
Webster, New York

ALISSA BURSE
Rocky River High School
Mint Hill, North Carolina

CAMERON BREWSTER
Cleveland State Community College
Cleveland, Tennessee

LAURIE FREE
Affton High School
St. Louis, Missouri

JEREMY MOHN
Blue Valley Northwest High School
Overland Park, Kansas

DANIELLE TOWNS-BELTON
Glencliff High School
Nashville, Tennessee

NERY VALDERRAMA
Nova High School
Davie, Florida

NATIONAL GEOGRAPHIC EXPLORERS

KATIE AMATO
Biological anthropologist

DIVA AMON
Deep-sea biologist

CHRISTOPHER C. AUSTIN
Biologist

CARTER CLINTON
Biological anthropologist

NICOLE COLÓN CARRIÓN
Biologist, ecologist

CAMILA ESPEJO
Veterinarian

HOLLY FEARNBACH
Zoologist

ANDRIAN GAJIGAN
Biochemist

PABLO GARCIA BORBOROGLU
Marine biologist and penguin conservationist

STANLEY GEHRT
Urban wildlife researcher

SHANE GERO
Marine biologist

DAVID GRUBER
Marine biologist

FEDERICO KACOLIRIS
Biologist

HEATHER KOLDEWEY
Marine biologist and environmentalist

BRENDA LARISON
Biologist

HEATHER JOAN LYNCH
Quantitative ecologist

EMILY OTALI
Primatologist

ROBERT A. RAGUSO
Biologist

KRISTEN RUEGG
Biologist

PARDIS SABETI
Computational biologist

ÇAĞAN ŞEKERCIOĞLU
Ornithology professor

ANUSHA SHANKAR
Biologist

DANIEL STREICKER
Infectious disease ecologist

VARUN SWAMY
Ecologist

GERARD TALAVERA
Evolutionary biologist

ROSA VÁSQUEZ ESPINOZA
Chemical biologist and Amazon conservationist

BRANWEN WILLIAMS
Oceanographer and climate scientist

KIMBERLY WILLIAMS-GUILLÉN
Conservationist

ROBERT WOOD
Roboticist

RAE WYNN-GRANT
Carnivore ecologist

STUDENT-CENTERED LEARNING IN BIOLOGY

DR. CATHERINE QUINLAN
Assistant Professor of Science Education
Howard University

National Geographic Biology is designed to facilitate more student-centered and cohesive ways of learning, while addressing the content-rich nature of biology. This is a shift away from centering the teacher to centering the learner. Each unit is built around an Anchoring Phenomenon that students revisit as they learn. At the chapter level, the program fosters curiosity and engagement for biology concepts using Case Studies, Connections, and Biotechnology Focus features that prompt reflections and lead to *Tying It All Together* at the end of each chapter. This integration is embedded in smaller yet connected sections that enable students to tap into relevant real-life phenomena that reflect the multidisciplinary and interdisciplinary nature of biology. With a balance between building skills and learning biology content, the program supports the development of College and Career Readiness skills and other essential skills for preparing the next generation to become scientifically literate citizens.

MAKING SENSE OF DATA

In today's rapidly changing world, students have access to more data than ever before. Understanding how to use data becomes more important than mere memorization of facts.

Now that I teach at the university level, I observe that students who enjoyed memorization because it was easier and did well because only memorization was required, struggle with the change to more meaningful learning. Think of the growing field of computational biology and other fields that continue to be created to address large amounts of data. That requires critical thinking and a certain level of comfort with uncertainty and exploration. National Geographic Biology provides you with opportunities for *Looking at the Data*, for *Math and English Language Arts Connections*, and *Minilabs* that hone in on skill building. Opportunities for skill building mean that students are more likely to carry over what they learn in biology into other contexts and into the subsequent years.

I had the unique opportunity to teach both tenth grade biology and then some of the same students in eleventh grade chemistry. Seeing what carried over from one year to the next led me to dramatically change the way I taught biology. For example, rather than place the photosynthesis equation on the board, I asked students to look at data and then to generate an equation showing what plants used and produced. I found that allowing students to take the time to construct meaning led them to remember what they learned, and their skills were more likely to carry over to the next year. If we only present information to students, the only skills they engage in is memorization and the only expectation we have of them is to remember or recall. If they don't make connections for themselves, then the information is only held in short-term memory and does not make it to long-term memory. In contrast, giving students the opportunity to engage with, derive, and construct meaning for themselves results in learning that lasts.

DIVERSITY AND INCLUSION

As biology teachers we're also called to address society's concerns for diversity, inclusion, and belonging. Our instructional materials should change to address the needs of all populations and to reflect the demographics of who will be tomorrow's scientists. Inclusion and Diversity is addressed through vignettes and illustrations from diverse scientists that are meaningfully integrated into the biology concepts. Students come to understand the variety of methods scientists use, and that scientists come from diverse backgrounds, which supports the NGSS *Understandings About the Nature of Science*. Vignettes such as *Blood Drive: Vaccinating Vampire Bats in the Peruvian Mountains*; *Under the Sea:*

Illuminating Unique Ecosystems in the Deep Ocean; and *Linking Healthy Chimpanzees to Healthy Children*, to name a few, provide engaging points of entry for exploring specific biology content—whether genetics, animal behavior, relationships in ecosystems, or diversity of living things. As students and teachers, we get pulled into their world in these features that give us a glimpse into what they are thinking—what led them to become interested in the work, why they are doing the work they're doing, the unique context and location, and questions they ask and pursue. One gets the feeling that we could also do what they're doing. This provides opportunity for all students, for diverse students, to see themselves in the field. Students also get to construct their own storylines as they themselves use science as a way of knowing. In the inquiry investigation labs students actively construct their own understandings by generating claims and gathering evidence using scientific practices. They not only collect and analyze data but use mathematical computations and reflect on their own thinking using the guided questions and self-reflective rubrics created specifically for students and teachers.

Imagine your students engaging in arguments about particular science concepts. Imagine all of these situations embedded in and facilitated by a well-organized text and program. Students can argue from evidence in small tasks or large tasks, or both. Students can discuss how they come to know ideas using evidence from the text and labs. In my own research into students thinking about science, I've learned that students not only connect differently with the information but also connect with different information. The benefits of confronting each other's thinking using evidence must be underscored. What's important about this is that as a biology teacher, you already have the buy-in and interest because your students are arguing in science. Students love to argue, but rarely get to do so in science, because of how we've taught science, and especially because of the content rich nature of biology. You'll be surprised to see that they will cover the content in their arguments and that your concern about covering the biology concepts required will be alleviated. Why? They look at the content to generate their claims, provide their warrants, and come up with their reasonings. When they construct their own meaning, they are more likely to retain the information. When they verbalize their thoughts, you have the opportunity to clarify and discuss.

The structure of the National Geographic program also scaffolds teacher facilitation of more student-centered approaches to teaching biology. This ranges from the inclusion of embedded questions for classroom discussions and arguments to opportunities for individual or small group research. The program provides teachers with ways to accommodate students while also accommodating teachers with limited access to lab resources. Questions and self-assessment rubrics help students to reflect on their own processes or prompt students to create their own questions, while providing teachers with alternate ways to assess progress and learning. In my 16 years of teaching ninth and tenth grade biology, I found that student-generated questions lead to very interesting discussions that reveal students' conceptions. These rich conversations not only provide opportunities to address students' understandings and misunderstandings but also encourage student engagement and active learning.

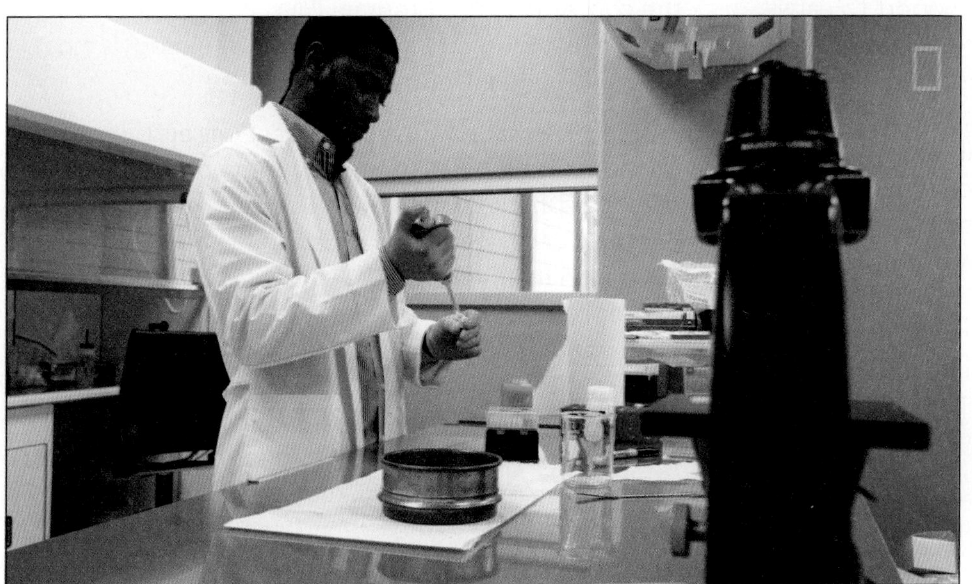

Explorer Dr. Carter Clinton, featured in Chapter 11, works to extract and analyze DNA samples from historical sites such as the African Burial Ground National Monument in New York City. His forensic analyses have helped shed light on the history of African Americans during colonial times.

EXPLORATION IN BIOLOGY

DR. CATHERINE WORKMAN
Vice President of Science
National Geographic Society

Hi! I'm Catherine Workman, Vice President of Science for the National Geographic Society. As part of the Science & Innovation division, I oversee our research, conservation, and technology grant-making. At the National Geographic Society, exploration of the planet is in our DNA, and we have awarded more than 15,000 grants since our founding in 1888. The recipients of these grants, who we call National Geographic Explorers, are a diverse community of changemakers from around the world (140 countries and counting!) working to support our mission to illuminate and protect the wonder of our world.

Explorers are the heart of the National Geographic Society. The National Geographic Society's Grants Program provides seed funding and support to early career individuals and more senior scientists who are working to address critical challenges, advance new solutions, and inspire positive transformation across all seven continents. The featured Explorers in this program—who I am excited for you to meet—received grants for biological science projects focused on research, conservation, or technology and aligned with one of the Society's focus areas: Ocean, Land, Wildlife, and Human Ingenuity. In this program, you'll learn that National Geographic Explorers are infinitely curious people who are passionate about our planet and making it a better place.

The National Geographic Society has a rich legacy of supporting biology and those studying Earth's living organisms. Some of our Explorers are iconic, like Jane Goodall and Jacques Cousteau, both of whom the Society supported in their earliest fieldwork before they were household names. Other Explorers you may not have heard of, but they are pushing the boundaries of botany, conservation, ecology, evolution, genetics, marine biology, microbiology, and physiology. I became an Explorer in 2007 when I was awarded a National Geographic Society grant to study the feeding ecology and adaptation of Delacour's langurs (*Trachypithecus delacouri*), a type of monkey living on rugged limestone karst mountains in northern Vietnam. My project investigated several hypotheses to explain the langurs' distribution on the karst. The grant funds paid for the analysis of soil and leaf samples that the monkeys ate, allowing me to better understand the chemical and nutritional ecology of their diet. Over the course of this series, you'll meet other Explorers and learn how their work is helping us to better understand our planet and also make it a better place.

THE BIOLOGY EXPLORER VIDEO SERIES

The *Explorers at Work* video series for this program helps set the stage for the concepts presented throughout the textbook. These videos, which I had the opportunity to host, illustrate key concepts for biology and touch on themes and ideas from several units. Due to COVID-19 travel and safety restrictions, each of the five units includes a *virtual* visit to the Explorers' fieldwork and lab locations around the world. The video series serves as a guide connecting the concepts in this program. Plus, it's a wonderful chance to hear firsthand from some of biology's most innovative and intrepid scientists.

For Unit 1, which focuses on interactions and relationships in ecosystems, Explorer and marine biologist Dr. Diva Amon takes us deep beneath the ocean to ecosystems and organisms that few people have seen firsthand. We cover topics such as how matter is transferred and energy flows in deep-sea ecosystems, how organisms interact in the deep sea, and how human activities disrupt deep-sea ecosystems. Amon describes what it's like to work nearly two miles beneath the surface of the ocean and her research methods, including the submersible in which she travels to the oceans' depths. She also shares what she has learned about deep-sea ecosystems and the unusual creatures that live there, as well as what sparked her original interest in marine biology.

In Unit 2, we cover cell systems and the important role of bacteria in our intestines. Explorer Dr. Katie Amato discusses the importance of the gut microbiome on human health, how some bacteria in the gut can destroy toxins and also supply essential nutrients that humans cannot make themselves, such as vitamin K. Drawing upon her research studying the gut microbiome of wild howler monkeys and baboons, Dr. Amato draws connections between human cells (such as intestinal cells), nutrition, and gut bacteria and describes how environmental factors and dietary changes affect the growth and composition of gut bacteria in both monkeys and humans. The interview wraps with Dr. Amato offering advice to high school students who are interested in becoming a microbiologist or other type of biologist.

Unit 3 takes us on a journey to a protected area of the Amazon rainforest in southeastern Peru, one of the most biologically diverse places on Earth where the treetop canopy reaches over 60 meters (200 ft.) high. Explorer Dr. Varun Swamy talks about the interconnectedness of organisms in tropical rainforests, specifically his research on plant-animal interactions and the processes by which tree diversity and regeneration are maintained in this ecosystem. Dr. Swamy explains how mammals like spider monkeys support tree diversity by spreading seeds far away from parent plants, how the use of drones and citizen scientists help him and his team to collect and analyze large amounts of rainforest canopy data, and the role of DNA barcoding in his work.

In Unit 4, the video explores how genetics can track the spread of viruses by using one of the most blood-curdling examples: vampire bats! Explorer and infectious disease ecologist Dr. Daniel Streicker takes us to the mountains of Peru where he researches how bat and human communities are regularly affected by rabies, a deadly disease that's incurable once symptoms take hold. Rabies is an example of a zoonotic pathogen, a disease-causing germ that spreads between animals and people. Dr. Streicker uses genetics from bat blood samples to learn how rabies spreads through the population. This was perhaps the most timely interview, as we talked about how his research can be applied to the COVID-19 pandemic—and to future pandemics.

Finally, in Unit 5 we explore how hummingbirds adapt to their environments, from the Andean highlands of Ecuador to the deserts of Arizona. Explorer Dr. Anusha Shankar explains why hummingbirds make such great subjects for studying topics such as metabolism, energy conservation, and bioenergetics, as well as torpor, which is when animals decrease their physiological activity by lowering their body temperature and metabolism. She also describes the role of genetics in hummingbirds' ability to fine-tune their energy expenditure and daily torpor. We learn that hummingbirds are evolutionarily adapted to survive in different environments, and climate change and other human-caused environmental changes might impact hummingbirds' energy budgets.

Our planet is teeming with complex and fascinating biological organisms, from microbes to green-blooded lizards to blood-sucking vampire bats to our own human species. Throughout this program, our aim is to sharpen students' analytical skills with exciting and interactive case studies so that they not only succeed in Biology but also become inspired to use these tools throughout their lives to better understand the world around them, their place in it, and how they can contribute to a more livable and sustainable world for all living creatures. Let's go!

National Geographic's Catherine Workman and Dr. Anusha Shankar discuss Shankar's research on daily torpor in hummingbirds.

PROGRAM COMPONENTS

ONLINE STUDENT EDITION

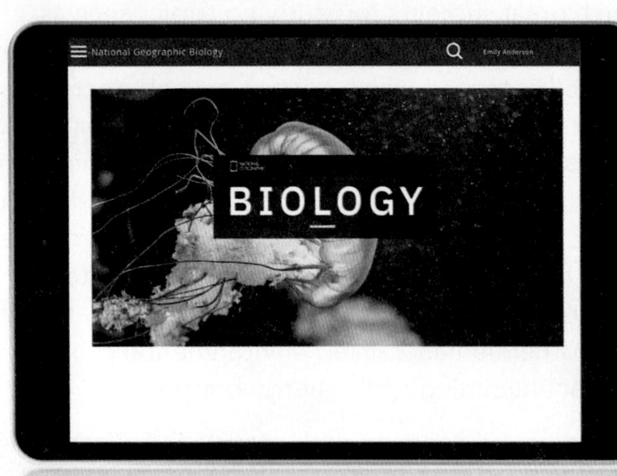

ONLINE TEACHER'S EDITION

STUDENT LAB MANUAL

TEACHER LAB MANUAL

ASSESSMENT HANDBOOK

STUDENT AND TEACHER RUBRICS

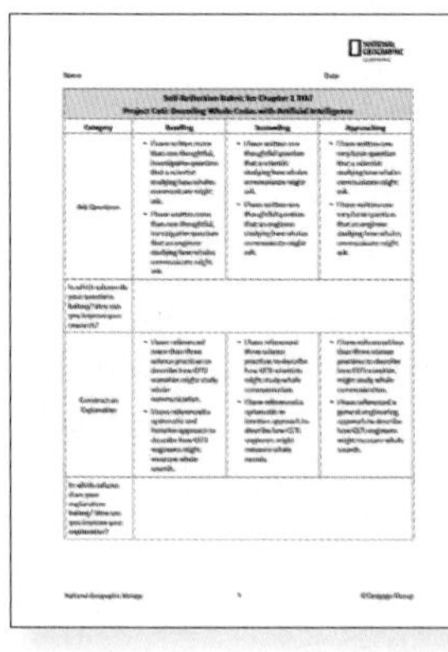

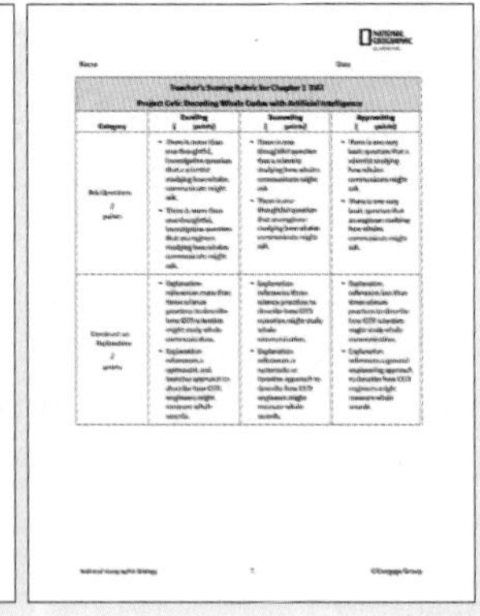

STUDENT EDITION

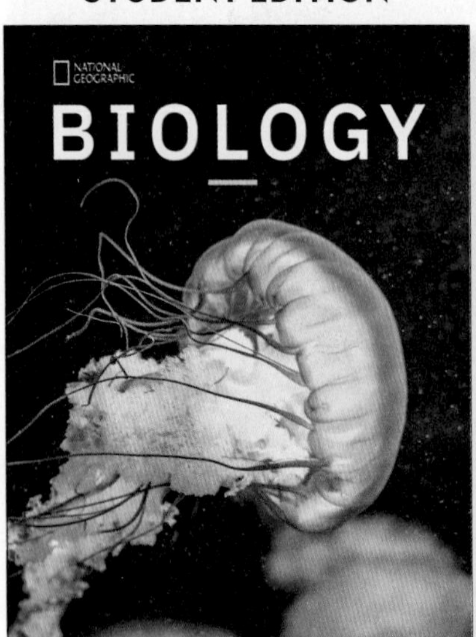

TEACHER'S EDITION

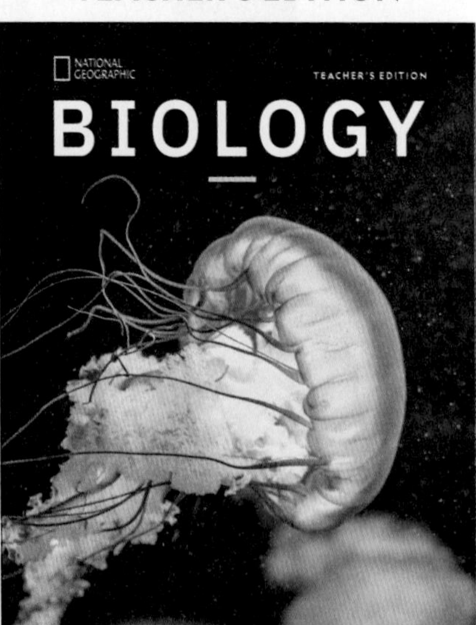

COGNERO TEST BANK

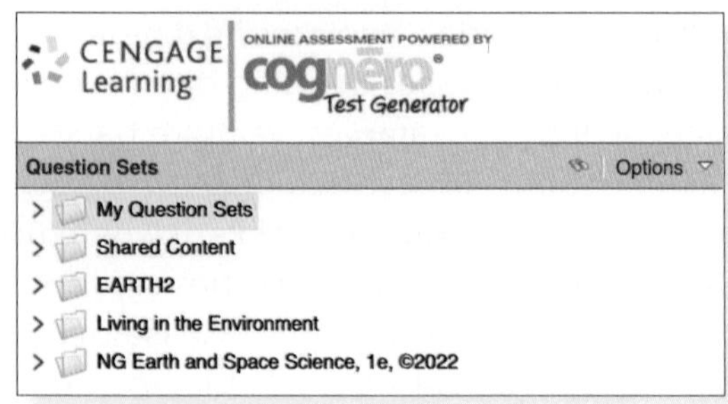

STUDENT EDITION OVERVIEW

EXPLORING BIOLOGY THROUGH NATIONAL GEOGRAPHIC

Capture students' imagination and inspire their curiosity with exciting National Geographic content and features that explore the vital issues of our times.

The **National Geographic Explorers** in *National Geographic Biology* are a diverse cross section of groundbreaking biologists, bioengineers, artists, and adventurers whose work ties to the biology concepts presented in the chapter. These relatable, passionate individuals are striving to improve our world.

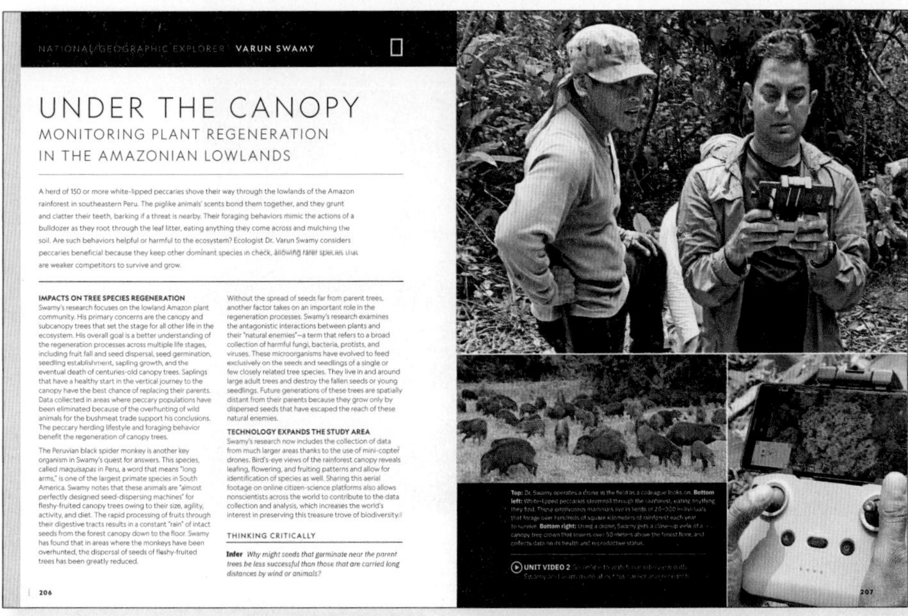

NATIONAL GEOGRAPHIC EXPLORER VARUN SWAMY

UNDER THE CANOPY
MONITORING PLANT REGENERATION IN THE AMAZONIAN LOWLANDS

A herd of 150 or more white-lipped peccaries shove their way through the lowlands of the Amazon rainforest in southeastern Peru. The piglike animals' scents bond them together, and they grunt and clatter their teeth, barking if a threat is nearby. Their foraging behaviors mimic the actions of a bulldozer as they root through the leaf litter, eating anything they come across and mulching the soil. Are such behaviors helpful or harmful to the ecosystem? Ecologist Dr. Varun Swamy considers peccaries beneficial because they keep other dominant species in check, allowing rarer species that are weaker competitors to survive and grow.

IMPACTS ON TREE SPECIES REGENERATION
Swamy's research focuses on the lowland Amazon plant community. His primary concerns are the canopy and subcanopy trees that set the stage for all other life in the ecosystem. His overall goal is a better understanding of the regeneration processes across multiple life stages, including fruit fall and seed dispersal, seed germination, seedling establishment, sapling growth, and the eventual death of centuries-old canopy trees. Saplings that have a healthy start in the vertical journey to the canopy have the best chance of replacing their parents. Data collected in areas where peccary populations have been eliminated because of the overhunting of wild animals for the bushmeat trade support his conclusions. The peccary herding lifestyle and foraging behavior benefit the regeneration of canopy trees.

The Peruvian black spider monkey is another key organism in Swamy's quest for answers. This species, called *maquisapas* in Peru, a word that means "long arms," is one of the largest primate species in South America. Swamy notes that these animals are "almost perfectly designed seed-dispersing machines" for fleshy-fruited canopy trees owing to their size, agility, activity, and diet. The rapid processing of fruits through their digestive tracts results in a constant "rain" of intact seeds from the forest canopy down to the floor. Swamy has found that in areas where the monkeys have been overhunted, the dispersal of seeds of fleshy-fruited trees has been greatly reduced.

Without the spread of seeds far from parent trees, another factor takes on an important role in the regeneration processes. Swamy's research examines the antagonistic interactions between plants and their "natural enemies"—a term that refers to a broad collection of harmful fungi, bacteria, protists, and viruses. These microorganisms have evolved to feed exclusively on the seeds and seedlings of a single or few closely related tree species. They live in and around large adult trees and destroy the fallen seeds or young seedlings. Future generations of these trees are spatially distant from their parents because they grow only by dispersed seeds that have escaped the reach of these natural enemies.

TECHNOLOGY EXPANDS THE STUDY AREA
Swamy's research now includes the collection of data from much larger areas thanks to the use of mini-copter drones. Bird's-eye views of the rainforest canopy reveals leafing, flowering, and fruiting patterns and allow for identification of species as well. Sharing this aerial footage on online citizen-science platforms also allows nonscientists across the world to contribute to the data collection and analysis, which increases the world's interest in preserving this treasure trove of biodiversity.

THINKING CRITICALLY

Infer *Why might seeds that germinate near the parent trees be less successful than those that are carried long distances by wind or animals?*

Top: Dr. Swamy operates a drone in the field as a colleague looks on. **Bottom left:** White-lipped peccaries steamroll through the rainforest, eating anything they find. These omnivorous mammals live in herds of 20–300 individuals that forage over hundreds of square kilometers of rainforest each year to survive. **Bottom right:** Using a drone, Swamy gets a close-up view of a canopy tree crown that towers over 50 meters above the forest floor, and collects data on its health and reproductive status.

UNIT VIDEO 2

206 207

ROSA VÁSQUEZ ESPINOZA

SURVIVAL MECHANISMS AND NEW MEDICINES

Dr. Rosa Vásquez Espinoza explores microorganisms in the Peruvian Amazon, studying their survival mechanisms to determine how those processes can be engineered to develop new natural medicinal products.

VIDEO 5-1 Go online to watch Vásquez Espinoza and a team of researchers gather sample organisms in the Boiling River ecosystem.

Inspired by her grandmother's knowledge of traditional medicines made from plants, herbs, and soils, chemical biologist Dr. Rosa Vásquez Espinoza grew up wondering how nature produces these powerful compounds. Her scientific search has led her to the Boiling River in the Peruvian Amazon. She steps carefully along the steamy banks of the river—one slip into its 99.1 °C waters could render third-degree burns.

Vásquez Espinoza stoops to scrape lichens off the side of a rocky ledge into a jar. Next, she dips a heat resistant test tube into the scalding water to gather cyanobacteria. She then moves upriver, where the water is a bit cooler, to harvest a sample of a microbial mat from the river bottom. She also takes samples of sediments that contain large populations of bacteria. After dinner, the team will set up a microscope and begin analyzing the samples. If one looks especially interesting, they will return to that area the next day to collect more samples.

Vásquez Espinoza thinks the bacteria produce unique molecules that protect them from the high temperatures of their habitats. She hypothesizes that because they enable microorganisms to survive under stressful conditions, these molecules may be engineered into medicines to fight bacterial, fungal, or viral infections, and even cancers. In the laboratory, Vásquez Espinoza also looks at enzymes, molecules that regulate chemical reactions in the organisms, to try to pinpoint substances that have therapeutic properties. Understanding how microorganisms produce these substances may allow scientists to develop sustainable processes for manufacturing and refining them for use as pharmaceutical drugs. In addition, these enzymes' ability to withstand a wide range of temperatures could make them useful for reducing negative environmental impacts of industrial processes that generate toxic waste and pollution.

"If this is possible now, what can we achieve in the next 10 to 20 years?" asks Vásquez Espinoza. "How many more bacteria can we find in our planet that can help make our lives better and also help us protect nature itself? How is it possible that we can go to the moon and don't know fully yet what's at the microbial level in our planet?"

THINKING CRITICALLY

Relate *Enzymes are made of amino acids. How does this information support the hypothesis that enzymes found in the Boiling River microbes could also work in medicines for humans or animals?*

136 CHAPTER 5 MOLECULES IN LIVING SYSTEMS

Science is a human endeavor, and students discover more about what scientists do from day-to-day in the **Explorers at Work** video series. What's it really like to collect (and dodge) monkey scat? capture vampire bats with nets? stumble upon a herd of white-lipped peccaries in a Peruvian rainforest? navigate a deep-sea submersible miles below the ocean surface? track tiny hummingbirds through a montane cloud forest? In these intimate interviews conducted by Dr. Catherine Workman of the National Geographic Society, Explorers describe details about their research and how they prepared for careers in STEM.

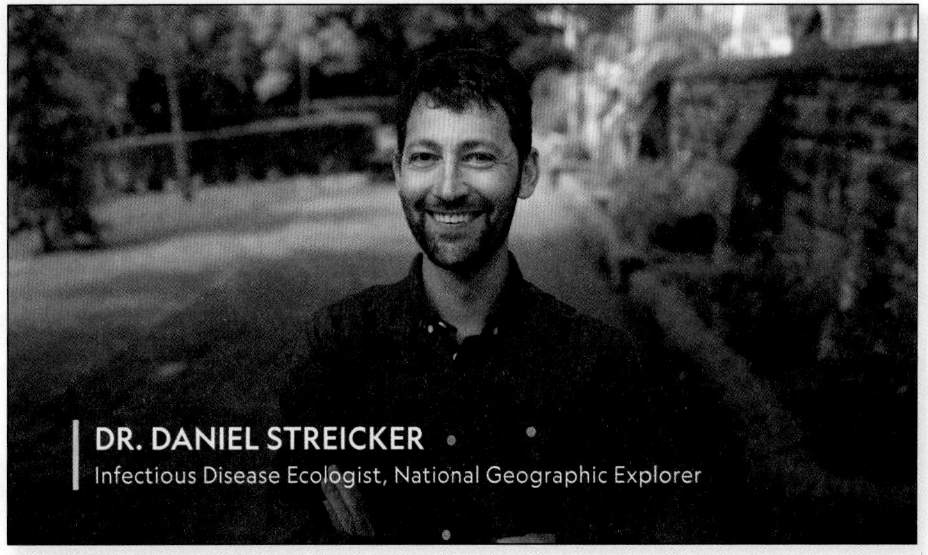

Storytelling through the work of National Geographic's world-class photographers brings nuanced ecological perspectives to biology topics.

ON ASSIGNMENT National Geographic photographer Ami Vitale began her career documenting conflict zones. An assignment in 2009 to photograph the transport and release of one of the world's last living male northern white rhinos named Sudan changed her focus from war zones to wildlife and environmental stories. In this photograph, taken in 2019, a wildlife ranger comforts Sudan moments before his death.

Like the On Assignments, **Connections** and **Biotechnology Focus** extend biology topics to real-world contemporary issues and applications. Use these features to segue into other local or global issues that are important to students, and practice competencies in Social-Emotional Learning (SEL) with support in the Teacher's Edition.

Ecological Succession

The species composition of a community will change over time. Often, some species change the habitat in ways that allow others to come in and replace them. This type of change, which takes place over a long interval of time, is called **ecological succession**.

When the concept of ecological succession was first developed in the late 1800s, it was thought to be a predictable and directional process. Which species are present at each stage in succession was thought to be determined primarily by physical factors such as climate, altitude, and soil type. In this view, succession ends in a "climax community," a collection of species that does not change over time and will be rebuilt in the event of a disturbance. Ecologists now know that the species composition of a community changes in unpredictable ways. Communities do not journey along a well-worn path to a predetermined climax state. Random events determine the order in which species arrive in a habitat, and thus affect the course of succession.

Ecological succession begins with the arrival of *pioneer species*, which are species whose traits allow them to colonize new or newly vacated habitats. Pioneer species have an opportunistic life history. This means they grow and mature quickly, and they produce many offspring capable of dispersing. Later, other hardier and slower-growing species replace the pioneers. Then the replacements are replaced, and so on. There are two types of succession: primary succession and secondary succession.

CHAPTER INVESTIGATION

Ecological Succession in an Aquatic Environment

How do biotic and abiotic factors influence succession in a freshwater pond community?

Go online to explore this chapter's hands-on investigation about ecological succession.

CONNECTIONS

PRAIRIE RESTORATION Prior to the arrival of Europeans, the Great Plains and Upper Midwest were covered by immense grasslands. Nearly 445,000 square kilometers of what is now Minnesota, North Dakota, and South Dakota were home to prairie ecosystems. Fire is the key to maintaining grassland ecosystems and preventing succession from taking hold. Historically, these grasslands burned on a regular basis, either caused naturally by lightning strikes or set intentionally by Native Americans to produce fertile land for agricultural purposes and to clear undergrowth to facilitate hunting and travel.

Today, fire is an important part of prairie restoration plans. The goals of prairie restoration are to reconnect isolated patches of prairie, reduce invasive species, and improve habitat for wildlife and grazing animals. Prescribed burns are a method of land management. The resulting nutrient-rich ash acts as a fertilizer and supports new growth of native vegetation. Fire also reduces the accumulation of brush, shrubs, and trees.

Every prescribed burn has a scientific plan that land managers have prepared in advance. It describes the objective of the fire, the fuels that will be used, the size of the fire, the precise environmental conditions under which it will burn, and conditions under which the fire will be suppressed if necessary.

Workers on a fire crew use a drip torch to ignite the prairie into flames as part of a prescribed fire.

3.3 ECOSYSTEM STABILITY AND CHANGE 75

In *National Geographic Biology,* students gain the knowledge and skills they will need to meet and exceed all of the NGSS Life Science Performance Expectations.

PHENOMENA-BASED LEARNING

Each unit of *National Geographic Biology* opens with an Anchoring Phenomenon. A Driving Question helps frame the phenomenon as something students can investigate throughout the unit.

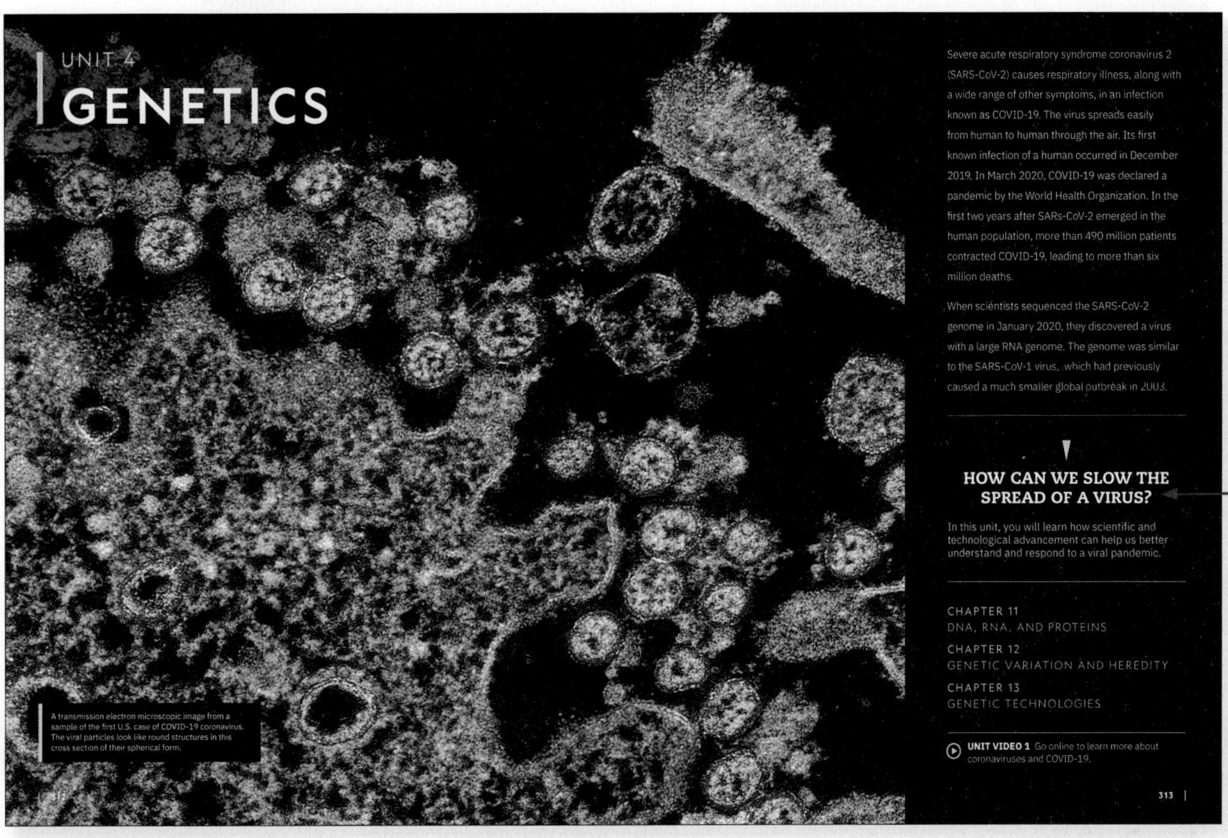

UNIT 4

GENETICS

A transmission electron microscopic image from a sample of the first U.S. case of COVID-19 coronavirus. The viral particles look like round structures in this cross section of their spherical form.

Severe acute respiratory syndrome coronavirus 2 (SARS-CoV-2) causes respiratory illness, along with a wide range of other symptoms, in an infection known as COVID-19. The virus spreads easily from human to human through the air. Its first known infection of a human occurred in December 2019. In March 2020, COVID-19 was declared a pandemic by the World Health Organization. In the first two years after SARs-CoV-2 emerged in the human population, more than 490 million patients contracted COVID-19, leading to more than six million deaths.

When scientists sequenced the SARS-CoV-2 genome in January 2020, they discovered a virus with a large RNA genome. The genome was similar to the SARS-CoV-1 virus, which had previously caused a much smaller global outbreak in 2003.

HOW CAN WE SLOW THE SPREAD OF A VIRUS?

In this unit, you will learn how scientific and technological advancement can help us better understand and respond to a viral pandemic.

CHAPTER 11
DNA, RNA, AND PROTEINS

CHAPTER 12
GENETIC VARIATION AND HEREDITY

CHAPTER 13
GENETIC TECHNOLOGIES

UNIT VIDEO 1 Go online to learn more about coronaviruses and COVID-19.

313

The **Driving Question** focuses students' observations into an investigable question they can answer at the end of the unit by using evidence and reasoning to apply biology concepts.

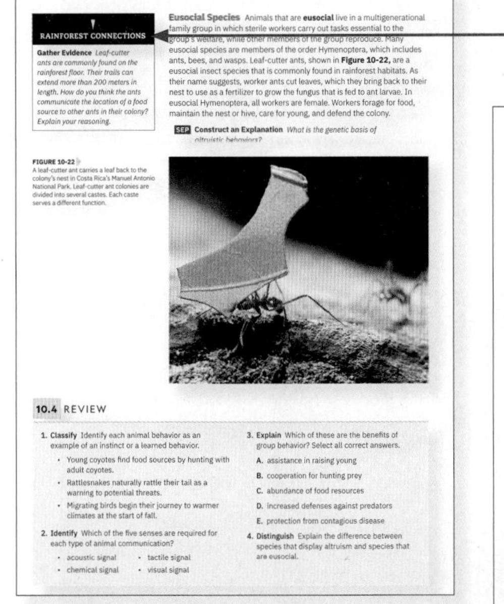

RAINFOREST CONNECTIONS

Gather Evidence *Leaf-cutter ants are commonly found on the rainforest floor. Their trails can extend more than 200 meters in length. How do you think the ants communicate the location of a food source to other ants in their colony? Explain your reasoning.*

Eusocial Species Animals that are **eusocial** live in a multigenerational family group in which sterile workers carry out tasks essential to the group's welfare, while other members of the group reproduce. Many eusocial species are members of the order Hymenoptera, which includes ants, bees, and wasps. Leaf-cutter ants, shown in **Figure 10-22,** are a eusocial insect species that is commonly found in rainforest habitats. As their name suggests, worker ants cut leaves, which they bring back to their nest to use as a fertilizer to grow the fungus that is fed to ant larvae. In eusocial Hymenoptera, all workers are female. Workers forage for food, maintain the nest or hive, care for young, and defend the colony.

SEP **Construct an Explanation** *What is the genetic basis of altruistic behaviors?*

FIGURE 10-22 A leaf-cutter ant carries a leaf back to the colony's nest in Costa Rica's Manuel Antonio National Park. Leaf-cutter ant colonies are divided into several castes. Each caste serves a different function.

10.4 REVIEW

1. **Classify** Identify each animal behavior as an example of an instinct or a learned behavior.
   - Young coyotes find food sources by hunting with adult coyotes.
   - Rattlesnakes naturally rattle their tail as a warning to potential threats.
   - Migrating birds begin their journey to warmer climates at the start of fall.
2. **Identify** Which of the five senses are required for each type of animal communication?
   - acoustic signal
   - tactile signal
   - chemical signal
   - visual signal
3. **Explain** Which of these are the benefits of group behavior? Select all correct answers.
   - A. assistance in raising young
   - B. cooperation for hunting prey
   - C. abundance of food resources
   - D. increased defenses against predators
   - E. protection from contagious disease
4. **Distinguish** Explain the difference between species that display altruism and species that are eusocial.

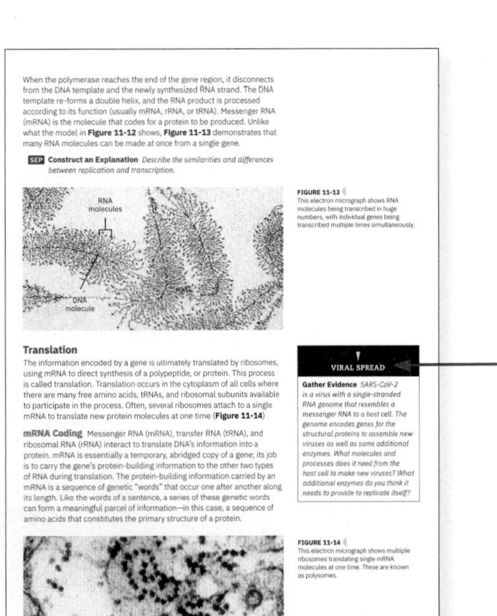

When the polymerase reaches the end of the gene region, it disconnects from the DNA template and the newly synthesized RNA strand. The DNA template re-forms a double helix, and the RNA product is processed according to its function (usually mRNA, rRNA, or tRNA). Messenger RNA (mRNA) is the molecule that codes for a protein to be produced. Unlike what the model in **Figure 11-12** shows, **Figure 11-13** demonstrates that many RNA molecules can be made at once from a single gene.

SEP **Construct an Explanation** *Describe the similarities and differences between replication and transcription.*

FIGURE 11-13 This electron micrograph shows RNA molecules being transcribed in huge numbers, with individual genes being transcribed multiple times simultaneously.

Translation

The information encoded by a gene is ultimately translated by ribosomes, using mRNA to direct synthesis of a polypeptide, or protein. This process is called translation. Translation occurs in the cytoplasm of all cells where there are many free amino acids, tRNAs, and ribosomal subunits available to participate in the process. Often, several ribosomes attach to a single mRNA to translate new protein molecules at one time (**Figure 11-14**)

mRNA Coding Messenger RNA (mRNA), transfer RNA (tRNA), and ribosomal RNA (rRNA) interact to translate DNA's information into a protein. mRNA is essentially a temporary, abridged copy of a gene; its job is to carry the gene's protein-building information to the other two types of RNA during translation. The protein-building information carried by an mRNA is a sequence of genetic "words" that occur one after another along its length. Like the words of a sentence, a series of these genetic words can form a meaningful parcel of information—in this case, a sequence of amino acids that constitutes the primary structure of a protein.

VIRAL SPREAD

Gather Evidence *SARS-CoV-2 is a virus with a single-stranded RNA genome that resembles a messenger RNA to a host cell. The genome encodes genes for the structural proteins to assemble new viruses as well as some additional enzymes. What molecules and processes does it need from the host cell to make new viruses? What additional enzymes do you think it needs to provide to replicate itself?*

FIGURE 11-14 This electron micrograph shows multiple ribosomes translating single mRNA molecules at one time. These are known as polysomes.

At multiple points throughout the units, students can look for evidence and connect concepts back to the **Driving Question**.

CRITICAL THINKING

14. **Explain** If DNA has only four different bases, how can it be used to produce so many different proteins?
15. NOS **Scientific Knowledge** Describe how the discoveries of Miescher, Griffith, and Avery built on each other to help advance our understanding of DNA.
16. **Explain** What happens when a ribosome reaches a stop codon?
17. **Contrast** Compare activators and repressors in gene regulation.

MATH AND ENGLISH LANGUAGE ARTS CONNECTIONS

1. **Synthesize Information** Describe what the words "transcription" and "translation" mean in other contexts outside of science and relate those meanings to what they mean in protein synthesis.
2. **Write a Function** Write an equation to describe the relationship between the number of nucleotides (n) in a DNA sequence and the number of amino acids (a) in the resulting polypeptide.
3. **Infer from Statistics** If a codon is randomly generated, what amino acids are most likely to be coded for? Which are least likely? Refer to **Figure 11-15**.
4. **Write Explanatory Text** Francis Crick said, "Rather than believe that Watson and Crick made the DNA structure, I would rather stress that the structure made Watson and Crick." Using what you know about the history of the discovery of DNA and what DNA does, how do you interpret this quote?
5. **Reason Abstractly** In geometry, complementary angles fit together to form an angle that measures exactly 90 degrees. Use this mathematical definition and **Figure 11-3** to help explain why base pairs in nucleotides are called complementary.

REVISIT VIRAL SPREAD

Gather Evidence In this chapter, you learned about how cells transfer information from their genomes to the RNA and protein products that are important for life. You have also learned here and in Section 8.4 that viruses have their own genomes. Viral genomes vary widely and require different help from their host cells to be expressed, but the overall goal remains the same: to express genes and replicate. Look at the structure of SARS-CoV-2, which was first presented in Section 8.4.

1. Observe the structures shown in the provided model of the SAR-CoV-2 viral particle. What are some of the genes that need to be encoded within the SARS-CoV-2 genome to make new virus particles? Which genes do you think code for proteins that are important for infecting a host cell?
2. What other questions do you have about how SARS-CoV-2 infects and replicates within host cells?

glycoprotein spike
membrane protein
envelope protein
lipid envelope
RNA and nucleocapsid protein

CHAPTER REVIEW 345

Each **Chapter Review** includes a section to revisit the unit's Anchoring Phenomenon, giving your students a low-stakes opportunity to apply certain Science and Engineering Practices in parallel with the main readings, activities, and assessments.

A culminating **Unit Activity** uses the **Claim, Evidence, and Reasoning (CER)** model to help students organize their evidence and thinking. Students practice writing a scientifically reasoned argument in response to the Anchoring Phenomenon's **Driving Question**.

UNIT 5 **ACTIVITY**

HOW DID HUMMINGBIRDS BECOME ADAPTED TO THEIR ENVIRONMENTS?

Argue From Evidence In this unit, you learned that evolution occurs in populations of species over time as they adapt to changes in their environment. The diversity of hummingbirds is one example of how certain species are able to rapidly expand into open ecological niches.

Using DNA data, scientists constructed a family tree for hummingbirds and their closest relatives. They determined that the branch leading to hummingbirds arose about 40 million to 50 million years ago when they split from their sister group, the swifts and treeswifts. Scientists think that this likely occurred in Europe or Asia. Hummingbird fossils that date back 30 million to 45 million years ago have been found in Germany, France, and Poland.

If hummingbirds first evolved in Eurasia, how did they get to South America? Scientists hypothesize that ancestral hummingbirds likely migrated across the Bering Land Bridge, which once connected Asia to Alaska, and traveled from there down to what is now South America. Over the course of 22 million years—a relatively short period of evolutionary time—a common ancestor gave rise to the 338 species of hummingbirds that exist today. Hummingbirds returned to North America around 12 million years ago, and others began to inhabit the Caribbean around five million years ago. Research indicates that new species of hummingbirds continue to arise, though at a much slower pace than they have in the past. Scientists estimate that over the next several million years, the total number of hummingbird species could double as they maximize the number of ecological niches available to them.

C Claim Make a claim about how climate change may affect hummingbird speciation in the future.

E Evidence Use the evidence you gathered throughout the unit to support your claim.

R Reasoning To help illustrate your claim, develop a model that illustrates how changes in a hummingbird's environment could affect its ability to survive and reproduce.

Two male booted racket-tail hummingbirds (*Ocreatus underwoodii*) engage in a dispute over feeding territory. The species, named after 20th-century naturalist Cecil Frank Underwood, are found at high altitudes in the Andean mountains of South America.

UNIT ACTIVITY 521

CHAPTER 5

MOLECULES IN LIVING SYSTEMS

Flamingos are born with a white or gray color. As they obtain carotenoids from their diet of shrimp and algae, their colors change.

Flamingos get their characteristic color from pigment molecules called carotenoids. These molecules give some animals and most plants their red, orange, and yellow colors. The structures of different carotenoid molecules determine which wavelengths of light they absorb. The types of molecules that make up living things share important structural characteristics. Scientists use these patterns to predict how a group of similar molecules functions within an organism or biochemical process.

Chapter 5 supports the NGSS Performance Expectation **HS-LS1-6**.

CASE STUDY
TURNING MOLECULES INTO MEDICINE

HOW ARE THE MOLECULES OF LIFE ASSEMBLED?

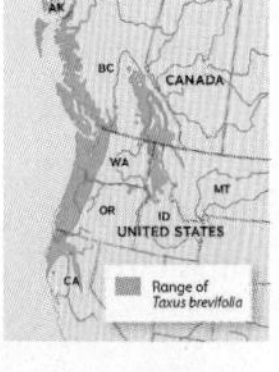

In many ways, Earth's living species are its most valuable resources. For example, many of the drugs used in a hospital or available in a pharmacy are derived from plants. Substances such as morphine and other pain relievers come from poppy plants. Colchicine from the meadow saffron treats a common form of arthritis called gout, and quinine from a tropical tree bark treats malaria. Madagascar periwinkle is used to treat cancers such as leukemia and Hodgkin disease.

The Pacific yew (*Taxus brevifolia*), shown in **Figure 5-1**, is a slow-growing tree native to the Pacific Northwest and the northern Rocky Mountains. The logging industry once considered it a "weed" tree with no commercial value. However, Native American people and specialty artisans have long used the Pacific yew's strong, decay-resistant wood for tools and decorative objects.

In the 1960s, scientists at the National Cancer Institute discovered that the thin bark of the Pacific yew contains a compound that inhibits the growth of tumors. They named this compound *paclitaxel* (**Figure 5-2**).

Clinical trials showed paclitaxel to be very effective against ovarian cancer compared with earlier forms of treatment. Unfortunately, the bark of a single yew contains only a tiny amount of paclitaxel, and removing the bark to extract the compound kills the tree. Hundreds of thousands of trees would need to be harvested to obtain enough paclitaxel to treat cancer patients. This fact sparked the search for other ways to synthesize the complex compound. Scientists have tried extracting compounds from yew needles and identifying species of fungi that can produce paclitaxel. They have also genetically engineered bacteria to make taxadiene, an intermediate compound needed to build a paclitaxel molecule. Paclitaxel can now be made in a laboratory, but significant challenges remain in making the process more sustainable and less expensive. Nevertheless, this naturally derived compound has been one of the most effective and versatile tools for cancer treatment to date.

Ask Questions *In this chapter, you will learn about patterns that occur in biological matter at the molecular level. As you read, generate questions you might need to ask if you were a biochemist who wanted to make naturally occurring compounds in a lab.*

FIGURE 5-1
Removing the thin purple bark from the Pacific yew tree is time-consuming and delicate work.

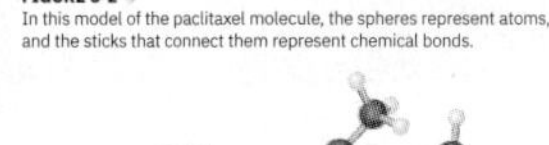

FIGURE 5-2
In this model of the paclitaxel molecule, the spheres represent atoms, and the sticks that connect them represent chemical bonds.

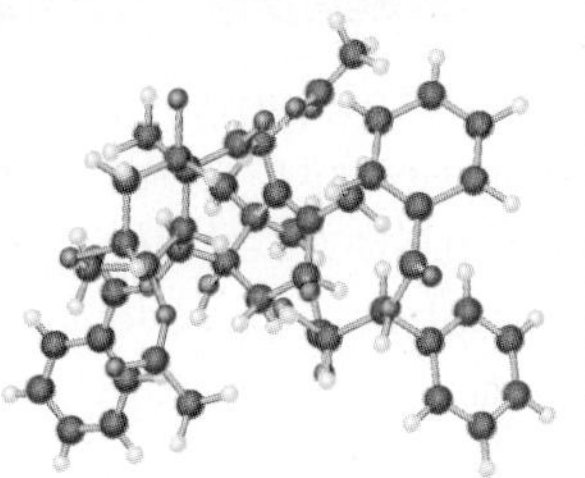

Each chapter also follows a phenomenon-based approach. The **Chapter Opener** and **Case Study** work together to introduce a phenomenon and related Driving Question, which are revisited at the end of the chapter in the Tying It All Together activity. Callouts within the chapters prompt students to connect concepts back to the Case Study as they read.

The **Tying It All Together** activity is a formative, guided research or engineering design project to address the Driving Question in the Case Study.

TYING IT ALL TOGETHER
TURNING MOLECULES INTO MEDICINE

HOW ARE THE MOLECULES OF LIFE ASSEMBLED?

Molecular substances in nature vary widely in complexity. By understanding patterns in the chemical structures of these substances, scientists can predict their functions within biological systems and organisms. The paclitaxel molecule introduced in the Case Study belongs to a category of organic molecules called alkaloids. Alkaloid molecules have at least one nitrogen atom that is typically part of a structure with multiple rings. Most alkaloids are found in plants, and scientists estimate that about 20 percent of plant species contain alkaloids. There are several hypotheses about the roles these compounds play: alkaloids may regulate plant growth, or protect plants from predators, or they may be waste products of plant metabolism.

Alkaloid names usually end in *-ine*. Many alkaloids, such as caffeine and nicotine, are known to have specific physiological effects on humans. Traditional medicines and healing practices are often derived from the therapeutic effects of alkaloids, some of which include pain relief, inflammation reduction, and antibacterial properties. As advances in biotechnology have allowed for detailed investigation of chemical compounds at the molecular level, scientists are increasingly motivated to study the structures and biological functions of alkaloids for their potential use as drugs to treat cancer and other conditions or diseases such as hypertension, diabetes, and asthma.

In this activity, you will investigate how a plant-derived alkaloid molecule is developed into a medical treatment. Work individually or in a group to complete these steps.

Gather Evidence

1. Research a specific plant alkaloid that is used in modern medicine. In your research, you should identify
   - the plant in which the alkaloid is found, and how the alkaloid is extracted from the plant
   - how the alkaloid affects the human body, and how these effects were discovered
   - the disease or condition the alkaloid is used to treat
   - the name of the medicine that is made from the alkaloid, and how it is made

Communicate Information

2. Construct a visual representation, such as a time line or flowchart, that describes the process by which the alkaloid you chose was developed into a medicine.

Argue From Evidence

3. Tens of thousands of plant alkaloids have been identified, but only a small percentage of these have been approved for pharmaceutical use. Use evidence from your research to support or refute the following claim: *Plant alkaloid research is an effective way to find new medical treatments.*

FIGURE 5-37
The Madagascar periwinkle (*Catharanthus roseus*) plant is a source of vincristine and vinblastine, alkaloids that are used to treat cancer and Hodgkin disease.

FIGURE 5-38
Claviceps purpurea, a fungus that infects rye, produces many medicinal alkaloids.

VIRTUAL LABS AND INTERACTIVE LEARNING

Transport students to places they have never been before with *National Geographic Biology*'s immersive and engaging **Virtual Labs**, **Simulations**, and **Interactive Figures**.

Students follow in the footsteps of National Geographic Explorers to conduct **Virtual Investigations** in the deep ocean, rainforest canopy, and other locations around the world.

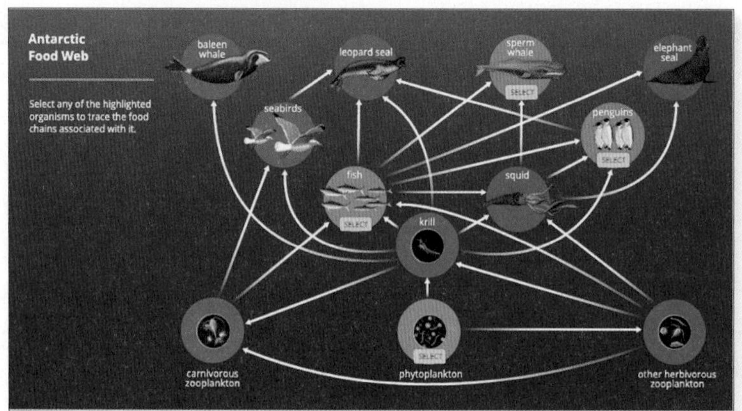

Simulations and **Interactive Figures** bring figures and concepts from the print book to life. For example, students can manipulate an Antarctic Food Web to see how a food web changes as organisms are removed. Other simulations allow students to use atoms to construct polymers such as DNA or watch a process unfold in an animation.

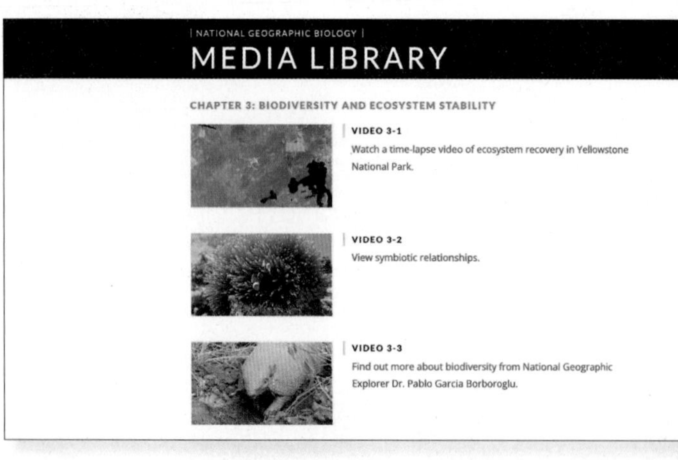

| NATIONAL GEOGRAPHIC BIOLOGY |

MEDIA LIBRARY

CHAPTER 3: BIODIVERSITY AND ECOSYSTEM STABILITY

VIDEO 3-1
Watch a time-lapse video of ecosystem recovery in Yellowstone National Park.

VIDEO 3-2
View symbiotic relationships.

VIDEO 3-3
Find out more about biodiversity from National Geographic Explorer Dr. Pablo Garcia Borboroglu.

Each chapter's **Media Library** lists the videos that appear at point of use throughout the chapter.

SUPPORT FOR ALL READERS WITH LEVELED TEXT

National Geographic Biology is available in two reading levels according to the "Stretch" Lexile® bands. The On Level version of the text matches the printed textbook for grades 9–10 (1050L–1335L). With the press of a button, however, students using MindTap can toggle to the simpler Modified Text version written to the Middle School grade band (925L–1185L).

ON-LEVEL TEXT

MODIFIED TEXT

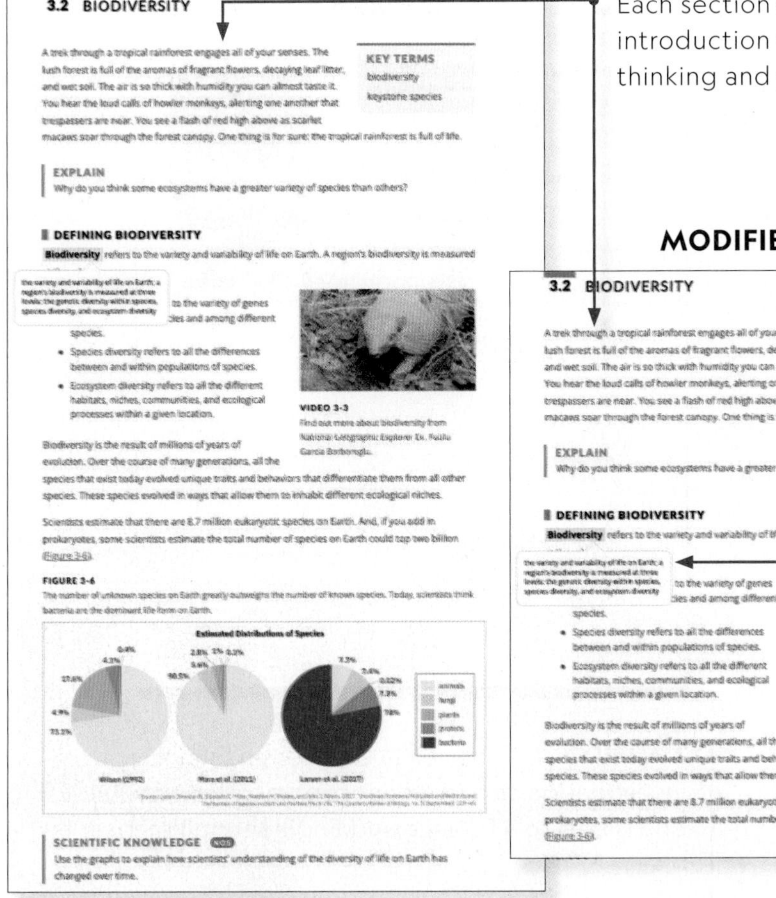

Each section opens with a brief, relatable introduction to the topic to get students thinking and primed for learning.

With **Modified Text** in the MindTap digital platform, students can simplify the text to a middle school reading level, reducing cognitive load and improving learning outcomes for struggling readers.

A definition of each **Key Term** gives students access to definitions at point of use. Other important terms are defined at point of use and with the Key Terms in the Glossary.

OPPORTUNITIES FOR THREE-DIMENSIONAL LEARNING

Questions interspersed throughout each chapter serve as 3D checkpoints, prompting students to engage with **Disciplinary Core Ideas** through the **Crosscutting Concepts** and **Science and Engineering Practices**. Checkpoints also foster greater metacognitive awareness so that students learn to gauge their comprehension as they read the text and visual supports.

Labels (SEP, CCC, NOS) indicate the NGSS dimension to which the concept or practice in red text pertains.

SEP **Evaluate Information** *What is a key characteristic of a biodiversity hotspot?*

CCC **Patterns** *Why do you think regions farthest from the Equator show the least amounts of biodiversity?*

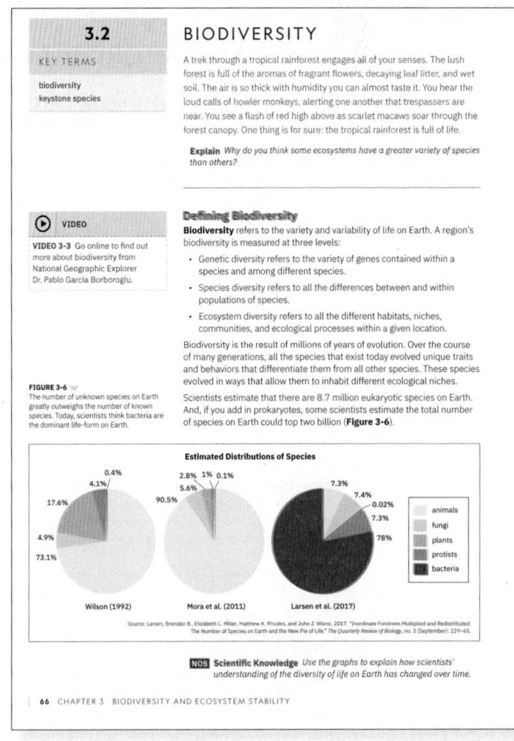

3.2

KEY TERMS

biodiversity
keystone species

BIODIVERSITY

A trek through a tropical rainforest engages all of your senses. The lush forest is full of the aromas of fragrant flowers, decaying leaf litter, and wet soil. The air is so thick with humidity you can almost taste it. You hear the loud calls of howler monkeys, alerting one another that trespassers are near. You see a flash of red high above as scarlet macaws soar through the forest canopy. One thing is for sure: the tropical rainforest is full of life.

Explain *Why do you think some ecosystems have a greater variety of species than others?*

VIDEO

VIDEO 3-3 Go online to find out more about biodiversity from National Geographic Explorer Dr. Pablo Garcia Borboroglu.

Defining Biodiversity

Biodiversity refers to the variety and variability of life on Earth. A region's biodiversity is measured at three levels:

- Genetic diversity refers to the variety of genes contained within a species and among different species.
- Species diversity refers to all the differences between and within populations of species.
- Ecosystem diversity refers to all the different habitats, niches, communities, and ecological processes within a given location.

Biodiversity is the result of millions of years of evolution. Over the course of many generations, all the species that exist today evolved unique traits and behaviors that differentiate them from all other species. These species evolved in ways that allow them to inhabit different ecological niches.

Scientists estimate that there are 8.7 million eukaryotic species on Earth. And, if you add in prokaryotes, some scientists estimate the total number of species on Earth could top two billion (**Figure 3-6**).

FIGURE 3-6
The number of unknown species on Earth greatly outweighs the number of known species. Today, scientists think bacteria are the dominant life-form on Earth.

Source: Larsen, Brendan B., Elizabeth C. Miller, Matthew K. Rhodes, and John J. Wiens. 2017. "Inordinate Fondness Multiplied and Redistributed: The Number of Species on Earth and the New Pie of Life." *The Quarterly Review of Biology*, no. 3 (September): 229–65.

NOS **Scientific Knowledge** *Use the graphs to explain how scientists' understanding of the diversity of life on Earth has changed over time.*

66 CHAPTER 3 BIODIVERSITY AND ECOSYSTEM STABILITY

HANDS-ON SCIENCE AND DATA LITERACY

Each chapter of *National Geographic Biology* provides multiple opportunities for hands-on learning. Quick labs designed for your classroom and full laboratory investigations give students practice with lab equipment and lab safety procedures. Data analysis activities give students practice reading data visualizations and identifying patterns in data sets.

In **Looking at the Data** activities, students apply Science and Engineering Practices by analyzing data in multiple formats and practicing mathematical and computational skills. Many activities are based on actual data about real-world topics that matter.

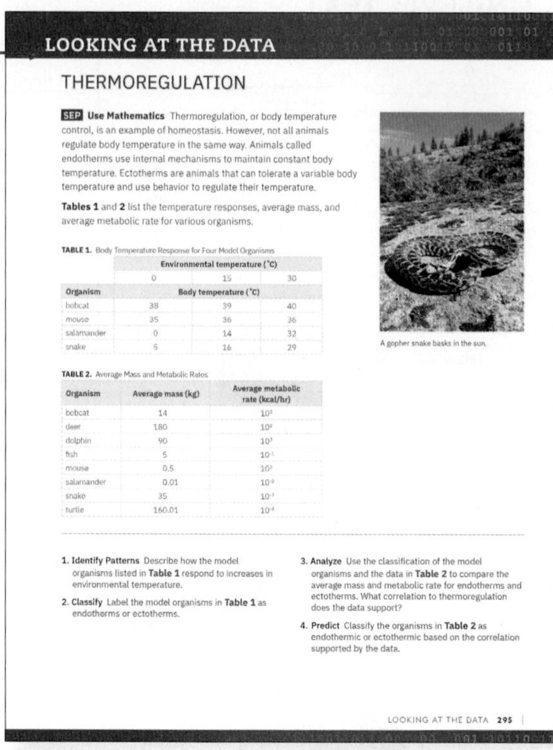

LOOKING AT THE DATA

THERMOREGULATION

SEP Use Mathematics Thermoregulation, or body temperature control, is an example of homeostasis. However, not all animals regulate body temperature in the same way. Animals called endotherms use internal mechanisms to maintain constant body temperature. Ectotherms are animals that can tolerate a variable body temperature and use behavior to regulate their temperature.

Tables 1 and **2** list the temperature responses, average mass, and average metabolic rate for various organisms.

TABLE 1. Body Temperature Response for Four Model Organisms

	Environmental temperature (°C)		
	0	15	30
Organism	**Body temperature (°C)**		
bobcat	38	39	40
mouse	35	36	36
salamander	0	14	32
snake	5	16	29

TABLE 2. Average Mass and Metabolic Rates

Organism	Average mass (kg)	Average metabolic rate (kcal/hr)
bobcat	14	10^1
deer	180	10^2
dolphin	90	10^3
fish	5	10^{-1}
mouse	0.5	10^3
salamander	0.01	10^{-2}
snake	35	10^{-1}
turtle	160.01	10^{-3}

A gopher snake basks in the sun.

1. **Identify Patterns** Describe how the model organisms listed in **Table 1** respond to increases in environmental temperature.
2. **Classify** Label the model organisms in **Table 1** as endotherms or ectotherms.
3. **Analyze** Use the classification of the model organisms and the data in **Table 2** to compare the average mass and metabolic rate for endotherms and ectotherms. What correlation to thermoregulation does the data support?
4. **Predict** Classify the organisms in **Table 2** as endothermic or ectothermic based on the correlation supported by the data.

LOOKING AT THE DATA 295

Two teacher-tested **Chapter Investigations** per chapter provide in-depth laboratory experiences. Find Guided Inquiry, Open Inquiry, and Design-Your-Own labs in the MindTap digital platform. Each unit also includes an **Engineering Design** activity. Supportive lab guides, worksheets, and rubrics are available in multiple file formats.

A **Minilab** built into each chapter enables a quick investigation as part of conceptual development. Each one is tied to a Science and Engineering Practice.

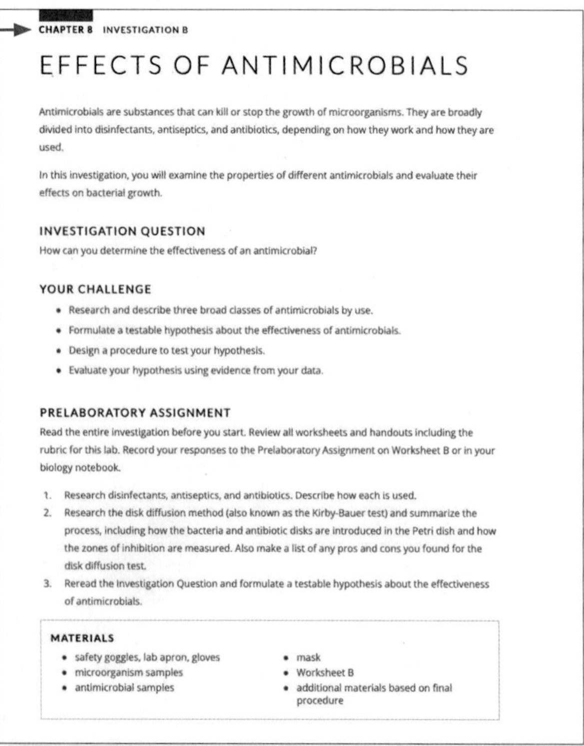

CHAPTER 8 INVESTIGATION B

EFFECTS OF ANTIMICROBIALS

Antimicrobials are substances that can kill or stop the growth of microorganisms. They are broadly divided into disinfectants, antiseptics, and antibiotics, depending on how they work and how they are used.

In this investigation, you will examine the properties of different antimicrobials and evaluate their effects on bacterial growth.

INVESTIGATION QUESTION

How can you determine the effectiveness of an antimicrobial?

YOUR CHALLENGE

- Research and describe three broad classes of antimicrobials by use.
- Formulate a testable hypothesis about the effectiveness of antimicrobials.
- Design a procedure to test your hypothesis.
- Evaluate your hypothesis using evidence from your data.

PRELABORATORY ASSIGNMENT

Read the entire investigation before you start. Review all worksheets and handouts including the rubric for this lab. Record your responses to the Prelaboratory Assignment on Worksheet B or in your biology notebook.

1. Research disinfectants, antiseptics, and antibiotics. Describe how each is used.
2. Research the disk diffusion method (also known as the Kirby-Bauer test) and summarize the process, including how the bacteria and antibiotic disks are introduced in the Petri dish and how the zones of inhibition are measured. Also make a list of any pros and cons you found for the disk diffusion test.
3. Reread the Investigation Question and formulate a testable hypothesis about the effectiveness of antimicrobials.

MATERIALS

- safety goggles, lab apron, gloves
- microorganism samples
- antimicrobial samples
- mask
- Worksheet B
- additional materials based on final procedure

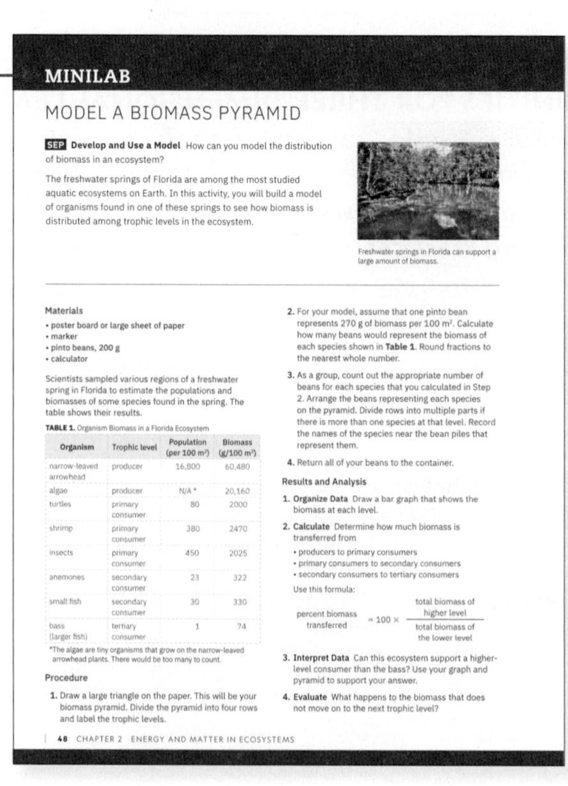

MINILAB

MODEL A BIOMASS PYRAMID

SEP Develop and Use a Model How can you model the distribution of biomass in an ecosystem?

The freshwater springs of Florida are among the most studied aquatic ecosystems on Earth. In this activity, you will build a model of organisms found in one of these springs to see how biomass is distributed among trophic levels in the ecosystem.

Freshwater springs in Florida can support a large amount of biomass.

Materials

- poster board or large sheet of paper
- marker
- pinto beans, 200 g
- calculator

Scientists sampled various regions of a freshwater spring in Florida to estimate the populations and biomasses of some species found in the spring. The table shows their results.

TABLE 1. Organism Biomass in a Florida Ecosystem

Organism	Trophic level	Population (per 100 m²)	Biomass (g/100 m²)
narrow-leaved arrowhead	producer	16,800	60,480
algae	producer	N/A *	20,160
turtles	primary consumer	80	2000
shrimp	primary consumer	380	2470
insects	primary consumer	450	2025
anemones	secondary consumer	23	322
small fish	secondary consumer	30	330
bass (larger fish)	tertiary consumer	1	74

*The algae are tiny organisms that grow on the narrow-leaved arrowhead plants. There would be too many to count.

Procedure

1. Draw a large triangle on the paper. This will be your biomass pyramid. Divide the pyramid into four rows and label the trophic levels.
2. For your model, assume that one pinto bean represents 270 g of biomass per 100 m^2. Calculate how many beans would represent the biomass of each species shown in **Table 1**. Round fractions to the nearest whole number.
3. As a group, count out the appropriate number of beans for each species that you calculated in Step 2. Arrange the beans representing each species on the pyramid. Divide rows into multiple parts if there is more than one species at that level. Record the names of the species near the bean piles that represent them.
4. Return all of your beans to the container.

Results and Analysis

1. **Organize Data** Draw a bar graph that shows the biomass at each level.
2. **Calculate** Determine how much biomass is transferred from
   - producers to primary consumers
   - primary consumers to secondary consumers
   - secondary consumers to tertiary consumers

   Use this formula:

$$\text{percent biomass transferred} = 100 \times \frac{\text{total biomass of higher level}}{\text{total biomass of the lower level}}$$

3. **Interpret Data** Can this ecosystem support a higher-level consumer than the bass? Use your graph and pyramid to support your answer.
4. **Evaluate** What happens to the biomass that does not move on to the next trophic level?

48 CHAPTER 2 ENERGY AND MATTER IN ECOSYSTEMS

TEACHER'S EDITION OVERVIEW

A thoughtfully designed Teacher's Edition for *National Geographic Biology* provides tools and strategies for impactful, effective classroom instruction.

The **Unit Overview** supports standards-based instructional planning with the NGSS progression from middle school to high school.

Information about the unit's National Geographic Explorer and Virtual Lab ties the unit's content to its Anchoring Phenomenon.

An overview of the unit-ending CER includes a summary of the main evidence-gathering activities within each of the chapters.

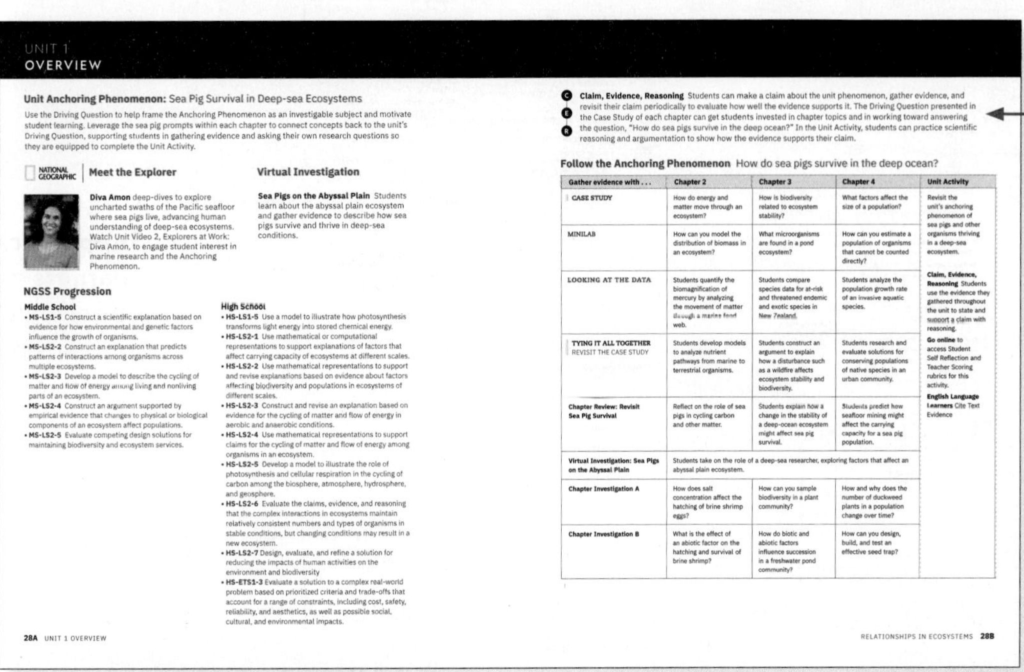

UNIT 1
OVERVIEW

Unit Anchoring Phenomenon: Sea Pig Survival in Deep-sea Ecosystems

Use the Driving Question to help frame the Anchoring Phenomenon as an investigable subject and motivate student learning. Leverage the sea pig prompts within each chapter to connect concepts back to the unit's Driving Question, supporting students in gathering evidence and asking their own research questions so they are equipped to complete the Unit Activity.

NATIONAL GEOGRAPHIC | **Meet the Explorer**

Diva Amon deep-dives to explore uncharted swaths of the Pacific seafloor where sea pigs live, advancing human understanding of deep-sea ecosystems. Watch Unit Video 2, Explorers at Work: Diva Amon, to engage student interest in marine research and the Anchoring Phenomenon.

Virtual Investigation

Sea Pigs on the Abyssal Plain Students learn about the abyssal plain ecosystem and gather evidence to describe how sea pigs survive and thrive in deep-sea conditions.

NGSS Progression

Middle School

- **MS-LS1-5** Construct a scientific explanation based on evidence for how environmental and genetic factors influence the growth of organisms.
- **MS-LS2-2** Construct an explanation that predicts patterns of interactions among organisms across multiple ecosystems.
- **MS-LS2-3** Develop a model to describe the cycling of matter and flow of energy among living and nonliving parts of an ecosystem.
- **MS-LS2-4** Construct an argument supported by empirical evidence that changes to physical or biological components of an ecosystem affect populations.
- **MS-LS2-5** Evaluate competing design solutions for maintaining biodiversity and ecosystem services.

High School

- **HS-LS1-5** Use a model to illustrate how photosynthesis transforms light energy into stored chemical energy.
- **HS-LS2-1** Use mathematical or computational representations to support explanations of factors that affect carrying capacity of ecosystems at different scales.
- **HS-LS2-2** Use mathematical representations to support and revise explanations based on evidence about factors affecting biodiversity and populations in ecosystems of different scales.
- **HS-LS2-3** Construct and revise an explanation based on evidence for the cycling of matter and flow of energy in aerobic and anaerobic conditions.
- **HS-LS2-4** Use mathematical representations to support claims for the cycling of matter and flow of energy among organisms in an ecosystem.
- **HS-LS2-5** Develop a model to illustrate the role of photosynthesis and cellular respiration in the cycling of carbon among the biosphere, atmosphere, hydrosphere, and geosphere.
- **HS-LS2-6** Evaluate the claims, evidence, and reasoning that the complex interactions in ecosystems maintain relatively consistent numbers and types of organisms in stable conditions, but changing conditions may result in a new ecosystem.
- **HS-LS2-7** Design, evaluate, and refine a solution for reducing the impacts of human activities on the environment and biodiversity
- **HS-ETS1-3** Evaluate a solution to a complex real-world problem based on prioritized criteria and trade-offs that account for a range of constraints, including cost, safety, reliability, and aesthetics, as well as possible social, cultural, and environmental impacts.

28A UNIT 1 OVERVIEW

Claim, Evidence, Reasoning Students can make a claim about the unit phenomenon, gather evidence, and revisit their claim periodically to evaluate how well the evidence supports it. The Driving Question presented in the Case Study of each chapter can get students invested in chapter topics and in working toward answering the question, "How do sea pigs survive in the deep ocean?" In the Unit Activity, students can practice scientific reasoning and argumentation to show how the evidence supports their claim.

Follow the Anchoring Phenomenon How do sea pigs survive in the deep ocean?

Gather evidence with . . .	Chapter 2	Chapter 3	Chapter 4	Unit Activity
CASE STUDY	How do energy and matter move through an ecosystem?	How is biodiversity related to ecosystem stability?	What factors affect the size of a population?	Revisit the unit's anchoring phenomenon of sea pigs and other organisms thriving in a deep-sea ecosystem. **Claim, Evidence, Reasoning** Students use the evidence they gathered throughout the unit to state and support a claim with reasoning. **Go online** to access Student Self Reflection and Teacher Scoring rubrics for this activity. **English Language Learners** Cite Text Evidence
MINILAB	How can you model the distribution of biomass in an ecosystem?	What microorganisms are found in a pond ecosystem?	How can you estimate a population of organisms that cannot be counted directly?	
LOOKING AT THE DATA	Students quantify the biomagnification of mercury by analyzing the movement of matter through a marine food web.	Students compare species data for at-risk and threatened endemic and exotic species in New Zealand.	Students analyze the population growth rate of an invasive aquatic species.	
TYING IT ALL TOGETHER REVISIT THE CASE STUDY	Students develop models to analyze nutrient pathways from marine to terrestrial organisms.	Students construct an argument to explain how a disturbance such as a wildfire affects ecosystem stability and biodiversity.	Students research and evaluate solutions for conserving populations of native species in an urban community.	
Chapter Review: Revisit Sea Pig Survival	Reflect on the role of sea pigs in cycling carbon and other matter.	Students explain how a change in the stability of a deep-ocean ecosystem might affect sea pig survival.	Students predict how seafloor mining might affect the carrying capacity for a sea pig population.	
Virtual Investigation: Sea Pigs on the Abyssal Plain	Students take on the role of a deep-sea researcher, exploring factors that affect an abyssal plain ecosystem.			
Chapter Investigation A	How does salt concentration affect the hatching of brine shrimp eggs?	How can you sample biodiversity in a plant community?	How and why does the number of duckweed plants in a population change over time?	
Chapter Investigation B	What is the effect of an abiotic factor on the hatching and survival of brine shrimp?	How do biotic and abiotic factors influence succession in a freshwater pond community?	How can you design, build, and test an effective seed trap?	

RELATIONSHIPS IN ECOSYSTEMS 28B

Planning Your Investigations supports hands-on investigations with details on advanced preparation, specific dimensions coverage, and time management.

Assessment Planning provides support for culminating Performance Tasks.

Additional Resources provides a curated list of videos and articles that complement unit content.

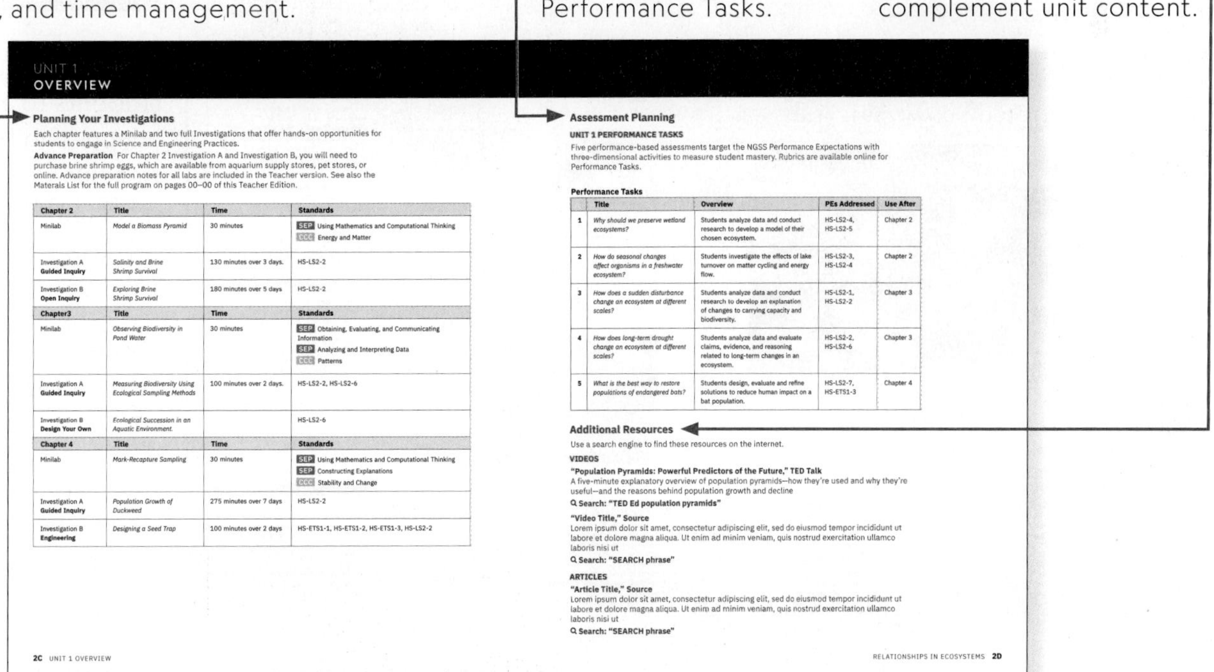

UNIT 1
OVERVIEW

Planning Your Investigations

Each chapter features a Minilab and two full Investigations that offer hands-on opportunities for students to engage in Science and Engineering Practices.

Advance Preparation For Chapter 2 Investigation A and Investigation B, you will need to purchase brine shrimp eggs, which are available from aquarium supply stores, pet stores, or online. Advance preparation notes for all labs are included in the Teacher version. See also the Materals List for the full program on pages 00–00 of this Teacher Edition.

Chapter 2	Title	Time	Standards
Minilab	*Model a Biomass Pyramid*	30 minutes	SEP Using Mathematics and Computational Thinking CCC Energy and Matter
Investigation A **Guided Inquiry**	*Salinity and Brine Shrimp Survival*	130 minutes over 3 days.	HS-LS2-2
Investigation B **Open Inquiry**	*Exploring Brine Shrimp Survival*	180 minutes over 5 days	HS-LS2-2
Chapter3	**Title**	**Time**	**Standards**
Minilab	*Observing Biodiversity in Pond Water*	30 minutes	SEP Obtaining, Evaluating, and Communicating Information SEP Analyzing and Interpreting Data CCC Patterns
Investigation A **Guided Inquiry**	*Measuring Biodiversity Using Ecological Sampling Methods*	100 minutes over 2 days.	HS-LS2-2, HS-LS2-6
Investigation B **Design Your Own**	*Ecological Succession in an Aquatic Environment.*		HS-LS2-6
Chapter 4	**Title**	**Time**	**Standards**
Minilab	*Mark-Recapture Sampling*	30 minutes	SEP Using Mathematics and Computational Thinking SEP Constructing Explanations CCC Stability and Change
Investigation A **Guided Inquiry**	*Population Growth of Duckweed*	275 minutes over 7 days	HS-LS2-2
Investigation B **Engineering**	*Designing a Seed Trap*	100 minutes over 2 days	HS-ETS1-1, HS-ETS1-2, HS-ETS1-3, HS-LS2-2

2C UNIT 1 OVERVIEW

Assessment Planning

UNIT 1 PERFORMANCE TASKS

Five performance-based assessments target the NGSS Performance Expectations with three-dimensional activities to measure student mastery. Rubrics are available online for Performance Tasks.

Performance Tasks

	Title	Overview	PEs Addressed	Use After
1	*Why should we preserve wetland ecosystems?*	Students analyze data and conduct research to develop a model of their chosen ecosystem.	HS-LS2-4, HS-LS2-5	Chapter 2
2	*How do seasonal changes affect organisms in a freshwater ecosystem?*	Students investigate the effects of lake turnover on matter cycling and energy flow.	HS-LS2-3, HS-LS2-4	Chapter 2
3	*How does a sudden disturbance change an ecosystem at different scales?*	Students analyze data and conduct research to develop an explanation of changes to carrying capacity and biodiversity.	HS-LS2-1, HS-LS2-2	Chapter 3
4	*How does long-term drought change an ecosystem at different scales?*	Students analyze data and evaluate claims, evidence, and reasoning related to long-term changes in an ecosystem.	HS-LS2-2, HS-LS2-6	Chapter 3
5	*What is the best way to restore populations of endangered bats?*	Students design, evaluate and refine solutions to reduce human impact on a bat population.	HS-LS2-7, HS-ETS1-3	Chapter 4

Additional Resources

Use a search engine to find these resources on the internet.

VIDEOS

"Population Pyramids: Powerful Predictors of the Future," TED Talk
A five-minute explanatory overview of population pyramids—how they're used and why they're useful—and the reasons behind population growth and decline

Search: "TED Ed population pyramids"

"Video Title," Source
Lorem ipsum dolor sit amet, consectetur adipiscing elit, sed do eiusmod tempor incididunt ut labore et dolore magna aliqua. Ut enim ad minim veniam, quis nostrud exercitation ullamco laboris nisi ut

Search: "SEARCH phrase"

ARTICLES

"Article Title," Source
Lorem ipsum dolor sit amet, consectetur adipiscing elit, sed do eiusmod tempor incididunt ut labore et dolore magna aliqua. Ut enim ad minim veniam, quis nostrud exercitation ullamco laboris nisi ut

Search: "SEARCH phrase"

RELATIONSHIPS IN ECOSYSTEMS 2D

Each **Chapter Planner** provides a section-by-section guide for classroom planning and time management. Each **Chapter Opener** previews the content through a lens of students' prior learning. It also includes information about the thought-provoking image and a **Case Study** overview.

A 5E instructional path frames each chapter.

Identifies the **PEs**, **SEPs**, **DCIs**, **and CCCs** addressed by the chapter content and resources.

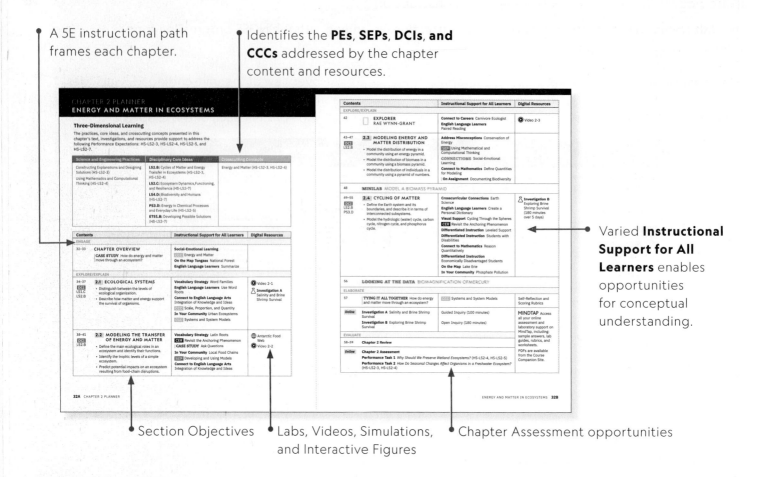

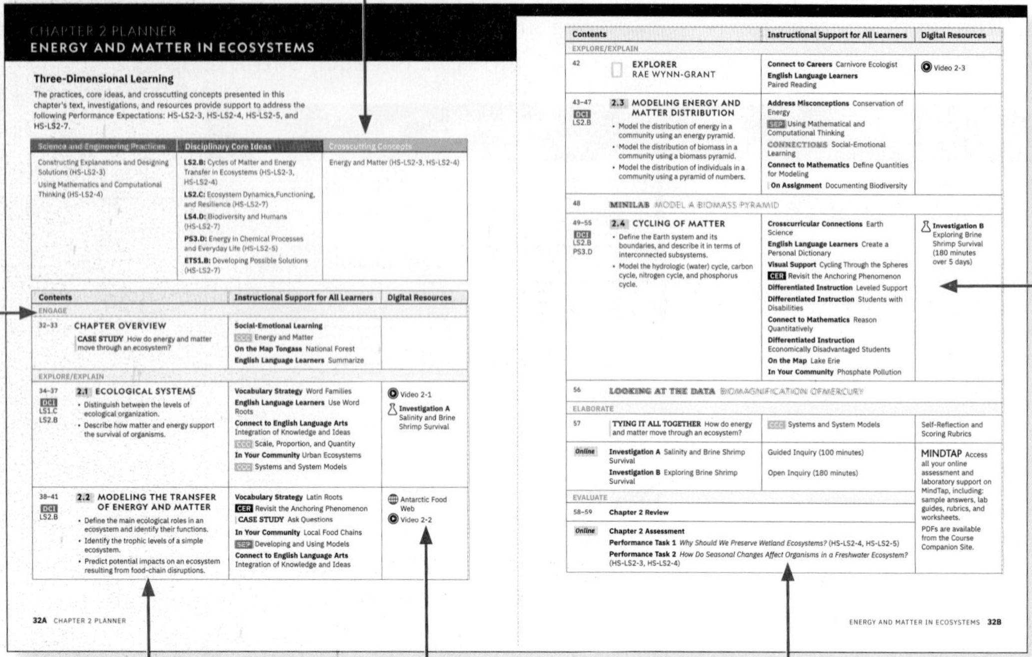

CHAPTER 2 PLANNER

ENERGY AND MATTER IN ECOSYSTEMS

Three-Dimensional Learning

The practices, core ideas, and crosscutting concepts presented in this chapter's text, investigations, and resources provide support to address the following Performance Expectations: HS-LS2-3, HS-LS2-4, HS-LS2-5, and HS-LS2-7.

Science and Engineering Practices	Disciplinary Core Ideas	Crosscutting Concepts
Constructing Explanations and Designing Solutions (HS-LS2-3) Using Mathematics and Computational Thinking (HS-LS2-4)	**LS2.B:** Cycles of Matter and Energy Transfer in Ecosystems (HS-LS2-3, HS-LS2-4) **LS2.C:** Ecosystem Dynamics,Functioning, and Resilience (HS-LS2-7) **LS4.D:** Biodiversity and Humans (HS-LS2-7) **PS3.D:** Energy in Chemical Processes and Everyday Life (HS-LS2-5) **ETS1.B:** Developing Possible Solutions (HS-LS2-7)	Energy and Matter (HS-LS2-3, HS-LS2-4)

Contents		Instructional Support for All Learners	Digital Resources
ENGAGE			
32–33	CHAPTER OVERVIEW **CASE STUDY** How do energy and matter move through an ecosystem?	**Social-Emotional Learning** CCC Energy and Matter **On the Map Tongass** National Forest **English Language Learners** Summarize	
EXPLORE/EXPLAIN			
34–37 DCI LS1.C LS2.B	**2.1** ECOLOGICAL SYSTEMS • Distinguish between the levels of ecological organization. • Describe how matter and energy support the survival of organisms.	**Vocabulary Strategy** Word Families **English Language Learners** Use Word Roots **Connect to English Language Arts** Integration of Knowledge and Ideas CCC Scale, Proportion, and Quantity **In Your Community** Urban Ecosystems CCC Systems and System Models	Video 2-1 **Investigation A** Salinity and Brine Shrimp Survival
38–41 DCI LS2.B	**2.2** MODELING THE TRANSFER OF ENERGY AND MATTER • Define the main ecological roles in an ecosystem and identify their functions. • Identify the trophic levels of a simple ecosystem. • Predict potential impacts on an ecosystem resulting from food-chain disruptions.	**Vocabulary Strategy** Latin Roots CER Revisit the Anchoring Phenomenon **CASE STUDY** Ask Questions **In Your Community** Local Food Chains SEP Developing and Using Models **Connect to English Language Arts** Integration of Knowledge and Ideas	Antarctic Food Web Video 2-2

32A CHAPTER 2 PLANNER

Contents		Instructional Support for All Learners	Digital Resources
EXPLORE/EXPLAIN			
42	EXPLORER RAE WYNN-GRANT	**Connect to Careers** Carnivore Ecologist **English Language Learners** Paired Reading	Video 2-3
43–47 DCI LS2.B	**2.3** MODELING ENERGY AND MATTER DISTRIBUTION • Model the distribution of energy in a community using an energy pyramid. • Model the distribution of biomass in a community using a biomass pyramid. • Model the distribution of individuals in a community using a pyramid of numbers.	**Address Misconceptions** Conservation of Energy SEP Using Mathematical and Computational Thinking CONNECTIONS Social-Emotional Learning **Connect to Mathematics** Define Quantities for Modeling **On Assignment** Documenting Biodiversity	
48	MINILAB MODEL A BIOMASS PYRAMID		
49–55 DCI LS2.B PS3.D	**2.4** CYCLING OF MATTER • Define the Earth system and its boundaries, and describe it in terms of interconnected subsystems. • Model the hydrologic (water) cycle, carbon cycle, nitrogen cycle, and phosphorus cycle.	**Crosscurricular Connections** Earth Science **English Language Learners** Create a Personal Dictionary **Visual Support** Cycling Through the Spheres CER Revisit the Anchoring Phenomenon **Differentiated Instruction** Leveled Support **Differentiated Instruction** Students with Disabilities **Connect to Mathematics** Reason Quantitatively **Differentiated Instruction** Economically Disadvantaged Students **On the Map** Lake Erie **In Your Community** Phosphate Pollution	**Investigation B** Exploring Brine Shrimp Survival (180 minutes over 5 days)
56	LOOKING AT THE DATA BIOMAGNIFICATION OF MERCURY		
ELABORATE			
57	**TYING IT ALL TOGETHER** How do energy and matter move through an ecosystem?	CCC Systems and System Models	Self-Reflection and Scoring Rubrics
Online	**Investigation A** Salinity and Brine Shrimp Survival **Investigation B** Exploring Brine Shrimp Survival	Guided Inquiry (100 minutes) Open Inquiry (180 minutes)	MINDTAP Access all your online assessment and laboratory support on MindTap, including: sample answers, lab guides, rubrics, and worksheets. PDFs are available from the Course Companion Site.
EVALUATE			
58–59	**Chapter 2 Review**		
Online	**Chapter 2 Assessment** **Performance Task 1** *Why Should We Preserve Wetland Ecosystems?* (HS-LS2-4, HS-LS2-5) **Performance Task 2** *How Do Seasonal Changes Affect Organisms in a Freshwater Ecosystem?* (HS-LS2-3, HS-LS2-4)		

ENERGY AND MATTER IN ECOSYSTEMS 32B

Varied **Instructional Support for All Learners** enables opportunities for conceptual understanding.

Section Objectives

Labs, Videos, Simulations, and Interactive Figures

Chapter Assessment opportunities

Social-Emotional Learning provides opportunities to help students develop and practice SEL competencies.

On the Map takes students to a location, connecting them to ecological or geographical features.

Video overviews offer time for planning purposes and strategies to help students extract information.

Wraparound teaching notes provide comprehensive support for instruction.

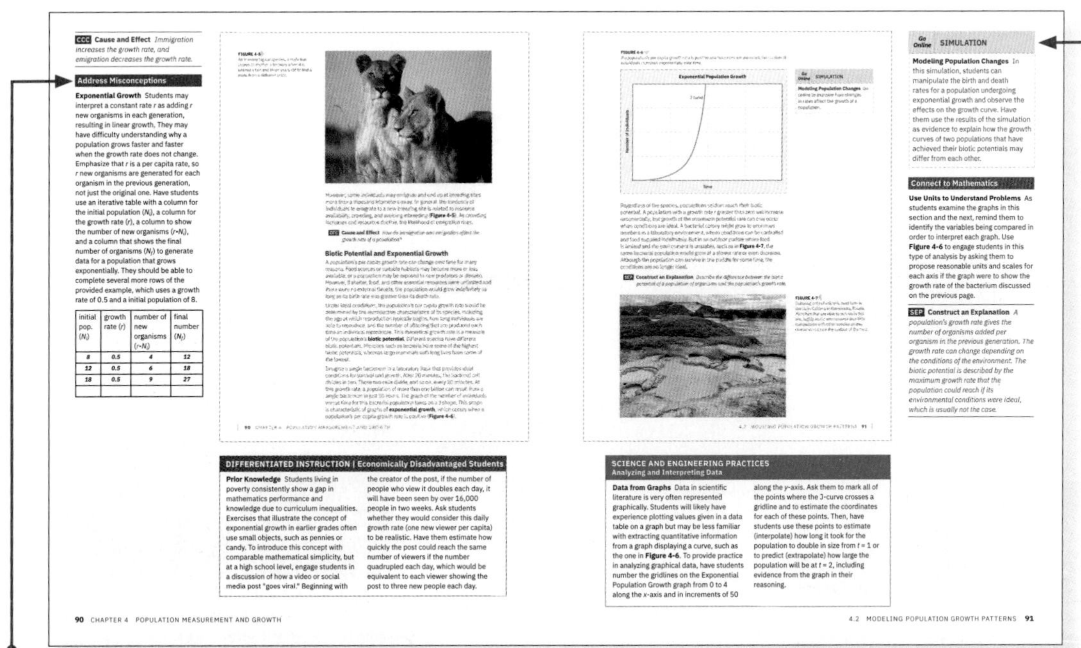

Go Online gives more background on what students do with Interactive Figures, Simulations, and Virtual Investigations.

Address Misconception notes alert teachers to common biology misconceptions and provide prompts to help students overcome the misconceptions.

Visual Support notes help students decode and analyze visuals for greater understanding of concepts.

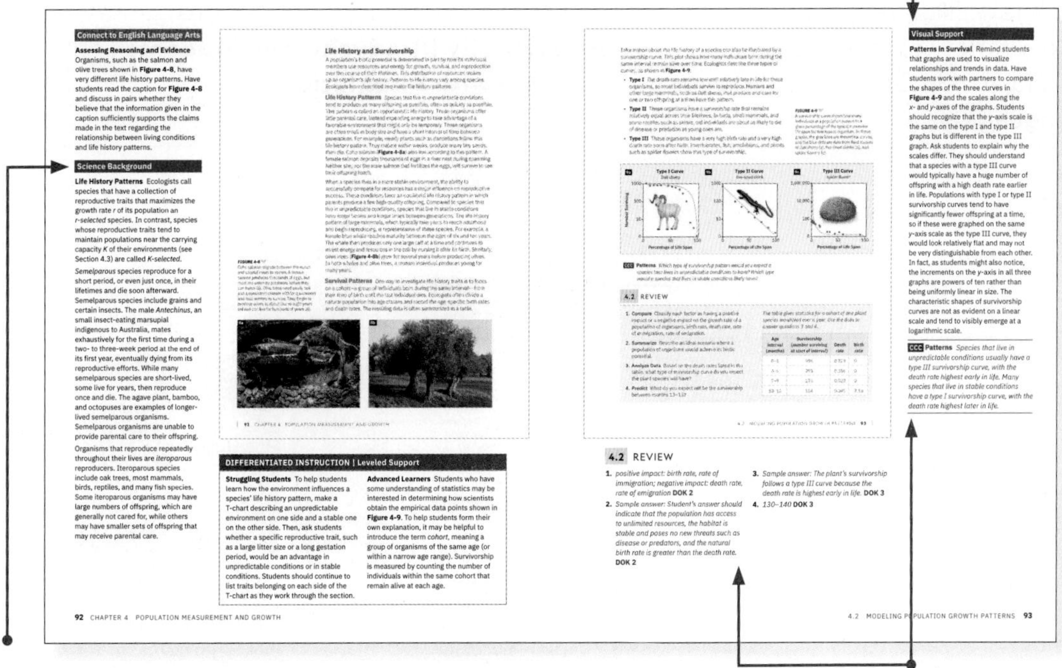

Science Background provides teachers with greater depth of information about the concepts and includes suggestions for further engaging students in discussion.

Answers to all Checkpoint, Section Review, and Chapter Review questions appear at point of use.

SUPPORT FOR DIVERSE LEARNERS

Differentiated Instruction notes offer methods and techniques for interacting with the content or participating in activities that add relevance and suggest modifications that support the work students are already doing.

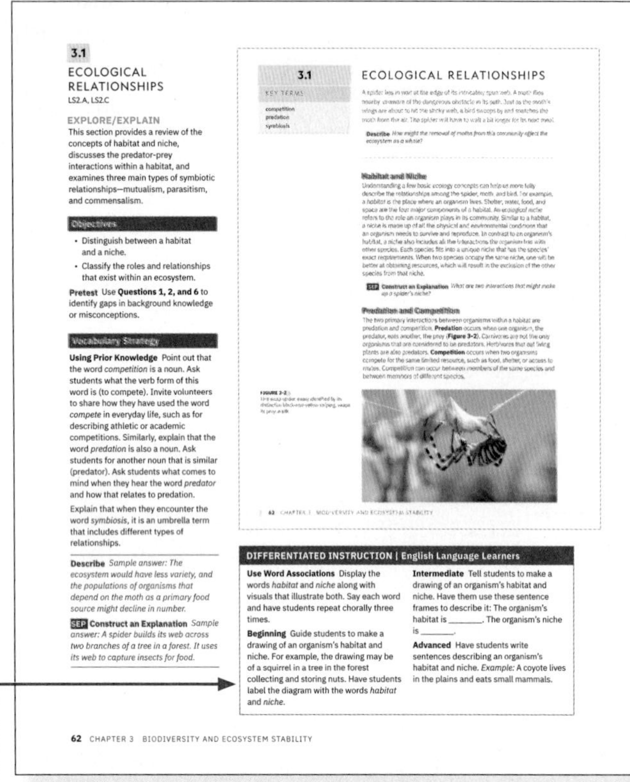

3.1
ECOLOGICAL RELATIONSHIPS
LS2.A, LS2.C

EXPLORE/EXPLAIN
This section provides a review of the concepts of habitat and niche, discusses the predator-prey interactions within a habitat, and examines three main types of symbiotic relationships—mutualism, parasitism, and commensalism.

Objectives
- Distinguish between a habitat and a niche.
- Classify the roles and relationships that exist within an ecosystem.

Pretest Use **Questions 1, 2, and 6** to identify gaps in background knowledge or misconceptions.

Vocabulary Strategy
Using Prior Knowledge Point out that the word *competition* is a noun. Ask students what the verb form of this word is (to compete). Invite volunteers to share how they have used the word *compete* in everyday life, such as for describing athletic or academic competitions. Similarly, explain that the word *predation* is also a noun. Ask students for another noun that is similar (predator). Ask students what comes to mind when they hear the word *predator* and how that relates to predation.

Explain that when they encounter the word *symbiosis*, it is an umbrella term that includes different types of relationships.

Describe *Sample answer: The ecosystem would have less variety, and the populations of organisms that depend on the moth as a primary food source might decline in number.*

SEP **Construct an Explanation** *Sample answer: A spider builds its web across two branches of a tree in a forest. It uses its web to capture insects for food.*

3.1 ECOLOGICAL RELATIONSHIPS

DIFFERENTIATED INSTRUCTION | English Language Learners

Use Word Associations Display the words *habitat* and *niche* along with visuals that illustrate both. Say each word and have students repeat chorally three times.

Beginning Guide students to make a drawing of an organism's habitat and niche. For example, the drawing may be of a squirrel in a tree in the forest collecting and storing nuts. Have students label the diagram with the words *habitat* and *niche*.

Intermediate Tell students to make a drawing of an organism's habitat and niche. Have them use these sentence frames to describe it: The organism's habitat is ______. The organism's niche is ______.

Advanced Have students write sentences describing an organism's habitat and niche. *Example:* A coyote lives in the plains and eats small mammals.

62 CHAPTER 3 BIODIVERSITY AND ECOSYSTEM STABILITY

Connect to English Language Arts
Assessing Reasoning and Evidence Organisms, such as the salmon and olive trees shown in **Figure 4-8**, have very different life history patterns. Have students read the caption for **Figure 4-8** and discuss in pairs whether they believe that the information given in the caption sufficiently supports the claims made in the text regarding the relationship between living conditions and life history patterns.

Science Background
Life History Patterns Ecologists call species that have a collection of reproductive traits that maximizes the growth rate *r* of its population an *r-selected* species. In contrast, species whose reproductive traits tend to maintain populations near the carrying capacity *K* of their environments (see Section 4.3) are called *K-selected*.

Semelparous species reproduce for a short period, or even just once, in their lifetimes and die soon afterward. Semelparous species include grains and certain insects. The male *Antechinus*, an small insect-eating marsupial indigenous to Australia, mates exhaustively for the first time during a two- to three-week period at the end of its first year, eventually dying from its reproductive efforts. While many semelparous species are short-lived, some live for years, then reproduce once and die. The agave plant, bamboo, and octopuses are examples of longer-lived semelparous organisms. Semelparous organisms are unable to provide parental care to their offspring.

Organisms that reproduce repeatedly throughout their lives are *iteroparous* reproducers. Iteroparous species include oak trees, most mammals, birds, reptiles, and many fish species. Some iteroparous organisms may have large numbers of offspring, which are generally not cared for, while others may have smaller sets of offspring that may receive parental care.

DIFFERENTIATED INSTRUCTION | Leveled Support

Struggling Students To help students learn how the environment influences a species' life history pattern, make a T-chart describing an unpredictable environment on one side and a stable one on the other side. Then, ask students whether a specific reproductive trait, such as a large litter size or a long gestation period, would be an advantage in unpredictable conditions or in stable conditions. Students should continue to list traits belonging on each side of the T-chart as they work through the section.

Advanced Learners Students who have some understanding of statistics may be interested in determining how scientists obtain the empirical data points shown in **Figure 4-9**. To help students form their own explanation, it may be helpful to introduce the term *cohort*, meaning a group of organisms of the same age (or within a narrow age range). Survivorship is measured by counting the number of individuals within the same cohort that remain alive at each age.

92 CHAPTER 4 POPULATION MEASUREMENT AND GROWTH

Strategies for **English Language Learners** provide instructional support to help ELLs understand biology concepts using a variety of approaches for reading, writing, speaking, and listening as well as utilizing their native language.

Varied strategies suggest how to provide **Leveled Support** as well as ideas to address **Economically Disadvantaged Students** and **Students with Disabilities**.

CONNECT TO ALL STUDENTS

Personalizing biology concepts makes them more tangible for students to grasp.

In Your Community connects concepts to students' local surroundings and direct experiences and observations.

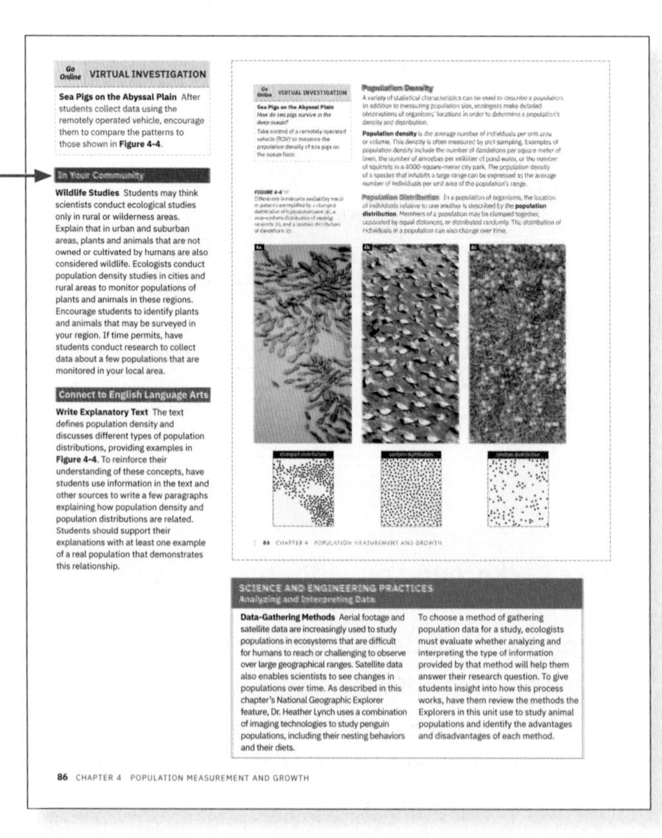

Go Online VIRTUAL INVESTIGATION
Sea Pigs on the Abyssal Plain After students collect data using the remotely operated vehicle, encourage them to compare the patterns to those shown in **Figure 4-4**.

In Your Community
Wildlife Studies Students may think scientists conduct ecological studies only in rural or wilderness areas. Explain that in urban and suburban areas, plants and animals that are not owned or cultivated by humans are also considered wildlife. Ecologists conduct population density studies in cities and rural areas to monitor populations of plants and animals in these regions. Encourage students to identify plants and animals that may be surveyed in your region. If time permits, have students conduct research to collect data about a few populations that are monitored in your local area.

Connect to English Language Arts
Write Explanatory Text The text defines population density and discusses different types of population distributions, providing examples in **Figure 4-4**. To reinforce their understanding of these concepts, have students use information in the text and other sources to write a few paragraphs explaining how population density and population distributions are related. Students should support their explanations with at least one example of a real population that demonstrates this relationship.

SCIENCE AND ENGINEERING PRACTICES
Analyzing and Interpreting Data

Data-Gathering Methods Aerial footage and satellite data are increasingly used to study populations in ecosystems that are difficult for humans to reach or challenging to observe over large geographical ranges. Satellite data also enables scientists to see changes in populations over time. As described in this chapter's National Geographic Explorer feature, Dr. Heather Lynch uses a combination of imaging technologies to study penguin populations, including their nesting behaviors and their diets.

To choose a method of gathering population data for a study, ecologists must evaluate whether analyzing and interpreting the type of information provided by that method will help them answer their research question. To give students insight into how this process works, have them review the methods the Explorers in this unit use to study animal populations and identify the advantages and disadvantages of each method.

86 CHAPTER 4 POPULATION MEASUREMENT AND GROWTH

SUPPORT FOR NEXT GENERATION SCIENCE STANDARDS (NGSS)

Green- and blue-boxed features throughout the Teacher's Edition provide strategies for integrating aspects of NGSS pedagogy in the biology classroom.

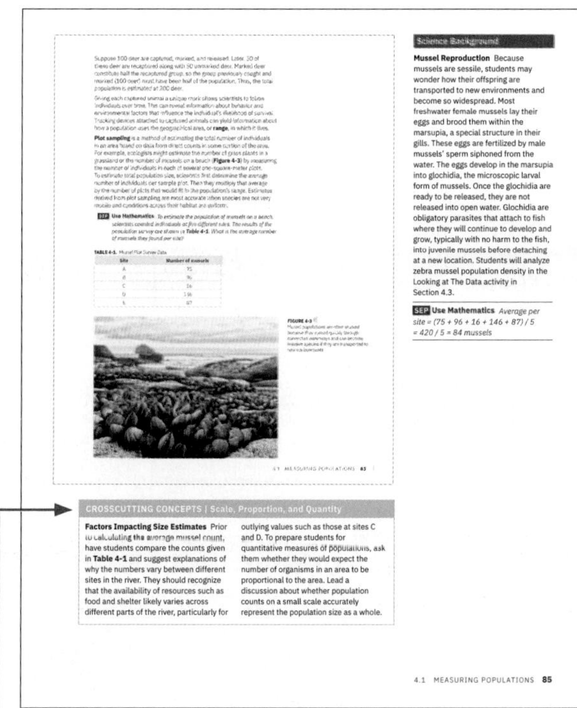

Science Background

Mussel Reproduction Because mussels are sessile, students may wonder how their offspring are transported to new environments and become so widespread. Most freshwater female mussels lay their eggs and brood them within the marsupia, a special structure in their gills. These eggs are fertilized by male mussels' sperm siphoned from the water. The eggs develop in the marsupia into glochidia, the microscopic larval form of mussels. Once the glochidia are ready to be released, they are not released into open water. Glochidia are obligatory parasites that attach to fish where they will continue to develop and grow, typically with no harm to the fish, into juvenile mussels before detaching at a new location. Students will analyze zebra mussel population density in the Looking at The Data activity in Section 4.3.

SEP Use Mathematics *Average per site = (75 + 96 + 16 + 146 + 87) / 5 = 420 / 5 = 84 mussels*

CROSSCUTTING CONCEPTS | Scale, Proportion, and Quantity

Factors Impacting Size Estimates Prior to calculating the average mussel count, have students compare the counts given in **Table 4-1** and suggest explanations of why the numbers vary between different sites in the river. They should recognize that the availability of resources such as food and shelter likely varies across different parts of the river, particularly for outlying values such as those at sites C and D. To prepare students for quantitative measures of populations, ask them whether they would expect the number of organisms in an area to be proportional to the area. Lead a discussion about whether population counts on a small scale accurately represent the population size as a whole.

4.1 MEASURING POPULATIONS **85**

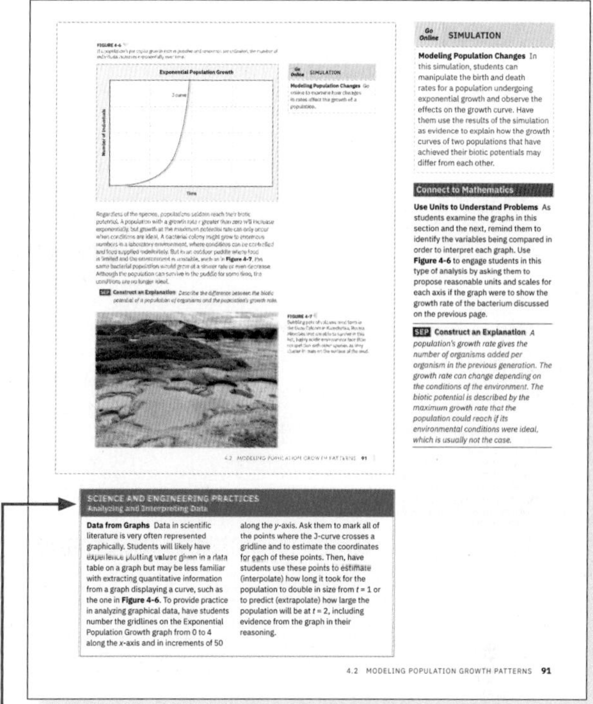

Go Online SIMULATION

Modeling Population Changes In this simulation, students can manipulate the birth and death rates for a population undergoing exponential growth and observe the effects on the growth curve. Have them use the results of the simulation as evidence to explain how the growth curves of two populations that have achieved their biotic potentials may differ from each other.

Connect to Mathematics

Use Units to Understand Problems As students examine the graphs in this section and the next, remind them to identify the variables being compared in order to interpret each graph. Use **Figure 4-6** to engage students in this type of analysis by asking them to propose reasonable units and scales for each axis if the graph were to show the growth rate of the bacterium discussed on the previous page.

SEP Construct an Explanation *A population's growth rate gives the number of organisms added per organism in the previous generation. The growth rate can change depending on the conditions of the environment. The biotic potential is described by the maximum growth rate that the population could reach if its environmental conditions were ideal, which is usually not the case.*

SCIENCE AND ENGINEERING PRACTICES
Analyzing and Interpreting Data

Data from Graphs Data in scientific literature is very often represented graphically. Students will likely have experience plotting values given in a data table on a graph but may be less familiar with extracting quantitative information from a graph displaying a curve, such as the one in **Figure 4-6**. To provide practice in analyzing graphical data, have students number the gridlines on the Exponential Population Growth graph from 0 to 4 along the x-axis and in increments of 50 along the y-axis. Ask them to mark all of the points where the J-curve crosses a gridline and to estimate the coordinates for each of these points. Then, have students use these points to estimate (interpolate) how long it took for the population to double in size from $t = 1$ or to predict (extrapolate) how large the population will be at $t = 2$, including evidence from the graph in their reasoning.

4.2 MODELING POPULATION GROWTH PATTERNS **91**

The green Crosscutting Concepts features support the use of one of the NGSS Crosscutting Concepts to help deepen students' understanding and connect with prior learning.

The blue **Science and Engineering Practices** features provide strategies and suggestions for using biological content in the text to engage students in one of the Science and Engineering Practices.

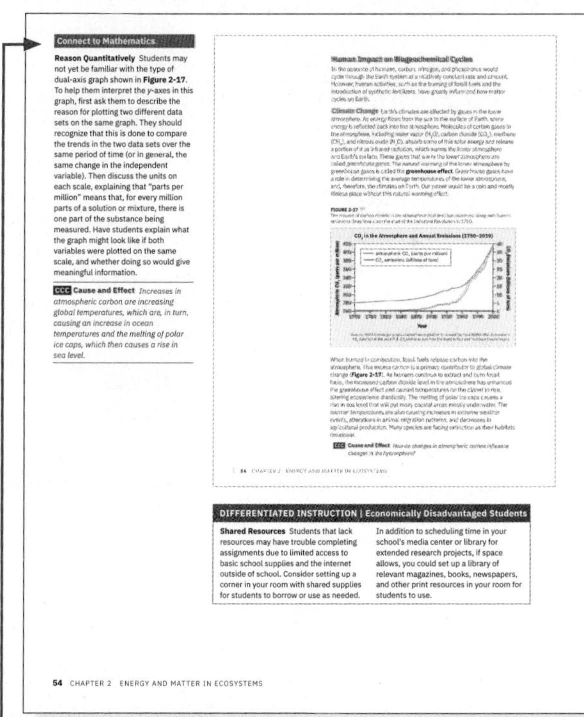

Connect to Mathematics

Reason Quantitatively Students may not yet be familiar with the type of dual-axis graph shown in **Figure 2-17**. To help them interpret the y-axes in this graph, first ask them to describe the reason for plotting two different data sets on the same graph. They should recognize that this is done to compare the trends in the two data sets over the same period of time (or in general, the same change in the independent variable). Then discuss the units on each scale, explaining that "parts per million" means that, for every million parts of a solution or mixture, there is one part of the substance being measured. Have students explain what the graph might look like if both variables were plotted on the same scale, and whether doing so would give meaningful information.

CCC Cause and Effect *Increases in atmospheric carbon are increasing global temperatures, which are, in turn, causing an increase in ocean temperatures and the melting of polar ice caps, which then causes a rise in sea level.*

DIFFERENTIATED INSTRUCTION | Economically Disadvantaged Students

Shared Resources Students that lack resources may have trouble completing assignments due to limited access to basic school supplies and the internet outside of school. Consider setting up a corner in your room with shared supplies for students to borrow or use as needed. In addition to scheduling time in your school's media center or library for extended research projects, if space allows, you could set up a library of relevant magazines, books, newspapers, and other print resources in your room for students to use.

54 CHAPTER 2 ENERGY AND MATTER IN ECOSYSTEMS

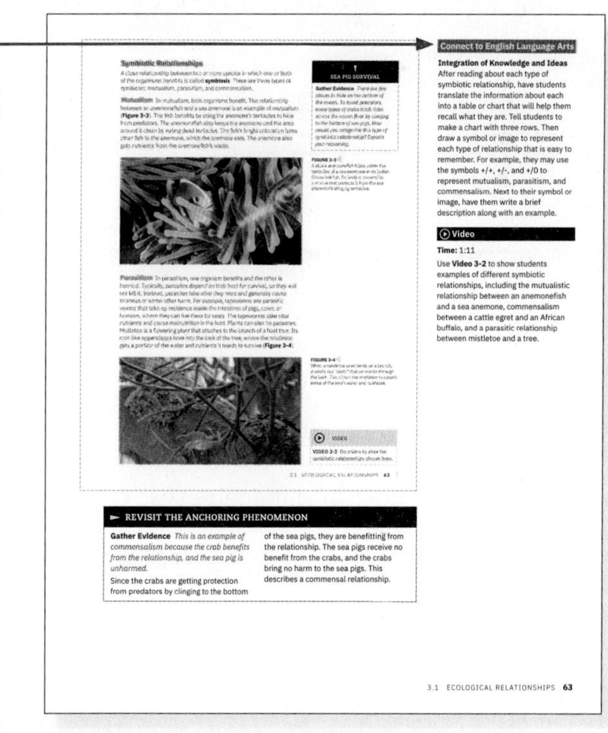

Connect to English Language Arts

Integration of Knowledge and Ideas After reading about each type of symbiotic relationship, have students translate the information about each into a table or chart that will help them recall what they are. Tell students to make a chart with three rows. Then draw a symbol or image to represent each type of relationship that is easy to remember. For example, they may use the symbols +/+, +/–, and +/0 to represent mutualism, parasitism, and commensalism. Next to their symbol or image, have them write a brief description along with an example.

Video

Time: 1:11

Use **Video 3-2** to show students examples of different symbiotic relationships, including the mutualistic relationship between an anemonefish and a sea anemone, commensalism between a cattle egret and an African buffalo, and a parasitic relationship between mistletoe and a tree.

REVISIT THE ANCHORING PHENOMENON

Gather Evidence *This is an example of commensalism because the crab benefits from the relationship, and the sea pig is unharmed.*

Since the crabs are getting protection from predators by clinging to the bottom of the sea pigs, they are benefiting from the relationship. The sea pigs receive no benefit from the crabs, and the crabs bring no harm to the sea pigs. This describes a commensal relationship.

3.1 ECOLOGICAL RELATIONSHIPS **63**

Connect to Mathematics and **Connect to English Language Arts** provide strategies for addressing the Common Core State Standards for Math and ELA that are aligned with each NGSS Performance Expectation.

3D ASSESSMENT RESOURCES

National Geographic Biology provides assessment in multiple formats throughout unit, chapter, and lab materials to support student learning and data-driven instructional strategies.

FORMATIVE ASSESSMENT

Formative assessments are comprised of Pretests, Checkpoint questions within the readings, Section Reviews, Looking at the Data, Tying It All Together projects, the CER-based Unit Activities, and Chapter Reviews. Assessment is incorporated into all Minilabs and Chapter Investigations.

Pretests provide diagnostic data for you to calibrate your lessons and customize individual assignments to address gaps and misconceptions. Teacher's Edition Section Openers provide alerts for which questions to use.

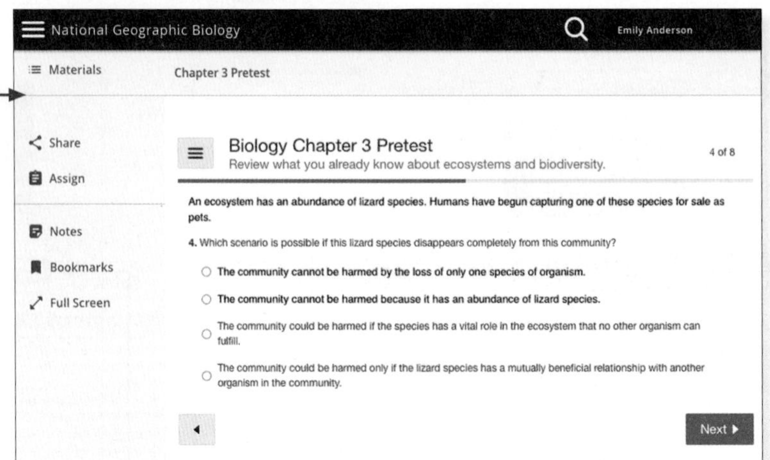

SUMMATIVE ASSESSMENT

Chapter Assessments offer a combination of open-response and machine-scored items carefully designed to measure students' understanding and retention of the content. **Unit Performance Tasks** assess bundled Performance Expectations.

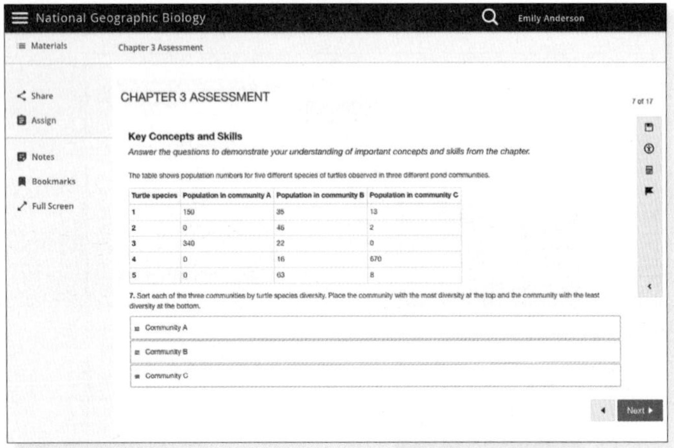

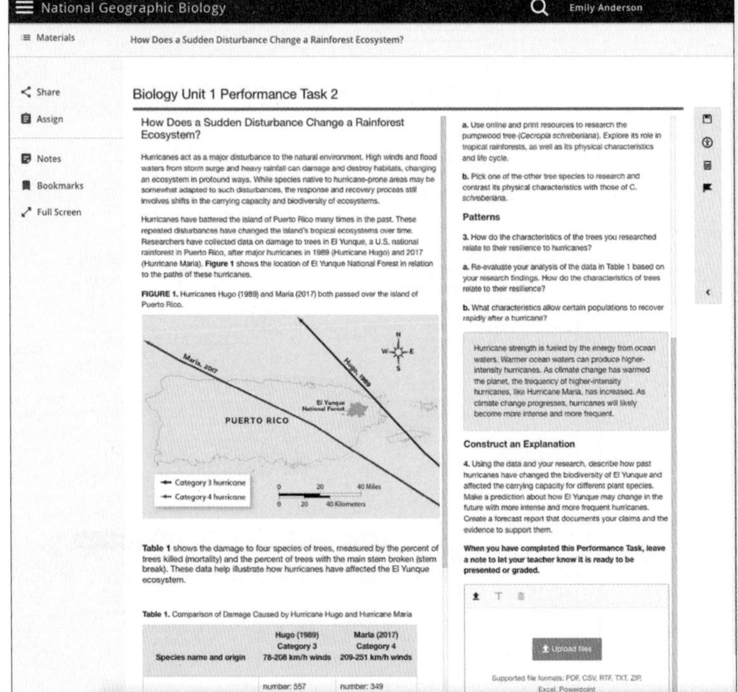

Cognero Test Bank is a flexible, online system that allows you to author, edit, and manage test content.

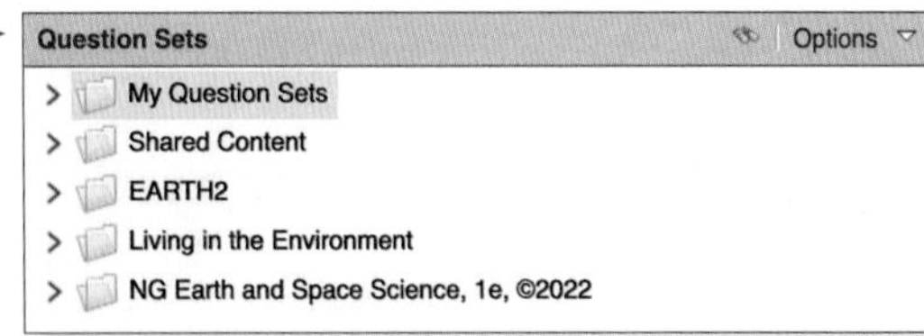

PACING GUIDE

Use the relative times shown as one tool to help prioritize segments of your course instruction and homework assignments in accordance with your curriculum needs. Timing is based on 50-minute periods or 90-minute blocks. Activities that do not require hands-on materials or advance preparation, such as Tying It All Together, have been built into the times for the sections they follow.

Chapter or Section Title	Periods	Blocks
Chapter 1 Introduction to Biology		
1.1 The Study of Life	1.0	0.5
Minilab Extracting DNA from Fruit	0.6	0.3
1.2 Constructing Explanations about the Natural World	1.0	0.5
1.3 Using Biology to Develop Solutions	1.0	0.5
Looking at the Data Mass Distribution of Mammals	0.5	0.3
Total Chapter 1	4.1	2.1
Unit 1 Relationships in Ecosystems	1.0	0.5
Chapter 2 Energy and Matter in Ecosystems		
2.1 Ecological Systems	1.0	0.5
2.2 Modeling the Transfer of Energy and Matter	1.0	0.5
2.3 Modeling Energy and Matter Distribution	1.0	0.5
Minilab Model a Biomass Pyramid	0.6	0.3
2.4 Cycling of Matter	1.0	0.3
Looking at the Data Biomagnification of Mercury	0.5	0.3
Total Chapter 2	5.1	2.4
Chapter 3 Biodiversity and Ecosystem Stability		
3.1 Ecological Relationships	1.0	0.5
3.2 Biodiversity	1.0	0.5
Looking at the Data The Biodiversity Conservation Paradox	0.5	0.3
3.3 Ecosystem Stability and Change	1.0	0.5
Minilab Observing Biodiversity in Pond Water	0.6	0.3
Total Chapter 3	4.1	2.1
Chapter 4 Population Measurement and Growth		
4.1 Measuring Populations	1.0	0.5
Minilab Mark-Recapture Sampling	0.6	0.3
4.2 Modeling Population Growth Patterns	1.0	0.5
4.3 Factors that Limit Population Growth	1.0	0.5
Looking at the Data Invasive Species Population Growth	0.5	0.3
Total Chapter 4	4.1	2.1
Unit 1 Activity	1.0	0.5

Chapter or Section Title	Periods	Blocks
Unit 2 Cell Systems	1.0	0.5
Chapter 5 Molecules in Living Systems		
5.1 Elements and Compounds	1.5	0.8
5.2 Water	1.0	0.5
Minilab Polar vs. Nonpolar Molecules	0.6	0.3
5.3 Carbon-Based Molecules	1.5	0.8
5.4 Chemical Reactions	1.5	0.8
Looking at the Data Digestive Enzymes and pH	0.5	0.3
Total Chapter 5	6.6	3.5
Chapter 6 Cell Structure and Function		
6.1 Cell Structures	3.0	1.5
Looking at the Data Microbiota of the Human Body	0.5	0.3
6.2 Cell Membranes	2.0	1.0
Minilab Selectively Permeable Membranes	0.6	0.3
6.3 Photosynthesis and Cellular Respiration	3.0	1.5
Total Chapter 6	9.1	4.6
Chapter 7 Cell Growth		
7.1 Cell Cycles	2.0	1.0
Looking at the Data Identifying Gene Mutations in Cancer Cells	0.5	0.3
7.2 Mitosis	1.5	0.8
Minilab Modeling Mitosis	0.6	0.3
7.3 Cell Differentiation	2.0	1.0
Total Chapter 7	6.6	3.4
Unit 2 Activity	1.0	0.5
Unit 3 Interactions in Living Systems	1.0	0.5
Chapter 8 Diversity of Living Systems		
8.1 Bacteria and Archaea	1.5	0.8
8.2 Protists	1.5	0.8
Minilab Features of Paramecium and Euglena	0.6	0.3
8.3 Fungi	1.0	0.5
Looking at the Data The C-Value Enigma	0.5	0.3
8.4 Viruses	1.5	0.8
Total Chapter 8	6.6	3.5

PACING GUIDE (continued)

Chapter or Section Title	Periods	Blocks
Chapter 9 Plant Systems		
9.1 Plant Origins	1.0	0.5
Minilab Investigating Leaf Stomata	0.6	0.3
9.2 Transport in Plants	1.0	0.5
9.3 Plant Growth and Reproduction	1.0	0.5
9.4 Plant Responses to the Environment	1.0	0.3
Looking at the Data Bud Burst and Flowering in a Changing Climate	0.5	0.3
Total Chapter 9	5.1	2.4
Chapter 10 Animal Systems		
10.1 Animal Diversity	1.0	0.5
10.2 Defining Animal Systems	1.0	0.5
Minilab Comparing Reaction Speed	0.6	0.3
10.3 Maintaining Homeostasis	1.0	0.5
Looking at the Data Thermoregulation	0.5	0.3
10.4 Animal Behavior	2.0	1.0
Total Chapter 10	6.1	3.1
Unit 3 Activity	1.0	0.5
Unit 4 Genetics	1.0	0.5
Chapter 11 DNA, RNA and Proteins		
11.1 Structure and Information	2.5	1.3
Minilab Modeling DNA Replication, Transcription, and Translation	0.6	0.3
11.2 Replication, Transcription, and Translation	3.0	1.5
11.3 Regulating Gene Expression	2.5	1.3
Looking at the Data Regulating Gene Expression	0.5	0.3
Total Chapter 11	9.1	4.7
Chapter 12 Genetic Variation and Heredity		
12.1 Meiosis	3.5	1.8
12.2 Mutations	1.5	0.8
12.3 Mendelian Inheritance	2.0	1.0
Looking at the Data Blood Type Compatibility	0.5	0.3
12.4 Other Patterns of Inheritance	2.0	1.0
Minilab Modeling Inheritance	0.6	0.3
Total Chapter 12	11.6	5.2
Chapter 13 Genetic Technologies		
13.1 Tools in Genetic Technology	1.5	0.8
13.2 Applications in Genetic Engineering	2.0	1.0
Looking at the Data Genetic Therapy Clinical Trials	0.5	0.3
13.3 Vaccine Development	2.0	1.0
Minilab Herd Immunity	0.6	0.3
Total Chapter 13	6.6	3.4
Unit 4 Activity	1.0	0.5

Chapter or Section Title	Periods	Blocks
Unit 5 Evolution and Changing Environments	1.0	0.5
Chapter 14 Evidence for Evolution		
14.1 Evolution of Life	2.5	1.3
Minilab Organizing Fossil Evidence	0.6	0.3
14.2 Fossil and Geological Evidence	2.5	1.3
Looking at the Data Forensic Radiometric Dating	0.5	0.3
14.3 Developmental, Anatomical, and Genetic Evidence	2.5	1.3
Total Chapter 14	8.6	4.5
Chapter 15 The Theory of Evolution		
15.1 Developing the Theory of Evolution by Natural Selection	3.0	1.5
Minilab Hawks and Mice	0.6	0.3
15.2 Evolution in Populations	2.0	1.0
Looking at the Data Tracking Evolution	0.5	0.3
15.3 Other Patterns in Population Genetics	1.5	0.8
Total Chapter 15	7.6	3.9
Chapter 16 Survival in Changing Environments		
16.1 Speciation	2.5	1.3
16.2 Extinction	2.0	1.0
16.3 Human Impact on the Environment	3.5	1.8
Looking at the Data Biodiversity and Deforestation	0.5	0.3
16.4 Reducing Human Impact on the Environment	2.5	1.3
Minilab Modeling Human-Caused Changes in the Environment	0.6	0.3
Total Chapter 16	11.6	6.0
Unit 5 Activity	1.0	0.5
Total for all Chapters	**121.1**	**61.9**

Chapter Investigations	Periods	Blocks
1A: Making Real-World Observations	1.0	0.5
1B: A Medicine Distribution Solution	2.0	1.0
2A: Salinity and Brine Shrimp Survival	3.0	1.5
2B: Exploring Brine Shrimp Survival	4.0	2.0
3A: Measuring Biodiversity Using Ecological Sampling Methods	2.0	0.0
3B: Ecological Succession in an Aquatic Community	3.0	1.5
4A: Population Growth of Duckweed	4.0	3.0
4B: Designing a Seed Trap	3.0	1.5
5A: Converting Carbohydrates	1.0	0.5
5B: Crime Scene Cleaners	4.0	2.0
6A: Factors Affecting Cellular Respiration	2.0	1.0
6B: Designing a Photobioreactor	4.0	2.0
7A: Plant Growth through Mitosis	0.5	0.3
7B: Cell Differentiation in Plant Leaves	1.0	0.5
8A: Classification Systems	2.0	1.0
8B: Effects of Antimicrobials	3.0	1.5

Chapter Investigations	Periods	Blocks
9A: Connecting Plant Structures with Their Functions	2.0	1.0
9B: Homeostasis in Plants	3.0	1.5
10A: The Effect of Exercise on Homeostasis	2.0	1.0
10B: Monitoring Animal Behavior	3.0	1.5
11A: Investigating the Building Blocks of Life	1.0	0.5
11B: Regulation of Gene Expression	2.0	1.0
12A: Design an Organism	1.0	0.5
12B: Mapping Fruit Fly Genes through Linkage	2.0	1.0

Chapter Investigations	Periods	Blocks
13A: DNA Evidence	2.0	1.0
13B: Fluorescent Genes	2.0	1.0
14A: Comparing Genetic Information Among Organisms	1.0	0.5
14B: What Lived Here?	1.0	0.5
15A: Genetic Drift	1.0	0.5
15B: Evolution of Antibiotic Resistance in Bacteria	2.5	1.5
16A: Modeling Speciation	2.0	1.0
16B: Wildlife Crossings and Corridors	2.0	1.0
Total for all Chapter Investigations	**69.0**	**34.8**

Unit	Virtual Investigations	Periods	Blocks
1	Sea Pigs on the Abyssal Plain	1.0	0.5
2	Bacteria in the Digestive System	1.0	0.5
3	Communication in the Rainforest	1.0	0.5
4	Fighting a Viral Pandemic	1.0	0.5
5	Hummingbirds on the Move	1.0	0.5
	Total for all Virtual Investigations	**5.0**	**2.5**

Unit	Performance Tasks	Periods	Blocks
1	**1:** Why Should We Preserve Wetland Ecosystems?	3.0	1.5
1	**2:** How Do Seasonal Changes Affect Organisms in a Freshwater Ecosystem?	3.0	1.5
1	**3:** How Does a Sudden Disturbance Change a Rainforest Ecosystem?	3.0	1.5
1	**4:** How Does Long-term Drought Change a Saltwater Ecosystem?	3.0	1.3
1	**5:** What Is the Best Way to Restore Habitat for Endangered Bats?	1.0	0.5
2	**1:** How Does Regenerative Medicine Reflect Nature?	3.0	1.5
2	**2:** What Are the Requirements for a Minimum Viable Ecosystem?	3.0	1.5
2	**3:** How Are Complex Carbon-based Molecules Built from Simple Atoms?	2.0	1.0
3	**1:** How Do Systems Interact to Maintain Homeostasis in Plants?	3.0	1.5
3	**2:** How Can We Test Systems that Interact to Maintain Homeostasis in Humans?	3.0	1.5
3	**3:** How Important Is Group Behavior to the Survival of Individuals in a Population?	3.0	0.0
4	**1:** How Does a Single-gene Trait Disappear and Reappear in a Subsequent Generation?	2.0	0.5
4	**2:** What Caused the Unusual Skin Discoloration in the People Living in a Rural Area?	2.0	1.0
4	**3:** What Are the Risks and Benefits of Genetically Engineered Food?	2.0	1.0
4	**4:** How Will We Curb the Spread of Mosquito-Borne Disease?	3.0	1.5
5	**1:** How Can We Determine Evolutionary Relationships?	3.0	1.5
5	**2:** How Does Bacterial Evolution Affect Public Health Globally?	2.0	1.0
5	**3:** How Is Climate Change Altering Species Evolution?	3.0	1.5
5	**4:**How Do Human-Induced Changes in the Environment Affect Different Species?	4.0	2.0
5	**5:** What Kind of Artificial Reef Is Most Effective at Preserving and Restoring Biodiversity?	2.0	1.0
	Total for all Performance Tasks	**53.0**	**24.3**

NEXT GENERATION SCIENCE STANDARDS

HS-LS1 From Molecules to Organisms: Structures and Processes

Performance Expectation	
HS-LS1-1. Construct an explanation based on evidence for how the structure of DNA determines the structure of proteins which carry out the essential functions of life through systems of specialized cells. **Assessment Boundary:** Assessment does not include identification of specific cell or tissue types, whole body systems, specific protein structures and functions, or the biochemistry of protein synthesis.	**Student Edition:** 343, 345 (#14) **Online:** Chapter Assessment 5 (#13), 7 (#12, 14) Unit 4 Performance Task 1
Disciplinary Core Ideas	
LS1.A: Structure and Function Systems of specialized cells within organisms help them perform the essential functions of life.	**Student Edition:** 104–105, 159, 193–196, 199–201, 203, 227, 234, 250–255, 257–261, 279, 308–309 **Online:** Chapter Investigation 7B Chapter Assessments 7, 10
All cells contain genetic information in the form of DNA molecules. Genes are regions in the DNA that contain the instructions that code for the formation of proteins, which carry out most of the work of cells.	**Student Edition**: 6, 26–27, 184–186, 193, 200–201, 316–324, 326–333, 335–341, 344–345, 382–389, 419 **Online:** Chapter Investigations 11A, 11B Chapter Assessments 1, 6, 7, 11
Science and Engineering Practices	
Constructing Explanations and Designing Solutions Construct an explanation based on valid and reliable evidence obtained from a variety of sources (including students' own investigations, models, theories, simulations, peer review) and the assumption that theories and laws that describe the natural world operate today as they did in the past and will continue to do so in the future.	**Student Edition:** 6, 70, 107, 119–123, 127, 130–131, 146–147, 192–193, 189–191, 200–201, 203, 229–231, 246–247, 348–350, 457–461, 465, 476–480 **Online:** Chapter Investigations 2A, 2B, 7A Chapter Assessments 5, 6, 7
Crosscutting Concepts	
Structure and Function Investigating or designing new systems or structures requires a detailed examination of the properties of different materials, the structures of different components, and connections of components to reveal its function and/or solve a problem.	**Student Edition:** 113, 131–132, 146–147, 194–196, 200–201, 203, 245, 318–319 **Online:** Chapter Investigations 7A, 7B, 8B Chapter Assessments 5, 7

HS-LS1 From Molecules to Organisms: Structures and Processes

Performance Expectation	
HS-LS1-2. Develop and use a model to illustrate the hierarchical organization of interacting systems that provide specific functions within multicellular organisms. **Clarification Statement:** Emphasis is on functions at the organism system level such as nutrient uptake, water delivery, and organism movement in response to neural stimuli. An example of an interacting system could be an artery depending on the proper function of elastic tissue and smooth muscle to regulate and deliver the proper amount of blood within the circulatory system. **Assessment Boundary:** Assessment Boundary: Assessment does not include interactions and functions at the molecular or chemical reaction level.	**Student Edition:** 203 **Online:** Chapter Investigations 9A, 10A Unit 3 Performance Task 1 Unit 3 Performance Task 2
Disciplinary Core Ideas	
LS1.A: Structure and Function Multicellular organisms have a hierarchical structural organization, in which any one system is made up of numerous parts and is itself a component of the next level.	**Student Edition:** 4–5, 178–179, 227, 250–256, 263–269, 271–273, 278–279, 282–284, 286–290, 308–309 **Online:** Chapter Investigations 9A, 10A Chapter Assessments 9, 10 Unit 3 Performance Task 1

Science and Engineering Practices	
Developing and Using Models Develop and use a model based on evidence to illustrate the relationships between systems or between components of a system.	**Student Edition:** 26–27, 37, 39–42, 46, 50, 55, 57–59, 107, 110–111, 117–118, 146–147, 169–175, 184–186, 192, 200–201, 245, 287–291, 307, 311, 334, 345, 376–377, 416, 474–475, 482–483, 516 **Online:** Chapter Investigations 6A, 6B, 7B, 9A, 10A, 16A Chapter Assessments 5, 7
Crosscutting Concepts	
Systems and System Models Models (e.g., physical, mathematical, computer models) can be used to simulate systems and interactions—including energy, matter, and information flows—within and between systems at different scales.	**Student Edition:** 12–14, 41, 46, 55, 58–59, 184–186, 200–201, 287–290, 311, 482–483, 516 **Online:** Chapter Investigations 1A, 9A, 10A, 16B Chapter Assessments 7, 13 Unit 1 Performance Task 1 Unit 3 Performance Task 1 Unit 5 Performance Task 5

HS-LS1 From Molecules to Organisms: Structures and Processes

Performance Expectation	
HS-LS1-3. Plan and conduct an investigation to provide evidence that feedback mechanisms maintain homeostasis. **Clarification Statement:** Examples of investigations could include heart rate response to exercise, stomate response to moisture and temperature, and root development in response to water levels. **Assessment Boundary:** Assessment does not include the cellular processes involved in the feedback mechanism.	**Student Edition:** 279 (#13) **Online:** Chapter Investigations 9B, 10A Unit 3 Performance Task 2
Disciplinary Core Ideas	
LS1.A: Structure and Function Feedback mechanisms maintain a living system's internal conditions within certain limits and mediate behaviors, allowing it to remain alive and functional even as external conditions change within some range. Feedback mechanisms can encourage (through positive feedback) or discourage (negative feedback) what is going on inside the living system.	**Student Edition:** 12–14, 162–168, 178–179, 256–261, 271–275, 278–279, 292–295, 308–309 **Online:** Chapter Investigations 9B, 10A Chapter Assessments 1, 6, 9, 10 Unit 3 Performance Task 1 Unit 3 Performance Task 2
Science and Engineering Practices	
Planning and Carrying Out Investigations Plan and conduct an investigation individually and collaboratively to produce data to serve as the basis for evidence, and in the design: decide on types, how much, and accuracy of data needed to produce reliable measurements and consider limitations on the precision of the data (e.g., number of trials, cost, risk, time), and refine the design accordingly.	**Student Edition:** 26–27, 57 **Online:** Chapter Investigations 1A, 3B, 5B, 8A, 8B, 9B, 10A Unit 3 Performance Task 2
Scientific Investigations Use a Variety of Methods Scientific inquiry is characterized by a common set of values that include: logical thinking, precision, open-mindedness, objectivity, skepticism, replicability of results, and honest and ethical reporting of findings.	**Student Edition:** 12–15, 462–464 **Online:** Chapter Investigation 10A
Crosscutting Concepts	
Stability and Change Feedback (negative or positive) can stabilize or destabilize a system.	**Student Edition:** 162–163, 257–261, 292–294 **Online:** Chapter Investigations 9B, 10A Unit 3 Performance Task 1 Unit 3 Performance Task 2

HS-LS1 From Molecules to Organisms: Structures and Processes

Performance Expectation	
HS-LS1-4. Use a model to illustrate the role of cellular division (mitosis) and differentiation in producing and maintaining complex organisms. **Assessment Boundary:** Assessment does not include specific gene control mechanisms or rote memorization of the steps of mitosis.	**Online:** Chapter Investigations 7A, 7B Unit 2 Performance Task 1
Disciplinary Core Ideas	
LS1.B: Growth and Development of Organisms In multicellular organisms individual cells grow and then divide via a process called mitosis, thereby allowing the organism to grow. The organism begins as a single cell (fertilized egg) that divides successively to produce many cells, with each parent cell passing identical genetic material (two variants of each chromosome pair) to both daughter cells. Cellular division and differentiation produce and maintain a complex organism, composed of systems of tissues and organs that work together to meet the needs of the whole organism.	**Student Edition:** 181–183, 189–191, 193–196, 200–201, 220–225, 263–269 **Online:** Chapter Investigations 7A, 7B Chapter Assessment 7 Unit 1 Performance Task 1
Science and Engineering Practices	
Developing and Using Models Use a model based on evidence to illustrate the relationships between systems or between components of a system.	**Student Edition:** 26–27, 37, 39–42, 46, 50, 55, 57–59, 107, 110–111, 117–118, 146–147, 169–175, 184–186, 192, 200–201, 245, 287–291, 307, 311, 334, 345, 376–377, 416, 474–475, 482–483, 516 **Online:** Chapter Investigations 6A, 6B, 7B, 9A, 10A, 16A Chapter Assessments 2, 5, 7
Crosscutting Concepts	
Systems and System Models Models (e.g., physical, mathematical, computer models) can be used to simulate systems and interactions—including energy, matter, and information flows—within and between systems at different scales.	**Student Edition:** 12–14, 41, 46, 55, 58–59, 184–186, 200–201, 287–290, 311, 482–483, 516 **Online:** Chapter Investigations 1A, 9A, 10A, 16B Chapter Assessments 5, 7, 13 Unit 1 Performance Task 1 Unit 3 Performance Task 1 Unit 5 Performance Task 5

HS-LS1 From Molecules to Organisms: Structures and Processes

Performance Expectation	
HS-LS1-5. Use a model to illustrate how photosynthesis transforms light energy into stored chemical energy. **Clarification Statement:** Emphasis is on illustrating inputs and outputs of matter and the transfer and transformation of energy in photosynthesis by plants and other photosynthesizing organisms. Examples of models could include diagrams, chemical equations, and conceptual models. **Assessment Boundary:** Assessment does not include specific biochemical steps.	**Online:** Chapter Investigation 6B Unit 2 Performance Task 2
Disciplinary Core Ideas	
LS1.C: Organization for Matter and Energy Flow in Organisms The process of photosynthesis converts light energy to stored chemical energy by converting carbon dioxide plus water into sugars plus released oxygen.	**Student Edition:** 120, 169–170, 174–177, 178–179, 212–214, 219–220 **Online:** Chapter Investigation 6B Chapter Assessments 2, 6 Unit 2 Performance Task 2

Science and Engineering Practices	
Developing and Using Models Use a model based on evidence to illustrate the relationships between systems or between components of a system.	**Student Edition:** 26–27, 37, 39–42, 46, 50, 55, 57–59, 107, 110–111, 117–118, 146–147, 169–175, 184–186, 192, 200–201, 245, 287–291, 307, 311, 334, 345, 376–377, 416, 474–475, 482–483, 516 **Online:** Chapter Investigations 6A, 6B, 7B, 9A, 10A, 16A Chapter Assessments 2, 5, 7
Crosscutting Concepts	
Energy and Matter Changes of energy and matter in a system can be described in terms of energy and matter flows into, out of, and within that system.	**Student Edition:** 34–37, 39–41, 46, 55, 58–59, 126–127, 146–147, 424–425 **Online:** Chapter Investigations 5A, 5B, 6A, 6B Chapter Assessments 2, 5

HS-LS1 From Molecules to Organisms: Structures and Processes

Performance Expectation	
HS-LS1-6. Construct and revise an explanation based on evidence for how carbon, hydrogen, and oxygen from sugar molecules may combine with other elements to form amino acids and/or other large carbon-based molecules. **Clarification Statement:** Emphasis is on using evidence from models and simulations to support explanations. **Assessment Boundary:** Assessment does not include the details of the specific chemical reactions or identification of macromolecules.	**Student Edition:** 147 (#14) **Online:** Chapter Investigations 5A, 5B Unit 2 Performance Task 3
Disciplinary Core Ideas	
LS1.C: Organization for Matter and Energy Flow in Organisms The sugar molecules thus formed contain carbon, hydrogen, and oxygen: their hydrocarbon backbones are used to make amino acids and other carbon-based molecules that can be assembled into larger molecules (such as proteins or DNA), used for example to form new cells.	**Student Edition:** 120, 128–134, 146–147, 344–345 **Online:** Chapter Investigations 5A, 5B Chapter Assessment 2
As matter and energy flow through different organizational levels of living systems, chemical elements are recombined in different ways to form different products.	**Student Edition:** 112, 126–127, 136, 146–147, 175, 212–214, 219–220, 229–231 **Online:** Chapter Investigations 2A, 2B, 7A Chapter Assessments 5, 6, 7
Science and Engineering Practices	
Constructing Explanations and Designing Solutions Construct and revise an explanation based on valid and reliable evidence obtained from a variety of sources (including students' own investigations, models, theories, simulations, peer review) and the assumption that theories and laws that describe the natural world operate today as they did in the past and will continue to do so in the future.	**Student Edition:** 6, 70, 107, 119–123, 127, 130–131, 146–147, 192–193, 189–191, 200–201, 203, 229–231, 246–247, 348–350, 457–461, 465, 476–480 **Online:** Chapter Investigations 2A, 2B, 7A Chapter Assessments 5, 6, 7
Crosscutting Concepts	
Energy and Matter Changes of energy and matter in a system can be described in terms of energy and matter flows into, out of, and within that system.	**Student Edition:** 34–37, 39–41, 46, 55, 58–59, 126–127, 146–147, 424–425 **Online:** Chapter Investigations 5A, 5B, 6A, 6B Chapter Assessments 2, 5 Unit 1 Performance Task 1 Unit 2 Performance Task 2

HS-LS1 From Molecules to Organisms: Structures and Processes

Performance Expectation	
HS-LS1-7. Use a model to illustrate that cellular respiration is a chemical process whereby the bonds of food molecules and oxygen molecules are broken and the bonds in new compounds are formed, resulting in a net transfer of energy. **Clarification Statement:** Clarification Statement: Emphasis is on the conceptual understanding of the inputs and outputs of the process of cellular respiration. **Assessment Boundary:** Assessment should not include identification of the steps or specific processes involved in cellular respiration.	**Online:** Chapter Investigation 6A Unit 2 Performance Task 2
Disciplinary Core Ideas	
LS1.C: Organization for Matter and Energy Flow in Organisms As matter and energy flow through different organizational levels of living systems, chemical elements are recombined in different ways to form different products.	**Student Edition:** 112, 146–147, 175, 212–214, 219–220, 229–231 **Online:** Chapter Investigation 5A Chapter Assessments 2, 5
As a result of these chemical reactions, energy is transferred from one system of interacting molecules to another. Cellular respiration is a chemical process in which the bonds of food molecules and oxygen molecules are broken and new compounds are formed that can transport energy to muscles. Cellular respiration also releases the energy needed to maintain body temperature despite ongoing energy transfer to the surrounding environment.	**Student Edition:** 169–175, 178–179 **Online:** Chapter Investigation 5A Chapter Assessment 6 Unit 2 Performance Task 2
Science and Engineering Practices	
Developing and Using Models Use a model based on evidence to illustrate the relationships between systems or between components of a system.	**Student Edition:** 26–27, 37, 39–42, 46, 50, 55, 58–59, 107, 110–111, 117–118, 146–147, 169–175, 184–186, 200–201, 287–291 **Online:** Chapter Investigations 6A 6B, 7B, 9A, 10A, 16A Chapter Assessments 2, 5, 7
Crosscutting Concepts	
Energy and Matter Energy cannot be created or destroyed—it only moves between one place and another place, between objects and/or fields, or between systems.	**Student Edition:** 36–37, 43–49, 58–59, 139, 146–147 **Online:** Chapter Investigation 6A Chapter Assessment 2 Unit 1 Performance Task 2

HS-LS2 Ecosystems: Interactions, Energy, and Dynamics

Performance Expectation	
HS-LS2-1. Use mathematical and/or computational representations to support explanations of factors that affect carrying capacity of ecosystems at different scales. **Clarification Statement:** Emphasis is on quantitative analysis and comparison of the relationships among interdependent factors including boundaries, resources, climate, and competition. Examples of mathematical comparisons could include graphs, charts, histograms, and population changes gathered from simulations or historical data sets. **Assessment Boundary:** Assessment does not include deriving mathematical equations to make comparisons.	**Student Edition:** 105 (#6, 15) **Online:** Chapter Investigation 4A Unit 1 Performance Task 3
Disciplinary Core Ideas	
LS2.A: Interdependent Relationships in Ecosystems Ecosystems have carrying capacities, which are limits to the numbers of organisms and populations they can support. These limits result from such factors as the availability of living and nonliving resources and from such challenges such as predation, competition, and disease. Organisms would have the capacity to produce populations of great size were it not for the fact that environments and resources are finite. This fundamental tension affects the abundance (number of individuals) of species in any given ecosystem.	**Student Edition:** 33, 58–59, 62, 64–65, 70, 77, 82–83, 86–87, 90–99, 103–105, 278–279, 500–501 **Online:** Chapter Investigations 2A, 2B Chapter Assessments 4, 12 Unit 1 Performance Task 3 Unit 1 Performance Task 4

Science and Engineering Practices	
Using Mathematics and Computational Thinking Use mathematical and/or computational representations of phenomena or design solutions to support explanations.	**Student Edition:** 44–46, 48, 50, 55, 58–59, 70–71, 80–81, 84–85, 88–96, 99, 102, 104–105, 114–116, 124, 144, 146–147, 159, 234, 378, 484–485, 517, 520 **Online:** Chapter Investigations 2A, 2B, 3A, 4A Chapter Assessments 2, 4, 5 Unit 1 Performance Task 1 Unit 1 Performance Task 2 Unit 1 Performance Task 3 Unit 4 Performance Task 3 Unit 5 Performance Task 2
Crosscutting Concepts	
Scale, Proportion, and Quantity The significance of a phenomenon is dependent on the scale, proportion, and quantity at which it occurs.	**Student Edition:** 86–87, 93, 99, 234, 346, 418–419, 502–50 **Online:** Chapter Investigation 1A Chapter Assessment 5 Unit 1 Performance Task 3

HS-LS2 Ecosystems: Interactions, Energy, and Dynamics

Performance Expectation	
HS-LS2-2. Use mathematical representations to support and revise explanations based on evidence about factors affecting biodiversity and populations in ecosystems of different scales. **Clarification Statement:** Examples of mathematical representations include finding the average, determining trends, and using graphical comparisons of multiple sets of data. **Assessment Boundary:** Assessment is limited to provided data.	**Student Edition:** 105 (#2, 6, 15) **Online:** Chapter Investigations 2A, 2B, 3A, 4A Unit 1 Performance Task 2 Unit 1 Performance Task 3 Unit 1 Performance Task 4
Disciplinary Core Ideas	
LS2.A: Interdependent Relationships in Ecosystems Ecosystems have carrying capacities, which are limits to the numbers of organisms and populations they can support. These limits result from such factors as the availability of living and nonliving resources and from such challenges such as predation, competition, and disease. Organisms would have the capacity to produce populations of great size were it not for the fact that environments and resources are finite. This fundamental tension affects the abundance (number of individuals) of species in any given ecosystem.	**Student Edition:** 33, 58–59, 62, 64–65, 70, 77, 82–83, 86–87, 90–99, 103–105, 278–279, 500–501 **Online:** Chapter Investigations 2A, 2B Chapter Assessments 4, 12
LS2.C: Ecosystem Dynamics, Functioning, and Resilience A complex set of interactions within an ecosystem can keep its numbers and types of organisms relatively constant over long periods of time under stable conditions. If a modest biological or physical disturbance to an ecosystem occurs, it may return to its more or less original status (i.e., the ecosystem is resilient), as opposed to becoming a very different ecosystem. Extreme fluctuations in conditions or the size of any population, however, can challenge the functioning of ecosystems in terms of resources and habitat availability.	**Student Edition:** 28–29, 41, 46, 58–5, 60–64, 66–67, 70, 72–78, 80–81, 86–99, 103, 176, 179, 203, 206–207, 248–249, 256, 270, 276–277, 311, 344–345, 426 **Online:** Chapter Investigations 3A, 3B, 4A Chapter Assessments 3, 4 Unit 5 Performance Task 5
Science and Engineering Practices	
Using Mathematics and Computational Thinking Use mathematical representations of phenomena or design solutions to support and revise explanations.	**Student Edition:** 44–46, 48, 50, 55, 58–59, 70–71, 80–81, 84–85, 88–96, 99, 102, 104–105, 114–116, 124, 144, 146–147, 159, 234, 378, 484–485, 517, 520 **Online:** Chapter Investigations 2A, 2B, 3A, 4A Chapter Assessments 2, 4, 5 Unit 1 Performance Task 1 Unit 1 Performance Task 2 Unit 1 Performance Task 3 Unit 4 Performance Task 3 Unit 5 Performance Task 2

Crosscutting Concepts	
Scale, Proportion, and Quantity Using the concept of orders of magnitude allows one to understand how a model at one scale relates to a model at another scale.	**Student Edition:** 37, 41, 46, 54–55, 58–59, 88, 104–105, 234 **Online:** Chapter Investigations 1A, 2A, 2B, 3A, 4A

HS-LS2 Ecosystems: Interactions, Energy, and Dynamics

Performance Expectation	
HS-LS2-3. Construct and revise an explanation based on evidence for the cycling of matter and flow of energy in aerobic and anaerobic conditions. **Clarification Statement:** Emphasis is on conceptual understanding of the role of aerobic and anaerobic respiration in different environments. **Assessment Boundary:** Assessment does not include the specific chemical processes of either aerobic or anaerobic respiration.	**Student Edition:** 107, 203 **Online:** Unit 1 Performance Task 2
Disciplinary Core Ideas	
LS2.B: Cycles of Matter and Energy Transfer in Ecosystems Photosynthesis and cellular respiration (including anaerobic processes) provide most of the energy for life processes.	**Student Edition:** 36–40, 58–59, 107, 169–175, 203, 220–225, 227 **Online:** Unit 1 Performance Task 2
Science and Engineering Practices	
Constructing Explanations and Designing Solutions Construct and revise an explanation based on valid and reliable evidence obtained from a variety of sources (including students' own investigations, models, theories, simulations, peer review) and the assumption that theories and laws that describe the natural world operate today as they did in the past and will continue to do so in the future.	**Student Edition:** 6, 70, 107, 119–123, 127, 130–131, 146–147, 192–193, 189–191, 200–201, 203, 229–231, 246–247, 348–350, 457–461, 465, 476–480 **Online:** Chapter Investigations 2A, 2B, 7A Chapter Assessments 5, 6, 7
Scientific Knowledge is Open to Revision in Light of New Evidence Most scientific knowledge is quite durable, but is, in principle, subject to change based on new evidence and/or reinterpretation of existing evidence.	**Student Edition:** 4–5, 14–15, 26–27, 66, 78–79,104–105, 210, 218, 246–247, 281,458–465 **Online:** Chapter Investigation 8A Chapter Assessment 8
Crosscutting Concepts	
Energy and Matter Energy drives the cycling of matter within and between systems.	**Student Edition:** 36–37, 39–41, 49, 58–59, 107, 164–167, 169–175, 203, 21–214 **Online:** Unit 1 Performance Task 2

HS-LS2 Ecosystems: Interactions, Energy, and Dynamics

Performance Expectation	
HS-LS2-4. Use mathematical representations to support claims for the cycling of matter and flow of energy among organisms in an ecosystem. **Clarification Statement:** Emphasis is on using a mathematical model of stored energy in biomass to describe the transfer of energy from one trophic level to another and that matter and energy are conserved as matter cycles and energy flows through ecosystems. Emphasis is on atoms and molecules such as carbon, oxygen, hydrogen and nitrogen being conserved as they move through an ecosystem. **Assessment Boundary:** Assessment is limited to proportional reasoning to describe the cycling of matter and flow of energy.	**Student Edition:** 56 (#6) **Online:** Chapter Assessment 2 (#14) Unit 1 Performance Task 1 Unit 1 Performance Task 2
Disciplinary Core Ideas	
LS2.B: Cycles of Matter and Energy Transfer in Ecosystems Plants or algae form the lowest level of the food web. At each link upward in a food web, only a small fraction of the matter consumed at the lower level is transferred upward, to produce growth and release energy in cellular respiration at the higher level. Given this inefficiency, there are generally fewer organisms at higher levels of a food web. Some matter reacts to release energy for life functions, some matter is stored in newly made structures, and much is discarded. The chemical elements that make up the molecules of organisms pass through food webs and into and out of the atmosphere and soil, and they are combined and recombined in different ways. At each link in an ecosystem, matter and energy are conserved.	**Student Edition:** 38–40, 43–46, 48–49, 55, 57–59, 204–205, 220–225 **Online:** Chapter Assessment 2 Unit 1 Performance Task 1 Unit 1 Performance Task 2

Science and Engineering Practices	
Using Mathematics and Computational Thinking Use mathematical representations of phenomena or design solutions to support claims.	**Student Edition:** 44–46, 48, 50, 55, 58–59, 70–71, 80–81, 84–85, 88–96, 99, 102, 104–105, 114–116, 124, 144, 146–147, 159, 234, 378, 484–485, 517, 520 **Online:** Chapter Investigations 2A, 2B, 3A, 4A Chapter Assessments 2, 4, 5 Unit 1 Performance Task 1 Unit 1 Performance Task 2 Unit 1 Performance Task 3 Unit 4 Performance Task 3 Unit 5 Performance Task 2
Crosscutting Concepts	
Energy and Matter Energy cannot be created or destroyed—it only moves between one place and another place, between objects and/or fields, or between systems.	**Student Edition:** 36–37, 43–49, 58–59, 139, 146–147 **Online:** Chapter Assessment 2

HS-LS2 Ecosystems: Interactions, Energy, and Dynamics

Performance Expectation	
HS-LS2-5. Develop a model to illustrate the role of photosynthesis and cellular respiration in the cycling of carbon among the biosphere, atmosphere, hydrosphere, and geosphere. **Clarification Statement:** Examples of models could include simulations and mathematical models. **Assessment Boundary:** Assessment does not include the specific chemical steps of photosynthesis and respiration.	**Student Edition:** 57, 179 (#15) **Online:** Unit 1 Performance Task 1
Disciplinary Core Ideas	
LS2.B: Cycles of Matter and Energy Transfer in Ecosystems Photosynthesis and cellular respiration are important components of the carbon cycle, in which carbon is exchanged among the biosphere, atmosphere, oceans, and geosphere through chemical, physical, geological, and biological processes.	**Student Edition:** 46, 51, 55, 58–59 **Online:** Chapter Assessment 2 Unit 1 Performance Task 1
PS3.D: Energy in Chemical Processes The main way that solar energy is captured and stored on Earth is through the complex chemical process known as photosynthesis. (secondary)	**Student Edition:** 37, 39, 169, 174 **Online:** Chapter Assessment 2
Science and Engineering Practices	
Developing and Using Models Develop a model based on evidence to illustrate the relationships between systems or components of a system.	**Student Edition:** 26–27, 37, 39–42, 46, 50, 55, 57–59, 107, 110–111, 117–118, 146–147, 169–175, 184–186, 192, 200–201, 245, 287–291, 307, 311, 334, 345, 376–377, 416, 474–475, 482–483, 516 **Online:** Chapter Investigations 6A ,6B, 7B, 9A, 10A, 16A Chapter Assessments 2, 5, 7
Crosscutting Concepts	
Systems and System Models Models (e.g., physical, mathematical, computer models) can be used to simulate systems and interactions—including energy, matter, and information flows—within and between systems at different scales.	**Student Edition:** 12–14, 41, 46, 55, 58–59, 184–186, 200–201, 287–290, 311, 482–483, 516 **Online:** Chapter Investigations 1A, 9A, 10A, 16B Chapter Assessments 5, 7, 13 Unit 1 Performance Task 1 Unit 3 Performance Task 1 Unit 5 Performance Task 5

HS-LS2 Ecosystems: Interactions, Energy, and Dynamics

Performance Expectation	
HS-LS2-6. Evaluate claims, evidence, and reasoning that the complex interactions in ecosystems maintain relatively consistent numbers and types of organisms in stable conditions, but changing conditions may result in a new ecosystem. **Clarification Statement:** Examples of changes in ecosystem conditions could include modest biological or physical changes, such as moderate hunting or a seasonal flood; and extreme changes, such as volcanic eruption or sea level rise.	**Student Edition:** 311 **Online:** Chapter Investigation 3B Unit 1 Performance Task 4
Disciplinary Core Ideas	
LS2.C: Ecosystem Dynamics, Functioning, and Resilience A complex set of interactions within an ecosystem can keep its numbers and types of organisms relatively constant over long periods of time under stable conditions. If a modest biological or physical disturbance to an ecosystem occurs, it may return to its more or less original status (i.e., the ecosystem is resilient), as opposed to becoming a very different ecosystem. Extreme fluctuations in conditions or the size of any population, however, can challenge the functioning of ecosystems in terms of resources and habitat availability.	**Student Edition:** 328–29, 41, 46, 58–5, 60–64, 66–67, 70, 72–78, 80–81, 86–99, 103, 176, 179, 203, 206–207, 248–249, 256, 270, 276–277, 311, 344–345, 426 **Online:** Chapter Investigations 3A, 3B, 4A Chapter Assessments 3, 4 Unit 5 Performance Task 5
Science and Engineering Practices	
Engaging in Argument from Evidence Scientific argumentation is a mode of logical discourse used to clarify the strength of relationships between ideas and evidence that may result in revision of an explanation.	**Student Edition:** 65, 78, 80–81, 296–300 **Online:** Chapter Investigations 3B, 10B, 16A
Scientific Knowledge is Open to Revision in Light of New Evidence Scientific argumentation is a mode of logical discourse used to clarify the strength of relationships between ideas and evidence that may result in revision of an explanation.	**Student Edition:** 12–15, 78, 518–519 **Online:** Chapter Investigation 3B
Crosscutting Concepts	
Stability and Change Much of science deals with constructing explanations of how things change and how they remain stable.	**Student Edition:** 30–31, 61, 65, 71–81, 124, 204–207, 424–425, 482–483, 518–519 **Online:** Chapter Investigations 3A, 3B Chapter Assessments 3, 16

HS-LS2 Ecosystems: Interactions, Energy, and Dynamics

Performance Expectation	
HS-LS2-7. Design, evaluate, and refine a solution for reducing the impacts of human activities on the environment and biodiversity.* **Clarification Statement:** Examples of human activities can include urbanization, building dams, and dissemination of invasive species.	**Student Edition:** 27 (#2), 417 **Online:** Chapter Investigation 16B Unit 1 Performance Task 5
Disciplinary Core Ideas	
LS2.C: Ecosystem Dynamics, Functioning, and Resilience Moreover, anthropogenic changes (induced by human activity) in the environment—including habitat destruction, pollution, introduction of invasive species, overexploitation, and climate change—can disrupt an ecosystem and threaten the survival of some species.	**Student Edition:** 30–31, 42, 54–55, 65, 71–73, 77, 80–81,102–103, 206–207, 277, 381, 492–493, 496–498, 500–508 **Online:** Chapter Investigation 16B Chapter Assessments 3, 16 Unit 5 Performance Task 5
LS4.D: Biodiversity and Humans Biodiversity is increased by the formation of new species (speciation) and decreased by the loss of species (extinction). (secondary)	**Student Edition:** 370, 486–488, 492, 517–519
Humans depend on the living world for the resources and other benefits provided by biodiversity. But human activity is also having adverse impacts on biodiversity through overpopulation, overexploitation, habitat destruction, pollution, introduction of invasive species, and climate change. Thus sustaining biodiversity so that ecosystem functioning and productivity are maintained is essential to supporting and enhancing life on Earth. Sustaining biodiversity also aids humanity by preserving landscapes of recreational or inspirational value. (secondary) (Note: This Disciplinary Core Idea is also addressed by HS-LS4-6.)	**Student Edition:** 65, 68, 70, 72–73, 232–233, 492–508, 510–515

Disciplinary Core Ideas	
ETS1.B: Developing Possible Solutions When evaluating solutions it is important to take into account a range of constraints including cost, safety, reliability and aesthetics and to consider social, cultural and environmental impacts. (secondary)	**Student Edition:** 18–21, 24, 314, 407–415, 413, 421 **Online:** Chapter Investigations 4B, 10B, 13A, 13B, 16B Chapter Assessment 13
Science and Engineering Practices	
Constructing Explanations and Designing Solutions Design, evaluate, and refine a solution to a complex real-world problem, based on scientific knowledge, student-generated sources of evidence, prioritized criteria, and tradeoff considerations.	**Student Edition:** 382–389, 513–514 **Online:** Chapter Investigations 4B, 10B, 13A, 13B
Crosscutting Concepts	
Stability and Change Much of science deals with constructing explanations of how things change and how they remain stable.	**Student Edition:** 30–31, 61, 65, 71–78, 80–81, 124, 204–207, 424–425 **Online:** Chapter Investigations 3A, 3B Chapter Assessment 3

HS-LS2 Ecosystems: Interactions, Energy, and Dynamics

Performance Expectation	
HS-LS2-8. Evaluate evidence for the role of group behavior on individual and species' chances to survive and reproduce. **Clarification Statement:** Emphasis is on: (1) distinguishing between group and individual behavior, (2) identifying evidence supporting the outcomes of group behavior, and (3) developing logical and reasonable arguments based on evidence. Examples of group behaviors could include flocking, schooling, herding, and cooperative behaviors such as hunting, migrating, and swarming.	**Student Edition:** 308 (#10, Revisit 1), 311 **Online:** Chapter Investigation 10B Unit 3 Performance Task 3
Disciplinary Core Ideas	
LS2.D: Social Interactions and Group Behavior Group behavior has evolved because membership can increase the chances of survival for individuals and their genetic relatives.	**Student Edition:** 296–304, 308–309, 311 **Online:** Chapter Investigation 10B Chapter Assessment 10 Unit 3 Performance Task 3
Science and Engineering Practices	
Engaging in Argument from Evidence Evaluate the evidence behind currently accepted explanations to determine the merits of arguments.	**Student Edition:** 65, 78, 80–81, 296–300 **Online:** Chapter Investigations 3B, 10B, 16A
Scientific Knowledge is Open to Revision in Light of New Evidence Scientific argumentation is a mode of logical discourse used to clarify the strength of relationships between ideas and evidence that may result in revision of an explanation.	**Student Edition:** 12–15, 78 **Online:** Chapter Investigation 3B
Crosscutting Concepts	
Cause and Effect Empirical evidence is required to differentiate between cause and correlation and make claims about specific causes and effects.	**Student Edition:** 54–55, 206–207, 295, 318–319, 364–365, 347, 378–379, 418–419, 458–464, 469–472, 476–479, 518–5196 **Online:** Chapter Investigations 9B, 10A, 11A, 11B, 15A, 15B, 16A Chapter Assessments 5, 12 Unit 4 Performance Task 2

HS-LS3 Heredity: Inheritance and Variation of Traits

Performance Expectation	
HS-LS3-1. Ask questions to clarify relationships about the role of DNA and chromosomes in coding the instructions for characteristic traits passed from parents to offspring. **Assessment Boundary:** Assessment does not include the phases of meiosis or the biochemical mechanism of specific steps in the process.	**Student Edition:** **Online:** Chapter Investigations 11A, 11B Unit 4 Performance Task 1

Disciplinary Core Ideas	
LS1.A: Structure and Function All cells contain genetic information in the form of DNA molecules. Genes are regions in the DNA that contain the instructions that code for the formation of proteins. (secondary) (Note: This Disciplinary Core Idea is also addressed by HS-LS1-1.)	**Student Edition:** 6, 26–27, 184–186, 200–201, 316–324, 326–333, 335–341, 344–345, 382–389, 419 **Online:** Chapter Investigations 11A, 11B Chapter Assessments 1, 6, 7, 11
LS3.A: Inheritance of Traits Each chromosome consists of a single very long DNA molecule, and each gene on the chromosome is a particular segment of that DNA. The instructions for forming species' characteristics are carried in DNA. All cells in an organism have the same genetic content, but the genes used (expressed) by the cell may be regulated in different ways. Not all DNA codes for a protein; some segments of DNA are involved in regulatory or structural functions, and some have no as-yet known function.	**Student Edition:** 8, 26–27, 187–188, 317–323, 326–334, 335–343, 344–345, 348–350, 379 **Online:** Chapter Investigations 11A, 11B Chapter Assessment 11
Science and Engineering Practices	
Asking Questions and Defining Problems Ask questions that arise from examining models or a theory to clarify relationships.	**Student Edition:** 26–27, 33, 41, 61, 81, 110–111, 113, 147, 168, 198, 209, 281, 306, 312–315, 343, 345, 347, 419, 455, 457, 483, 519 **Online:** Chapter Investigations 11A, 11B Chapter Assessment 1 Unit 3 Performance Task 1 Unit 3 Performance Task 2 Unit 5 Performance Task 1
Crosscutting Concepts	
Cause and Effect Empirical evidence is required to differentiate between cause and correlation and make claims about specific causes and effects.	**Student Edition:** 54–55, 206–207, 295, 318–319, 364–365, 347, 378–379, 418–419, 458–464, 469–472, 476–479, 518–519 **Online:** Chapter Investigations 9B, 10A, 11A, 11B, 15A, 15B, 16A Chapter Assessments 5, 12 Unit 4 Performance Task 2

HS-LS3 Heredity: Inheritance and Variation of Traits

Performance Expectation	
HS-LS3-2. Make and defend a claim based on evidence that inheritable genetic variations may result from (1) new genetic combinations through meiosis, (2) viable errors occurring during replication, and/or (3) mutations caused by environmental factors. **Clarification Statement:** Emphasis is on using data to support arguments for the way variation occurs. **Assessment Boundary:** Assessment does not include the phases of meiosis or the biochemical mechanism of specific steps in the process.	**Student Edition:** 377, 379 (Revisit #1) **Online:** Chapter 12 Assessment (#14) Unit 4 Performance Task 2
Disciplinary Core Ideas	
LS3.B: Variation of Traits In sexual reproduction, chromosomes can sometimes swap sections during the process of meiosis (cell division), thereby creating new genetic combinations and thus more genetic variation. Although DNA replication is tightly regulated and remarkably accurate, errors do occur and result in mutations, which are also a source of genetic variation. Environmental factors can also cause mutations in genes, and viable mutations are inherited.	**Student Edition:** 350–356, 358–361, 370, 376–379, 467–472, 476–479 **Online:** Chapter Investigation 12B Chapter Assessments 10, 12
Environmental factors also affect expression of traits, and hence affect the probability of occurrences of traits in a population. Thus the variation and distribution of traits observed depends on both genetic and environmental factors.	**Student Edition:** 365–375, 377–379, 467–472, 476–479 **Online:** Chapter Investigation 12B Chapter Assessments 11, 12 Unit 4 Performance Task 2

Science and Engineering Practices	
Engaging in Argument from Evidence Make and defend a claim based on evidence about the natural world that reflects scientific knowledge, and student-generated evidence.	**Student Edition:** 56, 246–247, 278–279, 311, 377, 398–402, 421, 509, 519, 521 **Online:** Chapter Investigations 3A, 8A, 8B, 12A, 12B, 14B, 16A Chapter Assessment 5 Unit 2 Performance Task 2 Unit 4 Performance Task 2
Crosscutting Concepts	
Cause and Effect Empirical evidence is required to differentiate between cause and correlation and make claims about specific causes and effects.	**Student Edition:** 354–55, 206–207, 295, 318–319, 364–365, 347, 378–379, 418–419, 458–464, 469–472, 476–479, 518–519 **Online:** Chapter Investigations 9B, 10A, 11A, 11B, 15A, 15B, 16A Chapter Assessments 5, 12 Unit 4 Performance Task 2

HS-LS3 Heredity: Inheritance and Variation of Traits

Performance Expectation	
HS-LS3-3. Apply concepts of statistics and probability to explain the variation and distribution of expressed traits in a population. **Clarification Statement:** Emphasis is on the use of mathematics to describe the probability of traits as it relates to genetic and environmental factors in the expression of traits. **Assessment Boundary:** Assessment does not include Hardy-Weinberg calculations.	**Student Edition:** 377 **Online:** Chapter Investigations 12A, 12B Chapter 12 Assessment (#8, 9, 13, 19, 20) Unit 4 Performance Task 3
Disciplinary Core Ideas	
LS3.B: Variation of Traits Environmental factors also affect expression of traits, and hence affect the probability of occurrences of traits in a population. Thus the variation and distribution of traits observed depends on both genetic and environmental factors.	**Student Edition:** 365–375, 377–379, 467–472, 476–479 **Online:** Chapter Investigation 12B Chapter Assessments 11, 12 Unit 4 Performance Task 2
Science and Engineering Practices	
Analyzing and Interpreting Data Apply concepts of statistics and probability (including determining function fits to data, slope, intercept, and correlation coefficient for linear fits) to scientific and engineering questions and problems, using digital tools when feasible.	**Student Edition:** 102, 369–370, 376, 467–472, 474–478, 482–483, 500–501 **Online:** Chapter Investigations 12A, 12B, 15A, 15B Unit 4 Performance Task 3 Unit 5 Performance Task 2
Crosscutting Concepts	
Scale, Proportion, and Quantity Algebraic thinking is used to examine scientific data and predict the effect of a change in one variable on another (e.g., linear growth vs. exponential growth).	**Student Edition:** 137–138, 369, 378–379, 418–419, 495–498 **Online:** Chapter Investigation 3A Chapter Assessment 5 Unit 4 Performance Task 3
Science Is a Human Endeavor Technological advances have influenced the progress of science and science has influenced advances in technology.	**Student Edition:** 3, 18–21, 177, 365, 378–379 **Online:** Chapter Assessment 8 Unit 4 Performance Task 3
Science and engineering are influenced by society and society is influenced by science and engineering.	**Student Edition:** 18–21, 102, 379 **Online:** Unit 4 Performance Task 3

HS-LS4 Biological Evolution: Unity and Diversity

Performance Expectation	
HS-LS4-1. Communicate scientific information that common ancestry and biological evolution are supported by multiple lines of empirical evidence. **Clarification Statement:** Emphasis is on a conceptual understanding of the role each line of evidence has relating to common ancestry and biological evolution. Examples of evidence could include similarities in DNA sequences, anatomical structures, and order of appearance of structures in embryological development.	**Student Edition:** 453 **Online:** Chapter Investigations 14A, 14B Unit 5 Performance Task 1
Disciplinary Core Ideas	
LS4.A: Evidence of Common Ancestry and Diversity Genetic information, like the fossil record, provides evidence of evolution. DNA sequences vary among species, but there are many overlaps; in fact, the ongoing branching that produces multiple lines of descent can be inferred by comparing the DNA sequences of different organisms. Such information is also derivable from the similarities and differences in amino acid sequences and from anatomical and embryological evidence.	**Student Edition:** 4–5, 210, 285, 426–434, 436–438, 442, 446–452, 454–455, 472, 480, 484, 517 **Online:** Chapter Investigations 14A, 14B Chapter Assessment 1
Science and Engineering Practices	
Obtaining, Evaluating, and Communicating Information Communicate scientific information (e.g., about phenomena and/or the process of development and the design and performance of a proposed process or system) in multiple formats (including orally, graphically, textually, and mathematically).	**Student Edition:** 58–59, 78, 145–147, 199, 215–216, 245, 307, 432–433 **Online:** Chapter Investigations 14A, 14B Unit 2 Performance Task 1 Unit 3 Performance Task 1 Unit 4 Performance Task 4
Science Models, Laws, Mechanisms, and Theories Explain Natural Phenomena A scientific theory is a substantiated explanation of some aspect of the natural world, based on a body of facts that have been repeatedly confirmed through observation and experiment and the science community validates each theory before it is accepted. If new evidence is discovered that the theory does not accommodate, the theory is generally modified in light of this new evidence.	**Student Edition:** 430–431, 442, 452 **Online:** Chapter Investigations 14A, 14B Chapter Assessment 8
Crosscutting Concepts	
Patterns Different patterns may be observed at each of the scales at which a system is studied and can provide evidence for causality in explanations of phenomena.	**Student Edition:** 8, 22–23, 35–36, 67, 74, 133–134, 136, 146–147, 246–247, 347, 357, 371–374, 418–419, 428–429, 452, 458–461, 469–472 **Online:** Chapter Investigations 8A, 14B, 15A, 15B Chapter Assessments 5, 13 Unit 5 Performance Task 1
Scientific Knowledge Assumes an Order and Consistency in Natural Systems Scientific knowledge is based on the assumption that natural laws operate today as they did in the past and they will continue to do so in the future.	**Student Edition:** 430–433, 436–442, 458–461, 465 **Online:** Chapter Investigations 8A, 14B, 15A, 15B Chapter Assessments 5, 13 Unit 5 Performance Task 1

HS-LS4 Biological Evolution: Unity and Diversity

Performance Expectation	
HS-LS4-2. Construct an explanation based on evidence that the process of evolution primarily results from four factors: (1) the potential for a species to increase in number, (2) the heritable genetic variation of individuals in a species due to mutation and sexual reproduction, (3) competition for limited resources, and (4) the proliferation of those organisms that are better able to survive and reproduce in the environment. **Clarification Statement:** Emphasis is on using evidence to explain the influence each of the four factors has on number of organisms, behaviors, morphology, or physiology in terms of ability to compete for limited resources and subsequent survival of individuals and adaptation of species. Examples of evidence could include mathematical models such as simple distribution graphs and proportional reasoning. **Assessment Boundary:** Assessment does not include other mechanisms of evolution, such as genetic drift, gene flow through migration, and co-evolution.	**Student Edition:** 481 **Online:** Chapter Investigation 15B Unit 5 Performance Task 2

Disciplinary Core Ideas	
LS4.B: Natural Selection Natural selection occurs only if there is both (1) variation in the genetic information between organisms in a population and (2) variation in the expression of that genetic information—that is, trait variation—that leads to differences in performance among individuals.	**Student Edition:** 458–465, 472, 481–483 **Online:** Chapter Investigations 15A, 15B
LS4.C: Adaptation Evolution is a consequence of the interaction of four factors: (1) the potential for a species to increase in number, (2) the genetic variation of individuals in a species due to mutation and sexual reproduction, (3) competition for an environment's limited supply of the resources that individuals need in order to survive and reproduce, and (4) the ensuing proliferation of those organisms that are better able to survive and reproduce in that environment.	**Student Edition:** 457–464, 469–472, 474–475, 481–483, 489–491, 517, 519 **Online:** Chapter Investigation 15B Chapter Assessment 15
Science and Engineering Practices	
Constructing Explanations and Designing Solutions Construct an explanation based on valid and reliable evidence obtained from a variety of sources (including students' own investigations, models, theories, simulations, peer review) and the assumption that theories and laws that describe the natural world operate today as they did in the past and will continue to do so in the future.	**Student Edition:** 6, 70, 107, 119–123, 127, 130–131, 146–147, 192–193, 189–191, 200–201, 203, 229–231, 246–247, 348–350, 457–461, 465, 476–480 **Online:** Chapter Investigations 2A, 2B, 7A Chapter Assessments 5, 6 ,7
Crosscutting Concepts	
Cause and Effect Empirical evidence is required to differentiate between cause and correlation and make claims about specific causes and effects.	**Student Edition:** 54–55, 206–207, 295, 318–319, 364–365, 347, 378–379, 418–419, 458–464, 469–472, 476–479, 518–519 **Online:** Chapter Investigations 9B, 10A, 11A, 11B, 15A, 15B, 16A Chapter Assessments 5, 12 Unit 4 Performance Task 2

HS-LS4 Biological Evolution: Unity and Diversity

Performance Expectation	
HS-LS4-3. Apply concepts of statistics and probability to support explanations that organisms with an advantageous heritable trait tend to increase in proportion to organisms lacking this trait. **Clarification Statement:** Emphasis is on analyzing shifts in numerical distribution of traits and using these shifts as evidence to support explanations. **Assessment Boundary:** Assessment is limited to basic statistical and graphical analysis. Assessment does not include allele frequency calculations.	**Student Edition:** 483 (#3, Revisit 1), 485 **Online:** Chapter Investigation 15A Unit 5 Performance Task 2 Unit 5 Performance Task 3
Disciplinary Core Ideas	
LS4.B: Natural Selection Natural selection occurs only if there is both (1) variation in the genetic information between organisms in a population and (2) variation in the expression of that genetic information—that is, trait variation—that leads to differences in performance among individuals.	**Student Edition:** 458–465, 472, 481–483 **Online:** Chapter Investigations 15A, 15B
The traits that positively affect survival are more likely to be reproduced, and thus are more common in the population.	**Student Edition:** , 26–27, 296–304, 462–466, 474–475, 481, 483, 485 **Online:** Chapter Investigation 15A Chapter Assessments 1, 15
LS4.C: Adaptation Natural selection leads to adaptation, that is, to a population dominated by organisms that are anatomically, behaviorally, and physiologically well suited to survive and reproduce in a specific environment. That is, the differential survival and reproduction of organisms in a population that have an advantageous heritable trait leads to an increase in the proportion of individuals in future generations that have the trait and to a decrease in the proportion of individuals that do not.	**Student Edition:** 7, 68, 422–423, 456, 458–466, 469–472, 481–483, 485, 489–491, 517, 519 **Online:** Chapter Investigation 15A Chapter Assessments 1, 16
Adaptation also means that the distribution of traits in a population can change when conditions change.	**Student Edition:** 424–425, 457, 472, 481, 483, 485, 489, 517, 519 **Online:** Chapter Investigations 15A, 15B Chapter Assessment 1

Science and Engineering Practices	
Analyzing and Interpreting Data Apply concepts of statistics and probability (including determining function fits to data, slope, intercept, and correlation coefficient for linear fits) to scientific and engineering questions and problems, using digital tools when feasible.	**Student Edition:** 369, 467–472, 476–478, 500–501 **Online:** Chapter Investigations 12A, 12B, 15A, 15B
Crosscutting Concepts	
Patterns Different patterns may be observed at each of the scales at which a system is studied and can provide evidence for causality in explanations of phenomena.	**Student Edition:** 8, 22–23, 35–36, 67, 74, 133–134, 136, 146–147, 246–247, 347, 357, 371–374, 418–419, 428–429, 452, 458–461, 469–472 **Online:** Chapter Investigations 8A, 14B, 15A, 15B Chapter Assessments 5, 13 Unit 5 Performance Task 1

HS-LS4 Biological Evolution: Unity and Diversity

Performance Expectation	
HS-LS4-4. Construct an explanation based on evidence for how natural selection leads to adaptation of populations. **Clarification Statement:** Emphasis is on using data to provide evidence for how specific biotic and abiotic differences in ecosystems (such as ranges of seasonal temperature, long-term climate change, acidity, light, geographic barriers, or evolution of other organisms) contribute to a change in gene frequency over time, leading to adaptation of populations.	**Student Edition:** 481, 483 (Revisit #1) **Online:** Chapter Investigation 15B Unit 5 Performance Task 3
Disciplinary Core Ideas	
LS4.C: Adaptation Natural selection leads to adaptation, that is, to a population dominated by organisms that are anatomically, behaviorally, and physiologically well suited to survive and reproduce in a specific environment. That is, the differential survival and reproduction of organisms in a population that have an advantageous heritable trait leads to an increase in the proportion of individuals in future generations that have the trait and to a decrease in the proportion of individuals that do not.	**Student Edition:** 7, 68, 422–423, 456, 458–466, 469–472, 481–483, 485, 489–491, 517, 519 **Online:**Chapter Investigation 15A Chapter Assessments 1, 16
Science and Engineering Practices	
Constructing Explanations and Designing Solutions Construct an explanation based on valid and reliable evidence obtained from a variety of sources (including students' own investigations, models, theories, simulations, peer review) and the assumption that theories and laws that describe the natural world operate today as they did in the past and will continue to do so in the future.	**Student Edition:** 6, 70, 107, 119–123, 127, 130–131, 146–147, 192–193, 189–191, 200–201, 203, 229–231, 246–247, 348–350, 457–461, 465, 476–480 **Online:** Chapter Investigations 2A, 2B, 7A Chapter Assessments 5, 6, 7
Crosscutting Concepts	
Cause and Effect Empirical evidence is required to differentiate between cause and correlation and make claims about specific causes and effects.	**Student Edition:** 54–55, 206–207, 295, 318–319, 364–365, 347, 378–379, 418–419, 458–464, 469–472, 476–479, 518–519 **Online:** Chapter Investigations 9B, 10A, 11A, 11B, 15A, 15B, 16A Chapter Assessments 5, 12 Unit 4 Performance Task 2
Scientific Knowledge Assumes an Order and Consistency in Natural Systems Scientific knowledge is based on the assumption that natural laws operate today as they did in the past and they will continue to do so in the future.	**Student Edition:** 430–433, 436–442, 465 **Online:** Chapter Investigations 14A, 14B

HS-LS4 Biological Evolution: Unity and Diversity

Performance Expectation	
HS-LS4-5. Evaluate the evidence supporting claims that changes in environmental conditions may result in (1) increases in the number of individuals of some species, (2) the emergence of new species over time, and (3) the extinction of other species. **Clarification Statement:** Emphasis is on determining cause and effect relationships for how changes to the environment such as deforestation, fishing, application of fertilizers, drought, flood, and the rate of change of the environment affect distribution or disappearance of traits in species.	**Student Edition:** 521 **Online:** Chapter Investigation 16A Unit 5 Performance Task 4

Disciplinary Core Ideas	
LS4.C: Adaptation Changes in the physical environment, whether naturally occurring or human induced, have thus contributed to the expansion of some species, the emergence of new distinct species as populations diverge under different conditions, and the decline—and sometimes the extinction—of some species.	**Student Edition:** 77, 308–309, 457, 484, 486–493, 495–498, 509, 516, 518–519, 528–519, 521 **Online:** Chapter Investigation 16A Chapter 16 Assessment
Species become extinct because they can no longer survive and reproduce in their altered environment. If members cannot adjust to change that is too fast or drastic, the opportunity for the species' evolution is lost.	**Student Edition:** 77, 492–493, 495–498
Science and Engineering Practices	
Engaging in Argument from Evidence Evaluate the evidence behind currently accepted explanations or solutions to determine the merits of arguments.	**Student Edition:** 65, 78, 80–81, 296–300 **Online:** Chapter Investigations 3B, 10B, 16A
Crosscutting Concepts	
Cause and Effect Empirical evidence is required to differentiate between cause and correlation and make claims about specific causes and effects.	**Student Edition:** 54–55, 206–207, 295, 318–319, 364–365, 347, 378–379, 418–419, 458–464, 469–472, 476–479, 518–519 **Online:** Chapter Investigations 9B, 10A, 11A, 11B, 15A, 15B, 16A Chapter Assessments 5, 12 Unit 4 Performance Task 2

HS-LS4 Biological Evolution: Unity and Diversity

Performance Expectation	
HS-LS4-6. Create or revise a simulation to test a solution to mitigate adverse impacts of human activity on biodiversity.* **Clarification Statement:** Emphasis is on testing solutions for a proposed problem related to threatened or endangered species, or to genetic variation of organisms for multiple species.	**Student Edition:** 27 (#2) **Online:** Chapter Investigation 16B Unit 5 Performance Task 5
Disciplinary Core Ideas	
LS4.C: Adaptation Changes in the physical environment, whether naturally occurring or human induced, have thus contributed to the expansion of some species, the emergence of new distinct species as populations diverge under different conditions, and the decline—and sometimes the extinction—of some species.	**Student Edition:** 77, 308–309, 457, 484, 486–493, 495–498, 509, 516, 518–519, 528–519, 521 **Online:** Chapter Investigation 16A Chapter 16 Assessment
LS4.D: Biodiversity and Humans Humans depend on the living world for the resources and other benefits provided by biodiversity. But human activity is also having adverse impacts on biodiversity through overpopulation, overexploitation, habitat destruction, pollution, introduction of invasive species, and climate change. Thus sustaining biodiversity so that ecosystem functioning and productivity are maintained is essential to supporting and enhancing life on Earth. Sustaining biodiversity also aids humanity by preserving landscapes of recreational or inspirational value. (Note: This Disciplinary Core Idea is also addressed by HS-LS2-7.)	**Student Edition:** 65, 68, 70, 72–73, 232–233, 492–508, 509–519
ETS1.B: Developing Possible Solutions When evaluating solutions, it is important to take into account a range of constraints, including cost, safety, reliability, and aesthetics, and to consider social, cultural, and environmental impacts. (secondary)	**Student Edition:** 18–21, 24, 314, 407–415, 413, 421 **Online:** Chapter Investigations 4B, 10B, 13A, 13B, 16B Chapter Assessment 13
Both physical models and computers can be used in various ways to aid in the engineering design process. Computers are useful for a variety of purposes, such as running simulations to test different ways of solving a problem or to see which one is most efficient or economical; and in making a persuasive presentation to a client about how a given design will meet his or her needs. (secondary)	**Student Edition:** 12–14, 18–21, 403 **Online:** Chapter Investigation 16B Chapter Assessments 1, 14
Science and Engineering Practices	
Using Mathematics and Computational Thinking Create or revise a simulation of a phenomenon, designed device, process, or system.	**Student Edition:** 434 **Online:** Unit 5 Performance Task 1

Crosscutting Concepts	
Cause and Effect Empirical evidence is required to differentiate between cause and correlation and make claims about specific causes and effects.	**Student Edition:** 54–55, 206–207, 295, 318–319, 364–365, 347, 378–379, 418–419, 458–464, 469–472, 476–479, 518–519 **Online:** Chapter Investigations 9B, 10A, 11A, 11B, 15A, 15B, 16A Chapter Assessments 5, 12 Unit 4 Performance Task 2

HS-ETS1 Engineering Design

Performance Expectation	
HS-ETS1-1. Analyze a major global challenge to specify qualitative and quantitative criteria and constraints for solutions that account for societal needs and wants.	**Online:** Chapter Investigation 1B Unit 4 Performance Task 4
Disciplinary Core Ideas	
ETS1.A: Defining and Delimiting Engineering Problems Criteria and constraints also include satisfying any requirements set by society, such as taking issues of risk mitigation into account, and they should be quantified to the extent possible and stated in such a way that one can tell if a given design meets them.	**Student Edition:** 18–21, 26–27
Humanity faces major global challenges today, such as the need for supplies of clean water and food or for energy sources that minimize pollution, which can be addressed through engineering. These global challenges also may have manifestations in local communities.	**Student Edition:** 26–27, 312–315, 403–405, 407–415, 421, 510–515 **Online:** Chapter Investigation 1B Unit 4 Performance Task 4
Science and Engineering Practices	
Asking Questions and Defining Problems Analyze complex real-world problems by specifying criteria and constraints for successful solutions.	**Student Edition:** 26–27, 103, 407–414 **Online:** Chapter Investigation 1B Chapter Assessment 1 Unit 5 Performance Task 5
Crosscutting Concepts	
Influence of Science, Engineering, and Technology on Society and the Natural World New technologies can have deep impacts on society and the environment, including some that were not anticipated. Analysis of costs and benefits is a critical aspect of decisions about technology.	**Student Edition:** 18–21, 392–402, 404–405, 407–409, 500–501, 511–515 **Online:** Chapter Investigations 1B, 4B, 13A, 13B Chapter Assessments 13, 14 Unit 1 Performance Task 5 Unit 4 Performance Task 4

HS-ETS1 Engineering Design

Performance Expectation	
HS-ETS1-2. Design a solution to a complex real-world problem by breaking it down into smaller, more manageable problems that can be solved through engineering.	**Online:** Chapter Investigation 16B Chapter 16 Assessment Unit 4 Performance Task 4
Disciplinary Core Ideas	
ETS1.C: Optimizing the Design Solution Criteria may need to be broken down into simpler ones that can be approached systematically, and decisions about the priority of certain criteria over others (trade-offs) may be needed.	**Student Edition:** 18–21, 513–515 **Online:** Chapter Investigation 6B Unit 4 Performance Task 4
Science and Engineering Practices	
Constructing Explanations and Designing Solutions Design a solution to a complex real-world problem, based on scientific knowledge, student-generated sources of evidence, prioritized criteria, and tradeoff considerations.	**Student Edition:** 25, 382–389, 509, 513–514, 518–519 **Online:** Chapter Investigations 4B, 10B, 13A, 13B Chapter Assessment 16 Unit 1 Performance Task 5 Unit 2 Performance Task 3 Unit 4 Performance Task 4 Unit 5 Performance Task 2

HS-ETS1 Engineering Design

Performance Expectation	
HS-ETS1-3. Evaluate a solution to a complex real-world problem based on prioritized criteria and trade-offs that account for a range of constraints, including cost, safety, reliability, and aesthetics as well as possible social, cultural, and environmental impacts.	**Student Edition:** 417 **Online:** Chapter Investigations 4B, 10B, 13A, 13B, 16B Chapter Assessment 13 (#14) Unit 1 Performance Task 5
Disciplinary Core Ideas	
ETS1.B: Developing Possible Solutions When evaluating solutions, it is important to take into account a range of constraints, including cost, safety, reliability, and aesthetics, and to consider social, cultural, and environmental impacts.	**Student Edition:** 18–21, 24, 314, 404–405, 407–415, 413, 421 **Online:** Chapter Investigations 4B, 10B, 13A, 13B, 16B Chapter Assessment 13 Unit 1 Performance Task 5
Science and Engineering Practices	
Constructing Explanations and Designing Solutions Evaluate a solution to a complex real-world problem, based on scientific knowledge, student-generated sources of evidence, prioritized criteria, and tradeoff considerations.	**Student Edition:** 25, 382–389, 509, 513–514, 518–519 **Online:** Chapter Investigations 4B, 10B, 13A, 13B Chapter Assessment 16 Unit 1 Performance Task 5 Unit 2 Performance Task 3 Unit 4 Performance Task 4 Unit 5 Performance Task 2
Crosscutting Concepts	
Influence of Science, Engineering, and Technology on Society and the Natural World New technologies can have deep impacts on society and the environment, including some that were not anticipated. Analysis of costs and benefits is a critical aspect of decisions about technology.	**Student Edition:** 18–21, 392–402, 404–405, 407–409, 500–501, 511–515 **Online:** Chapter Investigations 1B, 4B, 13A, 13B Chapter Assessments 13, 14 Unit 1 Performance Task 5 Unit 4 Performance Task 4

HS-ETS1 Engineering Design

Performance Expectation	
HS-ETS1-4. Use a computer simulation to model the impact of proposed solutions to a complex real-world problem with numerous criteria and constraints on interactions within and between systems relevant to the problem.	**Online:** Unit 5 Performance Task 5
Disciplinary Core Ideas	
ETS1.B: Developing Possible Solutions Both physical models and computers can be used in various ways to aid in the engineering design process. Computers are useful for a variety of purposes, such as running simulations to test different ways of solving a problem or to see which one is most efficient or economical; and in making a persuasive presentation to a client about how a given design will meet his or her needs.	**Student Edition:** 12–14, 18–21, 403 **Online:** Chapter Investigation 16B Chapter Assessments 1, 14
Science and Engineering Practices	
Using Mathematics and Computational Thinking Use mathematical models and/or computer simulations to predict the effects of a design solution on systems and/or the interactions between systems.	**Student Edition:** 378 **Online:** Chapter Investigations 3A, 16B Unit 1 Performance Task 4
Crosscutting Concepts	
Systems and System Models Models (e.g., physical, mathematical, computer models) can be used to simulate systems and interactions—including energy, matter, and information flows— within and between systems at different scales.	**Student Edition:** 12–14, 41, 46, 55, 58–59, 184–186, 200–201, 287–290, 311, 482–483, 516 **Online:** Chapter Investigations 1A, 9A, 10A, 16B Chapter Assessments 5, 7, 13 Unit 1 Performance Task 1 Unit 3 Performance Task 1 Unit 5 Performance Task 5

NEXT GENERATION SCIENCE STANDARDS BY CHAPTER

Chapter or Section	NGSS Standards and Dimensions
Chapter 1: Introduction to Biology, pages 2–27	**Performance Expectations** HS-ETS1-1; HS-ETS1-2 **Disciplinary Core Ideas** LS1.A, LS2.D, LS3.A, LS4.A, LS4.B, LS4.C, ETS1.A, ETS1.B, ETS1.C **Science and Engineering Practices** Asking Questions and Defining Problems; Developing and Using Models; Planning and Carrying Out Investigations; Constructing Explanations and Designing Solutions; Engaging in Argument from Evidence. **Connections to Nature of Science** Scientific Investigations Use a Variety of Methods; Scientific Knowledge is Based on Empirical Evidence; Scientific Knowledge is Open to Revision in Light of New Evidence; Science Models, Laws, Mechanisms, and Theories Explain Natural Phenomena. **Crosscutting Concepts** Patterns; Systems and System Models; Structure and Function. **Connections to Nature of Science** Science is a Human Endeavor. **Connections to Engineering, Technology, and Applications of Science** Influence of Engineering, Technology, and Science on Society and the Natural World.
Online Assessments and Investigations for Chapter 1	**Performance Expectations** HS-ETS1-1 **Disciplinary Core Ideas** ETS1.A; ETS1.B; LS1.A; LS4.A; LS4.B; LS4.C **Science and Engineering Practices** Asking Questions and Defining Problems; Planning and Carrying Out Investigations; Analyzing and Interpreting Data; Engaging in Argument from Evidence; Scientific Investigations Use a Variety of Methods. **Crosscutting Concepts** Scale, Proportion, and Quantity; Systems and System Models. **Connections to Engineering, Technology, and Applications of Science** Influence of Engineering, Technology, and Science on Society and the Natural World.
Unit 1: Relationships in Ecosystems	
Chapter 2: Energy and Matter in Ecosystems, pages 32–59	**Performance Expectations** HS-LS2-2; HS-LS2-3; HS-LS2-4; HS-LS2-5 **Disciplinary Core Ideas** LS2.A; LS2.B; LS2.C; LS2.C **Science and Engineering Practices** Asking Questions and Defining Problems; Developing and Using Models; Planning and Carrying Out Investigations; Analyzing and Interpreting Data; Using Mathematics and Computational Thinking; Constructing Explanations and Designing Solutions; Engaging in Argument from Evidence. **Connections to Nature of Science** Scientific Knowledge is Open to Revision in Light of New Evidence. **Crosscutting Concepts** Cause and Effect; Scale, Proportion, and Quantity; Systems and System Models; Energy and Matter; Stability and Change.
Online Assessments and Investigations for Chapter 2	**Performance Expectations** HS-LS2-2; HS-LS2-3; HS-LS2-4; HS-LS2-5; HS-LS2-6 **Disciplinary Core Ideas** LS1.C; LS2.A; LS2.B; ESS2.D; ESS3.C; PS3.D **Science and Engineering Practices** Developing and Using Models; Using Mathematics and Computational Thinking; Constructing Explanations and Designing Solutions. **Crosscutting Concepts** Scale, Proportion, and Quantity; Energy and Matter.
Chapter 3: Biodiversity and Ecosystem Stability, pages 60–81	**Performance Expectations** HS-LS2-1; HS-LS2-6; HS-LS2-7 **Disciplinary Core Ideas** LS2.A; LS2.C; LS4.C; LS4.D; ESS2.E; ESS3.C **Science and Engineering Practices** Asking Questions and Defining Problems; Using Mathematics and Computational Thinking; Constructing Explanations and Designing Solutions; Engaging in Argument from Evidence. **Connections to Nature of Science** Scientific Knowledge Is Open To Revision In Light Of New Evidence; Scientific Knowledge is Based on Empirical Evidence. **Crosscutting Concepts** Patterns; Cause and Effect; Systems and System Models; Stability and Change. **Connections to Nature of Science** Science Addresses Questions About the Natural and Material World **Connections to Engineering, Technology, and Applications of Science** Interdependence of Science, Engineering, and Technology.
Online Assessments and Investigations for Chapter 3	**Performance Expectations** HS-LS2-1, HS-LS2-2, HS-LS2-6 **Disciplinary Core Ideas** LS2.C **Science and Engineering Practices** Planning and Carrying Out Investigations; Analyzing and Interpreting Data; Using Mathematics and Computational Thinking; Engaging in Argument from Evidence. **Connections to Nature of Science** Scientific Knowledge is Open to Revision in Light of New Evidence. **Crosscutting Concepts** Scale, Proportion, and Quantity; Stability and Change.
Chapter 4: Population Measurement and Growth, pages 82–105	**Performance Expectations** HS-LS2-1; HS-LS2-2 **Disciplinary Core Ideas** LS1.A; LS2.C **Science and Engineering Practices** Analyzing and Interpreting Data; Using Mathematics and Computational Thinking; Engaging in Argument from Evidence. **Connections to Nature of Science** Scientific Knowledge is Open to Revision in Light of New Evidence. **Crosscutting Concepts** Scale, Proportion, and Quantity.
Online Assessments and Investigations for Chapter 4	**Disciplinary Core Ideas** HS-LS2-2; HS-LS2-7; HS-ETS1-3 **Science and Engineering Practices** Using Mathematics and Computational Thinking; Constructing Explanations and Designing Solutions. **Crosscutting Concepts** Influence of Engineering, Technology, and Science on Society and the Natural World.
Unit 1 Performance Tasks	**Performance Expectations** HS-LS2-1; HS-LS2-2; HS-LS2-3; HS-LS2-4; HS-LS2-5; HS-LS2-6; HS-LS2-7; HS-ETS1-3; HS-ESS2-6 **Disciplinary Core Ideas** LS2.A; LS2.B; LS2.C; ESS2.C; ESS2.D; ETS1.B **Science and Engineering Practices** Developing and Using Models; Using Mathematics and Computational Thinking; Constructing Explanations and Designing Solutions. **Crosscutting Concepts** Scale, Proportion, and Quantity; Systems and System Models; Energy and Matter. **Connections to Engineering, Technology, and Applications of Science** Influence of Engineering, Technology, and Science on Society and the Natural World.

Unit 2: Cell Systems	
Chapter 5: Molecules in Living Systems, pages 112–147	**Performance Expectations** HS-LS1-6 **Disciplinary Core Ideas** LS1.C; PS1.A; PS1.B; PS2.B **Science and Engineering Practices** Asking Questions and Defining Problems; Developing and Using Models; Using Mathematics and Computational Thinking; Constructing Explanations and Designing Solutions; Engaging in Argument from Evidence. **Connections to Nature of Science** Scientific Investigations Use a Variety of Methods; Scientific Knowledge is Based on Empirical Evidence. **Crosscutting Concepts** Patterns; Cause and Effect; Scale, Proportion, and Quantity; Energy and Matter; Structure and Function; Stability and Change.
Online Assessments and Investigations for Chapter 5	**Performance Expectations** HS-LS1-6 **Disciplinary Core Ideas** LS1.C **Science and Engineering Practices** Developing and Using Models; Planning and Carrying Out Investigations; Using Mathematics and Computational Thinking; Constructing Explanations and Designing Solutions; Engaging in Argument from Evidence. **Crosscutting Concepts** Scale, Proportion, and Quantity; Systems and System Models; Energy and Matter; Structure and Function.
Chapter 6: Cell Structure and Function, pages 148–179	**Performance Expectations** HS-LS1-5; HS-LS1-7; HS-LS2-3; HS-LS2-5 **Disciplinary Core Ideas** LS1.A; LS1.C; LS2.B; LS2.C **Science and Engineering Practices** Asking Questions and Defining Problems; Developing and Using Models; Using Mathematics and Computational Thinking. **Connections to Nature of Science** Scientific Investigations Use A Variety Of Methods. **Crosscutting Concepts** Energy and Matter; Structure and Function; Stability and Change. **Connections to Nature of Science** Science is a Human Endeavor.
Online Assessments and Investigations for Chapter 6	**Performance Expectations** HS-LS1-5; HS-LS1-7; HS-ETS1-2 **Disciplinary Core Ideas** LS1.A; LS1.C **Science and Engineering Practices** Developing and Using Models; Constructing Explanations and Designing Solutions. **Crosscutting Concepts** Energy and Matter.
Chapter 7: Cell Growth, pages 180–201	**Performance Expectations** HS-LS1-1; HS-LS1-4 **Disciplinary Core Ideas** LS1.A; LS1.B; LS3.A **Science and Engineering Practices** Asking Questions and Defining Problems; Developing and Using Models; Analyzing and Interpreting Data; Constructing Explanations and Designing Solutions; Engaging in Argument from Evidence. **Crosscutting Concepts** Systems and System Models; Structure and Function. **Connections to Nature of Science** Science Addresses Questions About the Natural and Material World.
Online Assessments and Investigations for Chapter 7	**Performance Expectations** HS-LS1-4 **Disciplinary Core Ideas** LS1.A; LS1.B **Science and Engineering Practices** Developing and Using Models; Constructing Explanations and Designing Solutions. **Crosscutting Concepts** Systems and System Models; Structure and Function.
Unit 2 Performance Tasks	**Performance Expectations** HS-LS1-4; HS-LS1-5; HS-LS1-6; HS-LS1-7 **Disciplinary Core Ideas** LS1.B; LS1.C; LS1.C; LS1.C **Science and Engineering Practices** Developing and Using Models; Constructing Explanations and Designing Solutions; Engaging in Argument from Evidence. **Crosscutting Concepts** Energy and Matter.
Unit 3: Interactions in Living Systems	
Chapter 8: Diversity of Living systems, pages 208–247	**Performance Expectations** HS-LS1-2 **Disciplinary Core Ideas** LS1.A; LS1.B; LS1.C; LS2.B; LS4.A; LS4.D **Science and Engineering Practices** Asking Questions and Defining Problems; Developing and Using Models; Analyzing and Interpreting Data; Using Mathematics and Computational Thinking; Constructing Explanations and Designing Solutions; Engaging in Argument from Evidence. **Connections to Nature of Science** Scientific Investigations Use a Variety of Methods; Scientific Knowledge is Open to Revision in Light of New Evidence. **Crosscutting Concepts** Patterns; Scale, Proportion, and Quantity; Energy and Matter; Structure and Function.
Online Assessments and Investigations for Chapter 8	**Science and Engineering Practices** Planning and Carrying Out Investigations; Constructing Explanations and Designing Solutions; Engaging in Argument from Evidence. **Connections to Nature of Science** Scientific Investigations Use a Variety of Methods; Scientific Knowledge is Open to Revision in Light of New Evidence; Science Models, Laws, Mechanisms, and Theories Explain Natural Phenomena. **Crosscutting Concepts** Patterns; Structure and Function. **Connections to Nature of Science** Science is a Human Endeavor.
Chapter 9: Plant Systems, pages 248–279	**Performance Expectations** HS-LS1-2, HS-LS1-3 **Disciplinary Core Ideas** LS1.A; LS1.B; LS2.A; LS2.C **Science and Engineering Practices** Engaging in Argument from Evidence. **Connections to Nature of Science** Science Models, Laws, Mechanisms, and Theories Explain Natural Phenomena. **Crosscutting Concepts** Systems and System Models; Stability and Change.

Unit 3: Interactions in Living Systems (continued)	
Online Assessments and Investigations for Chapter 9	**Performance Expectations** HS-LS1-2; HS-LS1-3 **Disciplinary Core Ideas** LS1.A **Science and Engineering Practices** Developing and Using Models; Planning and Carrying Out Investigations. **Crosscutting Concepts** Cause and Effect; Systems and System Models; Stability and Change.
Chapter 10: Animal Systems, pages 280–309	**Performance Expectations** HS-LS1-2; HS-LS1-3 **Disciplinary Core Ideas** LS1.A **Science and Engineering Practices** Developing and Using Models; Planning and Carrying Out Investigations. **Crosscutting Concepts** Cause and Effect; Systems and System Models; Stability and Change.
Online Assessments and Investigations for Chapter 10	**Performance Expectations** HS-LS1-2; HS-LS1-3; HS-LS2-8 **Disciplinary Core Ideas** LS1.A; LS2.D; LS3.B; ETS1.B **Science and Engineering Practices** Developing and Using Models; Planning and Carrying Out Investigations; Constructing Explanations and Designing Solutions; Engaging in Argument from Evidence. **Crosscutting Concepts** Systems and System Models; Stability and Change.
Unit 3 Performance Tasks	**Performance Expectations** HS-LS2-8 **Disciplinary Core Ideas** LS1.A; LS2.D **Science and Engineering Practices** Asking Questions and Defining Problems; Developing and Using Models; Planning and Carrying Out Investigations; Constructing Explanations and Designing Solutions; Engaging in Argument from Evidence. **Crosscutting Concepts** Systems and System Models; Stability and Change.
Unit 4: Genetics and Information	
Chapter 11: DNA, RNA and Proteins, pages 316–345	**Performance Expectations** HS-LS1-1; HS-LS3-1 **Disciplinary Core Ideas** LS1.A; LS1.C; LS2.C; LS3.A **Science and Engineering Practices** Asking Questions and Defining Problems; Developing and Using Models. **Connections to Nature of Science** Scientific Knowledge is Based on Empirical Evidence. **Crosscutting Concepts** Systems And System Models; Cause and Effect; Structure and Function.
Online Assessments and Investigations for Chapter 11	**Performance Expectations** HS-LS1-1, HS-LS3-1 **Disciplinary Core Ideas** LS1.A; LS3.A; LS3.B **Science and Engineering Practices** Asking Questions and Defining Problems. **Crosscutting Concepts** Cause and Effect.
Chapter 12: Genetic Variation and Heredity, pages 346–379	**Performance Expectations** HS-LS3-2; HS-LS3-3 **Disciplinary Core Ideas** LS3.A; LS3.B **Science and Engineering Practices** Asking Questions and Defining Problems; Developing and Using Models; Analyzing and Interpreting Data; Using Mathematics and Computational Thinking; Constructing Explanations and Designing Solutions; Engaging in Argument from Evidence. **Connections to Nature of Science** Scientific Investigations Use a Variety of Methods. **Crosscutting Concepts** Patterns; Cause and Effect; Scale, Proportion, and Quantity. **Connections to Nature of Science** Science is a Human Endeavor.
Online Assessments and Investigations for Chapter 12	**Performance Expectations** HS-LS3-2; HS-LS3-3 **Disciplinary Core Ideas** LS2.A; LS3.B **Science and Engineering Practices** Analyzing and Interpreting Data; Engaging in Argument from Evidence. **Crosscutting Concepts** Patterns; Cause and Effect.
Chapter 13: Genetic Technologies, pages 380–419	**Performance Expectations** HS-ETS1-3 **Disciplinary Core Ideas** LS1.A; LS2.C; ETS1.A; ETS1.B **Science and Engineering Practices** Asking Questions and Defining Problems; Developing and Using Models; Using Mathematics and Computational Thinking; Constructing Explanations and Designing Solutions; Engaging in Argument from Evidence. **Connections to Nature of Science** Scientific Investigations Use a Variety of Methods. **Crosscutting Concepts** Patterns; Cause and Effect; Scale, Proportion, and Quantity; Structure and Function. **Connections to Nature of Science** Science Addresses Questions About the Natural and Material World. **Connections to Engineering, Technology, and Applications of Science** Interdependence of Science, Engineering, and Technology; Influence of Engineering, Technology, and Science on Society and the Natural World.

Unit 4: Genetics and Information (continued)	
Online Assessments and Investigations for Chapter 13	**Performance Expectations** HS-LS3-3; HS-ETS1-1; HS-ETS1-2; HS-ETS1-3 **Disciplinary Core Ideas** ETS1.B **Science and Engineering Practices** Constructing Explanations and Designing Solutions. **Crosscutting Concepts** Patterns; Systems and System Models. **Connections to Engineering, Technology, and Applications of Science** Influence of Engineering, Technology, and Science on Society and the Natural World.
Unit 4 Performance Tasks	**Performance Expectations** HS-LS3-1; HS-LS3-2; HS-LS3-3; HS-ETS1-1; HS-ETS1-2 **Disciplinary Core Ideas** LS3.B; ETS1.A; ETS1.C **Science and Engineering Practices** Asking Questions And Defining Problems; Analyzing and Interpreting Data; Using Mathematics and Computational Thinking; Constructing Explanations and Designing; Engaging in Argument from Evidence. **Crosscutting Concepts** Cause and Effect; Scale, Proportion, and Quantity. **Connections to Nature of Science** Science is a Human Endeavor; Science Addresses Questions About the Natural and Material World. **Connections to Engineering, Technology, and Applications of Science** Influence of Engineering, Technology, and Science on Society and the Natural World.
Unit 5: Evolution and Changing Environments	
Chapter 14: Evidence for Evolution, pages 426–455	**Performance Expectations** HS-LS4-1 **Disciplinary Core Ideas** LS4.A; PS1.C; ESS2.A; ESS2.D **Science and Engineering Practices** Asking Questions and Defining Problems; Using Mathematics and Computational Thinking; Engaging in Argument from Evidence. **Connections to Nature of Science** Scientific Knowledge is Based on Empirical Evidence; Science Models, Laws, Mechanisms, and Theories Explain Natural Phenomena **Crosscutting Concepts** Patterns; Systems and System Models. **Connections to Nature of Science** Scientific Knowledge Assumes an Order and Consistency in Natural Systems.
Online Assessments and Investigations for Chapter 14	**Performance Expectations** HS-LS4-1 **Disciplinary Core Ideas** LS4.A, ETS1.B Science and Engineering Practices Engaging in Argument from Evidence. **Connections to Nature of Science** Science Models, Laws, Mechanisms, and Theories Explain Natural Phenomena. **Crosscutting Concepts** Patterns. **Connections to Nature of Science** Scientific Knowledge Assumes an Order and Consistency in Natural Systems. **Connections to Engineering, Technology, and Applications of Science** Influence of Engineering, Technology, and Science on Society and the Natural World.
Chapter 15: The Theory of Evolution, pages 456–483	**Disciplinary Core Ideas** HS-LS3-3; HS-LS4-2; HS-LS4-3; HS-LS4-4 **Science and Engineering Practices** Asking Questions and Defining Problems; Developing and Using Models; Analyzing and Interpreting Data; Constructing Explanations and Designing Solutions; Engaging in Argument from Evidence. **Connections to Nature of Science** Scientific Investigations Use a Variety of Methods; Scientific Knowledge is Based on Empirical Evidence; Scientific Knowledge is Open to Revision in Light of New Evidence. **Crosscutting Concepts** Patterns; Cause and Effect; Systems and System Models; Stability and Change. **Connections to Nature of Science** Scientific Knowledge Assumes an Order and Consistency in Natural Systems.
Online Assessments and Investigations for Chapter 15	**Performance Expectations** HS-LS4-2; HS-LS-4-3; HS-LS4-4 **Disciplinary Core Ideas** LS4.B, LS4.C **Science and Engineering Practices** Analyzing and Interpreting Data Constructing Explanations and Designing Solutions Engaging in Argument from Evidence. **Crosscutting Concepts** Patterns; Cause and Effect; Stability and Change. **Connections to Nature of Science** Scientific Knowledge Assumes an Order and Consistency in Natural Systems.
Chapter 16: Survival in Changing Environments, pages 484–519	**Performance Expectations** HS-LS2-7; HS-LS4-5; HS-LS4-6; HS-ETS1-4 **Disciplinary Core Ideas** LS2.A; LS2.C; LS4.A; LS4.B; LS4.C; LS4.D; ETS1.A; ETS1.C; ESS2.D ESS3.A; ESS3.C; ESS3.D **Science and Engineering Practices** Asking Questions and Defining Problems; Developing and Using Models; Analyzing and Interpreting Data; Using Mathematics and Computational Thinking; Constructing Explanations and Designing Solutions; Engaging in Argument from Evidence. **Connections to Nature of Science** Scientific Knowledge is Open to Revision in Light of New Evidence. **Crosscutting Concepts** Cause and Effect; Scale, Proportion, and Quantity; Systems and System Models; Stability and Change. **Connections to Nature of Science** Science Addresses Questions About the Natural and Material World. **Connections to Engineering, Technology, and Applications of Science** Influence of Engineering, Technology, and Science on Society and the Natural World.
Online Assessments and Investigations for Chapter 16	**Performance Expectations** HS-LS2-7; HS-LS4-5; HS-LS4-6; HS-ETS1-2; HS-ETS1-4 **Disciplinary Core Ideas** LS2.C ; LS4.A; LS4.C; LS4.D; ETS1.B; ESS3.A **Science and Engineering Practices** Developing and Using; Using Mathematics and Computational Thinking; Constructing Explanations and Designing Solutions; Engaging in Argument from Evidence. **Connections to Nature of Science** Scientific Knowledge is Open to Revision in Light of New Evidence. **Crosscutting Concepts** Cause and Effect; Systems and System Models; Stability and Change.
Unit 5 Performance Tasks	**Performance Expectations** HS-LS4-1; HS-LS4-2; HS-LS4-3; HS-LS4-4; HS-LS4-5; HS-LS4-6; HS-ETS1-2; HS-ETS1-4 **Disciplinary Core Ideas** LS2.C **Science and Engineering Practices** Asking Questions and Defining Problems; Developing and Using Models; Analyzing and Interpreting Data; Using Mathematics and Computational Thinking; Constructing Explanations and Designing Solutions. **Crosscutting Concepts** Patterns; Systems and System Models.

CONTENTS

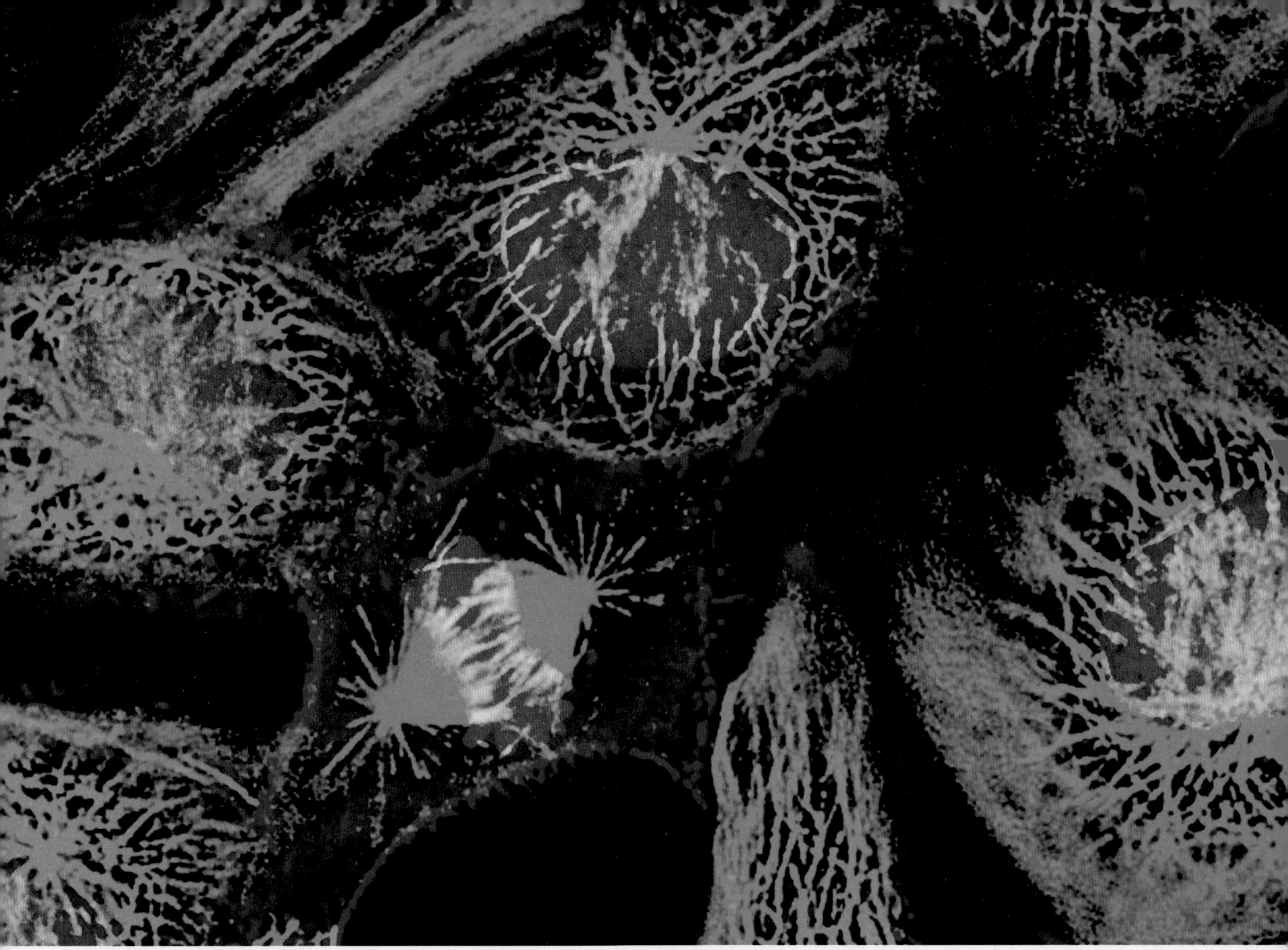

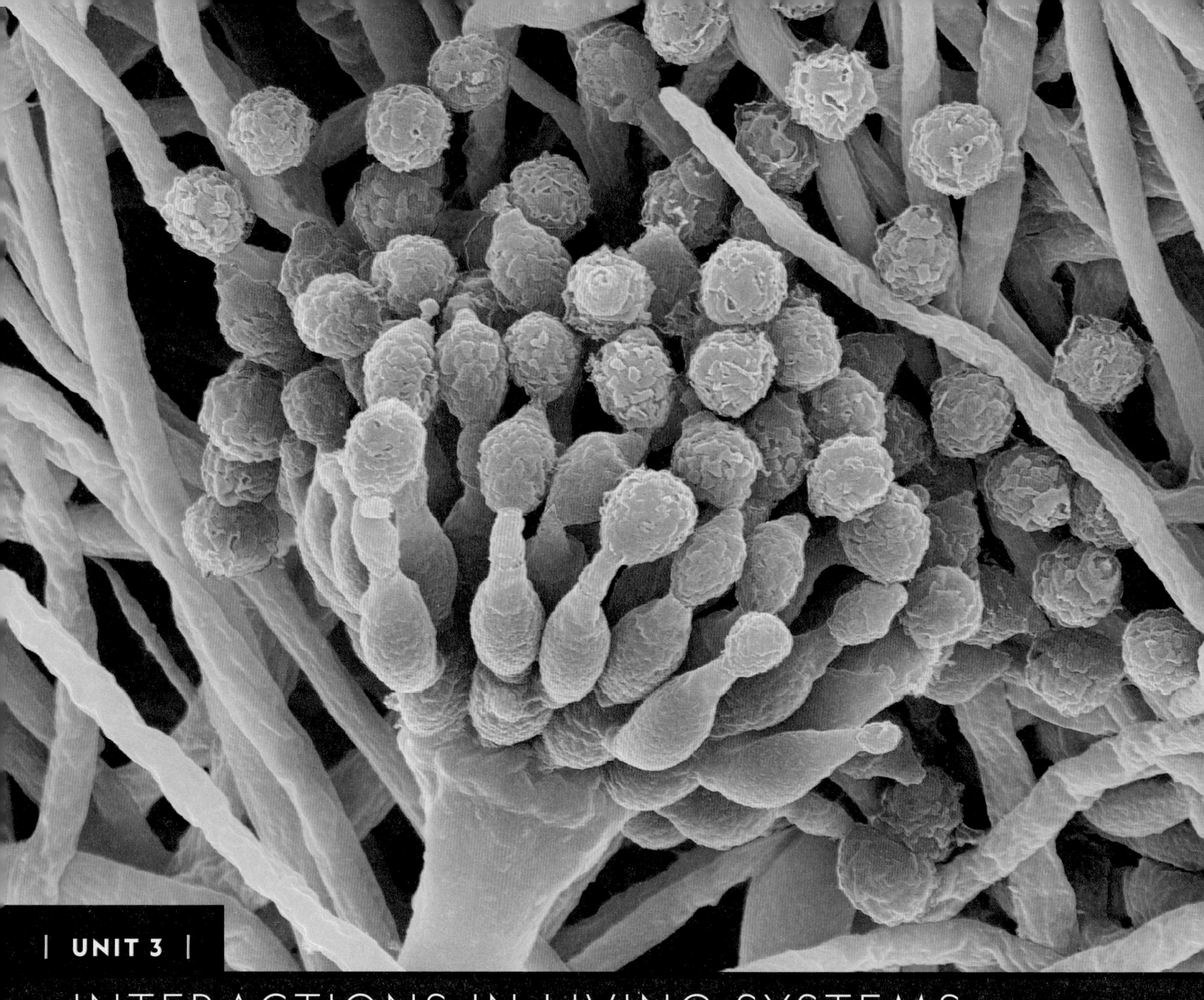

| UNIT 3 |

INTERACTIONS IN LIVING SYSTEMS 204

| UNIT 4 |

GENETICS

CHAPTER 12
GENETIC VARIATION AND HEREDITY **346**

CHAPTER 13
GENETIC TECHNOLOGIES **380**

ABOUT THE COVER Jellyfish are among the most ancient animals in the sea. They have swum gracefully through Earth's aquatic ecosystems for more than half a billion years. Today, hundreds of different species of jellyfish occupy vastly different marine environments, from just below the ocean surface down to the deepest depths.

Jellyfish such as these sea nettles drift with the current, sometimes propelling themselves with jets of water. These prolific carnivores catch food in their tentacles, which are lined with special toxin-producing cells that paralyze their prey.

INTRODUCTION TO BIOLOGY

Three-Dimensional Learning

The practices, core ideas, and crosscutting concepts presented in this chapter's text, investigations, and resources provide support to address the following Performance Expectations: **HS-ETS1-1** and **HS-ETS1-2**.

Science and Engineering Practices	Disciplinary Core Ideas	Crosscutting Concepts
Asking Questions and Defining Problems (HS-ETS1-1) Planning and Carrying Out Investigations Constructing Explanations and Designing Solutions (HS-ETS1-2) Engaging in Argument from Evidence Scientific Investigations Use a Variety of Methods Scientific Knowledge is Based on Empirical Evidence Scientific Knowledge is Open to Revision in Light of New Evidence Science Models, Laws, Mechanisms, and Theories Explain Natural Phenomena	**LS1.A:** Structure and Function **LS2.D:** Social Interactions and Group Behavior **LS4.A:** Evidence of Common Ancestry and Diversity **LS4.B:** Natural Selection **LS4.C:** Adaptation **ETS1.A:** Defining and Delimiting Engineering Problems (HS-ETS1-1) **ETS1.B:** Developing Possible Solutions **ETS1.C:** Optimizing the Design Solution (HS-ETS1-2)	Patterns Scale, Proportion, and Quantity Systems and System Models Structure and Function Science is a Human Endeavor Influence of Engineering, Technology, and Science on Society and the Natural World (HS-ETS1-1)

Contents		Instructional Support for All Learners	Digital Resources
ENGAGE			
002–003	**CHAPTER OVERVIEW** **CASE STUDY** How do we study the natural world?	**Social-Emotional Learning** **SEP** Planning and Carrying Out Investigations **On the Map** The Global Ocean **English Language Learners** Prereading	*Video 1-1*
EXPLORE/EXPLAIN			
004–008 **DCI** LS1.A LS4.A LS4.B LS4.C	**1.1 THE STUDY OF LIFE** • Explain life from a scientific perspective and how it is organized. • Relate DNA and inheritance to the diversity of life on Earth. • Describe how organisms can change over time.	**Differentiated Instruction** Leveled Support **Vocabulary Strategy** Word Wall **Visual Support** Classification Groups **In Your Community** Local Diversity **In Your Community** Genetic Modification **English Language Learners** Spelling and Memorizing Vocabulary **Address Misconceptions** The Pace of Evolution	**SIMULATION** Diversity of Life
008	**MINILAB** EXTRACTING DNA FROM FRUIT		

Contents		Instructional Support for All Learners	Digital Resources
EXPLORE/EXPLAIN			
009–017 DCI LS1.A ETS1.B	**1.2 CONSTRUCTING EXPLANATIONS ABOUT THE NATURAL WORLD** • Analyze the practices of science associated with the different parts of an investigation. • Explain what a system is and how science uses systems to answer questions. • Evaluate the role of bias in data collection and reporting.	**Vocabulary Strategy** Semantic Map **Visual Support** Butterfly Habitats **In Your Community** Monarch Butterflies, Online Reviews **English Language Learners** Graphic Organizers, Listening and Following Directions, Using Pronouns in Writing, Evaluate Arguments **On the Map** Journey of a Lifetime, Butterfly Origins CCC Patterns **Address Misconceptions** How Science Works, Model Limitations SEP Developing and Using Models **Connect to English Language Arts** Write an Argument, Evaluate a Science or Technical Text CONNECTIONS Social-Emotional Learning **On Assignment** Hercules	*Video 1-2* **Investigation A** Making Real-World Observations (50 minutes)
018–025 DCI ETS1.A ETS1.B ETS1.C	**1.3 USING BIOLOGY TO DEVELOP SOLUTIONS** • Describe the practices of engineering. • Differentiate between criteria and constraints and identify how they are utilized when defining a problem. • Design possible solutions to a problem that takes into account a range of criteria and constraints. • Explain why solutions are tested, evaluated, and modified for optimization.	**CASE STUDY** Ask Questions **English Language Learners** Concept Maps **Vocabulary Strategy** Knowledge Rating Scale **Connect to Mathematics** Reason Quantitatively SEP Influence of Science, Engineering, and Technology on Society and the Natural World **Address Misconceptions** Developing Solutions **Connect to English Language Arts** Use Sources in Multiple Formats **BIOTECHNOLOGY FOCUS** Biomimetics	**Investigation B** A Medicine Distribution Solution (110 minutes)
022	**LOOKING AT THE DATA** MASS DISTRIBUTION OF MAMMALS		
025	**EXPLORER** KIM WILLIAMS-GUILLÉN	**Connect to Careers** Biodiversity Conservationist Manager **English Language Learners** Sight Words	
ELABORATE			
025	**TYING IT ALL TOGETHER** How do we study the natural world?	SEP Asking Questions and Defining Problems	*Video 1-3* Self-Reflection and Scoring Rubrics
Online	**Investigation A** Making Real-World Observations **Investigation B** A Medicine Distribution Solution	Open Inquiry (50 minutes) Engineering (110 minutes)	MINDTAP Access all your online assessment and laboratory support on MindTap, including: sample answers, lab guides, rubrics, and worksheets. PDFs are available from the Course Companion Site.
EVALUATE			
026–027	**Chapter 1 Review**		
Online	**Chapter 1 Assessment**		

CHAPTER 1

ENGAGE

As the study of life, biology can be very engaging for students as they explore their diverse world and discover the relevance to their lives. In this chapter, they will look at this diversity of life and learn why it is so exceptional. Students will likely have some prior knowledge; however, they will take a deeper look at topics and apply skill sets to connect ideas, construct explanations, and propose solutions.

About the Photo This photo shows a pod, or group, of sperm whales in their ocean habitat. Sperm whales get their name from a waxy substance called *spermaceti,* which is formed in an oil sac inside their head cavity. The function of the spermaceti is to help a whale focus sounds. Prior to a 1986 moratorium on commercial whaling, sperm whales were nearly hunted to extinction. Though their populations are recovering, sperm whales are still listed as endangered under the Endangered Species Act.

Social-Emotional Learning

Students may have been introduced to many topics in earlier grades. However, at this level, topics will be more complex and expectations for the application of skills higher. You can help students develop a growth mindset by developing **self-awareness skills** that help individuals recognize strengths, as well as limitations, and pursue new ideas with confidence. **Self-management skills** will be important for when students encounter a concept they do not understand to avoid feelings of frustration. Encourage the use of **relationship skills** so students learn to take initiative in asking for help or giving help to another student in need.

CHAPTER 1

INTRODUCTION TO BIOLOGY

An adult female sperm whale, named Aurora by researchers, swims with two calves from her social unit. This type of whale has the biggest brain of any animal known to have lived.

Scientists study organisms such as sperm whales from many different perspectives. While some scientists may study the communication and ecological roles of sperm whales, others explore their physiological or genetic characteristics to better understand these animals.

Chapter 1 supports the NGSS Performance Expectations **HS-ETS1-1** and **HS-ETS1-2**.

SCIENCE AND ENGINEERING PRACTICES
Planning and Carrying out Investigations

Scientific Inquiry This chapter prepares students for approaching each lesson with a scientific mindset that is characterized by values such as open-mindedness, logical thinking, precision, and honest and ethical reporting of findings, among other things. Explain to students that these qualities are important, as science has a rich history of scientists building on and improving on the work of others. For this reason, work must be reported in a way that makes it replicable.

Planning and carrying out investigations is also central to the study of science, as scientific ideas require evidence. Discuss with students what the expectations will be as they carry out investigatory work, such as precision of measurements and reliability of data, as well as using results to refine their work.

CASE STUDY

PROJECT CETI: DECODING WHALE CODAS WITH ARTIFICIAL INTELLIGENCE

HOW DO WE STUDY THE NATURAL WORLD?

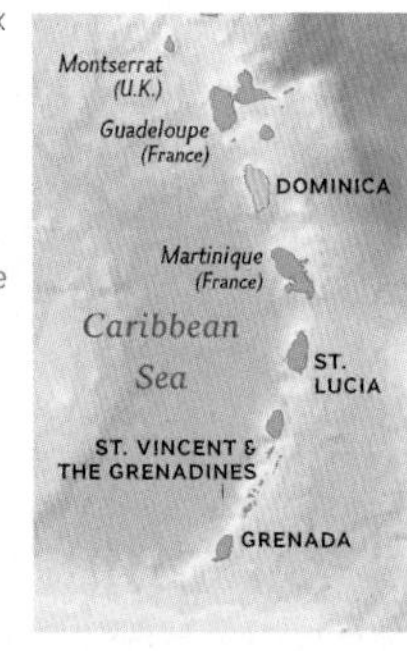

Sperm whales lead complex social lives. At the highest level, their society is organized into different clans, each numbering in the hundreds or thousands. They communicate with one another in short, patterned series of clicks called codas. Marine biologist and National Geographic Explorer Dr. Shane Gero has observed and analyzed thousands of codas from sperm whales around Dominica, an island country in the Caribbean. He has determined that different clans "speak" in different dialects just like American, Australian, and British are different dialects of English. Gero and others have shown that individual whales use specific click patterns to identify themselves as if by name. As calves, whales appear to babble-click much as human babies learn language by mimicking the sounds they hear at home.

Although plenty of animals communicate, the general consensus among scientists is that only humans use true language. Gero and another National Geographic Explorer, marine biologist Dr. David Gruber, are challenging this view. They want to leverage the latest advances in artificial intelligence (AI) to translate codas into something humans can understand. One day, we might even be able to talk to whales directly.

As an initial test of this idea, Gruber sent Gero's recordings to a team of computer scientists who specialize in machine learning, a subfield of artificial intelligence. The team applied AI techniques to Gero's recordings and trained a computer to recognize sperm whale clans based on their codas. The computer was able to correctly identify a whale's clan more than 93 percent of the time.

Excited by the result, Gruber founded Project CETI (**Ce**tacean **T**ranslation **I**nitiative) with a team of scientists to build on Gero's work. (Cetaceans are whales, dolphins, and porpoises.) Along with Gero, the team includes another National Geographic Explorer, roboticist Dr. Robert Wood, as well as experts in linguistics, machine learning, and camera engineering.

Ask Questions *In this chapter, you will learn about the practices scientists and engineers use to study the natural world. As you read, generate questions about how you think scientists and engineers work—including how they are similar and different.*

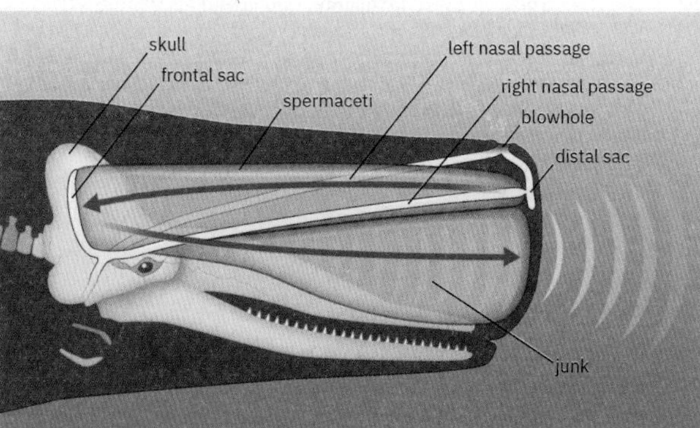

FIGURE 1-1 A sperm whale makes sound by snapping together two lip-like structures, called phonic lips, inside its head near the blowhole. The sound travels through the spermaceti organ, from the distal air sac to the frontal air sac near the whale's skull. From there, the sound reflects forward through an oil-filled organ, called the junk, to emerge from the front of the whale's head and travel through the water.

VIDEO 1-1 Go online to learn more from National Geographic Explorer Dr. Shane Gero about how families of sperm whales communicate.

DIFFERENTIATED INSTRUCTION | English Language Learners

Prereading Before students read the Case Study, explain that it is about how scientists research whale communication. Have students preview **Figure 1-1**. Then, preteach vocabulary such as *clan* ("family group"), *coda* ("a short, patterned series of clicks"), *dialect* ("a subgroup of a language"), and *machine learning*.

Beginning Have pairs point to the diagram and take turns reading the labels, sounding out words as needed. Then have them write the vocabulary words in their notebooks.

Intermediate Have pairs read the diagram and caption. Then have them write each vocabulary word and its definition in their notebooks.

Advanced Have students describe the diagram and restate the caption in their own words using each vocabulary word in a sentence.

CASE STUDY

ENGAGE

This Case Study focuses on the work of scientists who study communication among sperm whales by using technology to help decipher sperm whale sounds. Tell students that this is not the first time that humans have communicated with animals. Koko the gorilla was famous for communicating with researchers using sign language. These studies began in 1972 and continued until 2018, when Koko died. Discuss with students what advances in technology have enhanced scientists' ability to study animal communications. In the Tying It All Together activity, students will ask questions and construct an explanation of how science and engineering practices can be used to study whale communication.

Ask Questions Students should revisit the Case Study as they read the chapter to make connections with the content. See a specific suggestion in Section 1.3.

On the Map

The Global Ocean Have students study the map showing where the study of sperm whales is taking place. Ask them if they think sperm whales are found in other places around the world and to explain their reasoning. Tell students that sperm whales have one of the widest global distributions of any marine mammal. They can be found from the equator, where Dominica is located, to the waters of the Arctic and Antarctica. Use a globe or world map to show the size of the area.

Go Online VIDEO

Time: 1:49

Use **Video 1-1** to discuss how animals communicate. Ask students why it might be important to fund scientific research on animal communication. List their ideas on chart paper or the board. (Sample answers: *to learn more about nature, to extend the ability to communicate across organisms, to better understand emotions similar to humans*)

1.1

THE STUDY OF LIFE

LS1.A, LS4.A, LS4.B, LS4.C

EXPLORE/EXPLAIN

This section introduces students to biology, the study of living organisms. This is followed by a discussion of how scientists define life, as well as how life on Earth is organized using a system of taxonomy based on shared characteristics. Students are then introduced to DNA and inheritance and the flow of information that makes each organism unique, as well as how variations in DNA enable organisms to change over time.

Objectives

- Explain life from a scientific perspective and how it is organized.
- Relate DNA and inheritance to the diversity of life on Earth.
- Describe how organisms can change over time.

Pretest Use **Questions 1, 2,** and **3** to identify gaps in background knowledge or misconceptions.

Describe *Students might say that they interact with both humans and a dog or other pet regularly. Students may generally describe their characteristic body shapes and limbs, as well as any behaviors that are similar and different between the organisms they have chosen.*

Go Online SIMULATION

Diversity of Life Use the interactive version of **Figure 1-3** to have students compare and discuss each organism and its ecosystem. Then, have each student choose one particular structure or behavior of interest and explain to a partner how it is relevant to how the organism functions and survives in its ecosystem.

1.1

KEY TERMS

adaptation
biology
evolution
genome
genus
species
taxonomy

THE STUDY OF LIFE

On your way to school, maybe you pass by a grassy field, or walk under a tree, or step over weeds growing in a cracked sidewalk. No matter where you are, living things are all around you, on you, and even within you.

Describe *What living things can you think of that you observe or interact with regularly? Describe at least four different types of living things you are familiar with. What do they have in common? How are they different?*

The Diversity of Life

Life is incredibly diverse. A **species** is a group of closely related living things. On Earth, there are an estimated 8.7 million different species of plants and animals, and scientists have described only about 1.2 million of them. If you consider microbes, some scientists estimate that there may be as many as one trillion different species!

Go Online SIMULATION

FIGURE 1-3 Go online to begin exploring the Diversity of Life on Earth.

While some species are easy to find and observe, others are incredibly tiny and require special tools to see, and some can only be studied in the place where they live. Sometimes that place is difficult to reach, whether it is at the bottom of the ocean or inside another organism. **Biology** is the scientific study of living organisms. In this chapter and throughout this book, we will explore hundreds of organisms and learn how scientists study the world. Unit 1 explores the many ways in which organisms interact with each other and with their environment.

Defining Life Life has nested levels of organization. Cells are the building blocks of life. Cells include many parts that work together. These parts are made of molecules, and molecules are made up of atoms, which are the building blocks of all matter. A living organism consists of one or more cells. Some cells live and reproduce independently. Other cells live as part of a multicellular organism. Unit 2 describes cells and how they function, and Unit 3 explores how whole organisms are organized.

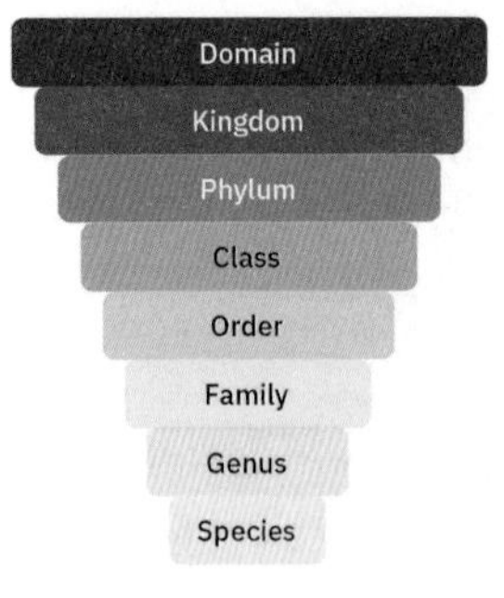

FIGURE 1-2 Moving from the top to the bottom of the pyramid, each taxonomic level becomes more specific and includes fewer organisms.

Classifying Life Each time a new species is discovered, it is named and classified, a practice called **taxonomy**. Taxonomy organizes organisms with shared characteristics into a series of eight progressively narrower categories, starting with the domains of life, then kingdoms, all the way down to species (**Figure 1-2**; see also Appendix E). Although biologists have used many of the same categories for over two centuries, taxonomy continually changes as scientific advances reveal increasingly detailed information about the similarities and differences between species. The largest category, the domain, emerged when advances in genetic technologies enabled scientists to classify all species into three groups based on molecular-level distinctions.

Each species is given a unique two-part scientific name. The first part of a scientific name is the **genus**, which is defined as a group of species that share a unique set of features. The second part of a scientific name is the name of the species. When both parts are combined, a scientific name identifies one unique organism. **Figure 1-3** shows only a tiny sample of the diversity of life found on Earth. The common name and scientific name are indicated for each species.

DIFFERENTIATED INSTRUCTION | Leveled Support

Use Mnemonics Explain to students that a mnemonic is a tool used to help people remember a long list of items in a series.

Struggling Students Provide a mnemonic to help students remember the levels of biological classification. For example, they may use **Dear King Philip Came Over For Good Spaghetti** to remember Domain, Kingdom, Phylum, Class, Order, Family, Genus, Species. To level up the thinking process, have students work in groups to develop their own mnemonic for the classification levels.

Advanced Learners You may provide the same mnemonic to students who have a solid understanding of the levels of classification, explaining that these terms will be used throughout their study of biology. To extend this, have students use the mnemonic to create a chart that includes pictures of organisms to represent each level.

FIGURE 1-3
The diversity of living organisms is demonstrated by these organisms: Australian-spotted jellyfish, Grandidier's baobab, salmonella, water bear, Philippine tarsier, and peacock spider.

Vocabulary Strategy

Word Wall Students are introduced to vocabulary words that may be used throughout their study of biology. Using a large display area, write the word "Biology" in the center. As students learn new vocabulary, have them add the word to the board. You can make this word wall interactive by having students add notes and other realia that will help them remember and be confident in using each term. English Language Learners may also write notes in their first language to help them build their language skills and biology knowledge.

Visual Support

Classification Groups As a class, examine each of the organisms in **Figure 1-3**. Then, have a discussion in which students describe what characteristics the organisms have in common and how they are different. Share with the class that the three domains of life are Archaea, Bacteria, and Eukarya. Encourage students to do quick Internet searches for photos of other organisms in each of these three groups to compare and contrast.

In Your Community

Local Diversity Students may think they have to go to distant areas to find great diversity among living things. Discuss with students that diversity of life can be found at different scales. For example, diverse populations exist under a fallen tree or in temporary vernal pools. Or, on a larger scale, parks and communities can be home to diverse organisms. Choose a practical location and take students outdoors with notebooks to record the living things they find. If an outdoor opportunity does not exist, search online for photos of local areas that show wildlife. Present these photos to students and ask them to record the living things they see. Encourage students to make their observations at different scales and to record details, such as where they observed each organism. Supply students with magnifyers or pocket microscopes to help them observe organisms and structures that are not visible to the naked eye.

In Your Community

Genetic Modification Ask students if they are familiar with labels that say GMO on packaging for produce or other food products. GMO stands for Genetically Modified Organism, or food that has been genetically modified by transferring genes from one organism to another organism. Point out that manipulation of plants and animals has occurred for thousands of years, as farmers used selective breeding to produce more robust crops and animals. However, genetic engineering is a relatively new practice that produces organisms with desired traits, such as disease- and drought-resistance or animals that are larger and more productive. This technique can also be used to develop organisms, such as fungi and bacteria, that produce medicines. Have students research genetically modified practices and guide them in an evidence-based debate about the pros and cons of GMOs.

Genetically Modified Organisms Students delve into this topic in depth in **Chapter 13** after learning more about genes and how they govern traits.

SEP Construct an Explanation *Students might say that they have green eyes, which are different from the blue eyes of their classmate. These characteristics are determined by small differences in the DNA sequences related to eye color.*

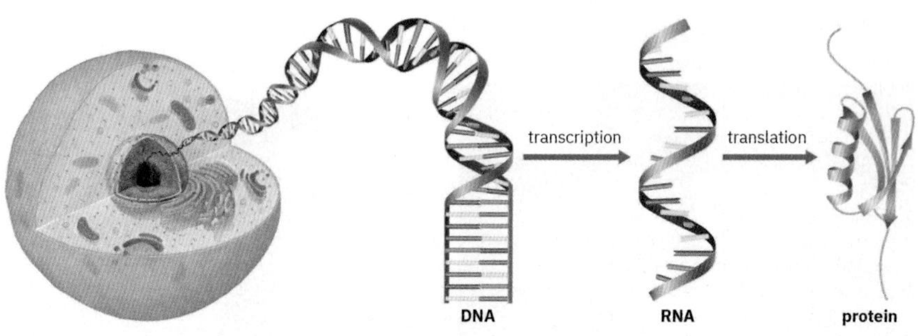

FIGURE 1-4 Cells, such as the animal cell modeled here, contain DNA. Each color bar on the DNA strand represents a different base, or character, of the four that make up the genetic code. These four bases code for all the animal's life functions and characteristics.

DNA and Genetic Variation

Every organism on Earth has a genome that is contained in each of its cells. A **genome** is the complete set of instructions that makes each organism different from other organisms. The genome is a sequence of DNA (deoxyribonucleic acid) that encodes the information for making RNA (ribonucleic acid) and protein molecules that perform all the functions required for life (**Figure 1-4**). The genome in each cell of an organism programs development and growth. It also guides ongoing activities that sustain an organism throughout its lifetime. Unit 4 details how genetic material codes for processes and characteristics that make each organism distinct.

Inheritance Inheritance, the passing of DNA and the traits it encodes from parent to offspring, occurs during reproduction. All organisms inherit their DNA from one or more parents. Individuals of the same species are alike in most aspects of body form, function, and behavior because their DNA is very similar. Humans look and act like humans and not like sunflowers because they inherited human DNA, which differs from sunflower DNA in the information it carries.

Genetic Variation Look at **Figure 1-3** and think about other organisms you know about. Organisms of different species have different sizes and shapes, and they use a variety of mechanisms and behaviors to survive and reproduce. However, these diverse organisms are unified by the four-base DNA code that all living organisms share. Each organism's unique genome sequence encodes information that leads to their particular characteristics.

In addition to the often obvious physical differences between species, individuals of almost every natural population (a group composed of a single species) also vary—just a little bit—from one another. For example, one human has green eyes, the next has brown eyes, and so on. Such variation arises from small differences in the sequences of DNA molecules. As you will learn in Units 4 and 5, differences among individuals of a species are the raw material of evolutionary processes.

SEP Construct an Explanation *Identify one physical trait that you and all of your classmates share, and another that differs between individual classmates. Explain why some traits are common to all individuals within a species, while others vary from one individual to another.*

DIFFERENTIATED LEARNING | English Language Learners

Spelling and Memorizing Vocabulary As students encounter new vocabulary, have them practice spelling the words and memorize the definitions. They could make flashcards or a personal dictionary to help them keep track of new words they learn.

Beginning Have students spell and sound out terms such as *genome*. Help them relate *gen-* to the word *gene*. In the second syllable, point out that the silent *e* at the end of the word makes the *o* a long-vowel sound.

Intermediate As students spell words such as *genetic* and *ribonucleic acid*, point out spelling patterns such as the adjective ending *-ic*.

Advanced Have students spell words such as *deoxyribonucleic acid* and *ribonucleic acid* and use word parts to explain how the meanings are different.

Evolution and Adaptation

You and every organism in this book are the result of millions of years of evolution. **Evolution** describes the ways in which a population of a species changes through time, over many generations. The variation in DNA among individuals is what enables species to evolve. Every time organisms reproduce, their offspring potentially inherit minor changes in DNA.

Some changes in DNA result in molecular- and cellular-level differences that lead to an advantage that helps an organism thrive in its environment. These inheritable advantages that are passed from generation to generation are called **adaptations**. For example, the kangaroo rat shown in **Figure 1-5** lives in the hot, dry deserts of North America. It has an unusual ability to extract water from the seeds it eats and can survive without drinking water because of its specialized kidneys. You will encounter examples of adaptations throughout this book. Unit 5 describes the mechanisms and evidence of evolution in detail.

FIGURE 1-5 Many adaptations are observable as an organism's physical characteristics or behaviors. This kangaroo rat moves at night to conserve water, and it has large hind legs and feet to help it escape predators.

CCC Structure and Function *Consider the variety of adaptations described for the kangaroo rat. Now look back at* ***Figure 1-3****. Can you identify any possible adaptations that might relate to how those organisms live in their habitats?*

1.1 REVIEW

1. **Sequence** List these terms by level of organization from simplest to most complex.
   - atoms
   - molecules
   - cells
   - organisms
2. **Describe** How could a population of a species change over a long period of time?
3. **Identify** What categories are used in taxonomy to name unique organisms? Select all correct answers.
   A. DNA
   B. genus
   C. species
   D. genome complexity
   E. number of adaptations
4. **Define** What is a genome?

Address Misconceptions

The Pace of Evolution When people think of evolution, they often think of a process that occurs over incredible time spans of millions of years. However, that is not always the case. Take, for example, the case of cichlids in Africa in which, through adaptive radiation, the fish developed into about 1,200 new species in less than 100,000 years. Adaptive radiation is the process by which multiple species evolve from a single ancestor. After being dispersed from the Nile River into three lakes in Africa, the cichlid population separated into new populations that were better adapted to the food resources and specific characteristics of their new environments. The adaptations were both physical and behavioral. One population became specialized in eating the scales of other fish. Another population developed behavioral adaptations that helped those fish survive by living among rocks and hunting prey. Another population's adaptation of carrying fertilized eggs in the mouth enabled those fish to live in turbulent waters. Cichlids demonstrate how evolution can occur over relatively short time scales. Connect the story of the cichlids being transported to new environments to the use of pesticides and herbicides to eradicate insects and weeds. The use of pesticides and herbicides has accelerated resistance to these substances in some populations. Organisms that have the resistance traits survive these chemicals and give rise to offspring with the similar adaptive traits.

CCC Structure and Function *Answers will vary. Students should describe characteristics that they can observe in one or more of the organisms. Students should plausibly explain how each characteristic may serve as an adaptation that helps the animal survive in its habitat. For example, students may point out how the tarsier's long fingers must be useful to help it grip and that its large eyes may be related to enhanced night vision.*

1.1 REVIEW

1. *atoms, molecules, cells, organisms* **DOK 1**
2. *Student answers may vary but should include that DNA can change slightly with each generation of organisms, and these minor changes can eventually change an entire population of a species.* **DOK 2**
3. *genus; species* **DOK 1**
4. *Sample answer: A genome is the complete set of DNA for an organism.* **DOK 1**

MINILAB
EXTRACTING DNA FROM FRUIT

In this activity, students use common materials to extract and observe DNA. Students will infer that fruit cells have DNA and will also observe the clumped strands of DNA. Students will then answer questions using evidence they gather.

Time: 30 minutes

Materials and Advance Preparation

- Strawberries are commonly used because they have relatively high amounts of DNA (7 chromosomes with 8 copies of each chromosome for a total of 56). Other options include bananas or wheat germ, though the sample amounts per student will need to be measured.
- Prepare the extraction mixture of water, alcohol, and salt before the activity to save time. Mix 1,000 mL of water with 10 g of salt and 100 mL of dish soap. This solution does not need to be exact.
- Use sturdy, sandwich-sized seal-top plastic bags. If the students vigorously squeeze a small bag, they can easily rupture it.
- The solubility of DNA in alcohol is lower when alcohol is cold. If you have a freezer available, chill the alcohol before use. *CAUTION: Students should be careful when handling isopropyl alcohol, which can be absorbed through skin contact. Have students wipe any spills and wash their hands.*

Procedure

In **Step 1**, students can cut the strawberries on the paper towel.

In **Step 2**, make sure students remove the air from the bag before kneading. Any air trapped inside the bag could cause the bag to burst.

In **Step 4**, the longer students knead, the more DNA will be extracted. If time allows, have students knead for up to five minutes.

In **Step 7**, be sure that students tilt the cup containing the strawberry mixture before they slowly pour in the alcohol.

MINILAB
EXTRACTING DNA FROM FRUIT

SEP Constructing Explanations What are some observable properties of DNA?

DNA is the basis of all living things. It is important to the study of life, which includes exploring such things as the structure and function of organisms and the systems in which they live. In this activity, you will use everyday materials to extract DNA from strawberries.

Strawberry fruits turn red when they ripen.

Materials

- strawberries (3)
- knife
- small seal-top plastic bag
- transparent cups or 250-mL beakers (3)
- dish soap, 10 mL
- salt, 1 g
- water, 100 mL
- small stirrer
- filter paper
- alcohol, 50 mL (pure ethanol or 91% isopropyl alcohol)
- paper towels

Safety     

Procedure

1. Use the knife to cut away any leaves from the strawberries. Then chop the strawberries into cubes. *CAUTION: Be careful when using sharp objects.*
2. Place the strawberries in the seal-top bag. Remove as much air from the bag as possible and seal it. Then gently knead the strawberries with your hand. Do not apply too much pressure, which can cause the bag to rupture. Continue until the strawberries are a fine paste.
3. In a cup, mix 100 mL of water with 1 g of salt and 10 mL of dish soap. Stir with the stirrer.
4. Open the seal-top bag and pour in the dish soap–salt solution. Seal the bag and gently mash it so that the solution and the strawberry puree are mixed well. Again, be careful not to rupture the bag. Continue kneading the mixture for at least two minutes.
5. Place the filter paper partially inside a cup. You can fold the extra paper outside the cup and hold it if necessary.
6. Open the seal-top bag. Carefully pour the mixture into the filter to filter out the solids. Dispose of the filter and the remaining strawberry mixture in the trash.
7. Pour 50 mL of alcohol into another cup. Tilt the cup with the strawberry mixture, then slowly pour the alcohol down the inside of the cup. There should be a thin layer of alcohol floating on top of the strawberry mixture.
8. The white substance that forms (precipitates) below the alcohol is the strawberry DNA. Gently wrap the white substance around the stirrer and remove from the cup.
9. Once you are done, clean up and return unused materials to the materials station.

Results and Analysis

1. **Describe** Describe the appearance of the DNA sample.
2. **Analyze** Does the DNA appear as you expected? Why or why not?
3. **Evaluate** Why did you need to observe a very large sample of DNA rather than a sample from a single cell? How would your observations differ if you could investigate at the microscopic level?

Results and Analysis

1. **Describe** *Sample answer: Inside the cup, the DNA looks like a white mass of threads. Outside the cup, the DNA looks like a lump of white jelly.* **DOK 1**
2. **Analyze** *Sample answer: The DNA does not appear as I thought it would. It was long and stringy, but I could not see the twisted ladder structure.* **DOK 2**
3. **Evaluate** *Sample answer: We had a very large sample size because we do not have the equipment to work at the level of a single cell. If we had a high-powered microscope, for example, we would be able to observe a smaller sample with more detail rather than a large a lump of DNA.* **DOK 3**

CONSTRUCTING EXPLANATIONS ABOUT THE NATURAL WORLD

You have likely used scientific thinking in your everyday life. Suppose you wanted to tend to an ailing houseplant. Was it getting too much sunlight or too little? Had it been watered too much or not enough? Could there be a problem with the soil? You could decide on one thing to change and observe the result. Biologists perform these types of activities all the time. Using an orderly, consistent approach to answer questions about the natural world helps assure that their conclusions are meaningful and reliable and can be verified by others.

Explain *Describe a question that you answered using scientific thinking. Explain the process you used to reach your conclusions.*

1.2

KEY TERMS

feedback
model
system

Using Scientific Thinking to Answer Questions

The author Vladimir Nabokov once said, "I cannot separate the aesthetic pleasure of seeing a butterfly and the scientific pleasure of knowing what it is." You have probably seen butterflies at some point or another in your life. Perhaps you have even wondered about their movements around the planet. National Geographic Explorer Dr. Gerard Talavera (**Figure 1-6**) studies exactly this question: Where does the vibrant, multicolored painted lady butterfly (*Vanessa cardui*) of Northern Europe migrate when the weather turns cold? What are its migration patterns?

FIGURE 1-6 Dr. Gerard Talavera surveys butterfly populations in southern Ghana.

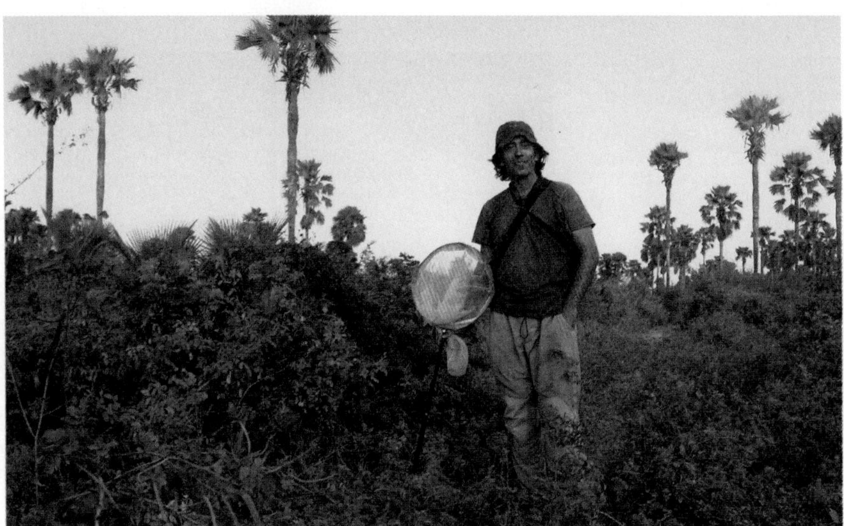

1.2

CONSTRUCTING EXPLANATIONS ABOUT THE NATURAL WORLD

LS1.A, ETS1.B

EXPLORE/EXPLAIN

This section provides a review of the role of science as a means for answering questions and solving problems, introduces eight practices of scientific inquiry, and explores how science is a human endeavor and subject to human error, biases, and misinterpretation.

Objectives

- Analyze the practices of science associated with the different parts of an investigation.
- Explain what a system is and how science uses systems to answer questions.
- Evaluate the role of bias in data collection and reporting.

Pretest Use **Questions 4** and **5** to identify gaps in background knowledge or misconceptions.

Vocabulary Strategy

Semantic Map Have students create a semantic map for the vocabulary in this section. Begin by choosing a key term that is central to the topic. Have students write that word at the top of a page. As they read through each section, encourage students to add arrows with new words to the map to demonstrate the relationship between the terms. For example, a student may begin with the term *model* at the top. Then, a box with several components or objects inside to represent a *system*. Students can place an arrow between the two terms to demonstrate how models can be used to make sense of a system.

Explain *Students should describe an example in which they have used deductive reasoning to come to a conclusion. Ideally, the students will explain how they made comparisons to make inferences (akin to using a control group versus an experimental group) or other scientific practices that validated their observations.*

Visual Support

Butterfly Habitats Direct students' attention to **Figure 1-7**. Guide them to the understanding that butterflies do not live in cold climates by having them look for similarities in the environment for each stage of the butterfly life cycle. Students should recognize that the environments shown in **Figure 1-7** show green leaves, indicating regions that are experiencing warm weather. Lead a class discussion about the different regions of the world where students think butterflies live and the corresponding season. Tie this information into the observation that multiple generations of butterflies take part in a migratory circuit.

Allow students to discuss and debate the hypothesis presented. Encourage students to brainstorm within their groups and provide possible alternative hypotheses. If time permits, allow students to research their hypothesis for evidence that supports it.

Go Online VIDEO

Time: 0:65

Use **Video 1-2** to elaborate on the painted lady butterflies at different stages of life.

In Your Community

Monarch Butterflies Students may be more familiar with monarch butterflies, which have a similar life cycle to painted ladies. Monarch butterflies have a typical life span of a few weeks. This means 3–4 generations of butterflies occur over the course of a single summer. In the fall, the last generation delays sexual maturity and migrates to their overwintering grounds. This generation lives about eight months. Once spring arrives, the butterflies become sexually mature and begin migrating north. It takes 3–5 generations to repopulate the monarchs' summer grounds.

Encourage students to research monarch migration routes and determine if they will be in your community. Students can also learn how to increase the amount of monarch-friendly habitat in your area.

The painted lady butterfly is found on nearly every continent. The species cannot survive for long in cold temperatures and is unable to hibernate, so it must migrate in response to seasonal changes. Scientists have debated how painted lady butterflies from as far north as Scandinavia are able to make the journey to warmer climates in Africa and back each year. Some argued that a single butterfly, with a life cycle of only 4–5 weeks (**Figure 1-7**), could not make so long a journey. They proposed that multiple generations of butterflies—perhaps up to 10—must take part in completing a single annual migratory circuit.

Talavera studies the migratory patterns of butterflies. A trip from Northern Europe to Africa includes a flight across the Mediterranean Sea, a distance of hundreds of kilometers over open water. When the butterflies reach Africa, they seek food and a safe place to rest. From this point they complete another incredibly long journey over Earth's largest hot desert, the Sahara. The butterflies reach their final destination, the lush African savanna, during the end of the rainy season. Is it possible that a single generation can complete this trip of more than 4000 kilometers? Based on field observations, Talavera hypothesized this was the case.

Talavera and the team he works with needed to develop a plan to investigate the migration phenomenon. They came up with multiple ways to test their predictions. One method was to use the pollen carried by individual butterflies to identify the plant species they visited and map the butterflies to specific regions along a migratory path. Another method was to analyze the chemistry of the butterflies' wings to model their migration patterns based on the regional chemical makeup of the food sources they would have eaten as caterpillars. Talavera also modeled the butterflies' ecological niches to predict their preferred breeding grounds over multiple generations. In addition, he used genetic information to understand the relationships between different populations and how these populations migrate all over the world. This effort required the collection of hundreds of specimens out in the field along the migratory path.

FIGURE 1-7 *Vanessa cardui* goes through four life stages: egg (a), caterpillar (b), chrysalis (c), and butterfly (d). Caterpillars are herbivores, so a large number of caterpillars hatching in an area can have an impact on the ecosystem in which they live.

VIDEO

VIDEO 1-2 Go online to observe painted lady butterflies at different stages of life.

DIFFERENTIATED INSTRUCTION | English Language Learners

Graphic Organizers Use a graphic organizer not only to take notes but also to understand what the text will be about. Before they read the section *Using Scientific Thinking to Answer Questions,* have students make a three-column chart with the headings: *Hypothesis, Ways to Test It,* and *Conclusions.* Point out that they will use paragraphs 2 and 3 to complete the left-hand column, paragraphs 4 and 5 to complete the middle column, and paragraphs 6, 7, and 8 to complete the right-hand column.

Beginning Have pairs work together to complete the chart in their first language or underline sentences from the text that fit each column.

Intermediate Have pairs complete the chart using phrases and short sentences. Encourage them to underline details in the text that helped them.

Advanced Have students individually complete the chart in their own words. Then have pairs compare notes and add any details they missed.

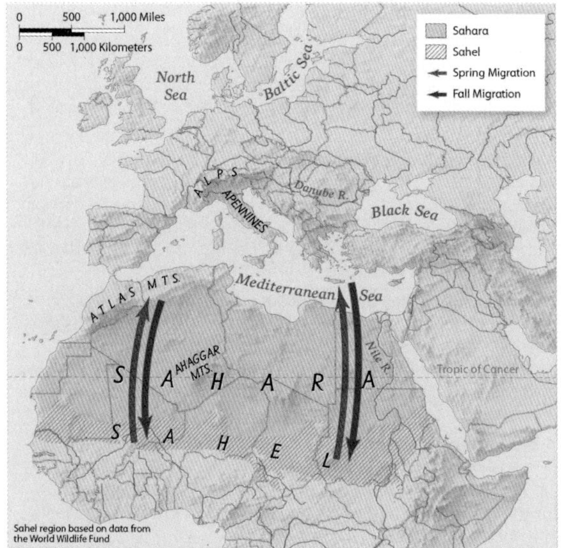

FIGURE 1-8 This map illustrates the journey taken by painted lady butterflies as they migrate from Europe and Northern Africa to central Africa in the fall, and vice versa in the spring. The butterfly's range extends across nearly this entire map, with the exception of the Sahara (and other desert regions), which the butterflies fly over. The Sahel is a transition region between the Sahara desert and the tropical savanna region to the south.

The scientists compared the data across multiple groups of butterflies to determine their migratory origins and destinations. Based on this evidence, Talavera and other scientists were able to model the migration patterns of painted lady butterflies on a map (**Figure 1-8**). They argued that the butterflies can indeed make the southward trans-Mediterranean journey in a single generation, flying from Northern Europe to the African savanna over one lifetime.

Researchers think the butterflies make use of patterns in wind gusts to move across long passages of the journey, sometimes flying as fast as 50 kilometers per hour at altitudes as high as 500 meters. Once they reach the African savanna, the butterflies produce several generations of offspring. These offspring progressively migrate further south, toward the tropics. When the land gets dry and food resources are in short supply, the butterflies migrate north again, reaching the Mediterranean in early spring.

The butterflies making the return journey to Northern Europe may be several generations removed from the original painted ladies that flew southward across the Mediterranean. Both of these long journeys, revealed by the work of Talavera and his colleagues, suggests that the butterflies' migration is truly remarkable from a human perspective.

Investigating the migratory patterns of *Vanessa cardui* and other butterflies is both fascinating and essential. Butterflies are important pollinators, helping plants to reproduce by spreading their pollen. Studying their behaviors and movements across continents increases our understanding of how pollinators sustain many other types of life on Earth.

CCC Patterns *Talavera was able to identify patterns in butterfly behavior and support his arguments using evidence. When have you observed a pattern and used it to draw a conclusion?*

On the Map

Journey of a Lifetime The butterfly migration route from Scandinavia to Africa is over 4,000 kilometers. This can be an abstract concept to students. To help students visualize the magnitude of this distance, share that a human walks an average of 5–6 kilometers per hour. At this rate, it would take a human 28–33 days to travel the butterfly migration route. Lead a class discussion on how humans could use resources to increase their walking speed. Then, relate this to the observations made by the scientists regarding the wind patterns used by the butterflies to increase their speed and shorten their migration time.

CCC Patterns *Answers will vary. Students should be able to describe a recurring observation and how it led them to a conclusion.*

CROSSCUTTING CONCEPTS | Patterns

Butterfly Migration Students explore the concept of identifying patterns as they learn how generations of butterflies complete an annual migration route and analyze the routes on a map. Emphasize to students that identifying trends and patterns in data is an important skill in science, as it can lead to evidence that either supports or does not support a proposed hypothesis. Focus student attention as they read about the migratory observations of the painted lady butterfly. Have students create a T-chart that lists evidence for or against the hypothesis.

Address Misconceptions

How Science Works Although students have been through several science courses by this age, they may still harbor misconceptions about the scientific process, such as that there is a single scientific method, that the process is purely analytical with no creativity, or that scientists' observations directly tell them how things work and that nothing is inferred. Set the tone for engaging students in the practices of science by reviewing the work of National Geographic Explorer Dr. Gerard Talavera through this lens. Have students look back on the information presented in the section *Using Scientific Thinking to Answer Questions* and categorize the actions Talavera took into the different scientific practices. Encourage discussion if students categorize the same action into different practices.

CHAPTER INVESTIGATION A

Open Inquiry *Making Real-World Observations*

Time: 50 minutes

Students collect data on the characteristics of an organism and its environment through observation and measurements. Then, they create graphs and other representations of the data to identify trends and analyze their findings to compare their organism to others.

Go online to access detailed teacher notes, answers, rubrics, and lab worksheets.

The Practices of Science

The central purpose of all science, including biology, is the gathering and sharing of information. For this information to be meaningful and reliable, scientists must approach their research as objectively as possible and with a shared set of values and practices. There are eight central practices of scientific inquiry.

Asking Questions Scientists ask questions about how the natural world works. Scientific questions can be investigated through observation and measurement. Scientists must aim to establish what is already known and determine what questions have not been answered. Scientists also ask questions to clarify the ideas of others. Scientists use this information to develop hypotheses to test.

Planning and Carrying Out Investigations Scientists plan and carry out investigations to test hypotheses in the field or laboratory. They must clearly identify all meaningful data and determine the variables and parameters they will set to obtain these data. Scientists use standardized methods, procedures, and measurement systems to minimize error and make valid comparisons between results.

Analyzing and Interpreting Data The data that result from scientific investigations must be interpreted to identify patterns and trends. Data can be qualitative, involving descriptions, or quantitative, involving measurements consisting of numbers and units. Scientists must identify sources of error in results.

Developing and Using Models A **model** is a tool for representing an idea or an explanation of observed phenomena. Models help scientists formulate questions as they proceed to a fuller understanding of the behavior of the world. A model can be a diagram, drawing, physical replica, mathematical representation, or computer simulation.

Constructing Explanations Scientists use evidence to construct explanations that fit their observations. New explanations are accepted when they are better able to explain phenomena than previous ideas.

Engaging in Argument from Evidence Scientists use evidence to defend and critique claims and explanations about the natural world. Scientists engage in argumentation at all stages of the scientific process.

Using Mathematics and Computational Thinking Scientists use mathematics to represent variables and their relationships. Mathematical models and computational tools help scientists make and test predictions. Statistical methods help scientists identify data patterns and correlations.

Obtaining, Evaluating, and Communicating Information Scientists must be able to communicate their findings clearly and persuasively. Information can be presented in a variety of ways, both written and oral. Scientists critique ideas to evaluate the validity of arguments from a variety of sources.

These eight practices show that science is a process, not a single set of facts or ideas. Scientists engage in all of the practices of science at different stages of their research, often more than one practice at a time, and not necessarily in any specific order. As scientists gain greater knowledge about the world, they use the evidence they gather to make further inferences. An inference is an interpretation of data based on information that is already known.

SCIENCE AND ENGINEERING PRACTICES
Developing and Using Models

System Models Help students understand how scientists use models to visualize and interpret data. Give small groups time to identify a system, such as a thermostat, ATM, or alarm clock, and its parameters and then determine how the group could create a model of it. Have students use their models to explain the boundaries that were applied to their system, how the components within the system interact, any system inputs and outputs, and how feedback from the system can affect how the system works. As students share their systems with other groups, allow them to look for similarities between the systems. Focus students' attention on models that are similar but with varying boundaries. Use these models to reinforce the importance of boundaries for focusing attention on specific information that can be determined from a system.

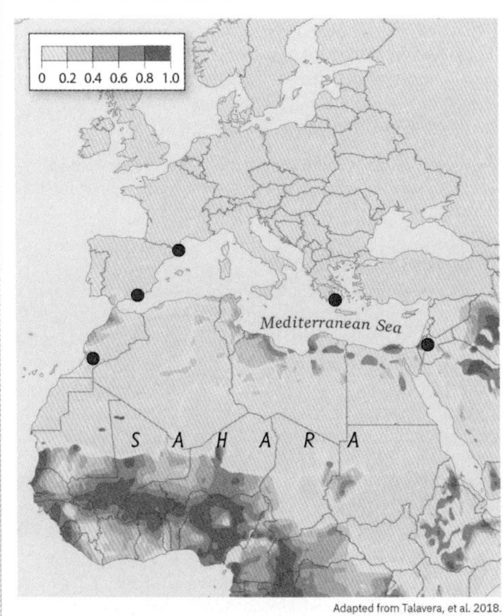

The red circles indicate five sites where Talavera and colleagues captured and analyzed butterflies. At each of these locations, butterflies with a common chemical composition were identified. These butterflies were considered to be one group in the study.

The scale and colors on the map approximately show the predicted probability of birthplace regions for the butterflies in this group. The darkest areas on the map represent the most likely birthplaces.

Adapted from Talavera, et al. 2018.

FIGURE 1-9 Talavera sampled the chemical composition of butterfly wings from many locations at different times of year. Combining these data with wind patterns and other observations, he was able to identify several groups of butterflies that shared probable birthplace regions. These regions have vegetation with a matching chemical composition, and the butterflies ate this vegetation as caterpillars. This map (simplified from a figure published in a peer-reviewed journal) models data for one of those butterfly groups, demonstrating how far those butterflies had traveled since their birth.

Models and Systems It is often easier to interpret data when they can be visualized using a model. A simple diagram can help you understand the structure of a cell. A graph can help correlate data, revealing a pattern or trend. Scientists often work with large sets of data that can only be analyzed using computer models. Working with other scientists, Talavera modeled the origins and movements of butterflies across the globe by interpreting chemical data he collected and correlating the data with locations on a map. These data have been simplified for **Figure 1-9**.

Another way that scientists model information is by defining and analyzing systems. We can define a system as the part of the universe on which a scientist's attention is focused when making observations. A **system** consists of components that may interact, though the system as a whole may behave very differently from each of its components. A scientist carefully defines the boundary of the system they are studying. For example, a cell biologist might choose to focus on a single cell as a system. Meanwhile, an ecologist might study an entire population of organisms and their environment as a single system.

In addition to the behavior of a system's parts, scientists are often interested in the cycling of matter and flow of energy into and out of the system across its boundaries. These are called the inputs and outputs of the system. For example, oxygen is an input to the cardiovascular system, and carbon dioxide is an output. **Feedback** is the process by which outputs from a system influence the behavior of the system—for example, a person's heart rate may increase due to feedback in the presence of excess carbon dioxide.

On the Map

Butterfly Origins Connect the map area shown in **Figure 1-9** to the physical map students observed in **Figure 1-8**. Have them note where mountains or other features are located on both maps. Then, remind them of Talavera's question: "What are the migration patterns of the painted lady butterfly?" Point out the scale shown in **Figure 1-9**, which refers to degrees of probability, with 0 being not probable and 1 being highly probable. Have students use information in the caption to argue how Talavera came to his conclusion.

Connect to English Language Arts

Write an Argument In this section, students learn how systems are used by scientists to focus observations within specific boundaries. After students read this page, have them discuss issues that could arise if a scientist did not define the boundaries of a system they are observing. Students should write an argument and support their argument using evidence from the text and prior knowledge to explain why it is essential for a system to have boundaries.

DIFFERENTIATED INSTRUCTION | English Language Learners

Listening and Following Directions Read the caption for **Figure 1-9** as students look at the map. Then check listening comprehension by directing them to point to the parts of the map described.

Beginning Have students point to the map in response. For example, say: *Point to five states where the scientists studied butterflies.*

Intermediate Have students point to the map and explain their answer. For example, say: *Point to the areas that represent the most likely places where the butterflies were born. How do you know?* (The areas are darker.)

Advanced Have students point to the map and explain their answer in detail. For example, say: *Point to the scale on the map. What does it represent? Point to and name the colors on the map. What do they represent?*

Address Misconceptions

Model Limitations Models provide a way to represent phenomena or objects in ways that make it easier to understand difficult scientific concepts. But there are limitations, or trade-offs, with this tool. For example, in order to understand the information contained within a model, it may be necessary to keep a model very basic or simple. But including only limited information can lead to a very narrow or even incorrect understanding of the concept being modeled.

SEP Systems and System Models *Answers will vary. Examples for a school include the school's walls or its perimeter as a boundary; inputs and outputs could include students and teachers, food or waste, thermal energy from the sun or a heating/cooling system, etc.*

A system can be part of a larger system or consist of smaller systems. Thinking in terms of a clearly defined system helps scientists organize and analyze their observations. One type of system that biologists often analyze is an ecosystem. An *ecosystem* consists of one or more communities of different species interacting with one another and with their nonliving environment. You will learn more about ecosystems in Unit 1.

SEP Systems and System Models *Consider a model of your home or school as a system. How would you define the system's boundaries? What are some inputs and outputs of the system?*

Science Is a Human Endeavor

Despite all efforts to standardize the practices of science and view data with objectivity, it is still a human endeavor, or activity. It is susceptible to human error, biases, and misinterpretation. This is why it is important that scientists actively engage in communicating and evaluating information. Scientific theories once thought flawless are proved wrong as our understanding of the universe increases. The entire scientific community benefits when their work is viewed with both impartiality and skepticism.

CONNECTIONS

DEVELOPING SCIENTIFIC EXPERIMENTS

Experimental and observational studies are key to the practice of science. To gather evidence and construct explanations, scientists must develop hypotheses and plan experiments with carefully controlled variables.

A researcher who wants to perform an experiment will first read about what others have discovered on the topic. They will then make a hypothesis, which is a testable explanation for a natural phenomenon. Next, they will test the hypothesis.

Some predictions are tested by systematic observation. Others require experimentation. Experiments are tests designed to determine whether a prediction is valid. If working directly with an object or event is not possible, experiments may be performed on a model system. For example, animal diseases are often used as models of similar human diseases.

A typical experiment explores a cause-and-effect relationship using variables. A variable is something that can differ, such as a characteristic that differs among individuals or an event that differs over time. An independent variable is defined or controlled by the person performing the experiment. A dependent variable is a factor that is influenced by the independent variable.

Biological systems are complex because they involve many interdependent variables. It can be difficult to study one variable separately from the rest. Because of this complexity, biology researchers often test two groups of individuals at the same time. For example, an experimental group may be a set of individuals that have a certain characteristic or receive a certain treatment. This group is tested side by side with a control group, which is identical to the experimental group except for one independent variable. Any differences in experimental results between the two groups is likely to be an effect of changing the independent variable.

Experimental results—data—offer a quantifiable way of evaluating a hypothesis. Data that validate the prediction are evidence in support of the hypothesis. Data that show the prediction is invalid are evidence that the hypothesis is flawed and should be revised.

National Geographic Explorer Dr. Rosa Vásquez Espinoza examines a sample gathered in the field. Learn more about her research in Chapter 5.

CONNECTIONS | DEVELOPING SCIENTIFIC EXPERIMENTS

Social-Emotional Learning A phenomenon known as the *placebo effect* can occur when participants of a control group start to experience improvement from an inactive treatment. Research shows that our thoughts can have a major influence on how our body processes illness, disease, treatments, and/or certain conditions. Support students in practicing self-awareness skills, such as linking feelings, values, and thoughts, as they consider how the placebo effect might impact scientific studies. For example, a 2014 study of a migraine medication showed that a placebo could demonstrate about 50 percent of the benefits of the actual medication and those benefits persisted even after being told the pill was a placebo. Have students privately ponder personal situations in which their desire for a positive outcome resulted in an experience that might have been greater than expected.

A key skill for scientific communication is the ability to provide and receive criticism in a clear, professional, and constructive manner. Scientific critique identifies how methods and procedures could be improved or gives a rationale for a different interpretation of results. All scientists should be prepared to receive critique openly and use this information to refine their processes and ideas.

Scientists are always updating and refining their ideas. Sometimes they disagree. These disagreements are ultimately healthy for science, as they force us to examine our preconceived notions and further advance our knowledge. For example, the COVID-19 pandemic exposed the public to an evolving set of interpretations about how the virus was spread. As scientists began to better understand the virus, our ability to protect ourselves and others increased.

We should also be aware of the limitations of science. Ethical issues often cannot be resolved using science. Misinformation and the misapplication of scientific facts have occasionally resulted in actions that have done harm to society or the environment. At the same time, scientists should not be afraid to stand firmly behind their principles.

Engaging with science is important for you as a student completing coursework. Beyond school, being a curious and well-informed member of society will help you when making decisions, some as basic as the products you buy, and others as exceptional as how you choose to view the world. Everyone benefits from a broad understanding of science and its practices, and the ability to use scientific thinking to solve problems both big and small.

SEP Argue From Evidence *Describe a time when you had to persuade another person using evidence that you could provide. How were you able to change their mind?*

1.2 REVIEW

1. **Analyze** Identify the practice(s) of science represented in each part of an investigation. There may be more than one practice for each part.
   - A researcher models migration patterns based on empirical data.
   - A science team comes up with multiple ways to test their predictions.
   - A scientist wonders where butterflies go when the weather turns cold.
   - Several scientists debate how European butterflies make the annual journey to Africa.
   - Scientists make sense of a statistical comparison of specimens taken from multiple locations.
   - At a national meeting, scientists share the evidence they collected in their investigation about butterfly migration.
   - Scientists publish a scientific journal article describing how butterflies make use of wind patterns to enable their migration.
2. **Justify** Describe the ways in which an ecosystem meets the criteria of a system.
3. **Describe** Use an example to support the claim that because science is a human endeavor, scientific disagreement is healthy for science.

In Your Community

Online Reviews Many people use online reviews. But online reviews do not have guidelines ensuring they are factual and fair. Have students find reviews for a favorite local store or restaurant and read them critically. Challenge them to identify evidence to justify the user's rating. Students should also note if different reviews have information that is repeated. Ask students if they would believe such a review. Then, explain that all human endeavors, including science, are affected by error, biases, and misinterpretation. For data to be useful, even online reviews, it must be gathered and interpreted with as much objectivity as possible.

SEP Argue from Evidence *Answers will vary. Look for use of factual data in students' persuasive responses.*

1.2 REVIEW

1. *Developing and using models; Planning and carrying out investigations; Asking questions; Engaging in argument from evidence; Analyzing and interpreting data, Using mathematics and computational thinking; Obtaining, evaluating, and communicating information; Constructing explanations* **DOK 2**
2. *Student answers may vary but should include the ideas that (1) scientists define the boundaries of an ecosystem, (2) scientists make observations within those boundaries, and (3) the boundaries contain interacting components, which are the species that interact with one another and with the nonliving parts of the environment* **DOK 2**
3. *Students answers may vary but should include the ideas that scientific disagreement encourages an examination of assumptions and methods used to find answers.* **DOK 3**

ON ASSIGNMENT

Hercules National Geographic photographer Thomas Peschak is known for his ability to tell stories about Earth's oceans, islands, and coasts. Over a number of months, Peschak was able to build a relationship with a pangolin known as Hercules in the Tswalu Kalahari Reserve. Hercules and other pangolins in the reserve are being studied to determine how they are being affected by climate change.

You can learn more about Hercules and Peschak at the National Geographic society website: www.nationalgeographic.org.

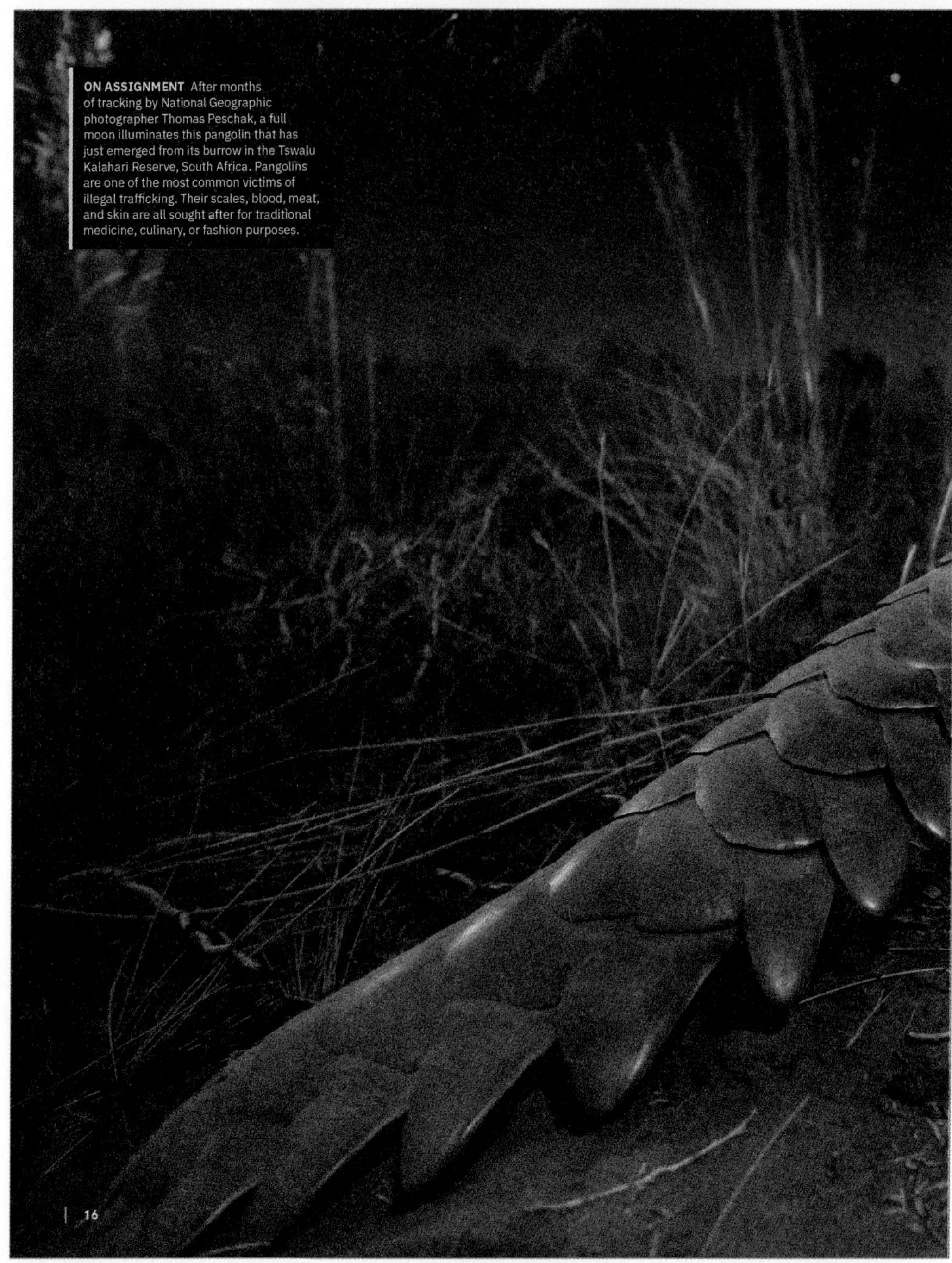

ON ASSIGNMENT After months of tracking by National Geographic photographer Thomas Peschak, a full moon illuminates this pangolin that has just emerged from its burrow in the Tswalu Kalahari Reserve, South Africa. Pangolins are one of the most common victims of illegal trafficking. Their scales, blood, meat, and skin are all sought after for traditional medicine, culinary, or fashion purposes.

16

Science Background

Pangolin There are eight species of pangolins, four species in Asia and four in Africa, and all are listed as critically endangered or vulnerable species. The scales that cover these animals are made of keratin and account for 20 percent of their weight. These shy animals are the world's most trafficked non-human mammal in the world. Poachers kill pangolins because of high demand for their scales, which some cultures believe have curative or magical properties. Some cultures consider pangolins to be a bad omen and kill them for this reason.

DIFFERENTIATED INSTRUCTION | English Language Learners

Using Pronouns in Writing As students write about the decline of the pangolin, have them check their writing to be sure they are using correct pronoun agreement and have a clear antecedent for each pronoun. Remind students that the pronoun takes the place of a noun. Review pronouns in English as needed. Point out that students should use a singular verb for a singular pronoun and a plural verb for a plural pronoun.

Beginning Have students underline each pronoun in their writing and draw an arrow to the antecedent, or the noun it refers to.

Intermediate Have pairs work together to check that they have used singular verb forms for singular pronouns and plural verb forms for plural pronouns.

Advanced Have partners exchange their writing and check for correct pronoun use. Then have them each correct their own work.

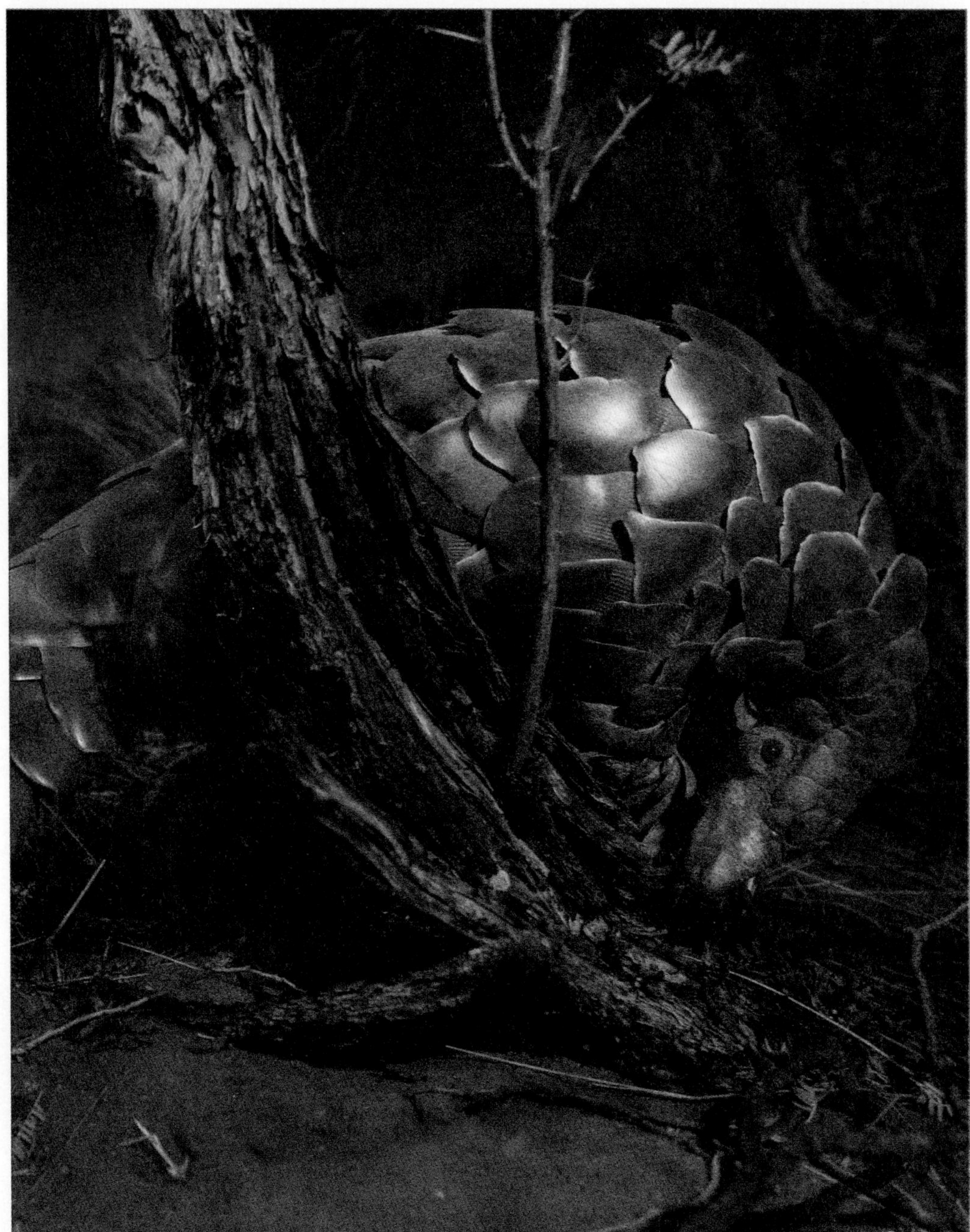

Connect to English Language Arts

Evaluate a Science or Technical Text After students learn about the decline of the pangolin due to trafficking, have them evaluate the claim made in the text using various sources to corroborate their findings. Have students keep a list of sources and cite the source for each fact in their writing. Students can write a brief paragraph describing which species of pangolins are most vulnerable and where most of them are at risk.

DIFFERENTIATED INSTRUCTION | English Language Learners

Evaluate Arguments Emerging language learners can expand their English comprehension and expand their reading skills by using secondary sources to evaluate the information in texts. Direct students to focus on the information on pangolins from the previous page and provide them with the time and access to resources that they can use to verify the information.

Beginning and Intermediate Students may benefit from doing their research and taking notes in their native language. Once students have completed their information gathering, they can write their paragraph using simple sentences appropriate to their English proficiency and content level.

Advanced Students should take their notes in English and use a variety of resources to support their findings. These students can write their paragraphs using complex sentences and more sophisticated vocabulary appropriate to their English proficiency level and content level.

1.3

USING BIOLOGY TO DEVELOP SOLUTIONS

ETS1.A, ETS1.B, ETS1.C

EXPLORE/EXPLAIN

This section walks students through the practices of engineering and uses a real-world problem to explain how engineers define a problem, develop possible solutions, and optimize solutions.

Objectives

- Describe the practices of engineering.
- Differentiate between criteria and constraints and identify how they are utilized when defining a problem.
- Design possible solutions to a problem that takes into account a range of criteria and constraints.
- Explain why solutions are tested, evaluated, and modified for optimization.

Pretest Use **Questions 6, 7,** and **8** to identify gaps in background knowledge or misconceptions.

CASE STUDY
PROJECT CETI: DECODING WHALE CODAS WITH ARTIFICIAL INTELLIGENCE

Ask Questions *To help students make connections with the content, have them refer back to the problem of communicating with whales that National Geographic Explorer Shane Gero is trying to solve. Have small groups compare and contrast the questions about medicines in the text. Then, ask them to brainstorm possible questions an engineer might ask when faced with the problem of understanding whale communication.*

Describe *Answers will vary. Students should be able to identify that even a tool as rudimentary as a pen is designed to be used in a specific way—for example, the shape/width of the barrel, the type of ink contained in the chamber, whether it is felt- or ballpoint-tipped, etc.*

1.3

KEY TERMS

biological engineer
constraint
criterion
engineering
trade-off

USING BIOLOGY TO DEVELOP SOLUTIONS

Have you ever considered the origins of various objects in your life? Who came up with them? What decisions went into their design? How were they made? What are they made from?

Even the simplest objects you interact with every day are the results of decisions, revisions, and improvements made over time. More sophisticated technologies require even greater thought and experimentation. In this section we will explore **engineering**, the application of science to design solutions to real-world problems.

Describe *Take a look around you and find a human-designed object. Describe how it is used and how its design suits its purpose.*

CASE STUDY
PROJECT CETI: DECODING WHALE CODAS WITH ARTIFICIAL INTELLIGENCE

Ask Questions *Generate questions about engineering approaches that could help scientists in decoding sperm whale communication.*

The Practices of Engineering

Scientists and engineers have many things in common, but often their goals are different. Scientists tend to seek explanations for the world around them, whereas engineers are concerned with using science to develop solutions. Instead of asking questions as a scientist would, an engineer identifies problems and tries to solve them using scientific knowledge. A **biological engineer**, or bioengineer, uses the principles of biology in particular to develop solutions. Bioengineers participate in a wide range of occupations, including the development of new drugs and medicinal therapies, improvements to agriculture, the design of safety equipment, and a host of other applications.

Consider, for example, the fact that many manufactured medicines and other useful compounds originated from naturally derived sources (**Figure 1-10**). A biologist might approach this fact and ask questions such as:

- What chemicals in this substance make it function as a medicine?
- Is its usefulness as a medicine due to a single chemical or a combination of chemicals?
- Can the organism that this substance is derived from be sustainably harvested?
- How does the drug work in the body?

The scientist is asking questions to construct explanations. An engineer faced with the same set of facts might ask questions such as:

- How can we improve the effectiveness of this medicine?
- Is there an easier way to obtain the compound(s) responsible for its usefulness as a medicine?
- Can the substance be synthesized in a lab setting?
- Are there better ways of storing or administering this medicine?

In order to answer any of these questions, it is important for the engineer to very clearly define the problem and the goals of its solution.

18 CHAPTER 1 INTRODUCTION TO BIOLOGY

DIFFERENTIATED INSTRUCTION | English Language Learners

Concept Maps Concept maps help language learners visualize relationships between new words they have learned. Have students make a concept map for the key terms in this lesson, starting with the term *biological engineer*.

Beginning Have students make a word web that includes the definition for each word. They could also write notes in their first language.

Intermediate Encourage students to think carefully about how to organize their concept map. They could use the headings in the text for inspiration. Have pairs compare concept maps and allow them to make adjustments if they wish.

Advanced Have students organize their concept map and draw arrows to show how the words are related. Then pair students, and have them explain their concept map to their partner.

FIGURE 1-10 *Penicillium chrysogenum* (a) is a source of antibiotics. Compounds from *Streptomyces coelicoflavus* (b) have shown promise as treatments for diabetes. The Madagascar periwinkle (c) produces substances that are used to treat cancer. Horseshoe crab (d) blood is used in tests to detect toxins.

Defining a Problem Now consider the real-world problem of designing a prosthetic leg for amputees. An engineer's goal is to make the best possible prosthesis that is reasonably achievable. However, simply stating that the goal is "to make a functional prosthetic leg" is somewhat ambiguous. It is much easier to imagine the requirements that the prosthesis will need to meet if we think about who it is being designed for and its intended purpose. Is it simply for walking on flat surfaces? Is it being designed for athletic activity? A carefully defined engineering problem allows engineers to develop and compare different solutions in a measurable way.

Suppose now that the prosthesis is being designed for a runner. This quickly gives us a much better picture of the user's requirements. The prosthesis should definitely be lightweight. It should be able to withstand high impact forces. It should probably be flexible, too. These requirements that a solution must meet are called criteria. A singular requirement is a **criterion**. Criteria are most easily compared when they are defined in a quantifiable way. Rather than saying the prosthesis should be lightweight, we could say the prosthesis must weigh no more than 1.5 kg. We could also say that it must be able to bear the force of a 73-kg (161-lb) person running at full speed while remaining flexible.

Once we have a clearer picture of the solution's criteria, we can also begin to think about its **constraints**, or the limitations that an engineer needs to consider when designing a solution. Constraints often include limiting factors such as cost, availability of materials, safety, aesthetics, and possible social, cultural, and environmental impacts. For the prosthetic leg, costs, safety, and the availability of materials are constraints we will have to take into account. Aesthetics might also be worth considering but probably is not as much of an issue.

Vocabulary Strategy

Knowledge Rating Scale Have students make a table of the key terms with four columns labeled 1, 2, 3, and 4. Before students begin reading, they should rate their knowledge of each term:

1 = I have never heard the word.

2 = I have heard the word, but do not know its meaning.

3 = I am familiar with the word.

4 = I know the word, its meaning, and could explain it to others.

Upon completing the reading, students should revisit the scale and re-rate the terms.

Connect to Mathematics

Reason Quantitatively Engineering mathematics is the branch of applied mathematics that covers mathematical methods and techniques used in engineering. While complex calculus is often used for building prototypes and final products, engineers still need to make basic calculations. Have students calculate the total weight of the person wearing the prosthesis in the example, then determine the proportion of that final weight that would be made up by the prosthesis. Students should explain the criteria in terms of proportion. *(74.5 kg; approximately 2%; The prosthesis should not weigh more than 2% of the body weight of the wearer.)*

SCIENCE AND ENGINEERING PRACTICES
Influence of Science, Engineering, and Technology on Society and the Natural World

Prosthetic Design There are a variety of different kinds of prostheses that are designed for different purposes. While the text brings students through an example of designing a prosthetic leg for a runner, a similar process can be followed if students were to design a prosthetic leg for a different purpose. Place students into groups and have them define the problem in other contexts. For example, students may want to solve this problem with a weightlifter or rock climber in mind, or even a surgeon, which would require different criteria. Once the group has defined the problem they want to solve, have them list criteria that their solutions must meet. Then, have them list out reasonable constraints that they should take into account if they were to design a solution.

Address Misconceptions

Developing Solutions A common misconception for students who are looking at an engineering challenge is that they may think the problem is easily solved with a single answer. However, engineers examine the problem from different angles and consider a variety of designs and materials that exist before choosing a solution. To rectify this misconception, students can revisit the prosthetic design problem in the previous spread and think about other considerations they may not have taken into account when they first looked at the problem. They can then further refine their criteria and constraints and develop several possible solutions that meet them.

Connect to English Language Arts

Use Sources in Multiple Formats Engineers must be able to obtain, evaluate, and share information in a variety of ways when working collaboratively. Students may think that all engineers are inventors, but guide them to understand that engineers do not begin the engineering process for a problem without first obtaining and evaluating solutions to similar problems. Have students use various sources of authoritative information to guide the refinement of their solution to the prosthetics design problem. They may use the Internet or library to access first person accounts, health and fitness journals, sports media, biomedical engineering documentation, or other resources to conduct their research into similar solutions.

FIGURE 1-11 Sprinters compete at the Tokyo 2020 Paralympic Games.

Developing Possible Solutions Once the problem is clearly defined and the solution's criteria and constraints are considered, engineers begin to brainstorm solutions. Brainstorming should be a judgment-free collaborative activity where new ideas are developed and shared. At this point, no idea should be off-limits. The goal is to come up with as many ideas as possible. Sometimes what seems like an impossible idea can help point in the direction of an even better solution.

With a set of possible solutions in mind, we can begin to compare how well they meet the criteria we have defined. This often requires balancing multiple criteria for the solution against one another while also keeping in mind its constraints.

BIOTECHNOLOGY FOCUS

TURNING TO NATURE FOR INSPIRATION Modern high-tech running blades were originally designed by American biomedical engineer Van Phillips, who was an amputee himself. Prior prosthesis designs were stiff and not useful for running. Inspired by the mechanics of cheetahs' legs, Phillips came up with a curved, springy design that allows users to run and jump. Running blades are made from carbon fiber, which is lightweight, flexible, and strong. The curved design stores the energy from a runner's stride and releases it like a spring, propelling the runner forward.

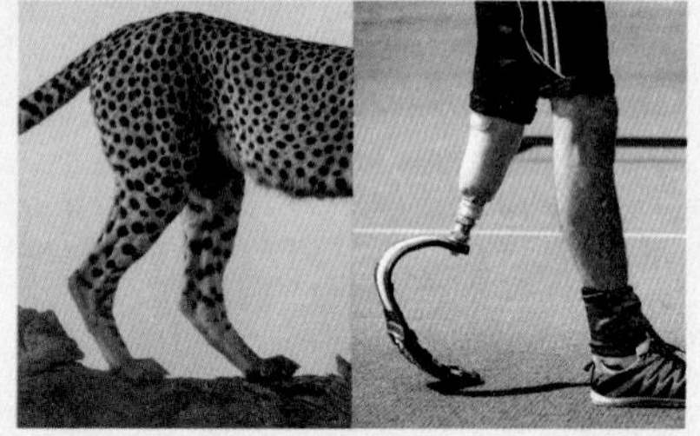

Some prosthetic devices take inspiration from nature.

BIOTECHNOLOGY FOCUS | Turning to Nature for Inspiration

Biomimetics Engineers often take inspiration from the natural world. The interdisciplinary field of biomimetics takes principles from biology, chemistry, and physics and applies them in engineering designs to solve human problems. To stimulate conversation about the topic of biomimetics, show students a strip of hook-and-loop tape and tell them how it was invented. While hunting, an engineer found that burs had attached to his pants and his dog's fur. Curious as to how they were attached, he took the burs home to investigate. Needing to study them at a greater scale, he observed them under a microscope where he saw that the burs had tiny hooks that enabled them to attach to hair and the fabric threads without falling off.

In Your Community Have students consider the products they use every day and determine if their design could be based on nature. Have them consider objects such as airplanes and wetsuits.

In the case of a prosthetic running leg such as in **Figure 1-11**, we might select a material that is a little more costly but is more lightweight. We might need to design the prosthesis with less flexibility but more ability to withstand the forces exerted on it while running. This prioritization of one criterion over another is called making a **trade-off**.

Optimizing the Solution Engineering design is an iterative process. This means that engineers will repeat the steps of the process over and over, gradually refining the solution each time. Sometimes a solution that seemed promising simply will not work and the engineers will have to go back to the drawing board. New criteria may come to light as solutions are tested. Or the entire problem may need to be redefined. Complex real-world problems are often solved by breaking them down into smaller, more manageable problems. Engineers should be prepared to treat what seems like a failure as an opportunity to learn more about the problem and better refine its solution.

When testing possible solutions, engineers often build test models called prototypes. They may rely on computer simulations to help envision how a solution will work. In situations when the solution will be a large-scale process, such as manufacturing a drug, engineers will often test the process on a smaller scale before "scaling up." As engineers repeat these iterative steps, an optimal solution is developed.

SEP Construct an Explanation *How are the practices of science and engineering similar? How are they different? For example, how might a scientist iterate to improve and optimize an experiment's design?*

1.3 REVIEW

1. **Classify** The practices of science and engineering are similar. Label each statement as an example of scientific practices, engineering practices, both, or neither.
   - Solutions are refined in iterative steps.
   - Failures can be used as learning opportunities.
   - Limits involve quantifiable criteria and specific constraints.
   - Goals are broken down into smaller, more manageable goals.
   - The answers to questions are obtained in a systematic process.
   - Restrictions involve defining system boundaries and appropriate data collection plans.
2. **Justify** Explain why quantifiable criteria are important for success in engineering projects. Include the concept of trade-offs in your answer.
3. **Describe** What are the benefits of an effective brainstorming process? Select all correct answers.
   A. Many solution ideas are generated.
   B. The problem will have an obvious solution.
   C. All proposed solutions are represented in the final solution.
   D. Multiple criteria and constraints are defined in the best solution.
   E. Solutions that may not be possible can point toward better solutions.

1.3 REVIEW

1. *engineering; both; engineering; engineering; scientific; scientific* **DOK 2**
2. *Students' answers may vary but should include the ideas that (1) quantifiable criteria remove subjective or biased judgment and that (2) prioritizing criteria to establish trade-offs is more straightforward when criteria are quantified.* **DOK 3**
3. *A, E* **DOK 1**

CHAPTER INVESTIGATION B

Engineering *A Medicine Distribution Solution*

Time: 110 minutes

Students will investigate the issue of medicine distribution in a small town to define an engineering problem and identify criteria and constraints. Then, they will develop and evaluate solutions and present a written proposal on their best solution for peer evaluation.

Go online to access detailed teacher notes, answers, rubrics, and lab worksheets.

Science Background

Inventing Drugs A global scientific and bioengineering challenge is developing drugs to treat the more than one billion people that have a "neglected" disease. To solve this problem for a specific disease, scientists must first identify a drug target. A drug target is a protein or other biological molecule in the human body that can interact with a drug to produce a therapeutic effect. Computer simulations can apply information about a drug target to identify promising drug candidates. A drug company can then carry out preclinical trials with the candidates using first cells and then animals. With each iteration, efficacy and safety are assessed and optimized. The next step in the drug development process is carrying out human clinical trials. Again, efficacy and safety are assessed and optimized with each iteration. If clinical trials are successful, the drug can be manufactured. This process is also optimized, with a goal of lowering costs by reducing the number of steps, increasing yield, and improving quality. Lower costs are important in solving the problem of neglected diseases. People with a neglected disease are either few in number or impoverished, which means there is often no financial incentive to develop drugs that can treat them.

SEP Construct an Explanation *The practices of science and engineering are nearly identical, except engineers define problems and develop solutions, while scientists ask questions and construct explanations.*

LOOKING AT THE DATA
MASS DISTRIBUTION OF MAMMALS

Connect this activity to the Science and Engineering Practices of using mathematics and computational thinking and analyzing and interpreting data. Students analyze and interpret data on the mass distribution of different mammals by organizing and comparing the data and identifying patterns. Students make inferences from the data and construct an explanation to explain the differences in the masses of terrestrial, marine, and aerial mammal species.

Connect to Mathematics

Infer from Statistics As students answer questions from the activity, ensure they are looking at the correct data set. Remind students to review column headings for key terms that match the key terms from the question. For example, **Question 6** has students analyzing data related to animal length and environment. A quick review of the column headings reveals that this information is not contained in **Table 1** as many students might think. Instead, students will find this information in the table provided in **Question 5**. Have students add the environment for each mammal listed in the table provided in **Question 5** on a sticky note or in their notebook.

LOOKING AT THE DATA

MASS DISTRIBUTION OF MAMMALS

SEP Use Mathematics To make it easier to compare numerical data, scientists have developed a standardized system of measurements known as the International System of units, or SI, from the French *Système Internationale*. The SI unit of mass is the kilogram, abbreviated kg. The masses of blue whales, the largest marine mammals, can be up to 180,000 kg, while the African elephant, the largest terrestrial mammal, can have a mass of more than 6000 kg. In comparison, the largest aerial mammals, such as a large flying fox bat, may have a mass of about 1 kg. **Table 1** shows an estimated mass distribution of terrestrial, marine, and aerial mammal species.

TABLE 1. Mass Distribution in Representative Terrestrial, Marine, and Aerial Mammals

Average estimated mass (kg)	Number of terrestrial mammal species	Number of marine mammal species	Number of aerial mammal species
0.001	140	0	360
0.01	1200	0	480
0.1	700	0	90
1	450	0	1
10	230	4	0
100	80	13	0
1000	5	27	0
10,000	0	9	0
100,000	0	3	0

Sources: Centre for Population Biology, Imperial College at Silwood Park

To compare the masses of mammal species, it is helpful to write and compare values in scientific notation. For example, to compare the mass of the smallest marine mammal to the mass of the smallest aerial mammal, we first write the ratio of the two masses to be compared as a fraction. Then, we convert the values to scientific notation and complete the calculation as shown:

$$\frac{10\text{ kg}}{0.001\text{ kg}} = \frac{1 \times 10^{1}\text{ kg}}{1 \times 10^{-3}\text{ kg}} = 1 \times 10^{(1-(-3))}\text{ kg} = 1 \times 10^{(4)}\text{ kg} = 10{,}000$$

The calculated value of 10,000 is one way to represent the ratio of the two masses. However, it can be easier to compare large and small values if we convert this ratio to an order of magnitude. To do this, we calculate the $\log_{10}$ of the value

$$\log_{10}(10{,}000) = 4$$

This means the average mass of the smallest marine mammals is four orders of magnitude larger than that of the smallest aerial mammals.

22 CHAPTER 1 INTRODUCTION TO BIOLOGY

SCIENCE AND ENGINEERING PRACTICES
Using Mathematics and Computational Thinking

Reflect on Data Have students read the first paragraph on the page. Draw their attention to the phrase "scientists have developed a standardized system of measurements known as the SI system." Have students make an assessment as to why a universal measurement system is necessary and lead a discussion about how scientific notation is helpful for comparing extremely large and small values. Students should understand that in order to share data, a standardized measurement system is required. To help reinforce this concept, ask students to reflect on how converting data from the U.S. measurement system to SI could lead to calculation errors. After students complete the reading, ask them to analyze and summarize the data represented in **Table 1**. Have students make a claim about the mass of mammals in different environments and use evidence from the table to support their claim.

FIGURE 1
The variety in size and shape among mammals is demonstrated by this manatee (a), red fox (b), giraffe (c), and fruit bat (d).

1. **Calculate** Convert the average masses in **Table 1** into scientific notation. Then determine the $\log_{10}$ of each value.
2. **Organize Data** Create a bar graph that shows the number of species in each mammal category using $\log_{10}$ average mass on the horizontal axis. The peak mass is the mass value that includes the most species. Record the peak mass of each distribution.
3. **Compare** Determine the differences in order of magnitude between the three mass distribution peaks.
4. **Interpret Data** What might explain the differences in the mass distributions of terrestrial, marine, and aerial mammals?
5. **Calculate** The SI unit for length is the meter (m). One meter is 100 centimeters (cm), and 1000 meters is 1 kilometer (km). Convert the estimated lengths of the mammals given in the following table to cm and km. Which unit would be most convenient for measuring and reporting the size of each mammal?

Mammal	Length (m)
manatee	3.0
red fox	1.0
giraffe	4.7
fruit bat	0.07

6. **Explain** What relationship, if any, would you expect to exist between an animal's length and its environment?

SAMPLE ANSWERS

1. *1×10^{-3}: -3; 1×10^{-2}: -2; : 1×10^{-1}: -1; 1×10^{0}: 0; 1×10^{1}: 1; 1×10^{2}: 2; 1×10^{3}: 3; 1×10^{4}: 4; 1×10^{5}: 5.* **DOK 2**
2. *-2; 3; -2* **DOK 2**
3. *Sample Answer: The peak mass of marine mammals is five orders of magnitude larger than the peak mass of terrestrial mammals and aerial mammals. Terrestrial mammals and aerial mammals have the same peak mass.* **DOK 2**
4. *Student's answer may include differences in the size of mammals, in their method of movement, or in the medium through which they move, including the effects of gravity, buoyancy, and friction. For example, marine mammals are supported by the water in which they move, so they can grow the largest, but aerial mammals must be able to fly, so they need to be smaller than terrestrial mammals.* **DOK 3**
5. *manatee: 0.0030 km, 300 cm; red fox: 0.0010 km, 100 cm; giraffe: 0.0047 km, 470 cm; fruit bat: 0.00007 km, 7 cm. Manatee, red fox, and giraffe length can all be easily reported in meters; whereas, fruit bat length is best reported in centimeters.* **DOK 3**
6. *Student's answer may mention that an organism's role in its environment may be related to its length, but it is not necessarily a consistent relationship. Giraffes and other animals may need a longer body length to reach particular foods, but other animals in that same environment may be shorter and smaller to evade predators.* **DOK 3**

DIFFERENTIATED INSTRUCTION | English Language Learners

Grammar and Usage As students answer the questions, have them check their work for correct grammar and spelling. Have them pay particular attention to negatives (for example, as they explain that an organism's role is not always related to its length) and spelling patterns (for example, when adding *-er* to form comparatives or *-est* to form superlatives).

Beginning Help students phrase their answer to question 6 as a negative statement. Have them underline any comparatives or superlatives, and help them check the spelling.

Intermediate Have students phrase their answer to question 6 as a negative statement. Have pairs work together to check spelling of comparatives and superlatives.

Advanced Have students write their answer to question 6 using a negative with and without a contraction (for example, *isn't* and *is not*). Have pairs exchange their work and check spelling.

EXPLORER
KIM WILLIAMS-GUILLÉN

About the Explorer

National Geographic Explorer Dr. Kim Williams-Guillén developed 3D-printed decoy sea turtle eggs that mimicked the look and feel of real sea turtle eggs to track poachers. Investigators were able to track a number of poachers in the illegal wildlife trade when some of the decoy eggs planted in nests were stolen. You can learn more about Kim Williams-Gullén and her research on the National Geographic Society's website: www.nationalgeographic.org.

Science Background

3D Printing of Eggs Williams-Gullén used a flexible plastic filament that could mimic the squishy texture of sea turtle eggs and specialized painting methods to create her decoys. While the decoy eggs can be distinguished from true eggs when set side by side, once the decoy eggs are buried in the sand with a hundred true eggs, they could not easily be distinguished.

Connect to Careers

Biodiversity Conservationist Manager Scientists who work toward decreasing illegal animal trade and trafficking are known as *biodiversity conservationist managers*. These scientists use practices to protect species while continuing to support ecotourism and responsible wildlife interactions. They generally need at least a master's degree in wildlife management or natural resources management and may work for government agencies, nonprofit conservation organizations, or environmental education agencies.

THINKING CRITICALLY

Define a Problem *Williams-Guillén wanted to track the whereabouts of illegally harvested eggs and thought a fake egg might be the right tool. The fake egg had to have the same characteristics as a real egg. One constraint was weight, which had to be just right. Williams-Guillén gauged success on the fact that at least some of the decoys yielded data that allowed tracking of eggs for at least some distance.*

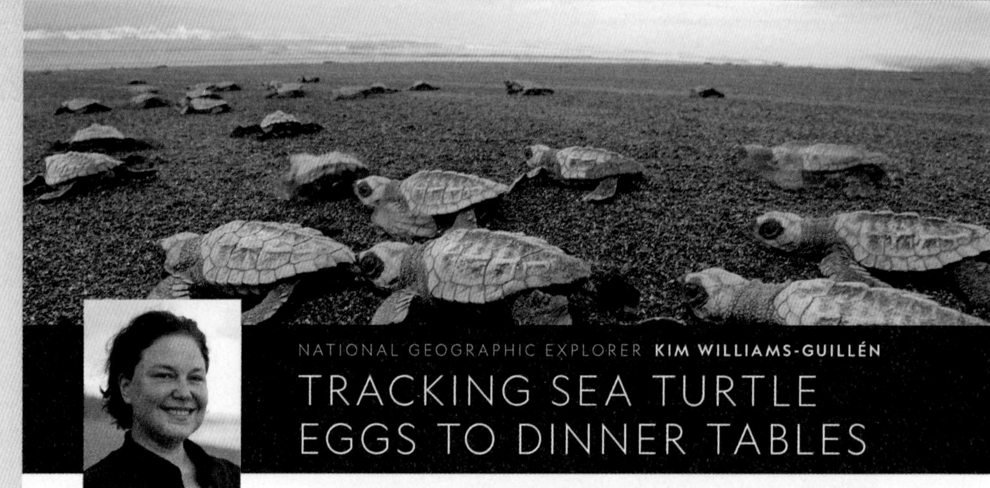

NATIONAL GEOGRAPHIC EXPLORER **KIM WILLIAMS-GUILLÉN**

TRACKING SEA TURTLE EGGS TO DINNER TABLES

Dr. Kim Williams-Guillén designed a decoy turtle egg with the aim of tracking poachers.

Conservationist Dr. Kim Williams-Guillén pondered how to track the locations of illegally harvested eggs of endangered and threatened sea turtles in Costa Rica. High in protein and said to taste richer than a chicken egg, sea turtle eggs are a delicacy served in some restaurants and bars. Each egg brings a poacher around one dollar, and a sea turtle lays about 100 leathery-shelled eggs in each nest.

The rainy season is especially busy for poachers around Ostional, Costa Rica. It begins in May and lasts through November. The olive ridley sea turtles' *arribadas*, Spanish for "arrival," begins in June. Once or twice a month, many thousands of them come ashore all at once, awkwardly clambering across the sand to dig nests about one meter deep in the sand. Olive ridleys lead a solitary life, migrating hundreds to thousands of kilometers in the open ocean. Every few years, each female returns to the same area where she hatched. The olives and their close relative, the Kemp's ridleys, synchronize those arrivals. The individuals of other species come ashore a few times during those months. The regular timing of sea turtle nesting opens the door for predators, both the natural and human kind.

The eggs can be collected in a box or bag without breaking. Williams-Guillén noted their resemblance to table tennis balls. Inspired by tracking devices seen on television shows, she concluded that a decoy egg could be made to mimic the texture, feel, size, and most importantly, the weight of a real egg. It would need to be fitted with electronics that could emit a strong signal for several days. The decoy would need to stay in the mix of the real ones and not drift to the top or sink to the bottom. And the fakes would have to look enough like the real thing to escape the notice of an experienced poacher.

Williams-Guillén's design solution was a 3D-printed decoy turtle egg embedded with a GPS-GSM transmitter. Her team placed a decoy in 101 sea turtle nests on four different beaches in Costa Rica. Twenty-five percent were later poached. Six decoys were discovered on the beach, likely identified and discarded by the collectors. The team also estimated a 32 percent failure rate based on their follow-up examination. Yet, the decoys allowed the team to track five of the illegally harvested nests, revealing one trail 137 kilometers long—nearly the entire length of the trade chain from beach collection to dinner table. The decoys were deemed a success for tracing the route of eggs—a positive step in curbing the illegal wildlife trade.

THINKING CRITICALLY

Define a Problem *What problem was Williams-Guillén trying to solve? What were her criteria and constraints? How could she determine whether her design solution was successful?*

DIFFERENTIATED INSTRUCTION | English Language Learners

Sight Words Language learners will better understand a text if they recognize some high-frequency words they can read on sight, rather than sounding them out. Preteach the following sight words in "Tracking Sea Turtle Eggs to Dinner Tables": *busy, come, entire, lead, ocean, through, weight.* Display the words and read them aloud for students to repeat.

Beginning Have students use sketches or gestures to express the meaning of each word.

Intermediate Have students write a definition or example of each word.

Advanced Have students use each word in a sentence that shows its meaning.

TYING IT ALL TOGETHER
PROJECT CETI: DECODING WHALE CODAS WITH ARTIFICIAL INTELLIGENCE

HOW DO WE STUDY THE NATURAL WORLD?

Translating an animal language into human language, as the Case Study describes, would have been impossible only a few years ago. Even today, it is a daunting goal that requires the cooperation of research scientists and engineers in multiple fields.

For Project CETI to be successful in deciphering whale codas, machine learning will require extremely large quantities of data—far more than the thousands of sperm whale recordings that have been collected during the past few decades. The Project CETI team must collect millions, and possibly tens of millions, of audio samples that include data about which whale was speaking, what other whales were present, and what they were doing at the time. Gathering this much detailed data requires an arsenal of acoustical sensors, underwater drones, and video cameras that can record whale sounds and document their activities wherever they go.

In this activity, you will analyze how the practices of science and engineering are involved in the Project CETI investigation. Work individually or in a group to complete these steps.

Ask Questions

1. Write a question that a scientist studying how whales communicate might ask.
2. Write a question that an engineer studying how whales communicate might ask.

Construct an Explanation

3. Describe how CETI scientists might use three of the science practices introduced in this chapter to study whale communication.
4. Describe the design process CETI engineers might use to measure whale sounds.

FIGURE 1-12 A specialized device that records and tracks sperm whales uses suction cups to stick to a whale's back.

SCIENCE AND ENGINEERING PRACTICES
Asking Questions and Defining Problems

Studying Whale Codas To help students prepare for this activity, begin by having them call out questions that they have about the Case Study to stimulate conversation in the classroom. Have students select their favorite questions and rewrite them in a way that a scientist might. In **Step 2,** have students rewrite their questions in a way that engineers might. List out the eight central science practices on the board from Section 1.2 and have students briefly describe each practice. Have them work individually or in groups to select the three practices they would like to focus on for their explanations developed for **Step 3**. Before beginning **Step 4**, have students identify the problem that engineers would be trying to solve. Students should focus on this problem and the design process to complete **Step 4**.

TYING IT ALL TOGETHER

ELABORATE

The Case Study examines how scientists are using artificial intelligence to attempt to decode whale codas. In the Tying It All Together activity, students will identify questions that scientists and engineers might ask and construct an explanation of how science and engineering practices can be used to study whale communication.

This activity is structured so that students may begin once they complete Section 2. **Step 1** has students identify questions that scientists studying whale communication might ask. **Step 2** asks students to identify questions an engineer might ask. **Step 3** connects to the science practices described in Section 1.2 and has students construct an explanation about how they can be used to study whale codas. Students will then consider the design process engineers would use to measure whale codas in **Step 4**. Alternatively, this can be a culminating project at the end of the chapter. The engineering practices defined here reflect the reasoning processes that students may use to find a solution for the Unit 1 Activity.

Go online to access the Student Self-Reflection and Teacher Scoring rubrics for this activity.

Go Online **VIDEO**

Time: 1:20

After listening **Video 1-3**, replay it so that students can see the audio bars. Have them discuss how the pattern of the bars follows the intensity of the sound and how that pattern can provide visual clues apart from the audio.

EVALUATE

REVIEW KEY CONCEPTS

1. *A* **DOK 2**
2. *D, E* **DOK 1**
3. *C, D, E* **DOK 1**
4. *Engineering Questions: How can bacteria containing human insulin genes be modified to increase insulin production? Can penicillin be modified in new ways to kill bacteria that are resistant to standard penicillin? Scientific Questions: How are bacteria affected by temperature changes? What types of bacteria are most likely to grow on human skin? Are beneficial intestinal bacteria harmed by antibiotics?* **DOK 2**
5. *observing bees visiting various flowers and wondering what factors affect their behavior: asking questions; calculating the average number of flowers of each color visited by bees over several days: analyzing and interpreting data; setting up patches of blue and yellow flowers that are otherwise identical and positioning cameras to record visits by bees: planning and carrying out investigations; devising an algorithm to predict the frequency of random bee visits versus visits influenced by color, odor, amount of nectar, and other factors: developing and using models* **DOK 2**
6. *Animalia; Chordata; Mammalia; Carnivora; Felidae; Felis; cactus* **DOK 2**
7. *A, B* **DOK 2**
8. *C, D, E* **DOK 3**

CHAPTER 1 **REVIEW**

REVIEW KEY CONCEPTS

1. **Relate** Which of the central practices of scientific inquiry is essential before scientists can plan and carry out investigations?
   - **A.** asking questions
   - **B.** constructing explanations
   - **C.** developing and using models
   - **D.** analyzing and interpreting data
2. **Identify** How could small physical differences between two individuals of the same species be explained? Select all correct answers.
   - **A.** The individuals have very dissimilar genomes.
   - **B.** The individuals do not belong to the same genus.
   - **C.** The individuals' parent(s) had slightly different DNA.
   - **D.** The individuals have nearly identical DNA with slight variations.
   - **E.** The individuals' DNA has been adapting and evolving throughout their lifetimes.
3. **Identify** A scientist is studying grasshoppers in local fields. Which statements describe ways the scientist could use models as part of the scientific process? Select all correct answers.
   - **A.** identifying the species of plants found in the local fields
   - **B.** observing the number of grasshoppers on different types of plants
   - **C.** creating a map to show predicted patterns of grasshopper distribution
   - **D.** graphing the number of grasshoppers observed at different times of day
   - **E.** developing data visualizations of the relationship between grasshopper density and rainfall
4. **Classify** Determine whether each question is an engineering question or a scientific question.
   - How are bacteria affected by temperature changes?
   - What types of bacteria are most likely to grow on human skin?
   - Are beneficial intestinal bacteria harmed when people take antibiotics?
   - Can penicillin be modified in new ways to kill bacteria that are resistant to standard penicillin?
   - How can bacteria containing human insulin genes be modified to increase insulin production?
5. **Classify** Label each example as a specific practice of science: asking questions, planning and carrying out investigations, analyzing and interpreting data, developing and using models.
   - observing bees visiting various flowers and wondering what factors affect their behavior
   - calculating the average number of flowers of each color visited by bees over several days
   - setting up patches of blue and yellow flowers that are otherwise identical and positioning cameras to record visits by bees
   - devising an algorithm to predict the frequency of random bee visits versus visits influenced by color, odor, amount of nectar, and other factors
6. **Sequence** The common house cat (*Felis catus*) is classified in the family Felidae, the order Carnivora, the kingdom Animalia, the species *catus*, the class Mammalia, the genus *Felis*, and the phylum Chordata. Place these levels of classification in order.
7. **Relate** A bioengineer has been hired by a drug company to design an over-the-counter oral antibiotic to treat a specific type of bacterial infection. Effective antibiotics for this disease already exist, but they can only be administered by injection in a hospital setting. Determine which statements represent *criteria* for an acceptable solution.
   - **A.** The drug must be delivered orally.
   - **B.** The drug must be effective in treating the disease.
   - **C.** The drug must be affordable for all socioeconomic groups.
   - **D.** The cost of developing the drug must be within the company's budget.
   - **E.** The drug must be more effective than existing drugs for the same disease.
8. **Evaluate** Which questions represent global challenges facing humanity that can be addressed by engineering? Select all correct answers.
   - **A.** How can we explain why species evolve over time?
   - **B.** How can we track the migration of a pink lady butterfly?
   - **C.** What can we do to improve the effectiveness of life-saving vaccines?
   - **D.** How can we increase agricultural production while limiting pollution?
   - **E.** How can we remove toxins and pollutants from sources of drinking water?

26 CHAPTER 1 INTRODUCTION TO BIOLOGY

CRITICAL THINKING

9. *Sample answer: This mutation is beneficial. It increases the salamander's chances of survival by helping it avoid predators. If the salamander survives to reproduce, it may pass on the genetic mutation to its offspring, which will also benefit from having light skin. So, it is likely that the mutation that causes lighter skin will become more common in future generations.* **DOK 2**
10. *Sample answer: Scientists recognize the limitations of human knowledge and reevaluate their ideas when new information becomes available. If scientists had clung to the hypothesis that proteins contained genetic information, they may not have realized that it was in DNA instead.* **DOK 4**
11. *Sample answer: One way is to describe the entire area of the proposed housing development as a single system from which animals can enter (inputs) and exit (outputs). Other possible inputs and outputs are related to human impact inputs, such as building materials, roads, houses; outputs such as trees, soil, and other natural habitats that humans change. In order to model the system, the ecologist would first need to survey the area, gather data on various forms of native wildlife, and map patterns in animal behavior. Then, the ecologist could compare this map to the proposed building plans and analyze the potential ecological impact.* **DOK 4**
12. *Sample answers: Do striped or unstriped salamanders produce more offspring in cool climates? Do striped and unstriped salamanders differ in body temperature when in the same place? Are striped salamanders becoming less common as average temperatures increase?* **DOK 2**

CRITICAL THINKING

9. **Explain** A genetic mutation causes an individual salamander to be a lighter shade of brown than other salamanders of the same species. This unique characteristic makes the light brown salamander harder for predators to spot. Explain how this mutation may or may not affect future generations of salamanders.

10. **NOS** **Scientific Knowledge** Scientists once hypothesized that proteins were responsible for inherited characteristics in organisms because proteins are very complex molecules. However, scientists later discovered evidence that DNA carries the genetic information organisms pass on to their offspring. How does this example illustrate the concept that science is a human endeavor?

11. **Synthesize** An ecologist has been hired to study how a proposed housing development may affect native wildlife and their habitats. How could the ecologist model the effects of different building plans on the area's biodiversity? In your response, address how the ecologist might define the system in order to study its inputs and outputs.

12. **Predict** Among eastern red-backed salamanders, some individuals have a stripe along their back and others do not. Researchers have found that in cooler climates, a higher percentage of salamanders are striped than in warmer climates. Earth's average temperature is currently increasing. What are three questions a scientist might ask in order to construct an explanation for this phenomenon?

MATH AND ENGLISH LANGUAGE ARTS CONNECTIONS

1. **Write Explanatory Text** Cacti, which are native to arid environments, have needles instead of conventional leaves to help them conserve water and defend themselves from herbivores. Explain how this adaptation is related to the genome of the cactus, how it is passed from one generation of cacti to the next, and what might cause this trait to change in future generations.

2. **Write Explanatory Text** A major highway is being built through an area where many animals are known to migrate regularly. Conservation groups are concerned that the highway will lead to large amounts of roadkill, habitat loss, and reduced biodiversity. Explain how a bioengineer might use the practices of engineering to develop a solution.

Use the information on painted lady butterfly migration on pages 9 to 11 to assist in answering the question.

3. **Reason Abstractly and Quantitatively** Based on information in the passage, estimate how many kilometers per day a painted lady butterfly would have to fly, on average, in order to complete the migration from Europe to the African savanna within its lifetime. What factors related to the butterfly's life cycle might affect this estimate?

4. **Write Explanatory Text** People who suffer from diabetes have trouble producing enough insulin, a hormone that helps regulate blood sugar levels. In 1982, genetic engineers modified the genome of a bacterium, essentially reprogramming bacterial cells to produce insulin for the treatment of patients with diabetes. Explain what makes this an example of bioengineering, and discuss how the practices of science and engineering likely contributed to this solution.

REVISIT THE ANCHORING PHENOMENON

1. *Sample answer: Changing environmental conditions may alter the migration patterns of some species of hummingbirds. Changing environmental conditions could also affect the availability of resources hummingbirds need to survive in their habitats and could result in their moving to different habitats or becoming extinct.* **DOK 3**

2. *Sample answer: What environmental factors most affect the speciation of hummingbirds? How have changes in the environment affected hummingbird speciation? How are hummingbird species changing today?* **DOK 3**

MATH AND ELA CONNECTIONS

1. *Sample answer: This characteristic is encoded in the cactus's DNA and is passed from parents to offspring. In future generations, a new adaptation may arise that becomes more common in the population if it is beneficial in some way.* **DOK 3**

2. *Sample answer: Student answers may vary substantially but should involve the three practices of engineering: defining the problem, developing possible solutions, and optimizing the solution. For example, the bioengineer might define the problem by identifying the geographical area impacted by the road and the animals likely to be affected. This might require gathering more information by, for example, studying animals and their migration patterns. Next, the bioengineer would need to develop possible solutions, such as rerouting the road or building paths to allow animals easier ways to cross the road. Finally, the engineer would need to select and optimize the solution, considering possible trade-offs that might need to be made.* **DOK 3**

3. *Sample answer: A painted lady butterfly's life cycle is between 15 and 29 days and the journey is roughly 4,000 kilometers. Assuming the butterfly spends its entire life flying, it would have to cover between 138 and 267 kilometers per day to travel 4,000 kilometers. However, the life cycle of a butterfly involves some phases when it cannot fly (egg, caterpillar, and chrysalis), which leaves fewer days for flying. So the butterflies would actually have to travel much farther per day than the estimates given above.* **DOK 3**

4. *Sample answer: This is an example of bioengineering because bacteria are being modified using knowledge of biology to solve a problem and benefit humans. This solution would not have been possible without prior research into the nature of DNA and how to modify the DNA of bacteria. Bioengineers probably tested and refined many possible solutions as they developed and optimized techniques for engineering bacteria to make insulin.* **DOK 3**

Unit Anchoring Phenomenon: Sea Pig Survival in Deep-sea Ecosystems

Use the Driving Question to help frame the Anchoring Phenomenon as an investigable subject and motivate student learning. Leverage the sea pig prompts within each chapter to connect concepts back to the unit's Driving Question, supporting students in gathering evidence and asking their own research questions so they are equipped to complete the Unit Activity.

NATIONAL GEOGRAPHIC

Meet the Explorer

Diva Amon deep-dives to explore uncharted swaths of the Pacific seafloor where sea pigs live, advancing human understanding of deep-sea ecosystems. Watch Unit Video 2, Explorers at Work: Diva Amon, to engage student interest in marine research and the Anchoring Phenomenon.

Virtual Investigation

Sea Pigs on the Abyssal Plain Students learn about the abyssal plain ecosystem and gather evidence to describe how sea pigs survive and thrive in deep-sea conditions.

NGSS Progression

Middle School

- **MS-LS1-5** Construct a scientific explanation based on evidence for how environmental and genetic factors influence the growth of organisms.
- **MS-LS2-2** Construct an explanation that predicts patterns of interactions among organisms across multiple ecosystems.
- **MS-LS2-3** Develop a model to describe the cycling of matter and flow of energy among living and nonliving parts of an ecosystem.
- **MS-LS2-4** Construct an argument supported by empirical evidence that changes to physical or biological components of an ecosystem affect populations.
- **MS-LS2-5** Evaluate competing design solutions for maintaining biodiversity and ecosystem services.

High School

- **HS-LS2-1** Use mathematical or computational representations to support explanations of factors that affect carrying capacity of ecosystems at different scales.
- **HS-LS2-2** Use mathematical representations to support and revise explanations based on evidence about factors affecting biodiversity and populations in ecosystems of different scales.
- **HS-LS2-3** Construct and revise an explanation based on evidence for the cycling of matter and flow of energy in aerobic and anaerobic conditions.
- **HS-LS2-4** Use mathematical representations to support claims for the cycling of matter and flow of energy among organisms in an ecosystem.
- **HS-LS2-5** Develop a model to illustrate the role of photosynthesis and cellular respiration in the cycling of carbon among the biosphere, atmosphere, hydrosphere, and geosphere.
- **HS-LS2-6** Evaluate the claims, evidence, and reasoning that the complex interactions in ecosystems maintain relatively consistent numbers and types of organisms in stable conditions, but changing conditions may result in a new ecosystem.
- **HS-LS2-7** Design, evaluate, and refine a solution for reducing the impacts of human activities on the environment and biodiversity.
- **HS-ETS1-3** Evaluate a solution to a complex real-world problem based on prioritized criteria and trade-offs that account for a range of constraints, including cost, safety, reliability, and aesthetics, as well as possible social, cultural, and environmental impacts.

Claim, Evidence, Reasoning Students can make a claim about the unit phenomenon, gather evidence, and revisit their claim periodically to evaluate how well the evidence supports it. The Driving Question presented in the Case Study of each chapter can get students invested in chapter topics and in working toward answering the question, "How do sea pigs survive in the deep ocean?" In the Unit Activity, students can practice scientific reasoning and argumentation to show how the evidence supports their claim.

Follow the Anchoring Phenomenon How do sea pigs survive in the deep ocean?

Gather evidence with . . .	Chapter 2	Chapter 3	Chapter 4	Unit Activity
CASE STUDY	How do energy and matter move through an ecosystem?	How is biodiversity related to ecosystem stability?	What factors affect the size of a population?	Revisit the unit's anchoring phenomenon of sea pigs and other organisms thriving in a deep-sea ecosystem. **Claim, Evidence, Reasoning** Students use the evidence they gathered throughout the unit to state and support a claim with reasoning. **Go online** to access Student Self Reflection and Teacher Scoring rubrics for this activity. **English Language Learners** Cite Text Evidence
MINILAB	How can you model the distribution of biomass in an ecosystem?	What microorganisms are found in a pond ecosystem?	How can you estimate a population of organisms that cannot be counted directly?	
LOOKING AT THE DATA	Students quantify the biomagnification of mercury by analyzing the movement of matter through a marine food web.	Students compare species data for at-risk and threatened endemic and exotic species in New Zealand.	Students analyze the population growth rate of an invasive aquatic species.	
TYING IT ALL TOGETHER REVISIT THE CASE STUDY	Students develop models to analyze nutrient pathways from marine to terrestrial organisms.	Students construct an argument to explain how a disturbance such as a wildfire affects ecosystem stability and biodiversity.	Students research and evaluate solutions for conserving populations of native species in an urban community.	
Chapter Review: Revisit Sea Pig Survival	Reflect on the role of sea pigs in cycling carbon and other matter.	Students explain how a change in the stability of a deep-ocean ecosystem might affect sea pig survival.	Students predict how seafloor mining might affect the carrying capacity for a sea pig population.	
Virtual Investigation: Sea Pigs on the Abyssal Plain	Students take on the role of a deep-sea researcher, exploring factors that affect an abyssal plain ecosystem.			
Chapter Investigation A	How does salt concentration affect the hatching of brine shrimp eggs?	How can you sample biodiversity in a plant community?	How and why does the number of duckweed plants in a population change over time?	
Chapter Investigation B	What is the effect of an abiotic factor on the hatching and survival of brine shrimp?	How do biotic and abiotic factors influence succession in a freshwater pond community?	How can you design, build, and test an effective seed trap?	

Planning Your Investigations

Each chapter features a Minilab and two full Investigations that offer hands-on opportunities for students to engage in Science and Engineering Practices.

Advance Preparation For Chapter 2 Investigation A and Investigation B, you will need to purchase brine shrimp eggs, which are available from aquarium supply stores, pet stores, or online. Advance preparation notes for all labs are included in the Teacher's version.

Chapter 2	Title	Time	Standards
Minilab	*Model a Biomass Pyramid*	30 minutes	**SEP** Using Mathematics and Computational Thinking **CCC** Energy and Matter
Investigation A **Guided Inquiry**	*Salinity and Brine Shrimp Survival*	130 minutes over 3 days	HS-LS2-2
Investigation B **Open Inquiry**	*Exploring Brine Shrimp Survival*	180 minutes over 5 days	HS-LS2-2
Chapter 3	**Title**	**Time**	**Standards**
Minilab	*Observing Biodiversity in Pond Water*	30 minutes	**SEP** Planning and Carrying Out Investigations, Engage in Argument From Evidence, Obtaining, Evaluating, and Communicating Information **CCC** Patterns
Investigation A **Guided Inquiry**	*Measuring Biodiversity Using Ecological Sampling Methods*	100 minutes over 2 days	HS-LS2-2, HS-LS2-6
Investigation B **Design Your Own**	*Ecological Succession in an Aquatic Environment.*	200+ minutes	HS-LS2-6
Chapter 4	**Title**	**Time**	**Standards**
Minilab	*Mark-Recapture Sampling*	30 minutes	**SEP** Using Mathematics and Computational Thinking **CCC** Scale, Proportion, and Quantity
Investigation A **Guided Inquiry**	*Population Growth of Duckweed*	190 minutes over 2 days plus observation time over 3 weeks	HS-LS2-2
Investigation B **Engineering**	*Designing a Seed Trap*	150 minutes over 3 days	HS-ETS1-3

Assessment Planning

UNIT 1 PERFORMANCE TASKS

Five performance-based assessments target the NGSS Performance Expectations with three-dimensional activities to measure student mastery. Rubrics are available online for Performance Tasks.

Performance Tasks

	Title	Overview	PEs Addressed	Use After
1	*Why Should We Preserve Wetland Ecosystems?*	Students analyze data and conduct research to develop a model of their chosen wetland ecosystem.	HS-LS2-4, HS-LS2-5	Chapter 2
2	*How Do Seasonal Changes Affect Organisms in a Freshwater Ecosystem?*	Students investigate the effects of lake turnover on matter cycling and energy flow.	HS-LS2-3, HS-LS2-4	Chapter 2
3	*How Does a Sudden Disturbance Change a Rainforest Ecosystem?*	Students analyze data and conduct research to explain how a hurricane changes carrying capacity and biodiversity in a rainforest.	HS-LS2-1, HS-LS2-2	Chapter 3
4	*How Does Long-term Drought Change a Saltwater Ecosystem?*	Students analyze data and evaluate how long-term drought changes the food web in a salt lake.	HS-LS2-2, HS-LS2-6	Chapter 3
5	*What Is the Best Way to Restore Habitat for Endangered Bats?*	Students design, evaluate, and refine solutions to reduce human impact on an endagered bat population.	HS-LS2-7, HS-ETS1-3	Chapter 4

Additional Resources

Use a search engine to find these resources on the internet.

VIDEOS

"Why is biodiversity so important?" TED-Ed
A four-minute animation gives an overview of the stability of diverse ecosystems and the importance of different biodiversity features for the survival of organisms on Earth.

Search: "TED-Ed biodiversity"

"What's in a Lichen?" National Geographic
The word *symbiosis* was invented by scientists to describe the relationship between a fungus and an algae. The third partner in this relationship took 150 years to discover. This four-minute National Geographic Short Film features the researchers who changed widely accepted scientific ideas using on new evidence.

Search: "What's in a lichen? Short film"

ARTICLES

"Using Strategy to Preserve Biodiversity While Saving Space," University of Hamburg
An article from the University of Hamburg, Cluster of Excellence: Climate, Climatic Change, and Society (CLICCS) discusses new strategy in defining protected areas in Latin America.

Search: "CLICCS strategy biodiversity"

ENGAGE

Unit Storyline

In this unit, students will investigate the question, "How do sea pigs survive in the deep ocean?" Each chapter provides information that will help answer this Driving Question.

Chapter 2 *How do matter and energy move throughout an ecosystem?*

Students learn about the movement of energy and matter through ecosystems, the modeling of this movement using ecological pyramids, and cycles of matter, such as the carbon cycle.

Chapter 3 *How is biodiversity related to ecosystem stability?*

Students are introduced to important ecological concepts, including competition, predation, and symbiosis. They also learn about ecosystem stability and different measures of biodiversity.

Chapter 4 *What factors affect the size of a population?*

Students learn about population dynamics, measurement, and modeling, as well as factors that limit population growth.

Unit Activity Students revisit the Anchoring Phenomenon to make a claim about the Driving Question and apply reasoning to the evidence they have gathered throughout the unit.

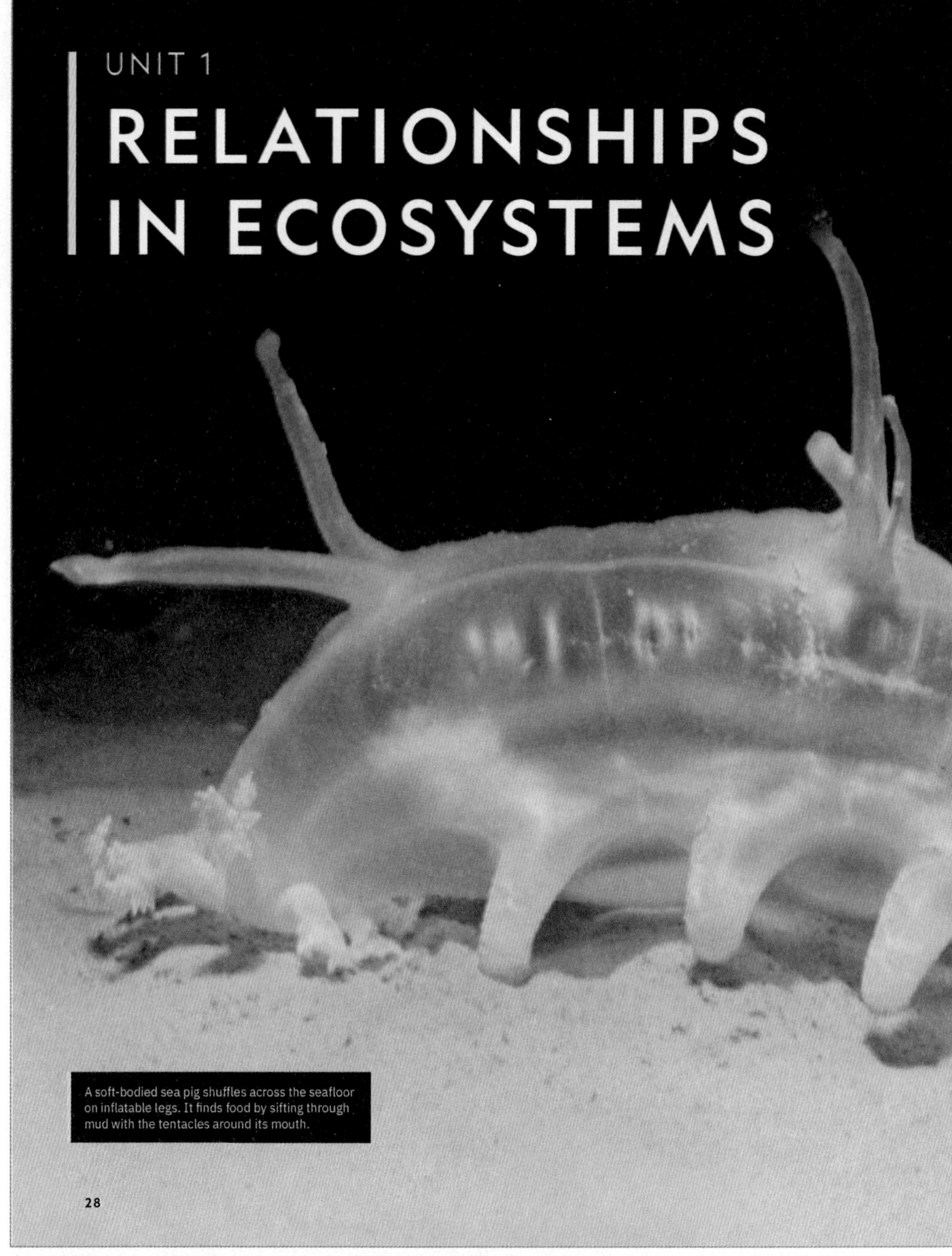

A soft-bodied sea pig shuffles across the seafloor on inflatable legs. It finds food by sifting through mud with the tentacles around its mouth.

SCIENCE BACKGROUND

Taxonomy of The Sea Pig phylum: Echinodermata; class: Holothuroidea ("sea cucumbers"); order: Elasipodida; family: Elpidiidae; genus: *Scotoplanes* ("sea pigs")

Sea Pigs The common name "sea pig" usually refers to *Scotoplanes globosa*, a translucent pink sea cucumber species found worldwide, but it is also applied to other species of similar appearance. Sea cucumbers are some of the most commonly observed abyssal megafauna.

The Abyssal Plain The abyssal zone of the ocean lies between 2 and 6 kilometers below the surface. Near continental shelves, a layer of muddy sediment, averaging 1 kilometer in depth, has buried most of the topography, forming uniformly flat plains. These plains make up more than half of Earth's total surface area and are its largest habitat. The abyssal plain's most biodiverse areas occur where rocky substrate is exposed, such as on volcanic seamounts and near tectonic plate boundaries.

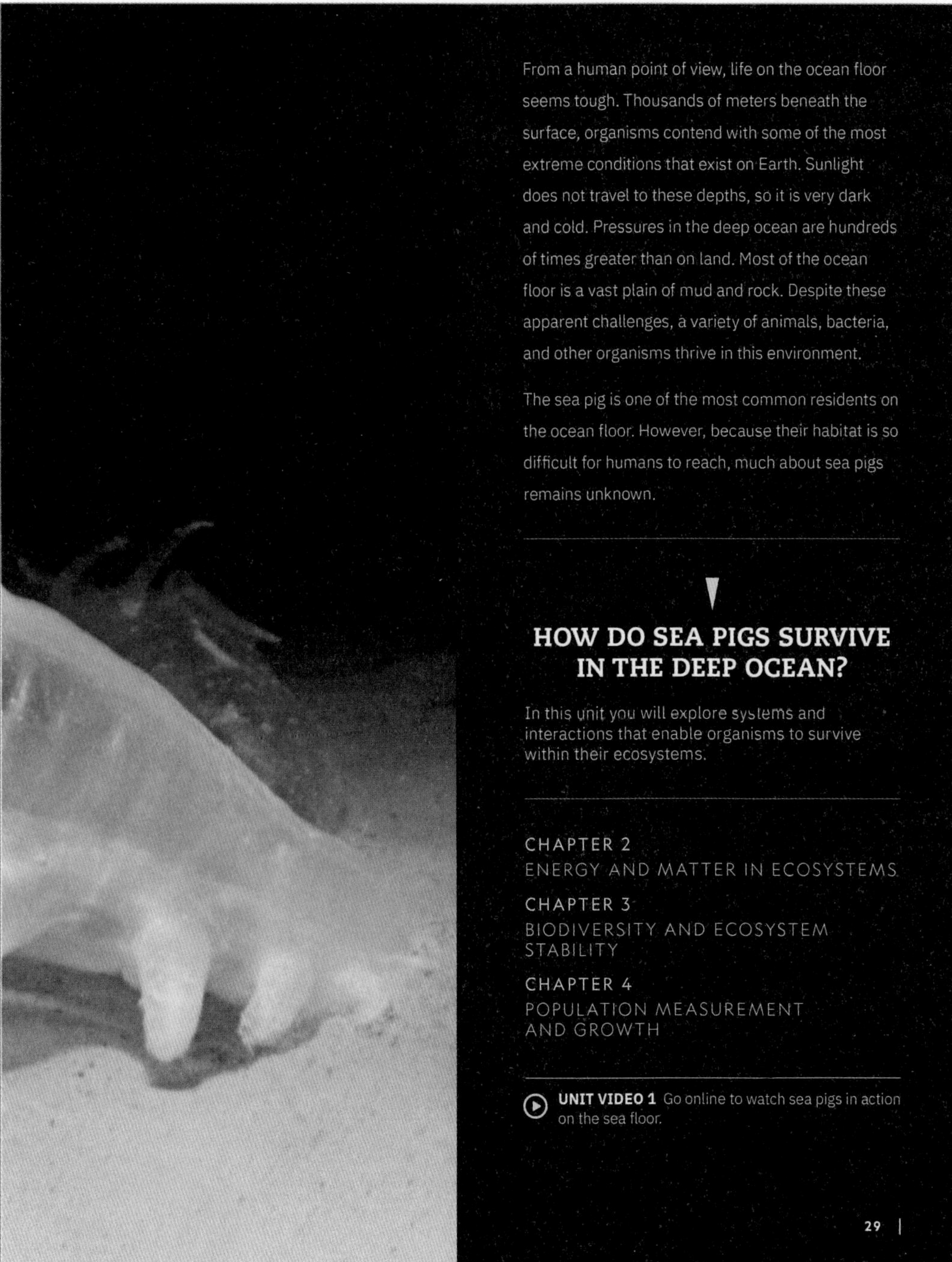
From a human point of view, life on the ocean floor seems tough. Thousands of meters beneath the surface, organisms contend with some of the most extreme conditions that exist on Earth. Sunlight does not travel to these depths, so it is very dark and cold. Pressures in the deep ocean are hundreds of times greater than on land. Most of the ocean floor is a vast plain of mud and rock. Despite these apparent challenges, a variety of animals, bacteria, and other organisms thrive in this environment.

The sea pig is one of the most common residents on the ocean floor. However, because their habitat is so difficult for humans to reach, much about sea pigs remains unknown.

HOW DO SEA PIGS SURVIVE IN THE DEEP OCEAN?

In this unit you will explore systems and interactions that enable organisms to survive within their ecosystems.

UNIT VIDEO 1 Go online to watch sea pigs in action on the sea floor.

29 |

Introduce the Anchoring Phenomenon

Ecosystems that exist on the ocean floor differ from terrestrial ecosystems. In the absence of sunlight, photosynthesis cannot drive energy flow and matter cycling. Examining this unusual ecosystem as an Anchoring Phenomenon encourages students to apply what they learn throughout the unit to a unique situation.

Driving Question Have students make a list of things they think they would need to know about sea pigs in order to answer the Driving Question. The photo and caption may prompt them to ask "What do sea pigs eat?" To analyze how a sea pig survives in its environment, students will need to focus on its most basic need: an energy source. What they learn in this unit about food webs, biodiversity, ecosystem stability, and factors that affect population size will inform their models of how sea pigs survive on the sea floor.

About the Photo This sea pig was photographed in the Monterey Bay Submarine Canyon by scientists on the research vessel *Point Sur.*

Video

Time: 1:34

Use **Unit Video 1**, from Ocean Networks Canada, to show a 15-centimeter long sea pig walking on the ocean floor using its hydraulically operated tube feet. Scientists think papillae on the upper body allow the sea pig to "smell" its way to food.

Go Online VIRTUAL INVESTIGATION

Sea Pigs on the Abyssal Plain

Time: about 50 minutes

Objectives Students will learn about the abyssal plain ecosystem and gather evidence to describe how sea pigs survive and thrive in deep-sea conditions.

Explore and Learn Students interact with the environment to obtain background information about the abyssal plain ecosystem and learn about the tools and techniques that scientists use to study organisms in the deep ocean.

Collect Data Students "drive" a remotely operated vehicle (ROV) to various locations on the abyssal plain to observe the organisms and environment, use a laser grid to count sea pigs per square unit, deploy a robotic arm to collect soil and environmental samples, and record their observations using tools available in the virtual laboratory.

Analyze and Report Students answer questions about the investigation, incorporating qualitative and quantitative data in a report to support their analyses.

EXPLORER
DIVA AMON

ENGAGE

About the Explorer

National Geographic Explorer Diva Amon investigates marine ecosystems that humans have only recently had the technology to observe directly. Her explorations have resulted in new information about animal species and their environments, which has furthered scientific understanding of deep-ocean ecosystems and how human activity affects them. You can learn more about Diva Amon and her research on the National Geographic Society's website: www.nationalgeographic.org.

On the Map

Mariana Trench Students may be familiar with the deepest known part of the ocean, Challenger Deep, which lies within the arc-shaped Mariana Trench in the Pacific Ocean. The Marianas Trench Marine National Monument, established in 2009 by President George W. Bush, encompasses the trench and surrounding areas. Although it lies nearly halfway across the globe, it includes the U.S. Commonwealth of the Northern Mariana Islands and the U.S. Territory of Guam, and is thus managed by U.S. agencies.

THINKING CRITICALLY

Analyze *Sample Answer: Mining equipment would scrape the seafloor, disturbing the habitats of organisms that live there. Removing minerals from the seafloor would also remove any organisms that live on these surfaces, which would decrease the food supply for species that eat those organisms.*

NATIONAL GEOGRAPHIC EXPLORER **DIVA AMON**

UNDER THE SEA

ILLUMINATING UNIQUE ECOSYSTEMS IN THE DEEP OCEAN

The deep ocean, the part of the ocean below 200 meters where sunlight fades rapidly, is one of the planet's last untouched wildernesses. This environment supports human life by absorbing carbon dioxide from the atmosphere and cycling nutrients into fisheries. Although less than 20 percent of the ocean floor has been mapped, it is estimated to contain vast deposits of natural resources, and countries and corporations are now exploring ways to extract them. To understand the effects of human activities and climate change on deep-ocean ecosystems, marine biologist and National Geographic Explorer Dr. Diva Amon seeks to answer the question, "What lives here?"

OTHERWORLDLY AND WEIRD Diva Amon has led scientific expeditions to explore the wide variety of largely unknown ecosystems in the Gulf of Mexico and the Marianas Trench Marine National Monument. These expeditions focused on characterizing deep-sea coral and sponge communities, bottom-fish habitats, and chemosynthetic environments. Chemosynthetic bacteria make food from chemicals seeping out of the seafloor. In the Gulf of Mexico, these bacteria colonize cold seeps, mud volcanoes, asphalt seeps, and brine pools. In the Marianas, Amon was among the first humans to observe recently formed hydrothermal vents that release plumes of chemical-filled underground water. These vents are hotbeds of chemosynthetic activity. Organisms such as tube worms and mussels that obtain food from the bacteria in these environments create habitats for many other species, which in turn become food for predators.

Of the hundreds of different species of organisms scientists observed on these research missions, most had not been seen before in these regions. Others were seen alive for the first time, and many were potentially undescribed species. For example, a slit shell snail observed in the Marianas in July 2016 was likely an undiscovered species. Scientists estimate that the deep ocean is home to one million species, two-thirds of which have not yet been discovered.

DEEP DIVE WITHOUT DEEP POCKETS Exploring the deep ocean is expensive. Remotely operated vehicles (ROVs) outfitted with machinery for collecting samples, high-resolution cameras, and other sensors are the workhorses of deep-sea research. These vehicles make measurements and take images that provide data about the types, populations, and behavior of organisms in their natural environment. They are deployed from ships that spend months at sea. During Amon's missions aboard the NOAA ship *Okeanos Explorer* and other ships, she described the rare and unique animals and ecosystems captured by the ROV *Deep Discoverer*. The video footage was live-streamed to the public over the internet. The same technology makes real-time data gathered by the ship's equipment available to collaborating scientists on shore.

This scale of investment in exploration is typically not accessible for developing countries, even though many manage deep-sea environments within their maritime zones. In her home country of Trinidad and Tobago, Amon's nonprofit organization, SpeSeas, has helped pilot first-time explorations using deep-sea cameras developed by National Geographic's Exploration Technology team and lower-cost ROVs that require minimal resources and expertise. The goal of this pilot project was to empower nations to develop and sustain the capacity to investigate their deep-sea backyards.

THINKING CRITICALLY

Analyze *Describe how a human activity such as seabed mining might affect organisms that live in a deep-ocean ecosystem.*

30

CROSSCURRICULAR CONNECTIONS

Human Geography The majority of the ocean floor lies beneath international waters and is subject to the same laws that govern human activity at the surface. The discovery of large reserves of polymetallic nodules scattered all over the ocean floor has spurred interest in further exploration of this environment for potential natural resource extraction. Some of the metals found in these nodules, such as manganese, copper, cobalt, and titanium, are projected to be more abundant in the deep ocean than in land reserves. The International Seabed Authority (ISA) organizes and controls all activities related to deep-sea mineral resources.

Scientists are working to characterize the unknown ecosystems of the deep ocean to understand how seabed mining would affect them. Tradeoffs between the economic potential of mining highly valuable substances from the ocean floor and the conservation of this unique environment are a viable real-world topic for student discussion and debate.

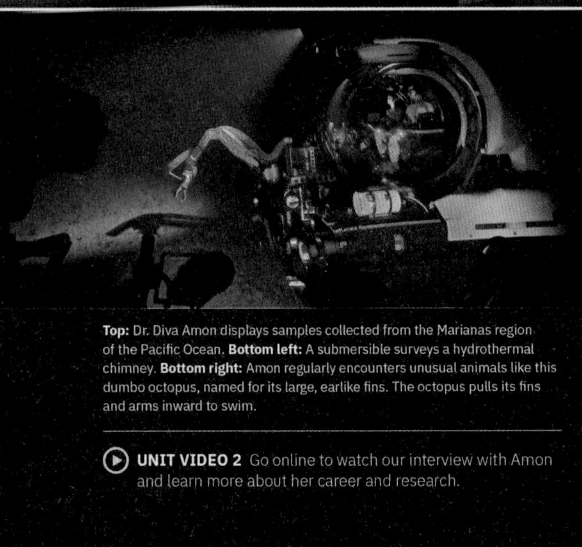

Top: Dr. Diva Amon displays samples collected from the Marianas region of the Pacific Ocean. **Bottom left:** A submersible surveys a hydrothermal chimney. **Bottom right:** Amon regularly encounters unusual animals like this dumbo octopus, named for its large, earlike fins. The octopus pulls its fins and arms inward to swim.

UNIT VIDEO 2 Go online to watch our interview with Amon and learn more about her career and research.

31

DIFFERENTIATED INSTRUCTION | English Language Learners

Taking Notes When Listening As students watch the video, encourage them to take notes about any words and concepts they do not understand. Invite students to watch the video again to grasp concepts and language they missed the first time.

Beginning students can take notes in their native language or jot down simple words in English.

Intermediate and Advanced students can take more complex notes as their ability allows. Then have students share their notes with a partner and help each other clarify concepts and discuss questions that came up.

Explorers at Work

Time: 8:53

Unit Video 2 interviews Diva Amon to allow students to hear first hand about her experiences inside the submersible and what it's like to explore at these depths. She describes the unusual fauna of the deep to help listeners visualize how it compares to what they see around them everyday. This lays the groundwork for students to apply what they learn throughout the unit to an ecosystem about which very little is known.

Connect to Careers

Marine Biologist Between 50 and 80 percent of all life on Earth is found in the oceans. Marine biologists have an enormous variety of organisms and natural habitats to study. They can specialize in studying animal behavior and survival techniques, conducting species inventories in new unexplored areas, or documenting the impact of climate change and human activity on marine ecosystems. Some marine biologists focus on marine biotechnology, which develops industrial processes and products that reflect nature.

Research assistants need a bachelor's degree in marine biology, but to lead research projects, a master's degree or PhD is required. Private research laboratories, governments, and universities employ marine biologists to conduct research in the field or to teach others about biology.

ROV Pilot Deep-ocean data are obtained using remote operated vehicles (ROVs) that are driven and maintained by pilots working aboard a research ship. ROV pilots have a wide variety of education and training. They may specialize in the electrical or mechanical aspects of robotics, or they may design and build scientific equipment that is mounted to the ROV. Pilots work for weeks or months at a time in teams that include other ROV pilots, scientists, and the ship's crew, so it is essential that they are able to cooperate with others under stressful conditions. ROV pilots regularly need to use their technical skills to solve problems rapidly and effectively.

Three-Dimensional Learning

The practices, core ideas, and crosscutting concepts presented in this chapter's text, investigations, and resources provide support to address the following Performance Expectations: **HS-LS2-2, HS-LS2-3, HS-LS2-4, HS-LS2-5, and HS-LS2-6.**

Science and Engineering Practices	Disciplinary Core Ideas	Crosscutting Concepts
Asking Questions and Defining Problems Developing and Using Models (HS-LS2-5) Using Mathematics and Computational Thinking (HS-LS2-2, HS-LS2-4) Constructing Explanations and Designing Solutions (HS-LS2-3)	**LS2.A:** Interdependent Relationships in Ecosystems (HS-LS2-2) **LS2.B:** Cycles of Matter and Energy Transfer in Ecosystems (HS-LS2-3, HS-LS2-4, HS-LS2-5) **LS2.C:** Ecosystem Dynamics, Functioning, and Resilience (HS-LS2-2, HS-LS2-6)	Patterns Cause and Effect Scale, Proportion, and Quantity (HS-LS2-2) Systems and System Models (HS-LS2-5) Energy and Matter (HS-LS2-3, HS-LS2-4) Stability and Change (HS-LS2-6)

Contents		Instructional Support for All Learners	Digital Resources
ENGAGE			
32–33	**CHAPTER OVERVIEW** **CASE STUDY** How do energy and matter move through an ecosystem?	**Social-Emotional Learning** CCC Energy and Matter **On the Map** Tongass National Forest **English Language Learners** Summarize	
EXPLORE/EXPLAIN			
34–37 DCI LS2.A LS2.B	**2.1 ECOLOGICAL SYSTEMS** • Distinguish between the levels of ecological organization. • Describe how matter and energy support the survival of organisms.	**Vocabulary Strategy** Word Families **English Language Learners** Use Word Roots **Connect to English Language Arts** Integration of Knowledge and Ideas CCC Scale, Proportion, and Quantity **In Your Community** Urban Ecosystems CCC Systems and System Models **Crosscurricular Connections** Chemistry	*Video 2-1* **Investigation A** Salinity and Brine Shrimp Survival (130 minutes over 3 days)
38–41 DCI LS2.B LS2.C	**2.2 MODELING THE TRANSFER OF ENERGY AND MATTER** • Define the main ecological roles in an ecosystem and identify their functions. • Identify the trophic levels of a simple ecosystem. • Predict potential impacts on an ecosystem resulting from food-chain disruptions.	**Vocabulary Strategy** Latin Roots CER Revisit the Anchoring Phenomenon **CASE STUDY** Ask Questions **In Your Community** Local Food Chains SEP Developing and Using Models **Connect to English Language Arts** Integration of Knowledge and Ideas	**Interactive Figure** Antarctic Food Web *Video 2-2*
42	**EXPLORER** **RAE WYNN-GRANT**	**Connect to Careers** Carnivore Ecologist **English Language Learners** Paired Reading	*Video 2-3*

Contents		Instructional Support for All Learners	Digital Resources
EXPLORE/EXPLAIN			
43–47 **DCI** LS2.B	**2.3 MODELING ENERGY AND MATTER DISTRIBUTION** • Model the distribution of energy in a community using an energy pyramid. • Model the distribution of biomass in a community using a biomass pyramid. • Model the distribution of individuals in a community using a pyramid of numbers.	**Address Misconceptions** Conservation of Energy **SEP** Using Mathematical and Computational Thinking **CONNECTIONS** Social-Emotional Learning **Connect to Mathematics** Define Quantities for Modeling **On Assignment** Documenting Biodiversity	
48	**MINILAB** MODEL A BIOMASS PYRAMID		
49–55 **DCI** LS2.A LS2.B LS2.C	**2.4 CYCLING OF MATTER** • Define the Earth system and its boundaries, and describe it in terms of interconnected subsystems. • Model the hydrologic (water) cycle, carbon cycle, nitrogen cycle, and phosphorus cycle.	**Crosscurricular Connections** Earth Science **English Language Learners** Create a Personal Dictionary **Visual Support** Cycling Through the Spheres **CER** Revisit the Anchoring Phenomenon **Differentiated Instruction** Leveled Support **Differentiated Instruction** Students with Disabilities **Connect to Mathematics** Reason Quantitatively **Differentiated Instruction** Economically Disadvantaged Students **On the Map** Lake Erie **In Your Community** Phosphate Pollution	**Investigation B** Exploring Brine Shrimp Survival (180 minutes over 5 days)
56	**LOOKING AT THE DATA** BIOMAGNIFICATION OF MERCURY		
ELABORATE			
57	**TYING IT ALL TOGETHER** How do energy and matter move through an ecosystem?	**CCC** Systems and System Models	Self-Reflection and Scoring Rubrics
Online	**Investigation A** Salinity and Brine Shrimp Survival **Investigation B** Exploring Brine Shrimp Survival	Guided Inquiry (130 minutes) Open Inquiry (180 minutes)	MINDTAP Access all your online assessment and laboratory support on MindTap, including: sample answers, lab guides, rubrics, and worksheets. PDFs are available from the Course Companion Site.
EVALUATE			
58–59	**Chapter 2 Review**		
Online	**Chapter 2 Assessment** **Performance Task 1** *Why Should We Preserve Wetland Ecosystems?* (HS-LS2-4, HS-LS2-5) **Performance Task 2** *How Do Seasonal Changes Affect Organisms in a Freshwater Ecosystem?* (HS-LS2-3, HS-LS2-4)		

CHAPTER 2

ENGAGE

In studying life science before this course, students are likely to have learned that food webs are used to trace the movement of matter and energy through an ecosystem. This chapter uses the food web model as an introduction to the concept of trophic levels, leading to a more quantitative analysis of energy flow and distribution.

Students will also learn about processes by which matter breaks down and recombines into different substances, enabling them to further examine the biogeochemical processes that drive the cycling of specific elements through an ecosystem.

About the Photo This photo was taken on Admiralty Island, which is one of three major islands in the Tongass National Forest in Southeast Alaska. The Tlingit name for the island is "Kootznoowoo," meaning "Fortress of the Bears." To prepare students for this chapter's content, ask them to use scientific terminology that they already know to describe the ecological relationships between organisms they see in the photograph.

Social-Emotional Learning

As students progress through the chapter, invite them to look for contexts and opportunities to practice some of the five social and emotional competencies. For example, as students participate in discussions about the content and the questions provided, support them in developing **relationship skills**. Help them practice effective communication and collaboration while answering questions and solving problems.

In addition, when students learn about relationships between humans and ecosystems (e.g., in "Using Data to Prevent Human–Bear Conflict" and "Human Impact on Biogeochemical Cycles"), remind them to practice **self-management** and **responsible decision-making**. Support students in managing their emotions, and challenge them to be curious and open-minded as they consider the consequences of human actions on the environment.

CHAPTER 2

ENERGY AND MATTER IN ECOSYSTEMS

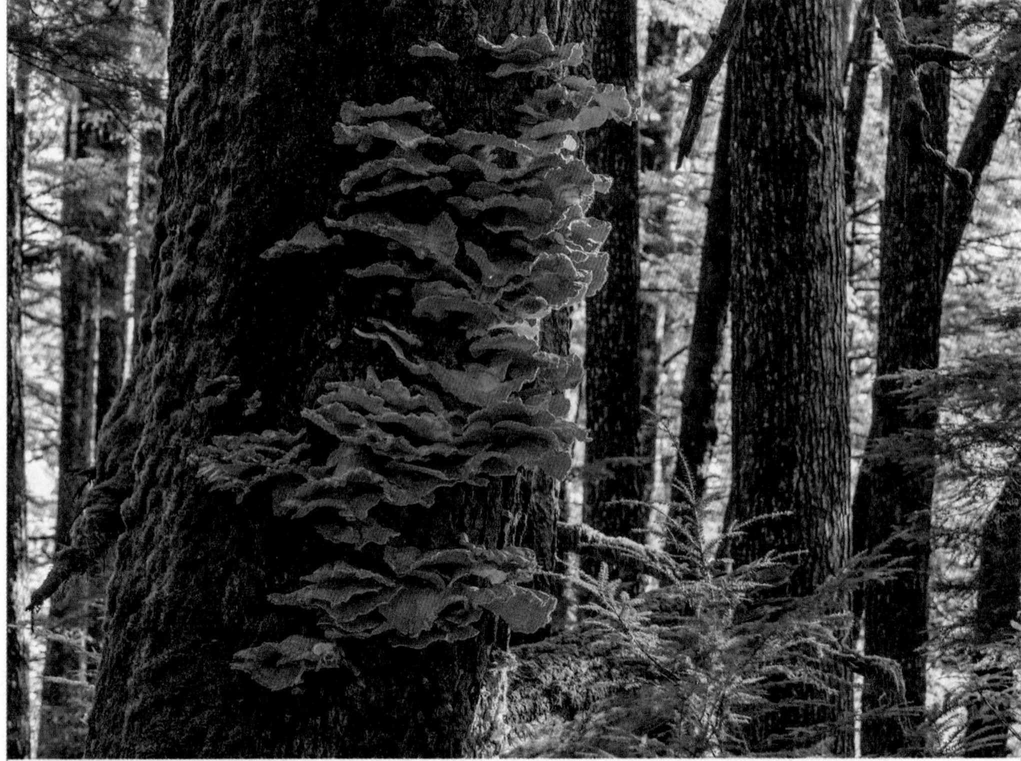

Hemlock, spruce, and cedar trees on Admiralty Island, Alaska, provide habitat and food for chicken mushrooms, lichens, and mosses.

Chapter 2 supports the NGSS Performance Expectations **HS-LS2-3**, **HS-LS2-4**, and **HS-LS2-5**.

The Tongass National Forest, which covers nearly 17 million acres in Southeast Alaska, is part of the largest temperate rainforest in the world. The forest has remained largely intact since glaciers retreated thousands of years ago. Fossil evidence shows that some animal species found in the forest today, such as bears and deer, also lived in the area at that time.

32 CHAPTER 2 ENERGY AND MATTER IN ECOSYSTEMS

CROSSCUTTING CONCEPTS | Energy and Matter

Modeling at Varied Scales This chapter focuses on modeling energy and matter transfer at ecological scales: between organisms in a community, between organisms and their environment, and among the biosphere, atmosphere, hydrosphere, and geosphere. Some fields of biology, such as physiology, cell biology, molecular biology, and biochemistry, essentially study how energy and matter enable life processes at various scales. Chapters 5 and 6 in Unit 2 addresses transformations of energy and matter at the molecular and cellular levels. Further reinforce this crosscutting concept throughout Unit 3 by having students organize information about living systems in terms of how they enable an organism to obtain energy and matter from its surroundings, transfer energy and matter within its body, and use energy and matter to survive.

CASE STUDY
SOMETHING FISHY IN THE FOREST

HOW DO ENERGY AND MATTER MOVE THROUGH AN ECOSYSTEM?

Biologists have identified several large populations of brown bears living in the temperate rainforest along the Pacific Ocean in Southeast Alaska. About 4300 brown bears inhabit Admiralty, Chichagof, and Baranof Islands in the Tongass National Forest. The bear population density on Admiralty Island is among the highest in the world, at one bear per square mile.

Bears are famous for their enormous appetites. They eat a variety of vegetation including berries, roots, and grasses. Bears also hunt large animals such as moose, deer, and caribou as well as smaller mammals. To obtain their primary source of protein, Alaskan coastal brown bears fish for salmon in the rivers that extend from the forest to the ocean (**Figure 2-1**). These species of salmon are born in freshwater streams and migrate to the ocean where they spend much of their lives. When it's time to reproduce, the salmon return to the same freshwater streams where they were born.

Typically, a bear eats only the most energy-rich parts of the fish, such as the brain and the eggs of female salmon that have not yet spawned. Brown bears gain just over one kilogram of fat per day to store energy to survive through the winter months, during which they lose 20 to 40 percent of their body mass. In particular, pregnant bears must build up ample fat reserves to gestate and feed cubs. Bears carry their meals away from the watershed, where they leave fish carcasses to decay or be scavenged by other animals.

Temperate rainforest soils naturally contain low levels of nitrogen, an element that is abundant in salmon and essential for plant growth. In a river ecosystem, plants and trees enhance conditions for the reproduction and growth of new fish. Thriving vegetation provides shade, stabilizes riverbank soil, filters sediment, and serves as a nutritious food source for other animals and microbes.

The relationship between bears and salmon affects many other organisms and is key to maintaining healthy coastal forest ecosystems. In turn, the forests provide suitable habitats for sustainable populations of bears and salmon. These relationships are characterized by the movement of both energy and matter throughout the ecosystem.

Ask Questions *As you read this chapter, generate questions about the ways in which matter cycles and energy flows between organisms and their environment.*

FIGURE 2-1 When salmon are plentiful in the late summer and fall, a brown bear can catch more than 30 fish per day.

DIFFERENTIATED INSTRUCTION | English Language Learners

Summarize Have students work in pairs. Assign each pair one paragraph from the Case Study to read together. Ask each pair to summarize their paragraph using sentence frames like the ones below. Suggest that students use a graphic organizer to help them find the main idea.

Beginning This is about _______.

Intermediate This paragraph tells how/what _______.

Advanced The main focus of this paragraph is _______. These details support the main focus: _______.

CASE STUDY

ENGAGE

The interactions described in the Case Study can be used to assess students' prior knowledge of relationships within ecosystems. Ask students to identify which organisms in this community exemplify producer and consumer roles, and which organisms might occupy decomposer roles. Students may also have experience constructing a food chain or basic food web that includes these organisms.

The temperate rainforest ecosystem provides an additional context in which students can practice modeling and refining their models as they read the chapter. In the Tying It All Together activity, students will use a model to explain a phenomenon that occurs in this ecosystem.

Ask Questions Students should revisit the Case Study as they read the chapter to make connections with the content. See a specific suggestion in Section 2.2.

On the Map

Tongass National Forest Use the map to discuss with students how geographic characteristics affect where, when, and how organisms obtain food in this environment. The densely packed islands and long coastlines, with rivers extending inland, result in some terrestrial consumers with greater access to freshwater and saltwater food sources. Likewise, aquatic organisms benefit from nearby land resources, such as falling plant matter.

Human Geography Native Alaskans have inhabited Southeast Alaska for thousands of years. What is today recognized as the Tongass National Forest is the ancestral lands of the Tlingit, Haida, and Tsimshian people.

Admiralty, Baranof, and Chichagof are colloquially referred to as the *ABC Islands*. Despite their large size (Admiralty and Chichagof are the seventh and fifth largest islands in the United States, respectively), relatively few people inhabit the islands. Sitka, which is located on Baranof Island, has the largest population, which was 8458 people in the 2020 U.S. census.

2.1

ECOLOGICAL SYSTEMS

LS2.A, LS2.B

EXPLORE/EXPLAIN

This section provides a review of Earth's interconnected systems, introduces the hierarchical organization of the biosphere, and describes the main processes through which energy and matter support organism survival.

Objectives

- Distinguish between the levels of ecological organization.
- Describe how matter and energy support the survival of organisms.

Pretest Use **Question 6** to identify gaps in background knowledge or misconceptions.

Vocabulary Strategy

Word Families The Greek root *bio-* (life) should be familiar to students from *biology* and other common words, such as *biography*. It is also the root of five Key Terms in this chapter: *biome, biosphere, biomass, biomass pyramid,* and *biogeochemical cycle*. Suggest that students add each of these terms to a word tree or other graphic organizer. Students can also add other terms, such as *biomagnification*, which they will see in the Looking at the Data feature, and *symbiosis*, which they will encounter in Chapter 3.

Video

Time: 1:45

Use **Video 2-1** to show students the differences between the four interacting systems (atmosphere, biosphere, geosphere, hydrosphere) that make up the Earth system.

Predict *Sample answer: If high tides ceased reaching the tidal pool, the community would not survive because seawater would not be replenished with nutrients from the ocean. Eventually, the seawater would evaporate.*

2.1 ECOLOGICAL SYSTEMS

KEY TERMS

atom, biome, biosphere, community, ecosystem, molecule, population

Walking along a rocky beach at low tide, you discover a shallow depression about the size of a bathtub. It is filled knee-deep with seawater. As you look closer, you notice clusters of seashells clinging to its upper edges. Lining the bottom of the pool are colonies of barnacles and purple mussels. Patches of algae sway among brightly colored sea anemones and sea stars.

You have stumbled upon a tide pool and a community of organisms specialized to survive there. It is like a strange world all its own. Soon, the next high tide will cover the beach and its tide pool communities, hiding them from view once more.

Predict *Consider the relationship between the organisms in a tide pool and the daily cycle of high and low tides. How might an overall increase in sea level affect tide pool organisms?*

VIDEO

VIDEO 2-1 Go online to learn more about Earth systems.

Earth Systems

You can think of a tide pool as a system that includes interacting living and nonliving things. Like all systems, it has inputs, such as nutrients from the ocean, and outputs, such as waste materials. It is just one example of the many interacting systems that make up the much larger Earth system. The Earth system is made up of four interconnected systems, or spheres: the atmosphere, hydrosphere, geosphere, and biosphere. These four systems are described in **Figure 2-2**.

FIGURE 2-2 Earth can be organized into four interacting systems.

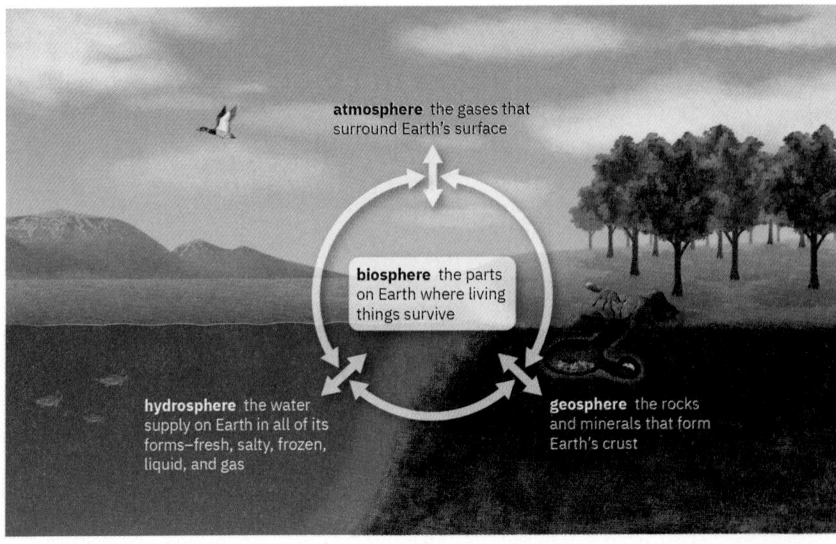
atmosphere the gases that surround Earth's surface
biosphere the parts on Earth where living things survive
hydrosphere the water supply on Earth in all of its forms–fresh, salty, frozen, liquid, and gas
geosphere the rocks and minerals that form Earth's crust

34 CHAPTER 2 ENERGY AND MATTER IN ECOSYSTEMS

DIFFERENTIATED INSTRUCTION | English Language Learners

Use Word Roots Display the word *biosphere* as two parts: *bio-* and *-sphere*, and guide understanding.

Beginning and **Intermediate** learners should sound out the parts. Explain that *bio-* means life, that a *sphere* is a ball shape, and that *biosphere* refers to the parts of Earth where living things survive. Have students identify the other words with *-sphere*, break them into parts, and sound them out.

Advanced students can read the meanings of *atmosphere, hydrosphere,* and *geosphere* in **Figure 2-2** as their mixed-level peers identify each sphere in the figure. Students can then infer the meanings of *atmo-*, *hydro-*, and *geo-*.

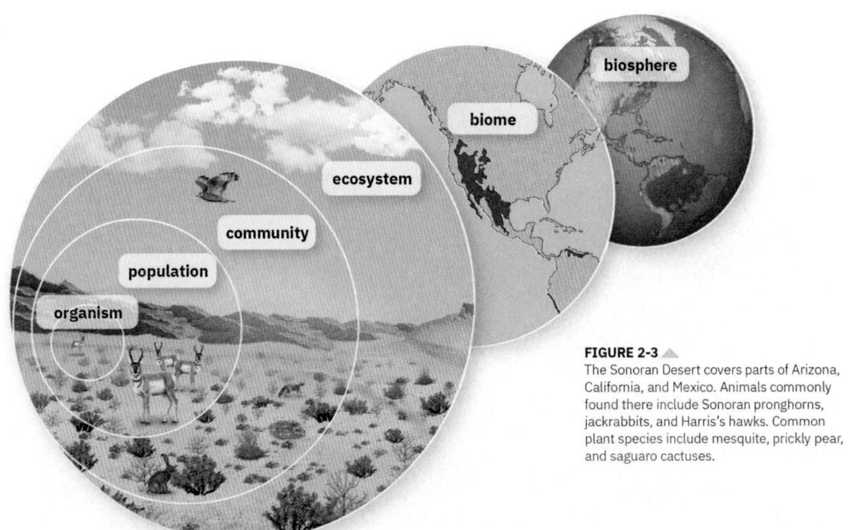

FIGURE 2-3 The Sonoran Desert covers parts of Arizona, California, and Mexico. Animals commonly found there include Sonoran pronghorns, jackrabbits, and Harris's hawks. Common plant species include mesquite, prickly pear, and saguaro cactuses.

Life in the Biosphere

The **biosphere** includes all the parts on Earth where life exists. It extends from the deep ocean to high mountaintops. Whether you are observing a tide pool, forest, desert, or other environment within the biosphere, you will find organisms suited to that environment interacting with one another.

Levels of Organization Ecologists organize organisms and their environments into a series of levels from smallest to largest.

- At the smallest scale is the individual organism, such as a single Sonoran pronghorn shown in **Figure 2-3**.
- A group of individual organisms of the same species forms a **population**. All of the pronghorns in this desert form a population.
- The pronghorn population is part of a larger **community** that includes the other organisms that pronghorns interact with, either directly or indirectly. This community includes the plants the pronghorns eat, as well as the other animals pronghorns compete with for access to resources. It also includes the predators that prey on pronghorns.
- A community exists within an ecosystem. An **ecosystem** includes all the living organisms in the community plus the nonliving environment in which the community exists. The nonliving environment includes things such as water, soil, and climate. The interactions between *biotic factors* (living parts) and *abiotic factors* (nonliving parts) allows for the transfer of matter and energy throughout the system.
- Ecosystems on land fit within larger **biomes**, which are regions characterized by similar plant communities and climate conditions. The ecosystem shown here is part of the desert biome.
- All of Earth's aquatic and land ecosystems make up the biosphere.

CHAPTER INVESTIGATION

Salinity and Brine Shrimp Survival

How does salt concentration affect the hatching of brine shrimp eggs?

Go online to explore this chapter's hands-on investigation about factors that affect survival.

2.1 ECOLOGICAL SYSTEMS 35

Connect to English Language Arts

Integration of Knowledge and Ideas System models introduced in Chapter 2, such as the food webs, ecological pyramids, and matter cycles, typically depict specific ecosystems as illustrative examples. When reading to understand how energy flows and matter transfers through ecosystems, students should be able to apply information from the model illustrations to apply the same concepts to different ecosystems.

Have students translate between specific visual information and general text by writing a label for each arrow in **Figures 2-2, 2-5, 2-7**, or **2-8**. Their labels should describe each transfer or transformation in terms of energy and matter.

CHAPTER INVESTIGATION A

Guided Inquiry *Salinity and Brine Shrimp Survival*

Time: 130 minutes over 3 days

Students will follow a step-by-step procedure to investigate how different salinities affect the hatching of brine shrimp.

Go online to access detailed teacher notes, answers, rubrics, and lab worksheets.

CROSSCUTTING CONCEPTS | Scale, Proportion, and Quantity

Levels of Organization Have students analyze **Figure 2-3** and consider the scale of each level of organization. Ask them to brainstorm phenomena that are significant at the different levels of hierarchical structure. Then discuss how a phenomenon or event that occurs at one scale may or may not have an impact at another scale. For example, an individual diseased organism might spread disease to the population with which it comes in contact, but this would not necessarily disrupt the local community if other organisms or species fill a similar role. In contrast, climate change at the biome or biosphere levels may have a large impact on all the levels below them.

Atoms, Molecules, and Compounds Students learn that matter can be broken down into atoms and compounds here, in the context of introducing matter transformations. **Section 5.1** provides more detail about subatomic structures and the formation of chemical bonds.

In Your Community

Urban Ecosystems Due to the prevalence of human infrastructure, urban ecosystems are often smaller in size or more fragmented than those typically found in less developed areas. However, students who live in cities can usually name at least a few organisms they encounter in environments near their homes, in city parks, or in and around local bodies of water. If possible, encourage them to safely observe animals in a given ecosystem and draft a model of what the community there might look like.

Elicit from volunteers some examples. Then hold a group discussion on how the density of the human population surrounding an ecosystem impacts the matter and energy flow within it.

FIGURE 2-4 This Cannon Beach tide pool can be studied as its own miniature ecosystem or as a small community within the larger Oregon coastal marine ecosystem.

Ecological Boundaries Like all systems, a community or an ecosystem has boundaries, but how ecologists define those boundaries depends on the relationships they are studying. For example, ecologists studying the organisms in a single tide pool may treat that tide pool as a single ecosystem, or they may consider the tide pool's organisms as just one community within a larger marine ecosystem that includes other communities (**Figure 2-4**). Regardless of their size, all communities depend on the transfer of matter and energy for their survival.

Matter and Energy in Ecosystems

Living things need the input of matter and energy to survive, grow, and reproduce. The movement of matter and energy supports life in Earth's ecosystems. Matter is cycled within ecosystems as the same atoms are used over and over again. Energy—typically from the sun—flows through ecosystems as organisms exchange matter.

Atoms, Molecules, and Compounds Every living and nonliving thing is made of matter. The smallest unit of matter is the **atom**. An element is a substance made from only one type of atom. Six elements essential for life are carbon (C), hydrogen (H), nitrogen (N), oxygen (O), phosphorus (P), and sulfur (S). These elements make up important biological molecules such as carbohydrates, amino acids, and DNA

Chemical Bonds Atoms exert forces on each other. Forces between two or more atoms can connect them to form a **molecule**. The forces that hold the atoms in a molecule together are called chemical bonds. These bonds connect atoms to form molecules. For example, when two oxygen atoms bond together, they form a molecule of oxygen, or O_2. A compound is a substance made up of two or more different elements bonded together. Carbon dioxide (CO_2), water (H_2O), and glucose ($C_6H_{12}O_6$) are examples of common compounds used by living things.

CROSSCUTTING CONCEPTS | Systems and System Models

Ecosystem Boundaries Understanding the inputs and outputs of a system is essential to interpreting scientific results. With regard to ecosystems, emphasize to students that ecological boundaries are not fixed geographical locations. To exemplify this concept, describe some well-studied ecosystems, such as Isle Royale in Lake Superior, the Sky Islands in southeastern Arizona and northern Mexico, and the Great Lakes, and show students the locations of these ecosystems on a map. Have students discuss how the boundaries of these ecosystems are related to their geographic characteristics. Describe a study that was performed in one of these ecosystems and then ask students to come up with a different scientific question that might require a scientist to redefine the ecosystem's boundaries in order to effectively investigate the question. Have students describe how the boundaries might change.

FIGURE 2-5 Sunlight is the ultimate source of energy in most ecosystems. In the process of photosynthesis, energy from sunlight is used to make glucose and oxygen from carbon dioxide and water.

Chemical Reactions In a chemical reaction, the atoms in one or more substances are rearranged into new substances with different chemical properties. Rearranging a system of atoms requires breaking chemical bonds and forming new chemical bonds. Though no new matter is created or destroyed during a chemical reaction, there is always an energy change: energy is either released or absorbed by the system. Chemical reactions are happening all the time in ecosystems. The majority of energy and matter transformations in living things occur in two chemical processes: photosynthesis and cellular respiration.

- **Photosynthesis** In this process, energy from the sun is transformed into chemical energy by rearranging the atoms in carbon dioxide and water into oxygen and sugar molecules (**Figure 2-5**). Photosynthetic organisms, such as plants, algae, and some bacteria, are the main source of chemical energy in most ecosystems.
- **Cellular Respiration** In this process, chemical energy is released by reactions that break down sugars and other carbon-based molecules. This energy is used to fuel cellular activities. Cellular respiration that occurs in the presence of oxygen is called aerobic respiration. Most organisms use aerobic respiration and need oxygen to live. In aerobic respiration, energy is released when sugar and oxygen are broken down to form carbon dioxide and water. Some organisms can live in low- or no-oxygen environments. These organisms rely on anaerobic respiration, which occurs without oxygen.

CCC Systems and System Models *What is the relationship between the inputs and outputs of photosynthesis and aerobic respiration?*

2.1 REVIEW

1. **Distinguish** Explain how a community is distinct from an ecosystem.
2. **Classify** Label each ecosystem component as a biotic or abiotic factor.
   - flowering plants
   - school of fish
   - willow tree
   - soil
   - pond water
   - air
3. **Sequence** Construct a model of the hierarchy of ecological organization. Your completed model should have six levels.
4. **Synthesize** Describe how chemical reactions are important to the movement of energy and matter in an ecosystem.

Chemical Reactions The main chemical compounds involved in photosynthesis and chemical respiration are introduced here. Students examine these reactions in further detail within cellular processes in **Section 6.3**.

Crosscurricular Connections

Chemistry Remind students of endo- and exothermic reactions, ones that absorb or release thermal energy. Show an instant hot pack, sealed in its package. When the package is opened and the pouch removed, the chemical inside, often iron, reacts with oxygen in the air to form iron (III) oxide, a reaction that releases heat. A simple demonstration of an endothermic reaction can be done by stirring baking sode into vinegar and measuring the temperature before and after.

CCC Systems and System Models *The inputs to photosynthesis are the outputs of cellular respiration.*

2.1 REVIEW

1. *Sample answer: An ecosystem consists of organisms that interact with each other and the physical environment. A community includes organisms that interact directly or indirectly with each other and does not include the environment.* **DOK 1**
2. *biotic: flowering plants, school of fish, willow tree; abiotic: pond water, soil, air* **DOK 2**
3. *largest to smallest: biosphere; biome; ecosystem; community; population; organism* **DOK 2**
4. *Sample answer: While matter is conserved in chemical reactions, breaking and forming new bonds causes a change in energy. Energy is either absorbed or released. Energy can be used by organisms for biological functions.* **DOK 1**

2.2

MODELING THE TRANSFER OF ENERGY AND MATTER

LS2.B, LS2.C

EXPLORE/EXPLAIN

This section provides a review of the ecological roles in an ecosystem, introduces trophic levels into models of the movement of energy and matter, and discusses the impact of disruptions to energy and matter pathways.

Objectives

- Classify organisms in an ecosystem according to their main ecological roles and their functions.
- Distinguish among the trophic levels of a simple ecosystem.
- Predict potential impacts on an ecosystem resulting from food-chain disruptions.

Pretest Use **Question 6** to identify gaps in background knowledge or misconceptions.

Vocabulary Strategy

Latin Roots Point out to students that most terms used to describe consumers come from Latin words. For example, the suffix *-vorae* means "to swallow or devour." *Herba* means "vegetation." *Carnus* means "flesh." *Omnis* means "all." *Detrere* means "to wear away."

Analyze *Energy is obtained from the breakdown of carbohydrates and other nutrients in an apple. The apple tree gets its energy from the sun.*

2.2

KEY TERMS

consumer
decomposer
food web
producer
trophic level

MODELING THE TRANSFER OF ENERGY AND MATTER

When you are feeling sluggish, a quick snack such as an apple will often give you the energy you need to get back to feeling "normal" in a short time. Like a cell phone, your body requires periodic "recharging" in order to function.

Explain *Where do you think the energy you get from eating an apple originally came from? Think about how an apple tree gets energy.*

SEA PIG SURVIVAL

Gather Evidence *Sea pigs digest organic matter from the ocean floor. Identify which ecological role sea pigs occupy in their community. Explain your answer.*

Ecological Roles in Ecosystems

Organisms are categorized by the way in which they get energy. The three major categories are producers, consumers, and decomposers.

Producers Organisms that make their own food are **producers**. They are also called *autotrophs*. (The prefix *auto-* means "self," and the suffix *-troph* means "nourishment.") Most producers, such as plants and algae, use photosynthesis to transform energy from the sun into sugars. Some bacteria and archaea obtain energy in the absence of sunlight. Many of these organisms live near deep-sea vents that spew superheated water full of energy-rich chemical compounds. These organisms use a process called *chemosynthesis* to convert the substances in the water into sugars.

Consumers Organisms that get their energy from eating other organisms are **consumers**. They are also called *heterotrophs*. (The prefix *hetero-* means "different.") Consumers can be classified by what they eat.

- Herbivores, such as grasshoppers, are organisms that eat producers.
- Carnivores, such as frogs, are organisms that eat other consumers.
- Omnivores, such as bears and foxes, are organisms that eat both producers and consumers.
- Detritivores, such as ravens and earthworms, are organisms that eat dead organic matter, which is called *detritus*.

Decomposers Organisms that break down decaying organic matter into simpler compounds that other organisms can use are **decomposers**. Most decomposers are bacteria, archaea, and fungi (**Figure 2-6**).

FIGURE 2-6 Wood bonnet mushrooms are commonly found growing on decaying logs and branches on the forest floor.

REVISIT THE ANCHORING PHENOMENON

Gather Evidence *Because sea pigs eat matter that is no longer living and has already been broken down, they are considered to be detritivores.*

Sea pigs consume particulate organic matter that consists of microorganisms such as bacteria, foraminifera (single-celled protists), and phytoplankton as well as wastes, secretions, and decayed body matter from other marine animals. They have been found to feed selectively on the most recently deposited particles on the seafloor, which have higher nutritional content than deeper sediments.

Communities are made up of a variety of producers, consumers, and decomposers. Producers and decomposers have indispensable roles in ecosystems. Producers provide both food and oxygen for all life. Consumers begin the process of breaking down the complex molecules that producers make, and the process is finished by decomposers so the materials can be recycled.

The Transfer of Energy and Matter in Ecosystems

Energy enters most ecosystems as sunlight, which is converted into chemical energy through organisms that carry out photosynthesis. This energy is used by animals and other consumers to fuel life processes. Thus, most organisms require a continuous source of sunlight to survive.

CASE STUDY
SOMETHING FISHY IN THE FOREST

Ask Questions *Generate questions about the trophic level a brown bear occupies as its diet changes throughout the year.*

Food Chains The movement of energy and matter within an ecosystem can be modeled using a food chain. Producers are the first link in the chain. Consumers eat the producers and are themselves eaten by other higher-level consumers.

A **trophic level** is the group of organisms in an ecosystem that occupy the same level in a food chain. There are five main trophic levels:

- Producers are at the base of every food chain.
- Primary consumers are herbivores that eat producers.
- Secondary consumers are carnivores and omnivores that eat herbivores.
- Tertiary consumers are carnivores and omnivores that eat secondary consumers.
- *Apex consumers* are the consumers at the top trophic level—those that eat other consumers but are not typically eaten by others.

For example, in the ocean, a food chain can be made up of microscopic floating algae at the producer level, which are eaten by krill at the primary consumer level. Penguins that eat the krill are secondary consumers. These organisms are then eaten by tertiary consumers, such as seals, and so on. A food chain can vary in length depending on the ecosystem it is modeling. In the case of this marine ecosystem, the food chain could grow very long, but will likely end with an apex predator such as a killer whale. A portion of a marine ecosystem food chain is illustrated in **Figure 2-7**.

As organisms are eaten along each link of a food chain, some of the energy they contain is used by the consumer at the next level. In this way, energy is transferred upward through each trophic level in the food chain.

FIGURE 2-7 A portion of a marine food chain includes phytoplankton at the producer level, krill at the primary consumer level, and penguins at the secondary consumer level.

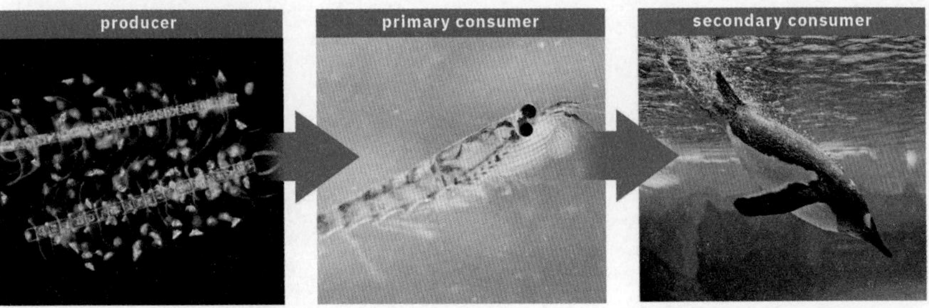

CASE STUDY
SOMETHING FISHY IN THE FOREST

Ask Questions To help students revisit the Case Study and make connections with the content, have students model a food chain that includes brown bears and salmon.

In Your Community

Local Food Chains Research typical species found in your local area, and prepare a listing of various plants and animals for display. Have pairs copy names onto stickies or index cards and create local food chains, trying to make them as long as possible.

Go Online INTERACTIVE FIGURE

Antarctic Food Web Students can remove or reinsert organisms in the interactive version of **Figure 2-8** to see how the food web relationships change. They should be able to recognize that the removal of some organisms significantly disrupts the food web, whereas other organisms do not have as much of an effect.

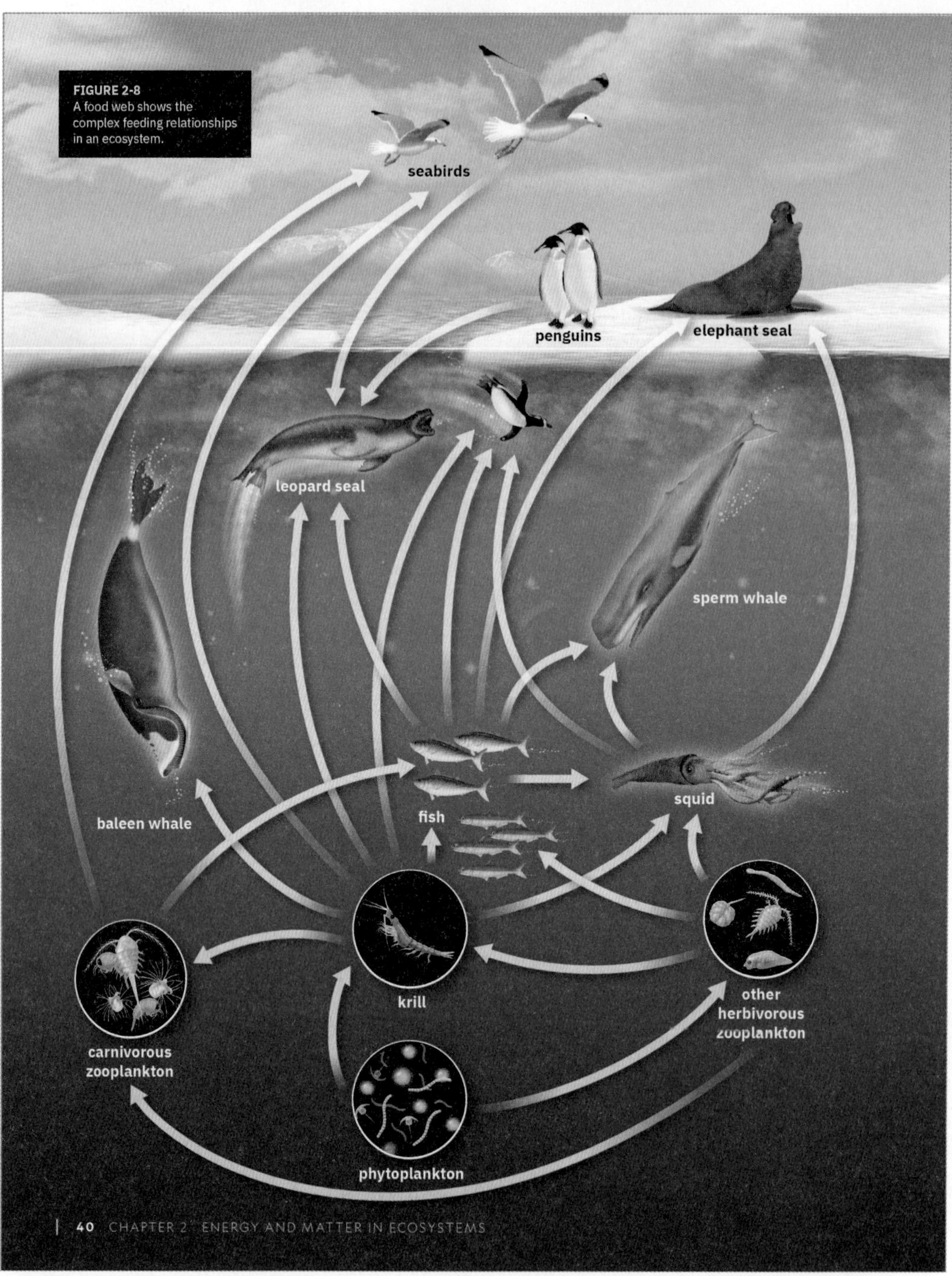

FIGURE 2-8
A food web shows the complex feeding relationships in an ecosystem.

SCIENCE AND ENGINEERING PRACTICES
Developing and Using Models

Limits of Models Students should recognize that food chains generally do not represent all members of a community and that they are subsets of food webs that can be constructed to represent the whole community (with more than one species at each trophic level). Students may notice that detritivores and decomposers are not represented in **Figure 2-8**. Ask students how they would refine the food web model shown here to include these types of organisms. You may wish to draw students' attention back to the Anchoring Phenomenon by encouraging them to build a food web based on the sea pig's deep-sea ecosystem.

Students can do a similar analysis of the limitations of the pyramid models presented in the next section.

Food Webs The flow of energy through an ecosystem is usually more complex than can be modeled using a simple food chain. Consider how different consumers can feed on many other types of organisms within a marine ecosystem. For example, a secondary-level fish feeds on primary-level krill. A leopard seal might eat the fish or the krill. When it eats the fish, the seal is a tertiary consumer, but when it eats the krill, it is a secondary consumer. Because many different organisms can eat and be eaten by other types of organisms within an ecosystem, food chains combine to form a complex system of feeding relationships known as a **food web**. The complex connections between organisms in an Antarctic food web are shown in **Figure 2-8**.

Go Online INTERACTIVE FIGURE

FIGURE 2-8 Go online to see how changes in an ocean community affect feeding relationships in an **Antarctic Food Web**.

SEP **Use a Model** *Identify organisms at each trophic level in this Antarctic food web. Why is a food web a more realistic model than a food chain?*

Food Web Stability

This Antarctic food web shows only a small number of the species in the ecosystem it represents. The stability of an ecosystem depends in part on the variety of species in the ecosystem. If a species in the food web is lost, other species may fill its role and maintain the food web's overall stability.

Disruptions to Energy and Matter Transfer A disruption in a food web due to changes in the environment or the loss of a species can have devastating consequences. As the impact of a change in the availability of energy on a lower level of a food web travels up the chain, consumers at higher levels are also affected. The loss of one link in a food chain can affect organisms at levels both above and below it.

VIDEO

VIDEO 2-2 Go online to learn more about the importance of krill to Antarctic food webs.

The complexity of a food web system often makes it difficult to predict how disruptions will affect other members of the community. For example, the human overfishing of krill, a small crustacean often used to produce nutritional supplements and animal feed, has significantly reduced the Antarctic krill population. This has led to concerns about the sustainability of other wildlife that feed on krill, such as fish, penguins, seals, and whales. As a result, the Scientific Committee on Antarctic Research (SCAR) has implemented the Krill Action Group to develop guidelines and regulations to manage the krill fishing industry.

2.2 REVIEW

1. **Analyze** Order the roles of producer, primary consumer, and secondary consumer from highest to lowest trophic level.
2. **Explain** Why are decomposers essential to the movement of matter in an ecosystem?
3. **Apply** Use **Figure 2-8** to support or refute the claim that an organism can feed at more than one trophic level.
4. **Identify** Which trophic level, if removed from a food web, would likely cause the most change in the ecosystem?
   A. producer
   B. primary consumer
   C. secondary consumer
   D. tertiary consumer

SEP **Use a Model** *Sample answer: Phytoplankton is a producer that is eaten by krill, a primary consumer. Krill get eaten by fish, which are a secondary consumer. Fish get eaten by seals, which are a tertiary producer. A food web is a more realistic model because it captures the complex feeding relationships in an ecosystem.*

Connect to English Language Arts

Integration of Knowledge and Ideas
The ability to translate information presented visually into written form is a required skill for many science and engineering practices. Have students write a paragraph, explaining in their own words, the feeding relationships depicted in the diagram of the aquatic food web.

Video

Time: 1:21

Use **Video 2-2** to emphasize to students the important role krill play in Antarctic food webs.

2.2 REVIEW

1. *highest to lowest: secondary consumer, primary consumer, producer* **DOK 1**
2. *Sample answer: Decomposers break down dead organic material so that it can be reused by other organisms, specifically producers. If there are no decomposers, dead organic matter will pile up, and nutrients would not cycle through an ecosystem.* **DOK 1**
3. *Sample answer: Organisms can be at different trophic levels in a food web. For example, krill are primary consumers when they eat phytoplankton and are secondary consumers when they eat carnivorous zooplankton.* **DOK 2**
4. *producer* **DOK 1**

EXPLORER
RAE WYNN-GRANT

About the Explorer

National Geographic Explorer Dr. Rae Wynn-Grant works to protect large carnivores, such as bears and lions. Her research sometimes involves mitigating conflicts between carnivore populations and landowners. You can learn more about Rae Wynn-Grant and her research on the National Geographic Society's website: www.nationalgeographic.org.

Connect to Careers

Carnivore Ecologist Because of their dominant positions in food webs, meat-eating predators are not often perceived by the public as needing protection. Carnivore ecologists study these organisms, their behaviors, and their ecological roles in different habitats and often serve as advocates for carnivore conservation. Universities with specialized ecology programs educate, train, and hire scientists to conduct research and teach. Ecologists in many different subfields also work with private organizations and public entities, such as the National Park Service, to share findings and to promote conservation efforts.

Video

Time: 2:10

Use **Video 2-3** to show students how Rae Wynn-Grant gathers and analyzes geographic information systems (GIS) data related to bear activity. With the use of statistical mapping methods, Wynn-Grant can identify the combinations of parameters that most reliably predict the occurrence of events, such as human-bear conflicts.

THINKING CRITICALLY

Predict *Sample answer: Grizzly bears would compete with other carnivores as predators for other prairie animals. However, as omnivores, they would also be able to eat plants.*

NATIONAL GEOGRAPHIC EXPLORER **RAE WYNN-GRANT**

USING DATA TO PREVENT HUMAN-BEAR CONFLICT

Dr. Rae Wynn-Grant studies spatial patterns in bear activity to reduce conflict between bears and humans.

VIDEO 2-3 Go online to see how Wynn-Grant collects and analyzes bear data.

By the early 20th century, humans drove grizzly and black bears out of most of their range in the United States. Today's bear populations are recovering thanks to decades of conservation efforts. But the growing overlap of human and bear habitats has made interactions between the two species more frequent. Carnivore ecologist Dr. Rae Wynn-Grant uses statistical models to identify patterns in bear activity to minimize the frequency and severity of human-bear conflict.

Vehicle collisions and illegal hunting pose serious risks to bears. In addition, as the climate changes, snow and seasonal temperature patterns that cue bears to hibernate no longer occur with the same regularity. Thus, some bears stay active later into the year when food is scarce, making them more likely to seek food sources associated with humans.

Wynn-Grant's statistical models map suitable bear habitats and predict areas of potential conflict with humans. This research involves boots-on-the-ground fieldwork to collect data on bears' health, whereabouts, and behaviors. As part of her research, she tagged black bears to locate their dens and track their survival rates in Nevada's Western Great Basin. By monitoring known "conflict bears," Wynn-Grant found "bear hotspots" correlated with behaviors such as raiding garbage cans and breaking into houses in search of food.

"My work might seem pretty specific to one animal and one place," Wynn-Grant says. "But it gives me an ability to make recommendations for any type of conservation, anywhere. Most conservation is being done in places where there are people, people who need to build towns and cities and highways and roads. So getting it right here means we can probably get it right somewhere else."

Bears can also eat slow-moving livestock and crops that provide them with high-calorie food for relatively little effort. This threat to ranchers' and farmers' livelihoods complicates habitat restoration efforts in the Great Plains of eastern Montana, where Wynn-Grant works to incentivize people to gather camera trap and field observation data. Ideally, the community's scientific engagement will help conserve safe ecological corridors for bears and other animals returning to prairie habitats.

THINKING CRITICALLY

Predict *The prairie is home to many species of birds, fish, snakes, and small mammals, as well as larger animals including bison, elk, bobcats, and coyotes. How might the return of grizzly bears affect food webs in prairie communities?*

DIFFERENTIATED INSTRUCTION | English Language Learners

Paired Reading Have students work in pairs to read this content. Tell students to take turns reading the paragraphs aloud. After each paragraph is read, have students try to use context clues or cognates to determine the meaning of unfamiliar words, then have them use a dictionary to confirm meanings. Note the vocabulary words that have Spanish cognates, including *rehabilitation, temperature, climate, hibernation*, and *conflict*.

MODELING ENERGY AND MATTER DISTRIBUTION

Less than 30 percent of the energy stored in gasoline is used by a gas-powered vehicle to move it down the road. Instead, most of the energy from the fuel is lost from the engine as heat to the environment. The transfer of energy between trophic levels is even more inefficient.

Explain *Why isn't all of the energy in one organism transferred to the organism that eats it?*

2.3

KEY TERMS

biomass
biomass pyramid
ecological pyramid
energy pyramid
pyramid of numbers

Ecological Pyramids

Food chains and food webs are a way to model the movement of energy and matter in an ecosystem. Each link shows a transfer of energy and matter from one trophic level to another. An **ecological pyramid** is a model used to compare an ecosystem's trophic levels in different ways.

Each tier of an ecological pyramid is one trophic level. Producers form the first level at the bottom of the pyramid. The trophic levels above the producer level are made up of consumers (primary consumers, then secondary consumers, and so on). Each consumer level obtains its energy from the level immediately below it. At every trophic level, some energy is used for metabolic processes and some is lost to the environment as heat. Also, many organisms die and decompose before they can be eaten. The small amount of remaining energy is passed up to the next trophic level. This results in the model's pyramid shape.

The number of trophic levels, and the number of organisms within each level, is limited by the amount of energy available to the ecosystem. Ecologists use three types of pyramid models: energy pyramids, biomass pyramids, and pyramids of numbers.

FIGURE 2-9 A kestrel, which is a type of falcon, commonly gets the energy it needs by eating primary consumers such as insects and mice.

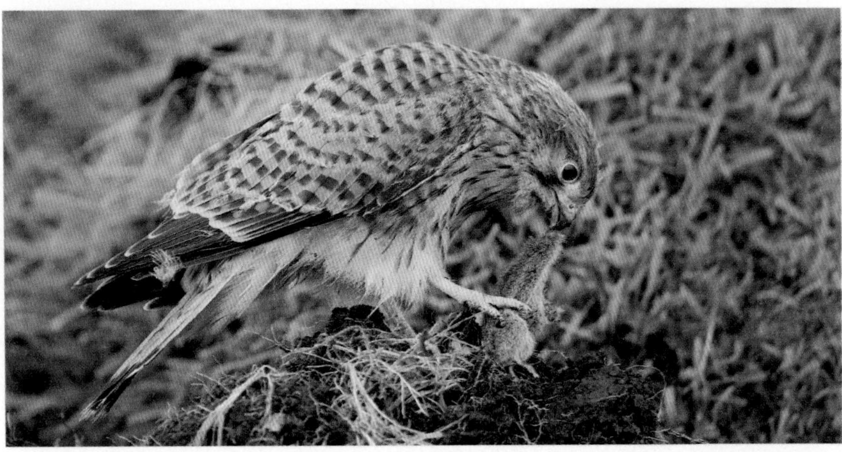

2.3

MODELING ENERGY AND MATTER DISTRIBUTION

LS2.B

EXPLORE/EXPLAIN

This section introduces pyramid models used to represent the distribution of energy and matter within ecosystems.

Objectives

- Model the distribution of energy in a community using an energy pyramid.
- Model the distribution of biomass in a community using a biomass pyramid.
- Model the distribution of individuals in a community using a pyramid of numbers.

Pretest Use **Question 2** to identify gaps in background knowledge or misconceptions.

Explain *Sample answer: The organism may not be able to digest all of the organism it eats, and some of the energy may also be lost to the environment as heat.*

Address Misconceptions

Conservation of Energy Because the word "conserve" means something similar to "keep" in many contexts, students may interpret the scientific concept of energy conservation as meaning that organisms in a higher trophic level retain all of the energy of the organisms at lower trophic levels, leading to the misconception that the higher an organism is in terms of its trophic level, the more energy it has.

To help students rectify their mental models, it may be useful to describe how specific forms of matter do accumulate at higher trophic levels, as exemplified in the Looking at the Data, and contrast this with energy, which is lost by an organism to the environment in the form of heat.

Emphasize that "lost" in the scientific context of energy is not the opposite of "conserved"; rather, we say that energy is conserved because we can account for all of the energy that has been transferred or transformed, even if it has changed to a form that cannot be used by the organism.

Elicit from students how energy loss in the form of heat can aid survival of individuals. At this time, expect responses about overheating or experiencing hypothermia and possible consequences. Students will learn in later chapters how enzymes, for example, work only in a certain core temperature range.

SEP Use Mathematics *700 kcal/m² per year*

Energy Pyramids

An **energy pyramid** models the distribution of energy across trophic levels in an ecosystem (**Figure 2-10**). It shows an ecosystem's productivity, or the efficiency with which energy is transferred up a food chain or food web. Energy is usually measured in joules (J) or kilocalories (kcal). Energy use in an ecosystem is inefficient because most of the energy produced is unable to be used by organisms.

Because producers absorb certain wavelengths of light, they convert only about one percent of the solar energy that reaches Earth into a usable form they can store in their cells. Yet they produce 100 percent of the energy available to primary consumers. As energy flows from one trophic level to the next, the amount of energy available at each level is only about 10 percent of the energy available at the previous trophic level.

For example, the producers in one community might produce 10,000 J per year of available energy. By feeding on the producers, the primary consumers would use 10 percent of that amount, or 1000 J. The secondary consumers would use 100 J per year, and the tertiary consumers would use 10 J per year, which is just 0.1 percent of the available energy at the base of the pyramid.

Although energy cannot be created or destroyed, clearly much of the energy available to each trophic level is not used. When a consumer such as a grasshopper eats part of a plant, some of the plant tissue cannot be digested and is excreted as waste. The grasshopper converts most of the remaining plant matter into energy that its own body can use for maintenance and movement. In doing so, some of the energy is lost to the environment as heat. On average, only about 10 percent of the energy from the grasshopper's food is stored as chemical energy in the cells of the grasshopper's body. This small amount is the chemical energy that is transferred to any predator that preys on the grasshopper.

SEP Use Mathematics *A population of top predators consumes 7 kcal/m² per year. If the ecosystem has four trophic levels, how much energy is available to the primary consumers?*

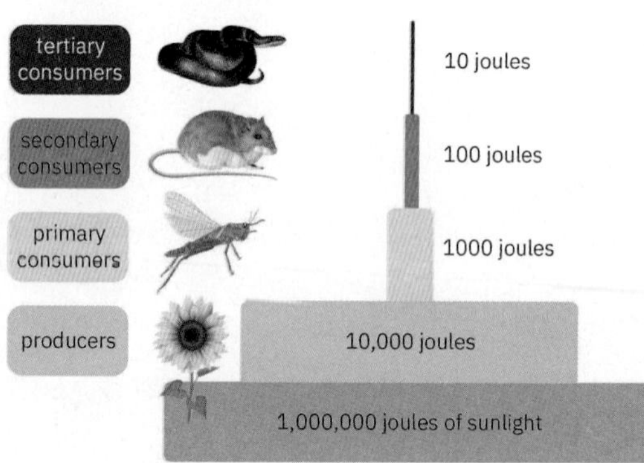

FIGURE 2-10 An energy pyramid shows the distribution of energy across trophic levels in an ecosystem. Only 10 percent of the available energy is transferred from one tier to the next. Most energy is lost to the environment as heat.

SCIENCE AND ENGINEERING PRACTICES
Using Mathematical and Computational Thinking

The "10 Percent Rule" This rule is often attributed to the work of prominent early 20th century ecologists Charles Elton and Raymond Lindeman. For discussion, ask students how they think scientists arrived at this particular ratio. They should recognize that how efficiently energy moves between trophic levels depends on the type of organisms involved, which differ by ecosystem. Thus, to model ecosystem energy transfer, scientists must combine data from indirect measurements of energy use, such as changes in the mass of an individual organism, with proportional reasoning.

Efficiency estimates for trophic levels within various ecosystems range from less than 1 percent up to 40 percent, where 10 percent is a rough average. To reinforce proportional thinking, have students apply some percentages in this range to the **Figure 2-10** pyramid to compare the energies available to the tertiary consumers.

tertiary consumers — 1.5 g/m²

secondary consumers — 11 g/m²

primary consumers — 37 g/m²

producers — 809 g/m²

FIGURE 2-11 A biomass pyramid shows the distribution of biomass at each trophic level in an ecosystem.

Biomass Pyramids

In contrast to an energy pyramid, a **biomass pyramid** models the total amount of biomass stored in living organisms at each trophic level in an ecosystem (**Figure 2-11**). **Biomass** is commonly defined as the total dry mass of all the organisms within a given area at a specific time. It is usually measured in grams of total biomass per unit area, such as g/m^2. The dry mass of an organism refers to its mass after all of the water has been removed. Biomass is also sometimes defined in terms of a single element, such as carbon. Like an energy pyramid, the largest tier in a biomass pyramid is typically the producer level. The shape of a biomass pyramid also indicates that at each trophic level, biomass decreases by approximately 90 percent. This corresponds to the 90 percent loss of energy between each level.

CONNECTIONS

EATING LOWER ON THE FOOD CHAIN People sometimes promote a vegetarian diet by touting the ecological benefits of "eating lower on the food chain." They are referring to energy that is lost in transfers between plants, livestock, and humans. When people eat plants, they get a larger proportion of the energy that the plant captured than they would if the plant was used to feed livestock. When plants are used to feed livestock, only a very small percentage of the energy that was stored in the plant body ends up in the meat a person can eat.

In 2019, the average American ate 100.8 kg (222 lb.) of meat. This amounts to about 0.3 kg (or 0.6 lb.) of meat per day. The majority of meat consumed in the United States is chicken at 50.1 kg per capita, followed by beef at 26.3 kg per capita, and then pork at 24 kg per capita.

As some Americans aim to decrease their overall meat consumption, several companies have introduced plant-based meat substitutes to the market. Plants such as grains and legumes are commonly used as protein sources for these plant-based meat substitutes.

Plant-based meals are typically more energy efficient than meat-based meals.

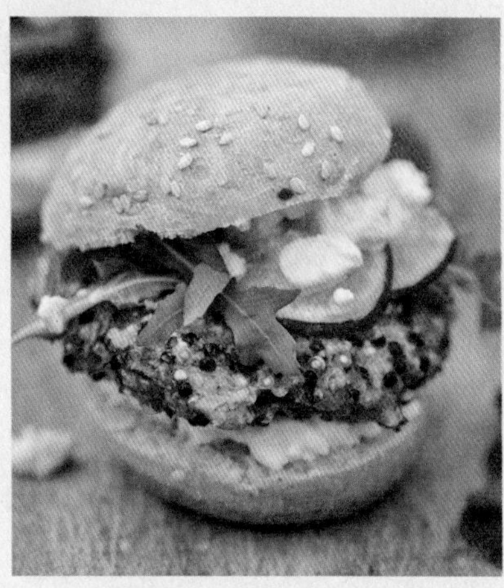

Science Background

Estimating Biomass Students will likely recognize that making a direct measurement of the biomass contained in a natural area over which community-scale ecology is studied is not feasible. Instead, scientists use average biomasses of representative organisms of the various species present to estimate the biomass of a given group of organisms.

In scientific literature, biomass is typically reported in terms of the mass of carbon because this element is common to all organisms. The amount of carbon in an organism is considered independently of its water content, which is much more variable. Using this convention, the mass of carbon in an organism is assumed to be one-half of its dry mass. (The dry mass of the organism is its mass after its water content has been removed.)

For example, a human body is about 60 percent water by mass, so the dry mass of a 50-kg human would be the remaining 40 percent, or 20 kg. The biomass of this human would then be one-half of the dry mass, or 10 kg carbon.

By assembling and analyzing data from a variety of biomass censuses for all the major taxa on Earth, scientists estimate the global biomass to be approximately 550 gigatons of carbon (1 Gt C = 10^{15} g of carbon), 450 Gt C of which are plants.

CONNECTIONS | Eating Lower on the Food Chain

Social-Emotional Learning Some students may not be familiar with the wide variety of dietary practices in the world. Others may also struggle with food security. Help students have a conversation about different food choices, such as those discussed in the Connections feature. They may have opportunities to demonstrate *responsible decision-making* by showing curiosity and *social awareness* through expressing empathy and respect for other students' perspectives during this discussion.

Connect to Mathematics

Define Quantities for Modeling Have students return to **Figure 2-8** and apply estimated quantities to a pyramid of biomass and a pyramid of numbers for an Antarctic food web. For example, students can research the average mass of an elephant seal and the number of elephant seals in an average Antarctic colony. They can then work backwards to estimate the average mass and numbers of squid, krill, and phytoplankton to support that food chain.

CCC Stability and Change *Sample answer: The number of organisms at each trophic level may differ across the seasons as the amount of available energy changes because plants go dormant and some animals migrate to different ecosystems during the colder months.*

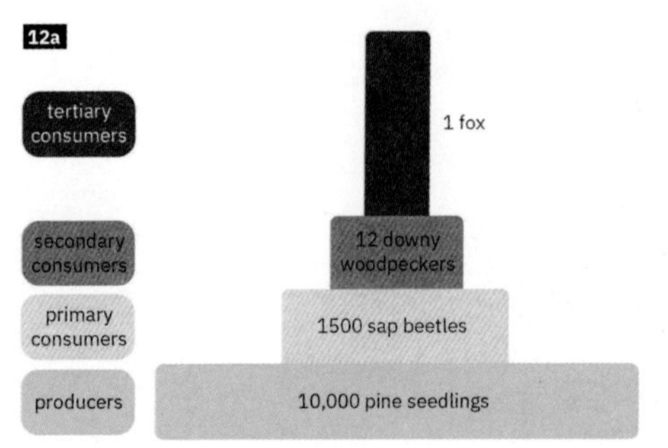

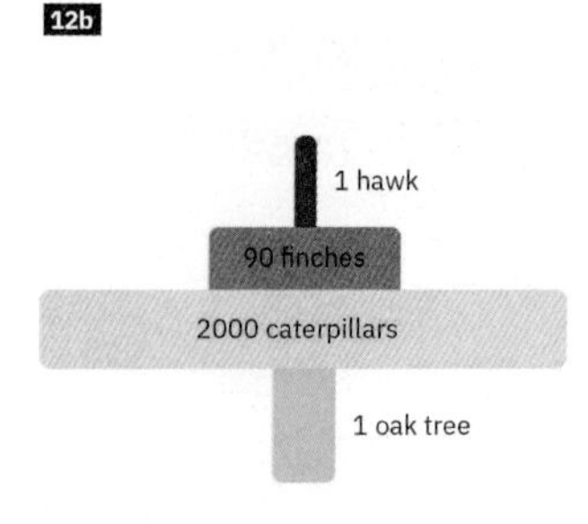

FIGURE 2-12 A pyramid of numbers shows the actual number of individual organisms at each trophic level in an ecosystem.

Pyramids of Numbers

A **pyramid of numbers** compares the relative number of individual organisms at each trophic level. For most communities, a pyramid of numbers will show that the number of producers required to support the other trophic levels is vastly greater in size (**Figure 2-12a**). The quantity of producers needed to support a community generally outnumbers all the consumer levels combined.

In some communities, however, a pyramid of numbers takes on a diamond shape (**Figure 2-12b**). A classic example of this is a pyramid in which the producers are individual trees that support a community of insects, birds, and possibly other predators. The number of trees is relatively small compared to the number of insects. For every tree, there may be thousands of insects that feed on it, and hundreds of predatory insects that feed on the primary consumers. A group of trees may support enough secondary consumers to support a few birds at the upper trophic levels.

CCC Stability and Change *Deciduous trees lose all their leaves each autumn. How might a pyramid of numbers for a deciduous forest ecosystem change across different seasons?*

2.3 REVIEW

1. **Sequence** List the types of organisms in order from the top to the bottom of an ecological pyramid.
   - tertiary consumer
   - producer
   - primary consumer
   - secondary consumer
2. **Explain** Why is less energy available to organisms higher in an energy pyramid than is available to organisms at the base of an energy pyramid?
3. **Predict** How might a fire affect the shape of a forest ecosystem's biomass pyramid?
4. **Represent Data** Build an energy pyramid using the data in the table.

Species	Energy (joules)
A	4,982,000
B	5200
C	500,738
D	47,033

5. **Identify** In your model, label each species with its ecological role in the ecosystem.

2.3 REVIEW

1. *top to bottom: tertiary consumer, secondary consumer, primary consumer, producer* **DOK 2**
2. *Sample answer: Only 10 percent of the energy available to organisms at one trophic level is available to the organisms in the trophic level above. Many chemical processes produce energy, but some of that energy is lost as heat in the environment and some is used by organisms for their own biological functions.* **DOK 2**
3. *Sample answer: A fire will wipe out producers, the trees and plants on the forest floor. The size of the base of the biomass pyramid will shrink. Other levels of the biomass pyramid may also shrink because consumers rely on producers, and trees are also the habitat and home for many organisms.* **DOK 3**
4. *top to bottom: B, D, C, A* **DOK 2**
5. *A: producer; B: tertiary consumer; C: primary consumer; D: secondary consumer* **DOK 2**

ON ASSIGNMENT Following a predawn ride up a Mexican mountain on a rented mule, National Geographic photographer Joel Sartore found himself surrounded by a swirling mass of monarch butterflies as soon as the sun rose. Every winter millions of monarch butterflies migrate to Sierra Chincua in central Mexico. What would a pyramid of numbers look like for this ecosystem?

ON ASSIGNMENT

Documenting Biodiversity National Geographic photographer Joel Sartore is known for his work on the Photo Ark. The goal of this multiyear project, which he began in 2005, is to document every species held by zoos, aquariums, captive breeders, and wildlife rehabilitation centers around the world. Why? He wants to inspire people to help save animal species at risk of extinction and document our planet's biodiversity. So far, Sartore has made portraits of more than 11,000 animal species; the Photo Ark archive contains over 38,000 images and videos.

You can learn more about the Photo Ark at the National Geographic Society's website: www.nationalgeographic.org/projects/photo-ark/.

Science Background

Monarch Butterfly Biosphere Reserve World Heritage Site Sierra Chincua lies within this 56,259 hectare area of rugged forested mountains about 100 kilometers northwest of Mexico City. It is one of five limited areas where tourists can visit. Nearly all populations of monarchs from east of the Mississippi overwinter across this site. The same individual does not make the trek from here to Canada and back; instead, four generations occur over that span of time. Scientists do not understand how individuals know how to migrate.

As an introduction to concepts presented in the next chapter, encourage interested students to research the impact of ecotourism on Sierra Chincua as well as the impact of logging by local residents.

MINILAB
MODEL A BIOMASS PYRAMID

Students use mathematical models to represent how energy stored in biomass moves through an ecosystem.

Time: 30 minutes

Advance Preparation

- One 16-ounce bag of dry pinto beans should be enough for each group. The beans may be replaced by another small object, such as beads. Provide a container for each group to store the beans in when finished. Each group will use about 350 pinto beans in the activity. The activity requires a large sheet of paper, approximately 75 × 75 cm^2.

Procedure

In **Step 4**, note that at higher trophic levels, biomass values become smaller, so a rounding error becomes more significant. This can be a discussion point related to mathematical representations and different scales.

Results and Analysis

1. **Organize Data** *Student bar graphs should show a large number of producers and decreasing amounts of primary, secondary, and tertiary consumers.* **DOK 3**
2. **Calculate** *The total mass of producers is 80,640 g. The total biomass of primary consumers is 2000 g + 2470 g + 2025 g = 6495 g. The percent biomass transferred from producers to primary consumers is (6495 / 80,640) × 100 = 8.1%. The total biomass of secondary consumers is 322 g + 330 g = 652 g. The percent biomass transferred from primary consumers to secondary consumers is (652 / 6495) × 100 = 10.0%. The total biomass of secondary consumers is 652 g. The biomass of tertiary consumers is 74 g. The percent biomass transferred from secondary consumers to tertiary consumers is (74 / 652) × 100 = 11.3%.* **DOK 2**
3. **Interpret Data** *The most likely answer is no. A quaternary consumer would need to reliably find and hunt the tertiary consumer, the seabass. Basses are relatively large and scarce. This ecosystem would not be able to support an organism that hunts the bass.* **DOK 2**
4. **Evaluate** *Answers will vary. At each level, some of the mass is converted to energy, which is used to fuel metabolic processes. Some of the energy is lost to the environment as heat. Some of the mass becomes waste products. Some of the mass cannot be transferred at all—shrimp have shells that cannot be easily digested, for example. The waste products and indigestible remains are consumed by decomposers over time.* **DOK 2**

MINILAB

MODEL A BIOMASS PYRAMID

SEP Develop and Use a Model How can you model the distribution of biomass in an ecosystem?

The freshwater springs of Florida are among the most studied aquatic ecosystems on Earth. In this activity, you will build a model of organisms found in one of these springs to see how biomass is distributed among trophic levels in the ecosystem.

Freshwater springs in Florida can support a large amount of biomass.

Materials

- **poster board or large sheet of paper**
- **marker**
- **pinto beans, 200 g**
- **calculator**

Scientists sampled various regions of a freshwater spring in Florida to estimate the populations and biomasses of some species found in the spring. The table shows their results.

TABLE 1. Organism Biomass in a Florida Ecosystem

Organism	Trophic level	Population (per 100 m^2)	Biomass (g/100 m^2)
narrow-leaved arrowhead	producer	16,800	60,480
algae	producer	N/A *	20,160
turtles	primary consumer	80	2000
shrimp	primary consumer	380	2470
insects	primary consumer	450	2025
anemones	secondary consumer	23	322
small fish	secondary consumer	30	330
bass (larger fish)	tertiary consumer	1	74

*The algae are tiny organisms that grow on the narrow-leaved arrowhead plants. There would be too many to count.

Procedure

1. Draw a large triangle on the paper. This will be your biomass pyramid. Divide the pyramid into four rows and label the trophic levels.
2. For your model, assume that one pinto bean represents 270 g of biomass per 100 m^2. Calculate how many beans would represent the biomass of each species shown in **Table 1**. Round fractions to the nearest whole number.
3. As a group, count out the appropriate number of beans for each species that you calculated in Step 2. Arrange the beans representing each species on the pyramid. Divide rows into multiple parts if there is more than one species at that level. Record the names of the species near the bean piles that represent them.
4. Return all of your beans to the container.

Results and Analysis

1. **Organize Data** Draw a bar graph that shows the biomass at each level.
2. **Calculate** Determine how much biomass is transferred from
   - producers to primary consumers
   - primary consumers to secondary consumers
   - secondary consumers to tertiary consumers

   Use this formula:

$$\text{percent biomass transferred} = 100 \times \frac{\text{total biomass of higher level}}{\text{total biomass of the lower level}}$$

3. **Interpret Data** Can this ecosystem support a higher-level consumer than the bass? Use your graph and pyramid to support your answer.
4. **Evaluate** What happens to the biomass that does not move on to the next trophic level?

CYCLING OF MATTER

Nutrition labels are a familiar part of most food packaging. The information shown in these labels helps consumers make healthful dietary decisions. These choices are often based on how energy efficient a food source is, as the body obtains energy from different types of compounds, such as carbohydrates or proteins, in different ways. Because your body is made up of elements obtained from the food you eat, another key to a wholesome diet is making sure to get these essential nutrients from a large variety of foods. But where does this matter come from? How does matter move through an ecosystem?

Predict *Where does the matter that makes up an organism's body come from?*

2.4

KEY TERMS

biogeochemical cycle
fossil fuel
greenhouse effect

Matter Cycles Through Earth's Systems

Earlier in this chapter you learned that energy from the sun provides the primary source of energy for producers at the base of a food chain. This energy flows through food webs as it moves from one trophic level to the next. Because the energy comes from an outside source (the sun), Earth is considered an open system in terms of energy. However, only a tiny amount of matter is lost from Earth at its atmospheric boundary with space. For this reason scientists model Earth as a closed system in terms of matter. Like energy, matter cannot be created or destroyed, so all of the substances that make up all living and nonliving objects on Earth come from already existing matter.

Because the amount of matter on Earth is limited, the cycling of life-sustaining elements is essential for the survival of living things on Earth. A **biogeochemical cycle** is the movement of matter through the biological and geological, or living and nonliving, parts of an ecosystem. Biogeochemical cycles include the water cycle, carbon cycle, nitrogen cycle, and phosphorus cycle. All of these cycles are driven by energy from the sun and Earth's core. They include a variety of interconnected processes:

- **Physical and Chemical Processes** These are the processes by which matter changes physical form or substances are transformed by chemical reactions.
- **Geological Processes** These are the processes that take place in the geosphere, such as the formation of rock or the movement of matter on and below Earth's surface.
- **Biological Processes** These are the processes performed by living organisms, such as photosynthesis and respiration.

All of the matter available to organisms on Earth is subject to these processes. Although matter used by organisms cannot be lost from the Earth system, matter can become unavailable or unusable for organisms in the biosphere.

SEP Construct an Explanation *Water is a substance essential to life on Earth. How might water or other substances be made unavailable to organisms in Earth's biosphere?*

2.4

CYCLING OF MATTER

LS2.A, LS2.B, LS2.C

EXPLORE/EXPLAIN

This section introduces the matter cycles for water, carbon, nitrogen, and phosphorus, and it discusses the importance of each matter cycle in supporting ecosystems. The effects of human activity on biogeochemical cycles are addressed at the end of the section.

Objectives

- Describe the Earth system in terms of its boundaries and interconnected subsystems.
- Model the hydrologic (water) cycle, carbon cycle, nitrogen cycle, and phosphorus cycle.

Pretest Use **Question 5** to identify gaps in background knowledge or misconceptions.

Predict *The matter that makes up an organism's body comes from the food it eats.*

SEP Construct an Explanation
Sample answer: Water and matter locked in nonbiological processes may become unavailable for use by organisms in the biosphere.

Crosscurricular Connections

Earth Science Students who have previously studied Earth science may recognize the water cycle shown in **Figure 2-13**. To prepare to use matter cycle models to account for the conservation of specific chemical elements within an ecosystem, have students describe why and how water changes during the evaporation, condensation, and precipitation processes labeled in the figure.

Stress the necessity of water for all life on Earth. Then stress the importance to life on land of recycling ocean salt water into freshwater rain in the hydrologic cycle. Display **Figure 2-13** and invite students to trace a path for a water molecule traveling through the cycle. Elicit that there is no single path water takes in the cycle; instead there is a network of interconnecting paths.

Similarly, in **Figure 2-14**, assess students' prior Earth science knowledge by asking them to describe what happens to matter during the processes of weathering, volcanic action, and uplifting over geologic time.

CCC Stability and Change *Sample answer: Water moves through the water cycle through the processes of condensation, precipitation, evaporation, and transpiration. A disruption to the cycle could limit the amount of water available to living organisms.*

The Water Cycle

The hydrologic cycle, or water cycle, circulates water through the environment (**Figure 2-13**). It begins as water evaporates from the surface of the ocean and other bodies of water and condenses to form clouds. Approximately 396,000 cubic kilometers of water evaporates each year, with about 78 percent of it condensing back to liquid form and falling on the ocean as precipitation.

On land, water enters streams and rivers as runoff within an area called a watershed. Some of the water seeps into the soil to become groundwater. The rest flows downhill through rivers and then enters the ocean via a coastal estuary, an area where salt and fresh water mix.

Water is also taken up by plants through their root systems and evaporates from their leaves through a process called *transpiration*. Approximately 90 percent of the water in the atmosphere is the product of evaporation from bodies of water, while the remaining 10 percent comes from plant-based transpiration.

CCC Stability and Change *How might a disruption to this cycle, such as a drought, affect living organisms?*

FIGURE 2-13
The water cycle circulates water on Earth.

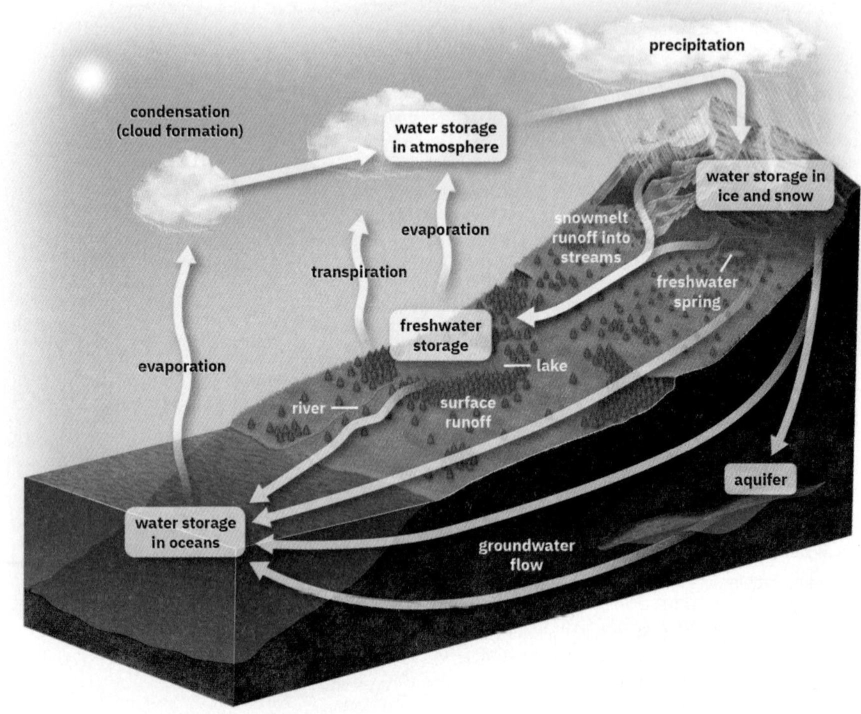

50 CHAPTER 2 ENERGY AND MATTER IN ECOSYSTEMS

DIFFERENTIATED INSTRUCTION | English Language Learners

Create a Personal Dictionary In this section, work with students to build a dictionary of science terms and grade-level vocabulary. **Beginning** and **Intermediate** students can use photos, drawings, and their native language to help them remember and review word meanings. **Advanced** students can write definitions in English. Have students focus on words that relate to cycles and processes, such as *condensation* and *precipitation*.

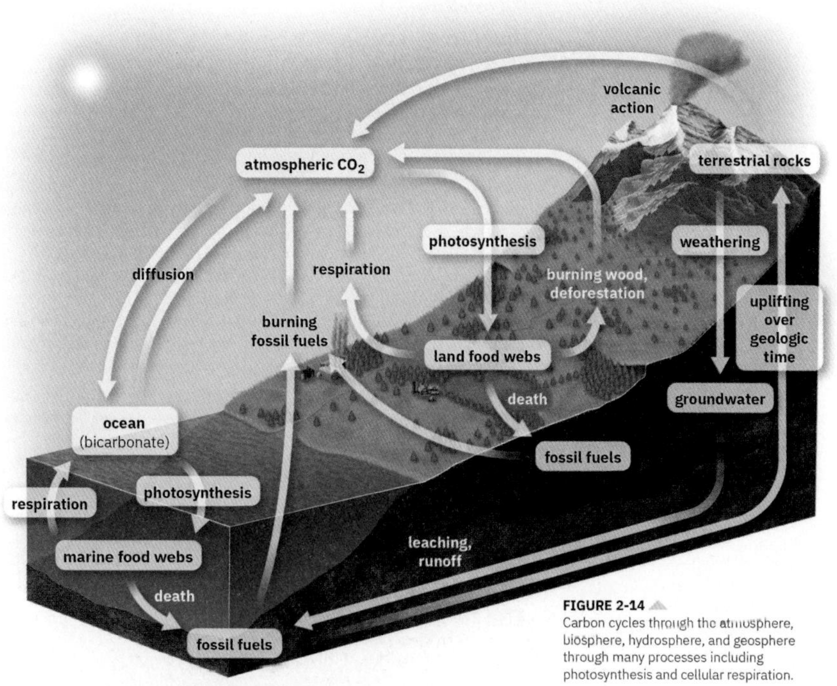

FIGURE 2-14 Carbon cycles through the atmosphere, biosphere, hydrosphere, and geosphere through many processes including photosynthesis and cellular respiration.

The Carbon Cycle

Carbon cycles through all of Earth's systems (**Figure 2-14**). Carbon enters land food webs when plants take up carbon dioxide from the air for use in photosynthesis. Carbon returns to the atmosphere as carbon dioxide when plants and other land organisms carry out aerobic respiration.

Carbon diffuses between the atmosphere and the ocean. Bicarbonate forms when carbon dioxide dissolves in seawater. Marine producers take up bicarbonate for use in photosynthesis, and marine organisms release carbon dioxide from aerobic respiration.

Earth's rocks and sediments are its single greatest reservoir of carbon. Marine organisms incorporate carbon into their shells. After they die, these shells become part of the sediments. Over time, the sediments become carbon-rich rocks such as limestone and chalk in Earth's crust. These rocks can be uplifted onto land by geological forces. However, producers take up carbon from the air rather than from rocks, so carbon in these reservoirs has little effect on ecosystems.

Fossil fuels such as coal, oil, and natural gas formed over hundreds of millions of years from the anaerobic decomposition of carbon-rich remains of ancient organisms. The burning of fossil fuels derived from these ancient remains puts additional carbon dioxide into the atmosphere.

SEP Develop a Model *Draw a model that shows the role of photosynthesis and cellular respiration in the cycling of carbon through Earth's systems.*

Visual Support

Cycling Through the Spheres Tell students that the carbon cycle can be thought of as the movement of carbon through Earth's four spheres. Break students into groups of four. Have groups assign each member a number from 1 to 4. Each number group will be responsible for carbon in one of the spheres: number 1 in the atmosphere; number 2 in the biosphere; number 3 in the hydrosphere; and number 4 in the geosphere. Groups will use **Figure 2-14** and the accompanying text to fill out an Idea Diagram for the carbon cycle. Each member will write the main ideas and details for their assigned sphere. Under "details," students should note what forms carbon takes in each sphere, how it is added and removed from the sphere, and any notable effects it has. Groups should write a one-sentence introduction and a one-sentence conclusion to bookend their findings.

SEP Use a Model *Drawings will vary. A model might show that carbon dioxide in the atmosphere is taken up by plants, which are part of the biosphere, during the process of photosynthesis. Living organisms release carbon dioxide into the atmosphere during the process of cellular respiration. Carbon diffuses from the atmosphere into the oceans, becoming part of the hydrosphere. Carbon becomes a part of the geosphere through the decomposition of living things and the formation of fossil fuels.*

REVISIT THE ANCHORING PHENOMENON

Sea Pig Survival Ask students to model the cycling of matter in the sea pig's deep-sea ecosystem by having them place their marine food webs in the larger context of the carbon cycle shown in **Figure 2-14**. The diagram shows that death is a process by which carbon moves from a land food web into the soil. This may help students recognize the need to refine their models by identifying a similar path by which the death of marine organisms returns carbon to the ocean.

Science Background

The Necessity of Nitrogen Nitrogen is a key component of living things and is vital to the construction of DNA and amino acids. In fact, the chemical bases that make up DNA, adenine (A), cytosine (C), guanine (G), and thymine (T), are collectively called *nitrogenous bases* because nitrogen is a key part of each.

Nitrogen is also a component of all amino acids, which form the molecular building blocks of proteins. Proteins provide some of the most important structures within cells and are necessary for numerous life-sustaining chemical reactions and processes.

Tell students that, like the carbon cycle, the nitrogen cycle is largely driven by the actions of organisms. In each process of the cycle, organisms break down nitrogenous compounds or combine them with other elements to obtain energy or nutrients or to build organic structures that they need. Have students work in pairs to read the text and study **Figure 2-15**.

SEP Energy and Matter *Sample answer: Without nitrogen-fixing bacteria, there wouldn't be enough nitrogen in the soil to support the growth of plants. With fewer plants, fewer animals would be supported in the ecosystem.*

The Nitrogen Cycle

Nitrogen gas (N_2) makes up 78 percent of Earth's atmosphere and is so stable that it does not readily combine with any other elements. That means living things cannot directly extract nitrogen from the atmosphere. Nitrogen must be broken down by the nitrogen cycle (**Figure 2-15**) before living things can use it to build molecules such as proteins and DNA.

In the process of nitrogen fixation, some nitrogen-fixing bacteria that live underground in the root nodules of legumes and a few other plants change gaseous N_2 to ammonia, NH_3. This ammonia is converted to ammonium (NH_4^+) during the process of ammonification. Nitrifying soil bacteria, archaea, and fungi convert ammonium to nitrates (NO_3^-) and nitrites (NO_2^-) in the process of nitrification. The bacteria gain energy from this process. Plant roots absorb nitrates and incorporate the nitrogen they contain into plant proteins and nucleic acids. Animals that eat the plants rearrange the chemicals into their own useful proteins and nucleic acids.

Waste products from the consumption of plant proteins are decomposed by ammonifying bacteria. The ammonia they release is ready to enter the nitrogen cycle again for nitrification. Finally, denitrifying bacteria and some fungi reduce nitrates back to their original gaseous N_2 form to complete the cycle.

Some nitrogen also enters the soil as a result of atmospheric nitrogen fixation by lightning. The energy of lightning causes nitrogen gas to react with oxygen in the air to form nitrogen oxides. When dissolved in rainwater, nitrates are formed, which are absorbed by the soil.

CCC Energy and Matter *How might a decrease in nitrogen-fixing bacteria affect the trophic levels in an ecosystem?*

FIGURE 2-15
The nitrogen cycle is the movement of nitrogen through Earth's air, water, and soil.

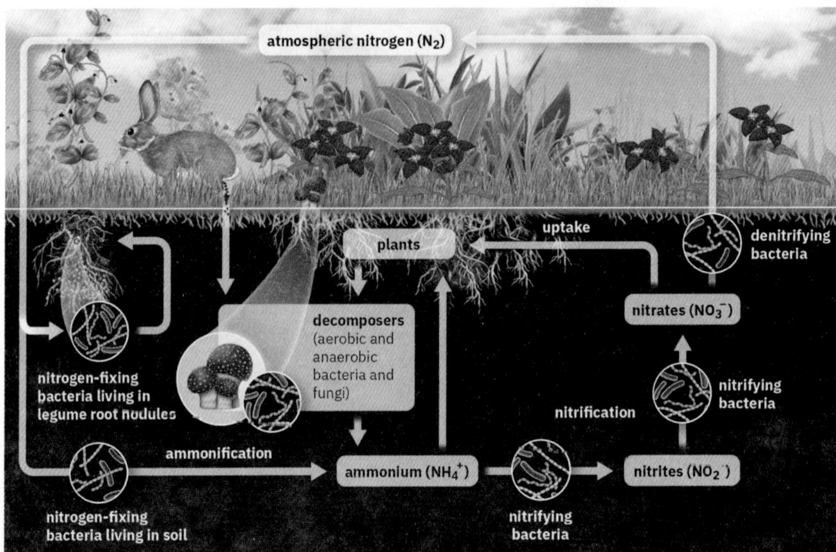

DIFFERENTIATED INSTRUCTION | Leveled Support

Struggling Students To aid in student understanding of the nitrogen cycle, have students generate their own graphic organizer, describing the flow of nitrogen through an ecosystem. Help them simplify some of the information provided in **Figure 2-15**. Students can make one cycle diagram that just shows how one atom of nitrogen might move through the atmosphere and populations of bacteria, then they can make a second diagram that just shows how nitrogen is used by plants and decomposers.

Advanced Learners Some students may have some knowledge about the use of nitrogen in agriculture. If any of your students are familiar with the application of nitrogen to field crops or the practice of crop rotation in which legumes are alternated with other crops, have them share their experiences with the class. Students can discuss why planting a legume (e.g., soybeans) in a field one year benefits a different type of crop planted in that same field the following year.

The Phosphorus Cycle

Phosphorus is part of several important biological molecules, including ATP, a compound that provides energy for life processes. Phosphorus is also an essential component in DNA and RNA. Unlike other matter cycles, phosphorus does not exist in the atmosphere in a gas phase. It is mainly cycled through the geosphere, hydrosphere, and biosphere. Most of Earth's phosphorus is bonded to oxygen as phosphate (PO_4^{3-}), which occurs in rocks and sediments. In the phosphorus cycle, phosphorus passes quickly through food webs as it moves from land to ocean sediments, then moves slowly back to land (**Figure 2-16**).

In the geochemical portion of the phosphorus cycle, weathering and erosion move phosphates from rocks into soil, lakes, and rivers. Leaching and runoff carry dissolved phosphate to the ocean. Here, most dissolved phosphorus comes out of solution and settles as rocky deposits. Movements of Earth's crust can uplift these deposits onto land, where weathering releases phosphates from rocks once again.

The biological portion of the phosphorus cycle begins when producers take up phosphate. Land plants take up dissolved phosphate from soil water. Animals on land get phosphates by eating the plants or one another. Phosphorus returns to the soil in the wastes and remains of organisms. In the seas, phosphorus enters food webs when producers take up phosphate dissolved in seawater. As on land, wastes and remains continually replenish the phosphorus supply.

FIGURE 2-16
Much of the phosphorus cycle takes place in the geosphere.

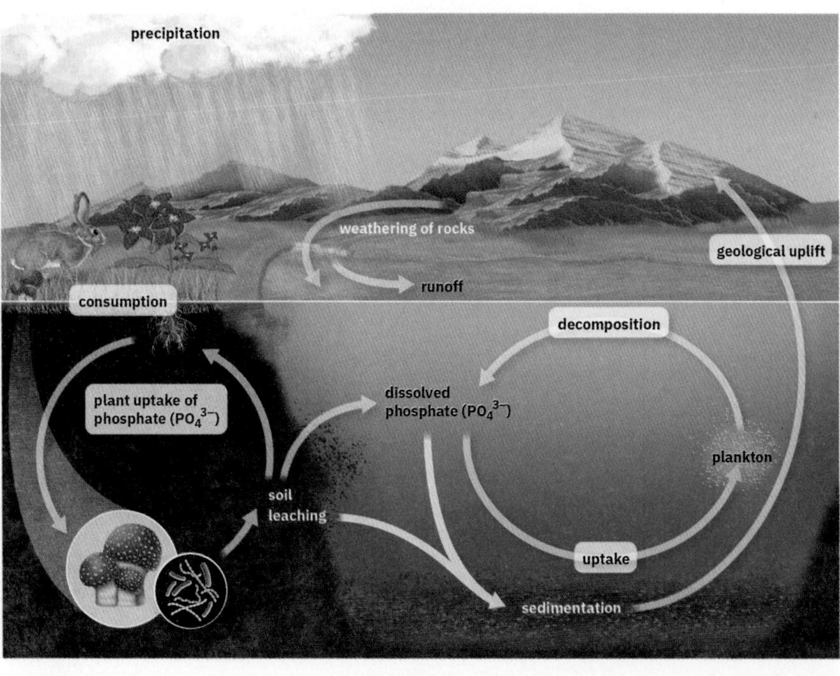

Science Background

The Phosphorus Cycle and Time
Phosphorus is an essential component of bones and teeth, brain tissue, nucleic acids, and the phospholipids that make up cell membranes.

Draw students' attention to the two smaller cycles within the larger phosphorus cycle shown in **Figure 2-16**. Ask students how this cycle differs from the water, carbon, and nitrogen cycles. Students should recognize that, unlike the other biogeochemical cycles, the phosphorus cycle lacks an atmospheric phase. This is because phosphorus is a solid or liquid at most temperature ranges on Earth. (For reference, the boiling point of phosphorus is 280.5 °C.) Instead, the majority of the phosphorus cycle takes place within the hydrosphere, the geosphere, and the biosphere.

Compared to other biogeochemical cycles, the phosphorus cycle is also the slowest cycle. While phosphorus passes through plants and animals relatively quickly via food webs, phosphate ions can be locked in ocean sediment for a period of 20,000 to 100,000 years.

Phosphorus is a limiting nutrient in both terrestrial and aquatic ecosystems. Phosphorus is commonly added as a fertilizer to agricultural fields to increase plant growth, as the quantities of naturally-occurring phosphorus in soil are limited.

Runoff from agricultural fields into bodies of water, such as rivers, lakes, and streams, is a common cause of harmful algal blooms.

DIFFERENTIATED INSTRUCTION | Students with Disabilities

Visual Impairment All students may benefit from a review of the matter cycles presented in this section. Have any visually-impaired students work with a small group of students to develop story-based descriptive summaries of each cycle. This task can also be divided among all students in the classroom, with each group focusing on a different matter cycle. Each summary should follow the movement of a single atom or molecule through the cycle. Encourage students to use storytelling skills when creating their summaries and to explain the movement of an atom or molecule as if it were on a journey.

Connect to Mathematics

Reason Quantitatively Students may not yet be familiar with the type of dual-axis graph shown in **Figure 2-17**. To help them interpret the *y*-axes in this graph, first ask them to describe the reason for plotting two different data sets on the same graph. They should recognize that this is done to compare the trends in the two data sets over the same period of time (or in general, the same change in the independent variable). Then discuss the units on each scale, explaining that "parts per million" means that, for every million parts of a solution or mixture, there is one part of the substance being measured. Have students explain what the graph might look like if both variables were plotted on the same scale, and whether doing so would give meaningful information.

CCC Cause and Effect *Increases in atmospheric carbon are increasing global temperatures, which are, in turn, causing an increase in ocean temperatures and the melting of polar ice caps, which then causes a rise in sea level.*

Human Impact on Biogeochemical Cycles

In the absence of humans, carbon, nitrogen, and phosphorus would cycle through the Earth system at a relatively constant rate and amount. However, human activities, such as the burning of fossil fuels and the introduction of synthetic fertilizers, have greatly influenced how matter cycles on Earth.

Climate Change Earth's climates are affected by gases in the lower atmosphere. As energy flows from the sun to the surface of Earth, some energy is reflected back into the atmosphere. Molecules of certain gases in the atmosphere, including water vapor (H_2O), carbon dioxide (CO_2), methane (CH_4), and nitrous oxide (N_2O), absorb some of this solar energy and release a portion of it as infrared radiation, which warms the lower atmosphere and Earth's surface. These gases that warm the lower atmosphere are called *greenhouse gases*. The natural warming of the lower atmosphere by greenhouse gases is called the **greenhouse effect**. Greenhouse gases have a role in determining the average temperatures of the lower atmosphere, and, therefore, the climates on Earth. Our planet would be a cold and mostly lifeless place without this natural warming effect.

FIGURE 2-17
The amount of carbon dioxide in the atmosphere (red line) has increased along with human emissions (blue line) since the start of the Industrial Revolution in 1750.

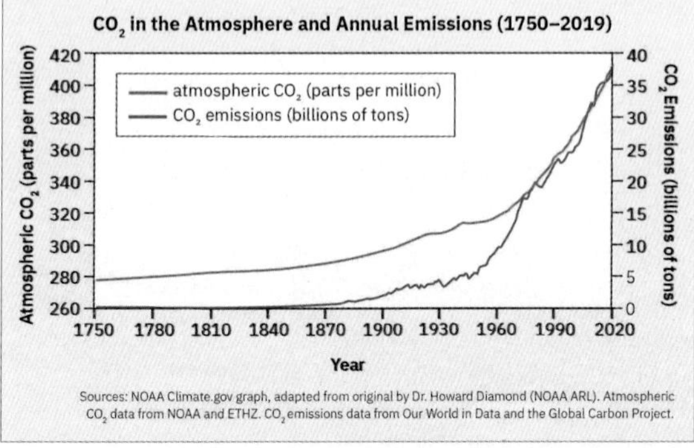

When burned in combustion, fossil fuels release carbon into the atmosphere. This excess carbon is a primary contributor to global climate change (**Figure 2-17**). As humans continue to extract and burn fossil fuels, the increased carbon dioxide level in the atmosphere has enhanced the greenhouse effect and caused temperatures on the planet to rise, altering ecosystems drastically. The melting of polar ice caps causes a rise in sea level that will put many coastal areas mostly underwater. The warmer temperatures are also causing increases in extreme weather events, alterations in animal migration patterns, and decreases in agricultural production. Many species are facing extinction as their habitats disappear.

CCC Cause and Effect *How do changes in atmospheric carbon influence changes in the hydrosphere?*

DIFFERENTIATED INSTRUCTION | Economically Disadvantaged Students

Shared Resources Students that lack resources may have trouble completing assignments due to limited access to basic school supplies and the internet outside of school. Consider setting up a corner in your room with shared supplies for students to borrow or use as needed. In addition to scheduling time in your school's media center or library for extended research projects, if space allows, you could set up a library of relevant magazines, books, newspapers, and other print resources in your room for students to use.

FIGURE 2-18
Large algal blooms (bright green) in Lake Erie result from agricultural and industrial runoff.

Nutrient Pollution In the early 1900s, scientists invented a method of fixing atmospheric nitrogen and producing ammonia on an industrial scale. This process allowed production of synthetic nitrogen fertilizers that have boosted crop yields.

The use of these fertilizers has helped feed a rapidly increasing human population, but it has also added large amounts of nitrogen-based compounds to our air and water. Nitrates from synthetic fertilizers can run off from agricultural fields and contaminate aquatic ecosystems, where it can encourage algal blooms (**Figure 2-18**).

Like nitrogen, phosphorus aids plant growth, so most fertilizers contain phosphorus as well as nitrogen. Phosphate-rich rocks are mined for use in the industrial production of fertilizer. Guano, which is phosphate-rich droppings from seabird or bat colonies, is also mined and used as fertilizer. Phosphates from fertilizers often run off from the site where they are applied and enter aquatic habitats. Other sources of aquatic phosphate pollution include animal waste from farms, sewage released from cities, and phosphate-rich detergents used to wash laundry and dishes. An influx of phosphorus can encourage the growth of aquatic producers, resulting in an algal bloom.

CHAPTER INVESTIGATION

Exploring Brine Shrimp Survival

What is the effect of an abiotic factor other than salinity on egg hatching and survival of brine shrimp?

Go online to explore this chapter's hands-on investigation and design your own investigation about abiotic factors.

2.4 REVIEW

1. **Identify** Which cycle most relies on the processes of photosynthesis and cellular respiration?
   A. carbon cycle C. nitrogen cycle
   B. phosphorus cycle D. water cycle
2. **Identify** Which of these processes are carried out by anaerobic bacteria? Select all correct answers.
   A. decomposition C. denitrification
   B. transpiration D. nitrification
3. **Analyze** Consider the cycles modeled in **Figures 2-13, 2-14, 2-15,** and **2-16** that show what happens to a water molecule, a carbon atom, a nitrogen atom, and a phosphorus atom, respectively. Describe two ways the water cycle is different from the other cycles.
4. **Design** Use **Figure 2-13** to make a simple model that follows an oxygen atom as it cycles through any two of these spheres: atmosphere, biosphere, hydrosphere.

2.4 REVIEW

1. *carbon cycle* **DOK 2**
2. *decomposition; nitrification* **DOK 1**
3. *Sample answer: Water molecules change state (solid, liquid, gas) but not chemical identity in the water cycle, while nitrogen and carbon go through chemical changes during their cycles. The water cycle shown does not include processes that occur within organisms (biosphere).* **DOK 3**
4. *Student model should identify that oxygen atoms are transferred from the atmosphere (in carbon dioxide and water) to the biosphere (photosynthesizing organisms) and back to the atmosphere (as oxygen molecules). The process is reversed by organisms through cellular respiration. The hydrosphere is used instead of the atmosphere for marine environments.* **DOK 3**

On the Map

Lake Erie Have students view Lake Erie's position within a larger-scale map of the Great Lakes that identifies details such as elevations, depths, and nearby towns and cities. Have students use the map to explain why the lake is highly susceptible to algal blooms. The U.S. Geological Survey (USGS) has a StoryMap on Harmful Algal Blooms in the Great Lakes, and they also map nitrogen and phosphorus levels in the Great Lakes watershed. Another useful resource is National Geographic's Mapmaker Interactive.

In Your Community

Phosphate Pollution Sources of phosphate pollution are not limited to agricultural runoff. By researching the local use and disposal of consumer products and other substances containing phosphates, students may be able to trace how these materials are able to travel from their communities to aquatic ecosystems and find out what measures are being taken to prevent phosphates from entering the local water supply.

CHAPTER INVESTIGATION B

Open Inquiry *Exploring Brine Shrimp Survival*

Time: 180 minutes over 5 days

Students will select a different abiotic factor and devise a step-by-step procedure similar to that of Investigation A in this chapter.

Go online to access detailed teacher notes, answers, rubrics, and lab worksheets.

LOOKING AT THE DATA
BIOMAGNIFICATION OF MERCURY

Connect this activity to the science and engineering practice of developing and using models and engaging in argument from evidence. Students identify patterns in the data and address matter cycling within a food chain. Students address the issue of seafood contamination by making sense of data that includes the physical characteristics and mercury concentration measured in four marine species.

SAMPLE ANSWERS

1. *Students' partial food webs should show swordfish at the highest trophic level connected directly to halibut. Halibut is connected directly to both herring and shrimp.* **DOK 2**
2. *shrimp* **DOK 2**
3. *swordfish* **DOK 2**
4. *swordfish; halibut; herring; shrimp* **DOK 2**
5. *The rankings are roughly opposite. The largest populations have the lowest mercury concentrations.* **DOK 3**
6. *Mercury levels are lowest at the lowest trophic level but increase at higher levels. Herring and shrimp at the lowest trophic level have the least mercury, while swordfish at the highest trophic level have more mercury. A large number of herring and shrimp are eaten by a smaller number of halibut. A large number of halibut are eaten by a smaller number of swordfish. The quantity of mercury absorbed by organisms at each higher trophic level increases.* **DOK 3**
7. *Two servings of halibut per week is considered safe.* **DOK 3**
8. *One way to decrease the mercury concentration of halibut is to farm them with low-mercury food sources, such as low-mercury shrimp.* **DOK 3**

LOOKING AT THE DATA

BIOMAGNIFICATION OF MERCURY

SEP Analyze and Interpret Data Seven of the most popular fish found on seafood menus are now considered unsafe to eat.

The Environmental Protection Agency (EPA) monitors mercury contamination in commercially fished species and their prey along the west coast of the United States. The data is used to provide seafood consumption warnings for the public.

Table 1 gives the average mercury concentration (in micrograms/gram) for four marine species monitored by the EPA. Herring and shrimp are two common food sources for halibut, while swordfish prey on halibut.

These spot shrimp were harvested in Prince William Sound, Alaska.

TABLE 1. Mercury Concentration in Marine Species

Species	Average mercury concentration (μg/g)	Species average mass (kg)	Average length (cm)
halibut	0.25	140	240
herring	0.078	0.2	25
shrimp	0.009	0.4	18
swordfish	0.99	350	300

Sources: EPA, NOAA

1. **Represent Data** Construct a partial marine food web that includes halibut, herring, shrimp, and swordfish.
2. **Infer** Based on the partial food web, which population do you expect to be the largest in any given community?
   A. halibut and herring
   B. herring and shrimp
   C. shrimp
   D. swordfish
3. **Infer** Based on the partial food web, which population do you expect to be the smallest in any given community?
   A. halibut and herring
   B. herring and shrimp
   C. shrimp
   D. swordfish
4. **Analyze Data** Rank the fish species by mercury concentration. List the species from highest to lowest.
5. **Identify Patterns** How does the ranking of species by population size compare to the ranking of species by mercury concentration levels?
6. **Formulate** State a claim about what happens to the concentration of mercury as it moves through trophic levels. Support your claim with evidence and propose an explanation.
7. **Apply** Halibut is popular on seafood menus. The Food and Drug Administration (FDA) reports that the maximum safe mercury consumption is at or below 0.46 μg/g per week. Based on this recommendation, how many servings per week of wild halibut would be safe?
8. **Design** Using only the data provided, propose one way that farming halibut could make it safer for human consumption.

56 CHAPTER 2 ENERGY AND MATTER IN ECOSYSTEMS

TYING IT ALL TOGETHER
SOMETHING FISHY IN THE FOREST

HOW DO ENERGY AND MATTER MOVE THROUGH AN ECOSYSTEM?

In this chapter, you learned about the transfer of energy and matter in an ecosystem. The Case Study described relationships among organisms in a Southeast Alaska coastal community (**Figure 2-19**). The rivers and streams in this ecosystem support five species of salmon, which spend up to seven years in the ocean where they gain more than 90 percent of their biomass. Largely due to this abundance of food, Alaskan coastal brown bears are some of the heaviest in the world. A male bear can weigh up to 680 kg after having feasted on salmon for a season.

Biologists studying the vegetation on Chichagof Island in the Tongass National Forest have found that plants that grow as far as 100 meters from rivers where salmon spawn contain higher levels of ocean-derived nitrogen than the same plant species at sites with no salmon (**Figure 2-20**). They estimated that 24 percent of the nitrogen in the needles of spruce trees near salmon streams comes from the ocean. The scientists also determined that trees near salmon habitats grow faster and larger than those farther away.

In this activity, you will develop models to analyze nutrient pathways from marine to terrestrial organisms. Work individually or in a group to complete these steps.

Gather Evidence

1. Research the types and populations of organisms in the Tongass National Forest ecosystem. Focus on defining the boundaries of an ecosystem with a community that includes brown bears, salmon, and trees.

Develop and Use a Model

2. Construct one or more models that represent both
   a. the flow of energy between organisms within the community
   b. the cycling of matter between these organisms and their environment

Your model(s) should identify all trophic levels present in the ecosystem you have defined and show the relative distribution of energy and biomass in the ecosystem.

Construct an Explanation

3. Use your models to explain how marine nitrogen travels from the ocean to the leaves of forest trees.

FIGURE 2-19
Sitka spruce trees grow three times faster on the banks of salmon streams than those that grow elsewhere.

FIGURE 2-20
Scientists measured the growth rings of spruce trees and the amount of marine nitrogen in their leaves to compare trees near salmon spawning streams to trees at reference sites with no salmon spawning.

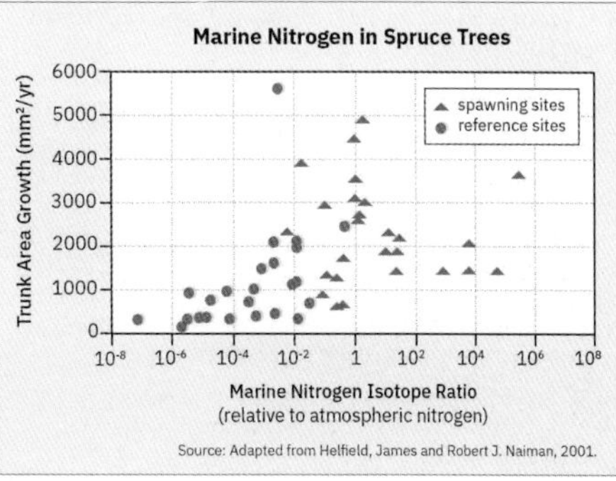

TYING IT ALL TOGETHER

ELABORATE

The Case Study described ecological relationships among brown bears, salmon, and plant life in Southeast Alaska. In the Tying It All Together activity, students define a specific ecosystem within the Tongass National Forest and apply what they have learned to trace the path of nitrogen through the ecosystem.

This activity is structured so that students may complete the steps as they study the chapter content. **Step 1** addresses Section 2.1 content. **Step 2a** corresponds to content in Sections 2.2 and 2.3, and **Step 2b** incorporates Section 2.4 content. Students will then need to synthesize ideas from all sections to complete **Step 3**.

Alternatively, this can be a culminating project at the end of the chapter. The modeling and analysis practiced here reflects a reasoning process that students may use to support a claim in the Unit 1 Activity.

Go online to access the Student Reflection and Teacher Scoring rubrics for this activity.

Science Background

Stable Isotope Analysis Chemical elements and their isotopes are addressed in **Section 5.1**. Isotopes of some elements occur in known proportions for certain environments in nature, which allows scientists to determine where the matter in an organism or material comes from. The ^{14}N isotope constitutes over 99 percent of the nitrogen on Earth. The rarer ^{15}N isotope is found primarily in marine environments. Thus, the elevated ratios of ^{15}N to ^{14}N in spruce tree leaves close to rivers where salmon spawn (**Figure 2-20**) support the hypothesis that matter enters this ecosystem via salmon migrating from the ocean and cycles through plants via the food web.

CROSSCUTTING CONCEPTS | Systems and System Models

Open and Closed Systems Help students define the boundaries of an ecosystem by discussing the difference between open and closed systems. In science, an open system is one in which both energy and matter can flow between the system and its surroundings, whereas in a closed system, only energy can enter or leave. Ecosystems, like all biological systems, are necessarily open. Students likely grasp the concept that within a community of organisms, matter and energy generally follow the same path. However, in the later parts of this activity, they will need to use abiotic factors in their ecosystem to accurately describe how energy and matter flow in and out, so students should prepare by identifying these factors as they define the ecosystem in step 1.

EVALUATE

REVIEW KEY CONCEPTS

1. *A, C, D* **DOK 2**
2. *D, E* **DOK 2**
3. *A, E* **DOK 2**
4. *carnivores: penguins; detritivores: worms; herbivores: zebras; omnivores: sea turtles, wasps* **DOK 2**
5. *energy pyramid: An ecosystem is not a closed system; Consumers obtain the energy they need from the trophic level immediately below it. Biomass pyramid: The amount of matter in living organisms at lower trophic levels is greater than at higher trophic levels. Numbers pyramid: The populations of producers are larger than the population of secondary consumers.* **DOK 2**
6. *D* **DOK 1**
7. *5* **DOK 2**
8. *Sample answer: Most transformations of energy and matter in living organisms occur in the chemical reactions of photosynthesis and cellular respiration. Organisms use photosynthesis to convert sunlight, carbon dioxide, and water into oxygen and sugar molecules. Organisms use cellular respiration to break down sugar and other carbon-based molecules to release chemical energy they need to live.* **DOK 2**
9. *Sample answer: All living organisms need energy for biological functions, growth, and movement. How organisms obtain the energy they need defines their main classification. Producers obtain energy entirely from nonliving sources. Consumers obtain energy from living organisms or waste matter. Decomposers obtain energy by breaking down dead organisms without consuming them.* **DOK 2**

CRITICAL THINKING

10. *Sample answer: The burning of fossil fuels releases excess carbon into the atmosphere. This carbon mixes with oxygen to form carbon dioxide, which is a greenhouse gas. The atmosphere warms due to increased carbon dioxide. When there is more carbon dioxide in the atmosphere, there is more available for photosynthesis. More carbon enters the biosphere. Carbon then accumulates in the geosphere via organic matter, such as dead organisms and waste. As global temperatures rise, ocean temperatures also rise, which can destroy ecosystems and ultimately the carbon in the hydrosphere decreases as it accumulates on the sea floor.* **DOK 3**
11. *Sample answer: Drought will decrease the number of producers in the grassy plains of Africa. Less grass and fewer plants and trees will directly impact the population of gazelles, which are primary consumers. Fewer primary consumers will impact secondary consumers, such as lions who feed on gazelles. The energy pyramid may keep the same number of trophic levels, but the energy available at all levels could decrease. Drought will kill off producers and gazelles, increasing the carbon stored in the geosphere and in the atmosphere but decreasing the carbon in the biosphere.* **DOK 3**

CHAPTER 2 **REVIEW**

REVIEW KEY CONCEPTS

1. **Relate** Which claims are true for all the organisms within a community? Select all correct answers.
   - **A.** They interact with each other.
   - **B.** They are part of the same population.
   - **C.** They are found within the same biome.
   - **D.** They have important roles in the same ecosystem.
   - **E.** They depend on the same biotic and abiotic factors.
2. **Compare** Why are food chains less effective than food webs for modeling ecosystem stability? Select all correct answers.
   - **A.** Energy flow can be shown in a food web but not in a food chain.
   - **B.** Unlike in food webs, all organisms in a food chain are connected.
   - **C.** A food chain shows a concise picture of how energy flows through a stable ecosystem.
   - **D.** Many complex organism interactions in a stable ecosystem are excluded in a food chain.
   - **E.** A disruption to a food chain may not impact the transfer of energy in other food chains within the same stable ecosystem.
3. **Compare** Which claims are most likely supported by the ecological pyramids for the two ecosystems described below? Select all correct answers.

   Consider two unrelated stable ecosystems. Ecosystem A has roughly 10,000 kcal/m^2 per year available for its primary consumers. Ecosystem B has only 5000 kcal/m^2 per year available for its primary consumers but contains the same number of trophic levels as Ecosystem A.

   - **A.** Ecosystem A has a greater variety of species than Ecosystem B.
   - **B.** In Ecosystem A, the populations of predators will be double the population of prey.
   - **C.** The trophic levels in Ecosystem A each contain half as many species as those in Ecosystem B.
   - **D.** Secondary consumers are not as well-supported in Ecosystem A as secondary consumers in Ecosystem B.
   - **E.** The amount of energy flowing from primary consumers to secondary consumers in Ecosystem A will be double that of Ecosystem B.
4. **Classify** Sort the organisms based on the given descriptions into four categories: *carnivores, detritivores, herbivores,* and *omnivores.*
   - Zebras feed on grass and small plants.
   - Sea turtles feed on corals, seaweed, and fish.
   - Wasps feed on insects and plants.
   - Penguins feed on fish.
   - Worms feed on decaying roots and leaves.
5. **Contrast** Select the best model to use to support or refute each of these claims. Choose from three models: *biomass pyramid, energy pyramid, pyramid of numbers.*
   - An ecosystem is not a closed system.
   - Ecosystems support fewer complex organisms than simple organisms.
   - Consumers obtain the energy they need from the trophic level immediately below them.
   - The populations of producers are larger than the populations of secondary consumers.
   - The amount of matter in living organisms at lower trophic levels is greater than at higher trophic levels.
6. **Identify** What happens to the matter that makes up an organism when it dies?
   - **A.** The matter is destroyed as the body decays.
   - **B.** The matter becomes trapped in the nonliving body.
   - **C.** The matter is converted into new types of matter by producers.
   - **D.** The matter is cycled back to the nonliving and living parts of the ecosystem.
7. **Analyze** A community of plants produces 50,000 kcal/m^2 per year of energy. The top predators in this ecosystem have only 5 kcal/m^2 per year available for consumption. How many trophic levels does this ecosystem have?
8. **Explain** Why are photosynthesis and cellular respiration important to the transfer of energy and matter on Earth? Explain your answer.
9. **Summarize** Justify the claim that all living organisms could be sorted into just three major categories: *producer, consumer,* and *decomposer.*

58 CHAPTER 2 ENERGY AND MATTER IN ECOSYSTEMS

CRITICAL THINKING

10. **Predict** Describe how the use of fossil fuels by humans might affect the carbon cycle and the relative amount of carbon in the atmosphere, hydrosphere, geosphere, and biosphere.

11. **Synthesize** Gazelles are herbivores that live on the grassy plains of Africa. How would a drought affect the food chain, energy pyramid, and carbon cycle within the ecosystem in which the gazelles live?

12. **Evaluate** Use **Figures 2-10** and **2-11** to evaluate the claim that Earth can be considered an open system and a closed system in terms of energy and matter.

13. NOS **Scientific Knowledge** Prior to the discovery of deep-sea thermal vent ecosystems, scientists thought all primary food sources in the deep ocean fell from the ocean surface. How did evidence of chemosynthetic organisms in these deep-sea ecosystems cause scientists to revise their understanding of how energy is produced in living systems?

MATH AND ENGLISH LANGUAGE ARTS CONNECTIONS

1. **Use Units to Understand Problems** When calculating biomass, biologists assume that about half of an organism's mass constitutes its biomass. In a 100-square-meter area, scientists count n plants. The average mass of a plant is m grams. Write an expression to estimate the mass of the plants in kilograms per square meter. Simplify your answer.

2. **Reason Abstractly and Quantitatively** Complete the table to show the amounts of energy that flow into and out of each trophic level in an ecosystem.

Trophic level	Energy available (J)	Energy lost (J)
producers		
primary consumers	10,000	
secondary consumers		

Read the following passage. Use evidence from the passage to answer the question.

Established in 1974, the Cabo Rojo National Wildlife Refuge serves as home to more than 300 different plant and bird species. Both native and migratory birds use the refuge, including two famous endemic species, the Puerto Rican emerald and the yellow-shouldered blackbird. Among the most populous plant species are the tamarind, an exotic species introduced to the island, the silk-cotton tree, and the guayacan, the official tree of the city of Cabo Rojo.

3. **Cite Textual Evidence** Does the passage describe a population, a community, or an ecosystem? Explain your answer.

4. **Write Explanatory Text** Choose two of the matter cycles and identify a place where parts of both cycles occur. Describe how the cycles are related in that location at the atomic or molecular level.

REVISIT SEA PIG SURVIVAL

Gather Evidence In this chapter, you have seen many examples of organisms and learned about their roles in the cycling of matter and flow of energy through their respective ecosystems. In Section 2, you learned that sea pigs scavenge organic matter from the ocean floor. They are part of a larger community of organisms that together cycle matter and transfer energy throughout the abyssal plain ecosystem.

1. As the sea pigs digest organic matter from the ocean floor, they play an important role in the cycling of matter. How does the sea pig's ecological role help cycle carbon and other matter in its ecosystem?

2. What other questions do you have about sea pigs? What else would you need to know to model the transfer of energy and matter throughout the sea pig's deep-ocean ecosystem?

CHAPTER REVIEW 59

REVISIT THE ANCHORING PHENOMENON

1. *As decomposers, sea pigs break down organic materials into their component parts and return them to the environment for other organisms to use.* **DOK 3**

2. *Sample answer: What organisms are the producers in this ecosystem? What are the primary consumers? Which matter cycles involve the deep ocean?* **DOK 3**

12. *Sample answer: Earth is an open system with respect to energy. Energy enters the system via solar radiation. All of that energy is used up in processes that occur in the atmosphere, biosphere, hydrosphere, and geosphere. Earth is a closed system with respect to matter. Matter is recycled through biogeochemical processes and biological functions.* **DOK 3**

13. *Sample answer: Food that falls to the deep ocean contains energy generated by photosynthesis. The discovery of chemosynthetic organisms that generate energy in deep-sea vent ecosystems showed scientists that not all energy in the deep sea comes from photosynthetic producers. Not all energy originates from the sun.* **DOK 3**

MATH AND ELA CONNECTIONS

1. *M•N/2000 km/m*^{2} **DOK 2**
2. *Producers: Energy available: 100,000 J; Energy lost: 90,000 J. Primary consumers: Energy lost: 9000 J. Secondary consumers: Energy available: 1000 J; Energy lost: 900 J.* **DOK 2**
3. *Sample answer: The passage describes a community that includes different species of birds and plants. It does not describe a single population since different species are mentioned, and it does not describe an ecosystem because there are probably other organisms that are not described.* **DOK 2**
4. *Sample answer: The water, nitrogen, and carbon cycles intersect in plants. Plant roots take up water, nitrates, nitrites, and ammonium from the soil. The plants then use the water for photosynthesis (to make sugar and oxygen) and use the nitrogen in proteins (assimilation).* **DOK 3**

BIODIVERSITY AND ECOSYSTEM STABILITY

Three-Dimensional Learning

The practices, core ideas, and crosscutting concepts presented in this chapter's text, investigations, and resources provide support to address the following Performance Expectations: **HS-LS2-1, HS-LS2-2, HS-LS2-6, HS-LS2-7**

Science and Engineering Practices	Disciplinary Core Ideas	Crosscutting Concepts
Asking Questions and Defining Problems Planning and Carrying Out Investigations Constructing Explanations and Designing Solutions (HS-LS2-7) Using Mathematics and Computational Thinking (HS-LS2-1, HS-LS2-2) Engaging in Argument from Evidence (HS-LS2-6) Obtaining, Evaluating, and Communicating Information Scientific Knowledge is Based on Empirical Evidence Scientific Knowledge is Open to Revision in Light of New Evidence (HS-LS2-2, HS-LS2-6)	**LS2.A** Interdependent Relationships in Ecosystems (HS-LS2-1, HS-LS2-2) **LS2.C:** Ecosystem Dynamics, Functioning, and Resilience (HS-LS2-2, HS-LS2-6, HS-LS2-7) **LS4.C** Biological Evolution: Unity and Diversity **LS4.D:** Biodiversity and Humans (HS-LS2-7)	Patterns Cause and Effect Stability and Change (HS-LS2-6, HS-LS2-7)

Contents		Instructional Support for All Learners	Digital Resources
ENGAGE			
60–61	**CHAPTER OVERVIEW** **CASE STUDY** Where there's smoke there's fire.	**Social-Emotional Learning** CCC Stability and Change **On the Map** Yellowstone National Park **English Language Learners** Ask and Answer Questions	*Video 3-1*
EXPLORE/EXPLAIN			
62–64 DCI LS2.A LS2.C	**3.1 ECOLOGICAL RELATIONSHIPS** • Distinguish between a habitat and a niche. • Classify the roles and relationships that exist within an ecosystem.	**Vocabulary Strategy** Using Prior Knowledge **English Language Learners** Use Word Associations **Connect to English Language Arts** Integration of Knowledge and Ideas CER Revisit the Anchoring Phenomenon **In Your Community** Symbiotic Relationships	*Video 3-2*
65	**EXPLORER** FEDERICO KACOLIRIS	**Connect to Careers** Conservation Scientist **Differentiated Instruction** Leveled Support	

Contents		Instructional Support for All Learners	Digital Resources
66–70 DCI LS2.C LS4.C LS4.D	**3.2 BIODIVERSITY** • Explain the three levels at which biodiversity is measured. • Distinguish between species richness and species evenness. • Identify the criteria for designation as a biodiversity hotspot. • Explain how keystone species differ from other species in a community.	**Vocabulary Strategy** Prefixes/Suffixes/ Root Relationships SEP Engaging in Argument from Evidence SEP Analyzing and Interpreting Data **Crosscurricular Connection** World Geography **In Your Community** Zoos and Conservation CCC Patterns **On Assignment** An Eye for the Elusive **Connect to English Language Arts** Assess Reasoning and Evidence **Vocabulary Strategy** Understanding in Context	*Video 3-3* **Investigation A** Measuring Biodiversity Using Ecological Sampling Methods (100 minutes over 2 days) **Interactive Figure** World's Biodiversity Hopspots
71	**LOOKING AT THE DATA** THE BIODIVERSITY CONSERVATION PARADOX		
EXPLORE/EXPLAIN			
72–77 DCI LS2.C LS4.D	**3.3 ECOSYSTEM STABILITY AND CHANGE** • Recognize natural and human-caused ecological disturbances. • Relate the effects of natural disturbance on ecosystems. • Compare primary and secondary succession.	CCC Scale, Proportion, and Quantity **In Your Community** Zoning Laws **Social-Emotional Learning** Social Awareness **CASE STUDY** Ask Questions **Connect to English Language Arts** Translate Quantitative or Technical Information **Differentiated Instruction** Leveled Support **CONNECTIONS** In Your Community **On The Map** Volcanic Islands in the South Pacific **English Language Learners** Paired Reading **Address Misconceptions** Forest Succession	**Investigation B** Ecological Succession in an Aquatic Environment (250+ minutes over 2 days plus observation time over 3 weeks)
78	**MINILAB** OBSERVING BIODIVERSITY IN POND WATER		
ELABORATE			
79	**TYING IT ALL TOGETHER** How do ecosystems change?	CCC Scale, Proportion, and Quantity	Self-Reflection and Scoring Rubrics
Online	**Investigation A** Measuring Biodiversity Using Ecological Sampling Methods **Investigation B** Ecological Succession in an Aquatic Environment	Guided Inquiry (100 minutes over 2 days) Design Your Own (250+ minutes over 2 days plus observation time over 3 weeks)	MINDTAP Access all your online assessment and laboratory support on MindTap, including: sample answers, lab guides, rubrics, and worksheets. PDFs are available from the Course Companion Site.
EVALUATE			
80–81	**Chapter 3 Review**		
Online	**Chapter 3 Assessment** **Performance Task 3** *How Does a Sudden Disturbance Change a Rainforest Ecosystem?* (HS-LS2-1, HS-LS2-2) **Performance Task 4** *How Does Long-term Drought Change a Saltwater Ecosystem?* (HS-LS2-2, HS-LS2-6)		

CHAPTER 3

ENGAGE

Students will likely bring prior knowledge of ecological relationships to this chapter from concepts learned in middle school and Chapter 2. Chapter 3 begins with a review of these relationships before exploring what biodiversity is, how it is measured, and why biodiversity differs among communities. Students will also explore both natural and human-caused disturbances and begin to see both the fragility and the resilience of ecosystems. The importance of biodiversity in maintaining ecosystem stability and adapting to disturbances provides the context for a broader application.

About the Photo Looking at images of wildfires often evokes negative feelings. However, encourage students to visualize the image from a different perspective, such as the forest floor. The mature trees form a canopy that blocks much of the sunlight, leaving plants at ground level struggling for resources. Ask students how a wildfire could benefit ground-level plants and the organisms that depend on them.

Social-Emotional Learning

As students progress through the chapter, invite them to look for contexts and opportunities to practice some of the five social and emotional competencies. For example, as students consider both human and natural causes of disturbances in ecosystems, there are ample opportunities to focus on **responsible decision-making**. Guide students as they learn to make reasonable, evidence-based judgments using information and data.

Learning about Earth's ecosystems and the organisms that live in them can reveal conflicting opinions. For example, in discussing "Human-Caused Disturbances," some students may have different opinions about decisions regarding land use or climate change. Practicing **relationship skills** and **social awareness** can help students communicate their opinions effectively, and listening to others' perspectives can lead to constructive dialogue.

CHAPTER 3

BIODIVERSITY AND ECOSYSTEM STABILITY

A disturbance, such as a wildfire in Yellowstone National Park, can actually help to maintain biodiversity in an ecosystem.

The mission of the U.S. National Park Service is to "preserve unimpaired the natural and cultural resources and values of the National Park System for the enjoyment, education, and inspiration for this and future generations." Parks are natural systems too, so the park service also endeavors to maintain each park's ecological processes, which can include major changes to the landscape, such as those caused by fire.

Chapter 3 supports the NGSS Performance Expectations **HS-LS2-6** and **HS-LS2-7**.

60 CHAPTER 3 BIODIVERSITY AND ECOSYSTEM STABILITY

CROSSCUTTING CONCEPTS | Stability and Change

Impact of Scale Throughout this chapter, students will see examples of stability and change at different scales within an ecosystem. A stable ecosystem is characterized by balance. Ecosystems that are considered stable may be able to return to a stable state shortly after an ecological disturbance. Ecological interactions, such as predation and competition, help maintain balance within an ecosystem. However, numerous natural and human-caused disruptions can produce change in even the most stable ecosystems. Biodiversity generally leads to greater ecosystem stability. Greater genetic and species diversity within an ecosystem makes the ecosystem more apt to withstand disturbances and changing conditions caused by disease or climate change. Reinforce the concept of stability and change as students learn about the role biodiversity plays in an ecosystem's ability to withstand disturbances at different scales.

CASE STUDY
WHERE THERE'S SMOKE THERE'S FIRE

HOW DO ECOSYSTEMS CHANGE?

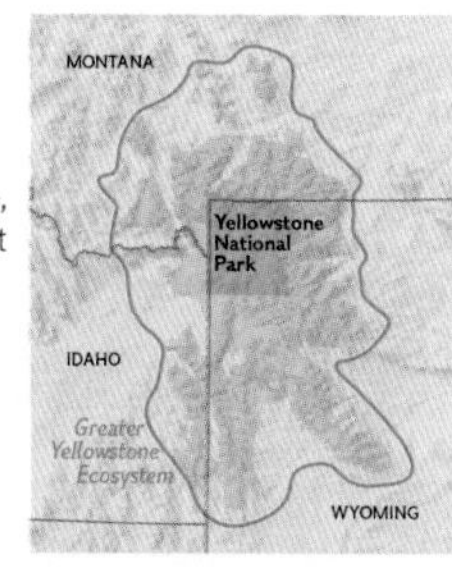

Yellowstone National Park, which extends across parts of Wyoming, Montana, and Idaho, is well known for its geysers, hot springs, and abundant wildlife. In the summer of 1988, the park made headlines around the country when wildfires raged unchecked across the region's unique landscape.

That summer, Yellowstone National Park and the Greater Yellowstone Ecosystem that surrounds it were in the midst of a severe drought. Dry conditions made the region ripe for a megafire. A lightning strike in late June sparked the first blaze. At the time, U.S. Park Service and Forest Service fire management policy was to allow natural fires to burn out on their own. However, by late July, fires were spreading at an alarming rate—as much as 6400 acres a day went up in smoke. Wind gusts would send hot cinders soaring more than a kilometer away from the main fire, where they seeded new fires that merged and grew. It was then decided to fight all fires. On August 20, strong winds caused the main fire to race across 150,000 acres, in what became known as "Black Saturday."

In the end, 1.4 million acres of the Greater Yellowstone Ecosystem had burned, including more than one-third of Yellowstone National Park's total area. In the summer of 1988 there were 51 fires within the park, nine of which were human-caused. The rest were started by lightning strikes. Although a snowfall in early September helped get the fires under control, the last fire was not extinguished until November.

People around the country worried that the national park had been destroyed by fire, and many were shocked that the nation's first national park had been allowed to burn for so long without any human intervention. However, fire has always been a force of change in the Greater Yellowstone Ecosystem. Tree ring studies indicate that major fire events have occurred there every 300 years or so. Prior to 1988, the last megafire was in the 1700s.

In fact, many ecosystems, including the Greater Yellowstone Ecosystem, need fire to function properly. The cones of lodgepole pine trees require the intense heat of a fire to open and disperse their seeds. Fire clears undergrowth from the forest, helps to spread seeds for new growth, and adds nutrients to the soil. In the year following the devastating fires, the blackened landscape was once again full of life, blanketed by wildflowers and signs of new tree growth.

Ask Questions *As you read the chapter, generate questions about the ways in which ecological disturbances affect an ecosystem's community of organisms.*

FIGURE 3-1
These satellite images show how the Greater Yellowstone Ecosystem has changed over time. The yellow zones indicate fire scars from the 1988 fire. Green indicates plant growth. Note that the 2018 image also shows damage from fires that occurred after 1988.

VIDEO 3-1 Go online to watch a time-lapse video of ecosystem recovery in Yellowstone National Park.

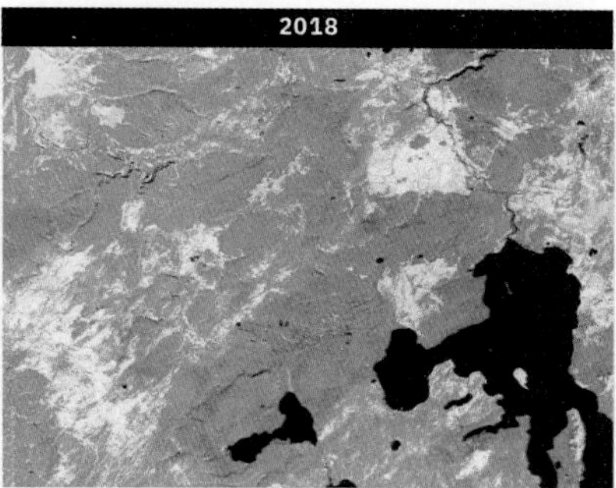

DIFFERENTIATED INSTRUCTION | English Language Learners

Ask and Answer Questions Working in pairs, have students take turns reading an assigned paragraph from the Case Study out loud. Explain to students that asking and answering questions can help them understand the main idea and supporting details in the article.

Beginning Have each student in a pair ask one question about the article. They can then work together to find the answer, conferring in their native language if necessary.

Intermediate Have each student ask one question and then swap with their partner. Tell them to work together to find the answers. Encourage the use of English, using their native language only when needed.

Advanced Have each student ask one question and then swap with their partner. Each partner should then find the answer and explain it to the other in English.

CASE STUDY

ENGAGE

The Case Study describes one example of a change in a specific ecosystem that can be used to gauge students' understandings about interactions within an ecosystem. Ask students to identify different habitats that may have been found in the forest prior to the fire, as well as the organisms found there. Students may be familiar with the idea of biodiversity, and through this discussion, they can begin to recognize the many different organisms living here that would be affected.

Students are likely aware of how disruptions such as a wildfire can impact an ecosystem and may have strong opinions. Guide the discussion so they can also recognize the benefits of such events.

In the Tying It All Together activity, students will research the effects of the 1988 wildfire and use the evidence found to evaluate the claim that wildfires are disturbances that affect biodiversity in places such as this.

Ask Questions Students should revisit the Case Study as they read the chapter to make connections with the content. See a specific suggestion in Section 3.3.

On the Map

Yellowstone National Park Use the map to discuss with students what geographic features affect the distribution of habitats for plants and animals. The Greater Yellowstone Ecosystem, including Grand Teton National Park, includes seven national forests and several mountain ranges. Habitats vary from mountains covered in snow year-round to grasslands. This ecosystem diversity supports the biodiversity found in the park.

Video

Time: 0:40

Use **Video 3-1** to show the landscape in Yellowstone National Park from 1987, just prior to the 1988 wildfire that devastated a third of the area, through the period of regrowth up to 2018. Have students make a timeline of the fire and recovery in the area to show the rate of progression.

3.1

ECOLOGICAL RELATIONSHIPS

LS2.A, LS2.C

EXPLORE/EXPLAIN

This section provides a review of the concepts of habitat and niche, discusses the predator-prey interactions within a habitat, and examines three main types of symbiotic relationships—mutualism, parasitism, and commensalism.

Objectives

- Distinguish between a habitat and a niche.
- Classify the roles and relationships that exist within an ecosystem.

Pretest Use **Questions 1, 2, and 6** to identify gaps in background knowledge or misconceptions.

Vocabulary Strategy

Using Prior Knowledge Point out that the word *competition* is a noun. Ask students what the verb form of this word is (to compete). Invite volunteers to share how they have used the word *compete* in everyday life, such as for describing athletic or academic competitions. Similarly, explain that the word *predation* is also a noun. Ask students for another noun that is similar (predator). Ask students what comes to mind when they hear the word *predator* and how that relates to predation.

Explain that when they encounter the word *symbiosis*, it is an umbrella term that includes different types of relationships.

Describe *Sample answer: The ecosystem would have less variety, and the populations of organisms that depend on the moth as a primary food source might decline in number.*

SEP Construct an Explanation *Sample answer: A spider builds its web across two branches of a tree in a forest. It uses its web to capture insects for food.*

3.1

KEY TERMS

competition
predation
symbiosis

ECOLOGICAL RELATIONSHIPS

A spider lies in wait at the edge of its intricately spun web. A moth flies nearby, unaware of the dangerous obstacle in its path. Just as the moth's wings are about to hit the sticky web, a bird swoops by and snatches the moth from the air. The spider will have to wait a bit longer for its next meal.

Describe *How might the removal of moths from this community affect the ecosystem as a whole?*

Habitat and Niche

Understanding a few basic ecology concepts can help us more fully describe the relationships among the spider, moth, and bird. For example, a *habitat* is the place where an organism lives. Shelter, water, food, and space are the four major components of a habitat. An *ecological niche* refers to the role an organism plays in its community. Similar to a habitat, a niche is made up of all the physical and environmental conditions that an organism needs to survive and reproduce. In contrast to an organism's habitat, a niche also includes all the interactions the organism has with other species. Each species fits into a unique niche that has the species' exact requirements. When two species occupy the same niche, one will be better at obtaining resources, which will result in the exclusion of the other species from that niche.

SEP Construct an Explanation *What are two interactions that might make up a spider's niche?*

Predation and Competition

The two primary interactions between organisms within a habitat are predation and competition. **Predation** occurs when one organism, the predator, eats another, the prey (**Figure 3-2**). Carnivores are not the only organisms that are considered to be predators. Herbivores that eat living plants are also predators. **Competition** occurs when two organisms compete for the same limited resource, such as food, shelter, or access to mates. Competition can occur between members of the same species and between members of different species.

FIGURE 3-2 This wasp spider, easily identified by its distinctive black-and-yellow striping, wraps its prey in silk.

DIFFERENTIATED INSTRUCTION | English Language Learners

Use Word Associations Display the words *habitat* and *niche* along with visuals that illustrate both. Say each word and have students repeat chorally three times.

Beginning Guide students to make a drawing of an organism's habitat and niche. For example, the drawing may be of a squirrel in a tree in the forest collecting and storing nuts. Have students label the diagram with the words *habitat* and *niche*.

Intermediate Tell students to make a drawing of an organism's habitat and niche. Have them use these sentence frames to describe it: The organism's habitat is ________. The organism's niche is ________.

Advanced Have students write sentences describing an organism's habitat and niche. *Example:* A coyote lives in the plains and eats small mammals.

Symbiotic Relationships

A close relationship between two or more species in which one or both of the organisms benefits is called **symbiosis**. There are three types of symbiosis: mutualism, parasitism, and commensalism.

Mutualism In mutualism, both organisms benefit. The relationship between an anemonefish and a sea anemone is an example of mutualism (**Figure 3-3**). The fish benefits by using the anemone's tentacles to hide from predators. The anemonefish also keeps the anemone and the area around it clean by eating dead tentacles. The fish's bright coloration lures other fish to the anemone, which the anemone eats. The anemone also gets nutrients from the anemonefish's waste.

SEA PIG SURVIVAL

Gather Evidence *There are few places to hide on the bottom of the ocean. To avoid predators, some types of crabs hitch rides across the ocean floor by clinging to the bottom of sea pigs. How would you categorize this type of symbiotic relationship? Explain your reasoning.*

FIGURE 3-3 A skunk anemonefish hides within the tentacles of a sea anemone in its Indian Ocean habitat. Its body is covered by a mucus that protects it from the sea anemone's stinging tentacles.

Parasitism In parasitism, one organism benefits and the other is harmed. Typically, parasites depend on their host for survival, so they will not kill it. Instead, parasites take what they need and generally cause sickness or some other harm. For example, tapeworms are parasitic worms that take up residence inside the intestines of pigs, cows, or humans, where they can live there for years. The tapeworms take vital nutrients and cause malnutrition in the host. Plants can also be parasites. Mistletoe is a flowering plant that attaches to the branch of a host tree. Its root-like appendages bore into the bark of the tree, where the mistletoe gets a portion of the water and nutrients it needs to survive (**Figure 3-4**).

FIGURE 3-4 When a mistletoe seed lands on a branch, it sends out "roots" that penetrate through the bark. This allows the mistletoe to absorb some of the tree's water and nutrients.

VIDEO

VIDEO 3-2 Go online to view the symbiotic relationships shown here.

Connect to English Language Arts

Integration of Knowledge and Ideas After reading about each type of symbiotic relationship, have students translate the information about each into a table or chart that will help them recall what they are. Tell students to make a chart with three rows. Then draw a symbol or image to represent each type of relationship that is easy to remember. For example, they may use the symbols +/+, +/-, and +/0 to represent mutualism, parasitism, and commensalism. Next to their symbol or image, have them write a brief description along with an example.

Video

Time: 1:11

Use **Video 3-2** to show students examples of different symbiotic relationships, including the mutualistic relationship between an anemonefish and a sea anemone, commensalism between a cattle egret and an African buffalo, and a parasitic relationship between mistletoe and a tree.

REVISIT THE ANCHORING PHENOMENON

Gather Evidence *This is an example of commensalism because the crab benefits from the relationship, and the sea pig is unharmed.*

Since the crabs are getting protection from predators by clinging to the bottom of the sea pigs, they are benefitting from the relationship. The sea pigs receive no benefit from the crabs, and the crabs bring no harm to the sea pigs. This describes a commensal relationship.

In Your Community

Symbiotic Relationships Students may think that symbiotic relationships are only found among exotic and unusual animals living in the wild. The fact is that these relationships are found close to home. Have students think of examples of parasitism, mutualism, and commensalism that can be observed in their home or community. Some students will be familiar with parasitic relationships, such as fleas or ticks on pets. Bees and flowers are an example of mutualism, as the benefit is to both, and burdock seeds on animals or birds nesting in trees can be classified as commensalism.

SEP Construct an Explanation *Sample answer: If there isn't another species that can fill the same ecological niche, the population of the other species in the relationship may also decline. This could affect the overall food web in the ecosystem and, in turn, affect the stability of the ecosystem.*

FIGURE 3-5 Cattle egrets and African buffalo are common sights in Kenya's Maasai Mara National Reserve.

Commensalism In commensalism, one of the organisms benefits from the relationship, while the other receives no benefit but is also not harmed. One example of commensalism is the relationship between African buffalo and cattle egrets. The cattle egrets eat the insects that are kicked up as the African buffalo grazes (**Figure 3-5**).

SEP Construct an Explanation *How might ecosystem stability be affected if one member of a symbiotic relationship was removed from the environment?*

3.1 REVIEW

1. **Distinguish** Use an example organism such as a bird, a fox, or a plant, to demonstrate the differences between an ecosystem, a habitat, and a niche.
2. **Predict** Which of these scenarios will increase competition between organisms in a community? Select all correct answers.
   - A. A new apex predator moves into the area.
   - B. A disease wipes out half the primary consumers.
   - C. Increased precipitation boosts vegetation growth.
   - D. Half the area is deforested for human development.
   - E. The number of consumers born each year equals the number that die each year.
3. **Classify** Categorize each symbiotic relationship as an example of mutualism, commensalism, or parasitism.
   - A. Birds eat disease-carrying pests such as ticks from the backs of rhinoceroses.
   - B. Some fish pick up scraps of food left behind as stingrays hunt for food.
   - C. Certain orchid species attach themselves harmlessly to the upper limbs of trees to obtain sunlight.
   - D. Fleas living off the blood of lions can cause illness in their hosts.
   - E. Ants eating peony flower nectar protect the plants from other herbivorous insects.

3.1 REVIEW

1. *Sample answer: A red fox's niche is as a predator that feeds on fish, small mammals, and fruits. Foxes help control rodent populations and disperse seeds from fruit. A fox's habitat can include riverbanks, mountains, meadows, and woodlands where it can dig burrows and hunt for food. Foxes exist in ecosystems that include other organisms, such as trees, fruit-bearing plants, small mammals, and fish, as well as abiotic factors, such as sunlight, water, and soil.* **DOK 2**
2. *A, B, D* **DOK 2**
3. *Mutualism: A, E; Commensalism: B, C; Parasitism: D* **DOK 2**

Dr. Federico Kacoliris is working to prevent critically endangered species in the Patagonian region of Argentina from going extinct.

The El Rincon stream frog (*Pleurodema somuncurense*) is found only in the headwaters of a single stream in Argentina. Threats to its already-small population include predation by non-native rainbow trout, damming, and habitat destruction by livestock. Biologist Dr. Federico Kacoliris is working to bring this critically endangered species back from the brink of extinction.

On Patagonia's Somuncura Plateau, temperatures frequently dip below freezing and the wind can whip by at speeds up to 60 km/hr. At first glance, this region does not seem to be ideal habitat for a frog species. The secret to the El Rincon stream frog's survival? It lives in hot springs, where the water temperature never drops below 18 °C (64 °F). However, given the species' precise habitat requirements and threats by outside forces, its range is limited to an area less than 5 km^2.

While the stream is located within the Somuncura Provincial Reserve, until 2012 the area was not actively managed. In 2013, Kacoliris established the Somuncura Foundation, with the goal of protecting the El Rincon stream frog and its habitat.

As part of the species' recovery plan, Kacoliris and a team of researchers collected 40 adult frogs from the wild and established a captive breeding population at La Plata Museum in Buenos Aires, Argentina. To protect the frogs' habitat, they built enclosures to exclude livestock from the stream and removed a small dam to restore water flow. Efforts were also made to educate the local people about the frog and to encourage them to help protect its habitat. In March 2017, 146 juvenile frogs and 50 tadpoles were released back into their native habitat. The following year, an additional 50 juveniles were released.

"My most rewarding experience was to have the opportunity of helping this species go home," Kacoliris said. "The releasing event was shared with children and teachers from a local school. A time after, we confirmed the frogs became established there. Having achieved a healthy habitat with frogs again, and sharing this action with children from the local school was a magic moment that I will always remember."

THINKING CRITICALLY

Argue From Evidence *Why was habitat restoration an important part of the El Rincon stream frog's species recovery plan?*

DIFFERENTIATED INSTRUCTION | Leveled Support

Struggling Students For students struggling with the concept of habitat destruction and recovery efforts for the El Rincon stream frog, have them look for local examples of habitat loss. Ask them to record what caused the habitat loss, such as clearing land for building, and what organisms were affected. They may then work in pairs to write their own species recovery plan.

Advanced Learners For students who easily grasp the concepts discussed here, assign them the role of an investigative reporter. Have them work in groups to develop a list of questions that they would like to pose to Dr. Kacoliris about his work, his career, or other topics related to what they are learning in this chapter about interactions and relationships in ecosystems.

EXPLORER
FEDERICO KACOLIRIS

About the Explorer

National Geographic Explorer Dr. Federico Kacoliris works to save endangered frog species from extinction. His work includes participation in the development of the Amphibians Rescue Center in Argentina, where he worked with captive breeding populations for reintroduction into their natural habitats. Part of his work is to identify, manage, and protect these habitats from disturbances that initially threatened the frogs. You can learn more about Federico Kacoliris and his work on the National Geographic Society's website: www.nationalgeographic.org.

Connect to Careers

Conservation Scientist Around the world, populations of amphibians are on the decline. The El Rincon stream frog is an endemic species that exists in only one geographic region. As such, they have unique characteristics that impact the survival of other species within that ecosystem. Conservation scientists who specialize in protecting these species are critical to their survival. These scientists also work in many other areas as they protect threatened or endangered species and their habitats. They may restore damaged natural areas or develop studies on the environmental impact of activities such as logging or fishing. Many universities offer programs that prepare students for work in this specialty while also working in conjunction with scientists in the field. Scientists may work in other settings, such as universities, museums, non-profit organizations, and conservation programs.

THINKING CRITICALLY

Argue From Evidence *Sample answer: If the habitat were not restored and plans not put in place to reduce the threats to the habitat, any returned population of frogs would still face the same threats and likely not have a chance of reestablishing a viable population.*

3.2

BIODIVERSITY

LS2.C, LS4.C, LS4.D

EXPLORE/EXPLAIN

This section defines biodiversity and the three levels at which it is measured, introduces the concept of measuring biodiversity within different communities, and discusses several different biodiversity hotspots. The section then introduces students to keystone species and their effect on the species found within communities.

Objectives

- Explain the three levels at which biodiversity is measured.
- Distinguish between species richness and species evenness.
- Identify the criteria for designation as a biodiversity hotspot.
- Explain how keystone species differ from other species in a community.

Pretest Use **Question 3** to identify gaps in background knowledge or misconceptions.

Vocabulary Strategy

Prefixes/Suffixes/Root Relationships Review with students that the prefix *bio-* means "life," and the word *biodiversity* is actually the short form of two words: *biological* and *diversity*. Tell students that *diversity* is the noun form of the adjective *diverse*. Ask for some examples of how the word *diverse* may be used. Guide them to recognize that the word means "of different kinds" or representing more than one group.

Explain *Students might say that areas with high numbers of species might have conditions that support a long growing season and have a large number of ecological niches that species can fill.*

SEP Nature of Science *Sample answer: Initially, scientists thought that bacteria made up a small amount of Earth's biodiversity, but more recent research indicates that bacteria are the major life form on Earth.*

3.2

KEY TERMS

biodiversity
keystone species

BIODIVERSITY

A trek through a tropical rainforest engages all of your senses. The lush forest is full of the aromas of fragrant flowers, decaying leaf litter, and wet soil. The air is so thick with humidity you can almost taste it. You hear the loud calls of howler monkeys, alerting one another that trespassers are near. You see a flash of red high above as scarlet macaws soar through the forest canopy. One thing is for sure: the tropical rainforest is full of life.

Explain *Why do you think some ecosystems have a greater variety of species than others?*

VIDEO

VIDEO 3-3 Go online to find out more about biodiversity from National Geographic Explorer Dr. Pablo Garcia Borboroglu.

Defining Biodiversity

Biodiversity refers to the variety and variability of life on Earth. A region's biodiversity is measured at three levels:

- Genetic diversity refers to the variety of genes contained within a species and among different species.
- Species diversity refers to all the differences between and within populations of species.
- Ecosystem diversity refers to all the different habitats, niches, communities, and ecological processes within a given location.

Biodiversity is the result of millions of years of evolution. Over the course of many generations, all the species that exist today evolved unique traits and behaviors that differentiate them from all other species. These species evolved in ways that allow them to inhabit different ecological niches.

Scientists estimate that there are 8.7 million eukaryotic species on Earth. And, if you add in prokaryotes, some scientists estimate the total number of species on Earth could top two billion (**Figure 3-6**).

FIGURE 3-6 The number of unknown species on Earth greatly outweighs the number of known species. Today, scientists think bacteria are the dominant life-form on Earth.

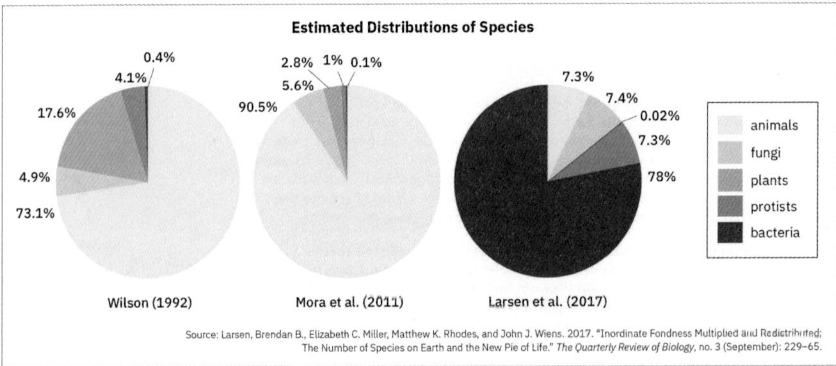

Source: Larsen, Brendan B., Elizabeth C. Miller, Matthew K. Rhodes, and John J. Wiens. 2017. "Inordinate Fondness Multiplied and Redistributed: The Number of Species on Earth and the New Pie of Life." *The Quarterly Review of Biology*, no. 3 (September): 229–65.

NOS Scientific Knowledge *Use the graphs to explain how scientists' understanding of the diversity of life on Earth has changed over time.*

66 CHAPTER 3 BIODIVERSITY AND ECOSYSTEM STABILITY

SCIENCE AND ENGINEERING PRACTICES
Engaging in Argument from Evidence

Influence of Technology Students should be aware of the idea that scientists often build on and improve upon the work of others as new insights and knowledge are gained. Technological advances over time in areas such as observing, collecting specimens, and genomic sequencing have enabled these new understandings. Students can see this idea in action in the graphs in **Figure 3-6** that show changes in data about the percentage of species on Earth. Have students compare and discuss the three graphs and where the changes occurred. They should analyze these differences and brainstorm ideas as to why such a significant change occurred over a relatively short time period, as well as how such new data would make a shift in thinking reasonable.

FIGURE 3-7
Red-shanked douc langur monkeys live in the tropical forest canopies of Vietnam and Laos.

Measuring Biodiversity

Communities can differ in species diversity even if they are of similar size. There are two components to diversity. The first, *species richness*, refers to the number of different species in a community. The second is *species evenness*, or the relative abundance of each species. For example, a pond that has five fish species in nearly equal numbers has a higher species diversity than a pond with one abundant fish species and four rare ones.

Community structure is dynamic, which means that in any community, the assortment of species and their relative abundances tend to change over time. Communities change over a long time span as they form and then age. For example, over a period of 100 years, changes in environmental conditions may cause an ecosystem to transition from a pond into a meadow. Ecosystems also change over the short term as a result of a sudden environmental change such as a wildfire or flood.

Each species can live only in a specific habitat (**Figure 3-7**). Thus, geography and climate affect community structure. Factors such as soil quality, sunlight intensity, rainfall, and temperature vary with latitude and elevation. Tropical (low latitude) regions receive the most energy from the sun, get ample rainfall, and have the most even temperature throughout the year. For the majority of plant and animal groups, the number of species is greatest near the Equator, and declines as you move toward the Poles. Tropical forest communities have more types of trees than temperate ones. Similarly, tropical reef communities are more diverse than marine communities found farther from the Equator.

CCC **Patterns** *Why do you think regions farthest from the Equator show the least amounts of biodiversity?*

CHAPTER INVESTIGATION

Measuring Biodiversity Using Ecological Sampling Methods
How can you measure biodiversity in a plant community?

Go online to explore this chapter's hands-on investigation about ecological sampling methods.

3.2 BIODIVERSITY **67**

Video

Time: 0:57

In **Video 3-3**, National Geographic Explorer Dr. Pablo Garcia Borboroglu defines biodiversity as a variety of different organisms living in a specific place. In this case, it is middle Patagonia where penguins, killer whales, right whales, and other animals live and interact together. As students view the video, ask them to note examples of different organisms and where they live. Discuss how these examples relate to Borboroglu's definition of biodiversity.

SEP **Patterns** *Sample response: Climate conditions in these regions do not support high rates of productivity and, therefore, can support only a small variety of organisms.*

CHAPTER INVESTIGATION A

Open Inquiry *Measuring Biodiversity Using Ecological Sampling Methods*

Time: 100 minutes over 2 days

Students will use two ecological sampling methods, quadrats and transects, to count plant species at a field site. They will then calculate the simplified diversity index and compare the two sampling methods.

Go online to access detailed teacher notes, answers, rubrics, and lab worksheets.

SCIENCE AND ENGINEERING PRACTICES
Analyzing and Interpreting Data

Diversity in Communities To develop a deeper understanding of species richness and species evenness, have students approach the concept quantitatively by analyzing data. Share the table below with students and discuss the differences between Habitat A and Habitat B.

Trees	Habitat A	Habitat B
pine species	550	1090
birch species	480	420
oak species	520	330

Using this data, have students complete the following sentence frame: Habitat A has greater species _______, and Habitat B has greater species _______. *(Answers: evenness / richness)*

Go Online INTERACTIVE FIGURE

World's Biodiversity Hotspots In the interactive version of **Figure 3-8**, students can find out more about the diversity of life in each of the 14 hotspots highlighted on the map.

Crosscurricular Connections

World Geography Have students practice their geography skills on a world map or globe. Provide the names of continents, major bodies of water, and countries on index cards. Working in small groups, students should draw a card from the box, locate it on the map, and tag it with a sticky note. As an option, you may choose to record how many locations each group could find in 10 minutes.

In Your Community

Zoo Conservation Programs Most students will not have experienced visiting a biodiversity hotspot; however, many will likely have visited a local zoo. Once considered solely as an attraction where families could spend the day viewing animals, zoos have taken on a much larger role in conserving biodiversity. For some species that are so severely threatened, the captive breeding programs at zoos may be their best chance at survival. Have students research the history and progression of zoos, including the work of William Hornaday, the planner for the Bronx Zoo and founder of the New York Zoological Society and, later, the Wildlife Conservation Society. Have students share what they find about the important contributions of zoos to conservation and public education.

SEP Evaluate Information *Sample response: A biodiversity hotspot has a relatively large number of species that are found nowhere else in the world and are threatened by human activities.*

Go Online INTERACTIVE FIGURE

FIGURE 3-8 Go online to find out more about what makes the **World's Biodiversity Hotspots** unique.

Biodiversity Hotspots

Biologists have identified biodiversity hotspots (**Figure 3-8**) as places rich in endemic species—that is, native species that exist nowhere else—that are threatened by humans. These hotspots, scattered throughout the world, make up about 2.5 percent of Earth's land but contain nearly half of all flowering plant species and more than a third of mammal, bird, reptile, and amphibian species. More than 20 percent of the global human population also lives within these regions. Biodiversity hotspots include

- **California Floristic Province** Located along the Pacific coast of North America, this hotspot is home to both giant sequoia and coastal redwood trees. The critically endangered California condor also lives here.
- **Atlantic Forest** Spread across portions of Brazil, Argentina, and Paraguay, this hotspot is home to 8000 endemic plant species.
- **Mountains of Southwest China** This hotspot is home to 12,000 plant species, 29 percent of which are found nowhere else in the world.
- **Madagascar** This island nation is home to a great diversity of chameleon species as well as hundreds of different lemur species, a unique primate found nowhere else in the world.

Conservation biologists and organizations have developed successful partnerships and educational programs with many countries where biodiversity hotspots exist. Economic solutions include fostering ecotourism and other alternative employment to relieve the human pressure on the native plants and animals found in these regions.

SEP Evaluate Information *What is a key characteristic of a biodiversity hotspot?*

FIGURE 3-8
Scientists have identified 36 biodiversity hotspots, 14 of which are identified here. To be designated as a hotspot, a region must have at least 1500 endemic plant species and have less than 30 percent of the original natural vegetation still intact.

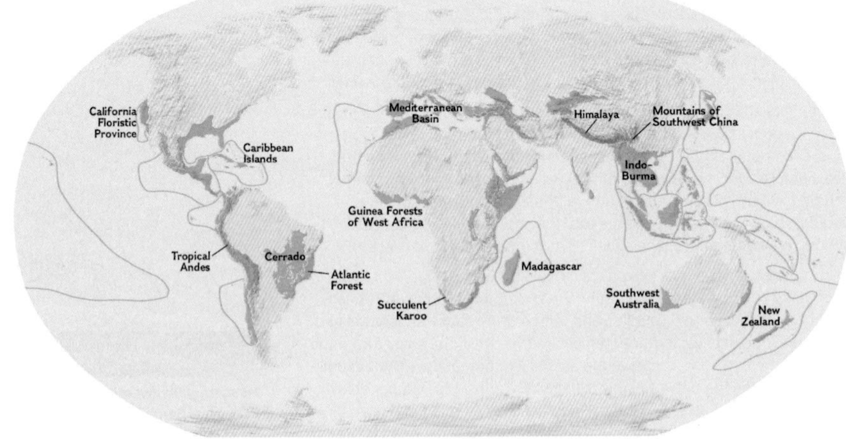

CROSSCUTTING CONCEPTS | Patterns

Biodiversity Hotspots Understanding how to identify patterns is essential to understanding and applying science concepts. Students have just learned about the connection between geography and climate in relation to biodiversity. Have students compare **Figure 3-8** to a globe or another world map and look for any patterns in the locations of biodiversity hotspots. About two-thirds of the world's species live in areas around the Equator where, with steady sunlight and plenty of precipitation, producers thrive in enough numbers to support a large number of species at higher trophic levels. However, quite a few hotspots lie outside the equatorial region. Have students research these hotspots to find out how they meet the criteria as a biodiversity hotspot and what other factors support species richness in these particular areas.

ON ASSIGNMENT National Geographic photographer Frans Lanting was one of the first photographers to document Madagascar's unique wildlife. To capture this photo, Lanting roamed along with troops of ring-tailed lemurs, fully immersing himself in their daily activities. The photographs Lanting sent back landed him his first major story for *National Geographic* magazine.

ON ASSIGNMENT

An Eye for the Elusive National Geographic photographer Frans Lanting has received many awards for his photography. In 2018, he was honored as the first recipient of the Wildlife Photographer of the Year competition's Lifetime Achievement Award. His photographic assignments have taken him across the globe, from the steamy Amazonian tropics to the frozen ice floes of Antarctica.

Lanting's work in Madagascar resulted in the first photographic images of the country's wildlife and tribal traditions to be shared with the world. Madagascar has mixed climates, including tropical rainforests, deserts, grasslands, and dry forests. For a photographer like Frans Lanting, Madagascar's landscape presents challenges as he tracks his subjects through all types of terrain, sometimes slithering on his stomach or following on foot to capture the perfect photograph of some of the planet's most elusive animals.

Frans Lanting has been described as having the "mind of a scientist, the heart of a hunter, and the eyes of a poet." You can learn more about Frans Lanting at the National Geographic Society's website: www.nationalgeographic.org/.

Connect to English Language Arts

Assess Reasoning and Evidence As a class, draw a simple food web made up of animals and plants. Have a class discussion about the food web and what happens when one part of the food web is disrupted or removed, which is that populations of one or more organisms within the web will be affected. Ask students to read the definition of a keystone species claiming that such organisms have a disproportionately large effect on a community, meaning a larger effect than those of other species in the food web. Then, ask if the text adequately supports this claim using the example of the beaver.

Vocabulary Strategy

Understanding in Context Show students an image of an architectural keystone. Ask students what might happen to the arch if the keystone was removed. (The arch would collapse.) Guide students toward developing a definition of keystone species based on this discussion.

SEP Construct an Explanation *Sample answer: A keystone species has an outsized impact on other species in its community.*

FIGURE 3-9 A beaver dam transforms a running stream into a calm pool of water in Tierra del Fuego, Argentina.

Keystone Species

A **keystone species** is a species that has a disproportionately large effect on a community relative to its size and abundance. For example, beavers are keystone species in some communities. These large, herbivorous rodents cut down trees by gnawing through their trunks. The beaver then uses the felled trees to build a dam, which forms a deep pool where a shallow stream would otherwise exist (**Figure 3-9**). By altering the physical conditions in a section of the stream, the beaver changes which species of fish and aquatic invertebrates can live there.

SEP Construct an Explanation *How is a keystone species different from other species in a community?*

3.2 REVIEW

1. **Identify** What are the characteristics of all biodiversity hotspots? Select all correct answers.
   A. They are dominated by invasive species.
   B. They have the greatest number of extinct species.
   C. They support a rich collection of endemic species.
   D. They exhibit extreme temperatures due to climate change.
   E. They provide habitats for organisms threatened by human activity.
2. **Describe** What is the relationship between biodiversity and ecological niche?
3. **Analyze Data** The table lists the number of different types of trees in three forests.

Type of tree	Forest 1	Forest 2	Forest 3
beech	10	10	14
pine	10	7	14
sycamore	0	3	14

Identify whether each forest has high *species richness*, high *species evenness*, or both.

3.2 REVIEW

1. *C, E* **DOK 1**
2. *Sample answer: The greater the number of ecological niches and the less overlap there is between them, the greater the biodiversity in an ecosystem.* **DOK 1**
3. *high species evenness: Forest 1; high species richness: Forest 2; both: Forest 3* **DOK 2**

LOOKING AT THE DATA

THE BIODIVERSITY CONSERVATION PARADOX

SEP **Analyze and Interpret Data** The overall biodiversity in New Zealand has increased over time despite an alarming species extinction rate—including the loss of nearly half its endemic bird species—since humans first settled there between 1200 and 1300 CE. Today, there are approximately 598 exotic animal species and 769 exotic plant species in New Zealand. These species were imported or migrated to the island.

The New Zealand Threat Classification System (NZTCS) database maintains a conservation assessment of native species. The classifications include threatened, at risk, not threatened, or data deficient. The data-deficient species have too little information available to determine their status. Some of these data are shown in **Figure 1**.

The endangered kea is a large parrot that lives in the mountains of New Zealand's South Island.

FIGURE 1
This bar graph shows the breakdown of select native organisms by conservation status.

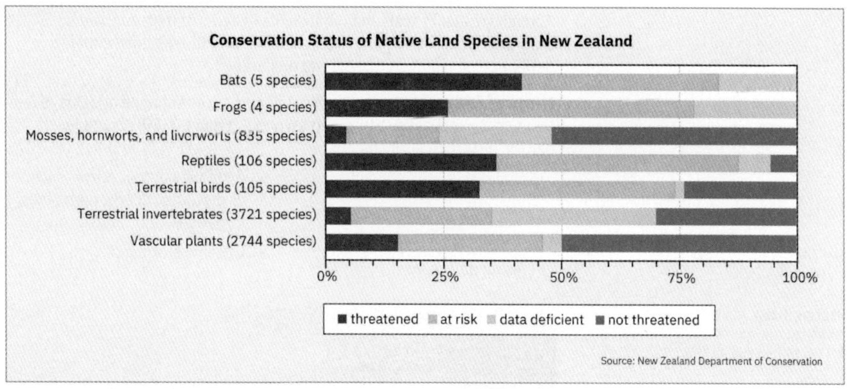

1. **Calculate** Use the data to estimate the current species richness of the endemic species in the study and exotic plants and animals in New Zealand.
2. **Estimate** The percentage of at-risk and threatened species can be used to estimate the number of species that need conservation efforts. For example, approximately four of five bat species and three of four frog species are threatened or at risk of extinction. Estimate the number of at-risk and threatened endemic species in the remaining categories.
3. **Calculate** Determine the ratio of the total number of at-risk and threatened endemic species studied to the total number of exotic species.
4. **Analyze Data** About 63 species have gone extinct since humans colonized New Zealand. How does the total number of currently at-risk and threatened species compare to the number of extinct species?
5. **Argue From Evidence** Use the data and your calculations to explain why biodiversity does not always reflect an ecosystem's overall health.

LOOKING AT THE DATA
THE BIODIVERSITY CONSERVATION PARADOX

Connect this activity to the Science and Engineering Practice of analyzing and interpreting data. Students use data to examine the conservation status of native organisms in New Zealand at one point in time. Students can compare the proportion of different types of organisms that are under threat and draw conclusions about the potential source of threats.

SAMPLE ANSWERS

1. *endemic species in study: 7520; exotic plants and animals: 1367 (598 + 769 from the introduction)* **DOK 2**
2. *mosses, hornworts, liverworts: 200 (.24 × 835); reptiles: 85 (.75 × 105); terrestrial birds: 79; terrestrial invertebrates: 1302 (.35 × 3721); vascular plants: 1235 (.45 × 2744)* **DOK 2**
3. *2:1 (4 + 3 + 299 + 85 + 79 + 1302 + 1235 = 2908; 2908 ÷ 1367 = 2)* **DOK 2**
4. *Sample answer: The number of at-risk and threatened species under study (2908) is 46 times greater than the number of extinct species (63).* **DOK 2**
5. *Sample answer: The biodiversity in New Zealand remains high due to the introduction of new species but at the cost of native species loss. At-risk endemic species outnumber the exotic species introduced to the island nation by a factor of two. The extinction rate has increased, as there are more than 40 times the number of species at risk of extinction as there are known extinct species. Invasive organisms compete for resources and introduce new predator-prey relationships that have disrupted ecosystems by eliminating endemic species.* **DOK 3**

3.3

ECOSYSTEM STABILITY AND CHANGE

LS2.C, LS4.D

EXPLORE/EXPLAIN

This section looks at ecological disturbances, both natural and human-caused, and the effects of disturbances on ecosystems. In discussing ecosystem stability, resilience and resistance are introduced, as well as ecological succession.

Objectives

- Recognize natural and human-caused ecological disturbances.
- Relate the effects of natural disturbance on ecosystems.
- Compare primary and secondary succession.

Pretest Use **Questions 4, 5, 7, and 8** to identify gaps in background knowledge or misconceptions.

Explain *Sample answer: Seeds might be blown in by wind or birds, and other small animals might drop seeds that are able to grow in the soil that accumulates in the cracks in the pavement.*

3.3

KEY TERMS

ecological disturbance
ecological succession
resilience
resistance

ECOSYSTEM STABILITY AND CHANGE

Have you ever noticed how an abandoned parking lot changes over time? At first, tiny plants begin to grow out of cracks in the pavement where new soil has accumulated. Eventually, grasses and small shrubs start growing there too. Plant roots widen existing cracks and form new ones, allowing more soil to build up and more plants to grow. If left alone for long enough, an entirely new community of organisms may become established in the once barren landscape.

Explain *How do you think new plants become established in an abandoned parking lot?*

Ecological Disturbances

An **ecological disturbance** refers to an event of intense environmental stress that occurs over a relatively short time period and causes a large change in the affected ecosystem. These disturbances can be natural or caused by humans. Disturbances can occur at different scales. For example, a tree that falls over in a forest is a relatively minor disturbance. Though smaller in scale than thousands of acres of trees toppled by a windstorm, the lone fallen tree still affects all the organisms that depended on it as a source of food or shelter.

Natural Disturbances A natural disturbance is change or damage to an ecosystem caused by a natural event (**Figure 3-10**). Examples of natural disturbances include

- avalanches and landslides
- fires
- disease epidemics
- insect outbreaks
- extreme weather events such as droughts, floods, hurricanes, tornadoes, and windstorms
- volcanic eruptions

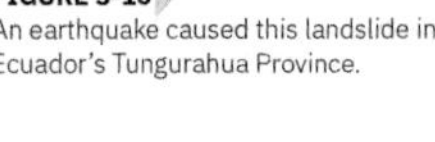

FIGURE 3-10 An earthquake caused this landslide in Ecuador's Tungurahua Province.

CROSSCUTTING CONCEPTS | Scale, Proportion, and Quantity

Disturbances and Scale The concept of scale is important to ecosystems, as they can exist in an infinite number of sizes. Ecosystems exist beneath fallen, rotting logs or even within small temporary vernal pools. Similarly, disturbances occur at different scales. An overflowing stream may be a small-scale change and easy to recover from with no long-term effects. Large-scale disturbances, such as wildfires, are widespread and ecosystems may take a long time to rebound.

Remind students that scale refers to things besides size, such as time. Guide a discussion in which students look at both natural and human-caused disturbances in terms of both size and time. Ask them to decide which would be more harmful, a small-scale disturbance such as paving a parking lot that will last many years or a volcanic eruption that happens quickly but recovery would start within a year.

FIGURE 3-11 Land clearing for human development projects is a major source of human-caused ecological disturbances.

Human-Caused Disturbances Humans are part of ecosystems. As such, human activities can also affect ecosystems in many ways, as shown in **Figure 3-11**. One way humans affect ecosystems is the introduction of non-native species. In the United States alone, humans have introduced approximately 50,000 non-native species. Not all non-native species are inherently "bad." In fact, many of the plant species we depend on as food sources are not native. For example, crop plants such as okra, millet, and yams all originated in Africa. Many insects, such as the European honeybee, perform an essential role in ecosystems as pollinators.

However, non-native species that outcompete native species can become a problem. These organisms are called *invasive species*. They wreak havoc in ecosystems by disrupting food webs and damaging habitats. They also cause trillions of dollars in economic damage. For example, invasive mussel species, which clog water intake pipes and distribution systems, are estimated to cause more than $1 billion in damages annually.

Other examples of human-caused ecological disturbances include

- habitat destruction that results from land clearing, such as for agriculture, mining, or urban development
- water, land, and air pollution
- overexploitation of species, such as by overhunting or overfishing vulnerable populations
- climate change

CCC Cause and Effect *How might human activities worsen the effects of a natural disturbance such as a hurricane?*

In Your Community

Zoning Laws Some students may not be aware of how different communities have zoning laws that regulate the use of land. Although these laws change from place to place, they determine what land can be used for residences, industry, commerce, or agriculture. They may also determine the scale at which changes can be made. Many communities are using zoning laws to preserve "green spaces" or other natural areas of a minimum size. Have students research their local zoning laws or other local laws intended to protect the natural environment.

Social-Emotional Learning

Social Awareness The image in **Figure 3-11** shows land that has been cleared for human development. While many students may support land conservation efforts and be opposed to land development, there may be others in the classroom with different opinions for a variety of reasons. This provides the opportunity to model and have students practice skills such as *taking others' perspectives* and *demonstrating empathy and compassion.*

CCC Cause and Effect *Sample answer: Destruction of wetlands and barrier islands as a result of land development by humans can result in more flooding and other storm-related damage from hurricanes and other severe storms.*

CASE STUDY
WHERE THERE'S SMOKE, THERE'S FIRE

Ask Questions To help students revisit the Case Study and make connections to the content, have students apply the intermediate disturbance hypothesis to what they know about the Yellowstone fire.

Connect to English Language Arts

Translate Quantitative or Technical Information The intermediate disturbance hypothesis is represented in the curved graphic model in **Figure 3-12**. This model can be used to describe and predict the impacts on biodiversity after different levels of environmental disturbance. Have student pairs use this model to describe when maximum species diversity is reached in an ecosystem.

CCC Patterns *Sample answer: Low-intensity disturbances result in a larger amount of competition, which leads to low diversity. High-intensity disturbances wipe out most organisms, so the only species present are colonizers, which also results in low diversity.*

CASE STUDY
WHERE THERE'S SMOKE
THERE'S FIRE

Ask Questions *Generate questions about how the intensity of a forest fire might affect ecosystem stability.*

Effects of Disturbance

The magnitude and frequency of disturbances affect communities. A variety of field studies support the *intermediate disturbance hypothesis*, which states that species richness is greatest when physical and biological disturbances are moderate in their intensity or frequency (**Figure 3-12**).

When disturbance is infrequent and of low intensity—meaning it does not remove many individual organisms—the most competitive species will exclude others, so diversity is low. By contrast, when disturbance occurs often or is of high intensity, most of the species present will be colonizers. With a moderate level of disturbance, the community will contain a mix of colonizer and mature species, and so will be most diverse. For example, a windstorm in a forest that knocks down a section of trees will result in a patch of younger plant species within a larger forest of mature trees.

FIGURE 3-12
According to the intermediate disturbance hypothesis, an ecosystem that experiences disturbances at a moderate frequency or intensity will have the highest biodiversity.

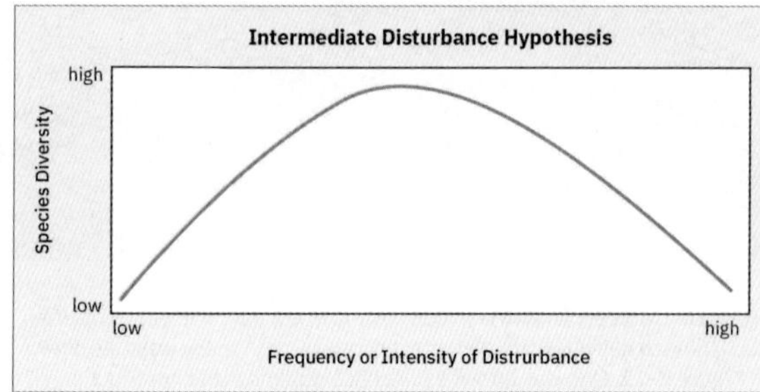

CCC Patterns *Why is species diversity greatest when ecological disturbances are moderate in intensity?*

Ecosystem Stability

Ecosystems are constantly changing. Environmental conditions, such as temperature, precipitation, and resource availability vary over time. Ecosystem stability refers to an ecosystem's ability to maintain consistent conditions. A stable ecosystem is able to withstand normal disturbances, maintaining a diverse population of species and supporting ecosystem functions such as the cycling of nutrients and energy. Two components of ecosystem stability are resilience and resistance.

- **Resilience** is the ability of an ecological community to regain its original structure and function following a disturbance. Ecosystems that are resilient show rapid recovery after a disturbance, quickly returning to their original state. Grasslands, which are adapted to frequent fire events, show quick recovery after a fire.
- **Resistance** is the ability of an ecological community to remain unchanged when affected by a disturbance. Ecosystems that are resistant show minimal change following a disturbance. A stable coral reef ecosystem with a high amount of biodiversity and minimal invasive species will likely show resistance after a hurricane, depending on the magnitude of the storm.

DIFFERENTIATED INSTRUCTION | Leveled Support

Struggling Students To help students understand **Figure 3-12**, have them work in groups to redraw the curve. Then, guide them in drawing pictures of organisms to represent each area of the graph. For example, at intermediate frequency and high species diversity, have them draw a mixed group of mature and immature trees. Similar drawings should be made each part of the graph. Students should then use their drawings to develop an explanation of how the magnitude and frequency of disturbances affect ecosystems.

Advanced Students For those students with a good understanding of **Figure 3-12**, have them work in groups to research studies that support it. One such study was conducted by Dr. Wayne Sousa. They should use the evidence to show how it supports the intermediate disturbance hypothesis and share their findings with the class.

Ecological Succession

The species composition of a community will change over time. Often, some species change the habitat in ways that allow others to come in and replace them. This type of change, which takes place over a long interval of time, is called **ecological succession**.

When the concept of ecological succession was first developed in the late 1800s, it was thought to be a predictable and directional process. Which species are present at each stage in succession was thought to be determined primarily by physical factors such as climate, altitude, and soil type. In this view, succession ends in a "climax community," a collection of species that does not change over time and will be rebuilt in the event of a disturbance. Ecologists now know that the species composition of a community changes in unpredictable ways. Communities do not journey along a well-worn path to a predetermined climax state. Random events determine the order in which species arrive in a habitat, and thus affect the course of succession.

Ecological succession begins with the arrival of *pioneer species*, which are species whose traits allow them to colonize new or newly vacated habitats. Pioneer species have an opportunistic life history. This means they grow and mature quickly, and they produce many offspring capable of dispersing. Later, other hardier and slower-growing species replace the pioneers. Then the replacements are replaced, and so on. There are two types of succession: primary succession and secondary succession.

CHAPTER INVESTIGATION

Ecological Succession in an Aquatic Environment
How do biotic and abiotic factors influence succession in a freshwater pond community?

Go online to explore this chapter's hands-on investigation about ecological succession.

CONNECTIONS

PRAIRIE RESTORATION Prior to the arrival of Europeans, the Great Plains and Upper Midwest were covered by immense grasslands. Nearly 445,000 square kilometers of what is now Minnesota, North Dakota, and South Dakota were home to prairie ecosystems. Fire is the key to maintaining grassland ecosystems and preventing succession from taking hold. Historically, these grasslands burned on a regular basis, either caused naturally by lightning strikes or set intentionally by Native Americans to produce fertile land for agricultural purposes and to clear undergrowth to facilitate hunting and travel.

Today, fire is an important part of prairie restoration plans. The goals of prairie restoration are to reconnect isolated patches of prairie, reduce invasive species, and improve habitat for wildlife and grazing animals. Prescribed burns are a method of land management. The resulting nutrient-rich ash acts as a fertilizer and supports new growth of native vegetation. Fire also reduces the accumulation of brush, shrubs, and trees.

Every prescribed burn has a scientific plan that land managers have prepared in advance. It describes the objective of the fire, the fuels that will be used, the size of the fire, the precise environmental conditions under which it will burn, and conditions under which the fire will be suppressed if necessary.

Workers on a fire crew use a drip torch to ignite the prairie into flames as part of a prescribed fire.

CHAPTER INVESTIGATION B

Open Inquiry *Ecological Succession in an Aquatic Environment*

Time: 250+ minutes over 2 days plus observation time over 3 weeks

Students will design their own procedure to investigate how a single abiotic or biotic factor could influence succession in a freshwater pond community. Students set up a model ecosystem and observe the changes over time, determining the effect of their chosen factor.

Go online to access detailed teacher notes, answers, rubrics, and lab worksheets.

Science Background

Pioneer Species Like early human settlers establishing settlements in an area, pioneer species are the first to colonize lifeless barren land, even those without soil. Lichens are a very common pioneer species. They are a symbiotic partnership of a fungus and an alga. Fungi are known for their ability to break down organic matter but cannot produce their own food, while algae are able to provide food through photosynthesis. Lichens are able to withstand harsh conditions and are found growing in places where few other organisms can survive, such as deserts or alpine tundra. As lichens grow, they produce an acid that breaks down rock. Over time, this broken down rock, mixed with dead lichens, produces soil that enables other plants, such as mosses, to become established. Thereafter, other plants and organisms become established.

CONNECTIONS | Prairie Restoration

In Your Community Some students may not be familiar with the history and contributions of Native Americans to sustainable land management practices. Long before land managers developed plans for prescriptive burns, Native Americans recognized the benefits of controlled fires to maintain grasslands and prevent their transformation into forests through succession. Have students research early land management practices used by Native Americans and how they relate to the sustenance of North American grasslands over thousands of years. Then, have students discuss the similarities and differences between those practices and the prescribed burns used today.

On the Map

Volcanic Islands in the South Pacific Hunga-Tonga-Hunga-Ha'apai formed in December 2014 following an explosive eruption of ash from an underwater volcano in the middle of the ocean. Have students locate the island on a map. The eruption's ejected material mixed with seawater, and within a month, a new island had formed. The speed at which the island formed was unusually fast. Within five years, it was home to flowering plants, sea birds, and even owls. Then, in January 2022, a massive volcanic eruption destroyed the island, leaving only two small landmasses separated by the sea. Have students learn more about this island and, working in groups, draw a series of 4–5 frames showing how primary succession could progress over time on the remaining volcanic islands.

FIGURE 3-13 Small plants have begun to colonize cracks that have formed within hardened lava on the Big Island of Hawaii.

Primary Succession This type of succession takes place in a barren habitat that lacks soil. The earliest pioneers to colonize such environments are often mosses and lichens, which are small, have a brief life cycle, and can tolerate intense sunlight, extreme temperature changes, and little or no soil. Some hardy annual flowering plants with wind-dispersed seeds, such as fireweed, are also frequent pioneers.

Primary succession occurs as land is exposed by the retreat of a glacier, on a newly formed volcanic island, or in a region where volcanic material has buried existing soil. On the Big Island of Hawaii, eruptions by the active volcano Kilauea provide a model example of primary succession. In 2018, volcanic activity added 3.5 km² of new land to the Big Island as the lava flowed offshore into the ocean. Additionally, 36 km² of existing land was covered by molten lava. Once the lava has cooled and hardened, the stage is set for primary succession to begin (**Figure 3-13**).

Most plant life is dependent on soil, which provides plants with the nutrients they need to grow and survive. Moss and lichen roots help to break apart lava and rock. When they die and decompose, nutrients are added to the soil. Many pioneer species have a symbiotic relationship with nitrogen-fixing bacteria that lets them grow in nitrogen-poor habitats. Seeds of later species can take root inside mats of mosses and lichens. Over time, organic wastes and remains accumulate and, by adding volume and nutrients to soil, help other species take hold. Later successional species often shade and eventually displace earlier ones. Typically, over a long period of time, smaller plants are replaced by shrubs and bushes, which are then replaced by trees and other large plants. As the plant community grows, animals begin to colonize the area, which helps to spread seeds and allows the plant community to continue to grow.

DIFFERENTIATED INSTRUCTION | English Language Learners

Paired Reading Working in pairs, have students take turns reading about primary and secondary succession.

Beginning As each paragraph is read, encourage students to look for words that are cognates, such as *primary* and *primario* or *secondary* and *secundario*.

Intermediate Provide copies of a blank Venn diagram, or draw one on the board for students to copy and use to diagram their comparison.

Advanced Have students first discuss with their partners the similarities and differences and then share with the class.

Address Misconceptions

Forest Succession When learning about secondary succession, students may think that all areas recover in the same way. Secondary succession in some forests occurs relatively quickly because soil and seeds are in place and, in some situations, nutrients may have been added back to the soil. However, sometimes a disturbance completely destroys a community, causing loss of soil nutrients. When rainforests are clear cut or slashed and burned, natural succession may not be possible.

CCC Stability and Change *Sample response: In primary succession, there is no soil present. In secondary succession, the soil is intact.*

Secondary Succession In this type of succession, a disturbed area within a community recovers. It commonly occurs in abandoned agricultural fields and burned forests (**Figure 3-14**). Because improved soil is present from the start, secondary succession usually occurs faster than primary succession. While primary succession can occur over a period of hundreds or even thousands of years, secondary succession can occur over a period of just 50 years.

FIGURE 3-14 Plants are beginning to regrow as a part of the secondary succession process following a forest fire in Ontario, Canada.

CCC Stability and Change *What is the main difference between primary and secondary succession?*

3.3 REVIEW

1. **Classify** Sort the ecological disturbances into natural and human-caused changes.
   - Water is contaminated by toxic chemicals near a hydraulic drilling site.
   - A prairie is converted to agricultural land.
   - A volcanic eruption forces the evacuation of nearby communities.
   - The emergence of cicadas in 15 states leaves exoskeletons on trees.
2. **Compare** How are the two components of ecosystem stability similar and different?
3. **Contrast** Which condition most determines whether succession is primary or secondary?
   A. lack of vegetation
   B. intensity of sunlight
   C. presence of soil
   D. human intervention
4. **Define** Which characteristics are ideal for a pioneer species? Select all correct answers.
   A. are mobile
   B. grow slowly
   C. photosynthetic
   D. resistant to toxins
   E. break rock into soil

3.3 ECOSYSTEM STABILITY AND CHANGE 77

3.3 REVIEW

1. *Human caused: Water is contaminated by toxic chemicals near a hydraulic drilling site. A prairie is converted to agricultural land. Natural: A volcanic eruption forces the evacuation of nearby communities. The emergence of cicadas in 15 states leaves exoskeletons on trees.* **DOK 2**
2. *Sample answer: The two components of ecosystem stability are ecosystem resistance and ecosystem resilience. Resistance is the ability to remain unchanged after an ecological disturbance. Resilience is the ability to recover from an ecological disturbance. Both resilience and resistance refer to how an ecosystem responds after a disturbance occurs.* **DOK 2**
3. *C* **DOK 2**
4. *C, E* **DOK 1**

MINILAB
OBSERVING BIODIVERSITY IN POND WATER

In this activity, students will use a microscope to observe some of the different microorganisms that are found in pond water and record their observations both as sketches and in words. They will use research materials to try to identify organisms they observed. Students will then share their results with other groups in the class and make a qualitative assessment about the level of biodiversity found in pond water.

Time: 30 minutes

Advance Preparation

- Gather pond water in a plastic container.
- Students sometimes have difficulty with the identification of organisms using a microscope. As an option, prepare a guide with drawings of different organisms they will be looking for, such as worms, algae, rotifers, or protozoa.

Procedure

In Steps 3 and 4, suggest that students use a Petri dish to outline four circles. Remind students to draw a different organism in each of the circles with labels to record the magnification, as well as the name of the organism.

Results and Analysis

1. **Summarize** *Answers will vary. Accept all reasonable observations and identifications.* **DOK 2**
2. **Compare** *Answers will vary. The most likely answer is that all students did not see the same organisms. This could be due to several factors, including different distributions of organisms with each water drop, difficulties locating and observing organisms using the microscope, and differences in identifying specific organisms.* **DOK 2**
3. **Analyze Data** *Specific answers will vary but should demonstrate an understanding that if multiple students or groups saw and identified the same organisms, those organisms are likely the most abundant, but if only one or two students observed a particular organism, then it is likely the least abundant.* **DOK 3**

MINILAB

OBSERVING BIODIVERSITY IN POND WATER

SEP Engage in Argument From Evidence What microorganisms are found in a pond ecosystem?

Pond water is filled with a diverse set of microscopic organisms, such as bacteria, protozoa, and algae. Pond water may also contain invertebrates, such as worms, crustaceans, and insects. In this activity, you will observe and identify some of the different microorganisms found in a sample of pond water.

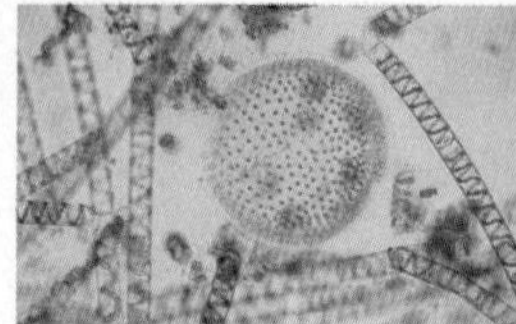

Common pond microorganisms include euglena, rotifer, and paramecium species.

Materials
- microscope
- microscope slide with depressed well
- coverslip
- pond water sample
- pipette or dropper
- pencil
- colored pencils (optional)
- computer or tablet with internet access

Safety  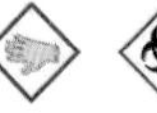    

Procedure

1. Obtain a sample of pond water from your teacher.
2. Place a drop or two of pond water in the well of the microscope slide. Then cover the slide with a coverslip. **CAUTION:** *Slides and coverslips are fragile, and their edges are sharp. Handle them carefully.*
3. Observe the slide using the low-power (4×) objective lens and scan for organisms. You are looking for tiny living organisms. They may look green, yellow-brown, or clear. Choose one to focus on and center it in your visual field.
4. Change your microscope to the medium (10×) or high power (40×) objective lenses to better view the organism. In your biology notebook, record a description of the organism and the magnification power you are using to view it. Do your best to draw the organism as you see it.
5. Repeat steps 3 and 4 until you have observed two to four different organisms.
6. Using internet resources, try to identify each organism you observed. Record the organism name next to the drawing in your biology notebook. If you cannot identify a specific organism, label it as "unidentified."
7. When you have completed the activity, dispose of the pond water as instructed by your teacher. Be sure to wash your hands with soap and water after you are done.

Results and Analysis

1. **Summarize** Make a summary of your observations and share your findings with the class.
2. **Compare** Compile a list of all the organisms observed by the class. Did you all observe the same organisms? How do you explain any differences?
3. **Analyze Data** Based on the class findings, which organisms were the most abundant? Which were the least abundant?
4. **Explain** What biotic and abiotic factors might influence the diversity of microorganisms found in pond water?
5. **Plan an Investigation** What changes could you make to the investigation to allow you to compare the biodiversity of different ponds?

4. **Explain** *In a typical pond ecosystem, abiotic factors that might influence the diversity of microorganisms include water, temperature/sunlight, salinity, nutrients, pH of soil, carbon dioxide, and oxygen. The biotic factors include such things as competition, predation, and symbiosis.* **DOK 3**
5. **Plan an Investigation** *Sample answer: I would collect pond water from two different sources. Then, I would examine several different samples from each pond water source and count the number of different types of organisms observed in each source. Finally, I would compare the numbers for the two different sources to determine which had the most biodiversity.* **DOK 4**

TYING IT ALL TOGETHER
WHERE THERE'S SMOKE THERE'S FIRE

HOW DO ECOSYSTEMS CHANGE?

In this chapter, you learned about the effects of ecological disturbances on ecosystem stability and change. The Case Study described the role fire plays in the Greater Yellowstone Ecosystem. The 1988 wildfire gave scientists the opportunity to study secondary succession as it has unfolded in real time.

Immediately after the fire, scientists made predictions about how the ecosystem might change over time. Some of these predictions were supported while others were only partially supported or did not occur at all. For example, scientists initially predicted that the abundance of native species would decrease within the burned zones and there would be a greater abundance of non-native species. However, they found that most native plant species reappeared within one to three years after the fire. And while populations of non-native species did increase in some areas, native plants were dominant in the burned zones.

Scientists made some surprising discoveries too. It was initially thought that aspens mainly reproduced from the sprouting roots of a parent tree. However, aspen seedlings were found to have established themselves throughout the burned conifer forests, far away from mature aspen stands. Genetic studies showed that these aspens had indeed grown from seeds and were not genetic clones.

Thirty years after the fire, as expected, lodgepole pines are again thriving in the region (**Figure 3-15**).

Research has shown that fire helps to increase diversity in a forest ecosystem by creating patches of new growth within older established forests (**Figure 3-16**). This dynamic is proving to be the case in Yellowstone National Park today.

In this activity, you will develop an argument to explain how an ecological disturbance such as a wildfire affects ecosystem stability and biodiversity. Work individually or in a group to complete these steps.

Gather Evidence

1. Research how the Greater Yellowstone Ecosystem has changed since the 1988 wildfire. How have its plant and animal communities changed over time? Have other fires impacted the park's ecosystems since 1988?

Scientific Knowledge Is Open to Revision in Light of New Evidence

2. How did a change in scientists' understanding of the role of fire in ecosystem function influence fire management policy in the United States?

Engage in Argument From Evidence

3. Use evidence from your research to evaluate the following claim: "Wildfire is a common ecological disturbance that affects biodiversity in Yellowstone National Park."

Extension Evaluate how climate change is affecting wildfire frequency and intensity. What effect does climate change have on ecosystem stability?

FIGURE 3-15
Lodgepole pine forest regeneration began soon after the Yellowstone fires of 1988.

FIGURE 3-16
Research indicates that biodiversity peaks around 25 years after a wildfire has occurred.

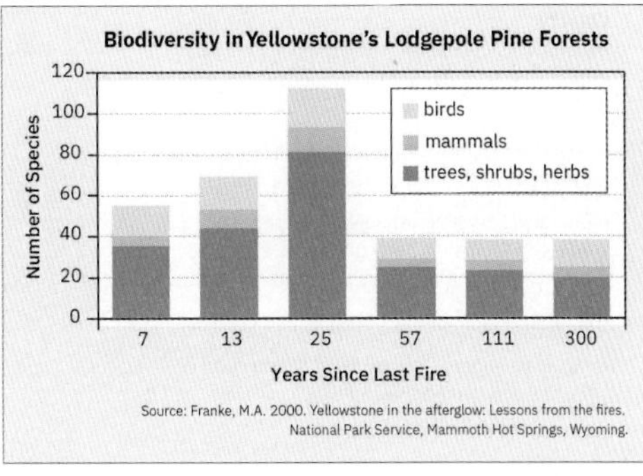

CROSSCUTTING CONCEPTS | Scale, Proportion, and Quantity

Some students may have visited Yellowstone National Park; however, even those students may not grasp the enormous scale of the 1988 wildfire. Tell students that the fire impacted 1.4 million acres of the park and discuss what that number means to them. Ask students to suggest comparisons to other areas that might be about the same size. To help them begin to recognize the scale of this area, tell them that the state of Delaware is about 1.6 million acres. Have them work in pairs to find out about the Desert National Wildlife Refuge in Southern Nevada, which is the largest wildlife refuge outside of Alaska.

TYING IT ALL TOGETHER

ELABORATE

The Case Study describes the role fire plays in the Greater Yellowstone Ecosystem and is a model for the study of secondary succession. In the Tying It All Together activity, students will develop an argument to explain how an ecological disturbance, such as the Yellowstone fire, affects ecosystem stability and biodiversity.

This activity is structured so that students may complete the steps as they study the chapter content. **Step 1** correlates with the chapter opener, as well as content in Sections 3.1 and 3.2, and content in Section 3.3 will be helpful in addressing **Step 2**. Students will then need to synthesize ideas from all sections and their research to complete **Step 3**.

Alternatively, this can be a culminating project at the end of the chapter. The gathering of evidence to support a claim or argument is a process that students may find useful in the Unit 1 Activity.

Go online to access the Student Self-Reflection and Teacher-Scoring rubrics for this activity.

Science Background

Biodiversity in Yellowstone's Lodgepole Pine Forest Historically, Yellowstone National Park has experienced severe fires every 100 to 300 years over the past 10,000 years. Within recorded history, trees have been killed, followed by a recovery. It seems as though Yellowstone's native plants and animals, as in other ecosystems, are well adapted to cycles of disturbance and recovery. As seen in **Figure 3-15**, lodgepole pine forests began regrowth shortly after the 1988 fires.

Ecological succession is the process through which an ecosystem moves toward the restoration of a stable community. Looking at the graph in **Figure 3-16**, it is possible to track the stages of regrowth, starting at year 7 after the fire, where trees, shrubs, and herbs began showing reestablishment, with birds and a few mammals returning. As plants become more established and productive, the trend is for more birds and mammals to return.

CHAPTER 3 REVIEW

REVIEW KEY CONCEPTS

1. *A* **DOK 1**
2. *A, B* **DOK 3**
3. *Mutualism: Algae avoid predators by living inside coral tissues and use photosynthesis to provide food for themselves and the corals; Commensalism: Barnacles attach themselves harmlessly to a humpback whale's body and receive a steady supply of food from the ocean as the whale swims about; Parasitism: Fleas live in a rat's fur and inject their mouthparts into the rat's skin to feed off the rat's blood; Cowbirds lay their eggs in other birds' nests.* **DOK 2**
4. *C, D, E* **DOK 2**
5. *D, E* **DOK 2**
6. *A, C, E* **DOK 1**
7. *Sample answer: As the number of niches in a natural area increases, the diversity of species would be expected to increase also. Each species has needs for resources that differ from other species. This allows many species to coexist without competing with one another for resources. Species' niches, therefore, differ from one another with each species requiring a unique niche. A greater number of niches in an area supports a greater variety of species.* **DOK 2**
8. *Sample answer: I agree with the claim because the grassland does not resist destruction by burning since many organisms are destroyed as it burns. However, resilience is demonstrated as grasses grow back quickly from roots that survived the fire. Grass growth is spurred by nutrients added to the soil during the fire and increased sunlight reaching the seedlings now that large trees and shrubs are gone.* **DOK 2**
9. *Sample answer: The field was undergoing secondary succession, as soil was already present. In this type of succession, soil is already present and pioneer species are not needed to make soil.* **DOK 3**

CHAPTER 3 REVIEW

REVIEW KEY CONCEPTS

1. **Contrast** Which statement correctly describes how a habitat differs from a niche?
   - **A.** A habitat is the place where an organism lives and a niche is the role filled by that organism along with its interactions within its community.
   - **B.** Habitats are the shelters used by animals for protection against predation and niches are the spaces plants use to put down roots for growth.
   - **C.** A habitat is the environment occupied by an entire ecosystem and niches are the individual environments occupied by organisms in that ecosystem.
   - **D.** Some organisms require large areas in which to live and these are called habitats. Other organisms require small areas and these are called niches.
2. **Analyze** Which ecological interactions are present in the community described below? Select all correct answers.

   Pitcher plants are carnivorous plants that grow in bogs, which are wet areas that have little water flow. Winged insects become trapped in the plants' long tube-like structures where they die and get absorbed by the plant. Bogs also provide habitat for frogs that feed on winged insects, snails, and slugs and grass snakes that feed on frogs, fish, and small mammals.
   - **A.** Pitcher plants are predators.
   - **B.** Frogs compete with pitcher plants.
   - **C.** Grass snakes and frogs are competitors.
   - **D.** Grass snakes have a symbiotic relationship with winged insects.
   - **E.** Winged insects have a symbiotic relationship with pitcher plants.
3. **Classify** Sort the organisms based on the given descriptions into three categories of symbiotic relationships: mutualism, commensalism, and parasitism.
   - Fleas live in a rat's fur and inject their mouthparts into the rat's skin to feed off the rat's blood.
   - Barnacles attach themselves harmlessly to a humpback whale's body and receive a steady supply of food from the ocean as the whale swims about.
   - Cowbirds lay their eggs in other birds' nests.
   - Algae avoid predators by living inside coral tissues and use photosynthesis to provide food for themselves and the corals.
4. **Explain** Which claims are true about biodiversity hotspots? Select all correct answers.
   - **A.** Human activities have little impact on biodiversity hotspots.
   - **B.** Biodiversity hotspots cover more than half of Earth's land surface.
   - **C.** Biodiversity hotspots contain species not found anywhere else on Earth.
   - **D.** Large numbers of species are threatened with extinction in biodiversity hotspots.
   - **E.** Educational programs are effective tools in reducing harm to biodiversity hotspots.
5. **Predict** Which characteristics would most likely be observed in an ecosystem that experiences a moderate level of disturbance? Select all correct answers.
   - **A.** nutrient-poor soil
   - **B.** lack of mature species
   - **C.** mostly colonizer species
   - **D.** relatively high species diversity
   - **E.** mix of colonizer and mature species
6. **Describe** What do pioneer species provide or make possible in an ecosystem? Select all correct answers.
   - **A.** organic matter
   - **B.** rocks
   - **C.** seeds
   - **D.** shade
   - **E.** soil
7. **Relate** How would you expect the diversity of species living in a natural area to be related to the number of niches present in that natural area? Explain.
8. **Explain** Grasslands can burn extensively after a lightning strike but return quickly to a stable state. An ecologist claims that these ecosystems are examples of resilience but not resistance. Explain whether you agree or disagree with this claim and why.
9. **Draw Conclusions** A plant ecologist observed changes in a field after it was no longer used for farming. Species A covered the field one year after abandonment. Species B replaced much of species A in the second year, and species C became the dominant plant species in the third year. What ecological process was the ecologist most likely observing? Explain your reasoning.

CRITICAL THINKING

10. *Sample answer: No. Two different species would compete for resources if they occupied the same niche and habitat in an ecosystem. Eventually one species would outcompete the other and drive it away.* **DOK 3**
11. *Sample answer: A keystone species supports many other species in the ecosystem. Therefore, if the keystone species begins to decline as the result of a disturbance, the entire ecosystem could be at risk of collapsing. Its well-being reflects the well-being of the entire ecosystem.* **DOK 3**
12. *Sample answer: The graph in Figure 3-12 shows that ecosystems that experience moderate disturbances have a high level of species diversity, while ecosystems with low- or high-intensity disturbances have low species diversity. The greater the species diversity, the more stable the ecosystem. Ecosystems that experience moderate disturbances are more stable than those that have low- or high-intensity disturbances.* **DOK 3**

CRITICAL THINKING

10. **Explain** Would it be possible for two different species to occupy the same niche and the same habitat in an ecosystem? Explain your answer.
11. **Synthesize** Why would it be important to track a keystone species in an ecosystem under threat of ecological disturbance?
12. **Argue From Evidence** Use evidence from **Figure 3-12** to support the idea that moderate disturbances actually enhance ecosystem stability.
13. **Explain** Why would it be unhealthy for an ecosystem to never undergo any disturbance of any kind?
14. NOS **Making Decisions** When making decisions about managing biodiversity hotspots, why is it important to take social and cultural contexts into consideration?

MATH AND ENGLISH LANGUAGE ARTS CONNECTIONS

Read the following passage. Use evidence from the passage to answer the question.

High biodiversity has been used as an indicator of ecosystem stability. Diversity provides more stability because diseases, invasive species, drought, fire, and other disturbances often harm certain species more than others. Having a diversity of species allows for greater chances of recovery after a disturbance.

1. **Cite Textual Evidence** How would you describe the stability of a forest with several different species but only a few members of each species? Cite information from the text to support your answer.

Read the following passages and examine the data table to answer questions 2 and 3.

An urban green space was divided into three zones of equal area. Species were counted weekly in each zone over a three-month period, then summed and divided by the number of weeks to produce the average values shown in the table.

Organisms	Zone A	Zone B	Zone C
large trees and tall shrubs	13.4	9.4	6
small plants	1.4	1	0.4
insects	8	6	2
birds	3.2	2.3	1
worms	2.2	2	1.3
mammals	1.3	0	0.2

Insect and bird diversity was higher in zones A and B but lower in zone C. This could be attributed to richness in large plant species within zones A and B in contrast to zone C. It has been shown that insects tend to inhabit areas of dense vegetation cover. Thick foliage provides protection from bird and mammalian predators. Birds also require large plant species and dense vegetation for nesting and protection. These results are consistent with large urban trees acting as keystone species that provide crucial habitat resources for many members of a single community.

2. **Evaluate a Science or Technical Text** Do the data support the conclusions presented in the passage regarding differences in biodiversity between zones? Explain your reasoning.
3. **Assess Reasoning and Evidence** Assess the extent to which the reasoning and evidence found in the passage support the claim that large tree species function as keystone species in urban green spaces.

REVISIT SEA PIG SURVIVAL

Gather Evidence In this chapter, you have seen how organisms interact together in different ecological relationships. In Section 1, you learned that sea pigs have a symbiotic relationship with some types of crabs. You also learned about ecosystem stability and change and the role of biodiversity in maintaining ecosystem stability.

1. How might a change in the stability of a deep-ocean ecosystem affect sea pig survival?
2. What other questions do you have about sea pigs? What else do you need to know to construct an explanation about how ecological disturbances affect their deep-ocean habitat?

REVISIT THE ANCHORING PHENOMENON

1. *Sample answer: If an ecological disturbance occurred in a deep-ocean ecosystem, the sea pig population would likely be adversely affected, especially if the disturbance reduced the number of food resources available to them.* **DOK 3**
2. *Sample answer: Does a deep-ocean ecosystem have low or high biodiversity? How often do ecological disturbances affect deep-ocean ecosystems?* **DOK 3**

13. *Sample answer: The most stable ecosystems are ones with high biodiversity. As succession proceeds over time, colonizer species are replaced by a uniform community of species. A certain amount of disturbance results in the formation of new habitats that support colonizer species, which increases the overall biodiversity of the ecosystem.* **DOK 3**
14. *Sample answer: Humans also live in and use resources from biodiversity hotspots. Considering social and cultural contexts helps to provide a balance between protecting the ecosystem and allowing for the livelihoods and well-being of human populations.* **DOK 3**

MATH AND ELA CONNECTIONS

1. *Sample answer: The forest would be fairly stable because there would be many different species to respond in different ways to a disturbance. According to the text, some species would survive while others would die, allowing the community to rebuild. However, having low populations of each species would be a weakness because there would be greater chances that an entire species could be wiped out. The forest would be more stable if it had larger populations of each species.* **DOK 3**
2. *Sample answer: The data support the conclusions. For every organism type except mammals, the biodiversity value was highest in zone A and lowest in zone C, with values for zone B in between. This indicates that a greater variety of species are typically found in zones A and B than in zone C.* **DOK 3**
3. *Sample answer: The data show a pattern that supports this claim. A similar pattern is seen across each of the three zones. As the diversity value for large trees and tall shrubs decreases, the diversity values for insects and birds also decrease. This suggests that trees are essential for birds and insects to survive in urban green spaces. However, this is a correlation and not proof of this claim. More research is needed to better test whether certain species of trees function as keystone species in this type of ecosystem.* **DOK 3**

POPULATION MEASUREMENT AND GROWTH

Three-Dimensional Learning

The practices, core ideas, and crosscutting concepts presented in this chapter's text, investigations, and resources provide support to address the following Performance Expectations: **HS-LS2-1**, **HS-LS2-2**, **HS-LS2-7**, and **HS-ETS1-3**.

Science and Engineering Practices	Disciplinary Core Ideas	Crosscutting Concepts
Analyzing and Interpreting Data Using Mathematics and Computational Thinking (HS-LS2-1, HS-LS2-2) Constructing Explanations and Designing Solutions (HS-LS2-7, HS-ETS1-3)	**LS2.A:** Interdependent Relationships in Ecosystems (HS-LS2-1, HS-LS2-2) **LS2.C:** Ecosystem Dynamics, Functioning, and Resilience (HS-LS2-2, HS-LS2-7) Developing Possible Solutions (HS-LS2-7, HS-ETS1-3)	Scale, Proportion, and Quantity (HS-LS2-1, HS-LS2-2) Science is a Human Edeavor Influence of Engineering, Technology, and Science on Society and the Natural World (HS-ETS1-3)

Contents		Instructional Support for All Learners	Digital Resources
ENGAGE			
82–83	CHAPTER OVERVIEW **CASE STUDY** What factors influence the size of a population?	**Social-Emotional Learning** CCC Scale, Proportion, and Quantity **On the Map** Coyote Range Changes **English Language Learners** Summarize	*Video 4-1*
EXPLORE/EXPLAIN			
84–87 DCI LS2.A LS2.C	**4.1** MEASURING POPULATIONS • Identify methods used by scientists to measure populations. • Describe how mathematical representations can be used to estimate populations without having to count each individual organism.	**Vocabulary Strategy** Graphic Organizer **English Language Learners** Make Comparisons CCC Scale, Proportion, and Quantity **In Your Community** Wildlife Studies **Connect to English Language Arts** Write Explanatory Text SEP Analyzing and Interpreting Data	**Virtual Investigation** Sea Pigs on the Abyssal Plain **Investigation B** Designing a Seed Trap (150 minutes over 3 days) *Video 4-2*
88	**MINILAB** MARK-RECAPTURE SAMPLING		
89–93 DCI LS2.A LS2.C	**4.2** MODELING POPULATION GROWTH PATTERNS • Calculate growth rates given data. • Explain how immigration and emigration affect population growth patterns. • Use mathematical and computational models to describe how populations grow in an environment. • Analyze survivorship curves to identify life history patterns of organisms.	**Connect to Mathematics** Use Units to Understand Problems SEP Using Mathematics and Computational Thinking **Address Misconceptions** Exponential Growth **Differentiated Instruction** Economically Disadvantaged Students **Connect to Mathematics** Use Units to Understand Problems SEP Analyzing and Interpreting Data **Connect to English Language Arts** Assessing Reasoning and Evidence **Differentiated Instruction** Leveled Support **Visual Support** Patterns of Survival	**SIMULATION** Modeling Population Changes

Contents		Instructional Support for All Learners	Digital Resources
EXPLORE/EXPLAIN			
94	EXPLORER HEATHER J. LYNCH	**Connect to Careers** Quantitative Ecologist **English Language Learners** Ask and Answer Questions	Video 4-3
95–101 DCI LS2.A LS2.C	**4.3 FACTORS THAT LIMIT POPULATION GROWTH** • Explain how limiting factors constrain the carrying capacity of an environment. • Explain how density-dependent factors can slow population growth rates. • Understand how density-independent factors differ from density-dependent factors and how they affect populations.	**Vocabulary Strategy** Categorize and Label **CASE STUDY** Ask Questions **Crosscurricular Connections** Algebra I **Differentiated Instruction** Leveled Support **Visual Support** Double-Line Graphs **Connect to Mathematics** Write a Function **CER** Revisit the Anchoring Phenomenon **BIOTECHNOLOGY FOCUS** Aquatic DNA Sampling **On Assignment** A Changing Earth **English Language Learners** Use Dialogue to Argue	**Investigation A** Population Growth of Duckweed (190 minutes over 2 days plus observation time over 3 weeks) **SIMULATION** Modeling Population Changes
102	**LOOKING AT THE DATA** INVASIVE SPECIES POPULATION GROWTH		
ELABORATE			
103	**TYING IT ALL TOGETHER** What factors influence the size of a population?	**Social-Emotional Learning** Social Awareness and Responsible Decision-Making **SEP** Constructing Explanations and Designing Solutions	Self-Reflection and Scoring Rubrics
Online	**Investigation A** Designing a Seed Trap **Investigation B** Population Growth of Duckweed	Guided Inquiry (190 minutes over 2 days plus observation time over 3 weeks) Open Inquiry (150 minutes over 3 days)	MINDTAP Access all your online assessment and laboratory support on MindTap, including: sample answers, lab guides, rubrics, and worksheets. PDFs are available from the Course Companion Site.
EVALUATE			
104–105	**Chapter 4 Review**		
Online	**Chapter 4 Assessment** **Performance Task 5** *What is the Best Way to Restore Habitat for Endangered Bats?* (HS-LS2-7, HS-ETS1-3)		

CHAPTER 4

ENGAGE

In middle school, students are likely to have studied living and nonliving factors that affect biodiversity and populations in ecosystems. This chapter expands on that prior knowledge by exploring different methods used to measure populations and to model population growth patterns to learn how factors affect ecosystems at different scales.

Students will learn more about how scientists measure populations and population densities, as well as the models used to explore population growth. Students will also explore the density-dependent factors and density-independent factors that affect population growth.

About the Photo This photo is of the ruins of the Temple of Zeus outside of the city of Agrigento, Sicily. Humans have often settled in regions where their needs for food could be met, whether around rich agricultural soils or near rivers and other bodies of water that held fish. To prepare students for this chapter's content, ask them to use scientific terminology that they already know to discuss how the availability of resources affects where people live.

Social-Emotional Learning

As students progress through the chapter, invite them to look for contexts and opportunities to practice some of the five social and emotional competencies. For example, as students participate in discussions about the content and the questions provided, support them in developing **relationship skills**. Help them practice effective communication and to collaborate while answering questions and solving problems.

In addition, when students learn about how human-caused events can affect ecosystems (e.g., in "Sharing Our Cities" and "Density-Independent Factors"), remind them to practice **social awareness** and **responsible decision-making**. Support students in managing their emotions, and challenge them to be curious and open-minded as they consider the role of humans in nature.

CHAPTER 4

POPULATION MEASUREMENT AND GROWTH

The city of Agrigento, Sicily, overlooks the Temple of Olympian Zeus. Rich agricultural soils and sulfur mines have supported the city since it was founded in 580 B.C.E. by Greek colonists.

A study of human history shows that for many thousands of years, humans have moved across the globe to gain access to the natural, societal, and technological resources that sustain them. Ecologists find that the same is true of any species in an ecosystem. As the availability of resources in a given environment changes, the number of individuals the environment can support changes as well.

Chapter 4 supports the NGSS Performance Expectations **HS-LS2-1** and **HS-LS2-2**.

82 CHAPTER 4 POPULATION MEASUREMENT AND GROWTH

CROSSCUTTING CONCEPTS | Scale, Proportion, and Quantity

Sampling Populations In this chapter, students apply the concepts of scale and proportion as they examine how sampling done in a smaller area can be used to model population size and population growth patterns over a larger area. Some fields of biology, especially conservation biology and environmental biology, study how factors affect populations and allow for the growth and stabilization of different population sizes. Chapter 4 in Unit 1 of this course focuses on how orders of magnitude can be used to extrapolate population densities using smaller sample sizes. You can further reinforce this crosscutting concept in Unit 2 by helping students understand how chemical reactions in the cell are generally sampled at very small scales but that collected data can be used to predict how chemical reactions occur at a larger scale in an organism.

CASE STUDY
SHARING OUR CITIES

WHAT FACTORS INFLUENCE THE SIZE OF A POPULATION?

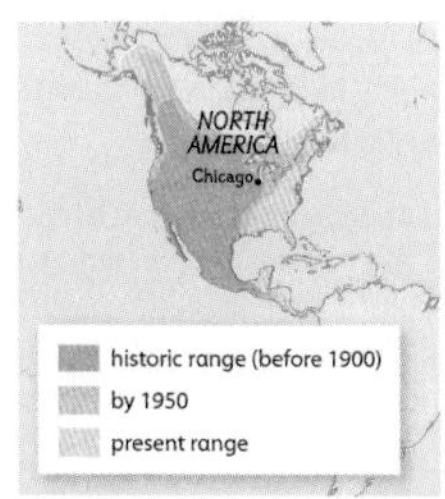

The city of Chicago, Illinois, and its surrounding suburbs form one of the most populous metropolitan areas in the United States. Along with more than five million human residents, the region hosts a population of nearly 4000 coyotes. Hundreds of coyotes live within the city of Chicago itself, often making dens in public parks and cemeteries. They have also been spotted walking down residential streets and past upscale hotels and shops in the downtown shopping district. Researchers even tracked a mating pair of coyotes to the parking lot of Soldier Field Stadium, where they had built a den in the concrete for their litter of five pups.

Coyotes are native to North American prairies and deserts, which provide open spaces for hunting their preferred prey of rabbits, rodents, and other small animals. While humans have disturbed these natural habitats, human activity has also contributed to expanding the coyote's range in many ways. Despite having been perceived as a pest species and targeted for eradication over the past century, coyote populations have thrived, spreading eastward to the Atlantic coast and as far south as Panama. Urban development is thought to encourage coyote habitat expansion by clearing forested areas for agricultural use, removing larger predators such as wolves and cougars, and providing resource-rich suburban landscapes.

Chicago is not the only major U.S. city where coyotes live successfully. Thousands of coyote sightings have also been reported in Washington, D.C., and Denver, Colorado, in recent years. To help coyote and human populations coexist peacefully, wildlife ecologists have closely studied coyote behavior in these urban environments. They have found that city and suburban coyotes are larger and live longer than rural coyotes. Near dense human populations, coyotes use wooded habitats and are more active at night in order to avoid contact with people (**Figure 4-1**). For the most part, urban coyotes continue to hunt their natural prey species and do not consume human-sourced food. As top predators, they perform an essential ecological role by controlling rodent, deer, and Canada goose populations in areas inhabited by humans.

Ask Questions *In this chapter, you will learn how various factors influence organism populations. Generate questions about the effects of human activity on populations of native species.*

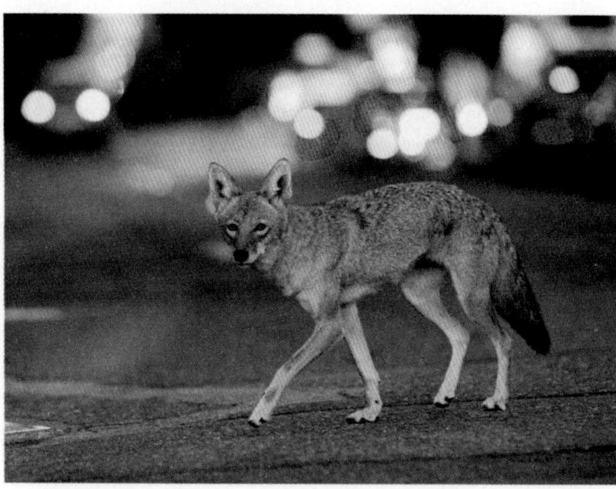

FIGURE 4-1 Shifting its activities to the evening hours helps minimize human interactions for this urban-dwelling coyote.

VIDEO 4-1 Go online to see the streets of Chicago from a coyote's point of view, as filmed for National Geographic Explorer Dr. Stanley Gehrt's urban wildlife studies.

DIFFERENTIATED INSTRUCTION | English Language Learners

Summarize Have students work in pairs. Assign each pair one paragraph from the Case Study to read together. Ask each pair to summarize their paragraph using sentence frames like the ones provided. If students struggle, suggest that they use a graphic organizer to help them find the main idea.

Beginning This is about ________.

Intermediate This paragraph tells how/what ________.

Advanced The main focus of this paragraph is ________. These details support the main focus: ________.

CASE STUDY

ENGAGE

The situation described in the Case Study can be used to assess prior knowledge of how human activities affect organism populations. Elicit how human population growth has affected organisms. Assess prior understanding of how the concepts of population size, population density, and range differ and how these are related. In the Tying it All Together activity, students will research and evaluate solutions for conserving populations of native species within an urban community.

Ask Questions Students should revisit the Case Study as they read the chapter to make connections with the content. See a specific suggestion in Section 4.3.

On the Map

Coyote Range Changes Use the map of North America to elicit discussion about how the range of coyotes has changed over the last 200 years. Have students engage with prior knowledge by asking them to describe the coyote's historical range of prairies and deserts and their present ranges of the more densely human-populated areas east and south.

Human Geography Have students compare changes in human population densities across North America over the past few centuries to the shift in the range of the coyote population. They may recognize that in general, although human populations have expanded westward, coyote populations have move eastward toward urban areas.

Video

Time: 0:59

Use **Video 4-1** to show the coyote's perspective of the urban environment. As part of the Urban Coyote Project, National Geographic Explorer Stanley Gehrt's research team attached Crittercam collars to coyotes living in and around Chicago to track their movements and obtain video footage of their activities. Discuss with students how coyote populations may be affected by human development. Elicit discussion of both positive and negative impacts of human development on coyote populations.

4.1

MEASURING POPULATIONS

LS2.A, LS2.C

EXPLORE/EXPLAIN

This section provides an overview of how scientists characterize populations of organisms in nature, including the concepts of population size, density, and distribution patterns.

Objectives

- Identify methods used by scientists to measure populations.
- Describe how mathematical representations can be used to estimate populations without having to count each individual organism.

Pretest Use **Question 1** to identify gaps in background knowledge or misconceptions.

Vocabulary Strategy

Graphic Organizer Students may have studied *population density* and *population size* in previous science or social studies courses; however, there are likely other terms in the lesson they are unfamiliar with. Suggest that students use a Frayer model graphic organizer to keep track of the definition, characteristics, examples, and non-examples of each term as they work through the lesson.

Compare and Contrast *Sample answer: Birds make nests, so you could count the number of nests and observe how many birds use each nest as you would count members of human households. Birds cannot fill out questionnaires (or tell you that they've already done so), so you would need a way to distinguish between individual birds to make sure to not count the same bird multiple times as they move around the park.*

4.1 MEASURING POPULATIONS

KEY TERMS

- mark–recapture sampling
- plot sampling
- population density
- population distribution
- population size
- range

By law, the United States government conducts a census every ten years to measure the population of each community in the nation. To do so, the Census Bureau builds an accurate list of housing unit addresses and asks each household to respond to a short questionnaire about its occupants. This population data is used to determine how to apportion seats in the House of Representatives, draw Congressional districts, and distribute federal funding. According to the 2020 census, there were 331,449,281 residents of the United States in that year.

Compare and Contrast *Suppose you want to count the number of birds in a park. How would this procedure be similar to a census? How would it be different?*

Counting Organisms

A *population* is a group of organisms of the same species that interbreed with one another. Population ecologists are scientists who investigate the factors that affect populations. Data from population ecology studies can be used to make decisions about how to manage a species and to assess the effects of human activities on wild species.

Population size is the total number of individuals in a population. It is often impractical to count all members of a population, so biologists frequently use sampling techniques to estimate population size.

Mark–recapture sampling is used to estimate the population size of mobile animals. With this technique, animals are captured, marked with a unique identifier of some sort, and released (**Figure 4-2**). At a later time, scientists capture another group of individuals from the same population. The proportion of marked animals in the second sample is taken to be representative of the proportion marked in the population as a whole.

FIGURE 4-2 To study the migration of a population of rare Kirtland's warblers, scientists attached tiny radio tags to birds in their winter home in the Bahamas. They then used radio telemetry to detect and track individual birds' movements in their Michigan breeding grounds.

DIFFERENTIATED INSTRUCTION | English Language Learners

Make Comparisons In this section, students encounter the scientific terms *population density, population distribution,* and *population size*. These three terms are similar in that they all refer to the characteristics of a population, but the differences may not be apparent. Support English Language Learners and other students by using analogies and comparisons to help clarify and differentiate between these important concepts. Student analogies should vary depending on their proficiency levels. Examples include:

Beginning These students can use sentence frames such as:

A population is ______.

Population density is ______.

Intermediate Students can make simple comparisons between the terms and ask for clarification as necessary.

Advanced Students have a strong grasp of the differences between the words and are able to make more sophisticated comparisons and analogies.

Suppose 100 deer are captured, marked, and released. Later, 50 of these deer are recaptured along with 50 unmarked deer. Marked deer constitute half the recaptured group, so the group previously caught and marked (100 deer) must have been half of the population. Thus, the total population is estimated at 200 deer.

Giving each captured animal a unique mark allows scientists to follow individuals over time. This can reveal information about behavior and environmental factors that influence the individual's likelihood of survival. Tracking devices attached to captured animals can yield information about how a population uses the geographical area, or **range**, in which it lives.

Plot sampling is a method of estimating the total number of individuals in an area based on data from direct counts in some portion of the area. For example, ecologists might estimate the number of grass plants in a grassland or the number of mussels on a beach (**Figure 4-3**) by measuring the number of individuals in each of several one-square-meter plots. To estimate total population size, scientists first determine the average number of individuals per sample plot. Then they multiply that average by the number of plots that would fit in the population's range. Estimates derived from plot sampling are most accurate when species are not very mobile and conditions across their habitat are uniform.

SEP Use Mathematics *To estimate the population of mussels on a beach, scientists counted individuals at five different sites. The results of the population survey are shown in **Table 4-1**. What is the average number of mussels they found per site?*

TABLE 4-1. Mussel Plot Survey Data

Site	Number of mussels
A	75
B	96
C	16
D	146
E	87

FIGURE 4-3 Mussel populations are often studied because they spread quickly through connected waterways and can become invasive species if they are transported to new environments.

Science Background

Mussel Reproduction Because mussels are sessile, students may wonder how their offspring are transported to new environments and become so widespread. Most freshwater female mussels lay their eggs and brood them within the marsupia, a special structure in their gills. These eggs are fertilized by male mussels' sperm siphoned from the water. The eggs develop in the marsupia into glochidia, the microscopic larval form of mussels. Once the glochidia are ready to be released, they are not released into open water. Glochidia are obligatory parasites that attach to fish where they will continue to develop and grow, typically with no harm to the fish, into juvenile mussels before detaching at a new location. Students will analyze zebra mussel population density in the Looking at The Data activity in Section 4.3.

SEP Use Mathematics *Average per site = (75 + 96 + 16 + 146 + 87) / 5 = 420 / 5 = 84 mussels*

CROSSCUTTING CONCEPTS | Scale, Proportion, and Quantity

Factors Impacting Size Estimates Prior to calculating the average mussel count, have students compare the counts given in **Table 4-1** and suggest explanations of why the numbers vary between different sites in the river. They should recognize that the availability of resources such as food and shelter likely varies across different parts of the river, particularly for outlying values such as those at sites C and D. To prepare students for quantitative measures of populations, ask them whether they would expect the number of organisms in an area to be proportional to the area. Lead a discussion about whether population counts on a small scale accurately represent the population size as a whole.

Go Online VIRTUAL INVESTIGATION

Sea Pigs on the Abyssal Plain After students collect data using the remotely operated vehicle, encourage them to compare the patterns to those shown in **Figure 4-4**.

In Your Community

Wildlife Studies Students may think scientists conduct ecological studies only in rural or wilderness areas. Explain that in urban and suburban areas, plants and animals that are not owned or cultivated by humans are also considered wildlife. Ecologists conduct population density studies in cities and rural areas to monitor populations of plants and animals in these regions. Encourage students to identify plants and animals that may be surveyed in your region. If time permits, have students conduct research to collect data about a few populations that are monitored in your local area.

Connect to English Language Arts

Write Explanatory Text The text defines population density and discusses different types of population distributions, providing examples in **Figure 4-4**. To reinforce their understanding of these concepts, have students use information in the text and other sources to write a few paragraphs explaining how population density and population distributions are related. Students should support their explanations with at least one example of a real population that demonstrates this relationship.

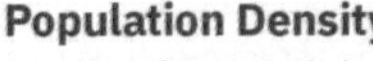

Go Online VIRTUAL INVESTIGATION

Sea Pigs on the Abyssal Plain
How do sea pigs survive in the deep ocean?
Take control of a remotely operated vehicle (ROV) to measure the population density of sea pigs on the ocean floor.

Population Density

A variety of statistical characteristics can be used to describe a population. In addition to measuring population size, ecologists make detailed observations of organisms' locations in order to determine a population's density and distribution.

Population density is the average number of individuals per unit area or volume. This density is often measured by plot sampling. Examples of population density include the number of dandelions per square meter of lawn, the number of amoebas per milliliter of pond water, or the number of squirrels in a 4000-square-meter city park. The population density of a species that inhabits a large range can be expressed as the average number of individuals per unit area of the population's range.

Population Distribution In a population of organisms, the location of individuals relative to one another is described by the **population distribution**. Members of a population may be clumped together, separated by equal distances, or distributed randomly. The distribution of individuals in a population can also change over time.

FIGURE 4-4 Differences in resource availability result in patterns exemplified by a clumped distribution of hippopotamuses (a), a near-uniform distribution of nesting seabirds (b), and a random distribution of dandelions (c).

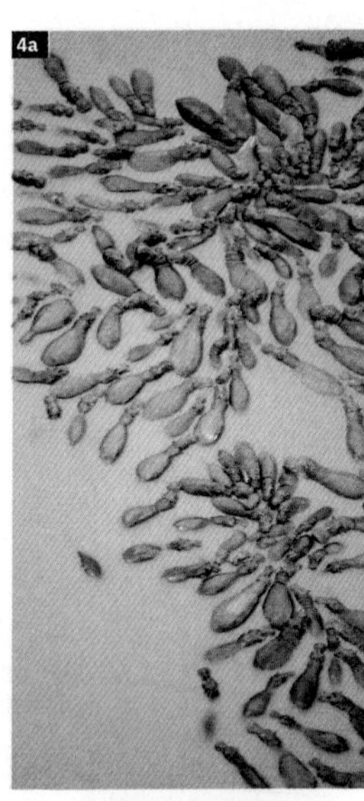

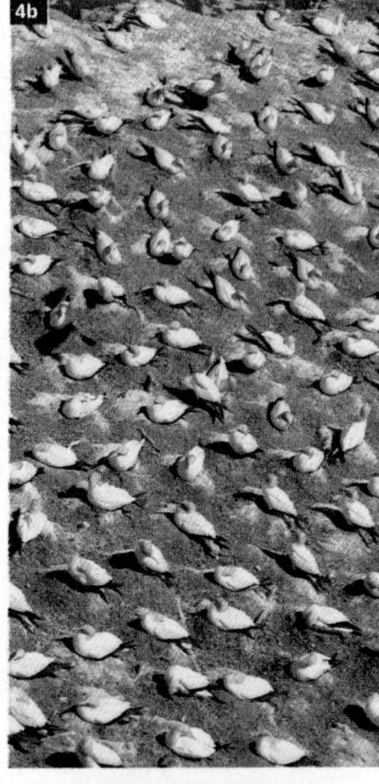

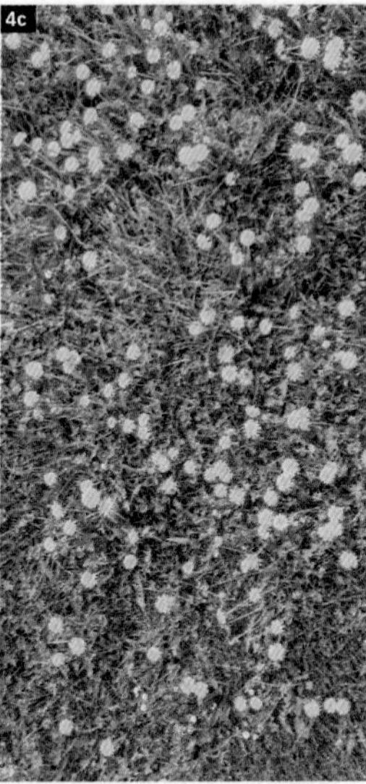

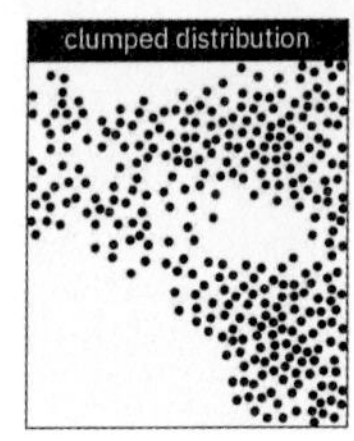

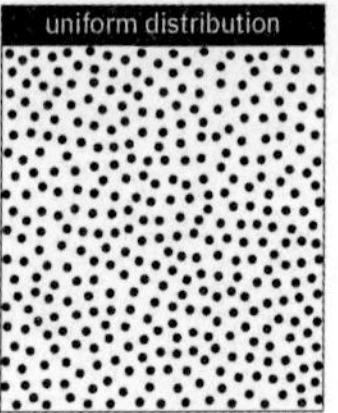

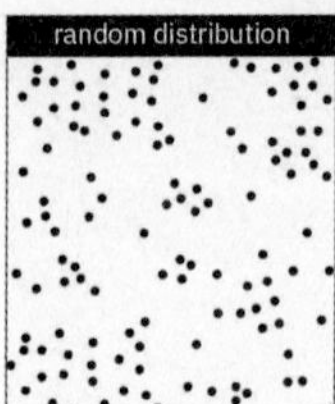

SCIENCE AND ENGINEERING PRACTICES
Analyzing and Interpreting Data

Data-Gathering Methods Aerial footage and satellite data are increasingly used to study populations in ecosystems that are difficult for humans to reach or challenging to observe over large geographical ranges. Satellite data also enables scientists to see changes in populations over time. As described in this chapter's National Geographic Explorer feature, Dr. Heather Lynch uses a combination of imaging technologies to study penguin populations, including their nesting behaviors and their diets.

To choose a method of gathering population data for a study, ecologists must evaluate whether analyzing and interpreting the type of information provided by that method will help them answer their research question. To give students insight into how this process works, have them review the methods the Explorers in this unit use to study animal populations and identify the advantages and disadvantages of each method.

- **Clumped Distribution** This is the most common distribution pattern. In a *clumped distribution*, members of a population are closer to one another than would be predicted by chance alone. A patchy distribution of resources encourages clumping. Hippopotamuses clump in muddy river shallows (**Figure 4-4a**). Moisture-loving ferns may cover a damp, north-facing slope and be absent from an adjacent drier and sunnier south-facing slope. A limited ability to disperse also increases the likelihood of a clumped distribution—the acorn really does not fall far from the tree.
- **Near-Uniform Distribution** This distribution type results from intense competition for limited resources. In a *near-uniform distribution*, individuals are more evenly spaced than would be expected by chance alone. Creosote bushes in some parts of the American Southwest grow in this pattern. Competition for limited water among the root systems keeps the bushes from growing close together. Similarly, seabirds in breeding colonies often show a near-uniform distribution. To defend their nesting sites, Australian gannets aggressively lunge and strike at others that come within reach of their beaks (**Figure 4-4b**).
- **Random Distribution** When resources are distributed uniformly through the environment, and proximity to others neither benefits nor harms individuals, a *random distribution* occurs. For example, when the wind-dispersed seeds of dandelions land on the uniform environment of a suburban lawn, the dandelion plants grow in random locations without any particular pattern (**Figure 4-4c**).

The scale of the sampled area and the timing of a study can influence observed population densities and distributions. For example, seabirds are spaced almost uniformly at a nesting site, but the nesting sites are clumped along a shoreline. The birds crowd together during the breeding season but disperse when breeding is over.

CCC Scale, Proportion, and Quantity *In the plot sampling method, the area where organisms are counted directly, such as a 1-m² plot of grassland, is called a quadrat. Which type(s) of population distribution would be most effectively surveyed using a quadrat? Explain your answer.*

CHAPTER INVESTIGATION

Designing a Seed Trap
How can you build and test an effective seed trap?
Go online to explore this chapter's hands-on investigation and build your own seed trap prototype.

VIDEO

VIDEO 4-2 Go online to get an aerial view of penguins at their Antarctic breeding grounds.

4.1 REVIEW

1. **Calculate** What is the population density of 10,000 amoebas in a 25-milliliter sample of pond water?
2. **Analyze** A population of acacia trees grows in a savanna at a density of 100 trees per hectare in the east and 200 trees per hectare in the west. What type of distribution does this population have?
   A. clumped
   B. dense
   C. near-uniform
   D. random
3. **Calculate** Scientists capture a sample of 36 individuals from an estimated population of 180 foxes. Twelve of the captured foxes are red and 24 are gray. Based on this sample, how many foxes in the whole population would they expect to be red?
4. **Justify** Would mark–recapture sampling or plot sampling be a more suitable method for estimating the population of clams in a shallow river? Explain your choice.

4.1 REVIEW

1. *400 amoebas per milliliter* **DOK 2**
2. *A* **DOK 2**
3. *60 foxes* **DOK 2**
4. *Sample answer: Plot sampling would be more suitable. Clams do not move very much, so they can be counted wherever they live. If they are not evenly distributed, you could sample multiple plots along the river at about the same time without the risk of the same clams traveling between different plots and being counted twice.* **DOK 3**

CHAPTER INVESTIGATION B

Guided Inquiry *Designing a Seed Trap*

Time: 150 minutes over 3 days

Students will develop, build, and test an effective seed trap targeting a specific type of seed.

Go online to access detailed teacher notes, answer, rubrics, and lab worksheets.

CCC Scale, Proportion, and Quantity *A quadrat sample could be a good model for uniformly distributed or randomly distributed populations because the density of these distributions is about the same throughout their ranges. A quadrat might not be a good model for a population with a clumped distribution because the population density differs from place to place, so the organism count in the quadrat might give too low or too high of a population estimate, depending on what part of the range was sampled.*

Video

Time: 0:49

Use **Video 4-2** to ask students to identify the type of penguin population distribution they see in the video. By pausing the video when the penguins, as viewed from above, appear to be near-uniformly distributed, students can use the average height of an adult penguin (70 cm) to estimate the distance between nests and the size of the area visible onscreen. Counting the penguins will enable students to estimate the population density. Ask them to evaluate the validity of their estimate based on how the overall population appears in the video.

MINILAB
MARK–RECAPTURE SAMPLING

Students use averages and proportions to analyze data they obtain from modeling the mark-recapture method. By examining the experimental design aspects of this method, students engage in thinking about factors that affect the size of a natural population and how these factors might contribute to data variation or inaccuracy.

Time: 30 minutes

Advance Preparation

- Any small objects that students can quickly mark individually can be used for this activity. Containers with approximately 150 to 250 objects should give useful results when 20 are marked in Step 2 and 20 are randomly drawn in Step 3.
- Students should not count or be given the total number of objects in their containers before doing the activity. Using opaque containers may help minimize selection bias when sampling at random.

Procedure

In Steps 3 and 4, make sure students understand that they should only be counting how many objects in their sample of 20 are marked and not marking any additional objects.

Sample Data

Trial	Animals trapped	Tagged animals recaptured	Percent of trapped animals tagged
1	20	3	3/20=15%
2	20	4	20%
3	20	0	0%
4	20	1	5%
5	20	3	15%

MINILAB

MARK–RECAPTURE SAMPLING

SEP Use Mathematics How can you estimate a population of organisms that cannot be counted directly?

One of the methods biologists use to estimate population size is the mark–recapture method. This method involves capturing some of the animals in a population, marking them in a way that does not hurt them, and releasing them back into their environment. Later, a random sample of animals from the same population is captured. Biologists can use the proportion of recaptured marked animals in the sample to estimate the size of the population.

This monarch butterfly is marked with a lightweight tag.

Materials

- unknown number of beans in a container
- marker

Procedure

1. The beans in the container represent all of the animals in a population you are studying. Suppose that because their range is very large, it is not feasible to count each individual animal in the population. However, you are able to safely trap and tag some of the animals for later identification. Remove 20 beans from the container. These represent the animals you have trapped during a particular month of your investigation.
2. Make a mark on each bean to tag each animal. Release all of the animals back into their environment by putting the beans back into the container. Shake the container to simulate the animals moving to unknown locations within their range.
3. Again, remove a group of 20 beans from the container at random. Count how many of these animals have tags. In a data table, record the number of animals trapped and the number of tagged animals recaptured. Release the animals by returning the beans to the container.
4. Repeat step 3 several more times. Record your results in the table.
5. For each trial, calculate the percentage of trapped animals that had tags and record your results.

Results and Analysis

1. **Model** Compare the data you obtained for each trial. Do your results vary widely, or are they generally consistent between trials? If these were the results of an actual mark–recapture investigation, what factors in nature could explain the differences in results between trials?
2. **Interpret Data** How do the percentages you calculated in step 5 relate to the number of objects you originally tagged and the total number of objects in the container?
3. **Estimate** Use your data to estimate the total number of beans in the container. Show your calculations.
4. **Calculate** Count all of the beans in the container. How close was your estimate to the actual value? Calculate the percent error in your result.
5. **Evaluate** What changes could you make in your mark–recapture sampling method to improve the accuracy of your estimate?

Results and Analysis

1. **Model** *Sample answer: If the population within the area of study changed between counts (for example, due to birth, seasonal migration, or death), the results might vary widely. If animals were recaptured in a different location than where they were marked and the different locations have different densities of animals, this could also cause variation in the results.* **DOK 3**
2. **Interpret Data** *The percentage of tagged animals in a captured group gives an estimate of the percentage of the total population of animals that are tagged.* **DOK 2**
3. **Estimate** *Students should use average data over multiple trials to make their estimates. For the Sample Data, the average percentage of tagged animals is 55%/5 trials = 11%, so 20 is 11% of the total population N: 20 = 0.11 N, or N = 20/0.11 ≈ 182 animals.* **DOK 2**
4. **Calculate** *For example, if students were to obtain an actual count of 157 objects, using the Sample Data and the formula yields a percent error of about 14%.* **DOK 2**
5. **Evaluate** *Sample answer: Animals should be trapped during the same season and at the same location, using multiple trials under the same conditions. Trapping and tagging a larger number of animals would make the percent of tagged recaptured animals represent the total population more closely.* **DOK 3**

MODELING POPULATION GROWTH PATTERNS

Populations of organisms, whether they are sunflowers, eagles, or humans, change over time. This change is ultimately because of two factors: the number of births and the number of deaths in the population. For humans, the birth rate is often expressed as the number of live births per 1000 people per year, and the death rate is expressed as the number of deaths per 1000 people per year. **Table 4-2** gives birth and death rates in 2019 for several countries.

4.2

KEY TERMS

biotic potential
exponential growth
per capita growth rate
survivorship curve

TABLE 4-2. World Bank Population Statistics

Country	Birth rate (per 1000 people)	Death rate (per 1000 people)
Czech Republic	10.5	10.5
Japan	7.0	11.1
Kazakhstan	21.73	7.19
Ukraine	8.1	14.7
United States	11.4	8.7

Source: World Bank 2019.

Predict *In which countries would you expect the population to increase in the coming years? Are there any countries where the population is expected to decrease?*

Growth Rates

We can measure births and deaths in terms of rates per individual, or per capita. *Capita* means head, as in a head count. Imagine 1200 mice living in the same field. If 600 mice are born each month, then the birth rate *b* is

(600 births/1200 mice) = 0.5 births per mouse per month.

Similarly, if 120 mice die each month, then the death rate *d* is

(120 deaths/1200 mice) = 0.1 deaths per mouse per month.

To calculate the overall **per capita growth rate** *r* for a population, subtract the population's per capita death rate from its per capita birth rate:

$$r = b - d$$

SEP Use Mathematics *Find the mouse population's per capita growth rate.*

Immigration and Emigration

Population size is also affected by immigration and emigration. Immigration is the movement of individuals into a population, increasing its size. Emigration is the movement of individuals out of a population, decreasing its size. In many animal species, young of one or both sexes leave the area where they were born to breed in a different place. For example, young freshwater turtles typically emigrate from their parental population and become immigrants at another pond some distance away. By contrast, some species of seabirds breed where they were born.

4.2 MODELING POPULATION GROWTH PATTERNS 89

SCIENCE AND ENGINEERING PRACTICES
Using Mathematics and Computational Thinking

Growth Rate Explain to students that an initial population with a constant per capita growth rate will generally double at a predictable time. At a 5 percent growth rate, it takes about 14 years for a population to double. At a growth rate of 6 percent, the population doubles in less than 12 years. At a 10 percent growth rate, the population doubles in only 7 years. Have students use these values to try to find the relationship between the growth rate and doubling time. They may recognize that if they multiply one by the other in each case, the product is approximately the same.

The doubling time is estimated by dividing the growth rate, expressed as a percentage, into the number 70. This "rule of 70" is also commonly applied in finance to estimate the amount of time in which an investment will double. The value of 70 is derived by solving the exponential growth function $N = N_0e^{rt}$ for t when $N = 2N_0$, giving $t = (\ln 2)/r \approx 0.69/r$.

4.2

MODELING POPULATION GROWTH PATTERNS

LS2.A, LS2.C

EXPLORE/EXPLAIN

This section provides a review of growth rates, immigration and emigration, and introduces the concepts of biotic potential and exponential growth. Students will also connect life history patterns of organisms to survivorship curves.

Objectives

- Calculate growth rates given data.
- Explain how immigration and emigration affect population growth patterns.
- Use mathematical and computational models to describe how populations grow in an environment.
- Analyze survivorship curves to identify life history patterns of organisms.

Pretest Use **Question 3** to identify gaps in background knowledge or misconceptions.

Predict *The population should increase when the birth rate is greater than the death rate, as in Kazakhstan and the United States. The population should decrease when the death rate is greater than the birth rate, as in Japan and Ukraine.*

SEP Use Mathematics *For $b = 0.5$ and $d = 0.1$, $r = b - d = 0.5 - 0.1 = 0.4$. The population increases by 0.4 mice per mouse per month.*

Connect to Mathematics

Use Units to Understand Problems The subject of this section requires students to understand rate units. Review what *per* means, both conceptually and mathematically, by discussing the units used on this page. Ask students why "per 1000 people" is typical for human population rates, and have them identify which of the two *pers* in a "mice per mouse per year" growth rate corresponds to *per capita*.

CCC **Cause and Effect** *Immigration increases the growth rate, and emigration decreases the growth rate.*

Address Misconceptions

Exponential Growth Students may interpret a constant rate r as adding r new organisms in each generation, resulting in linear growth. They may have difficulty understanding why a population grows faster and faster when the growth rate does not change. Emphasize that r is a per capita rate, so r new organisms are generated for each organism in the previous generation, not just the original one. Have students use an iterative table with a column for the initial population (N_i), a column for the growth rate (r), a column to show the number of new organisms ($r \cdot N_i$), and a column that shows the final number of organisms (N_f) to generate data for a population that grows exponentially. They should be able to complete several more rows of the provided example, which uses a growth rate of 0.5 and a initial population of 8.

initial pop. (N_i)	growth rate (r)	number of new organisms ($r \cdot N_i$)	final number (N_f)
8	*0.5*	*4*	*12*
12	*0.5*	*6*	*18*
18	*0.5*	*9*	*27*

FIGURE 4-5 As in many big cat species, a male lion leaves its mother's territory when it is between two and three years old to find a mate from a different pride.

However, some individuals may emigrate and end up at breeding sites more than a thousand kilometers away. In general, the tendency of individuals to emigrate to a new breeding site is related to resource availability, crowding, and avoiding inbreeding (**Figure 4-5**). As crowding increases and resources decline, the likelihood of emigration rises.

CCC **Cause and Effect** *How do immigration and emigration affect the growth rate of a population?*

Biotic Potential and Exponential Growth

A population's per capita growth rate can change over time for many reasons. Food sources or suitable habitats may become more or less available, or a population may be exposed to new predators or disease. However, if shelter, food, and other essential resources were unlimited and there were no external threats, the population would grow indefinitely as long as its birth rate was greater than its death rate.

Under ideal conditions, the population's per capita growth rate would be determined by the reproductive characteristics of its species, including the age at which reproduction typically begins, how long individuals are able to reproduce, and the number of offspring that are produced each time an individual reproduces. This theoretical growth rate is a measure of the population's **biotic potential**. Different species have different biotic potentials. Microbes such as bacteria have some of the highest biotic potentials, whereas large mammals with long lives have some of the lowest.

Imagine a single bacterium in a laboratory flask that provides ideal conditions for survival and growth. After 20 minutes, the bacterial cell divides in two. Those two cells divide, and so on, every 20 minutes. At this growth rate, a population of more than one billion can result from a single bacterium in just 10 hours. The graph of the number of individuals versus time for this bacterial population takes on a J shape. This shape is characteristic of graphs of **exponential growth**, which occurs when a population's per capita growth rate is positive (**Figure 4-6**).

90 CHAPTER 4 POPULATION MEASUREMENT AND GROWTH

DIFFERENTIATED INSTRUCTION | Economically Disadvantaged Students

Prior Knowledge Students living in poverty consistently show a gap in mathematics performance and knowledge due to curriculum inequalities. Exercises that illustrate the concept of exponential growth in earlier grades often use small objects, such as pennies or candy. To introduce this concept with comparable mathematical simplicity, but at a high school level, engage students in a discussion of how a video or social media post "goes viral." Beginning with the creator of the post, if the number of people who view it doubles each day, it will have been seen by over 16,000 people in two weeks. Ask students whether they would consider this daily growth rate (one new viewer per capita) to be realistic. Have them estimate how quickly the post could reach the same number of viewers if the number quadrupled each day, which would be equivalent to each viewer showing the post to three new people each day.

FIGURE 4-6 If a population's per capita growth rate is positive and resources are unlimited, the number of individuals increases exponentially over time.

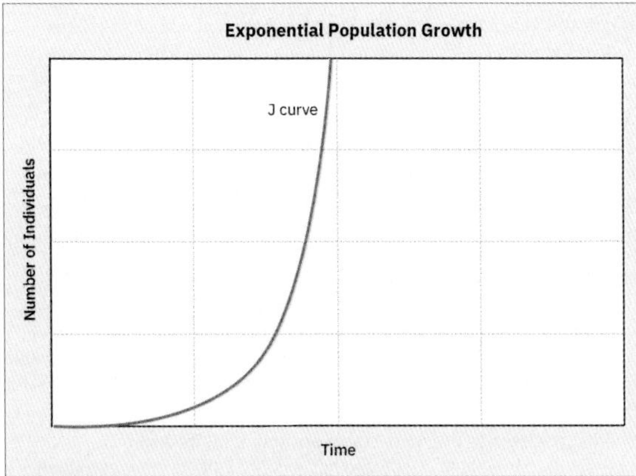

Go Online SIMULATION

Modeling Population Changes Go online to examine how changes in rates affect the growth of a population.

Regardless of the species, populations seldom reach their biotic potential. A population with a growth rate *r* greater than zero will increase exponentially, but growth at the maximum potential rate can only occur when conditions are ideal. A bacterial colony might grow to enormous numbers in a laboratory environment, where conditions can be controlled and food supplied indefinitely. But in an outdoor puddle where food is limited and the environment is unstable, such as in **Figure 4-7**, the same bacterial population would grow at a slower rate or even decrease. Although the population can survive in the puddle for some time, the conditions are no longer ideal.

SEP **Construct an Explanation** *Describe the difference between the biotic potential of a population of organisms and the population's growth rate.*

FIGURE 4-7 Bubbling pots of volcanic mud form in the Uzon Caldera in Kamchatka, Russia. Microbes that are able to survive in this hot, highly acidic environment face little competition with other species as they cluster in mats on the surface of the mud.

Go Online SIMULATION

Modeling Population Changes In this simulation, students can manipulate the birth and death rates for a population undergoing exponential growth and observe the effects on the growth curve. Have them use the results of the simulation as evidence to explain how the growth curves of two populations that have achieved their biotic potentials may differ from each other.

Connect to Mathematics

Use Units to Understand Problems As students examine the graphs in this section and the next, remind them to identify the variables being compared in order to interpret each graph. Use **Figure 4-6** to engage students in this type of analysis by asking them to propose reasonable units and scales for each axis if the graph were to show the growth rate of the bacterium discussed on the previous page.

SEP **Construct an Explanation** *A population's growth rate gives the number of organisms added per organism in the previous generation. The growth rate can change depending on the conditions of the environment. The biotic potential is described by the maximum growth rate that the population could reach if its environmental conditions were ideal, which is usually not the case.*

SCIENCE AND ENGINEERING PRACTICES
Analyzing and Interpreting Data

Data from Graphs Data in scientific literature is very often represented graphically. Students will likely have experience plotting values given in a data table on a graph but may be less familiar with extracting quantitative information from a graph displaying a curve, such as the one in **Figure 4-6**. To provide practice in analyzing graphical data, have students number the gridlines on the Exponential Population Growth graph from 0 to 4 along the *x*-axis and in increments of 50 along the *y*-axis. Ask them to mark all of the points where the J-curve crosses a gridline and to estimate the coordinates for each of these points. Then, have students use these points to estimate (interpolate) how long it took for the population to double in size from $t = 1$ or to predict (extrapolate) how large the population will be at $t = 2$, including evidence from the graph in their reasoning.

Connect to English Language Arts

Assessing Reasoning and Evidence Organisms, such as the salmon and olive trees shown in **Figure 4-8**, have very different life history patterns. Have students read the caption for **Figure 4-8** and discuss in pairs whether they believe that the information given in the caption sufficiently supports the claims made in the text regarding the relationship between living conditions and life history patterns.

Science Background

Life History Patterns Ecologists call species that have a collection of reproductive traits that maximizes the growth rate *r* of its population an *r-selected* species. In contrast, species whose reproductive traits tend to maintain populations near the carrying capacity *K* of their environments (see Section 4.3) are called *K-selected*.

Semelparous species reproduce for a short period, or even just once, in their lifetimes and die soon afterward. Semelparous species include grains and certain insects. The male *Antechinus*, an small insect-eating marsupial indigenous to Australia, mates exhaustively for the first time during a two- to three-week period at the end of its first year, eventually dying from its reproductive efforts. While many semelparous species are short-lived, some live for years, then reproduce once and die. The agave plant, bamboo, and octopuses are examples of longer-lived semelparous organisms. Semelparous organisms are unable to provide parental care to their offspring.

Organisms that reproduce repeatedly throughout their lives are *iteroparous* reproducers. Iteroparous species include oak trees, most mammals, birds, reptiles, and many fish species. Some iteroparous organisms may have large numbers of offspring, which are generally not cared for, while others may have smaller sets of offspring that may receive parental care.

Life History and Survivorship

A population's biotic potential is determined in part by how its individual members use resources and energy for growth, survival, and reproduction over the course of their lifetimes. This distribution of resources makes up an organism's *life history*. Patterns in life history vary among species. Ecologists have described two major life history patterns.

Life History Patterns Species that live in unpredictable conditions tend to produce as many offspring as possible, often as quickly as possible. This pattern is called an opportunistic life history. These organisms offer little parental care, instead expending energy to take advantage of a favorable environment that might only be temporary. These organisms are often small in body size and have a short interval of time between generations. For example, weedy plants such as dandelions follow this life history pattern. They mature within weeks, produce many tiny seeds, then die. Coho salmon (**Figure 4-8a**) also live according to this pattern. A female salmon deposits thousands of eggs in a river nest during spawning. Neither she, nor the male salmon that fertilizes the eggs, will survive to see their offspring hatch.

When a species lives in a more stable environment, the ability to successfully compete for resources has a major influence on reproductive success. These conditions favor an equilibrial life history pattern in which parents produce a few high-quality offspring. Compared to species that live in unpredictable conditions, species that live in stable conditions have larger bodies and longer times between generations. The life history pattern of large mammals, which typically take years to reach adulthood and begin reproducing, is representative of these species. For example, a female blue whale reaches maturity between the ages of six and ten years. The whale then produces only one large calf at a time and continues to invest energy and resources in the calf by nursing it after its birth. Similarly, olive trees (**Figure 4-8b**) grow for several years before producing olives. In both whales and olive trees, a mature individual produces young for many years.

Survival Patterns One way to investigate life history traits is to focus on a cohort—a group of individuals born during the same interval— from their time of birth until the last individual dies. Ecologists often divide a natural population into age classes and record the age-specific birth rates and death rates. The resulting data is often summarized in a table.

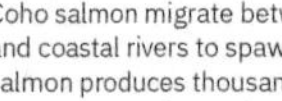

FIGURE 4-8 Coho salmon migrate between the ocean and coastal rivers to spawn. A female salmon produces thousands of eggs, but most are eaten by predators before they can hatch (a). Olive trees need sandy soil and a consistent climate with long summers and mild winters to survive. They begin to produce olives at about four to eight years old and can live for hundreds of years (b).

DIFFERENTIATED INSTRUCTION | Leveled Support

Struggling Students To help students learn how the environment influences a species' life history pattern, make a T-chart describing an unpredictable environment on one side and a stable one on the other side. Then, ask students whether a specific reproductive trait, such as a large litter size or a long gestation period, would be an advantage in unpredictable conditions or in stable conditions. Students should continue to list traits belonging on each side of the T-chart as they work through the section.

Advanced Learners Students who have some understanding of statistics may be interested in determining how scientists obtain the empirical data points shown in **Figure 4-9**. To help students form their own explanation, it may be helpful to introduce the term *cohort*, meaning a group of organisms of the same age (or within a narrow age range). Survivorship is measured by counting the number of individuals within the same cohort that remain alive at each age.

Information about the life history of a species can also be illustrated by a survivorship curve. This plot shows how many individuals born during the same interval remain alive over time. Ecologists describe three types of curves, as shown in **Figure 4-9**.

- **Type I** The death rate remains low until relatively late in life for these organisms, so most individuals survive to reproduce. Humans and other large mammals, such as Dall sheep, that produce and care for one or two offspring at a time have this pattern.
- **Type II** These organisms have a survivorship rate that remains relatively equal across their lifetimes. In birds, small mammals, and some reptiles such as skinks, old individuals are about as likely to die of disease or predation as young ones are.
- **Type III** These organisms have a very high birth rate and a very high death rate soon after birth. Invertebrates, fish, amphibians, and plants such as spider flowers show this type of survivorship.

FIGURE 4-9 A survivorship curve shows how many individuals in a population survive to a given percentage of the typical maximum life span for that type of organism. In these graphs, the gray lines are theoretical curves, and the blue dots are data from field studies of Dall sheep (a), five-lined skinks (b), and spider flowers (c).

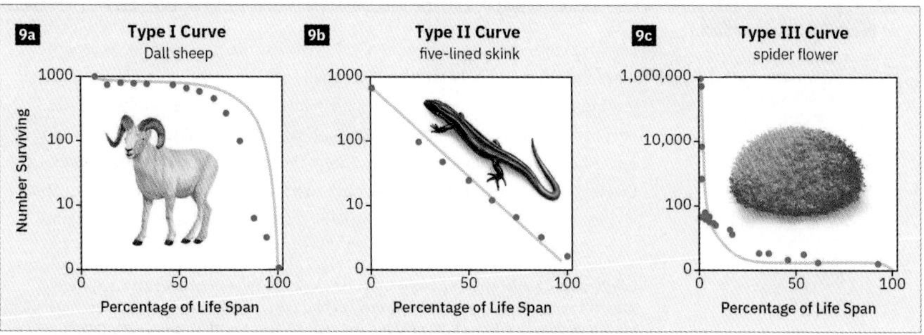

CCC Patterns *Which type of survivorship pattern would you expect a species that lives in unpredictable conditions to have? Which type would a species that lives in stable conditions likely have?*

4.2 REVIEW

1. **Compare** Classify each factor as having a positive impact or a negative impact on the growth rate of a population of organisms: birth rate, death rate, rate of immigration, rate of emigration.
2. **Summarize** Describe an ideal scenario where a population of organisms would achieve its biotic potential.
3. **Analyze Data** Based on the death rates listed in the table, what type of survivorship curve do you expect the plant species will have?
4. **Predict** What do you expect will be the survivorship between months 13–15?

The table gives statistics for a cohort of one plant species monitored over a year. Use the data to answer questions 3 and 4.

Age interval (months)	Survivorship (number surviving at start of interval)	Death rate	Birth rate
0–3	996	0.329	0
4–6	295	0.356	0
7–9	176	0.023	0
10–12	154	0.045	3.13

Visual Support

Patterns in Survival Remind students that graphs are used to visualize relationships and trends in data. Have students work with partners to compare the shapes of the three curves in **Figure 4-9** and the scales along the *x*- and *y*-axes of the graphs. Students should recognize that the *y*-axis scale is the same on the type I and type II graphs but is different in the type III graph. Ask students to explain why the scales differ. They should understand that a species with a type III curve would typically have a huge number of offspring with a high death rate earlier in life. Populations with type I or type II survivorship curves tend to have significantly fewer offspring at a time, so if these were graphed on the same *y*-axis scale as the type III curve, they would look relatively flat and may not be very distinguishable from each other. In fact, as students might also notice, the increments on the *y*-axis in all three graphs are powers of ten rather than being uniformly linear in size. The characteristic shapes of survivorship curves are not as evident on a linear scale and tend to visibly emerge at a logarithmic scale.

CCC Patterns *Species that live in unpredictable conditions usually have a type III survivorship curve, with the death rate highest early in life. Many species that live in stable conditions have a type I survivorship curve, with the death rate highest later in life.*

4.2 REVIEW

1. *positive impact: birth rate, rate of immigration; negative impact: death rate, rate of emigration* **DOK 2**
2. *Sample answer: Student's answer should indicate that the population has access to unlimited resources, the habitat is stable and poses no new threats such as disease or predators, and the natural birth rate is greater than the death rate.* **DOK 2**
3. *Sample answer: The plant's survivorship follows a type III curve because the death rate is highest early in life.* **DOK 3**
4. *130–140* **DOK 3**

EXPLORER
HEATHER J. LYNCH

About the Explorer

National Geographic Explorer Dr. Heather J. Lynch studies animal populations in Antarctic regions and has developed mathematical tools and automated methods to estimate penguin populations in the Antarctic. You can learn more about Lynch's research on the National Geographic Society's website: www.nationalgeographic.org

Video

Time: 0:39

Use **Video 4-3** to show drone footage of a penguin colony on an Antarctic island. Penguin guano, stained pink with krill, is visible on the rocky ground.

Connect to Careers

Quantitative Ecologist Ecological studies often generate highly complex datasets that incorporate information in multiple dimensions, including spatial information, time series, climate and weather data, and electronic signals generated by advanced sensor technology. Quantitative ecologists apply statistics and mathematical modeling to analyze data from diverse sources. These analyses are used to answer scientific questions or inform public policy in fields such as natural resource management and population ecology. Students wishing to pursue a career in quantitative ecology will typically need to earn a doctoral degree with a strong background in biology, mathematics, statistics, and programming.

THINKING CRITICALLY

Identify Cause and Effect *Sample answer: As the climate changes, species such as the Gentoo penguins that are able to adapt to unpredictable conditions (such as by having a more varied diet, more flexible breeding requirements, and tolerance for warmer temperatures) will maintain or increase in population, but species that cannot adapt to new conditions will decrease in population.*

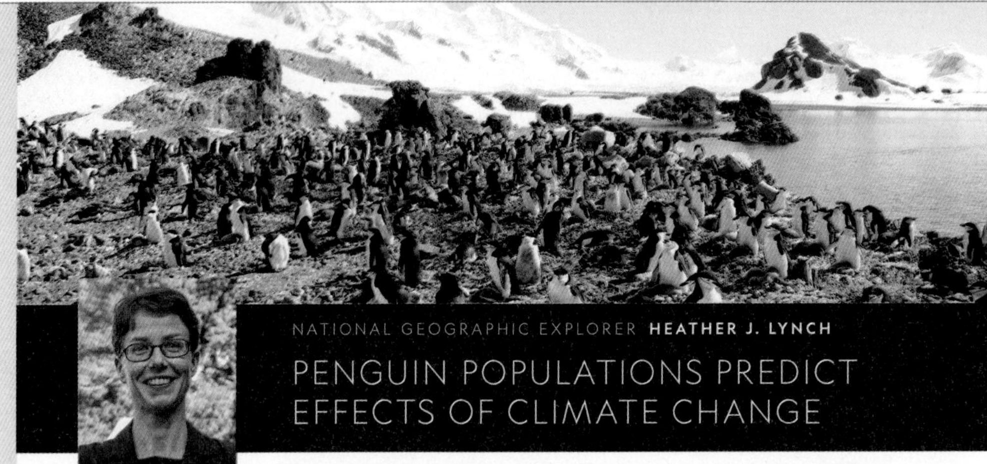

NATIONAL GEOGRAPHIC EXPLORER **HEATHER J. LYNCH**

PENGUIN POPULATIONS PREDICT EFFECTS OF CLIMATE CHANGE

Dr. Heather J. Lynch works to develop mathematical models and apply statistical analyses to conservation biology.

VIDEO 4-3 Go online to observe a penguin colony that Lynch's group found.

Scientists monitor penguin populations to assess the health of the Southern Ocean ecosystem in one of the most rapidly warming places on Earth. However, visiting penguins in their far-flung, inhospitable habitats is often difficult. Until very recently, many colonies had not been visited for decades and some had never been properly surveyed. To address this problem, quantitative ecologist Dr. Heather J. Lynch has pioneered methods of combining high-resolution satellite imagery with high-performance computing to study polar animal populations.

Lynch is a professor at Stony Brook University whose research focuses on a long-term project to map Antarctica's penguin and seabird populations. Like many seabirds, penguins are highly sensitive to local conditions. Antarctica's five penguin species migrate to specific breeding sites that are suited to their varied nesting and feeding habits. Penguins eat fish and krill, which are small pink crustaceans that feed on phytoplankton under the sea ice. Pink penguin guano stains the areas where penguins nest, and those guano stains are visible by satellite. Using images of penguin colonies and their characteristic guano staining, Lynch and her research group discovered a series of penguin "supercolonies" in the remote Danger Islands. In all, Lynch and her colleagues have discovered over a million penguins through satellite imagery. In addition, they figured out how to use the shade of pink in the guano to analyze the proportion of krill and fish in the penguins' diets. This information is critical to understanding the threats penguins may be facing.

A number of different stakeholders benefit from accurate information on penguin populations. For example, scientists need to understand the effectiveness of Marine Protected Areas, which are designed to conserve and protect marine habitats and the resources within them. Commercial interests, such as krill harvesters and tour operators, also need to know where penguin populations exist to avoid disrupting their food sources and habitats. To address this broad variety of needs, Lynch's group created the Mapping Application for Penguin Populations and Projected Dynamics (MAPPPD). This internet-based tool uses a globally sourced dataset to estimate current Antarctic penguin populations and forecast population trends.

THINKING CRITICALLY

Identify Cause and Effect *Lynch's population models integrated 41 years of data from 70 penguin breeding sites on the Antarctic Peninsula. These studies show that Adélie and chinstrap penguin populations are decreasing, and gentoo penguin populations are increasing. What factors might explain why population trends differ for different penguin species living in the same region?*

DIFFERENTIATED INSTRUCTION | English Language Learners

Ask and Answer Questions Have students work in mixed groups. Students will form questions and answer them based on the work by Dr. Lynch. Questions should vary depending on students' proficiency levels. Examples include:

Beginning These students can use sentence frames such as:

What does __________ study?

What does __________ mean?

Intermediate What kinds of animals does Dr. Lynch study?

Advanced How does Dr. Lynch study penguins?

FACTORS THAT LIMIT POPULATION GROWTH

4.3

KEY TERMS

carrying capacity
density-dependent factor
density-independent factor
logistic growth

The city of Detroit, Michigan, was settled by French traders in 1701 and incorporated as a city in 1815. Its position on the Detroit River that links Lake Erie and Lake St. Clair made it ideal for transporting goods by ship. The city's population grew exponentially, adding tens of thousands of people per year during the early 20th century. In 1950, Detroit was the fourth largest city in the United States and the epicenter of the automobile industry, with a population of more than 1.8 million people. However, the city's population has significantly declined since that era (**Figure 4-10**).

> **Predict** *What change in resources might have caused Detroit's population decline over the last 70 years? What do you think will happen to the city's population in the future? How might the population of your town or city change in the future?*

FIGURE 4-10
The city's population increased over its first 150 years, but has been decreasing since 1950.

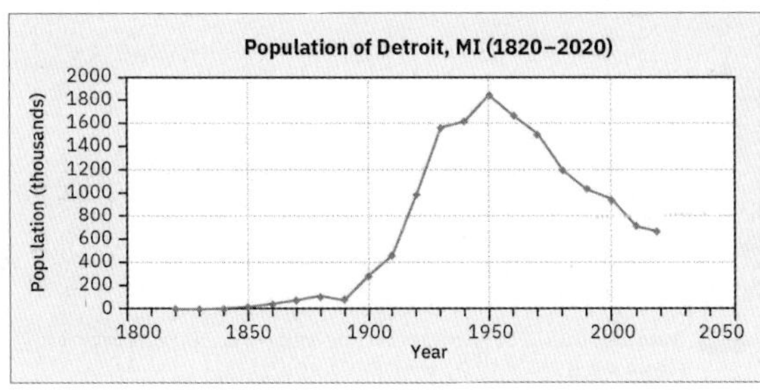

Limiting Factors and Carrying Capacity

Even in a stable environment with favorable conditions, there are limits to how much a population can grow. For a population of humans, some of these limits are imposed by natural boundaries such as the bodies of water that surround Detroit. In addition, humans draw borders and write laws that regulate the number of people that can live in a city, much like coyotes establish territories inhabited only by specific packs.

As in a city, circumstances that curb exponential growth exist in any natural environment. For example, consider a population of water fleas that live in a small pond with plenty of algae and bacteria to eat and no fish or other predators. Even in these conditions, the water flea population could not keep growing exponentially for an extended period of time because they would run out of food and living space. A larger pond, or one with more food, might sustain a larger population of water fleas. However, the population would still not be able to increase indefinitely because the environment's resources are finite.

4.3

FACTORS THAT LIMIT POPULATION GROWTH

LS2.A, LS2.C

EXPLORE/EXPLAIN

This section explains how density-dependent factors limit population growth to the carrying capacity of the environment and compares the effects of density-dependent and density-independent factors on population size.

Objectives

- Explain how limiting factors constrain the carrying capacity of an environment.
- Explain how density-dependent factors can slow population growth rates.
- Understand how density-independent factors differ from density-dependent factors and how they affect populations.

Pretest Use **Questions 2, 4, 5,** and **6** to identify gaps in background knowledge or misconceptions.

Vocabulary Strategy

Categorize and Label As they work through the section, students should identify factors as density-dependent and as density-independent. Have students discuss the following examples with a partner:

- A region experiences a long-term drought.
- Scorpionfish wipe out small fish in a reef.
- A typhoon washes out an island.
- A wasp lays its eggs inside of caterpillars.

They should label the limiting factor in each example as a *density-dependent factor* or as a *density-independent factor* and explain their reasoning.

Predict *Sample answer: Lower job availability, such as in auto manufacturing, caused many people to leave the city.*

CHAPTER INVESTIGATION A

Guided Inquiry *Population Growth of Duckweed*

Time: 190 minutes over 2 days plus observation time over 3 weeks

Students will follow a step-by-step procedure to investigate how a duckweed population changes over time.

Go online to access detailed teacher notes, answer, rubrics, and lab worksheets.

SEP Interpret Data *The carrying capacity is about 150 individuals.*

CASE STUDY
SHARING OUR CITIES

Ask Questions To help students make connections with the content, have them draw a mathematical model showing how an S-curve might be affected by density-dependent factors that affect an urban coyote population.

Crosscurricular Connections

Algebra I Students who are taking or have taken Algebra I courses will likely have some experience with exponential growth functions. The S-shaped curve that characterizes growth limited by density-dependent factors is a logistic function that is typically introduced in advanced math courses. However, students who understand the concept of slope can analyze a limited population growth function by drawing tangent lines to the curve at various locations. Have students do this for the graph in **Figure 4-11** and ask them to use the slopes of the tangent lines to describe how the growth rate changes over time. They should recognize that the maximum growth rate occurs when the population reaches half of the carrying capacity *K*.

In the "Population Growth of Duckweed" Chapter Investigation, students will further distinguish between the growth rate and the per capita growth rate.

CHAPTER INVESTIGATION

Population Growth of Duckweed
How and why does the number of duckweed plants in a population change over time?

Go online to explore this chapter's hands-on investigation about factors that affect population growth.

Carrying Capacity When limiting factors affect population size over time, **logistic growth** occurs, yielding an S-shaped curve. The graph shows exponential growth at first. Then, as the growth rate decreases, the population levels off at a constant value (**Figure 4-11**). This **carrying capacity** is the maximum number of individuals that the population's environment can support indefinitely. Carrying capacity is species-specific, environment-specific, and can be affected by changes in the environment.

FIGURE 4-11
In nature, the increase in a real population may resemble exponential growth at first, but many factors contribute to slowing this growth as time passes.

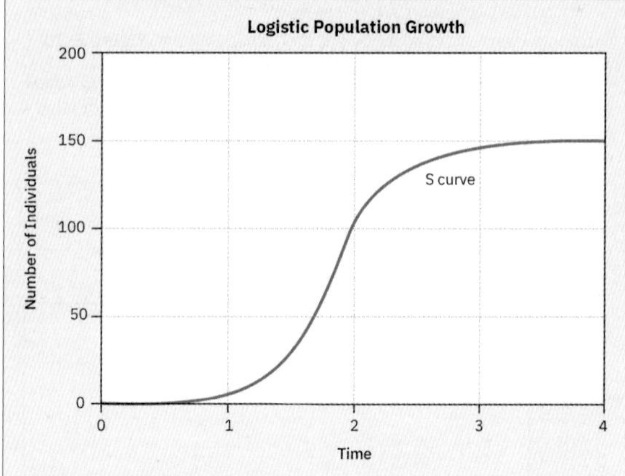

SEP Interpret Data *Estimate the carrying capacity of the population in Figure 4-11.*

As time goes on, the growth rate for most organisms stabilizes near zero for many different reasons. Factors that limit the growth of a population can be categorized as density dependent or density independent.

Density-Dependent Factors

A **density-dependent factor** is a limiting factor that is affected by the number of individuals in a given area. The denser the population becomes, the more its growth rate will tend to decrease. Competition, parasitism, disease, and predation are all examples of density-dependent factors that can slow population growth rates and decrease the carrying capacity of an environment.

Competition As population density increases, so does competition for resources such as food and shelter. Eventually, the population reaches the point where there is no longer enough of a resource to sustain it. For example, when small sapling trees in a forest compete for sunlight, many trees are able to grow in a small space (**Figure 4-12**). Eventually, as they grow, the trees crowd each other out, blocking off the sunlight so only a few can survive, and the carrying capacity is reached. Competition has a detrimental effect even on the survivors, because energy they use to compete for resources is not available for reproduction.

CASE STUDY
SHARING OUR CITIES

Ask Questions *Generate questions about how human activity might influence the density-dependent factors that affect an urban coyote population.*

FIGURE 4-12 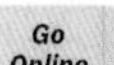
Plants in the same environment compete for light and space. This competition occurs among plants of the same species or with plants of different species.

Parasites and Pathogens The negative effects of diseases transmitted by parasites or pathogens may increase with population density, which most likely occurs when the disease is transmitted through air and water sources used by the population. In humans, airborne viruses such as severe acute respiratory syndrome (SARS) occur at higher rates in densely populated regions. Chytridiomycosis, a disease that affects hundreds of frog species worldwide, is caused by a fungus that eats the proteins in the frog's skin. This results in infections that may eventually lead to the frog's death by cardiac arrest. The fungus can infect frogs as it floats in the water where the frogs live. It also spreads by contact between individual frogs (**Figure 4-13**). This disease has, to date, no effective control methods in the wild, so it poses a serious threat to biodiversity. Scientists estimate that since the 1990s, more than 200 frog species have declined or become extinct due to chytridiomycosis.

Go Online SIMULATION

Modeling Population Changes Go online to investigate how density-dependent factors affect the size of a population.

FIGURE 4-13
The American bullfrog (*Lithobates catesbeianus*) tolerates the fungus that causes chytridiomycosis and can carry it to habitats where it infects other frog species.

Go Online **SIMULATION**

Modeling Population Changes In this simulation, students will modify a number of density-dependent factors to manipulate an environment. This will allow them to see how density-dependent factors can affect a population size.

Spreading Pathogens in Human Populations In **Section 13.3**, students will analyze methods for mitigating viral transmission within a population and from reservoir species to humans.

DIFFERENTIATED INSTRUCTION | Leveled Support

Struggling Students Use a simplified example of an aquarium to help students understand density-dependent factors. Have students consider an aquarium containing one species of fish and discuss what would happen if one mating couple of fish lived in the aquarium as opposed to many fish living in the aquarium. Have students identify the limiting factors of the example and how they affect the fish population.

Advanced Learners Have students work in groups to develop a simple game that can be used to learn how to identify whether a limiting factor is density-dependent or density-independent. Students should build their game and share it with the class.

Visual Support

Double-Line Graphs Students may not recall how to read a double-line graph like the one shown in **Figure 4-14**. To help them interpret the graph, first ask them to describe the reason for plotting two different data sets on the same graph. They should recognize that this is done to compare the trends in the two data sets over the same period of time (or in general, the same change in the independent variable). Have students explain the patterns they see and use the data to make a claim about how the hare and lynx populations interact.

SEP Construct an Explanation *Sample answer: Yes. Humans introducing a new organism to an area could increase the population of organisms that are its parasites or that prey on the new organism.*

Connect to Mathematics

Write a Function Have students use the grid lines and the shapes of the data sets plotted in the **Figure 14-4** graph to estimate how often population data was recorded. Then, ask them to identify a pattern in the relationship between the times at which the hare and lynx populations reach maximum values. Have students write a function to represent this relationship. Student answers will vary but should show that for each time the hare population spikes, the lynx population spikes about a year or two later.

Predation In any community, the number of predators and the number of prey are connected. Generally, an increase in the size of a predator population results in a decrease in the abundance of its prey. Predator and prey populations sometimes rise and fall in a predictable cyclical fashion. **Figure 4-14** shows historical data for the numbers of Canadian lynx and their main prey, the snowshoe hare. Both populations rise and fall over an approximately ten-year cycle, with the number of predators lagging behind the number of prey. Field studies indicate that lynx population numbers increase or decrease mainly in response to hare population numbers. However, the size of the hare population is affected by the availability of the hare's food as well as the number of lynx. As a result, hare populations continue to rise and fall even when predators are experimentally excluded from areas.

FIGURE 4-14
Canadian lynx population estimates in the late 19th century are based on counts of lynx pelts that fur trappers sold to the Hudson's Bay Company.

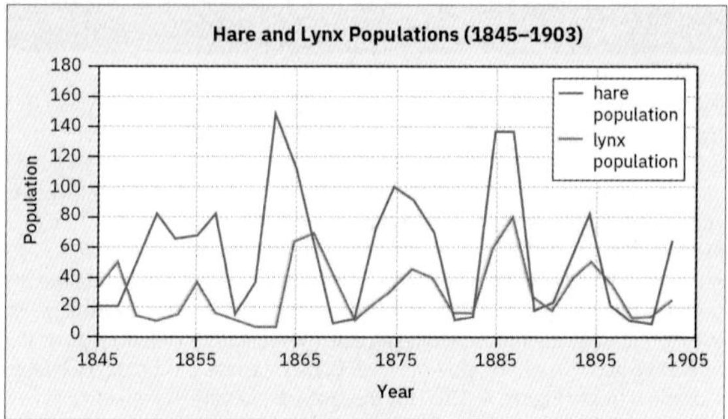

Density-Independent Factors

Sometimes, natural disasters or weather-related events affect population size. A volcanic eruption, hurricane, or flood can greatly diminish the size of a population in an instant, whether it is high or low in density. These events are called **density-independent factors**, because the density of the affected population does not influence the likelihood of the factor's occurrence or the magnitude of its impact. Human-caused events, such as oil spills or clearing land for development or agriculture, can also act as density-independent factors.

In nature, density-dependent and density-independent factors interact to determine a population's size. The occurrence of a density-independent event may intensify the effect of an existing density-dependent factor or introduce new density-dependent factors. For example, in addition to killing some portion of the animals and plants that inhabit a forest, a major forest fire reduces the amount of suitable habitat and food for the remaining populations, and surviving organisms may be less healthy and more vulnerable to predation and disease.

SEA PIG SURVIVAL

Gather Evidence *What density-dependent factors could limit the populations of organisms that live on the ocean floor? What density-independent factors could limit these populations?*

SEP Construct an Explanation *Can a density-independent factor lead to an increase in the size of a population? Explain your answer.*

REVISIT THE ANCHORING PHENOMENON

1. *Sample answer: Competition for food is the main density-dependent factor that limits ocean floor populations. Density-independent disturbances to the seafloor that would occur during a human activity, such as exploration or mining, could also limit these populations.* **DOK 3**

There are a variety of sea pigs that live in the ocean, and some species, such as *Scotoplanes globosa*, can be found in every ocean of the world. Sea pigs have soft, fleshy bodies without an exoskeleton or thick skin. This makes them easily susceptible to the density-dependent factor of parasites and pathogens. There are a number of sea pig parasites that eat or bore through the sea pig's external body walls to eat them from the inside out.

BIOTECHNOLOGY FOCUS

ENVIRONMENTAL DNA Canadian lynx are now very rarely seen and are thus listed as a threatened species according to the Endangered Species Act. Because their tracks are very similar to those of bobcats, it is difficult to accurately determine their populations. Carnivore researchers have collaborated with aquatic scientists to develop a reliable technique for identifying lynx tracks.

An organism's DNA is found in all of its cells, and each species has a unique genome. Animals shed surface cells and waste products as they move through their environments. Cellular DNA can survive in water for a few weeks before degrading. This allows scientists to extract and analyze DNA from a water sample taken from an aquatic environment, giving them information about the species of organisms that exist in the water. The DNA analysis techniques they use are sensitive enough to detect rare species that have not otherwise been observed in the sampled environment.

Animal tracks made in snow become water samples when the snow melts. Using carefully designed methods to avoid contamination, scientists scooped up snow near and underneath fresh tracks they encountered in the northern Rocky Mountains. By applying the same techniques used to analyze environmental DNA for the presence of aquatic species, they were able to identify lynx tracks and differentiate them from tracks made by several carnivore species that share a habitat with the Canadian lynx.

Camera traps occasionally capture lynx in the wild.

4.3 REVIEW

1. **Classify** Label each scenario as a factor that will increase or decrease the carrying capacity for a species.
   A. The most populous predator of the species emigrates.
   B. Favorable environmental conditions boost the birth rate.
   C. All the organisms infected with a contagious disease die off.
   D. A new parasite for the species immigrates into the community.

The table gives the population sizes in one food chain in a community. Use the data to answer questions 2 and 3.

Organism	Population size (number of organisms)
primary consumers	7×10^5
producers	6×10^6
secondary consumers	3×10^5
tertiary consumers	1×10^1

2. **Analyze Data** A new tertiary consumer predator moves into the community. Sketch the numbers pyramid for the community before and after the arrival of the new predator.
3. **Predict** An infectious disease decimates a population of organisms in a community. For each trophic level, predict what might happen to the numbers pyramid if the disease were to kill species of organisms at that level.
4. If the death rate of a population increases and the birth rate stays the same, in what ways might the graph of population vs. time change? Select all correct answers.
   A. The carrying capacity could decrease if the increase in the death rate is due to a change in a density-dependent factor.
   B. The carrying capacity could increase because there would be room for more organisms than before.
   C. The population would reach the carrying capacity more slowly.
   D. The growth rate of the population could decrease temporarily if the increase in death rate is due to a density-independent factor.

Organism DNA Scientists use DNA to identify which organisms exist in an environment. **Chapter 13** discusses tools for analyzing a genome to identify DNA sequences that can be traced to a particular species or individual.

Biotechnology Focus

Aquatic DNA Sampling Although students will learn more about DNA in later chapters, they likely know from middle school that DNA is present in the cells of all living things and codes for traits. The term *environmental DNA* refers to DNA that organisms shed into the environment, such as skin cells, feces, hair, and reproductive secretions. In terrestrial ecosystems, such eDNA generally ends up in the soil. In aquatic ecosystems, eDNA can be present in relatively high densities. Multiple species can be identified from a sample as small as 250 milliliters. Typically, aquatic field surveys collect samples from 250 milliliters to 1 liter to analyze back in the lab. Students might calculate how much snow around the footprints would need to be collected to yield that much water. They easily can find snow-to-water ratios on the Internet.

4.3 REVIEW

1. *increases CC: the most populous predator of the species emigrates, all the organisms infected with a contagious disease die off; decreases CC: favorable environmental conditions boost the birth rate, a new parasite for the species immigrates into the community.* **DOK 2**
2. *Sample answer: The number pyramid before the arrival of the predator proportionally matches the data in the table. The number pyramid after the arrival of the predator shows a decrease in secondary consumers and an increase in primary consumers, producers, and tertiary consumers.* **DOK 3**
3. *Sample answer: If the disease impacts producers, the entire ecosystem could collapse. If it impacts the primary or secondary consumers, the population of organisms at lower levels would increase, while those above could collapse. If the disease impacts the tertiary consumers, the ecosystem may continue unharmed since there are so few in this trophic level.* **DOK 3**
4. *A, C, D* **DOK 3**

Science Background

Decomposition When a region is subjected to severe drought, populations of organisms that are highly dependent on water will move away, decrease, or die. When organisms die, decomposers such as fungi and bacteria break them down. However, decomposers require water as well. Acacia trees, such as the ones shown in the On Assignment photograph, are drought tolerant, meaning they can live in very dry conditions. Have students discuss why decomposers have not broken down these acacia trees. Students might conclude that the organisms that would have decomposed the dead trees did not survive the level of drought in which the trees still live.

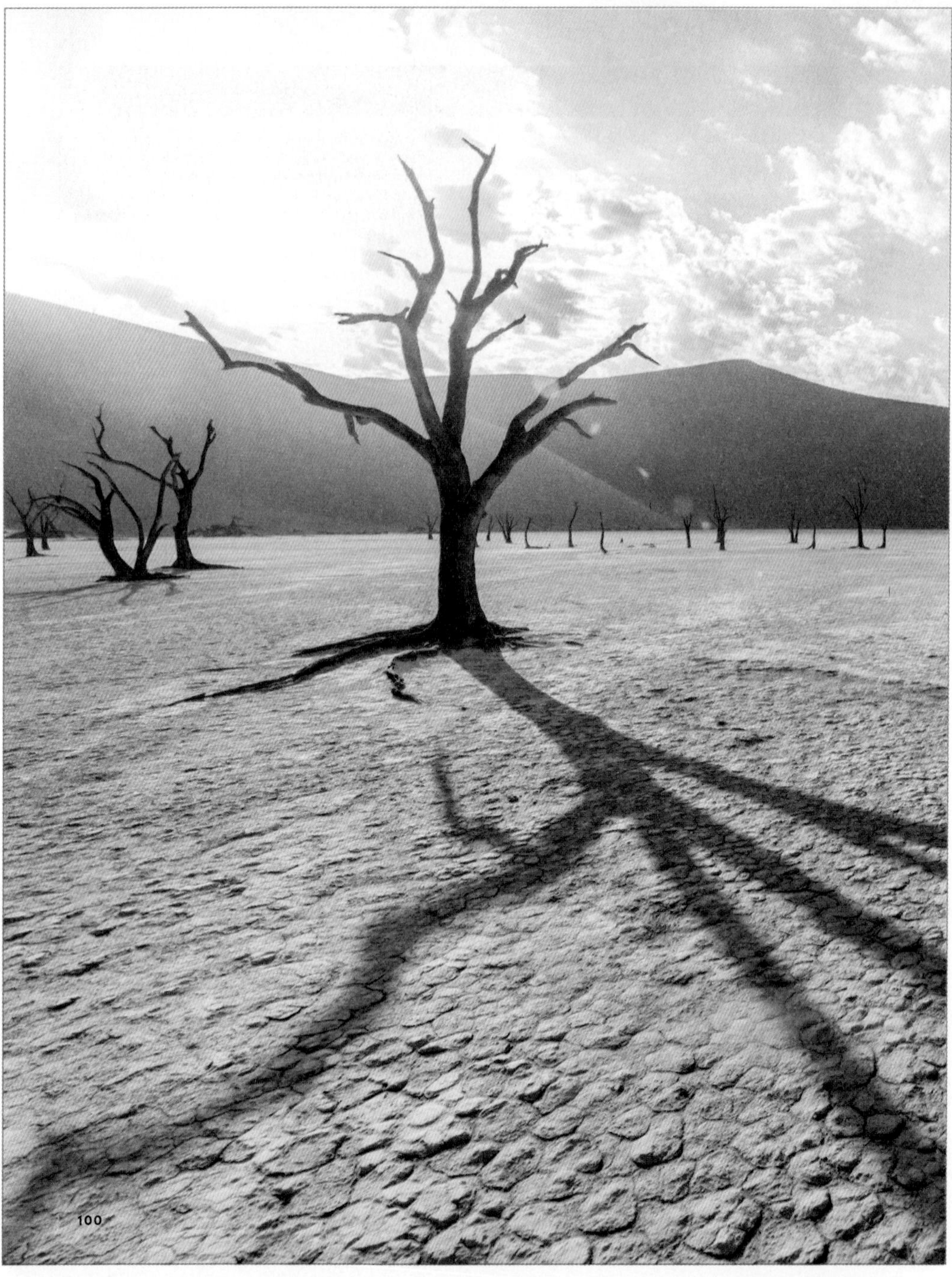

DIFFERENTIATED INSTRUCTION | English Language Learners

Use Dialogue to Argue Pair students with varying levels of language proficiency and have them engage in a role-playing exercise. Both students should imagine that they live in a region suffering from drought. One student will play the role of a citizen arguing for greater regulation of water use. The other student will play the role of a citizen who believes environmental regulations will harm the local economy.

Beginning/Intermediate Give both students time to plan and write out the evidence they will use to support their position. Then, have pairs engage in a respectful dialogue, with each partner making their case.

Advanced After pairs have concluded their debates, invite a few pairs to summarize their dialogue for the rest of the class.

ON ASSIGNMENT The Deadvlei, or "dead marsh," is located in Namibia's Namib-Naukluft National Park. Commonly referred to as "the forest frozen in time," this unique desert ecosystem, which features some of the tallest sand dunes in the world, was once a river marsh. A change in climate over 900 years ago caused the marsh to dry up. Subsequent drought conditions prevented the acacia trees from decomposing. This image was taken by National Geographic photographer Edson Vandeira.

ON ASSIGNMENT

A Changing Earth National Geographic photographer Edison Vandeira is known for his work as a photojournalist who captured images of the fires in the Brazilian Pantanal, a once lush wetland. However, he got his start as a travel photographer who traveled to remote regions and photographed them.

Through his work, Vandeira aims to draw attention to the state of the planet and provide a sense of urgency to protect the environment. He hopes that his photographs highlight how the decisions people make every day affect the environment, for good or bad. You can learn more about Edison Vandeira on his website: https://www.edsonvandeira.com

LOOKING AT THE DATA
INVASIVE SPECIES POPULATION GROWTH

Connect this activity to the Science and Engineering Practices of analyzing and interpreting data and using mathematics and computational thinking. Students analyze data shown in a chart to identify patterns in the zebra mussel population density. They interpret the data in light of the problem of invasive species population growth and use it to defend a potential solution.

SAMPLE ANSWERS

1. *The highest exponential growth rate occurred between Years 4 and 5, and the population peaked in Year 6.* **DOK 2**
2. *A disease that kills zebra mussels may have been introduced in Year 6, or Minnesota's boating regulations may have taken effect in that year.* **DOK 2**
3. *Zebra mussel populations can grow exponentially because there are few factors that control their population. When they enter an ecosystem, they quickly dominate and force out native species. If the species spreads to other aquatic ecosystems, more native species can be wiped out. Because the zebra mussels attach themselves to solid objects, they need to be removed from boats before they accidentally get transferred into a new lake. Eliminating humans as a cause for the spread of the invasive species is one way to protect native species.* **DOK 3**
4. *Divide the density in Year 3 by the density in Year 2.* **DOK 3**
5. *Sample answer: The largest per capita growth was between Year 1 and Year 2, when zebra population increased by a factor of 155 mussels/(mussel•m^2•year).* **DOK 2**

LOOKING AT THE DATA

INVASIVE SPECIES POPULATION GROWTH

SEP Analyze and Interpret Data In many states, new environmental regulations require recreational boaters to drain boats and liquid containers on land after using public freshwater lakes. Boats moving to a different lake must be fully decontaminated and inspected.

Scientists are concerned about zebra mussels, an invasive species that first appeared in Lake Erie in 1988 and now exists in at least 32 states. The mollusks attach themselves firmly to solid objects and accumulate on the surfaces of docks and boats. They harm native species, compete for food and space, and quickly overwhelm stable aquatic food webs.

A study of an inland lake in Minnesota recorded the zebra mussel population over a period of 10 years. **Figure 1** shows the combined population density of zebra mussels over time.

FIGURE 1
The density of zebra mussels in one inland lake is reported as a number of organisms per square meter.

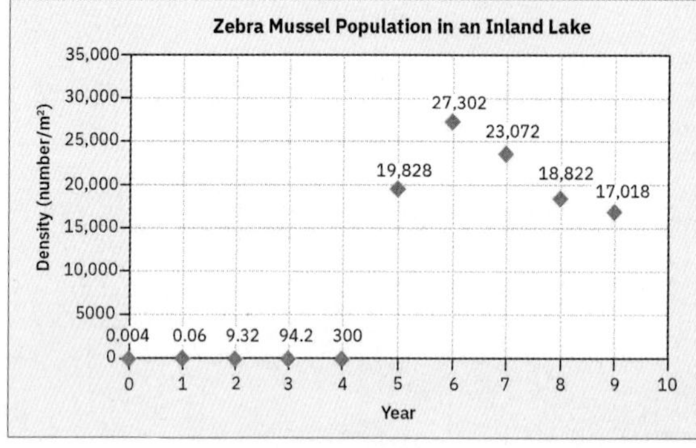

1. **Identify Patterns** During which time period did the population grow at the highest rate, and in which year did it peak?
2. **Predict** What are potential reasons for the population decline in the later years of the study?
3. **Justify** Use the historical population data and details about the role of zebra mussels in the ecosystem to justify the environmental regulations for recreational boaters.
4. **Formulate** A per capita change in a population is the amount of change "per head," or per organism in the initial population. How would you calculate the per capita change in the zebra mussel population density between year 2 and year 3?
5. **Compare** Between which two consecutive years of the study was the per capita increase in zebra mussel population density the largest?

102 CHAPTER 4 POPULATION MEASUREMENT AND GROWTH

TYING IT ALL TOGETHER ENGINEERING
SHARING OUR CITIES

WHAT FACTORS INFLUENCE THE SIZE OF A POPULATION?

In this chapter, you learned about how the characteristics of a species and its environment affect its population. The Case Study described a coyote population living in and around the city of Chicago. Human populations anywhere on Earth cohabit with populations of other native species. We share Los Angeles with mountain lions, Mumbai with leopards (**Figure 4-15**), and Hong Kong and Barcelona with wild boars. Endangered salamanders swim in spring-fed pools with the citizens of Austin, and penguins flock to the beaches of Cape Town.

When developing urban areas, humans make substantial changes to the environment that influence whether and how other species can survive there. Urbanization alters local conditions such as temperature, air and water quality, and light and noise levels. The placement of infrastructure such as roads and dams can lead to habitat fragmentation, cutting off organisms' access to vital resources. In turn, the presence of native species in urban areas can affect human life to varying degrees. While some species may be dangerous to humans, others only act as a nuisance or property hazard, and still others go nearly unnoticed.

In this activity, you will research and evaluate solutions for conserving populations of native species within an urban community. Work individually or in a group to complete these steps.

Obtain Information

1. Choose a city and research the effects of urban development or human interactions on the population of a native species that lives and interacts with humans there.

Propose Solutions

2. Describe at least two actions humans have taken, or could potentially take, to conserve the population you researched.

Evaluate Solutions

3. Compare each of the actions you described in terms of meeting constraints such as safety, cost-effectiveness, environmental impact, and social acceptability. Write an argument to support the solution you think would be most effective based on its ability to operate within these constraints.

FIGURE 4-15
Leopards visit a residential area near Mumbai, India, to prey on dogs that forage for garbage left out by humans.

SCIENCE AND ENGINEERING PRACTICES
Constructing Explanations and Designing Solutions

Solutions Involving Behavior Change Consider reviewing the engineering process that was introduced in Chapter 2 to help students navigate this activity. In the Tying It All Together Activity, rather than evaluating various designs of a physical object, students will likely be generating or addressing solutions that involve changing the behavior of people or animals. In addition to evaluating these solutions based on constraints such as cost, safety, and environmental impacts, students should also consider defining specific constraints related to public acceptance of the solutions. For example, some native species may have significance to indigenous human populations in your area, or they may play a role in the local economy, perhaps by drawing tourists or for recreational or subsistence hunting or foraging.

TYING IT ALL TOGETHER

ELABORATE

The Case Study describes a coyote population living in and around the city of Chicago and how human populations and coyotes cohabit other regions of North America. In the Tying It All Together activity, students research the effects of urbanization on a native species and argue for actions to help conserve that population.

This activity is structured so that students may complete the steps using a simplified design process. **Step 1** has students obtaining the information about the organisms they wish to study. In **Step 2**, students will propose solutions to conserve the population they are studying. In **Step 3**, students evaluate their solutions in terms of constraints.

While this activity may be performed at any point during the lesson, it would be best used as a culminating project at the end of the chapter.

Go online to access the Student Self-Reflection and Teacher-Scoring rubrics for this activity.

Social-Emotional Learning

Social Awareness and Responsible Decision-Making Some students may have strong feelings about the presence of coyotes or other displaced animal species in urban or suburban settings to a greater degree than in rural areas where those species might be expected. Use this as an opportunity to pair students with differing opinions and guide them to listen to their partners and try to *take others' perspectives*. Then, in class discussion, have students *identify possible solutions for this social problem* of human impact on native species.

EVALUATE

REVIEW KEY CONCEPTS

1. *C* **DOK 2**
2. *A* **DOK 2**
3. *D* **DOK 2**
4. *D* **DOK 1**
5. *density-dependent: pine bark beetles eating pine trees, lack of food resources due to drought; density-independent: flooding after a heavy rain, lava flow covering a forest* **DOK 2**
6. $b < d$ **DOK 2**
7. *Sample answer: It depends on the scale at which the observations are made. A population may be uniformly distributed within an area that has good resources, but a wider perspective shows that the population is clumped around those resources.* **DOK 2**
8. *Sample answer: If the organisms do not reach reproductive age quickly, it will take a long time for the population to grow.* **DOK 3**
9. *A, D* **DOK 2**
10. *uniform* **DOK 2**
11. *D, E* **DOK 3**
12. *B* **DOK 1**

CRITICAL THINKING

13. *Sample answer: Plot sampling would work well because the trees do not move, and the different areas of the forest could be evaluated for distribution of resources.* **DOK 3**
14. *Sample answer: A mark-recapture project assumes that the capture randomly samples from the population. If the animals being captured have a feature that makes them more likely to be captured, then it is not a random sample of the population. Similarly, if marking the animal makes it more or less likely to be captured, then the second sample will not be random.* **DOK 3**
15. *Sample answer: The frog population is likely close to carrying capacity when the fungus is introduced. After the fungus is introduced, the frog population would rapidly decrease. The pond might reach a new lower carrying capacity of frogs, or the frogs may become extinct if they generally die before they can reproduce. The initial fungus population is small but will initially experience rapid growth as it spreads between frogs.* **DOK 3**
16. *Sample answer: The exponential growth illustrated by the J-curve never reaches a limit, while the S-curve levels off. The presence of density-dependent factors, such as limited resources, increased competition, and disease, cause the S-curve to level off.* **DOK 2**
17. *Sample answer: Organisms with a type I survivorship curve produce relatively few offspring and invest time and energy into each one. This approach works best when available resources and challenges are predictable. If environmental conditions change unexpectedly, then the likelihood of any of those offspring surviving is very low. In circumstances with wildly fluctuating conditions, producing a large number of offspring increases the chances that a few may survive the varying conditions, so in those circumstances, a type II approach is better.* **DOK 3**
18. *Some possible answers include: How many of their offspring usually survive to reproductive age? How does the death rate change over the course of the life of the hamster? Does the hamster*

CHAPTER 4 **REVIEW**

REVIEW KEY CONCEPTS

1. **Calculate** In a meadow, 10 rabbits are captured, marked, and released. The next day, 10 rabbits are captured. Of the recaptured rabbits, 2 have been marked. What is a reasonable estimate for the size of the rabbit population?
   - **A.** between 2 and 10
   - **B.** 10
   - **C.** 50
   - **D.** 100
2. **Interpret** In the sample described in question 1, half of the rabbits had dark gray fur and the rest had brown fur. What can be assumed about the coloration of the population of rabbits?
   - **A.** About half will have dark gray fur.
   - **B.** Gray is the most common fur color.
   - **C.** Gray fur increases the likelihood of being captured.
   - **D.** There are likely many other colors of rabbits in the meadow.
3. **Predict** Which factor would likely result in a clumped population distribution?
   - **A.** flat topography
   - **B.** competition for space
   - **C.** long-distance dispersal of seeds
   - **D.** unequal distribution of resources
4. **Define** How do density-dependent factors affect a population?
   - **A.** They cause a sudden decline in the population.
   - **B.** They cause exponential growth of the population.
   - **C.** They cause the carrying capacity to increase as the population grows.
   - **D.** They slow exponential growth until the population levels out at carrying capacity.
5. **Classify** Categorize the following examples as density-dependent or density-independent factors.
   - lava flow covering a forest
   - flooding after a heavy rain
   - pine bark beetles eating pine trees
   - lack of food resources due to drought
6. **Relate** Construct a mathematical expression that relates birth rate b and death rate d when a population with no migration is shrinking.
7. **Explain** How can a population be described as having a uniform and a clumped distribution at the same time?
8. **Identify Cause and Effect** A population of organisms lives in ideal environmental conditions. However, the population is not growing rapidly. What life history factors might cause this?
9. **Classify** Which reason(s) would justify the use of a plot sampling method to estimate the size of a population? Select all correct answers.
   - **A.** The organism being sampled has limited mobility.
   - **B.** The distribution of resources within the area is unknown.
   - **C.** There are large differences in population density within the area.
   - **D.** You are able to distinguish organisms that have previously been counted.
   - **E.** Rapid and extreme changes in conditions are occurring throughout the area.
10. **Classify** A species of plants produces toxic chemicals that prevent any seeds from germinating in the soil nearby. What word best describes the population distribution that will likely result?
11. **Analyze** How does the ability to attach tracking devices to animals help a scientist study the population to which they belong? Select all correct answers.
    - **A.** It increases the animal's ability to find a mate.
    - **B.** It increases the likelihood of the animals surviving.
    - **C.** It allows the scientist to change the way the animals behave.
    - **D.** It allows the scientist to learn about the range of the population.
    - **E.** It allows the scientist to mark the animals for estimating population size.
12. **Identify** What kinds of changes can lead to an increase in carrying capacity?
    - **A.** More organisms immigrate into the area.
    - **B.** More resources become available in the area.
    - **C.** More waste by-products accumulate in the area.
    - **D.** Carrying capacity is fixed and cannot be changed.

CRITICAL THINKING

13. **Justify** What method would you use to estimate the population of maple trees in a forest? Explain your reasoning.

14. **Evaluate** What factors could make a population estimate from a mark–recapture project inaccurate?

15. **Analyze** In a pond, a species of frogs lives at a high population density. A fungus that causes chytridiomycosis is then introduced into the pond. Describe how a population vs. time graph would look for

    a. the frogs, before and after the introduction of the fungus.

    b. the fungus.

    What circumstances cause the two graphs to look different?

16. **Contrast** What environmental circumstances result in a J-shaped population curve? What circumstances result in an S-shaped curve?

17. **Analyze** Why are stable environmental conditions necessary for the survival of organisms that have a type I survivorship curve?

18. **Ask Questions** Evaluate the following claim: *Hamsters can have large litters of young, sometimes more than 20, and this means hamsters have a type III survivorship curve.* What kinds of questions should you ask about the life history of the hamster that might help you support or refute this claim?

19. **NOS Scientific Knowledge** Critter cams, radio tags, and environmental DNA help scientists track organisms in the wild. How can evidence from new technologies help confirm or change our scientific knowledge about organism populations?

MATH AND ENGLISH LANGUAGE ARTS CONNECTIONS

1. **Write an Argument** Use your knowledge of population growth factors to explain why introducing a few individuals of a new species into a new location could be problematic. Use your understanding of biodiversity to frame your argument.

2. **Write a Function** Develop a formula to calculate the size of a population based on mark–recapture data. Use *P* for the estimated population size, *c* for the number of organisms captured the first time, *r* for the number of organisms captured the second time, and *m* for the number of marked organisms that were captured the second time.

3. **Reason Quantitatively** A population of organisms grows so that its population vs. time graph forms an S-shaped curve. Describe the slope of a line that is parallel to the curve

    a. at the bottom part

    b. in the middle part

    c. at the top part.

4. **Graph Functions** Graph the organism data given in the table. Describe the graph and classify the survivorship curve of this organism.

Number surviving	Percent of life span
100	5
91	75
82	93
45	96
10	99

5. **Write Informative Text** Choose an organism. Describe its life history and determine the type of survivorship curve it likely has.

6. **Model with Mathematics** The expression $r = b - d$ describes the per capita growth rate *r* of a population with birth rate *b* and death rate *d*. How could emigration (*e*) and immigration (*i*) be included in the expression for *r*?

REVISIT SEA PIG SURVIVAL

Gather Evidence In this chapter, you have learned about factors that affect population size. Many useful mineral resources exist on the ocean floor, so several nations and companies have been exploring the potential of seabed mining.

1. Would human attempts to explore and obtain minerals from the seafloor act as a density-dependent or a density-independent factor? Explain your answer.

2. How might mining the ocean floor affect the carrying capacity for a sea pig population?

REVISIT THE ANCHORING PHENOMENON

1. *Sample answer: These activities would be density-independent factors because they would affect all populations that live in the regions that are explored or mined, regardless of how many organisms form the population.* **DOK 2**

2. *Sample answer: If mining operations removed sediments containing food from the sea pigs' habitat, the carrying capacity of their environment would decrease.* **DOK 3**

care for its young? How often do hamsters reproduce? At what age do they reproduce? **DOK 3**

19. *Sample answer: New tracking technologies give information about how specific individuals in a population behave and where they go. Genetic technology can identify unexpected species that live in or use an environment. Both of these types of information would update our knowledge about species interactions in an ecosystem, which is then used to make more effective decisions about ecological conservation.* **DOK 4**

MATH AND ELA CONNECTIONS

1. *Sample answer: Even though there are only a few individuals, if the conditions are ideal, they could reproduce rapidly and quickly become a large population, growing exponentially. A large population has the potential to alter the ecosystem through disrupting food webs or the environment. The biodiversity of the area would be altered by having a new organism and by the impact the new organism would have on the native organisms. If the new organism displaces native organisms, the biodiversity would decrease.* **DOK 3**

2. $P = (c \cdot r) / m$ **DOK 2**

3. *At the bottom of the S-curve, the growth is very slow, and the slope of the line is positive but close to zero. At the middle of the S-curve, the slope is steeper, as the population grows rapidly; the slope is a very large number. At the top of the S-curve, the slope is nearly flat, close to zero; it may be negative if the population is shrinking.* **DOK 2**

4. *The graph should show a large number of organisms surviving until late in life. This is a type I survivorship curve.* **DOK 2**

5. *Answers will vary based on the organism picked. Sample answer: Elephants reach reproductive age after several years and produce just one calf at a time. They care for that calf. Most elephants reach adulthood, and mortality only becomes high at the end of the life span. These traits are frequently seen in organisms with a type I survivorship curve.* **DOK 3**

6. $r = b + i - d - e$ **DOK 3**

Class Discussion

Encourage students to practice speaking, listening, and collaborative skills when discussing these questions as a class or in smaller groups as they review the topics in Unit 1.

1. How can a food web model help represent the biodiversity and resilience of an ecosystem? What are the limitations of the model?
2. Consider a population of apex predators and their prey that thrive in an old growth forest. If that forest was destroyed in a forest fire, what do you think would happen to the populations of the predators and the prey over time—immediately after the fire and as the ecosystem goes through succession?
3. A change in the phosphorus cycle leads to an influx of phosphorus into a lake ecosystem. What are some potential effects on the carrying capacity of the ecosystem in the short term and in the long term?

Go Online

Five Performance Tasks for Unit 1 are available on MindTap to assess students' mastery of the following NGSS Performance Expectations:

- Task 1: HS-LS2-4 and HS-LS2-5
- Task 2: HS-LS2-3 and HS-LS2-4
- Task 3: HS-LS2-1 and HS-LS2-2
- Task 4: HS-LS2-6 and HS-LS2-2
- Task 5: HS-LS2-7 and HS-ETS1-3

Go Online

The Virtual Investigation, **Sea Pigs on the Abyssal Plain**, includes observational evidence that students may find useful for supporting their claims about how sea pigs survive in the deep ocean.

UNIT 1 **SUMMARY**

RELATIONSHIPS IN ECOSYSTEMS

CHAPTER 2
ENERGY AND MATTER IN ECOSYSTEMS

How do energy and matter move through an ecosystem?

- Living things need the input of matter and energy to survive, grow, and reproduce. The interaction between biotic and abiotic components of an ecosystem allows for the transfer of matter and energy throughout the system.
- Both matter and energy are conserved in Earth's ecosystems.
- Energy and matter transformations occur through chemical reactions. Two important chemical reactions in ecosystems are photosynthesis and cellular respiration.
- The movement of matter and energy through an ecosystem can be modeled using food webs, ecological pyramids, and matter cycles.

CHAPTER 3
BIODIVERSITY AND ECOSYSTEM STABILITY

How is biodiversity related to ecosystem stability?

- Ecosystems that have more biodiversity are more stable than ecosystems with little biodiversity.
- Biodiversity can be measured in terms of species richness and species evenness.
- Natural and human-caused disturbances affect ecosystem stability.
- Resistant ecosystems experience minimal change after a disturbance. Resilient ecosystems recover rapidly after a disturbance and quickly return to their original state.
- Ecosystem recovery after a disturbance occurs in predictable ways.

CHAPTER 4
POPULATION MEASUREMENT AND GROWTH

What factors affect the size of a population?

- A population with a constant growth rate greater than zero and access to unlimited resources will increase exponentially.
- The carrying capacity of an environment is the maximum number of individuals of a species than an environment can support indefinitely.
- Density-dependent factors, such as competition and parasitism, are limiting factors that are affected by the number of individuals in an area.
- Density-independent factors, such as natural disasters and human activity, are limiting factors that affect a population regardless of its size.

VIRTUAL INVESTIGATION

Sea Pigs on the Abyssal Plain
How do sea pigs survive in the deep ocean?
Take control of a remotely operated vehicle (ROV) to observe organisms on the ocean floor.

HOW DO SEA PIGS SURVIVE IN THE DEEP OCEAN?

Argue From Evidence In this unit, you learned that the energy for life processes originates from the sun. Nevertheless, diverse ecosystems thrive on the ocean floor at sunless depths between 3 and 6 kilometers, in generally flat regions called abyssal plains. These plains make up approximately 70 percent of the global seabed.

The Monterey Bay Aquarium Research Institute (MBARI) has monitored an abyssal ecosystem off the central coast of California for 30 years. Their ROV takes frequent high-resolution images of the seafloor and its inhabitants and has measured carbon and oxygen levels to determine what the animals and bacteria there eat and produce. Sea pigs are abundant in this ecosystem. Their main food source is fresh "marine snow," the dead organic matter and waste that sinks from shallow waters to the ocean floor.

Although they share their habitat with many other species of sea cucumbers, octopuses, marine worms, mollusks, crabs, fish, and coral, sea pigs are not usually a food source for other organisms.

MBARI's data indicate that a typical slow and steady supply of marine snow would not provide enough food to support the number of consumers they find in the abyss, but occasional large bursts of matter falling from events at or near the ocean surface can sustain populations deep below for long periods of time.

 **Claim** Make a claim about the role of sea pigs in an abyssal plain ecosystem.

 **Evidence** Use the evidence that you gathered throughout the unit to support your claim.

 **Reasoning** To help illustrate your reasoning, you can develop a model that explains how abyssal plain organisms obtain the energy and matter they need to survive.

In 2019, MBARI located the remains of a baleen whale on the ocean floor. A fallen whale attracts a changing assortment of organisms: larger scavengers feed on the whale's soft parts, smaller organisms look for decomposing tissue in the surrounding sediment, and bacteria and other microbes access the fat trapped inside the whale's bones.

DIFFERENTIATED INSTRUCTION | English Language Learners

Cite Text Evidence Support students by allowing them to discuss their ideas verbally before beginning to write and by allowing different degrees of detail in students' responses.

Beginning Encourage students to construct all or part of their CER in their native language and use a translator tool to compare it with the English version.

Intermediate Point students to their personal dictionaries and the class word wall to add details to their evidence statements.

Advanced Suggest pairs first discuss their reasoning before they write individual paragraphs stating it.

REVISIT THE ANCHORING PHENOMENON

Claim

Evidence

Reasoning

This Unit Activity asks students to formulate a claim to answer the question posed at the beginning of the unit, cite evidence they obtained throughout the unit, and use the concepts they learned to explain how that evidence supports their claim.

C Formulating a Claim Students should write their answer to the question "How do sea pigs survive in the deep ocean?" in the form of a single sentence. Examples of claims that students might make using terms and concepts from the unit chapters include: "Sea pigs get the matter and energy they need to survive from other dead organisms," or "Sea pigs depend on consumers and decomposers for the food they need to survive," or "Sea pig populations are sustained by the remains of large marine animals."

E Citing Evidence Have students review their answers to the Sea Pig Survival questions in Sections 2.2, 3.1, and 4.3 and in each Chapter Review as they determine what evidence to use to support their claim. If you wish to have students look for evidence from sources other than this text, they can find useful information on the internet about the MBARI Pelagic-Benthic Coupling expeditions and the NOAA Discovering the Deep: Exploring Remote Pacific Marine Protected Areas expedition.

R Explaining Reasoning In addition to their explanatory writing, models such as food webs and matter cycle diagrams are convenient structures around which students may build their explanations.

Go online to access the Student Self Reflection and Teacher Scoring rubrics for this activity.

Unit Anchoring Phenomenon: Bacteria in Your Intestines

Use the Driving Question to help frame the Anchoring Phenomenon as an investigable subject and motivate student learning. Leverage the bacteria in your gut prompts within each chapter to connect concepts back to the unit's Driving Question, supporting students in gathering evidence and asking their own research questions so they are equipped to complete the Unit Activity.

NATIONAL GEOGRAPHIC

Meet the Explorer

Katie Amato is a biological anthropologist who studies the symbiotic relationship between the human gut and the microbes that live there by studying a similar relationship in howler monkeys. Watch Unit Video 2, Explorers at Work: Katie Amato to engage student interest in microbiology research and the Anchoring Phenomenon.

Virtual Investigation

Bacteria in the Digestive System Students learn about roles of the microbiome in human health and gather evidence to describe how a variety of microorganisms live and interact inside the human gut

NGSS Progression

Middle School

- **MS-LS1-3** Use argument supported by evidence for how the body is a system of interacting subsystems composed of groups of cells.
- **MS-LS1-7** Develop a model to describe how food is rearranged through chemical reactions forming new molecules that support growth and/or release energy as this matter moves through an organism.
- **MS-LS2-3** Develop a model to describe the cycling of matter and flow of energy among living and nonliving parts of an ecosystem
- **MS-LS1-6** Construct a scientific explanation based on evidence for the role of photosynthesis in the cycling of matter and flow of energy into and out of organisms.

High School

- **HS-LS1-1** Construct an explanation based on evidence for how the structure of DNA determines the structure of proteins, which carry out the essential functions of life through systems of specialized cells.
- **HS-LS1-4** Use a model to illustrate the hierarchical organization of interacting systems that provide specific function within multicellular organisms.
- **HS-LS1-5** Use a model to illustrate how photosynthesis transforms light energy into stored chemical energy.
- **HS-LS1-6** Construct and revise an explanation based on evidence of how carbon, hydrogen, and oxygen from sugar molecules may combine with other elements to form amino acids and/or other large carbon-based molecules.
- **HS-LS1-7** Use a model to illustrate that cellular respiration is a chemical process whereby the bonds of food molecules and oxygen molecules are broken and the bonds in new compounds are formed resulting in a net transfer of energy.
- **HS-LS2-3** Construct and revise and explanation based on evidence for the cycling of matter and flow of energy in aerobic and anaerobic conditions.
- **HS-LS2-5** Develop a model to illustrate the role of photosynthesis and cellular respiration in the cycling of carbon among the biosphere, atmosphere, hydrosphere, and geosphere.

C E R **Claim, Evidence, Reasoning** Students can make a claim about the unit phenomenon, gather evidence, and revisit their claim periodically to evaluate how well the evidence supports it. The Driving Question presented in the Case Study of each chapter can get students invested in chapter topics and in working toward answering the unit's Driving Question, "What do bacteria do in your intestines?" In the Unit Activity, students can practice scientific reasoning and argumentation to show how the evidence supports their claim.

Follow the Anchoring Phenomenon What do bacteria do in your gut?

Gather evidence with . . .	Chapter 5	Chapter 6	Chapter 7	Unit Activity
CASE STUDY	How are the molecules of life assembled?	How do structures work together in cell systems?	How do cells divide and grow?	Revisit the unit's Anchoring Phenomenon of bacteria that live in the human gut and how that relationship benefits both. **Claim, Evidence, Reasoning** Students use the evidence they gathered throughout the unit to support their claim with reasoning. **Go online** to access Student Self Reflection and Teacher Scoring rubrics for this activity. **English Language Learners** Retelling with Visuals
MINILAB	How can you determine if a substance is polar or nonpolar?	How do molecules move through the cell membrane?	How can cell growth be modeled?	
LOOKING AT THE DATA	Students will analyze and interpret data to learn about lactose intolerance.	Students will calculate and use data to compare bacterial cell counts in the human body.	Students will simulate identifying cancerous cells in a DNA microarray.	
TYING IT ALL TOGETHER REVISIT THE CASE STUDY	Students will research a specific plant alkaloid to determine how it was developed and its effects on the human body.	Students will research how artificial photosynthesis works and two approaches being used to develop it.	Students will research what makes HeLa cells so unique that they make powerful research tools.	
Chapter Review: Revisit Bacteria in the Gut	Reflect on how bacteria synthesize and break down chemical compounds in the gut.	Consider how antibiotics can affect the gut and lead to further illness.	Reflect on how bacteriophages affect growth of bacteriophages in the gut.	
Virtual Investigation: Bacteria in the Digestive System	Students take on the role of a microbiologist to gather evidence to describe how a variety of microorganisms live and interact inside the human gut.			
Chapter Investigation A	How does salivary amylase affect the carbohydrates in our food?	What are factors that affect cellular respiration in yeast?	How do rates of cell division relate to plant growth?	
Chapter Investigation B	Which type of detergent is most effective at breaking down protein-based substances?	How can we build and test a photobioreactor that maximizes the production of oxygen and algae on a space station?	Can leaf structures be identified based on the appearance of differentiated cells?	

Planning Your Investigations

Each chapter features a Minilab and two full Investigations that offer hands-on opportunities for students to engage in Science and Engineering Practices.

Advance Preparation For Chapter 5, you will need to collect five different polar and non-polar substances, as well as prepare solutions of amylase, starch, and maltose. For Chapter 6, you will need to prepare solutions and obtain *Chlorella* and alga growth mediums. For Chapter 7, you will need to gather common flowering plants. Advance preparation notes for all labs are included in the Teacher's version.

Chapter 5	Title	Time	Standards
Minilab	*Polar vs. Nonpolar Molecules*	30 minutes	SEP Plan and Conduct an Investigation
Investigation A **Guided Inquiry**	*Converting Carbohydrates*	50 minutes plus advance preparation	HS-LS1-6
Investigation B **Open Inquiry**	*Crime Scene Cleaners*	200 minutes plus advance preparation	HS-LS1-6
Chapter 6	**Title**	**Time**	**Standards**
Minilab	*Selectively Permeable Membranes*	30 minutes	CCC Stability and Change
Investigation A **Guided Inquiry**	*Factors Affecting Cellular Respiration*	100 minutes over 2 days plus advance preparation	HS-LS1-7
Investigation B **Open Inquiry**	*Designing a Photobioreactor*	160 minutes over 4 days	HS-LS1-5, HS-ETS1-2
Chapter 7	**Title**	**Time**	**Standards**
Minilab	*Modeling Mitosis*	30 minutes	SEP Developing and Using Models CCC Systems and System Models
Investigation A **Guided Inquiry**	*Plant Growth through Mitosis*	50 minutes plus advanced preparation	HS-LS1-4
Investigation B **Open Inquiry**	*Cell Differentiation in Plant Leaves*	50 minutes	HS-LS1-4

Assessment Planning

UNIT 2 PERFORMANCE TASKS

Three performance-based assessments target the NGSS Performance Expectations with three-dimensional activities to measure student mastery. Rubrics are available online for Performance Tasks.

Performance Tasks

	Title	Overview	PEs Addressed	Use After
1	*How Does Regenerative Medicine Reflect Nature?*	Students conduct research to learn how select organisms undergo regeneration of body parts and how that understanding can be applied in human medicine.	HS-LS1-6	Use after Chapter 7
2	*What Arc the Requirements for a Minimum Viable Ecosystem?*	Students design a paludarium and use the blueprints to model the energy and matter flows in an ecosystem.	HS-LS1-5 HS-LS1-7	Use after Chapter 6
3	*How Are Complex Carbon-Based Molecules Built from Simple Atoms?*	Students design a game where players build up or break down biological molecules using criteria and constraints they identify.	HS-LS1-4	Use after Chapter 5

Additional Resources

Use a search engine to find these resources on the internet.

VIDEOS

"How our microbes make us who we are." TED Talk
A 17-minute talk on how the community of bacteria living inside our bodies have a huge, and largely unexplored, role in our health.

Search: microbes human TED

ARTICLES

"How the Microbiome Could Be the Key to New Cancer Treatments" Smithsonian
An article from Smithsonian Magazine discusses new research into how cancer drugs interact with the gut microbiome.

Search: microbiome cancer smithsonian

"Biologists Discover Unknown Powers in Mighty Mitochondria" Quanta
An article that describes research into how mitochondria regulate cell activities.

Search: mighty mitochondria quanta

"New Clues to Chemical Origins of Metabolism at Dawn of Life" Quanta
An article describing how primordial cells may have developed metabolisms.

Search: chemical metabolism smithsonian

"Solving Peto's Paradox to better understand cancer" Proceedings of the National Academy of Sciences
An article discussing how different animal species have evolved to deal with cancer.

Search: peto cancer pnas

Unit Storyline

In this unit, students will answer the question, "What do bacteria do in your intestines?" Each chapter provides part of the answer to this Driving Question.

Chapter 5 *How are molecules of life assembled?"*

Students learn about the different types of chemical bonds between atoms that form molecules and compounds and the forces that determine the strength of those bonds. They also learn what happens in a chemical reaction.

Chapter 6 *How do structures work together in cell systems?*

Students explore prokaryotic and eukaryotic cells and the structures found within each. The focus then turns to the structure and function of the cell membrane, including movement of materials in and out of the cell and finally, an exploration of photosynthesis and cellular respiration.

Chapter 7 *What factors affect the size of a population?*

Students learn about cell growth and division with the prokaryotic and eukaryotic cell cycles, with focus on the stages of mitosis, followed by cell differentiation in both animals and plants.

Unit Activity Students revisit the Anchoring Phenomenon to make a claim about the Driving Question and apply reasoning to the evidence they have gathered throughout the unit.

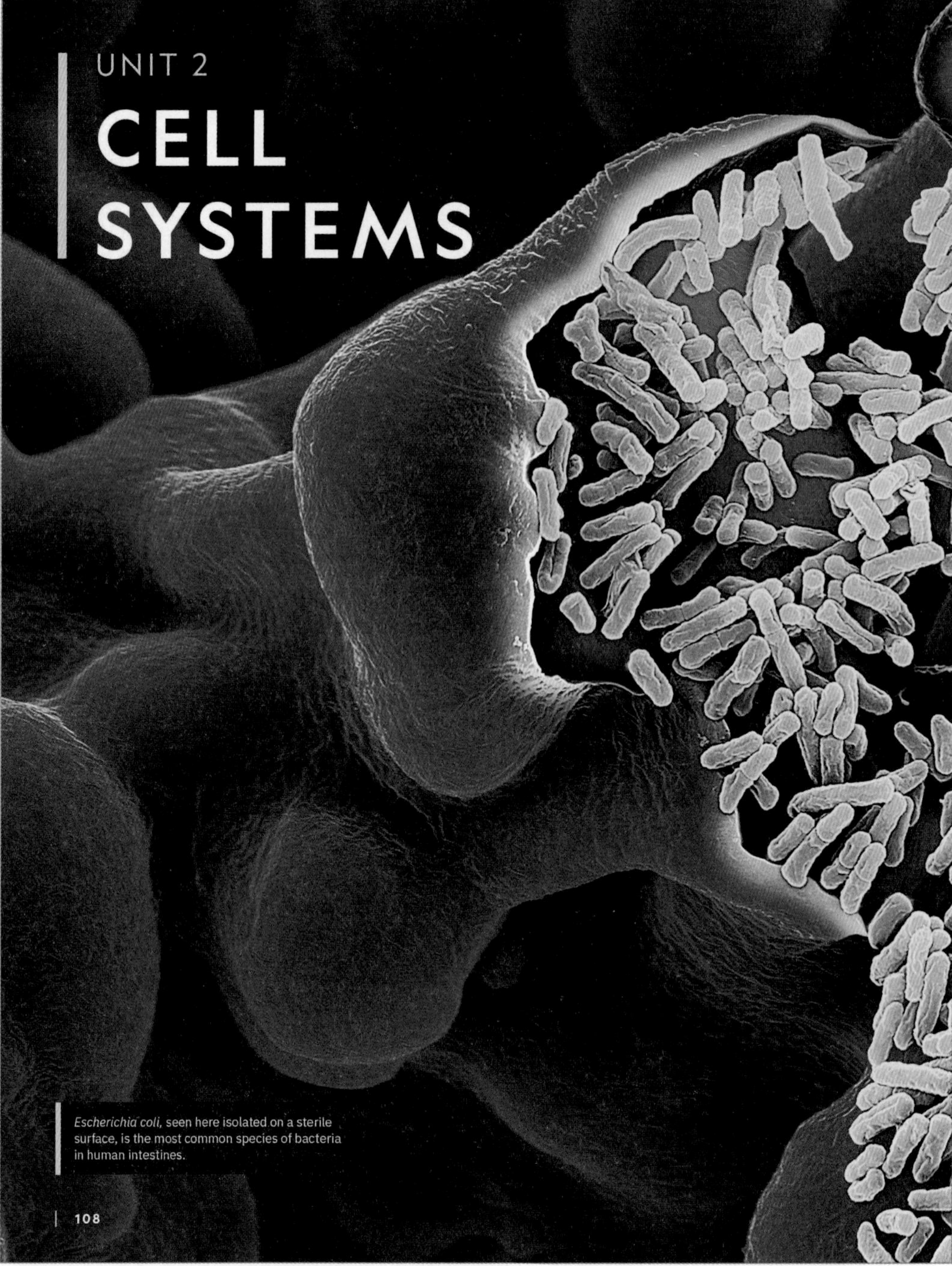

Escherichia coli, seen here isolated on a sterile surface, is the most common species of bacteria in human intestines.

SCIENCE BACKGROUND

Large Intestines Few microorganisms are found in the stomach and small intestine due to the effects of gastric acids and other secretions that are antimicrobial. However, this changes dramatically in the transition to the large intestine where bacterial populations increase significantly as the environment changes.

Bacteria Bacteria are single-celled prokaryotes that are found almost everywhere on Earth, even hundreds of meters below the ice of Antarctica. The bacteria that reside in the human gut are primarily anaerobes, outnumbering aerobic bacteria by 100 to 1000:1. These bacteria help with the final stages of digestion, as well as the production of vitamins, such as vitamin K, that are absorbed across the intestinal wall into the bloodstream. The type of bacteria found in the colon is influenced by age, diet, and other factors. Antibiotics also have an effect on the microbiota, including those that are ingested by eating products in which antibiotics have been given to animals.

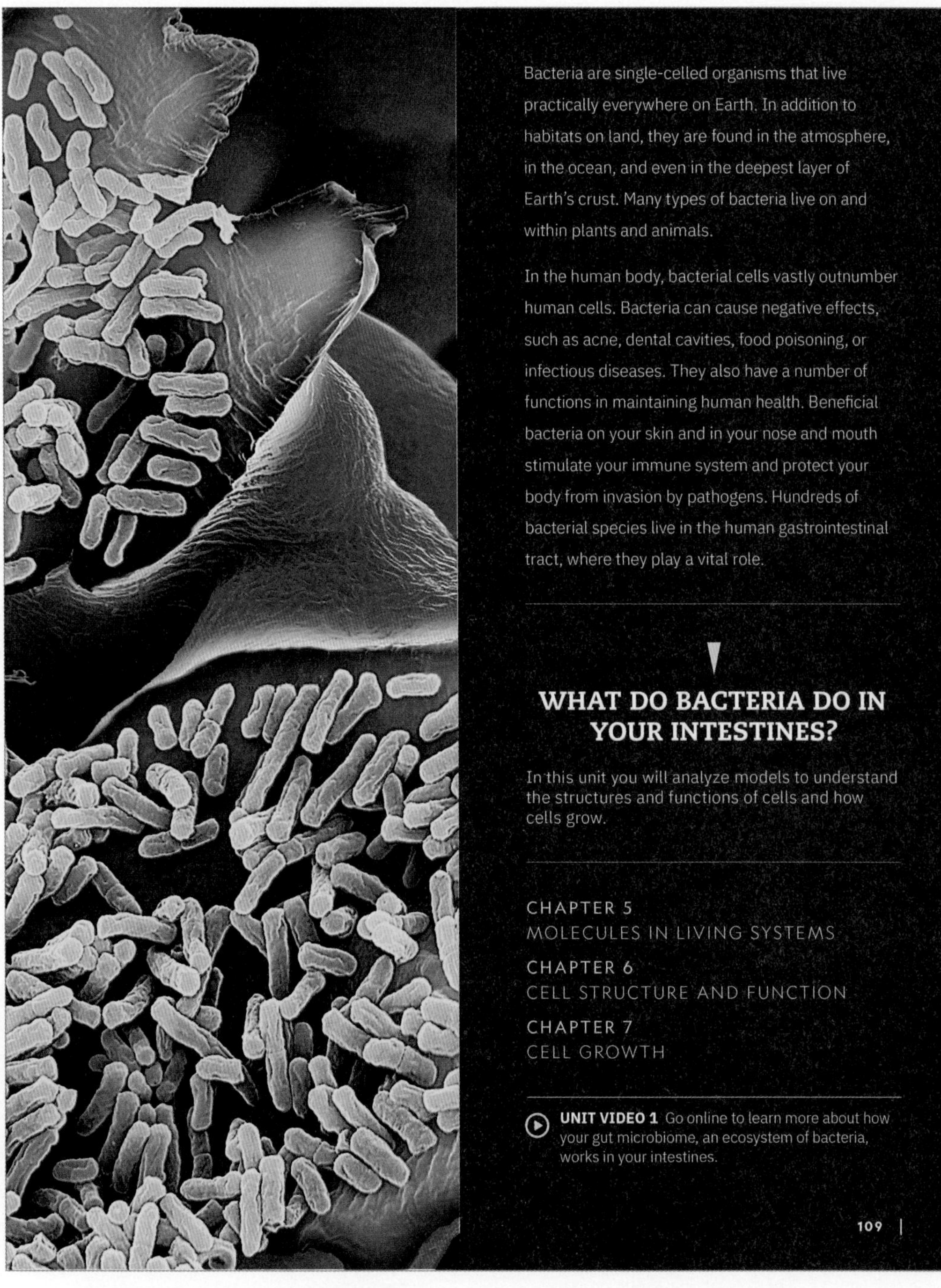

Bacteria are single-celled organisms that live practically everywhere on Earth. In addition to habitats on land, they are found in the atmosphere, in the ocean, and even in the deepest layer of Earth's crust. Many types of bacteria live on and within plants and animals.

In the human body, bacterial cells vastly outnumber human cells. Bacteria can cause negative effects, such as acne, dental cavities, food poisoning, or infectious diseases. They also have a number of functions in maintaining human health. Beneficial bacteria on your skin and in your nose and mouth stimulate your immune system and protect your body from invasion by pathogens. Hundreds of bacterial species live in the human gastrointestinal tract, where they play a vital role.

WHAT DO BACTERIA DO IN YOUR INTESTINES?

In this unit you will analyze models to understand the structures and functions of cells and how cells grow.

UNIT VIDEO 1 Go online to learn more about how your gut microbiome, an ecosystem of bacteria, works in your intestines.

109 |

Introduce the Anchoring Phenomenon

Bacteria and Human Health This text introduces the phenomenon of bacteria in the human body. There are many well-known negative effects of bacteria in the body, often resulting in a visit to a doctor and a prescription for antibiotics. However, many bacteria function to maintain our health, including those that reside in the intestinal tract and benefit human hosts by helping to break down food, make important vitamins, and build immunity.

Driving Question Prompt students to think of thier own experiences with bacteria by mentioning a visit to a doctor to get a prescription for an antibiotic to kill harmful bacteria causing an illness. Elicit from students if they are aware of any ways bacteria are helpful to the human body. Point out that many helpful bacteria reside in your intestines and elsewhere on and in the body. What students learn in this unit will inform their analysis of models of chemical reactions, cellular processes, and cell structure and growth.

About the Photo This is a false-color scanning electron micrograph of *Escherichia coli* bacteria (yellow) living in the intestine of a mammal. *E. coli* are common inhabitants of the gut of all mammals, including humans.

Video

Time: 2:39

Use **Unit Video 1**, from National Public Radio (NPR) to give students an overview of microbes in the human body, from their origins to their locations to their influence on human health.

Go Online VIRTUAL INVESTIGATION

Bacteria in the Digestive System

Time: about 50 minutes

Objectives Students learn about role of the microbiome in human health and gather evidence to describe how a variety of microorganisms live and interact inside the human gut.

Explore and Learn Students explore a microbiology laboratory to discover equipment and techniques scientists use to study bacterial growth.

Collect Data Students use a variety of laboratory equipment and techniques to study bacterial growth. They will compare results from microscope studies, growth experiments, and microfluidic arrays to develop an understanding of the factors that affect the growth of bacteria, and the roles of those bacteria in human health.

Analyze and Report Students answer questions about the investigation, incorporating qualitative and quantitative data in a report to support their analyses.

EXPLORER
KATIE AMATO

ENGAGE

About the Explorer

National Geographic Explorer Katie Amato always liked observing things. Today, she is learning more about the human gut microbiome by studying primates, specifically "New World" monkeys, such as howler monkeys, and "Old World" monkeys, such as baboons. Baboons share 99% of their DNA with humans, making them good models for learning about bacteria in the human gut. Dr. Amato is interested in discovering the effect of bacterial populations in the human gut to nutrition and health in different global populations, some of which may have severely limited access to nutritious foods. With this information, along with data about other factors, Dr. Amato's goal is to understand the impact of diet on the gut.

On the Map

Palenque National Park Dr. Amato conducted field studies of howler monkeys in the lowland rainforest of Palenque National Park, near the Mayan ruins in Chipas, Mexico. Although this area was disturbed during the 2,000 year occupation by the Maya, much of the land recovered and remains undisturbed, particularly around the remaining ruins. The maintenance of protected areas has been critically important in efforts to conserve species and ecosystems. Among the ruins of ancient temples, free-ranging groups of howler monkeys are divided into separate social groups.

THINKING CRITICALLY

Model *Sample answer: As primates, human and baboon physiology is similar in many ways, as is their diet. Because baboons in the wild naturally experience similar nutritional pressures, such as limited food access and pregnancy, they could be studied instead of humans.*

NATIONAL GEOGRAPHIC EXPLORER **KATIE AMATO**

SOMETHING GOOD IN YOUR GUT

CONNECTING GUT MICROBIOMES TO HUMAN HEALTH

The 30–40 square meters of surface area in your gut interacts with the outside environment every time you take a bite of food. Your gut is home to as many microorganisms as cells that make up your body. The collection of gut microbes forms a symbiotic relationship with you. The microbes thrive in your gut's environment while your body benefits from their metabolic activities. Biological anthropologist and National Geographic Explorer Dr. Katie Amato first started taking a close look at this relationship in howler monkeys in 2010, and she is now trying to answer questions related to the gut microbiome's impact on human nutrition and brain growth.

A WORLD UNIQUELY YOUR OWN By about age three, the composition and function of the human gut's microbiome look much like those of an adult. They stay that way into an individual's 40s, when they begin to naturally shift due to changes associated with aging. That is, unless the microbiome is disrupted by illness, antibiotic use, or other disturbances. Surprisingly, because of these environmental impacts, each person's gut microbiome is unique, like a fingerprint, even though similarities exist among healthy individuals.

Amato focuses her research on the microbe community in the large intestine. She's very interested in humans and the role of the gut's microbiota when food is scarce or when nutritional needs increase, such as during pregnancy. Carrying out research in humans can be difficult, so she uses other primates as models. By collecting data from fecal samples of primates in different environments, she can determine how microbes interact with host nutrition and health.

USING SCIENCE TO EFFECT SOCIAL CHANGE Amato's research on howler monkeys indicates that the gut microbiome can help protect hosts against periods of food shortage by digesting fiber to produce more energy. Amato also gained insights by comparing gut microbiota data of nonhuman primates with data from humans. These data showed that the human gut microbiome is more similar to that of baboons than to that of other apes. These patterns likely reflect similarities between the diets and habitats of baboons and human ancestors. The baboons' gut microbiomes, however, showed less variation among individuals than humans did. The greater variation in human gut microbiomes suggests to Amato that humans may be more influenced by environmental changes in location or diet than other primates.

In her research, Amato observes the influence of environmental factors on the gut microbiome, including diet, antibiotics, social interactions, and outdoor exposure. She also sees a relationship between these factors and human health. If she can link certain gut microbes to certain traits and determine how the environment impacts the growth of those microbes, then she can identify which environmental factors are most likely to affect health. By doing so, Amato can help people survive short-term food shortages and better understand the impact of long-term food insecurity on the gut microbiome as well as conditions such as malnutrition, growth stunting, obesity, and diabetes.

THINKING CRITICALLY

Model *Why could baboon gut microbiomes be a good model for studying human gut microbiomes?*

110

CROSSCURRICULAR CONNECTIONS

Human Nutrition and Health
Nutritionists look at the relationships between diet and health and wellness. Organizations such as the USDA provide guidelines from birth to adulthood, as the needs of the human body change at different phases of life. However, without a healthy gut, even with the best diet, a person can have health issues. The gut microbiome is very important to overall health and preventing different chronic diseases. Studies have shown that eating particular types of food produce predictable changes in the host bacterial populations. The relationships between diet and diseases such as inflammatory bowel disease (IBD), inflammatory skin diseases, autoimmune arthritis, and atherosclerosis have also been shown. For example, patients with IBD tend to have less bacterial diversity in their gut. Although a one-size-fits-all model does not exist, nutritionists recommend diets high in fiber, fermented and prebiotic foods, whole grains, and rich in polyphenols for a healthy gut microbiome.

Top: For her Ph.D. dissertation, Dr. Katie Amato spent a year doing field research on wild howler monkeys in Mexico. **Bottom left:** In the wild, baboons eat a variety of plant parts and sometimes meat. Their diet is more like that of humans than that of the more closely related chimpanzees, which primarily eat fruit. **Bottom right:** Amato analyzes fecal samples in her lab at Northwestern University.

UNIT VIDEO 2 Go online to watch our interview with Amato and learn more about her career and research.

111

DIFFERENTIATED INSTRUCTION | English Language Learners

Listening and Taking Notes Have students draw a graphic organizer with a *Central Idea* box across the top and three *Detail* boxes below it. Have them use it to take notes as they watch the video. Allow them to pause and rewatch parts of the video as needed.

Beginning Have students write *Amato's research* in the *Central Idea* box and *howler monkeys/baboons, diet,* and *gut microbiome* in the *Detail* boxes. As they watch, have them write down what they learn about each term, using their first language or English.

Intermediate Have students write words and short phrases in their graphic organizer as they watch the first time. Then, have them watch again and work in small groups to write sentences for the central idea and details.

Advanced Have students individually complete their graphic organizers as they watch the video, using complete sentences. Then, have them summarize the information.

Explorers at Work

Time: 7:08

Unit Video 2 interviews Katie Amato who takes students inside the human body to understand how our cells are affected by the myriad microbes living inside us. She also discusses how howler monkeys and baboons can serve as models for assessing the impact of disturbances in the human gut microbiome on human health.

Connect to Careers

Biological Anthropologist
Anthropology is the study of humans over time, including physiology, culture, and social relationships. Biological anthropologists share these interests but specialize in how humans have adapted to different environments, causes of disease and death, and evolution. To accomplish their goals, biological anthropologists study humans and their ancestors through fossils. They may also study primates, including their evolutionary history, as well as current social behaviors in their natural environment and their diet, health, and life spans. Careers in biological anthropology require an advanced degree.

Primatologist Primatology is a career that focuses on non-human primates, such as gorillas and chimpanzees. Primatologists may choose to pursue specific interests, such as genetics, behavior, evolution, or pursue veterinary medicine. Some primatologists choose to work in the field, observing primates in their natural habitats and recording data on the animals' behaviors. Others may be interested in education, research, or conservation and rescue efforts. Careers in primatology require a four-year college degree and, most often, graduate studies are required.

MOLECULES IN LIVING SYSTEMS

Three-Dimensional Learning

The practices, core ideas, and crosscutting concepts presented in this chapter's text, investigations, and resources provide support to address the following Performance Expectation: **HS-LS1-6.**

Science and Engineering Practices	Disciplinary Core Ideas	Crosscutting Concepts
Constructing Explanations and Designing Solutions (HS-LS1-6)	**LS1.C:** Organization for Matter and Energy Flow in Organisms (HS-LS1-6)	Energy and Matter (HS-LS1-6)

Contents		Instructional Support for All Learners	Digital Resources
ENGAGE			
112–113	CHAPTER OVERVIEW **CASE STUDY** How are the molecules of life assembled?	**Social-Emotional Learning** CCC Energy and Matter **On the Map** The Pacific Northwest **English Language Learners** Prereading	
EXPLORE/EXPLAIN			
114–120 DCI LS1.C PS1.A	**5.1** ELEMENTS AND COMPOUNDS • Distinguish the parts of an atom and their properties. • Explain different kinds of bonds between atoms. • Characterize different kinds of attractions that can exist between molecules.	**Vocabulary Strategy** Word Wall SEP Developing and Using Models **On the Map** ^{2}H Isotopes **BIOTECHNOLOGY FOCUS** In Your Community **Address Misconceptions** How Atoms Become Charged **Crosscurricular Connections** Physical Science CCC Energy and Matter **Connect to Mathematics** Model with Mathematics **English Language Learners** Language Patterns **Connect to English Language Arts** Write Informative Text	
121–125 DCI PS1.A PS2.B	**5.2** WATER • Describe the unique properties of water that make life on Earth possible. • Explain the differences between acids and bases and how pH is used to measure substances.	**Vocabulary Strategy** Sentence Starters **Visual Support** Molecules in Water **English Language Learners** Prior Knowledge and Experience **In Your Community** Acid Rain CCC Structure and Function **Crosscurricular Connections** Earth Science **Connect to Mathematics** Use a Number Line	
125	**MINILAB** POLAR VS. NONPOLAR MOLECULES		

Contents		Instructional Support for All Learners	Digital Resources
EXPLORE/EXPLAIN			
126–136 **DCI** LS1.C PS2.B	**5.3 CARBON-BASED MOLECULES** • Relate monomers, polymers, and hydrocarbons. • Describe the structure of carbohydrates and their functions in organisms. • Describe the structure of lipids and their functions in organisms. • Explain the relationship between nucleotides and nucleic acids. • Explain the relationship between amino acids and proteins.	**Vocabulary Support** Word Webs **English Language Learners** Answering a Question about a Text **SEP** Constructing Explanations and Designing Solutions **In Your Community** Table Sugar **CCC** Energy and Matter **Visual Support** Complex Carbohydrates, Structure of a DNA Molecule, Protein Structures **CER** Revisit the Anchoring Phenomenon **Connect to English Language Arts** Draw Evidence from Texts, Summarize a Text **Address Misconceptions** Use a Model **Differentiated Instruction** Leveled Support, Economically Disadvantaged Students, Leveled Support **SEP** Constructing Explanations and Designing Solutions **On Assignment** *Oxalis tuberosa*	**Investigation A** Converting Carbohydrates (50 minutes plus advance preparation) **SIMULATION** *Building Complex Molecules*
136	**EXPLORER** ROSA VÁSQUEZ ESPINOZA	**Connect to Careers** Extremophile Biologist	*Video 5-1*
137–145 **DCI** LS1.C PS1.A PS2.B	**5.4 CHEMICAL REACTIONS** • Explain the process of chemical reactions in biological systems. • Describe how energy in organic molecules is released during chemical reactions and used for cellular processes. • Characterize a reaction reaching equilibrium. • Summarize how catalysts affect the rate of reactions.	**In Your Community** Greenhouse Gases **English Language Learners** Expressing Opinions and Ideas **Connect to Mathematics** Model with Mathematics **SEP** Using Mathematics and Computational Thinking **CASE STUDY** Ask Questions **Connect to English Language Arts** Draw Evidence from Texts **CCC** Stability and Change **Visual Support** Enzymes and Catalysts **Differentiated Instruction** Leveled Support **BIOTECHNOLOGY FOCUS** Bioremediation of Plastic Waste	*Video 5-2* *Video 5-3* **Investigation B** Crime Scene Cleaners (200 minutes plus advance preparation)
144	**LOOKING AT THE DATA** DIGESTIVE ENZYMES AND pH		
ELABORATE			
145	**TYING IT ALL TOGETHER** How are the molecules of life assembled?	**SEP** Constructing Explanations and Designing Solutions	Self-Reflection and Scoring Rubrics
Online	**Investigation A** Converting Carbohydrates **Investigation B** Crime Scene Cleaners	Guided Inquiry (50 minutes plus advance preparation) Open Inquiry (200 minutes plus advance preparation)	MINDTAP Access all your online assessment and laboratory support on MindTap, including: sample answers, lab guides, rubrics, and worksheets. PDFs are available from the Course Companion Site.
EVALUATE			
146–147	**Chapter 5 Review**		
Online	**Chapter 5 Assessment Performance Task 3** *How Are Complex Carbon-based Molecules Built from Simple Atoms?* (HS-LS1-6)		

CHAPTER 5

ENGAGE

In middle school, students will have been introduced to molecules in organisms and learned about the role of photosynthesis and how elements are cycled as matter in the environment. This chapter expands on that prior knowledge by exploring the molecules of living systems and their interactions.

Students will be reminded about the parts of atoms and learn how bonds between atoms form. They will explore how properties of water support life and how organic molecules function in organisms, and they will see how chemical reactions in biological systems allow for, and support, life.

About the Photo Flamingos get their color from the foods that they eat, including plants, algae, and brine shrimp. These food sources produce natural red, yellow, or orange pigments, called carotenoids. In zoos, flamingos are typically provided these foods to maintain their pink color.

Some organisms can experience a color change if they consume a large quantity of certain molecules. For example, humans who eat a large amount of carotenoid-containing foods may experience yellow-orange skin coloration. To prepare students for this chapter's content, ensure they understand that different molecules can have different effects on organisms.

Social-Emotional Learning

As students progress through the chapter, invite them to look for different contexts and opportunities to practice some of the five social and emotional competencies. For example, support students in their **self-awareness** by helping them find ways to connect the content to their own personal experiences. When students learn about how scientists find molecules in plants (e.g., "Turning Molecules into Medicine"), remind them to practice **responsible decision-making** and **social awareness**. Support students in recognizing how an increased understanding of societal issues, such as the COVID-19 pandemic or climate change, can affect the funding and speed with which topics are researched.

CHAPTER 5

MOLECULES IN LIVING SYSTEMS

Flamingos are born with a white or gray color. As they obtain carotenoids from their diet of shrimp and algae, their colors change.

Flamingos get their characteristic color from pigment molecules called carotenoids. These molecules give some animals and most plants their red, orange, and yellow colors. The structures of different carotenoid molecules determine which wavelengths of light they absorb. The types of molecules that make up living things share important structural characteristics. Scientists use these patterns to predict how a group of similar molecules functions within an organism or biochemical process.

Chapter 5 supports the NGSS Performance Expectation **HS-LS1-6**.

CROSSCUTTING CONCEPTS | Energy and Matter

Energy and Molecular Bonds In this chapter, students explore the concept of energy and matter as they learn how energy is involved in the formation and breaking of bonds. Molecular biology in particular explores these reactions to examine the molecules and molecular structures that are required for life. The chemical reactions that occur in living organisms can be described by how the energy and matter flow into and out of organisms, and how molecules within organisms interact with each other.

Energy and matter were first introduced in **Unit 1**, **Chapter 2** in the context of how energy drives the cycling of matter in ecosystems. **Chapter 6** in this unit further explores energy and matter in relation to the reactions that occur during photosynthesis and cellular respiration. You can reinforce the crosscutting concept in this lesson by helping students understand that energy is not only stored in the bonds between molecules, but is also used to form and break the bonds.

CASE STUDY
TURNING MOLECULES INTO MEDICINE

HOW ARE THE MOLECULES OF LIFE ASSEMBLED?

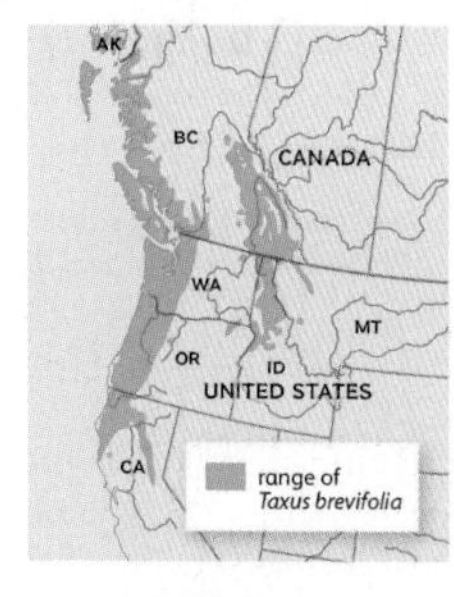

In many ways, Earth's living species are its most valuable resources. For example, many of the drugs used in a hospital or available in a pharmacy are derived from plants. Substances such as morphine and other pain relievers come from poppy plants. Colchicine from the meadow saffron treats a common form of arthritis called gout, and quinine from a tropical tree bark treats malaria. Madagascar periwinkle is used to treat cancers such as leukemia and Hodgkin disease.

The Pacific yew (*Taxus brevifolia*), shown in **Figure 5-1**, is a slow-growing tree native to the Pacific Northwest and the northern Rocky Mountains. The logging industry once considered it a "weed" tree with no commercial value. However, Native American people and specialty artisans have long used the Pacific yew's strong, decay-resistant wood for tools and decorative objects.

In the 1960s, scientists at the National Cancer Institute discovered that the thin bark of the Pacific yew contains a compound that inhibits the growth of tumors. They named this compound *paclitaxel* (**Figure 5-2**).

Clinical trials showed paclitaxel to be very effective against ovarian cancer compared with earlier forms of treatment. Unfortunately, the bark of a single yew contains only a tiny amount of paclitaxel, and removing the bark to extract the compound kills the tree. Hundreds of thousands of trees would need to be harvested to obtain enough paclitaxel to treat cancer patients. This fact sparked the search for other ways to synthesize the complex compound. Scientists have tried extracting compounds from yew needles and identifying species of fungi that can produce paclitaxel. They have also genetically engineered bacteria to make taxadiene, an intermediate compound needed to build a paclitaxel molecule. Paclitaxel can now be made in a laboratory, but significant challenges remain in making the process more sustainable and less expensive. Nevertheless, this naturally derived compound has been one of the most effective and versatile tools for cancer treatment to date.

Ask Questions *In this chapter, you will learn about patterns that occur in biological matter at the molecular level. As you read, generate questions you might need to ask if you were a biochemist who wanted to make naturally occurring compounds in a lab.*

FIGURE 5-1
Removing the thin purple bark from the Pacific yew tree is time-consuming and delicate work.

FIGURE 5-2
In this model of the paclitaxel molecule, the spheres represent atoms, and the sticks that connect them represent chemical bonds.

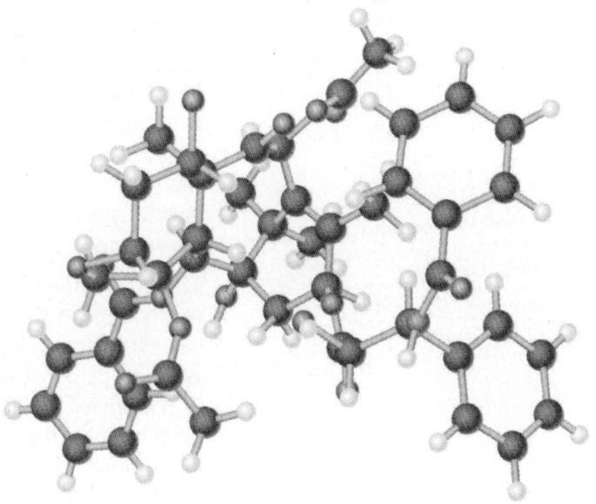

DIFFERENTIATED INSTRUCTION | English Language Learners

Prereading Have students read the title and headings and preview the illustrations and captions to help them understand what the text is about.

Beginning Explain that range in the label on the map means "where something lives." Have students point to the bark in the photo. Ask them to describe the compound the model molecule shows. Have them say where they think it comes from.

Intermediate Have students answer these questions: *Where is Taxus brevifolia found? Why might people need bark from the Pacific yew tree? What do you think paclitaxel is?* Point out the similar letters in *Taxus brevifolia* and *paclitaxel.*

Advanced Have students use the illustrations to explain how they think scientists "turn molecules into medicine."

CASE STUDY

ENGAGE

The situation described in the Case Study can be used to assess students' prior knowledge of how molecules can impact system functions in organisms. Ask students to identify common molecules humans ingest, such as penicillin, aspirin, caffeine, and ethyl alcohol. Elicit the effect of these molecules on system functions.

Generalize the discussion to include how molecules found in various sources are being used to develop new medicines. This provides additional context in which students can begin to develop a deeper understanding of molecular impact on systemic functions. In the Tying it All Together activity, students will research how a plant-derived alkaloid molecule is developed for medical treatments.

Ask Questions Students should revisit the Case Study as they read the chapter to make connections with the content. See a specific suggestion in Section 5.4.

On the Map

The Pacific Northwest Use the map of the range of *Taxus brevifolia* to discuss with students how geographic features impact environmental factors that determine where plants grow. Suggest to students to research the Pacific northwest biome in which the Pacific yew thrives, specifically the tree's role in the local ecosystems. Students may also discuss how a slow-growing tree is impacted by human encroachment and logging practices, and suggest steps to ensure the species' survival. This discussion could then be generalized to a biome, such as the Amazonian rainforest, where untold numbers of beneficial molecules may be sourced in the biodiversity of the region.

Human Connection Indigenous people of the Pacific northwest where this species thrived used the wood for making tools for hunting, fishing, and household utensils because it is very hard and resists decay. Many indigenous peoples also used it as medicine to treat internal injuries and lung diseases among other maladies.

5.1

ELEMENTS AND COMPOUNDS

LS1.C, PS1.A

EXPLORE/EXPLAIN

This section provides an overview of the structure of the atom, introduces the concept of isotopes, and reviews different kinds of bonds that exist between atoms. It also covers the weaker attractions that exist between molecules.

Objectives

- Distinguish the parts of an atom and their properties.
- Explain different kinds of bonds between atoms.
- Characterize different kinds of attractions that can exist between molecules.

Pretest Use **Questions 1, 6,** and **7** to identify gaps in background knowledge or misconceptions.

Vocabulary Strategy

Word Wall As students work through the key terms in this section, have them create word posters. Students should include the key term, its definition, and an example showing the image. You can have students do these individually or in pairs. As students work through the section, have the class vote on the posters that they think best embody the terms and hang these on the word wall.

Predict *Sample answer: Gold atoms have different properties than oxygen atoms. For example, a gold atom has more protons and neutrons than an oxygen atom and therefore more mass.*

5.1

ELEMENTS AND COMPOUNDS

KEY TERMS

- chemical bond
- covalent bond
- element
- hydrogen bond
- ion
- ionic bond
- isotope
- nonpolar
- polar

An Olympic gold medal is required to contain at least 6 grams of pure gold—slightly more than the mass of a U.S. quarter. Imagine 6 grams of gold being divided into smaller and smaller pieces. The smallest possible particle that acts like and appears as gold is a gold atom. Atoms are almost unimaginably tiny. They are much smaller than the smallest particle we can see under a light microscope. The gold in an Olympic medal would consist of about 1.8×10^{22}, or 18,000,000,000,000,000,000,000, gold atoms.

Predict *Gold and oxygen are very different substances. What are some ways in which a gold atom and an oxygen atom might differ?*

The Parts of an Atom

Though an atom is extremely small, it is composed of even smaller components called subatomic particles. The three main types of subatomic particles are *protons, neutrons,* and *electrons*. As shown in **Figure 5-3**, an atom is arranged with protons and neutrons in a central nucleus that is surrounded by a cloud of much smaller electrons.

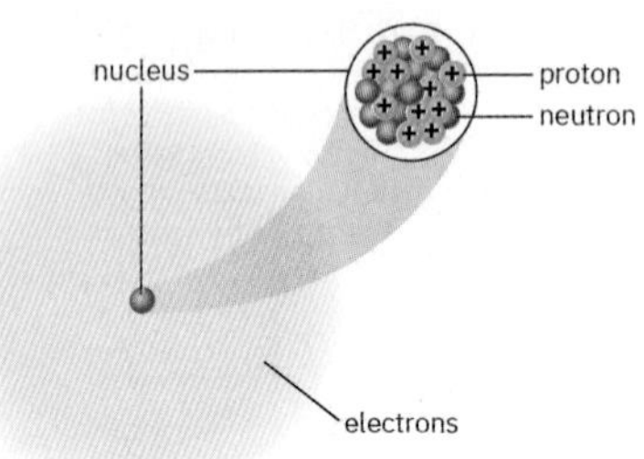

FIGURE 5-3 Protons and neutrons make up the nucleus in the center of an atom. Electrons move rapidly around the nucleus. The positions of the electrons are represented by a "cloud" that is darker where the electrons are more likely to be found and lighter where they are less likely to be found.

Protons and Elements The number of protons in an atom's nucleus is called its atomic number, and the atomic number of an atom determines its element. An **element** is a pure substance that consists only of atoms with the same characteristic number of protons. For example, the atomic number of carbon is 6. All atoms with six protons are carbon atoms, no matter how many electrons or neutrons they have. Elemental carbon (the substance) consists only of carbon atoms, and all of those atoms have six protons.

Carbon is a major component of all living things, including humans. When estimating the amount of biomass contained in a trophic level, ecologists assume that, on average, half of the mass of an organism (not including water) is carbon. Even beyond Earth, carbon is considered to likely be important for life. Scientists searching for life across the universe look for evidence of carbon-containing molecules such as carbon dioxide (CO_2) and methane (CH_4). **Table 5-1** lists six elements that have significant roles in biological, physical, and geological systems that support life on Earth.

SCIENCE AND ENGINEERING PRACTICES
Developing and Using Models

Atomic Models Students should recognize that the models in the pages that follow are not exactly what atoms look like. Note that the models are an interpretation of how the particles within atoms relate to one another. Ask students why a model is helpful when learning about objects that are either very large or very small in scale.

Guide students to examine **Figure 5-3**. There, students can see that the protons and neutrons occur together to form the nucleus of an atom. Ensure students understand that that the nucleus of an atom is not the same thing as the nucleus of a cell. Have students read the caption and discuss why the electrons are illustrated as a cloud as opposed to each electron being in a fixed location.

Have students refer to **Table 5-1**. Group students in pairs and have each pair select one of the six elements. Then have each pair draw a similar model to the one in **Figure 5-3** for the element they chose from the table.

TABLE 5-1. Six Elements Essential for Life

Element	Symbol	Atomic number	Occurrences in nature
carbon	C	6	all organisms and fossil fuels, carbon dioxide gas, Earth's crust, the ocean
hydrogen	H	1	all organisms, water
oxygen	O	8	in water, Earth's atmosphere (as oxygen and in carbon dioxide), Earth's crust
nitrogen	N	7	Earth's atmosphere (as nitrogen gas), soil, plants, proteins
phosphorus	P	15	all organisms, minerals, animal waste
sulfur	S	16	minerals, geothermal emissions

Isotopes Different forms of the same element that have the same number of protons but a different number of neutrons are called **isotopes**. For example, one carbon atom may have six neutrons, and another may have seven. An isotope is named using the total number of protons and neutrons in its nucleus. The carbon isotope with six neutrons is carbon-12 (abbreviated ^{12}C). The one with seven neutrons is carbon-13 (^{13}C). Isotopes of the same element are interchangeable in a biological system.

Some isotopes, such as ^{14}C, are radioactive—their nuclei break up spontaneously over time to form other elements. The time it takes for half of a radioactive isotope to decay is called its half-life. Radioactive isotopes exist in tiny amounts in any material. For example, a very small percentage (about 0.012 percent) of the potassium in your body is radioactive ^{40}K.

BIOTECHNOLOGY FOCUS

STABLE ISOTOPE ANALYSIS Not all isotopes are radioactive. Some of the natural isotopes of carbon, hydrogen, oxygen, and nitrogen do not decay. Some of these stable isotopes occur in animal tissues that grow quickly, such as hair and feathers. The isotope ratios of hydrogen and oxygen in these materials reflect the water and atmospheric conditions the animal has experienced, so isotope analysis can give clues as to where the animal has been over a period of time. Scientists use this information to track migratory birds.

In ecology, stable isotope analysis follows the principle that "you are what you eat." Isotopes of some elements make their way through food webs in a consistent pattern as matter is eaten by consumers and incorporated into their bodies. For example, ^{13}C can be used to identify the producer level in a food web. Nitrogen-15 is typically biomagnified, so higher ^{15}N ratios appear in higher trophic levels. When direct observation of food web interactions is not feasible, stable isotope analysis can help ecologists infer an organism's niche within its ecosystem.

The relative amounts (ratios) in which stable isotopes are present in any geographic location comprise that location's isotopic "signature." The ratio of ^2H increases from north to south across North America.

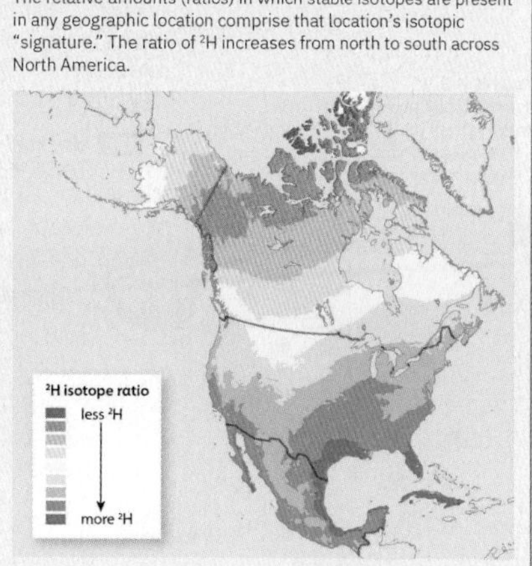

On the Map

^2H Isotopes Hydrogen has two stable isotopes, ^1H and ^2H. ^2H, also known as deuterium, has one proton and one neutron in the nucleus while the more common ^1H has one proton and no neutrons. Use the map of North America and the key to explain to students that the further south they move, more deuterium exists in the environment. Have students postulate how they think scientists use isotopes, like deuterium, to analyze where organisms have been over time.

BIOTECHNOLOGY FOCUS | Stable Isotope Analysis

In Your Community Like deuterium, the further south you move, the more of the isotope of oxygen-18, or ^{18}O, exists. Draw a molecule of water, H_2O, on the board. Explain to students that hydrogen of both ^1H and ^2H and oxygen of both ^{16}O and ^{18}O can be components of H_2O. Project the map and have students roughly identify the location of your community. Ask students to identify the relative isotope ratio of your area. Choose 2–3 other cities or states for students to locate. Have students identify if a location has a relatively higher or lower isotope ratio than your area and postulate why. Students may surmise that climate factors are at play. Generally, warmer temperatures result in more rainfall and higher humidity and isotope ratios are higher. Other factors such as altitude and distance inland also impact the ratios.

Address Misconceptions

How Atoms Become Charged Some students may think that atoms become charged when they gain or lose protons. Guide students to **Figure 5.4**. Ask them to explain what would happen to the carbon atom if it gained a proton (it would become a nitrogen atom) or lost a proton (it would become a boron atom). Elicit how the number of protons is a defining characteristic of a given kind of atom. Then ask students to explain what would happen if the same carbon atom gained or lost an electron. Encourage students to ponder the balance of charges within the atom. They should understand that the carbon atom would have a positive charge if it loses an electron and a negative charge should it gain one. Atoms change charge by the loss or gain of electrons, not protons. Because the neutrons carry no charge, they are not involved in an atom's relative charge.

Science Background

Ions in the Human Body Many atoms exist as ions in the human body and are necessary for normal systemic functioning. In addition to potassium, as identified in **Figure 5-5**, other common atoms that exist as ions include chloride, calcium, and sodium. Ions cannot cross the plasma membranes in cells because of their charge and require special proteins that are activated to enter or leave a cell. Because of this, ions are used to maintain the membrane potentials that control nerve signaling and muscle contractions. Pair students and have them select an ion to briefly research. Have each pair share the ion they selected with the class and identify its function within the human body.

SEP Use Mathematics *−3. From Table 5-1, a nitrogen atom has 7 protons: (+1) 7 + (−1) 10 = 7 − 10 = −3*

Charge The property of electric *charge* distinguishes protons, neutrons, and electrons from each other. Protons carry one unit of positive electric charge (+1). Electrons carry one unit of negative electric charge (−1). Neutrons have no charge. The negative charge of an electron is the same magnitude as the positive charge of a proton, so these two charges cancel one another. Thus, an atom with exactly the same number of electrons as protons is *neutral*—it carries no charge (**Figure 5-4**).

FIGURE 5-4
A neutral carbon atom has six protons, six neutrons, and six electrons.

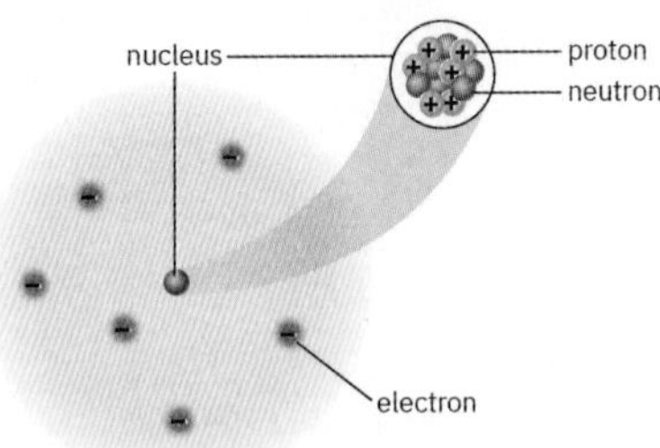

The strength of the force between charged particles depends on the size of the charges and the distance between them. The direction of the forces is determined by the signs of the charges. Charged particles of the same sign repel each other, and charged particles of opposite signs attract each other. An atom's negatively charged electrons are drawn to its positively charged nucleus by these attractive forces, but the electrons also repel each other as they move in a cloud around the nucleus.

An **ion** is an atom that has a different number of electrons than protons. Ions play an important role in cellular functions. For example, the body uses potassium ions to regulate the concentrations of fluids inside cells. All potassium atoms have 19 protons, so a potassium atom that has 19 electrons has no charge (**Figure 5-5a**). However, a potassium atom that has only 18 electrons is an ion with a charge of +1. The 18 +1 charges on 18 of the protons are canceled out by the 18 −1 charges on all 18 electrons, leaving one proton with a charge of +1. The overall effect is an ion with a +1 charge (**Figure 5-5b**).

SEP Use Mathematics *What is the charge of a nitrogen atom that has 10 electrons?*

FIGURE 5-5
A neutral potassium atom (a) and a potassium ion (b) both have 19 protons, but the potassium ion has one less electron, giving it a charge of +1.

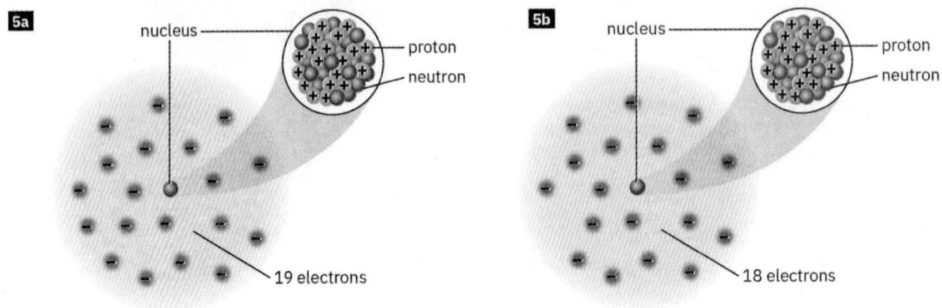

TABLE 5-2. Important Ions in the Human Body

Ion	Function
Ca^{2+}	formation of bones and teeth; muscle contraction; blood clotting
Cl^-	component of gastric juices; regulation of body fluids
Fe^{2+}	component of hemoglobin; role in oxygen transfer
H^+	blood pH control; formation of ATP
K^+	nerve impulses
Mg^{2+}	component of bones and teeth; muscle and nerve function
Na^+	water balance; nerve impulses
Zn^{2+}	role in cell division; growth; healing

Ions have a major role in living systems. **Table 5-2** lists some of the important ions in the human body and provides a short summary of their functions.

Bonds Between Atoms

In Chapter 2, you learned that atoms bond to form molecules such as water and carbon dioxide, and that **chemical bonds** are formed or broken in chemical reactions. The bonding properties of an atom arise from the behavior of its charged particles in the presence of other atoms.

The electrons moving around the nucleus of an atom have energy. The amount of energy an electron has depends on how far it is from the nucleus. When atoms are close to each other, the outermost electrons participate in forming chemical bonds.

Covalent Bonding Two atoms form a **covalent bond** by sharing electrons. In this type of bond, an outermost electron spends some time in the cloud around its nucleus and some time in the cloud around the other nucleus. The two atoms in hydrogen gas (H_2), nitrogen gas (N_2), or oxygen gas (O_2) are bound together covalently to form molecules, as shown with hydrogen in **Figure 5-6**.

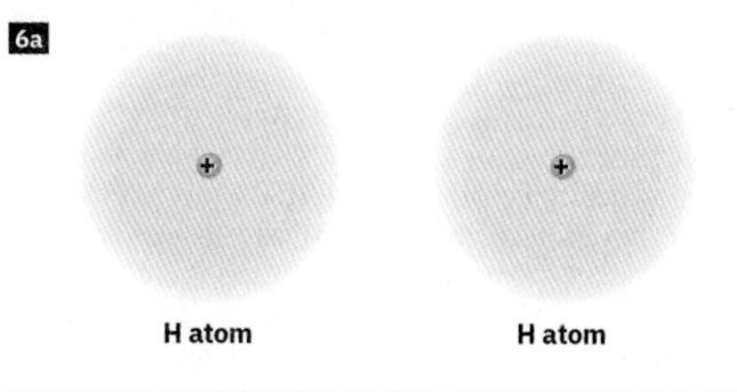

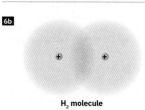

6b

H_2 molecule

FIGURE 5-6 In individual hydrogen atoms, each atom's electron stays close to its nucleus (a). When two hydrogen atoms are close to one another, the attractive forces between each positively charged nucleus and the other atom's negatively charged electron become strong enough to bind the atoms together (b). As they move around, the two electrons spend most of their time in the space between the two nuclei, forming a covalent bond.

5.1 ELEMENTS AND COMPOUNDS 117

Crosscurricular Connections

Physical Science Students often struggle with the subtle differences between covalent and ionic bonds. To help students solidify their mental models, it may be useful to have them model covalent and ionic bonds. To model a covalent bond, have groups of students hold, or share, two "electrons" made of balled up paper. To model an ionic bond, one student should hand an "electron" to the another student.

Atoms and Molecules This section gives an in-depth explanation on atoms and the bonds they form. **Section 6.3** discusses atoms and chemical reactions in the context of photosynthesis and cellular respiration.

CROSSCUTTING CONCEPTS | Energy and Matter

The Role of Energy at the Atomic Level Understanding the relationship between energy and matter is essential to interpreting the bonds between atoms. Emphasize to students that energy is required to bond atoms into molecules, which are larger building blocks of matter. Energy is also required to break bonds apart. To exemplify this concept, provide some real-life examples. Have students think about putting together a bicycle or a piece of modular furniture. Describe how when building either the bicycle or the piece of furniture, energy must be applied by a person or machine building the object. In addition, students should understand that to dismantle the object, energy must also be applied by the person or machine to take the object apart. Have students discuss and come up with other examples that can be used to model how energy is used to combine or separate matter.

Connect to Mathematics

Model with Mathematics Direct students to **Table 5-2.** Focus on the number of bonds each of these four molecules can form. Have students replicate the table, but add an additional column labeled *Number of electrons*. Then, direct students back to **Table 5-1**. Remind them that in an uncharged atom, the number of electrons equals the atomic number of the atom. Have students record the number of electrons in the appropriate column. Elicit from students that the number of electrons does not equal the number of bonds an atom can form.

Science Background

The Electron Cloud and Bonding Explain that even though electrons cannot be pinpointed within the cloud, they tend to stay in certain areas. In unreactive noble gases, two electrons remain close to the nucleus while the outer region from the nucleus can include up to eight electrons. Hydrogen, which has only one electron has the potential to form a bond. Carbon has two electrons near the nucleus and four outer electrons. This allows for four potential bonding sites. Encourage students to talk with a partner to determine why oxygen and nitrogen can form two and three bonds respectively.

SEP Develop a Model *Sample answer: One carbon atom could bond with four hydrogen atoms to form a neutral molecule,* CH_4.

```
     H
     |
H —  C — H
     |
     H
```

Two carbon atoms could bond with each other leaving three electrons from each carbon atom (total of 6) free to bond with atoms such as hydrogen.

```
     H    H
     |    |
H —  C —  C — H
     |    |
     H    H
```

Atoms of different elements have different tendencies to attract electrons to themselves, so the electron will spend more time around the nucleus of the atom that is more attractive. **Figure 5-7** shows how atoms of two different elements share electrons.

FIGURE 5-7 Fluorine is more attractive to electrons than hydrogen, so the electrons spend more time close to the fluorine atom. This is represented by the darker part of the electron cloud.

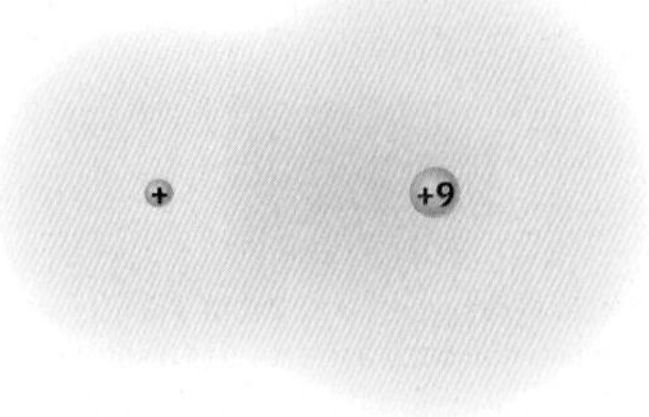

H atom F atom

Elements also differ in the number of outermost electrons they have available to form bonds. **Table 5-3** shows the number of covalent bonds the most biologically important atoms can form with other atoms in a molecule.

TABLE 5-3. Number of Bonds Formed by Essential Elements

Atom	Symbol	Number of bonds
carbon	C	4
hydrogen	H	1
oxygen	O	2
nitrogen	N	3

SEP Develop a Model *According to* ***Table 5-3****, a carbon atom forms four bonds and a hydrogen atom forms one bond. How many of each of these atoms could form a neutral molecule? Sketch the molecule.*

Ionic Bonding If two atoms differ very greatly in their tendency to attract electrons, one or more of the outermost electrons in the less attractive atom's cloud can transfer entirely to the more attractive atom's cloud. This results in two ions: the atom that lost electrons now has a positive charge, and the atom that gained electrons has a negative charge. The force of attraction between these oppositely charged ions forms an **ionic bond**. For example, the compound NaCl, or table salt, is held together by ionic bonds between a positive sodium ion (Na^+) and a negative chloride ion (Cl^-).

An ionic bond can also form between an atom and a group of atoms, or between two groups of atoms. Some common ions formed by groups of atoms are ammonium (NH_4^+) and phosphate (PO_4^{3-}).

Chemical Bonds and Energy When the electrons in two atoms move to form a chemical bond, energy is released. Conversely, energy is required to break a chemical bond. The amount of energy needed to break a bond depends on the type of bond and which elements form the bond.

DIFFERENTIATED INSTRUCTION | English Language Learners

Language Patterns Have students describe covalent and ionic bonds using similar patterns of language.

Beginning Have students classify these sentences as *covalent bonding* or *ionic bonding*: Two atoms share electrons (*covalent*). One atom loses an electron, and one atom gains an electron (*ionic*). An outermost electron spends time in both clouds (*covalent*). An outermost electron transfers entirely to one cloud. One atom has a positive charge, and one atom has a negative charge (*ionic*).

Intermediate Provide these sentence frames: *In (a covalent bond / an ionic bond), atoms ______. An outermost electron ______.*

Advanced Have pairs describe how each kind of bond forms, explain what happens to the outermost electrons, and give examples of each.

Attractions Between Molecules

Molecules can be attracted to each other as well. Attractive forces between molecules vary in strength. You have seen changes in these forces when you observe water melting or freezing. At low temperatures, water molecules are held close together in a repeating structure to form ice. As the temperature increases, the forces between the molecules become weaker, allowing them to move around as a liquid. Increasing the temperature even more makes the molecules move faster and farther apart until the forces between them are insignificant. The water thus takes the form of a gas.

Polarity Electronegativity is a measure of an atom's attraction for the electrons in covalent bonds. When the atoms in a molecule have similar electronegativity, the electrons are shared equally and the molecule is referred to as **nonpolar**. The electrons are shared equally in the covalent bonds of the hydrogen gas H_2 and oxygen gas O_2, for example, so both molecules are completely nonpolar.

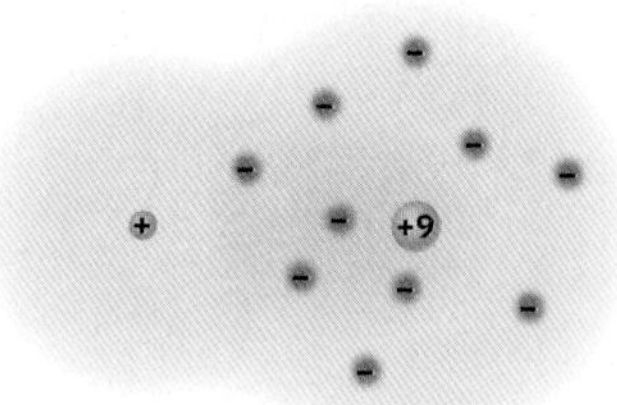

FIGURE 5-8 The hydrogen end of a hydrogen fluoride molecule has a net positive charge due to the relative lack of electrons around the positively charged H nucleus. Likewise, the fluorine end of the molecule has a net negative charge.

Many other atoms form similar bonds. However, they share unequally, with one atom having a greater "pull" than the other and thus an unequal pull on shared electrons. When two different elements, such as hydrogen and fluorine, have different electronegativity values, they do not share electrons equally, as shown in **Figure 5-8**. The part of the molecule that is more attractive to electrons will act as a center of negative charge, and the part that is less attractive to electrons will act as a center of positive charge. Such a molecule is referred to as being **polar**.

Water molecules are polar due to their shape. **Figure 5-9** shows the bent shape of a water molecule (H_2O), in which the two hydrogen atoms are closer together on one side of the oxygen atom. The oxygen atom attracts the hydrogen electrons toward itself so that, effectively, the hydrogen ends of the molecule are slightly positively charged and the oxygen end is slightly negatively charged.

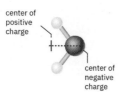

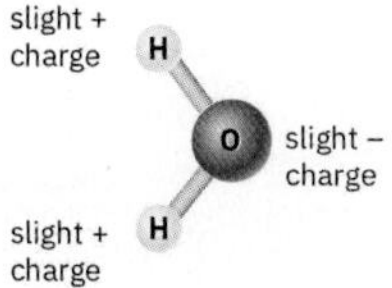

FIGURE 5-9 The bent shape of a water molecule results in a center of positive charge toward the hydrogen atoms and a center of negative charge toward the oxygen atom.

Polarity Students learn that polarity can affect how electrons move within a molecule and that some atoms have a stronger attraction to electrons. **Chapter 6** covers how polarity is involved in controlling how molecules interact with the cellular membrane.

Science Background

Polarity and Water The unique characteristics and interactions of water molecules are what give water its unique properties. Show students a glass that is so full of water that they can see that the water level is higher than edges of the glass itself. Ask if any students can name the phenomena that they are observing. (*surface tension*) Direct students to **Figure 5-9** and have them postulate how the charge distribution in a water molecule contributes to this phenomenon.

Connect to English Language Arts

Write Informative Text This section covers a variety of different kinds of bonds that atoms can form between each other, and between molecules. Have students write an informative text, in the form of a pamphlet, booklet, or using digital media that could be used to teach about each type of bond. The text should include the names of the bonds, descriptions, examples, and a simple model of each type, such as what is shown in **Figure 5-10**. Encourage students to use headings to help organize their information and captions or labels to clarify their models.

SEP Construct an Explanation
Yes. A water molecule consist of two hydrogen atoms bonded to oxygen. The electron in each of the hydrogen atoms can form a bond with an oxygen atom in another water molecule. Therefore, one water molecule can form two hydrogen bonds.

Hydrogen Bonding The elements fluorine, oxygen, and nitrogen are the most highly attractive to electrons. When a polar molecule includes hydrogen bonded to one of these atoms, the F, O, or N atom also attracts the H electrons in nearby molecules to hold the substance together. The force of attraction between a hydrogen atom in one molecule and a highly attractive atom in another molecule is called a **hydrogen bond**. Because hydrogen atoms are very small compared to other elements, the hydrogen end of a molecule can come very close to the positively charged side of the other molecule to which its electron is attracted. The small distance between atoms makes the forces between their molecules stronger (**Figure 5-10**).

SEP Construct an Explanation *Would you expect water molecules to form hydrogen bonds? Explain your answer.*

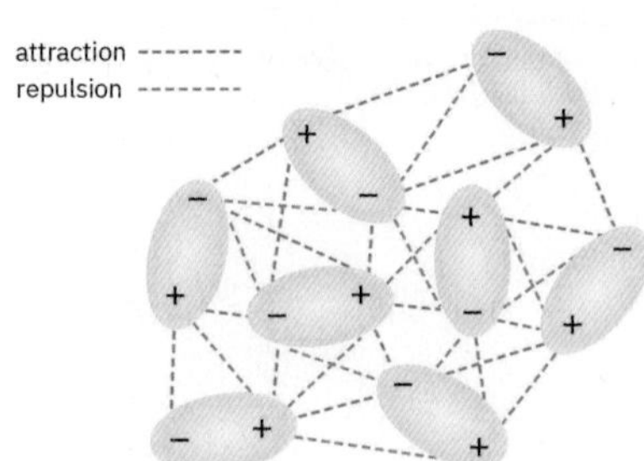

FIGURE 5-10 Hydrogen bonds form between atoms of different polar molecules.

5.1 REVIEW

1. **Calculate** An oxygen atom has 8 protons, 8 electrons, and 8 neutrons. What is the charge of an oxygen atom? What charge does the nucleus have?
2. **Compare** In the nitrogen cycle, bacteria in plant roots convert atmospheric nitrogen (N_2) into ammonia (NH_3), which is then converted into ammonium (NH_4^+) by bacteria living in the soil. Which statement explains a difference between ammonia and ammonium?
   A. Ammonia has fewer protons than ammonium.
   B. Ammonia has more electrons than ammonium.
   C. Ammonia has fewer neutrons than ammonium.
   D. Ammonia has more bonds than ammonium.
3. **Relate** What specific role does the energy from sunlight play in rearranging atoms during photosynthesis?
4. **Model** A molecule of pentane is made up of 5 carbon atoms and 12 hydrogen atoms. Use the information in **Table 5-3** to sketch a model of a pentane molecule.

5.1 REVIEW

1. *The charge of an oxygen atom is (+1) 8 protons + (–1) 8 electrons = 0. The charge of the nucleus is (+1) 8 protons + (0) 8 neutrons = +8.* **DOK 2**
2. *A* **DOK 2**
3. *Sample answer: Energy from sunlight breaks apart the bonds between carbon and oxygen in CO_2 and between hydrogen and oxygen in H_2O.* **DOK 2**
4. 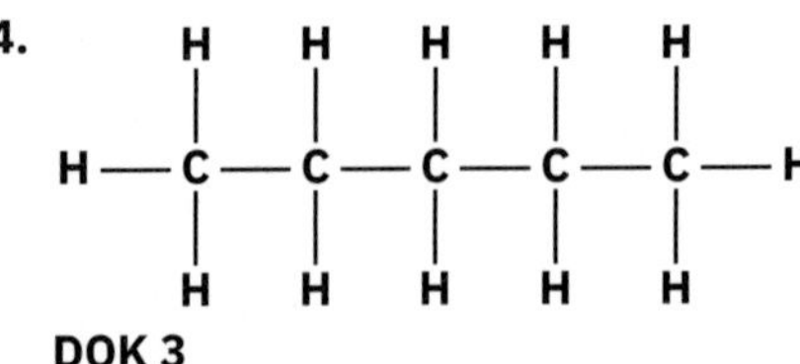

DOK 3

WATER

Scientists looking for life on planets, asteroids, and moons often search for signs of water. For example, they have found frozen water and evidence of past water flow on Mars. They study these sites in hopes of finding additional evidence that organisms lived there in the past or exist there now.

Infer *Why do you think finding water on other worlds would be an important sign of potential life?*

5.2

KEY TERMS

acid
base
pH
solution

Properties of Water

The unique properties of water make life on Earth possible. Scientific theories of how the first cells formed all involve water, which is present in all living organisms and is thus a fundamental resource in any ecosystem. The hydrogen bonds between its molecules give water many characteristics that are crucial to sustaining life.

FIGURE 5-11 A seal pup pokes its head into the water below a layer of ice.

States of Water Water is the only substance that is present in solid, liquid, and gas states under the temperatures and pressures typically found on Earth. Due to strong hydrogen bonds, the temperature of water changes much more slowly than almost all other common substances when heated. You see this on a lake in the summer. The air temperature will be hot, but the lake will remain cool, and vice-versa in the winter. Water's ability to resist rapid changes in temperature maintains stable conditions in aquatic ecosystems and within the cells of organisms. A lot of energy is needed to break the hydrogen bonds between molecules as water changes in state from liquid to gas.

5.2

WATER

LS1.C

EXPLORE/EXPLAIN

This section explains the properties of water. It also covers the relationship between polarity and solubility before introducing the differences between acids and bases.

Objectives

- Describe the unique properties of water that make life on Earth possible.
- Explain the differences between acids and bases and how pH is used to measure substances.

Pretest Use **Question 9** to identify gaps in background knowledge or misconceptions.

Infer *Sample answer: Water is essential to life on Earth, so it would probably also be needed for life anywhere else in the universe. If we found water on another world, there might be life there as well.*

Vocabulary Strategy

Sentence Starters As students reach the appropriate points in this section, use sentence starters such as the following to help them reframe the text into their own words. Then make comparisons with the defining sentence in the text.

A solution forms when ________.

A nonpolar molecule does ________.

Acids form when ________.

Bases form when ________.

The value pH measures ________.

Visual Support

Molecules in Water Explain to students that the state of water changes depending on how much energy is present. Break students into groups of three to model the movement of water molecules in each state. Groups can use **Figure 5-12** and the accompanying text to determine the degree of motion they should use in each state. Then, have the groups fill out an Idea Diagram for the states of water. Each member will write the main idea and details for their assigned or chosen state of water. Under "details" students should draw a model of their state of water. Groups should then write a one-sentence introduction and a one-sentence conclusion to bookend their findings.

SEP Construct an Explanation *Sample answer: During cold seasons, aquatic organisms still live in the water below the ice, which is insulated from the cold air. If ice sank rather than floating on the surface of a body of water, eventually the body of water would freeze along with the organisms in it.*

Cohesion and Adhesion Students learn about the properties of cohesion and adhesion. **Chapter 9.2** includes a more in-depth discussion on how these properties function in plant structures.

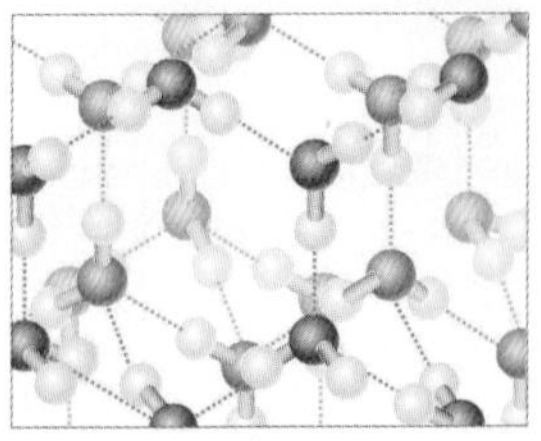

ice

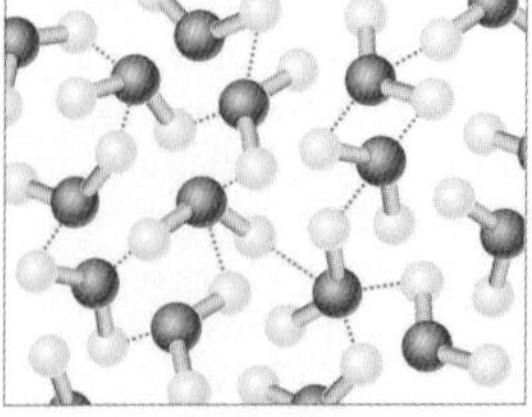

liquid water

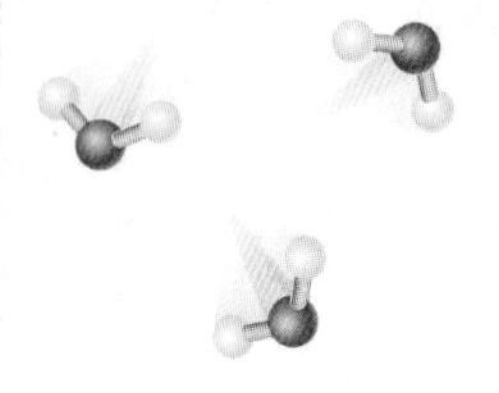

water vapor

FIGURE 5-12 The molecules in ice are farther apart than they are in liquid water. In water vapor, the gaseous form of water, the water molecules are very much farther apart than they are in liquid water.

Density Unlike almost all other substances, liquid water decreases in density as the temperature drops. In the liquid state, the small size of the hydrogen atoms allows water molecules to be close together, as shown in **Figure 5-12**, but the molecules move without any particular pattern. At 0 °C, the molecules settle into a crystal structure in which the hydrogen bonds hold them farther apart. Thus, the solid form of water is less dense than the liquid form, which is why ice floats on water.

SEP Construct an Explanation *Why would the facts that water changes temperature slowly and that ice floats in water be essential to the survival of aquatic organisms?*

Cohesion and Adhesion When it rains on a leaf, the water tends to bead up and stick together (**Figure 5-13**). Molecules of some substances resist separating from one another, a property called *cohesion*. Water has cohesion because hydrogen bonds collectively exert a continuous pull on its individual molecules. You can observe cohesion as surface tension, where the surface of liquid water behaves a bit like an elastic sheet.

Similarly, water molecules are attracted to molecules of other substances. This property is called *adhesion*. Cohesion and adhesion drive many processes that sustain multicellular bodies. Inside a plant, water molecules pull on each other and adhere to the walls of narrow pipelines of tissue, enabling columns of water to rise from the roots to the leaves. In some trees, these pipelines extend more than 50 meters above the soil.

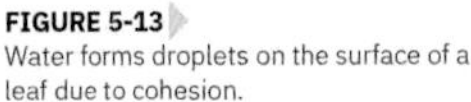

FIGURE 5-13 Water forms droplets on the surface of a leaf due to cohesion.

DIFFERENTIATED INSTRUCTION | English Language Learners

Prior Knowledge and Experience Have students discuss real-life experiences with the states of water, such as skating or playing hockey on an ice, swimming in a pool, washing a car, using a water hose, or steam coming up from a bowl of soup.

Beginning Give students time to write or draw their experience before sharing it. Ask questions such as: *Where were you? What did you do? Was it ice, water, or steam?*

Intermediate Have partners take turns sharing experiences and building on each other's ideas. For example, if one student mentions a steaming bowl of soup, the other might describe a steaming cup of tea.

Advanced Have students describe experiences with all three states of water and give specific details about how they know which state of water it was.

Polarity and Solubility The fact that water molecules are highly polar gives rise to another property that is important in life processes. Opposite charges attract, so if another substance made of polar molecules is placed in water, the positively charged (H) ends of the water molecules attract the negatively charged ends of the other polar molecules. Likewise, the negatively charged (O) ends of the water molecules attract the positively charged ends of the other polar molecules. In water, this attraction is often strong enough to pull those other polar molecules apart, dissolving the substance. A *solvent* is a substance in which another substance, called the *solute*, is dissolved. The mixture of the two substances is called a **solution**. Water is called the "universal solvent" because it dissolves so many other substances (**Figure 5-14**).

Nonpolar molecules, however, do not have different charge distributions at the ends, so they cannot be easily pulled apart by water molecules. To remember this, we say that "like dissolves like." Polar substances can dissolve other polar substances, but a polar substance cannot dissolve a nonpolar substance. For example, oil separates in a layer on top of water.

FIGURE 5-14 A butterfly perches on a black caiman's eyelid to drink its tears, which contain nutrients dissolved in their water.

Acids and Bases

Recall that when a hydrogen atom is covalently bonded to another atom that strongly attracts electrons, the hydrogen's electron is pulled slightly away from its proton. Hydrogen bonding in water can pull that proton right off the molecule. The detached proton is called a hydrogen ion (H^+). The electron stays with the rest of the molecule, making it a negatively charged ion. For example, a water molecule that loses a proton becomes a hydroxide ion (OH^-). The loss is more or less temporary, because these two ions easily get back together to form a water molecule.

An **acid** is a substance that gives up hydrogen ions in water. By contrast, a **base** is a substance that accepts hydrogen ions in water. Many organisms and biological processes are sensitive to how acidic or basic the surrounding solution is.

Polarity and Solubility Here, students learn how differences in charge distribution result in polar and nonpolar molecules. Students explore the implications of polarity on cell membranes in **Chapter 6**.

In Your Community

Acid Rain Acid rain is precipitation that has a lower pH than typical water. It is caused when certain oxides are released into the atmosphere and then combine with precipitation before it reaches Earth's surface. It can lead to erosion of building materials and other materials that are exposed to the rain. It can also have harmful effects on organisms and the soil in regions that receive high amounts of acid rain. If possible, direct students to collect rainwater and test its pH. Acid rain generally has a pH of between 4.2 and 4.4 while regular rain has a pH of about 5.6. Have students draw conclusions about the acidity of the water compared to regular rainwater and tap water. Have students postulate about the source of the oxides contributing to the acid rain in your community.

CROSSCUTTING CONCEPTS | Structure and Function

Polarity and Solubility Encourage students to keep in mind how the structures of molecules regulate how they interact with other molecules. Provide groups of students with small plastic bottles that are half full of clear isopropanol (rubbing alcohol). Have students add a few drops of food coloring into the isopropanol and shake the bottles. Then, have them use a funnel to fill up the rest of the bottles with baby oil. Students should notice that the oil floats on top. Ask why the oil does not dissolve into the isopropanol. They should be able to deduce that the isopropanol is polar and therefore it does not dissolve in the oil. Have students discuss whether the food coloring contains polar or nonpolar molecules. The food coloring dissolves in the isopropanol because it is a polar substance.

Crosscurricular Connections

Earth Science Some students may have been exposed to the idea of ocean acidification in an Earth Science or Environmental Science class. Share with students that before the industrial revolution, ocean pH was about 8.2, but it is now, on average, 8.1. Point them to **Figure 5-15** for comparisons.

To demonstrate the effects, obtain mussel or clam shells from a local grocery and clean thoroughly. Prepare small containers of solutions more acidic than seawater, or if time allows, have small groups do so using common liquids from **Figure 5-15** or various dilutions of vinegar. Place one or two shells in each container. Be sure the containers are filled to the brim and capped tightly. Then, after a few days, carefully extract the shells from the liquids and compare their quality with reserved shells that were kept in a "seawater" control solution of filtered water and salt with a salinity of 35 parts per thousand.

Remind students that the exoskeletons of shellfish and coral contain calcium as their own endoskeleton does. Have students discuss possible reasons why marine organisms cannot maintain their skeletons and shells as the ocean becomes more acidic.

Connect to Mathematics

Use a Number Line Direct students to the pH scale shown in **Figure 5-15** and guide them to understand that it is a type of number line. Lead a discussion to help students understand that the numbers shown on this line are averages, and that not all of these solutions will have these exact pH values. Have students make their own pH scale number line and identify the pH of other substances using pH paper or another indicator.

We use a value called **pH** as a measure of the relative amounts of H^+ and OH^- ions present in a water-based solution. If a solution has an equal number of H^+ and OH^- ions, its pH is 7 (or neutral). Pure water is neutral. A solution that has more H^+ ions than OH^- ions is acidic, and its pH is less than 7. Conversely, a solution with fewer H^+ ions than OH^- ions is basic, with a pH greater than 7. The pH values of several common substances are shown in **Figure 5-15**.

Acids can be strong or weak. Strong acids ionize completely in water to give up all of their H^+ ions. Hydrochloric acid (HCl) is an example. When HCl dissolves in water, all of the molecules give up H^+ ions, leaving Cl^- ions behind. The H^+ released from HCl makes gastric fluid in your stomach very acidic (pH 1–2). By contrast, weak acids do not ionize completely in water. Carbonic acid (H_2CO_3) is an example. It forms when carbon dioxide gas dissolves in plasma, the fluid portion of human blood. Carbonic acid is a weak acid, so only some of its molecules give up a hydrogen ion in water. When carbonic acid loses a hydrogen ion, it becomes an ionic molecule called bicarbonate.

Together, carbonic acid and bicarbonate form a buffer. A buffer is a set of chemicals that can keep the pH of a solution stable by alternately donating and accepting ions that contribute to pH. The addition of too much acid or base can overwhelm a buffer's capacity to stabilize pH. This outcome is catastrophic in a cell or body because most biological molecules function properly only within a narrow range of pH. A slight deviation from that range can halt cellular processes.

FIGURE 5-15 As shown on this pH scale, acidic substances have pH values less than 7, and basic substances have pH values greater than 7.

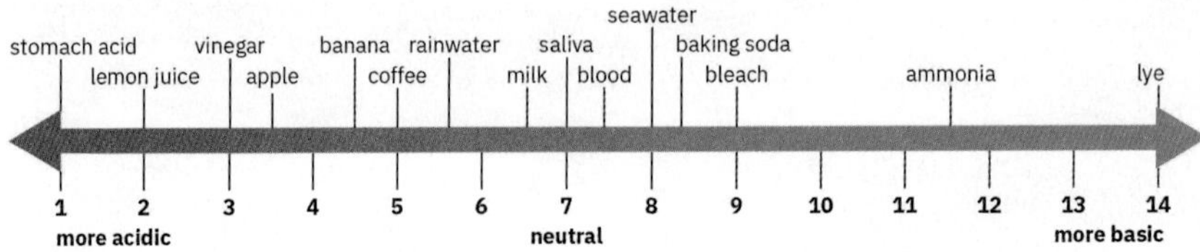

5.2 REVIEW

1. **Sequence** List the three states of water in order from least dense to most dense.
2. **Infer** Sugar dissolves easily in water. What property does this fact demonstrate?
   - A. Sugar is acidic.
   - B. Sugar has ionic bonds.
   - C. Sugar molecules are polar.
   - D. Sugar is more dense than water.
3. **Explain** Use the properties of water to explain why when it rains on a road, a layer of gasoline from the road's surface will float on top of a puddle of rainwater.
4. **Infer** Which property of water allows it to be sucked up from a glass into a straw?
   - A. It has cohesion.
   - B. Its pH is slightly basic.
   - C. It resists rapid changes in temperature.
   - D. It is less dense as a gas than as a liquid.

5.2 REVIEW

1. *gas, solid, liquid* **DOK 2**
2. *C* **DOK 2**
3. *Water is a polar substance, so because the gasoline forms a layer rather than dissolving in the water, the gasoline must be nonpolar. Also, water must be denser than gasoline for gasoline to float on top of it.* **DOK 3**
4. *A* **DOK 2**

MINILAB

POLAR VS. NONPOLAR MOLECULES

SEP Plan and Conduct an Investigation How can you determine if a substance is polar or nonpolar?

Water is often called the "universal solvent." In this activity, you will conduct an experiment that uses this property of water to determine whether the molecules of other common substances are polar or nonpolar. You will determine each substance's polarity by mixing it with water and observing the result.

Materials

- test tubes with rubber stoppers or caps (5)
- test tube rack
- labels (5)
- marker
- distilled water, 50 mL
- 10-mL graduated cylinder
- substances to test for polarity (5)
- pipet or dropper for each liquid substance
- spoon spatula for each solid substance
- timer

Safety      

Procedure

1. Obtain the distilled water and the substances to test from your teacher. Label each test tube with the name of each substance.
2. Determine how you will be able to tell whether each substance is polar or nonpolar. Record your criteria.
3. Predict whether each substance is polar or nonpolar. Record your predictions.
4. Use the graduated cylinder to measure and pour 10 mL of water into each test tube.
5. To test a liquid substance, add 10 drops of the substance to the water in one of the test tubes. Place the stopper on the test tube. Mix the contents by shaking or swirling the tube for 20 seconds. Observe the results. Return the test tube to the rack.
6. To test a solid substance, use a spatula to add a pea-sized sample to the water in one of the test tubes. Place the stopper on the test tube. Mix the contents by shaking or swirling the tube. Observe the results. Place the test tube back in the holder.
7. When you have tested all five substances, dispose of the mixtures as instructed by your teacher. Be sure to wash your hands with soap and water after you are done.

Results and Analysis

1. **Interpret Data** How do the results of your tests indicate whether a substance is polar or nonpolar? Did your results match your predictions? Justify your answer.
2. **Explain** Use what you know about the structure of water molecules to explain why water is useful for testing the polarity of other substances.
3. **Apply** Which of your substances are nonpolar? How would you confirm this result using a different solvent?
4. **Draw Conclusions** Some substances only partially dissolve in water. For example, when whole milk is mixed with water, it appears diluted and lighter in color. What might this tell you about this substance?

MINILAB
POLAR VS. NONPOLAR MOLECULES

Students make predictions and then test to determine whether select substances are polar or nonpolar and describe why the data collected about ability to dissolve in water would provide information about the polarity of a substance.

Time: 30 minutes

Advance Preparation

• Collect at least five different polar and nonpolar substances for testing. Common polar substances include salt, sugar, vitamin C, isopropyl alcohol (rubbing alcohol), ethanol (grain alcohol), glycerin, and corn syrup. Common nonpolar substances include flour, talcum powder, petroleum jelly, vitamin D, vegetable oil, and mineral spirits. **CAUTION:** *Isopropyl alcohol, ethanol, talcum powder, and mineral spirits are both poisonous and flammable. These materials should be used only in a well-ventilated room and away from open flame. Caution students not to drink or taste any of the materials used in the lab, even if they are commonly used in foods.*

Procedure

In **Step 3**, remind students to read all the Procedure steps before making their predictions.

In **Step 5**, review how to safely remove the test tube cover in order to prevent any substances from spraying out. Alternatively, provide a stirring rod for each test tube to eliminate the need for stoppers or caps.

In **Step 6**, remind students to use only a small amount of solid. If they use too much, the water may need to be heated for the substance to fully dissolve. You may also want to encourage students to wait a minute or two after shaking the tube before they decide whether the substance has dissolved.

Results and Analysis

1. **Interpret Data** *Sample answer: Salt and isopropyl alcohol dissolved in water, so these substances are polar. Flour, vitamin D, and vegetable oil did not dissolve in water, so these substances are nonpolar. No, I predicted that flour and vitamin D would dissolve in water because they were solids like salt, but they did not.* **DOK2**
2. **Explain** *Sample answer: Water is a polar molecule because the oxygen atom has a slightly negative charge, and the hydrogen atoms have slightly positive charges. This causes water molecules to attract each other and other polar molecules. Therefore, many polar substances dissolve in water. Nonpolar molecules do not dissolve in polar solvents, such as water.* **DOK3**
3. **Apply** *Sample answer: I would expect that the nonpolar substances would dissolve in a nonpolar solvent, such as vegetable oil, while the polar substances would not dissolve. For example, vitamin D should dissolve in vegetable oil, but isopropyl alcohol should not.* **DOK4**
4. **Draw Conclusions** *Sample answer: Substances that only partially dissolve in water are probably a mix of polar and nonpolar molecules.* **DOK3**

5.3

CARBON-BASED MOLECULES

LS1.C, PS2.B

EXPLORE/EXPLAIN

This section introduces the different building blocks of organic molecules, and covers the topics of carbohydrates, lipids, nucleotides in nucleic acids, and the amino acids that make up proteins.

Objectives

- Relate monomers, polymers, and hydrocarbons.
- Describe the structure of carbohydrates and their functions in organisms.
- Describe structure of lipids and their functions in organisms.
- Explain the relationship between nucleotides and nucleic acids.
- Explain the relationship between amino acids and proteins.

Pretest Use **Questions 2** and **5** to identify gaps in background knowledge or misconceptions.

Infer *Sample answer: The elements alone must not be what determines the properties of different substances. Different combinations of elements form different molecules, such as sugar and plastic, that have different properties.*

Vocabulary Support

Word Webs As students work through this section, suggest they begin a word web with the term *organic molecule* at the top or center. Then, encourage them to create branches for *carbohydrates, lipids, nucleic acids,* and *proteins,* and group related terms.

5.3 CARBON-BASED MOLECULES

KEY TERMS

amino acid	lipid
ATP	monomer
carbohydrate	nucleotide
cellulose	nucleic acid
DNA	polymer
glucose	protein
glycogen	starch

The tastes and textures of food that you eat depend largely on the presence of certain molecules. Fruit tastes sweet because it contains sugar molecules, and cellulose molecules make raw vegetables crunchy. The molecular makeup of a substance also determines whether or not an animal can digest it. Cows and goats can digest grass, but humans cannot. Many animals try to eat plastic, but their digestive systems cannot break it down.

Infer *Both sugar and plastic consist mainly of carbon, hydrogen, and oxygen atoms. Why do you think sugar and plastic have very different properties even though they are made of the same elements?*

Elements of Life

Organisms exchange matter and energy with the environment as they grow, maintain themselves, and reproduce. All organisms take up carbon compounds from the environment and use them to build the molecules of life. These molecules are eventually broken down, and their parts are cycled back to the environment in by-products, wastes, and remains. Biological molecules are made of carbon, hydrogen, nitrogen, and other elements, with the key element being carbon. Its ability to form four bonds makes carbon an ideal building block for a vast assortment of molecules. In many ways, the chemistry of carbon is the chemistry of life itself.

Compounds that consist mainly of carbon and hydrogen are called *organic compounds*. Except for water, virtually all of the important compounds in the body of any organism are organic compounds.

FIGURE 5-16 Open-air landfills in the Iberian Peninsula attract migrating white storks during winter. Although these dump sites provide ample food, some storks also ingest plastic,. Although made mainly of carbon, plastic supplies little to no energy and can damage their internal organs.

126 CHAPTER 5 MOLECULES IN LIVING SYSTEMS

DIFFERENTIATED INSTRUCTION | English Language Learners

Answering a Question About a Text Pause after each section to have students answer questions to demonstrate comprehension of the text.

Beginning Ask questions that can be answered yes or no or in short phrases, such as *What are organic compounds made of? Are hydrocarbon chains reactive?*

Intermediate Ask Wh-questions that can be answered in sentences, such as *What are hydrocarbon chains? Why are monomers and polymers different?*

Advanced Ask questions that prompt students to connect ideas in the text, such as *Why can carbon atoms form hydrocarbon chains? What makes a hydrocarbon chain more reactive? Why?*

Building Blocks of Organic Molecules

The molecules of life are assembled from simpler organic subunits called **monomers**. These building blocks combine to form sugars, fatty acids, amino acids, and nucleotides. A molecule that consists of repeated monomers is a **polymer**.

Many biological molecules, such as carbohydrates, lipids, proteins, and nucleic acids, are very large polymers consisting of thousands of subunits. The structure and function of a biological molecule arises from (and depends on) the order, orientation, and interactions of its subunits.

Hydrocarbons Some of the most common organic polymers are *hydrocarbons*, substances made entirely of carbon and hydrogen. Because a carbon atom can form four covalent bonds, it can join atoms into long chains of repeating structures that feature single, double, or triple covalent bonds. Petroleum, a fossil fuel, is made of hydrocarbons. Gasoline is made from petroleum. Octane, shown in **Figure 5-17**, is one of the compounds in gasoline.

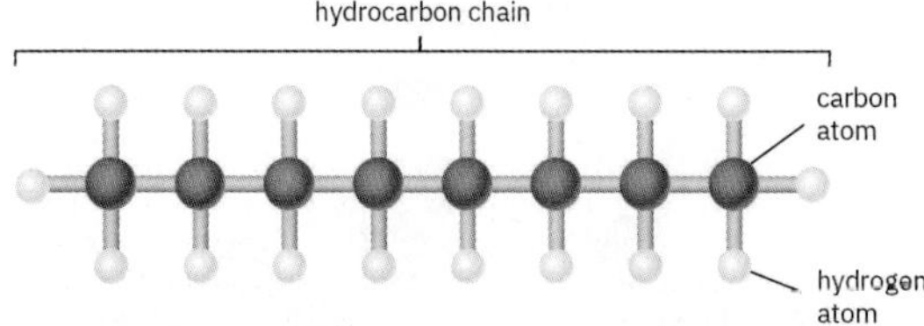

FIGURE 5-17 An octane molecule consists of eight carbon atoms linked in a chain, with 18 hydrogen atoms completing the remaining available bonds.

By themselves, hydrocarbon chains are not very reactive, but they can be made into much more reactive substances by adding functional groups. A *functional group* is a group of atoms that, when added to a hydrocarbon, gives the compound a particular property. Most of the functional groups found in biological molecules contain oxygen or nitrogen, which makes them polar. For example, adding a hydroxyl (—OH) group to the end of a hydrocarbon chain turns it into an alcohol. Carboxyl (—COOH) groups make a molecule into an acid (**Figure 5-18**).

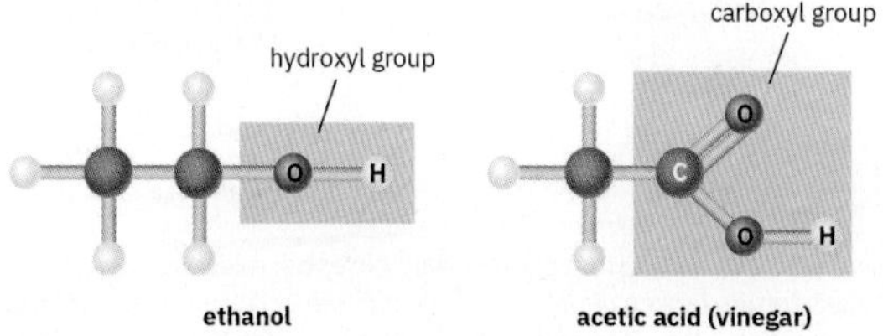

FIGURE 5-18 Replacing a hydrogen atom at the end of a short hydrocarbon chain with an oxygen and hydrogen (hydroxyl) group gives the two ends of the molecule different structures. The resulting alcohol molecule is polar. A carboxyl group attached to a hydrocarbon chain makes it polar and acidic.

SEP Construct an Explanation *Would you expect a hydrocarbon such as octane to dissolve in water? Why or why not?*

Science Background

Hydrocarbons Two classes of hydrocarbons exist. Saturated hydrocarbons have only single bonds between carbon atoms and each carbon atom has the maximum number of hydrogen atoms bound to it, hence being saturated with hydrogen. Unsaturated hydrocarbons have at least one double or triple bond between carbon atoms but the remaining available outer electrons of the carbon atoms are not bound to hydrogen.

SEP Construct an Explanation
Hydrocarbons do not dissolve in water because they are nonpolar.

SCIENCE AND ENGINEERING PRACTICES
Constructing Explanations and Designing Solutions

Sources of Hydrocarbons Hydrocarbons are most often associated with an important natural resource called fossil fuels—oil, natural gas, and coal. Fossil fuels contain varying kinds of hydrocarbon chains. All are formed naturally in the earth's crust from the remains of plants and animals that lived millions of years ago, thus the name *fossil fuels*. Group students and have them briefly research the sources of coal, petroleum, and natural gas and the processes under which they form. Then have each group construct an explanation about how the molecules in prehistoric organisms have been converted over time into the hydrocarbons found in these fossil fuels.

Carbohydrates Students learn that carbohydrates are fuels that organisms use to grow, live, and survive. **Chapter 6** discusses how plants use energy from the sun to make carbohydrates from carbon dioxide and water during photosynthesis.

In Your Community

Table Sugar Table sugar is extracted from both sugar cane and sugar beets and is known as sucrose, or $C_{12}H_{22}O_{11}$. Guide students to **Figure 5-19** and explain that sucrose is made of one molecule of glucose and one molecule of fructose that are bonded together. Share with students that in 2019, the United States was the 5th largest worldwide consumer of sugar although it outranks second-place Germany in *per capita* consumption by more than 20 grams per person. India is the leading consumer as a country but does not make the top 10 in per capita consumption. India is also the largest producer of sugar either in cane or beet form. The US ranks 8th in cane sugar production and 3rd in beet sugar production. Given the high production of table sugar in the United States, have students research to find if there is a local producer of table sugar in your community or if there are sugarcane or sugar beet farmers in your region.

Carbohydrates

Carbohydrates are organic compounds that consist of carbon, hydrogen, and oxygen. Cells use various types of carbohydrates as structural materials and for storing energy. These compounds are oxidized (burned) in the process of cellular metabolism, which releases the energy that organisms need to survive. Think of carbohydrates as fuels that, like gasoline for a car, keep an organism going. Carbohydrates also make up the backbone of **DNA**, the hereditary material in every cell.

Simple Sugars **Glucose** and fructose molecules are the building blocks of the most common carbohydrates. Monomers such as these two molecules are *monosaccharides*. Monosaccharides are often called simple sugars because they are the simplest carbohydrate, and many of them have a sweet taste. Simple sugars have a backbone of five or six carbon atoms and include two or more hydroxyl (—OH) groups.

The —OH functional groups make sugar molecules polar, so they can dissolve and move easily through the water-based internal environments of all organisms. In water, simple sugar molecules form ring-shaped structures such as the ones in **Figure 5-19**.

FIGURE 5-19 Sugar monomers have rings of six atoms, like glucose (a), or five atoms, like fructose (b). When a sugar monomer is incorporated into the structure of a larger molecule, it is often represented using a hexagon or a pentagon.

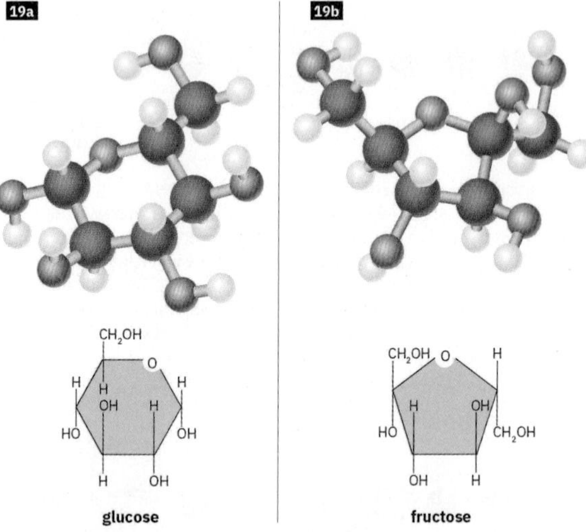

Simple sugars have extremely important biological roles. Bonds in glucose molecules are broken during chemical processes that cells use to release energy. That energy is then used to power other reactions. Simple sugars are also remodeled into other molecules or used as structural materials to build organic polymers.

SEP **Construct an Explanation** *Why are simple sugars a good source of quick energy?*

CROSSCUTTING CONCEPTS | Energy and Matter

Factors that Impact Energy Release Complex carbohydrates are typically compared in terms of how much energy each type of molecule stores. Students often develop a mental model that breaking bonds releases this stored energy in the molecule, and that breaking individual bonds equates to releasing more energy. In fact, breaking bonds absorbs energy. The details of breaking down glucose are described in **Section 6.3**, which shows the pathway and the reactions involved. That pathway shows that in the breakdown of a carbohydrate molecule such as glucose, molecules of ATP are formed which provide the usable energy for many chemical reactions in the cell. To help students adjust their mental models, point out here and in the next section that to break down molecules, such as cellulose and starch, requires energy. Then, emphasize that bonds in these molecules are broken during chemical processes that cells use to release energy that then powers other reactions.

Complex Carbohydrates Complex carbohydrates are chains of hundreds to thousands of sugar monomers. The most common *polysaccharides*—cellulose, starch, and glycogen—all consist only of glucose monomers, but as substances their properties are very different because of differences in the bonding patterns that link their monomers.

Cellulose is the most abundant organic molecule on Earth. It is the major structural material of plants (Chapter 9), forming tough fibers that act like reinforcing rods inside stems and other plant parts (**Figure 5-20**). Cellulose does not dissolve in water, and it is not easily broken down. Some bacteria and fungi can break it apart into its component sugars, but humans and other mammals cannot. Dietary fiber, or "roughage," usually refers to the indigestible cellulose in our vegetable foods.

BACTERIA IN YOUR GUT

Gather Evidence *Why might the relationship between bacteria and humans be described as symbiotic?*

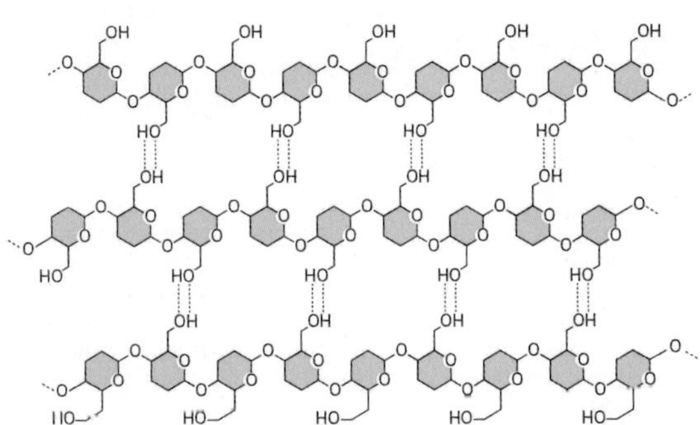

FIGURE 5-20 Cellulose is a polymer made of long straight chains of glucose monomers. The chains are joined parallel to each other by hydrogen bonds.

In **starch**, different bonding patterns between glucose monomers can make a chain that coils up into a spiral (**Figure 5-21**) or form a branched pattern of chains. Starch does not dissolve readily in water, but it is easier to break down than cellulose. These properties make starch ideal for storing sugars in the watery interiors of plant cells.

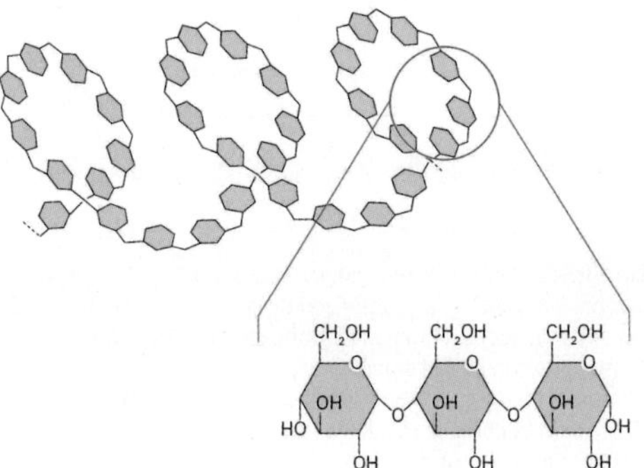

FIGURE 5-21 Amylose, a type of polymer in starch, forms tightly packed spirals that resist being broken down.

CHAPTER INVESTIGATION

Converting Carbohydrates
How does salivary amylase affect the carbohydrates in our food?

Go online to explore this chapter's hands-on investigation about simple sugars and starches.

Visual Support

Complex Carbohydrates Cellulose, as seen in **Figure 5-20** is a complex carbohydrate. Pair students to work through the concept of complex carbohydrates, and draw their attention to **Figure 5-20, Figure 5-21**, and **Figure 5-22.** Have students identify visible similarities and differences between cellulose, starch, and glycogen. Student groups can then make charts to list the three different types of complex carbohydrates, note the similarities and differences they identified during their discussions, and draw a simple model of each type. They should understand that all three types of complex carbohydrates are made up of chains of sugar monomers, but that the way that they are bound together differs between each type.

CHAPTER INVESTIGATION A

Guided Inquiry *Converting Carbohydrates*

Time: 50 minutes plus advance preparation

Students will follow a step-by-step procedure to investigate how starch can be broken down into smaller sugars and construct an explanation of their findings.

Go online to access detailed teacher notes and answers, rubrics, and lab worksheets.

REVISIT THE ANCHORING PHENOMENON

Gather Evidence *Bacteria can break down cellulose but humans cannot, so having bacteria in the gut could help humans break down the cellulose they eat into smaller molecules for energy.*

While many people may think of bacteria as germs that can cause disease and illnesses, bacteria are found in high volumes in most animals, and in humans can exist in nearly equal numbers to the number of cells in a human body.

Generally, bacteria live in harmony with their human host, providing support in digestion and upkeep of the skin. For example, certain bacteria in the gut have an enzyme that humans do not have that helps synthesize the essential nutrient vitamin B12. In general, the host-microbe relationship in humans is both complex and critical to meeting the nutritional and metabolic needs of humans, especially when treating for certain illnesses.

Connect to English Language Arts

Draw Evidence from Texts Once students complete the section on lipids, have them reflect on what they have read and draw evidence from the text. Then, have them write a paragraph explaining in their own words the difference between a non-reactive hydrocarbon and a lipid.

Science Background

Lipids Stearic acid, shown in **Figure 6-23,** has all single bonds. All the bonds are the same, so the chain is straight. Fats with these kinds of chains are called saturated fats, which are solid at room temperature. Some chains have double bonds in some places that cause the chains to bend. Unsaturated fats contain bent chains, and are liquid at room temperature. Elicit from students the names of some solid and liquid fats and ask whether each is saturated or unsaturated. Ask them to explain their answers using what they now know about the structures of their hydrocarbon chains.

Phospholipids This section introduces the concept of phospholipids and how they behave in water. **Chapter 6.2** discusses these interactions in more depth and explains how their structures allow for the fluid movement of phospholipid molecules in the fluid mosaic model.

Animals store sugars in the form of **glycogen**, a polysaccharide that consists of highly branched chains of glucose monomers (**Figure 5-22**). Muscle and liver cells contain most of the body's glycogen.

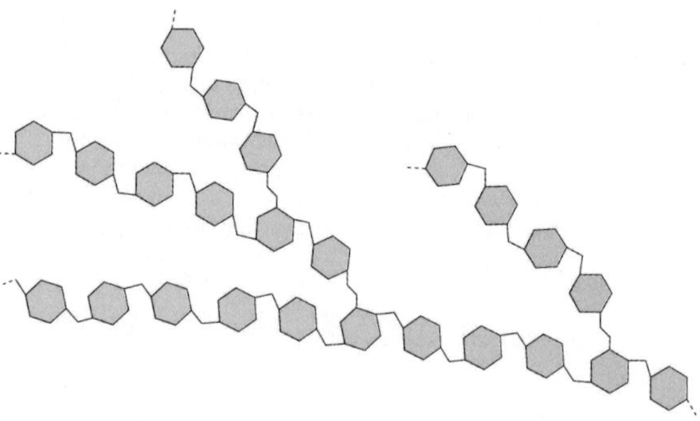

FIGURE 5-22 Chains of glucose form branches in a glycogen structure. Although it is commonly called "animal starch," glycogen is also found in bacteria and fungi.

Lipids

Lipids are a group of fatty, oily, or waxy organic compounds that are similar to carbohydrates in many ways. Like carbohydrates, lipids are composed mainly of carbon, hydrogen, and oxygen atoms. Energy storage is one of their primary functions. Lipid molecules vary in structure, but all are partly or completely nonpolar. They are typically made up of one or more hydrocarbon chains that attach to a single, small functional group that includes oxygen.

Fatty Acids Many lipids incorporate *fatty acids*—organic molecules with a carboxyl (—COOH) functional group "head" and a long hydrocarbon "tail" that varies in length (**Figure 5-23**). The tail is nonpolar (fatty), and the head is polar (acid). This gives the two ends of a fatty acid molecule different properties. You are already familiar with these properties, because these molecules are the main component of soap. The nonpolar tails attract oily dirt, and the polar heads attract the dirt to water.

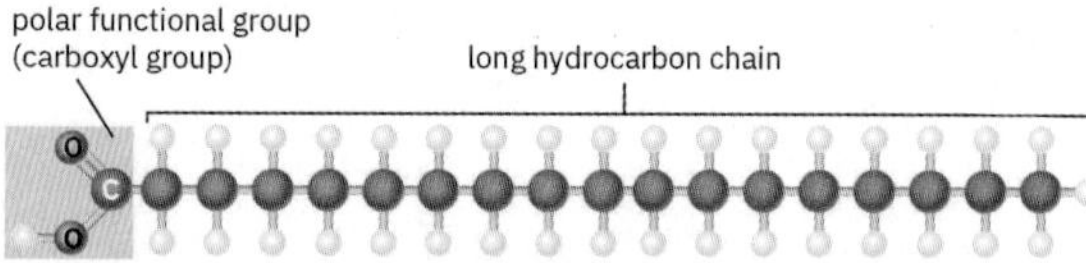

FIGURE 5-23 Stearic acid is a fatty acid that is commonly found in animal fats.

Phospholipids One class of biologically essential lipid molecules is the phospholipid, which is characterized by two fatty acid tails joined by a polar head that includes a functional group called a *phosphate group*. The phosphate group consists of a phosphorus atom bonded to four oxygen atoms. Some of the oxygen atoms are not bonded to anything else, making this end of the phospholipid molecule polar and negatively charged. Phosphate groups are found in several types of biological molecules.

In water, phospholipids tend to arrange themselves in layers in which the fatty acid ends face toward each other and the heads face away from each other. This configuration, shown in **Figure 5-24**, is called a phospholipid bilayer. Phospholipid bilayers play an important role in cell structures.

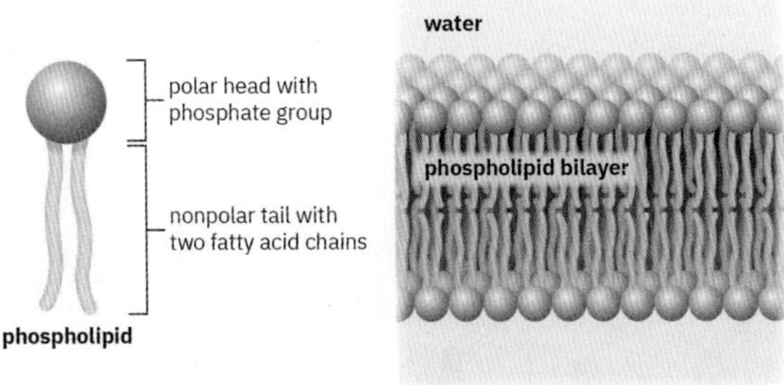

FIGURE 5-24 The nonpolar tails of a phospholipid molecule are hydrophobic, meaning they repel water. The polar head is hydrophilic. It attracts water.

SEP Construct an Explanation *Why do you think phospholipid molecules form a phospholipid bilayer in water?*

Nucleotides and Nucleic Acids

You have learned that deoxyribonucleic acid (DNA) molecules carry the genetic instructions for life. The building blocks of these molecules are monomers called nucleotides.

Nucleotide Structure A **nucleotide** consists of a monosaccharide (simple sugar) ring bonded to a nitrogen-containing base and one, two, or three phosphate groups. Nucleotides can act as individual molecules or as monomers that link to form **nucleic acids**.

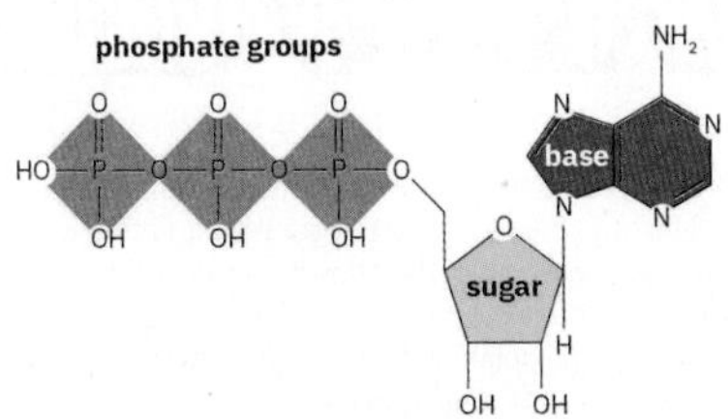

FIGURE 5-25 In a nucleotide, the phosphate groups are bonded to one of the carbon atoms in the sugar. The base, a group of atoms that form one or two flat rings, is bonded to another of the sugar's carbon atoms.

The **ATP** (adenosine triphosphate) nucleotide, shown in **Figure 5-25**, has a simple sugar component of ribose, three phosphate groups, and a base called adenine. This important molecule carries energy during cellular processes by transferring a phosphate group to another molecule. Removing one of the three phosphate groups from an ATP nucleotide turns it into an ADP (adenosine diphosphate) nucleotide.

SEP Construct an Explanation
The polar heads, which are attracted to the surrounding water, face outward toward the water, and the nonpolar ends face away from the water, toward the inside of the layer.

Address Misconceptions

Use a Model Students begin to understand the usefulness of structural models as they analyze **Figure 5-25.** Although introduced to them in **Figure 5-19**, here they can see how the display of the bonding arrangement of the atoms in the molecule are more apparent than in the geometry of a ball-and-stick representation. Encourage them to look back at **Figure 5-25** as they analyze how nucleotides bond with each other to form a DNA molecule in **Figure 5-26.**

DIFFERENTIATED INSTRUCTION | Leveled Support

Struggling Students Read aloud the description of nucleotides and nucleic acids. Emphasize the differences between the two and direct students to **Figure 5-26**. Have students work in groups to identify the parts of the structure of a DNA molecule. Students should understand that the sugar-phosphate backbone is the same for each monomer, but that the base pairs may differ for each monomer.

Advanced Learners Students may be familiar with the nucleotides—adenine, cytosine, guanine, thymine, and uracil—that make up DNA and RNA. However, they may not be familiar with ATP and that it is also a nucleotide. Ask these students to draw an ATP molecule, using a structure similar to the one shown in **Figure 5-25** and a molecule of adenosine. Then, students should identify the differences between the two structures.

Deoxyribose Nucleic Acid Students learn about the nucleic acids in DNA and how it consists of a sugar-phosphate backbone and a base pair. **Chapter 11** provides more detail about how DNA functions to code for genetic sequences within organisms.

Visual Support

Structure of a DNA Molecule Students may struggle to differentiate between a nucleotide and a nucleic acid, such as DNA. Group students and direct them to **Figure 5-26**. Have each group use digital media or paper to replicate the model. They may choose to make a 2D or 3D model, but they should use their own language to identify which part of the structure is the monomer (nucleotide) and what makes up the entire nucleic acid, which in this case is a strand of DNA. Students should understand that a nucleotide consists of a sugar phosphate backbone and a base pair that are covalently bonded to the next nucleotide on the same strand, but that the two strands are connected via hydrogen bonds.

CCC Structure and Function *Sample answer: Complex carbohydrates and DNA are both polymers. Their monomers both have simple sugars, but the monomer for a complex carbohydrate is just a simple sugar, whereas the monomer for DNA is a nucleotide consisting of a simple sugar bonded to a base and some phosphate groups.*

Go Online SIMULATION

Building Complex Molecules
Simulation 5.3 allows students to build organic molecules out of C, H, O, N, P, and other basic units. The simulation begins with the breakdown of a sugar molecule to C, H, O.

Nucleic Acids The molecules that contain the genetic information that determines the makeup of the proteins in all of an organism's cells are polymers of nucleotides. In an individual DNA (deoxyribonucleic acid) nucleotide, the sugar component is deoxyribose, and there are three phosphate groups. The nitrogenous base can be one of four types

- adenine (A)
- cytosine (C)
- thymine (T)
- guanine (G)

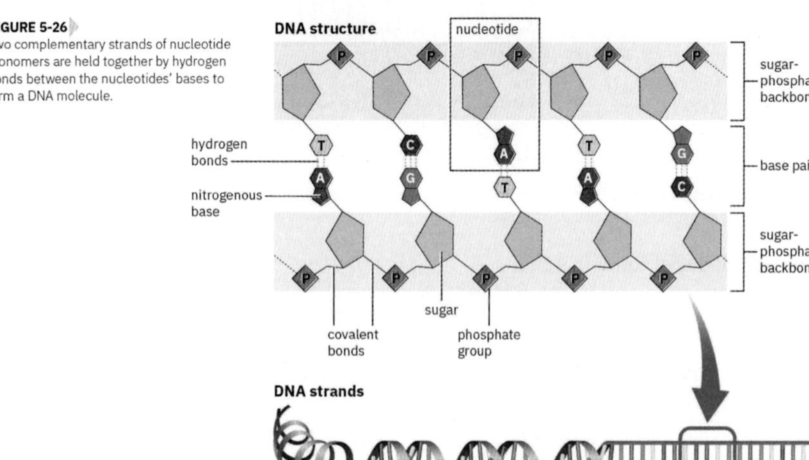

FIGURE 5-26 Two complementary strands of nucleotide monomers are held together by hydrogen bonds between the nucleotides' bases to form a DNA molecule.

Go Online SIMULATION

Building Complex Molecules Go online to use atoms to assemble polymers such as DNA.

In a DNA molecule, adenine pairs with thymine, and cytosine pairs with guanine. Millions of DNA nucleotides linked together comprise a strand of DNA. The backbone of the strand is made of the sugars and phosphate groups in the nucleotides, as shown in **Figure 5-26**. Two such strands coiled in a shape called a double helix form a DNA molecule. The information coded in a DNA molecule lies in the sequence of the four nucleotide bases.

RNA (ribonucleic acid) is the nucleic acid that translates the DNA code into amino acids (and ultimately proteins). It is similar to DNA but it has a different sugar, ribose, rather than deoxyribose, and it uses the nitrogenous base uracil (U) instead of thymine. In an RNA molecule, adenine pairs with uracil, and cytosine pairs with guanine.

CCC Structure and Function *How are a complex carbohydrate molecule and a DNA molecule similar? How are they different?*

132 CHAPTER 5 MOLECULES IN LIVING SYSTEMS

DIFFERENTIATED INSTRUCTION | Leveled Support

Struggling Students Some students may not be able to visualize three-dimensional versions represented by the flat models shown in **Figure 5-26**. Point out that these are just visual representations of DNA and are not to scale, nor do they reflect what DNA actually looks like at a molecular level. Support these students by showing them a labeled blueprint of a car, a small model car, and an actual car. Help these students understand that these three examples show an actual item and that the blueprint and model car are different representations of the same thing. Guide them in a discussion as to why different representations can be used to model something that is either too large or too small to easily see or manipulate.

Advanced Learners Encourage students to use various materials such as straws and clay to create three-dimensional versions of the two-dimensional model shown in **Figure 5-26.**

Amino Acids and Proteins

Proteins perform thousands of different functions in living things. Some proteins are enzymes, molecules that drive all chemical reactions that take place inside a cell. Structural proteins support cell parts and multicellular bodies. Muscle tissue, for example, is made almost entirely of protein. Proteins transport substances, help cells communicate, and defend the body. They also regulate genes. Every cell in your body contains exactly the same genes, but only a subset of genes are active in certain cell types at specific points in time. Proteins turn these genes on and off to control the processes and behaviors that a cell exhibits.

Amino Acids An **amino acid** is a small organic compound with an amine group (—NH_2), a carboxyl group (—COOH, the acid), and a side chain called an R group. A few amino acids also include a sulfhydryl group (—SH). In most amino acids, all three groups are attached to the same carbon atom (**Figure 5-27**). Cells make the thousands of different proteins they need from only 20 types of amino acid monomers. Each type of amino acid has a different R group.

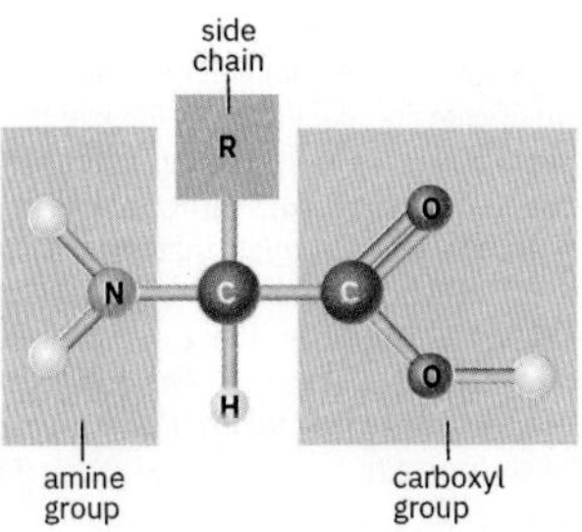

FIGURE 5-27 Amino acids vary by the side chain they contain, but all amino acids have an amine group and a carboxyl group.

Protein Structure A **protein** is made up of a chain of amino acids. Proteins are built brick by brick, amino acid by amino acid. The covalent bond that links amino acids in a protein is called a peptide bond. The order of the amino acids is called the protein's primary structure (**Figure 5-28**). Each type of protein has a unique primary structure.

CCC **Patterns** *A gene is a DNA sequence that carries instructions for making a protein. What characteristic do DNA and protein structures have in common that could explain this relationship?*

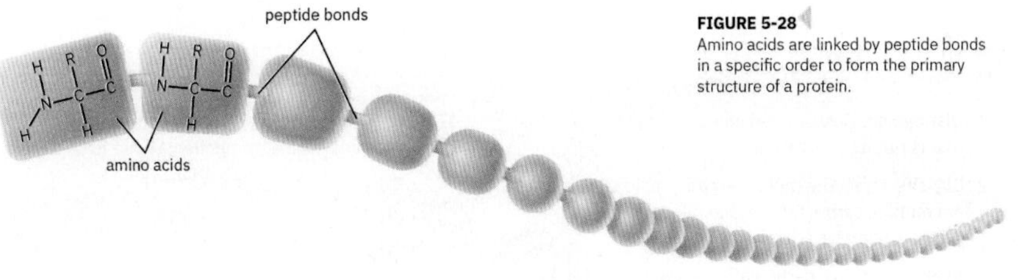

FIGURE 5-28 Amino acids are linked by peptide bonds in a specific order to form the primary structure of a protein.

SCIENCE AND ENGINEERING PRACTICES
Constructing Explanations and Designing Solutions

How Amino Acids Join While students may understand the parts of the generic amino acid shown in **Figure 5-27**, they may be confused as to how this amino acid would join with the next. Explain to students that when peptide bonds form, a water molecule is lost. Allow students to form groups to examine **Figure 5-27** and **Figure 5-28** to determine where the bond between two amino acids will form. If students are stuck, let them know that the white atoms in **Figure 5-27** represent hydrogen molecules. Once students have determined that the single bonded OH portion of the carboxyl group would bind to one of the hydrogens on the N on the amine group, have students construct a sentence that explains what happens when a peptide group between amino acids forms.

Science Background

Amino Acids and R Groups Generally, the R groups among the 20 amino acids differ in their polarity or distribution of charge. Amino acids with nonpolar R groups will repel water. In an aqueous solution, certain proteins will fold nonpolar amino acids toward the protein's interior. Polar but uncharged amino acids form proteins that will bond with water and other molecules. Polar but charged amino acids may be acidic (or easily give up a proton) or basic (easily accepting a proton) in protein formation. Students may research which amino acid fall into each category.

CCC **Patterns** *Sample answer: The sequence of monomers is important in both DNA and proteins. DNA carries information in the sequence of its nucleotides, and a protein's primary structure is determined by the sequence of its amino acids.*

Visual Support

Protein Structures Provide groups of students with 30 small, 2 cm x 2 cm, colored squares of construction paper. Explain to students that each of these squares represents an amino acid. Have students make two strips of amino acids by connecting 15 squares together in a row using tape. Ask them which level of a protein structure these two strips represent. Students should understand that the connected amino acids form the primary structure of a protein. As students move through the next page, have them manipulate the strips to form the secondary, tertiary, and quaternary protein structures.

Connect to English Language Arts

Summarize a Text As students work through the primary, secondary, and tertiary structures of proteins, have them write a brief summary of the properties of each structure level. Once they have completed the section, have students revise their summaries to include only the most important ideas and leave out minor details.

A long chain of amino acids joined by peptide bonds is known as a polypeptide chain. As amino acids are added to a polypeptide chain, the chain curls up, bending and twisting as hydrogen bonds form between the amino acids. The hydrogen bonds pull sections of the polypeptide chain into characteristic patterns such as coils, as shown in **Figure 5-29,** or sheets connected by flexible loops and tight turns. These patterns are the protein's secondary structure. Almost all proteins have similar patterns in their secondary structures.

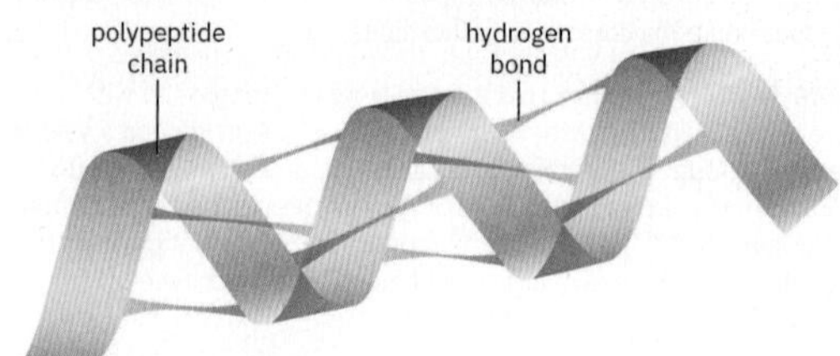

FIGURE 5-29 Hydrogen bonds occur between parts of a growing polypeptide chain to form its secondary structure.

Hydrogen bonding and other interactions between coils and sheets can make them fold up into compact shapes. Each of these shapes has a particular function that is more or less separate from the rest of the protein. For example, some barrel-like shapes rotate like motors in small molecular machines. Others form tunnels that allow substances to move into and out of a cell. These functional shapes are called the protein's tertiary structure (**Figure 5-30a**). They make a protein into a working molecule. Proteins that consist of two ore more polypeptide chains have a quaternary structure. Most enzymes are like this, with multiple polypeptide chains that collectively form a roughly spherical shape (**Figure 5-30b**).

30a

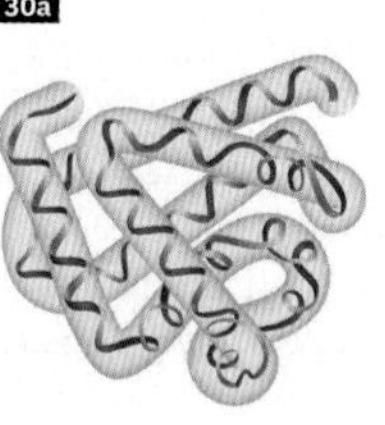

30b

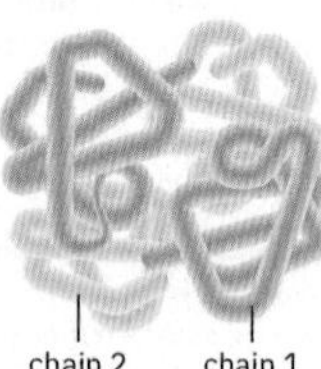

FIGURE 5-30 Proteins with compact tertiary structures have shapes that perform specific functions (a). Proteins made from two or more polypeptide chains have a quaternary structure (b).

5.3 REVIEW

1. **Distinguish** Give an example of an organic molecule that is not a hydrocarbon.
2. **Classify** Label the molecules as simple sugars or complex carbohydrates: cellulose, fructose, glucose, sucrose.
3. **Explain** Why are carbohydrates soluble in water while lipids are not?
4. **Model** Draw one amino acid using the given components. Not all components need to be used.
   - CH
   - CH_2OH
   - CH_3
   - COOH
   - NH_2

5.3 REVIEW

1. *Student answer should be any molecule that contains hydrogen, carbon, and at least one atom of any other element. For example, ethanol C_2H_5OH.* **DOK 2**
2. *simple sugars: fructose, glucose; complex carbohydrates: cellulose, sucrose* **DOK 2**
3. *Carbohydrates have a greater proportion of oxygen than lipids. Oxygen makes molecules polar and thus soluble in water.* **DOK 2**
4. *Student answer should include NH_2, CH, COOH, and either CH_3 or CH_2OH.* **DOK 3**

ON ASSIGNMENT For people in the Peruvian Andes, tuber crops are an essential source of carbohydrates. Farmers plant dozens of varieties together in the same plot. Specific types of tubers are used at weddings, festivals, or funerals, and others are eaten with everyday meals. This farmer showed National Geographic photographer Jim Richardson a harvest of potatoes and bright yellow oca (*Oxalis tuberosa*), a tuber native to Peru and rich in amino acids.

ON ASSIGNMENT

Oxalis tuberosa National Geographic photographer Jim Richardson is known for being able to tell a story with the pictures he takes, and he specializes in photographs around the topics of food production, water, and soil conservation. He has spent over 35 years traveling internationally to capture the stories about how food is produced around the world.

This photograph, taken in the Peruvian Andes, shows a farmer holding out a basket of oca, one of over 1,300 varieties of tubers that exist in the region. While students are likely familiar with potatoes, share that oca are the second most highly cultivated tubers on the planet and are high in protein and antioxidants, molecules that have protective properties.

You can learn more about the photographer's work at https://www.jimrichardsonphotography.com.

DIFFERENTIATED INSTRUCTION | Economically Disadvantaged Students

Food Insecurity Though not always apparent, there are many students who have experienced food insecurity or have had few experiences with the wide variety of different kinds of plants and vegetables that are available across the world. Share with students images of a number of different root and tuber vegetables and explain that while tuber vegetables may look like stems, tubers like potatoes, oca, jicama, and yams are actually modified stems of a plant, and not a part of the root. Have students compare and contrast the differences between what a tuberous plant looks like under the soil versus a root plant, like carrot, beet, or garlic.

EXPLORER
ROSA VÁSQUEZ ESPINOZA

About the Explorer

National Geographic Explorer Rosa Vásquez Espinoza studies the organisms of the Boiling River in the Amazon. She studies the enzymes that the organisms in the Boiling River produce in order to determine what molecules protect them in the harsh environment. She hopes these molecules may be engineered into medicines. You can learn more about Rosa Vásquez Espinoza and her research on the National Geographic Society's website: www.nationalgeographic.org.

Connect to Careers

Extremophile Biologist While many extremophiles would not survive in the average conditions that other organisms thrive in, there are some that can. Tardigrades are a microscopic animal that can survive 30 years without food or water, in freezing and boiling temperatures, and in space. They also live in regular water. A protein has been found in their body that protects their DNA from damage and scientists are looking for other protective molecules that allow them to survive as they do.

Video

Time: 1:03

Video 5-1 shows Dr. Vásquez Espinoza and a team of researchers gather sample organisms in the Boiling River ecosystem. Allow students time to imagine the remoteness of the area and what the air around the Boiling River might feel like.

THINKING CRITICALLY

Relate *Because all living things share the same small set of amino acids, those found in the microbes would be the same as found in humans so it would be possible to insert them into human processes or model their properties.*

NATIONAL GEOGRAPHIC EXPLORER **ROSA VÁSQUEZ ESPINOZA**

SURVIVAL MECHANISMS AND NEW MEDICINES

Dr. Rosa Vásquez Espinoza explores microorganisms in the Peruvian Amazon. She is studying their survival mechanisms to determine how those processes can be engineered to develop new natural medicinal products.

VIDEO 5-1 Go online to watch Vásquez Espinoza and a team of researchers gather sample organisms in the Boiling River ecosystem.

Inspired by her grandmother's knowledge of traditional medicines made from plants, herbs, and soils, chemical biologist Dr. Rosa Vásquez Espinoza grew up wondering how nature produces these powerful compounds. Her scientific search has led her to the Boiling River in the Peruvian Amazon. She steps carefully along the steamy banks of the river—one slip into its 99.1 °C waters could render third-degree burns.

Vásquez Espinoza stoops to scrape lichens off the side of a rocky ledge into a jar. Next, she dips a heat resistant test tube into the scalding water to gather cyanobacteria. She then moves upriver, where the water is a bit cooler, to harvest a sample of a microbial mat from the river bottom. She also takes samples of sediments that contain large populations of bacteria. After dinner, the team will set up a microscope and begin analyzing the samples. If one looks especially interesting, they will return to that area the next day to collect more samples.

Vásquez Espinoza thinks the bacteria produce unique molecules that protect them from the high temperatures of their habitats. She hypothesizes that because they enable microorganisms to survive under stressful conditions, these molecules may be engineered into medicines to fight bacterial, fungal, or viral infections, and even cancers. In the laboratory, Vásquez Espinoza also looks at enzymes, molecules that regulate chemical reactions in the organisms, to try to pinpoint substances that have therapeutic properties. Understanding how microorganisms produce these substances may allow scientists to develop sustainable processes for manufacturing and refining them for use as pharmaceutical drugs. In addition, these enzymes' ability to withstand a wide range of temperatures could make them useful for reducing negative environmental impacts of industrial processes that generate toxic waste and pollution.

"If this is possible now, what can we achieve in the next 10 to 20 years?" asks Vásquez Espinoza. "How many more bacteria can we find in our planet that can help make our lives better and also help us protect nature itself? How is it possible that we can go to the moon and don't know fully yet what's at the microbial level in our planet?"

THINKING CRITICALLY

Relate *Enzymes are made of amino acids. How does this information support the hypothesis that enzymes found in the Boiling River microbes could also work in medicines for humans or animals?*

136 CHAPTER 5 MOLECULES IN LIVING SYSTEMS

CHEMICAL REACTIONS

Humans have discovered that when we cook food, its flavor and texture change in ways that make it tastier and easier to digest. Many fruits and vegetables can be eaten raw, but some foods should not be eaten without cooking them. For example, plants such as green potatoes, lima and red kidney beans, and cassava naturally contain toxic compounds that are broken down by heat.

Explain *How do you think cooking a food source changes it into a different substance?*

5.4

KEY TERMS

- activation energy
- catalyst
- chemical reaction
- enzyme
- equilibrium
- product
- reactant
- substrate

Chemical Reactions in Biological Systems

All biological systems and processes involve changes in matter and energy. At a molecular level, these changes occur during chemical reactions. A **chemical reaction** is a process in which substances change into different substances through the breaking and forming of chemical bonds. Depending on the specific substances involved, this whole process absorbs or releases energy.

VIDEO

VIDEO 5-2 Go online to see how bonds in glucose and fructose molecules are broken to recombine atoms in a chemical reaction.

Reactants and Products **Reactants** are substances that are the inputs of a chemical reaction. The outputs of the reaction are its **products**. Chemical bonds in the reactants are broken and new bonds form as the atoms are rearranged to make the products. For example, as shown in **Figure 5-31**, by breaking the bonds between carbon and oxygen atoms, the atoms from glucose and fructose molecules can be reassembled to produce sucrose and water.

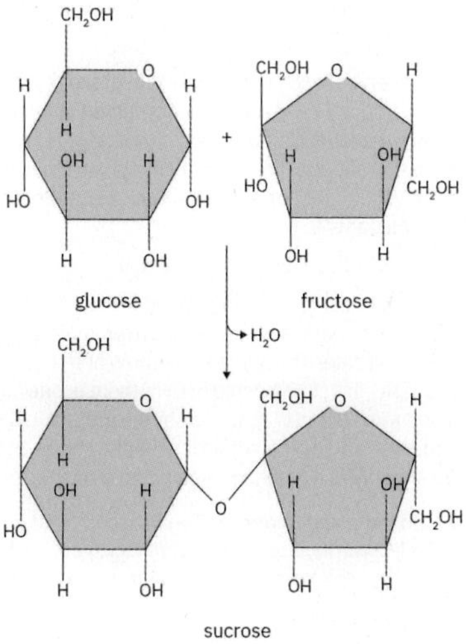

FIGURE 5-31 The simple sugars glucose and fructose react to form a sucrose molecule and a water molecule. Sucrose is found in the roots, fruits, and nectar of plants. It is also known as table sugar.

5.4

CHEMICAL REACTIONS

LS1.C, PS1.A, PS2.B

EXPLORE/EXPLAIN

This section introduces chemical reactions and how they occur in biological systems, the role of energy in chemical processes, the concept of system equilibrium, and how catalysts, enzymes, and other substances can affect the rates of biochemical reactions.

Objectives

- Explain the process of chemical reactions in biological systems.
- Describe how energy in organic molecules is released during chemical reactions and used for cellular processes.
- Characterize a reaction reaching equilibrium.
- Summarize how catalysts affect the rate of reactions.

Pretest Use **Questions 3, 4,** and **8** to identify gaps in background knowledge or misconceptions.

Explain *Sample answer: The heat that you add to raw food when you cook it would break down the molecules in the original substance so that they can be recombined into different molecules, making the cooked food into a different substance.*

Video

Time: 20 seconds

Video 5-2 shows the formation of a disaccharide molecule from a glucose and fructose molecule. Share with students that the carbon molecules are gray, the oxygen molecules are red, and the hydrogen molecules are white. Have students identify the molecule that is removed when the two sugars combine. (*water*)

Video

Time: 30 seconds

Video 5-3 shows a methane combustion reaction where a methane molecule reacts with oxygen to release carbon dioxide. Share with students that the carbon molecules are gray, the oxygen molecules are red, and the hydrogen molecules are white. Ask students to identify how the reaction shown supports the concept of conservation of matter.

In Your Community

Greenhouse Gases Methane is a powerful greenhouse gas that is often released into the atmosphere by human activities or agricultural practices. Guide students to the EPA website (https://www.epa.gov) to identify other greenhouse gases that exist in the atmosphere. Have students select one of the greenhouse gases and find the chemical reactions that it is derived from. As students share their findings with the class, brainstorm how industrial or other practices in your community contribute to them.

CCC Scale, Proportion, and Quantity *6 CO_2 and 6 H_2O*

VIDEO

VIDEO 5-3 Go online to see how methane and oxygen react to form carbon dioxide and water.

Conservation of Matter By now, you have seen the chemical formulas for several compounds, such as water (H_2O), carbon dioxide (CO_2), and oxygen gas (O_2). The letters in these chemical formulas indicate the elements that make up a compound, and the subscript numbers to the right of each element are the number of atoms of that element in the compound. (If there is no number after an element, there is only one atom of that element.) Not only do chemical formulas describe exactly what constitutes a specific compound, but these formulas also provide a useful way to keep track of what happens to matter that undergoes a chemical reaction.

In a chemical reaction, matter is conserved. Although the products are different compounds from the reactants, no new matter is created, and no matter is destroyed. The total number of atoms of each element in the products is the same as the total number of atoms of each element in the reactant. This fact enables us to predict what the products of a reaction between specific reactants might be.

When a substance such as wood or coal is burned, a chemical reaction occurs between the oxygen in the air and one or more of the compounds in the substance, releasing energy in the form of light and heat. Methane (CH_4) is a hydrocarbon compound found in natural gas. When a compound that consists of only carbon and hydrogen is burned in oxygen (O_2), the products must be made of only carbon, hydrogen, and oxygen atoms. As shown in **Figure 5-32**, these products are water (H_2O) and carbon dioxide (CO_2).

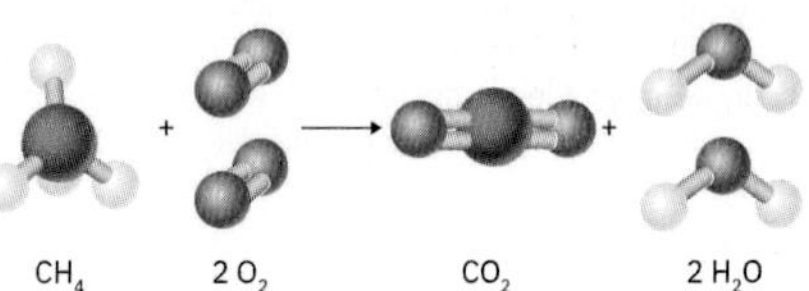

FIGURE 5-32 One methane molecule reacts with two oxygen molecules to form one carbon dioxide molecule and two water molecules. An arrow pointing from the reactants to the products is used to indicate that a chemical reaction changes the reactants to the products.

For matter to be conserved, each individual atom of the reactants must become part of the products, and the products cannot include atoms that were not part of the reactants. During a reaction, one or more reactants become one or more products. Intermediate molecules may form between reactants and products. A reaction is shown as an equation in which an arrow points from reactants to products:

$$CH_4 + 2O_2 \rightarrow CO_2 + 2H_2O$$

A number before a chemical formula in such equations indicates the number of molecules. (If there is only one of a given molecule, the 1 is not shown.) A subscript indicates the number of atoms of that element per molecule. For example, in one molecule of CH_4, there is one carbon atom and four hydrogen atoms. Note that atoms shuffle around in a reaction, but they never disappear. A balanced chemical equation shows that the same number of atoms that enter a reaction remain at the reaction's end.

CCC Scale, Proportion, and Quantity *The chemical formula for glucose is $C_6H_{12}O_6$. If a glucose molecule reacts with six oxygen (O_2) molecules, how many carbon dioxide molecules and how many water molecules are produced?*

DIFFERENTIATED INSTRUCTION | English Language Learners

Expressing Opinions and Ideas Have students discuss possible sources of pollution and greenhouse gases in their community. Explain that greenhouse gases got this name because they warm Earth like a glass greenhouse warms plants.

Beginning Have students taking turns expressing their ideas. Provide sentence frames: *I think ______ in our area cause pollution. There is/are ______. I see/smell ______.*

Intermediate Have pairs express their opinions and agree or disagree. Provide sentence frames: *I think ______ causes pollution because ______. I agree/disagree, because ______.*

Advanced Have students express their opinions and support them with reasons and evidence. Encourage group members to ask: *Why do you think that? Provide sentence frames: One reason is ______. I* see/read/know that ______.

Energy in Chemical Processes

Cells store energy by building large organic molecules such as polysaccharides, lipids, and proteins from smaller components. The energy of these molecules is released in chemical reactions and used to power life processes such as cell division and growth.

Activation Energy All chemical reactions need an input of energy to break bonds in the reactants. The amount of energy required to start a chemical reaction is called its **activation energy**. For example, in the process of photosynthesis, sunlight provides the activation energy to begin a series of reactions that convert reactants, including carbon dioxide and water, into the products, sugar and oxygen. Similarly, to burn a fuel such as wood, you first need to add energy in the form of heat to ignite the wood.

FIGURE 5-33 Whether they release or absorb energy, all chemical reactions require an input of activation energy to proceed.

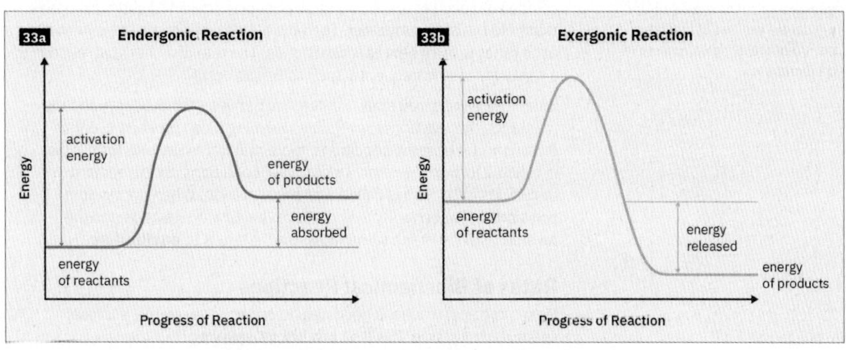

Endergonic and Exergonic Reactions An *endergonic* reaction or process is one in which energy is absorbed. In an endergonic chemical reaction, the total energy of the products is greater than the total energy of the reactants, as illustrated by the graph in **Figure 5-33a**. Photosynthesis is an endergonic process in which the leaves of a plant absorb solar energy to build sugar molecules (Section 6.3). Similarly, when you bake a cake, you add energy in the form of heat, which is absorbed by the reactants (the ingredients) to form a delicious new substance.

In an *exergonic* reaction or process, energy is released. The total energy of the products of an exergonic reaction is less than the total energy of the reactants, as illustrated by the graph in **Figure 5-33b**. Exergonic reactions occur spontaneously once their activation energy is reached. Burning wood is an exergonic process that releases energy in the form of heat and light. Cells use exergonic reactions to release the energy stored in molecules in the process of cellular respiration (Section 6.3).

CCC Energy and Matter *The chemical reaction in which glucose and fructose molecules combine to form sucrose is an example of a condensation reaction. Condensation reactions occur when a cell assembles a polysaccharide, such as cellulose, from glucose monomers. Would you expect the process of building a polysaccharide to be endergonic or exergonic? Explain.*

Connect to Mathematics

Model with Mathematics Chemical reactions require energy to break bonds so that new bonds can form. Both endergonic and exergonic reactions require that the activation energy level is met to start the reaction, but the two types of reactions have very different amounts of energy in their products once the reaction is complete. These reactions can be modeled by using graphs, as shown in **Figure 5-33**. Have students work with a partner to analyze the graphs to understand the differences between endergonic and exergonic reactions.

Chemical Reactions Students learn that endergonic reactions absorb energy and that exergonic reactions release energy in this section. **Chapter 13** explores how scientists use chemical reactions in biotechnology.

SEP Construct an Explanation *Building a polysaccharide with condensation reactions would be an endergonic process. If cells release energy by breaking up a large molecule, then the reaction to form that molecule would have had to absorb energy to store it in chemical bonds.*

SCIENCE AND ENGINEERING PRACTICES
Using Mathematics and Computational Thinking

Graphs as Visual Representations In **Figure 5-33**, have students identify the *x*-axis and the *y*-axis in the graphs. Ensure that students understand that the progress of reaction is shown on the *x*-axis and that the amount of energy is shown on the *y*-axis. Ask students to discuss why these two graphs were used to represent the concepts of endergonic reactions and exergonic reactions. Have students discuss how the graphs would change if something were added to the reaction that lowered the activation energy of the reaction. Students should understand that the peak in both reactions would decrease, but that the other points of the graphs would remain the same.

CASE STUDY
TURNING MOLECULES INTO MEDICINE

Ask Questions To help students revisit the Case Study and make connections with the content, have them make a model that would show in which direction a reaction would move if more reactants were added to a reaction.

Connect to English Language Arts

Draw Evidence from Texts The rates of biochemical reactions can be affected by a number of different factors. As students complete the section on the Rates of Biochemical Reactions, have them analyze the information and images provided in the text to explain the differences in the effects of using heat or pH to speed up a chemical reaction, or using enzymes.

Conservation of Energy Energy is neither created nor destroyed during a chemical reaction. The difference between the energy of the products and the energy of the reactants is equal to the amount of energy absorbed from some source outside the reaction system or released into the surroundings. During a reaction, energy can be transferred from one place to another or converted from one form to another. For example, some of the solar energy absorbed in a photosynthesis reaction is converted into chemical energy that is stored in glucose molecules.

CASE STUDY
TURNING MOLECULES INTO MEDICINE

Ask Questions *Generate a question about how equilibrium affects the process of making a naturally occurring compound in a laboratory.*

Equilibrium

A mixture of two substances consists of many of both types of molecules randomly moving around near each other in the same space. For a chemical reaction to occur between the substances, the molecules must collide with each other with enough energy to break the chemical bonds that hold the atoms together. Then the atoms must come close enough to each other to form new chemical bonds. The rate of a chemical reaction is a measure of how frequently such collisions occur.

When there is a small energy difference between the products and the reactants, the reaction is reversible, meaning it can proceed in either direction. Under some conditions, more product molecules form. This is called a forward reaction. Under other conditions, more reactant molecules form. This is called a reverse reaction. When the random collisions between molecules produce the forward reaction and the reverse reaction at the same rates, the system is at **equilibrium**.

Rates of Biochemical Reactions

Most biological reactions need help to occur. For example, sucrose seems highly stable. It will sit in a box for years without breaking down spontaneously into glucose and fructose in an exergonic reaction. Cells store energy in organic molecules that are even larger than sucrose, but an organism does not have the time for the spontaneous breakdown of an organic molecule to occur when it needs this energy.

Catalysts A chemical reaction can be sped up by raising the temperature dramatically or changing the pH. However, these conditions are often unsafe or toxic to the organism or cells in which they would need to occur. Instead, cells use **catalysts**, molecules that increase the rate of a chemical reaction without being used up in the process. Catalysts do so by lowering the activation energy needed in the reaction (**Figure 5-34**).

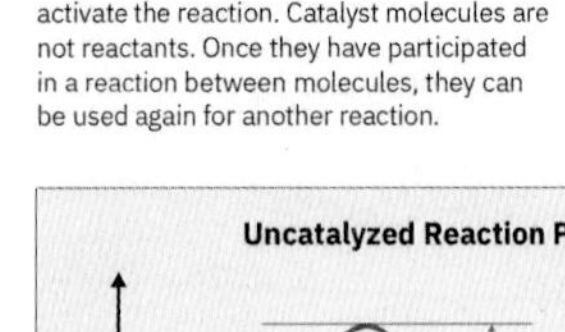

FIGURE 5-34 A catalyst speeds up a chemical reaction by lowering the amount of energy needed to activate the reaction. Catalyst molecules are not reactants. Once they have participated in a reaction between molecules, they can be used again for another reaction.

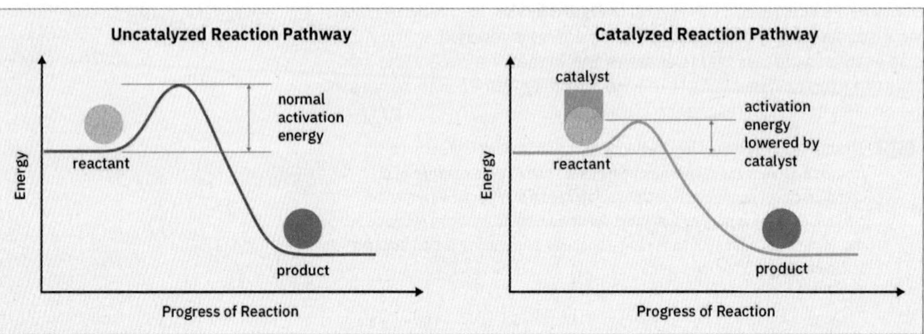

CROSSCUTTING CONCEPTS | Stability and Change

Equilibrium in Systems Chemical equilibrium as described in the text is common in chemical systems such as the movement of carbon dioxide between the water and air in a closed bottle of carbonated water. This occurs with no additional input of energy. It also occurs in biological systems, but more rarely. One example is the ability of hemoglobin to bind with oxygen in the lungs and release it elsewhere in the body. A steady state or state of dynamic equilibrium occurs more frequently in biological systems where a continual input of energy enables a continual reactive state, such as the movement of ions across the cell membrane. Encourage students to look up the meaning of the word *dynamic* and then construct an explanation for the phrase "stability can mean constant or continual change."

Enzymes and Substrates Biological catalysts are called **enzymes**. Some enzymes are nucleic acids, but most are proteins. The structure of an enzyme molecule enables its function.

Enzymes lower the activation energy of a reaction by providing pathways for molecules to interact more easily than they would have otherwise. An enzyme has one or more *active sites* on its surface that change their shape so that they can only bind to molecules that have a specific shape, size, polarity, and charge. A molecule that fits an enzyme's active site is called its **substrate**. Together, an enzyme and its substrate(s) form an *enzyme-substrate complex*.

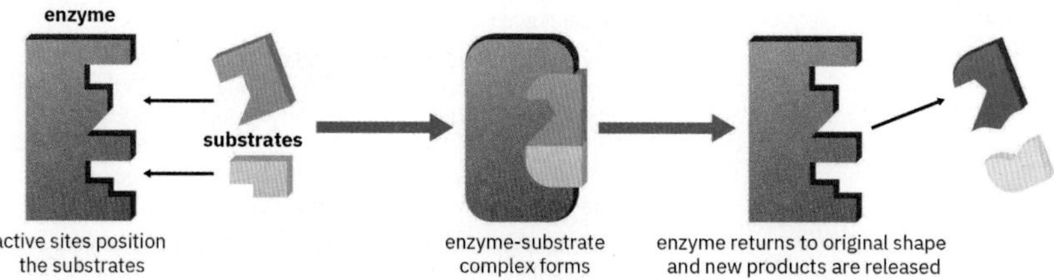

enzyme
substrates
active sites position the substrates
enzyme-substrate complex forms
enzyme returns to original shape and new products are released

FIGURE 5-35 An enzyme catalyzes the reaction between two substrates by positioning the substrates when they bind to its active sites, moving them together so that they can react, and releasing the product.

In a solution containing randomly-moving molecules, the flexible shape of the enzyme-substrate complex can lower the energy required for substrates to react by bringing them closer to each other and orienting them in positions that favor the reaction, as shown in **Figure 5-35**. The active sites can also work to deform substrates in a way that breaks the bonds between their atoms. Enzymes can also control the rate or direction of a reaction by removing product or reactant molecules, specific substrates, or other molecules present in the solution, that may interfere with the reaction. Once an enzyme's active site releases its substrate, it returns to its original shape so that it can bind to another substrate.

Most enzymes are highly specific and catalyze only a few closely related chemical reactions or, in many cases, only one particular reaction. For example, the enzyme sucrase catalyzes only the breakdown of sucrose. It will not break down similar sugars. Enzymes are often named by changing a suffix *-ose* to *-ase*. Thus, the enzyme maltase breaks down the sugar maltose; the enzyme lactase breaks down the sugar lactose. **Table 5-4** lists a few other important enzymes and their functions.

TABLE 5-4. Enzymes and Their Functions

Enzyme	Function
ATP synthase	synthesizes ATP from ADP and phosphate
nitrogenase	converts atmospheric nitrogen to ammonia
pepsin	breaks down proteins during digestion in the stomach
polymerase	makes RNA and DNA strands in transcription and translation
rubisco	catalyzes the first major step in carbon fixation
trypsin	breaks down proteins during digestion in the small intestine

CCC **Structure and Function** *Use the structural characteristics of proteins and the function of enzymes to explain why it makes sense that many enzymes are proteins.*

Enzymes Students learn about how enzymes lower the activation energy of a reaction and act as biological catalysts in this section. Students will find more details on how the body builds proteins and enzymes through the process of translation in **Section 11.2**.

Visual Support

Enzymes as Catalysts Use **Figure 5-35** to guide student to focus on the fact that enzymes do not *provide* energy, they only *lower* the activation energy of a reaction. Enzymes have a variety of pathways that they can use to control a reaction. Have students compare the shapes of the substrates and products before and after the reaction and compare the enzyme from before the reaction and after the reaction. Students should recognize that the new product that was formed has a different shape, but that matter was conserved, while the enzyme returned to its original state. Lead a class discussion to examine how the enzyme returning to the original state is useful in biological reactions.

DIFFERENTIATED INSTRUCTION | Leveled Support

Struggling Students After viewing **Figure 5-35**, some students may think that enzymes only exist to form new products from multiple substrates to build larger molecules. Explain to students that enzymes can work in either direction. They are not limited to only building molecules but can also be involved in breaking down molecules. Have students work in pairs to develop an analogy that shows how enzymes function in biological systems.

Advanced Learners Students who have an understanding of how enzymes function as catalysts may be interested in learning about some of the enzymes that function within their own bodies. Have these students identify at least 5 enzymes in the human body and what happens when these enzymes misfunction. Students may choose any enzymes they would like, but possible enzymes include lactase, carbohydrase, lipase, and protease. Have students share their findings with the class.

Systems in Change Students learn that a variety of factors can influence enzymes to either increase or decrease their efficacy and change how they affect reactions, which occur at a small scale. **Chapter 16** discusses how changes in larger systems, like ecosystems, affect the species that live in the environment.

CHAPTER INVESTIGATION B

Guided Inquiry *Crime Scene Cleaners*

Time: 200 minutes plus advance preparation

Students will design an experiment to compare the effectiveness of enzyme-based detergents on protein-based waste in this forensics-oriented investigation.

Go online to access detailed teacher notes, answers, rubrics, and lab worksheets.

Construct an Explanation *To avoid having too much* C. difficile, *you would need to prevent the bacteria from getting enough of the proline that they need to grow. To do this, you could disable the HypD enzyme by changing the conditions in the gut so it breaks apart or doesn't work as well, or blocking its active site with some other molecule so it can't make an enzyme-substrate complex with the hydroxy-L-proline molecule.*

CHAPTER INVESTIGATION

Crime Scene Cleaners
Which type of detergent is most effective at breaking down protein-based substances?

Go online to explore this chapter's hands-on investigation about digestive enzymes.

Factors That Influence Enzymes Environmental factors such as temperature, salt concentration, and pH influence an enzyme's shape, which in turn influences its function. Each enzyme works best in a particular range of conditions. With some exceptions, an organism's enzymes work best under its typical physiological conditions: a pH of about 7, normal body temperature, and a salt concentration that matches that of a typical cell.

In general, enzymes increase in efficiency as the temperature rises, but only up to a point. When the temperature rises too much, enzymes begin to shut down. Human body temperatures above 42 °C (107.6 °F) adversely affect the function of many of our enzymes, which is why severe fevers are dangerous.

The activity of enzymes is also influenced by the amount of salt in the surrounding fluid. If there is too little salt, the polar parts of the enzyme attract one another so strongly that the enzyme's shape changes. Too much salt interferes with the hydrogen bonds that hold the enzyme in its characteristic shape, and the enzyme *denatures* (unfolds) and can no longer form a complex with its substrates.

Extreme changes in pH change the proportion of hydrogen ions in an enzyme's environment. The presence of an unusual amount of charged particles affects hydrogen bonds and ionic bonds, and can break the bonds that shape the enzyme and its active site. Almost all biological processes involve chemical reactions that use enzymes. Thus, in order to survive, cells must maintain the optimal conditions that enzymes need to operate.

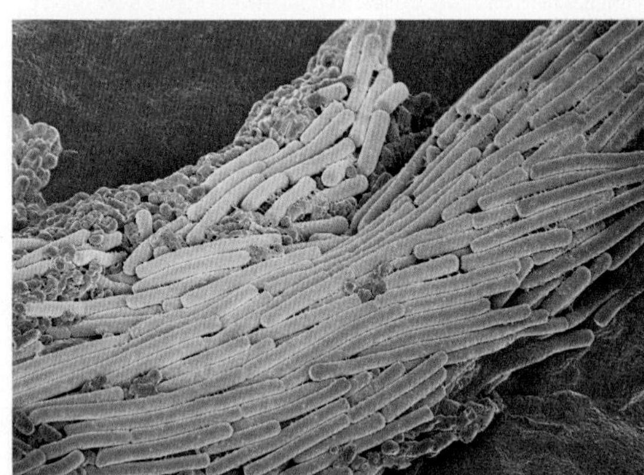

FIGURE 5-36 The long rod-like organisms in this electron microscope image are *Clostridium difficile*, a species of bacteria that lives in the human gastrointestinal tract. An excess of *C. difficile* can lead to life-threatening inflammation of the colon or intestines.

SEP **Construct an Explanation** Clostridium difficile, *shown in* ***Figure 5-36****, makes an enzyme called HypD that helps break down hydroxy-L-proline, a molecule that forms the structure of collagen. The bacteria then use the proline (an amino acid) from this molecule for their own growth. Humans do not produce HypD. How might the HypD enzyme be targeted to fight a* C. difficile *infection?*

BIOTECHNOLOGY FOCUS
Bioremediation of Plastic Waste

In Your Community The well-known Great Pacific Garbage Patch is just one of five garbage patches—huge areas of floating debris that collect in large systems of circulating ocean currents. Much of the debris is made up of microplastics, tiny pieces of plastic that don't degrade, but collect in the ocean. Scientists have had luck in isolating an enzymes that can break down some of these plastics and are hoping that they will be a tool in remedying the collection of plastic wastes that are collecting in environments around the world.

Take Action Have students do research to find volunteer opportunities in their area to support environmental issues. Their actions could be as simple as organizing a trash pickup drive around their school or more involved, such as campaigning against single-use plastic bags.

BIOTECHNOLOGY FOCUS

BIOREMEDIATION OF PLASTIC WASTE Plastics are human-made organic polymers. Plastic molecules are engineered to produce materials with numerous useful properties, one of which is high durability under a wide variety of conditions. This characteristic, combined with the low cost of manufacturing plastic goods, results in a vast global accumulation of plastic waste. One method that scientists are exploring to solve this problem involves identifying and studying organisms that are able to metabolize tough polymers that occur in nature. The chemical mechanisms these organisms use to turn natural polymers into biomass, carbon dioxide, and water may also be applicable to speeding up the degradation of molecularly similar plastics.

Lignin, the polymer that gives wood and bark cells their strength, does not degrade easily. However, many mushrooms, often seen attached to the trunks of trees, can break down lignin. These species are part of a group known as white-rot fungi. Scientists have found that nylon, a strong synthetic material, is also susceptible to degradation by enzymes extracted from white-rot fungi.

Cutin is a fatty acid polymer that makes the surfaces of plants waterproof. Scientists have isolated specific cutinases, or enzymes that break down cutin, in species of bacteria that decompose plants. These enzymes have been shown to be effective at degrading polyethylene terephthalate (PET), a transparent plastic commonly used in food and beverage packaging. Some bacteria have evolved to thrive in plastic-rich environments such as recycling plants and plastic waste deposits by using a cutinase-like enzyme called PETase to digest PET.

Waxworms, the larvae of wax moths, are parasitic caterpillars that eat honey and beeswax from the beehives in which they live. They can also rapidly digest polyethylene, a plastic that is highly resistant to degradation. Polyethylene has strong molecular bonds similar to bonds in beeswax compounds. Scientists are now studying the enzymes and pathways that enable waxworms to break down this ubiquitous plastic.

Darkling beetle larvae, also called mealworms, eat polystyrene foam. Because of its low density, this type of plastic is costly to melt down for transport to recycling facilities.

5.4 REVIEW

1. **Classify** The energy of the products in a chemical reaction is 6 kJ. The energy of the reactants is 12 kJ. Is this reaction exergonic or endergonic?
2. **Justify** Is energy considered a reactant in a chemical reaction?
   - A. No. All reactants are matter.
   - B. No. In an exergonic reaction, energy is a product.
   - C. Yes. In an endergonic reaction, energy is a required reactant.
   - D. Yes. Activation energy is needed for a chemical reaction to occur.
3. **Analyze** In which chemical equation is matter conserved?
   - A. $2\ H + 2\ O \rightarrow H_2O$
   - B. $HCO_3^- + 2\ H^+ \rightarrow H_2CO_3$
   - C. $6\ CO_2 + 6\ H_2O \rightarrow C_6H_{12}O_6 + 6\ O_2$
   - D. $6\ C_6H_{12}O_6 + 6\ O_2 \rightarrow 6\ CO_2 + 6\ H_2O$
4. **Classify** Tryptophan is an amino acid found in many foods. In the small intestine, a chymotrypsin molecule attaches to tryptophan to help break down the amino acid. Identify the substrate and the enzyme in this example.

Science Background

Breaking Down Plastics In 2018 in the United States, nearly 27 million tons of plastics were disposed of in landfills, as opposed to only 3 million tons of plastics being recycled. This is despite huge national and local initiatives to inspire the public to recycle plastics. While many consumers may think they are recycling their plastic waste, complex practices in sorting and handling plastics of unknown or mixed origin often lead to plastics being diverted to landfills. Much research is being conducted to improve recycling processes and to create biodegradable plastics that can be broken down by enzymes, microbes, or environmental factors such as heat and water. Encourage students to find out more about this research and share the findings with the class.

5.4 REVIEW

1. *endergonic* **DOK 2**
2. *A* **DOK 2**
3. *C* **DOK 2**
4. *tryptophan: substrate; chymotrypsin: enzyme* **DOK 2**

LOOKING AT THE DATA
DIGESTIVE ENZYMES AND pH

Connect this activity to the science and engineering practice of analyze and interpret data and engaging in argument from evidence. Students analyze patterns in the data and identify the pH level at which different enzymes are most active. Students address the issue of lactose intolerance and propose a solution or treatment for someone experiencing the condition.

SAMPLE ANSWERS

1. *amylase: 7; lactase: 6; pepsin: 2.5; trypsin: 8* **DOK 2**
2. *amylase (pH 7) – mouth; lactase (pH 6) – small intestine; pepsin (pH 2.5) – stomach; trypsin (pH 8) – small intestine* **DOK 2**
3. *stomach, mouth* **DOK 2**
4. *small intestine* **DOK 2**
5. *Sample answer: A person who experiences lactose intolerance should avoid foods that contain the disaccharide sugar found in dairy and milk-based products or add lactase enzyme drops or tablets when they consume those products.* **DOK 3**

LOOKING AT THE DATA

DIGESTIVE ENZYMES AND pH

SEP Use Mathematics Scientists estimate that over two-thirds of the world's population is lactose intolerant. People with this condition commonly develop digestive difficulties in early adulthood. Lactose is a disaccharide found in dairy and milk-based products. It is broken down by an enzyme called lactase.

Lactase is one of several enzymes responsible for digestion. Enzymes accelerate the chemical reactions that break down complex biological molecules in food to release energy and smaller molecules. **Table 1** lists the digestive breakdown processes for four digestive enzymes. An enzyme's activity depends on its concentration and environmental conditions such as pH. **Figure 1** shows the dependency of enzyme activity on pH for the four digestive enzymes.

The organs in the human digestive system provide environments with different acidity. For example, the mouth and colon typically have a pH of 7, while the stomach's pH range is 1–3 and the small intestine's pH range is 6–8.

TABLE 1. Digestive Enzyme Roles

Digestive enzymes	Digestive breakdown process
amylase	polysaccharides ↓ disaccharides
lactase	disaccharides ↓ monosaccharides
pepsin	proteins ↓ peptides, amino acids
trypsin	peptides ↓ amino acids

FIGURE 1
A graph shows the relative enzyme activity over a range of pH.

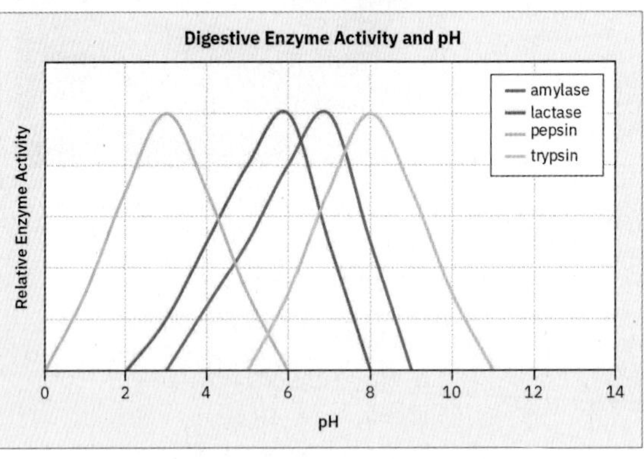

1. **Organize Data** At what pH is each digestive enzyme the most active?
2. **Interpret Data** Match the digestive enzymes with the human organ in which they are most active.
3. **Relate** Identify the two organs in which proteins and carbohydrates begin the digestive process.
4. **Identify** Which organ is most responsible for lactose intolerance?
5. **Design** Propose a solution or treatment for someone experiencing lactose intolerance.

144 CHAPTER 5 MOLECULES IN LIVING SYSTEMS

TYING IT ALL TOGETHER
TURNING MOLECULES INTO MEDICINE

HOW ARE THE MOLECULES OF LIFE ASSEMBLED?

Molecular substances in nature vary widely in complexity. By understanding patterns in the chemical structures of these substances, scientists can predict their functions within biological systems and organisms. The paclitaxel molecule introduced in the Case Study belongs to a category of organic molecules called alkaloids. Alkaloid molecules have at least one nitrogen atom that is typically part of a structure with multiple rings. Most alkaloids are found in plants, and scientists estimate that about 20 percent of plant species contain alkaloids. There are several hypotheses about the roles these compounds play: alkaloids may regulate plant growth, or protect plants from predators, or they may be waste products of plant metabolism.

Alkaloid names usually end in *-ine*. Many alkaloids, such as caffeine and nicotine, are known to have specific physiological effects on humans. Traditional medicines and healing practices are often derived from the therapeutic effects of alkaloids, some of which include pain relief, inflammation reduction, and antibacterial properties. As advances in biotechnology have allowed for detailed investigation of chemical compounds at the molecular level, scientists are increasingly motivated to study the structures and biological functions of alkaloids for their potential use as drugs to treat cancer and other conditions or diseases such as hypertension, diabetes, and asthma.

In this activity, you will investigate how a plant-derived alkaloid molecule is developed into a medical treatment. Work individually or in a group to complete these steps.

Gather Evidence

1. Research a specific plant alkaloid that is used in modern medicine. In your research, you should identify
   - the plant in which the alkaloid is found, and how the alkaloid is extracted from the plant
   - how the alkaloid affects the human body, and how these effects were discovered
   - the disease or condition the alkaloid is used to treat
   - the name of the medicine that is made from the alkaloid, and how it is made

Communicate Information

2. Construct a visual representation, such as a time line or flowchart, that describes the process by which the alkaloid you chose was developed into a medicine.

Argue From Evidence

3. Tens of thousands of plant alkaloids have been identified, but only a small percentage of these have been approved for pharmaceutical use. Use evidence from your research to support or refute the following claim: *Plant alkaloid research is an effective way to find new medical treatments.*

FIGURE 5-37 The Madagascar periwinkle (*Catharanthus roseus*) plant is a source of vincristine and vinblastine, alkaloids that are used to treat cancer and Hodgkin disease.

FIGURE 5-38 *Claviceps purpurea*, a fungus that infects rye, produces many medicinal alkaloids.

TYING IT ALL TOGETHER 145

TYING IT ALL TOGETHER

ELABORATE

The Case Study discusses how scientists use organic molecules, like alkaloids, to develop medical treatments. In the Tying It All Together activity, students research an alkaloid that is used in modern medicine and use evidence to argue whether plant alkaloid research is an effective way to find new medical treatments.

This activity is structured so that students may complete the steps as soon as they have completed Section 5.3. **Step 1** has students gather evidence on a specific plant alkaloid that is used in modern medicine and identify its effects. In **Step 2**, students construct a visual representation that describes the process through which the alkaloid was developed into medicine. In **Step 3**, they use the evidence they have gathered to argue whether plant alkaloids are a good source for medical treatments.

Alternatively, this can be a culminating project at the end of the chapter. The analysis practiced here reflects processes that students may use to support a claim in the Unit 2 activity.

Go online to access the Student Self-Reflection and Teacher Scoring rubrics for this activity.

SCIENCE AND ENGINEERING PRACTICES
Constructing Explanations and Designing Solutions

Alkaloids in Medicines To help students construct explanations and discuss how scientists use plant alkaloids as sources for medical treatments, it may be useful to discuss examples of alkaloids that are used in medicine. Students may think the alkaloids in plants must be purified from the plants to be used for medicinal purposes. Guide students to understand that once scientists can isolate the active compounds that have specific functions that lead to treatments, they can create artificial compounds at greater concentrations than those that can typically be found in nature. It may be useful to share examples of plant alkaloids that students can use as a source to start their research. Share with students that a few examples of plant alkaloids include atropine, pilocarpine, morphine, quinine, and ephedrine.

CHAPTER 5 **REVIEW**

EVALUATE

REVIEW KEY CONCEPTS

1. *A, D* **DOK 2**
2. *C, E* **DOK 2**
3. *D* **DOK 1**
4. *Sample answer: Covalent bonds occur when two atoms share the electrons contributed by each atom. Electrons are not shared between two atoms in ionic bonds. Instead, the electrons move from one atom to another. The atom gaining electrons becomes negatively charged and the atom losing the electrons becomes positively charged. The two ions bond through the force of attraction between these opposite charges* **DOK 2**
5. *acids: Gives up H+ ions in water; Produces a solution with excess free H+ ions; Lowers pH to less than 7 when dissolved in water; bases: Accepts free H+ ions in water; Produces a solution with excess free OH– ions; Raises pH to greater than 7 when dissolved in water* **DOK 2**
6. *B* **DOK 1**
7. *5 or five* **DOK 1**
8. *Yes, carbon-12 and carbon-14 differ by two neutrons, which have no charge. So carbon-12 and carbon -14 have the same charge.* **DOK 2**
9. *Sample answer: The reactions are exergonic. They involve breakdown of large molecules to smaller molecules, and they release of energy in the form of heat that warms the runners' skin.* **DOK 2**
10. *Reactants: glucose and oxygen. Products: carbon dioxide and water.* **DOK 1**
11. *A, B, C, E* **DOK 2**

CHAPTER 5 **REVIEW**

REVIEW KEY CONCEPTS

1. **Contrast** Based on **Figures 5-19** and **5-25**, which functional groups are found in nucleic acids but not in carbohydrates? Select all correct answers.
   - A. amine (—NH_2)
   - B. carboxyl (—COOH)
   - C. hydroxyl (—OH)
   - D. phosphate (—PO_4^{3-})
   - E. sulfhydryl (—SH)
2. **Classify** Which substances are polymers? Select all correct answers.
   - A. amino acids
   - B. nucleotides
   - C. phospholipids
   - D. polysaccharides
   - E. proteins
3. **Identify** A chemist purifies a single substance from an organism and finds that it contains phosphate groups. In which category could this biological molecule belong?
   - A. carbohydrate
   - B. fatty acid
   - C. hydrocarbon
   - D. nucleic acid
4. **Explain** How do covalent bonds and ionic bonds differ?
5. **Classify** Match each description with acids or bases.
   - gives up H^+ ions in water
   - accepts free H^+ ions in water
   - produces a solution with excess free H^+ ions
   - produces a solution with excess free OH^- ions
   - lowers pH to less than 7 when dissolved in water
   - raises pH to greater than 7 when dissolved in water
6. **Identify** A scientist determines that a certain substance has no net charge. Which of these statements must be true?
   - A. The number of neutrons and protons in the substance must be equal.
   - B. The number of protons and electrons in the substance must be equal.
   - C. The number of electrons and neutrons in the substance must be equal.
   - D. The number of protons, neutrons, and electrons in the substance must be equal.
7. **Calculate** How many molecules of oxygen (O_2) are required to react with one molecule of propane (C_3H_8) to produce four molecules of water (H_2O) and three molecules of carbon dioxide (CO_2)?
8. **Compare** Do atoms of carbon-12 and carbon-14 have the same charge? Explain why or why not.
9. **Explain** A large carbohydrate molecule can react with water to break it down into smaller sugar molecules. For example, sucrose and water react to form the simple sugars glucose and fructose, releasing energy. If energy must be absorbed to break a chemical bond, how can a reaction that breaks a larger molecule down into smaller molecules be exergonic?
10. **Identify** When glucose is oxidized to release energy, glucose reacts with oxygen to produce carbon dioxide and water. What are the reactants and what are the products in this reaction?
11. **Predict** The exergonic breakdown of starch into glucose is catalyzed by an enzyme called amylase. Which statements are true about this reaction? Select all correct answers.
    - A. Some bonds in starch are broken during the reaction.
    - B. A starch molecule can bind to the active site of amylase.
    - C. Glucose is released from the active site after the reaction.
    - D. The bonds of glucose have more energy than the bonds of starch.
    - E. The activation energy of the reaction is lower with amylase than without amylase.

CRITICAL THINKING

12. *Sample answer: Most of the matter in an organism is water, so a lipid molecule that repels water would be useful for preventing things from interacting with water, such as by separating the inside of a cell from the surrounding water.* **DOK 3**
13. *Sample answer: In a carbon dioxide molecule, the negative charge will tend to move equally toward opposite sides of the molecule. The molecule is linear, so the charge distribution on one end of the molecule is the same as on the other, making the molecule nonpolar.* **DOK 2**
14. *Sample answer: Glucose would be able to contribute carbon, oxygen, and hydrogen atoms to nucleic acid synthesis. Nitrogen and phosphorus atoms would need to come from another source. The law of conservation of matter states that matter must come from other matter; therefore, the elements making up nucleic acids must come from other molecules.* **DOK 2**
15. *Sample answer: The activation energy required to break down starch is high, so it occurs very slowly. When ingested, enzymes in the body help accelerate the process.* **DOK 3**

CRITICAL THINKING

12. **Draw Conclusions** The long hydrocarbon chains that make up lipid molecules are nonpolar. Nonpolar molecules are usually *hydrophobic*, meaning they repel water. How would this property of lipids be useful in an organism?

13. **Explain** Carbon dioxide has the chemical formula CO_2. It is a linear molecule with a carbon atom in the center that is double-bonded to an oxygen atom on either side. Oxygen tends to attract electrons more strongly than carbon. Would you expect carbon dioxide to be polar or nonpolar? Explain your answer.

14. **Explain** Many organisms use glucose as a source of matter to build other molecules they need for cell metabolism and growth. Based on **Figures 5-19** and **5-25**, what elements can glucose contribute to the synthesis of nucleic acids? What elements would need to be obtained from another source? Explain.

15. **Justify** Corn starch can often sit in a pantry for years without breaking down. Yet, when ingested, corn starch can be digested very quickly by the body. How can these two observations be reconciled?

MATH AND ENGLISH LANGUAGE ARTS CONNECTIONS

1. **Reason Quantitatively** A baker mixes 3 g of yeast with 1000 g of flour and 500 g of water. After allowing the dough to rise, the dough weighs 1450 g. Quantify the change in mass. What can best explain the change in mass of the dough?

2. **Write Informative Text** Enzymes are sometimes included in detergents to help clean stains from clothing. What types of enzymes are used in such applications? How do the enzymes contribute to stain removal? In your answer, include the terms *activation energy* and *catalyst*.

3. **Model with Mathematics** The chemical reaction that links amino acids together to make a protein requires a net consumption of energy. Which of the two models in **Figure 5-33** best represents the energy changes taking place during this reaction, and how does the law of conservation of energy apply to this case? Explain your reasoning.

4. **NOS Explaining Natural Phenomena** When molecules react inside cells, the laws of conservation of energy and conservation of matter hold true. These laws also hold true at the level of whole ecosystems. What evidence supports these laws at the molecular level and the ecosystem level? Explain.

BACTERIA IN YOUR GUT

Gather Evidence In this chapter, you learned about how bacteria can break down certain compounds that humans cannot, such as cellulose. Some bacteria in the gut have also been shown to destroy toxins and mutagens. In addition, bacteria also supply essential nutrients that humans cannot make themselves, such as vitamin K.

1. The variety of bacteria in the human gut have many different mechanisms to synthesize or break down chemical compounds. How do human hosts benefit from having a healthy and diverse bacteria population in their digestive systems?

2. What other questions do you have about the role of the gut bacteria in promoting human health?

REVISIT THE ANCHORING PHENOMENON

1. *Sample answer: Different bacteria have different capabilities. If the host is not able to consume enough of a vitamin like vitamin K, it helps that the gut bacteria can synthesize it. Bacteria that can break down food material such as cellulose help the host extract more substances from the food they eat.* **DOK 2**

2. *Sample answer: How many different vitamins and other nutrients are provided by bacteria? Do they provide other nutrients that supplement and help the host, even if the host can also produce them? I've also heard that healthy gut bacteria contribute to mental health. How does that work? Is it because of chemicals they make?* **DOK 3**

MATH AND ELA CONNECTIONS

1. *Sample answer: The dough weighs 53 grams less than the total mass of the ingredients used to make the dough. This suggests that chemical reactions caused the bread to rise, and some of the products of the chemical reactions were gases that escaped from the dough into the atmosphere.* **DOK 3**

2. *Sample answer: Enzymes are biological catalysts that speed chemical reactions. One way to remove stains is to use enzymes such as proteases, lipases, and amylases to break down biological molecules. These enzymes lower the activation energy needed to chemically break down these molecules, speeding up the breakdown process. When the molecules are broken down into their component parts, they separate from the fabric.* **DOK 3**

3. *Sample answer: Since energy is conserved, a net consumption of energy for this reaction means that the products have more energy than the reactants. Therefore, the endergonic reaction best represents the reaction.* **DOK 2**

4. *Sample answer: The evidence at both levels is similar. At the molecular level, evidence supporting the law of conservation of matter and energy includes the observation that all atoms in products come from atoms in reactants and that all energy before and after a reaction can be accounted for. Similarly, evidence in support of these laws at the ecosystem level includes the observation that all matter in an ecosystem comes from previous matter and that all energy moving through an ecosystem can be accounted for before, during, and after it moves through the ecosystem.* **DOK 2**

CELL STRUCTURE AND FUNCTION

Three-Dimensional Learning

The practices, core ideas, and crosscutting concepts presented in this chapter's text, investigations, and resources provide support to address the following Performance Expectations: **HS-LS1-5, HS-LS1-7, HS-LS2-3,** and **HS-LS2-5**.

Science and Engineering Practices	Disciplinary Core Ideas	Crosscutting Concepts
Developing and Using Models (HS-LS1-5, HS-LS1-7, HS-LS2-5) Constructing Explanations and Designing Solutions (HS-LS2-3, HS-LS2-7) Using Mathematics and Computational Thinking Connections to Nature of Science: Scientific Investigations Use a Variety of Methods	**LS1.A:** Structure and Function **LS1.C:** Organization for Matter and Energy Flow in Organisms (HS-LS1-5, HS-LS1-7) **LS2.B:** Cycles of Matter and Energy Transfer in Ecosystems (HS-LS2-3, HS-LS2-5) **LS2.C:** Ecosystem Dynamics, Functioning, and Resilience	Energy and Matter (HS-LS1-5, HS-LS1-7, HS-LS2-3) Structure and Function Stability and Change

Contents		Instructional Support for all Learners	Digital Resources
ENGAGE			
148–149	**CHAPTER OVERVIEW** **CASE STUDY** How do structures work together in cell systems?	**Social-Emotional Learning** CCC Stability and Change **Historical Connection** Argonne National Laboratory **English Language Learners** Cognates	
EXPLORE/EXPLAIN			
150–161 DCI LS1.A LS2.C	**6.1 CELL STRUCTURES** • Identify components of all cells. • Describe the structure and function of components in prokaryotic cells. • Describe the structure and function of components in eukaryotic cells. • Explain the Endosymbiont theory.	**Vocabulary Strategy** Word Web **English Language Learners** Taking Notes on a Text, Monitoring and Self-Correcting **Address Misconceptions** Cell Size, Mitochondria and Chloroplasts CONNECTIONS Discovery of Cells and Cell Theory **Visual Support** Components of Prokaryotes, Comparing Plant and Animal Cells, Organelles in Plant and Animal Cells **Crosscurricular Connections** Health CER Revisit the Anchoring Phenomena SEP Developing and Using Models CCC Structure and Function, Systems and System Models, Structure and Function **Connect to Mathematics** Model with Mathematics **Differentiated Instruction** Students with Disabilities **Connect to English Language Arts** Integration of Knowledge and Ideas **Differentiated Instruction** Leveled Support **In Your Community** Local Snakes **On Assignment** Chromatophores	**Virtual Investigation** Bacteria in the Digestive System *Video 6-1*

	Contents	Instructional Support for all Learners	Digital Resources
EXPLORE/EXPLAIN			
159	**LOOKING AT THE DATA** MICROBIOTA OF THE HUMAN BODY		
162–168 DCI LS1.A	**6.2 CELL MEMBRANES** • Describe the structure of the cell membrane and how proteins function within it. • Explain homeostasis and the involvement of the cell membrane in its maintenance. • Summarize the different modes of transport across the cellular membrane.	**CASE STUDY** Ask Questions **Connect to English Language Arts** Use Digital Media SEP Developing and Using Models **Differentiated Instruction** Leveled Support **Address Misconceptions** Passive Transport **Differentiated Instruction** English Language Learners **In Your Community** Vesicle Movement **Visual Support** Exocytosis and Endocytosis	**SIMULATION** *Crossing Membranes*
168	**MINILAB** SELECTIVELY PERMEABLE MEMBRANES		
169–177 DCI LS1.C	**6.3 PHOTOSYNTHESIS AND CELLULAR RESPIRATION** • Explain the importance of energy usage by cells. • Describe the processes involved in cellular respiration and ATP synthesis. • Summarize the stages of photosynthesis and the light-dependent and light-independent reactions. • Identify other energy conversion pathways available to cells.	**Vocabulary Strategy** Using Prior Knowledge SEP Developing and Using Models CER Revisit the Anchoring Phenomena **Connect to English Language Arts** Use Digital Media **Visual Support** Electron Transport Chain **In Your Community** Fermented Foods CCC Energy and Matter **Connect to Mathematics** Reason Quantitatively **Differentiated Instruction** Leveled Support **Crosscurricular Connections** Physical Science	**Investigation A** Factors Affecting Cellular Respiration (100 minutes over 2 days plus advance preparation) *Video 6-2* **Investigation B** Designing a Photobioreactor (160 minutes over four days) *Video 6-3*
176	**EXPLORER** ANDRIAN GAJIGAN	**Connect to Careers** Marine Virologist **English Language Learners** Making Connections Between Ideas	
ELABORATE			
177	**TYING IT ALL TOGETHER** How do structures work together in cells?	SEP Constructing Explanations and Designing Solutions	Self-Reflection and Scoring Rubrics
Online	**Investigation A** Factors Affecting Cellular Respiration **Investigation B** Designing a Photobioreactor	Guided Inquiry (100 minutes over two days plus advance preparation) Guided Inquiry (160 minutes over four days)	MINDTAP Access all your online assessment and laboratory support on MindTap, including: sample answers, lab guides, rubrics, and worksheets. PDFs are available from the Course Companion Site
EVALUATE			
178–179	**Chapter 6 Review**		
Online	**Chapter 6 Assessment** **Performance Task 2** *What Are the Requirements for a Minimum Viable Ecosystem?* (HS-LS1-5, HS-LS1-7)		

CHAPTER 6

ENGAGE

In middle school, students will have learned that living things are made of cells, and that different structures in cells contribute to their functions. Students will have been exposed to the basic concepts of photosynthesis and cellular respiration. This chapter expands on that prior knowledge by exploring structures in eukaryotic and prokaryotic cells and how organelles in the cell function together.

Students will learn about cell membranes and how they allow for the transport of materials. Students will also learn more about the molecules and specific pathways cells utilize to harness and use energy.

About the Photo This photo shows the alga *Acetabularia calculus.* While students may think these organisms look like plants or fungi, each umbrella and stalk is actually one individual cell. The structures have different functions. The caps of the alga capture light energy, while the stalks hold the alga to the ground. Umbrella algae have unusually large nuclei, and are used to study how nuclei function in the cell. To prepare students for this chapter's content, ensure that they understand that structures in cells function together to support life.

Social-Emotional Learning

As students progress through the chapter, invite them to look for contexts and opportunities to practice some of the five social and emotional competencies. For example, as students work through the content in the chapter, support them in their **self-management**. Help them set personal goals as they work toward understanding the content.

In addition, when students learn about how scientists are attempting to construct artificial cells (e.g., in "Artificial Cell Technology"), remind them to practice **responsible decision-making** and **self-awareness**. Support students in managing prejudices and biases and challenge them to make a reasoned judgement after analyzing information and data.

CHAPTER 6

CELL STRUCTURE AND FUNCTION

Acetabularia caliculus, also known as umbrella algae, have stalk and radial cap structures that consist of single cells. These algae grow in mudflats, coral reefs, and mangroves.

All living things are made of one or more cells. While most cells are microscopic, some individual cells are visible to the unaided eye, such as this algae. At any size, every cell is a system that performs specific functions in an organism. Understanding how cells work is helping scientists to develop new technologies.

Chapter 6 supports the NGSS Performance Expectations **HS-LS1-5**, **HS-LS1-7**, **HS-LS2-3**, and **HS-LS2-5**.

148 CHAPTER 6 CELL STRUCTURE AND FUNCTION

CROSSCUTTING CONCEPTS | Stability and Change

Role of Feedback Mechanisms In this chapter, students explore the concept of stability and change as they learn how cellular membranes use feedback mechanisms to manage a cell's interactions with its environment. Some fields of biology, especially microbiology and molecular biology, study how feedback mechanisms regulate the function of a system. Feedback mechanisms involved in stability and change are also explored in Earth science topics that examine how changes on the planet's surface can cause changes to other systems.

Chapter 9 of this course explores how feedback mechanisms involved with stability and change function in material transport within plants. You can reinforce this crosscutting concept in Chapter 10 by helping students understand that feedback mechanisms maintain homeostasis in animals as well.

CASE STUDY

ARTIFICIAL CELL TECHNOLOGY

HOW DO STRUCTURES WORK TOGETHER IN CELL SYSTEMS?

Cells make up all living things, from the smallest single-celled bacteria to the largest multicellular organisms. Scientists around the world have been trying a variety of approaches to build functioning artificial cells (**Figure 6-1**). Artificial cells that can grow and divide on their own could lead to a better understanding of life and have many useful technological applications.

The quest to develop artificial cells began in the late 1950s. Scientists used synthetic and natural materials to mimic cell membranes that could selectively let certain materials in and out. In the following decades, scientists were able to encapsulate proteins and other small molecules into artificial cell-like capsules. Artificial cell membranes have been constructed of lipids, polymers, and proteins, in a variety of combinations.

Some medical applications for this technology involve using cell-inspired lipid membrane capsules to deliver medicinal molecules to patients. Although these lipid capsule structures used for medicine delivery are not exactly artificial cells, they are an important step in the progress of this field of research. These capsules can be designed to target specific cells within the body. The molecules within are delivered when the capsule is absorbed by the patient's cells. In 2020, some of the vaccines to protect against COVID-19 were successfully developed for delivery using a lipid capsule technology.

While some scientists have used synthetic materials to construct artificial cells, others have used a combination of synthetic materials and parts harvested from other cells. Some projects have approached the development of artificial cells by making them from scratch, whereas others have attempted to strip away parts of existing cells to discover the minimal viable parts needed for a functional cell. The variety of approaches reflects the diversity of goals for applications of artificial cell technology.

In recent years scientists have attempted to generate cells that can replicate themselves and communicate with each other. In 2010, scientists built a cell with a synthetic genome by replacing an existing cell's genome with synthetic DNA. Several years later, a self-replicating cell with a minimum genome of fewer than 500 genes was developed (a relatively simple *Escherichia coli* bacterium contains about 4000 genes). As the research continues to progress, scientists' goals are to make these cells more stable and identify critical roles of each gene.

Ask Questions *As you read, generate questions about components and processes of cells that you think could be important and useful in the application of artificial cell technologies.*

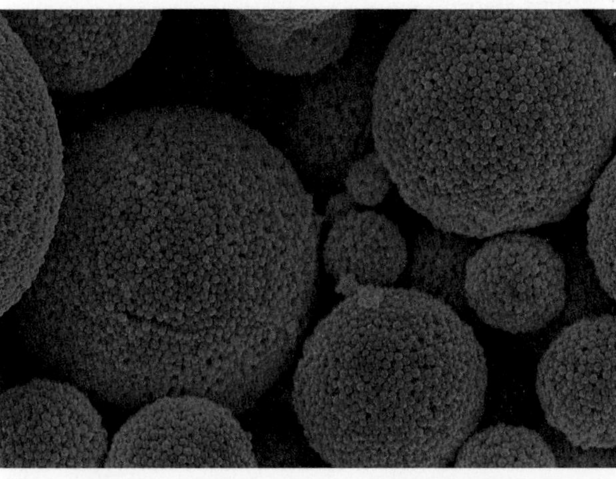

FIGURE 6-1 Scientists at Argonne National Laboratory developed these artificial "protocells" that can harvest solar energy.

CASE STUDY 149

DIFFERENTIATED INSTRUCTION | English Language Learners

Cognates Cognates can help students decode words in English using sound-letter relationships. Have students underline and define the following Spanish cognates as they read the case study: *bacteria, artificial, synthetic, natural,* and *proteins.*

Beginning Have pairs work together to list matching cognates in their native language and in English and use a dictionary to define them.

Intermediate Have pairs use each cognate in a written sentence that shows its meaning. Have them trade sentences with another pair to check.

Advanced Have students explain what they learned about each cognate from reading the case study. Challenge students to identify other cognates in the case study, such as *generate* and *critical.*

CASE STUDY

ENGAGE

The scenario described in the Case Study can be used to assess students' prior knowledge of cellular structure and function. Students should have learned about cell membranes in middle school. Ask students to consider how the structure of lipid membranes used to encapsulate medicines might allow for their targeted delivery to cells.

The artificial cell technology provides additional context in which students can construct explanations about how cells maintain homeostasis. In the Tying it All Together activity, students will research artificial photosynthesis and explain how it can be used to improve existing energy technologies.

Ask Questions Students should revisit the Case Study as they read the chapter to make connections with the content. See a specific suggestion in Section 6.2.

Historical Connection

Argonne National Laboratory Students may wonder about other research conducted at the Argonne National Laboratory. This U.S. Department of Energy laboratory, located west of Chicago, Illinois, was established in 1946 to create the world's first self-sustaining nuclear reaction as part of the Manhattan Project. Today, the facility is the largest research facility in the Midwest and its mission has evolved into a broad range of research to benefit humankind. Argonne scientists focus on a variety of topics in science and engineering, including nuclear physics, renewable energy, supercomputing, medical research, and environmental sustainability.

The scientists at Argonne National Laboratory conduct research to make new discoveries to change the way we live. In 2019, these scientists were able to produce ATP by designing and connecting artificial cells that could communicate with each other. One set of cells generated protons under light while the other set of cells used those protons to generate ATP. Have students identify possible questions that the researchers at the laboratory may ask in their research.

6.1

CELL STRUCTURES

LS1.A, LS1.C

EXPLORE/EXPLAIN

This section describes the structures found in all cells, covers the different components found in prokaryotic and eukaryotic cells, and introduces the endosymbiont theory.

Objectives

- Identify components of all cells.
- Describe the structure and function of components in prokaryotic cells.
- Describe the structure and function of components in eukaryotic cells.
- Explain the Endosymbiont theory.

Pretest Use **Questions 1, 2,** and **3** to identify gaps in background knowledge or misconceptions.

Vocabulary Strategy

Word Web Students may already be familiar with the term “cell membrane” but they may not have been exposed to the other key terms, such as prokaryote and eukaryote, introduced in this section. Have students work together to make a physical or digital word web that connects key terms and other new vocabulary from this section to each other. You may consider starting with “cell membrane” in the center of the word web and asking students to add new terms as they read.

Describe *Students may describe cells as miniature organisms, as a factory, or they may simply describe the parts of the cell they can see.*

6.1

KEY TERMS

cell membrane
chloroplast
cytoskeleton
endosymbiont theory
eukaryote
mitochondria
nucleus
prokaryote

CELL STRUCTURES

Look at your surroundings. Do you see a friend, a pet, a tree, or perhaps wooden furniture? You see them as a whole person or object, possibly with legs or leaves or other parts. Now imagine looking at their finger, or paw, or a leaf through a microscope, not knowing what you will see. For most of human history, we did not have microscopes. Scientists could only guess at what made up the living objects around them.

Describe *Imagine you are the first scientist to see a living object through a microscope, and you are astounded to see that each item is made up of countless tiny pieces. How would you describe the small pieces you see to other scientists?*

Components of Cells

A cell is the smallest unit of life. All cells share certain organizational and functional features such as a cell membrane, cytoplasm, and DNA (**Figure 6-2**). The **cell membrane** (sometimes called a plasma membrane) is a lipid bilayer that separates the cell from the external environment and controls exchanges between the cell and its environment. The cell membrane encloses *cytoplasm*, a jellylike mixture of water, sugars, ions, and proteins. A cell's internal components, including organelles, are suspended in the cytoplasm. *Organelles* are structures that carry out special functions inside a cell. Organelle membranes separate certain substances and activities.

Nearly every cell contains a DNA genome. In nearly all prokaryotes, that DNA is suspended in the cytoplasm. By contrast, the DNA of a eukaryotic cell is contained in a **nucleus** (plural *nuclei*), an organelle with a double membrane. Some eukaryotes are independent, free-living cells; others consist of many cells working together in multicellular organisms.

Cell systems have traditionally been classified as either prokaryote or eukaryote. Prokaryotes (bacteria and archaea) have simpler cell structures than eukaryotes (protists, fungi, plants, and animals). Appendix E and Unit 3 describe more about the diversity of bacteria, archaea, and eukaryote systems. In this section we will focus on model cell systems of bacteria, animal, and plant cells.

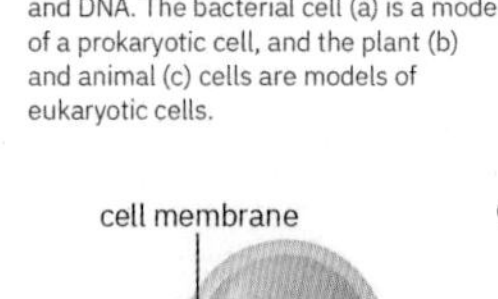

FIGURE 6-2
Cells include a cell membrane, cytoplasm, and DNA. The bacterial cell (a) is a model of a prokaryotic cell, and the plant (b) and animal (c) cells are models of eukaryotic cells.

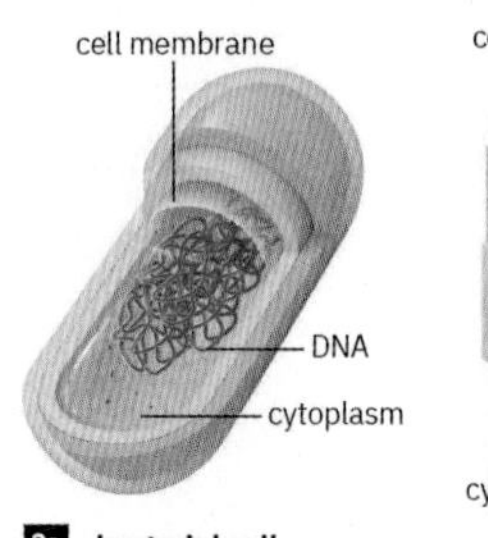

2a bacterial cell

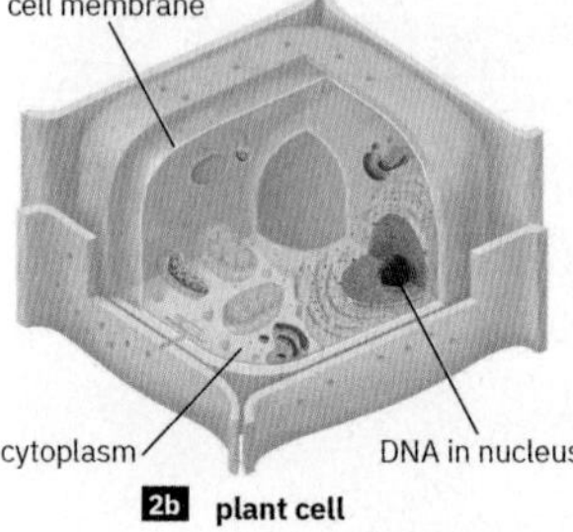

2b plant cell

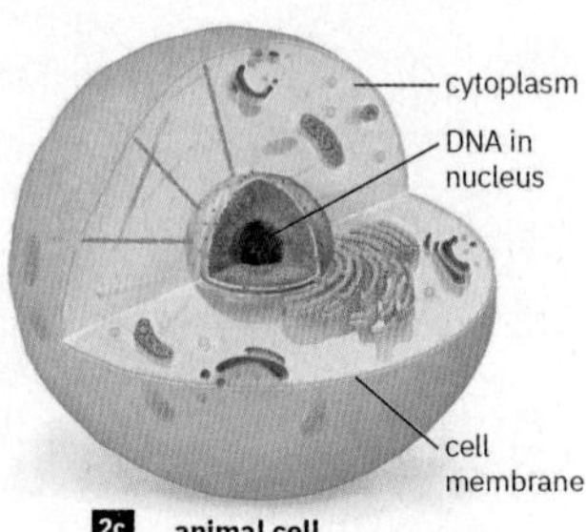

2c animal cell

DIFFERENTIATED INSTRUCTION | English Language Learners

Taking Notes on a Text Have students create a three-column chart with columns titled Organelle, Structure, and Function. Point out the headings in the text that name organelles, and have students add these to their chart. Have students complete the chart as they read.

Beginning Have students draw the structure and write a word or two to describe the function. After reading, have students compare charts with a partner.

Intermediate Have students identify the organelle and write a short phrase in the structure and function columns.

Advanced Have students write detailed notes about the structure and function of each organelle. After reading, have them work in small groups, taking turns using their chart to explain the structure and function of one organelle.

CONNECTIONS

DISCOVERY OF CELLS AND CELL THEORY Before microscopes were invented, no one knew that cells existed because nearly all cells are invisible to the unaided eye. Italian astronomer Galileo Galilei designed the first compound microscope using a concave lens and convex lens in the early 1600s. The design was improved by English physicist Robert Hooke 65 years later. Using a three-lens microscope, Hooke observed tiny hairs on a plant stem and the intricate pores in a piece of cork. He coined the term "cells" to describe the smallest components he observed.

A decade later, the first living cells were seen by Dutch scientist Antoni van Leeuwenhoek using a single-lens simple microscope he developed. He noted tiny organisms in rainwater, insects, fabric, sperm, feces, and other samples. In scrapings of tartar from his teeth, Leeuwenhoek saw what he described as "many very small animalcules, the motions of which were very pleasing to behold." He (incorrectly) assumed that movement defined life, and (correctly) concluded that the moving "beasties" he saw were alive. Leeuwenhoek might have been less pleased to behold his animalcules if he had grasped the implications of what he saw: Our world, and our bodies, teem with microbial life.

As described in this chapter, today we know that a cell carries out metabolism, maintains homeostasis, and reproduces either on its own or as part of a larger organism. By this definition, each cell is alive even if it is part of a multicellular body, and all living organisms consist of one or more cells. We also know that cells reproduce by dividing, so it follows that all existing cells must have arisen by division of other cells. As a cell divides, it passes its hereditary material—DNA—to offspring. Taken together, these generalizations constitute the cell theory, which is one of the foundations of modern biology.

Cell Theory

1. Each organism consists of one or more cells.
2. The cell is the structural and functional unit of all organisms. A cell is the smallest unit of life, individually alive even as part of a multicellular organism.
3. All living cells arise by division of preexisting cells.
4. Cells contain hereditary material (DNA), which they pass to their offspring when they divide.

Prokaryotic Cells

All bacteria and archaea are single-celled organisms (**Figure 6-3**), but individual cells of many species cluster together. Outwardly, cells of the two groups appear to be so similar that archaea were once presumed to be a type of bacteria. Both were classified as **prokaryotes**, a word that means "before the nucleus." By 1977, it had become clear that archaea are more closely related to eukaryotes than to bacteria, so they were classified as their own separate domain. The term "prokaryote" is now only an informal designation.

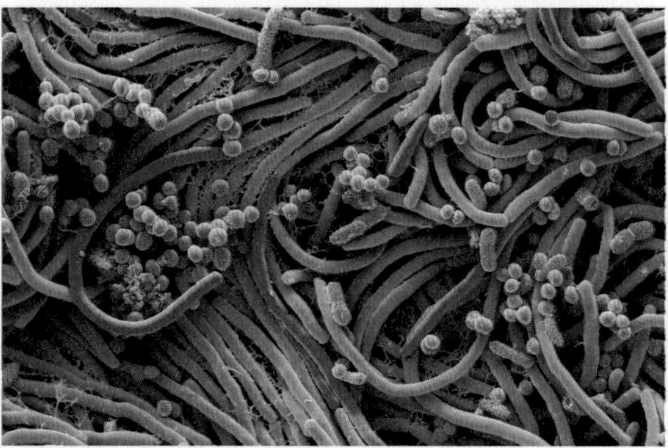

FIGURE 6-3 A scanning electron microscope image shows bacteria from the surface of a mobile phone. Bacteria come in a variety of shapes and sizes.

Address Misconceptions

Cell Size When comparing prokaryotic and eukaryotic cells, students may have the misconception that they are approximately the same size, given that models, such as the one in **Figure 6-2** are not typically shown to scale. Ask students to find the average diameter of a eukaryotic cell and a prokaryotic cell. Students can use the information to develop a to-scale model for the size of the two cell types.

Bacteria, Archaea, and Eukaryote Systems Students learn about the structures and functions of cell systems in bacteria, animal, and plant cells. **Chapter 8** provides more detail about the diversity of living systems while **Chapters 9** and **10** introduce how cells function within entire plant and animal systems.

CONNECTIONS | Discovery of Cells and Cell Theory

Social-Emotional Learning Share how Leeuwenhoek's contemporaries viewed his work with skepticism. Although he produced numerous detailed accounts of his observations, his work was largely overlooked and forgotten. Scientists began to articulate cell theory almost 150 years after Leeuwenhoek first observed single-celled organisms. His desire to protect his reputation as a scientist may have contributed to his obscurity. Leeuwenhoek refused to share details for his single-lens microscope, and for many years the compound microscope remained the preferred instrument. The early compound microscopes, however, lacked the magnification power of Leeuwenhoek's design. Scientists who may have wanted to reproduce his observations simply lacked the tools to do so. Ask students to discuss the value of sharing information, or how they might react to a challenging claim without any way of verifying it. They should get opportunities to practice **self-awareness** by reflecting on what they take for granted and reflect on **responsible decision-making** as they consider how people contribute to their communities.

Visual Support

Components of Prokaryotes The model in **Figure 6-4** shows components of bacterial cells, some of which are also found in eukaryotic cells. As students progress through the rest of the section, they should be able to identify which components are unique to bacterial cells and which are also found in eukaryotic cells. A Venn diagram may help students organize the cell components. You may wish to have students research how structures that are shared by bacterial and eukaryotic cells, such as ribosomes, differ in their structure. Assign groups of students different cell components and present their findings to the class.

Crosscurricular Connections

Health Share with students that many illnesses, such as strep throat, are caused by bacteria. Some antibiotics used to treat these conditions work by disrupting bacterial cell walls. Ask students why those antibiotics are harmful to prokaryotes, but do not necessarily affect their own eukaryotic body cells.

Point out that after antibiotic treatment of such illnesses, many people experience intestinal distress due to the disruption of the microbiome in the human gut. Encourage interested students to look into the use of probiotics and how they can mitigate that effect.

FIGURE 6-4 Some common features of bacterial cells are shown. Different bacteria and archaea have a variety of other shapes and structures.

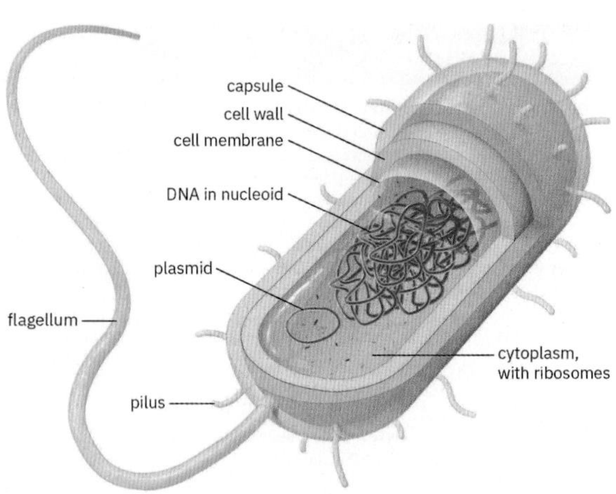

Components of Prokaryotes Prokaryotic cells perform many of the same functions as eukaryotic cells, but with fewer intracellular structures and a less complicated internal framework. Protein filaments under the cell membrane reinforce the cell's shape and provide a scaffolding for internal components (**Figure 6-4**):

- **Cytoplasm** The cytoplasm is a jellylike mixture of water, sugars, ions, and proteins that fills a cell. Ribosomes and DNA plasmids are suspended there. Most metabolism occurs in cytoplasm.
- **Ribosomes** Ribosomes are complex structures made of many protein and RNA molecules. Ribosomes build proteins.
- **Plasmid** Plasmids are small circular molecules of DNA that carry a few genes. Plasmid genes can provide advantages such as resistance to antibiotics. Transfer of plasmids between bacterial cells is an important mechanism of genetic exchange.
- **Nucleoid** Irregularly shaped regions of cytoplasm that contain the cell's essential genetic information are called nucleoids.
- **Cell Membrane** Cell membranes are lipid bilayers with integrated proteins that control the exchange of molecules with the surrounding environment. The structures and functions of cell membranes are described in detail in Section 6.2.
- **Cell Wall** Crystalline material makes up the cell wall that encloses the cell membrane in most prokaryotes. A cell wall protects the cell and supports its shape.
- **Capsule** Many species of bacteria have a capsule, which is a thick, gelatinous enclosure and/or a loosely attached layer of slime. These sticky structures help the cells adhere to each other as well as to many types of surfaces, and they also offer protection against predators and toxins.

BACTERIA IN YOUR GUT

Gather Evidence *Many bacteria in your intestines include a capsule structure. Why would a capsule be a useful feature for bacteria living inside the intestines?*

REVISIT THE ANCHORING PHENOMENON

Gather Evidence *Students might say that capsules help the bacteria to cling together and to the surfaces in the intestine so they are not all carried away every time food or waste comes through.*

Bacterial capsules consist of a polysaccharide layer, or a layer made of a complex carbohydrate. Most types of bacteria have very specific polysaccharides in their capsules. While many bacterial capsules are not recognized by the human immune system, these polysaccharides can be a target for medical treatment. Challenge students to gather evidence online to explain how components of the bacterial capsule might be used to develop medicines or vaccines. Group students together and have them make a poster describing how this structure, unique to bacteria, can be targeted by science.

- **Pilus** A protein filament that projects from the surface of some prokaryotes is called a pilus (plural *pili*). Pili help cells move across or cling to surfaces. A "sex" pilus attaches to another cell and then shortens itself. The attached cell is reeled in, and DNA is transferred from one cell to the other through the pilus.
- **Flagellum** A flagellum (plural *flagella*) is a long, slender structure used for movement. Many prokaryotes have one or more flagella projecting from their surface. A prokaryotic flagellum rotates like a propeller that drives the cell through fluids.

CCC Structure and Function *Pick two of the prokaryotic cell components described, and explain how the structure of the component helps it perform its intended function.*

Go Online VIRTUAL INVESTIGATION

Bacteria in the Digestive System
What do bacteria do in your intestines?
Put on your lab coat to grow some bacteria and study how they live and interact.

Eukaryotic Cells

Every cell in a **eukaryote** contains a nucleus and typically many additional organelles. Organelles enclosed by lipid bilayer membranes, including endoplasmic reticulum, Golgi bodies, chloroplasts, and mitochondria, are characteristic of eukaryotes. An enclosing membrane allows these structures to regulate the types and amounts of substances that enter and exit an organelle. Through this type of control, an organelle can maintain a specialized environment that allows it to carry out a particular function or store molecules that need to be isolated from the rest of the cell. Ribosomes are found in the cytoplasm of both prokaryotic and eukaryotic cells, and ribosomes are also attached to some parts of the endoplasmic reticulum in eukaryotes.

The Nucleus The cell nucleus (**Figure 6-5**) serves two important functions. First, the nucleus keeps the cell's genetic material isolated and safe from metabolic processes that might damage it. The DNA is suspended in nucleoplasm, a viscous fluid similar to cytoplasm. Second, the nucleus controls the movement of molecules through the nuclear envelope, a specialized membrane composed of two lipid bilayers. Embedded in the two lipid bilayers are nuclear pore complexes, tiny portals that regulate the passage of molecules between the cytoplasm and nucleoplasm. Proteins making up nuclear pores are anchored by a dense mesh of fibrous proteins that supports the inner surface of the membrane. The nucleus also contains one or more nucleoli (singular *nucleolus*), the structure where ribosomes are synthesized.

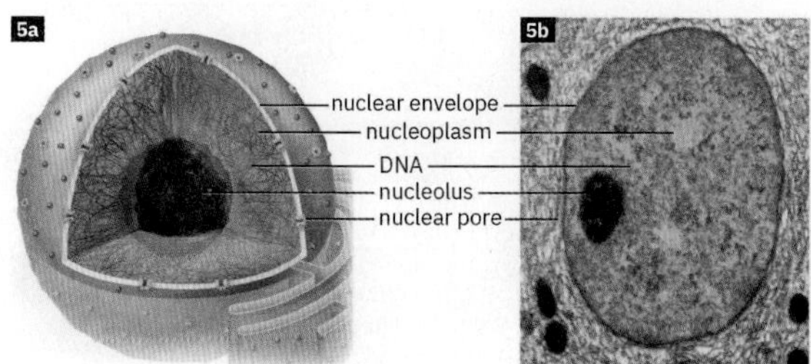

FIGURE 6-5 A model of an animal cell nucleus (a) is shown alongside a transmission electron microscope image of the nucleus in a pancreas cell from a rat (b).

SCIENCE AND ENGINEERING PRACTICES
Developing and Using Models

Cell Models Students should recognize that the 3D illustrations that show cellular components and what a cell looks like under a microscope do not exactly match. They should understand that both images show the same components, but a model is often easier to interpret than a microscopic photograph. Draw students' attention to **Figure 6-5**, which shows a model of a nucleus on the left and a transmission electron microscope image on the right. Have student discuss why a model is used to identify the structures in a cell. Ask students to explain why a model is helpful when learning about items that are either very large or very small in scale. Ensure students can identify other figures in this section as combinations of illustrations and photos of microscopic structures.

Go Online VIRTUAL INVESTIGATION

In the Virtual Investigation, *Bacteria in the Digestive System,* students can use a microscope to view several species of intestinal bacteria and parasites.

CCC Structure and Function *Sample answer: The cell wall includes crystalline structures, which are strong enough to support the shape of the cell and protect its contents. A flagellum is long and extends out from the cell, so it can move and propel the cell through its environment.*

Science Background

The Nucleus The nucleus was the first organelle observed in the cell, due to its size. However, its function remained unknown until 1838 when it was suggested that the nucleus was involved in cellular replication. This concept was heavily debated. In 1877, a German biologist, Oscar Hertwig, provided evidence that the nucleus is involved in the fertilization of sea urchins. He showed that the nucleus of sperm would fuse with the nucleus of an egg.

Not all cells have nuclei. Mature red blood cells in mammals do not have nuclei. Ask students to explain why red blood cells must have a nucleus at some point in development. (*to replicate during cell division*). Tell students that during cell division in red blood cells in mammals, one daughter cell receives all the DNA and the other receives no DNA. Only the cell without the DNA survives. Losing the nucleus enables the red blood cell to carry and transport more oxygen in the blood.

Visual Support

Comparing Plant and Animal Cells Pair or group students to examine **Figure 6-6**. Remind students that these images are actually models of cells that they are viewing, and do not reflect what a specific plant or animal cell would look like under a microscope. Have the groups of students make their own three-dimensional models of plant cells and animal cells. Students should ensure that each organelle and structure shown in the visual are included in their models and that the structures are labeled. As students complete the section, they should put together a labeled guide that explains the function of each organelle in their cellular model. Student organelle guides may be a booklet where each organelle is described, or a presentation that they share with the class, however, the guides should align to their models.

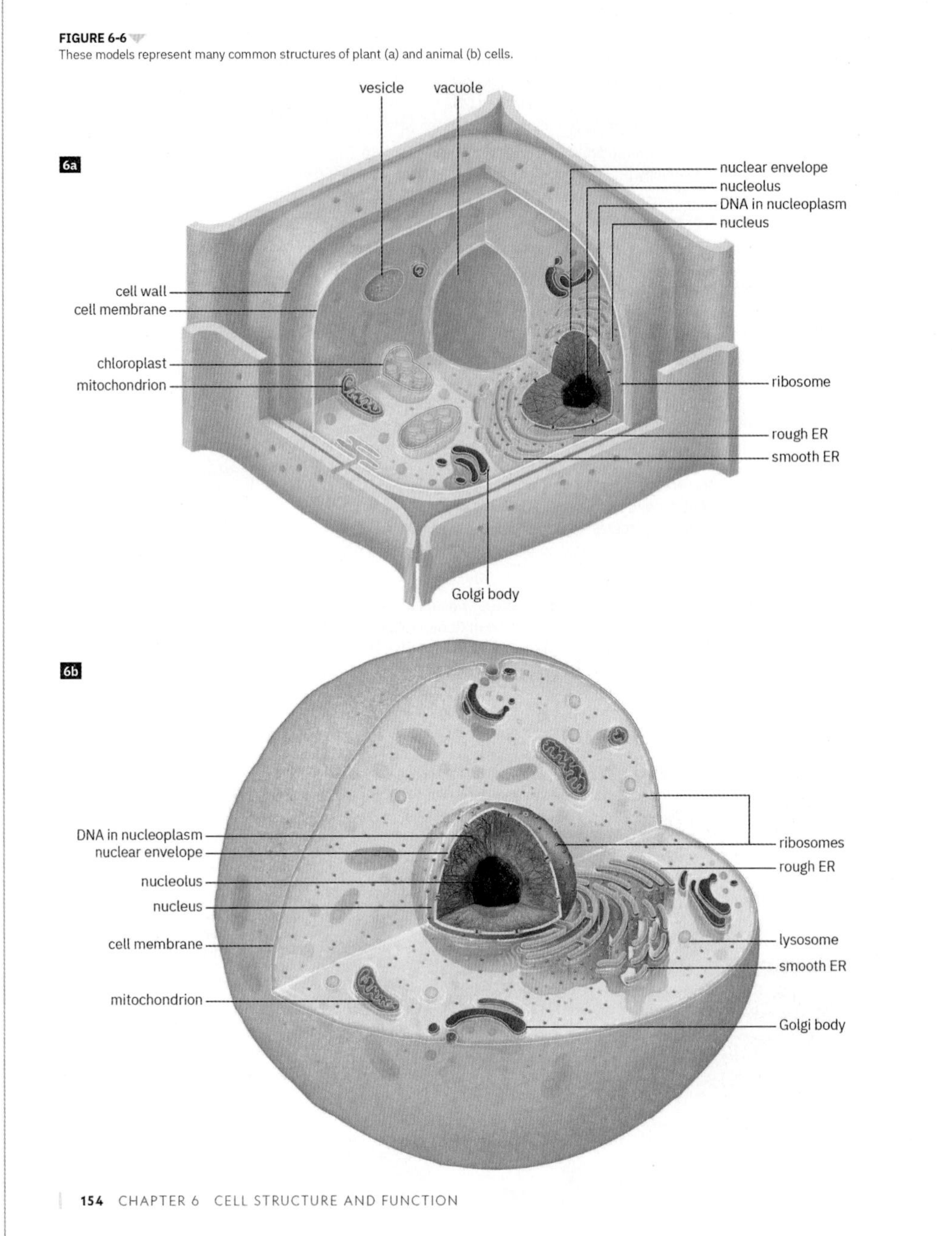

FIGURE 6-6 These models represent many common structures of plant (a) and animal (b) cells.

CROSSCUTTING CONCEPTS | Structure and Function

Structural Similarities Eukaryotic cells are grouped together because they all have membrane-bound organelles. These shared structures have a similar function in both plant and animal cells. The nucleus contains the genetic information for both cell types, and ribosomes are involved in protein synthesis for both cell types. Guide students to review the models of the plant cell and the animal cell in **Figure 6-6**. Have them call out the structures found in the plant cell as you list them for the class. Then, have them call out the structures for the animal cell. If the structure in the animal cell is also found in the plant cell, write the letter "B" for both, next to the structure. If the structure is only found in the animal cell, list the structure and write the letter "A" next to it. Once you have gone through all of the animal cell structures, go back through the list and write the letter "P" for the organelles only found in plant cells.

The Cytoskeleton Between the nucleus and cell membrane of all eukaryotic cells is a system of interconnected protein filaments called the **cytoskeleton**. Elements of the cytoskeleton reinforce, organize, and move cell structures (including the nucleus), and sometimes even whole cells. Some are permanent. Others form only at certain times. Motor proteins that associate with cytoskeletal elements move cell parts. The cytoskeleton is made up of three different kinds of filaments. These include microtubules, microfilaments, and intermediate filaments. All are built of many individual subunits that can be lengthened or shortened by adding or subtracting subunits (**Figure 6-7**). Each type of filament has a different structure specific to its function and relative stability.

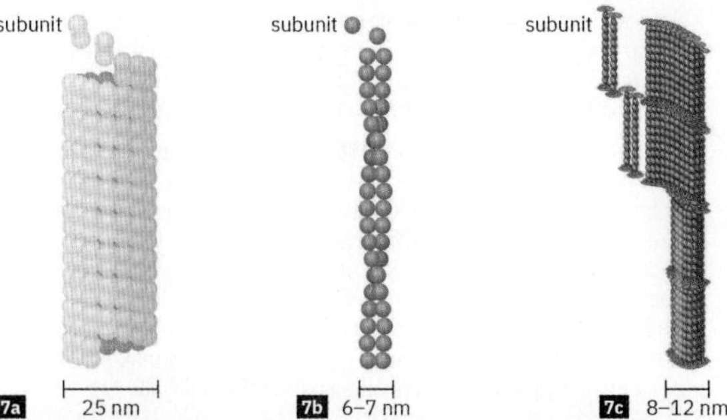

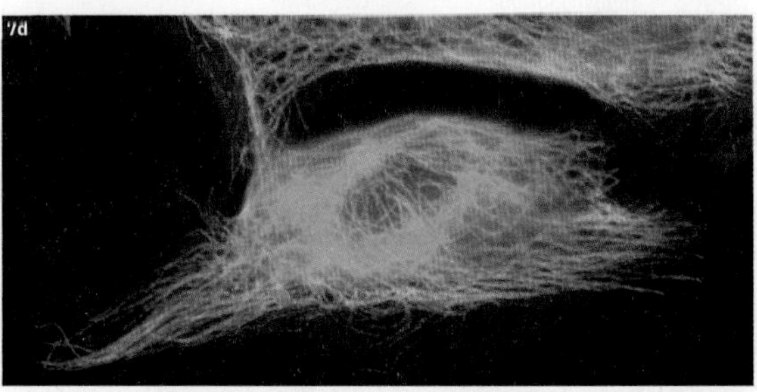

FIGURE 6-7 Microtubules (a) are involved in moving cell parts or moving the whole cell. The ability to add or remove microtubule subunits to lengthen or shorten microtubules is important in mitosis (Section 7.2). Microfilaments (b) reinforce cell membranes and function in muscle contractions. Intermediate filaments (c) structurally support cell membranes and tissues and are the most stable of these three parts of the cytoskeleton. A fluorescence microscopy image (d) of epithelial cells isolated from a kangaroo rat kidney shows microtubules in green and microfilaments in red. DNA appears in blue.

Other Membrane-Enclosed Organelles in the Cytoplasm

A variety of membrane-enclosed organelles do jobs in the cytoplasm of cells. Some of them bud from other membranes and others are networked to each other and the nuclear membrane.

- **Vesicles** These structures bud from the cell membrane or from membrane-enclosed organelles. Many types of vesicles carry substances between or within an organelle, or to and from the cell membrane. Small organelles called *peroxisomes* are vesicles containing enzymes that break down molecules such as amino acids and fatty acids (for metabolism) or hydrogen peroxide and ammonia (because they are toxic). Enzymes in vesicles called *lysosomes* break down and recycle macromolecules and cellular debris, repair the cell membrane, and process foreign particles such as bacteria and viruses.

Science Background

The Cytoskeleton The cytoskeleton in eukaryotes has the homologues of tubulin and actin in bacteria. These molecules were discovered in 1992, and it is now believed that the complexity of the cytoskeleton in eukaryotes evolved before the common ancestor of eukaryotes.

In eukaryotes, the cytoskeleton reinforces, organizes, and moves cell structures, including the cell itself. Challenge students to use outside resources to develop an explanation of how the cytoskeleton moves a cell.

Connect to Mathematics

Model with Mathematics Have students examine **Figure 6-7** and point out that the numbers underneath each part of the illustration. Ask students why scaled images are used in science texts to show the relative sizes or distances between objects. Ask them to determine if the images in 7a, 7b, and 7c are drawn to scale. Provide students rulers if they wish to measure and compare the actual sizes of the images on the page.

DIFFERENTIATED INSTRUCTION | Students with Disabilities

Visual Impairments Students who are visually impaired may not observe the differences between the microtubules, microfilaments, and intermediate filaments models shown in **Figure 6-7**. Have students work in small groups to identify differences among the structures beyond the color in this representation. Project an enlarged image to aid comparison. Also, consider supplying students with manipulatives such as pop beads or interlocking bricks to model structural differences to show how adding or subtracting units can facilitate movement. Encourage students to discuss reasons for the names of the different structures.

Video

Time: 39 seconds

Video 6-1 shows a microscopic time lapse of organelles moving within a plant cell. Ask students to identify structures they see including the nucleus, cytoskeleton, and other organelles.

Membranes Students learn that organelles are membrane bound and have different functions. **Chapter 5** provides more detail about the carbon-based molecules that make up membranes.

Visual Support

Organelles in Plant and Animal Cells
To help students identify the structures shown in **Figure 6-8**, project **Figure 6-6** so they can make a comparison between the models and photos. Have students compare and contrast the structures they see in both sets of images. Ask students to identify the structure in **8b** that indicates this cell is a plant cell. Students should be able to point at the rectangular cell wall. Have them take turns at identifying other unlabeled structures in the images. They should be able to point out additional mitochondrion, vacuoles, and Golgi bodies in the images by looking for similar shapes between the labeled and unlabeled structures.

VIDEO

VIDEO 6-1 Go online to observe how organelles move within a cell.

- **Vacuoles** Large, fluid-filled vesicles called *vacuoles* store or break down waste, debris, toxins, or food. In plants, lysosome-like vesicles fuse to form a very large central vacuole that makes up most of the volume of the cell. Fluid pressure in a central vacuole keeps plant cells plump, so stems, leaves, and other plant parts stay firm. The central vacuole has additional functions in some cells.
- **Endoplasmic Reticulum** The system of sacs and tubes that make up *endoplasmic reticulum* (ER) extends out from the outer lipid bilayer of the nuclear envelope. Two kinds of ER, rough and smooth, are named for their appearance in electron microscope images. Smooth ER is involved in the synthesis and storage of lipid molecules. The membrane of rough ER is typically folded into flattened sacs and has thousands of attached ribosomes that give it a "rough" appearance. These ribosomes make protein molecules that thread into the ER's interior as they are assembled. Proteins that are incorporated into the cell membrane or exported from the cell are constructed here.
- **Golgi Body** A *Golgi body* (also called a Golgi apparatus) has a folded membrane that often looks like a stack of pancakes. Enzymes inside Golgi bodies put finishing touches on proteins and lipids that have been delivered from ER. The finished products—membrane proteins and lipids, proteins for secretion, and enzymes—are sorted and packaged in new vesicles. Some of the new vesicles deliver their cargo to the cell membrane; others become lysosomes. In plant cells, Golgi bodies have an additional, major function: They make complex, branched polysaccharides that are a part of the cell wall.

FIGURE 6-8 Some of the membrane-enclosed structures in a white blood cell from a guinea pig (a) and in a cell from a root of a thale cress plant (b) are shown.

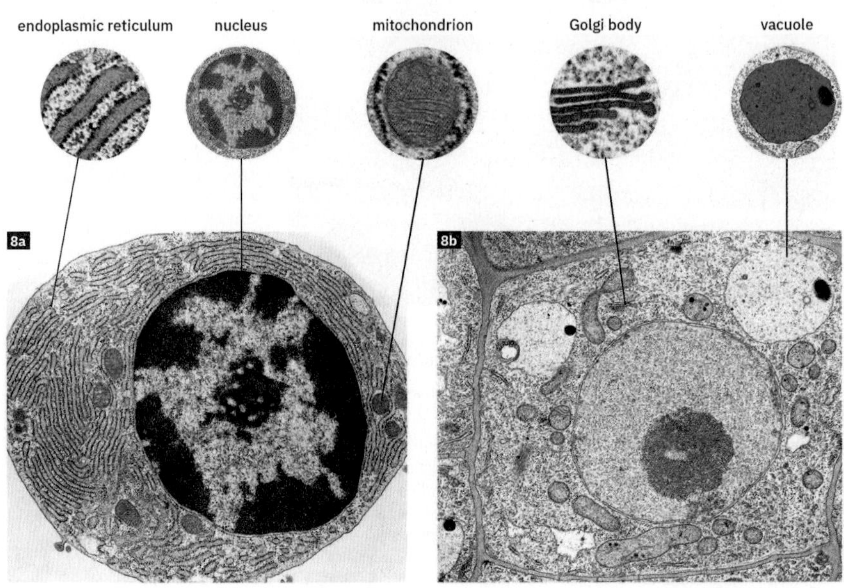

DIFFERENTIATED INSTRUCTION | English Language Learners

Monitoring and Self-Correcting
Encourage students to correct their own speech and writing if they mispronounce new vocabulary. If they are unsure how to say a word, they can slow down and sound it out. If they know they pronounced a word incorrectly, they can pause and repeat it correctly. As needed, say the word for students to repeat.

Beginning Model struggling to pronounce *mitochondria* and self-correcting by pausing to say it slowly, syllable-by-syllable. Have students repeat. As students read aloud, have them follow a similar process for other words they struggle with.

Intermediate Have students read aloud in pairs. If they struggle with a word, have them reread the sentence with the word more slowly.

Advanced Have students evaluate their own pronunciation during and after reading using these questions: *What did I pronounce correctly? Which words did I struggle with?* Have students work in pairs to practice the words they struggled with.

Mitochondria The structure of a **mitochondrion** (plural *mitochondria*) is specialized for carrying out reactions of aerobic cellular respiration, which generates the ATP used by cells as an energy source (**Figure 6-9**). The inner membrane has many folds that increase the surface area where some reactions of cellular respiration occur. Mitochondria have their own DNA and ribosomes within the matrix. Nearly all eukaryotic cells (including plant cells) have mitochondria, but the number of mitochondria varies by the type of cell and by the organism. For example, single-celled eukaryotes such as yeast often have only one mitochondrion, but human skeletal muscle cells have a thousand or more. In general, cells that have the highest demand for energy tend to have the most mitochondria. Typical mitochondria are between 1 and 4 micrometers in length.

9a

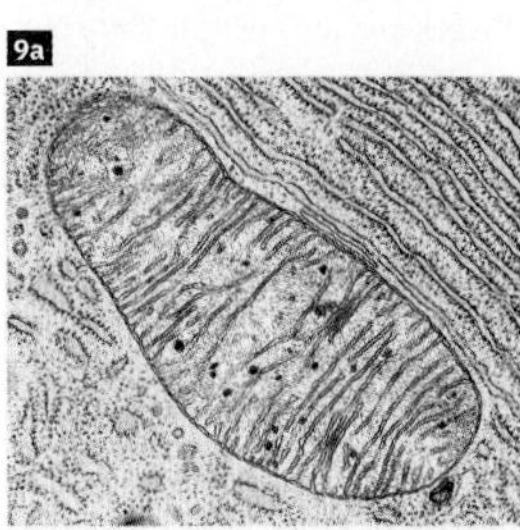

9b

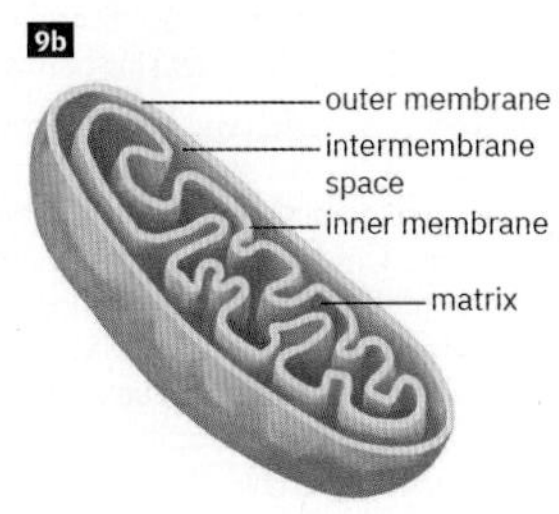

FIGURE 6-9 A microscope image of a mitochondrion in a cell from a bat pancreas (a). Each mitochondrion has two membranes, one highly folded inside the other. The inner compartment formed by these membranes is called the matrix (b). The outer compartment is called the intermembrane space.

Chloroplasts Photosynthesis in plants and in photosynthetic protists takes place in **chloroplasts** (**Figure 6-10**). Plant chloroplasts are oval or disk-shaped. Each has two outer membranes enclosing a fluid interior, the *stroma*, that contains enzymes as well as the chloroplast's own DNA and ribosomes. In the stroma, a third, highly folded membrane forms a single, continuous compartment that appears as stacks of discs. Each disc is called a *thylakoid*. Photosynthesis occurs at the thylakoids, which incorporate many pigments, including chlorophyll (the substance that makes most plants green). During photosynthesis, these pigments capture energy from sunlight to generate other molecules that can be stored and used to drive reactions in cells. In general, specialized cells within plant leaves have the most chloroplasts, because plant leaves receive more sunlight than other parts of plants.

10a

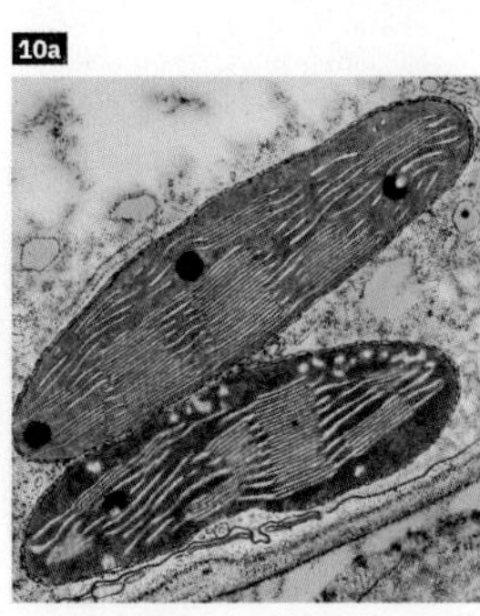

10b

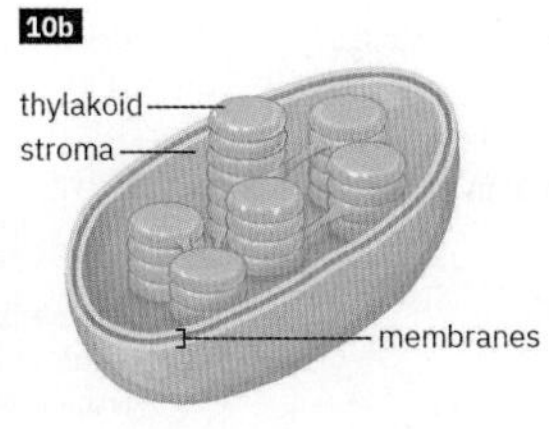

FIGURE 6-10 The transmission electron microscope image shows a chloroplast from a leaf of corn (a). Each chloroplast has two outer membranes (b). Photosynthesis occurs in the thylakoids, which are formed within a third, highly folded inner membrane that is called the thylakoid membrane.

Address Misconceptions

Mitochondria and Chloroplasts Students are likely familiar with the terms "mitochondria" and "chloroplasts," but it is possible that they have the common misconceptions that mitochondria are found only in animal cells and chloroplasts are found only in plant cells. Students may think plant cells do not contain mitochondria because they use their chloroplast to capture energy from the sun during photosynthesis.

To address this misconception, it may be useful to have a general class discussion on the topic of cellular respiration, which is covered in **Section 6.3**. Students should understand that both plant and animal cells have mitochondria. Share with them that some animals that consume algae may also contain chloroplasts, but that this is very rare. Ask students to discuss how they think having chloroplasts can be beneficial to these animals.

CROSSCUTTING CONCEPTS | Systems and System Models

Determine System Type Mitochondria and chloroplasts are components of a system, but they are each systems themselves. Have students examine **Figures 6-9** and **6-10**. Discuss the differences between open and closed systems. Have students construct an argument about the type of system mitochondria and chloroplasts represent. Help students who struggle with the concept understand that both are open systems because they are sharing matter and energy with their surroundings.

Connect to English Language Arts

Integration of Knowledge and Ideas From **Section 3.1**, students should be familiar with the ecological concepts of mutualism, a relationship where both species benefit, and commensalism, a relationship where one organism benefits and the other is neither harmed nor benefits from the relationship. Group students together and have them draw evidence from the text to help them generate questions about the different types of relationships endosymbionts share with their hosts. Students should relate their ideas to the endosymbiont theory covered in class. Have students conduct a short research project to answer their questions and share their findings with the class.

SEP Construct an Explanation
Metabolic requirements, variety of functions (e.g., more mitochondria in cells that need more energy vs those that need less), structural integrity and variety (e.g., cell walls and support vacuoles in plants)

Endosymbiont Theory

According to the **endosymbiont theory** for the origin of mitochondria and chloroplasts, these organelles descended from bacteria that entered a host cell and lived inside it. Organisms that live inside another organism are called *endosymbionts*. The term is usually reserved for cells that help their host, or at least do not harm it.

Mitochondria have their own DNA, which is circular and otherwise similar to bacterial DNA. They also divide independently of the cell and have their own ribosomes. Mitochondria are thought to have arisen when an anaerobic host cell engulfed heterotrophic bacteria capable of aerobic cellular respiration. The endosymbionts lived inside their host, where they continued to carry out aerobic cellular respiration and to reproduce. When the host cell divided, it passed on "guest" cells and the ability to carry out this efficient energy-releasing pathway.

As the two species lived together over many generations, the endosymbiont lost some genes that duplicated the function of host genes and donated other genes to the chromosome of the host cell. At the same time, the host became dependent on the ATP produced by its endosymbiont. Eventually, the host and endosymbiont became incapable of living independently. The endosymbiont had evolved into the organelle we call a mitochondrion. The high degree of similarity among the genomes of all mitochondria indicates that these organelles descended from one species of bacteria.

The first chloroplasts evolved after cyanobacteria were engulfed and became endosymbionts for an early eukaryote. Like mitochondria, chloroplasts have many similarities with their ancestral bacteria, including their own circular DNA and their own ribosomes. Genetic similarities between cyanobacteria and chloroplasts have confirmed their close evolutionary relationship, and both perform photosynthesis using the oxygen-producing pathway that evolved in cyanobacteria.

SEP Construct an Explanation *As described in this section, eukaryotic cells (**Figure 6-6**) have many more complicated internal components than prokaryotic cells (**Figure 6-4**). Why do you think multicellular organisms are made up of eukaryotic cells?*

6.1 REVIEW

1. **Describe** What is the cytoskeleton and how does it help cell components?
2. **Classify** Label each cell component as a part of a prokaryotic cell, eukaryotic cell, or both.
   - capsule
   - cytoplasm
   - cytoskeleton
   - ER
   - nucleus
   - ribosomes
3. **Identify** Which cell component is responsible for protecting genetic material from chemical processes that occur in the cell?
   A. DNA
   B. Golgi body
   C. nucleus
   D. vesicle
4. **Summarize** Describe three observations that support the hypothesis that mitochondria descended from bacteria that became incorporated into the cytoplasm of eukaryotic cells.

6.1 REVIEW

1. *The cytoskeleton is made up of interconnected filaments found between the cell membrane and the nucleus of eukaryotic cells. The cytoskeleton helps cells and cell parts move and it structurally supports cell membranes.* **DOK 1**
2. *prokaryote: capsule; eukaryote: nucleus, ER, cytoskeleton; both: cytoplasm, ribosomes* **DOK 2**
3. *C.* **DOK 2**
4. *Students' answers may include any two of the following: (1) Mitochondria resemble bacteria in size, form, and biochemistry. (2) Mitochondria have their own DNA that is similar to bacterial DNA. (3) Mitochondria divide independently of the cell. (4) Mitochondria have their own ribosomes.* **DOK 2**

LOOKING AT THE DATA

MICROBIOTA OF THE HUMAN BODY

SEP Use Mathematics Bacteria have been observed in nearly every organ of the human body.

The largest organ of the human body is the skin, which covers a surface area of about 2 square meters for the average adult. The number of bacteria, also known as the bacterial cell count, is estimated to be about 2×10^{11} on the skin surface of healthy adults. Other body parts and bodily fluids contain measurable amounts of bacteria. **Table 1** lists the typical volumes of organs and fluids in an adult human body and representative estimates of the concentration of bacterial cells.

Staphylococcus epidermis is a type of bacterium that lives on human skin.

TABLE 1. Bacteria in Human Organs and Fluids

Organ or fluid	Estimated volume (cm^3)	Estimated concentration of bacteria ($cells/cm^3$)
large intestine	400	3×10^{10}
lower small intestine	400	1×10^{8}
saliva	100	1×10^{9}
stomach	250	1×10^{3}
upper small intestine	400	1×10^{3}

Source: U.S. National Library of Medicine

On average, human cells are larger and less dense than bacterial cells. The largest mammalian cells have a volume of 10,000 μm^3 with density comparable to that of water, 1 g/cm^3. In comparison, typical bacterial cells have a volume of 0.7 μm^3 with a density of about 1.1 g/cm^3.

1. **Compare** Use the data to determine the bacterial cell count in human organs and fluids. Rank the organs and fluids from highest to lowest bacterial cell count.
2. **Estimate** Based on your calculations and assuming the rest of the human body has negligible bacterial counts, roughly how many bacterial cells in total are found in an average adult human body?
3. **Calculate** Determine an estimate for the mass of a single human cell in kilograms. Then determine the human cell count in an average 10^2-kg adult.
4. **Interpret** Based on the data provided about where bacteria are found in the human body, what is at least one important role bacteria play in the normal function of the human body?

DIFFERENTIATED INSTRUCTION | Leveled Support

Struggling Students To aid students who struggle with the calculations, provide a brief overview of how to multiply and divide numbers with exponents. It may help to work through an example of one of the organs or fluids. Encourage students to work in groups as they complete their calculations and review each other's work. If a student finds that a classmate's calculation is incorrect, have them guide each other to find the correct answers.

Advanced Learners If students quickly complete the calculations, have them estimate the cell count of humans of different size. For example, students may estimate the human cell count in 25 kg, 50 kg, and 70 kg individuals. Students may also calculate the estimated concentration of bacteria ($cells/cm^3$) on skin if the estimated volume is 2,600 cm^3.

LOOKING AT THE DATA
MICROBIOTA OF THE HUMAN BODY

Connect this activity to the Science and Engineering Practices of using mathematics and computational thinking and analyzing and interpreting data. Students analyze the estimated volume of bacteria found in organs in the human body and use mathematical principals to calculate the number of bacterial cells. They then interpret their data to identify at least one role that bacteria play in the function of the body.

SAMPLE ANSWERS

1. *large intestine: 1.2×10^{13}; saliva: 1×10^{11}; lower small intestine: 4×10^{10}; upper small intestine: 4×10^{5}; stomach: 2.5×10^{5}* **DOK 2**
2. *1.2×10^{13}* **DOK 2**
3. *estimated mass of single human cell = density x volume = $1\ g/cm^3 \times 10^{-8}\ cm^3 = 10^{-8}\ g = 10^{-11}\ kg$; human cell count in 10^2 kg adult = $10^2\ kg / 10^{-11}\ kg = 1 \times 10^{13}$; human cell count is about the same as the bacterial cell count, 1.2 : 1* **DOK 2**
4. *Sample answer: Bacteria are found in largest numbers in the digestive tract so they must play an important role in digestion and excretion.* **DOK 3**

Science Background

Specialized Cells Write the phrase "specialized cell" on the board and have students read the caption. Ask students to use their own words to define what they think the term means. Guide them to understand that many different types of organisms have specialized cells, including for example, humans. Red blood cells and muscle cells perform very specific functions in the body. Mammalian red blood cells are specialized to carry oxygen around the body and lack both a nucleus and mitochondria. Muscle cells are packed with mitochondria because they function to produce and distribute energy. The pit viper's skin contains specialized cells. Pair students in groups and have them research and identify at least two other specialized cells in another kind of animal. They might refer to **Chapter 10** for ideas. They should note at least one unique quality about each specialized cell and share their findings with the class.

In Your Community

Local Snakes Share with students that all 48 states in the contiguous United States have endemic snakes. While some snakes are venomous, most are not, and they are typically beneficial to the environment. Have students make a guide of the most common snakes that can be found in your community. Students should identify the structures that are unique to the species, whether or not they are venomous, and how they are beneficial to the region. Allow students to share their guides in small groups or with the class.

ON ASSIGNMENT This close up of a venomous eyelash pit viper, *Bothriechis schlegelii*, was taken by National Geographic photographer Charlie Hamilton James while in Costa Rica's Osa Peninsula. Special pigment-bearing cells known as chromatophores are a part of the skin and are responsible for the distinctive colors and patterns of each snake.

ON ASSIGNMENT

Chromatophores National Geographic photographer Charlie Hamilton James is known for his work as a photojournalist, and he specializes in photographs that showcase topics around conservation, natural history and anthropology. He has spent decades photographing the rainforests of Central and South America where he has captured the images of many different species of animals and recorded the lives of many indigenous peoples.

This photograph, taken for his Costa Rica OSA Peninsula collection Featured in February 2021 shows a venomous eyelash pit viper which has specialized skin cells that lend it its color. This species of snake is considered to be one of the most widespread arboreal vipers and their habitat ranges from Costa Rica to Peru. Their specialized chromatophores are highly variable, and the species has a wide variety of coloration patterns.

Encourage students to hypothesize how the chromatophores help the snake to survive in a changing environment.

You can learn more about Charlie Hamilton James on his website: http://charliehamiltonjames.com.

CROSSCUTTING CONCEPTS | Structure and Function

Toxins in Snake Venom The eyelash pit viper has a number of specialized structures that function to improve its chance of survival. In addition to the chromatophores, these snakes, like all vipers have specialized organs that form venom. Venom is essentially specialized spit, as it is made up of about 90% water and only a few types of enzymatic and nonenzymatic proteins. These proteins can be highly toxic. Eyelash pit viper venom functions as a hemotoxin—it destroys red blood cells.

Have students synthesize information from multiple sources to identify the differences between hemotoxins, cytotoxins, and neurotoxins—all toxins found in venom of different species—and their functions. Have students explain how they affect the cells of an animal bitten by a snake. Students can make a poster modeling how the toxins interact with cells and share their posters with the class.

Alternatively, encourage interested students to research how antivenoms work and why it is important for them to be administered soon after being bitten.

6.2

CELL MEMBRANES

LS1.A

EXPLORE/EXPLAIN

This section provides an overview of the cell membrane and the structure and function of the components within. It also covers the different processes of transport available to molecules, such as diffusion, osmosis, passive transport, and active transport.

Objectives

- Describe the structure of the cell membrane and how proteins function within it.
- Explain homeostasis and the involvement of the cell membrane in its maintenance.
- Summarize the different modes of transport across the cellular membrane.

Pretest Use **Question 4** to identify gaps in background knowledge or misconceptions.

Analyze *Students might say that cells need to keep out pathogens, take in food, or expel wastes.*

Phospholipids Students learn about the interactions of phospholipids in the fluid mosaic model. They read that the fluidity of the membrane is caused by the lack of chemical bonds between phospholipids. **Chapter 5** discusses the structures of phospholipids and their chemical properties in greater detail.

6.2

KEY TERMS

endocytosis
exocytosis
facilitated diffusion
fluid mosaic model
homeostasis
osmosis
selective permeability

CELL MEMBRANES

Opening a window on a nice day may seem like a good idea. An open window lets in fresh air, but you might also end up with a room full of houseflies and other insects. A window screen can let in the air and sounds while keeping out the things you would rather stay outside.

Analyze *Like a room with a window and window screen, what are some reasons a cell might need to control what materials can enter and exit?*

Cell Membrane Structure

Cell systems need to maintain optimal conditions while also communicating and exchanging material with nearby cells and the surrounding environment. A cell membrane regulates what materials can enter or exit the cell, allowing the cell to maintain homeostasis within its environment and communicate with surrounding cells. Cellular **homeostasis** is the maintenance of optimal chemical and physical conditions within a cell, even if the external environment changes.

Fluid Mosaic Model The foundation of all cell membranes is a lipid bilayer that consists mainly of phospholipids. As described in Section 5.3, a phospholipid has a phosphate-containing head and two fatty acid tails. The head is polar and hydrophilic (water-loving), so it interacts with water molecules. The two long hydrocarbon tails are nonpolar and hydrophobic (water-fearing), so they do not interact with water molecules. As a result of these opposing properties, phospholipids mixed with water will spontaneously organize themselves into two phospholipid sheets (bilayer), with hydrophobic tails facing each other and hydrophilic heads facing the aqueous extracellular and intracellular environments (**Figure 6-11**).

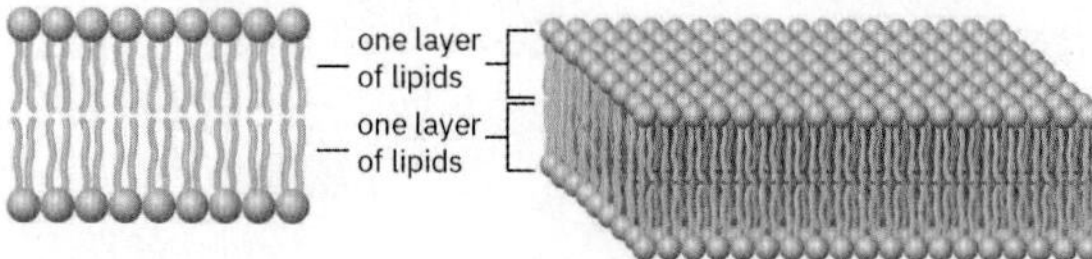

FIGURE 6-11 Phospholipids are organized in cell membranes with the hydrophobic tails facing each other and the hydrophilic heads facing the environments inside and outside the cell.

Proteins and other molecules are embedded in or attached to the lipid bilayer of a cell membrane. Many of these molecules move around the membrane more or less freely. The **fluid mosaic model** describes a membrane as a two-dimensional liquid made up of different substances. The "mosaic" part of the name comes from the many different types of molecules in the membrane.

Membrane fluidity occurs because phospholipids in the bilayer are not chemically bonded to one another. They stay organized as a result of collective hydrophobic and hydrophilic attractions. Individual phospholipids in the bilayer drift sideways and spin around their long axis, and their tails wiggle.

Proteins in the Cell Membrane Many types of proteins are associated with a cell membrane (**Figure 6-12**). Some are temporarily or permanently attached to one of the lipid bilayer's surfaces. Others have a hydrophobic part that anchors the protein permanently in the bilayer. Filaments inside the cell fasten some membrane proteins in place. Each type of protein in a membrane gives a specific function to it. Different cell membranes can carry out different tasks depending on which proteins they include. A cell membrane contains certain proteins that no internal membrane has, so it has functions that no other membrane does.

- **Adhesion Proteins** Cells stay organized in animal tissues because adhesion proteins in their cell membranes fasten them together and hold them in place. Adhesion proteins also provide a cell with information about its position relative to other cells or structures.
- **Receptor Proteins** Cell membranes and some internal membranes incorporate receptor proteins, which trigger a change in the cell's activities in response to a stimulus or signaling molecule. Each type of receptor protein receives a particular stimulus such as a hormone binding to it. The response triggered by a receptor may involve metabolism, movement, division, or even cell death.
- **Enzymes** All cell membranes incorporate *enzymes*, which are proteins that help catalyze chemical reactions in cells (Section 5.4). Some of these enzymes act on proteins or lipids that are part of the membrane. Others work in series and use the membrane as a scaffold. Enzymes associated with the membranes of both mitochondria and chloroplasts, such as ATP synthase, are important parts of cellular respiration and photosynthesis.
- **Transport Proteins** All membranes also have transport proteins, which move specific substances across the bilayer. Transport proteins are important because lipid bilayers are impermeable to most substances, including the ions and polar molecules that cells must take in and expel on a regular basis.

SEP **Construct an Explanation** *Explain why lipid molecules come together to form a double membrane. Why is this structure important for the integration of proteins that function in cell membranes?*

CASE STUDY
ARTIFICIAL CELL TECHNOLOGY

Ask Questions *Generate questions about possible similarities and differences between cell membranes and proteins in artificial cells and natural cells.*

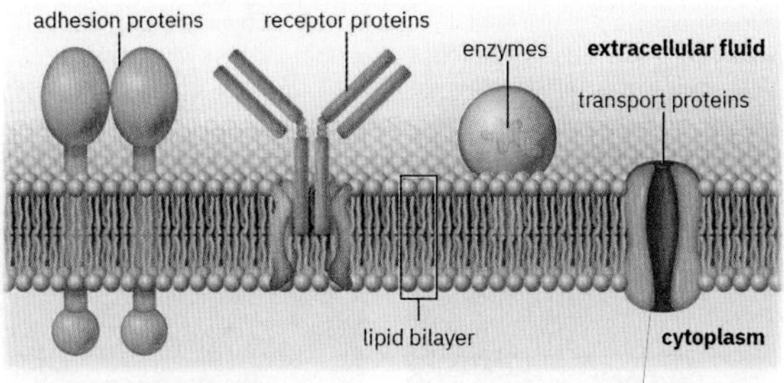

FIGURE 6-12 Proteins embedded in the cell membrane perform a variety of functions. Enzymes are enclosed within a membrane that connects to the lipid bilayer.

CASE STUDY
ARTIFICIAL CELL TECHNOLOGY

Ask Questions To help students revisit the Case Study and make connections with the content, have them generate a list of the components that would be required for an artificial cell membrane. Students should compare possible structures in artificial cell membranes with the structures shown in the model of a natural cell membrane.

Connect to English Language Arts

Use Digital Media Students are introduced to common types of proteins that are embedded in the cell membrane in this section and in **Figure 6-12**. Place students in small groups and assign each group one of the four protein types identified on the page. Allow students to do research on their proteins and put together a brief presentation using digital media to share with the class. Student presentations should include the protein type that they were assigned, the function of the protein in the cell membrane, and a brief description of at least one example of the protein that can be found in a plant or animal cell.

Construct an Explanation *Lipids have one end that is hydrophobic and one end that is hydrophilic. The hydrophobic ends come together as they move away from water, forming the membrane. The flexibility of the membrane allows many proteins to be embedded in the membrane and perform important functions.*

SCIENCE AND ENGINEERING PRACTICES
Developing and Using Models

Schematic Models Students should recognize that the models that are used to visualize the different kinds of membrane proteins are not replicas of what these proteins look like in the cell. The models are used as general schematics to help visualize how the proteins are seated in the cell membrane and how they function. Have students make a chart where they identify each class of cellular membrane protein, draw their own model, and describe how it functions with the membrane. Students should describe the limitations of their models. If students work in groups, you consider assigning each group one type of protein and make a large, class chart that students can refer to as they move through the lesson.

Cell Membrane Students learn that materials can cross the cell membrane due to selective permeability or through transport proteins, allowing molecules to enter and leave the cell as necessary. **Chapter 11** provides detail on how proteins are made and processed in the cell.

Science Background

Diffusion and Osmosis Diffusion can easily be modeled in the air by using a room spray on one side of the room and having students raise their hands as they begin to smell it. It can also be modeled in liquids using a glass of water and food dye. When first dropped in water, the food coloring will be highly concentrated in one area. However, it will quickly diffuse and color the entire volume of water, given time.

Osmosis requires the movement of water across a membrane. One way to model this is to use gummy bear candies and a container of water. Share that during production, gummy bear candies have most of their water removed. Ask students what they think will happen if a gummy bear is set in water overnight. Place one gummy bear in tap water and leave the other in air as the control. Have students explain why the gummy bear in the water expanded.

Movement of Material Across Cell Membranes

Lipid bilayers have **selective permeability**, which means that some substances can move across them and others cannot. Permeability is determined by the physical size and charge of molecules. Hydrophobic molecules, gases, and small, uncharged polar molecules can easily move through the hydrophobic core of the lipid bilayer. However, ions and large polar molecules cannot.

Diffusion Molecules are constantly moving into, throughout, and out of cells. An important part of this movement is *diffusion*, the spreading of atoms or molecules through a fluid or gas. The constant jiggling of particles within a fluid or gas causes particles to randomly bounce off each other or nearby structures. Molecules diffuse from areas of high concentration to areas of low concentration.

The concentration of molecules in a solution is known as the solute concentration. A solution with more concentrated solute is referred to as *hypertonic*, relative to a *hypotonic* solution with a lower concentration of solute.

Osmosis When a lipid bilayer separates two fluids with differing solute concentrations, water will diffuse across it. **Osmosis** is the diffusion of water across membranes, and it is driven by differences in the total solute concentration on opposite sides of the membrane. Water diffuses from a hypotonic fluid into a hypertonic one to balance the solute concentration across the lipid bilayer. This continues until the two fluids are *isotonic*, which means they have the same overall solute concentration. Another way to consider this is that hypotonic solutions have a high concentration of free water molecules, that is, molecules that are not bound to solutes. Those water molecules move to a hypertonic solution that has a lower concentration of free water molecules.

Osmosis can be demonstrated with red blood cells (**Figure 6-13**). If a cell's cytoplasm becomes hypertonic with respect to the fluid outside of its cell membrane, water will diffuse into the cell. If the cytoplasm becomes hypotonic, water will diffuse out. In either case, the solute concentration of the cytoplasm will change. If it changes significantly, the cell's enzymes will stop working, with lethal results. Most cells have mechanisms that compensate for osmosis when the solute concentration of cytoplasm differs from extracellular fluid. In cells with no such mechanism, the volume and solute concentration of cytoplasm changes when water diffuses into or out of the cell.

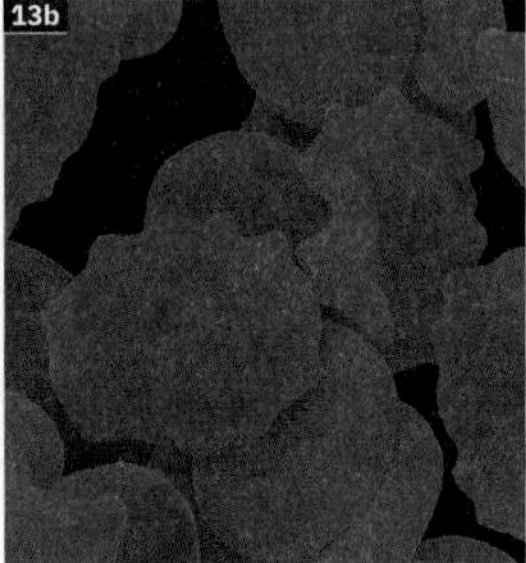

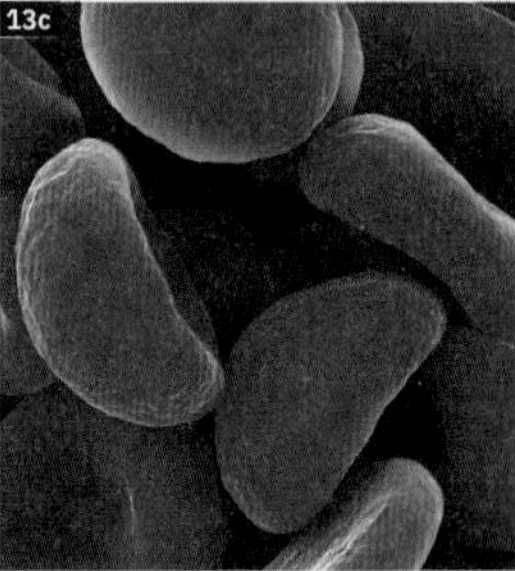

FIGURE 6-13 Red blood cells in an isotonic solution (such as the fluid portion of blood) have an indented disc shape (a). Red blood cells in a hypertonic solution shrivel up because water diffuses out of them (b). Red blood cells in a hypotonic solution swell up because water diffuses into them (c).

DIFFERENTIATED INSTRUCTION | Leveled Support

Struggling Students Students may struggle differentiating between the terms *hypertonic* and *hypotonic*. To help students differentiate between the terms, ask students what they think the term *hyper* means. Students will likely equate the term to high energy. Guide them to understand that the prefix *hyper* means *higher* or *above normal*. Ask students to now consider what a *hypertonic* solution is. They should understand that it is a solution that has a high solute concentration. Explain to students that *hypo* is the opposite of *hyper*.

Advanced Learners Have students find common synthetic or artificial membranes that have uses in different industries. Ask students to determine why different membranes might be used for the same purpose and to identify advantages and disadvantages to each membrane. Students may also choose to determine the effectiveness of each membrane type. Students should summarize how semipermeable membranes are similar to the cell membrane and share their findings with the class.

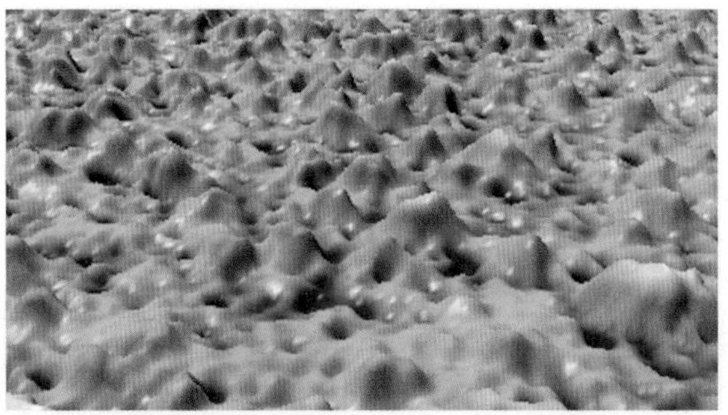

FIGURE 6-14 Cell membrane proteins, including transport proteins, form dense clusters on the cell membrane. Atomic force microscopy allows scientists to scan a membrane surface and detect proteins sticking up just a few micrometers from the surface. Each peak is a protein in this cell membrane.

Transport Proteins Substances that do not diffuse directly through lipid bilayers can cross a cell membrane only through transport proteins embedded in the membrane (**Figure 6-14**). Each type of transport protein allows a specific substance to cross. For example, calcium pumps pump only calcium ions, and glucose transporters transport only glucose. The ability to move only specific molecules across the cell membrane is an important part of homeostasis. The cell has to regulate what can move into and out of the cytoplasm to maintain its composition.

Some transport proteins form pores, which are open channels through a membrane. Gated transport proteins open and close in response to a stimulus such as a shift in electric charge or binding to a signaling molecule. Other types of transport proteins change shape upon binding and releasing their specific molecule on the other side of the membrane.

Passive Transport Osmosis and diffusion of solutes through the membrane are examples of passive transport. *Passive transport* does not require any energy because the movement of the solute and the direction of its movement are driven entirely by the solute's concentration gradient. **Facilitated diffusion** is the passive transport of solutes through the cell membrane with the help of a transport protein. **Figure 6-15** demonstrates one example of facilitated diffusion, in which glucose is transported with the help of a glucose transport protein.

FIGURE 6-15 A glucose molecule in extracellular fluid binds to a glucose transporter embedded in the cell membrane (a). Binding causes the transport protein to change shape (b). The transport protein releases the glucose in the cytoplasm and returns to its original shape (c).

15a

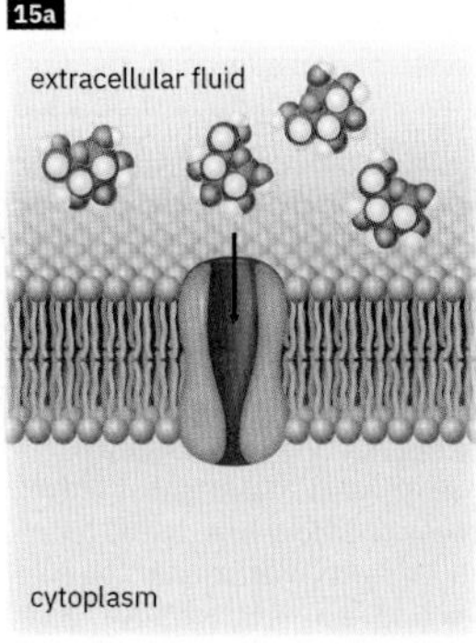

15b

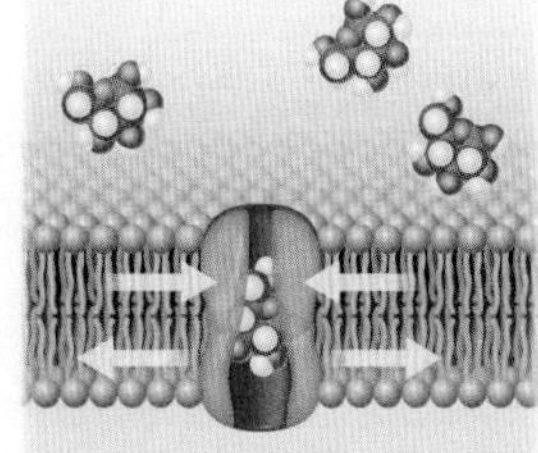

15c

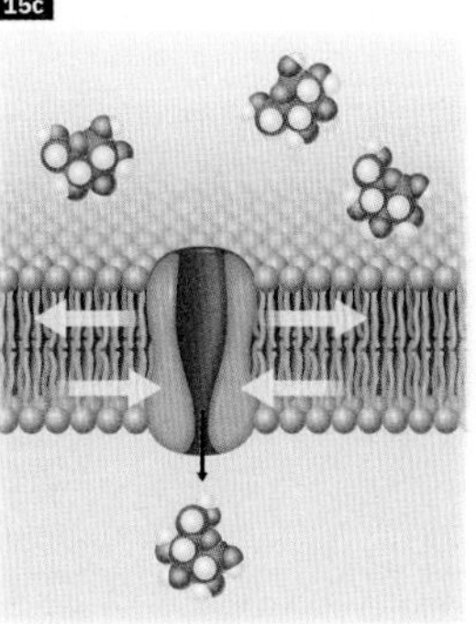

Address Misconceptions

Passive Transport Students may think that everything enters the cell by passive transport. Only the smallest molecules can freely diffuse across cell membranes. These molecules include water, carbon dioxide, and oxygen. Larger molecules can also be transported into the cell using facilitated diffusion, which is also passive transport but require the help of a transport protein.

Have students examine **Figure 6-15** which shows an example of facilitated diffusion. In this example, glucose is being moved across the cell membrane by transport proteins. Have students discuss the difference between facilitated diffusion and active transport. Confirm they understand that while active transport and facilitated diffusion both use proteins to assist in transport, active transport works against the concentration gradient, moving substances from areas of low concentration to areas of high concentration, which requires energy.

Science Background

Active Transport Common examples of active transport proteins that are found in humans are calcium pumps that are highly concentrated in muscle cells and sodium-potassium pumps which are found in nearly all cell membranes. Project **Figure 6-15** and **Figure 6-16** for students. Have them identify the differences they see between the two kinds of transport proteins. Students should recognize the need for ATP as an energy source to move the calcium ions from the area of low concentration (cytoplasm) to the area of high concentration (extracellular fluid).

Some proteins involved in active transport are integral to maintaining a cell's membrane potential. The membrane potential is the difference in the electrical potential between the inside and outside of a cell, but the cytoplasm is always more negative than the extracellular fluid.

Use a Model *Each protein has a structure that makes it specific to a certain molecule or function. Each type of protein and its function can be regulated separately to be sure that the right molecules enter or exit the cell as needed.*

Go Online **SIMULATION**

Crossing Membranes After observing the different types of transport across the cell membrane, ask students why there are multiple modes of transport. Reinforce the idea of controlling the variety of types and sizes of molecules that enter and exit the cell.

Active Transport Many cellular processes require transporting solutes across a membrane into a hypertonic solution. Moving a solute against its concentration gradient requires energy. In *active transport*, a transport protein uses energy to pump a solute against its gradient across a cell membrane (**Figure 6-16**). Typically, energy from ATP (Section 6.3) is used to change the shape of the protein. The shape change causes the protein to release a bound solute to the other side of the membrane. Each solute moved by active transport has a specific transport protein. One example of active transport occurs with calcium ions, which act as messengers that trigger various processes inside cells. The concentration of these ions in cytoplasm must be kept thousands of times lower than in extracellular fluid. This gradient is maintained by calcium pumps, which export calcium ions from a cell by active transport.

FIGURE 6-16
Two calcium ions bind to the specific transport protein (a). Energy from ATP helps the protein to change shape so that the calcium ions are ejected to the opposite side of the membrane (b). After the calcium ions are released, the transport protein resumes its original shape (c).

16a

extracellular fluid

cytoplasm

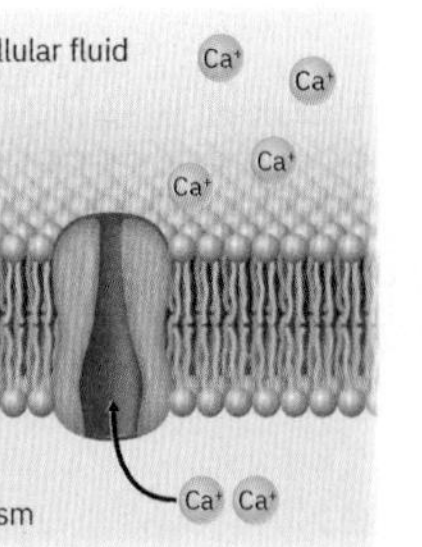

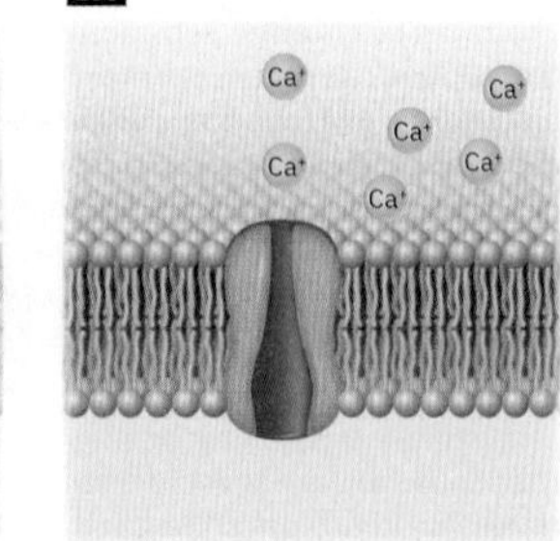

SEP **Use a Model** *Using the figures and examples in this section, describe why it is important to have different, specific proteins for transporting different materials across the cell membrane.*

Go Online **SIMULATION**

Crossing Membranes Go online to transport different substances into and out of a cell.

Vesicle Movement When a membrane is disrupted, the fatty acid tails of the phospholipids in the bilayer become exposed to their watery surroundings. Recall that in water, hydrophobic phospholipids spontaneously rearrange themselves so that their nonpolar tails stay together. Because of this, a membrane tends to seal itself after a disruption. Vesicles form the same way. When a patch of membrane bulges into the cytoplasm, the hydrophobic tails of the lipids in the bilayer are repelled by the watery fluid on both sides. The fluid "pushes" the phospholipid tails together, which helps round off the bud as a vesicle and seals the rupture in the membrane.

Vesicles are constantly carrying materials to and from the cell membrane. This movement requires energy because it involves motor proteins that drag the vesicles along cytoskeletal elements. During **exocytosis**, substances are released out of a cell. In this process, a vesicle in the cytoplasm moves to the cell's surface and fuses with the cell membrane. As the fusion occurs, the contents of the vesicle are released to the surrounding fluid outside the cell (**Figure 6-17a**). During **endocytosis**, large molecules or liquids are taken into a cell by an inward folding of the cell membrane. There are multiple pathways of endocytosis, but all take up substances near the cell's surface in bulk, as opposed to one molecule or ion at a time via transport proteins.

DIFFERENTIATED INSTRUCTION | English Language Learners

Monitoring Understanding During the class discussion on vesicle movement, have students monitor their own understanding. Pause to allow students to ask for repetition, restate to confirm understanding, or ask classmates to define terms or clarify ideas.

Beginning Encourage students to ask for repetition to help them understand: *Could you say that again?* or *Could you say that more slowly?*

Intermediate Have students restate in their own words to confirm understanding. Provide a sentence frame such as the following: *So, you're saying that ________. Is that right?*

Advanced Have students ask their classmates to define terms or clarify ideas. Model some examples: *What does ________ mean? Is that similar to ________? How is that different from ________?*

Cells can bring in fluids and molecules from outside the cell in a nonspecific way, or they can ingest molecules that bond to specific receptors that trigger endocytosis. For endocytosis that is triggered by specific molecules, as shown in **Figure 6-17b**, receptor proteins in the cell membrane bind to a substance such as a hormone, or a particle such as a bacterium. The binding triggers a shallow pit to form in the membrane, just under the receptors. The pit sinks into the cytoplasm and traps the targeted molecules in a vesicle as it closes back on itself.

FIGURE 6-17 In exocytosis (a), a vesicle in the cytoplasm fuses with the cell membrane. Lipids and proteins of the vesicle's membrane become part of the plasma membrane as its contents are expelled. In endocytosis (b), shown here with receptors that match specific molecules, a pit forms in the cell membrane. The target molecules are trapped in a vesicle as the pit deepens and sinks into the cell's cytoplasm.

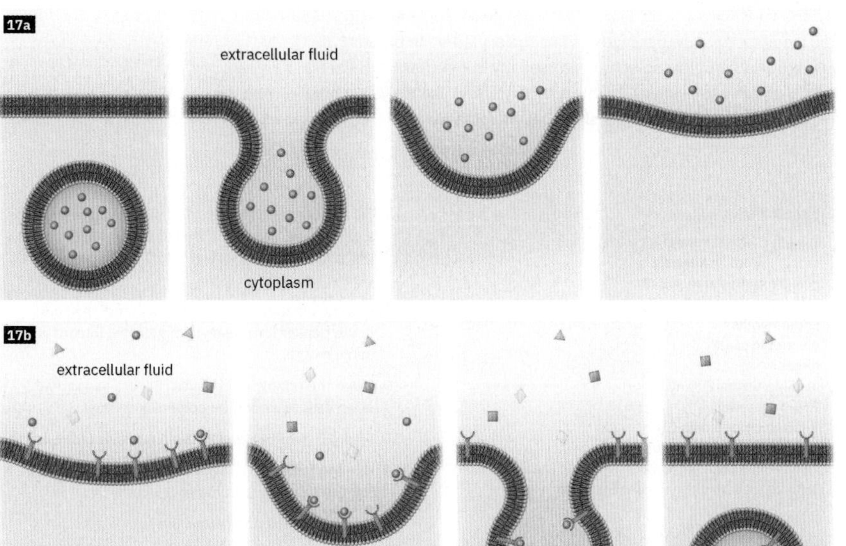

6.2 REVIEW

1. **Relate** In each scenario described, identify whether water must move into or out of the cell to reach equilibrium.
   - The solute concentration outside the cell is higher than inside the cell.
   - The concentration of water molecules outside the cell exceeds the number inside the cell.
   - There is a higher concentration of solute particles inside the cell than outside the cell.
2. **Define** What is the difference between diffusion and osmosis?
3. **Identify** Which of the following types of transport require no energy input? Select all correct answers.
   A. vesicle exocytosis
   B. pumping of calcium
   C. facilitated diffusion of glucose
   D. osmosis of water into a red blood cell

In Your Community

Vesicle Movement Vesicles allow larger particles or substances to enter or exit the cell. These are different from transport proteins or channels which do not greatly disrupt the cell. It might help students to draw analogies to structures or services in their homes or communities. Have a class discussion where students make comparisons between transport proteins and vesicles and structures or services with which they are familiar. Students may compare passive transport proteins with open doorways, active transport proteins with closed and locked doors that require energy to open and close them. They may compare vesicle movement with a delivery service. Guide the class discussion to ensure that students have a strong understanding of the similarities and differences between the transport mechanisms and structures.

Visual Support

Exocytosis and Endocytosis Have students consider the prefixes for *exocytosis* and *endocytosis*. Ask students to recall what the prefixes *exo-* and *endo-* mean. Project **Figure 6-17** and ask students to use clues in the illustrations to identify the location of the cell membrane, the cytoplasm, the vesicles, and the molecules that move into and out of the cell. Draw arrows to indicate the motion of the vesicles and confirm that **Figure 6-17a** shows exocytosis and **Figure 6-17b** shows endocytosis.

6.2 REVIEW

1. *Water moves out of the cell: The solute concentration outside the cell is higher than inside the cell. Water moves into the cell: There is a higher concentration of solute particles inside the cell than outside the cell. Water moves into the cell: The concentration of water molecules outside the cell exceeds the number inside the cell* **DOK 2**
2. *Sample answer: Diffusion is the spontaneous movement of atoms or molecules through a fluid or a gas. Materials diffuse from higher to lower concentrations and across membranes. Osmosis is a special case describing the movement of water molecules across a boundary such as a cell membrane.* **DOK 2**
3. *C, D* **DOK 2**

MINILAB
SELECTIVELY PERMEABLE MEMBRANES

Students use a model to investigate how cells maintain internal conditions to stay alive. The dialysis tubing will represent the selectively permeable cell membrane.

Time: 30 minutes

Advance Preparation

All the materials needed for this minilab can be purchased online. The dialysis tubing should be cut before beginning the lab. Each group should receive a 20 cm piece of dialysis tubing and two 15 cm pieces of string soaked in water.

Glucose-and-starch solution can be prepared manually. Because starch is not soluble in cold water, place 1000 mL of distilled or deionized water in a beaker on a hot plate and boil it. Weigh 10 g of starch and mix it with 30–50 mL of water to make a paste. Pour this paste into the boiling water and stir. A 15% glucose solution can be prepared by mixing 15 g of glucose with 85 mL of water (final volume should be 100 mL). Glucose is often sold under the name dextrose.

Procedure

In **Step 5**, students are asked to wash off the bag in case any glucose or starch contamination was smeared outside the tubing. Since the measured mass difference between the initial and final weight is expected to be about 1 g in ideal conditions, drying the tubing before weighing improves the accuracy of the measurement.

In **Step 7**, expect the water outside the tube to turn lighter over time. Small iodine ions will gradually diffuse into the dialysis tube. The solution inside the tubing will eventually turn darker.

For **Step 8**, different brands of glucose strips can use different units. Remind the students to use units from glucose test strips when they record their data.

In **Step 9**, discuss the relative sizes of the substances involved. Starch forms comparatively long chains while glucose forms a 6-carbon molecule. Water molecules and iodine ions are smaller than glucose and starch molecules.

If possible, extend the wait time in **Step 9** to 25 or 30 minutes.

MINILAB

SELECTIVELY PERMEABLE MEMBRANES

SEP Develop and Use a Model How do molecules move through the cell membrane?

Cell membranes allow passage of some molecules while blocking passage of others. This selective permeability is important for helping cells maintain homeostasis. In this activity you will use dialysis tubing to model some of the functions of the cell membrane.

Suggested Materials

- 2.5-cm diameter dialysis tubing, 20 cm
- string, 15 cm (2 pieces)
- glucose-and-starch solution
- scissors
- paper towels
- electronic scale
- forceps
- distilled water, 200 mL
- 250-mL beaker
- 5-mL transfer pipets (2)
- potassium iodide solution
- glucose indicator strips (2)
- color chart with glucose concentration

Safety      

Procedure

1. Obtain a piece of the dialysis tubing and the two pieces of string that are soaking in water.
2. Open the tubing to a cylinder shape. Use one piece of string to tie one end of the tube to make a bag.
3. Pour the glucose-and-starch solution until the bag is about half full.
4. Tie off the open end of the tubing with the other piece of string. Cut off any extra string material from both ends.
5. Wash the outside of the tubing gently but thoroughly under running water. Then pat it dry with a paper towel.
6. Weigh the tube and record its initial mass.
7. Fill the beaker with 200 mL distilled water, then use a pipet to add several drops of the starch indicator, potassium iodide solution. **CAUTION:** *Iodine stains clothing and skin.*
8. Use the glucose indicator strips to test the solution in the beaker for presence of glucose. Record your initial results.
9. Place the tubing in the water. Record your initial observations about the color of the water.
10. After 15 minutes, record the final color of the water.
11. Use the forceps to take the tube out of the water. Pat the tube dry with paper towels.
12. Weigh the tube and record the final mass.
13. Using glucose indicator strips, test the solution in the beaker for presence and concentration of glucose. Record your results.
14. Once you have all the data you need, move to the cleanup stage.

Results and Analysis

1. **Analyze** Consider the flow of glucose through the membrane. Would the rate of this flow be faster at the start or at the end of the experiment?
2. **Predict** Consider how the glucose concentration changed in the beaker. Describe a setup where the glucose would move in the opposite direction.
3. **Explain** How does this model demonstrate selective permeability? Use your results, including any changes in the mass of the tube and in glucose and starch concentrations, to support your answer.
4. **Argue From Evidence** Most animal cells will swell and burst if they are placed in pure water. Explain why this would happen.

Results and Analysis

1. **Analyze** *Sample answer: There was a higher concentration of glucose inside the dialysis tubing initially, and there was no glucose outside. This most likely means the glucose would diffuse faster at the start and gradually slow down.* **DOK 2**
2. **Predict** *Sample answer: If the outside solution had a higher concentration of glucose, glucose would move from outside to inside the tubing.* **DOK 2**
3. **Argue from Evidence** *Sample answer: A cell has water with many different dissolved substances in it. Pure water does not have any dissolved substances. If a cell is placed in pure water, water will flow toward the higher concentration of dissolved substances—into the cell. This would cause the cell to swell.* **DOK 3**

PHOTOSYNTHESIS AND CELLULAR RESPIRATION

Think about how energy and matter move through ecosystems. For example, when one organism in a food web eats another, it acquires energy and carbon for its own life processes. It uses that energy to carry out cell functions, and the carbon atoms become part of molecules important for each individual cell.

Explain *Think of the cell structures and processes you've learned about in this chapter. Describe how one of those structures or processes uses carbon and energy.*

6.3

KEY TERMS

cellular respiration
electron transport chain
fermentation
photosynthesis

Energy in Cells

All life is sustained by inputs of energy, but not all forms of energy can sustain life. Sunlight, for example, is abundant here on Earth, but it cannot directly power protein synthesis or other energy-requiring reactions that all organisms need to perform to stay alive. To do so, sunlight must first be converted to chemical bond energy. Unlike light, chemical energy can power the reactions of life, and it can be stored for later use. This section will provide a brief overview of the processes that most often are used to convert energy into a usable chemical form in organisms. Details of the chemical reactions and molecules involved in each step of the processes of photosynthesis and cellular respiration are described in the Appendix.

Photosynthesis and cellular respiration are a critical part of the cycling of matter and flow of energy in ecosystems (Section 2.4). **Figure 6-18** illustrates that the process of photosynthesis converts carbon dioxide, water, and energy from the sun into stored energy in the form of glucose. Aerobic respiration then converts sugar and oxygen into usable energy in the form of ATP.

FIGURE 6-18 Photosynthesis produces sugars such as glucose and, in most organisms, it also releases oxygen. The breakdown of glucose in aerobic cellular respiration requires oxygen, and it produces carbon dioxide and water (a). The stable structure of glucose stores energy from the sun until it is converted to usable ATP in cellular respiration (b).

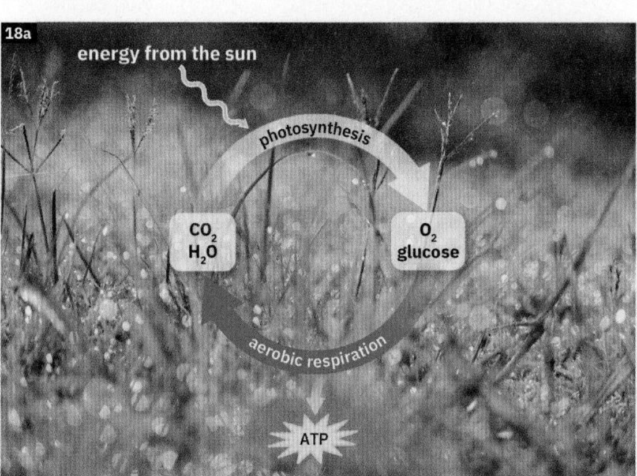

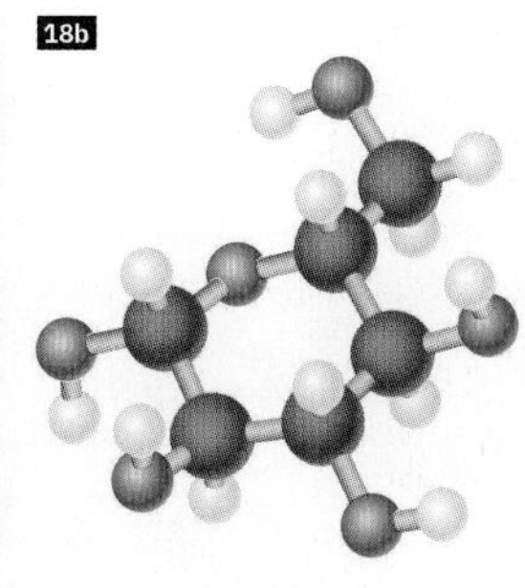

6.3

PHOTOSYNTHESIS AND CELLULAR RESPIRATION

LS1.C

EXPLORE/EXPLAIN

This section provides an overview of how energy is utilized in cells, the metabolic pathways involved in cellular respiration, how energy is captured in the process of photosynthesis, and other energy conversion pathways used by the cell.

Objectives

- Explain the importance of energy usage by cells.
- Describe the processes involved in cellular respiration and ATP synthesis.
- Summarize the stages of photosynthesis and the light-dependent and light-independent reactions.
- Identify other energy conversion pathways available to cells.

Pretest Use **Questions 5** and **6** to identify gaps in background knowledge or misconceptions.

Explain *Students might say that oxygen from photosynthesis must have allowed more energy to be made so aerobic respiration was advantageous. And/or that oxygen was poisonous for anaerobic organisms.*

Vocabulary Strategy

Using Prior Knowledge Students should have learned about cellular respiration and photosynthesis in their middle school science classes. **Section 2.1** also provided a short overview of the processes. As students work through this section, suggest they use a graphic organizer, such as a KWL chart, to identify the terms they are familiar with and capture new information around those terms and link these to new terms and their related information.

SCIENCE AND ENGINEERING PRACTICES
Developing and Using Models

Varied Forms of Molecular Models The chemical reactions described in this section happen at the molecular level so they are represented using different types of models. Remind students that scientists develop and use models to describe phenomena that are difficult to observe and measure. Some models are chemical reaction equations that show reactants and products represented with chemical formulae. Some models, like **Figure 6-18b**, use three-dimensional shapes to represent molecular structures like those seen in Chapter 5. Other models use shapes and arrows to represent processes. As students work through this section, have them spend time with each illustration and read the captions carefully. They can practice explaining the models to each other to strengthen their understanding.

Science Background

Energy from Food Energy from food is not released all at once in the body, but instead goes through a series of oxidation reactions where electrons are transferred from one molecule to the next. After each reaction, the resulting product has a lower amount of energy than the initial molecule did because it handed off electrons to electron acceptor molecules. The energy from these reactions is not used immediately, but is instead converted to energy rich molecules, like NADH, NADPH, $FADH_2$, and ATP. ATP is the most common of these in the human body and is discussed here as one of the products of cellular respiration. As students move through the lesson, have them make a chart to identify the different energy storing molecules, where they form, and their function.

CCC Energy and Matter *Sample answer: ADP combines with a phosphate group, which adds matter and energy to the molecule, forming ATP. When ATP is broken down, energy is released, and a phosphate group (matter) is removed from the molecule to form ADP.*

Energy from Food Students learn about how sugar molecules are broken down at the molecular level during cellular respiration. **Chapter 10** defines how animal systems interact to use the energy from food to maintain homeostasis.

BACTERIA IN YOUR GUT

Gather Evidence *There is no sunlight or oxygen inside your intestines. How do the bacteria and other organisms living there obtain energy and make ATP to perform cellular reactions?*

Molecules Involved in Respiration and Photosynthesis

There are many important molecules involved in cellular respiration and photosynthesis. Some may be familiar, such as CO_2, H_2O, and O_2, while others may seem more complex. Use the following list as a reference to help identify the roles of other molecules described in this section:

- **Energy Storage** Sugar molecules, such as the glucose produced by photosynthesis, store energy in stable forms.
- **Energy for Other Reactions** A phosphate group added to ADP (adenosine diphosphate) generates ATP (adenosine triphosphate) in the cycle shown in **Figure 6-19**. ATP provides usable energy for many chemical reactions within the cell, such as the reactions required in active transport.
- **Electron Carriers** NADH, NADPH, and $FADH_2$ carry high-energy electrons from one part of the photosynthesis and cellular respiration processes to another. These molecules have a cyclical pattern similar to ATP, except they gain and lose electrons and hydrogen ions (H^+). For instance, NADPH molecules become $NADP^+$, and vice versa.
- **Electron Acceptors** Electrons and hydrogen ions that are delivered by electron carriers ultimately combine with electron acceptor molecules. For example, O_2 accepts electrons and combines with hydrogen ions to form H_2O in aerobic cellular respiration.

FIGURE 6-19
A phosphate group is added to ADP to make ATP, which can be used to provide energy in a reaction. After ATP is used as part of a reaction, the ADP can be recycled and used again. P_i represents the free inorganic phosphate group that is added to ADP to make ATP (a). An ATP molecule has three phosphate groups, which reflect the "tri" in the molecule name. ADP only has two phosphate groups (b).

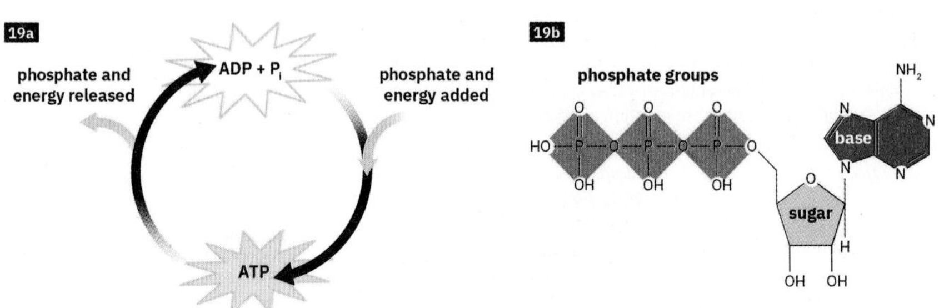

CCC Energy and Matter *Observe the ATP cycle represented in* ***Figure 6-19a****. Describe what is happening to the matter and energy in this system throughout the cycle.*

Cellular Respiration

All organisms use energy stored in sugars to power the reactions that sustain life. However, to use the energy stored in sugars, cells must first transfer it to ATP or other molecules that can participate directly in chemical reactions. In addition to generating ATP to power reactions in the cell, the process of cellular respiration also releases energy as heat. Because organisms lose energy to the environment constantly, these reactions are important to maintain ideal temperatures for all organisms.

Cellular respiration is the process that harvests energy from an organic molecule to store it in the usable form of ATP. Cells harvest energy from an organic molecule, such as glucose, by breaking it down one reaction at a time. This releases the energy of the molecule in small amounts that can be captured.

REVISIT THE ANCHORING PHENOMENON

Gather Evidence *Students might say that bacteria obtain energy from food products and waste that your body does not fully digest, or from other organisms. These organisms must process those food products through anaerobic respiration pathways to make ATP.*

Scientists have found nearly 2000 different species of bacteria in the human digestive system. Bacteria in the gut are typically bathed in a nutrient rich soup that consists of probiotics that increase the bacterial gut population, and prebiotics, molecules from undigested food that bacteria feed upon. However, even when a person is ill and may not have an appetite, the bacteria in their gut are still fed. The cells in intestines can release a certain sugar to keep the bacteria in the gut fed. The bacteria can use this sugar during their own anaerobic respiration process to make ATP. Ask students why intestinal cells would make nutrients for bacteria.

Cellular respiration that requires oxygen is called *aerobic cellular respiration*. With each breath, you take in oxygen for your trillions of cells. You ultimately exhale the products: carbon dioxide and water. Aerobic respiration harvests energy from glucose in a process that can be described in three major stages: glycolysis, the Krebs cycle, and the electron transport chain (**Figure 6-20**). In eukaryotes, glycolysis occurs in the cytoplasm, while the Krebs cycle and electron transport chain take place within the mitochondria.

The overall pathway can be modeled with this chemical equation:

$$\text{glucose} + \text{oxygen} \rightarrow \text{carbon dioxide} + \text{water} + \text{energy}$$
$$C_6H_{12}O_6 + O_2 \rightarrow CO_2 + H_2O + ATP$$

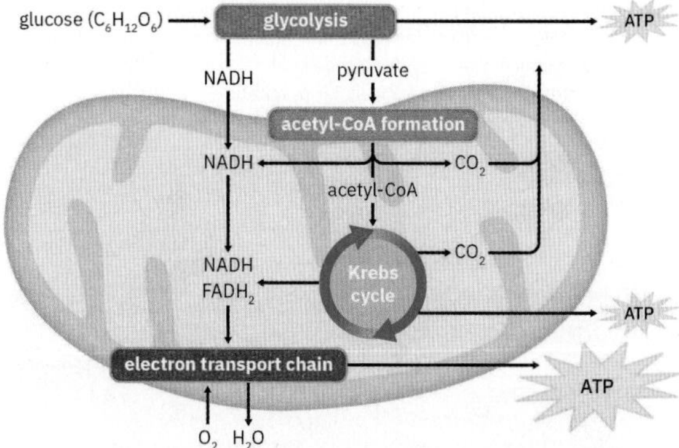

FIGURE 6-20 Trace the arrows to follow the movement of molecules through the steps of aerobic cellular respiration, noticing what reactants go into each step, and what products are generated.

SEP **Use a Model** *Compare the inputs and outputs of the chemical equation to the model shown in **Figure 6-20**. What are similarities and differences between the two models of aerobic cellular respiration?*

Glycolysis Glycolysis is a series of anaerobic reactions that produce ATP by splitting glucose (a six-carbon molecule) into two three-carbon molecules called *pyruvate*. The term *glycolysis* comes from *glyco*, which means "glucose," and *lysis*, which means "split." The pathway occurs in the cytoplasm of all cells, and it is the first step in both aerobic respiration and fermentation. The energy from this reaction is captured in electrons carried by NADH, and in high-energy phosphate bonds of ATP.

Acetyl-CoA Formation The two pyruvate molecules from glycolysis are not yet prepared for the next major stage. First the two pyruvate molecules move into a mitochondrion, where they are converted to acetyl-CoA, a molecule important in many reactions.

The Krebs Cycle The next stage of aerobic respiration, the *Krebs cycle*, releases energy from the acetyl-CoA molecule in a cyclical series of reactions. Energy released in these reactions is captured in electrons carried by NADH and $FADH_2$, and in high-energy phosphate bonds of ATP. All of the carbon atoms that were once part of glucose end up in CO_2 molecules, which diffuse out of the cell.

CHAPTER INVESTIGATION

Factors Affecting Cellular Respiration

What are some factors that affect cellular respiration in yeast?

Go online to explore this chapter's hands-on investigation to measure cellular respiration in yeast and observe how different factors affect that process.

Connect to English Language Arts

Use Digital Media Aerobic cellular respiration is a very complex process that takes place in the cytoplasm and mitochondria of the cell, but students may struggle with following the process using only **Figure 6-20**. Have them work in groups and use digital media to make an interactive or step by step model of the process shown in the image. Students should include a brief description of each of the steps in their models and share them with the class.

SEP **Use a Model** *Sample answer: Both models show inputs and outputs of aerobic cellular respiration. The chemical equation shows the general inputs and outputs for the entire process. The model in Figure 6-20 shows more detail, with inputs and outputs for each individual stage in the cellular respiration process.*

CHAPTER INVESTIGATION A

Guided Inquiry *Factors Affecting Cellular Respiration*

Time: 100 minutes over two days plus advance preparation

Students will follow a step-by-step procedure to investigate what happens when amylase is added to food containing starch.

Go online to access detailed teacher notes, answers, rubrics, and lab worksheets.

Visual Support

Electron Transport Chain ATP synthase is made up of a protein complex and studs the entire inner mitochondria. It acts like a reverse ion pump. The function of the electron transport chain is to generate and maintain the hydrogen gradient between the intermembrane space and the matrix, so that it can be used to generate ATP. To facilitate the learning of this complex process shown in **Figure 6-21**, group students into teams of 4-6 and have them write a short script where they act out the electron transport chain. Each student should have a speaking role and explain at least one portion of the process. Have each group act out their version of the electron transport chain for the class.

In Your Community

Fermented Foods Students may not realize that they have likely eaten foods produced through fermentation, including sourdough bread, some cheeses, yogurt, kimchi, and sauerkraut. Ask students whether they have eaten any of these products and have them describe the taste. Compare the descriptions of the tastes of different fermented food products. Share with students that byproducts of the fermentation process are what give these foods their sour tastes.

Electron Transport Chain and ATP Synthase The last step of aerobic respiration is the **electron transport chain** (**Figure 6-21**). In eukaryotes, this step occurs at the inner membrane of mitochondria. In prokaryotes, it occurs at infoldings of the inner cell membrane. Electrons and hydrogen ions are provided by the NADH and $FADH_2$ molecules generated by glycolysis and the Krebs cycle. Electrons flow through the electron transport chain and set up a hydrogen ion gradient that drives ATP synthesis. As the electrons move through the electron transport chain, they give up energy little by little. Molecules in the chain harness that energy to actively transport the hydrogen ions across the inner membrane, from the matrix to the intermembrane space.

The hydrogen ions pumped across the membrane by the electron transport chain pass back through a membrane channel that is part of the ATP synthase enzyme. As the hydrogen ions pass back along the gradient from high concentration to low concentration, they drive the formation of most of the ATP molecules formed during cellular respiration. Electrons and hydrogen ions in the intermembrane space combine with the electron acceptor molecule (O_2 in aerobic cellular respiration) to form water.

FIGURE 6-21 NADH and $FADH_2$ deliver electrons that provide energy to move hydrogen ions into the intermembrane space. Oxygen accepts the electrons and some hydrogen ions at the end of the chain, forming water. The concentration gradient drives hydrogen ions to pass back into the matrix, through the ATP synthase enzyme, which generates large amounts of ATP for the cell.

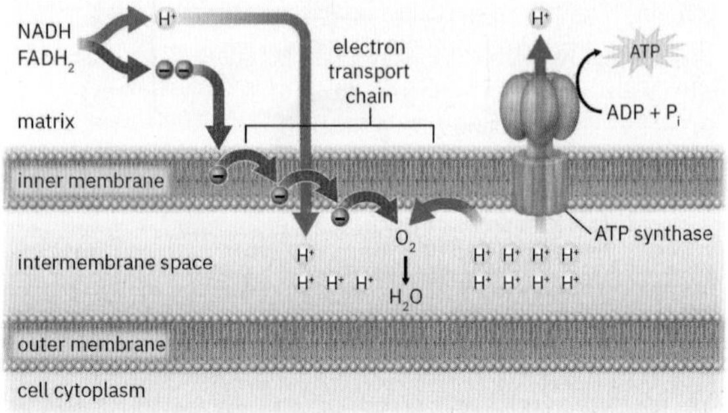

Other Energy Conversion Pathways

Fermentation Almost all cells can carry out **fermentation**, which is an anaerobic process similar to respiration that converts glucose to ATP that can be used by the cell. In fermentation, the overall gain in energy comes from glycolysis, and the rest of the process primarily serves to return molecules such as NAD^+ to glycolysis so more ATP can be generated. Fermentation generates much less ATP than respiration processes. Many cells can switch between aerobic respiration and fermentation as needed. For example, when muscle cells run out of oxygen, they switch to a fermentation pathway that generates lactic acid. As lactic acid accumulates (often during exercise), muscles feel fatigued.

Fermentation pathways vary greatly and are named after their end product. Lactate fermentation performed by beneficial bacteria is used to prepare many foods that involve lactic acid. Yogurt, for example, is made by allowing bacteria such as *Lactobacillus bulgaricus* and *Streptococcus thermophilus* to grow in milk. Alcoholic fermentation converts glucose to ethyl alcohol. A yeast species called *Saccharomyces cerevisiae* (**Figure 6-22**) helps produce bread and other products through alcoholic fermentation.

CROSSCUTTING CONCEPTS | Energy and Matter

Tracing Movement of Matter Have students analyze **Figure 6-21** and consider how each stage of the aerobic cellular respiration process begins with matter. Tracing the movement of matter helps us understand how different processes utilize the energy stored in the molecules of that matter. Remind students that while the energy reactions described in the text occur at a cellular level, they provide enough energy for organisms to move, breathe, and grow.

As students consider each part of the reactions that occur during aerobic respiration, encourage them to keep in mind the relationships between matter and energy that are occurring. Have students brainstorm the initial source of the energy found in the glucose. As students move through the section, they will learn about photosynthesis and how the sun is the initial source of energy that is captured by plants to make food.

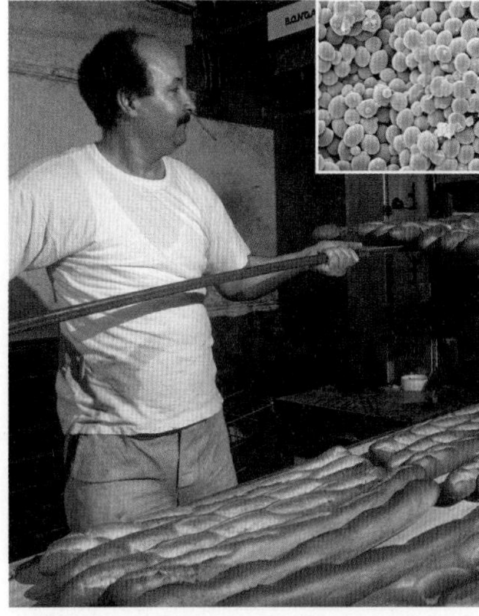

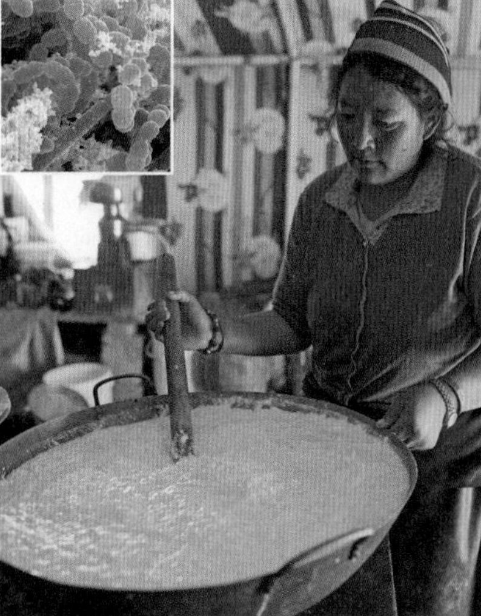

Anaerobic Cellular Respiration Early in the development of life on Earth, prior to photosynthesizing organisms, there was not an abundance of oxygen, and cellular respiration was all anaerobic. Now that Earth is an oxygen-rich environment, most organisms use aerobic respiration, but many bacteria and some other organisms still live in conditions with low or no oxygen. Anaerobic cellular respiration is cellular respiration that uses a molecule other than oxygen at the end of the electron transport chain.

FIGURE 6-22
Saccharomyces cerevisiae cells release CO_2 during fermentation, which forms pockets in bread. Yogurt is a product formed through lactate fermentation by *Lactobacillus bulgaricus* in milk. The scanning electron micrographs show these bacteria magnified by more than 2000 times.

Chemosynthesis In some ecosystems where there is no sunlight (**Figure 6-23**) photosynthetic organisms are not the primary producers of energy (Section 2.2). Chemosynthetic organisms get energy for assembly of their food molecules by removing electrons from inorganic substances such as hydrogen sulfide. Many of these organisms are bacteria and archaea that live in the deep ocean at hydrothermal vents or cold seeps.

 VIDEO

VIDEO 6-2 Go online to observe an example of an ecosystem that relies on chemosynthesis.

FIGURE 6-23
Bubbles of methane gas feed chemosynthetic bacteria at a cold seep at the Pascagoula Dome in the Gulf of Mexico. These chemosynthetic organisms support the growth of these mussels and other organisms, which then attract other animals to this deep-sea habitat.

Science Background

Anaerobes Students may wonder how ecosystems can exist in places where there is no sunlight. Organisms that can survive in these environments are typically known as extremophiles. Some of these organisms are anaerobes and are able to grow and flourish in environments that lack oxygen. These organisms go on to form the producer level of the ecosystem. There are many different species of anaerobes, and some can be found in the human body. Facultative anaerobes can grow in the presence or absence of oxygen. They can either grow aerobically, using oxygen, or anaerobically. Obligate anaerobes cannot survive in the presence of oxygen. Most obligate anaerobes are found deep in the ocean near hydrothermal vents and cold seeps. Instead of using oxygen as an electron acceptor, these organisms use other molecules at the end of the electron transport chain. Anaerobes near hydrothermal vents and cold seeps are also often chemosynthetic because these environments lack sunlight.

Chemosynthetic Organisms Students learn about the primary producers in ecosystems that lack light. **Chapter 2** introduced how energy and matter move through ecosystems.

6.3 PHOTOSYNTHESIS AND CELLULAR RESPIRATION 173

CHAPTER INVESTIGATION B

Guided Inquiry *Designing a Photobioreactor*

Time: 160 minutes over four days

Students will design an experiment to compare the effectiveness of enzyme-based detergents on protein-based waste in this forensics-oriented activity.

Go online to access detailed teacher notes, answers, rubrics, and lab worksheets.

SEP Use a Model *Sample answer: The model in Figure 6-24 shows that the light-dependent reactions have inputs of water and energy from light, and it has outputs of oxygen, NADPH, and ATP. The light-independent reactions have inputs of carbon dioxide, as well as the NADPH and ATP that come from the light-dependent reactions. Outputs of light-independent reactions include glucose as well as the NADP+ and ADP molecules that cycle back to the light-dependent reactions.*

Connect to Mathematics

Reason Quantitatively Have students review the chemical equation shown above **Figure 6-24**.

Write the equation on the board.

$$CO_2 + H_2O \rightarrow C_6H_{12}O_6 + O_2 + H_2O$$

Explain to students that this equation shows the overall process of photosynthesis, but it isn't a balanced chemical equation. Balanced chemical equations should have the same number of molecules of each element on either side of the reaction. Ask students to identify how the equation should be written so that it is balanced. Help students understand that since there are 6 carbon molecules on the product on the right that there should be 6 carbon dioxide molecules for the reactant. Help students reason quantitatively and arrive at the balanced equation,

$$6CO_2 + 6H_2O \rightarrow C_6H_{12}O_6 + 6O_2.$$

CHAPTER INVESTIGATION

Designing a Photobioreactor

How can we maximize oxygen and algae production in a photobioreactor?

Go online to explore this chapter's hands-on investigation to design a photobioreactor that can generate energy and oxygen for use during space travel.

Photosynthesis

Photosynthesis is a pathway that uses the energy of sunlight to drive the synthesis of sugar molecules such as glucose from carbon dioxide and water. The sugars can be stored as polysaccharides for later use, remodeled into other compounds, or broken down to release the energy in their bonds through cellular respiration. Plants and most other autotrophs that harvest energy from the sun by photosynthesis are known as *producers* (Section 2.2).

In plants, photosynthetic protists, and cyanobacteria, the light-dependent reactions are carried out by molecules embedded in a thylakoid membrane. The light-independent reactions run in stroma, the thick, cytoplasm-like fluid that fills the chloroplast. The thylakoid membrane is suspended in the stroma. Similar to the membranes involved in the electron transport chain of cellular respiration, these highly folded membrane structures in the chloroplast are critical for generating ATP.

VIDEO

VIDEO 6-3 Go online to find out more about photosynthesis from National Geographic Explorer Dr. Branwen Williams.

The structures of chloroplasts mentioned above are described in detail in Section 6.1. Chloroplasts are descendants of ancient cyanobacteria, so it is unsurprising that photosynthesis in eukaryotes is similar to photosynthesis in cyanobacteria. Modern cyanobacteria have thylakoid membranes that carry out the light-dependent reactions, and the light-independent reactions occur in the cytoplasm of these cells.

Photosynthesis is a complex pathway with many reactions. The details of the process are described in the Appendix. Photosynthesis can generally be represented by two major stages: light-dependent reactions and light-independent reactions (**Figure 6-24**).

The overall process of photosynthesis can be modeled by the following equation:

$$\text{carbon dioxide} + \text{water} \xrightarrow{\text{light}} \text{glucose} + \text{oxygen}$$

$$CO_2 + H_2O \xrightarrow{\text{light}} C_6H_{12}O_6 + O_2$$

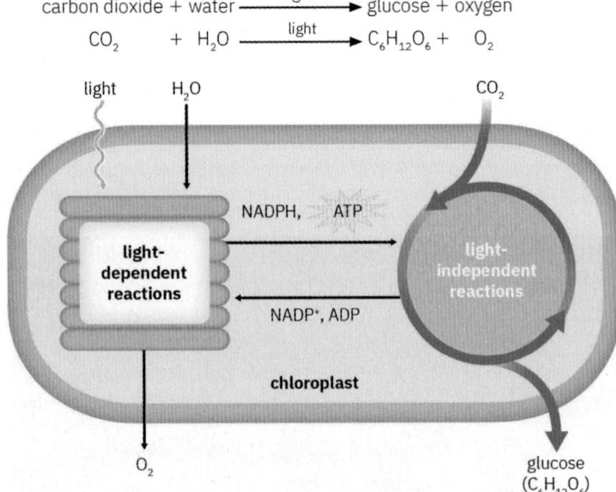

FIGURE 6-24 Light drives the production of ATP during the light-dependent reactions in the thylakoid membranes. NADPH and oxygen are also produced. In the light-independent reactions that take place in the stroma, NADPH and ATP from the light-dependent reactions drive sugar production.

SEP Use a Model *Use* ***Figure 6-24*** *to describe the inputs and outputs for each of the two major stages of photosynthesis.*

DIFFERENTIATED INSTRUCTION | Leveled Support

Struggling Students Some students may struggle with the differences between light-dependent reactions and light-independent reactions. To aid these students in understanding the differences, split the class into groups. Half of the groups should create posters identifying the reactants and products involved in the light-dependent reaction. These groups should also identify that these reactions can only occur under light because light is required to split hydrogen from oxygen. The other groups should each make a poster identifying the reactants and products involved in light-independent reactions. They should also identify the energy source for these reactions and explain why these reactions do not require light.

Advanced Learners Challenge students to identify the importance of the Calvin Cycle in carbon fixation. These students can make a digital media presentation or a video of their findings and share it with the class.

Light-Dependent Reactions The reactions of the first stage of photosynthesis are driven by light and called the *light-dependent reactions*. Visible light is a small part of all the electromagnetic energy radiating from the sun. Visible light powers photosynthesis, which begins when light is absorbed by photosynthetic pigments. Different pigments absorb different colors of light. The main photosynthetic pigment, chlorophyll a, absorbs violet and red light, so it appears green. Most photosynthetic organisms use a combination of pigments to capture more light for photosynthesis.

The light-dependent pathway splits water molecules and releases oxygen. Hydrogen ions and electrons from the water molecules are loaded onto $NADP^+$ to form NADPH. Similar to cellular respiration, the light-dependent reactions of photosynthesis use the harvested energy to move hydrogen ions across a membrane with an electron transport chain. The gradient of hydrogen ions powers the enzyme ATP synthase to make ATP. ATP and NADPH are used to power the light-independent reactions that follow.

Light-Independent Reactions The second stage of photosynthesis builds sugar molecules such as glucose from carbon dioxide and water. They are collectively called the *light-independent reactions* because light energy does not power them. Instead, they run on energy delivered by NADPH and ATP that formed during the first stage. At the end of the second stage, $NADP^+$ and ADP are recycled to work again in the reactions of the first stage. Like many chemical reactions, this process is aided extensively by enzymes, such as rubisco, which helps convert carbon dioxide to glucose. Rubisco is one of the most abundant proteins on Earth.

SEP Construct an Explanation *Why are organisms capable of photosynthesis known as "producers" within their ecosystems?*

6.3 REVIEW

1. **Define** Link the processes involving matter and energy to photosynthesis or cellular respiration.
   - Glucose molecules are produced.
   - Glucose molecules are broken down.
   - Light is converted to chemical energy.
   - Stored energy is converted into chemical energy.
2. **Describe** Identify and describe the roles of two important enzymes involved in photosynthesis.
3. **Define** Which processes can proceed in the absence of oxygen? Select all correct answers.
   A. chemosynthesis
   B. alcoholic fermentation
   C. lactic acid fermentation
   D. aerobic cellular respiration
   E. anaerobic cellular respiration

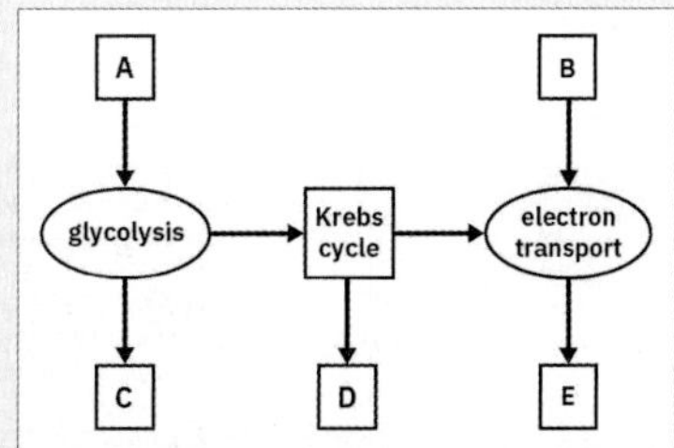

4. **Model** Complete a flowchart model of aerobic cellular respiration by labeling the inputs and outputs to each part of the process. You may use labels ATP, glucose, O_2, and CO_2 more than once.

Crosscurricular Connections

Physical Science Students should be familiar with the electromagnetic spectrum. Ask them to identify colors that they have seen in plants. Students may identify a variety of colors as some plant leaves may be green, yellow, pink, and purple, but the most common color called out will likely be green. Refer students to the chloroplast model in **Figure 6-24**. They should understand that chloroplasts are certain colors they are because they only absorb certain wavelengths of light that provide the energy for photosynthesis These pigments can change with the seasons. If time permits, share with students different graphs of the absorption spectra of chlorophyll. Ask students to use the graphs to determine the most likely colors of plant leaves.

SEP Construct an Explanation *Photosynthetic organisms produce the sugar molecules (glucose) that are needed for all organisms to run processes requiring energy in their cells.*

6.3 REVIEW

1. *Photosynthesis: Glucose molecules are produced. Light is converted to chemical energy. Cellular respiration: Glucose molecules are broken down. Stored energy is converted into chemical energy.* **DOK 1**
2. *ATP synthase enzyme is involved in the synthesis of ATP during photosynthesis. Rubisco is an enzyme that helps convert CO_2 into glucose.* **DOK 1**
3. *A, B, C, E* **DOK 2**
4. *A: glucose; B: O_2; C: ATP; D: CO_2 and ATP; E: ATP* **DOK 3**

EXPLORER
ANDRIAN GAJIGAN

About the Explorer

National Geographic Explorer Andrian Gajigan works to study algal blooms and the involvement of viruses in the population of dinoflagellates. His research involves determining which viruses keep dinoflagellate populations in control and which viruses are beneficial to dinoflagellates and other marine algae and encourage their growth. You can learn more about Andrian Gajigan and his research on the National Geographic Society's website: www.nationalgeographic.org.

Science Background

Dinoflagellate Dinoflagellates are single-celled eukaryotes that exist in aquatic environments. Dinoflagellates make up a portion of the phytoplankton population in the ocean, an important food source for many animals. These organisms have characteristics of both plant and animal cells, but aren't a part of either group, as they are considered a kind of algae. When they grow uncontrollably, they can negatively impact the local ecosystem.

Connect to Careers

Marine Virologist Most people are familiar with infectious disease scientists who study the viruses that cause Ebola and HIV but not all viruses infect humans. Marine virologists may study infectious diseases that can infect animal, plant, and algal populations in the ocean. They may also study rare viruses that are found in the ocean. This is a growing field of study.

THINKING CRITICALLY

Explain *Through photosynthetic processes, algae capture the sun's energy, passing it along to the other single-celled organisms and animals that eat them. While blooms can be beneficial, harmful ones are detrimental to people and ocean life, and lead to death through toxins or oxygen depletion due to decomposition.*

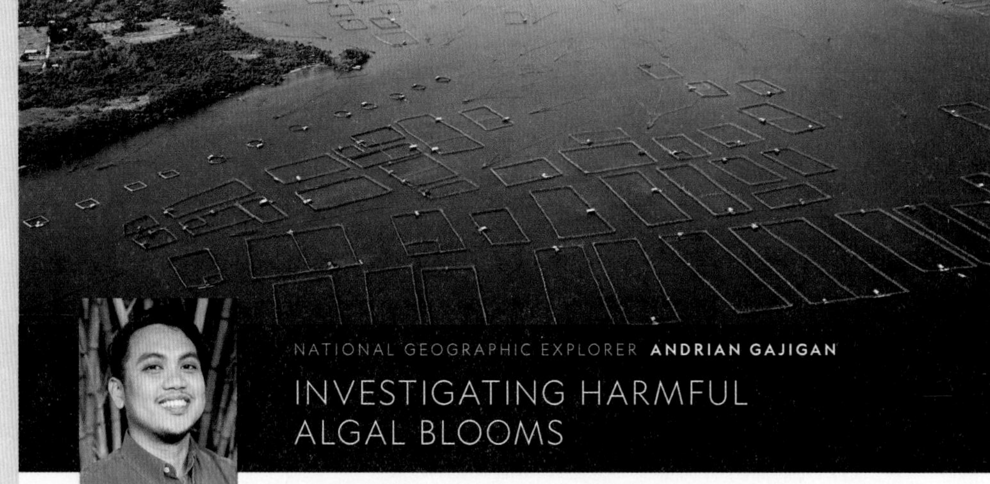

NATIONAL GEOGRAPHIC EXPLORER **ANDRIAN GAJIGAN**

INVESTIGATING HARMFUL ALGAL BLOOMS

Red tides occur worldwide, but Dr. Andrian Gajigan is concentrating on those in the Asia-Pacific region.

In the first half of 2021, the Philippines Bureau of Fisheries and Aquatic Resources issued more than 20 shellfish alerts, recommending people avoid eating shellfish due to red tide toxins. If ingested, these toxins can be debilitating or even fatal for humans and ocean life. How can such red tide events be reduced or stopped? Biochemist Dr. Andrian Gajigan is trying to determine the answer.

Growing up in the Philippines, Gajigan witnessed firsthand the damage caused by harmful algal blooms (HABs), sometimes known as red tides. HABs form when rapid growth, or blooming, of single-celled organisms turns water red. In the Philippines, single-celled dinoflagellates and diatoms are responsible for HABs. Of the 2000 species of known dinoflagellates, roughly one-half are photosynthetic but only about 20 produce toxins.

Algal blooms can be beneficial as the photosynthesizing organisms fuel the ocean food web. When toxic species bloom, however, the toxins move through the food chain causing fish to die and shellfish to become dangerous to eat. The amount of photosynthesis occurring during blooms of toxic and nontoxic algal species does increase oxygen in the water. However, when large quantities of photosynthetic organisms die and decompose, so much oxygen is depleted from the water that many fish and other animals either leave the area or die because not enough oxygen is available to carry out cellular respiration.

Gajigan is investigating the role of viruses in red tides, focusing on those that infect dinoflagellates and other marine algae. Often, viruses keep populations of dinoflagellates in check. But viruses can also form symbiotic relationships with cells and give them a survival advantage. By studying the viruses, Gajigan is trying to figure out what encourages blooming algal growth and how it might be slowed or stopped.

Gajigan sees his research as an intersection of his fascination with science and community activism in his home country. Outside the lab, Gajigan encourages equity and inclusion in STEM fields. He wants to establish a research program in the Philippines to train the next generation of explorers and lead projects across the Asia-Pacific region as a way of staying connected to his roots.

THINKING CRITICALLY

Explain *Why are marine algae important in the ocean food web? Why would finding a way to slow or stop harmful algal blooms be important to society as well as to the environment?*

DIFFERENTIATED INSTRUCTION | English Language Learners

Making Connections Between Ideas Have students explain the connection between Andrian Gajigan's firsthand observations of algal blooms as a child and his work today.

Beginning Have students work in pairs to answer these questions: *What did Gajigan see as a child? What does he study today? What does he want to learn?* They can denote phrases in the text that helped them answer.

Intermediate Have pairs complete this sentence frame: *Gajigan most likely studies ___ today because he ___ as a child.* Have partners take turns reading aloud sentences from the text that support their ideas.

Advanced Have students explain why Gajigan studies the role of viruses in red tides today, drawing evidence from the text to support their answer. Encourage them to elaborate on why red tides are a problem.

TYING IT ALL TOGETHER ENGINEERING
ARTIFICIAL CELL TECHNOLOGY

HOW DO STRUCTURES WORK TOGETHER IN CELL SYSTEMS?

In this chapter you learned about parts of cells and how those structures carry out particular functions in both prokaryotic and eukaryotic cells. Many scientists, inspired by the natural process of photosynthesis and its mechanisms in nature, have started to develop different methods of artificial or synthetic photosynthesis.

Photovoltaic cell technology has been used to build solar panels and solar arrays for decades (**Figure 6-25**). Over time, there have been major improvements in efficiency. However, there are still significant costs and material wastes associated with the technology. Researchers think artificial photosynthetic approaches can lead to more efficient and flexible modes of converting sunlight to usable energy by using materials that are readily available as parts of living cells. Although there are other carbon-based molecules that may be made by artificial photosynthesis pathways, the primary goal of many researchers is to use light energy to split water molecules and make hydrogen fuel.

Like other aspects of artificial cell research, studies of artificial photosynthesis include a wide variety of approaches. These approaches include harvesting chloroplasts for use in the construction of artificial cells and building synthetic cells that use photosynthetic pathways not present in nature (**Figure 6-26**).

In this activity you will research artificial photosynthesis and explain how it could be used to improve existing energy technologies.

Analyze the Problem

1. Scientists think artificial photosynthesis could be the key to the next generation of renewable energy technologies. Research how artificial photosynthesis works. Research at least two different approaches used by scientists to develop artificial photosynthesis systems.

Evaluate Solutions

2. Describe some current limitations of artificial photosynthesis. Compare artificial photosynthesis with other current renewable energy technologies, such as photovoltaic cells used in solar arrays. Describe some advantages that might come from further developments in artificial photosynthesis.

Science, Technology, and Society

3. Write an opinion article about why continually developing new and more efficient energy technologies is important for society and the environment.

FIGURE 6-25
An aerial photograph shows the extensive layout of photovoltaic cells built into rugged terrain in China.

FIGURE 6-26
Synthetic cells from the U.S. Department of Energy Joint Genome Institute contain tiny chloroplasts.

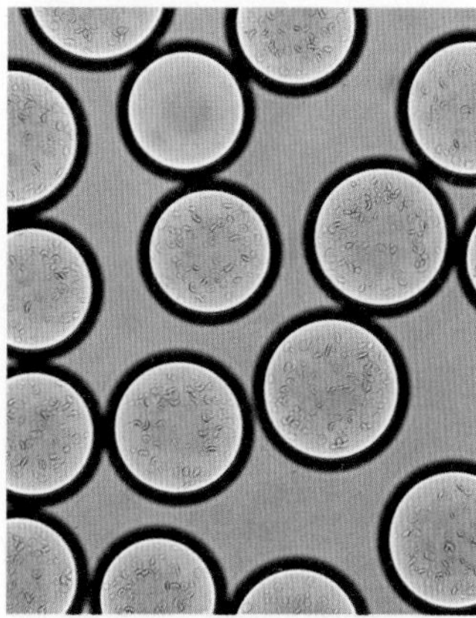

TYING IT ALL TOGETHER **177**

TYING IT ALL TOGETHER

ELABORATE

The Case Study discusses how scientists are working on developing artificial cells and how they hope to use them to develop different methods to harness the power of photosynthesis. In the Tying It All Together activity, students research artificial photosynthesis and are expected to explain how it can be used to improve existing energy technologies.

This activity is structured so that students may complete the steps once they have completed the lesson. **Step 1** has students obtaining information on how artificial photosynthesis works and two approaches used by scientists to develop artificial photosynthetic systems. In **Step 2**, students will describe current limitations in artificial photosynthetic systems and compare it with other renewable energy technologies that are currently in use. In **Step 3**, students write an opinion article on the importance of developing more efficient energy technologies on the environment and society.

This activity would be best used as a culminating project at the end of the chapter.

Go online to access the Student Self-Reflection and Teacher Scoring rubrics for this activity.

SCIENCE AND ENGINEERING PRACTICES
Constructing Explanations and Designing Solutions

Reliable Energy Information Resources To aid students with the Case Study, encourage them to collect a few resources that they can use to start their own research. Stress to students that when they are conducting research to construct explanations, they should make sure to always use reliable sources. Students might begin their research on artificial cells and photosynthesis using the National Institute of Standards and Technology (www.nist.gov), Nature (www.nature.com), Brookhaven National Laboratory (www.bnl.gov), and the National Science Foundation (www.nsf.gov). Useful sources for information on renewable energy technologies include U.S. Department of Energy (www.energy.gov), U.S. Energy Information Administration (www.eia.gov), National Renewable Energy Laboratory (www.nrel.gov), and the International Renewable Energy Agency (www.irena.org).

CHAPTER 6 REVIEW

EVALUATE

REVIEW KEY CONCEPTS

1. *B, E* **DOK 1**
2. *nucleus, ER, chloroplast, cytoskeleton, vesicle* **DOK 1**
3. *Eukaryotic: animals, fungi, plants; Prokaryotic: archaea, bacteria* **DOK 2**
4. *C* **DOK 3**
5. *D* **DOK 2**
6. *Sample answer: Both NADPH and NADH act as electron carriers between different stages.* **DOK 2**
7. *Sample answer: They will reform a lipid bilayer with the phosphate heads oriented toward the aqueous environment and the lipid tails oriented together in the middle, away from the aqueous environment.* **DOK 2**
8. *C* **DOK 1**
9. *enzyme: facilitates chemical reactions, transport protein: controls entry of substances into the cell, receptor protein: senses signals from the extracellular environment, adhesion protein: attaches a cell to surrounding cells* **DOK 1**
10. *Sample answer: The molecules that make up the membrane are able to move around freely within the membrane, which is why they are described as being fluid. The membrane also has a variety of molecules embedded in it, including proteins, cholesterol, glycoproteins, glycolipids, and phospholipids. The diversity of the different molecules form a mosaic – lots of different pieces coming together to form a whole.* **DOK 3**
11. *B* **DOK 2**

CHAPTER 6 REVIEW

REVIEW KEY CONCEPTS

1. **Identify** Select the characteristics that describe a typical eukaryotic cell but NOT a typical prokaryotic cell. Select all correct answers.
   A. capable of dividing
   B. contains a nucleus
   C. has a cell membrane
   D. contains DNA in the form of a plasmid
   E. has complex membrane-enclosed organelles

2. **Identify** Name each of the described cellular structures.
   - contains a eukaryotic cell's DNA
   - has two forms: rough and smooth
   - is an organelle where the reactions of photosynthesis take place
   - is made up of microfilaments, intermediate filaments, and microtubules
   - is a membrane-enclosed sac used to transport materials from one organelle to another

3. **Classify** Identify each as prokaryotic or eukaryotic.
   - animals
   - archaea
   - bacteria
   - fungi
   - plants

4. **Draw Conclusions** With the help of transport proteins, many glucose molecules produced in a leaf cell during photosynthesis are moved into other cells of the plant. What type of membrane transport would be used to do this?
   A. endocytosis
   B. exocytosis
   C. facilitated diffusion
   D. osmosis

5. **Predict** Which stage of cellular respiration or photosynthesis would be unable to begin without carbon dioxide?
   A. glycolysis
   B. Krebs cycle
   C. light-dependent reaction
   D. light-independent reaction

6. **Compare** How is the role of NADPH in photosynthesis similar to the role of NADH in cellular respiration?

7. **Predict** What will happen to the phospholipid molecules in a lipid bilayer if they are disrupted and broken apart within an aqueous environment?

8. **Describe** Which stage of cellular respiration takes place in the cytoplasm of the eukaryotic cell?
   A. acetyl-CoA formation
   B. electron transport chain
   C. glycolysis
   D. Krebs cycle

9. **Label** Determine which type of membrane-embedded protein—enzyme, adhesion protein, receptor protein, or transport protein—is described by each phrase below.
   - facilitates chemical reactions
   - attaches a cell to surrounding cells
   - controls entry of substances into the cell
   - senses signals from the extracellular environment

10. **Explain** Why is the term *fluid mosaic* used to describe cellular membranes?

11. **Analyze** In the absence of oxygen, a certain organism is still able to produce ATP from glucose. Which process is the organism likely utilizing?
    A. chemosynthesis
    B. fermentation
    C. Krebs cycle
    D. light-independent reaction

CRITICAL THINKING

12. *Sample answer: The ability to have separate compartments allows the cell to carry out chemical reactions and store substances in specialized areas. This can increase efficiency and allows eukaryotic cells to diversify their activities.* **DOK 3**
13. *A folded membrane creates more surface area. With a folded membrane, there is more room for the important enzymes and structures involved in photosynthesis and cellular respiration. This allows the organelle to produce more.* **DOK 3**
14. *Increased quality of microscopic images has allowed us to observe much more detail and thus increased our understanding of the diversity of cells and the complexity of their structure. In addition, the ability for many scientists to make observations over hundreds of years has greatly increased the amount of knowledge we have about cells.* **DOK 3**
15. *Both breathing and cellular respiration use oxygen and release carbon dioxide.* **DOK 3**
16. *Both structures are membrane-bound organelles that contain substances at a higher concentration than is found in the cytoplasm. They can be distinguished by function. Vesicles play a key role in transport of substances between organelles and to the exterior of the cell, while vacuoles are more important for storage. Vacuoles are often larger and longer lasting than vesicles. Large vacuoles storing water can be found in many plant cells.* **DOK 2**

CRITICAL THINKING

12. **Explain** What advantage do eukaryotic cells derive from having membrane-enclosed organelles, which are lacking in prokaryotic cells?

13. **Explain** Both chloroplasts and mitochondria have complex inner membranes with a high number of folds. What is the advantage of having folded, convoluted inner membranes?

14. **NOS Scientific Knowledge** Antoni van Leeuwenhoek made incorrect assumptions as he observed microscopic organisms for the first time. How have advancements in microscopy increased our understanding of cells?

15. **Explain** Respiration can refer to gas exchanges that occur when breathing air in and out of the lungs, as well as to the chemical reactions that take place in mitochondria. What similarities are there between these two processes that make the use of the word *respiration* appropriate for both?

16. **Compare** What distinguishes vacuoles from other types of vesicles?

17. **Explain** Describe how carbon cycling in ecosystems relates to the processes of photosynthesis and cellular respiration.

MATH AND ENGLISH LANGUAGE ARTS CONNECTIONS

1. **Synthesize Information** The words *hypotonic* and *hypertonic* are used in this chapter to describe solute concentration. Name some other words that use the prefixes *hypo-* and *hyper-*. How do the definitions of those words relate to the meanings of hypotonic and hypertonic?

2. **Model With Mathematics** Sketch a graph to show how the amount of glucose produced during photosynthesis would change over the course of a year in a tree that loses its leaves in the winter. Describe the shape of the graph.

3. **Reason Quantitatively** The cellular structures described in this chapter are very small and could not be observed until microscopes were invented. Cells vary in size, but many eukaryotic cells are about 10 micrometers (µm) in diameter. The human eye can only see objects as small as about 0.1 millimeter (mm) in size. Assuming that the average size of a eukaryotic cell is 10 µm, how many eukaryotic cells, end to end, could fit onto a 0.1-mm line?

Read the following excerpt from the writings of Antoni van Leeuwenhoek about his early observations of microorganisms. Use evidence from the passage to answer the question.

I then most always saw, with great wonder, that in the said matter there were many very little living animalcules, very prettily a-moving. The biggest sort ... had a very strong and swift motion, and shot through the water ... like a pike does through the water. The second sort ... oft-times spun round like a top ... the other animalcules were in such enormous numbers, that all the water ... seemed to be alive.

4. **Draw Evidence From Texts** What did van Leeuwenhoek find most interesting about his observations? How does the focus of his observations differ from what a microbiologist might focus on today?

REVISIT BACTERIA IN YOUR GUT

Gather Evidence In this chapter, you learned bacterial cells have structures and mechanisms that allow them to survive and obtain energy inside the intestines. However, dietary changes and medications can drastically change the types and amounts of bacteria growing in the human gut and can sometimes lead to illness.

1. Sometimes people taking antibiotics experience an increase in pathogenic bacteria populations in their gut and develop gastrointestinal illness as a result. How do you think antibiotics might cause this?

2. What other questions do you have about how changes in the variety of gut bacteria can negatively affect human health?

REVISIT THE ANCHORING PHENOMENON

1. *Sample answer: Antibiotics kill some bacteria better than they kill others. If an antibiotic kills part of the gut bacteria population, other species that cause illness might grow more and make the person sick.* **DOK 2**

2. *Sample answer: Do pathogens come into the gut from outside, or are they a part of the normal bacterial populations? How many of the bacteria in the gut would be pathogens if they were on their own? Why do they not cause illness under normal circumstances?* **DOK 3**

17. *Sample answer: Vesicles are membrane-enclosed organelles that contain substances at a higher concentration than is found in the cytoplasm. Generally, vesicles play a key role in transport of substances between organelles and to the exterior of the cell. A vacuole is a type of vesicle that is specifically important for storage. Vacuoles are often larger and longer lasting than other vesicles. Large vacuoles storing water can be found in many plant cells.* **DOK 3**

MATH AND ELA CONNECTIONS

1. *Hypothermia and hypodermic are similar to hypotonic in that they refer to being less or below. Hypothermia means a lower body temperature, hypodermic means beneath the skin, and hypotonic means there is a lower solute concentration. Hyperactive and hyperbole are similar to hypertonic in that they refer to being more than or above. Hyperactive means very active, hyperbole means exaggerating to more than the truth, and hypertonic means higher solute concentration.* **DOK 4**

2. *Sample answer: The graph shows low production of glucose in the winter months when the tree does not have leaves and high glucose production in the summer when the tree has leaves.* **DOK 2**

3. *Ten 10-µm cells would fit into 0.1 mm* **DOK 3**

4. *van Leeuwenhoek seems most interested in the large number of organisms he observes, and how they are moving. He repeatedly describes their motion and speed. He says 'an unbelievably great company,' indicating he is amazed by the number of organisms. A microbiologist today might be interested in those things, but might also focus on the shape and structures of the cells, in an effort to identify them.* **DOK 3**

Three-Dimensional Learning

The practices, core ideas, and crosscutting concepts presented in this chapter's text, investigations, and resources provide support to address the following Performance Expectations: **HS-LS1-1** and **HS-LS1-4.**

Science and Engineering Practices	Disciplinary Core Ideas	Crosscutting Concepts
Asking Questions and Defining Problems Developing and Using Models (HS-LS1-4) Constructing Explanations and Designing Solutions (HS-LS1-1)	**LS1.A:** Structure and Function (HS-LS1-1) **LS1.B:** Growth and Development in Organisms (HS-LS1-4) **LS3.A:** Inheritance of Traits	Systems and System Models (HS-LS1-4) Structure and Function (HS-LS1-1)

Contents		Instructional Support for All Learners	Digital Resources
ENGAGE			
180–181	CHAPTER OVERVIEW **CASE STUDY** How do cells divide and grow?	**Social-Emotional Learning** CCC Systems and System Models **On the Map** Baltimore, Maryland **English Language Learners** Summarize	
EXPLORE/EXPLAIN			
182–187 DCI LS1.A LS1.B	**7.1** CELL CYCLES • Compare and contrast the prokaryotic and eukaryotic cell cycle. • Explain the importance of eukaryotic cell cycle regulation.	**Vocabulary Strategy** Prefixes/Suffixes/Root Relationships CER Revisit the Anchoring Phenomena **Connect to Mathematics** Exponential Growth SEP Using Mathematics and Computational Thinking **English Language Learners** Roots and Affixes **In Your Community** Cancer Prescreening **Social-Emotional Learning** Social Awareness **Differentiated Instruction** Leveled Support	*Video 7-1* **SIMULATION** Eukaryotic Cell Cycle *Video 7-2*
187	**LOOKING AT THE DATA** IDENTIFYING GENE MUTATIONS IN CANCER CELLS		
188–192 DCI LS1.B LS3.A	**7.2** MITOSIS • Describe chromosomes and the structure of DNA. • Sequence the stages of mitosis and cytokinesis.	CCC Scale, Proportion, and Quantity **Vocabulary Strategy** Prefixes/Suffixes/Root Relationships **CASE STUDY** Ask Questions **Differentiated Instruction** Leveled Support **Vocabulary Strategy** Word Wall **Differentiated Instruction** Students with Disabilities **Address Misconceptions** Phases of Mitosis	*Video 7-3* **Investigation A** Plant Growth Through Mitosis (50 minutes plus advance preparation)

Contents		Instructional Support for All Learners	Digital Resources
EXPLORE/EXPLAIN			
192	**MINILAB** MODELING MITOSIS		
193–197 **DCI** LS1.A LS1.B	**7.3 CELL DIFFERENTIATION** • Explain how cell differentiation results in different morphological structures in multicellular organisms. • Discuss the role of pluripotent stem cells in cell differentiation. • Apply the concept of cell differentiation to three types of tissues in plants. conversion pathways available to cells.	**Vocabulary Strategy** Prefixes/Suffixes/ Root Relationships **Visual Support** Specialized Cells **English Language Learners** Using Academic Language **BIOTECHNOLOGY FOCUS** Stem Cell Research **Connect to English Language Arts** Draw Evidence from Text **In Your Community** Pinching Plants **On Assignment** Capturing the Eyes of Organisms	**Chapter Investigation B** Cell Differentiation in Plant Leaves (50 minutes)
198	EXPLORER CAMILA ESPEJO	**Connect to Careers** Wildlife Veterinarian	
ELABORATE			
199	**TYING IT ALL TOGETHER** How do structures work together in cells?	**Differentiated Instruction** Economically Disadvantaged Students	Self-Reflection and Scoring Rubrics
Online	**Investigation A** Plant Growth Through Mitosis **Investigation B** Cell Differentiation in Plant Leaves	Guided Inquiry (50 minutes plus advance preparation) Guided Inquiry (50 minutes)	MINDTAP Access all your online assessment and laboratory support on MindTap, including: sample answers, lab guides, rubrics, and worksheets. PDFs are available from the Course Companion Site
EVALUATE			
200–201	**Chapter 7 Review**		
Online	**Chapter 7 Assessment** **Performance Task 1** *How Does Regenerative Medicine Reflect Nature?* (HS-LS1-4)		

CHAPTER 7

ENGAGE

Students should have an understanding of cell structure and function from Chapter 6, as well as a basic understanding of mitosis from middle school. This chapter builds on that knowledge with a focus on cell division and its regulation through the cell cycle. Through an understanding of cell cycle regulation, students can begin to process how cells, given the same genetic information, can develop into different systems with different functions.

About the Photo This photo was taken using fluorescence microscopy where parts of the cell are molecularly labeled with a fluorescent dye, in this case DNA and microtubules. Encourage students to use the caption to identify these structures and say whether they are in the cells' nuclei or cytoplasm. Then, engage students in this chapter's content by eliciting comparisons between the center cell and those surrounding it and asking what the center cell may have to do with cell growth.

Social-Emotional Learning

As students progress through the chapter, invite them to look for contexts and opportunities to practice some of the five social and emotional competencies. For example, as students expand their knowledge of cell division, the concepts become more complex. Students can practice **self-management skills** when encountering a concept they do not understand to avoid feelings of frustration. Encourage the use of **relationship skills** so students learn to take initiative in asking for help, or giving help to another student in need.

In addition, as students learn about disruptions in the cell cycle that can cause cancer, remain thoughtful that some students may be experiencing similar situations in their own family or social circles. Remind students to practice **social awareness** by demonstrating concern for the feelings of others through empathy and compassion.

CHAPTER 7

CELL GROWTH

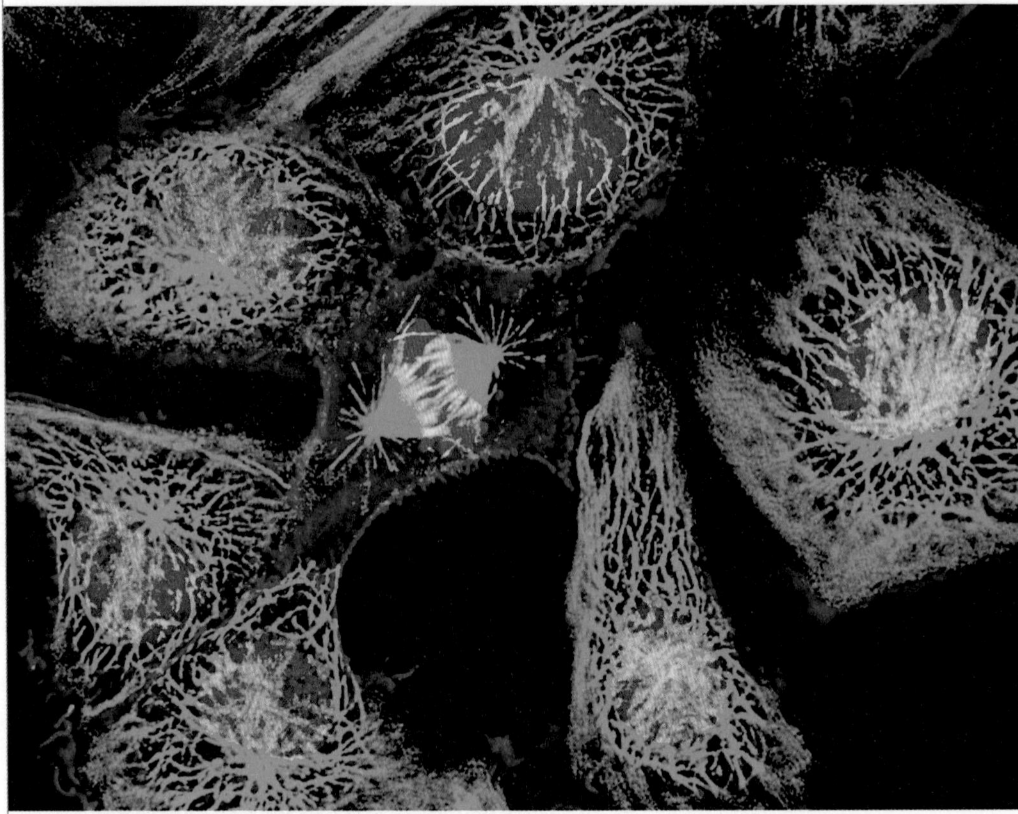

Fluorescent dyes brightly highlight DNA (yellow/orange) and microtubules (green) in cells originally from the kidney of an African clawed frog. The center cell is undergoing mitosis.

For multicellular organisms to grow and survive, their cells must continually replicate and divide. But when this process doesn't occur in an orderly manner, things can quickly go awry.

Chapter 7 supports the NGSS Performance Expectations **HS-LS1-1** and **HS-LS1-4**.

180 CHAPTER 7 CELL GROWTH

CROSSCUTTING CONCEPTS | Systems and System Models

Cellular Systems This chapter provides opportunities to apply the concepts of systems and system models. Students should be focusing on the functions of different cell parts that work together as a whole to keep the cell alive. Have students work in groups and think of one familiar object that is a system and answer questions such as: What parts make up the object and what is their function, does each part have its own properties and are they the same as the whole object, or can any one part carry out the job of the whole object?

When learning about prokaryotic and eukaryotic cell cycles, students will focus on a single-celled organism as a system. Ask them to relate a single-celled organism to the everyday object they just discussed using the same questions as guiding points. When moving on to eukaryotic cells, point out that cells in multicellular organisms perform specialized functions and operate as part of a larger system.

CASE STUDY

IMMORTAL CELLS: THE STORY OF HENRIETTA LACKS

HOW DO CELLS DIVIDE AND GROW?

In February 1951, the onset of puzzling symptoms led 31-year-old Henrietta Lacks to visit a doctor at Johns Hopkins Hospital in Baltimore, Maryland. Lacks was diagnosed with what would turn out to be an aggressive form of cervical cancer. She would be dead within the year, leaving behind her husband and their five children.

Without obtaining Lacks's consent—a common practice at the time—a biopsy of her cervical tissue was given to Dr. George Gey's tissue lab at Johns Hopkins. There, the scientists discovered that, unlike most other cells, the cells from her tissue sample doubled every 20–24 hours. This discovery led to the cultivation of the first line of living human cells. Although Lacks died on October 4, 1951, her immortal cells continue to live on today.

These cells were named "HeLa" cells, using the first two letters of her first and last names, although her true identity was not released to the public until many years later. Many scientists consider this line of cells to be one of the most important biomedical breakthroughs of the 20th century. In addition to an integral role in the development of the polio vaccine, HeLa cells have also been used to study a variety of viruses and diseases, including human immunodeficiency virus (HIV) and tuberculosis. They have also been used to study the effects of hormones, toxins, and drugs on the growth of cancer cells. The discovery that a certain stain could be used on HeLa cells to make their chromosomes visible led to breakthroughs in genetic medicine. This technique allowed scientists to find genetic links to diseases such as Down syndrome and led to the development of genetic screening tests such as amniocentesis.

Two Nobel Prizes have been awarded to scientists who used HeLa cells in their scientific investigations, including research on human papillomavirus (HPV), a virus that causes cervical cancer. HeLa cells have been used in tens of thousands of published studies. Trillions of HeLa cells are in use as a part of scientific research around the world. HeLa cells have even been sent into space to help determine the effects of low gravity and radiation on cell growth.

Ask Questions *As you read, generate questions about why controlled cell division is important for the growth and maintenance of organisms. Consider what makes HeLa cells different from other cells.*

FIGURE 7-1
Cells taken from Henrietta Lacks (1920–1951) were used to develop HeLa cell lines.

FIGURE 7-2
A scientist at the Curie Institute in Paris, France, observes HeLa cells using fluorescent microscopy to visualize the location of a specific protein, shown in red on the computer screen.

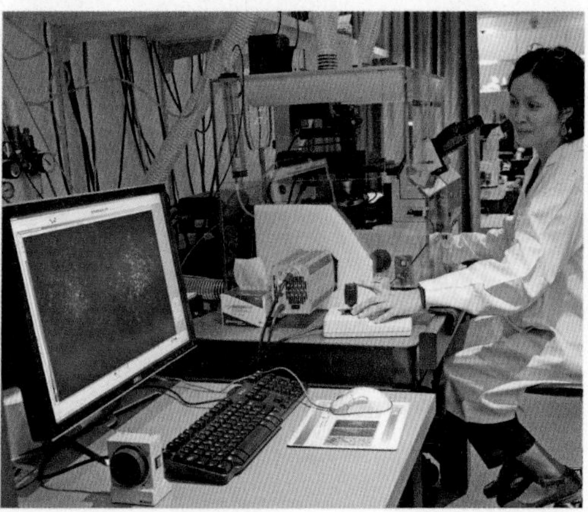

DIFFERENTIATED INSTRUCTION | English Language Learners

Summarize Have students work in small groups to read and summarize the article.

Beginning Provide sentence frames to help students write a brief summary: *Scientists studied Henrietta Lacks's ______. The cells ______ every 20-24 hours. HeLa cells help scientists study ______ and ______.* If needed, provide a word bank (*cancer cells, doubled, viruses, diseases*).

Intermediate Provide sentence frames to help students summarize each paragraph: *Henrietta Lacks died from ______. A lab ______ and discovered that ______. HeLa cells helped scientists study ______ and find ______. Today, scientists ______.*

Advanced Provide questions to guide students' summary: *Where did HeLa cells come from? Why are they important?* Encourage students to answer each question with at least two sentences.

CASE STUDY

ENGAGE

This Case Study, focusing on HeLa cells, relays a story about a person who made significant contributions to science without even knowing it. It serves as an anchor for content development in this chapter as students will use this Case Study as a springboard for learning about the cell cycle and its regulation, cell division, and medical advancements using HeLa cells that thrived because of unchecked cell division.

The story of Henrietta Lacks also provides the context for discussions about medical ethics then and now, including informed consent. Students may place this Case Study in the context of the early 1950s when cell biology was very much a work in progress and the issue of consent was not a legal question. Encourage students to ask questions about how consent laws have changed and how that has affected generations of Lacks's family.

Ask Questions Students should revisit the Case Study as they read the chapter to make connections with the content. See a specific suggestion in Section 7.2.

On the Map

Baltimore, Maryland The Johns Hopkins Hospital was established by a wealthy philanthropist, Johns Hopkins, as an institution that would provide care for anyone, regardless of gender, age, or race. Today, it has hospitals located throughout Maryland and is world renowned for its biomedical research facility and patient care.

Human Connection Ask students if they know someone affected by cancer, Parkinson's disease, HIV, or COVID-19. If so, they have been touched by the contribution of Henrietta Lacks. Lacks's cells have led to many life-saving medical breakthroughs such as the 1954 Nobel-Prize winning polio vaccine. HeLa cells have also contributed to the study of treatments for hemophilia, leukemia, and Parkinson's disease. They were instrumental in identifying the infectiousness of the SARS-CoV-2 (COVID-19) virus. Suggest that interested students research and create a time line of the breakthroughs attributed to HeLa cells.

7.1

CELL CYCLES

LS1.A, LS1.B

EXPLORE/EXPLAIN

This section begins with a look at the prokaryotic cell cycle and the conditions for growth rate before introducing the eukaryotic cell cycle, including an overview of each part of the cycle and regulatory mechanisms. Finally, apoptosis is discussed as part of the process used to replace or remove cells.

Objectives

- Compare and contrast the prokaryotic and eukaryotic cell cycle.
- Explain the importance of eukaryotic cell cycle regulation.

Pretest Use **Questions 1, 2,** and **4** to identify gaps in background knowledge or misconceptions.

Analyze *Students might say that people grow 12 inches in 4 years before they start high school, but the size of their ears or fingers does not change very much. Or they may point out how quickly their hair grows. Cells in each of these parts of the body are not dividing at the same rate.*

Vocabulary Strategy

Prefixes/Suffixes/Root Relationships Knowing recurring affixes in biology terms can help students figure out meaning. For example, the suffix *–sis* means an action or process. The prefix *cyto-* refers to the cell.

Suggest that students make a chart with affixes they find in the vocabulary and list the meaning beside each. They may add to this chart as new affixes are encountered through the chapter. For example, the suffix *–phase* means a distinct stage in a series of events and is used often in this chapter.

Video

Time: 0:35

Use **Video 7-1** of *E. coli* to discuss the prokaryotic cell cycle. Tell students that the footage has been speeded up 440 times and that *E. coli* can divide every 20–30 minutes in ideal conditions. Have students posit explanations for why this species can divide so quickly.

7.1

KEY TERMS

apoptosis
binary fission
cytokinesis
interphase
mitosis
oncogene

CELL CYCLES

The human body grows by dividing its cells to make new ones. Think about the average size of a fifth grade school student and how much taller they will grow in the four years until they start high school. Now think about an adult you know and whether they have grown any taller over a four-year time frame.

Analyze *Consider the types of cells in a human, such as blood cells and skin cells. Do you think cells in your body all grow and divide at the same rate?*

VIDEO

VIDEO 7-1 Go online to see how quickly *E. coli* grow and divide.

Life Cycles

In biology, a life cycle is a sequence of distinct stages that occur during an organism's lifetime, from a single cell, through development, until that individual's death. Consider the life cycle of a butterfly, for example: egg, larva (caterpillar), pupa (chrysalis), and adult (butterfly). At each stage, the butterfly looks and acts differently. Individual cells also have a life cycle and a life span.

Prokaryotic Cell Cycle Prokaryotic cell cycles are generally described in three parts that include (1) the time between reproduction events, (2) the replication of DNA, and (3) the completion of cell duplication. Bacteria typically reproduce by **binary fission**, a type of asexual reproduction (**Figure 7-3**). The process begins when the cell replicates its single chromosome. Addition of new membrane and wall material elongates the cell to nearly double its size and moves the two DNA molecules apart. Membrane and cell wall material are deposited across the cell's midsection, yielding two identical descendant cells.

FIGURE 7-3 The bacterial cell (a) duplicates its chromosome (b), then divides (c) after nearly doubling in size. The new cells (d) can then begin the cycle and duplicate again.

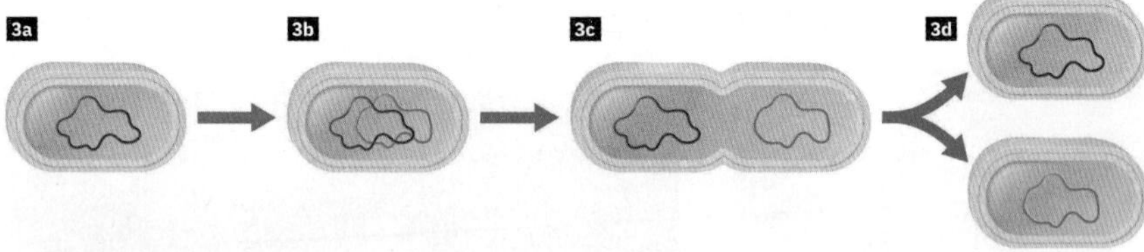

BACTERIA IN YOUR GUT

Gather Evidence E. coli *and other bacterial populations can grow incredibly quickly. What factors help keep the different types of bacteria in the human gut in balance so no one type of bacteria takes over?*

The growth rate of bacteria is affected by the availability of resources. When environmental conditions are optimal and enough resources are available to double their mass and their DNA, bacteria will rapidly proceed to the reproductive part of their cell cycle. If resources are scarce, they will divide more slowly.

Under optimal conditions in a laboratory setting, *Escherichia coli* can divide every 20 minutes and quickly grow exponentially into a huge number of new organisms. The *E. coli* that are widely present in the human digestive system have access to an abundance of carbohydrates and a favorable growth temperature, but they also have to deal with competition from other species and rapid changes in conditions and nutrition provided by their host.

REVISIT THE ANCHORING PHENOMENON

Gather Evidence *Students might say that limitations in the amount of nutrients and competition between the different types of bacteria keep any one species from taking over.*

The growth rate of bacteria is dependent on the availability of resources, optimal conditions such as temperature, and level of competition from other species for the same resources. In the human gut, *E.coli* have access to plenty of nutrients and ideal temperatures for growth; however, there is competition from other intestinal species, as well as constant changes in internal conditions resulting from diet changes.

Eukaryotic Cell Cycle Dividing eukaryotic cells pass through a series of intervals and events shown in **Figure 7-4**. A typical cell spends most of its time in interphase. During **interphase**, the cell increases its mass, roughly doubles the number of its cytoplasmic components, and copies its chromosomes in preparation for division. Interphase is typically the longest part of the cycle, and it consists of three stages: G1, S, and G2. G1 and G2 were named "Gap" intervals because outwardly they seem to be periods of inactivity, but they are not. Most cells are building molecules and performing other functions during G1.

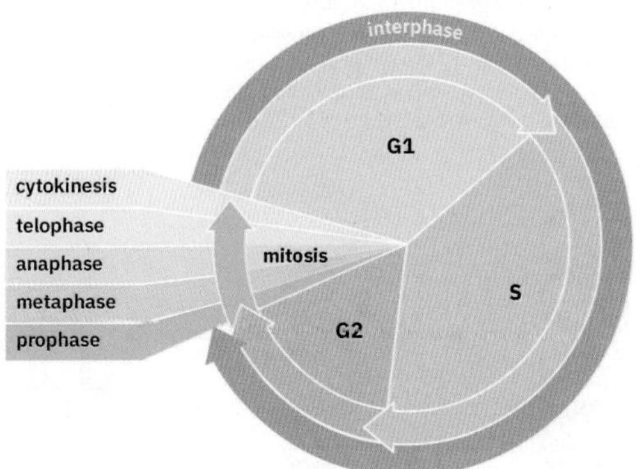

FIGURE 7-4 G1, S, and G2 are part of interphase, and the length of these intervals differs among cells. Mitosis includes several stages that describe how a cell's DNA is divided into two nuclei.

Go Online SIMULATION

Go online to examine how a cell grows and divides during the **Eukaryotic Cell Cycle**.

Cells preparing to divide enter S ("synthesis") phase, when DNA replication occurs (**Figure 7-5**). In G2, cells continue to grow and make proteins.

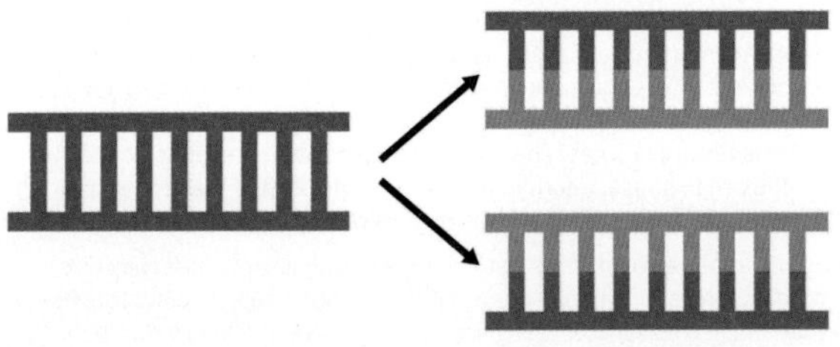

FIGURE 7-5 DNA is replicated in the S phase of the cell cycle. DNA replication is described as *semiconservative* because each strand of the original DNA is paired with a new DNA strand to form two copies. Details of the mechanism of DNA replication are described in Chapter 11.

The remainder of the cell cycle consists of the cell division process. **Mitosis** is a mechanism for cellular division that maintains the correct number of chromosomes. The cytoplasm usually undergoes **cytokinesis** after mitosis, dividing the cytoplasm and pinching the cell membrane to produce two new descendant cells. The steps of mitosis will be described in detail in the next section.

SEP Construct an Explanation *How do you think the cell cycles might be different for a fast-growing bacterium, compared with a cell in a fully grown multicellular organism?*

Connect to Mathematics

Exponential Growth After watching the video, explain that bacterial reproduction is an example of exponential growth. In this situation, the individual reproductive rate stays the same, however the population grows faster as it increases. Work through an example with students starting with a single bacterium that reproduces at a rate of 30 minutes per reproductive cycle to show how many bacteria would exist after 10 hours. Doubling at a rate of 30 minutes, reproduction would occur 20 times in 10 hours, making the number of bacteria 2^{20}, or 1,048,576. Remind students that this is dependent on resource availability.

Go Online SIMULATION

Eukaryotic Cell Cycle After observing the interactive cell cycle, explain to students that, although the process seems to flow without pause, there is a pause, as shown in **Figure 7-4**.

SEP Construct an Explanation
Sample answer: A bacterial life cycle duplicates an entire organism and primarily depends on having conditions favorable for growth. Most cells in an adult multicellular organism only grow to form the structures the organism needs, rather than dividing as quickly as possible.

SCIENCE AND ENGINEERING PRACTICES
Using Mathematics and Computational Thinking

Surface-to-Volume Ratio Ask students to consider the question of why cells might divide rather than just growing larger. Remind them that everything a cell needs enters through the cell membrane, and waste products leave through the same structure. The rate at which this happens depends on the total area of the cell membrane. The rate at which food and oxygen are used up depends on the cell volume, or space within the cell. Show students what happens when the length of each side of a cube doubles from 1 to 2 to 4. For example, a cube with 1-inch sides would have a volume of 1 cm^3, a surface area of 6 cm^2, and a surface-to-volume ratio of 6. A cube with 4-inch sides would have a volume of 64 cm^3, a surface area of 96 cm^2, and a surface-to-volume ratio of 1.5. The takeaway is that the surface-to-area ratio gets smaller as the cell gets bigger, making it much more difficult to get needed materials into the cell and waste products out.

Science Background

mRNA and Vaccines Share with students that DNA in eukaryotic cells never leaves the nucleus, but remains protected within the nucleus where the information is copied onto mRNA. Messenger RNA travels out of the nucleus to where it provides instructions for the production of proteins.

The virus that causes COVID-19 is recognizable by its spike proteins that allow it to penetrate host cells and infect them. Some COVID-19 vaccines that were developed were made using mRNA. The mRNA vaccine provides the instructions for cells in the body to make copies of the spike protein. If a person is exposed to the virus, cells in the body can immediately recognize the COVID-19 virus and attack the virus. Discuss with students whether such a vaccine would affect a person's DNA, and explain why or why not. (*an mRNA vaccine cannot alter the DNA for three reasons: the DNA is in the nucleus not the cytoplasm where the mRNA vaccine would be, mRNA cannot be reverse transcribed back to DNA, and mRNA cannot be integrated into DNA)*

SEP Use a Model *Sample answer: Genes and their protein products involved in DNA synthesis would be expressed and active during S phase. Checkpoint genes would be performing important functions at the transitions between phases.*

Eukaryotic Cell Cycle Control

Cell division is tightly regulated via changes in gene expression. Like the accelerator of a car enables it to move forward, some changes cause the cell cycle to advance. Other mechanisms are like car brakes, pausing the cell cycle. In the adult human body, brakes on the cell cycle normally keep the vast majority of cells in a resting state known as G0. Most of your nerve cells, skeletal muscle cells, heart muscle cells, and fat-storing cells spend most of your life resting.

Genes are functional units of DNA. When genes are expressed, an RNA copy of the gene is generated and used to build a protein (**Figure 7-6**). The genes expressed and timing of gene expression not only control the cell cycle but also ultimately shape the identity of each cell in a multicellular organism. This process, called *cell differentiation*, is discussed in Section 7.3. Details of gene expression are described in Section 11.3.

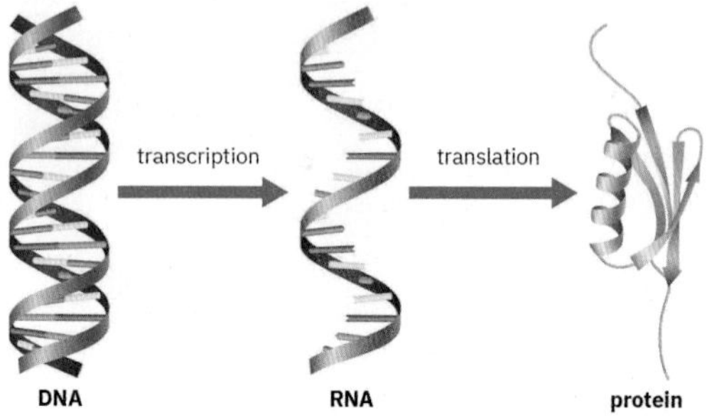

FIGURE 7-6 Genes are functional units of DNA that carry information. That information is typically transcribed into an RNA molecule that is used by ribosomes to instruct the synthesis of proteins. Proteins perform most of the work in the cell, including those functions that regulate the cell cycle.

Built-in checkpoints in the cell cycle ensure that the cell does not progress from one phase to another unless certain criteria are met. For example, adequate nutrients are available to support cell growth, cellular damage is identified and repaired, and the entire genome has been completely and accurately copied. Protein products of "checkpoint genes" interact to carry out this control process. The primary checkpoints represent an opportunity for the cell to pause or terminate cell division—each of these checkpoints is regulated by a different subset of checkpoint genes.

For example, a checkpoint that operates in S phase monitors the cell's chromosomes for damage during DNA replication. Checkpoint proteins that function as sensors recognize damaged DNA and bind to it. Upon binding, they trigger other events that stall the cell cycle, and also enhance expression of genes involved in DNA repair. After the problem has been corrected, the brakes are lifted and the cell cycle proceeds. If the problem remains uncorrected, other checkpoint proteins may initiate a series of events that eventually cause the cell to self-destruct.

SEP Use a Model *Using the steps shown in the eukaryotic cell cycle model (**Figure 7-4**), explain when you think checkpoint genes would be expressed. When do you think genes involved in replicating and repairing DNA might be expressed?*

DIFFERENTIATED INSTRUCTION | English Language Learners

Roots and Affixes Have students use the prefix *trans-* ("across"), the root scribe ("write"), and the suffix *-tion* (which changes a verb to a noun) to understand the meanings of *transcribe/transcription* and *translate/translation* in the context of cell division.

Beginning Have pairs use the word parts to write simple definitions in their own words. Then have them check their work in a glossary or dictionary.

Intermediate Have pairs define the words using the word parts. Then have them look up the words and discuss how the everyday meanings ("to write down what is heard" and "to say the same thing in a different language") differ from the science meanings.

Advanced Have pairs use the word parts to explain transcription and translation in cell division. Partners can take turns, each describing one step.

Cancer Disruption of cell cycle checkpoints can have negative consequences. Checkpoint genes (and their protein products) are often involved in mechanisms that cause cancer. Sometimes a random or inherited mutation in a checkpoint gene results in a dysfunctional protein product. In other cases, the expression of a checkpoint gene is wrong, and a cell makes too much or too little of its product. When checkpoint mechanisms fail, a cell loses the ability to repair damage and/or replicate the genome accurately. The problem is compounded when checkpoint malfunctions are passed along to the cell's descendants. Over time, this uncontrolled cell growth can lead to tumor formation.

 VIDEO

VIDEO 7-2 Go online to learn more about how cancer begins.

A gene that has a tumor-causing mutation is called an **oncogene**. An oncogene can transform a normal cell into a tumor cell. Oncogene mutations in reproductive cells can be passed to offspring, which is a reason that some types of tumors run in families. Genes that are involved in cell growth that can mutate to become oncogenes are called *proto-oncogenes*.

One proto-oncogene (*EGFR*) encodes an epidermal growth factor (EGF) receptor that binds EGF molecules. When a normal EGF receptor binds an EGF molecule, it triggers the cell to begin mitosis (**Figure 7-7a**). Mutations in the *EGFR* gene can change the EGF receptor so that it stimulates mitosis even in the absence of EGF molecules (**Figure 7-7b, c**). Many tumors have mutations in the *EGFR* gene.

FIGURE 7-7 Under normal circumstances, an EGF receptor stimulates mitosis only when it binds an EGF molecule (a). However, a mutated EGF receptor can stimulate mitosis constantly, leading to excess cell division and cancer (b). The photo (c) shows human breast tissue labeled with a brown tracer molecule that marks active EGF receptors. These cells are dividing uncontrollably.

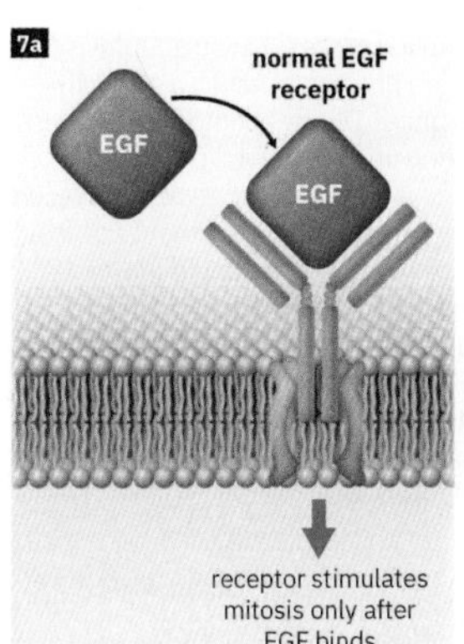

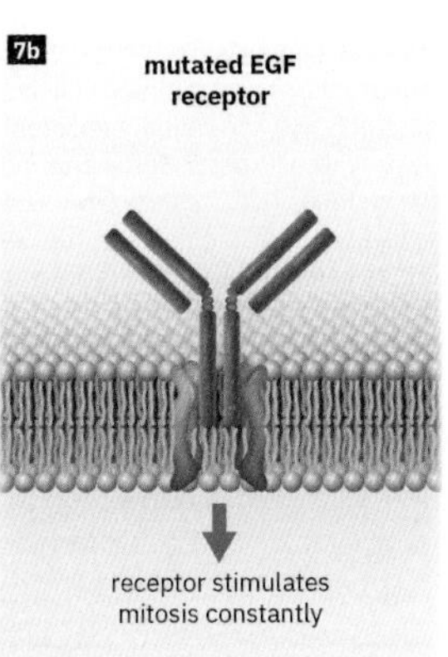

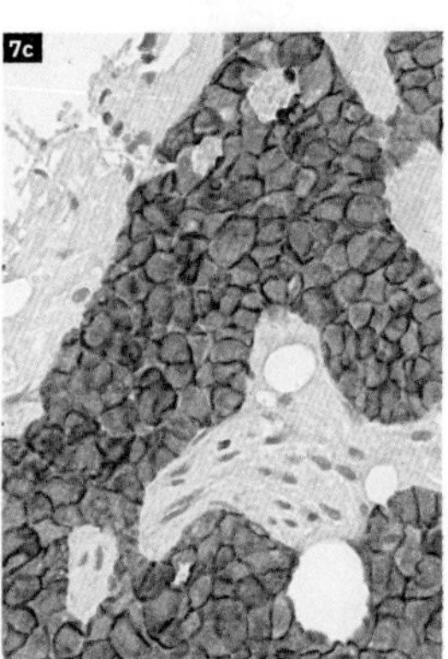

Other genes may cause cancer when mutations ruin the function of their protein products. Tumor suppressor genes code for proteins that are related to DNA repair and other regulatory functions. When these proteins are missing, tumors form. The protein products of checkpoint genes called *BRCA1* and *BRCA2* are tumor suppressors. The *BRCA* gene products help maintain the structure and number of chromosomes during mitosis and repair damaged DNA. They also play a role in regulating growth factors for cells of breast and ovarian tissues. Mutations in these genes lead to tumors in the breast, prostate, ovary, and other tissues.

Video

Time: 1:35

Use **Video 7-2** to discuss how cancer begins. Explain that genetic changes that disrupt normal cell growth and division cause the cells to grow and divide uncontrollably. Abnormal cell growth leads to tumors. Discuss the drop in cancer mortality rates in the U.S. and have students posit explanations for why this is the case.

In Your Community

Cancer Prescreening Many advances have been made in the treatment of different cancers. One of the most effective ways to battle cancer is through early screening and lifestyle choices. Early screenings vary with age, but include those for cancers of the cervix, breasts, skin, prostate, and colon. When caught early, most cancers are easier to treat successfully. Ask students to find out about cancer screenings and when they should begin. Many communities may provide support for these screenings. Suggest students create a public service announcement poster educating people about cancer screening. Encourage students to make sidebars about how lifestyle choices impact the potential to develop cancer.

Social-Emotional Learning

Social Awareness Cancer is a diagnosis that touches many families and friends of students. It also is a very feared disease. In discussing this topic, students should practice **social awareness** skills, such as showing concern for the feelings of others and demonstrating compassion. If classmates are aware of a student's situation, or the student has shared their experience, it would be appropriate to talk about demonstrating these skills.

Self-awareness skills can also help students to identify their own emotions, and **self-management** can help them in learning how to manage those emotions and any stressors that accompany those feelings.

DIFFERENTIATED INSTRUCTION | Leveled Support

Checkpoints Students routinely go through class periods each day that sometimes include "checkpoints." For example, to move forward to a science lab, a student must meet certain criteria, such as wearing safety goggles, tying back hair, or having lab materials ready.

Struggling Students Have students create a model that depicts how their day is similar to the cell cycle. Tell them to include one place in the day where there might be a "checkpoint," such as being prepared for gym class.

Advanced Learners Have students create a similar model. When they are finished, ask them to describe the possible consequences of not having checkpoints. For example, if safety precaution are not checked, an accident may occur that could cause harm. If the problems are corrected, then the student may continue on in the lab.

Science Background

Good and Bad of Apoptosis The word *apoptosis* comes from Latin roots meaning "to fall off." This happens somewhat literally in the amphibian shown in **Figure 7-8**. In humans, apoptosis is what prevents them from having webbed fingers and toes, as these cells are eliminated during fetal development. Apoptosis also helps the body to rid itself of cells that are damaged beyond repair.

A rate of apoptosis that is too high or too low can contribute to an array of human diseases. Apoptosis is a mechanism for eliminating damaged and dysfunctional cells. Many kinds of cancer have defects in apoptosis, allowing damaged cells to continue to divide in an unregulated manner. On the other hand, apoptosis can contribute to some diseases like neurodegenerative diseases because the cells that are damaged and lost cannot be replaced.

Relate apoptosis malfunction to diseases such as lung cancer, melanoma, Alzheimer's, Parkinson's, stroke, heart failure, and rheumatoid arthritis. Have students suggest which groups of cells are involved and what might have triggered the malfunction.

FIGURE 7-8 Tadpoles develop back legs and then front legs as they turn into adult frogs. The cells of the tadpole's tail are eliminated through apoptosis during development.

Cell Death and Apoptosis As organisms grow and develop, sometimes cells need to be replaced or removed, ending the cell cycle. One way that cells die is through a controlled process called apoptosis. **Apoptosis**, or programmed cell death, is a process by which a cell self-destructs in a controlled, predictable manner. During apoptosis, enzymes destroy a cell's parts. Some enzymes chop up cytoskeletal proteins and the histones that organize DNA. Others snip apart nucleic acids. As a result of this self-inflicted damage, the cell shrivels up and dies without spewing out its contents. These cells are engulfed by other cells, and many of the remaining parts are broken down and recycled. In a mature organism, apoptosis plays a role in normal cell turnover and in defense against cancer. During development, apoptosis shapes structures by eliminating excess cells, removing those cells that do not contribute to the expected final form. For example, in the development of an adult frog, a tadpole's tail is eliminated through apoptosis (**Figure 7-8**).

7.1 REVIEW

1. **Calculate** Under ideal conditions, how long would it take a colony of 1000 *E. coli* bacteria to reach a population size of 64,000 cells?
2. **Model** Create a simple flow chart for the three stages of interphase as a cell prepares to divide.
3. **Summarize** At the most basic level, what is the difference between a cancer cell and a normal cell? Why do cancer cells result in tumors?
4. **Define** Which processes are affected by checkpoint genes during the cell cycle? Select all correct answers.
   - A. cell differentiation
   - B. repair of damaged DNA
   - C. recognition of DNA replication errors
   - D. self-destruction of cells with certain problems
   - E. monitoring for favorable conditions for duplication

7.1 REVIEW

1. *2 hours* **DOK 2**
2. *G1 stage: cells build molecules and conduct normal functions; S stage: cells synthesize DNA and prepare to divide; G2 stage: cells make proteins needed for division.* **DOK 2**
3. *Sample answer: Problems with the checkpoint genes in a normal cell can cause the cell to divide abnormally and become a cancer cell. The cell may divide too quickly or not die off when it is supposed to. These problems are transferred to the new cells after the cell divides, so the new cells divide abnormally. When these cancer cells accumulate, they form cancerous tumors.* **DOK 2**
4. *B, C, D, E* **DOK 2**

LOOKING AT THE DATA

IDENTIFYING GENE MUTATIONS IN CANCER CELLS

SEP Analyze and Interpret Data Gene mutations that cause cancerous tumors can be identified in tissue samples.

Cancerous tumors are often the result of gene mutations that cause some genes to be expressed too much or cause other genes to shut down and not function. One technique called a DNA microarray can detect the expression of genes in tissue samples. In a DNA microarray, the DNA for each gene being tested is attached to individual spots in an array of many wells.

Test samples are generated by making DNA that matches all the genes expressed in the cells of a tissue sample. Healthy cell DNA is marked with one color of fluorescent dye, and the cancer cell DNA is marked with another color of fluorescent dye. The dyes fluoresce or "light up" when the test DNA matches up with the DNA attached to the well. When the samples are added together to the microarray, the resulting color of fluorescent light tells scientists whether the genes are expressed in normal cells, cancer cells, or both. Microarrays in the lab are designed to test the expression of thousands of genes at once.

Figure 1 is a representation of the results of a partial DNA microarray performed on two tissue samples, one with normal cells and one with cancer cells, collected from the same type of tissue. **Table 1** identifies the genes being tested in each well.

DNA Microarray Results

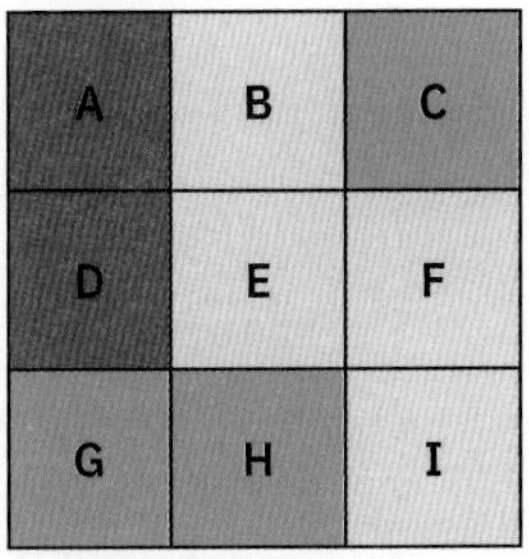

A	B	C
D	E	F
G	H	I

FIGURE 1 Results of a DNA microarray are shown. Each microarray well tests one specific gene. The fluorescence color in each well of the microarray indicates which cells in the samples (normal or cancer) are expressing each gene. The colors have the following meanings:

If the well is red, the gene is expressed at a higher level in cancer cells.

If the well is green, the gene is expressed at a higher level in normal cells.

If the well is yellow, the gene is expressed equally in both types of cells.

TABLE 1. The Roles of the Genes Tested in the DNA Microarray

A	supports cellular structure and shape
B	encourages cell growth
C	stops cells from dividing too fast
D	encourages cell growth
E	provides strength to cells in skin, hair, and nails
F	provides strength to cells in skin, hair, and nails
G	secretes mucus to line air pathways
H	enables lung cells to expand and contract
I	stops cells from dividing too fast

1. **Organize Data** Interpret the results of the DNA microarray to identify how genes are expressed in normal and cancer cells.
2. **Classify** Using the data given in **Figure 1** and **Table 1**, identify the oncogenes and mutated tumor suppressor genes.
3. **Analyze** Based on the results, how do you think the cancerous cells compare with normal cells? Explain your answer.
4. **Infer** Based on the genes being tested, what type of tissue sample was most likely used in this DNA microarray test?

   A. hair

   B. lung

   C. nail

   D. skin

LOOKING AT THE DATA
IDENTIFYING GENE MUTATIONS IN CANCER CELLS

Connect this activity to the Science and Engineering practice of developing and using models and engage in argument from evidence. Students will be simulating the use of DNA microarrays that can be used to analyze the difference in gene expression between cancerous and noncancerous tissues. Students will begin their analysis by familiarizing themselves with the meaning of each color in the array and then matching each letter to the role of the gene being tested. They can then analyze the array and interpret the results.

SAMPLE ANSWERS

1. **Organize Data** *Genes expressed more in cancer cells: A, D. Genes expressed more in normal cells: G, H, C. Genes expressed similarly in both normal and cancer cells: B, E, F, I.* **DOK 2**
2. **Classify** *oncogenes: D, tumor suppressor genes: C,I* **DOK 2**
3. **Analyze** *The mutated genes A and D are responsible for the structure and shape of the cells and for encouraging cell growth. The cancer cells have an unusual structure or shape; they also go through the cell cycle too fast.* **DOK 3**
4. **Infer** *lung* **DOK 2**

7.2

MITOSIS

LS1.B, LS3.A

EXPLORE/EXPLAIN

This section reviews the basic structure of chromosomes and DNA, including tightly wound DNA-histone spools that can be packed into a cell nucleus. This is followed by a look at the activities leading up to cell division and a walkthrough of each mitotic phase and the complete cell cycle.

Objectives

- Describe chromosomes and the structure of DNA.
- Sequence the stages of mitosis and cytokinesis.

Pretest Use **Question 3** to identify gaps in background knowledge or misconceptions.

Infer *Students might say it is important for each cell to have the same contents so it can reproduce itself and faithfully produce additional cells that have all the same components needed for life.*

Science Background

Chromosome Numbers in Different Species Each species has a characteristic number of chromosomes; humans have 46. Share the table below with students or display one from an online source using a keyword search such as "organisms number of chromosomes." Have students compare the number of chromosomes among species. Ask students to make a claim about the number of chromosomes and the size or complexity of an organism. Encourage them to support their claims with evidence.

Organism	Number of chromosomes
Adder's tongue fern	1,262
Amoeba	50
Chimpanzee	48
Goldfish	94
Dog	78
Carrot	18

7.2 MITOSIS

KEY TERMS

homologous chromosome
sister chromatid

Imagine you are preparing for a classroom crafting project and are responsible for making sure each of your classmates all have the same materials available for the project. Each craft station needs to have the exact same items, such as glue, paper, string, and wire, so every student can complete the task successfully. Similarly, when cells divide, they need to ensure new cells all end up with the correct materials to live.

Infer *Why do you think it would be important for a dividing cell to make sure all of the new cells have the same contents?*

Chromosomes

Most prokaryotes have a single circular chromosome. By contrast, most eukaryotic cells have multiple chromosomes that differ in length and shape. Human cells have 46 chromosomes. Stretched out end to end, the double-stranded DNA in a single human cell would be about 2 m (6.5 ft) long, but all that DNA fits into a nucleus less than 10 micrometers in diameter. Proteins that associate with DNA help keep it organized and can help to pack DNA very tightly when needed. In eukaryotes, a DNA molecule wraps at regular intervals around proteins called *histones*. These DNA-histone spools look like beads on a string in micrographs (**Figure 7-9a**). Interactions among histones coil the DNA into a tight fiber. A typical eukaryotic chromosome consists of one DNA molecule and its many associated proteins.

In preparation for mitosis, a eukaryotic cell will duplicate its chromosomes through DNA replication (Section 11.2). After replication, the two identical DNA molecules are referred to as **sister chromatids**, attached to one another at a constricted region called the *centromere*. The duplicated chromosomes condense fully into their familiar "X" shapes (**Figure 7-9b**) before the cell divides.

9a

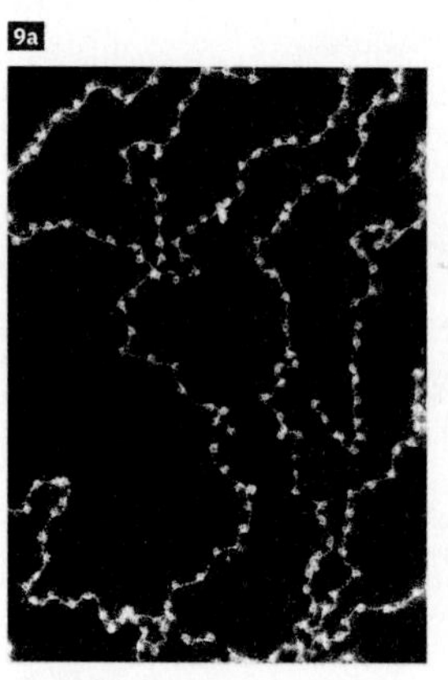

9b

FIGURE 7-9 At regular intervals, the double-stranded DNA molecule is wrapped around a core of histone proteins (a). The DNA and proteins associated with it twist tightly into a fiber. The fiber coils multiple times and at its most condensed, a duplicated eukaryotic chromosome has an X shape (b). This structure consists of two identical DNA molecules (sister chromatids) attached at a centromere.

CROSSCUTTING CONCEPTS | Scale, Proportion, and Quantity

Size Comparisons Students have learned that double-stranded DNA in a single human cell can be about 2 meters (6.5 feet) long and that this fits into a cell nucleus that is less than 10 micrometers in diameter. To help students recognize the significance of this difference in size, display a thread that is 2 meters long. Explain that a micrometer is one millionth of a meter. Ask students to describe their thoughts on what size that might be. For comparison, use a grain of salt, which is about 1 millimeter (one thousandth of a meter), or the width of a strand of hair that averages about 70 micrometers.

Mitosis

In the human body, somatic cells are those that are not involved in reproduction. These cells are diploid, which means they contain two copies of each chromosome. One set of chromosomes is contributed by an egg and the other set is contributed by a sperm in the process of fertilization. Except for a pairing of different sex chromosomes (XY) in males, the chromosomes of each pair are homologous, or the same. Females have matching sex chromosomes (XX). **Homologous chromosomes** have the same length, shape, and centromere location, and they carry the same genes. However, they do not have the exact same DNA sequence, unlike the sister chromatids produced by DNA replication.

The two homologous chromosomes of a pair are not identical. They are inherited from the gametes (egg, sperm) of two genetically distinct parents. Cell division by mitosis produces two cellular descendants, each with the same number of chromosomes as the original cell (**Figure 7-10**). In addition to total number, each descendant cell must inherit the full complement of chromosomes, without duplicates or missing chromosomes. A cell cannot function properly without a copy of each chromosome.

CASE STUDY
IMMORTAL CELLS: THE STORY OF HENRIETTA LACKS

Ask Questions *Although Henrietta Lacks's original cells had 46 chromosomes, most HeLa cell lines now have between 70 and 90 chromosomes, among other changes. What other questions do you have about HeLa cells and how they are different from the somatic cells in your body?*

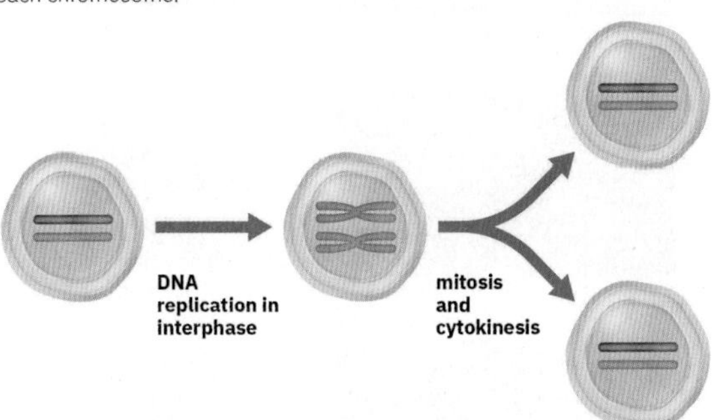

FIGURE 7-10 Mitosis maintains the correct number of chromosomes. One homologous chromosome is inherited from each parent. DNA replication results in sister chromatids, identical copies of each chromosome. Then, mitosis and cytokinesis separate the sister chromatids into two new cells. Each descendant cell is genetically identical to the parent cell.

Compare **Figure 7-10** with the cell cycle diagram shown in Section 7.1 (**Figure 7-4**). When a cell is in G1, each of its chromosomes is made of one double-stranded DNA molecule. DNA replication occurs during the synthesis, or S phase, so by G2 each chromosome has two double-stranded DNA molecules.

These sister chromatids stay attached to one another until mitosis is almost over, and then they are pulled apart and packaged into two separate nuclei. When sister chromatids are pulled apart, each becomes an individual chromosome that consists of one double-stranded DNA molecule. Thus, each of the two new nuclei that form by mitosis contains the same number and types of chromosomes as the parent cell. During cytokinesis, these nuclei are packaged into separate cells. Each new cell starts its life in G1 of interphase.

SEP Construct an Explanation *What are the differences between homologous chromosomes and sister chromatids?*

Vocabulary Strategy

Prefixes/Suffixes/Root Relationships
Write the word *chromosome* on the board, underlining *chromo-* . Explain to students that the affix *chromo-* means "color" and is often associated with photography. While it may seem odd that this affix is associated with cells and genetics, it was so used because chromosomes contain a substance that stains easily. Another common affix used in biology is *–somes*, which means "the body." Under a microscope, a chromosome might appear as a red or purple body.

CASE STUDY
IMMORTAL CELL: THE STORY OF HENRIETTA LACKS

Ask Questions To help students revisit the Case Study and make connections with the content, prompt them to think about in which part of the cell cycle would additional chromosomes occur and how that might have happened. You can also lead a short discussion about whether these additional chromosomes are completely new, coding for traits not in the original cells, or if they are duplicates of the original 46. Discuss the impact of either answer on the resulting new cells.

SEP Construct an Explanation
Sample answer: Homologous chromosomes are matching chromosomes from each parent but are not exact copies. Sister chromatids are formed just for the process of mitosis and are exact copies that can be split up so each nucleus and cell gets one of every homologous chromosome.

DIFFERENTIATED INSTRUCTION | Leveled Support

Students may have difficulty distinguishing between homologous chromosomes and sister chromatids.

Struggling Students Working in pairs, have student use craft materials such as clay, yarn, and beads to create a model of a homologous pair of chromosomes and sister chromatids. Using the models, ask students to create a 2-column chart and list similarities and differences between homologous pairs and sister chromatids. They may then use resources to add to their chart.

Advanced Students Have students compare and contrast homologous chromosomes and sister chromatids using a Venn diagram. Students should note that sister chromatids and homologous chromosomes are the same size and have the same genes. However, in homologous chromosomes, some genes are in the same location, but they have different alleles (different version of the same gene) and, unlike sister chromatids, they are not attached by a centromere.

Vocabulary Strategy

Word Wall To help students remember the phases of mitosis, create an interactive word wall. Outline a graphic organizer with "Mitosis" at the top with a definition and a branched diagram to each word describing a phase in mitosis. Allow students to add to the word wall with definitions, meanings of prefixes, visuals, or graphics that will help them connect words to their meanings. Students for whom English is not their primary language may also post notes in their native language.

Science Background

Microtubules Formed from molecules of the protein tubulin, microtubules form the spindle fibers in the cell during cell division. They also function by forming the foundation of the cytoskeleton that helps maintain cell shape. Discuss with students how the structure of microtubules matches their function as they can shorten or lengthen by the addition of tubulin molecules at the ends of the tubes. This allows them to move chromosomes during cell division.

FIGURE 7-11
Although mitosis is a continual process, it is often described in stages. Micrographs show plant cells (onion root, *left*) and animal cells (fertilized eggs of a roundworm, *center*) at each stage. The illustrations (*right*) demonstrate mitosis for a diploid cell with two chromosome pairs. The nucleus and chromosomes are enlarged to show each stage.

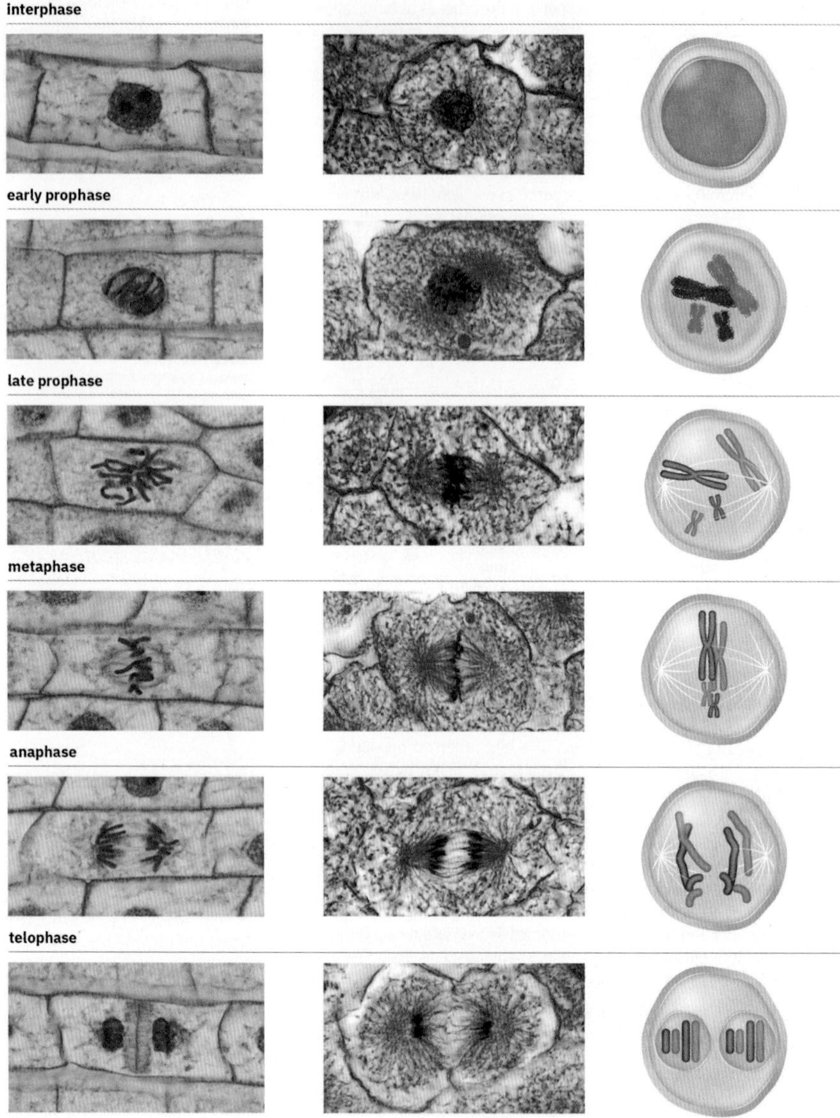

190 CHAPTER 7 CELL GROWTH

DIFFERENTIATED INSTRUCTION | Students with Disabilities

Visual Impairments After instruction on the stages of mitosis, guide students who have a visual impairment to build a two-dimensional model of each phase using different textured objects, such as string, tubing, paper plates, textured paper, glue, and toothpicks.

Guide students to prepare their model using a separate paper plate for each phase. For the first plate, interphase, a cut-out circle of textured paper can be used to represent the nucleus. Using the string as the chromosomes, guide students to begin making the model for interphase, before and after DNA replication. In prophase, the nucleus dissolves so the circle of textured paper would not be used and the string chromosomes are glued directly to the plate. During prophase, the chromosomes are held together by a centromere, so assistance may be offered to attach them with a toothpick. Continue guidance as needed for the remaining phases.

Stages of Mitosis and the Cell Cycle

Mitosis can be divided into separate stages (**Figure 7-11**). These stages include:

- **Interphase** Cells prepare for mitosis during this stage of the cell cycle. When a cell is in interphase, its chromosomes are relaxed in many regions to allow gene expression and DNA replication.
- **Early Prophase** Chromosome condensation begins. A tightly condensed structure keeps the chromosomes from getting tangled and breaking during division. As prophase continues, the nuclear envelope breaks up. Fibers form the *mitotic spindle*, a specialized structure of microtubules and other molecules.
- **Late Prophase** This stage is sometimes called prometaphase. As the nuclear envelope completely breaks down and chromosomes fully condense, spindle fibers made of microtubules extend from opposite poles, attaching to chromosomes at the centromere. The individual sister chromatids are each attached to a spindle fiber (one from each pole). The spindle fibers begin moving the chromosomes to the center.
- **Metaphase** During this stage, a fully formed mitotic spindle with sister chromatids all attached to the spindle aligns at the center of the cell.
- **Anaphase** During anaphase, spindle microtubules separate the sister chromatids of each duplicated chromosome and move them toward opposite poles of the cell.
- **Telophase** This stage begins when one set of chromosomes reaches each spindle pole. Each set consists of the same number and kinds of chromosomes that the parent cell nucleus had before replication. The nuclear envelope reassembles around each set of chromosomes as they decondense. At this point, mitosis is finished.
- **Cytokinesis** After mitosis is completed, the cell membrane pinches to form a cleavage furrow to separate the cell into two new cells (in animals) or forms a cell plate for the separation (in plants).

CCC Structure and Function *How is the structure of sister chromatids important in ensuring that each resulting nucleus gets one of each chromosome during the stages of mitosis?*

VIDEO

VIDEO 7-3 Go online to watch ovary cells from a hamster undergo mitosis.

CHAPTER INVESTIGATION

Plant Growth Through Mitosis

How do rates of cell division relate to plant growth?

Go online to explore this chapter's hands-on investigation about mitosis in plants.

7.2 REVIEW

1. **Model** Describe how you would construct a simple structural model of a single eukaryotic chromosome using large beads, string, and any other materials you might need. Identify the role of each material in the model.
2. **Relate** Human cells have 46 chromosomes. How many of these chromosomes come from the father and how many come from the mother?
3. **Sequence** Describe mitosis by ordering processes that occur during different stages.
   - Nuclear envelope breaks up completely.
   - Nuclear envelope forms around chromosomes.
   - Sister chromatids are aligned in the middle of the cell.
   - Chromosomes in the nucleus begin to condense around centromeres.
   - Sister chromatids are pulled apart toward opposite poles of the cell.

Video

Time: 0:43

Use **Video 7-3** of a cell from a hamster ovary undergoing mitosis. Students can visibly follow the phases of mitosis, including prophase, anaphase, metaphase, as well as cytokinesis. Students can follow along with the diagram in **Figure 7-10** to identify and discuss each of the phases seen in the video.

Address Misconceptions

Phases of Mitosis Students commonly use cell division and mitosis interchangeably. It is important that they understand the distinction between the two. Mitosis refers strictly to the division of the nucleus, and cytokinesis refers to the division of the cytoplasm. Students should understand that prophase through telophase are the steps in mitosis, while interphase and cytokinesis are steps of cell division before and after mitosis.

CHAPTER INVESTIGATION A

Guided Inquiry: *Plant Growth through Mitosis*

Time: 50 minutes plus advance preparation

Students will observe plant cells to explain the role of mitosis in plant growth and calculate the mitotic index in a plant.

Go online to access detailed teacher notes, answers, rubrics, and lab worksheets.

CCC Structure and Function *Sample answer: Sister chromatids keep duplicated chromosomes together as the nucleus gets organized during prophase and lines up the chromosomes in the middle during metaphase. Then sister pairs can get pulled apart without any risk of pulling 2 of the same chromosome to one nucleus.*

7.2 REVIEW

1. *The student should describe a model that represents how DNA is condensed into chromosomes with the DNA wrapping into tighter and tighter coils, with the help of histones. Beads can represent histones and centromeres (if students describe a model of sister chromatids). The string can wrap around the histones (beads). A small rod or pencil could be used to wrap the histone/string combination into a right coil that represents a condensed section of chromosomes.* **DOK 3**
2. *father: 23; mother: 23* **DOK 1**
3. *Order: 4, 1, 3, 5, 2* **DOK 2**

MINILAB
MODELING MITOSIS

Students will develop models to represent each phase of mitosis in order to visualize the role of cellular division in producing complex organisms.

Time: 30 minutes

Advanced Preparation

- Gather a variety of materials for students to use, such as craft paper, poster board, beans, clay, yarn, pipe cleaners, or bendable wire. Alternatively, a day or two before, ask students to bring materials from home.

Procedure

In **step 1**, if students are given the option to use a camera or create a digital model, have them list what technology or programs they are using.

In **step 5**, be sure to emphasize that mitosis is only nuclear division. It is not the entire cell cycle.

Results and Analysis

1. **Evaluate** *Sample answer: The condensed and uncondensed chromosomes were made from different materials. A better model could use the same material for both. It was also difficult to build the desired shapes using the bendable wire. Another material, like string, could be used instead.* **DOK 2**
2. **Compare** *Sample answer for cells starting with three pairs of duplicated chromosomes: The cell started with three pairs of duplicated chromosomes, or six chromosomes, and in telophase each new nucleus now has three duplicated chromosomes. The chromosomes were replicated before mitosis so each of the new cells would have the correct number of chromosomes.* **DOK 2**
3. **Draw Conclusions** *Sample answer: The spindle fibers work to align and then pull apart chromatids. If spindle fibers are not formed, mitosis would not be able to proceed into metaphase and anaphase.* **DOK 3**
4. **Explain** *Sample answer: The chromatids are pulled apart during anaphase. Therefore, a lagging chromatid is a problem that occurs in that phase. A lagging chromatid can cause daughter cells with different chromosome counts to form: One daughter cell can have a missing chromosome, and the other daughter cell can have an extra copy of the chromosome.* **DOK 3**

MINILAB

MODELING MITOSIS

SEP Develop a Model How can cell growth be modeled?

Cell division allows organisms to make more cells and grow. Before cells can divide from one cell into two descendant cells, the replicated chromosomes in the cell must separate into two new nuclei. In this activity, you will build models showing what happens during each step of mitosis in an animal cell.

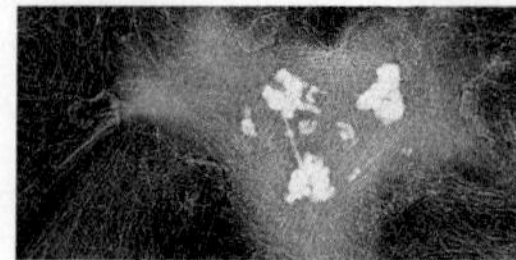

This cell is dividing abnormally, with the DNA (yellow) separating into three new cells.

Suggested Materials

- camera, tablet, computer
- craft paper, poster paper
- beans, moldable clay, pasta, yarn
- pipe cleaners, bendable wire
- scissors, glue, transparent tape

Procedure

1. Decide whether you will make a digital or physical model.
2. List the materials you will use to build your model.
3. Discuss with your teacher how you will make and present your model.
4. If building a physical model, determine what materials will represent each of these parts: cell membrane, chromosomes (sister chromatids), nuclear envelope, and spindle fibers.
5. Build and label a model of the cell that has three to four pairs of chromosomes going through mitosis. Your model should represent what happens in the cell during early prophase, late prophase (or prometaphase), metaphase, anaphase, and telophase.
6. Once you are done, clean up and return unused materials to the materials station.

Results and Analysis

1. **Evaluate** Consider the materials you selected to represent various components of the cell during mitosis. Were these materials good choices for representing the structure and function of each cell component? Why or why not?
2. **Compare** How many chromosomes were in the nucleus at the start of prophase compared with the number of chromosomes in each of the two new nuclei at the end of telophase? Explain why there is a difference.
3. **Draw Conclusions** Some herbicides work by preventing the formation of spindle fibers in plant cells. Using your models, explain how these herbicides would affect mitosis.
4. **Explain** In some cases, the mitotic spindles cannot attach properly to some of the chromosomes. As the chromosomes are pulled apart to the poles of the cell, some chromatids cannot be pulled at the same speed and start to lag. Using your models, explain which stage of mitosis would be affected and the likely result of this situation.

192 CHAPTER 7 CELL GROWTH

CELL DIFFERENTIATION

Think about how different parts of your school are organized to match their function. Classrooms may be full of chairs and desks, while a gym or other large room might have extra space for students to move. A cafeteria might have long tables for eating. The cells in multicellular organisms are in some ways like the different rooms and areas that make up your school.

Explain *Similar to different rooms in your school, multicellular organisms have different cell types. Why do you think organisms need a variety of cells?*

7.3

KEY TERMS

cell differentiation
meristem
stem cell

Multicellular Organisms

All cells in sexually reproducing organisms are derived from a single cell, the *zygote*. Consider the anatomy of the organisms in **Figure 7-12**. Each of their very distinct structures is made up of cells, and those cells in each organism all contain the same exact *genome*. How can these different structures develop if the cells all contain the same "instructions"? As the cells divide and the organism develops, different cell types express different subsets of genes. This is called *selective gene expression*. It is the key to **cell differentiation**, the process by which cells become specialized to serve different functions in a multicellular living system (Sections 9.1 and 10.2). Cell differentiation can also be affected by factors in the environment that cause changes in gene expression, such as nutrient availability and temperature.

FIGURE 7-12 What do an ocean sunfish in the Pacific Ocean, an armadillo in the United States, a South American green mantis, and a dragon blood tree in Yemen have in common? Each of these organisms started from a single cell that led to many differentiated cell types.

7.3

CELL DIFFERENTIATION

LS1.A, LS1.B

ENGAGE

This section builds on the idea of cell differentiation that was introduced in Section 7.1 in the context of the eukaryotic cell cycle. Here, students can think in terms of a regulated cell cycle as they develop an understanding of cell differentiation in which only specific subsets of genes are expressed, making it possible for many different morphological structures to develop from identical genetic information.

Objectives

- Explain how cell differentiation results in different morphological structures in multicellular organisms.
- Discuss the role of pluripotent stem cells in cell differentiation.
- Apply the concept of cell differentiation to three types of tissues in plants.

Pretest Use **Questions 5, 6,** and **7** to identify gaps in background knowledge or misconceptions.

Vocabulary Strategy

Prefixes/Suffixes/Root Relationships Review with students that the word *differentiation* is the noun form of the verb *differ*. Explain that the verb *differ* means to be different from, or unlike, something. The word *differentiation* means the process of becoming different. The *–ion* suffix is a tip that means "the action or process of."

Science Background

Levels of Organization Engage students in a discussion to assess what they know about levels of organization and about groups of specialized cells in multicellular organisms. Ask them to name types of cells or tissues that would be found in the organisms pictured in **Figure 7-12**.

Explain *Students might say that, like a school, various parts of an organism have different functions so different cell types are needed to carry them out.*

Visual Support

Specialized Cells Use the visual to engage students in a discussion about different types of cells in their bodies. Ask students to explain how all of the cells pictured can have the same DNA, yet be so different. They should understand that they all have the exact same genetic instructions, but different segments of the DNA are expressed. Have students identify the types of tissues and organs each of these cells are a part of. Students might then work in small groups to argue about how the structure of each cell is adapted to its function, such as how the dendrites and axons of a nerve cell function to transmit signals. Consider having students share arguments with the class as they all take notes.

Stem Cells

Stem cells are special undifferentiated cells. Their descendants can be either stem cells or they can become different specialized cell types that make up the tissues and organs of a complex organism. The cells shown in **Figure 7-13**, as well as many other cell types, make up the tissues and organs of multicellular organisms (Section 10.2).

Most multicellular organisms replace or repair damaged cells and tissues lost to injury, but some organisms have a unique capacity for regeneration. Some sea stars, for example, can regrow an entire body from a single arm and a bit of the central disk. No vertebrate can grow a body from a limb, but many salamanders and lizards can replace a lost tail.

FIGURE 7-13 Each cell in an organism has the same set of DNA instructions, but different genes are expressed. A small selection of human cell types is shown here.

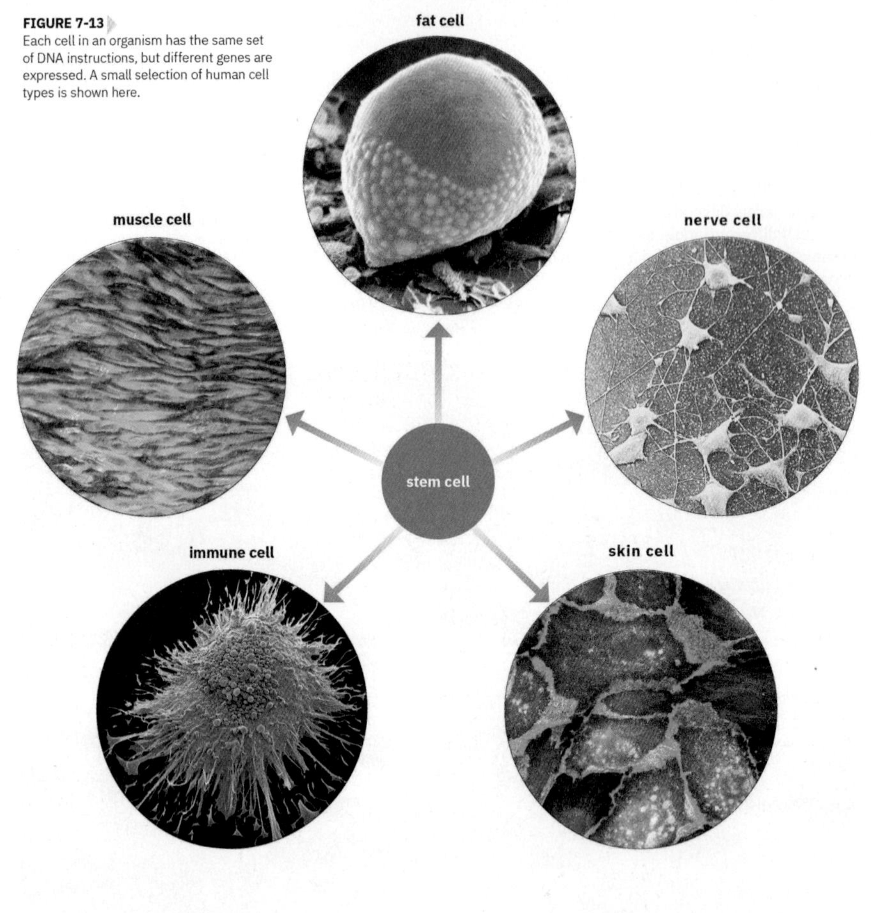

DIFFERENTIATED INSTRUCTION | English Language Learners

Using Academic Language Have pairs discuss the visual on this page using the following terms: *stem cell, DNA, segments, expressed, tissues*, and *organs*.

Beginning Provide sentence frames: *In the middle is a ___. It can become a cell in different ___ and ___. It depends on what ___ segments are ___.* Have students read the labels and point to where each of these cell types might be in their own bodies.

Intermediate Have pairs describe how a stem cell becomes a fat cell and what structures it develops. Provide a word bank with the academic words for students to use.

Advanced Have pairs explain why a stem cell can become any of the types of cells pictured. Provide a list of academic words. Encourage them to describe the structures of some of the different cell types.

BIOTECHNOLOGY FOCUS

STEM CELL RESEARCH Researchers use stem cells to investigate normal cell differentiation and disease progression, to produce healthy cells to replace damaged or diseased cells that the body cannot replace on its own, and to test the safety and efficacy of drugs.

Stem cells with pluripotency have the most promise in regenerative medicine, because they can give rise to all cell types in the human body. One source of pluripotent human embryonic stem cells is from donated or nonviable human embryos (blastocyst stage). Another source is to use "reprogrammed" cells—cells obtained from adult tissue that have been genetically engineered for pluripotency. These cells, called induced pluripotent stem (iPS) cells, behave like embryonic stem cells. To produce iPS cells, for example, fibroblasts from skin are converted to iPS cells by introducing synthetically modified RNAs. The RNAs are taken up by fibroblasts, then translated into proteins that induce dedifferentiation, which is the opposite of differentiation.

There are many remaining questions and technological hurdles for using stem cells in the treatment of human disease, but some results have been promising. An iPS cell–based therapy for a degenerative form of blindness known as age-related macular degeneration has improved vision in a small number of patients. Stem cell–based treatments for a variety of conditions have also proven effective in animal models. For example, researchers have used embryonic stem cells to cure diabetes in mice.

Axolotls, which are a type of salamander, can regenerate limbs, spinal cord, and parts of the brain and heart. This makes them a popular model for researchers who study regeneration. Mammals cannot regenerate entire structures. However, they do have the ability to regenerate tissues over time, including the skin, blood, intestinal lining, and liver. Stem cells and other similar cells are key to producing new or replacement tissues.

Human embryonic stem cells form soon after fertilization, when mitotic divisions produce a small cluster of cells. Each cell in the cluster is *pluripotent,* meaning it can develop into any of the cell types in a human body. Stem cells become more specialized as development continues. In an adult, for example, blood stem cells located in bone marrow give rise to both red and white blood cell types but not muscle cells or nerve cells.

Skin cells, blood cells, and other cell types that your body replaces on an ongoing basis arise continually from adult stem cells. Skin or blood lost to injury can be replaced. However, organs or tissues such as the heart and brain, for example, do not have a robust adult stem cell population. Thus, some tissues are not replaced if they are damaged or lost. Death of cardiac muscle cells during a heart attack permanently weakens the heart, and spinal cord injuries can result in permanent paralysis.

CCC Structure and Function *Observe the different cell types shown in* ***Figure 7-13****. Describe how you think two of the cell type structures are likely to support their functions.*

Meristem Cells A young plant grows mainly by lengthening. An older woody plant grows by lengthening and thickening. During both processes, tissues arise from the activity of meristems. A **meristem** is a region of undifferentiated cells that divides continually during the growing season. Meristem cells in plants are comparable to stem cells in animals. When they divide, some meristem cells differentiate while others remain undifferentiated. In plants, the differentiating cells give rise to specialized tissues as they enlarge and mature.

CHAPTER INVESTIGATION

Cell Differentiation in Plant Leaves

Can leaf structures be identified based on the appearance of differentiated cells?

Go online to explore this chapter's hands-on investigation on how cells in a leaf differentiate.

Science Background

Producing Liver Cells The human liver plays a key role in helping our bodies process drugs. For this reason, liver cells are used to test the safety and efficacy of new drugs. However, these cells, called hepatocytes, are not readily available. Researchers are now using stem cell technology to generate liver cells that have been found to be equally effective, if not better, than live harvested cells to test drug safety. While human liver cells can vary a lot due to donor differences, those generated using stem cells are consistent.

CHAPTER INVESTIGATION B

Guided Inquiry: *Cell Differentiation in Plant Leaves*

Time: 50 minutes

Students will observe leaf tissue to determine which cells are undifferentiated and which cells are most specialized.

Go online to access detailed teacher notes, answers, rubrics, and lab worksheets.

CCC Structure and Function

Students might note the unique structure and many connections of a nerve cell and think that will enable the cell to make many connections in the brain. They may see that the fat cell looks like it is a good structure for storage rather than a specific activity. The elongated muscle cell might be useful for contracting and tightening muscles.

BIOTECHNOLOGY FOCUS | Stem Cell Research

Social-Emotional Learning Some students may come to class with prior ideas about stem cell research. They may have strong feelings for or against the use of cells from human embryonic tissue. Emphasize that science class is a place to teach and learn about facts; however, it is important to acknowledge and respect individual opinions. Before engaging in a discussion remind students of skills such as **social awareness** and understanding others' perspectives, as well as making reasoned judgments after analyzing information, data, and facts. Encourage students with questions to research the subject using authoritative resources such as those from the National Institutes of Health or respected hospitals such as the Mayo Clinic or Cedars Sinai.

Connect to English Language Arts

Draw Evidence from Text Have students write a short essay to compare and contrast meristems in plants and stem cells in animals. Have them organize their ideas by describing similarities and then differences. Remind students to support their points with specific evidence from the text, including facts and examples. They should end with a concluding statement that follows from the information they have presented.

In Your Community

Pinching Plants When most people visit a local nursery to purchase plants for a garden or as a gift, they look for full, healthy-looking plants and pass by those that are described as "leggy" with long, stalky stems. Nursery workers help plants grow fuller and bushier by pinching the plant. Discuss with students where the best place to pinch a plant might be based on what they have learned. Different from pruning, a plant should be pinched at the plant's apical meristem where most meristematic tissue is located.

Figure 7-14 shows some of the cell types found in plants. Plants have two primary organs: roots (typically underground) and shoots (stems and leaves). These organs are made up of three different tissue types, including dermal, vascular, and ground tissue. The dermal tissue system consists of cells that cover and protect the plant's surfaces. The vascular tissue system includes the cells that carry water and nutrients from one part of the plant to another. The ground tissue system is essentially everything that is not part of the dermal or vascular tissue systems. Cells that carry out the most basic processes for survival (growth, photosynthesis, and storage) are part of the ground tissue system. More details of plant tissues, organs, and systems are described in Section 9.1.

FIGURE 7-14
Primary (apical) meristem growth is shown in a shoot tip of privet (a) and a root tip of corn (b). Meristem tissues differentiate into complex structures demonstrated by the vascular tissue cells in a stem cross section (c) and dermal, vascular, and ground tissue cells in a barley leaf cross section (d).

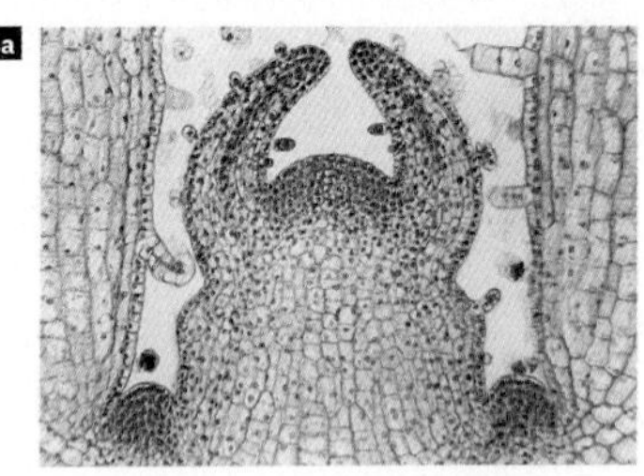

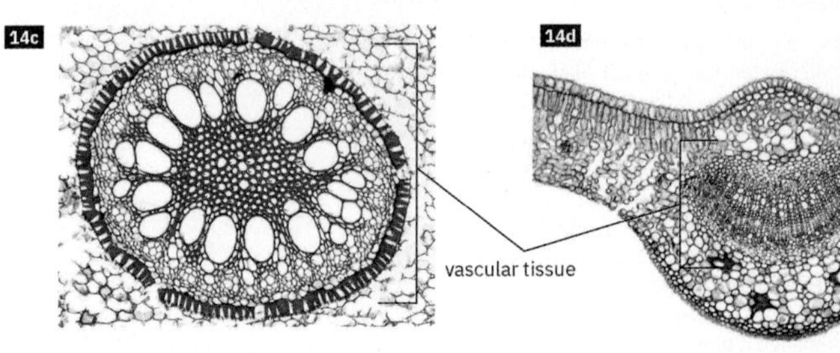

7.3 REVIEW

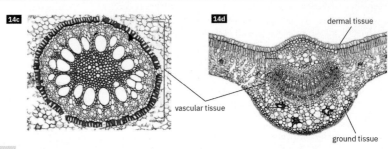

1. **Define** What causes cells to become specialized as parts of different tissues in a multicellular organism?
   - A. Cells in different tissues have unique DNA.
   - B. Cells can randomly differentiate during mitosis.
   - C. Selective gene expression causes cells to become specialized.
   - D. All cells with different functions are inherited from the parents.
2. **Explain** Give one example of why living organisms need to have stem cells, even after they are fully developed.
3. **Relate** Meristem can differentiate into which types of tissues? Select all correct answers.
   - A. dermal tissue
   - B. ground tissue
   - C. muscle tissue
   - D. nervous tissue
   - E. vascular tissue

7.3 REVIEW

1. *C* **DOK 2**
2. *Students' answers may include the ongoing replacement of skin cells and blood cells, the replacement of tissue lost to injury, or the regeneration of body parts in animals.* **DOK 3**
3. *A, B, E* **DOK 2**

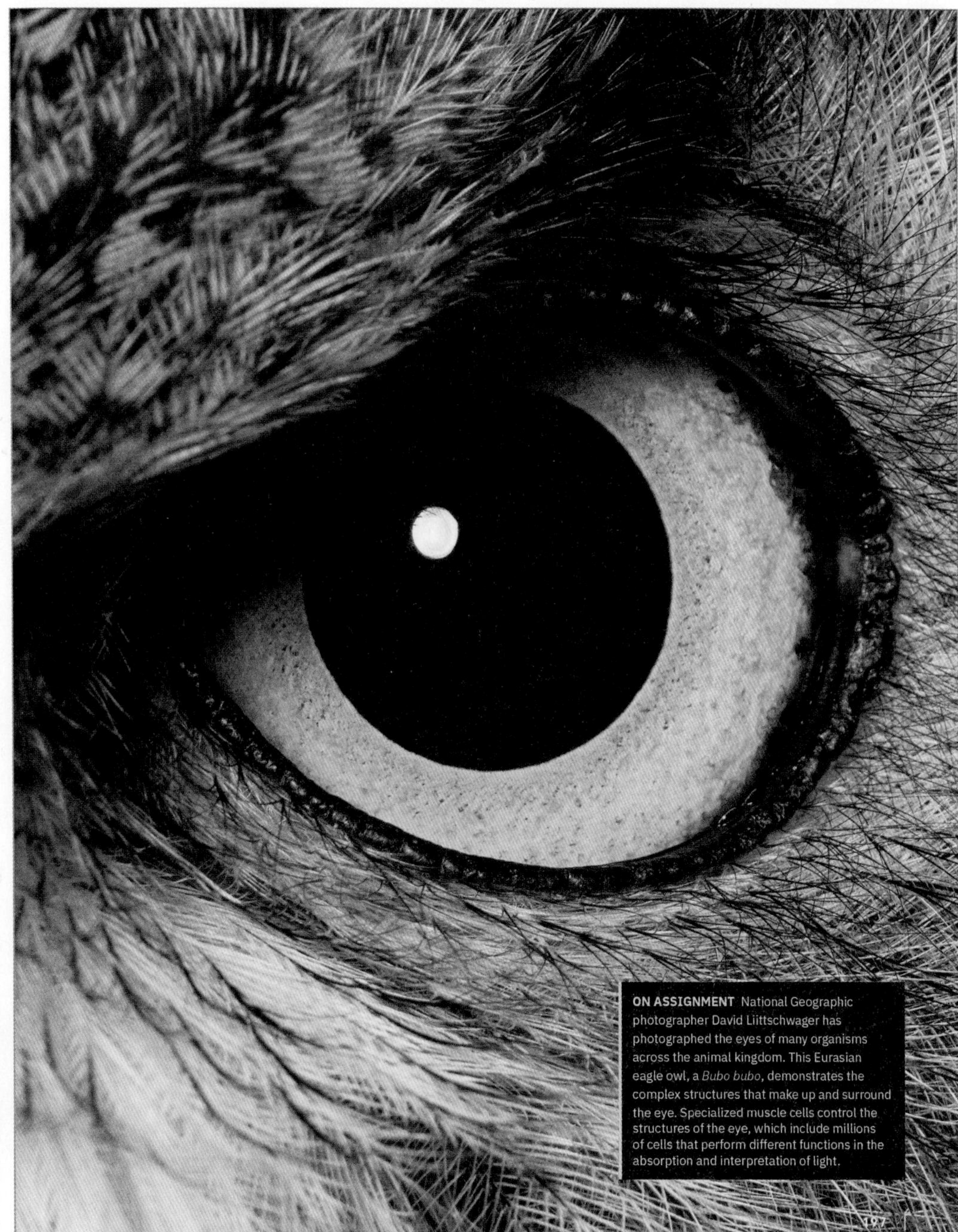

ON ASSIGNMENT National Geographic photographer David Liittschwager has photographed the eyes of many organisms across the animal kingdom. This Eurasian eagle owl, a *Bubo bubo*, demonstrates the complex structures that make up and surround the eye. Specialized muscle cells control the structures of the eye, which include millions of cells that perform different functions in the absorption and interpretation of light.

ON ASSIGNMENT

Capturing the Eyes of Organisms
David Liittschwager is a contributing photographer for National Geographic, as well as other magazines. His photographs have been exhibited at many museums, such as the American Museum of Natural History in New York City, and the National Academy of Sciences in Washington, D.C. He is the recipient of the Endangered Species Coalition Champion Award for Education and Outreach.

His work on a series of photographs that capture the detailed structure of various organisms' eyes are featured in a National Geographic article, "Inside the Eye: Nature's Most Exquisite Creation," by Ed Yong, which discusses the anatomy and physiology of different eyes and how they perceive their environments very differently.

You can learn more about David Liittchwager at the National Geographic Society's website: www.nationalgeographic .org/projects/photo-ark.

EXPLORER
DR. CAMILA ESPEJO

About the Explorer

National Geographic Explorer Dr. Camila Espejo works to protect Tasmanian Devils from a contagious cancer. As a veterinarian, Dr. Espejo brings a background in wildlife medicine and disease ecology to her research on extracellular vesicles (EV) and the progression of tumor formation. You can learn more about Dr. Espejo and her research on the National Geographic Society's website: www.nationalgeographic.org.

Science Background

Tassie populations have declined in some areas by as much as 90 percent since this cancer emerged about 20 years ago. In a population, transmission typically occurs when the mating animals bite each other and can be deadly within 6 months.

Extracellular vesicles (EV) are very small, lipid-covered particles that are secreted into the blood and other fluids. The vesicles carry RNA, DNA, and proteins. As EVs transport products between cells, they transport the biomarkers that serve as an early warning for eventual metastasis.

Connect to Careers

Wildlife Veterinarian Veterinarians are trained to diagnose and treat healthcare issues in animals, as well as establish well-care programs for their patients and owners. Specialties include domestic pets, agricultural animals, exotic animals, and wildlife animals. A wildlife veterinarian often designs and conducts field studies and leads a team to collect samples, such as blood, to assess diseases and health threats that can affect whole populations. Most have doctorates and additional training.

THINKING CRITICALLY

Infer *Sample answer: It is leading to vaccines and therapies to help cure Tassies. The same could happen with humans if this protein marker is found in humans or another is discovered.*

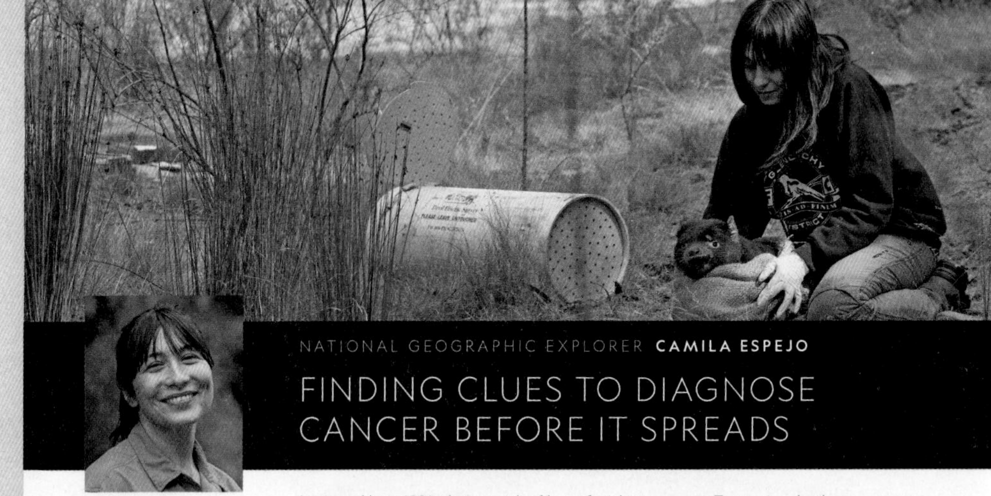

NATIONAL GEOGRAPHIC EXPLORER **CAMILA ESPEJO**

FINDING CLUES TO DIAGNOSE CANCER BEFORE IT SPREADS

Dr. Camila Espejo's background in wildlife medicine and disease ecology helps to inform her research with Tasmanian devils.

Intrigued by a 1996 photograph of large facial tumors on a Tasmanian devil, veterinarian Dr. Camila Espejo set out to learn more about the contagious cancers threatening the devils. Could studying them help diagnose cancer in humans?

Found only in Tasmania, these large marsupials were hunted nearly to extinction in the 1800s. Today, Tasmanian devils face a new threat: two forms of a fatal cancer that causes disfiguring facial tumors. Researchers originally thought the cancers were caused by a virus, because very few transmissible cancers exist. Espejo had another hypothesis. She began taking blood samples to test for extracellular vesicles (EVs). EVs are tiny membrane-enclosed structures that store and transport products between cells. They are thought to be key to the spread of cancer throughout the body.

To capture Tasmanian devils, the research team sets out special traps baited with lamb. At night, the devils go into the traps to eat and fall asleep. The researchers return in the early morning to weigh and microchip the animals. They also take biological samples such as blood. Espejo tests the blood serum for EVs and a common protein biomarker that she first discovered by growing tumor cells in the lab. By testing individual devils' blood, she confirmed that devils show this biomarker in their EVs three to six months before the appearance of tumors. She thinks the protein is the host's response to the spreading cancer cells, and "mark" that cancer is present. The biomarker is the same in all devils that have one of the facial tumor cancers. Humans with cancer also have similar biomarkers. EVs made by tumor cells advance the spread of cancer by communicating with other cells throughout the body. The presence or absence of the biomarker proteins allows the researchers to identify healthy Tasmanian devils to put into wildlife reserves.

"Currently, I am working to develop a more scalable early diagnostic test for devil facial tumor disease," Espejo says. "This will provide hope to this iconic species and boost conservation strategies to preserve devils from their possible extinction. In the long term, I hope to keep helping other species under threat, especially in the development of new diagnostic platforms."

With more conclusive data, the potential exists for using a similar biomarker test in humans to diagnose cancers before they spread.

THINKING CRITICALLY

Infer *Think about the cell cycle and how it is changed by cancer. How could identifying a biomarker related to the spread of cancer cells impact health care?*

198 CHAPTER 7 CELL GROWTH

TYING IT ALL TOGETHER
IMMORTAL CELLS: THE STORY OF HENRIETTA LACKS

HOW DO CELLS DIVIDE AND GROW?

Unlike normal cells, HeLa cells do not have the mechanism that limits the number of times a cell can divide. In 1989, scientists at Yale University discovered that HeLa cells have high levels of telomerase, an enzyme required for elongation of telomeres. Telomeres are specialized structures at the end of linear chromosomes, and they get shorter with each cell division, leading to aging and eventual cell death. HeLa cells can divide continuously in part because the telomeres are continually elongated.

HeLa cells have become an invaluable resource to biomedical researchers. However, the history of these cells, in light of the historical treatment of African Americans and concerns raised by the Lacks family, poses several complex ethical issues.

In 2013, Henrietta Lacks's genome was published online by a molecular biology lab in Europe, and was subsequently taken down because of privacy concerns. Today, members of the Lacks family sit on the board of a National Institutes of Health (NIH) working group that determines who is allowed to access HeLa genetic sequence information. Although millions of dollars have been made through the commercialization of patented HeLa cells, the family has only recently been compensated. Most of this compensation has gone to the Henrietta Lacks Foundation, which was established in 2010 to raise awareness about ethical issues in health care access and scientific research, with a particular focus on minorities and underserved communities.

In this activity, you will research the biology of HeLa cells and examine ethical issues related to their use. Work individually or in a group to complete these steps.

Gather Evidence

1. Research what makes HeLa cells unique and different from the cells you learned about in this chapter. Describe why those unique qualities have made HeLa cells such a powerful research tool.

Nature of Science

2. Consider and write brief statements about the following ethical issues presented by this case study:
   A. informed consent
   B. anonymity of samples collected from patients
   C. right to privacy regarding genetic information

Present Your Findings

3. Make a digital presentation featuring a variety of media, including images or videos, to present your research and findings to your class.

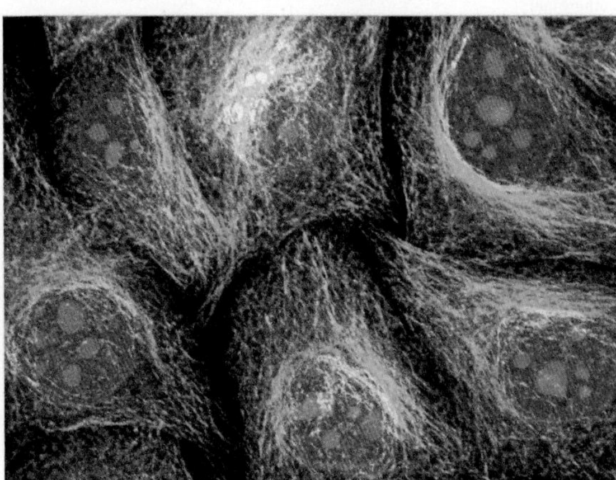

FIGURE 7-15 This confocal light micrograph of HeLa cells shows the cell nuclei in pink, mitochondria in red, and microfilaments of the cytoskeleton in green.

DIFFERENTIATED INSTRUCTION | Economically Disadvantaged Students

Make sure all students have equal access to resources, such as books and the Internet. Toward this goal, set up some time for student groups to work together with shared supplies in the classroom, or in the school's media center or library.

Check with your librarian to see if there are copies available of *The Immortal Life of Henrietta Lacks* by Rebecca Skloot, winner of multiple awards including the National Academy of Science, National Academy of Engineering, and Institute of Medicine's 2011 Communication Award for Best Book.

Additionally, encourage students working within groups to be aware of their teammates' strengths and talents. Those with talents in illustration, graphic design, or computer design may take the lead with the media. Students with great research skills can take the lead in gathering information.

TYING IT ALL TOGETHER

ELABORATE

The Case Study described unique characteristics of HeLa cells, so named after Henrietta Lacks, who died from aggressive cervical cancer in 1931. Although the medical breakthroughs that resulted from the use of these cells in research are significant, this case also brings attention to the issue of informed consent, as the patient was not informed at the time the tissue was taken. In the Tying It All Together activity, students will gather evidence as to the significance of HeLa cells from a biomedical perspective, as well as consider patient rights around informed consent and privacy.

This activity is structured so that students may complete the steps as they study the chapter content. **Step 1** correlates with the chapter opener, as well as content in Sections 7.1 and 7.2. Content in Section 7.3 may bring in other considerations as student look at the benefits of biomedical research in terms of patient rights in **Step 2**. Students will then need to synthesize ideas from all sections and their research to complete **Step 3**.

Alternatively, this can be a culminating project at the end of the chapter. The gathering of evidence here to support a claim or argument is a process that students may find useful in the Unit 1 Activity.

Science Background

Immortal Cell and Henrietta Lacks It wasn't until 1971 that Henrietta's identity was revealed. Seventy years after Henrietta's death, the family and children have received recognition, compensation, and control over the use of her cells. Henrietta's story has also resulted in changes to the law regarding patient rights and informed consent. Today, the Henrietta Lacks Foundation, which provides assistance to people and families in similar situations, has received donations from some of the labs that used and profited from her immortal cells.

CHAPTER 7 **REVIEW**

EVALUATE

REVIEW KEY CONCEPTS

1. *A, C, D* **DOK 2**
2. *G1: The cell prepares to replicate its DNA; S: Chromosomes are duplicated; G2: Proteins needed for chromosomal movement are synthesized; mitosis: Duplicate chromosomes move apart.* **DOK 2**
3. *bacterial cell division: binary fission; eukaryotic cell division: cancer, oncogenes, mitosis, cell differentiation* **DOK 2**
4. *Checkpoint genes code for proteins that are involved in regulating the cell cycle; If newly synthesized DNA is damaged and cannot be repaired, tumor formation may result; A tumor may develop in a multicellular eukaryotic organism as the result of a mutation in a proto-oncogene.* **DOK 3**
5. *differentiated cells: skin cells, immune cells, vascular tissue cells; undifferentiated cells: embryonic stem cells, meristem cells, stem cells* **DOK 2**
6. *Microtubules attach to the centromere region that connects duplicate chromosomes.* **DOK 1**
7. *Sister chromatids will not be able to separate from each other and move to opposite ends of the dividing cell.* **DOK 2**
8. *Once a cell differentiates, it is no longer pluripotent; Undifferentiated cells can divide to replace injured cells; Differentiated cells selectively express genes needed for one type of cell.* **DOK 2**
9. *Sample answer: Chemical change in DNA could cause a proto-oncogene to become an oncogene. This mutation would cause the cell to lose its ability to regulate its own cell division process and cause it to continue dividing uncontrollably.* **DOK 3**
10. *Sample Answer: Each of the two descendant cells that form as a result of cell division must receive exact copies of the DNA present in the original cell.* **DOK 2**

CHAPTER 7 **REVIEW**

REVIEW KEY CONCEPTS

1. **Compare** Which activities occur during cellular division in both bacteria and eukaryotes? Select all correct answers.
   - **A.** DNA is replicated.
   - **B.** Chromosomes condense around histones.
   - **C.** Proteins are synthesized as a cell prepares to divide.
   - **D.** New plasma membrane forms as cell division proceeds.
   - **E.** The cell nucleus divides followed by division of the cell cytoplasm.
2. **Relate** Match each statement with one stage of the cell cycle: G1, S, G2, or mitosis.
   - Chromosomes are duplicated.
   - Duplicate chromosomes move apart.
   - The cell prepares to replicate its DNA.
   - Proteins needed for chromosomal movement are synthesized.
3. **Contrast** Which terms are relevant to bacterial cell cycles, and which are relevant to eukaryotic cell cycles?
   - binary fission
   - mitosis
   - cancer
   - oncogenes
   - cell differentiation
4. **Explain** Which statements are true concerning cell cycle control in eukaryotes? Select all correct answers.
   - **A.** A tumor is a group of cells that has stopped dividing.
   - **B.** Checkpoint genes code for proteins that are involved in regulating the cell cycle.
   - **C.** Most cells in multicellular organisms divide constantly throughout the organism's life.
   - **D.** If newly synthesized DNA is damaged and cannot be repaired, tumor formation may result.
   - **E.** A tumor may develop in a multicellular eukaryotic organism as the result of a mutation in a proto-oncogene.
5. **Classify** Sort each cell type into one of two categories: differentiated cells, undifferentiated cells.
   - embryonic stem cells
   - skin cells
   - immune cells
   - stem cells
   - meristem cells
   - vascular tissue cells
6. **Describe** Which statement describes an event that occurs during mitosis?
   - **A.** DNA unwinds from histones to assume an elongated shape.
   - **B.** Sister chromatids remain paired, moving together as new descendant cells form.
   - **C.** Microtubules attach to the centromere region that connects duplicate chromosomes.
   - **D.** Homologous chromosomes are pulled apart, each descendant cell receiving a different set of chromosomes.
7. **Predict** Suppose a mutation event prevents a cell from producing tubulin subunits that can assemble to form microtubules. Which outcome would most likely occur as this cell undergoes cell division?
   - **A.** The cell membrane will not be able to break down to allow chromosomal movement.
   - **B.** Replication of chromosomes will not occur to produce genetically identical sister chromatids.
   - **C.** Chromosomes will not be able to fully condense and will tangle and break as division proceeds.
   - **D.** Sister chromatids will not be able to separate from each other and move to opposite poles of the dividing cell.
8. **Relate** Which claims are true about differentiated and undifferentiated cells? Select all correct answers.
   - **A.** Once a cell differentiates, it is no longer pluripotent.
   - **B.** Undifferentiated cells can divide to replace injured cells.
   - **C.** As a cell undergoes differentiation, it loses genetic material.
   - **D.** Undifferentiated cells express all genes encoded in their DNA.
   - **E.** Differentiated cells selectively express genes needed for one type of cell.
9. **Identify Cause and Effect** Certain chemical substances have been shown to react with DNA to change its chemical structure. As a result, these substances have been labeled as carcinogens, or substances that cause cancer. Why might chemical changes in DNA lead to cancer?
10. **Explain** Why does DNA have to be accurately duplicated during cell division?

200 CHAPTER 7 CELL GROWTH

CRITICAL THINKING

11. *Sample answer: Cells would most likely increase synthesis of DNA repair enzymes during S phase, when they are replicating their DNA. Cells would need DNA repair enzymes available at this stage to correct any mistakes before the DNA is packaged into chromosomes.* **DOK 3**
12. *Sample answer: So far, the student has only the homologous pairs of each chromosome for her model. She will need to cut four more pieces of yarn as she did earlier to make exact duplicates of each homologous pair. This will be necessary to model the cell after it has passed through the S phase when DNA is replicated.* **DOK 3**
13. *Sample answer: Both processes occur during cell division with mitosis first followed by cytokinesis. Mitosis involves division of the nucleus and cytokinesis involves division of the cytoplasm. If mitosis occurred without cytokinesis, the cell would end up with two nuclei. If cytokinesis occurred without mitosis, one descendant cell would end up with no nucleus.* **DOK 2**

CRITICAL THINKING

11. **Infer** Certain enzyme levels increase during certain phases of the cell cycle and decrease during others. In which phase of cell division—G1, S, G2, or mitosis—would you expect to see an increase in the level of DNA repair enzymes? Explain your reasoning.

12. **Model** A student wants to model cell division for a diploid cell having two chromosomes. To represent the two chromosomes, she has cut two short pieces of yarn of equal length and two long pieces of yarn of equal length. What will the student need to do to continue making her model? Explain your reasoning.

13. **Relate** How are cytokinesis and mitosis related? Why are both necessary for a cell to successfully divide?

14. **NOS Science, Technology, and Society** In 2006, researchers at Kyoto University in Japan developed a technique for inducing mouse skin cells to "dedifferentiate" back to a pluripotent stem cell state. They did this by changing gene expression in the cells so that the cells reverted to their original undifferentiated form. These cells are called induced pluripotent stem cells, or iPS cells. Since then, researchers have produced human iPS cells from human skin cells. Theoretically, human iPS cells could be used to produce any type of human cell, including sperm and egg cells. What ethical questions does this situation raise? Pose at least two questions.

MATH AND ENGLISH LANGUAGE ARTS CONNECTIONS

1. **Reason Abstractly and Quantitatively** Complete the table to show the total number of cells resulting after each round of cell division.

Generation	Total number of cells
0	1
1	2
2	
4	

2. **Infer From Statistics** A section of a root tip was prepared for observation under a microscope. One thousand cells were analyzed to determine the number of cells at each stage of mitosis. The results of this analysis are summarized in the table.

Stage of mitosis	Number of cells
early prophase	26
late prophase	4
metaphase	24
anaphase	19
telophase	14

What can you infer from this information about the amount of time that any given cell spends in mitosis? In interphase? Explain your reasoning.

Read the following passage. Use evidence from the passage to answer the question.

Planarian flatworms are small, soft-bodied invertebrates usually less than 3 cm in length. They have wide, flat bodies and two eyes on the top of their heads. If you slice a planarian flatworm in half, each half will grow back into a healthy flatworm. In fact, if you cut a flatworm into 200 pieces, each piece will regenerate into an entirely new and healthy worm. Most other animals lack planarians' ability to regrow lost body parts, but they can regrow cells to some degree, such as when a wound heals. The processes that allow a wound to heal are similar to those that enable a flatworm to regenerate lost parts of its body.

3. **Draw Evidence From Texts** Use evidence from the text to contrast stem cell locations and capabilities in planarian flatworms versus humans.

4. **Write Explanatory Text** A diagram in a textbook depicts an embryo of a frog with an arrow pointing toward a tadpole. Another arrow points from the tadpole to an adult frog. Write text to explain how this series of illustrations involves both cell division and differentiation.

► REVISIT BACTERIA IN YOUR GUT

Gather Evidence In Section 1 of this chapter, you learned about factors affecting the growth of bacteria. In addition, there is an extensive population of viruses, called bacteriophages, that can infect and kill bacteria.

1. How do you think bacteriophages in your gut might affect bacteria populations there?

2. What other questions do you have about the bacteria and viruses that live in your digestive system?

► BACTERIA IN YOUR GUT

1. *Sample Answer: Bacteriophages can kill bacteria and keep the population in check. In normal conditions, bacteriophages should help stop any particular type of bacteria from growing too much.* **DOK 2**

2. *Sample Answer: How many viruses are there in the human gut? Is there a different bacteriophage or other virus for each type of bacteria? Why do bacteria have a different and specific type of virus that infects them?* **DOK 3**

14. *Sample answer: Should iPs cells be used to produce human embryos? Should individuals be able to decide whether tissue removed from their bodies will be used to produce iPS cells?* **DOK 3**

MATH AND ELA CONNECTIONS

1. *Generation 2: 4 cells; Generation 4: 16 cells.* **DOK 2**

2. *Sample answer: A cell spends a very short amount of time in mitosis and most of its time in interphase. This can be concluded from the low number of cells observed to be in mitosis (87) and the large number of cells observed to be in interphase.* **DOK 3**

3. *Sample answer: Because planarian flatworms can regrow their entire body from any small body part, they must have pluripotent stem cells spread throughout their bodies. Humans, on the other hand, must not have these type of stem cells because they cannot regrow lost body parts this way. However, humans must have stem cells that allow wounded tissue to heal and new skin to grow to replace damaged skin.* **DOK 2**

4. *Sample answer: Embryonic stem cells in the frog embryo undergo cell division to add new cells in order to grow into a tadpole. These cells undergo differentiation to form structures including eyes, fins, and tails. Later, as the tadpole changes into an adult frog, more new cell types are formed through cell division of stem cells, and these differentiate to give rise to another set of new structures, including legs and feet.* **DOK 3**

Class Discussion

Use each of the following questions to engage students in a collaborative discussion relevant to each chapter.

1. How can you use the structural characteristics of molecules to predict their functions within a living organism? Provide specific examples.
2. Consider the mechanisms and reactions within the chloroplasts of a plant cell during photosynthesis. How can an understanding of these structures and processes help scientists develop artificial photosynthesis? What would be the benefits to that technology?
3. How does the structure of a cell relate to how it grows and reproduces? How does the structure of a bacterium help it grow and thrive in the gut environment?

Go Online

Three Performance Tasks for Unit 2 are available on MindTap to assess students' mastery of the following NGSS Performance Expectations:

- Task 1: HS-LS1-4
- Task 2: HS-LS1-5 and LS1-7
- Task 3: HS-LS1-6

Go Online

The Virtual Investigation, **Bacteria in the Digestive System**, includes observational evidence about the role of the microbiome in human health that students may find useful in supporting their claims about what bacteria in the do in the intestines.

UNIT 2 **SUMMARY**

CELL SYSTEMS

CHAPTER 5
MOLECULES IN LIVING SYSTEMS

How are the molecules of life assembled?

- Atoms join by chemical bonds to form molecules and compounds. In a chemical reaction, energy is used to break these bonds and the atoms are recombined into new substances.
- Life depends on the unique properties of water. The solubility, high cohesion, and ability to regulate temperature are related to the fact that water molecules are highly polar and tend to form hydrogen bonds with each other.
- Many organic molecules are assembled from simpler carbon-based subunits. The major groups of organic compounds are carbohydrates, lipids, proteins, and nucleic acids.
- Chemical reactions can absorb or release energy, but they require activation energy to proceed. Enzymes lower the required energy and increase the rate of a reaction.

CHAPTER 6
CELL STRUCTURE AND FUNCTION

How do structures work together in cell systems?

- Eukaryotic cells have more complex structures than prokaryotic cells. These include a nucleus to protect DNA and membrane-enclosed organelles that perform specific functions within cells.
- The lipid bilayer structure of membranes and associated proteins allows cells to control what materials enter and exit cells and organelles.
- Aerobic cellular respiration uses glucose and oxygen to generate ATP that can be used in reactions. Other respiration and fermentation pathways are used to generate ATP in the absence of oxygen.
- Photosynthesis uses light and carbon dioxide to make glucose that cells can use to store energy.

CHAPTER 7
CELL GROWTH

How do cells divide and grow?

- Cells duplicate DNA, grow, and divide through the cell cycle. This cycle is influenced by both genetic and environmental factors.
- Mitosis organizes chromosomes to ensure cells end up with the correct number of chromosomes.
- Mutations in genes that control the cell cycle can lead to the development of cancerous tumors.
- Different cell types in a multicellular organism contain the same DNA, but differences in gene expression allow stem cells and meristems to form different cell types with specific functions.

Go Online **VIRTUAL INVESTIGATION**

Bacteria in the Digestive System
What do bacteria do in your intestines?
Put on your lab coat to grow some bacteria and study how they live and interact.

WHAT DO BACTERIA DO IN YOUR INTESTINES?

Develop a Model In this unit you learned how molecules interact to form cell systems and how cell systems form the tissues that make up complex multicellular organisms. You learned in Unit 1 how many different types of organisms exist together to maintain balanced ecosystems.

Human intestines enclose a complex ecosystem that hosts many species of bacteria, archaea, viruses, and fungi. Together, all these microscopic organisms are known as the gut microbiome. Due to its complexity, and because many of these organisms cannot live outside the intestine, it has been difficult to determine all the types of organisms and their roles. Recently, scientists have been able to use genetic technology to assess the microbiome's diversity of DNA. They have identified hundreds of previously unidentified species of bacteria and viruses that routinely colonize human intestines. These organisms maintain a balanced ecosystem within the intestines, resisting infection and contributing to the health of the human host.

Scientists have developed many new methods to learn more about the symbiotic relationship between hosts and their gut microbiomes and to study the interactions between the colonizing organisms. Some studies attempt to understand differences in human microbiomes, as influenced by factors such as region and diet. Other laboratory studies use "gut-on-a-chip" models to mimic the complexity of the intestinal system by directly studying how bacterial species interact with intestinal cells in very small, controlled quantities. Our understanding of the intestinal microbiome is constantly changing and improving.

Claim Make a claim about the role of *E. coli* and other bacteria in the intestines of humans.

Evidence Use the evidence that you gathered throughout the unit to support your claim.

Reasoning To help illustrate your reasoning, develop a model that describes the structures of the bacteria and how they function within the host intestine, as well as how the diversity of organisms in the intestine affects the bacteria.

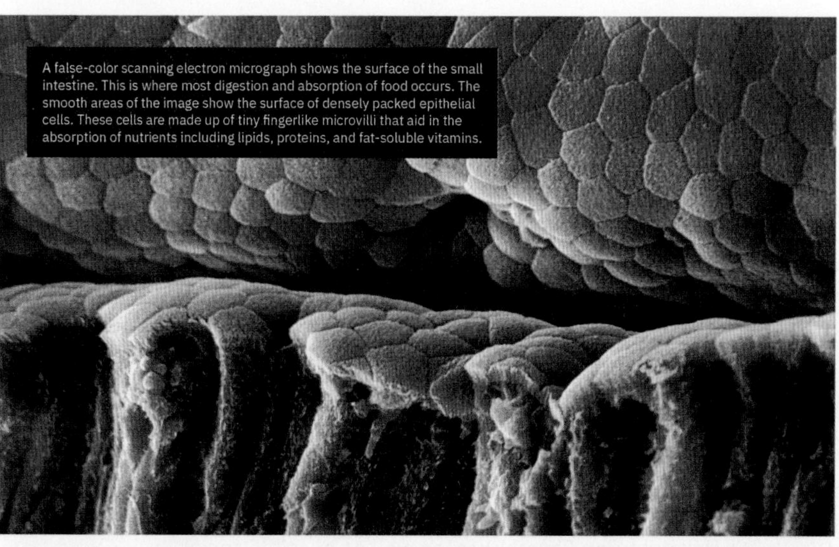

A false-color scanning electron micrograph shows the surface of the small intestine. This is where most digestion and absorption of food occurs. The smooth areas of the image show the surface of densely packed epithelial cells. These cells are made up of tiny fingerlike microvilli that aid in the absorption of nutrients including lipids, proteins, and fat-soluble vitamins.

DIFFERENTIATED INSTRUCTION | English Language Learners

Subject-Verb Agreement Have students edit their writing for subject-verb agreement.

Beginning Have students choose the correct verb form in this sentence: Bacteria help/helps in digestion. Point out that *bacteria* is plural, although it does not end in -s or -es, so the correct form is *help*. Have students check for this error in their writing.

Intermediate Have students choose three sentences from their writing and identify the subject and verb in each. Then, have them tell a partner whether the subject-verb agreement is correct. Provide a sentence frame: *The subject is _______, which is singular/plural. The verb is _______, which is a singular/plural form.*

Advanced Have students read this sentence from the caption: *The smooth areas of the image show the surface of densely packed columnar epithelial cells.* Have students underline the subject and explain why the verb form is *show* rather than *shows*. (*The subject is areas, which is plural, not image, which is singular.*)

UNIT 2 **ACTIVITY**

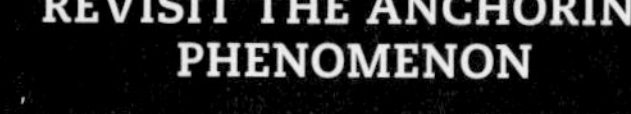

REVISIT THE ANCHORING PHENOMENON

 Claim

 Evidence

 Reasoning

This Unit Activity asks students to formulate a claim to answer the question posed at the beginning of the unit, cite evidence they have obtained throughout the unit, and use the concepts they have learned to explain how this evidence supports their claim.

C Formulating a Claim Students should write their answer to the question "What do bacteria do in your intestines?" in the form of a single sentence. Examples of claims that students might make using terms and concepts from the unit chapters include: "Bacteria help in the final stages of digestion of food" or "Bacteria produce some vitamins needed by the body."

E Citing Evidence Have students review their answers to the Bacteria in Your Gut questions in Sections 5.3, 6.1, and 7.1 and in each Chapter Review as they determine what evidence to use to support their claim. The Unit Video 1 and the online Virtual Investigation, Bacteria in Your Gut, may provide additional evidence. If you wish to have students look for evidence from sources other than this text, they can find useful information from National Geographic.

R Explaining Reasoning Have students work in groups to draw a model of the prokaryotic cell cycle. They should then describe the conditions that would promote optimal cell division and represent mathematically how many cells would arise from one bacterial cell division if it occurred every 20 minutes.

Go online to access the Student Self-Reflection and Teacher Scoring rubrics for this activity

Unit Anchoring Phenomenon: Rainforest Connections

Use the Driving Question to help frame the Anchoring Phenomenon as an investigable subject and motivate student learning. Leverage the rainforest connections prompts within each chapter to connect concepts back to the unit's Driving Question, supporting students in gathering evidence and asking their own research questions so that they are equipped to complete the Unit Activity.

NATIONAL GEOGRAPHIC | Meet the Explorer

Varun Swamy is an ecologist who studies the interactions that exist between organisms in tropical rainforests. Watch **Unit Video 2**, Explorers at Word: Varun Swamy, to engage student interest in tropical ecosystem interactions and the Anchoring Phenomenon.

Virtual Investigation

Communication in the Rainforest Students learn about the many different organisms that make up the El Yunque Rainforest ecosystem in Puerto Rico and gather evidence to describe how those organisms communicate.

NGSS Progression

Middle School

- **MS-LS1-1** Conduct an investigation to provide evidence that living things are made of cells; either one cell or many different numbers and types of cells.
- **MS-LS1-2** Develop and use a model to describe the function of a cell as a whole and ways parts of cells contribute to the function.
- **MS-LS1-3** Use argument supported by evidence for how the body is a system of interacting subsystems composed of groups of cells.
- **MS-LS1-4** Use argument based on empirical evidence and scientific reasoning to support an explanation for how characteristic animal behaviors and specialized plant structures affect the probability of successful reproduction of animals and plants respectively.
- **MS-LS1-8** Gather and synthesize information that sensory receptors respond to stimuli by sending messages to the brain for immediate behavior or storage as memories.
- **MS-LS2-2** Construct an explanation that predicts patterns of interactions among organisms across multiple ecosystems.

High School

- **HS-LS1-2** Develop and use a model to illustrate the hierarchical organization of interacting systems that provide specific functions within multicellular organisms.
- **HS-LS1-3** Plan and conduct an investigation to provide evidence that feedback mechanisms maintain homeostasis.
- **HS-LS2-8** Evaluate the evidence for the role of group behavior on individuals' and species' chances to survive and reproduce.
- **HS-ETS1-3** Evaluate a solution to a complex real-world problem based on prioritized criteria and trade-offs that account for a range of constraints, including cost, safety, reliability, and aesthetics, as well as possible social, cultural, and environmental impacts.

Claim, Evidence, Reasoning Students can make a claim about the unit phenomenon, gather evidence, and revisit their claim periodically to evaluate how well the evidence supports it. The driving question presented in the Case Study of each chapter can get students invested in chapter topics and in working toward answering the unit's driving question, "How do rainforest species communicate?" In the Unit Activity, students can practice scientific reasoning and argumentation to show how the evidence supports their claim.

Follow the Anchoring Phenomenon How do rainforest species communicate?

Gather evidence with . . .	Chapter 8	Chapter 9	Chapter 10	Unit Activity
CASE STUDY	How are the characteristics of an organism related to its place in an ecosystem?	How do plants respond to changes in the environment?	How do animals respond to changes in the environment?	Revisit the unit's Anchoring Phenomenon of the systems in a rainforest that allow organisms to survive, grow, reproduce, and respond to the environment. **Claim, Evidence, Reasoning** Students use the evidence they gathered throughout the unit to support their claim with reasoning. **Go online** to access Student Self Reflection and Teacher Scoring rubrics for this activity. **English Language Learners** Varying Sentences
MINILAB	What features of paramecia or euglenas can you identify?	What can the stomata of a plant tell you about its environmental conditions?	Do body systems react to visual stimuli faster than audio stimuli?	
LOOKING AT THE DATA	Students will analyze and interpret data to learn about the C-value enigma.	Students will analyze and use data to predict when the first bud burst and first flowering days will be in the future.	Students will analyze the metabolic rate of different species to identify if they are endotherms or ectotherms.	
TYING IT ALL TOGETHER REVISIT THE CASE STUDY	Students will research the characteristics of *Aspergillus niger* that enable it to thrive in space.	Students will research how plant systems interact in the removal of harmful substances from contaminated water or soil.	Students will research the body system interactions involved in bird migration.	
Chapter Review: Revisit Rainforest Connections	Reflect on the roles of microscopic organisms in rainforests.	Students explain responses caused by stimuli that affect plant root systems.	Consider how organisms communicate in social groups.	
Virtual Investigation: Communication in the Rainforest	Students learn about the many different organisms that make up the El Yunque Rainforest ecosystem in Puerto Rico and gather evidence to describe how those organisms communicate.			
Chapter Investigation A	How can you determine the effectiveness of an antimicrobial?	How is plant structure connected to function?	How does exercise affect homeostasis?	
Chapter Investigation B	What classification system should be used to sort different algae?	How do the roots of a plant develop in response to moisture?	How can you design, build, and test an effective device to monitor animal behavior?	

Planning Your Investigations

Each chapter features a Minilab and two full Investigations that offer hands-on opportunities for students to engage in Science and Engineering Practices.

Advance Preparation For Chapter 8, you will need to order live specimens of *Paramecium* or *Euglena*, prepared slides of ten different kinds of algae, and growth medium. For Chapter 9, you will need to prepare leaf impressions, obtain prepared premade flower slides, and obtain plant seeds or seedlings. For Chapter 10, you will need to produce a list of endangered species that exhibit group behaviors. Advance preparation notes for all labs are included in the Teacher's version.

Chapter 8	Title	Time	Standards
Minilab	Features of *Paramecium* and *Euglena*	30 minutes	SEP Developing and Using Models SEP Constructing Explanations and Designing Solutions
Investigation A **Guided Inquiry**	Classification Systems	100 minutes or 2 days	SEP Engaging in Argument from Evidence SEP Scientific Knowledge is Open to Revision in Light of New Evidence CCC Patterns
Investigation B **Design Your Own**	Effects of Antimicrobials	150 minutes over 3 days	SEP Constructing Explanations and Designing Solutions SEP Engaging in Argument from Evidence CCC Structure and Function
Chapter 9	**Title**	**Time**	**Standards**
Minilab	Investigating Leaf Stomata	30 minutes	SEP Developing and Using Models CCC Systems and System Models CCC Stability and Change
Investigation A **Guided Inquiry**	Connecting Plant Structures with Their Functions	100 minutes over 2 days	SEP Developing and Using Models CCC Systems and System Models
Investigation B **Open Inquiry**	Homeostasis in Plants	205–255 minutes over 1–3 weeks	SEP Planning and Carrying Out Investigations CCC Cause and Effect CCC Stability and Change
Chapter 10	**Title**	**Time**	**Standards**
Minilab	Comparing Reaction Speed	30 minutes	SEP Developing and Using Models CCC Systems and System Models
Investigation A **Guided Inquiry**	The Effect of Exercise on Homeostasis	80 minutes over 2 days	SEP Developing and Using Models SEP Planning and Carrying Out Investigations SEP Scientific Investigations Use a Variety of Methods CCC Cause and Effect CCC Systems and System Models
Investigation B **Open Inquiry**	Monitoring Animal Behavior	150 minutes over 3 days	HS-ETS1-3

Assessment Planning

UNIT 3 PERFORMANCE TASKS

Three performance-based assessments target the NGSS Performance Expectations with three-dimensional activities to measure student mastery. Rubrics are available online for Performance Tasks.

Performance Tasks

	Title	Overview	PEs Addressed	Use After
1	*How Do Systems Interact to Maintain Homeostasis in Plants?*	Students research how plants respond to water stress and build a model to illustrate the response.	HS-LS1-2	Chapter 9
2	*How Can We Test Systems That Interact to Maintain Homeostasis in Humans?*	Students research and model one mechanism the human body uses to maintain homeostasis and design an experiment to test how human body systems interact during exercise.	HS-LS1-2 HS-LS1-3	Chapter 10
3	*How Important Is Group Behavior to the Survival of Individuals in a Population?*	Students research group behavior in a selected species and compose a multimedia story about the life of a population of this species faced with environmental or ecological threats.	HS-LS2-8	Chapter 10

Additional Resources

Use a search engine to find these resources on the Internet.

VIDEOS

"We're covered in germs. Let's design for that." TED

Jessica Green, engineer and biodiversity scientist, answers the question, "Can we design buildings that encourage happy, healthy microbial environments?" in a nine-minute segment.

Search: "microbes design TED"

"You Didn't Know Mushrooms Could Do All This" National Geographic

This video is a four-minute short on how mushrooms are being tested in innovative and imaginative ways to help society by utilizing the natural abilities of various purposes.

Search: "you didn't know mushrooms"

ARTICLES

Why Australia's Trash Bin-Raiding Cockatoos Are the 'Punks of the Bird World' Smithsonian Magazine

The article describes a research study about how cockatoos learn garbage-foraging behavior within their social groups.

Search: "Smithsonian trash cockatoos"

Chimpanzees more likely to share tools, teach skills when task is complex Washington University in St. Louis

A new study led by researchers at three universities finds that chimpanzees that use a multi-step process and complex tools to gather termites are more likely to share tools with novices.

Search: "chimpanzees share tools"

ENGAGE

Unit Storyline

In this unit, students will answer the question, "How do rainforest species communicate?" Each chapter provides part of the answer to this Driving Question.

Chapter 8 *How are the characteristics of an organism related to its place in an ecosystem?*

Students learn about the different classes of microorganisms and viruses. The lesson first introduces bacteria and archaea, then moves on to protists and fungi. Students then learn about viruses and how they are different from microorganisms.

Chapter 9 *How do plants respond to changes in the environment?*

Students learn about plant origins and diversity, as well as how materials are transported in plants. Students then learn about the factors that allow for plant growth and reproduction before moving on to how plants respond to the environment.

Chapter 10 *How do animals respond to changes in the environment?*

Students learn about animal origins and diversity before learning how animal systems function to maintain homeostasis. The focus then shifts to how animal behaviors support their reproductive success.

Unit Activity Students revisit the Anchoring Phenomenon to make a claim about the Driving Question and apply reasoning to the evidence they have gathered throughout the unit.

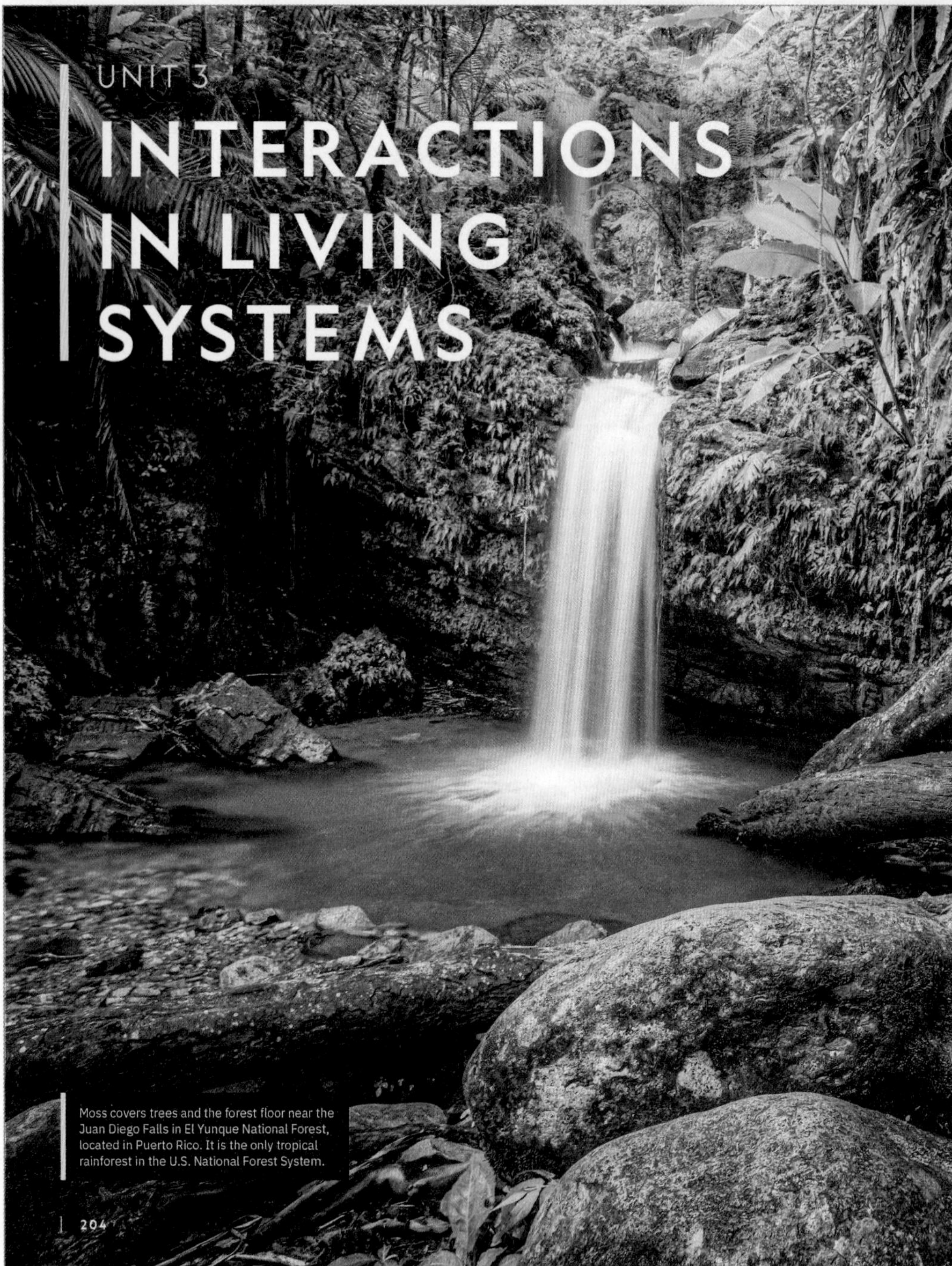

Moss covers trees and the forest floor near the Juan Diego Falls in El Yunque National Forest, located in Puerto Rico. It is the only tropical rainforest in the U.S. National Forest System.

SCIENCE BACKGROUND

Rainforests Rainforests can be found on every continent on the planet, except Antarctica. These regions host 50% of all animal and plant species but make up less than 6% of Earth. They are characterized by receiving more than 200 cm of rain annually and consisting of mostly evergreen plants. They also typically have very dense vegetation across four layers.

Tropical Rainforest El Yunque National Rainforest is the only tropical rainforest in the United States. In addition to high amounts of rain, tropical rainforests are warm year-round and fall between the Tropic of Cancer and the Tropic of Capricorn. These diverse regions have such high humidity that they produce up to 75% of their own rainfall, which can reach 1000 cm annually.

Temperate Rainforest These rainforests are found in coastal and mountainous regions in the areas lying between 30 and 60 degrees north or south of the equator. While temperate rainforests receive less rain than tropical rainforests, they do experience up to 500 cm of rain annually.

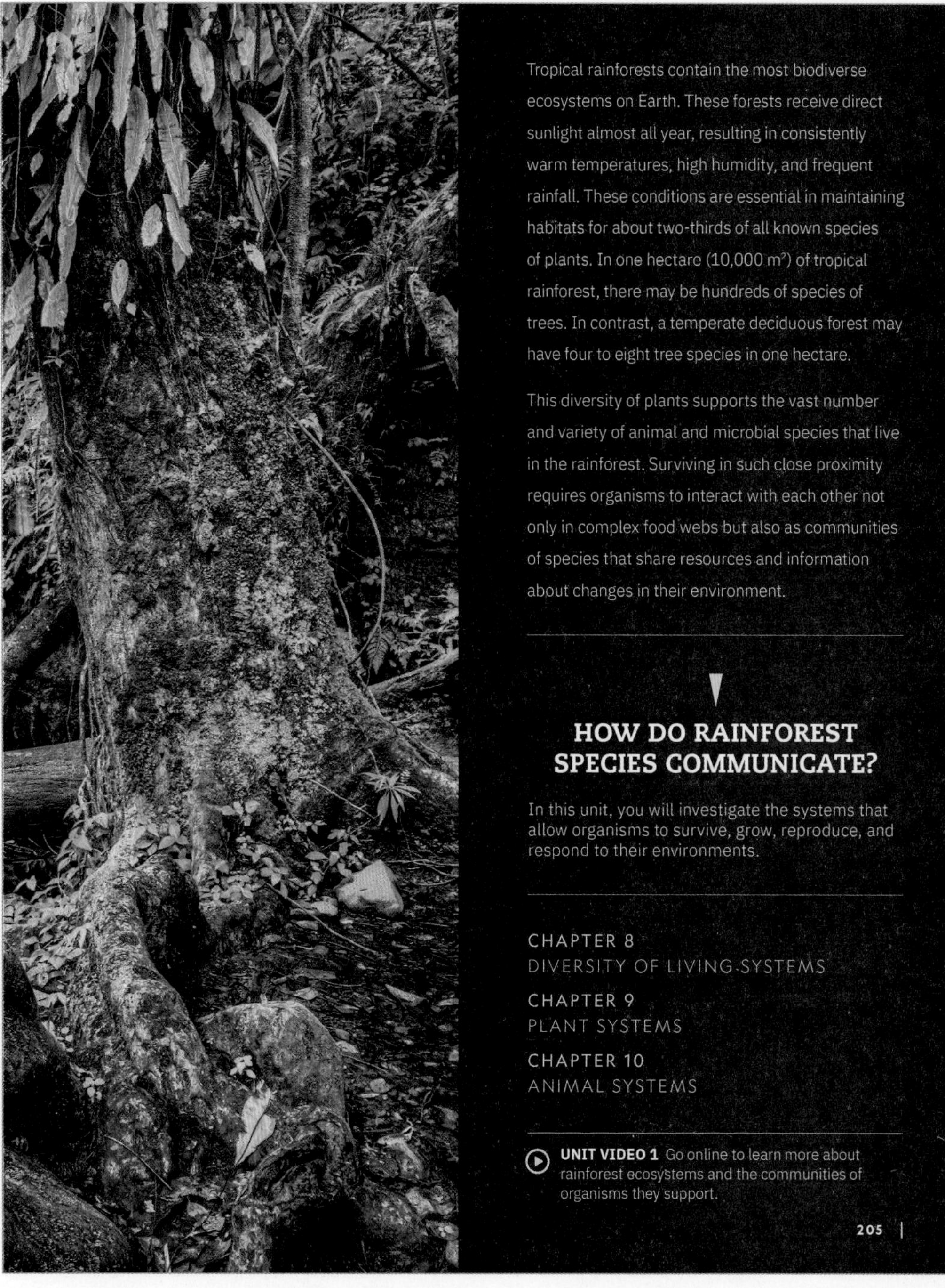

Tropical rainforests contain the most biodiverse ecosystems on Earth. These forests receive direct sunlight almost all year, resulting in consistently warm temperatures, high humidity, and frequent rainfall. These conditions are essential in maintaining habitats for about two-thirds of all known species of plants. In one hectare (10,000 m^2) of tropical rainforest, there may be hundreds of species of trees. In contrast, a temperate deciduous forest may have four to eight tree species in one hectare.

This diversity of plants supports the vast number and variety of animal and microbial species that live in the rainforest. Surviving in such close proximity requires organisms to interact with each other not only in complex food webs but also as communities of species that share resources and information about changes in their environment.

HOW DO RAINFOREST SPECIES COMMUNICATE?

In this unit, you will investigate the systems that allow organisms to survive, grow, reproduce, and respond to their environments.

UNIT VIDEO 1 Go online to learn more about rainforest ecosystems and the communities of organisms they support.

205

Introduce the Anchoring Phenomenon

Rainforests consist of four different layers that develop smaller biomes with different conditions within the larger ecosystem. The variety of organisms that exist between the emergent layer, the canopy, the understory layer, and the forest floor layer share resources and information that lead to a healthy ecosystem. Examining the interactions that exist between the organisms across the layers encourages students to apply what they learn throughout the unit to a complex situation.

Driving Question Have students make a list of things they think they would need to know about rainforests to answer the Driving Question. The photo and caption may prompt them to ask, "What organisms live in a rainforest?" To analyze the systems that allow organisms to communicate in a rainforest, students will need to focus on the systems that exist in organisms that live in a rainforest. What they learn in this unit about microorganisms, plants, and animals will inform their understanding of how rainforest species communicate.

About the Photo Juan Diego Falls consists of two waterfalls that run through El Yunque National Rainforest, Puerto Rico. The one seen in this image is approximately 2 m tall; however, a 12 m tall waterfall exists upstream. The trees in the foreground of the image are covered in moss.

Video

Time: 1:50

Unit Video 1 from National Geographic provides background on rainforests, shows a variety of organisms that can be found in them and identifies some of the foods and medicines that are sourced from them.

Go Online VIRTUAL INVESTIGATION

Communication in the Rainforest

Time: about 50 minutes

Objectives Students will learn about the diversity of the El Yunque Rainforest ecosystem and gather evidence to describe how organisms in the rainforest communicate.

Explore and Learn Students interact with the environment to obtain background information about the rainforest ecosystem and learn about the tools scientists use to study them.

Collect Data Students visit different locations in the El Yunque Rainforest to make observations and measurements of the forest organisms using a variety of scientific tools. They will analyze samples from the field, and they will fly a drone over parts of the forest as they learn about its recovery from damage caused by Hurricane Maria.

Analyze and Report Students answer questions about the investigation, incorporating qualitative and quantitative data in a report to support their analyses.

ENGAGE

About the Explorer

National Geographic Explorer Dr. Varun Swamy is a tropical rainforest ecologist. He has primarily focused on the plant-animal interactions that exist in the rainforests of the Madre de Dios River Basin of the Peruvian Amazon. Swamy has a growing interest in crowdsourcing data by citizen scientists and hopes that the information collected can help improve the understanding of systems and cycles that exist in tropical rainforests. You can learn more about Varun Swamy on the National Geographic Society's website: www.nationalgeographic.org.

On the Map

Students should understand that rainforests are found in both the tropics around the equator and in some temperate regions on Earth. The Madre de Dios River basin is the largest watershed of the area and a part of the drainage basin for the Amazon River. It extends across parts of Peru, Bolivia, and Brazil. The river and its tributaries are important waterways for transportation. Several industries are located on the banks of the river, including rubber harvesting, gold mining, and farming. There are several national parks and reserves in the region. The Huarayo peoples have traditionally occupied parts of the region in scattered settlements.

THINKING CRITICALLY

Infer *Sample answer: Not all seeds will germinate even in the right environmental conditions. Those germinating in the vicinity of the parent will have a better chance of experiencing the right conditions than those dispersed to other places.*

NATIONAL GEOGRAPHIC EXPLORER **VARUN SWAMY**

UNDER THE CANOPY

MONITORING PLANT REGENERATION IN THE AMAZONIAN LOWLANDS

A herd of 150 or more white-lipped peccaries shove their way through the lowlands of the Amazon rainforest in southeastern Peru. The piglike animals' scents bond them together, and they grunt and clatter their teeth, barking if a threat is nearby. Their foraging behaviors mimic the actions of a bulldozer as they root through the leaf litter, eating anything they come across and mulching the soil. Are such behaviors helpful or harmful to the ecosystem? Ecologist Dr. Varun Swamy considers peccaries beneficial because they keep other dominant species in check, allowing rarer species that are weaker competitors to survive and grow.

IMPACTS ON TREE SPECIES REGENERATION

Swamy's research focuses on the lowland Amazon plant community. His primary concerns are the canopy and subcanopy trees that set the stage for all other life in the ecosystem. His overall goal is a better understanding of the regeneration processes across multiple life stages. These include fruit fall and seed dispersal, seed germination, seedling establishment, sapling growth, and the eventual death of centuries-old canopy trees. Saplings that have a healthy start in the vertical journey to the canopy have the best chance of replacing their parents. Data collected in areas where peccary populations have been eliminated because of the overhunting of wild animals for the bushmeat trade support his conclusions. The peccary herding lifestyle and foraging behavior benefit the regeneration of canopy trees.

The Peruvian black spider monkey is another key organism in Swamy's quest for answers. In Peru, this species is called *maquisapas*, a word that means "long arms." It is one of the largest primate species in South America. Swamy notes that these animals act as "seed-dispersing machines" for fleshy-fruited canopy trees due to their size, agility, activity, and diet. The rapid processing of fruits through their digestive tracts results in a constant "rain" of intact seeds from the forest canopy down to the floor. Swamy has found that in areas where the monkeys have been overhunted, the dispersal of seeds of fleshy-fruited trees has been greatly reduced.

Without the spread of seeds far from parent trees, another factor takes on an important role in regeneration processes. Swamy's research examines the antagonistic interactions between plants and their "natural enemies"—a term that refers to a broad collection of harmful fungi, bacteria, protists, and viruses. These microorganisms have evolved to feed exclusively on the seeds and seedlings of a single or a few closely related tree species. They live in and around large adult trees and destroy the fallen seeds or young seedlings. Future generations of these trees are spatially distant from their parents because they grow only by dispersed seeds that have escaped the reach of these natural enemies.

TECHNOLOGY EXPANDS THE STUDY AREA

Swamy's research now includes the collection of data from much larger areas thanks to the use of mini-copter drones. Bird's-eye views of the rainforest canopy reveals leafing, flowering, and fruiting patterns and allow for identification of species as well. Sharing this aerial footage on online citizen-science platforms also allows nonscientists across the world to contribute to the data collection and analysis, which increases the world's interest in preserving this treasure trove of biodiversity.

THINKING CRITICALLY

Infer *Why might seeds that germinate near the parent trees be less successful than those that are carried long distances by wind or animals?*

206

CROSSCURRICULAR CONNECTIONS

World History Humans have been living in tropical rainforests for about 45,000 years, and as long as we have inhabited them, we have been changing them. Early modern humans used fire and other tools to gather what they needed to survive and are thought to have drastically changed the flora and fauna that lived in the rainforests. However, it wasn't until about 10,000 years that humans started farming. Ancient, cultivated plants, like the Brazil nut trees, and other fruit plants are overrepresented in over 3,000 pre-Columbian settlement areas. It is thought that human activities have influenced about 20% of the species in the Amazon, both from cultivating native plants and the introduction of other species that may have been able to outcompete other native plants. Archeologists are using remote sensing technologies to learn more about the ancient people who coexisted with the rainforests. They hope to find clues to their agricultural and architectural practices that can be used in modern times to conserve the biodiversity found in rainforests.

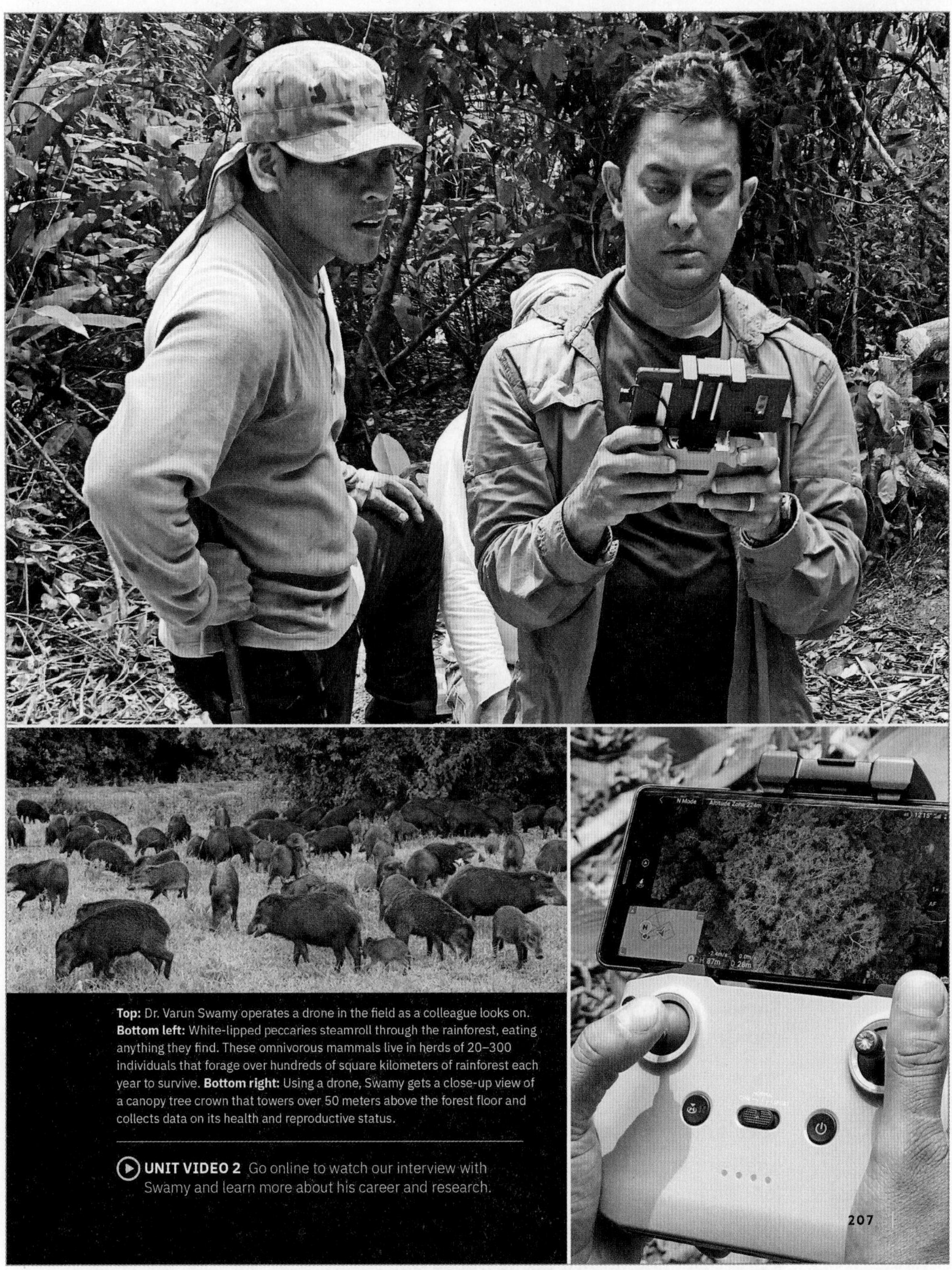

Top: Dr. Varun Swamy operates a drone in the field as a colleague looks on. **Bottom left:** White-lipped peccaries steamroll through the rainforest, eating anything they find. These omnivorous mammals live in herds of 20–300 individuals that forage over hundreds of square kilometers of rainforest each year to survive. **Bottom right:** Using a drone, Swamy gets a close-up view of a canopy tree crown that towers over 50 meters above the forest floor and collects data on its health and reproductive status.

UNIT VIDEO 2 Go online to watch our interview with Swamy and learn more about his career and research.

Explorers at Work

Time: 7:44

Unit Video 2 interviews Dr. Varun Swamy who likens the interconnectedness of rainforest species to a piece of "communal artwork that is the result of thousands of individual artists." He goes on to talk about how he uses drones to collect data in the rainforest canopy and how he enlists the help of the public in analyzing that data. This lays the groundwork for students to apply what they learn throughout the unit about the vast array of life on Earth to rainforest communication.

Connect to Careers

Tropical Rainforest Ecologist Ecology is the study of the interactions of the plants and animals that live in an environment. Tropical rainforest ecologists focus on the species that exist within the specific rainforest ecosystems. They may use traditional field biology techniques and chemical testing of collected materials, but they may also use technology, such as drones, satellite photography, and field cameras, to collect data. Research technicians typically need an associate's degree in ecology, but to lead research projects, a bachelor's degree or higher is required. A master's degree may be required for specialized positions. Governments, universities, and environmental consulting agencies who are surveying rainforest regions hire ecologists to survey areas and determine the health of the ecosystem.

Forest Ecology Drone Pilot Data about the diversity in a forest can be collected by individuals operating drones from the ground. Forest ecology drone pilots plan and execute drone missions in different forest regions. They often camp near the sites where data is to be collected and typically work independently or in close collaboration with a researcher. Depending on their location, drone pilots may need a certification in remote piloting, using geospatial software and know-how to navigate using maps and handheld GPS.

DIFFERENTIATED INSTRUCTION | English Language Learners

Retelling or Summarizing Explain that retelling or summarizing what they heard in their own words will help students better understand and remember the content. Encourage students to take notes as they watch. Allow them to pause and rewatch parts as needed.

Beginning Provide guiding questions to small groups to summarize the most important ideas: *What animals help spread seeds? What are plants' "natural enemies," and where are they? How does Swamy study how seeds spread?*

Intermediate Provide guiding questions to help pairs retell the ideas in the video: *What does Swamy research? Why is it important? How does he study it? What other information did you learn?*

Advanced Have students work in pairs to retell as much of the information from the video as they can. Remind students to use their own words to explain the ideas, rather than repeating phrasing from the video.

Three-Dimensional Learning

The practices, core ideas, and crosscutting concepts presented in this chapter's text, investigations, and resources provide support to address the following Performance Expectation: **HS-LS1-2**.

Science and Engineering Practices	Disciplinary Core Ideas	Crosscutting Concepts
Asking Questions and Defining Problems Planning and Carrying Out Investigations Constructing Explanations and Designing Solutions Engaging in Argument from Evidence Obtaining and Communicating Information Scientific Investigations Use a Variety of Methods Scientific Knowledge is Open to Revision in Light of New Evidence	**LS1.A** Structure and Function (HS-LS1-2) **LS1.B** Growth and Development of Organisms **LS1.C** Organization for Matter and Energy Flow in Organisms **LS2.B** Cycles of Matter and Energy Transfer in Ecosystems **LS4.A** Evidence of Common Ancestry and Diversity **LS4.D** Biodiversity and Humans	Patterns Energy and Matter Structure and Function

Contents		Instructional Support for All Learners	Digital Resources
ENGAGE			
208–209	**CHAPTER OVERVIEW** **CASE STUDY** How are the characteristics of an organism related to its place in an ecosystem?	**Social-Emotional Learning** CCC Structure and Function **On the Map** ISS Orbit **English Language Learners** Details in Text and Graphics	
EXPLORE/EXPLAIN			
210–217 DCI LS1.A	**8.1 BACTERIA AND ARCHAEA** • Relate the metabolism of an organism to the environments in which it might be able to live. • Compare and contrast the structural differences between archaea and bacteria. • Examine ways that bacterial pathogens are transmitted and cause infection in new hosts.	**Vocabulary Strategy** Word Definition Map **English Language Learners** Summarizing for Comprehension **Visual Support** Cladograms SEP Scientific Knowledge is Based on Empirical Evidence **Address Misconceptions** Beneficial Bacteria **Differentiated Instruction** Leveled Support **Connect to English Language Arts** Draw Evidence From Texts **English Language Learners** Reading a Table **Vocabulary Strategy** Understanding in Context CCC Structure and Function **In Your Community** Food Poisoning **On the Map** Lyme Disease Across the U.S. SEP Constructing Explanations and Designing Solutions **Vocabulary Strategy** Prefixes/Suffixes/ Root Relationships	**Investigation B** Effects of Antimicrobials (150 minutes over 3 days)

Contents		Instructional Support for All Learners	Digital Resources
EXPLORE/EXPLAIN			
218–225 DCI LS2.C	**8.2 PROTISTS** • Evaluate how the structures of some protists are related to how they function in their environment. • Explain why protists are described as a kingdom. • Identify characteristics common to protists.	**SEP** Scientific Knowledge is Open to Revision in Light of New Evidence **Address Misconceptions** Protist Size **Visual Support** Common Eukaryote Structures **English Language Learners** Using Visuals **Connect to English Language Arts** Compare and Contrast **Differentiated Instruction** Students with Disabilities **Visual Support** Movement-Related Structures **Differentiated Instruction** Leveled Support **Differentiated Instruction** Students with Disabilities **Visual Support** SAR Group **CCC** Structure and Function **Connect to English Language Arts** Summarize **English Language Learners** Vocabulary Strategies **In Your Community** Food Additives	**Investigation A** Classification Systems (100 minutes or 2 days) Video 8-1
EXPLORE/EXPLAIN			
226	**MINILAB** FEATURES OF PARAMECIA AND EUGLENA		
227–233 DCI LS2.C	**8.3 FUNGI** • Describe the structures that help fungi to produce and disperse spores. • Explain the symbiotic relationship between fungi and plants. • Identify characteristics that are common to all fungi.	**English Language Learners** Spellings for Long *i* **CASE STUDY** Ask Questions **CCC** Structure and Function **In Your Community** Mutualisms **Differentiated Instruction** Leveled Support **Visual Support** Lichen **CER** Revisit the Anchoring Phenomenon **In Your Community** Fungal Diseases **SEP** Constructing Explanations and Designing Solutions **CONNECTIONS** Social-Emotional Learning **Visual Support** Fungal Diversity	*Video 8-2* *Video 8-3*

Contents		Instructional Support for All Learners	Digital Resources
EXPLORE/EXPLAIN			
234	**LOOKING AT THE DATA** THE C-VALUE ENIGMA		
236–243 **DCI** LS2.C	**8.4 VIRUSES** • Characterize the four major groups of viral genomes and where they replicate. • Compare and contrast bacteriophage and HIV replication. • Identify the characteristics that are common to all viruses.	**SEP** Constructing Explanations and Designing Solutions **Vocabulary Strategy** Knowledge Rating Scale **CCC** Structure and Function **Visual Support** Viral Replication **Vocabulary Strategy** Word Definition Map **Differentiated Instruction** Leveled Support **Connect to English Language Arts** Synthesize Information **Social-Emotional Learning** Social Awareness **English Language Learners** Linguistic Support **Address Misconceptions** Herd Immunity **On the Map** Infectious Diseases Worldwide **Differentiated Instruction** Economically Disadvantaged Students **Visual Support** Virus Reassortment **On Assignment** Documenting Biodiversity **Connect to English Language Arts** Write Informative Text	**Interactive Figure** HIV Replication *Video 8-4*
244	**EXPLORER** DR. EMILY OTALI	**Connect to Careers** Primatologist	

<table>
<tr><th colspan="2">Contents</th><th>Instructional Support for All Learners</th><th>Digital Resources</th></tr>
<tr><td colspan="4">ELABORATE</td></tr>
<tr><td>245</td><td>TYING IT ALL TOGETHER How are the characteristics of an organism related to its place in an ecosystem?</td><td>Connect to Mathematics Model with Mathematics</td><td>Self-Reflection and Scoring Rubrics</td></tr>
<tr><td>Online</td><td>Investigation A Classification Systems
Investigation B Effects of Antimicrobials</td><td>Guided Inquiry (100 minutes or 2 days)
Design Your Own (150 minutes over 3 days)</td><td rowspan="3">MINDTAP Access all your online assessment and laboratory support on MindTap, including: sample answers, lab guides, rubrics, and worksheets.
PDFs are available from the Course Companion Site.</td></tr>
<tr><td colspan="3">EVALUATE</td></tr>
<tr><td>246–247</td><td colspan="2">Chapter 8 Review</td></tr>
</table>

CHAPTER 8

ENGAGE

In middle school, students will have learned about cells and that all living things are made up of one cell or many and varied cells. Students will have been exposed to the role of individual parts of a cell, specifically the nucleus, chloroplasts, mitochondria, cell membrane, and cell wall. They will also have an understanding of the primary differences between plant and animal cells. This chapter expands on that prior knowledge by exploring features and characteristics of bacteria, archaea, protists, fungi, and viruses.

Students will explore the environmental interactions of these organisms and how the metabolism of an organism determines where it can thrive. They will also examine various structures and learn how they function in organisms.

About the Photo This photo shows a magnified view of the hyphae and fruiting structure of the fungus *Aspergillus ustus.* The technology used to produce this magnified view is an electron microscope. The images are produced by scanning the sample with a focused beam of electrons. To prepare students for this chapter's content, ensure that they understand the role that advancements in technology has played in observing the features and characteristics of the miniscule microorganisms and their cellular components.

Social-Emotional Learning

As students progress through the chapter, invite them to look for contexts and opportunities to practice some of the five social and emotional competencies. For example, as they learn about the emergence and spread of new diseases, support students in practicing **social awareness** and **relationship skills** when they encounter different perspectives. As they learn about the treatment of diseases, remind students to practice **responsible decision-making** to protect themselves and the health of the people in their community.

CHAPTER 8

DIVERSITY OF LIVING SYSTEMS

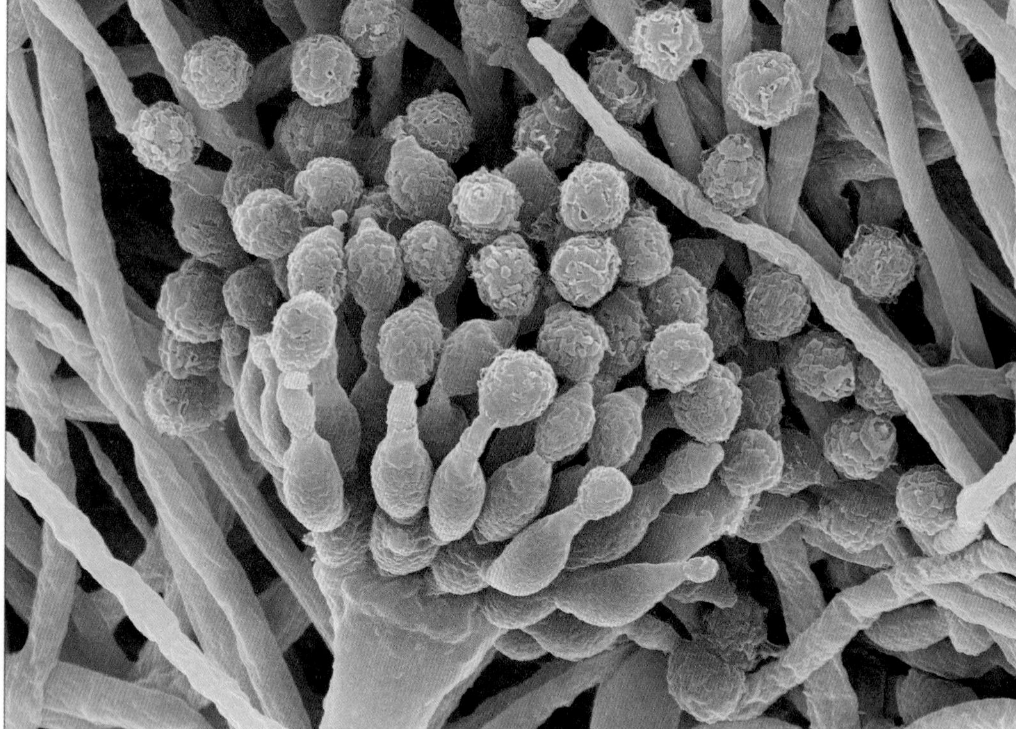

The hyphae and fruiting structure of the fungus *Aspergillus ustus* is magnified 600 times in this scanning electron micrograph.

Most organisms are invisible to the unaided eye. Although we can often see large groups of microorganisms when they grow together, most individual microbes can only be observed directly using a microscope. Magnification has revealed that these tiny creatures live in practically every habitat on Earth.

Chapter 8 supports the NGSS Performance Expectation **HS-LS1-2.**

CROSSCUTTING CONCEPTS | Structure and Function

Organism Interactions In this chapter, students explore the concept of structure and function as they learn about the varied characteristics of organisms and how they interact in our world. Students learn that microorganisms are typically too small to be visible with the naked eye, and scientists observe them with different types of high-powered magnification technologies.

Organism are made up of structures that play a role within the organism and in the organism's interactions with the environment.

As students explore how the features of an organism allow it to thrive in its environment, reinforce this crosscutting concept by helping students understand that the environment must also be conducive to an organism's survival.

CASE STUDY
SPACE STATION STOWAWAYS

HOW ARE THE CHARACTERISTICS OF AN ORGANISM RELATED TO ITS PLACE IN AN ECOSYSTEM?

The International Space Station (ISS) is a laboratory hovering in low Earth orbit, operated by several multinational space agencies. Scientists investigate numerous phenomena in the microgravity environment. In recent missions, biologists have performed experiments in microbiology, plant growth, and the physiological effects of space travel. They have also used the ISS's satellite imaging capabilities to assess the productivity of Earth ecosystems. To ensure that experiments function properly, everything that enters or is attached to the space station is decontaminated before it is brought on board. Dust and debris from Earth are removed from all cargo and scientific equipment, and even the astronauts are quarantined pre-flight to avoid transporting infectious microbes to the station.

However, all humans host essential and diverse communities of microbes that cannot be removed. Living on the ISS under unusual conditions such as microgravity, high exposure to radiation, and psychological stress can weaken the immune system. All the air and water on the ISS is recycled through closed supply systems that can harbor microbial contaminants. Thus, the ISS crew thoroughly cleans the station every week, and NASA closely monitors the microbiome aboard the station to determine what types of microbes exist there (**Figures 8-1** and **8-2**).

More than 200 species of bacteria and fungi have been identified in air, water, and surface samples collected on the ISS. The most abundant types of microorganisms found are associated with human microbiomes. Scientists have also identified some previously unknown bacterial strains that thrive on the ISS.

Studying how microbes survive in space not only helps scientists understand potential risks to astronauts' health but also uncovers possibilities for using microbe-based biotechnology in space applications. For example, bacteria or fungi that can survive space travel could potentially be useful for producing food and medicine for long-term missions. Scientists are also investigating the ability of some microbial species to make fertile soil by breaking up rock. Other species might form building materials on other planets.

Ask Questions *In this chapter, you will study many diverse organisms, some closely related to those that live on the ISS. As you read, generate questions about how characteristics of these organisms relate to where and how they live.*

FIGURE 8-1
Dr. Don Pettit prepares biological samples for storage in a freezer on the ISS.

FIGURE 8-2
A diverse collection of organisms can be isolated from different locations on the ISS. The majority of identified species are bacteria and fungi (eukaryota).

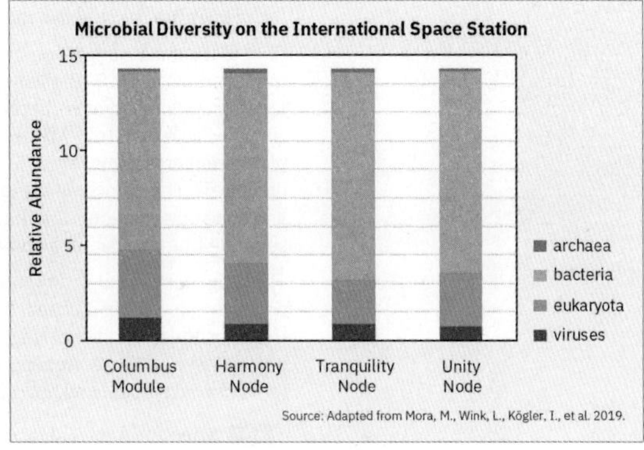

DIFFERENTIATED INSTRUCTION | English Language Learners

Details in Text and Graphics Have mixed groups reread the text and discuss connections with the graphics. Have them read the photo caption and find related details in the first paragraph of the text then examine the graph and find related details in the second and third paragraphs.

Beginning Students Have students use the text and graphics to complete these sentence frames: ISS stands for _______. The biological samples on the ISS are for _______. There are microbes in different places on the ISS because _______. Most of them are _______.

Intermediate Have students use the text and graphics to answer these questions: Why are there biological samples on the ISS? Why are there microbes on the ISS? Where are they found? What are most of them classified as?

Advanced Have students explain why the author most likely included the photo of a scientist working with biological samples on the ISS. Then, have them explain the information in the graph and describe how it supports the ideas in the text.

CASE STUDY

ENGAGE

The situation described in the Case Study can be used to assess students' prior knowledge of how organisms thrive in their environment. Ask students to identify a structure that they know helps an organism to survive.

Revisit the Case Study after each section in the chapter to have students identify the relationship between the reproductive structures they have explored and how those structures might function in the given environment. Remind students that all living things reproduce either asexually or sexually and the environment is a factor in how an organism reproduces. In the Tying it All Together activity, students will research how an organism grows and reproduces.

Ask Questions Students should revisit the Case Study as they read the chapter to make connections with the content. See a specific suggestion in Section 8.3.

On the Map

ISS Orbit Share with students that at certain times, the ISS may be visible in your location due to reflected sunlight, just as the moon is. Find your location on spotthestation.nasa.gov to see when it will be visible in your area. The ISS orbits Earth approximately 16 times per day, with the orbital track shifting to the west with each pass. The orbit tracks over the same area of Earth's surface every three days.

Human Connection Researchers reported in a 2021 study of bacterial strains found on the ISS the discovery of three new strains, which were named *Methylobacterium ajmalii*. Species of this genus are often involved in plant processes such as nitrogen fixation and stress tolerance. The scientists think the new strains may help astronauts grow their own food while in space.

8.1

BACTERIA AND ARCHAEA

LS1.A

EXPLORE/EXPLAIN

This section provides an overview of the classification system for organisms, introduces the concept of bacterial and archaea kingdoms, and reviews the different structures and characteristics associated with the organisms found in these kingdoms. It also covers the different ways these groups of organisms obtain and use energy and matter (metabolic diversity).

Objectives

- Relate the metabolism of an organism to the environments in which it might be able to live.
- Compare and contrast the structural differences between archaea and bacteria.
- Examine ways that bacterial pathogens are transmitted and cause infection in new hosts.

Pretest Use **Questions 1** and **5** to identify gaps in background knowledge or misconceptions.

Vocabulary Strategy

Word Definition Map As students work through the key terms in this section, have them create word definition maps. Students should include the key term, its definition, and an example and non-example.

As students work through the section, have them include any additional relevant information, such as images, related words, and page number information found in the text for reference.

Predict *Students might mention advances in microscopy, or they might know something about DNA sequencing technology and figure out that organisms might be sortable based on molecular characteristics.*

SEP Scientific Knowledge *Sample answer: Bacteria and archaea were long considered to be part of the same group, but advances in DNA sequencing and phylogenetic analysis have enabled scientists to identify many more species and understand that bacteria and archaea are distinct domains of organisms.*

8.1 BACTERIA AND ARCHAEA

KEY TERMS

- endospore
- halophile
- methanogen
- phylogenetic analysis
- thermophile

Imagine you are sorting a box full of school supplies into groups based on similar characteristics. You might sort them by size or material, or possibly by how they are used. For years, scientists sorted and classified organisms in similar ways, based on comparing observable characteristics. However, that does not work very well when the organisms are incredibly tiny.

Predict *How do you think scientists might classify different species of bacteria that are tiny and appear to have similar characteristics?*

Classifying Prokaryotes

When the modern system of naming organisms was developed in the 1700s, there were just two kingdoms, Plantae (plants) and Animalia (animals). In the 1860s, a third kingdom was introduced, Protista, which included all microorganisms. Microscopes continued to improve, and scientists continued learning more about cells and microorganisms.

Historically, scientists classified unidentified single-celled organisms by comparing them against a known group based on shape, metabolism, and properties of the cell wall. The more traits the organism shared with the known group, the closer they were thought to be related. The wide variety of single-celled organisms, in particular, were difficult to classify but clearly did not fit into a single kingdom. By the 1950s, scientists suggested there were five kingdoms.

In the 1970s, scientists began studying the sequences of RNA in the ribosomes of organisms. Then, in the 1990s, technology advanced to allow scientists to begin studying entire genomes of organisms. These developments changed the way organisms could be classified and led to the three-domain system. These uses of DNA and RNA sequence comparison are a type of **phylogenetic analysis**, which is a constantly improving group of methods in which genetic sequences are compared to estimate relatedness between organisms. Phylogenetic analysis techniques are described in more detail in Section 14.1 and Appendix E.

Prior to the initial suggestion of the three-domain system in 1977 and its formal proposal in 1990, scientists often grouped bacteria and archaea together as "prokaryotes" because they are both single celled and have similar visible features. However, sequencing of RNA in the ribosomes of different organisms revealed that archaea were a distinct group of organisms. The branching in **Figure 8-3** reflects the hypothesis that archaea and eukaryotes are thought to have a common ancestor that is separate from the bacteria domain.

The three domains are commonly divided further into six kingdoms of organisms, including archaea, bacteria, plants, fungi, animals, and protists. However, these traditional kingdom divisions are not based on modern phylogenetic analyses. Appendix E provides more information on the three-domain model and how scientists classify organisms.

NOS Scientific Knowledge *How have advancements in technology and scientific techniques affected how scientists classify microorganisms?*

210 CHAPTER 8 DIVERSITY OF LIVING SYSTEMS

DIFFERENTIATED INSTRUCTION | English Language Learners

Summarizing for Comprehension Summarizing shorter sections of text can help students identify and understand the most important ideas. Have students summarize one or two paragraphs at a time to demonstrate comprehension before moving on.

Beginning Have students work in pairs to identify key terms in the paragraph. Have them use short phrases to explain these in their own words.

Intermediate Have pairs write a one-sentence summary for each paragraph. At the end, have them combine their sentences into a summary for the section.

Advanced Have students write a few sentences to summarize a section of text. Remind them to include only the most important ideas and leave out minor details. They should explain ideas in their own words.

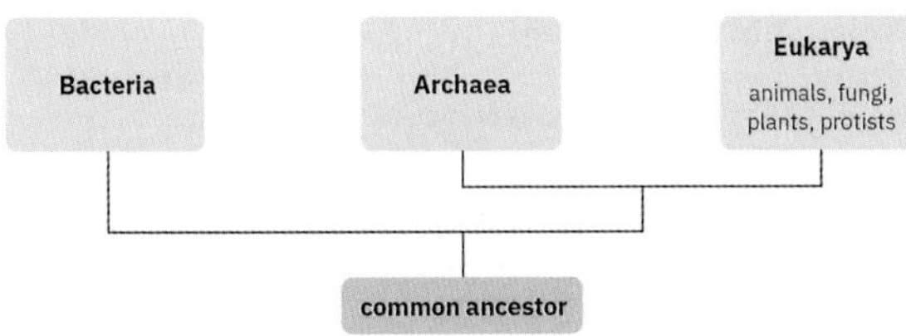

FIGURE 8-3 Bacteria, Archaea, and Eukarya encompass the three domains of life, all of which descended from a common ancestor. Archaea and Eukarya are thought to have a common ancestor that is separate from the Bacteria domain. On cladograms such as this one, a longer distance between groups indicates a more distant evolutionary relationship. Shorter distances indicate that groups are more closely related.

Prokaryote Diversity

Scientists' understanding of the diversity and number of bacteria and archaea has also changed dramatically with improvements in DNA sequencing and phylogenetic analyses. Many bacteria and archaea are difficult to isolate or grow outside of their native environments, so they are difficult to study. New techniques allow DNA in samples collected from an environment to be sequenced and analyzed. Many thousands of new species of bacteria and archaea have been identified in environments ranging from hot springs to human intestines.

Prokaryote Structures **Figure 8-4** shows structures that are typical of bacteria. These structures are discussed in detail in Chapter 6. Bacteria are often shaped like rods (bacilli) and spheres (cocci), but others are shaped like commas (vibrios) or spirals (spirilla or spirochetes). In addition to those shapes typically found among bacteria, archaea species with square, triangle, and other irregular shapes have also been discovered.

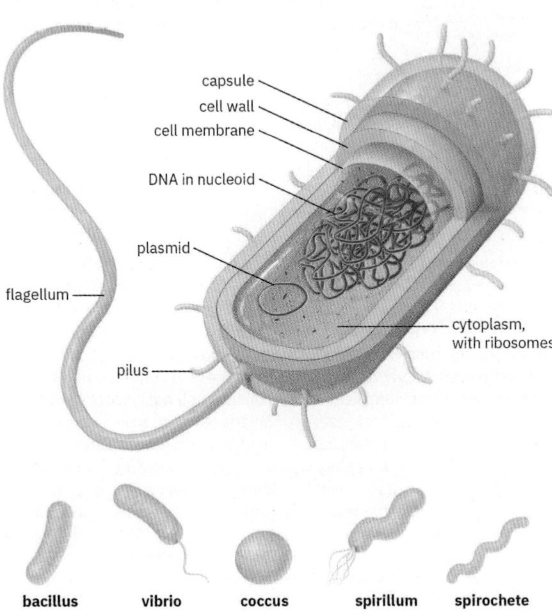

FIGURE 8-4 Most prokaryotes have similar basic structures. Bacteria and archaea come in many shapes, including bacillus, vibrio, coccus, spirillum, and spirochete.

Visual Support

Cladograms A cladogram, sometimes called a phylogenetic tree, is a diagram that shows evolutionary relationships among organisms with a shared common ancestor. Use **Figure 8-3** to explain to students how a cladogram is similar to a family tree that shows familial relationships over generations. Have students follow the lines from the common ancestor to see that species in aacteria are more closely related to the common ancestor of archaea and eukaryotes than to species of archaea and eukaryotes.

Science Background

Prokaryote Cells Prokaryotic cells differ structurally from eukaryotic cells and are much smaller. In addition to the basic structures identified in **Figure 8-4,** some bacteria can form resistant endospores when necessary nutrients are lacking in an environment. The process begins with asymmetrical cell division and over several hours, a cell within the cell forms, and eventually the dehydrated endospore is released into the environment when the parent cell lyses. The endospore is resistant to trauma and can survive extreme temperatures, high UV radiation, and other environmental stressors. Endospores can remain dormant for an extended period of time until the environment again becomes favorable. Then the endospore absorbs water and resumes normal growth. Challenge students to predict how the presence of this feature might affect the ability of scientists to make vaccines or provide cures for certain infectious bacteria.

SCIENCE AND ENGINEERING PRACTICES
Scientific Knowledge Is Based on Empirical Evidence

Technology Development Encourage students to reflect on how their ability to collect empirical evidence has changed as they were introduced to new technologies.

Students should recognize that when they were younger, their tools were very basic, such as a hand magnifier or a ruler. Now that they are older, many students may have used a microscope capable of viewing items at 10x, 40x, and 100x the actual size. They may have measured mass with a triple beam balance and volume with a graduated cylinder.

Encourage students to think about how increasingly technologically advanced tools aid scientists in their research. What discoveries were made when microscopes were first used to observe living organisms? Students should relate their own experiences with advancements in scientific tools to how advancements in technology affect how scientists classify organisms.

Address Misconceptions

Beneficial Bacteria If students have not yet been exposed to the Anchoring Phenomenon of Unit 2, some students may think all bacteria are harmful. Remind students that the human body is host to millions of beneficial bacteria. Bacterial cells can be found in the nose, mouth, gut, and even skin. One milliliter of saliva can contain 40 million bacterial cells!

Beneficial bacteria species in the human body release nutrients by breaking down food, neutralizing toxins, providing protection from harmful microorganisms, and supporting the immune system. Bacteria species even synthesize some vitamins that humans need.

Have students discuss the different environments that host bacterial cells and consider if bacteria species benefit by living in the human body.

SEP Construct an Explanation
Sample answer: Photoheterotrophs and photoautotrophs need to live in environments where there is sunlight, whereas chemoautotrophs and chemoheterotrophs can extract energy from a host or chemicals in the environment.

Prokaryote Metabolism

Metabolic diversity—the different ways organisms obtain and use energy and matter—gives certain species of bacteria and archaea the ability to thrive in a variety of environments. Many bacteria are anaerobic, which means they can (or must) live where there is no oxygen. Others are aerobic, which means they require (or at least tolerate) oxygen. There are four possible modes of obtaining energy and nutrients (**Table 8-1**) and, as a group, bacteria and archaea use them all.

- *Photoautotrophs* carry out photosynthesis to build their own food from an inorganic source of carbon such as carbon dioxide. Like plants, they assemble their food using light as an energy source.
- *Chemoautotrophs* fuel the assembly of their food by using energy from inorganic substances such as hydrogen sulfide. These bacteria and archaea are producers in dark places such as the seafloor.
- *Photoheterotrophs* use light energy to fuel the breakdown of organic molecules they obtain from the environment.
- *Chemoheterotrophs* obtain both energy and carbon through the breakdown of organic compounds. Pathogenic bacteria are chemoheterotrophs that extract the organic compounds they need to live from a host. Other bacterial chemoheterotrophs are decomposers.

SEP Construct an Explanation *Describe how the metabolism of an organism affects what environments it might be able to live in.*

TABLE 8.1. Modes of Obtaining Energy and Carbon by Different Groups of Organisms

Type of metabolism	Carbon source	Energy source	Organisms
chemoautotrophs	inorganic source such as CO_2	chemicals	archaea, bacteria
chemoheterotrophs	organic source such as glucose	chemicals	animals, archaea, bacteria, fungi, nonphotosynthetic protists
photoautotrophs	inorganic source such as CO_2	light	archaea, bacteria, photosynthetic protists, plants
photoheterotrophs	organic source such as glucose	light	archaea, bacteria

Bacterial Diversity and Ecology

There are thousands of species of bacteria, and new ones are constantly being discovered. In addition to genetic, shape, and metabolic differences, bacteria are often distinguished based on their cell walls, which affects their ecological roles. *Gram-positive* bacteria have cell walls with a thick outer layer of peptidoglycan, which is a molecule made of sugar and amino acids. This layer retains purple stain when a technique called Gram staining is used to prepare samples for viewing under a microscope. In *Gram-negative* bacteria, the cell wall is surrounded by an outer membrane and has a thinner peptidoglycan layer that retains very little dye during Gram staining. Outer membranes in some Gram-negative bacteria generate vesicles to help the bacteria adhere to surfaces or evade immune systems of their hosts.

DIFFERENTIATED INSTRUCTION | Leveled Support

Struggling Students After viewing **Table 8-1**, help students connect the type of metabolism and the ability of bacteria and archaea to thrive in a variety of environments. Students can extend **Table 8-1** by adding an additional column for Environments. Help students look for patterns across the types of organisms and the environments by looking for similarities and differences among the organisms. Direct students to look at the energy sources that organisms use and connect those to the type of environment in which they survive and thrive. Support students with sentence stems such as: Photoautotrophs use _______ as an energy source to make their own food and would thrive in _______ environment.

Advanced Learners Students who have an understanding of the metabolic diversity of bacteria and archaea may be interested in learning about some of the specific species within these domains. Have these students describe the actual conditions of the environments in which they thrive.

While there is no clear consensus for the classification of bacteria, it is helpful to look at some traditional informal groups. Cyanobacteria, Gram-positive bacteria, and proteobacteria are a few of the groups that show the diversity and ecological importance of bacteria.

Cyanobacteria Cyanobacteria are a group of Gram-negative bacteria that perform photosynthesis. Photosynthesis evolved in many bacteria, but only cyanobacteria release oxygen as a by-product. Evidence suggests that chloroplasts in plants evolved from ancient cyanobacteria. This is described in the *endosymbiont theory* detailed in Section 6.1. Some cyanobacteria also carry out nitrogen fixation, which means they incorporate nitrogen from the air into ammonia (NH_3). Nitrogen fixation is an important ecological service. Photosynthetic eukaryotes need nitrogen but they cannot use the gaseous form (N_2). Plants and photosynthetic protists can, however, take up ammonia released by nitrogen-fixing bacteria. The nitrogen cycle is described in Section 2.4.

Some cyanobacteria partner symbiotically with fungi to form *lichens*. Others grow on the surface of soil, but most are aquatic. Aquatic cyanobacteria grow as single cells or as long filaments of cells arranged end to end (**Figure 8-5**). When conditions become unfavorable for growth, some filamentous cyanobacteria produce thick-walled resting cells. These cells can be dispersed and remain dormant until conditions improve.

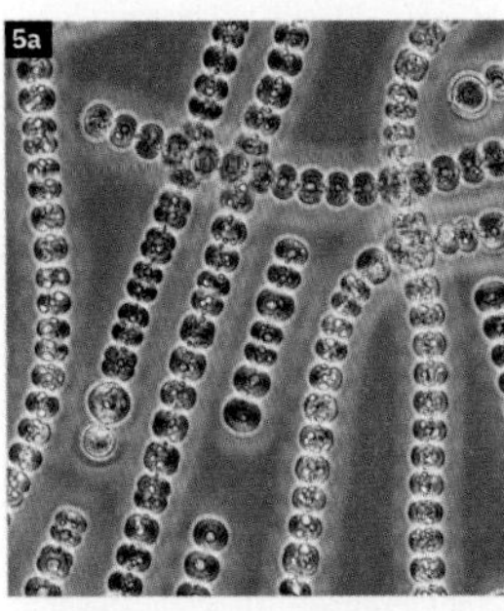

FIGURE 8-5 This micrograph reveals chains of cyanobacteria, including larger, circular resting cells (a). In large quantities, cyanobacteria can form blooms that harm aquatic ecosystems, such as this bloom in Lake Okeechobee, Florida (b).

Gram-Positive Bacteria Most Gram-positive bacteria are chemoheterotrophs, and many are decomposers. Decomposers break down complex organic molecules in the wastes and remains of other organisms, leaving inorganic leftovers that serve as nutrients for plants. Decomposers also break down pesticides and pollutants, improving the environment for other organisms. *Actinomycetes* species are Gram-positive decomposers that grow through soil as long, branching chains of cells. Their presence gives freshly exposed soil its distinctive "earthy" smell. Many antibiotics, including streptomycin and vancomycin, were first isolated from *Actinomycetes* bacteria.

Lactic acid bacteria are another Gram-positive subgroup. These cells break down sugars in milk and produce lactic acid as a by-product of fermentation (Section 6.3). *Lactobacillus* species can spoil milk, but they are also used to produce cheese and yogurt. Lactic acid released by these bacteria produces a sour taste and changes the shape of milk proteins, causing the milk to thicken.

Connect to English Language Arts

Draw Evidence From Texts This section introduces three groups of bacteria, cyanobacteria, Gram-positive bacteria, and proteobacteria that highlight the diversity and ecological importance of bacteria. Have students write an informative summary of the information presented for each of the bacteria groups. To help them organize the information for their summary, have students create a three-column chart with one column for each group of bacteria and row headings for structures, metabolism type, environment, and ecological role.

DIFFERENTIATED INSTRUCTION | English Language Learners

Reading a Table Have students use the information in **Table 8.1** to help them understand the bulleted text. Point out that in English we read left to right, top to bottom, so the information in each row tells about the type of metabolism listed in the left column.

Beginning Provide sentence frames for pairs to restate the information in the table: _______ *gets carbon from a(n)* _______. *Its energy comes from* _______. *Some examples of* _______ *are* _______. Then, have them reread the corresponding bullet point in the text and compare.

Intermediate Have pairs read each phrase in a bullet point and find the corresponding information in the table, for example: "carry out photosynthesis to build their own food" corresponds to "energy source: light" in the table.

Advanced Have students work in pairs or groups of four and take turns explaining the information in the table using details in the text.

Vocabulary Strategy

Understanding in Context Support students in recognizing context clues that can be used to help develop vocabulary comprehension. For example, surrounding words can provide readers with hints to the meaning of a new word. As students work through this section, have them first identify occurrences of key terms. Then, have students take a second read through the sentences before and after where the key term appears. On the second read through, students should be looking for areas where the word meaning is directly stated, a similarity term, such as like, or an example is provided. These are often clues to the meaning of vocabulary terms.

CCC Structure and Function *Sample answer: Gram-positive organisms have a thick outer layer of peptidoglycan. Some of these species can form endospores that let them survive in poor conditions in the soil. Gram-negative bacteria have an outer membrane. For some species, that enables them to adhere to each other and to hosts and sometimes to evade host immune responses.*

Soil bacteria in the *Clostridium* genus and the *Bacillus* genus are Gram-positive cells that form endospores when conditions are unfavorable. An **endospore** contains the cell's DNA and a bit of cytoplasm in a protective coat (**Figure 8-6**). It is functionally similar to the resting cells cyanobacteria use to survive difficult conditions, but it is much tougher. Unlike active cells, endospores withstand exposure to extreme heat, freezing temperatures, drought conditions, disinfectants, and ultraviolet radiation. Many human pathogens produce endospores.

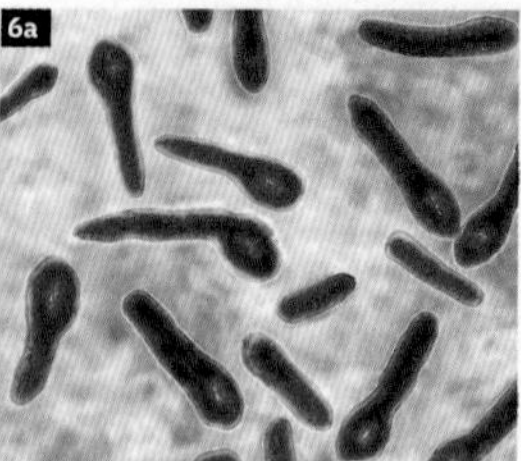

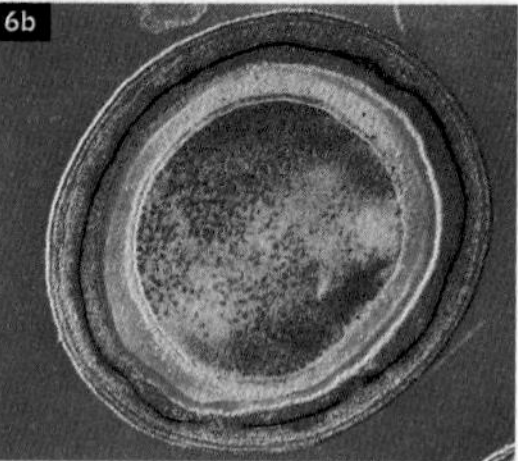

FIGURE 8-6 *Clostridium* bacteria take on a drum-stick shape (a) when endospores (b) are forming. Endospores can withstand unfavorable conditions.

Proteobacteria Proteobacteria is the largest bacterial group. This diverse group of Gram-negative bacteria is linked based on similarity in their ribosomes' RNA sequences, but they do not necessarily have similar structural or metabolic characteristics. Most proteobacteria are aerobic. Some are photosynthetic, but they do not produce oxygen. Other proteobacteria are chemoautotrophs or chemoheterotrophs. The chemoautotrophic species *Thiomargarita namibiensis* is among the largest prokaryotes. It can be up to 0.75 mm wide (**Figure 8-7a**).

Many chemoheterotrophic proteobacteria live inside other organisms. Members of the genus *Rhizobium* live in the roots of legumes such as peas and beans. The bacteria receive sugars from the plant and, in turn, provide the plant with ammonia they produced by fixing nitrogen. The most well-studied prokaryote is *Escherichia coli* (**Figure 8-7b**), a bacillus that lives in mammalian intestines and is easily grown in the laboratory. *Rickettsias* are very tiny proteobacteria that live inside cells. The genomes of mitochondria are most similar to those of *Rickettsias*. This indicates that the bacterial ancestors of mitochondria were related to proteobacteria.

CCC Structure and Function *Describe the structural differences between Gram-positive and Gram-negative organisms. How do these differences affect the ecological roles of some of these organisms?*

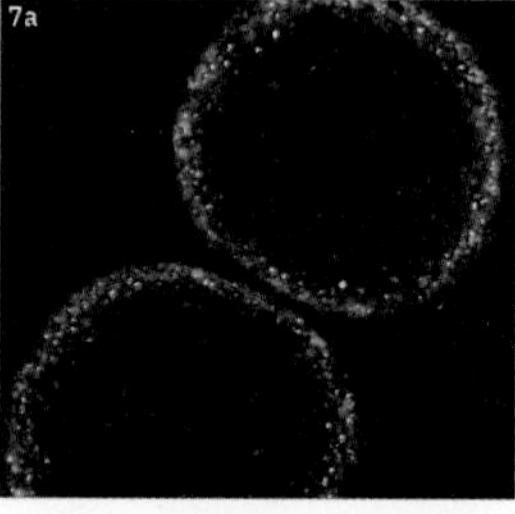

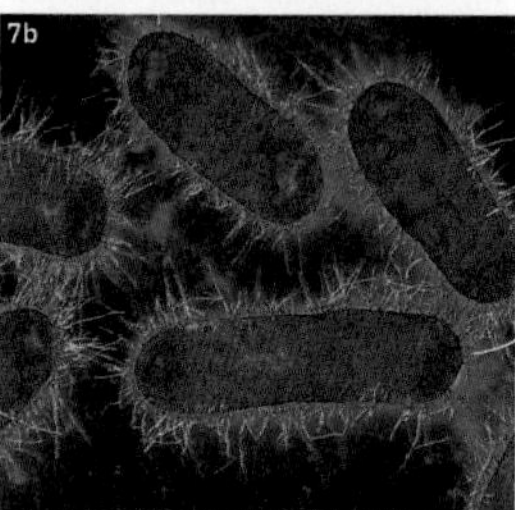

FIGURE 8-7 *T. namibiensis* cells (a) are large enough for individual cells to be seen with the unaided eye. *E. coli* (b) is the most-studied prokaryote, common in laboratories and important in mammalian digestion.

CROSSCUTTING CONCEPTS | Structure and Function

Operating Systems Emphasize to students that to understand how a system works, they must first examine what it is made of, the shapes of its parts, and the role they play within the system. To reinforce this concept, provide some real-life examples of how features within a system work together that might be more tangible for students. Have them visualize the process for designing an app or computer program. Describe how when building an app, each piece of coding plays an important role in the successful operation of the app once it is fully created. One incorrectly placed binary code, backslash, or tilde and the app becomes inoperable. Remind students that this coding is not visible while working in the app but rather it is running behind the scenes. Have students discuss and come up with other examples that can be used to model how the varying structures within a system are critical to the smooth operation of the system.

Bacteria and Human Health

Bacteria and other microorganisms that normally live in or on our body are our normal microbiota (explored in Unit 2). These cells outnumber our own cells, and they are our first defense against pathogens. For example, lactic acid produced by bacteria in the human mouth, gut, and vagina help keep acid-intolerant pathogenic bacteria from taking hold. Some bacteria that live in our gut also benefit us by producing essential vitamins. Our intestines are home to lactic acid bacteria that synthesize some B vitamins that we cannot make, and also to *E. coli* that produce vitamin K.

Toxins and Disease Bacteria cause many common diseases (**Table 8-2**). Most pathogenic bacteria harm us by way of toxins that disrupt our health. The toxin may be a substance that bacteria release into their environment or a molecule that is part of the cell's wall. Botulinum toxin released by *Clostridium botulinum* disables nerve cells, which is why a dangerous paralyzing food poisoning occurs after ingesting the bacteria. Toxins from bacteria may not harm cells directly, but instead elicit an immune response that produces symptoms such as fever and aches.

Whooping cough (pertussis) and tuberculosis are bacterial respiratory diseases spread by coughs or sneezes. *Mycobacterium tuberculosis*, the cause of tuberculosis, infects about a quarter of the human population. In most people, the infection is inactive and does no harm. If the infection becomes active, bacteria grow in the lungs and cause severe lung damage. Each year, more than a million people die as a result of active tuberculosis.

Streptococcus and *Staphylococcus* are Gram-positive cocci that infect the outer layers of the skin, often causing infection when an injury occurs. *Streptococcus* also causes strep throat and scarlet fever. Gonorrhea, syphilis, and chlamydia are bacterial infections transmitted by sexual contact.

TABLE 8-2. Examples of Bacterial Diseases

Disease	Typical bacterial cause	Description
botulism, tetanus	*Clostridium botulinum, Clostridium tetani*	muscle paralysis by bacterial toxin
chlamydia	*Chlamydia trachomatis*	sexually transmitted infection
cholera	*Vibrio cholerae*	diarrheal illness
gonorrhea	*Neisseria gonorrhoeae*	sexually transmitted infection
impetigo, boils	group A *Streptococcus, Staphylococcus aureus*	blisters, sores on skin
Lyme disease	*Borrelia burgdorferi*	rash, flulike symptoms, spread by ticks
Rocky Mountain spotted fever	*Rickettsia rickettsii*	rash, flulike symptoms, spread by ticks
strep throat	*Streptococcus pyogenes*	sore throat, can damage heart
syphilis	*Treponema pallidum*	sexually transmitted infection
tuberculosis	*Mycobacterium tuberculosis*	respiratory disease
whooping cough	*Bordetella pertussis*	childhood respiratory disease

In Your Community

Food poisoning Food poisoning is a generic name for any foodborne illness. Many foodborne illnesses are caused by bacteria. One common cause of food poinsoning is the bacteria *Esherichia coli,* which typically lives in the intestine of healthy people and animals. When *E. coli* is ingested in contaminated food, it enters the small intestine and produces a toxin called Shiga. This toxin damages the lining of your small intestine, causing vomiting and diarrhea. According to the Centers for Disease Control, about 265,000 infections occur in the United States each year. *E. coli* are found in all areas of the United States and are able to thrive both inside and outside of a living organism.

Have students assess the environments in their homes and neighborhoods for risk of *E. coli* infection. Challenge them to look into the various ways that *E. coli* is spread through a community. Examples include undercooked meats, contaminated vegetables, raw milk, soft cheeses, or contaminated water. Have students connect these causes to cases that have occurred either at a local level or on a national level, such as food recalls or restaurants that have had breakouts. Help students to understand that even though *E. coli* is a naturally occurring and beneficial gut bacterium, it is dangerous when removed from its intended environment.

CHAPTER INVESTIGATION B

Design Your Own *Effects of Antimicrobials*

Time: 150 minutes over 3 days

Students investigate the effectiveness of different antimicrobials on bacterial growth and construct an explanation of their findings.

Go online to access detailed teacher notes, answers, rubrics, and lab worksheets.

SEP Communicate Information

Sample answer: Bacteria can be transmitted through aerosol transmission, entrance through an injury, tainted food or water, and insect vectors.

On the Map

Lyme Disease Across the U.S. The blacklegged tick (or deer tick, *Ixodes scapularis*) spreads Lyme disease in the northeastern, mid-Atlantic, and north-central United States. The western blacklegged tick (*Ixodes pacificus*) spreads Lyme disease on the Pacific Coast. Direct students to use an internet keyword search such as "range map lyme disease CDC" to find the CDC's interactive map of Lyme disease incidence for the most recent year. Data from 2019 shows that only Hawaii had zero cases of Lyme disease. Students can use the map to assess their relative risk of contracting Lyme disease, which can lead to serious health issues. They can also learn more about transmission and symptoms of the disease at the CDC site.

CHAPTER INVESTIGATION

Effects of Antimicrobials

How can you determine the effectiveness of an antimicrobial?

Go online to explore this chapter's hands-on investigation to design an experiment to test the effectiveness of different antimicrobial substances.

Bacterial pathogens sometimes enter the body in tainted food or water. Most cases of bacterial food poisoning occur after bacteria-rich animal feces contaminate food. In regions where safe drinking water is not readily available, cholera sickens millions of people each year and kills an estimated 100,000 people. Cholera-causing bacteria (*Vibrio cholerae*) produce a toxin that causes intestinal cells to malfunction, and this results in a potentially fatal diarrhea. Tainted water also spreads *Helicobacter pylori*, which causes stomach ulcers. *H. pylori* is the only bacterial species known to cause cancer; a long-term infection increases the risk of stomach cancer.

Lyme disease (**Figure 8-8**) is a bacterial disease that is transmitted by a *vector*, which carries the bacteria between hosts. Ticks move the spirochete bacteria (typically *Borrelia burgdorferi*) that cause the disease between vertebrate hosts. Infection may initially cause a bull's-eye-shaped rash at the site of the tick bite. A long-term infection can harm joints and the nervous system.

SEP Communicate Information *Describe at least two ways that bacterial pathogens are transmitted to cause infection in new hosts.*

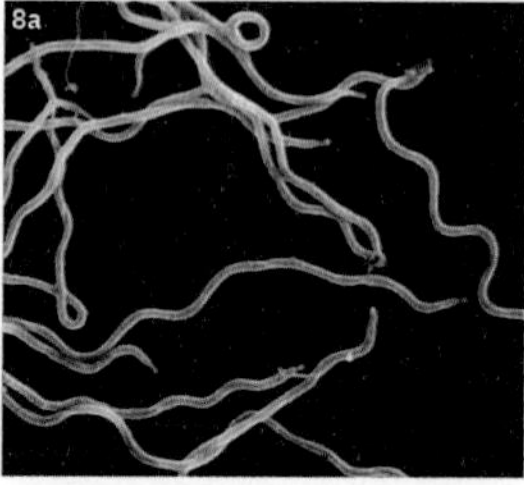

FIGURE 8-8 The spirochete bacteria (a) that cause Lyme disease are transmitted to humans by vector insects such as this deer tick (b).

Archaea

Archaea and bacteria often live alongside each other and have many physical similarities, but they have many differences that are not visually obvious. Archaea cell walls include different molecules that help them live in what we would consider extreme environments. Also, their cell membrane contains lipids that are different from bacteria. Like eukaryotes, archaea organize their DNA around histone proteins, which bacteria do not have. The enzymes and signals that archaea use to transcribe RNA and translate proteins are more similar to those of eukaryotes than bacteria.

Archaeal Habitats and Diversity Some archaea species can live in places where few or no other cells survive (**Figure 8-9**). For example, they are the most abundant cells in the ocean's deepest, darkest waters. Archaea were first discovered in hot springs, and many are extreme **thermophiles**, meaning they grow at a very high temperature. Some archaea that live near deep-sea hydrothermal vents grow even at 110 °C (230 °F). Other archaea are extreme **halophiles**, meaning they live in highly salty water. Salt-loving archaea live in the Dead Sea, the Great Salt Lake, and smaller brine-filled lakes. Most are photoheterotrophs that capture light energy using a red pigment.

SCIENCE AND ENGINEERING PRACTICES
Constructing Explanations and Designing Solutions

Environment Interactions As students observed from **Figure 8-7** and **Figure 8-8,** archaea and bacteria look relatively similar. Students also learned from analysis of **Table 8-1** that both organisms live in many of the same environments. Lead a discussion with students that highlights the differences between the two cell types. Point students to the difference in archaea cell wall and cell membrane structure and the presence of histones in their DNA. Guide students to think of this through the lens of structure and function. As you discuss with the class, place relevant information within a large Venn diagram on the board. Then, guide students to use the information to begin an explanation of how the structures of archaea and bacteria compare and relate to function. Students should include how different structures might influence the different environments where certain species may be able to live and thrive. They could start the explanation by stating "The _______ structures help archaea to function in specific environments because _______."

FIGURE 8-9 Microbial communities around deep-sea hydrothermal vents include both bacteria and archaea, which perform chemosynthesis and serve as producers for other organisms that can flourish in this ecosystem.

Many archaea, including some extreme thermophiles and extreme halophiles, are **methanogens**. While some bacterial species can live in most of the environments where archaea are found, the only known methanogens are species of archaea. Methanogens are chemoautotrophs, and their ATP-producing reactions also produce methane (CH_4), an odorless gas. Methanogenic archaea are strict anaerobes, meaning they cannot live in the presence of oxygen. They abound in sewage, marsh sediments, and the animal gut. Cattle have methanogens in their stomachs and release methane gas by belching. About a third of humans have significant numbers of methanogens in their intestine. So far, no archaea have been found to be human pathogens.

SEP Construct an Explanation *Explain generally how the structures of archaea and bacteria compare. How do these structures affect where different species live?*

8.1 REVIEW

1. **Evaluate** Which questions are useful in defining the characteristics of a new species of bacteria? Select all correct answers.
   - A. Do the bacteria produce oxygen?
   - B. What is the shape of the bacteria?
   - C. How do the bacteria obtain energy?
   - D. How many cells do the bacteria have?
   - E. How thick is the layer of peptidoglycan in the bacterial cell wall?
2. **Define** List some of the ways that different types of bacterial infections can spread.
3. **Distinguish** Sort the characteristics according to which group(s) of bacteria they apply to: cyanobacteria, Gram-positive bacteria, or proteobacteria. Characteristics may be related to more than one group.
   - have an outer cell membrane
   - perform photosynthesis
   - form endospores
   - have a thick cell wall
4. **Describe** In what type of environments have archaea typically been found?

Vocabulary Strategy

Prefixes/Suffixes/Root Relationships Review thermophiles and halophiles with students. Explain that the suffix *philes* means "to love" and the prefix *thermo* means "heat." So, a thermophile is an organism that loves heat. The prefix *halo* means "salt." So, a halophile is an organism that loves salt.

Encourage students to apply this understanding to other biology terms, such as acidophile, cryophile, and psammophile.

SEP Construct an Explanation
Sample answer: Archaea and bacteria look relatively similar and live in many of the same environments, but they have important differences in their cell walls, cell membranes, and other cellular machinery, which make archaea more similar to eukaryotes. Archaea cell walls and other proteins are structured in a way that allows them to live in more extreme environments.

8.1 REVIEW

1. *Do the bacteria produce oxygen? What is the shape of the bacteria? How do the bacteria obtain energy? How thick is the layer of peptidoglycan in the bacterial cell wall?* **DOK 3**
2. *Students' answer may include any of the following: via air, via sexual contact, via food or water, and via insect bites.* **DOK 2**
3. *cyanobacteria: have an outer cell membrane, all perform photosynthesis; Gram-positive bacteria: some form endospores, have a thick cell wall; proteobacteria: have an outer cell membrane.* **DOK 1**
4. *Sample answer: First discovered in hot spring environments, archaea are often thermophiles or halophiles that live in hot or high-salt conditions, respectively. Some archaea also live in environments with no oxygen and produce methane.* **DOK 1**

8.2

PROTISTS

LS2.C

EXPLORE/EXPLAIN

This section provides an overview of the classification of eukaryotes and protists, introduces the concept of protist diversity, and reviews the different structures and characteristics associated with the organisms found in some protist eukaryotic supergroups. It also covers the ecological diversity of protists.

Objectives

- Evaluate how the structures of some protists are related to how they function in their environment.
- Explain why protists are described as a kingdom.
- Identify characteristics common to protists.

Pretest Use **Question 6** to identify gaps in background knowledge or misconceptions.

Analyze *Students might say that scientists classified many organisms together because they did not fit into other kingdoms. So many different types of organisms have been described as protists.*

SEP Scientific Knowledge *Sample answer: Phylogenetic analysis has allowed scientists to understand the evolutionary relationship between organisms and find similarities in DNA sequences between protists and other organisms. While previously some protists were sorted together based on observable characteristics, we now know that some species are more closely related to animals or plants than they are to other protists.*

8.2

KEY TERMS

colonial organism
phytoplankton
protist
pseudopod
zooplankton

PROTISTS

Do you have a junk drawer or a box where you throw things you are not sure what to do with? For years, scientists classified protists together in their own kingdom because they were organisms that did not quite fit into any other categories. Now, however, scientists have much more evidence about how protists are related to each other and other organisms.

Analyze *How has the kingdom of protists been like a "junk drawer" for scientists?*

Classification of Eukaryotes and Protists

The kingdoms of the eukaryote domain that are likely the most familiar to you are plants, animals, and fungi. However, eukaryotes also include a fourth diverse kingdom of organisms commonly referred to as **protists**. Despite the designation as a kingdom in many classification systems, protists are no longer considered a valid biological group because no single trait unites them. Some protists are more closely related to plants, fungi, or animals than to other protists.

Scientists continue to investigate exactly how all the various eukaryotes are related, and phylogenetic analyses have led to systems that differ from the traditional kingdoms. One widely used organization distributes eukaryotes into six supergroups based on evolutionary relationships, or how closely scientists think they are related. Supergroups are a taxonomic level of organization between domain and kingdom (Appendix E).

Four of the supergroups include protists exclusively. A fifth group includes plants (Chapter 9) and their protist relatives. Animals (Chapter 10), fungi, and their protist relatives are members of the sixth group. **Figure 8-10** shows the five most important supergroups. This section will review characteristics and examples of some protists that are in each of these five eukaryotic supergroups. You will notice that even within each supergroup, the organisms described are incredibly diverse, and there are not necessarily any uniting characteristics to easily describe each supergroup.

NOS Scientific Knowledge *How have advances in scientific knowledge helped scientists better understand the relationships between the many different organisms called protists?*

FIGURE 8-10 Eukaryotes are often divided into supergroups for ease of discussion. However, scientists' understanding of the evolutionary relationships between organisms is constantly changing and improving.

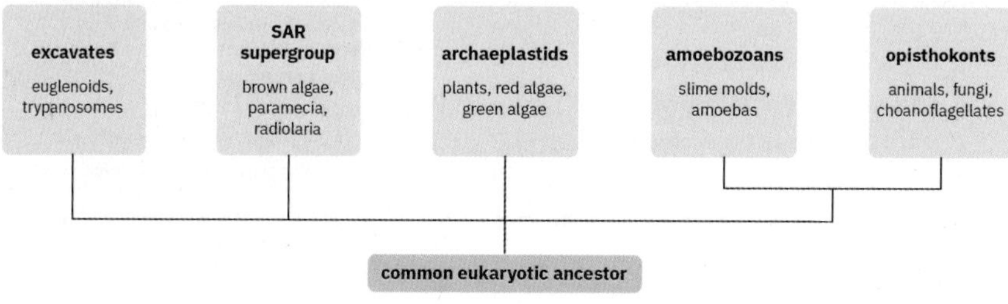

SCIENCE AND ENGINEERING PRACTICES
Scientific Knowledge Is Open to Revision in Light of New Evidence

New Species Remind students that classification systems in science are fluid and change as new information becomes available. To reinforce this concept, play a classification game with them. Provide students with categories, such as reptiles, birds, amphibians, and fish. Explain to them that based on the available information, they must select a category for an unknown animal species. Provide students with an initial observation, such as the species lays eggs. Students should recognize that based on this information, the species could go in any of the provided categories. Then, share that the organism is cold blooded. Since birds are warm blooded, the unknown species cannot be a bird. Continue sharing new information until there is only one possible category remaining. Relate activity back to eukaryotes classification and the role of phylogenetic analyses in helping to better classify organisms from the protist kingdom.

Protist Characteristics

Most protists live as single cells, but some are colonial or members of multicellular organisms. A **colonial organism** consists of cells that live together but remain self-sufficient. Each cell retains the traits required to survive and reproduce on its own. By contrast, the cells of a multicellular organism have a division of labor and are interdependent.

Given the differences in organization, it is not surprising that protists vary widely in size. The smallest known free-living protist is a single-celled, green alga about 1 micrometer in diameter. The largest protists are giant kelps that regularly grow more than 30 meters in length.

Metabolism Most protists need oxygen to live. However, some are adapted to living in places with little or no oxygen. These anaerobic protists have modified mitochondria that allow them to make ATP without oxygen.

Protists include *autotrophs* and *heterotrophs*. Autotrophs, such as various types of algae, carry out photosynthesis by the same oxygen-producing photosynthetic pathway as cyanobacteria. They produce much of the oxygen we breathe and serve as food for other organisms, including humans. Heterotrophic protists obtain carbon and energy by breaking down organic materials. Some absorb organic molecules directly from their environment. Others capture and ingest smaller organisms such as bacteria. A few are able to change how they obtain energy depending on their environment.

Habitats The overwhelming majority of protists are free-living and aquatic in seawater or freshwater habitats. However, some live in moist soils, and others live on or in other organisms. Protists can benefit their hosts. For example, some photosynthetic protists live in the tissues of corals and supply their animal host with sugars. Other protists are pathogens, including the one that causes malaria.

Common Structures Protists are eukaryotes, so they all have a nucleus. In addition, most have other standard eukaryotic organelles such as an endoplasmic reticulum and Golgi body. All protists have at least one mitochondrion. Some typical organelles are shown in **Figure 8-11**, which illustrates the body systems of two different single-celled, freshwater protists called *Euglena* and *Paramecium*.

FIGURE 8-11 These diagrams highlight interesting features of euglena and paramecium systems. Euglena and paramecium both have typical features of eukaryotes that are not shown, including endoplasmic reticula, Golgi bodies, and mitochondria.

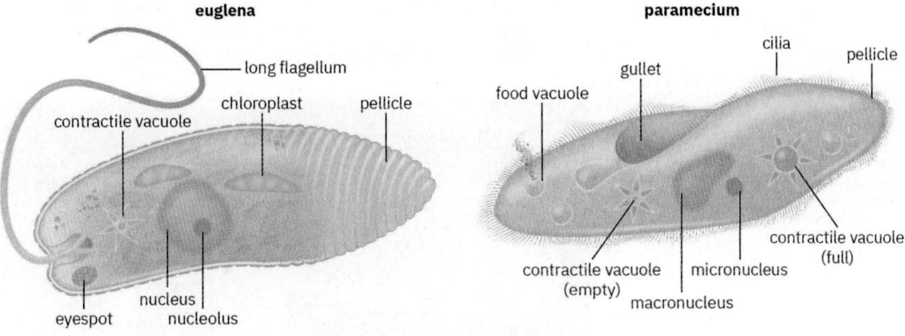

Address Misconceptions

Protist Size Students may incorrectly think that all protists are microscopic and single-celled. Kelp, for example, is the largest known protist. Refer students to **Figure 8-15** to visually reinforce the size of and magnitude of kelp. Students may also confuse kelp for a plant since it resembles tall grass. Remind them kelp is actually a brown algae seaweed.

Visual Support

Common Eukaryote Structures Students may look at **Figure 8-11** and assume these are the only structures common to the organisms being modeled. To help students reconcile the information in **Figure 8-11** with the actual structures in common, it may be useful to have them review the complete list of eukaryote features from Chapter 6. Have students work together in groups to add any common features from Chapter 6 that are not present in **Figure 8-11.**

DIFFERENTIATED INSTRUCTION | English Language Learners

Using Visuals Photos and diagrams can help students develop new vocabulary and enhance and confirm understanding of the text. To support these students, refer them to **Figure 8-11.** Have them review the labels from the image. Then, have students skim the text looking for the same key terms. As they encounter the terms in their reading, they should pause and locate the term in **Figure 8-11**.

Beginning Have students use the information from the text and **Figure 8-11** to create a list of terms in common and terms that vary.

Intermediate Have students use the information from the text and **Figure 8-11** to identify the function of key features.

Advanced Have students use the information from the text and **Figure 8-11** to describe how certain features enable protist survival based on their environment.

Connect to English Language Arts

Compare and Contrast In this section, students learn about protist structures that help them survive in their habitats. After students read this page, have them compare and contrast protist characteristics that enable them to survive and reproduce in specific environments. Students should discuss both similarities and differences in features. Encourage them to evaluate how the features that are different can be beneficial to the protist based on their environment.

Structures That Vary Protists have a wide variety of characteristics that help them survive in their particular habitats.

- **Outer Coverings** Most single-celled protists have a protective layer at their cell surface. *Radiolaria* and diatoms make a glassy silica shell. *Foraminifera* build a calcium carbonate shell. By contrast, the outer layer of *Euglena* is a *pellicle* that constrains the cell's shape, but flexes enough to allow the organism to bend and squeeze through openings.
- **Movement-Related Structures** Most single-celled protists are able to move from place to place during part or all of their life cycle. Some protists, such as *Euglena*, propel themselves by moving one or more flagella. Others, such as *Paramecium*, move by means of cilia. Others crawl along by extending lobes of cytoplasm.
- **Contractile Vacuoles** The interiors of freshwater protists such as *Euglena* and *Paramecium* are saltier than their freshwater habitat, so water tends to diffuse into the cell. To keep from bursting, these protists operate one or more contractile vacuoles. These are organelles that collect excess water from the cytoplasm and then contract and expel it out of the cell through a pore.
- **Chloroplasts** Many protists have chloroplasts. Chloroplasts work to convert light energy into chemical energy that can be stored and ultimately used for cell processes.

Protist Diversity

As indicated, the term *protists* includes a wide variety of different organisms. The following survey of the five most important supergroups represents only a part of the structural and ecological diversity of protists.

Excavates All excavates are unwalled cells with one or more flagella. The supergroup name refers to an "excavation," a groove that functions in feeding in some members of this group.

- *Giardia intestinalis* is anaerobic and has two nuclei and multiple flagella. It causes giardiasis, a common waterborne disease. A person becomes infected by ingesting cysts that were passed in the feces of a previous host. When a cyst reaches the host's intestine, the cell inside becomes active and attaches itself to the intestinal wall (**Figure 8-12**). The resulting symptoms include cramps, nausea, and diarrhea. Giardia cysts often enter streams and lakes via the feces of wild animals, which is why hikers are advised to treat water from such sources before drinking it. Boiling kills the cysts. Microfilters are also effective.

FIGURE 8-12 The *Giardia lamblia* in this scanning electron micrograph are another species that causes giardiasis. The active, feeding stage is shown, with flagella that help the organism move.

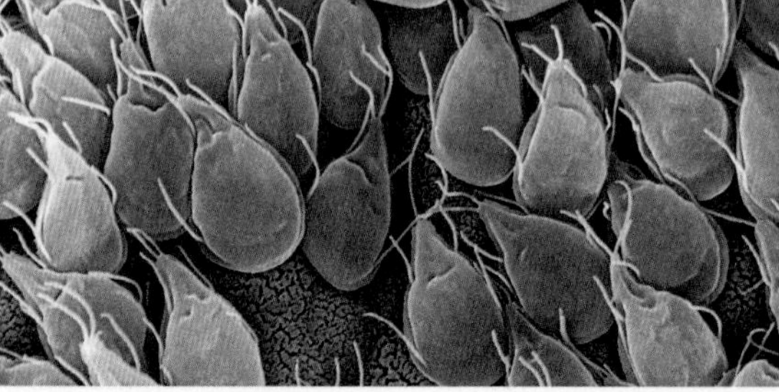

DIFFERENTIATED INSTRUCTION | Students with Disabilities

Auditory Impairments Students with auditory impairments benefit from visual supports to help encourage communication of concepts. When developing an understanding of structures that vary and structures that are common, refer students to **Figure 8-11**. Have students work in groups to look for structures that are present on both models. Have them highlight these terms on both models in the same color. Each new common structure should be highlighted in a new color. This process will help students visually see which structures are in common and which structures vary. Have them reference their highlighted **Figure 8-11** to help them evaluate any differences in structures and how the differences are beneficial to the protist.

- Trypanosomes are long, tapered cells with a flagellum that emerges from the back of the cell and extends toward the front of the cell (**Figure 8-13**). Movement of the flagellum causes the membrane to wiggle in a wavelike manner that propels the cell. All trypanosomes are parasites. Biting insects transmit those that infect humans. For example, tsetse flies spread *Trypanosoma brucei*, the agent of African sleeping sickness. *T. brucei* lives in blood and other body fluids, where it absorbs nutrients across its body wall. Infected people feel drowsy during daytime and often cannot sleep at night. An untreated infection results in coma and, eventually, death.

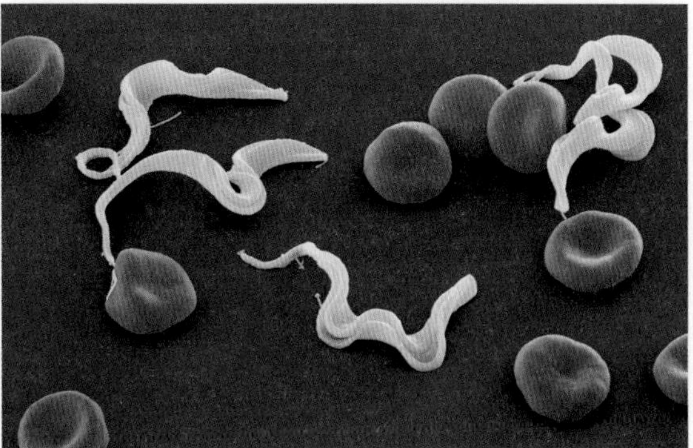

FIGURE 8-13 This scanning electron micrograph shows *Trypanosoma brucei* alongside red blood cells from a human host.

- Euglenoids such as *Euglena* (**Figure 8-14**) are free-living, primarily freshwater protists closely related to trypanosomes. Depending on their environment, some are heterotrophs that engulf bacteria and small protists. Others are autotrophs. Euglenoids have chloroplasts that evolved from a green alga. They carry out photosynthesis when there is light but can become heterotrophs in the dark. An eyespot near the base of a long flagellum helps the cells detect light.

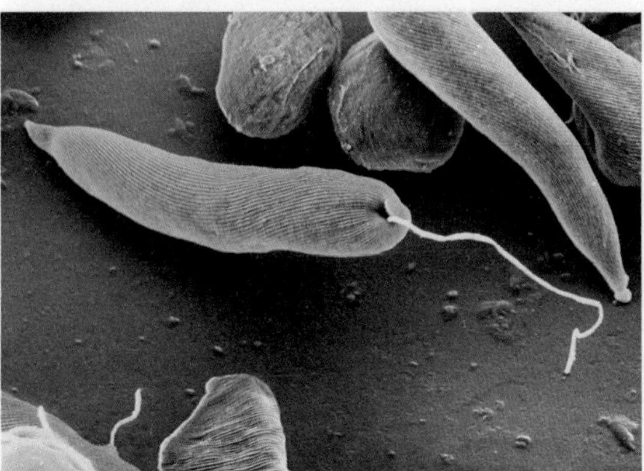

FIGURE 8-14 This color-enhanced scanning electron micrograph shows the outside of *Euglena granulata*, which is covered by a protective pellicle and has a flagellum for movement.

Visual Support

Movement-Related Structures Help students process the different mechanisms that help protists survive in their particular environment by having them identify the flagella in **Figure 8-13** and **Figure 8-14**. Have students looks for any similarities. Then, as they read, have students pay attention to the text for any reference to their observed similarities. Each figure has flagella that are noticeably different in shape and size. Students should recognize that the flagellum is unique to the needs of each specific protist.

DIFFERENTIATED INSTRUCTION | Leveled Support

Struggling Students Some students may have a difficult time keeping track of variations among the protist supergroups. Help students make sense of this information with a 5 x 5 chart graphic organizer. Label columns: Protist Name, Example, Unicellular/Multicellular, Feeding, and Unique Characteristics. Students work together to complete the informational chart. Then, have students use their chart to discuss how the structures of some protists are related to how they function in their environment.

Advanced Learners Students who have an understanding of protist diversity may be interested in creating a Habitat For Rent poster. The poster should include an image of the habitat, selling points for the habitat (i.e., located near a big tourist attraction would appeal to *Giardia intestinalis* as they require a host to attach to), and a list of organisms that might already be living in the "neighborhood," or local restaurants that would make the area desirable to the organism.

CHAPTER INVESTIGATION A

Guided Inquiry *Classification Systems*

Time: 100 minutes or 2 days

Students observe, identify, and classify characteristics of 10 algae samples to develop their own dichotomous key. Then, they test their key using the algae samples to categorize and name them, compare their key to those of other group's keys to identify similarities and differences.

Go online to access detailed teacher notes, answers, rubrics, and lab worksheets.

CHAPTER INVESTIGATION

Classification Systems

What classification system should be used to sort different algae?

Go online to explore this chapter's hands-on investigation to design your own classification system for algae.

SAR Supergroup The SAR supergroup is made up exclusively of protists. It is an incredibly diverse grouping of organisms.

- Brown algae are multicellular "seaweeds" of temperate or cool seas. Most grow along rocky seashores. They range in size from microscopic filaments to the largest protists—the giant kelps (**Figure 8-15**). Like trees in a forest, kelps shelter a wide variety of other organisms. Kelps are also the source of algins (or alginates), which are polysaccharides used to thicken foods, drinks, cosmetics, and personal care products.
- Diatoms are single-celled or colonial protists that secrete a transparent silica shell. The shell consists of two parts that fit one inside the other, like a shoe box and its lid. Diatoms are a component of the **phytoplankton**, the community of photosynthetic cells that live in the upper sunlit waters of lakes and seas. Phytoplankton is the base for aquatic food chains and the source of about half of Earth's photosynthetically produced oxygen. When marine diatoms die, their remains fall to the seafloor.
- *Paramecium* is a single-celled organism common in ponds (**Figure 8-16a**). Its covering of cilia allows it to move and helps it feed. The cilia sweep water laden with food particles into an oral groove at the cell surface and then into a gullet where vesicles digest the food.
- Malaria is a potentially deadly infectious disease caused by four species in the genus *Plasmodium* (**Figure 8-16b**). Different species infect different hosts. In 2020, 241 million cases of malaria in humans led to more than 600,000 deaths. A human infection begins after a mosquito carrying an infectious form of the *Plasmodium* bites a person. The organism completes its life cycle inside the human host, and the next mosquito that bites that human can pick up the *Plasmodium* and continue to spread the disease.

FIGURE 8-15
This kelp forest, located on Cortes Bank in the North Pacific Ocean, is home to many other organisms.

222 CHAPTER 8 DIVERSITY OF LIVING SYSTEMS

DIFFERENTIATED INSTRUCTION | Students with Disabilities

Visual Impairments Students who have visual impairments may struggle to make connections between the text and the information in **Figure 8-15**. Students who have partial visual impairments can be supported by allowing groups to discuss images prior to reading. Encourage groups to paint a mental picture of the image displayed on the page by describing colors and organisms observed. As students read, encourage them to note any references to color and/or organisms. For example, in the first bulleted information of the SAR Supergroup, seaweed is described as brown algae. Have students pause in their reading to synthesize this information with the visual imagery of **Figure 8-15**.

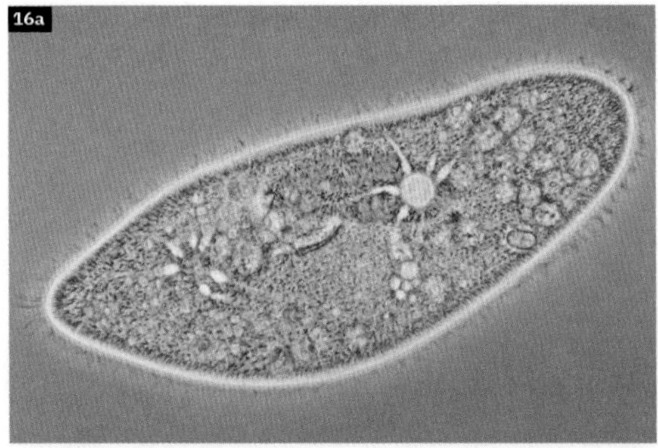

- Radiolaria have a silica shell (**Figure 8-17**) and many gas-filled vacuoles that keep the cell afloat. Radiolaria are part of plankton and are most abundant in nutrient-rich tropical waters.
- Foraminifera have a chalky calcium carbonate shell. Most species live on the seafloor, where they probe the sediments for prey. Others are part of the marine **zooplankton**, a collection of tiny heterotrophs that drift or swim in the sea. Foraminifera and radiolaria have lived in the seas for millions of years, so remains of countless cells have fallen to the seafloor. Over time, geologic processes can transform accumulated foraminifera shells into chalk and limestone, two types of sedimentary rock.

CCC Structure and Function *Describe how the structures of some protists in the SAR supergroup are related to how they function in their particular environments.*

FIGURE 8-16
Paramecium and *Plasmodium* species are vastly different, despite being part of the same supergroup. Notice the full contractile vacuole in this freshwater *Paramecium caudatum* (a). These *Plasmodium gallinaceum* are shown invading the midgut of a mosquito (b). This species of *Plasmodium* causes malaria in poultry.

FIGURE 8-17
This light micrograph reveals the variety of silica structure shapes that enclose radiolaria. Living radiolaria push lobes of cytoplasm (pseudopods) through pores in the shells to obtain food.

Visual Support

SAR Group Have students look at **Figure 8-16** and examine the images of the paramecium and plasmodium cells. Explain that while the two species are classified in the SAR group, they have different structural characteristics that allow them to thrive. Ask students to identify some of these characteristics and explain their function. (*The cilia and contractile vacuole of the paramecium help it move and feed in water. The narrow shape and twisting of the plasmodium help it invade cells.*)

CCC Structure and Function *Sample answer: Flagella and cilia help protists in aqueous environments move. Gas-filled vesicles can help radiolaria stay afloat in water. Amoeba and slime molds stream cytoplasm into pseudopod shapes to pull themselves across surfaces.*

CROSSCUTTING CONCEPTS | Structure and Function

SAR Variations Emphasize to students that the structures within the organism that differ often offer the greatest insight into the organism. Focus student attention as they read about some of the protists in the SAR supergroup on information related to movement-related structures, outer coverings, contractile vacuoles, and chloroplasts as these are the features with the greatest variation among them. Have students create a three-column T-chart to guide their reading of the SAR supergroup. Label the column headings: Movement-Related Structures, Outer Coverings, and Chloroplasts. As they encounter information related to each heading, they should place the name of the protist and the function of the feature within the column. Then, encourage students to use their T-charts to describe how the structures of some protists in the SAR supergroup are related to how they function in their environments.

Video

Time: 11 seconds

Video 8-1 shows how slime molds move across a surface. Share with students that slime molds move by changing their shape. Have students identify the structure that is used to allow the slime mold to change shape *(cell wall)*.

Connect to English Language Arts

Summarize This section introduces several protist categories from the SAR supergroup. Have students write an informative summary of the information presented for each protist subgroup. Encourage students to use a graphic organizer as they read to ensure they are collecting similar key information for each classification type.

 VIDEO

VIDEO 8-1 Go online to see how slime molds move across a surface.

Amoebozoans Amoebozoans are shape-shifting heterotrophic cells that typically lack a cell wall or pellicle. They move and capture smaller cells by extending thick lobes of cytoplasm called **pseudopods**. There are two groups of amoebozoan protists: amoebas and slime molds.

- Amoebas are single cells (**Figure 8-18a**) that live in marine or freshwater sediments or in damp soil. They are important predators of bacteria. A few amoebas are human pathogens. *Entamoeba* infects the human gut, causing cramps, aches, and bloody diarrhea. Infections are most common in developing nations where amoebic cysts contaminate drinking water. The freshwater species *Naegleria fowleri* can enter the nose and cause a rare but deadly brain infection.
- Slime molds are sometimes described as "social amoebas." They are common on the floor of temperate forests. Cellular slime molds spend most of their existence as individual amoeba-like cells. Each cell ingests bacteria and reproduces by mitosis. If food runs out, thousands of cells stream together to form a cohesive multicellular unit that can move to a suitable area. Plasmodial slime molds spend most of their life cycle as a plasmodium, a mass with many nuclei that forms when a cell undergoes mitosis many times without dividing its cytoplasm. A plasmodium streams along surfaces, feeding on microbes and organic matter (**Figure 8-18b**).

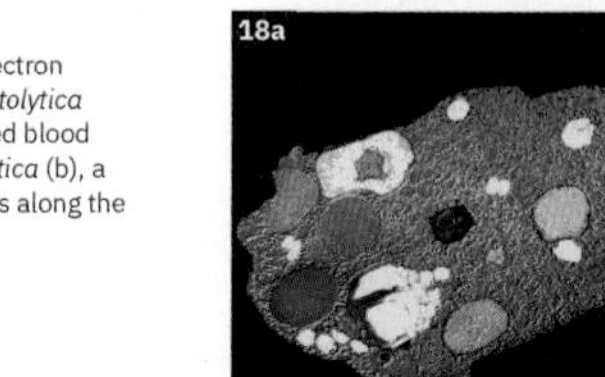

FIGURE 8-18 A false-color transmission electron micrograph of *Entamoeba histolytica* (a) reveals ingested human red blood cells (green ovals). *Fuligo septica* (b), a plasmodial slime mold, moves along the bark of a dead pine tree.

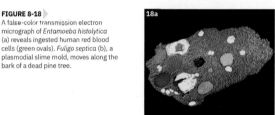

Archaeplastids The archaeplastid supergroup includes land plants, green algae, and the protists that are most closely related to plants. All archaeplastids have cells with cellulose walls and chloroplasts that evolved from cyanobacteria. Green algae (**Figure 8-19a**) were long considered to be protists but scientists now classify many of them as plants. Green algae share many features with land plants, including the characteristic green chlorophyll pigments.

Red algae are photosynthetic, generally multicellular protists that live in clear, warm seas. Most grow either as thin sheets or in a branching pattern (**Figure 8-19b**). One subgroup, the coralline algae, deposits calcium carbonate in its cells. The rigid material produced by these algae contributes bulk to some tropical reefs. Two gelatinous polysaccharides, agar and carrageenan, are extracted from red algae for use as thickeners or stabilizers in foods, cosmetics, and personal care products. Agar is also important in microbiology. When mixed with appropriate nutrients, it forms a semisolid culture medium that can be used for growing bacteria or fungi. Another red algal product is nori, the sheets of dried seaweed used to wrap sushi. Nori is derived from the red alga *Porphyra*.

DIFFERENTIATED INSTRUCTION | English Language Learners

Vocabulary Strategies Have students use strategies such as drawing or making physical models to help them learn and remember new vocabulary. Review amoebozoans and archaeplastids. Explain that the term *amoebozoans* comes from the Greek root *amoibe*, which means "to change." Show a piece of clay being molded into different shapes. The archaeplastids include land plants and others organisms closely related to plants. Draw a basic image with water near land. Include plants on the land.

Beginning Working in pairs, have students create a model of each term.

Intermediate Working in pairs, have students use the information provided about each new word and write a definition using their own words.

Advanced Working in pairs, have students write a short paragraph using the words *amoebozoans* and *archaeplastids*.

In Your Community

Food Additives Carrageenan and algin are substances extracted from the cell walls of red algae and are used in many products as food additives. They are used as thickeners and can be found in thousands of food products. Chocolate milk, yogurt, candy bars, ice cream, and even baby formula all contain algae-based food additives. Have students make a list of their favorite foods. Then, have them research the ingredient list to see which items contain food additives with a protist origin.

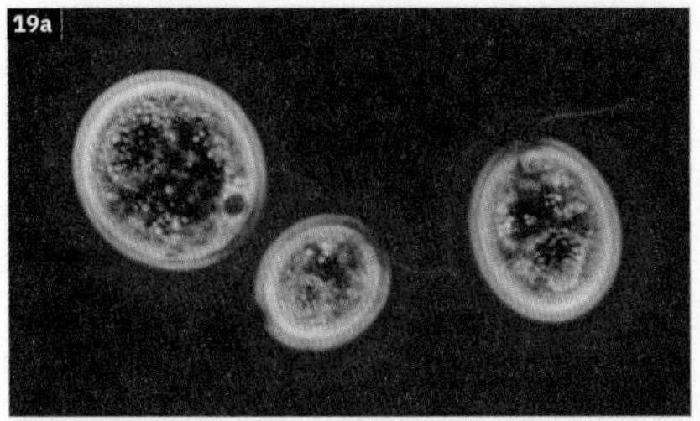

FIGURE 8-19 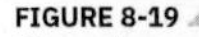
Chlamydomonas (a) is one genus of green algae that lives in soil and water and includes approximately 150 species. Light microscopy reveals that these single-celled algae each have two flagella. *Eucheuma* (b) is a genus of algae that includes species that are brown, green, and red in color. The multicellular red algae shown is being harvested for food at a seaweed farm in Malaysia.

Opisthokont Opisthokonts include fungi, animals, and the protist relatives of both groups. The protist group most closely related to animals is the *choanoflagellates*. Each choanoflagellate cell has a flagellum surrounded by a "collar" of thread-like projections (**Figure 8-20**). Movement of the flagellum sets up a current that draws food-laden water through the collar. Some sponge cells have a similar structure and function.

Most choanoflagellate species live as single cells, but some are colonial. The colonies arise by mitosis, when descendant cells do not separate after division, but rather stick together with the help of adhesion proteins. The adhesion proteins of choanoflagellates are similar to those of animals, and researchers have discovered that even solitary choanoflagellates have these proteins. Researchers continue to study choanoflagellates to learn more about what the earliest animals may have been like.

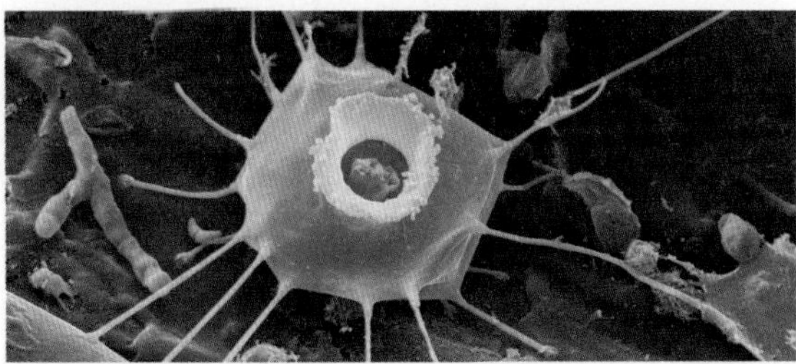

FIGURE 8-20
This choanoflagellate was isolated from wet soil. Its long flagellum is not visible here.

8.2 REVIEW

1. **Describe** Identify the characteristics common to nearly all protists. Select all correct answers.
   - A. have cells that have a nucleus
   - B. contain chloroplasts
   - C. have mitochondria
   - D. live as isolated single cells
   - E. populate aquatic habitats
   - F. require oxygen to live
2. **Explain** Why are protists often described as a kingdom even though they are not considered a valid biological group?
3. **Identify** Match the organisms to the structure that helps it move: flagella, cilia, or pseudopods.
   - paramecia
   - amoebas
   - trypanosomes
4. **Distinguish** Which supergroups include protists most closely related to animals, and which include those most closely related to plants?

8.2 REVIEW

1. *have cells that have a nucleus; have mitochondria* **DOK 1**
2. *Sample answer: Protists have no collective traits that distinguish them or unite them into a biological group. Traditionally, organisms that did not fit into other groups were lumped into the kingdom of protists. Phylogenetic analyses have revealed that protists are all over the board, with some species being more closely related to animals or plants than to other protists.* **DOK 2**
3. *flagella: trypanosomes; cilia: paramecia; pseudopods: amoebas* **DOK 2**
4. *related to animals: opisthokonts; related to plants: archaeplastids* **DOK 2**

MINILAB
FEATURES OF PARAMECIA AND EUGLENA

Students explore structure and function by observing either paramecia or euglena under a microscope to identify the features of the organism and evaluate the structural organization to make comparisons to other organisms.

Time: 30 minutes

Materials and Advance Preparation

- Premade paramecia or euglena slides can be ordered from an online lab supply store. As an alternative, live specimens of paramecia or euglena, which can also be ordered from an online lab supply store, can be used. If live specimens are used, be sure to loosen the lid on the container and aerate the culture using a pipette to replace oxygen depleted during shipping. Replace the lid loosely and store the specimens in a dark cupboard, away from direct sunlight.
- Instruct students on how these specimens should be placed on slides as well as how to properly dispose of materials after the lab is completed.

Procedure

In **Step 1**, if necessary, review how to use the microscope.

In **Step 5**, encourage students to complete their diagrams first. They can then go back and add more detail and labels as time permits. Encourage students to complete as much as they can on their own before referencing diagrams in the textbook or from other sources.

In **Step 6**, remind students to record the structures that they are able to identify under the microscope, as well as those they are unable to identify but are shown in the reference diagrams.

Results and Analysis

1. Identify *Sample answer: Using the microscope, I was able to identify the following features of the paramecium: oral groove, food vacuole, contractile vacuole, and cilia. I was not able to identify the radiating canals.* **DOK 1**

MINILAB

FEATURES OF PARAMECIA AND EUGLENAS

SEP Construct an Explanation What features of paramecia or euglenas can you identify?

Observing paramecia or euglenas under a microscope will provide the opportunity to recognize an organism's structures by looking at a real-world example rather than viewing a diagram in a text. This will allow you to see the structural organization within an organism. You will construct an explanation about how paramecium or euglena systems and structures compare with both simpler bacteria and more complex multicellular organism systems.

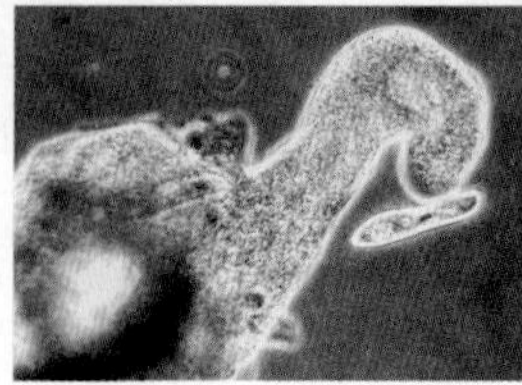
An amoeba extends a psuedopod to engulf a paramecium.

Materials
- compound microscope
- prepared slides of paramecia or euglenas
- reference diagrams of paramecia or euglenas (**Figure 8-11**) and bacteria (**Figure 8-4**)

Safety    

Procedure

1. Place your prepared slide of either paramecia or euglenas on the microscope stage.
2. Use the low-power objective (4× magnification, then 10× magnification) to locate several paramecia or euglenas.
3. With the organisms in focus, use the high-power objective (40× magnification) to make observations.
4. In your biology notebook or a digital device such as a laptop or tablet, create a diagram that reflects the organism you see under the microscope.
5. Label the structures of the organism based on what you know about protists such as paramecia or euglenas. If you know the name of the structure, label it. Also include what you think the structure's function might be, whether you label the structure or not.
6. Compare your observations and diagram to the diagram provided by your teacher or from the textbook. Identify structures in the provided diagrams that match the structures from your observations. Update your diagram with additional labels and functions of those structures and make any corrections.
7. Return your materials to their stations and clean up.

Results and Analysis

1. **Identify** What structures of the organism were you able to identify under the microscope? Were there any structures in the labeled diagram from the textbook or provided by your teacher that you were unable to identify using your microscope?
2. **Compare** How do the structures of protists, such as paramecia or euglenas, compare with the structures of bacteria? What are the similarities and differences?
3. **Infer** How are the structures that make up the living systems of paramecia or euglenas the same as those of more complex multicellular systems, such as humans? How are they different?

2. Compare *Sample answer: Protists have a well-defined cell structure that contains a nucleus as well as several specialized organelles, such as a food vacuole, a contractile vacuole, and cilia. Bacteria do not have a nucleus or specialized organelles. A similarity is that bacteria are single-celled organisms and protists can be single-celled organisms. However, protists can also be multicellular.* **DOK 2**

3. Infer *Sample answer: Protists such as paramecia and euglenas contain a nucleus as well as specialized smaller organelles. Their cellular structure is similar to that of more complex multicellular organisms such as humans. One significant difference between the cells of protists and more complex multicellular organisms is that single-celled protists are capable of moving on their own, whereas individual cells within a complex multicellular organism cannot move on their own. Additionally, some protists are single-celled organisms, whereas all plants and animals are multicellular.* **DOK 3**

FUNGI

An iceberg in the ocean can appear to be enormous, but 90 percent of its ice is below the surface where you cannot easily see it. Some fungi are similar to icebergs. Above ground, you may see a mushroom. But the mushroom is connected to a vast underground network far larger than what you see above the ground.

Infer *For the type of fungus described, what do you think might be the function of the part of a fungus that sticks up above the ground? The part below ground?*

8.3

KEY TERMS

chitin	lichen
endophyte	mycelium
fungus	mycorrhiza
hypha	spore

Characteristics of Fungi

A **fungus** (plural, *fungi*) is a eukaryotic heterotroph that feeds by secreting digestive enzymes onto organic material and then absorbing the resulting breakdown of products. Like plants, fungi have cell walls and a life cycle that involves producing **spores**—small reproductive units that can be dispersed and grow new organisms. **Chitin**, a nitrogen-containing polysaccharide (Section 5.3), is the main component of fungal cell walls. Like animals, to whom they are more closely related than plants, fungi are heterotrophs. Unlike either plants or animals, fungi include both single-celled and multicellular species.

Organization and Structure Single-celled fungi are commonly referred to as yeasts. We use some yeasts to bake bread and make alcoholic beverages. Others cause human yeast infections. Yeasts most often reproduce asexually by budding. During budding a descendant cell with a small amount of cytoplasm pinches off from its parent.

VIDEO

VIDEO 8-2 Go online to see yeast reproduce by budding.

Most fungi are multicellular. Powdery mildews, shelf fungi, and mushrooms are examples. A multicellular fungus consists of a mesh of microscopic thread-like filaments collectively referred to as a **mycelium** (plural, *mycelia*). Each filament in the mycelium is a **hypha** (plural, *hyphae*). Look for these features in the figures throughout this section.

Metabolism, Nutrition, and Dispersal Most fungi require oxygen, but some species are full-time or part-time anaerobes. For example, yeasts that are used in baking bread and making wine can switch between aerobic respiration and fermentation depending on the availability of oxygen in their environment (Section 6.3).

Most fungi are *decomposers* (Section 2.2). Decomposers are organisms that feed on organic remains by breaking down molecular components into smaller subunits that can be absorbed. A smaller number of fungi live on or inside living organisms and draw nourishment from their host.

Fungi disperse by producing spores. Fungal spores can be single celled or multicellular, but all are microscopic. Many spores cannot move on their own, but some types of fungi produce spores with flagella.

DIFFERENTIATED INSTRUCTION | English Language Learners

Spellings for Long *i* Point out that the words in this lesson have different spellings for the long-*i* and short-*i* sounds. Say each sound for students to repeat and then have them identify the sounds in the vocabulary words.

Beginning Read the vocabulary words and have students raise a hand when they hear long *i*. Point out the spelling for each (*i* in *chitin*, *fungi*, and *lichen*; *y* in *mycelium*; both *y* and *i* in *mycorrhiza*).

Intermediate Read pairs of vocabulary words and have students identify the one with one or more long-*i* sounds. Then have students think of other examples where *y* has the long-*i* sound (for example, cytoplasm).

Advanced Read the vocabulary words and have students say the letter that makes the long-*i* sound. Then have pairs take turns using the words in sentences, being sure to pronounce the long-*i* sound in each.

8.3

FUNGI

LS2.C

EXPLORE/EXPLAIN

This section provides a review of fungi characteristics, introduces types of fungi common to the five major subgroups, and reviews the ecological roles of fungi. It also covers some of the different uses for fungi, from food to medicine.

Objectives

- Describe the structures that help fungi to produce and disperse spores.
- Explain the symbiotic relationship between fungi and plants.
- Identify characteristics that are common to all fungi.

Pretest Use **Questions 2, 4,** and **7** to identify gaps in background knowledge or misconceptions.

Infer *Students might say that the fungus might stick out of the ground to reproduce, whereas the underground portion of the fungus might gather nutrients and water from the environment.*

Video

Time: 2:09

Video 8-2 shows yeast reproducing by budding. *Saccharomyces cerevisiae*, commonly known as baker's yeast or brewer's yeast, is a model organism. This single-celled organism can divide as rapidly as once every 90 minutes under optimal laboratory conditions through budding, in which smaller daughter cells pinch, or bud, off the mother cell. Researchers can move genes in and out of yeast cells, either on plasmids or within the yeast chromosomes, allowing them to easily mutate or manipulate yeast gene expression and study the resulting phenotypic effects. Ask students why such research in yeast is relevant. *(to better understand the structure and function of chromosomes)*

CASE STUDY
SPACE STATION STOWAWAYS

Ask Questions To help students make connections with the content, have them make a diagram to show the life cycle of a fungus. Then, have them describe the environment in space and compare this to what they know about fungal habitats. Encourage students to refer to their life cycle diagrams to help identify possible environmental features in space that could be advantageous to the life cycle of fungi. For example, the air supply system onboard the space shuttle might make it easier for spore transportation, which could result in increased reproduction.

Video

Time: 24 seconds

Video 8-3 shows the formation of fruiting bodies from a mushroom and how this enables it to disperse spores. Share with students that, through the process of sexual reproduction, the mushroom forms the fruiting body structures that produce and disperse spores. Have students identify the structure that allows fungi to disperse. *(spores)*

CCC Structure and Function *Sample answer: Hyphae extend out and produce spores asexually from their tips. Spores with flagella are able to swim in aquatic environments. Sac and club fungi both produce fruiting bodies that extend above ground and distribute spores.*

CASE STUDY
SPACE STATION STOWAWAYS

Ask Questions *Generate questions about the characteristics and structures that might be critical for fungal species to live and thrive in their particular ecosystems, such as on the International Space Station.*

VIDEO

VIDEO 8-3 Go online to see how some mushrooms form fruiting bodies to disperse spores.

Types of Fungi

As of 2022, scientists have used phylogenetic analyses to identify eight phyla in the Fungi kingdom (Appendix E). The different types of fungi can also be sorted into five major subgroups based on the structure of their spores and how the fungi produce them (**Figure 8-21**).

- There are four phyla of fungi that make spores that have flagella and can move. The most commonly known of these fungi are called *chytrids*. Most chytrids are aquatic decomposers, but some live inside plants or animals. Chytrids usually reproduce asexually, producing spores at the tips of hyphae.
- Two phyla of fungi have traditionally been called *zygote fungi*. Zygote fungi typically produce thick-walled resting spores during sexual reproduction. However, they most often exist as a mold, which is a mass of hyphae that grows rapidly over organic material and produces spores asexually at the tips of specialized hyphae.
- Although they are not always considered a separate phylum, *glomeromycetes* are a group of soil fungi that form a mutually beneficial partnership with plant roots. Their hyphae extend inside the cell wall of plant root cells. Glomeromycetes produce spores asexually at the tips of hyphae that extend out of the host's root.
- *Sac fungi* (one phylum) reproduce sexually by producing spores in sac-shaped cells. In some species such as morels, these cells form on a fruiting body. Sac fungi include yeasts, molds, parasitic species, and species that partner with photosynthetic cells to form lichens.
- *Club fungi* (one phylum) reproduce sexually by producing spores on club-shaped cells. Mushrooms are the fruiting bodies of club fungi. Shelf fungi, puffballs, and stinkhorns are other examples of club fungi. The group also includes plant pathogens such as smuts and rusts.

CCC Structure and Function *Describe the structures that help fungi produce and distribute spores.*

FIGURE 8-21 Different fungi have different spore-bearing structures.

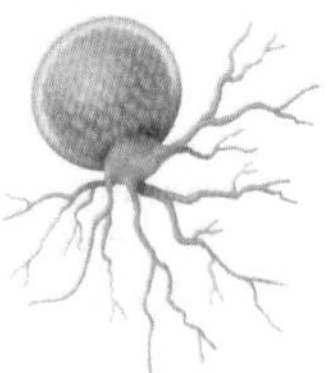
chytrids

zygote fungi

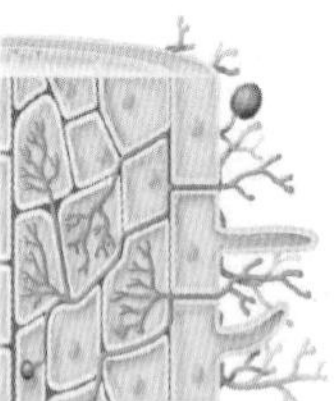
glomeromycetes

sac fungi

club fungi

CROSSCUTTING CONCEPTS | Structure and Function

System Organization Emphasize to students that to begin understanding how a system works, they must first thoroughly examine what it is made of, the shapes of its parts, and the role these parts play within the organization of the system. To reinforce this concept, remind students that fungi are sorted based on the *structure of their spores* and how the fungi *produce them*. Encourage students to focus on these specific features throughout their reading. As students observe **Figure 8-21**, have them identify the different spore features from each fungus shown.

Ecological Roles of Fungi

Fungi provide an important ecological service by breaking down complex compounds in organic wastes and remains. Some soluble nutrients released by this process enter soil or water. Plants and other producers can then take up the nutrients to meet their own needs. Bacteria also serve as decomposers, but they tend to grow mainly on surfaces. By contrast, fungal hyphae can extend deep into a decaying log or other bulky food source.

Mutualisms Many fungi take part in mutualisms, which are beneficial partnerships between different species (Section 3.1). Some fungi act as **endophytes**, living within different parts of plants, including the roots, stems, and leaves. While not all endophytes are mutualistic, many provide benefits for their plant host and exchange materials with the plant. For the plants they inhabit, endophytes have been shown to improve stress tolerance, promote growth, and help protect against pathogens.

A **mycorrhiza** (plural, *mycorrhizae*) is a type of mutualism in which a soil fungus forms a partnership with the root of a vascular plant. An estimated 80 percent of the vascular plants form mycorrhizae with a glomeromycetes fungus. Hyphae of these fungi enter root cells and branch in the space between the cell wall and the plasma membrane (**Figure 8-22**). Some sac fungi and club fungi form mycorrhizae in which hyphae surround a root and grow between its cells. Most forest mushrooms are fruiting bodies of mycorrhizal club fungi.

Hyphae of all mycorrhizal fungi functionally increase the absorptive surface area of their plant partner. Hyphae are thinner than even the smallest roots and can grow between soil particles. The fungus shares water and nutrients taken up by its hyphae with plant root cells. In return, the plant supplies the fungus with sugars.

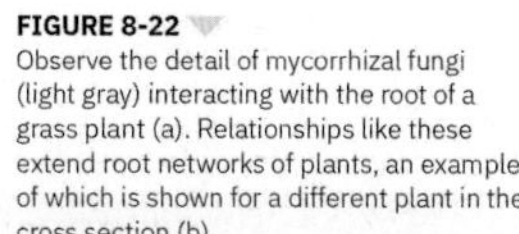

FIGURE 8-22 Observe the detail of mycorrhizal fungi (light gray) interacting with the root of a grass plant (a). Relationships like these extend root networks of plants, an example of which is shown for a different plant in the cross section (b).

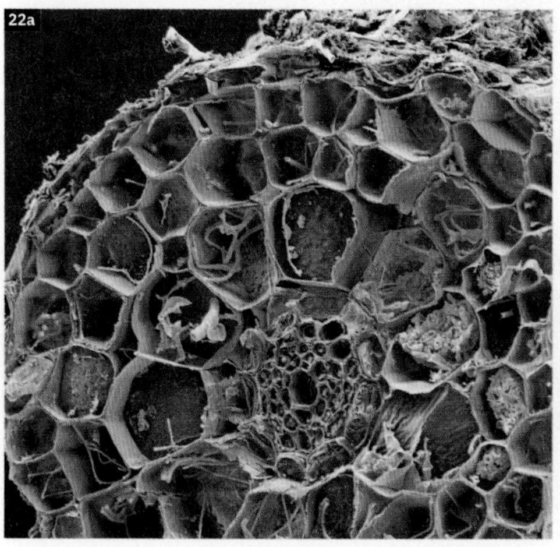

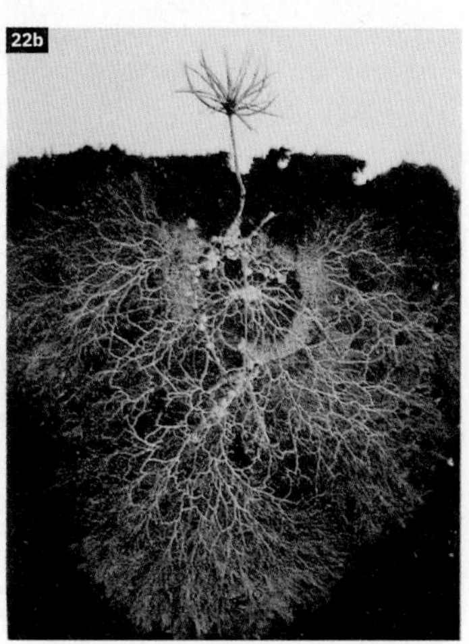

In Your Community

Mutualisms Horticulturists use scientific knowledge to improve the growth of plants and flowers. They use mycorrhizal applications to treat seeds, apply to unrooted cuttings, or integrate into growing plants. Mycorrhizae products help to trigger germination of dormant propagules that, in turn, kick start the symbiotic colonization of the root system. Encourage students to find out if there are any local garden centers or nurseries that use mycorrhizae products to treat their plants. Have students evaluate the application used to promote plant growth.

DIFFERENTIATED INSTRUCTION | Leveled Support

Struggling Students Encourage students to create a T-chart to organize what they learn about fungi. Have them label the column headings: Fungi Characteristics, Fungus Types, Ecological Role, and Uses. Then, as they read, have students write down facts that they learn about each and examples. For example, in the Uses column, they might write about food, medicines, and scientific research. In the Fungus Types column, they might about any of the fungi found in the five major subgroups. As students move through the section, have them refer to their chart to add additional facts and examples.

Advanced Learners Students who have an understanding of fungi may be interested in learning about some of the everyday uses for fungus. Encourage students to explore their home environment for example of fungi. For example, students could explore their home pantry for common foods that are made with fungi, such as olives, chocolate, bread, or coffee.

Visual Support

Lichen Students may look at **Figure 8-23a** and **Figure 8-23b** and not make the connection that both images are of the same organism. Explain to students that **Figure 8-23b** is a magnified view of **Figure 8-23a**. Encourage them to identify the algae cells and fungal threads shown in **Figure 8-23b**.

SEP Construct an Explanation *Sample answer: Hyphae and mycelium appear to be suited to making connections between fungi and other organisms, and they are able to transfer materials between them.*

RAINFOREST CONNECTIONS

Gather Evidence *How do mycorrhizal networks connecting trees help maintain the health of the rainforest?*

Mycorrhizal networks transfer water, carbon, nitrogen, and other materials between different plants. Other chemicals that can be transferred between plants serve as signals, effectively letting plants communicate when nutritional conditions change or when there is a threat that is affecting a plant in the network. In response to a defense signal, other plants in the network can increase production of toxins and other compounds that help them resist infestation or other attacks.

A **lichen** is a composite organism that consists of a sac fungus and either cyanobacteria or green algae (**Figure 8-23**). Some types of lichen include more than two species. Fungus makes up the bulk of a lichen's mass, with its hyphae providing structural support to the photosynthetic cells. These cells provide the fungus with sugars and, if the cells are cyanobacteria, with fixed nitrogen. Lichens play an important ecological role by colonizing places too hostile for most organisms, such as exposed rocks. They break down the rock and produce soil by releasing acids and by holding water that freezes and thaws. Once soil forms, plants can move in and take root. Long ago, lichens may have lived on land before plants.

Fungal partners also enhance the nutrition of some animals. Chytrid fungi that live in the stomachs of grazing hoofed mammals such as cattle, deer, and moose aid their hosts by breaking down otherwise indigestible cellulose. Similarly, fungal partners of some ants and termites serve as an external digestive system. Leaf-cutter ants gather bits of leaves to feed a fungus that lives in their nest. The ants cannot digest leaves, but they do eat some of the fungus.

SEP Construct an Explanation *Explain how the structures of fungi help them in establishing symbiotic relationships with other species.*

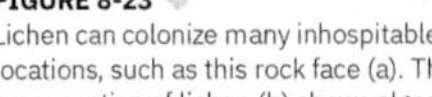

FIGURE 8-23 Lichen can colonize many inhospitable locations, such as this rock face (a). The cross section of lichen (b) shows algae cells in dark red near the surface of the lichen. The threadlike hyphae of the fungus are stained blue and red.

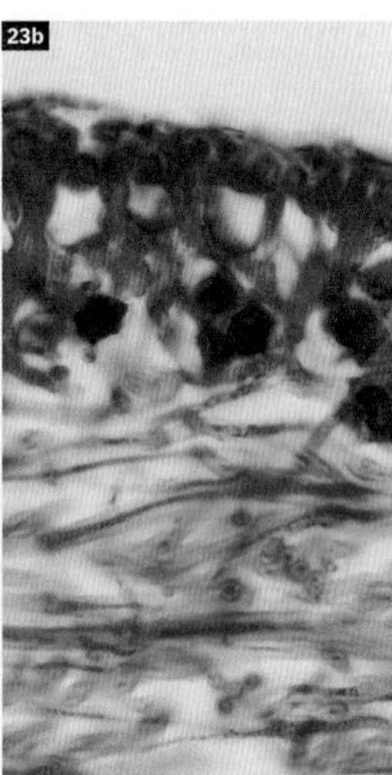

REVISIT THE ANCHORING PHENOMENON

Gather Evidence *Sample Answer: Mycorrhizal networks help distribute important nutrients and signals between trees in the rainforest, allowing more trees to respond to threats and obtain what they need to grow.*

Fungi are well known for their symbiotic relationship with plants. As the fungus colonizes the roots of a host plant, it enables the root network to expand. This, in turn, allows the plant to receive increased levels of water and nutrient absorption, while the plant provides the fungus with carbohydrates from photosynthesis. With rainforests being dominant in plant species, it makes sense that some plants would compete for food and water. Mycorrhizal networks connecting trees in the rainforest could help expand root systems and increase the ability of the tree to receive water and nutrients.

Parasites and Pathogens Many sac fungi and club fungi are plant parasites. Powdery mildews (sac fungi) and rusts and smuts (club fungi) are parasites that grow in living plants. Their hyphae extend into cells of stems and leaves, where they suck up photosynthetically produced sugars. The resulting loss of nutrients stunts the plant, minimizes seed production, and may eventually kill it. However, the plant usually does not die before the fungus has produced spores on the surface of its infected parts.

Other fungi are pathogens that kill plant tissue with their toxins and then feed on the remains. The club fungus *Armillaria* causes root rot by infecting trees and woody shrubs in forests worldwide. Once an infected tree dies, the fungus decomposes the stumps and logs left behind. In one forest in Oregon, the mycelium of a single honey mushroom (*Armillaria solidipes*) extends across nearly 10 square kilometers. It has been growing for an estimated 2400 years.

Many more fungi infect plants than animals. Among animals, those that do not maintain a high body temperature are most vulnerable to fungal infections. White-nose bat syndrome (caused by *Pseudogymnoascus destructans*) has caused widespread problems in North American bat populations since 2007. Hundreds of fungal species infect insects, and some turn their hosts into zombies that no longer control themselves, forcing the insects to disperse fungal spores (**Figure 8-24**).

Most human fungal infections involve sac fungi on body surfaces. Infected areas are raised, red, and itchy. For example, several species of sac fungi infect skin of the feet, causing "athlete's foot." Sac fungi also cause skin infections misleadingly known as "ringworm." No worm is involved. A ring-shaped lesion forms as fungal hyphae grow out from the initial infection site. Life-threatening fungal infections are rare and most often occur in people whose immune response is weak because of other health issues.

FIGURE 8-24
An ant infected by a *Cordyceps* fungus stops performing its normal activities as it is taken over by the fungus. After it dies, hyphae burst out of the ant to release spores.

In Your Community

Fungal Diseases Some fungal species can cause illnesses. Fungal nail infections and ringworm are two of the more common fungal diseases. Both are examples of superficial infections of skin and nails and are relatively minor. Fungal diseases can also occur inside the body, such as lung and bloodstream infections, which are more serious in nature. Recognizing the signs and symptoms of serious fungal diseases early is critical in providing treatment. Encourage students to help bring awareness to common fungal diseases by participating in Fungal Disease Awareness Week sponsored by the Center for Disease Control. Have students create informational pamphlets and/or posters that can be displayed at appropriate locations throughout the school.

SCIENCE AND ENGINEERING PRACTICES
Constructing Explanations and Designing Solutions

Symbiotic Relationships As students observed from **Figure 8-22b** and **Figure 8-23b**, many fungi have a symbiotic relationship with different species. Lead a discussion with students that highlights the key benefits of the hyphae from the fungus to its host plant. Provide students with sticky notes. Allow students to work in groups to list as many structural benefits as they would like. Then, have groups place their sticky notes in a central location. As a class, organize the different benefits listed on the notes into groups, such as structure (hyphae and mycelium) and then subgroups within the two structures. Then, guide students to use the different groupings to create an outline that can be used to begin an explanation of how the structures of fungi help them in establishing symbiotic relationships with other species.

Science Background

Fungi History The oldest known mummified body, Otzi, or "The Ice Man," had mushrooms in his digestive system when he died in 3300 BCE. Scientists found that Otzi had intestinal parasites and stomach ulcers along with other physical ailments when he died. It is thought that he may have consumed birch polypore muchrooms (*Fomitopsis betulina*) prior to his death due to their anti-inflammatory and fever-reducing properties. It might interest students to learn that the mushrooms were still present in his stomach when his body was discovered in 1991. Have students discuss how the mushrooms could still be there thousands of years later. Encourage them to identify and describe the significance of mushrooms in human history.

CONNECTIONS

HUNTING FOR TRUFFLES Truffle-producing fungi are highly prized by gourmet food consumers. While prices vary depending on the market, Piedmont white truffles (*Tuber magnatum*) typically cost more than $3300 per kilogram. Burgundy black truffles (*Tuber uncinatum*) average more than $870 per kilogram. Because they cannot be readily cultivated and demand a high price, truffle hunters are very secretive about their favorite hunting areas. Interestingly, the secret to hunting truffles is related to a dispersal strategy that has evolved among this group of fungi.

Truffles are the fruiting bodies of some mycorrhizal fungi that partner with tree roots. Truffles form underground near their host trees and, when mature, produce an odor similar to that of an amorous male wild pig. Female wild pigs detect the scent and root through the soil in search of their seemingly subterranean suitor. When they unearth the truffles, they eat them and then disperse fungal spores in their feces.

Because humans would rather eat the truffles themselves instead of feeding them to pigs, dogs are often used to hunt for truffles. Dogs have a sense of smell thousands of times better than humans, and they can be trained to sniff out only ripe truffles. Truffles take years to develop and ripen.

This man in France hunts truffles with his pig. The pig can detect the truffles where they grow underground.

Uses of Fungi

Many fungal fruiting bodies serve as human food. Button, shiitake, and oyster mushrooms decompose organic matter, so these species are easily cultivated. By contrast, other edible mycorrhizal fungi such as chanterelles, porcini mushrooms, morels, and truffles need a living plant partner, so they are typically gathered from the wild. Each year thousands of people become ill after eating poisonous mushrooms they mistook for edible ones (**Figure 8-25**).

Fungal fermentation reactions help us make a variety of products. Fermentation by one mold (*Aspergillus*) helps make soy sauce. Another mold (*Penicillium*) produces the tangy blue veins in cheeses. A packet of baker's yeast holds spores of the sac fungus *Saccharomyces cerevisiae*. When these spores grow in bread dough, they ferment sugar and produce carbon dioxide that causes the dough to rise. Other strains of *S. cerevisiae* are used to make wine and beer, and they are also used in the production of ethanol biofuel.

Geneticists and biotechnologists make use of yeasts for research. For example, checkpoint genes that regulate the eukaryotic cell cycle (Section 7.1) were first discovered in the yeast *S. cerevisiae*. This discovery was the first step toward our current understanding of how mutations in these genes cause human cancers. Genetically engineered *S. cerevisiae* and other yeasts are now used to produce proteins that serve as vaccines and medicines.

232 CHAPTER 8 DIVERSITY OF LIVING SYSTEMS

CONNECTIONS | Hunting for Truffles

Social-Emotional Learning Some students from urban areas may find it odd to go on a walk with a pig in search of mushrooms. Guide students in a conversation about diverse social norms, such as the one described in the Connections feature. Promote **social awareness** by encouraging students to learn about different regions of the world and unique agricultural food-related animal relationships in different regions. For example, birds are a common household pet in Turkish, Arabic, and Persian cultures, whereas dogs are generally not pets. Allow students with different backgrounds to discuss their cultural experiences.

Some naturally occurring fungal compounds have medicinal or psychoactive properties. The initial source of the antibiotic penicillin was a soil fungus (*Penicillium chrysogenum*). Another soil fungus gave us cyclosporin, an immune suppressant used to prevent rejection of transplanted organs. Ergotamine, a compound used to relieve migraines, was first isolated from ergot (*Claviceps purpurea*), a club fungus that infects rye plants.

FIGURE 8-25 Fungal fruiting bodies take many forms. Identifying whether a mushroom is useful, edible, or poisonous is critical. Fly agaric (a) and porcini (b) mushrooms have some visual similarities, but the first is poisonous and the second is edible. Chicken of the woods (c) is edible, while turkey tail mushrooms (d) have medicinal purposes.

8.3 REVIEW

1. **Identify** Which characteristics are common to **ALL** fungi? Select all correct answers.
   - A. reproduce asexually
   - B. require oxygen to live
   - C. digest organic material
   - D. have a cell wall made of chitin
   - E. form a mesh of filaments
2. **Describe** What is one way that fungi and plants benefit in a mutualistic relationship?
3. **Identify** Which type of spores are involved in the asexual reproduction of fungi?
   - A. spores with flagella at tips of hyphae
   - B. spores in sac-shaped cells
   - C. thick-walled resting spores
   - D. spores on club-shaped cells
4. **Describe** Consider one example of a parasitic relationship between a fungus and a plant. What happens to the fungi and plants in this relationship?

Visual Support

Fungal Diversity Explain to students that while scientific estimates of the number of fungal species varies widely, from two to a few million, only about 150,000 species have been identified. Remind students that fungal spores are microscopic particles that allow fungi to reproduce, and the hyphae are the branching fibers that make up the root-like structure of a fungus. Refer students to **Figure 8-25.** Have them conduct research on the mushrooms shown in the images and allow them to work in groups to identify where the spores and hyphae from each image shown could be located. Ask students to draw a diagram of one of the mushrooms researched that shows the location of the spores and hyphae. Encourage them to draw multiple views, such as a vertical cross section, the underside of the cap alone, and hyphae in the ground.

8.3 REVIEW

1. *C, D* **DOK 1**
2. *Sample answer: Fungi obtain nutrients from plants, and plants gain the ability to absorb more water from the environment.* **DOK 2**
3. *A* **DOK 2**
4. *Sample answer: Plants die due to the toxins in fungi, and then the fungi obtain energy by decomposing the dead plant tissue.* **DOK 2**

LOOKING AT THE DATA
THE C-VALUE ENIGMA

Connect this activity to the science and engineering practice of using computational thinking. Students will be analyzing the C-value against the number of genes in various organisms. The C-value is the amount of nuclear DNA in a haploid gamete, measured in base pairs, that is the species' genome size. Student will compare the genome size and number of coding genes of different organisms to explain the observations that simple organisms can sometimes have more complex genomes than complex organisms.

SAMPLE ANSWERS

1. *C-value increases with number of genes, but this is not a linear relationship or there are a few outliers like rice and the amoeba.* **DOK 2**
2. *The C-values span a wide range of number sizes from thousandths to hundreds while the range of number of genes is nearly all in the tens of thousands. A log-log plot allows us to compare wide ranges of numbers to each other that would otherwise be difficult to place on a single graph.* **DOK 3**
3. *humans; axolotl; lungfish; rice; mosquito; amoeba; malaria parasite; yeast; bacteria* **DOK 2**
4. *Larger genome than humans: amoeba, lungfish, axolotl; smaller genome than humans: rice, mosquito, yeast, malaria parasite, bacteria.* **DOK 2**
5. *Organism complexity does not follow genome size. More complex organisms do not always have the largest genome. The number of protein-encoding genes may be a more useful indicator of complexity.* **DOK 3**

LOOKING AT THE DATA

THE C-VALUE ENIGMA

SEP Use Computational Thinking The genome of an axolotl salamander is ten times larger than the genome of a human and more than 100,000 times larger than the genome of a bacterium.

Genome size, represented by the total amount of DNA in any cell of an organism, is often reported as the *C-value*, measured in base pairs. It is a constant among all organisms of the same species. The small sections of the total DNA that encode for proteins are called genes. The human genome consists of 3.2 billion base pairs with about 25,000 genes. **Table 1** lists the C-value and approximate number of protein-encoding genes for a selection of organisms. **Figure 1** shows the values in the table on a log-log plot.

TABLE 1. C-value and Number of Genes for a Selection of Organisms

Organism	C-Value (10^9 base pairs)	Approximate number of protein-encoding genes
Ambystoma mexicanum (axolotl)	320.000	23,000
Anopheles gambiae (mosquito)	0.3	13,500
Escherichia coli (bacteria)	0.005	3200
Homo sapiens (human)	3.2	25,000
Neoceratodus forsteri (Australian lungfish)	43	31,120
Oryza sativa (rice)	0.47	51,000
Plasmodium falciparum (malaria parasite)	0.023	5000
Polychaos dubium (amoeba)	670	15,000
Saccharomyces cerevisiae (yeast)	0.012	6000

FIGURE 1
This log-log plot of C-value and number of genes shows a wide range of values.

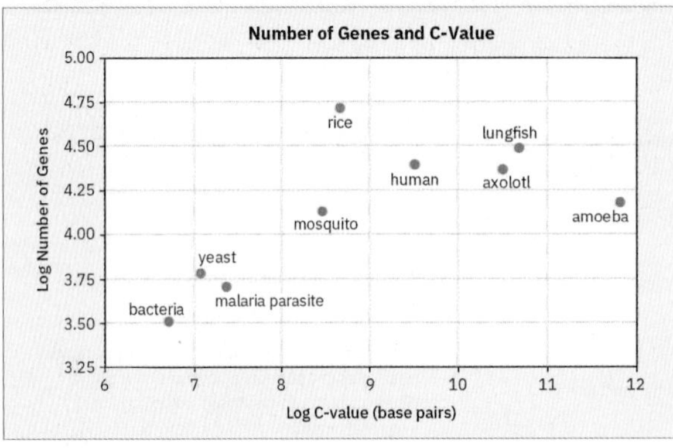

234 CHAPTER 8 DIVERSITY OF LIVING SYSTEMS

FIGURE 2
Organisms such as axolotls (a), rice plants (b), yeast (c), and Australian lungfishes (d) do not all necessarily have the number of genes you would expect, relative to their apparent complexity.

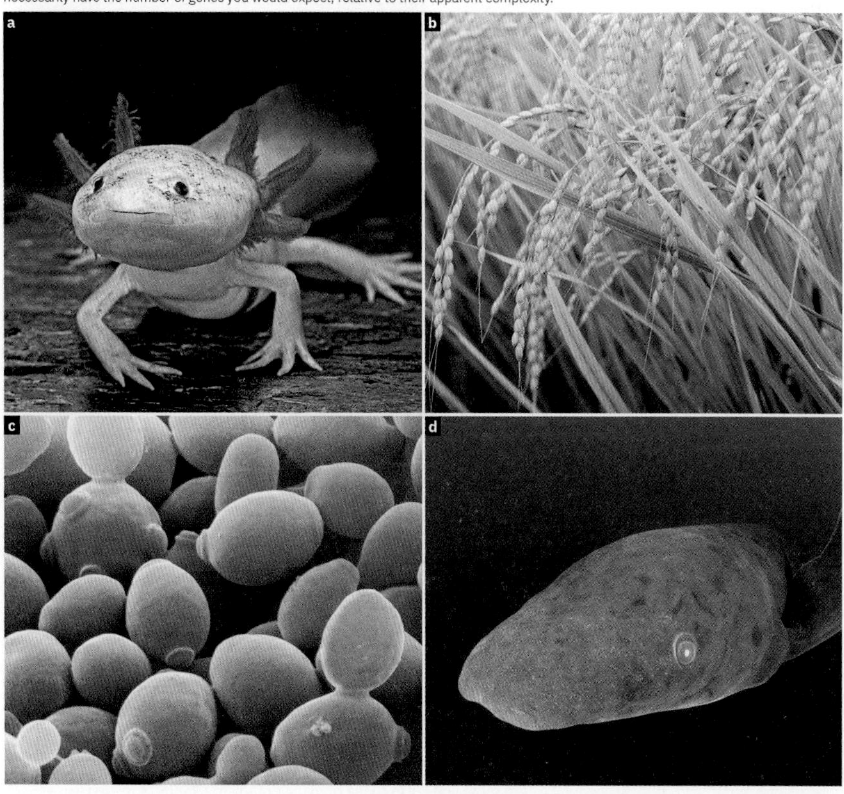

1. **Identify Patterns** Describe the general trend between number of genes and C-value.
2. **Justify** Explain why plotting a log-log plot helped identify patterns in the data.
3. **Compare** Based on what you know about them, rank the organisms by their complexity.
4. **Organize Data** Group organisms by relative genome size compared to humans.
5. **Synthesize** Use your analysis of the data to support a claim about C-value and organism complexity.

Connect to Mathematics

Model with Mathematics Trends in data can be useful for making predictions or identifying possible outlier data. As students analyze **Figure 1**, remind them that an increasing line represents a positive trend, a decreasing line a negative trend, and a horizontal line is a neutral trend. First, have students identify the type of trend displayed in **Figure 1**. Allow them to discuss what this information tells us about the data being graphed. Students should recognize that the graph shows a positive trend, which indicates that as the log C-value increases so does the log number of genes. Then have students look for possible outliers. They should recognize that the data for rice and amoeba fall outside of the trend line modeled in the graph. Lead a class discussion about how the data for rice and amoeba follow a similar, but not identical, trend to the other data. Students should recognize that the values increase at a different rate than the other items displayed in the graph.

8.4

VIRUSES

LS2.C

EXPLORE/EXPLAIN

This section provides an overview of virus classification and structures, introduces the concept of viral replication, and reviews the difference between common, emerging, and influenza viral diseases. It also covers the development and use of vaccines in minimizing the spread of viral diseases.

Objectives

- Characterize the four major groups of viral genomes and where they replicate.
- Compare and contrast bacteriophage and HIV replication.
- Identify the characteristics that are common to all viruses.

Pretest Use **Questions 3** and **8** to identify gaps in background knowledge or misconceptions.

Explain *Students might say that viruses will need the equipment necessary to make proteins (they may remember ribosomes from Unit 2). Students may know that cells have special protein complexes to replicated DNA and RNA.*

SEP Construct an Explanation *Sample answer: Inside a eukaryotic cell, the cell machinery that replicates DNA and transcribes mRNA is inside the nucleus. The DNA viruses need the same machinery to build new viral genomes and proteins.*

8.4 VIRUSES

KEY TERMS

bacteriophage
capsid
lysogenic pathway
lytic pathway
retrovirus
viral envelope

Have you ever been performing a lab activity and were missing the lab equipment you needed? Or maybe you could not find the material that was best for building a model of a scientific process, and you had to borrow from another group. Every time a virus infects a host cell, it has to borrow equipment from the host to reproduce its own proteins and genome.

Explain *What kind of "equipment" do you think a virus might need to replicate itself?*

Viruses and Their Classification

A virus is a noncellular infectious particle that can replicate only inside a living cell. By many definitions, viruses are not alive. Unlike cells, viruses do not maintain homeostasis or engage in metabolic processes such as ATP production, and they cannot reproduce on their own. A virus that is not inside a host cell is referred to as a *viral particle*. With rare exceptions, viral particles can only be viewed with an electron microscope.

Viruses outnumber the stars in the universe and exist everywhere on Earth. For every bacterium, there are about 10 virus particles. Viruses infect every kind of known organism, yet each virus is highly specialized. A few viruses, such as the mimivirus, that are large enough to see with a light microscope can themselves become infected by other viruses.

A viral particle always includes a viral genome inside a protein shell called a **capsid**. During infection, some proteins of the capsid must bind to proteins at the surface of a host cell. Because of this binding, each type of virus can only infect specific kinds of cells. Viruses are classified by the structure of their capsid, their genome, and how they replicate.

Virus Genomes The genetic material of a virus particle can be either DNA or RNA. In cells, RNA serves many important functions, including as an intermediate message (a messenger RNA, or mRNA) between the DNA genome and the translation of proteins the cell needs (Chapter 11). DNA replication and transcription of DNA into mRNA occurs in the nucleus. Translation of mRNA to proteins occurs in the cytoplasm. Because viruses rely on these same processes when infecting a host cell, the type of virus genome affects how it behaves and replicates. Virus genomes generally fall into the four following categories:

- **Double-Stranded DNA (dsDNA)** dsDNA viruses have genomes that are most similar to those of living organisms. Many of these viruses replicate in the host cell's nucleus and rely on its cell machinery. Some dsDNA viruses are able to replicate their genome and capsids on their own in the host cell's cytoplasm. Herpesviruses, papillomaviruses, poxviruses, and adenoviruses are examples of dsDNA viruses.
- **Single-Stranded DNA (ssDNA)** ssDNA viruses of eukaryotes mostly replicate in the nucleus. Parvoviruses, which infect cats and dogs and cause fifth disease in humans, are ssDNA viruses.

SEP Construct an Explanation *Why do you think most DNA viruses replicate inside the nucleus of their host cells?*

236 CHAPTER 8 DIVERSITY OF LIVING SYSTEMS

SCIENCE AND ENGINEERING PRACTICES
Constructing Explanations and Designing Solutions

Viral Genomes As students learn about genomes, they discover that DNA viruses need the same machinery as a eukaryotic cell to build new viral genomes and proteins. Students also learn that the cell machinery that replicates DNA and transcribes mRNA is inside the nucleus. Lead a discussion with students that highlights the machinery that replicates DNA. As you discuss with the class, place relevant information on the board. Then, guide students to organize information on the board in order of relevance. Once the key or top point has been identified from all of the ideas on the board, have students focus only on this point to develop their explanations.

- **Single-Stranded RNA (ssRNA)** Some RNA genomes are very similar to mRNA, and the viral RNA can be directly translated into proteins by a host cell, while others undergo additional steps to replicate. Coronaviruses and influenza viruses are both important groups of viruses with ssRNA genomes. Most ssRNA viruses replicate in the cytoplasm. Among the exceptions are human immunodeficiency virus (HIV), which has a replication phase that occurs in the nucleus.
- **Double-Stranded RNA (dsRNA)** dsRNA viruses evolved from ssRNA viruses. Different members of this group infect a wide variety of hosts, including bacteria, fungi, plants, and animals. The rotaviruses that cause gastroenteritis have dsRNA genomes. RNA virus genomes typically encode their own genome replication enzymes and replicate in the cytoplasm of host cells.

Virus Capsid Structures A virus capsid consists of a few unique protein subunits arranged in a repeating pattern. When viral proteins assemble as a helix around a strand of nucleic acid, the result is a rod shape or a filamentous, fiber-like shape. Many plant viruses, including the tobacco mosaic virus (ssRNA), have this type of structure (**Figure 8-26a**).

Bacteriophages are viruses that infect bacteria, and most have a complex structure. **Figure 8-26b** includes the T4 bacteriophage, a dsDNA virus common in the human gut. The viral genome is enclosed within a polyhedral (many-sided) "head." Other protein components of the T4 viral particle include a hollow helical sheath through which viral DNA is injected into a bacterial cell, and "tail fibers" that attach the virus to that cell.

The tobacco mosaic virus and the T4 bacteriophage are called *nonenveloped viruses* because their outermost layer is the protein capsid. By contrast, the nearly spherical (often icosahedral, or 20-sided) capsid of most animal viruses is covered by a **viral envelope**, a layer of cell membrane derived from the cell in which the virus formed. SARS-CoV-2 (severe acute respiratory syndrome coronavirus 2)—the cause of COVID-19—is an enveloped ssRNA virus (**Figure 8-26c**).

CCC Structure and Function *Consider that most virus genomes need to be small. Why do you think virus capsid structures might use many repeating structures instead of a large number of unique proteins?*

FIGURE 8-26 These are just a few examples of the different shapes and structures that can make up viral particles.

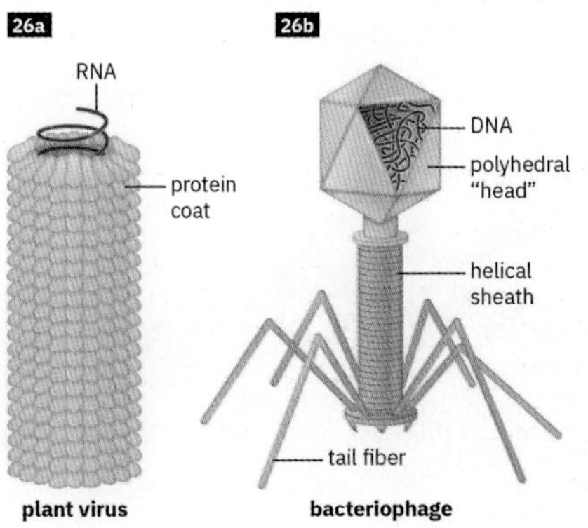

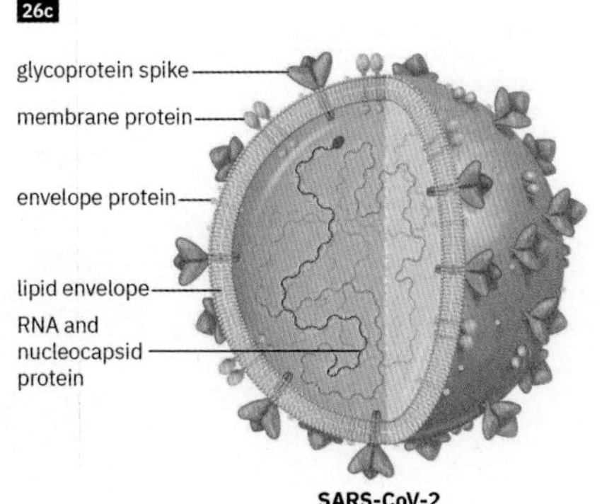

26c

glycoprotein spike
membrane protein
envelope protein
lipid envelope
RNA and nucleocapsid protein
SARS-CoV-2

CROSSCUTTING CONCEPTS | Structure and Function

Virus Structure Remind students that viruses do not meet the full definition of a living organism and that viruses carry out certain functions, like reproduction, by utilizing the host cells to perform these basic functions. Have students relate this to why a virus only requires a few types of proteins to build large enough shapes to hold their genomes, and the genome only needs a few genes to code for all the many proteins that form the entire capsid. Encourage students to focus on these specific structures throughout their reading. As students observe **Figure 8-26**, have them identify the different DNA and RNA features from each virus shown.

Vocabulary Strategy

Knowledge Rating Scale Have students make a table of the key terms with four columns labeled 1, 2, 3, and 4. Before students begin reading, they should rate their knowledge of each term:

1 = I have never heard the word.

2 = I have heard the word, but do not know its meaning.

3 = I am familiar with the word.

4 = I know the word, its meaning, and could explain it to others.

Students should use their prior knowledge to write a definition for words listed in categories 3 and 4. As students complete the reading, they should add definitions and examples for terms in categories 1 and 2. After students have completed reading the section, have them revisit the chart and re-rank terms.

CCC Structure and Function *Sample answer: Repeating structures that only require a few types of proteins allow viruses to build large enough shapes to hold their genomes and other proteins, but the genome only needs a few genes to code for all the many proteins that form the entire capsid.*

Visual Support

Virus Replication Explain to students that virus replication is unique in that viruses have different genome structures, unlike living organisms that all have double-stranded DNA genomes. The methods of genome replication in viruses can vary. Each virus depends on the infected hosts in different ways. **Figure 8-27** shows how a bacteriophage can replicate using two different methods.

Have students work in groups to compare and contrast both methods. Encourage them to identify any potential benefits of using one method over the other for replication.

Virus Replication

Virus replication is particularly interesting and varied, because each virus depends on the infected hosts in different ways. Typically, when a virus encounters a host cell it can infect, viral capsid proteins bind proteins of a host's cell membrane. Then viral genes enter the cell. Viral genes hijack the cell's internal machinery, causing it to make viral components that self-assemble as viral particles. The particles are then released into the environment. Two life cycle examples are those of bacteriophages and human immunodeficiency virus (HIV).

FIGURE 8-27 Bacteriophages can undergo lytic or lysogenic pathways inside a host bacterium.

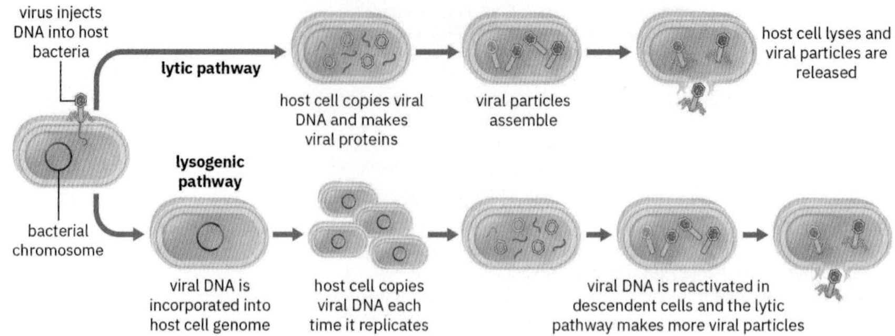
virus injects DNA into host bacteria
lytic pathway
host cell copies viral DNA and makes viral proteins
viral particles assemble
host cell lyses and viral particles are released
lysogenic pathway
bacterial chromosome
viral DNA is incorporated into host cell genome
host cell copies viral DNA each time it replicates
viral DNA is reactivated in descendent cells and the lytic pathway makes more viral particles

Bacteriophage Replication A bacteriophage such as T4 replicates using two pathways (**Figure 8-27**). Both pathways start when the virus attaches to a bacterium and injects DNA (**Figure 8-28**).

In the **lytic pathway**, viral DNA is immediately replicated and transcribed, causing the host cell to produce new viral DNA and proteins. These viral components assemble as virus particles, which escape when the cell breaks open (lyses) and dies.

In the **lysogenic pathway**, viral DNA becomes integrated into the host cell's genome. However, viral genes are not expressed, so the host cell remains healthy. When this cell reproduces, a copy of the viral DNA is passed on with the host's genome. The DNA remains in the genome of the cell's descendants until it is reactivated and starts the lytic pathway.

FIGURE 8-28 A bacteriophage starts an infection by injecting its DNA genome into a host cell. This colored transmission electron micrograph shows a T4 bacteriophage injecting its DNA into *E. coli*.

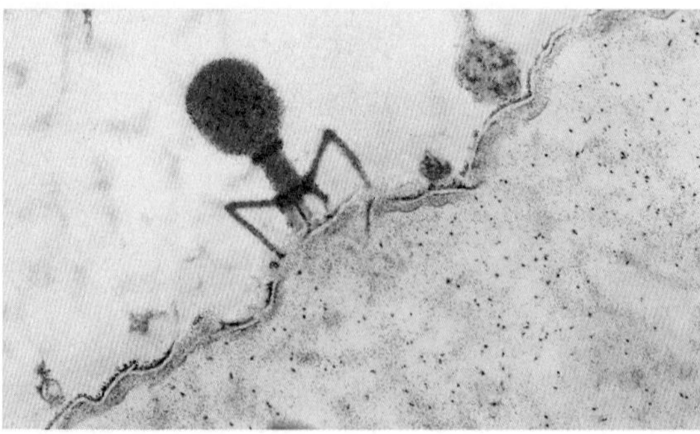

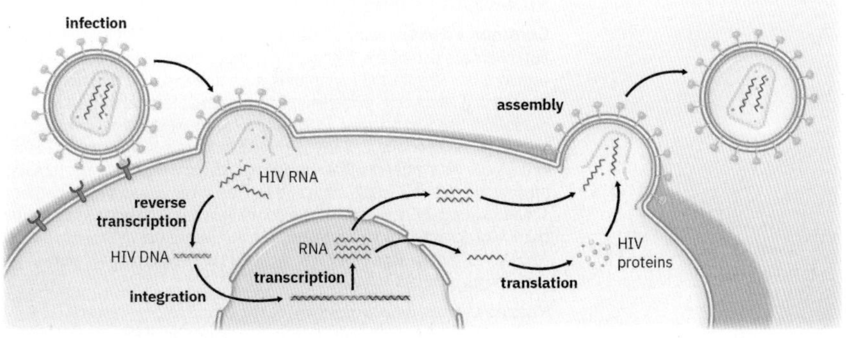

HIV Replication RNA viruses, such as HIV, that use reverse transcriptase to produce viral DNA are called **retroviruses**. This means that HIV's ssRNA genome is "reverse transcribed" into DNA that can be incorporated into the host genome where it can code for viral particles (**Figure 8-29**). Drugs that fight HIV infection interfere with different steps of the replication cycle. HIV replication can be described by the following steps:

- **Infection** During infection, spikes of viral glycoprotein attach to proteins in the cell's plasma membrane. The host cell's plasma membrane and the viral envelope fuse, allowing viral enzymes and RNA to enter the cell.
- **Reverse Transcription and Integration** The viral enzyme reverse transcriptase converts viral RNA into double-stranded DNA. Once viral DNA has been made, it is moved into the nucleus, along with another viral enzyme that integrates the DNA into the host chromosome.
- **Transcription** Viral DNA is transcribed into RNA. Some of that transcribed RNA is the genome of new HIV particles.
- **Translation** Some of the transcribed RNA encodes viral proteins and is translated to make the proteins of the HIV capsid.
- **Assembly** The viral proteins and RNA assemble at the host cell's plasma membrane. New viruses gain a viral envelope when they bud off from the cell (**Figure 8-30**).

FIGURE 8-29
The human immunodeficiency virus brings its own enzymes to a human host cell, which allow it to reverse transcribe its RNA genome into DNA that it integrates into the host genome.

Go Online INTERACTIVE FIGURE

FIGURE 8-29 Go online to learn more about virus structure and step through the process of **HIV Replication**.

FIGURE 8-30
A colored transmission electron micrograph shows an HIV particle budding off a cell (a). The key structures of an HIV particle include several structural proteins, enzymes, and an RNA genome (b).

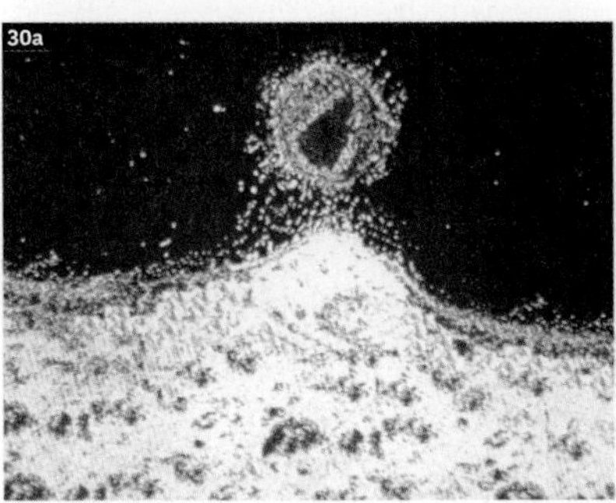

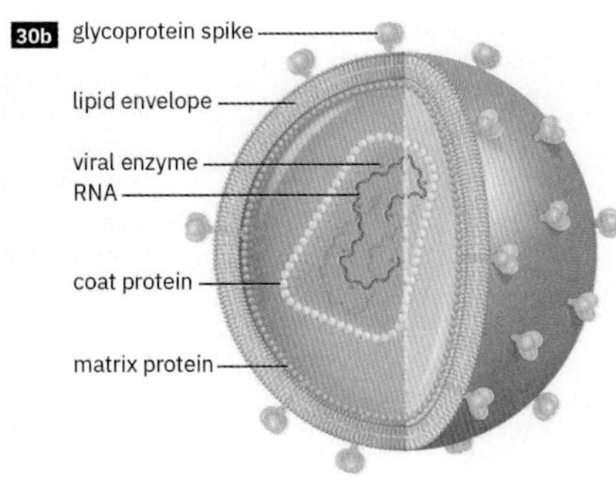

Vocabulary Strategy

Word Definition Map As students work through the key terms in this section, have them create word definition maps. Students should include the key term, its definition, and an example and non-example. As students work through the section, encourage them to use synonyms, antonyms, or a picture when possible to help develop an understanding of the new term.

Go Online **INTERACTIVE FIGURE**

HIV Replication In the interactive version of **Figure 8-29,** students can learn about the structure of the human immunodeficiency virus and explore its replication pathway.

DIFFERENTIATED INSTRUCTION | Leveled Support

Struggling Students After viewing **Figure 8-29**, some students may not fully understand the reproduction process involved with RNA viruses. Help students understand the process of reverse transcription by having them add sticky notes with numbering labels to each step on **Figure 8-29**. Then, have students identify the first and last step using the same color note. All the steps in between should be a different colored note. This will help students to visualize what happens during the beginning, middle, and end of RNA replication.

Advanced Learners Students who have an understanding of reverse transcription can develop a fictitious virus and corresponding drug or vaccine. They should focus on how the virus reproduces. Students should use this information to create a drug or vaccine that aims to interfere with one or more steps of the replication cycle. Students may present their virus and drug/vaccine to the FDA (classroom) for approval.

Connect to English Language Arts

Synthesize information In the wake of the global pandemic known as COVID-19, viral diseases and how they evolve and reproduce have become a major interest for all. The quest to better understand germs and how to minimize our exposure to them is evident in the number of hand sanitizers, disinfectants, and cleaning supplies available for human use. Have students write a one-sentence informative summary for each virus presented. Encourage students to use a graphic organizer as they read to ensure they are collecting similar key information for each virus.

Social-Emotional Learning

Social Awareness Students may have developed negative opinions about the impact of various viral diseases or reflect the negative opinions of their friends and families. Approach the presentation of viruses and human health from a scientific standpoint, focusing on the transmission and positive role of vaccinations. Remind students to *demonstrate empathy and compassion* for anyone suffering from such diseases. Guide them to also consider the *influences of organizations and systems on their own behavior* and how those might have contributed to any negative opinions.

Video

Time: 1:07

Video 8-4 shows the structure of the Ebola virus and provides information about its deadly effects. Though rare, Ebola has a fatality rate of 90 percent. Discuss with students the societal impacts of an Ebola outbreak, such as have occurred in some African countries. Ask students to identify some of the affects on society. *(apart from increased overall death rates, communities may experience resource shortages, emotional unrest, and quarantine)*

Viruses and Human Health

Common Viruses Many viruses live in our body without any ill effects, but others are pathogens, meaning they cause disease. Viral diseases usually produce brief, mild symptoms, such as by causing common colds. Viruses that infect cells in the lining of our gastrointestinal tract may cause vomiting and diarrhea. Infections by some viral pathogens can persist for long periods, or even for life. An initial infection by one of these viruses causes symptoms that subside quickly. However, the virus remains present in a latent (resting) state and can reawaken. For example, herpes simplex virus 1 (HSV-1) can remain dormant in nerve cells for years before becoming activated by stress. When this virus replicates, it causes painful "cold sores" at the edge of the lips. Another herpesvirus causes similar sores on the genitals.

Measles, mumps, rubella (German measles), and chicken pox are potentially deadly viral diseases of childhood that, until recently, were common worldwide. Today, most children in developed countries have been vaccinated against these illnesses. A vaccine primes the body to fight a pathogen.

A few viruses can cause cancer. Infection by certain strains of sexually transmitted human papillomaviruses (HPV) is the main cause of cervical cancer. Similarly, infection by some hepatitis viruses raises the risk of liver cancer. Epstein-Barr virus, the cause of infectious mononucleosis, also raises the risk of lymphoma (a blood cell cancer).

VIDEO

VIDEO 8-4 Go online to learn more about the viruses that cause Ebola.

Emerging Viral Diseases An emerging disease is a disease that has only relatively recently been detected in humans, or has recently expanded its range. These examples are considered to be emerging diseases in terms of human history.

- **AIDS** Acquired immunodeficiency syndrome (AIDS) is a viral disease that was first identified in humans in 1981. Since then, it has caused about 39 million deaths. AIDS is a communicable disease, meaning the HIV infection that causes AIDS spreads from one infected person to another. Scientists now know that HIV was circulating in parts of Africa since the beginning of the 20th century before it became a widespread disease in other parts of the world.
- **West Nile** The virus that causes West Nile disease cannot spread directly from person to person. Mosquitoes carry this virus from host to host, so these insects are called vectors. Birds are a "reservoir" for West Nile virus, which means that the virus can replicate in birds, and that birds serve as a source of virus for new human infections. West Nile virus was unknown in North America until 1999, when it emerged in New York. It has since spread across the continent. From 1999 through 2019, there were nearly 52,000 reported human cases of West Nile disease in the United States and 2390 deaths.
- **Zika** The Zika virus is a relative of West Nile virus. It is transmitted by mosquitoes and also by sexual contact. Zika has been known in Africa since 1947, but was recently introduced to South America. Most people have mild symptoms, but some develop a temporary paralysis. An infection during pregnancy raises the risk of miscarriage and of microcephaly (an unusually small brain and head) in the developing child.

240 CHAPTER 8 DIVERSITY OF LIVING SYSTEMS

DIFFERENTIATED INSTRUCTION | English Language Learners

Linguistic Support Have students read chunks of the text in pairs and then discuss it. Preteach key vocabulary and high-frequency words students will encounter in the text and provide support for discussions as needed.

Beginning Have students underline key terms they are familiar with and write a question mark beside words they need clarification on. Provide dictionaries or online translators, or clarify meaning with visual supports such as gestures and sketches.

Intermediate Have pairs define key terms and identify the most important ideas in the text. Allow them to discuss in their native language as needed. Encourage them to refer to the visuals in the text for support.

Advanced Have pairs explain key terms and ideas in the text. Encourage students to discuss the ideas in the text using their own words. Provide minimal linguistic support as needed.

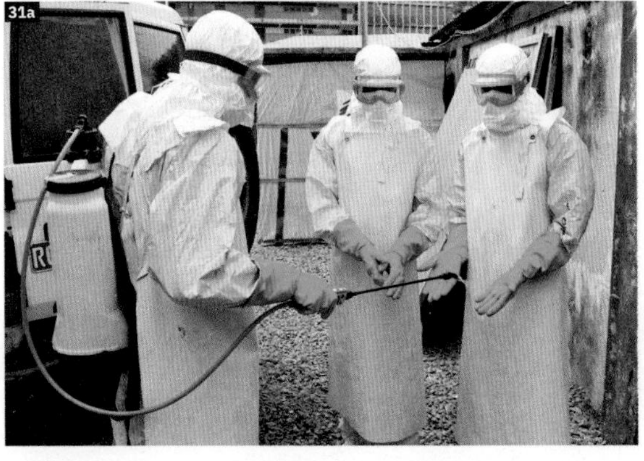

- **Ebola** Ebola is caused by enveloped ssRNA viruses first identified in Africa in 1976. Within three weeks of infection, a person develops flulike symptoms, followed by a rash, vomiting, diarrhea, and bleeding from all body openings. The virus is transmitted by contact with body fluids, so caregivers must wear protective gear (**Figure 8-31**). While most outbreaks are small, an outbreak that began in Guinea killed more than 11,000 people between the end of 2013 and 2016.
- **SARS and COVID-19** SARS (prevalent in 2002–2004) and COVID-19 (starting in 2019) are caused by similar coronaviruses, named for their crown or "corona"-shaped capsid. Both viruses are thought to have jumped from bat species to become infectious in humans and were first detected in China. These coronavirus infections are transmitted through the air and cause respiratory and other symptoms. Through May 2022, more than 522 million cases of COVID-19 were recorded worldwide, with more than 6 million reported deaths.

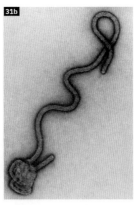

FIGURE 8-31 Medical personnel must take extreme caution when treating patients or performing any tasks associated with the highly contagious and deadly Ebola virus (a). A colored transmission electron micrograph shows the unique structure of an Ebola virus particle (b).

Influenza Viruses and Viral Reassortment Influenza viruses are enveloped ssRNA viruses. Influenza strains are defined by and named for the varying structures of two viral proteins—hemagglutinin (H) and neuraminidase (N)—that extend through the viral envelope (**Figure 8-32**).

FIGURE 8-32 A colored transmission electron micrograph of an avian influenza virus (a) shows many of the features demonstrated in the influenza virus model (b). The hemagglutinin and neuraminidase proteins are important for infection.

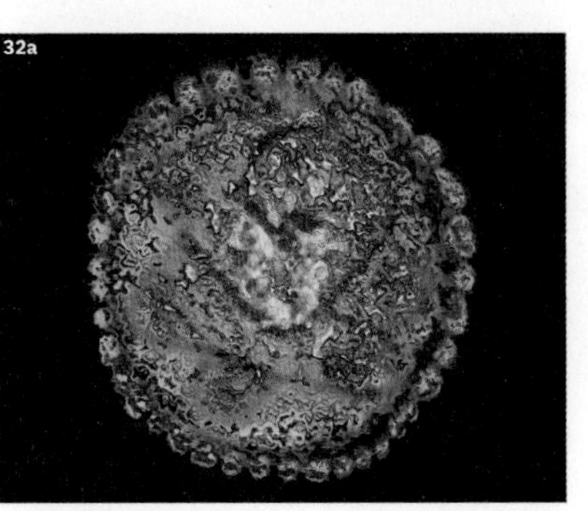

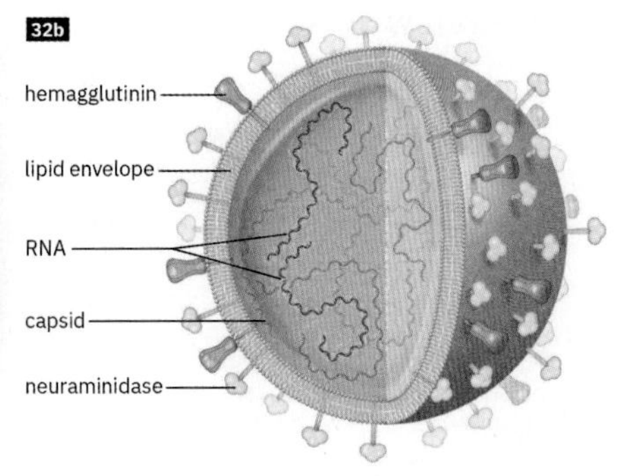

DIFFERENTIATED INSTRUCTION | Economically Disadvantaged Students

Housing Insecurity Though not always apparent, there are many students who have experienced or currently experience housing insecurity. Their living condition may be one in which they must share a small space with a number of other family members or other, unrelated people. Tight living quarters are not conducive to outbreak or pandemic control strategies, which are generally based on social distancing, hygiene, and contact tracing. Help students to navigate related human health conversations, such as herd immunity, as they arise by having them practice social awareness and demonstrating empathy and compassion.

Address Misconceptions

Herd Immunity The idea of herd immunity came back into the foreground during the COVID-19 pandemic, yet many misunderstand the concept. In herd immunity, unvaccinated individuals are protected by the herd or people around who either were vaccinated or survived the disease and thus developed immunity because the virus has fewer ways to pass to the unvaccinated. Both the World Health Organization (WHO) and the Centers for Disease Control (CDC) recommend vaccinations as the surest way to reach herd immunity because allowing a disease to spread naturally can result in many unnecessary cases and deaths, as evidenced by tracking COVID-19 cases and vaccination rates.

The percentage of the population that need to be immune varies with the disease. Measles requires about 95% of a population to be vaccinated, while polio requires only about 80%. The percentage to reach herd immunity from COVID-19 is not yet known but expected to be less than that of polio given the general infection rate. Encourage students to track vaccination rates for various diseases through websites such as those of WHO and CDC.

On the Map

Infectious Diseases Worldwide An Internet keyword search such as "infections diseases world map" results in several sources that show the prevalence of certain diseases in certain parts of the world. Relate such a graphic to the need for certain vaccinations before traveling to other parts of the world.

Visual Support

Virus Reassortment Students may have a difficult time making sense of the process being modeled in **Figure 8-33**. To help students make sense of the process, have them add additional labels to the diagram using sticky notes. Have them label the purple and blue circles on the left, *Influenza A* and *Influenza B*. Carry labels down to the half circles on the host cell. Label the half circle on the right of the host cell, *Influenza C*. Carry labels up to full circle on the top right.

SEP Construct an Explanation
Sample answer: Viral reassortment means that the membrane proteins of influenza can change very easily, so each year new strains of influenza are circulating, with different surface proteins that need different vaccines.

Influenza genomes are made up of eight genome segments that can be swapped between related viruses that infect a host at the same time (**Figure 8-33**). Viral reassortment occurs all the time, and it can be dangerous. For example, the H5N1 strain is a bird flu that has a high mortality rate in humans but is not easily transmitted. H1N1 (the swine flu) is rarely deadly, but it is transmitted easily. If both strains were to infect the same individual at the same time, the result could be a new strain that has a high mortality rate and is also highly transmissible.

To keep up with changes in the prevalence of different H and N proteins in the viral population, scientists develop a new flu vaccine every year. However, determining which flu strains will be around in the future is not an exact science. Even after a flu shot, a person is susceptible to a virus that differs from the strains targeted by the vaccine. New influenza strains arise through mutation and also by viral reassortment.

SEP Construct an Explanation *Why do new vaccines have to be developed every year for influenza viruses?*

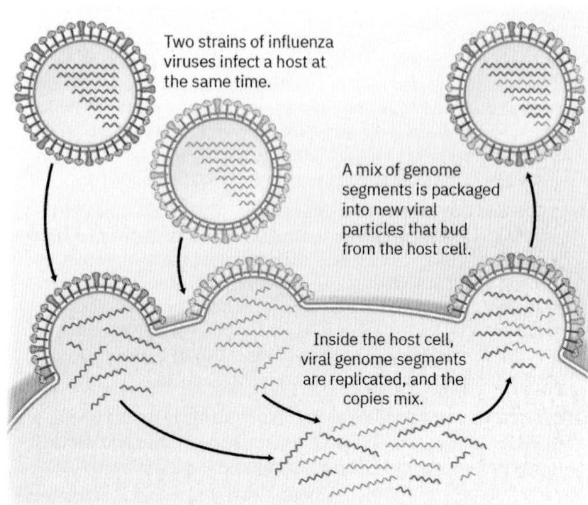

Two strains of influenza viruses infect a host at the same time.
A mix of genome segments is packaged into new viral particles that bud from the host cell.
Inside the host cell, viral genome segments are replicated, and the copies mix.

FIGURE 8-33 Viral reassortment can cause rapid changes in the influenza virus population. Changes in H and N proteins, in particular, affect how the virus interacts with and infects host cells.

8.4 REVIEW

1. **Identify** Which characteristics are common to **ALL** virus particles? Select all correct answers.
   - A. contain DNA or RNA
   - B. have a protein-based capsid
   - C. are single-celled with nucleus
   - D. replicate within a host cell
   - E. have double-stranded genetic material
2. **Describe** What are the four major groups of viral genomes, and where do they typically replicate?
3. **Describe** Which shape describes most of the capsids of viruses that infect animals?
   - A. filamentous fiber
   - B. nonenveloped helix
   - C. enveloped icosahedral
   - D. polyhedral head with helix
4. **Contrast** In what ways are bacteriophage and HIV replication similar and different?

8.4 REVIEW

1. *contain DNA or RNA, have a protein-based capsid, replicate within a host cell.* **DOK 1**
2. *Both double-stranded and single-stranded DNA viruses tend to replicate in the cell nucleus. For the most part, both double-stranded and single-stranded RNA viruses replicate in the cytoplasm.* **DOK 2**
3. *C* **DOK 2**
4. *Sample answer: One pathway of bacteriophage replication is similar to HIV replication. In the lysogenic pathway, bacteriophage DNA becomes integrated into the host cell's genome. This is how HIV replicates as well. Replication is different between the two viruses since bacteriophage have two pathways and are based on DNA while HIV uses reverse transcription from RNA to DNA in the replication process.* **DOK 3.**

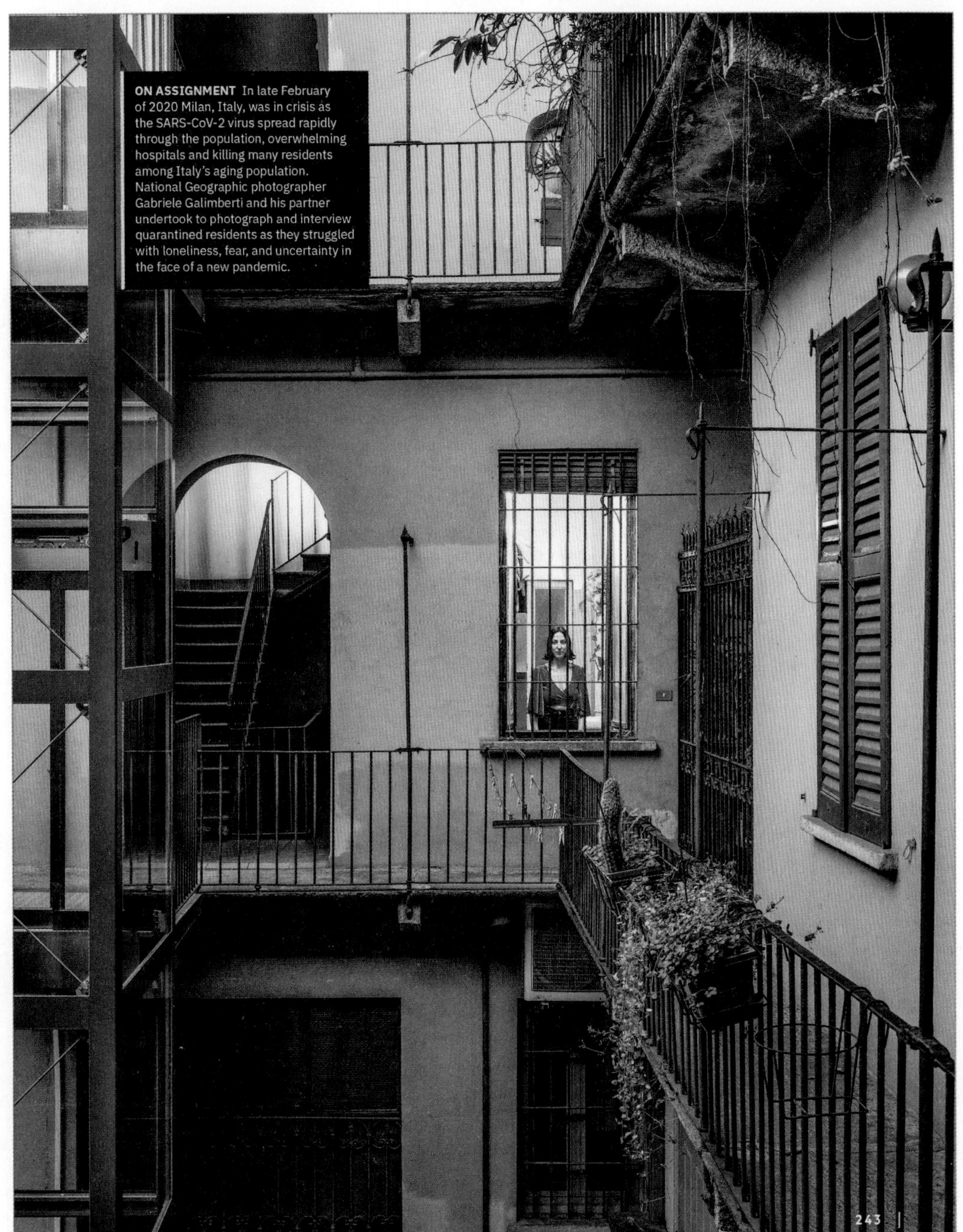

ON ASSIGNMENT In late February of 2020 Milan, Italy, was in crisis as the SARS-CoV-2 virus spread rapidly through the population, overwhelming hospitals and killing many residents among Italy's aging population. National Geographic photographer Gabriele Galimberti and his partner undertook to photograph and interview quarantined residents as they struggled with loneliness, fear, and uncertainty in the face of a new pandemic.

ON ASSIGNMENT

Mental Health Awareness Gabriele Galimberti is an Italian photographer located in Tuscany, Italy. He is passionate about documentary photography and uses his lens to tell people's story. This photo was taken in Milan, Italy, during the height of the COVID-19 pandemic. The image is intended to shed light on the mental exhaustion suffered by many during the pandemic as a result of government and self-imposed quarantines. During this period, mental health issues skyrocketed. In the United States alone, approximately 4 in 10 adults reported systems of isolation, fear, depression, and anxiety. This is a significant increase from pre-pandemic reporting of only about 1 in 10 adults with similar symptoms.

Connect to English Language Arts

Write Informative Text Many students will recall the impact of the COVID-19 pandemic on their daily lives and may have experienced the feeling of isolation depicted by the photograph. Invite students to reflect on their own experiences during that time or to interview older family members or friends to write a historical narrative about the impact of the event. Encourage them to incorporate what they have learned about the science of viruses, vaccinations, and COVID-19 in particular.

EXPLORER
DR. EMILY OTALI

About the Explorer

National Geographic Explorer Dr. Emily Otali studies the behavior, ecology, and physiology of chimpanzees in their natural habitat of Kibale National Park in Uganda. Her research involves studying viruses that could be harmful to the chimpanzees that inhabit the national park. She routinely collects clinical samples at local schools for virus identification to predict potential viral outbreaks. She hopes that by predicting viral outbreaks, she may be able to reduce harmful exposure to the chimpanzees. You can learn more about Emily Otali and her research on the National Geographic Society's website: www.nationalgeographic.org.

Connect to Careers

Primatologist Scientists who study both living and extinct primates in their natural habitats are referred to as primatologists. *Primatologists* combine their habitat observations with experiments to develop an understanding of species evolution and behavior. This is especially helpful when encountering endangered species. Have students discuss how a primatologist might find information about microorganisms beneficial when working to preserve endangered species. *(Microorganisms can be symbiotic or harmful to the host. In harmful instances, the scientist would seek to find ways to minimize exposure.)*

THINKING CRITICALLY

Argue from Evidence *Look for logic in students responses, such as documenting the health of workers entering the park and making connections to the health of their children. Doing this over time would show trends.*

NATIONAL GEOGRAPHIC EXPLORER **EMILY OTALI**

LINKING HEALTHY CHIMPANZEES TO HEALTHY CHILDREN

Dr. Emily Otali works with communities surrounding the Kibale National Park in Uganda to protect the chimpanzees that live there.

The chimpanzees were making strange grunting calls, a signal that a hunt was about to begin. One clung to the base of a tree, beating it with his feet and a stick. Others gathered, focused on another huge tree. The next thing primatologist Dr. Emily Otali saw were monkeys raining down from the tree and racing away in all directions. By the time she reached the tree, the chimpanzees were feasting on five monkeys. The hunt was over.

Otali was exhilarated by this experience of observing her first hunt. She knew at age nine that she wanted to work with animals as an adult. As the Field Director of the Kibale Chimpanzee Project in Uganda, she celebrates the fact that she lives her childhood dream. The 1400 or so chimpanzees in Kibale National Park, divided into about seven different communities of 50–200 members each, comprise one of the largest populations in East Africa. Even though they live in a protected area, and are generally increasing in population, they are not without threats from the humans that surround them.

Of great interest to Otali is the threat of respiratory diseases moving into the chimpanzee population from the humans who work as guides, trail cutters, foresters, and researchers or who visit as tourists. Since 1987, the rate of respiratory infections has increased yearly and has become the leading cause of death in the Kibale chimpanzees, a statistic repeated in other East African populations. In 2013 alone, five chimpanzees from one Kibale community died during a single outbreak attributed to a human rhinovirus-C. This was the first time that this virus was identified in chimpanzees.

The rhinoviruses that cause respiratory problems invade and destroy tissues in the body. After lifelong exposure to such viruses, the human body's immune system can often fight off an infection after a few days. However, the chimpanzees, lacking this long-term exposure, have not developed similar protective responses. Knowing that schools often act as sources of respiratory infections, Otali's project "Healthy Children—Healthy Chimps," focuses on using outbreaks in the area's youth as an early warning system of sorts. She visits forest edge schools to monitor the respiratory health of the students, collect clinical samples for virus identification, and educate children on their role in keeping chimpanzees healthy.

THINKING CRITICALLY

Argue From Evidence *What evidence would you look for to support Otali's claim that the respiratory diseases enter the park through humans going there?*

TYING IT ALL TOGETHER
SPACE STATION STOWAWAYS

HOW ARE THE CHARACTERISTICS OF AN ORGANISM RELATED TO ITS PLACE IN AN ECOSYSTEM?

The Case Study described why microorganisms are found even on the International Space Station (ISS), and how scientists study the effects of space travel on these organisms. In some cases, the presence of bacteria and fungi in space can have unfortunate consequences. In the late 1990s, the Russian space station Mir was inundated with black and green mold growing behind control panels and forming slimy films on numerous surfaces. The mold posed a serious safety risk. It corroded windows, rubber gaskets, electrical components, space suits, and other materials. Even on Earth, indoor mold and mildew are difficult to remove; astronauts on the ISS must combat the same kinds of hardy fungal formations to protect their health, their equipment, and the station itself (**Figure 8-34**).

Aspergillus niger is a species of fungus commonly found on the ISS. Normally harmless to humans, it can cause respiratory disease in people with weakened immune systems. The fungus is also of particular interest to space scientists because it is widely used in the biotechnology industry to produce a variety of acids, proteins, enzymes, and other useful substances. Studies have found that *A. niger* can withstand extreme conditions of the kind that occur in space, including prolonged microgravity, massive doses of radiation, and low temperatures, making it a promising candidate for manufacturing materials needed on long missions (**Figure 8-35**).

In this activity, you will investigate how the characteristics of *Aspergillus niger* enable it to thrive in space. Work individually or in a group to complete these steps.

Obtain and Evaluate Information

1. Research the fungus *Aspergillus niger*. In your research, identify its physical qualities and the systems it uses to grow, form colonies, and reproduce.

Develop a Model

2. Construct a model that illustrates how at least two different characteristics of *Aspergillus niger* would help it survive in space.

Communicate Information

3. Use digital media to make a presentation that explains your model.

FIGURE 8-34 This door shows one surface where fungal growth occurs regularly on the ISS.

FIGURE 8-35 *Aspergillus niger* spores are very resistant to radiation, with significant rates of survival after extremely high doses.

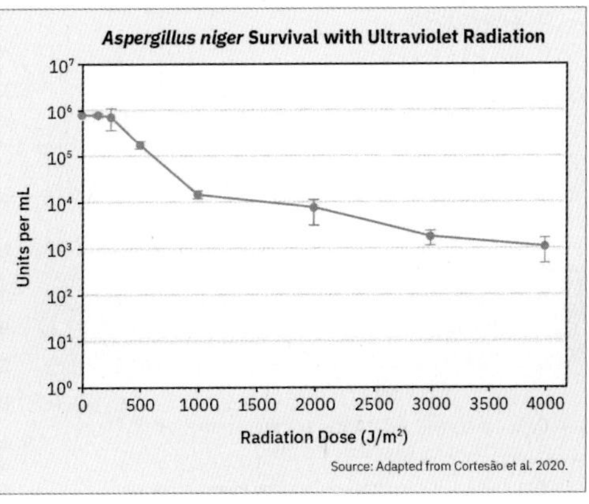

TYING IT ALL TOGETHER

ELABORATE

The Case Study discusses why microorganisms are found even on the International Space Station (ISS) and how scientists study the effects of space travel on these organisms. In the Tying It All Together activity, students research the fungus *Aspergillus niger* and develop a model to explain the characteristics of this fungus that allow it to survive in space.

This activity is structured so that students may complete the steps as soon as they have completed Section 8.3. **Step 1** has students gather evidence on a specific fungus to identify structures used to grow and reproduce. In **Step 2**, students develop a model that shows how different characteristics help the fungus survive in space. In **Step 3**, students communicate information from their models by making a digital presentation.

Alternatively, this can be a culminating project at the end of the chapter. The analysis practiced here reflects processes that students may use to support a claim in the Unit 2 activity.

Go online to access the Student Self-Reflection and Teacher Scoring rubrics for this activity.

Connect to Mathematics

Model with Mathematics As students observe **Figure 8-34**, remind them that an increasing line represents a positive trend, decreasing line a negative trend, and a horizontal line is a neutral trend. Students should also be aware that the labels of a figure contain important information. Encourage students to read the labels prior to starting data analysis. Students will also need to be familiar with the basic principles of graphing points on a coordinate plane. Review the *x*-axis, *y*-axis, and how to locate an ordered pair (*x*, *y*) with students.

Have students break the data on the graph into positive, negative, and neutral trends. Then, have them discuss and determine, based on the trends in the graph, if *Aspergillus niger* spores are resistant to radiation.

CHAPTER 8 **REVIEW**

EVALUATE

REVIEW KEY CONCEPTS

1. *Analyze the sequence of its genome.* **DOK 1**
2. *photoautotroph: C. tepidum; chemoautotroph: S. denitrificans; photoheterotroph: R. palustris; chemoheterotroph: N. eutropha* **DOK 2**
3. *produce toxins; tolerate acidic conditions; obtain nutrients from other organisms* **DOK 2**
4. *by increasing a plant's ability to absorb water and nutrients from soil; by allowing chemical compounds to move between plants as a form of communication* **DOK 2**
5. *Colonial single-celled organism can live independently, whereas cells from a multicellular organism cannot. Each cell in a colonial single-celled organism can reproduce to form a new organism; only certain cells in a multicellular organism can carry out this function.* **DOK 1**
6. *genomes; proteins* **DOK 2**
7. *Step 1: HIV infects the host cell; Step 2: Viral RNA is reverse transcribed into DNA; Step 3: Viral DNA is integrated into the host genome; Step 4: Viral DNA is transcribed into RNA; Step 5: Viral RNA is translated into proteins; Step 6: New viruses are assembled* **DOK 2**
8. *enzymes for RNA replication* **DOK 2**
9. *Sample answer: A mycelium has many filamentous hyphae that create a large surface area available for spreading out over and through dead organic matter. This allows the fungus to rapidly and efficiently absorb nutrients in its decomposition role.* **DOK 3**
10. *Sample answer: Viruses that follow a lytic pathway enter host cells, replicate, and leave the host cell by causing it to rupture, thus destroying the host cell. Viruses that follow a lysogenic pathway follow a similar path but have an extra step in which they insert their DNA into the host's genome before they replicate and exit the host cell. This extra step allows them to remain dormant inside the host cell for extended periods of time.* **DOK 2**

CHAPTER 8 **REVIEW**

REVIEW KEY CONCEPTS

1. **Identify** Which approach would be most useful in determining whether a newly discovered unicellular organism should be classified as bacteria, archaea, or eukaryotes?
   - **A.** Define its cellular shape.
   - **B.** Test its metabolic requirements.
   - **C.** Analyze the sequence of its genome.
   - **D.** Observe how it responds to Gram staining.
2. **Classify** Use the characteristics of four bacterial species described in the table to identify each species as a photoautotroph, chemoautotroph, photoheterotroph, or chemoheterotroph.

Bacteria species	Energy source	Carbon source
Rhodopseudomonas palustris	sunlight	organic acids
Sterolibacterium denitrificans	sulfur compounds	carbon dioxide
Nitrosomonas eutropha	ammonia	organic animal waste
Chlorbium tepidum	sunlight	carbon dioxide

3. **Predict** Some bacterial pathogens enter the human body via ingested food or water, multiply, and attack the human intestines. Which characteristic(s) would you expect to observe in these species? Select all correct answers.
   - **A.** require oxygen
   - **B.** produce toxins
   - **C.** tolerate acidic conditions
   - **D.** convert sunlight into energy
   - **E.** obtain nutrients from other organisms
4. **Explain** How do mycorrhizae assist plants? Select all correct answers.
   - **A.** by inhibiting plant pathogens from entering plant cells
   - **B.** by breaking down and removing the decaying parts of plants
   - **C.** by increasing a plant's ability to absorb water and nutrients from soil
   - **D.** by increasing the number of photosynthetic cells that produce sugars for the plant
   - **E.** by allowing chemical compounds to move between plants as a form of communication
5. **Contrast** How do colonial single-celled organisms differ from multicellular organisms? Select all correct answers.
   - **A.** Colonial single-celled organisms do not contain mitochondria; cells of multicellular organisms do.
   - **B.** Colonial single-celled organisms can live independently, whereas cells from a multicellular organism cannot.
   - **C.** Colonial single-celled organisms are always much smaller than multicellular organisms, not reaching more than a few centimeters in size.
   - **D.** Colonial single-celled organisms differentiate to perform many different functions, whereas cells in multicellular organisms all express the same genes.
   - **E.** Each cell in a colonial single-celled organism can reproduce to form a new organism; only certain cells in a multicellular organism can reproduce.
6. **Contrast** What components are present in both viruses and cells? Select all correct answers.
   - **A.** cytoplasm
   - **B.** genomes
   - **C.** organelles
   - **D.** proteins
   - **E.** ribosomes
7. **Sequence** Place each step in the correct sequence to describe how a retrovirus such as HIV infects its host cell.
   - Viral DNA is integrated into the host genome.
   - Viral RNA is translated into structural proteins.
   - HIV infects the host cell.
   - New viruses are assembled.
   - Viral DNA is transcribed into RNA.
   - Viral RNA is reverse transcribed into DNA.
8. **Predict** What do you think the genome of an ssRNA virus needs to encode if the virus replicates in a eukaryotic host cell's cytoplasm?
   - **A.** enzymes for DNA replication
   - **B.** enzymes for reverse transcription of RNA to DNA
   - **C.** enzymes for RNA replication
   - **D.** ribosomes
9. **Infer** How is the structure and shape of a mycelium well adapted to the role of decomposition that many fungi have in their ecosystems?
10. **Compare** How do viral lytic and lysogenic pathways compare?

11. *Sample answer: Historically, scientists used the physical characteristics of eukaryotes to group them into categories based on similar body shapes and body parts. Later, the development of computer technologies and DNA sequencing techniques allowed scientists to analyze genetic similarities and differences between organisms. This allowed scientists to identify evolutionary relationships between eukaryotes and group them into the six supergroups used at present.* **DOK 2**

CRITICAL THINKING

12. *Sample answer: G. intestinalis cells leave the body of one host in the host's feces. Outside the host's body, the cells must form cysts in order to survive. Once the cysts are ingested, they can re-enter an active state and infect the new host. T. brucei are not transmitted in a similar way. They live in the body fluids of their hosts and of their vector, the tse-tse fly. T. brucei can absorb nutrients from blood and other body fluids while it is being passed between hosts; thus it does not need to form cysts at any point in its life cycle.* **DOK 3**

CRITICAL THINKING

11. NOS **Scientific Knowledge** What evidence did scientists historically use to classify eukaryotes, and how did new technologies provide evidence that led to the more recent classification of eukaryotes into six supergroups?

12. **Identify Patterns** *Giardia intestinalis* and *Trypanosoma brucei* are both parasitic protists. *G. intestinalis* typically enters a new host in the form of cysts before the cells become active in the host's digestive tract. *T. brucei* does not form cysts before infecting a new host. Explain why *G. intestinalis* takes the form of cysts, whereas *T. brucei* does not.

13. **Justify** An effective antifungal drug can kill fungal pathogens without harming the host's cells. A scientist is developing a new antifungal drug for use in humans. She proposes using a chemical compound known to inhibit chitin synthase, an enzyme that is important in the biosynthesis of chitin. What argument can the scientist use to justify her proposal?

14. **Synthesize Information** The fungus that causes white-nose syndrome in North American bats was likely transported from Europe to China. Given that bats do not migrate between these locations, how do you think the infectious fungus could have been transported?

MATH AND ENGLISH LANGUAGE ARTS CONNECTIONS

1. **Write an Argument** The Outer Space Treaty of 1967 established a set of principles governing space exploration. One principle reads: "States shall avoid harmful contamination of space and celestial bodies." Some scientists worry that microorganisms from Earth could colonize other planets if carried there by human-made spacecraft. Is this a valid concern? Write an argument explaining your reasoning.

2. **Reason Quantitatively** Researchers used light microscopy to observe the pumping action of contractile vacuoles in living cells of the protist *Chlamydomonas*. The researchers measured the time interval between contractions as a function of solute concentration in the external medium surrounding the cells. Describe the relationship between pumping rate and solute concentration. What does this relationship suggest about the role of the contractile vacuole in these freshwater organisms?

External medium	Solute concentration (mole/kg water)	Time between contractions (seconds)
distilled water	0.000	15.8
distilled water: tap water (1:1)	0.032	22.0
tap water	0.065	28.9

Read the following passage. Cite evidence from the passage to answer the question.

By the late 1880s, a variety of pathogenic bacteria, fungi, and protists had been identified using a combination of microscopic observations and controlled infection experiments. However, viruses were completely unknown, mainly because microscopes were not powerful enough to view them. In 1892, scientists discovered that a disease affecting tobacco plants was caused by an agent that passed through pores too small to allow bacteria through. Many scientists thought that the agent must be a very small bacterium. Others thought that chemical compounds excreted by bacteria must cause the disease. It wasn't until virus particles could be viewed using an electron microscope in 1941 that scientists agreed that the tobacco mosaic virus was the agent responsible for the disease.

3. **Cite Textual Evidence** How does this passage illustrate the idea that new technologies advance scientific knowledge? Cite information from the text to support your answer.

REVISIT RAINFOREST CONNECTIONS

Gather Evidence In this chapter, you learned how many tiny organisms, and some viruses that infect them, live in all different environments. The rainforest is an incredibly biodiverse habitat with many interacting organisms.

1. What are some roles that you think bacteria might play in the rainforest ecosystem?
2. What other questions do you have about how microscopic organisms interact in the rainforest?

REVISIT THE ANCHORING PHENOMENON

1. *Sample answer: Bacteria can be decomposers, cycling nutrients and other matter in an environment, so they break down dead matter in the rainforest. Bacteria probably cause disease in some organisms in the rainforest, but they will also have beneficial roles such as helping digestion in the intestines of rainforest mammals.* **DOK 3**

2. *What kind of bacterial diseases are important in rainforest ecosystems? With the wide diversity of other organisms in the rainforest ecosystem, are there also more unique types of bacteria there?* **DOK 3**

13. *Sample answer: All fungi contain chitin in their cell walls, whereas human cells lack chitin. An inhibitor of chitin synthase would stop growth of a fungus infecting human tissue while not harming human tissue. Such a drug could be both effective and safe to use.* **DOK 3**

14. *Sample answer: Travelers visiting caves in Europe or China could have accidently transported the spores to North America because the spores could have survived for long periods in their dormant state. Once in North America, some spores likely became airborne and landed on hibernating bats.* **DOK 4**

MATH AND ELA CONNECTIONS

1. *Sample answer: Yes. Many bacteria and archaea are anaerobic and could survive the lack of oxygen in space. As well, some archaea are extremophiles that can survive in the extreme low temperatures they would encounter in space. Also, many bacteria form endospores that are very resistant to harsh environmental conditions. If they travel through space in this form, these bacteria might be able to grow and reproduce if they land in a favorable environment. This would create a contamination problem that space agencies should try to avoid.* **DOK 3**

2. *The data show that the lower the solute concentration, the faster the the contractile vacuole pumps. This suggests that the role of contractile vacuoles is to rid organisms of excess water that diffuses across the cell membrane. The cytoplasm of the cell has a higher solute concentration than the surrounding fresh water, so water diffuses into the cell. Water would move in faster as the solute concentration in the surrounding medium decreases, so the contractile vacuole would have to pump faster when the solute concentration is lower.* **DOK 3**

3. *Sample answer: Scientists draw conclusions from the information they are able to collect. Scientists had no method for directly identifying viruses until the electron microscope was invented. This technology allowed scientists to directly observe viruses for the first time.* **DOK 2**

PLANT SYSTEMS

Three-Dimensional Learning

The practices, core ideas, and crosscutting concepts presented in this chapter's text, investigations, and resources provide support to address the following Performance Expectations: **HS-LS1-2** and **HS-LS1-3.**

Science and Engineering Practices	Disciplinary Core Ideas	Crosscutting Concepts
Engaging in Argument from Evidence	**LS1.A** Structure and Function (HS-LS1-2, HS-LS1-3) **LS1.B** Growth and Development in Organisms **LS2.C** Ecosystem Dynamics, Functioning, and Resilience	Stability and Change (HS-LS1-3)

Contents		Instructional Support for All Learners	Digital Resources
ENGAGE			
248–249	CHAPTER OVERVIEW **CASE STUDY** How do plants respond to changes in the environment?	**Social-Emotional Learning** CCC Structure and Function **On the Map** Chernobyl **English Language Learners** Cognates	
EXPLORE/EXPLAIN			
250–255 DCI LS1.A	**9.1** PLANT ORIGINS AND DIVERSITY • Illustrate some of the adaptations plants need to survive on land. • Differentiate the characteristics of the four major plant groups. • Indicate the relationship between structure and function of plant organs.	**Vocabulary Strategy** Word Wall CCC Stability and Change **Connect to English Language Arts** Develop and Strengthen Writing **Differentiated Instruction** Leveled Support **Visual Strategy** Using Cladograms **English Language Learners** Roots and Affixes **Visual Strategy** Using Diagrams SEP Developing and Using Models **Connect to Mathematics** Choose Levels of Accuracy **In Your Community** Observing Plant Parts **Differentiated Instruction** Students with Disabilities **Visual Support** Stem and Root Variations	

Contents		Instructional Support for All Learners	Digital Resources
EXPLORE/EXPLAIN			
256	**MINILAB** INVESTIGATING LEAF STOMATA		
257–262 DCI LS1.A	**9.2 TRANSPORT IN PLANTS** • Relate the structures involved in plant transport to their functions. • Describe the process and mechanics of both fluid and nutrient transport. • Construct an explanation of how plants control the conservation and release of water.	**Vocabulary Strategy** Using Prior Knowledge **On the Map** Redwood National Forest **SEP** Developing and Using Models **Crosscurricular Connections** Chemistry **Visual Support** Water Adhesion and Cohesion **English Language Learners** Paired Writing **CASE STUDY** Ask Questions **In Your Community** Tapping Maple Syrup **Differentiated Instruction** Leveled Support **On Assignment** Night Sky Photographer **In Your Community** Light Pollution	**INTERACTIVE FIGURE** Movement of Water and Nutrients Through a Plant **Investigation A** Connecting Plant Structures with Their Functions (100 minutes or 2 days)
263–269 DCI LS1.A, LS1.B, LS2.C	**9.3 PLANT REPRODUCTION** • Sequence the steps in the life cycle of a gymnosperm and an angiosperm. • Diagram the relationship between the structure of flowering plant parts and their function in reproduction. • Compare the different methods of asexual reproduction.	**Vocabulary Strategy** Prefixes/Suffixes/ Root Relationships **Visual Support** Alternation of Generations **Differentiated Instruction** Leveled Support **In Your Community** Ferns **Connect to English Language Arts** Draw Evidence From Texts **SEP** Developing and Using Models **Connect to English Language Arts** Integration of Knowledge and Ideas. **CONNECTIONS** Saving Seeds **Address Misconceptions** Trees and Flowers **Vocabulary Strategy** Word Wall **Differentiated Instruction** Students with Disabilities **Connect to Careers** Beekeeper **CCC** Structure and Function **Differentiated Instruction** Leveled Support **Visual Support** Plant Reproduction	*Video 9-1* *Video 9-2*
270	**EXPLORER** ROBERT RAGUSO	**Connect to Careers** Chemist **English Language Learners** Speaking Strategies	

Contents		Instructional Support for All Learners	Digital Resources
EXPLORE/EXPLAIN			
271–275 DCI LS1.A	**9.4 PLANT RESPONSES TO THE ENVIRONMENT** • Summarize the four main types of plant hormones that regulate specific plant behaviors. • Describe different environmental stimuli and the type of responses they trigger. • Assess plant responses specific to daily and seasonal changes in a plant's environment.	**Vocabulary Strategy** Semantic Map **Differentiated Instruction** Leveled Support **Address Misconceptions** Positive and Negative Tropisms **English Language Learners** Using New Vocabulary **Crosscurricular Connections** Organic Chemistry **In Your Community** Around-the-Clock Plants **CER** Revisit the Anchoring Phenomena **Crosscurricular Connections** Chemistry **English Language Learners** Writing About Possibilities	*Video 9–3* **Investigation B** Homeostasis in Plants (250 minutes or three days plus observation time over 2 weeks)
276	**LOOKING AT THE DATA** BUD BURST AND FLOWERING IN A CHANGING CLIMATE		

Contents		Instructional Support for All Learners	Digital Resources
ELABORATE			
277	**TYING IT ALL TOGETHER** How do plants respond to changes in the environment?	**In Your Community** In Your Backyard **Differentiated Instruction** Leveled Support	Self-Reflection and Scoring Rubrics
Online	**Investigation A** Connecting Plant Structures with Their Functions **Investigation B** Homeostasis in Plants	Guided Inquiry (100 minutes or 2 days) Design Your Own (250 minutes or three days plus observation time over 2 weeks)	MINDTAP Access all your online assessment and laboratory support on MindTap, including: sample answers, lab guides, rubrics, and worksheets. PDFs are available from the Course Companion Site.
EVALUATE			
278–279	**Chapter 9 Review**		
Online	**Chapter 9 Assessment** **Performance Task 1** *How Do Systems Interact to Maintain Homeostasis in Plants?* (HS-LS1-2)		

CHAPTER 9

ENGAGE

Students may have an understanding of plant structure and function from middle school, as well as the basics of some plant responses. This chapter builds on that knowledge as students take a deeper look at plant origins, adaptations, and how plant systems function, such as water transport. They will also explore different methods of plant reproduction and the adaptations that have allowed plants to be so successful.

About the Photo The Pando aspen grove consists of over 40,000 individual trees. However, as the grove that sprouted from a single seed thousands of years ago shares a single root system, it is considered a single organism that weighs about 6000 metric tons (13 million pounds). There are signs that the Pando is starting to decline, due to a lack of regeneration. The causes of the decline include overgrazing by mule deer, insects, and disease.

Social-Emotional Learning

As students progress through the chapter, invite them to look for contexts and opportunities to practice some of the five social and emotional competencies. For example, students may develop **responsible decision-making skills** by making reasoned judgments after learning and analyzing new information, data, and facts about different plants and how they function. Plants have a place in the history of different cultures. They have been used as resources to meet the needs of people around the world, such as food, medicines, and shelter. Students can develop **social awareness and relationship skills** by demonstrating a willingness to learn about cultural differences in this area or recognize and share such experiences from their own culture.

CHAPTER 9

PLANT SYSTEMS

This grove of aspen trees in Utah is thought to be the largest and most dense living organism on Earth.

This aspen grove, named Pando, a Latin word meaning "I spread," covers 106 acres and consists of more than 40,000 trees. Aspen is a clonal species, which means that genetically identical trees can sprout from its roots. Though current trees are only around 130 years old, scientists think the aspen grove got its start near the end of the last ice age, about 11,700 years ago.

Chapter 9 supports the NGSS Performance Expectations **HS-LS1-2** and **HS-LS1-3.**

248 CHAPTER 9 PLANT SYSTEMS

CROSSCUTTING CONCEPTS | Structure and Function

Plant Cells This chapter provides many opportunities to teach the concepts of structure and function. The structure of a plant cell is linked to its function and varies in specialization. Plant cells are organized into specialized tissues, such as the xylem and phloem that function to transport materials.

The structures of a plant are tied to how the plant grows, survives, and reproduces.

Throughout this lesson, encourage students to make the connection between structure and function and model the use of those terms. Have students take notes and participate in discussions using sentence frames that vocalize the structure and function relationship such as: *A leaf is a structure of the plant. The plant needs food to survive. The function of the leaf is to make food for the plant.*

CASE STUDY
FLOWER POWER

HOW DO PLANTS RESPOND TO CHANGES IN THE ENVIRONMENT?

Sunflowers (*Helianthus annuus*) are common sights in agricultural fields where they are grown for food and oil production. But farms are not the only places where fields of sunflowers grow. These easily identifiable flowers have also been planted after environmental disasters such as the nuclear accidents at the Chernobyl Nuclear Power Plant in Ukraine in 1986 and at the Fukushima Daichi Nuclear Power Plant in Japan in 2011.

Planting sunflowers in contaminated areas is an example of *phytoremediation*, which is the use of plants to extract chemical contaminants from soil or water. The goal of phytoremediation is to recover polluted soil and help restore soil fertility.

Sunflowers grow quickly and can grow nearly anywhere without much intervention. These popular plants are *hyperaccumulators*, meaning they are able to take up and store high concentrations of metals from the soil or water where they grow. These substances are stored in the plants' tissues, including their roots, stems, and leaves. Sunflowers are particularly adept at absorbing zinc, copper, lead, and radioactive wastes.

Phytoremediation of radioactive materials works because some radioactive isotopes are chemically similar to nutrients that the plants normally absorb. For example, cesium behaves like potassium, a critical nutrient for plant growth and development. Strontium is similar to calcium, which plants require for structural support. As part of a Chernobyl remediation project, sunflowers were grown in rafts floating in a contaminated pond. Within two weeks, the amounts of radioactive cesium and strontium in the water had decreased by 90 percent.

Once the sunflowers absorb the materials, the plants are harvested and safely disposed of. Phytoremediation is a particularly attractive solution for soil contamination (**Figure 9-1**), as it is significantly less expensive to grow and dispose of plants than it would be to remove the soil itself. However, while the use of sunflowers has gained popularity as a remediation technique, several factors can affect its success. For example, not all species of sunflowers are effective hyperaccumulators, so planting the right species is key. The type of soil can also affect how well the plants are able to extract toxic materials. While the sunflowers have been effective at Chernobyl, projects at Fukushima did not have the same results. Researchers continue to study how to make the most effective use of these plants in the reclamation of polluted sites.

Ask Questions *As you read this chapter, generate questions about systems and processes that would enable a plant to transport toxic substances from soil or water into its tissues.*

FIGURE 9-1 Coal-fired electrical power plants produce fly ash that contains heavy metal pollutants. The heavy metals accumulate in the surrounding soil and water. Phytoremediation can be used to remove these substances.

DIFFERENTIATED INSTRUCTION | English Language Learners

Cognates The use of cognates can be useful in helping students comprehend meanings of new English words.

Beginning Have students read through the article and identify words that are cognates, or English words that are similar to those in their native language. You can then ask them to make a chart with both the English word and the cognate. For example in Spanish:
agriculture→agricultura
accident→accidente
radioactive→radiactivo
contamination→contaminación

Intermediate Have students identify cognates from the Case Study. Ask them to write sentences using the English word and placing the cognate immediately after in parentheses.

Advanced/Advanced High Working in pairs, have students ask and answer questions using cognates that they have identified in the Case Study.

CASE STUDY

ENGAGE

This Case Study focuses on sunflowers and looks beyond their beauty to their use in remediating contaminated areas. The Chernobyl disaster killed two people on the night of the event and 28 people died within a few weeks from acute radiation syndrome. Many others suffered from thyroid cancers attributed to the disaster. The thyroid was affected largely through the ingestion of food that had been contaminated with radioactive iodine, including children who drank milk from grass-fed cows after the grass was contaminated.

In the Tying It All Together activity, students will analyze data related to different plant species and their effectiveness in removing specific metals from the soil.

Ask Questions Students should revisit the Case Study as they read the chapter to make connections with the content. See a specific suggestion in Section 9.2.

On the Map

Chernobyl The Chernobyl Nuclear Power Plant is located in northern Ukraine, about 130 km north of the capital city of Kyiv. Located about 3 km from the power plant is the small town of Pripyat, which was built to house Chernobyl's workers and their families, totaling about 50,000 people. The day after the disaster, the town was evacuated and families were relocated.

Today, a 1239 km Chernobyl Exclusion Zone exists around the site of the power plant that has been classified as uninhabitable. Have students locate Chernobyl on a map and circle the uninhabitable zone around the area.

Human Connection Relatively few major nuclear disasters have occurred, but they affected large numbers of people who lived nearby. Coal-fired power plants emit hazardous materials during regular operation. Fly ash, a fine silica-based powdery by-product can travel up to 30 km through the air and is risky to lung health.

As a class, have students create a risk versus benefits chart of power plants based on facts. Then, guide a reasoned debate on the use of nuclear or coal-fired power.

9.1

PLANT ORIGINS AND DIVERSITY

LS1.A

EXPLORE/EXPLAIN

This section begins by introducing the transition of plants from water to land. This helps students better understand the structural and reproductive adaptations found in land plants today. Some groups met the demands of life on land sooner than others. Students will learn about nonvascular plants that are still dependent on moist environments and vascular plants that adapted to life on land with specialized tissues to transport water and food.

Objectives

- Illustrate some of the adaptations plants need to survive on land.
- Differentiate the characteristics of the four major plant groups.
- Indicate the relationship between structure and function of plant organs.

Pretest Use **Questions 1** and **2** to identify gaps in background knowledge or misconceptions.

Vocabulary Strategy

Word Wall Create an interactive word wall with vocabulary listed so that items can be attached. As students learn each new word, invite them to attach items to the board that will help them connect with the word. For example, they may attach a helpful prefix or a picture. Non-native speakers may post items in their native language. Encourage students to draw arrows along with notes to show how words are related.

Explain *Sample answer: Plants produce most of the food for other organisms on Earth because they photosynthesize.*

9.1

KEY TERMS

cuticle
lignin
nonvascular plant
organ
organ system
phloem
stomata
vascular plant
vascular tissue
xylem

PLANT ORIGINS AND DIVERSITY

What color is Earth? In images of the sunny side of the planet taken from space, white clouds swirl above blue oceans, and land masses appear in shades of brown. Of the many forms of life on Earth, only one can be directly observed at this scale–the patches of green visible on almost all continents are plants. There are about 435,000 unique species of land plants, and plants constitute over 80 percent of all the biomass on Earth.

Explain *Why do you think plants are such a dominant life-form on Earth?*

Plant Evolution

Plants are multicellular eukaryotes that carry out photosynthesis and typically live on land. Fossil evidence indicates that between 500 million and 400 million years ago, land plants evolved from a freshwater green alga that grew at the water's edge. Like their green algal relatives (Section 8.2), plants have cells with walls made of cellulose, they have chloroplasts that contain chlorophyll, and they store sugars as starch.

A green alga absorbs all the water, dissolved nutrients, and gases it needs through its outer surface. Water also keeps the alga afloat, supporting its weight and maximizing its exposure to sunlight for photosynthesis. But to survive on land, plants needed to adapt to very different conditions than their aquatic ancestors faced.

Structural Adaptations Land plants are exposed to air, so the water they need for photosynthesis readily evaporates into their surroundings. To reduce water loss, plants secrete a waxy **cuticle** that seals their outermost tissue layer, the epidermis (**Figure 9-2**). Pores called **stomata** (singular, *stoma*) span the cuticle and epidermis. Stomata can open to allow carbon dioxide into the plant or close to conserve water.

FIGURE 9-2 A cross section of a leaf shows some plant structures that evolved as adaptations for surviving on land.

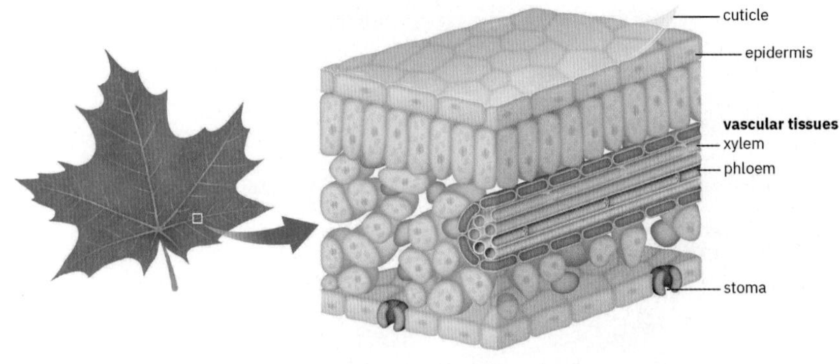

CROSSCUTTING CONCEPTS | Stability and Change

Modeling Plant Adaptations To develop students' awareness and interest in plants and their evolutionary history, have them design a garden that showcases the key adaptations that enabled plants to make the move from water to land.

Working in groups, students should design their gardens in sections that represent the key stages in plant evolution. A time-saving option is to assign each group one stage that can later be combined into one drawing.

As an example, the first section of the garden can be 500 million years ago (mya) when plants lived only in water. The next section can be 500–400 mya, when plants developed a waterproof cuticle, and continuing on with the appearance of vascular tissue, seeds, and flowers.

Students should include a script for their garden design that explains each adaptation and construct an explanation as to how it enabled plants to move from water to land.

On land, a plant must obtain water and nutrients from soil and transport these materials to wherever they are needed in the plant's body. Some land plants accomplish this by direct absorption, similarly to green algae. In contrast, more than 90 percent of modern plant species have internal pipelines made of **vascular tissue** to perform transport functions. **Xylem** is the vascular tissue that distributes water and mineral ions taken up by roots. **Phloem** is the vascular tissue that distributes sugars made by photosynthetic cells.

A polymer called **lignin** stiffens cell walls in vascular tissue. Lignin-reinforced vascular tissue not only distributes materials. It also helps land plants stand upright and supports their branches. Branches with many leaves increase the amount of the plant's surface area that is available for capturing light and exchanging gases with the atmosphere.

Reproductive Adaptations Like related algae, plants produce two types of reproductive cells. *Spores* are single cells that develop into a new organism by dividing. *Gametes* are sex cells that fuse together and develop into an embryo. Plants reproduce by releasing spores or gametes into their environment.

New plant structures evolved to enable reproductive cells to disperse in dry habitats, such as by wind or land animals. For example, a *pollen grain* is typically a one- or two-celled structure that contains a male gamete. The outer layer of a pollen grain is highly resistant to abrasion, heat, decay, and other environmental dangers it may encounter on the way from one plant to another (**Figure 9-3**).

The development of the *seed* was an enormous advancement for land plants. Seeds are multicellular structures that include a preformed embryonic root, stem, and leaf as well as a food supply to increase the embryo's chance of survival. A protective seed coat allows water and oxygen to reach the growing embryo until it is ready to germinate.

CCC Structure and Function *What are two structural advantages that enable plants to survive on land?*

FIGURE 9-3 Pollen grains vary widely in size, ranging in diameter from a few to hundreds of micrometers in land plants. Pollen grains are so indestructible that they commonly appear in the fossil record. The pollen grains in this scanning electron microscope image come from various types of plants. A plant's species can often be identified by the markings and shape of its pollen grains.

Science Background

Plant Fossils and the Geologic Time Scale The fossil record shows evidence that the first land plants appeared about 450 million years ago. Initially, these plants were small and similar to today's bryophytes.

Provide copies of the geologic time scale to students and have them discover and discuss changes over time in plants, as well as corresponding geologic events and conditions on Earth.

Connect to English Language Arts

Develop and Strengthen Writing Ask students to write an explanatory paragraph about how plants were able to make the move from water to land. Their paragraph should identify the challenges presented by life on land (*inconsistent supply of water, possible overexposure to sunlight, reproduction*) and adaptations in plant structures and functions that made the transition possible. Encourage students to read and edit their paragraphs for clarity and to make sure their content addresses the topic.

CCC Structure and Function *Sample answer: The cuticle and vascular tissue allow plants to survive on land.*

DIFFERENTIATED INSTRUCTION | Leveled Support

Key Terms For students to successfully understand plant reproduction, it is important for all students to understand the terms *spore, gamete, pollen,* and *seed* and the differences among them. Review the words with students.

Struggling Learners Ask for student volunteers to explain what a spore and a gamete are and then do the same for pollen and seed. Working in pairs, have students create comparison charts for both sets of words. For example, "Gametes are sex cells that must fuse together to form an embryo; Spores are single cells that can develop into a new organism by dividing."

Advanced Learners Have students create two comparison charts, one for spore and gamete and a second for pollen and seeds. Then, use the comparison to write a descriptive paragraph about each pair of words as they relate to plant reproduction.

Science Background

Pioneer Species American history has many examples of exploring pioneers who paved the way through uncharted territories and made it possible for others to survive and thrive. This took resilience and the ability to work with the resources available.

Engage students in a discussion as to why bryophytes are considered a pioneer species. Bryophytes are typically the first plants to appear following a natural disaster, such as a volcanic eruption, that scrubs the soil surrounding it of all living things. Bryophytes absorb water and nutrients from barren, rocky terrain. In doing so, rock breaks into pieces that form soil. As more soil forms, different plants can grow. As more plants grow, a greater diversity of animals and other organisms can thrive. Eventually, a community of living things takes hold.

Visual Strategy

Using Cladograms Briefly review what students learned in Chapter 8 about how diagrams such as **Figure 9-4** depict relatedness. Ask them what the diagram tells them about nonvascular plants and seedless vascular plants. Have them continue identifying similarities and differences between the remaining plant groups.

CCC Structure and Function
Nonvascular plants have no transport system to carry water throughout their bodies, so all parts of the plant need to be in contact with a water source.

Major Plant Groups

Modern plants can be divided into two main groups based on whether they have a system for moving water and nutrients within the plant. The **vascular plants** have transport systems made of vascular tissue. Although some **nonvascular plants** known as bryophytes do not have xylem or phloem, they do have internal conducting vessels. They absorb nutrients through their surface rather than withdrawing them from soil, so they can colonize rocky sites where vascular plants cannot take root. Nonvascular plants are the only plant life in some parts of the Arctic and Antarctic. Evolutionary relationships between the nonvascular and vascular plants remain a matter of investigation.

Vascular plants can be further categorized by their reproductive systems. Seedless vascular plants include the ferns and other similar plants that reproduce by dispersing spores. The seed plants include gymnosperms, which produce unprotected seeds on a cone or stem, and angiosperms (flowering plants), which produce seeds enclosed within a fruit. **Figure 9-4** summarizes the relationships among the major plant groups and the traits of each group.

FIGURE 9-4 The first land plants to evolve from an ancient alga were nonvascular. Seedless vascular plants emerged later, followed by vascular plants with seeds.

CCC Structure and Function *Many tiny individual moss plants clump together to form the blanket of moss shown in **Figure 9-4**. In general, nonvascular plants tend to be very small in size relative to vascular plants. Why do you think this is?*

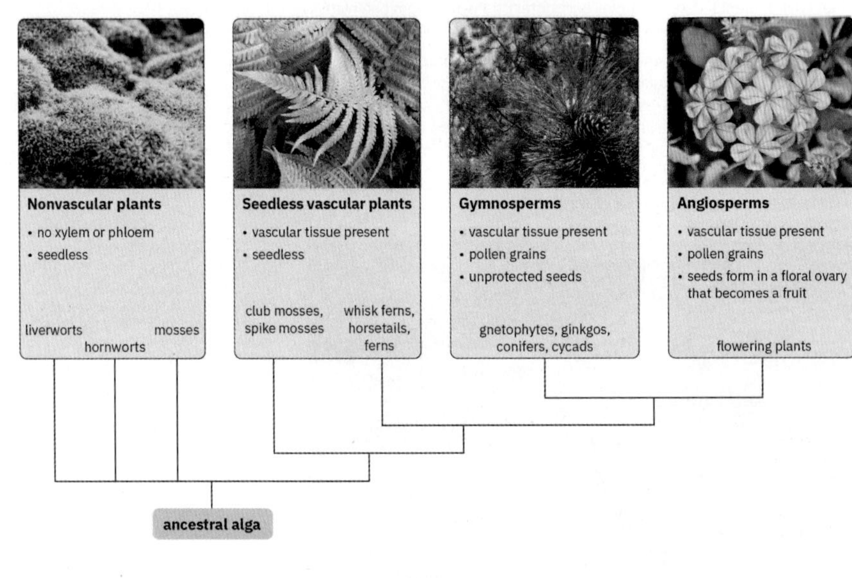

DIFFERENTIATED INSTRUCTION | English Language Learners

Roots and Affixes Learning roots, as well as associating scientific names with more familiar terms, is a helpful tool for language learners at all levels. Share the following roots: *sperm – seed, gymno- naked, angio – receptacle, vascular – of or pertaining to vessels or tubes,* and *non – not, or reverse of.* As a class, use the roots to define *gymnosperm*. Response: naked seed, or unprotected seed. Have students write each word root on the front of an index card and its meaning on the back.

Beginning and Intermediate Working in pairs, have students use their cards to practice recalling each root and its meaning. Then, say words such as *angiosperm* and *nonvascular* and have students use their cards to form the words from the roots.

Advanced Tell students to practice recalling each root and its meaning. Then, ask them to use the roots to define words such as *angiosperm* and *nonvascular* and use each one in a complete sentence.

Plant Tissues

The body of a plant is organized into tissues, organs, and organ systems. *Meristem* tissue is made up of undifferentiated cells that divide and develop into more complex tissues that provide specialized functions. Plants are made up of three general types of complex tissue:

- **Dermal Tissue** The dermal tissue system consists of tissues that cover and protect the plant's exposed surfaces. The epidermis is dermal tissue.
- **Vascular Tissue** The vascular tissue system includes the tissues that carry water and nutrients from one part of the plant body to another (**Figure 9-5**). Xylem tubes have two structures, called vessels and tracheids. Phloem is made up of sieve tubes. Vascular tissue also provides plants with structural support.
- **Ground Tissue** The ground tissue system is essentially everything that is not part of the dermal or vascular tissue systems. Cells that carry out the most basic processes necessary for survival are part of the ground tissue system. Flexible, thin-walled *parenchyma* cells give ground tissue a variety of properties that are essential for growth, photosynthesis, and storage functions.

FIGURE 9-5 In this cross section of a *Clematis* stem, bundles of vascular tissue are surrounded by ground tissue. The stem's outer surface is made of dermal tissue.

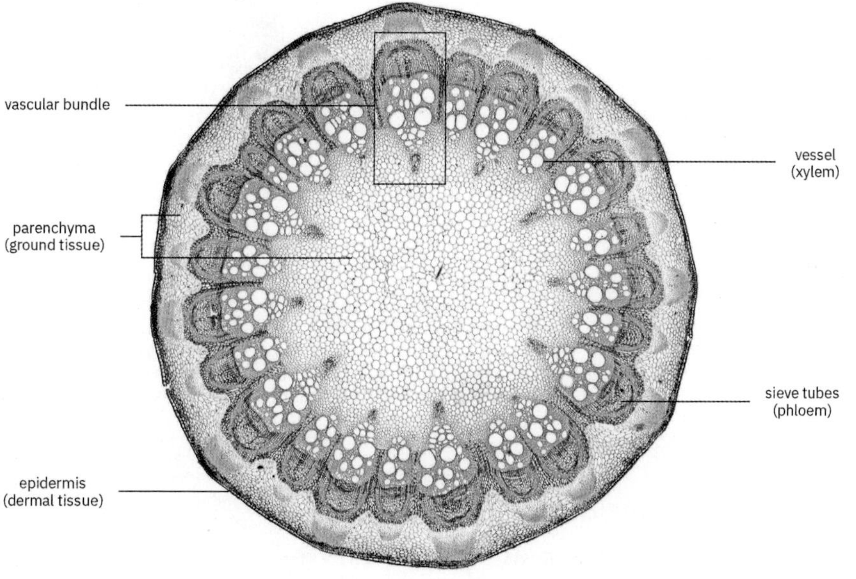

9.1 PLANT ORIGINS AND DIVERSITY 253

Science Background

Specialized Cells in Plants

Parenchyma is one of three types of ground tissue found in plants. The other two types are collenchyma and sclerenchyma. Cells making up all three types of tissue share the unique structures found in plants, including a central vacuole, plastids, and a cell wall surrounding the cell membrane. However, small differences between them are related to their different functions. Parenchyma cells have a large central vacuole with thin, flexible cell walls. They are involved in functions such as photosynthesis and storage of water and nutrients. They also make up the fleshy part of fruits. Collenchyma cells have thicker walls and function to support parts of the plant. Sclerenchyma cells have rigid walls and provide a frame to support the plant in parts that are no longer growing, such as woody stems.

Visual Strategy

Using Diagrams Provide pairs of students with a microscope and prepared slide of a dicot stem cross-section, such as a *Ranunculus*. Prior to viewing, have students study **Figure 9-5** and become familiar with the labeled parts. Direct students to make a drawing of what they see on the slide and label each part and then compare their drawing to **Figure 9-5**. Ask volunteers to use their drawings to describe the location of the xylem and phloem (*the xylem and phloem are bundled together with the xylem toward the center of the stem and the phloem toward the outside of the stem).*

SCIENCE AND ENGINEERING PRACTICES
Developing and Using Models

Modeling Plant Parts To help students demonstrate the relationships among the parts in a plant's vascular system, have them develop and construct a model using **Figure 9-5** as a guide. Provide the following materials to each group: tissue roll or paper towel roll cut in half (epidermis); colored drinking straws cut in 10-cm sections (phloem); smaller coffee stirrer straws (xylem); card stock paper (vascular cambium); scissors, tape, and cotton or other filler (parenchyma).

Provide small labels so students can label the parts of their model. You may save each group's model for the next section where more content about transport will be presented.

Connect to Mathematics

Choose Levels of Accuracy Have students ponder the surface area of a tree's leaves compared to their volume by considering the size of a mature red oak tree: 15–23 m (50–75 ft.) tall with a canopy width of 12–18 m (40–60 ft.). The tree is covered in lobed, oval leaves, each of which is 12–25 cm (5–10 in.) by 10–15 cm (4–6 in.). If balled up, an oak leaf would have a volume of about 50–60 mL (1/4 c.).

To roughly estimate the number of leaves on a red oak, students can first calculate the size of the circular area beneath the canopy and then determine the number of single leaves that would cover that area. Then, students can mulitply that result by the leaf area index (LAI), which ranges from 4–8 for deciduous trees. LAI is unitless because it is a ratio of areas. Using these data, students can estimate the number of leaves on a mature red oak tree by multiplying the LAI by the number of single leaves beneath the crown.

Have students then compare the total surface area of the leaves with their volume. After sharing results, guide them to discuss the relative accuracy of the data, their results, and whether they can consider their results valid.

In Your Community

Observing Plant Parts Arrange for students to observe live plants at a garden center, university greenhouse, or conservatory, or have students research plants online. Alternatively, you can bring a variety of different types of plants to display, preferably with all parts intact. Guide a discussion of the similarities and differences that they found. Discuss how local conditions affect the plants that are grown. If viewing exotic plants, discuss how the growing conditions are controlled so that the plants can survive.

CCC Systems and System Models *Students' sketches should show the parts of the shoot system above the ground and roots underground. Water and nutrients from the soil flow from the root system upward through a plant. Sugar moves from areas where photosynthesis occurs, such as leaves, to other parts of the plant.*

Plant Organs and Organ Systems

An **organ** is a structure made up of different types of tissue that work together to perform a task. Plant organs includes stems, leaves, roots, and reproductive structures. An **organ system** is a set of interacting organs that carry out a particular body function. A vascular plant's body has two organ systems: shoots and roots (**Figure 9-6**). The *shoot system* consists of stems, leaves, and reproductive organs such as flowers. The *root system* absorbs water and dissolved minerals, and it often serves to anchor the plant in soil. Most shoots grow above the ground and most roots grow below it, but there are many exceptions to this general rule.

FIGURE 9-6 A plant's vascular system contains a hierarchy of systems and organs.

Stems In addition to water and nutrient transport, the stems of a vascular plant provide structural support and keep leaves positioned for photosynthesis. Stems can grow above or below the soil, and in many species they are specialized for storage or asexual reproduction. They characteristically have nodes, which are regions of the stem where leaves attach. New shoots (and roots) can form at nodes. A young plant grows mainly by lengthening of its stems. Older woody plants grow by stems lengthening and thickening.

Leaves The main organs of photosynthesis in most plant species are leaves. Leaves contain the stomata that allow carbon dioxide to enter the plant and allow water to leave. Leaves are typically thin, with a high ratio of surface area to volume.

Roots The primary function of a plant's roots is to take up water from the soil. As roots grow, they stabilize the plant and provide more surface area to absorb water. The roots of a typical plant are as extensive as its shoots, and often more so. In many species, roots also store nutrients.

CCC Systems and System Models *Draw a simple diagram that models the root and shoot systems of a plant that is common in the area where you live. In your drawing, use arrows to show the flow of water and sugar molecules.*

DIFFERENTIATED INSTRUCTION | Students with Disabilities

Visual Impairment Provide visually impaired learners with a board with the headings "Root System" and "Shoot System" in large print and/or braille. Preview the concepts taught in the lesson, describing the characteristics of each plant part in detail in terms of texture, size, and shape: e.g., *the leaf is thin like a sheet of paper, the stem is long and thin and supports the leaf,* or *the roots are made of many long, thin, and fibrous shoots.* Then, allow students extra time to observe plants that have been brought in to the classroom using their available senses. If assistance is needed, guide them through each part of the plant.

After observing the intact plant, assist in dissecting the plant into roots, stems, and leaves. Have students categorize each part by placing them under the correct heading on the board. Discuss as a class how the characteristics of each plant part help it perform its task.

Stem and Root Variations Many types of cacti and succulents have fleshy photosynthetic stems specialized to store water. Flowers, spikes, small leaves, or entire plants may form at nodes. The triangular stems of the cactus in **Figure 9-7a** perform photosynthesis for this plant, which has no leaves.

New roots typically branch from existing roots. However, in some plants, roots can form on stems or leaves. In a few species, such as the banyan tree in **Figure 9-7b**, roots form above ground at nodes on trunks or lower branches and then grow into the ground. In some plants that grow on other plants for support, such as the orchids in **Figure 9-7c**, roots develop from nodes in the stem and hang in the air. These roots absorb moisture from the surrounding air.

FIGURE 9-7 The green stems of *Ariocarpus fissuratus* (a) look like leaves, but they store water and are photosynthetic. The roots of a banyan tree (b) hang down from its branches. Orchids (c) grow in tropical climates where their roots can obtain water from humid air.

9.1 REVIEW

1. **Describe** Early plants developed an adaptation that allowed them to flourish on land. What are the two key properties of this structure?
2. **Distinguish** Which characteristics occur in seeds but not in spores? Select all correct answers.
   A. multicellular
   B. unprotected
   C. does not require germination
   D. embryonic root, stem, and leaf
   E. easily dispersed in wind and water
3. **Distinguish** Match the terms to describe the functions of these plant tissues: *dermal*, *vascular*, and *ground*.
   - stores nutrients
   - plant growth
   - transports water
   - photosynthesis
   - protects plant surfaces
   - transports nutrients

Visual Support

Stem and Root Variations Discuss each of the examples of stem and root variations and where they are located. Cacti generally are found in hot, arid climates, such as deserts, that receive less than 25 centimeters of rain per year. Epiphytes, such as orchids, are found in tropical rain forests that are characterized by dense treetops that form a canopy, which blocks sunlight from reaching the forest floor. Banyan trees begin as epiphytes when their seeds land on the branches of a host tree and their roots grow downward to the ground. The roots form columns that support the banyan tree.

Have students choose one of these examples and construct an explanation as to how the adaptation of their selected plant helps it survive in its environment. Remind students that they must include evidence to support their claims.

9.1 REVIEW

1. *Plants are covered with a wavy cuticle that prevents water from evaporating from the body. The cuticle has tiny pores called stomata that allow carbon dioxide and oxygen to be transferred between the plant body and atmosphere.* **DOK 2**
2. *A, D.* **DOK 2**
3. *dermal: protects plant surfaces; vascular: transports water, transports nutrients; ground: stores nutrients plant growth, photosynthesis.* **DOK 2**

MINILAB
INVESTIGATING LEAF STOMATA

Students will identify stomata on plant leaves and distinguish between those that are open and closed. They will then use the data collected to support their claims about the type of environment where each plant is found.

Time: 30 minutes

Advanced Preparation

- Gather locally available succulent and non-succulent dicots. To prepare leaf impressions, apply a thin layer of clear, quick-dry nail polish to the leaf undersides, avoiding the veins. After drying 5–8 minutes, place a piece of clear tape on the polished side and gently peel it off. Place the tape with the nail polish adhered to it on a blank microscope slide. Both the leaf impressions and the actual leaves will be observed.

Procedure

In **Step 1**, students will be able to see a close-up view of stomata using the 40X objective lens. When using this lens, students will be able to observe a number of stomata with less detail.

In **Step 3**, it may be difficult for students to determine if a stomata is open or closed. In this case, have students count the total number of stomata and record that number.

In **Step 5**, encourage students to think about the leaf shape, size, and number of stomata and what can be inferred from these observations. Leaves with a small surface area and few stomata, common in succulent plants, are adaptations to prevent moisture loss. These plants are commonly found in arid environments. On the other hand, a plant with broad leaves and large numbers of stomata is more likely to be found in an area where precipitation occurs more frequently.

Results and Analysis

1. **Describe** *Sample answer: The stomal opening and the guard cells form the stomata. The stomata is part of the dermal tissue of the plant. The dermal tissue is part of the leaf structure. The leaf is one of the main parts of the plant.* **DOK 1**

MINILAB

INVESTIGATING LEAF STOMATA

SEP Conduct an Investigation What can the stomata of a plant tell you about its environmental conditions?

Plants have pores on their leaves called stomata (singular, *stoma*). In most plants, each stoma is surrounded by a pair of guard cells that are responsible for opening and closing the pore. Stomata play a critical role in gas exchange through the leaves.

In this activity, you will study the stomata from two different plants. You will count the number of open and closed stomata in each plant and then make inferences about the plants and the environment in which they live.

Leaf stomata respond to changes in the humidity of the surrounding air.

Materials

- compound microscope
- leaf impression slides (2)
- leaf samples (2)

Safety 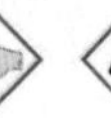   

Procedure

1. Place the first leaf impression slide under the microscope. View the sample using the 4×, 10×, and 40× objective lenses. The stomata are the circular structures seen throughout the leaf. They will appear to be more circular in shape if they are open.
2. Observe a single stoma using the 40× objective lens and sketch it in your biology notebook. Label the stomal opening and guard cells.
3. Switch back to the 10× objective lens. Count the number of open stomata you see in your field of view and record it in your biology notebook. Next, count and record the number of closed stomata you see in your field of view.
4. Repeat Steps 1 through 3 for the second leaf impression slide.
5. Study the leaf samples from both plants whose leaves you viewed under the microscope. Based on your observations of the stomata and your observations of the leaves, write a prediction in your biology notebook about the environment each plant lives in.
6. Return materials to their stations and clean your work area.

Results and Analysis

1. **Describe** What is the hierarchical relationship among the guard cells, stomata, and the leaf?
2. **Infer** Based on the stomata data you collected, which plant is likely to lose less water to the environment? What does this tell you about the environment in which the plant normally lives?
3. **Infer** What characteristics of the succulent leaf enable the plant to retain water and maintain homeostasis in a desert environment?
4. **Predict** If one of the plants were experiencing an extended dry period, how would you expect its stomata to react?

2. **Infer** *Sample answer: The spurge has fewer stomata, so it is likely to lose less water to the environment. The spurge may be native to a warmer climate compared to the apple.* **DOK 2**
3. **Infer** *Sample answer: I noticed that the leaf is thick, which may be to store additional water. The leaf also feels waxy, and this feature may help the plant retain water.* **DOK 2**
4. **Predict** *Sample answer: In an extended dry period, the plant would be likely to minimize its water loss. This means most of its stomata would be closed.* **DOK 3**

TRANSPORT IN PLANTS

Your heart is a muscular pump that pushes blood through the vessels of your circulatory system to deliver oxygen, nutrients, and specialized cells throughout your body. Similarly, a plant's vascular system transports the substances the plant needs to survive. Unlike your circulatory system, a plant has no "heart" to pump liquids.

Explain *What do you think drives the movement of water through a plant?*

9.2

KEY TERMS

feedback mechanism
transpiration

Water Transport in Plants

In vascular plants, water travels inside long, narrow xylem cells called *tracheids*. Lignin strengthens tracheid cell walls, lending structural support to stems and other organs that contain xylem. In addition to tracheids, flowering plants have xylem *vessels*, which are tubes made from shorter, wider cells stacked end to end. Vessel walls are reinforced by rings, spirals, or networks of lignin, as shown in **Figure 9-8**. Water flows lengthwise through tracheids and vessels. Vessels have openings in their walls that allow water to flow laterally from one vessel to another as well.

Like drinking straws, xylem tubes must conduct water upward from the roots of a plant to its leaves. However, tracheids and vessels are made of dead cells that cannot use energy to pump water against gravity. Instead, the upward movement of water in vascular plants occurs as a result of evaporation and cohesion.

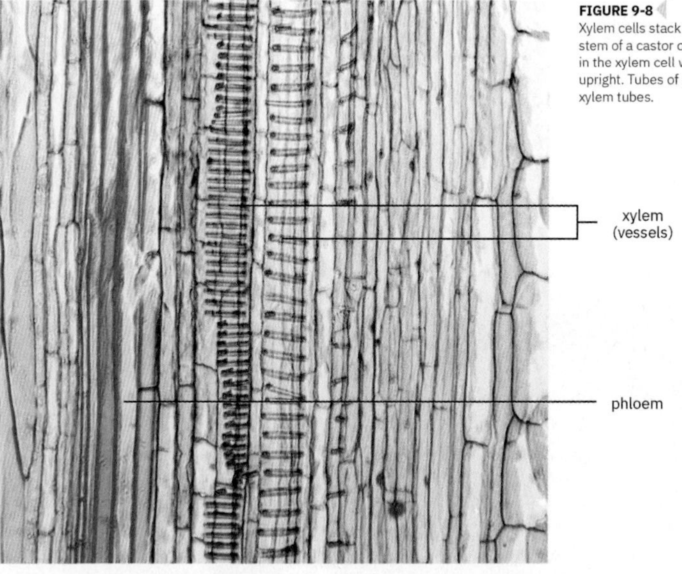

FIGURE 9-8 Xylem cells stack to form vessels in the stem of a castor oil plant. Rings of stiff lignin in the xylem cell walls help hold the stems upright. Tubes of phloem run alongside the xylem tubes.

SCIENCE AND ENGINEERING PRACTICES
Developing and Using Models

Model Water Movement Discuss with students what they know so far about the structure and function of vascular tissues in plants. Encourage use of the terms *xylem, tracheid cells, phloem*, and *lignin*. Refer to **Figure 9-8** as you discuss how the structure of xylem tubes functions in a way similar to drinking straws.

Tell students that they will be modeling the way water moves through a plant.

Provide students with 250 mL of water in a beaker or glass jar, food coloring, and several stalks of celery with about 1 cm freshly cut off the bottom. Have them predict what they will observe after placing the celery stalks in the water with food coloring added. After 24 hours, observe the celery and have students explain what they learned from the model.

9.2

TRANSPORT IN PLANTS

LS1.A

EXPLORE/EXPLAIN

This section reviews the structures and mechanics involved in transporting both water and nutrients in plants. In that context, the cohesion-tension theory and the pressure flow theory are introduced to explain how the mechanics of transport in a plant work.

Objectives

- Relate the structures involved in plant transport to their functions.
- Describe the process and mechanics of both fluid and nutrient transport.
- Construct an explanation of how plants control the conservation and release of water.

Pretest Use **Questions 3 and 4** to identify gaps in background knowledge or misconceptions.

Vocabulary Strategy

Using Prior Knowledge Feedback is information that triggers a response. Discuss how thermostats in different areas of the school work to keep the temperature constant. Organisms use feedback mechanisms to maintain balance within a system.

On the Map

Redwood National Forest Nestled in the Redwood National Forest in California is the world's tallest tree, known as Hyperion. It has been measured at 116.07 meters (380.81 feet), about the height of a 35-story building. Structures in the tree transport water and nutrients to support its growth. Hyperion's age is between 600 and 800 years. Have students locate Redwood National Forest in northern California and look for geographic features that might contribute to the growth of trees (*the Pacific coast has high levels of moisture and moderating temperatures).*

Explain *Sample answer: Water moves through a plant because of evaporation and cohesion.*

Crosscurricular Connections

Chemistry Transpiration is the evaporation of water inside the plant through the stomata. Evaporation is the change in water from liquid to gas that occurs at a lower temperature than water's boiling point. Water molecules in the liquid state have a wide range of kinetic energy, and those that have kinetic energy above a threshold amount can overcome intermolecular bonds and escape into the atmosphere in the gas state. The result of transpiration is a force that pulls water and its dissolved nutrients upward from the soil to all parts of the plant.

Visual Support

Water Adhesion and Cohesion
Students can see the movement of water in the visual; however, they often have difficulty in understanding the concepts related to the movement. These can be overcome by relating these concepts to familiar experiences. For example, cohesion can be demonstrated by using a dropper to place water on a coin until it forms a slight arc over the edge. Adhesion can be demonstrated using a piece of paper towel, which is made of cellulose and mimics the cell wall. Dipping the piece of paper in water demonstrates the movement of water as the molecules stick to the cellulose. This movement is called *capillary action* and is important in understanding how water moves from roots to leaves.

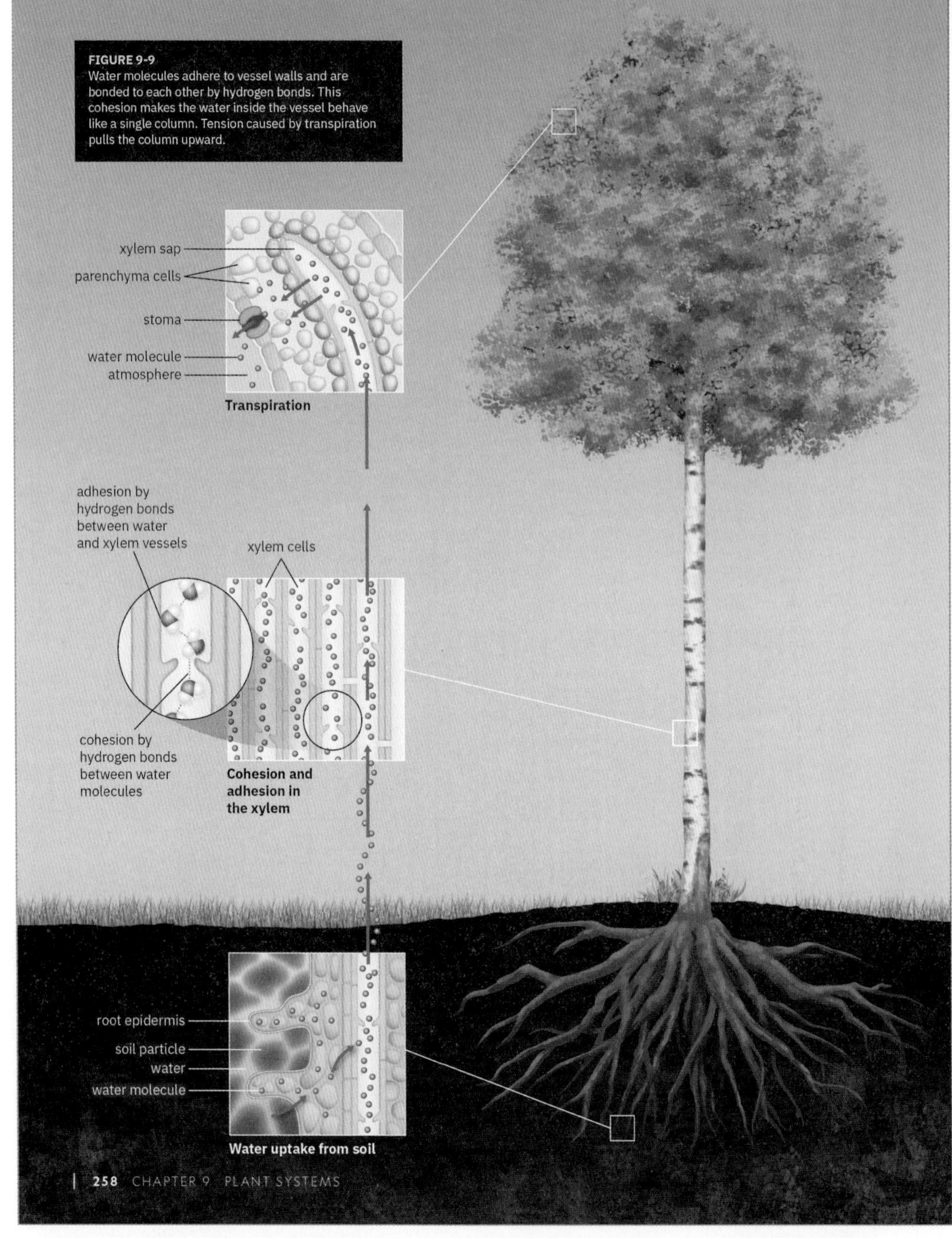

FIGURE 9-9 Water molecules adhere to vessel walls and are bonded to each other by hydrogen bonds. This cohesion makes the water inside the vessel behave like a single column. Tension caused by transpiration pulls the column upward.

DIFFERENTIATED INSTRUCTION | English Language Learners

Paired Writing Review the words *adhesion, cohesion,* and *transpiration*. Explain that the verb *adhere* means to stick to. Show a piece of *adhesive* tape and how it sticks to surfaces. The verb *cohere* means to be united or form a whole. A *cohesive* group works closely together. Finally, the prefix *–trans* means "across" and *–spirare* means "to breathe."

Beginning Working in pairs, have students complete the sentence frames: *When water molecules stick together, it is _______. When water molecules stick to another surface, it is _______. When water vapor moves across the leaf surface from the plant to the air, it is _______.*

Intermediate Working in pairs, have students use the information provided about each new word and write a definition using their own words.

Advanced Working in pairs, have students write a short paragraph using the words *adhesion, cohesion,* and *transpiration*.

Cohesion–Tension Theory The *cohesion–tension theory* explains that water in xylem is pulled upward as water molecules move out of the plant into the surrounding air. This creates a continuous negative pressure called *tension*. **Figure 9-9** illustrates this mechanism, which begins with water evaporating from leaves and stems. The evaporation of water from a plant's aboveground parts is called **transpiration**.

Transpiration's effect on the water inside a plant is a bit like what happens when you suck a liquid through a straw. Because the water molecules are attracted to each other by hydrogen bonds, a pull on one molecule tugs on the others to which it is attracted (cohesion). As a result, the force (tension) applied by transpiration pulls on entire columns of water that fill xylem tubes. The tension extends all the way from leaves that may be more than 20 meters in the air, down through the branches and stem, and down into the roots where water is being taken up from soil.

Water is pulled upward in continuous columns because xylem tubes are narrow, and water moving through a narrow channel (such as a straw or a xylem tube) resists breaking into droplets. This phenomenon is partly an effect of water's cohesion and partly because water molecules are attracted to hydrophilic materials, such as the cellulose in the walls of xylem tubes (Chapter 5).

SEP **Construct an Explanation** *What is the role of transpiration in the movement of fluids through xylem?*

Mechanisms for Water Conservation Plants open and close their stomata to regulate transpiration rates. A pair of guard cells borders each stoma. When these two specialized epidermal cells swell with water, they bend slightly so a gap (the stoma) forms between them. When the guard cells lose water, they collapse against one another, closing the stoma. **Figure 9-10** shows open and closed stomata on the surface of a leaf.

Open stomata allow water to leave the plant. Even under conditions of high humidity, the interior of a leaf or stem contains more water than air. Thus, water vapor diffuses out of a stoma whenever it is open. As much as 99 percent of the water a plant takes up from the soil can be lost by transpiration from open stomata.

Go Online INTERACTIVE FIGURE

FIGURE 9-9 Go online to trace the **Movement of Water and Nutrients Through a Plant**.

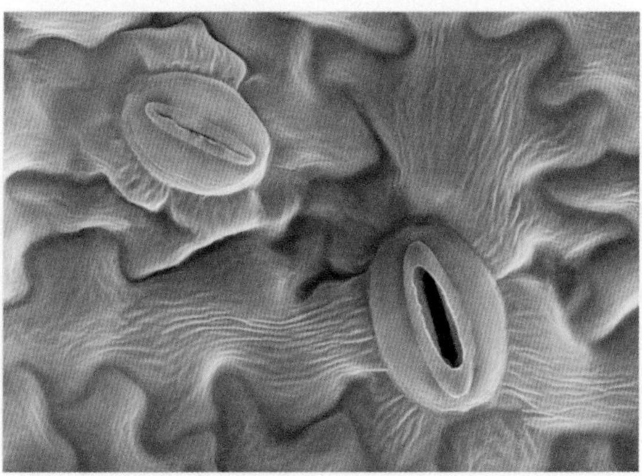

FIGURE 9-10 The guard cells on this lavender leaf are about 14 micrometers long. The stoma on the left is closed, while the one on the right is open.

Go Online **INTERACTIVE FIGURE**

Movement of Water and Nutrients Through a Plant

Students can place either a water or a sugar molecule at various (ordered) locations to follow the path of the molecule through a plant.

Science Background

Adaptations that Slow the Rate of Transpiration Plants that grow in dry environments have several kinds of leaf adaptations that slow the rate of transpiration. These include reduced surface area, fewer stomata, only opening stomata at night, small hairs, and stomata sunk in depressions. Hairs and sunken stomata slow the rate of transpiration because they increase the thickness of the layer of still air that hugs the surfaces of the leaves. The layer of still air is more saturated with water vapor than air farther away from the leaves because of transpiration.

SEP Construct an Explanation
Sample answer: Transpiration, which is the evaporation of water at the leaf, makes a vacuum that pulls water up through the plant.

CHAPTER INVESTIGATION A

Connecting Plant Structures with Their Functions

Guided Inquiry: *How is plant structure connected to function?*

Time: 100 minutes or two days

Students will dissect a flowering plant and prepare a labeled diagram of the different parts.

Go online to access detailed teacher notes, answers, rubrics, and lab worksheets.

CCC Cause and Effect *Sample answer: A plant species population might evolve over time to have smaller or fewer stomata on their leaf surfaces to prevent the loss of moisture in response to an environment's sustained increase in air temperature.*

CASE STUDY
FLOWER POWER

Ask Questions To help students make connections with the content, share with them the "About the Photo" information from the chapter opener. Point out that the Pando is in decline mainly because regeneration cannot keep up with grazing pressure from deer, elk, and livestock. Discuss this topic in the context of clones, pointing out that clones are genetically identical to the parent stock. Although the aspens can reproduce by seed, they usually reproduce asexually. Factors such as disease, blight, climate change, and wildfire suppression also negatively impacts the grove.

Photosynthesis Students summarize the process of photosynthesis in **Chapter 6** to learn about sugar production in plants. Sudents who need more detail can be directed to **Appendix D**.

CHAPTER INVESTIGATION

Connecting Plant Structures With Their Functions

How is plant structure connected to function?

Go online to explore this chapter's hands-on investigation about plant anatomy.

Carbon dioxide also enters a plant through the stomata. A lot more water is lost through open stomata than molecules of carbon dioxide are gained. During the day, each open stoma loses about 400 molecules of water for every molecule of carbon dioxide it takes in. When a plant's stomata are closed, it cannot exchange enough carbon dioxide and oxygen with the surrounding air to support critical metabolic processes.

Recall that homeostasis refers to the maintenance of stable internal conditions (Section 6.2). **Feedback mechanisms** allow a plant to maintain homeostasis by balancing the loss of water to the atmosphere with its need to bring in carbon dioxide for photosynthesis. Stomata open or close based on cues that include humidity, light intensity, the level of carbon dioxide inside the leaf, and hormonal signals from other parts of the plant. For example, the roots of some plants release a hormone when soil moisture is low. This stimulus triggers the guard cells to release water in response. As a result, the guard cells collapse and the stomata close.

CCC Cause and Effect *Over evolutionary time, the size of a plant species' stomata may change, or the plant species may develop fewer or more stomata in a given area, in response to environmental changes. How might a sustained increase in air temperature affect the size or number of stomata on a leaf?*

CASE STUDY
FLOWER POWER

Ask Questions *Generate questions about the roles of a plant's water and nutrient transport mechanisms in phytoremediation.*

Nutrient Transport in Plants

Phloem is the vascular tissue that conducts sugars and other organic solutes through the plant. Conducting tubes of phloem are called *sieve tubes*. Each sieve tube is made up of a stack of living cells called *sieve elements*. Unlike in xylem tubes, fluid flows only lengthwise through sieve tubes.

Pressure Flow Theory A plant makes sugar molecules during photosynthesis. Some of these molecules are used by or stored in the cells that make them. The rest of the sugar molecules are moved through sieve tubes to other parts of the plant.

Inside sieve tubes, fluid rich in sugars—mainly sucrose—flows from a source to a sink. A source is a region of plant tissue where sugars are being made or released from storage. Photosynthetic tissues are typical source regions. A sink is a region where sugars are being broken down or put into storage for later use. Tissues of developing roots and fruits are typical sink regions. The *pressure flow theory* explains how fluid pressure pushes sugar-rich fluid inside a sieve tube (**Figure 9-11**).

A pressure gradient between the two regions drives the movement of sugar-rich fluid from a source to a sink. For example, consider mesophyll cells in a leaf, which make far more sugar than they use for their own metabolism. Photosynthesis in mesophyll cells sets up the leaf as a source region. Excess sugar molecules made by these cells diffuse into adjacent companion cells through the cell wall.

Once inside a companion cell, sugar molecules diffuse through the cell wall into associated sieve elements. The sugar increases the solute concentration of the sieve tube's cytoplasm, so it becomes hypertonic with respect to the surrounding cells. Water follows the sugar by osmosis, moving into the sieve tube from surrounding cells.

DIFFERENTIATED INSTRUCTION | Leveled Support

As students learn about the movement of nutrients through the phloem, it is important that they understand the concepts of diffusion and osmosis. Ask students what they recall about diffusion and osmosis from their study of transport across cell membranes.

Struggling Students Explain to students that diffusion is the movement of molecules from an area of higher concentration to an area of lower concentration. You might offer the analogy of people in a very crowded area randomly moving to an area where it is less crowded. Osmosis is the diffusion of water from regions of high concentration to regions of low concentration.

Advanced Learners Review the terms *diffusion* and *osmosis* with students. Have them provide a description of what each term means. Remind students that diffusion and osmosis are forms of passive transport and require no chemical energy. Discuss with them how materials in plants move along a pressure gradient.

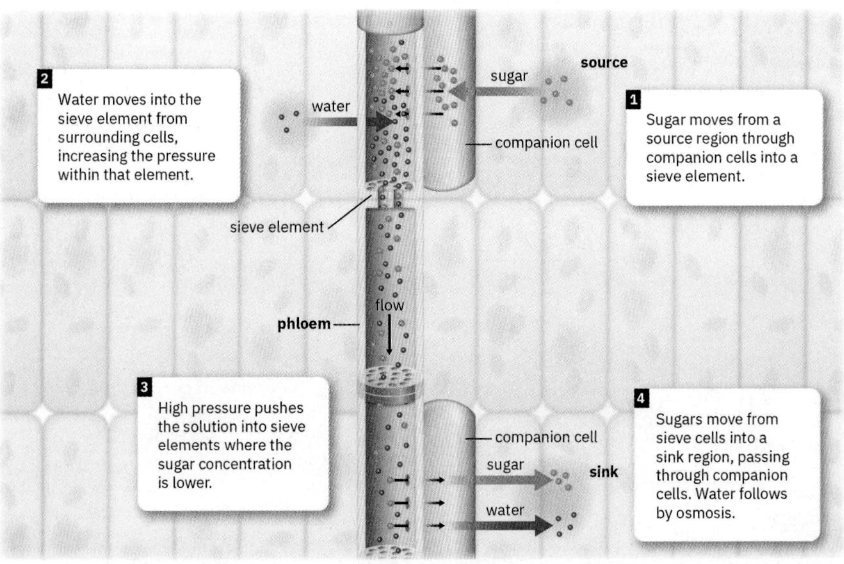

source
sugar
water
companion cell
2
Water moves into the sieve element from surrounding cells, increasing the pressure within that element.
1
Sugar moves from a source region through companion cells into a sieve element.
sieve element
flow
phloem
3
High pressure pushes the solution into sieve elements where the sugar concentration is lower.
4
Sugars move from sieve cells into a sink region, passing through companion cells. Water follows by osmosis.
companion cell
sugar
sink
water

FIGURE 9-11 The movement of water by osmosis into regions of high concentrations of sugar drives the pressure changes in phloem.

The rigid cell walls of mature sieve elements cannot expand very much, so the inflow of water raises the pressure in the tube. Pressure inside a sieve tube can be very high—up to five times higher than that of an automobile tire. High fluid pressure inside the tube pushes sugar-rich cytoplasm from one sieve element to the next, toward a sink region where the fluid pressure is lower. Pressure inside a sieve tube decreases at a sink because sugars leave the tube in this region. Sugar molecules diffuse into companion cells and then move into adjacent sink cells. Water follows, again by osmosis, so the fluid pressure inside the sieve elements decreases in sink regions.

CCC **Structure and Function** *Explain how sieve element structure helps in transporting the products of photosynthesis.*

9.2 REVIEW

1. **Explain** Which condition forms the basis of the pressure flow theory?
   A. Water free of nutrients diffuses easily by osmosis.
   B. The tissues of roots and fruits serve as storage for nutrients.
   C. The cell walls of sieve tubes are flexible and expand with the inflow of fluid.
   D. Nutrient-rich fluid in photosynthetic tissues has higher pressure than fluid elsewhere.
2. **Relate** Which plant structures are necessary for transpiration? Select all correct answers.
   A. flowers C. leaves E. phloem
   B. stems D. xylem
3. **Relate** Explain the vital role stomata play in regulating plant metabolic processes such as photosynthesis.
4. **Contrast** Describe the similarities and differences between xylem and phloem.

9.2 TRANSPORT IN PLANTS 261

In Your Community

Tapping Maple Syrup When the leaves of sugar maple trees make food through photosynthesis, the sugars are stored in the form of starch in the tree's roots. In the spring, the starch is converted back into sugar that moves upward through the vascular tissue to provide the new buds with energy to grow. As the sugars moves through the vascular system, the trees can be tapped to collect the sap. Maple trees yield an average of 70 liters (18 gallons) of sap.

Science Background

Cells that Stack Together Sieve tube cells are specialized plant cells that include a sieve plate and are lined end on end to create a tube-like structure for the transport of nutrients through the phloem. The sieve plate is a set of minute pores found on the transverse ends of the elongated phloem cell. The structure of the elongated cells, in addition to the presence of sieve plates, allows for the phloem cells to stack and transport nutrients between cells without the need for movement through the cell wall. Sieve cells are an example of the specialization of cells and tissues in angiosperms that enabled these plants to adapt to a terrestrial habitat.

CCC **Structure and Function** *A sieve element's rigid walls do not allow it to expand very much, so when water enters them, the pressure increases. This fluid pressure transports sugar molecules (the products of photosynthesis) to sink regions where the fluid pressure is lower.*

9.2 REVIEW

1. *D* **DOK 2**
2. *B, D, E* **DOK 2**
3. *Sample answer: Photosynthesis requires the input of carbon dioxide from the atmosphere, which enters the leaf through the open stomata. However, when the stomata are open, water is lost from the plant. Opening and closing stomata allows a plant to maintain homeostasis and balance its need for water with its need to exchange gases.* **DOK 3**
4. *Sample answer: Xylem and phloem are tube-like internal structures that transport substances to different parts of a plant. Xylem are made of dead cells that move water entering a plant's roots upward. Water can flow lengthwise and laterally across xylem tubes. Phloem are made of living cells that move nutrients, sugars, and organic molecules to other plant parts. Nutrients can only flow lengthwise across phloem tubes.*

ON ASSIGNMENT

Night Sky Photographer Babak Tafreshi is a National Geographic photojournalist who specializes in night sky photography. His images and time-lapse videos juxtapose the night sky with iconic landmarks. In 2007, Babak founded The World at Night (TWAN) nonprofit program and worked as its director. The program includes an elite group of about 40 photographers in 25 countries who use their work to reconnect people with the beauty of the night sky in different places on Earth. In 2009, Babak received the Lennart Nilsson Award, the world's most recognized award for scientific photography, for his global contributions.

In Your Community

Light Pollution The use of artificial light has created a problem called *light pollution*. Light pollution affects the health of people, animals, and plants, including trees. Trees evolved to respond to the transition of day to night, and they have mechanisms to measure the amount and length of time of the light they receive. Trees use this information to guide their seasonal changes. Where there is light pollution, trees receive information about the length of the day that is incorrect and they may not thrive as a result. Communities have been taking steps to reduce light pollution by installing artificial lights on roads and in public areas that are less harmful because they use certain kinds of bulbs, have fixtures that direct all the light downward, and are automatically dimmed during off-peak hours. At home, people can reduce their contribution to light pollution by turning off lights when they are not in use.

ON ASSIGNMENT National Geographic photographer Babak Tafreshi captured this image of giant sequoia trees in Yosemite National Park. They are among the largest and tallest trees on Earth. An opening in the canopy between these 80-meter tall trees gives a glimpse of the Milky Way galaxy in a clear sky.

262

PLANT REPRODUCTION

9.3

KEY TERMS

carpel
pollination
stamen

After a walk through a grassy field, you notice both you and your dog are covered in tiny pointed objects, each with a few thin, sharp spikes. These objects, called spikelets, contain grass seeds and can easily poke into the body of a moving host. With your help, some of the spikelets may end up in a favorable location for the seeds to grow, ensuring a future for the next generation of plants.

Infer *A single grass plant releases hundreds of spikelets per year. Why do you think a plant would need to produce such an abundance of seeds?*

Plant Life Cycles

Plants have a clearly defined life cycle in which they alternate between producing spores and producing gametes (**Figure 9-12**). In the *gametophyte generation*, the plant produces gametes, or sex cells, by mitosis. At the end of this generation, two gametes fuse to form a zygote that contains genetic information from both of its parent cells.

During the next stage, called the *sporophyte generation*, the zygote undergoes mitosis to develop into a multicellular plant. To complete this stage, the plant's cells undergo meiosis (Section 12.1) to make *spores*, which are single reproductive cells that have half the genetic information of the parent plant. Spores are capable of developing into a new individual without combining with another reproductive cell. A spore undergoes mitosis to become a gametophyte plant, and the cycle repeats.

Plant cells belonging to each generation exist within specific plant body structures. For example, in a giant sequoia tree, the sporophyte consists of the parts we can see—roots, trunk, branches, leaves, and cones—which all grew from a zygote. The gametophyte takes the form of cells found within pollen grains and in tiny structures deep inside the cones. In all plants, one generation is dominant over the other, and the nondominant generation may rely on the dominant generation for survival.

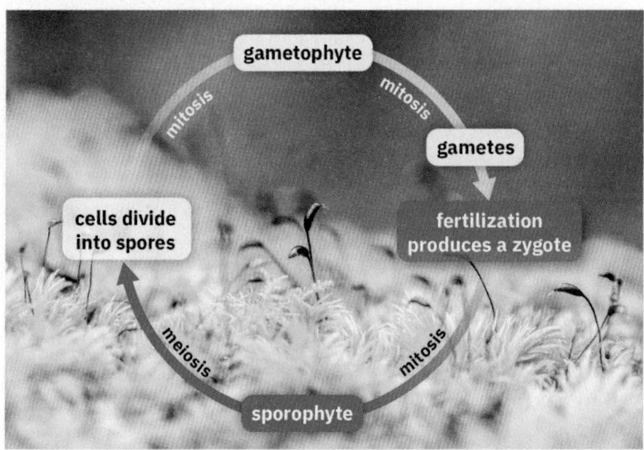

FIGURE 9-12 A gametophyte produces zygotes, and a sporophyte produces spores. Both generations can coexist as parts of the same plant.

DIFFERENTIATED INSTRUCTION | Leveled Support

Struggling Students Write the words *gamete* and *spore* on the board. Tell students to make a two-column chart with one column heading "Gamete" and the other column heading "Spore." As you review the terms, have students list the distinguishing features in their chart, for example, "Spores grow by mitosis into a new individual—a gametophyte" and "A gamete is a cell that must fuse with another gamete to form a new individual."

Advanced Learners Students who demonstrate a clear understanding of the terms *gamete* and *spore* may be asked to synthesize this information in a short paragraph explaining the difference between *gametophyte generation* (*the plant produces gamete by mitosis*) and *sporophyte generation* (*spores grow into new individuals by mitosis*).

9.3

PLANT REPRODUCTION

LS1.A, LS1.B, LS2.C

ENGAGE

This section develops one of the key concepts related to the movement of plants from water to land, which is how plants reproduce when out of a water environment. Students will develop an understanding of plant life cycles and distinguish between the gametophyte generation and the sporophyte generation in the context of sexual and asexual reproduction.

Objectives

- Sequence the steps in the life cycle of a gymnosperm and an angiosperm.
- Diagram the relationship between the structure of flowering plant parts and their function in reproduction.
- Compare the different methods of asexual reproduction.

Pretest Use **Questions 6 and 7** to identify gaps in background knowledge or misconceptions.

Vocabulary Strategy

Prefixes/Suffixes/Root Relationships Review with students that the word *alternation* is the noun form of the verb *alternate.* Ask students if they have used the word *alternate,* meaning to "occur or exist one after the other repeatedly." The word *alternation* means the repeated occurrence of two things in turn. Have students write their own definition of *alternation of generations*.

Visual Support

Alternation of Generations Review **Figure 9-12** with students and give them the opportunity to ask questions, such as what the different colored arrows mean. Have students work in pairs to write their own description of the sequence of events in a plant life cycle, beginning with the gametophyte.

Infer *Students might say that producing a large amount of seeds ensures that at least a few may have a chance to grow into mature plants.*

In Your Community

Ferns Ferns are fairly prolific in some communities, particularly in damp wooded areas. Discuss with students how ferns differ from plants, such as moss. Use diagrams and photos from Internet research to display the parts of a fern—the fronds, rhizomes (horizontal stems), and roots—as well as sori. The U.S. Forest Service (www.fs.fed.us/wildflowers/beauty/ferns/fern structure) explores the structures of a variety of ferns from across the country. A simple image search will also yield a variety of the approximately 380 fern species in North America.

Have students write a description of a frond and make diagrams of the sori on the underside. Encourage them to identify features common to all ferns.

Connect to English Language Arts

Draw Evidence From Texts It is important for students to learn to recognize relationships between related ideas as they read. Have them write a short paragraph that analyzes the similarities and differences in reproduction between nonvascular and vascular plants, using mosses and ferns as examples. Students should include which generation, sporophyte or gametophyte, is dominant over the other in each situation, as well as comparisons of structures that affect their reproductive processes.

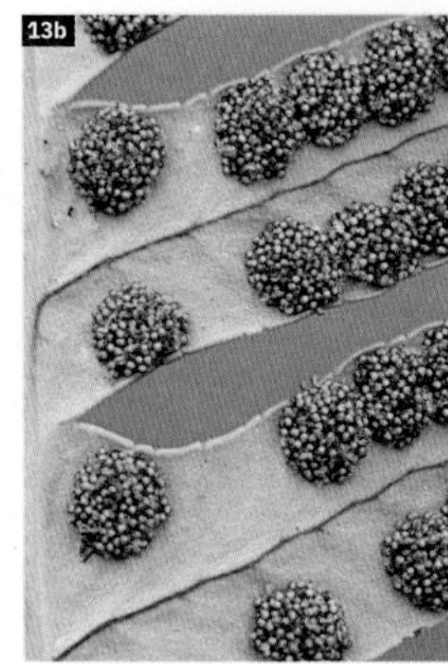

FIGURE 9-13 Sporophytes take a variety of forms. The thin stalks of moss sporophytes (a) aid in spore dispersal. Fern sporangia (b) cluster on fronds of the dominant sporophyte. Pine tree cones (c) produce spores.

Sexual Reproduction

Plant sexual reproduction requires fertilization of a female gamete (egg) by a male gamete (sperm). The offspring plant inherits genes from both parent cells. An individual plant may produce gametes of only one type so that another plant is needed for sexual reproduction to occur. A single plant may also produce both male and female gametes. In some species, this type of plant can fertilize itself, but in other species, male and female gametes of the same plant are incompatible with each other.

Nonvascular Plants Nonvascular plants are the only modern plants in which the gametophyte generation is dominant. However, even the gametophytes tend to be small and low growing due to the absence of a vascular system. All nonvascular plants disperse by releasing spores. For example, in **Figure 9-13a**, moss sporophyte capsules rise above the green leaf-like gametophyte generation. The sporophyte's tall, thin shape positions the spores to be easily carried away by wind. After dispersal by the wind, a spore germinates and grows into a filament of cells from which one or more gametophytes develop. Closer to the ground, rain stimulates moss gametophytes to release flagellated sperm. In order to reach and fertilize the plant's eggs, the sperm require a film of water in which to swim. Mosses also reproduce asexually by fragmentation when a bit of gametophyte breaks off and develops into a new plant.

Seedless Vascular Plants The key difference between ferns and other vascular plants is their production of spores rather than seeds. Spore production usually occurs on certain areas of the fronds (leaves), which develop sporangia in clusters called *sori* (**Figure 9-13b**). Inside the sporangia, meiosis occurs to form spores. The spores disperse and grow into tiny, heart-shaped gametophytes. Though ferns are vascular plants, fertilization in ferns is much like that of mosses. Sperm from a male gametophyte must be carried by water to the female plant. The sporophyte generation is dominant in ferns. The sporophyte is larger than the gametophyte and it has roots and large fronds that persist for an extended time. In contrast, the gametophyte dies soon after reproducing.

SCIENCE AND ENGINEERING PRACTICES
Developing and Using Models

Fern Life Cycle Model Working in small groups, provide students with a mature fern plant along with other materials, such as white card stock, glue, and markers. Try to provide plants that have sori available on the underside of the fronds. If not available, you may have students use alternate materials to represent the sori, such as small peppercorns.

Tell students that they will create a model of the life cycle of a fern using multiple mediums. Wherever possible, they should use parts from the fern plant; otherwise, they may draw any parts or stages that are not available. They should make sure all plant parts and processes are clearly labeled. When they have completed their model, have them create a flowchart that describes the steps in the life cycle of a fern, from new sporophyte to mature sporophyte, on a separate paper.

Gymnosperms Gymnosperms feature two evolutionary advances for plants. First, the sporophyte, rather than the gametophyte generation, is dominant in gymnosperms. Second, gymnosperms carry out fertilization without being surrounded by water. They are the first plants to be completely adapted to a terrestrial lifestyle.

VIDEO

VIDEO 9-1 Go online to see how a cedar tree disperses pollen.

Pine trees have a life cycle typical of gymnosperms. Like most gymnosperms they are wind-pollinated. The pine tree is a sporophyte that produces spores on specialized structures called cones (**Figure 9-13c**). Conifers produce both male and female cones. In most pine species, each tree makes one type of cone: either small, soft pollen (male) cones or large, woody seed (female) cones. Both types of cones have scales arranged around a central axis. Each scale of a pollen cone contains pollen sacs in which pollen grains develop. Pollen cones release these tiny grains to drift with the winds. Fertilization takes place within the female cone. After fertilization, the zygote grows into an embryo within the seed. The surrounding cone provides nutrition for the developing embryo. The seed is released, germinates, then develops into a new sporophyte.

CCC Structure and Function *What are the primary differences between moss and fern reproduction and reproduction in other types of plants?*

CONNECTIONS

SAVING SEEDS In the past, food crops were location-specific. People developed and grew plant varieties that did well in their region, and they saved seeds from their crops to plant the following year. Now, many traditional varieties of crop plants are disappearing as a few large companies supply farmers with the same strains of seeds worldwide. Different varieties of plants are resistant to different diseases, so widespread planting of one variety increases the risk that a disease could decimate the global supply of a crop. The more varieties of a crop we plant, the more likely it is that some will resist a particular disease.

Sustaining the wild relatives of crop plants provides a form of insurance because it maintains a reservoir of genetic diversity. Plant breeders can draw upon that diversity to meet future challenges. Storing seeds in seed banks is one way to ensure the survival of potentially useful plants. Today, there are more than 1700 seed banks around the world. The most ambitious, the Svalbard Global Seed Vault, was built in 2008 on a Norwegian island about 1100 kilometers from the North Pole. This so-called doomsday vault serves as a backup for other seed banks. The seeds are stored deep inside a mountain, in a permanently chilled, earthquake-free zone 130 meters above sea level. The location was chosen to keep the seeds high and dry even if global climate change causes the polar ice caps to melt. The vault also has an advanced security system and has been engineered to withstand any nearby explosions.

The Svalbard vault now holds the world's most diverse collection of seeds, with more than one million samples and material from nearly every nation. Under the deep-freeze conditions in Svalbard, the seeds are expected to survive for about 100 years.

The Svalbard Global Seed Vault is capable of storing 4.5 million seed samples.

Video

Time: 0:24 seconds

Use **Video 9-1** to discuss how pollen can be dispersed in the wind. Discuss other methods of pollen and seed dispersal, such as animal scat containing seeds from fruit the animal has eaten and pollen and seeds clinging to fur, or on the legs of flying insects. Discuss why these modes of dispersal are important for plant reproduction. (*plants cannot move*).

Connect to English Language Arts

Integration of Knowledge and Ideas Working in small groups, have students translate the text information about the life cycle of conifers into a diagram that shows the alteration of generations beginning with the mature plant. Guide them to read the paragraph closely and take notes on the key details that they think should be included in their diagram. Remind them to label each part of the diagram and use arrows to show the correct order of events.

CCC Structure and Function *Mosses and ferns require water to reproduce.*

CONNECTIONS | Saving Seeds

Social-Emotional Learning Some students may have strong feelings about issues such as population growth, hunger, and climate change. This article provides an opportunity for students to think about solutions for such problems.

Students can demonstrate *responsible decision-making* by approaching the article through the lens of curiosity and open-mindedness about learning the risk involved with not maintaining plant diversity.

Discuss an example of the consequences of limited diversity. One that may be part of some students' family history is the Irish potato famine. Farmers consistently planted a single variety, the Irish Lumper potato, and eliminated genetic diversity from the local area. When the potato crops were infected with a fungus, they all died because no other varieties existed that may have had some resistance to the disease. The resulting famine caused starvation and death.

Address Misconceptions

Trees and Flowers It is a common misconception that trees do not produce flowers. Some species of trees, such as gymnosperms, do not have flowers but trees that are angiosperms do. Examples of angiosperm trees include magnolias and tulip trees that have very large and showy flowers. In other angiosperm trees, the flowers are relatively small and pollen is carried by the wind.

Vocabulary Strategy

Word Wall Students often get confused and have trouble remembering the difference between the anther and filament or the stigma and style. Start an interactive word wall just for terminology related to flower parts. List each word in **Figure 9-14** and allow students to add definitions, images, or phrases that will help them remember the words. Make it possible for students to also indicate the relationships between the structures and how they work together. Encourage students less proficient in English to post notes in their primary language.

Flowering Plant Reproduction

Flowering plants, with more than 300,000 living species, are the world's most successful plants, having adapted to almost every habitat on Earth. The success of the flowering plants can be attributed to their flowers and fruits. A bright, colorful flower attracts animals such as birds, bats, bees, and other insects that obtain nectar and, in the process, deliver pollen to other flowers. Different animals eat the fruit that develops from a flower, dispersing the seeds far and wide. Both pathways represent major improvements in ensuring genetic diversity and reproductive efficiency for flowering plants over gymnosperms and other groups that rely on more passive forms of reproductive cell distribution.

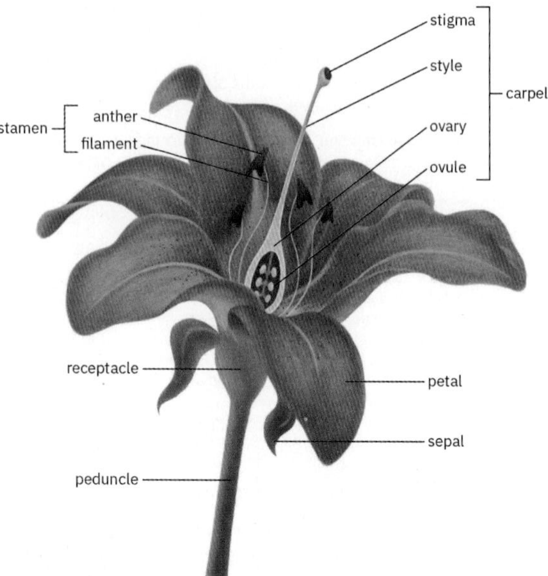

FIGURE 9-14 A flowering plant's reproductive system is part of its shoot system and contains groups of reproductive organs that produce gametes, promote their fertilization, and feed and protect the zygote.

Reproductive Parts Like all vascular plants, the sporophyte is the dominant phase in flowering plants. Gametophytes are produced within the flowers of the sporophyte. A *flower* is a specialized reproductive shoot. Floral structure varies, but most flowers include the parts shown in **Figure 9-14**. *Sepals*, which usually have a green leaf-like appearance, ring the base of a flower and enclose it until it opens. Inside the sepals are *petals*, which are often colored and scented. The petals surround the stamens. **Stamens** are organs that produce pollen. A typical stamen consists of a tall stalk (the filament), topped by an anther that holds four pollen sacs. The innermost part of the flower is the **carpel**, the organ that captures pollen and produces eggs. The carpel consists of a stigma, style, and ovary. The sticky or hairy *stigma*, which is specialized for receiving pollen, tops a stalk called the *style*. At the base of the style is an *ovary*, a chamber that contains one or more ovules. After fertilization, an *ovule* matures into a seed and the ovary becomes a fruit.

DIFFERENTIATED INSTRUCTION | Students with Disabilities

Visual Impairments This activity allows students to tactilely study plant parts. Prior to the activity, discuss the plant parts and describe them in detail, including size, shape, color, and how structures are connected on the flower.

Provide students with different kinds of flowers, avoiding those with thorns. For low vision students, white provides a better contrast so flowers such as a white lily work well. Other options include tulips, daffodils, or gladiolus.

Encourage students with visual impairments to describe what they observe using other available senses; e.g., fragrance, texture, size, or shape. Have them observe the whole flower structurally. Beginning with the peduncle guide students through a "tour" of the flower. Explain that in a flowering plant, what they might call the stem is actually the peduncle as it is the stalk to which the flower is attached. Use **Figure 9-14** as a guide to walk students through the remaining parts of the flower, allowing them to experience each structure.

Pollination The fertilization process in flowering plants, called double fertilization, is unique in that two sperm are involved. One sperm fertilizes the egg, forming a zygote that develops into the embryo in the seed. The other sperm fuses with two cells in the female gametophyte to form a special nutritive tissue in the seed called endosperm.

Pollination is the act of transferring pollen grains from the male anther of a flower to the female stigma. Some flowering plants are self-pollinators. That is, the pollen produced in one anther of a flower pollinates the stigma of the same flower. Other flowering plants are cross-pollinators. In this process, the pollen from a flower in one plant is used to pollinate a flower of a different plant of the same species.

Pollen is transferred from one flower in a variety of ways, including dispersal by wind, water, or animals. Grasses and other plants pollinated by wind do not benefit by attracting animals, so they do not expend resources making scented, brilliantly colored, or nectar-laden flowers. Wind-pollinated flowers are typically small, nonfragrant, and green, with insignificant petals and large stamens that produce a lot of pollen.

Passive pollination by animals, as shown in **Figure 9-15**, occurs as a pollinator moves from one flower to another to obtain the nectar that is typically stored in the female part of the plant. If it comes into contact with the flower's anther, the pollinator's body picks up pollen that may be transferred to the stigma of the same flower or another flower. The flowers of a given species bloom only during specific and limited times of the year. This limits the pollinator to seeking nectar from one species at a time and helps to ensure that pollen picked up from one flower will be brought to a flower of the same species.

CCC **Structure and Function** *How do flower structures facilitate animal-driven pollination?*

FIGURE 9-15 
Crocus flowers bloom in the spring and are visited by bees that carry their pollen from flower to flower.

VIDEO

VIDEO 9-2 Go online to learn how bees' buzzing helps some plants reproduce.

CROSSCUTTING CONCEPTS | Structure and Function

Pollinators There are a variety of species with unique structures that help them function as pollinators. Among these are butterflies, moths, bats, and, of course, birds and bees. Provide images that show some of these pollinators and their unique structures up close.

Butterflies and moths do not have mouths. Instead, they have *proboscises*—long tube-like mouthparts. In butterflies and moths, the proboscis is like a long straw through which they drink nectar. Butterflies feed on flowers by day, while moths are more active at night and rely on scent instead of color to find sources of food.

Some bats use their sense of vision, hearing, and smell to find nectar-bearing flowers. With long noses and tongues, they can drink the nectar. The pollen sticks to the bat's fur and is transported to the next flower the bat visits.

Have students investigate other interesting pollinators and research the unique structures that allow them to function in this role.

Connect to Careers

Beekeeper Beekeepers work in the agricultural industry. They maintain beehives, harvest the honey, and bring their products to market. In addition to harvesting honey, beekeepers also use the wax produced by the hive and even royal jelly that is used as a supplement. Some beekeepers work on a local community level, while others work for larger commercial farms. Local beekeepers may also rent out their beehives to orchards that use the bees to pollinate their fruit trees.

It is the responsibility of the beekeeper to help monitor the health of the bee species. This is particularly true as bee populations suffer from parasitic diseases, colony collapse disorder (CCD), and other challenges that threaten the species. Beekeepers maintain the health of their hives while also making a commitment to bee welfare, biodiversity practices, and conservation efforts. Some beekeepers help educate the public about bees by visiting local schools or printing labels and materials that teach people about these important pollinators.

Video

Time: 1:10 seconds

Video 9-2 explains how bumble bees use the buzzing sound to get pollen. This is called *sonication*, or buzz pollination. This is an adaptation that bumble bees have to get pollen that other bees cannot. Ask students why it can be beneficial for organisms to have adaptations, or the ability to do something, such as gathering food, in a different way than other organisms do.

CCC **Structure and Function** *Sample answer: Brightly colored petals help to attract pollinators like insects and birds. Pollen grains stick to the legs of insects and other pollinating animals.*

Science Background

Kalanchoë daigremontiana

Researchers at the University of California Davis studied why this species of kalanchoë cannot produce viable seeds and reproduces only through the tiny plantlets growing on the edge of its leaves. They looked at two genes in this plant and in close relatives, some of which make viable seeds. One gene was involved in shoot growth while the other gene was involved in making seeds. In *Kalanchoë daigremontiana*, a variant of the gene involved in making seeds was expressed in leaves as well. When that variant was introduced in other plants, those plants could no longer make viable seeds. So, although *Kalanchoë daigremontiana* has lost the ability to reproduce sexually via seeds, some of the embryo-making process has been maintained in the leaves. The researchers think the findings could be useful in manipulating plant reproduction. Elicit from students how the number of plantlets on the plant's leaves could make it as successful as reproduction with seeds.

Asexual Reproduction

Many flowering plants can reproduce asexually. Each new plant is a clone, or genetic replica, of its parent. Entire forests of quaking aspen trees (*Populus tremuloides*) are actually stands of clones that grew from root suckers—shoots that sprout from shallow lateral roots. The quaking aspen grove in Utah that was described at the beginning of this chapter consists of about 47,000 genetically identical shoots. The plant shown in **Figure 9-16**, *Kalanchoë daigremontiana*, is commonly known as "mother of thousands." It reproduces asexually by growing tiny plantlets on its leaves.

FIGURE 9-16 Each tiny plant along the leaf edges of the *Kalanchoë daigremontiana* plant is a clone of the parent plant.

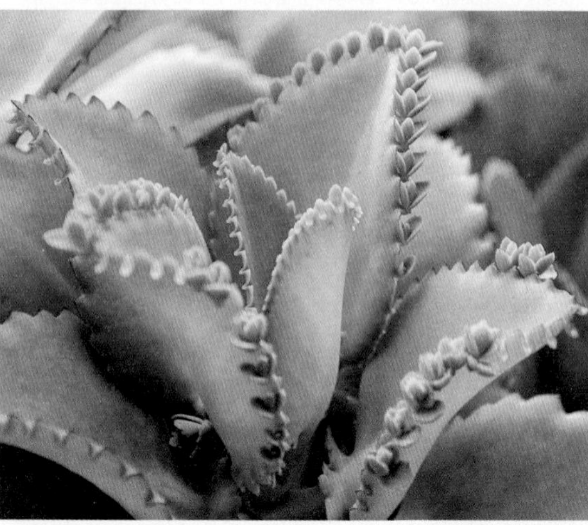

Vegetative Reproduction Vegetative reproduction is a form of asexual reproduction in which a new plant grows from a stem, leaf, or root of a parent plant. In many plant groups, a new plant can regenerate from just a fragment of tissue from its parent. Some plants form specialized structures for vegetative reproduction.

- *Stolons* are stems that branch from the main stem of the plant and grow horizontally on the ground or just under it. They are commonly called runners because in many plants they "run" along the surface of the soil. Stolons have nodes from which roots and shoots can sprout and develop into new plants. Strawberries reproduce using stolons (**Figure 9-17a**).
- *Rhizomes* are fleshy stems that typically grow underground in a horizontal direction. In many plants, the main stem is a rhizome, and it often serves as the primary region for storing food. Shoots that sprout from nodes grow aboveground for photosynthesis and flowering. Irises are examples of rhizomes (**Figure 9-17b**).
- *Tubers* are thick, fleshy storage structures made from the stem of a plant. Tiny buds that grow on a tuber's surface potentially develop into a new plant. Potatoes are examples of tubers (**Figure 9-17c**).

DIFFERENTIATED INSTRUCTION | Leveled Support

Specialized Plant Structures Allow time for students to observe each plant in the images provided in **Figure 9-17** and make notes on the different specialized structures they see. Ask students to infer from the structure the role it plays in plant reproduction and to discuss their ideas with a partner.

Struggling Students Have students create a chart with a column for each type of plant they observe. At the top of each column, they should identify the specialized structure. In the rows beneath, they should include a one-sentence definition or description of the structure along with a drawing made from their observations. Encourage them to record any questions they may have and provide an opportunity for discussion.

Advanced Learners Have students create a three-column chart that includes the name of each specialized structure, a drawing made from their observations, and a short paragraph describing the unique characteristics of each structure.

- *Bulbs* consist of a short, flattened stem covered in overlapping layers of thickened modified leaves called scales. The dry, paper-like outer scale of many bulbs serves as a protective covering. The lower portion of a bulb gives rise to roots. The upper portion produces scales and new shoots. Bulb scales contain starch and other substances that a plant holds in reserve during times when conditions in the environment are unfavorable for growth. When favorable conditions return, the plant uses these stored substances to sustain rapid growth. Onions and lilies are examples of bulbs (**Figure 9-17d**).

SEP Construct an Explanation *What are two advantages and two disadvantages of asexual reproduction in plants?*

FIGURE 9-17
Specialized structures for vegetative reproduction include strawberry stolons (a), iris rhizomes (b), potato tubers (c), and onion bulbs (d).

9.3 REVIEW

1. **Distinguish** Match each term to mosses, ferns, or conifers to distinguish the fertilization requirements for each plant group. Each term may be related to one or more plant groups.
   - cones
   - spores
   - pollen
   - water
2. **Explain** What two features have made angiosperms the most successful plants on Earth, and why?
3. **Classify** Match the flower structures to the stamen or carpel. Some flower structures may not have a match.
   - anther
   - pollen sacs
   - filament
   - sepals
   - ovary
   - stigma
   - petals
   - style
4. **Explain** Why is it beneficial for a flowering plant species to bloom only during a specific and limited time of the year?

Visual Support

Plant Reproduction Have students use **Figure 9-17** to create a table that compares the different types of vegetative reproduction. Guide students to set up a three-column table with the first column listing the type of reproduction. The second column should then include a detailed description and the third should include at least three examples. You can also opt to have students include a drawing to help with comprehension.

SEP Construct an Explanation
Sample answer: advantages: only requires one parent, the process is relatively quick; disadvantages: offspring is identical to parent plant, making it more susceptible to disease/ environmental change; lack of dispersal means offspring will compete with parent plant for access to water and nutrients.

9.3 REVIEW

1. *conifers: cones, pollen; mosses: spores, water; ferns: spores, water* **DOK 2**
2. *Flowers and fruits attract animals that participate in pollination and dispersal of seeds. These processes help maximize genetic diversity.* **DOK 2**
3. *stamen: anther, filament, pollen sacs; carpel: ovary, stigma, style* **DOK 2**
4. *Sample answer: When flowering plants bloom during specific and limited times of the year, pollinators are more likely to cross-pollinate with the same species. Cross-pollination is advantageous because of the variety of offspring that it produces.* **DOK 3**

EXPLORER
DR. ROBERT RAGUSO

About the Explorer

National Geographic Explorer Dr. Robert Raguso specializes in the study of floral chemistry, studying plant-insect interactions at all levels from molecules to ecosystems. He has played a significant role in establishing chemical ecology as a field. You can learn more about Dr. Raguso and his research on the National Geographic Society's website: www.nationalgeographic.org.

Science Background

Scent Attraction When thinking of plant-pollinator interactions, most scientists have focused on physical attributes of plants, such as flower color or shapes. Brightly colored flowers are known to attract pollinators, as are ultraviolet markings on plants that act as "runways." However, many organisms that interact with plants are colorblind or have poor sight.

As part of Dr. Raguso's work, he and his colleagues have found that populations of the same species that have been geographically separated evolved to produce pollinator-specific scents. In another example of co-evolution, woodland star plants depend on a specific type of moth for pollination. After collecting samples from all recognized species of woodland star plants in 94 separate populations, it was found that almost every population studied was different in its scent chemistry.

Connect to Careers

Chemist Most careers in chemistry require a four-year degree. This degree can lead to many opportunities ranging from agriculture to manufacturing to pharmaceuticals. With additional education, there are opportunities in education, engineering, medicine, research, and the environment.

THINKING CRITICALLY

Explain *A standard reference material serves as a control for an analytical test, such as determining the chemical components of an aroma compound.*

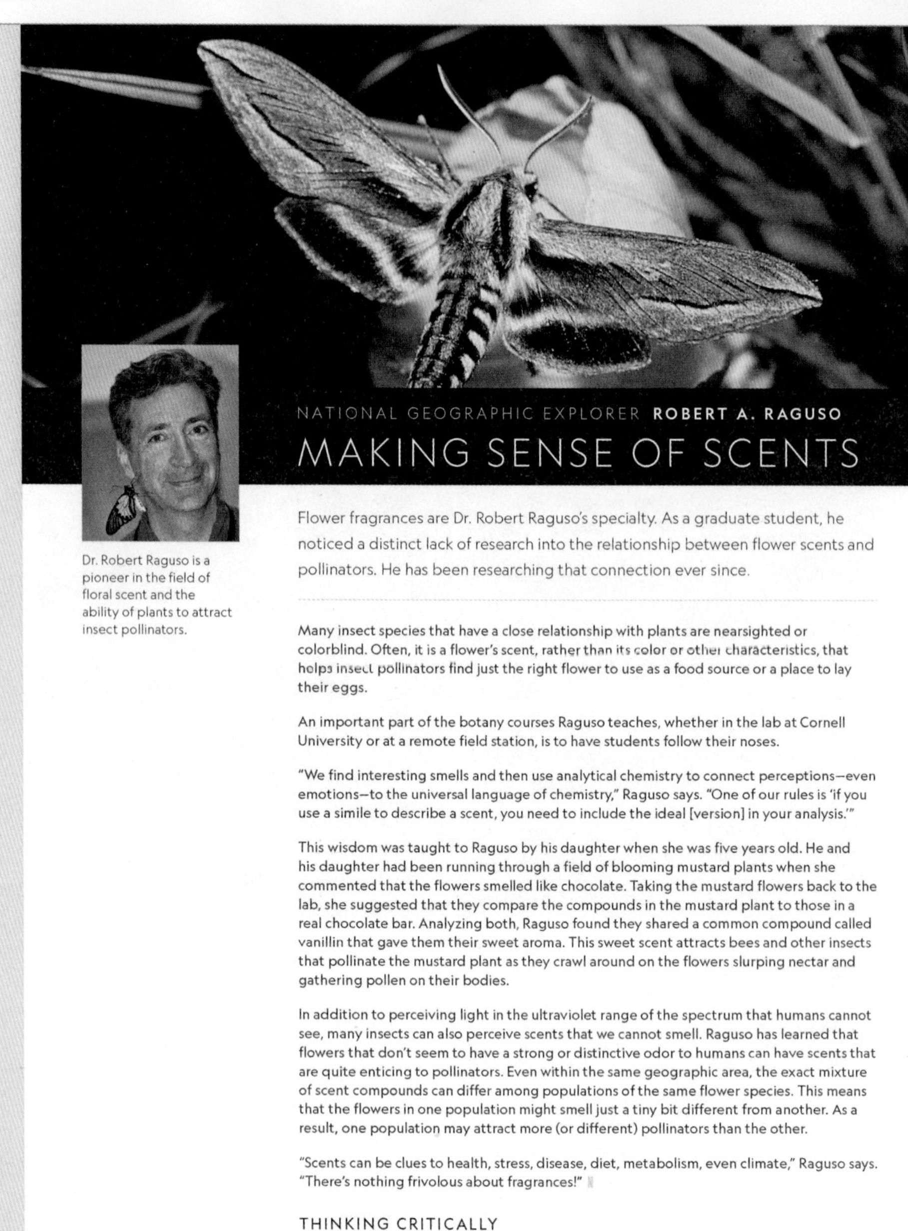

NATIONAL GEOGRAPHIC EXPLORER **ROBERT A. RAGUSO**

MAKING SENSE OF SCENTS

Dr. Robert Raguso is a pioneer in the field of floral scent and the ability of plants to attract insect pollinators.

Flower fragrances are Dr. Robert Raguso's specialty. As a graduate student, he noticed a distinct lack of research into the relationship between flower scents and pollinators. He has been researching that connection ever since.

Many insect species that have a close relationship with plants are nearsighted or colorblind. Often, it is a flower's scent, rather than its color or other characteristics, that helps insect pollinators find just the right flower to use as a food source or a place to lay their eggs.

An important part of the botany courses Raguso teaches, whether in the lab at Cornell University or at a remote field station, is to have students follow their noses.

"We find interesting smells and then use analytical chemistry to connect perceptions—even emotions—to the universal language of chemistry," Raguso says. "One of our rules is 'if you use a simile to describe a scent, you need to include the ideal [version] in your analysis.'"

This wisdom was taught to Raguso by his daughter when she was five years old. He and his daughter had been running through a field of blooming mustard plants when she commented that the flowers smelled like chocolate. Taking the mustard flowers back to the lab, she suggested that they compare the compounds in the mustard plant to those in a real chocolate bar. Analyzing both, Raguso found they shared a common compound called vanillin that gave them their sweet aroma. This sweet scent attracts bees and other insects that pollinate the mustard plant as they crawl around on the flowers slurping nectar and gathering pollen on their bodies.

In addition to perceiving light in the ultraviolet range of the spectrum that humans cannot see, many insects can also perceive scents that we cannot smell. Raguso has learned that flowers that don't seem to have a strong or distinctive odor to humans can have scents that are quite enticing to pollinators. Even within the same geographic area, the exact mixture of scent compounds can differ among populations of the same flower species. This means that the flowers in one population might smell just a tiny bit different from another. As a result, one population may attract more (or different) pollinators than the other.

"Scents can be clues to health, stress, disease, diet, metabolism, even climate," Raguso says. "There's nothing frivolous about fragrances!"

THINKING CRITICALLY

Explain *Why is it useful to compare analytical results to standard reference materials, as Raguso does with aroma compounds?*

270 CHAPTER 9 PLANT SYSTEMS

DIFFERENTIATED INSTRUCTION | English Language Learners

Speaking Strategies Point out the word simile in paragraph 3. Explain that a *simile* is a way to describe something by comparing it to something else using the word *like* or *as*: "the flowers smelled like chocolate." Explain that students can use a similar strategy if they do not know an English word; they can use other words to define or describe it. Allow time for students to practice this skill in small groups. They can take turns describing a word for group members to guess.

Beginning Provide a sentence frame: *It smells/sounds/looks like ______.*

Intermediate Have students describe a special scent or flower using similes and other descriptions. They could use this sentence frame: *It reminds me of ______ because ______.*

Advanced Have students choose a word from the article to describe, such as pollinator or population. Have them describe the word and give examples.

PLANT RESPONSES TO THE ENVIRONMENT

When it gets too hot, an animal such as a lizard can scurry underneath a rocky outcrop to beat the heat in the shade. However, while plants don't have the luxury of movement, they do have a number of mechanisms in place that let them respond to changes in their environment and maintain homeostasis.

Analyze *Consider the structures of a plant. What are some ways a plant might cope with hot and dry conditions?*

9.4

KEY TERMS

hormone
tropism

Plant Responses

Just like animals, plants have mechanisms to maintain the internal conditions required for survival. Additionally, plants have adaptations that allow them to survive in a variety of environments with differing conditions. For example, the shape of a cactus's spines helps to reduce water loss in its arid habitat.

Plant Hormones Most animal development occurs before adulthood. By contrast, plant development continues throughout the plant's lifetime. A plant adjusts to external conditions by changing growth patterns. Changes in form and function are triggered by environmental cues such as temperature, gravity, night length, the availability of water and nutrients, and the presence of pathogens or herbivores. This developmental flexibility requires extensive coordination among plant cells. Cells in different tissues or even different parts of a plant coordinate activities by communicating with one another through hormones. A **hormone** is a chemical substance that is produced in one part of an organism and acts as a messenger to either activate or suppress activity in another part of the organism. For example, plant cells are stimulated to undergo mitosis by the release of hormones that circulate within the plant's tissues.

All cells in a plant have the ability to make and release hormones. Plant hormones can be released far from the tissue they affect. For example, cells in an actively growing root release a hormone that also keeps shoot tips actively growing. **Table 9-1** lists examples of plant hormones and their functions.

TABLE 9-1. Plant Hormones and Their Functions

Hormone	Functions
gibberellins	Gibberellins stimulate cell division and elongation in stems. They also promote germination and stimulate flowering in some plants.
ethylene	Ethylene stimulates fruit ripening. It is also involved in abscission, which is the process by which a plant sheds leaves or other parts.
cytokinins	Cytokinins stimulate differentiation of cells in the growing tip of a plant's roots. They also stimulate cell division in shoots and prevent the formation of lateral roots.
auxins	Auxins are involved in the formation of plant parts, differentiation of vascular tissues, formation of lateral roots, and shaping the plant body in response to environmental stimuli.

9.4

PLANT RESPONSES TO THE ENVIRONMENT

LS1.A

EXPLORE/EXPLAIN

Plants have responses to changing conditions in their environment that enable them to survive. This section presents plant responses to hormones, environmental triggers, and to injury. Building on this, plant responses to circadian rhythms, seasonal changes, and climatic conditions are discussed.

Objectives

- Summarize the four main types of plant hormones that regulate specific plant behaviors.
- Describe different environmental stimuli and the type of responses they trigger.
- Assess plant responses specific to daily and seasonal changes in a plant's environment.

Pretest Use **Question 5** to identify gaps in background knowledge or misconceptions.

Vocabulary Strategy

Semantic Map When used in combination with a prefix, the word *tropism* describes many different plant responses. Have students build a semantic map, beginning with the word *tropism* in the center and adding each specific type after they encounter it.

Analyze *Sample answer: Plants have a thick waxy cuticle that prevents moisture loss; some plants close the stomata on their leaves during hot and dry conditions to prevent moisture loss.*

DIFFERENTIATED INSTRUCTION | Leveled Support

Plant Hormones To help students retain these hormone names and promote understanding over memorization, provide some clues and associations. For example, the word *auxin* is derived from the Greek word *auxein*, which means "to grow." The prefix *cyto-* means "cell." Students may recall that cytokinesis is cell division that occurs at the end of mitosis. Cytokinins promote cytokinesis.

Struggling Students Working in groups, have students write the names of plant hormones on one card. On separate cards, have them write a clue or an effect of each hormone. Turn the cards upside down in two piles, one with the names and one with the clues. Students can take turns correctly matching each name with a clue card.

Advanced Learners Working in groups, have students predict what the uses of plant hormones in horticulture might be, based on their effects. For example, gibberellins may be used to increase the size of certain fruits.

Video

Time: 0:21 seconds

Video 9-3 shows a bean plant bending toward the light as it grows due to the release of auxin hormones. Have student identify on which side of the plant the cells are shortening and on which side the cells are lengthening. Ask them to predict why this is happening.

CHAPTER INVESTIGATION B

Homeostasis in Plants

Design Your Own: *How do the roots of a plant develop in response to moisture?*

Time: 250 minutes or three days plus observation time over 2 weeks

Students will design an experiment to demonstrate the relationship between plant seedling root growth and different soil moisture conditions.

Go online to access detailed teacher notes, answers, rubrics, and lab worksheets.

Address Misconceptions

Positive and Negative Tropisms Plants can exhibit both positive and negative tropisms, and these can occur at the same time. A positive tropism is movement or growth toward a stimulus, and a negative tropism is movement or growth away from the stimulus. When stems grow toward the light, it is a positive phototropism; however, when roots grow away from light, it is a negative phototropism. Similarly, roots are positively gravitropic and stems are negatively gravitropic. Ask students for examples of how positive and negative tropisms can benefit a plant. (*Roots experiencing negative phototropism will grow away from light; roots experiencing positive gravitropism grow toward soil, water, and minerals.*)

CCC Cause and Effect *In phototropism, the hormone auxin stimulates cells on the dark side of the plant to grow longer. As a result, the plant bends toward the light source.*

VIDEO

VIDEO 9-3 Go online to watch a plant respond to gravity and a light source as it grows.

Environmental Triggers Environmental cues—such as light, gravity, and contact—can trigger movements in plants. The movement of a plant in response to an environmental stimulus is called a **tropism**.

- **Light** A directional response to light is called a *phototropism*. Many plants can change the orientation of their leaves or other parts in a direction that depends on a light source. For example, an elongating stem curves toward a light source, thus optimizing the amount of light captured by leaves for photosynthesis in low-light environments (**Figure 9-18a**). Leaves or flowers of some plants even change position in response to the changing angle of the sun throughout the day, a phototropic response called *heliotropism*.
- **Contact** A directional response to contact with an object is called *thigmotropism*. A vine's tendrils coiling around an object is an example of this type of tropism (**Figure 9-18b**).
- **Gravity** Even if a seedling is turned upside down just after germination, its primary root and shoot will curve so the root grows down and the shoot grows up. A directional response to Earth's gravitational pull is called *gravitropism*. The hormone auxin is involved in gravitropism. Low levels of auxin stimulate root growth, but high levels of auxin inhibit root growth. In horizontally growing roots, auxin builds up on the bottom side of the root so that the upper side grows faster. As a result, the root tip grows downward in the direction of the force of gravity.
- **Water** The directional growth of plant roots in response to water levels is called *hydrotropism*. Though less understood than other types of tropisms, evidence suggests that plant roots can detect water in their environment and modify the direction of root tip growth.

CHAPTER INVESTIGATION

Homeostasis in Plants

How do the roots of a plant develop in response to moisture?

Go online to explore this chapter's hands-on investigation about feedback mechanisms in plants.

CCC Cause and Effect *Explain how the movement of a plant toward a light source is an example of a feedback mechanism.*

FIGURE 9-18

The plant hormone auxin, which promotes cell growth along the length of a stem, causes a calla lily (a) to bend toward a sunny window by acting only on cells on the shady side of the stem. Thigmotropism makes a pumpkin vine (b) coil around a rod by continually bending toward it.

272 CHAPTER 9 PLANT SYSTEMS

DIFFERENTIATED INSTRUCTION | English Language Learners

Using New Vocabulary Have students use new vocabulary, such as the word *tropism* and related words, in writing to help them remember the spelling and meaning.

Beginning Have students draw and label each type of tropism. Students should also label the environmental cue the plant is moving toward.

Intermediate Have students write a sentence or two to describe each type of tropism. They could also add a simple sketch for each.

Advanced Have students write an explanation of each type of tropism. They should describe an example of each, based on the text and/or the photos.

FIGURE 9-19 Plants detect threats and communicate with other organisms using chemical signals. This wasp was drawn to a plant by volatile compounds the plant released. The compounds indicate the presence of a predatory caterpillar that became the wasp's prey.

Injury Responses Volatile compounds are chemical compounds that evaporate readily into the air. Plants release volatile chemicals that can be used for communication with other plants and other organisms. The plant's internal conditions and environmental cues determine what compounds are released. For example, when a tomato plant detects the chemical signature of substances that a hornworm secretes from its mouth, the plant releases volatile compounds that attract parasitic wasps that feed on the worms (**Figure 9-19**). Neighboring plants also detect volatile chemicals released in response to herbivory and respond by increasing production of chemicals that make them less palatable.

RAINFOREST CONNECTIONS

Gather Evidence *Unlike animals, plants do not have specialized organ systems for producing and receiving sound. How do plants communicate with other species?*

Daily and Seasonal Responses

Plants respond to cycles of environmental change with daily and seasonal cycles of activity.

Circadian Rhythms A circadian rhythm is a cycle of biological activity that repeats every 24 hours or so. For example, a bean plant holds its leaves horizontally during the day but folds them close to its stem at night. Bean plants exposed to constant light or constant darkness for a few days will continue to move their leaves in and out of the "sleep" position at the normal time of sunset and sunrise. Similar mechanisms cause flowers of some plants to open only at certain times of day. For example, the flowers of plants pollinated mainly by night-flying bats open, secrete nectar, and release fragrance only at night. Periodically closing flowers protects the delicate reproductive parts of the plant at times when the likelihood of pollination is lowest.

REVISIT THE ANCHORING PHENOMENON

Gather Evidence *Sample answer: Plants use chemical signals to communicate with one another.*

As discussed in the text, it is well known that plants can warn one another about insect infestations via chemical signals sent through the air. In addition to communicating via these volatile chemicals, or VOCs, research has also shown that plants communicate with one another by sending chemical signals through a connected underground network of fungi, or mycorrhizae. In one study, scientists covered plants to prevent communication by airborne chemical signals. However, when some of the plants were infested with aphids, an uninfested plant connected to the others via a mycorrhizae network also began to mount a defense against the pests. Plants that were not connected to the network were not alerted of the potential threat and did not mount a similar defense.

Crosscurricular Connections

Organic Chemistry Volatile organic compounds (VOC) are released by plants into the atmosphere from leaves, flowers, and fruits, as well as into the soil from roots. They are used as a defense mechanism to protect plants against herbivores and to attract pollinators. VOCs are not limited to use by plants. Humans have used these carbon-based compounds for centuries as perfumes, as well as food flavorings, preservatives, and herbal remedies.

VOCs are a very large group of chemicals and are present in many different products that people use, particularly in their work or home. Some can create health problems even with short-term exposure. The most common forms are benzene, ethylene glycol, formaldehyde, and more. Common sources are paints and adhesive, carpets, air fresheners and other cleaning products, gasoline, and even dry cleaning.

In Your Community

Around-the-Clock Plants Encourage students to look for examples of plants showing daily responses to their environment. Brightly-colored four o'clock flowers (*Mirabilis jalapa*) get their name from the fact that they typically open their trumpet-shaped flowers around 4:00 p.m. each day and stay open until the next morning. Moths that are only active at night pollinate the flowers.

Nyctinastic movements are movements in response to daily cycles of night and dark. They can be observed in many plants, such as bean plants. A common houseplant, the prayer plant, gets its name from the way in which the leaf blades move to a vertical position at night, resembling praying hands. During the day, the leaves return to a horizontal position. The botanist Linnaeus made a "flower clock" by placing different species of plants in a circle, with the movements of each species occurring at a specific time of day.

The movement of four o'clock flowers ensures pollination of the plant by moths. Discuss with students how other movements of plants may benefit them.

Crosscurricular Connections

Chemistry Chemical compounds called *phytochromes* give plants the ability to sense the changes in day length. Changes in the length of light and dark periods throughout the year cause a change in the chemistry of these photoreceptors. Phytochrome exists in two forms based on the wavelength of light it absorbs, either red or far-red (infrared). Phytochrome includes parts (chromophores) that absorb light signals and convert them into biochemical signals. Besides photoperiodism, phytochrome is also involved in bud dormancy and seed generation.

Science Background

Fall Colors Seasonal responses of trees change the landscape from the green of spring to the reds, oranges, and gold of fall. Warm days and crisp nights are both responsible for the color change in leaves. Warm days allow for chlorophyll to produce sugars, but as night temperatures drop, leaf veins narrow. The narrowing of leaf veins keeps the majority of the sugar produced during the day trapped in the leaves. The abundance of sugar in the leaves leads to a production of anthocyanin, the deep-red pigment seen in fall foliage. Longer nights also impact the pigment in leaves. The shortened amount of daylight degrades the chlorophyll in the leaves and allows other leaf pigments, such as the carotenoids (orange carotenes and yellow xanthophylls), to become visible. These red, orange, and yellow pigments support chlorophyll in the photosynthetic process as well as protect chlorophyll against harmful effects of light energy.

FIGURE 9-20 In temperate climate regions, temperature falls and the length of night increases in autumn. Trees respond by ceasing food production, and the unused chlorophyll in their leaves breaks down. In the absence of the green pigment of chlorophyll, the red and yellow pigments remaining in the leaf become the most visible.

Seasonal and Climatic Responses Except at the Equator, the amount of daylight varies with the season. Days are longer in summer than in winter, and the difference increases with latitude. These seasonal changes in light availability trigger seasonally appropriate responses in plants. *Photoperiodism* refers to an organism's response to changes in the length of day relative to night.

Flowering is a photoperiodic response in many species. Such species are called long-day or short-day plants depending on the season in which they flower. These names are somewhat misleading, as the main trigger for flowering is actually the length of night, not the length of day. Irises and other long-day plants flower only when the hours of darkness fall below a critical value. Chrysanthemums and other short-day plants flower only when the hours of darkness are greater than a critical value.

Yearly cycles of leaf loss and dormancy are also photoperiodic responses. Plants that drop their leaves before dormancy are typically native to regions too dry or too cold for optimal growth during part of the year (**Figure 9-20**). For example, deciduous trees in the northeastern and midwestern regions of the United States lose their leaves around September or October. The trees remain dormant during the months of harsh weather that would otherwise damage tender leaves and buds. Growth resumes when milder conditions return in spring.

On the opposite side of the world, many tree species native to tropical monsoon forests of South Asia lose their leaves during the dry season. The region receives a lot of rain annually, but almost none falls between October and April. Dormancy helps the trees survive the seasonal drought-like conditions. New growth appears near the beginning of the monsoon season, just in time to be watered by ample rains.

DIFFERENTIATED INSTRUCTION | English Language Learners

Writing About Possibilities As students construct their explanation, support them in using verbs correctly to talk about what *may* or *might* happen. Explain that in English we use *may* or *might* before the base verb to talk about possibilities: *It might rain tomorrow. We may stay home and play board games.*

Beginning Have students write one sentence with *may* or *might* in their explanation.

Intermediate Have students include a conditional statement with *may* or *might* using this sentence frame: *If people _______, then the plants might _______.*

Advanced Have students write a detailed explanation with *may* or *might* and conditional statements.

FIGURE 9-21 Cacti such as these saguaro, cholla, and barrel cacti in the Sonoran Desert only open their stomata at night when the temperature decreases. This adaptation reduces the amount of water lost to the environment due to transpiration.

Climate also drives plant responses (**Figure 9-21**). The seeds of desert annuals, plants that live for only one season, avoid drought by sprouting only when the soil is moist. The new plants make seeds quickly before they, too, die. Desert perennials make seeds over several seasons. In deserts where rains fall seasonally, perennials conserve water by making leaves only after a rain and shedding them when dry conditions return.

SEP Construct an Explanation *In regions of the world that are hit by tropical cyclones, high winds damage and uproot plants, and coastal plant habitats may be flooded with seawater for extended periods. How might plant structures and systems withstand and respond to periods of high winds and excess water?*

9.4 REVIEW

1. **Label** Match the functions to the plant hormone, *ethylene*, *gibberellin*, *cytokinin*, or *auxin*, that triggers them.
   - breaking dormancy
   - cell division
   - differentiation of root cells
   - formation of vascular tissue
   - body responses to stimuli
   - ripening of fruit
   - stem elongation

2. **Classify** In some plants, what type of response is flowering?

   A. gravitropic
   B. phototropic
   C. photoperiodic
   D. thigmotropic

3. **Distinguish** Which scenarios demonstrate plant tropisms? Select all correct answers.

   A. The stamen of a flower bends toward an insect that lands on it.
   B. The continually growing leaves of a plant move in a rhythmic daily pattern.
   C. The roots of plant growing on the edge of a cliff extend straight downward.
   D. The temperature of the flowers of a plant increases when the flowers turn toward the sunlight.
   E. The flowers of a plant open in the early morning, wilt by the afternoon, and fall off the stem within a few days.

Science Background

Desert Annuals Some desert plants, such as mesquite and creosote, are perennials, meaning they can live for many years. Annuals, on the other hand, live a single year during which they produce seeds before they die. However, in the desert, this can be risky to the survival of the species. If a desert annual produced seeds and they all germinated just before a dry spell, the plant would miss its chance to reproduce. However, researchers have shown that desert annuals have a way to anticipate this problem. When the plants produce seeds, only a portion of them germinate. If the year is particularly dry, the plants will likely die. However, there is still another large portion that wait—sometimes for up to a decade—for the rains to come.

SEP Construct an Explanation
Sample answer: If plant cells are damaged by wind or excessive water, they could respond by releasing a hormone to repair the damage. If their leaves are submerged, they might close the stomata to avoid taking in too much water.

9.4 REVIEW

1. *ethylene: breading dormancy, ripening of fruit; gibberellin: cell division, stem elongation; cytokinin: differentiation of cells; auxin: formation of vascular tissue, body responses to stimuli* **DOK 2**
2. *C* **DOK 2**
3. *A, B, C* **DOK 2**

LOOKING AT THE DATA
BUD BURST AND FLOWERING IN A CHANGING CLIMATE

Connect this activity to the Science and Engineering practice of mathematical thinking. Students will be analyzing data based on recorded data, beginning in 1856 and through subsequent years, of observable changes in plants. In this case, the data represents dates of first bud burst and flowering. Students will analyze recorded data and make determinations about the effects of temperature increases over time and the appearance of first bud bursts and flowering. Using these patterns, they can then predict any changes expected in the future, assuming all other conditions remain the same.

SAMPLE ANSWERS

1. *7* **DOK 2**
2. *first bud burst = −5X + 150; first flowering: y= −3x + 155* **DOK 2**
3. *First bud burst is more affected by changes in temperature than flowering. A change in temperature will cause a greater change in first bud burst date than first flowering date.* **DOK 3**
4. *In 1856, first bud burst was May 16, and first flowering was May 19; 3 days between responses. In 2012, first bud burst was April 14, and first flowering was April 28; 14 days between responses.* **DOK 2**
5. *The mean spring temperature in 2030 is expected to be approximately 12.5 °C. First bud burst will occur on day 94, around April 5, and first flowering will occur on day 114, around April 25. The difference is 20 days.* **DOK 3**
6. *As global temperature continues to increase, there is more time between the warm temperatures that cause trees to come out of dormancy and the start of spring. Trees become more stressed, leading to less flowering and less fruit over time.* **DOK 3**

LOOKING AT THE DATA

BUD BURST AND FLOWERING IN A CHANGING CLIMATE

SEP Mathematical Thinking Changing climate conditions affect the biological responses of plants to the beginning of spring. Nature field notes taken by Henry David Thoreau, 19th-century American naturalist and philosopher, are among the earliest records of the unfurling of green leaves and flowering as winter turns into spring. A research site in Concord, Massachusetts, where Thoreau lived, continues to monitor plant and tree responses to both seasonal and long-term climate change. **Figure 1** documents springtime shifts between 1856 and 2012, based on combined data from the late 1800s and a recent study.

As temperatures warm, leaves burst from plant buds and flowers begin to bloom.

FIGURE 1
Temperature dependence is shown for two types of observable changes in plants as spring begins: first bud burst (emergence of new leaves) and first flowering.

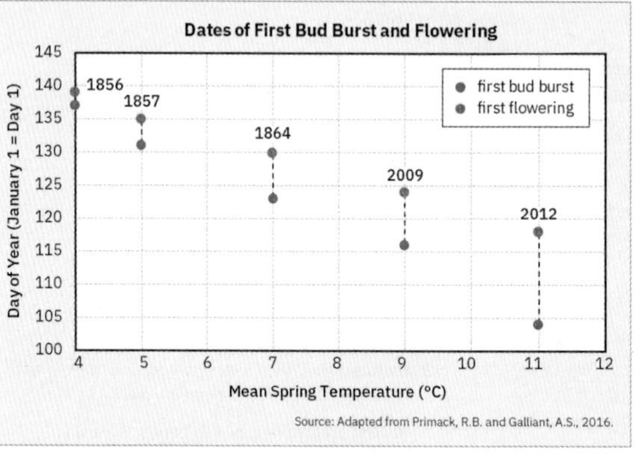

1. **Analyze Data** What is the change in mean spring temperature between 1856 and 2012?
2. **Model** Use the graph to develop linear functions that model the dependence of first bud burst day and first flowering day on mean spring temperature.
3. **Justify** Use the mathematical models to state a claim that compares the plant responses to changes in temperature.
4. **Calculate** Day 1 corresponds to January 1. Use this to determine the approximate dates of first bud burst and first flowering in 1856 and in 2012. About how many days pass between the plant responses?
5. **Apply** The mean spring temperature in Concord, MA, during 2020, one of the warmest years on record, was about 15°C. Estimate the first bud burst and first flowering dates in 2020. How many days passed between plant responses?
6. **Predict** What will happen to the number of days between first bud burst and first flowering if the trends in global climate change continue?
7. **Predict** How might the observed pattern in the number of days between first bud burst and first flowering affect the reproductive cycle of flowering plants in this region?

276 CHAPTER 9 PLANT SYSTEMS

TYING IT ALL TOGETHER ENGINEERING
FLOWER POWER

HOW DO PLANTS RESPOND TO CHANGES IN THE ENVIRONMENT?

In this chapter, you learned about plant structures and how they function to help plants grow, survive, reproduce, and respond to changes in their environment. The Case Study described the process of phytoremediation, which takes advantage of a plant's selective intake of water and nutrients to remediate soil and water that has been contaminated with pollutants. Factors including soil type, the amount and type of contamination, and the length of the growing season affect a plant's efficiency at cleaning contaminated sites.

Sunflowers are not the only plant species used in phytoremediation projects. Other plants used to remediate contaminated areas include Indian mustard, willow, poplar, various types of grasses, and legumes such as mung beans and horse gram (**Figure 9-22**). Some remediation projects have specifically incorporated plants for their aesthetic qualities. For example, willow trees were incorporated into the design of Amsterdam's Westergasfabriek Park in the Netherlands (**Figure 9-23**). This former gas facility was repurposed as a large public recreational park.

In this activity, you will develop a model to show how plant systems interact in the removal of harmful substances from contaminated water or soil. Work individually or in a group to complete these steps.

Gather Evidence

1. Research the process of phytoremediation. What plant systems are involved in the removal of harmful materials from the soil or water the plant is grown in? What plant species are the most successful at removing contaminants?

Develop and Use a Model

2. Choose a plant from your research and use digital media to construct and present a model that shows how the plant could be used to remove harmful substances from the soil or water where it grows. Label the plant systems involved and indicate how the substance moves through the plant.

Evaluate Solutions

3. Use your research and model to explain why cultivating certain plants might be an effective solution to remediate contaminated areas. What are some constraints related to using plants in remediation projects?

FIGURE 9-22
Plant species differ in their effectiveness at removing specific metals from soil.

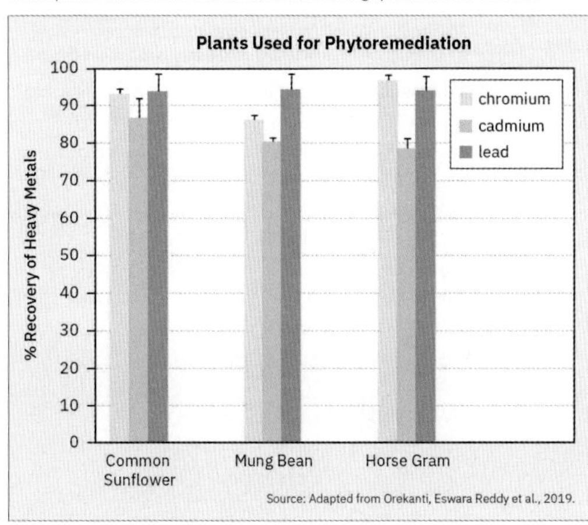

FIGURE 9-23
Phytoremediating plants are included in attractive parklike landscapes at former coal and gas processing sites.

DIFFERENTIATED SUPPORT | Leveled Support

Using Data from Graphs Support students in analyzing the data found in the graphs as they use them to cite evidence for their claims.

Struggling Students Students should understand that bar graphs are used to represent lots of data in a visual way. Using **Figure 9-23,** help students recognize data patterns and how they can be interpreted. Tell them to look at each type of metal contaminant individually across species to see which is better at reducing levels of each contaminant. Point out the error bars, which show a degree of uncertainty about the top range of the data.

Advanced Learners Have students look at the data patterns and use them to justify predictions. Have them practice writing statements such as: "Based on the data, horse gram is more effective at reducing levels of chromium; however, the common sunflower is the most effective in cleaning up cadmium."

TYING IT ALL TOGETHER

ELABORATE

The Case Study at the beginning of the chapter discussed the use of sunflowers as useful tools in the clean up of contaminated areas, such as sites of nuclear disasters. In the sections that followed, students learned more about plant structures and functions, including those that would allow plants to take up contaminants from soil or water and store them in tissues. Students will now apply this understanding as they research different plant species and their effectiveness in phytoremediation.

This activity is structured so that students may complete the steps as they study the chapter content. **Step 1** correlates with the chapter opener, as well as content in Sections 9.1 and 9.2. Section 9.2 will be of particular help as students construct models of how materials move through their selected plant in **Step 2**. Students will then need to synthesize ideas from all sections and their research to complete **Step 3**.

Alternatively, this can be a culminating project at the end of the chapter. The gathering of evidence here to support a claim or argument is a process that students may find useful in the Unit 3 Activity.

Go online to access the Student Self-Reflection and Teacher Scoring rubrics for this activity.

In Your Community

In Your Backyard In the early 2000s, people living in an area of Washington, DC, learned that their neighborhood had once been a World War I-era weapons testing site. But it was discovered that while the military-grade weapons were gone, unsafe levels of arsenic remained.

Some of the families with high levels of contamination had the soil in their yards dug up and removed. Others were given the option to plant ferns that would extract the arsenic. Of the 22 families that chose this method, 16 had clean soil after one growing season. Engage students in a discussion about the potential advantages of using phytoremediation in this case, such as reduced costs, preservation of soil, ease of management.

CHAPTER 9 REVIEW

EVALUATE

REVIEW KEY CONCEPTS

1. *B, C, D* **DOK 2**
2. *A* **DOK 2**
3. *A, B, E* **DOK 1**
4. *roots detect low soil moisture, release of hormone, guard cells release water, guard cells collapse, stomata close* **DOK 2**
5. *D* **DOK 1**
6. *Endosperm* **DOK 1**
7. *B, C, D* **DOK 1**
8. *gravitropism – gravity; heliotropism – light or sun; hydrotropism – water; phototropism – light; thigmotropism – contact or touch* **DOK 1**
9. *B, E* **DOK 1**
10. *xylem, phloem, vascular tissue* **DOK 2**
11. *source* **DOK 1**
12. *mitosis* **DOK 1**
13. *A spore is a single cell that will grow into a gametophyte. All plants produce spores. A seed is a multicellular structure that will grow into a sporophyte. It is only produced by conifers and flowering plants. A fruit is a structure that surrounds the seed and plays a role in dispersal. Only flowering plants produce fruits.* **DOK 2**

CRITICAL THINKING

14. *Sample answer: Ethylene is an important plant hormone that can act as a volatile chemical and trigger ripening, both in the fruit producing the ethylene and also in fruits nearby. By increasing air circulation and ventilation, the ripening of fruit can be delayed by preventing the effect of ethylene building up.* **DOK 3**
15. *Sample answer: Algae living in an environment surrounded by water did not need to adapt to avoid drying out. Land plants need water to survive but can easily lose a large amount of water in the dry air. Adaptations such as the cuticle prevent water loss. Stomata allow the plant to control the exchange of gases and water with the air; they open when the plant needs to bring in carbon dioxide, and they close if the plant is losing too much water.* **DOK 3**
16. *Answers will vary. Accept all reasonable answers and experimental designs that logically match with student hypotheses.* **DOK 3**
17. *Xylem is used for transporting water through the plant. Xylem is made of vessel elements and tracheids that are stacked on top of each other to form a narrow tube. These cells are dead but still function to conduct water through the plant. Phloem is used for transporting nutrients through the plant. Phloem is made up of sieve tubes stacked on top of each other. Next to sieve tubes are cells called sieve elements that load the nutrients into the tubes to be transported. The sieve tubes and elements are living cells that can pump nutrients to different parts of the plant.* **DOK 2**
18. *Sample answer: Photoperiodism and other adaptations can help a species flower at a specific time of year. If a species flowers at a specific time, then the pollen released is more likely to reach another flower of the same species that is flowering at the same time. Flowers can also be adapted to*

CHAPTER 9 REVIEW

REVIEW KEY CONCEPTS

1. **Compare** In what ways are plants that grow on land similar to green algae? Select all correct answers.
   - A. They produce seeds.
   - B. They contain chloroplasts.
   - C. They have cellulose in their cell walls.
   - D. They store sugars in the form of starch.
   - E. They use vascular tissue to transport nutrients and water.
2. **Classify** Which statement describes how flowering plants and vascular plants are related?
   - A. All flowering plants are vascular plants.
   - B. All vascular plants are flowering plants.
   - C. Flowering plants cannot be vascular plants.
   - D. Flowering plants can be vascular or nonvascular.
3. **Identify** Which are roles of roots? Select all correct answers.
   - A. absorb water from the soil
   - B. anchor the plant
   - C. initiate sexual reproduction
   - D. protect the plant from extreme temperatures
   - E. store nutrients
4. **Sequence** After a long drought, the soil around a plant has become dry. Sequence the steps in the process that helps the plant conserve water.
   - Stomata close.
   - Guard cells collapse.
   - Hormone is released.
   - Guard cells release water.
   - Roots detect low soil moisture.
5. **Identify** Which plant group has a dominant gametophyte stage?
   - A. conifers
   - B. ferns
   - C. flowering plants
   - D. mosses
6. **Define** In double fertilization, one sperm fertilizes the egg. What structure is formed when the other sperm fuses with two cells in the gametophyte?
7. **List** Which are the typical characteristics of a flower that relies on wind pollination? Select all correct answers.
   - A. full of nectar
   - B. green
   - C. large amounts of pollen
   - D. small
   - E. strong fragrance
8. **Identify** Name the tropism each stimulus causes.
   - gravity
   - light
   - sunlight
   - touch
   - water
9. **Identify** Which are examples of photoperiodism? Select all correct answers.
   - A. a vine coiling around a branch
   - B. a tree losing its leaves in autumn
   - C. a flower growing toward the brightest source of light
   - D. a plant closing its stomata when conditions are dry
   - E. a plant flowering only when the length of darkness is short enough
10. **Predict** Which two labeled parts of the leaf cross section in **Figure 9-2** would contain the most lignin?
11. **Identify** As a tree starts to leaf out in the spring, are the roots acting as a source or sink for the plant?
12. **Identify** What cell division process allows a gametophyte to produce gametes?
13. **Compare** Differentiate between a spore, a seed, and a fruit. Which plant groups produce which?

278 CHAPTER 9 PLANT SYSTEMS

CRITICAL THINKING

14. **Construct an Explanation** After being harvested, fresh fruit is often shipped to markets thousands of miles away, a process that can take days or weeks. To prevent fruit from overripening during transport, shipping companies use ventilation to limit the buildup of gases emitted by the fruit. Explain how this technique can help slow ripening.

15. **Analyze** Explain how the development of stomata allowed early plants to adapt to life on land.

16. **NOS Scientific Investigations** A student wants to examine how a disruption in homeostasis affects plant growth. Identify a hypothesis and describe a possible experimental setup the student could use to test that hypothesis.

17. **Contrast** How do xylem and phloem differ? Include their functions and structures in your description.

18. **Explain** What are some typical adaptations of flowering plants that increase the chances that the pollen from one flower reaches another flower of the same species?

MATH AND ENGLISH LANGUAGE ARTS CONNECTIONS

1. **Reason Quantitatively** Transporting water from the roots to the top of a tall tree requires a large difference in pressure. Compare the pressure at the top of the tree to the pressure at the bottom of the tree. Explain why this difference occurs and how it enables water to travel upward.

2. **Write Explanatory Text** Why do the asexual reproductive structures of plants often make good sources of food for humans? Give an example.

3. **Synthesize Information** The Irish Potato Famine in the 1840s was due, in part, to the planting of a single variety of potato. This variety was very susceptible to potato late blight, a disease that was introduced into Ireland in the 1840s and quickly decimated the entire potato crop. How could planting multiple varieties of potatoes have helped prevent the famine?

Consider the following paragraph on how to propagate strawberries. Use evidence from the passage to answer the question.

A strawberry plant can reproduce by sending horizontal stems, or stolons, outward along the ground. When a stolon finds nutrient-rich soil, it sends roots down into the soil, and a new strawberry plant begins to grow. Eventually the new plant may disconnect from the original plant. One plant has now become two separate—but genetically identical—plants.

4. **Cite Textual Evidence** What type of reproduction is described in this passage? What evidence supports your claim?

5. **Reason Quantitatively** In a certain greenhouse, short-day plants are grown in the winter. The short winter days mean the plants receive about 10 hours of natural light each day. A nearby greenhouse grows the same plants. The conditions are the same except that, during the night, maintenance workers inadvertently expose the plants to artificial light as they come and go, turning the lights on and off as they complete their work each night. Describe how the night length experienced by the plants differs between the first and second greenhouse, and predict which greenhouse will have plants that flower.

REVISIT RAINFOREST CONNECTIONS

Gather Evidence Above ground, plants have various mechanisms for responding to stimuli and changes in their environment. One type of response, releasing volatile organic compounds, signals other organisms with information about the stimulus. The transfer of information enables interactions between individual plants and between plants and other species.

1. What types of stimuli might the root system of a rainforest plant encounter, and what types of responses might the plant have to these stimuli?

2. What other questions do you have about the ways in which rainforest plants interact?

REVISIT RANFOREST CONNECTIONS

1. *In the soil, plant roots would encounter water or solid objects such as rocks. The roots might respond by changing their direction of growth toward or away from the stimulus.*

2. *Sample answers: What role does fungi play in plant communication? What are different types of ecological relationships that plants have? How do plants interact with other organisms?*

specific pollinators, and these pollinators will then move pollen between flowers that are of the same species. Adaptations such as shape, size, smell, and color of the flower can be specific to certain pollinators and help encourage pollinators to visit many flowers of the same species. **DOK 3**

MATH AND ELA CONNECTIONS

1. *Sample answer: The pressure at the top of the tree is lower than at the bottom of the tree. Negative pressure is created at the top of the tree as water evaporates from the leaves during transpiration. Water flows from areas of high pressure (in this case, the roots) to areas of low pressure (in this case, the leaves); thus, transpiration from the leaves helps pull water up from the roots.* **DOK 2**

2. *The plant produces the reproductive structures with reserves of nutrients for the new plant that will grow. These reserves of nutrients also make good food for humans. For example, the food stored in the potato tuber allows the potato plant to propagate asexually but also provides a food source for people.* **DOK 3**

3. *Sample answer: Potatoes are propagated through asexual reproduction. The tubers produced are clones of the parent. Because all of the potatoes being grown were genetically identical, they were all susceptible to the disease. If different varieties of potato had been grown, some of them might have been more resistant to potato light blight.* **DOK 4**

4. *This is asexual reproduction. The new clone plant is genetically identical, indicating asexual reproduction. In addition, the description of horizontal stems, runners, or stolons is a method of plant propagation that is asexual.* **DOK 3**

5. *Sample answer: The plants in the second greenhouse will experience a much shorter night length due to exposure to artificial light. Because the night length does not exceed the critical night length, the plants in the second greenhouse will not flower.* **DOK 2**

ANIMAL SYSTEMS

Three-Dimensional Learning

The practices, core ideas, and crosscutting concepts presented in this chapter's text, investigations, and resources provide support to address the following Performance Expectations: **HS-LS1-2, HS-LS1-3, HS-LS2-8,** and **HS-ETS1-3.**

Science and Engineering Practices	Disciplinary Core Ideas	Crosscutting Concepts
Asking Questions and Defining Problems Developing and Using Models (HS-LS1-2) Planning and Carrying Out Investigations (HS-LS1-3) Constructing Explanations and Designing Solutions (HS-ETS1-3) Engaging in Argument from Evidence (HS-LS2-8) NOS Scientific Knowledge is Open to Revision in Light of New Evidence (HS-LS2-8)	**LS1.A** Structure and Function (HS-LS1-2, HS-LS1-3) **LS2.D** Social Interactions and Group Behavior (HS-LS2-8) **LS4.A** Evidence of Common Ancestry and Diversity **LS4.B** Natural Selection **ETS1.B** Developing Possible Solutions (HS-ETS1-3)	Patterns Systems and System Models (HS-LS1-2) Structure and Function Stability and Change (HS-LS1-3) Cause and Effect (HS-LS2-8)

Contents		Instructional Support for All Learners	Digital Resources
ENGAGE			
280–281	CHAPTER OVERVIEW **CASE STUDY** How do animals respond to changes in the environment?	**Social-Emotional Learning** CCC Systems and System Models **On the Map** Pacific Ocean **English Language Learners** Sentence Structure	
EXPLORE/EXPLAIN			
282–285 DCI LS1.A, LS4.A	**10.1** ANIMAL DIVERSITY • Summarize the shared characteristics in animals. • Compare the similarities and differences of the structures present within the major animal groups. • Interpret the differences between fossils found from the Precambrian era and the Cambrian period.	**English Language Learners** Retelling with Visuals **Connect to English Language Arts** Use Digital Media **Visual Support** Phylogenetic Tree SEP Developing and Using Models **Crosscurricular Connections** Earth Science	
286–290 DCI LS1.A	**10.2** DEFINING ANIMAL SYSTEMS • Describe the levels of organization found in vertebrates. • Model the interactions between the four major organ systems of vertebrates: skeletal, muscular, nervous, and cardiovascular. • Summarize how the organ systems function together to support nutrient absorption and immunity and defense against pathogens.	**Vocabulary Strategy** Root Relationships CCC Systems and System Models **Visual Support** Body System Functions **Differentiated Instruction** Leveled Support **Connect to English Language Arts** Research SEP Developing and Using Models **In Your Community** Stray Animals **Address Misconceptions** Innate Immunity **English Language Learners** Supporting Text Evidence	

Contents		Instructional Support for All Learners	Digital Resources
EXPLORE/EXPLAIN			
291	**MINILAB** COMPARING REACTION SPEED		
292–294 **DCI** LS1.A	**10.3 MAINTAINING HOMEOSTASIS** • Give examples of how homeostasis is maintained in an organism. • Describe how negative feedback mechanisms are involved in homeostasis. • Compare the difference between negative and positive feedback mechanisms.	**CASE STUDY** Ask Questions **CCC** Stability and Change **Visual Support** Negative Feedback Mechanisms **English Language Learners** Using Non-Verbal Clues **Vocabulary Strategy** Semantic Map **Connect to Mathematics** Model with Mathematics	**Interactive Figure** Body Systems and Homeostasis **Investigation A** The Effect of Exercise on Homeostasis (100 minutes or 2 days)
295	**LOOKING AT THE DATA** THERMOREGULATION		
296–305 **DCI** LS1.A, LS2.D, LS4.B	**10.4 ANIMAL BEHAVIOR** • Differentiate between instinctive behaviors and behaviors that are learned. • Construct an explanation of the different ways that animals communicate with other members of their species and other species. • Summarize how parenting behavior affects offspring survival. • Distinguish how group living and social behavior can increase individual survival.	**Vocabulary Strategy** Using Prior Knowledge **CCC** Cause and Effect **Address Misconceptions** Learned Behavior **English Language Learners** Talking about Media **Connect to English Language Arts** Use Sources in Multiple Formats **CONNECTIONS** Can You Hear Me Now? **In Your Community** Local Predators, Multigenerational Families **Connect to Mathematics** Model with Mathematics, Graph a Function **Differentiated Instruction** Leveled Support **CCC** Cause and Effect **Social-Emotional Learning** Relationship Skills **CER** Revisit the Anchoring Phenomena **On Assignment** Kangaroo Joeys	**Virtual Investigation** Communication in the Rainforest Video 10-1 Video 10-2 **Investigation B** Monitoring Animal Behavior (250 minutes or three days plus observation time over 2 weeks) Video 10-3
306	**EXPLORER** HOLLY FEARNBACH	**Connect to Careers** Cetologist **English Language Learners** Distinguishing /wh/, /w/, and /h/	Video 10-4
ELABORATE			
307	**TYING IT ALL TOGETHER** How do animals respond to changes in the environment?	**SEP** Developing and Using Models	Self-Reflection and Scoring Rubrics
Online	**Investigation A** The Effect of Exercise on Homeostasis **Investigation B** Monitoring Animal Behavior	Design Your Own (100 minutes or 2 days) Engineering Design (150 minutes or 3 days)	MINDTAP Access all your online assessment and laboratory support on MindTap, including: sample answers, lab guides, rubrics, and worksheets. PDFs are available from the Course Companion Site.
EVALUATE			
308–309	**Chapter 10 Review**		
Online	**Chapter 10 Assessment** **Performance Task 2** *How Can We Test Systems That Interact to Maintain Homeostasis in Humans?* (HS-LS1-2, HS-LS1-3) **Performance Task 3** *How Important Is Group Behavior to the Survival of Individuals in a Population?* (HS-LS2-8)		

CHAPTER 10

ENGAGE

In middle school, students will have learned that the bodies of multicellular organisms are systems made of interacting subsystems and that cells form tissues, which form organs that make up organ systems. Students will have also learned about the evidence of common ancestry and diversity in the animal kingdom, as well as how animal behaviors increase their reproductive odds. This chapter expands on that prior knowledge by examining specific functions of animal systems and how animal bodies maintain homeostasis.

Students will dive deeper into the diversity that exists in the animal kingdom and will learn about the differences between instinctual and learned behaviors in animals.

About the Photo This photo shows bar-tailed godwits, a breed of bird that scientists suggest take advantage of specific weather conditions to make their migratory flights. Scientists hypothesize that when bar-tailed godwits migrate south, they take advantage of tailwinds to decrease the amount of effort needed to fly, allowing them to make the flight to Australia and New Zealand nonstop. To prepare students for this chapter's content, ensure that they understand that animal systems and behaviors function to increase the chances of reproduction and survival.

Social-Emotional Learning

As students progress through the chapter, invite them to look for contexts and opportunities to practice some of the five social and emotional competencies. For example, as students work through the content in the chapter, support them in their **self-management**. Help them set personal goals as they work toward understanding the content.

In addition, as students explore the effects of homeostasis, (Chapter Investigation A), remind them to practice **relationship skills**. Support students in communicating effectively and practicing teamwork and collaborative problem-solving.

CHAPTER 10

ANIMAL SYSTEMS

Bar-tailed godwits are large wading birds that belong to the sandpiper family. They often gather in large flocks at their wintering grounds or when preparing to migrate.

Nearly 40 percent of the world's bird species migrate. Many of these species spend a portion of the year at their breeding grounds in far northern climates. Then, before winter sets in, they migrate south in groups to warmer climates. Migratory behaviors are a dramatic example of the ways in which animals respond to changes in their environment.

Chapter 10 supports the NGSS Performance Expectations **HS-LS1-2, HS-LS1-3,** and **HS-LS2-8.**

CROSSCUTTING CONCEPTS | Systems and System Models

Animal Systems In this chapter, students apply the concept of systems and system models as they explore the subsystems that make up the larger system known as an organism. Students will examine the levels of organization found in animals, including the major body systems, as well as how the structures in the body are maintained through homeostasis. Some fields of biology, especially anatomy and physiology, study how body systems support survival and reproduction.

Chapter 2 in Unit 1 related systems and system models to the cycling of matter and energy within an ecosystem, and Chapter 7 in Unit 2 related the concept to cell growth and differentiation. Chapters 9 and 10 of this unit explore the relationships between systems and how system models can be used to illustrate interacting systems within multicellular organisms. This crosscutting concept is reinforced in Chapter 16 of Unit 5 where students study how humans interact with the systems in their environment.

CASE STUDY

THE INCREDIBLE NONSTOP FLIGHT OF THE BAR-TAILED GODWIT

HOW DO ANIMALS RESPOND TO CHANGES IN THE ENVIRONMENT?

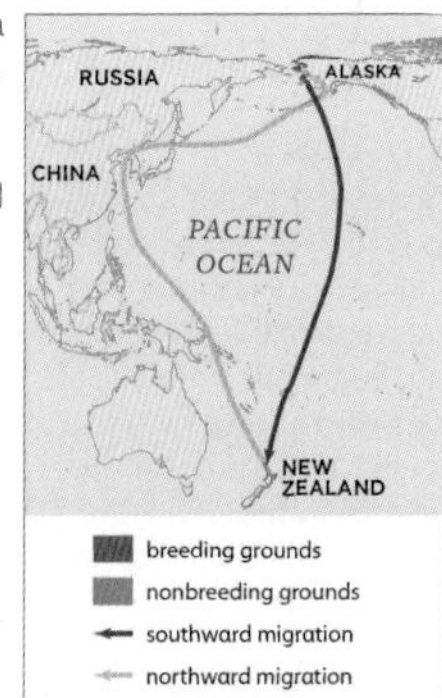

In early February 2007, a female bar-tailed godwit, named E7 after its leg tag, was fitted with an internal battery-operated satellite transmitter, similar to the one shown in **Figure 10-1**. In March, E7 took off from its wintering grounds in New Zealand and landed eight days later in Yalu Jiang, China, in the Yellow Sea. E7 spent five weeks at this coastal wetland refueling before taking off again on May 1 for its breeding grounds in Alaska, where it landed on May 6 after flying nonstop for five days.

Although researchers only expected the bird's satellite transmitter to send out data for the northbound flight, the battery lasted long enough to record the bird's southbound flight back to New Zealand as well.

The researchers were surprised when the data showed that the bird made the flight from Alaska to New Zealand in one fell swoop. From August 29 to September 7, E7 flew 11,700 km (7270 miles)—the equivalent of flying round trip from New York City to San Francisco and then back to San Francisco again—without stopping once. E7 held the record for the longest nonstop migratory bird flight until 2020, when a male bar-tailed godwit (named 4BBRW after the color of bands on its legs) made a nonstop 12,000-km journey from Alaska to New Zealand over a period of 11 days. Scientists estimated that the bird flew at speeds up to 88.5 km/h (55 mph), aided by strong easterly winds.

Bar-tailed godwits prepare for their nonstop journey south by gorging on mollusks, worms, and crustaceans in their Alaskan habitat in the Yukon-Kuskokwim Delta. By doing so, they dramatically increase their fat reserves, doubling their mass. Their bodies are able to store more fat due to the shrinkage of their intestines and gizzard during their migration. Other characteristics that make this bird well adapted for such a long flight are its aerodynamically shaped body and long tapered wings.

Ask Questions *In this chapter, you will learn about animal systems. As you read, generate questions about how a bird's body systems interact to allow it to make its annual migration.*

FIGURE 10-1
This bar-tailed godwit, identified as Z0 from its leg tag, is also outfitted with a satellite transmitter (note the antenna extending beyond its tail feathers). Data sent back from this transmitter allowed researchers to track this individual bird's migratory flight path from New Zealand to Alaska and back.

DIFFERENTIATED INSTRUCTION | English Language Learners

Sentence Structure Help students break sentences into parts to focus on each idea in the text. Have them explain in their own words to check understanding.

Beginning Help students unpack the last sentence in paragraph 1. Have them divide it into three parts, dividing the sentence before the words *before* and *where*. For each part, ask where E7 was, for how long, and what it was doing.

Intermediate Have students underline the time phrases in paragraph 1: early February 2007, March, eight days later, five weeks later, May 1, five days, May 6. Have them use short phrases to explain what happened on each date or during each time period.

Advanced Have students rewrite the second and third sentences in paragraph 3 without the information set off by em-dashes and parentheses, so they can focus on the most important details. Have them compare the two migratory flights.

CASE STUDY

ENGAGE

The situation described in the Case Study can be used to assess students' prior knowledge of animal systems and behaviors and whether they are aware of how the systems interact to maintain homeostasis within an organism. Ask students to identify behaviors that animals may use to increase their chances for survival and reproduction, such as hibernation or complex dances and mating rituals.

The bar-tailed godwit is unique in that it makes one of the longest known nonstop journeys of all organisms. In the Tying It All Together activity, students will conduct research to identify the interactions that occur within birds' body systems that allow for migration.

Ask Questions Students should revisit the Case Study as they read the chapter to make connections with the content. See a specific suggestion in Section 10.3.

On the Map

Pacific Ocean Use the map of the bar-tailed godwits' migratory path to discuss with students why the birds migrate from New Zealand to Alaska and back each year. Remind students that the Northern and Southern Hemispheres experience opposite seasons, so it is warm in the Southern Hemisphere during the winters in the Northern Hemisphere and vice versa. Encourage students to investigate the different environments that exist in the islands that make up New Zealand and determine if the bar-tailed godwit tends to winter in a specific location.

Human Connection New Zealand is a country that is made up of over 600 islands that have various environments. The Maori, the indigenous Polynesian people of New Zealand, are believed to have arrived on the island in the 1300s from the eastern Polynesian islands by canoe. Given the distance between New Zealand and the other Polynesian islands, the Maori culture advanced independently from other Polynesian cultures.

10.1

ANIMAL DIVERSITY

LS1.A, LS4.A

EXPLORE/EXPLAIN

This section provides an overview of characteristics shared by animals, discusses the nine major animal groups, identifies similarities and differences between them in animal structures, and provides an introduction to animal evolution.

Objectives

- Summarize the shared characteristics in animals.
- Compare the similarities and differences of the structures present within the major animal groups.
- Interpret the differences between fossils found from the Precambrian era and the Cambrian period.

Pretest Use **Questions 1, 3, 4, 5,** and **6** to identify gaps in background knowledge or misconceptions.

Identify *Students might say all animals are multicellular and must consume food to get the energy they need to survive.*

10.1

KEY TERMS

chordate
invertebrate
vertebrate

ANIMAL DIVERSITY

You are an animal. So are gigantic blue whales larger than three buses stacked end to end and tiny rotifers smaller than the period at the end of this sentence. Despite their huge variety of shapes and sizes, all animals have a shared set of characteristics and common ancestry.

Identify *What are two characteristics common to all animals?*

Animal Characteristics

Scientists have identified over 1 million species of animals. However, the actual number of animal species on Earth is estimated to be closer to 7.7 million. Known animal species are categorized into more than 30 different phyla. You and some of the animals that are likely most familiar to you—dogs, birds, fish, frogs—are members of the **chordate** phylum. These animals (including you) all belong to a subgroup of the chordates called **vertebrates**, which have backbones. Animals that do not have a backbone are called **invertebrates**. The invertebrates make up more than 95 percent of known species and include diverse forms such as corals, sea stars, spiders, and insects. A variety of animals is shown in **Figure 10-2**.

FIGURE 10-2 Animals vary greatly in size and shape. Brown tube sponges (a) are simple animals. Rosemary beetles (b) have an iridescent exoskeleton. Panther chameleons (c) exhibit bright shades of color. Red pandas (d) are fur-covered mammals.

DIFFERENTIATED INSTRUCTION | English Language Learners

Retelling with Visuals Photos and diagrams give students concrete support as they retell information from the text. As students read each section, have pairs take turns retelling the information using the visuals associated with the text.

Beginning Have students identify examples of key terms in the photos. For example, have them identify two chordates (a vertebrate and an invertebrate) in **Figure 10-2**.

Intermediate Have students explain key terms using the photos. For example, have them point to parts of the photos in **Figure 10-3** to explain asymmetry, radial symmetry, and bilateral symmetry.

Advanced Have students put details from the text in their own words as they explain the visuals. For example, have them use **Figure 10-4** to compare characteristics of the major animal groups.

3a

3b

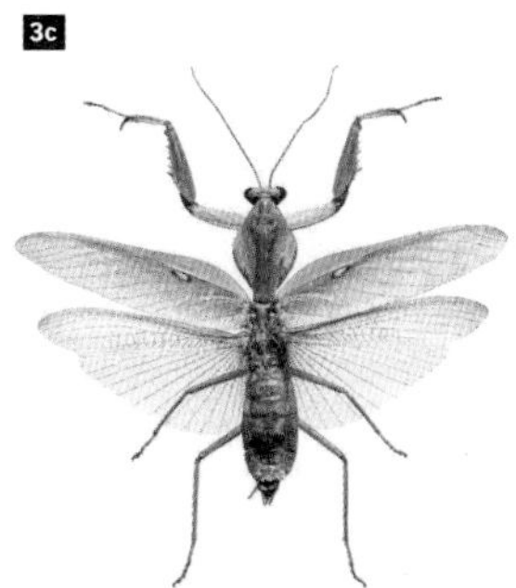
3c

FIGURE 10-3 A branching vase sponge (a) is a simple animal that lacks tissues and has an asymmetrical body plan. A white-spotted jellyfish (b) has two tissue layers and a radial body plan. A praying mantis (c) has three tissue layers and a bilateral body plan.

Tissue Organization The earliest animals were probably clumps of cells, and this level of organization continues to exist in sponges. However, most other modern animals have cells organized as tissues. Tissue organization begins during early development in animal embryos. Embryos of jellies and other cnidarians have two tissue layers: an outer ectoderm and an inner endoderm. In other modern animals, embryonic cells typically rearrange themselves to form a middle tissue layer called the mesoderm. Evolution of a three-layer embryo allowed for an important increase in structural complexity. Most internal organs in animals develop from embryonic mesoderm.

Body Symmetry Animals may be asymmetrical, radial, or bilateral. Sponges, which are animals with the simplest structural organization, are asymmetrical (**Figure 10-3a**). This means their body cannot be divided into halves that are mirror images. Jellies, sea anemones, and other cnidarians have radial symmetry (**Figure 10-3b**). Their body parts are repeated around a central axis, like spokes of a wheel. Radial animals attach to an underwater surface or drift along in the water. A radial body plan lets them capture food that can arrive from any direction. Animals that have a three-layer body plan typically have bilateral symmetry (**Figure 10-3c**). This means the body's left and right halves are mirror images. Bilateral animals such as worms, snails, praying mantises, and lobsters have a distinctive head end with a concentration of nerve cells. They usually move headfirst through their environment.

Developmental Differences The two lineages of bilateral animals are defined in part by developmental differences. In protostomes, the first opening of the digestive cavity that forms in the embryo is the mouth. The anus is formed second. (*Proto–* means "first" and *stoma* means "opening.") In deuterostomes, the anus forms from the first opening and the mouth forms from the second opening. (*Deutero–* means "second.")

Body Cavities Some animals, such as jellyfish and sea anemones, digest food in a saclike cavity with a single opening. Most animals, however, have a tubular gut, with a mouth at one end and an anus at the other. Parts of the tube are typically specialized for taking in food, digesting it, absorbing nutrients, or compacting wastes. While a tubular gut can carry out these tasks at the same time, a saclike cavity cannot.

Connect to English Language Arts

Use Digital Media The ability to communicate information using digital media is a key skill for many scientists and engineers. Have students work in groups to select a topic covered in this section. Students should research the topic of their choosing and determine how it compares between vertebrates and invertebrates. They should use evidence from the section and research to develop a digital media presentation to communicate their findings.

Visual Support

Phylogenetic Trees Students should understand that **Figure 10-4** shows an evolutionary tree built to make hypotheses about the evolutionary relationship among animals. For example, roundworms are more closely related to arthropods than annelids. Students may be unfamiliar with a number of terms shown in the model. Help students define each term in the blue lozenges. If they struggle to understand the differences between radial and bilateral symmetry, share an aboral image of a jellyfish and a ventral image of a horse. Help students find the lines of symmetry in each animal. Guide them to **Figure 10-2** and challenge them to identify to which group each animal belongs.

Science Background

Echinoderms and Body Symmetry Echinoderms include organisms such as starfish, sea urchins, sand dollars, and sea cucumbers. Students may look at these organisms and wonder why they are not categorized with cnidarians because these organisms appear to have radial symmetry. However, while these echinoderms exhibit radial symmetry as adults, during development, their larval forms have bilateral symmetry. In fact, echinoderms are thought to be more closely related to chordates than any other group of animal due to the development of the blastocyst into the deuterostome.

CCC Structure and Function

Echinoderms and chordates are both deuterostomes that exhibit bilateral symmetry.

Animals share a set of common characteristics:

- All animals are multicellular eukaryotes. Unlike plant cells, animal cells do not have cell walls. Animals use a protein called collagen for structural support.
- All animals have specialized cells, and most are organized into tissues, organs, and organ systems with which they carry out specific functions.
- Animals are heterotrophs that eat food and digest it inside their body, usually within a digestive system.
- Animals have diverse body plans that are adapted to their methods of obtaining food and reproducing.
- Most animals can move from one location to another during some stage of their lives.
- Most animals reproduce sexually.
- Most animals have well-developed sense organs, nervous systems, and muscle systems that allow them to respond rapidly to stimuli and changing conditions.

Major Animal Groups

An evolutionary tree for the nine major animal groups is shown in **Figure 10-4**. These groupings are based on structural, developmental, and genetic comparisons. Like all evolutionary trees, this one is a hypothesis and is open to revision if new data or new interpretations arise.

CCC Structure and Function *Use **Figure 10-4** to determine which characteristics echinoderms and chordates share.*

FIGURE 10-4 This evolutionary tree shows the evolutionary relationships among the nine major animal groups. Of these groups, eight are invertebrates. Chordates are the only vertebrates.

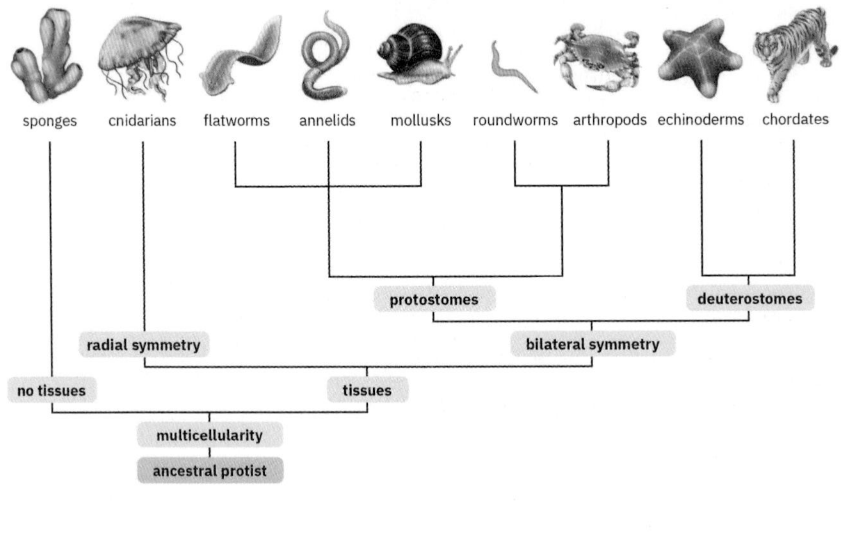

SCIENCE AND ENGINEERING PRACTICES
Developing and Using Models

Animal Characteristics Students should understand that the model in **Figure 10-4** is drawn to help people understand the relationships between the nine different animal groups. Help students understand the relationships by learning about the characteristics of each animal group. Break them out into nine groups. Assign each group one animal group and have each student in the group find one or two characteristics that are unique to each animal group. List the animal groups across the board in the same order as they are shown in the figure. Then, have the student groups capture the characteristics for their animal group under its name. Once the chart is complete, have them compare and contrast the characteristics of the animals found in each group. Students should be able to connect that as they move from left to right on the list, the grouping becomes more complex. Share with students that more complex models of phylogenetic trees exist and can even show the believed relationships between animals and plants.

FIGURE 10-5 These tubular fossils are of *Cloudina carinata*. It was a worm-shaped organism that lived about 550 million years ago. These fossils were discovered at the archaeological site El Membrillar in Badajoz, Spain.

Animal Evolution

The oldest widely accepted evidence of animals dates to the final period of the Precambrian era (635 million–541 million years ago). Evidence from this time period includes many trace fossils—rocks with trails or burrows made when unidentified early animals moved across or tunneled through sediments on the ancient seafloor. There are also a variety of body fossils from this time period (**Figure 10-5**). All known animals from this period were soft bodied. Some were unlike any modern animal. Others resemble and may be related to modern invertebrates. Various late Precambrian fossils are considered possible relatives of sponges, cnidarians, annelids, mollusks, or arthropods. Thus, the origin of bilateral body plans and the divergence between protostomes and deuterostomes are thought to have occurred during this time period.

The abundance and types of animal fossils left behind rose dramatically with the onset of the Cambrian period (541 million–485 million years ago). The increase in the number and diversity of fossil animals resulted in part because the first animals with hard parts evolved. Remains of these animals were more likely to become fossils than soft-bodied animals. By the end of the Cambrian, representatives of every major animal phylum now known were present in the seas.

10.1 REVIEW

1. **Classify** Sort these animals by their symmetry: radial or bilateral.
   - anemone
   - hammerhead shark
   - butterfly
   - zebra mussel
2. **Describe** Which characteristics are common to ALL vertebrates? Select all correct answers.
   - **A.** are autotrophs
   - **B.** cells have nucleus
   - **C.** early development occurs in stages
   - **D.** have backbone
   - **E.** multicellular
3. **Compare** Use **Figure 10-4** to identify the characteristics that are similar and different between mollusks and chordates.
4. **Explain** Why is there more evidence of animal life from the Cambrian period than from the end of the Precambrian era?

10.1 ANIMAL DIVERSITY **285**

Crosscurricular Connections

Earth Science Students who have previously studied Earth Science may be familiar with the fossil record and the principles of relative and absolute age. Share with students that prior to the mid-1900s, geologists generally used observations to determine the age of rocks and fossils based on the characteristics of the rocks and the principles of original horizontality and superposition to determine which layers of rock were older. Indeed this gradualist view of geologic features was very influential in the thought processes of Charles Darwin while journeying on the *HMS Beagle*. Many of the books he took with him were geology books, including Charles Lyell's *Principles of Geology*.

Have students identify the likely differences between the fossils from the Precambrian era and the Cambrian period. They should understand that the fossils from the Precambrian era would not have bones as the divergence into organisms with hard parts did not occur until the Cambrian period.

10.1 REVIEW

1. *radial symmetry: anemone; bilateral symmetry: butterfly, zebra mussel, hammerhead shark* **DOK 2**
2. *B, C, D, E* **DOK 1**
3. *Both mollusks and chordates are multicellular animals with tissues that have a bilateral body plan. Unlike mollusks, which are protostomes, chordates are deuterostomes.* **DOK 2**
4. *The first animals with hard parts evolved during the Cambrian, meaning they were more likely to be fossilized than the soft-bodied animals of the Precambrian.* **DOK 2**

10.2

DEFINING ANIMAL SYSTEMS

LS1.A

EXPLORE/EXPLAIN

This section provides an introduction to the levels of organization within animal systems, covers the major body systems found in vertebrates, and explains how the body systems interact.

Objectives

- Describe the levels of organization found in vertebrates.
- Model the interactions between the four major organ systems of vertebrates: skeletal, muscular, nervous, and cardiovascular.
- Summarize how the organ systems function together to support nutrient absorption and immunity and defense against pathogens.

Pretest Use **Question 2** to identify gaps in background knowledge or misconceptions.

Explain *Students might say that their skeletal and muscular systems interact together when they move.*

Vocabulary Strategies

Prefixes/Suffixes/Root Relationships
The word *inflammation* comes from the Latin prefix *in-*, which students should be familiar with, and the root *flammare,* which means "to flame." Suggest that students add key terms to a word tree or graphic organizer as they work through the chapter. Students may look up the etymology of key terms in the lesson as they fill in their graphic organizers.

10.2

KEY TERMS

immunity
inflammation

DEFINING ANIMAL SYSTEMS

Even the simplest animals—the sponges—are made up of specialized cells. More complex animals, such as you, are composed of a system of interacting cells, tissues, and organs.

Explain *What are two systems that interact in your body? How do they interact?*

Levels of Organization

In all animals, development produces a body with multiple types of cells. For most animals, these cells are organized into tissues. All vertebrates have four types of tissue: epithelial, connective, muscle, and nervous.

Typically, animal tissues are organized into organs. Recall that an organ is made up of two or more tissues that are arranged in a way that allows the organ to carry out specific tasks. Consider the cat heart, an organ that includes all four tissue types. The heart wall is made up mostly of cardiac muscle tissue. Epithelial tissue covers the heart's internal and external surfaces, heart valves are made of stretchy connective tissue, and nervous tissue delivers signals to and from the heart. Coordinated activities of these tissues let the heart undergo rhythmic contractions that move blood through the body at an appropriate pace.

As with plants, two or more organs in an organ system interact physically, chemically, or both to perform a common task. For example, in a cat's circulatory system, the pressure generated by a contracting heart (an organ) moves blood (a tissue) through blood vessels (organs), which transport gases and solutes to and from body cells (**Figure 10-6**).

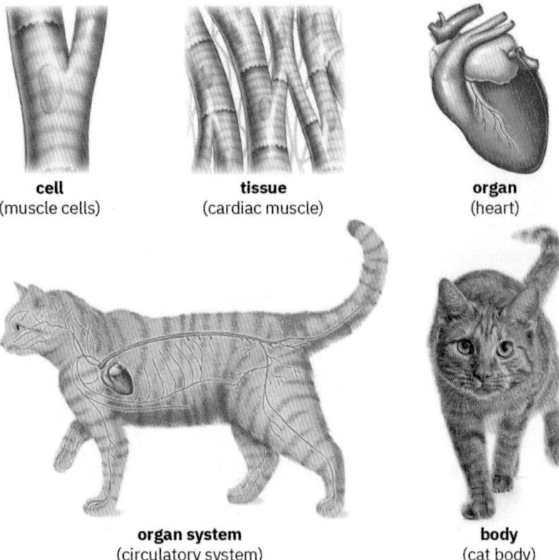

FIGURE 10-6 Most animals can be organized into five levels. From smallest to largest, these levels are: cell, tissue, organ, organ system, and whole body.

CROSSCUTTING CONCEPTS | Systems and System Models

Organ Systems Models can be used to simulate systems and interactions that may be hard to see. For example, you cannot easily see the levels of organization that exist within an organism. Guide students to observe the models and photo shown in **Figure 10-6**. Point out to students that when they interact with cats, they only interact with the full system, the organism. However, the organism is made up of interacting subsystems, which are identified in the models shown in the figure. In the figure, students can see a simplified model of the circulatory system, a subsystem in the cat that interacts with other systems. The heart is an organ, which is a smaller subsystem within the circulatory system. Challenge students to select a different organ system listed in **Table 10-1** found within a cat and identify the subsystems that make up the organ system. Have them lead a discussion about the systems with the class and identify how the systems interact.

Major Body Systems

Tissues interact structurally and functionally in organs. Organs, in turn, interact in organ systems. **Table 10-1** lists the major vertebrate body systems and their functions.

TABLE 10-1. Major Vertebrate Body Systems and Their Functions

Body system	Major organs and other parts	Major functions
circulatory	blood, blood vessels, heart, lymph nodes, lymphatic vessels	distributes materials and heat throughout the body
digestive	esophagus, gallbladder, large intestine, liver, mouth, pancreas, pharynx, small intestine, stomach	takes in food and water; breaks food down and absorbs nutrients, then eliminates food residues
endocrine	adrenal glands, hypothalamus, pancreas, parathyroids, pituitary, ovaries, testes, thyroid	secretes hormones that control the activities of other organ systems
integumentary	hair, nails, sweat and oil glands	protects body from injury, dehydration, pathogens; moderates temperature; excretes some wastes; detects external stimuli
lymphatic	lymph nodes, spleen, thymus	collects and returns excess tissue fluid to the blood; defends the body against infection and cancers
muscular	cardiac, skeletal, and smooth muscles	moves the body and its parts; maintains posture; produces heat to maintain body temperature
nervous	brain, peripheral nerves, spinal cord	detects external and internal stimuli; coordinates responses to stimuli; integrates organ system activities
reproductive	female: fallopian tubes, ovaries, uterus, vagina male: ducts and glands, penis, testes	female: produces eggs; nourishes and protects developing offspring male: produces and transfers sperm to a female
respiratory	nasal cavity, nose, lungs, pharynx, trachea	takes in oxygen needed for aerobic respiration; expels carbon dioxide; helps maintain pH
skeletal	bones, cartilage, ligaments, tendons	supports and protects body parts; site of muscle attachment; produces blood cells; stores minerals
urinary	bladder, kidneys	maintains volume and composition of blood; excretes excess fluid and wastes; helps maintain pH

Body System Functions

Body systems interact together to accomplish tasks. For example, as shown in **Figure 10-7**, the circulatory system works with the respiratory, digestive, and urinary systems in delivering oxygen and nutrients to cells throughout the body, and clearing away their wastes. In this section, we will focus on a few vertebrate body systems and their interactions.

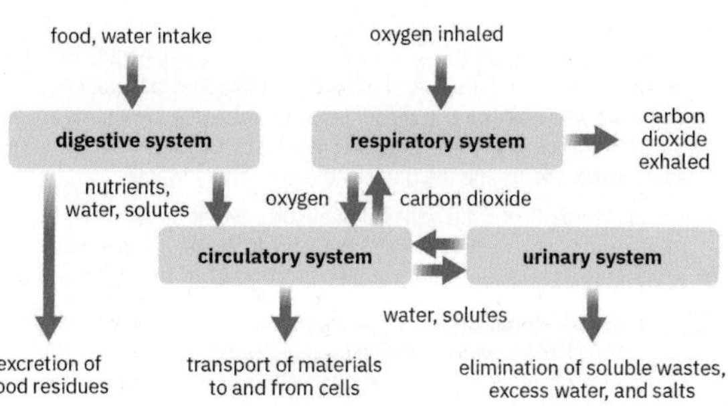

FIGURE 10-7 The organ system interactions shown here keep the body supplied with essential substances and eliminate unwanted wastes.

Visual Support

Body System Functions Body systems function to maintain homeostasis within an organism. **Figure 10.7** identifies four body systems and the main inputs and outputs of those systems. Have students identify the systems that are involved with taking matter in and those that are directly involved in releasing waste matter into the environment. Students should recognize that the circulatory system is primarily involved in moving matter to and from cells. Other systems are engaged in taking in matter in the forms of food and oxygen or releasing matter as food residues from the digestive system, soluble wastes, water and salt from the urinary system, and carbon dioxide that is exhaled as a product of cellular respiration. Have students discuss in groups how the different systems function together to maintain homeostasis within an organism and what happens if feedback mechanisms fail.

DIFFERENTIATED INSTRUCTION | Leveled Support

Struggling Students Work through **Figure 10-7** in sections. Guide students to look at one section of the model at a time in order to determine how the different systems work together. It may help to have them translate the visual into a different media or to talk out the interactions. Students may benefit from making their own chart where they list the systems in the first column, the inputs in the second column, the outputs in the third column, and identify other systems they directly interact with in the fourth.

Advanced Learners Students who show exceptional interest in body systems and their functions may find it stimulating to learn about how the organs in one or more of the body systems work together to complete a specific function. Allow these students to explore the body system of their choice and present their findings using models or other visuals. Encourage these students to identify how each organ contributes to the feedback mechanisms in the body.

Connect to English Language Arts

Research Ask students to research the differences that exist in the digestive or immune systems of two different kinds of animals. Then, have them write a short research paper where they synthesize their findings from multiple sources and demonstrate that they understand the differences between the organ systems in the two different animals. You may choose to encourage students to select animals from different classes of vertebrates, for example, a cow and a shark. Despite being vertebrates, these two organisms have very different digestive systems.

Develop a Model *Sample answer: A simple flow chart might show the digestive system connected to the circulatory system in the absorption and transportation of nutrients.*

Nutrient Absorption Body systems involved with nutrient absorption include the digestive, muscular, and circulatory systems. Vertebrates have a complete digestive tract (**Figure 10-8**). Its various regions specialize in digesting food, absorbing nutrients, and concentrating and storing unabsorbed wastes.

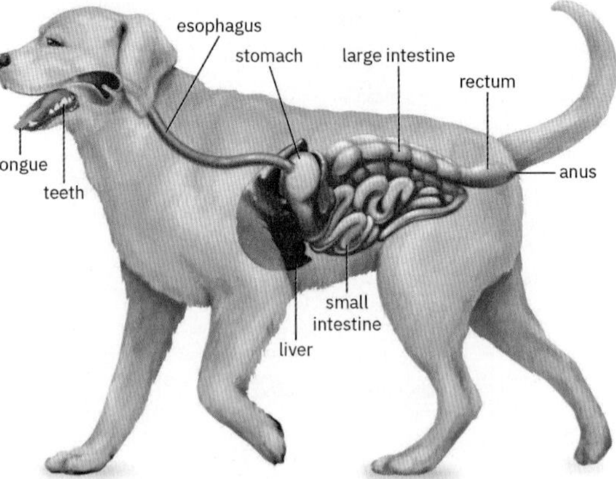

FIGURE 10-8 Vertebrates, such as this dog, have a complete digestive tract. The processing of food begins in the mouth and ends when waste exits the body through the anus.

- Processing of food begins inside the mouth. Salivary glands make saliva that moistens food and begins the process of chemical digestion. Swallowing moves the food–saliva mixture into the pharynx (throat), from which it continues into the esophagus.
- The wall of the esophagus and all portions of the digestive tract beyond it contain smooth muscle, part of the muscular system. By a process called *peristalsis*, waves of smooth muscle contraction propel material through the digestive tract.
- The esophagus extends into the abdominal cavity, where it connects to the stomach. The stomach has three main functions. First, it stores food and controls the rate at which food continues on to the small intestine. Second, it mechanically breaks down food. Third, it secretes substances that aid in chemical digestion.
- The stomach empties into the intestine, which has two regions that differ in their structure and function. The small intestine is the region that completes the process of chemical digestion and absorbs most nutrients. Secretions from the liver, gallbladder, and pancreas assist the small intestine in these tasks. The absorbed nutrients are transferred from the small intestine into the bloodstream, where they are transported around the body by the circulatory system.
- The large intestine is the region that absorbs water and salts and concentrates waste. At the end of the large intestine is the rectum, which stores waste until it is expelled from the body through the anus.

SEP **Develop a Model** *Draw a simple model that shows how the digestive system interacts with another body system during the process of nutrient absorption.*

SCIENCE AND ENGINEERING PRACTICES
Developing and Using Models

Vertebrate Digestive System Use **Figure 10-8** to help students follow along the bulleted list that describes the process of nutrient absorption in a vertebrate. To make this section more interactive, provide students with print outs of the dog. Have them read the section in pairs. As they move through each bullet, they should identify where the process is happening in the dog's body. If an organ or tissue mentioned in the list is not identified in the model, have students use external sources, as necessary, to add the missing organ label. Students may then make a chart of all the organs found in the model and description. They should identify the organs in the first column and their structures in the second column. Have them identify the functions of the tongue, teeth, and spleen, which are included in the model but not described in the text.

Immunity and Defense The body's ability to resist and fight infections is called **immunity**. The immune system includes interactions among the integumentary, lymphatic, and circulatory systems. Major parts involved in the immune system are shown in **Table 10-2**.

TABLE 10-2. Immune System Parts and Their Functions

Part	Function
bone marrow	produces red blood cells, plasma cells, and white blood cells
lymph node	filters and destroys pathogens
skin, mucous membrane	serves as first-line defense against pathogens
spleen	stores white blood cells; filters blood
thymus	maturation site for specialized white blood cells
white blood cell	searches for, attacks, and destroys pathogens

Surface barriers usually prevent microorganisms from entering the body's internal environment. The tough outer layer of skin is an example. Microorganisms flourish on skin's waterproof, oily surface, but they rarely penetrate its thick layer of dead cells (**Figure 10-9**). The thinner epithelial tissues that line the body's interior tubes and cavities also have surface barriers. Sticky mucus secreted by cells of these linings can trap microorganisms. The mucus contains an enzyme that kills bacteria. In the sinuses and respiratory tract, the coordinated beating of cilia sweeps trapped microorganisms away before they have a chance to breach the delicate walls of these structures. These anatomical barriers are the body's first line of defense against pathogens.

A pathogen that breaches these barriers activates mechanisms of innate immunity, the second line of defense. *Innate immunity* is a set of immediate, general responses that help protect all multicellular organisms from infection.

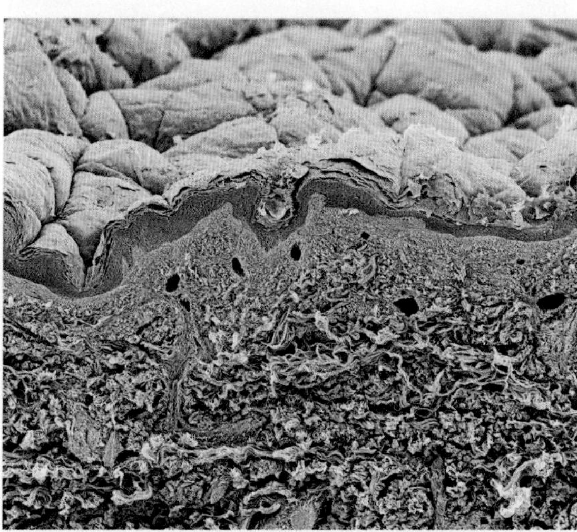

FIGURE 10-9 This color-enhanced scanning electron microscope (SEM) image shows the layers of human skin. A thick, waterproof layer of dead skin cells at and just below skin's surface usually keeps microorganisms such as bacteria, fungi, and viruses from penetrating internal tissues.

In Your Community

Stray Animals While many pets are kept on preventative treatments to reduce their chances of being infected by a variety of pathogens, stray animals are often a common reservoir for disease. They can carry ticks, fleas, a variety of worms, parainfluenza, and even rabies.

Take Action Encourage students to find out about the common diseases carried by stray and wild animals in your area. They may choose to reach out to veterinarians or your local animal shelter and interview an employee. They should also find out how a pathogen can get past the immune system barriers of the animals it infects.

Address Misconceptions

Innate Immunity Students may conflate *innate immunity* with *natural immunity*. Innate immunity refers to the innate immune system, which includes the surface barriers that prevent pathogens from entering the body. It also includes a number of proteins and phagocytotic cells that provide a nonspecific line of defense to protect organisms from infection. These reactions are fast, inflammatory, and nonspecific and will often recognize the chemical properties of the pathogens. However, natural immunity is a type of acquired immunity that comes from being infected by a pathogen and having recovered. Once exposed to the pathogen again, the body actively and specifically targets the pathogen for destruction. To help students rectify this misconception, guide them through an activity where they compare and contrast the differences between innate and natural immunity.

DIFFERENTIATED INSTRUCTION | English Language Learners

Supporting Text Evidence Have students reread the second paragraph on page 289. Point out that the first sentence is the central idea. Have students find text evidence in the rest of the paragraph that supports this idea.

Beginning Have students identify two examples of surface barriers that prevent microorganisms from entering (*outer layer of skin, epithelial tissues in interior tubes, and cavities such as the sinuses and respiratory tract*).

Intermediate Have students answer these questions: *Why is the outer layer of skin a surface barrier? What are two ways epithelial tissues prevent microorganisms from entering?*

Advanced Have students explain how the skin and epithelial cells keep microorganisms from entering the body's internal environment.

Science Background

Adaptive Immunity While the immune system consists of innate immunity and adaptive immunity, the latter can be further split into humoral immunity and cell-mediated immunity. Humoral immunity is driven by B lymphocytes, cells that produce antibodies. These antibodies bind to antigens on microbes, which signal other immune cells to eliminate the pathogens. However, antibodies cannot access the pathogens that may already be inside of a cell. This is when cell-mediated immunity must come into play. Cell-mediated immunity is driven by two different kinds of T lymphocytes. Helper T lymphocytes activate macrophages to destroy pathogens that have been ingested. Cytolytic T lymphocytes can kill any kind of cell that is infected with the pathogen. One of the main driving factors between these two cell types is where the pathogen exists when the immune system destroys it.

HIV is a virus that gets around the adaptive immune system by destroying helper T lymphocytes. Guide students through a discussion on how the different parts of the immune system function together to fight disease. Tie in feedback mechanisms by talking about what happens when one part of the immune system stops functioning.

Inflammation is a fast, local response that both destroys affected tissues and jump-starts the healing process. The primary function of inflammation is to form a physical barrier that prevents the spread of infection. Examples of inflammatory responses include swelling, redness, warmth, and pain.

The activation of innate immunity triggers mechanisms of adaptive immunity, which is the third line of defense. *Adaptive immunity* tailors an immune response to a vast array of specific pathogens that an individual encounters during their lifetime. For example, when a person recovers from chicken pox, the body develops a memory of the infection. That memory specifically protects the body from the virus that causes chicken pox if it is exposed to the virus again in the future.

White blood cells are active in both innate and adaptive immune responses. These cells circulate in blood and lymph and many populate the lymph nodes, spleen, and other solid tissues. White blood cells communicate by secreting and responding to chemical signaling molecules. These molecules, which include proteins and amino acids, allow cells throughout the body to coordinate activities during an immune response. As part of an immune response, white blood cells engulf and digest pathogens (**Figure 10-10**).

FIGURE 10-10 In this color-enhanced SEM image, a white blood cell (brown) is attacking a group of bacteria cells (blue).

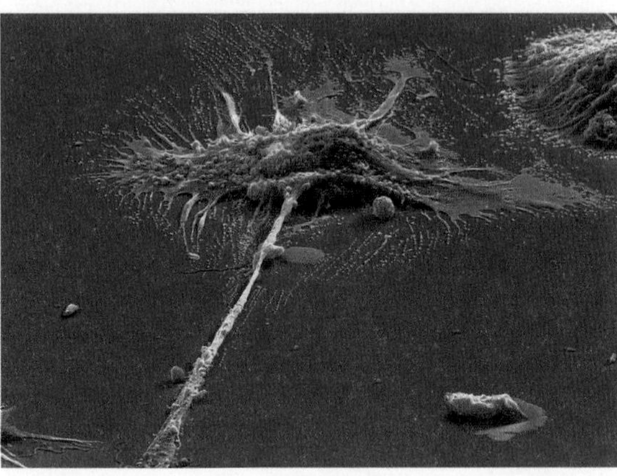

10.2 REVIEW

1. **Sequence** Order the terms to make a model of the levels of organization in animal bodies. List the terms from largest scale (top) to smallest scale (bottom).
   - body
   - organ system
   - cell
   - tissue
   - organ
2. **Distinguish** Describe an example of two organs working together in the human digestive, respiratory, circulatory, or urinary systems.
3. **Explain** Describe in your own words how the integumentary, lymphatic, and circulatory systems interact together to protect the body from pathogens.

10.2 REVIEW

1. *body, organ system, organ, tissue, cell* **DOK 2**
2. *Students' answers should include two organs in the digestive, respiratory, or cardiovascular system and how they work together to accomplish a goal. Sample answer: The mouth and the esophagus work together to break down food and bring food and water into the body for digestion.* **DOK 3**
3. *Sample answer: The integumentary system acts as a barrier to prevent pathogens from entering the body. The lymphatic and circulatory systems circulate lymph and blood, which contains white blood cells that engulf and digest pathogens.* **DOK 3**

MINILAB

COMPARING REACTION SPEED

SEP Analyze and Interpret Data Do body systems react to visual stimuli faster than auditory stimuli?

Many everyday activities you do, such as throwing a basketball or catching a flying disc, involve multiple body systems. To shoot a basket, the nervous system gauges the distance to the hoop and sends information to the musculoskeletal system, which then performs the throw. Catching an object, such as a flying disc, is even more complicated. The nervous system estimates where the disc will be and sends out information to the musculoskeletal system. This allows the catcher to put their hand in the path of the disc and gets them to close their fingers at the right time to grasp it.

In this activity, you will test the body's reaction time in catching a falling object. You will also determine whether the reaction time is faster with a visual stimulus compared to an auditory stimulus.

Athletes react to an auditory signal at the start of a race.

Materials
- meter stick

Procedure

1. Work in pairs for this experiment. To begin, Student A will release the meter stick and Student B will catch it. Develop a standard method for holding and dropping the meter stick. For example, the student dropping the meter stick could hold the meter stick from its top edge, while the student catching the meter stick could hold their finger and thumb loosely around the 10 cm mark of the meter stick.
2. To start the visual reaction test, Student A should indicate that a five-second window is starting. At some random time during that five-second window, Student A releases the meter stick.
3. In your biology notebook, record the distance the meter stick fell before being caught. Depending on where Student B initially placed their fingers around the meter stick, subtract any distance as necessary. Record the distance as the place just above where the fingers caught the meter stick.
4. Repeat Steps 2–3 three times and record the results in your biology notebook.
5. To start the auditory reaction test, use the same hand positioning you decided on in Step 1. This time, Student B will close their eyes.
6. Student A calls "drop" at the same time as they release the meter stick.
7. In your biology notebook, record the distance the meter stick fell before being caught.
8. Repeat Steps 5–7 three times and record the results in your biology notebook.
9. Switch roles to allow Student A to test their reaction time.
10. Use the formula $t = 0.045\sqrt{d}$, where d is distance in centimeters and t is time in seconds, to determine the reaction time for each trial.

Results and Analysis

1. **Sequence** Draw a model or flow chart showing the body systems involved in processing information for this experiment.
2. **Identify Patterns** Did you observe a pattern in the repeated trials of the visual part of this activity? For the auditory part? Explain possible reasons for whether a pattern was observed for either stimulus.
3. **Evaluate** Which stimulus, visual or auditory, had a faster reaction time? Why do you think this is the case?

Results and Analysis

1. **Sequence** *Sample answer: My eyes received the signal that the ruler was dropping. That information went to my brain, which sent a signal through my body to my fingers telling me to grab the ruler.* **DOK 2**
2. **Identify Patterns** *Sample answer: Generally, our response times were better in trials two through four. This is probably because we got better at catching the meterstick by repeating the action over and over.* **DOK 2**
3. **Evaluate** *Sample answer: We found the auditory stimulus resulted in faster reaction times compared to the visual stimulus. We think people react faster to audio stimulus for basic actions like closing their fingers.* **DOK 3**

MINILAB
COMPARING REACTION SPEED

In this activity, students will model interactions of the nervous, skeletal, and muscular systems. Although likely a familiar exercise to many, this activity adds a visual and auditory component and mathematical computations.

Time: 30 minutes

Materials and Advance Preparation

Consider starting the lab with a brief discussion about whether catching a dropped object is a reflex action. Ask: *Is it possible to get better at catching a dropped object with practice?* Students' answers may vary. Generally, reflex responses cannot be trained to be faster, while non-reflex responses can. Ask the students to think about which body systems would be involved in a reflex and a non-reflex response.

Procedure

For **Step 1**, make sure students follow the instructions on how and where the meterstick should be held by Student A and how and where Student B should position their hand. If necessary, pick one spot on the meterstick where Student B in all groups positions their hand so everyone starts in the same spot. Keep in mind that catching the meterstick is not a reflex action. Students may think this is a reflex because they are trying to respond as fast as possible.

For **Steps 4** and **8**, students will do a total of four trials for visual stimulus and four trials for audio stimulus. Students should create a data table in their biology notebook to record their results for each trial.

For **Step 6**, make sure students drop the meterstick as soon as they call drop. There should not be any time difference between the call and the actual drop.

For **Step 10**, explain to students that the equation they are using is based on the formula

$d = \frac{1}{2}at^2$, where d is distance, a is the acceleration due to gravity (981 cm/s^2), and t is time.

10.3

MAINTAINING HOMEOSTASIS

LS1.A

EXPLORE/EXPLAIN

This section provides an introduction of how feedback mechanisms function to maintain stability in an organism while its environment changes.

Objectives

- Give examples of how homeostasis is maintained in an organism.
- Describe how negative feedback mechanisms are involved in homeostasis.
- Compare the difference between negative and positive feedback mechanisms.

Pretest Use **Questions 7** and **8** to identify gaps in background knowledge or misconceptions.

Describe *Sample answer: Enzymes stop functioning correctly when body temperatures are too low or too high.*

CCC Stability and Change *Sample answer: When you touch a hot surface, sensory receptors in your fingers send information to the control center (your brain), which sends a message to the body's effectors (muscles) to move your fingers away from the hot surface.*

CASE STUDY
THE UNFATHOMABLE NONSTOP FLIGHT OF THE BAR-TAILED GODWIT

Ask Questions To help students revisit the Case Study and make connections with the content, have them generate model feedback loops to show the body system interactions for homeostasis that occur in the migrating birds.

Go Online **INTERACTIVE FIGURE**

Body Systems and Homeostasis
Students identify the receptor, control center, and effector in a bird that fluffs its feathers in response to cold.

10.3 MAINTAINING HOMEOSTASIS

KEY TERMS

control center
effector
negative feedback mechanism
positive feedback mechanism
sensory receptor

Internal and external factors, such as temperature, stress, and infection, can affect an animal's ability to maintain stable conditions in its body. For example, in humans, enzymes that control the chemical reactions that sustain life work best around 37 °C (98.6 °F).

Describe *What do you think happens to enzyme function when body temperatures are too high or too low?*

CASE STUDY
THE INCREDIBLE NONSTOP FLIGHT OF THE BAR-TAILED GODWIT

Ask Questions *Generate questions about the body system interactions involved in the maintenance of homeostasis in migrating birds.*

Feedback Mechanisms

In animals, homeostasis involves interactions among sensory receptors, the brain, and muscles and glands. **Figure 10-11** shows the control systems involved in maintaining homeostasis.

- A **sensory receptor** is a cell or cell component that responds to a change in a specific stimulus such as temperature or blood pressure. Sensory receptors involved in homeostasis function like internal guards. They monitor the body for changes. When conditions within the body change, an imbalance occurs. This imbalance is a stimulus that signals to the body that conditions have changed.
- Information from sensory receptors throughout the body flows to a **control center**. The control center processes the incoming information. It compares this information to a set point, or normal condition in the body.
- If the information differs from the set point, the control center sends a message to the body's **effectors**—muscles and glands—to carry out the required response to return the body to the set point, or normal conditions.

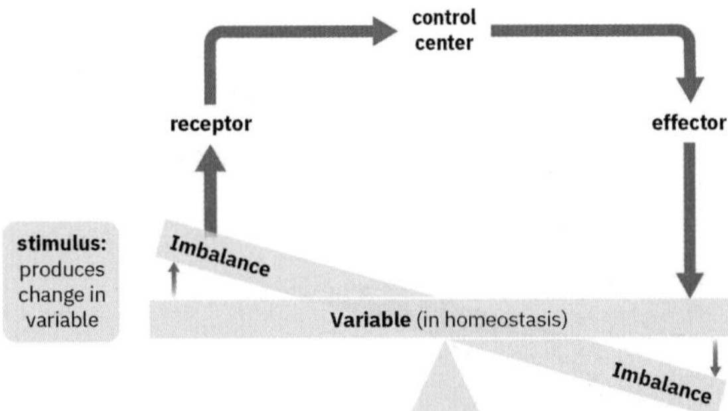

FIGURE 10-11 When an imbalance occurs in an animal's body, information from sensory receptors is sent to the control center. The control center in turn sends a message that activates the body's effectors (muscles and glands) to return the body to stable, or balanced, conditions.

Go Online **INTERACTIVE FIGURE**

FIGURE 10-11 Go online to see how an imbalance affects **Body Systems and Homeostasis**.

CCC Stability and Change *Use the terms shown* ***Figure 10-11*** *to describe how your body responds when you accidentally touch a hot surface such as a stovetop.*

CROSSCUTTING CONCEPTS | Stability and Change

Feedback for Homeostasis It is important for students to understand that changes occur within the body to maintain homeostasis, but what changes occur depend on the imbalance. To emphasize this point, have student groups analyze **Figure 10-12** to identify a variable that can cause an imbalance. They may identify hunger, pain, cold, or any change of their choosing. Then, have them identify the sensory receptor that would signal to the body that the conditions changed, the control center that would process the information from the sensory receptor, and effectors that would respond to return the body back to homeostasis. In many cases, the nervous system would be the main control center; however, there are some cases where other systems would be the control center. If necessary, allow student groups time to determine which system would be the control center.

In vertebrates, the nervous system is the body's main control center. The brain, spinal cord, and nerves are part of this system, as are sensory organs such as the eyes. The nervous system gathers information about the internal and external environment, integrates this information, and issues commands to muscles and glands. Neurons make up the communication lines of a nervous system.

The endocrine system consists of all the body's hormone-secreting glands and cells. Portions of the endocrine system and nervous system are so closely linked that scientists sometimes refer to the two systems collectively as the neuroendocrine system. Glands in the endocrine system and neurons in the nervous system receive signals from the hypothalamus, a structure that is the center for homeostatic control of the internal environment. Most organs respond to both hormones and signals from the nervous system.

Negative Feedback Mechanisms Homeostasis typically involves a **negative feedback mechanism**, in which a change brings about a response that reverses the change. An air conditioner with a thermostat provides a familiar nonbiological example of a negative feedback mechanism. A person sets the air conditioner to a desired temperature. When the temperature rises above this preset point, a sensor in the air conditioner detects the change and turns the unit on. When the temperature decreases to the desired level, the thermostat detects the change and turns off the air conditioner. (Note that the use of "negative" in this context does not mean the response is bad or undesirable, but rather that it negates the stimulus that caused it.)

A similar negative feedback mechanism operates when you exercise on a hot day (**Figure 10-12**). Muscle activity generates heat, and your body's internal temperature rises. Sensory receptors in the skin detect the increase and signal your brain. In response, the brain signals effectors. These signals increase the flow of blood from the body's hot interior to the skin. The shift maximizes the amount of heat given off to the surrounding air. At the same time, sweat glands in the skin increase their output. Evaporation of sweat helps cool the body surface. You breathe faster and deeper, speeding the transfer of heat from the blood in your lungs to the air. Your rate of heat loss increases, and your body's internal temperature decreases back to normal.

CHAPTER INVESTIGATION

The Effect of Exercise on Homeostasis
How does exercise affect homeostasis?
Go online to explore this chapter's design-your-own investigation about the feedback mechanisms that maintain homeostasis.

FIGURE 10-12
This diagram shows the negative feedback mechanism that maintains human body temperature under hot conditions.

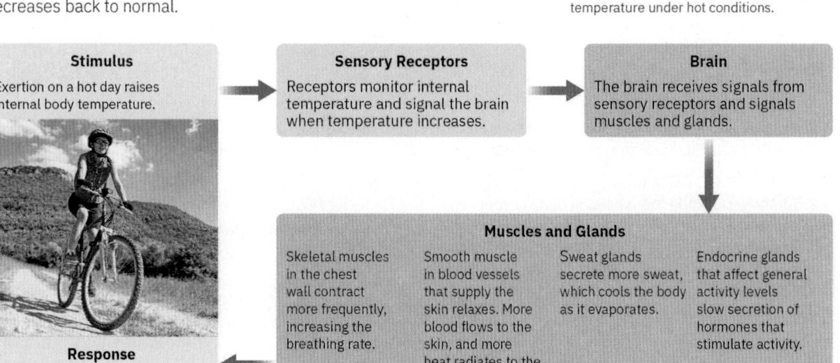

Stimulus
Exertion on a hot day raises internal body temperature.
Sensory Receptors
Receptors monitor internal temperature and signal the brain when temperature increases.
Brain
The brain receives signals from sensory receptors and signals muscles and glands.
Muscles and Glands
Skeletal muscles in the chest wall contract more frequently, increasing the breathing rate.
Smooth muscle in blood vessels that supply the skin relaxes. More blood flows to the skin, and more heat radiates to the surrounding air.
Sweat glands secrete more sweat, which cools the body as it evaporates.
Endocrine glands that affect general activity levels slow secretion of hormones that stimulate activity.
Response
Body temperature declines.

CHAPTER INVESTIGATION A

Design Your Own *The Effect of Exercise on Homeostasis*

Time: 100 minutes or 2 days

Students will design their own investigation to determine how exercise affects feedback mechanisms that maintain homeostasis.

Go online to access detailed teacher notes, answers, rubrics, and lab worksheets.

Energy from Food Students learn how animal systems interact to use the energy from food to maintain homeostasis. **Chapter 6** summarizes how sugar molecules are broken down at the molecular level during cellular respiration. **Appendix D** explores this process in greater detail.

Visual Support

Negative Feedback Mechanisms
Direct students to **Figure 10-12** and have them follow the path from the stimulus to the sensory receptors, to the brain, and then to the muscle glands. Have students discuss how the mechanisms adjust if the person exercises harder or when they have completed their bike ride. Students can consider what the loop would look like if someone was sitting outside on a cold day. Students should identify that the sensory receptors would signal that the internal temperature decreased. The brain would interpret the signals and signal instructions to muscles and glands. Students should be able to explain that the effectors would function in a different manner if the body was trying to keep warm instead of cool. Ensure that they understand that these are negative feedback loops because these systems are constantly adjusting to maintain homeostasis.

DIFFERENTIATED INSTRUCTION | English Language Learners

Using Non-Verbal Cues Before students read the last paragraph following the blue Feedback Mechanisms heading and examine **Figure 10-12**, have them explain to a classmate what happened when they rode a bike or otherwise exercised on a hot day. Encourage them to use gestures and act out what happens, especially if they do not know the English words. After describing their own experience, they will be better equipped to understand the scientific explanation of what happens.

Beginning Have students act out getting warmer, sweating, breathing deeper, and cooling off as they pedal a bike on a hot day.

Intermediate Have students use phrases or short sentences accompanied with gestures.

Advanced Have students give a detailed explanation with some gestures to clarify meaning if they are unsure of any vocabulary.

Vocabulary Strategy

Semantic Map Students may struggle differentiating between negative and positive feedback mechanisms. Suggest they make a semantic map for each term to visually connect the meanings. Students should place each word in its own map, define the term, include examples of systems they each control, and use symbols. They may show that negative feedback mechanisms typically work by reversing a change, while positive feedback mechanisms work by maintaining the direction of the change.

Connect to Mathematics

Model with Mathematics Students should understand that the messages sent to the effectors are chemical reactions. Remind students that the rate of reactions can be modeled by using graphs. Allow them time to work in pairs and develop a model of what the rate of the signals would look like for a negative feedback mechanism versus what it would look like for a positive feedback mechanism. Make sure that students understand that positive feedback mechanisms function until the change is no longer needed and then stop, while a negative feedback loop is continuous because it is constantly working to maintain homeostasis. In **Figure 10-13**, once the damaged site is sealed, the chemicals that attract more platelets stop.

CCC Stability and Change *In a negative feedback mechanism, a response reverses the direction of the stimulus. In a positive feedback mechanism, the response maintains the direction of the stimulus.*

FIGURE 10-13 The formation of a blood clot is an example of a positive feedback mechanism. Rather than returning the body to a set point, a positive feedback loop enhances the change away from a set point.

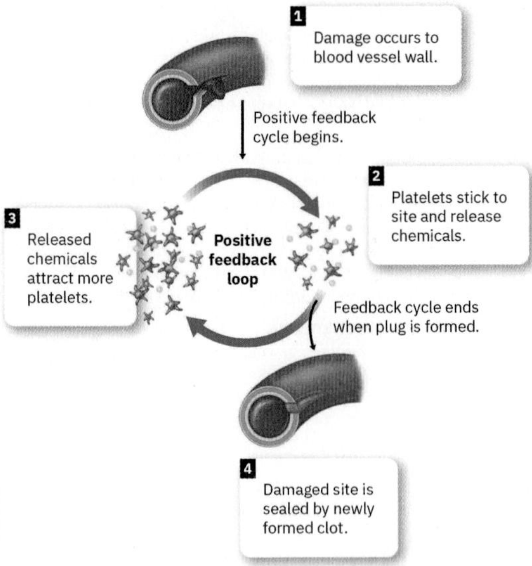

1 Damage occurs to blood vessel wall.
Positive feedback cycle begins.
2 Platelets stick to site and release chemicals.
Positive feedback loop
3 Released chemicals attract more platelets.
Feedback cycle ends when plug is formed.
4 Damaged site is sealed by newly formed clot.

Positive Feedback Mechanisms Unlike the negative feedback mechanisms that help to maintain homeostasis, a **positive feedback mechanism** maintains the direction of a stimulus, rather than stopping it. One example is blood clotting, shown in **Figure 10-13**. Once a clotting factor is triggered, it triggers the next factor in the sequence until a clot is formed. In the case of blood clotting, chemicals that attract platelets continue to be released by the body until a plug is formed by the accumulated platelets at the site of blood vessel damage.

CCC Stability and Change *What is the main difference between negative and positive feedback mechanisms?*

10.3 REVIEW

1. **Explain** Use the example of body temperature regulation to illustrate the role of a negative feedback mechanism.
2. **Identify** Describe the functions of a sensory receptor, control center, and effector in maintaining homeostasis. How are these three parts of a control system related?
3. **Distinguish** Identify each of the following response outcomes as part of a positive feedback or negative feedback mechanism.
   - response reverses a change
   - response enhances a change
   - response reinforces stability
   - response increases instability

10.3 REVIEW

1. *Sample answer: To maintain a body's internal temperature, a negative feedback mechanism reverses changes that the external environment introduces. If the temperature increases, the negative feedback mechanism triggers sweating, an action which helps to cool the body. If the temperature decreases, the negative feedback mechanisms trigger shivering, which helps to warm the body.* **DOK 3**
2. *Sample answer: The sensory receptors, control center, and effectors work together to maintain homeostasis. Sensory receptors detect a change in internal conditions and send a signal to the control center. The control center processes the information and sends a message to the effectors to carry out the required response to return the body to stable conditions.* **DOK 2**
3. *negative: response reverses a change, response reinforces stability; positive: response enhances a change, response increases instability* **DOK 2**

LOOKING AT THE DATA

THERMOREGULATION

SEP **Use Mathematics** Thermoregulation, or body temperature control, is an example of homeostasis. However, not all animals regulate body temperature in the same way. Animals called endotherms use internal mechanisms to maintain constant body temperature. Ectotherms are animals that can tolerate a variable body temperature and use behavior to regulate their temperature.

Tables 1 and **2** list the temperature responses, average mass, and average metabolic rate for various organisms.

A gopher snake basks in the sun.

TABLE 1. Body Temperature Response for Four Model Organisms

	Environmental temperature (°C)		
	0	15	30
Organism	**Body temperature (°C)**		
bobcat	38	39	40
mouse	35	36	36
salamander	0	14	32
snake	5	16	29

TABLE 2. Average Mass and Metabolic Rates

Organism	Average mass (kg)	Average metabolic rate (kcal/hr)
bobcat	14	10^2
deer	180	10^3
dolphin	90	10^3
fish	5	10^{-1}
mouse	0.5	10^2
salamander	0.01	10^{-3}
snake	35	10^{-3}
turtle	160.01	10^{-3}

1. **Identify Patterns** Describe how the model organisms listed in **Table 1** respond to increases in environmental temperature.
2. **Classify** Label the model organisms in **Table 1** as endotherms or ectotherms.
3. **Analyze** Use the classification of the model organisms and the data in **Table 2** to compare the average mass and metabolic rate for endotherms and ectotherms. What correlation to thermoregulation does the data support?
4. **Predict** Classify the organisms in **Table 2** as endothermic or ectothermic based on the correlation supported by the data.

LOOKING AT THE DATA
THERMOREGULATION

Connect this activity to the Science and Engineering Practices of using mathematics and computational thinking and analyzing and interpreting data. Students analyze two sets of data: the body temperature of four organisms at three different environmental temperatures and the average mass and metabolic rates of eight other organisms. Students use mathematical principles to analyze the patterns and are expected to predict whether the organisms are endothermic or ectothermic based on the data.

SAMPLE ANSWERS

1. *As environmental temperature increases, the body temperature remains about the same in the bobcat and mouse but increases in the salamander and snake.* **DOK 2**
2. *bobcat and mouse: endotherms; salamander and snake: ectotherms* **DOK 2**
3. *Endotherms (bobcat and mouse) have different mass but the same average metabolic rate. Ectotherms (salamander and snake) have a range of mass but the same average metabolic rate. There is a correlation between thermoregulation and metabolic rate. Endotherms have a high metabolic rate, while ectotherms have a low metabolic rate.* **DOK 2**
4. *endotherms: bobcat, deer, dolphin, mouse; ectotherms: fish, salamander, snake, turtle* **DOK 3**

10.4

ANIMAL BEHAVIOR

LS1.A, LS2.D, LS4.B

EXPLORE/EXPLAIN

This section introduces different kinds of animal behaviors, including instinct and learning, as well as how animals communicate with each other. The section also covers how parenting and group behavior can lead to increased survival for individuals.

Objectives

- Differentiate between instinctive behaviors and behaviors that are learned.
- Construct an explanation of the different ways that animals communicate with other members of their species and other species.
- Summarize how parenting behavior affects offspring survival.
- Distinguish how group living and social behavior can increase individual survival.

Pretest Use **Question 9** to identify gaps in background knowledge or misconceptions.

Analyze *Students might say that by alerting others, the prairie dog helps to ensure that other members of the colony will survive any potential attack by a predator. Another example of an alarm call is a deer snorting/huffing in response to a predator.*

Vocabulary Strategy

Using Prior Knowledge Students will have been introduced to the idea that animal behaviors affect survival and reproduction in their elementary and middle school science classes. As they are introduced to the key terms in this section, have them make connections with their prior knowledge and identify other examples of the terms that they are familiar with.

10.4

KEY TERMS

altruism
eusocial
instinctive behavior
learned behavior
pheromone

ANIMAL BEHAVIOR

In a colony of prairie dogs, one individual stands on its hind legs and scans the area for predators as the rest of the group forages for food. When a hawk soars above, the prairie dog issues a loud alarm call to warn the others. The rest of the colony members scurry into their burrow for safety.

Analyze *What is the function of an alarm call? Can you think of other examples of this type of behavior?*

Instinct and Learning

All animals have some **instinctive behaviors**—innate responses to a specific and usually simple stimulus. The life cycle of the cuckoo bird provides several examples of instinct at work. The common cuckoo is a brood parasite, meaning it lays its eggs in the nests of other birds. The female cuckoo scatters her eggs among the nests, leaving just one egg in each. A newly hatched cuckoo is blind, but contact with an object beside it stimulates an instinctive response. The cuckoo moves the object, which is usually one of its foster parents' eggs, onto its back, then shoves it out of the nest. The cuckoo will also shove its foster siblings over the side, if they hatched before it did. Instinctively shoving these objects out of the nest ensures that the cuckoo receives its foster parents' undivided attention.

Instinctive responses are adaptive, or beneficial for survival, when the triggering stimulus almost always signals the same situation and requires the same response. Getting rid of an egg or unrelated sibling always benefits a cuckoo chick because it rids the nest of future competitors for food. Instinctive responses sometimes open the way to exploitation. For example, most adult birds instinctively fill any wide-open mouth they see in their nest, a response that is clearly adaptive when those mouths belong to their chicks. However, this parental instinct opens the way to exploitation by cuckoos, which often look nothing like the chicks they displace (**Figure 10-14**).

FIGURE 10-14 A reed warbler feeds a dragonfly to a much larger common cuckoo chick. This is a behavioral response to a simple cue: the chick's gaping mouth.

CROSSCUTTING CONCEPTS | Cause and Effect

Unexpected Effects Understanding cause-and-effect relationships is essential to understanding how to interpret scientific observations. Most animal behaviors and traits can be linked back to how they increase an animal's ability to survive and reproduce. However, as seen in **Figure 10-14**, occasionally the behaviors do not work out for the animal as expected. Have students make a chart. As they work through the section, have them list behaviors described in the text. They should then identify the cause-and-effect relationship each behavior has on the survival or reproductive success for the animal. Make sure that students are making correct correlations between the causes and effects and that they are making specific claims in their charts that are supported by the text.

FIGURE 10-15 Chimpanzees use sticks as tools for extracting tasty termites from a nest. The method of shaping the stick and catching termites is learned by imitation.

A **learned behavior** is a behavior that occurs only as a result of practice or experience. Some instinctive behaviors can be modified with learning. For example, a newly hatched toad instinctively tries to capture and eat any insect-sized moving object it sees. This behavior is modified after encounters with bees or wasps. The toad quickly learns that insect-sized objects with certain color patterns provide a sting rather than a meal. A genetic capacity to learn, combined with experiences in the environment, shapes most forms of behavior.

Some animals learn a behavior by imitation. Consider how a chimpanzee strips leaves from branches to make a "fishing stick" for use in capturing termites as food (**Figure 10-15**). The chimpanzee inserts a fishing stick into a termite mound, then withdraws the stick and eats the termites that cling to it. Different groups of chimpanzees use different methods of tool shaping and insect fishing. In each group, young chimpanzees learn by imitating the behavior of others.

SEP Argue From Evidence *What is the difference between a learned behavior and an instinctive behavior?*

Address Misconceptions

Learned Behavior Students may think that animals purposefully try to learn new skills to improve their chances of survival or reproduction. Help students understand that when animals learn new skills, it is typically as a reaction to a stimulus. For instance, the text calls out the example of a toad learning that certain color patterns will result in a sting. The toad is associating the colors of the insect with pain, which is why it learns to avoid eating wasps and bees. While innate behaviors are passed on from offspring to parents, learned behaviors are obtained after the organism experiences some sort of stimulus from the action itself, be it immediate gratification from feeding, like the chimpanzees in **Figure 10-15**, or pain from touching or eating something that is harmful.

SEP Argue from Evidence *An instinctive behavior is one that is inborn and often only requires a simple stimulus to trigger it. A learned behavior is a behavior that must be taught through experience.*

Go Online **VIRTUAL INVESTIGATION**

Communication in the Rainforest
Students learn about the many different organisms that make up the El Yunque rainforest ecosystem in Puerto Rico and gather evidence to describe how those organisms communicate.

Video

Time: 0:31 seconds

Video 10-1 shows the mating rituals of a frigatebird in the Caribbean. Have students identify the different signals that frigatebirds use to communicate when trying to find a mate. You may choose to pause the video and replay at the midpoint and have students identify the obvious signals observable in this segment (*sight and sound*).

Go Online VIRTUAL INVESTIGATION

Communication in the Rainforest
How do rainforest species communicate?
Explore how animals communicate in a Puerto Rican rainforest.

Animal Communication

Communication signals are cues that transmit information from one member of a species to another. A communication signal arises and persists only if it benefits both the signal sender and the signal receiver.

Chemical Signals A **pheromone** is a chemical communication signal that is secreted by one individual and brings about a change in another member of the same species. In many insects, a pheromone helps individuals locate a suitable mate. For example, a female silk moth releases a sex pheromone that a male silk moth can detect from as far as a kilometer away.

Sound Acoustic signals, or sounds, often announce the presence of an animal or group of animals. Many male vertebrates, some fish, and many insects make sounds to attract prospective mates. In numerous species, sounds made by males also function in territoriality. Some birds and mammals give alarm calls that inform others of potential threats. In many cases, the calls convey more information than simply "Danger!" A ground squirrel makes one type of bark when it detects an eagle and another when it sees a coyote. Upon hearing the call, other ground squirrels respond appropriately. They either dive into burrows (to escape an eagle's attack) or stand tall (to spot the coyote).

VIDEO

VIDEO 10-1 Go online to watch the courtship display of a male magnificent frigatebird.

Sight Visual communication is most widespread in animals that have good eyesight and are active during the day. In most birds, mating is preceded by a courtship display that involves specific movements and actions (**Figure 10-16**). A courtship display assures a prospective mate that the displayer is of the correct species and in good health. When two potential rivals meet, a threat display demonstrates each individual's strength and how well armed it is. If the rivals are not evenly matched, the weaker individual retreats. Threat displays benefit both participants by allowing them to avoid an energetically costly fight that could lead to injury.

FIGURE 10-16 The male magnificent frigatebird inflates its bright red throat pouch and clacks its slender beak to attract a mate. Large populations of these seabirds are found in the Galápagos Islands.

DIFFERENTIATED INSTRUCTION | English Language Learners

Talking about Media Have students use **Figure 10-16** and **Video 10-1** to explain concepts from the sections on "Sound" and "Sight" in the text. Have students take notes on the photo caption and what they observe in the photo and video before discussing.

Beginning Allow students to make notes in their native language if needed. Have them write down words in two categories: Sounds and Sights. Then, have them share and compare words with a partner.

Intermediate Have students use phrases and short sentences to describe the sights and sounds they observe. Encourage them to make drawings. Then, have them explain their notes to a classmate.

Advanced Have students take notes and then explain to a classmate how the sights and sounds they observed illustrate sentences in the text (for example, "Many male vertebrates . . . make sounds to attract prospective mates.").

Touch With tactile communication, touch transmits information. For example, after discovering food, a foraging honeybee worker returns to the hive and dances on the honeycomb surrounded by a crowd of workers (**Figure 10-17**). The interior of the hive is dark, so the workers follow the dancer's progress by touch. The speed and orientation of a successful forager's dance provides information about the location of the food.

VIDEO

VIDEO 10-2 Go online to watch a honeybee perform a waggle dance.

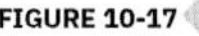

FIGURE 10-17 The angle of the waggle run, in which the bee wiggles back and forth in a straight line, gives the direction of the food. The faster the dance is performed, the closer the food.

CONNECTIONS

CAN YOU HEAR ME NOW? Right whales, like other whales, use sound to communicate. The sound they make most frequently is a whoop-like contact call that informs other right whales of their presence.

Sound-recording devices attached to individual right whales show that the level of shipping noise influences the whales' calling patterns. Whales call more loudly and are shriller in the presence of a ship. The louder the ship noise, the louder the whales call. Such vocal alterations improve sound transmission, but vital information can still be drowned out. Scientists do not know if the changes the whales make to their voices are sufficient to let their calls be understood despite the noise.

For humans, chronic exposure to noise is stressful, and the same may be true for whales. It would be unrealistic to ask tankers to stop their engines for days to monitor how noise affects whales. However, a quiet period did occur immediately after the terrorist attacks of September 11, 2001, when all shipping traffic was halted for a short period of time. At the time, researchers were collecting whales' feces to measure hormone secretions, including glucocorticoids, a class of hormones excreted in response to stress.

When the data about ship-related noise and hormone secretions were combined, scientists found that the brief decline in shipping noise was accompanied by a significant drop in the whales' stress hormone level. There was no drop in stress hormone levels immediately after September 11 of other years. These data suggest that the normally high noise level related to shipping causes a chronic stress response in right whales. We know that, in other mammals, chronic stress dampens the immune response and impairs reproduction. Thus, noise-induced stress may slow the recovery of this highly endangered species.

Noise from cargo ships and other ocean traffic affects the ability of whales to communicate with one another.

Video

Time: 0:18 seconds

Video 10-2 shows a bee communicating with other members of its hive using a waggle dance. Replay the video a few times and have students identify the bee that is doing the waggle dance and the bees that are touching it to determine where they can find food.

Connect to English Language Arts

Use Sources in Multiple Formats In this section, students learn about frigatebirds and honeybees from both the text and the videos. Allow them to select an animal of their choice that is not identified in the text but communicates with other organisms. Examples include dolphins, ostriches, and crickets. Have students gather information that includes data, video, text, and other multimedia sources to identify how their animal of choice communicates. Most animals will use more than one modality. Then, have them write a brief essay about how the animal's communication method increases the likelihood of survival.

CONNECTIONS | Can You Hear Me Now?

Social-Emotional Learning Students who live in high-stress environments often struggle to self regulate when compared to students who experience a manageable amount of stressors. Some students may live in very high-stress environments, such as poverty or homelessness, that they cannot control. Others who may not live in high-stress homes may still struggle with handling stressful situations. Help students have a conversation about stress, how they handle it, manage it, and let go of it. Students may identify that removing the stressor, such as what happened to the whales, has an immediate effect on their emotions and mental state. Other students may be able to share the tools they use to handle stressful situations. Students may have the opportunities to demonstrate *social awareness* by expressing empathy and respect for other students' situations and *self-management* by identifying the stress management strategies that they use to handle such situations.

In Your Community

Local Predators Have students identify a local species that has a behavior to "trick" other species to their benefit. Also known as *tactical deception*, some organisms have behaviors that protect them in the face of predators. For example, opossums have an involuntary reaction to fear that makes them appear dead when threatened. They will freeze and release a foul odor. This behavior can lead predators to avoid the animal. Have students discuss the species they selected and identify the behavior(s) of the species that increases its chances for survival.

SEP Construct an Explanation *Animal communication allows animals to transmit information from one member of a species to another.*

Eavesdroppers and False Signals Predators can benefit by intercepting signals sent by their prey. For example, frog-eating bats locate a male tungara frog by listening for his mating call. This poses a dilemma for the male frog. Female tungara frogs prefer complex calls, but complex calls are easier for bats to locate. Thus, when bats are near, male frogs call less often and with less flair. This subdued signal is a trade-off between competing pressures to attract a mate and to avoid being eaten. Other predators lure prey with counterfeit signals. Fireflies are nocturnal beetles that attract mates with flashes of light. When a predatory female firefly sees the flash from a male of the prey species, she flashes back as if she were a female of his own species.

SEP Construct an Explanation *What is the function of animal communication?*

Parenting Behavior

Parental care requires time and energy that an individual could otherwise invest in reproducing again. It arises only if the genetic benefit of providing care (an increased chance of offspring survival) offsets the cost (reduced opportunity to produce more offspring). If the parental care provided by a single parent is sufficient to successfully rear an offspring, individuals who spare themselves this cost by leaving parental duties to their mate are at an advantage. Thus, whether parental care occurs and, if so, who provides it varies among species.

Most fish provide no parental care, but in those that do, this duty usually falls to the male. In many species, males guard eggs until they hatch. In some cases, males retain eggs or young on or in their body. Male seahorses have a special brood pouch. Mouth brooding, which refers to holding developing eggs in an organism's mouth, occurs in a variety of fish species, including the jawfish (**Figure 10-18**). The prevalence of male parental care in fish may be related to sex differences in how fertility changes with age. A female fish's fertility increases with age, whereas a male's does not. Thus, a female fish who invests energy in parental care forfeits more reproductive potential than does a male fish.

FIGURE 10-18 After a female jawfish lays an egg mass, the male fertilizes the eggs. The male then takes the eggs into its mouth, where the eggs are incubated until they hatch.

In birds, two-parent care is most common. Chicks of birds that cooperate in care of their young tend to hatch while in a relatively helpless state, so the chicks require the care of two parents to survive. In about 90 percent of mammals, the female rears young. In the remaining 10 percent, both sexes participate. Female mammals sustain developing young in their body, so they do not have the option of pursuing a new mate rather than providing care. Females also have a greater investment in newborns than males, and this investment continues with lactation.

Group Living and Social Behavior

Some animals live solitary lives, coming together only to mate. However, individuals of many species spend time in groups. Some groups are temporary, such as schools of fish. Others, such as prairie dog colonies or wolf packs, are more permanent. A tendency to group together, whether briefly or for the longer term, will occur only if the benefits of being near to others outweigh the costs.

Benefits of Group Behavior Whenever animals cluster together, individuals at the margins of the group inadvertently shield others from predators. The selfish herd effect describes how groups can emerge from the tendencies of individuals to reduce their risk of danger by drawing near one another. For example, starlings form large swooping flocks called murmurations in part to confuse and disorient a predator such as a falcon.

Once a group forms, multiple individuals can be on the alert for predators. In some cases, an animal that spots a threat will warn others of the threat's approach. Birds, monkeys, prairie dogs, and ground squirrels are among the animals that make alarm calls in response to a predator. Even in species that do not make alarm calls, individuals can benefit from the vigilance of others. Often, when an animal notices another group member beginning to flee, it will do likewise. In what is called the "confusion effect," a predator has more difficulty choosing an individual to pursue when a group of prey is scattering. Some animals stand their ground to present a united defense. For example, a group of threatened musk oxen forms a tight circle so a predator faces their sharp horns (**Figure 10-19**).

CHAPTER INVESTIGATION

Monitoring Animal Behavior
How can you design, build, and test an effective device to monitor animal behavior?

Go online to explore this chapter's hands-on investigation about animal behavior.

FIGURE 10-19 When threatened by predators, musk oxen stand shoulder to shoulder to form a defensive line or circle to protect their more vulnerable young.

Connect to Mathematics

Model with Mathematics Students are likely familiar with emperor penguins and how both parents raise the chick until it is old enough to survive on its own. This sort of parental care is common in birds but not as much in other animal species. The text identifies that 90 percent of mammals are raised by only their mothers. Have students determine the percentages of other animal types, such as reptiles, fish, insects, and birds, that are either raised by one or two parents. Then, have them make a stacked bar graph where they model the percentage of each class of animal that does not provide parental care, provides parental care by one parent, or provides parental care by both parents.

CHAPTER INVESTIGATION B

Engineering Design *Monitoring Animal Behavior*

Time: 150 minutes or 3 days

Students will design, build, and test a prototype of a device that can be used to monitor behaviors exhibited by endangered species.

Go online to access detailed teacher notes, answers, rubrics, and lab worksheets.

DIFFERENTIATED INSTRUCTION | Leveled Support

Struggling Students To aid students in understanding how parenting behavior and living in groups increases the likelihood that an individual will survive to reproduce, have them consider the relative size of adults and offspring. Have students consider why grown prey are much harder to catch and eat than young prey. When young individuals are protected by adults, the predators have a harder time targeting them. This makes it more likely that the young will grow old and large enough to protect themselves.

Advanced Learners Students can relate the content in this section to prior knowledge. Have them review life history patterns and survivorship curves in Section 4.2. Then, have students list the animals discussed in the Parenting Behavior content of this section. They should identify the most likely survivorship curves associated with the species. If time permits, allow students to select other animals in this section and identify the survivorship curves of those species.

Connect to Mathematics

Graph a Function Guide students to **Figure 10-20** and have them consider the cooperative hunting efficiency that is described in the lion example. Ask students to think about what a graph of cooperative hunting efficiency will look like if the wolves have the same success rate as the lions. Remind them that the success rate increases by the same factor for every additional hunter but that the food will also need to be divided further. Have students write a function to describe the relationship, and then draw a graph of the function. Student functions should show that as the number of lions increases, the percentage of a successful hunt increases by 15 percent. Students should have a linear graph with a slope of 0.15. $f(n) = 0.15n$.

Science Background

Lions Students are likely aware that lions often live and hunt in packs called *prides*. Prides break apart and come back together but generally consist of a number of related females, their offspring, and a coalition of males. While a female lion is taking care of her offspring, she will typically not mate until the young are at least 18 months old, but she will mate if she has no young. When a new coalition of males takes over a pride, they often attempt to kill off all of the cubs produced by the males in the previous coalition, but the female members of the pride will attempt to protect each other's cubs. Lead students in a discussion on why this phenomenon occurs. Guide students to understand that the males are attempting to pass on their genetic information, which will only happen when they produce offspring with the females. However, the related females have shared genetic information with the offspring of the other females in their packs.

FIGURE 10-20 An Arctic wolf pack works together to attack a herd of musk oxen on Ellesmere Island in Nunavut, Canada.

Wolves and lions are social animals that live in multigenerational groups and cooperate in hunting prey (**Figure 10-20**). Cooperative hunting allows a group of predators to capture larger or faster prey. However, on a per-hunter basis, cooperative hunting is usually no more efficient than hunting alone. In one study, researchers observed that a solitary lion catches prey about 15 percent of the time. Two lions hunting together catch prey twice as often as a solitary lion, but having to share the spoils of the hunt means the amount of food per lion is the same. Among carnivores that hunt cooperatively, hunting success is not the major advantage of group living. Individuals hunt together but also cooperate in fending off scavengers, caring for one another's young, and defending the group's territory.

Costs of Group Behavior Individuals that group together are easier for predators to locate. In addition, parasites and contagious diseases spread more quickly when individuals come into frequent close contact. Grouping also increases competition for food and mates, and the resulting conflict increases stress.

In most social groups, costs and benefits are not equally distributed among the group members. Many animals that live in groups form a hierarchy, a social system in which dominant animals get a greater share of resources and breeding opportunities than subordinates. Typically, dominance is established by physical confrontation. In wolf packs, the dominant male and dominant female are usually the only breeders.

CROSSCUTTING CONCEPTS | Cause and Effect

Benefits and Drawbacks Students should have a general understanding of the cause-and-effect relationships that exist in group living. In order to support their understanding, allow students to work through this section as a group. After they have completed the reading, have them consider the effect of animals living in groups, for both predators and prey. Allow them to discuss why the animals discussed in the lesson live in groups. Then, have them identify possible effects, both good and bad, of living in a group, and connect how effects at a small scale can lead to changes in the entire group. Students should use evidence from the text and can use outside sources to support their arguments, but they should carefully assess whether or not the evidence supports their claim. After the groups have completed making their claims, bring the class back together for a full class discussion.

All members of a wolf pack hunt and carry food back to individuals that guard the young in their den. Why would a wolf stay in a pack where it is subordinate? Belonging to a group enhances the subordinate's chance of survival, and some subordinates do get a chance to reproduce in periods when food is abundant. A subordinate may also one day move to the dominant reproductive role.

SEP Argue From Evidence *What is one cost and one benefit of social behavior?*

Evolution of Cooperation In many animal groups, individuals display **altruism**, which is behavior that decreases an individual's likelihood of survival and reproduction but increases the likelihood of survival and reproduction of other group members. Giving an alarm call is an example of an altruistic act (**Figure 10-21**). It puts the caller at an increased risk of predation but reduces the risk to others.

Field studies of social animals confirm that many altruistic behaviors are preferentially directed at relatives. For example, ground squirrels and prairie dogs are more likely to give an alarm call if they have close relatives living nearby. In hyenas, which often have intragroup fights, dominant females are more likely to risk injury by joining a fight to assist a relative than to assist a nonrelative.

Cooperative behavior can also arise through reciprocity, in which an individual assists group members who have previously assisted it or may assist it in the future. Consider vampire bats, which roost during the day in large groups. Hungry bats who have not successfully fed the night before are offered food (regurgitated blood) by other bats. Close kinship increases the likelihood of food sharing, but sharing also occurs among unrelated bats. Bats in need who previously offered food to many different recipients receive more offers of food than bats who were less generous.

VIDEO

VIDEO 10-3 Go online to learn more about prairie dog alarm calls.

FIGURE 10-21 A black-tailed prairie dog makes a loud alarm call in response to an intruder. Such calls alert the prairie dog's relatives to take cover from potential danger, which aids in their survival.

Video

Time: 1:11

Video 10-3 shows a prairie dog and a coyote that hunts it. Play the video before students read the content covered in *Evolution of Cooperation* and ask them to discuss why they think the risky behavior of calling out when predators are nearby would be helpful for the group while dangerous for the individual.

SEP Argue from Evidence *Animals that exhibit social behaviors may have a better chance of survival by having a social structure in which some individuals look out for predators or other dangers; a cost is that individuals that group together are more easily spotted by predators.*

Social-Emotional Learning

Relationship Skills Cooperative behaviors are observed in many species, but the scope of human cooperation is considered to be higher than most other organisms. Invite students to consider how they interact in groups. For example, they could think about how they work on group projects. Students should consider how they *practice teamwork and collaborative problem-solving*. They should also think about how they might recognize when they should *seek or offer support and help when needed*. As collaborative projects are assigned, support students in *showing leadership skills* as they move from project to project and team members shift around.

In Your Community

Multigenerational Families Have students consider the similarities that exist between social animal species and multigenerational families of humans that live together. Students should understand that every member of either group has a job that they may do individually or collectively but that overall is expected to help the entire group. Allow students that live in multigenerational families that would like to share their experiences discuss the roles that the different family members have.

SEP Construct an Explanation *Organisms that perform altruistic behaviors do so because the behavior typically aids close relatives, thus ensuring that their genes are maintained in a population.*

REVISIT THE ANCHORING PHENOMENON

Gather Evidence *Sample answer: Leaf-cutter ants likely leave a trail of pheromones for other worker ants to follow. Pheromones persist in the environment for a longer period of time than other communication signals.*

Ants from different species and from different groups use pheromones to communicate with each other. In addition to guiding each other to a food source, ants can also find their way back to their colony by following their own pheromones. As other ants make their way to the food source, they drop additional pheromones, making the trail stronger. This allows ants to make group decisions as to which food sources to focus on. The stronger the trail of pheromones, the more profitable the food source. In addition to food, ants also use pheromones to identify dead ants, which can be moved away from the active colony. By removing dead ants, the likelihood of disease decreases and the hive is kept healthy.

RAINFOREST CONNECTIONS

Gather Evidence *Leaf-cutter ants are commonly found on the rainforest floor. Their trails can extend more than 200 meters in length. How do you think the ants communicate the location of a food source to other ants in their colony? Explain your reasoning.*

Eusocial Species Animals that are **eusocial** live in a multigenerational family group in which sterile workers carry out tasks essential to the group's welfare, while other members of the group reproduce. Many eusocial species are members of the order Hymenoptera, which includes ants, bees, and wasps. Leaf-cutter ants, shown in **Figure 10-22,** are a eusocial insect species that is commonly found in rainforest habitats. As their name suggests, worker ants cut leaves, which they bring back to their nest to use as a fertilizer to grow the fungus that is fed to ant larvae. In eusocial Hymenoptera, all workers are female. Workers forage for food, maintain the nest or hive, care for young, and defend the colony.

SEP Construct an Explanation *What is the genetic basis of altruistic behaviors?*

FIGURE 10-22 A leaf-cutter ant carries a leaf back to the colony's nest in Costa Rica's Manuel Antonio National Park. Leaf-cutter ant colonies are divided into several castes. Each caste serves a different function.

10.4 REVIEW

1. **Classify** Identify each animal behavior as an example of an instinct or a learned behavior.
   - Young coyotes find food sources by hunting with adult coyotes.
   - Rattlesnakes naturally rattle their tail as a warning to potential threats.
   - Migrating birds begin their journey to warmer climates at the start of fall.
2. **Identify** Which of the five senses are required for each type of animal communication?
   - acoustic signal
   - chemical signal
   - tactile signal
   - visual signal
3. **Explain** Which of these are the benefits of group behavior? Select all correct answers.
   A. assistance in raising young
   B. cooperation for hunting prey
   C. abundance of food resources
   D. increased defenses against predators
   E. protection from contagious disease
4. **Distinguish** Explain the difference between species that display altruism and species that are eusocial.

10.4 REVIEW

1. *instinct: Rattlesnakes naturally rattle their tail as a warning to potential threats. Migrating birds begin their journey to warmer climates at the start of fall. learned behavior: Young coyotes find food sources by hunting with adult coyotes.* **DOK 2**
2. *acoustic signal: hearing; chemical signal: smell, taste; tactile signal: touch; visual signal: sight* **DOK 2**
3. *A, B, D* **DOK 2**
4. *Altruistic species put the needs of the social group ahead of their own needs, sometimes to their own disadvantage. A eusocial species is one that has many nonreproducing individuals that instead carry out tasks essential for the group's welfare.* **DOK 2**

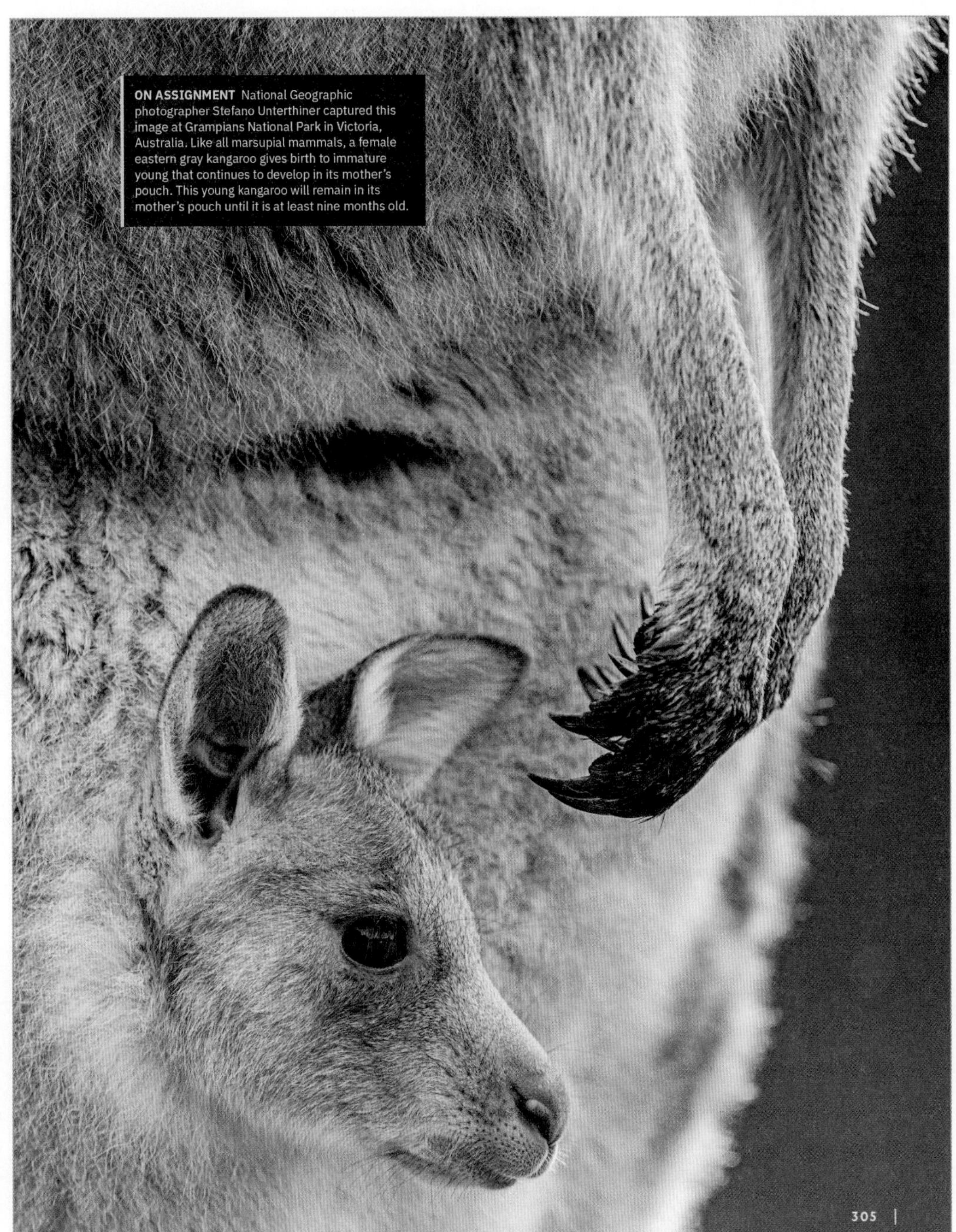
ON ASSIGNMENT National Geographic photographer Stefano Unterthiner captured this image at Grampians National Park in Victoria, Australia. Like all marsupial mammals, a female eastern gray kangaroo gives birth to immature young that continues to develop in its mother's pouch. This young kangaroo will remain in its mother's pouch until it is at least nine months old.

305

ON ASSIGNMENT

Kangaroo Joeys National Geographic photographer Stefano Unterthiner is a zoologist who travels the world to photograph animals. He spends long periods of time in close contact with the animals he is telling the stories of in order to capture the nuances of animal life and bring people into nature. He has spent over twenty years photographing animals and has a strong commitment to environmental and wildlife conservation. This photo, taken while on assignment for a February 2019 National Geographic issue on the conflicts between kangaroos and humans, shows a mother kangaroo with a joey in her pouch. Marsupials like kangaroos, koalas, and opossums are the only mammals that have pouches within which young develop after birth. You can learn more about Stefano Unterthiner on his website: http://www.stefanounterthiner.com.

Science Background

Marsupials Students are likely aware that kangaroos are marsupials and carry their young in their pouch, but they may not be aware that there are over 250 marsupial species that share this trait. Marsupials are mammals that give birth to young that partially develop in the uterus. Once the young are born prematurely, they travel to the mother's pouch where they attach to a nipple to feed. They remain there until the completion of their development. Over 200 of the more than 250 known marsupials are native to Australia, but approximately 70 species span the Americas, primarily in Central and South America. Students might be familiar with opossums, the only known marsupial to range across North America.

EXPLORER
HOLLY FEARNBACH

About the Explorer

National Geographic Explorer Holly Fearnbach studies whale populations around the world. She uses drones to photograph whales to study their health. You can learn more about Holly Fearnbach and her research on the National Geographic Society's website: www.nationalgeographic.org.

Science Background

Killer Whales While colloquially known as killer whales, *Orcinus orca* is actually the largest species of dolphin. They are apex predators and live in family groups called *pods*. Pods can be relatively small but can also be as large as 40 individuals. Female orcas generally nurse their young for up to two years. Some young orcas may leave the pod and join a new one after they mature, but others remain for their entire lives.

Connect to Careers

Cetologist Because of the intelligence, of whales and dolphins, the public is often intrigued by them. A cetologist is a marine biologist who focuses on the study of cetaceans: whales, dolphins, and porpoises. These researchers study the behaviors and ecological role of these animals. They are often also involved in conservation efforts to support the preservation of these organisms that have been affected by human activities, such as commercial fishing and oil mining.

Video

Time: 2:03

Video 10-4 shows how the Octocopter uses photogrammetry to collect quantitative data on whales' health.

THINKING CRITICALLY

Identify Patterns *Sample answer: Biometric measurements collected from the aerial images can be used to detect fat loss and assess the overall body condition of individual whales in the population and thus help scientists to determine the health of the whale population as a whole.*

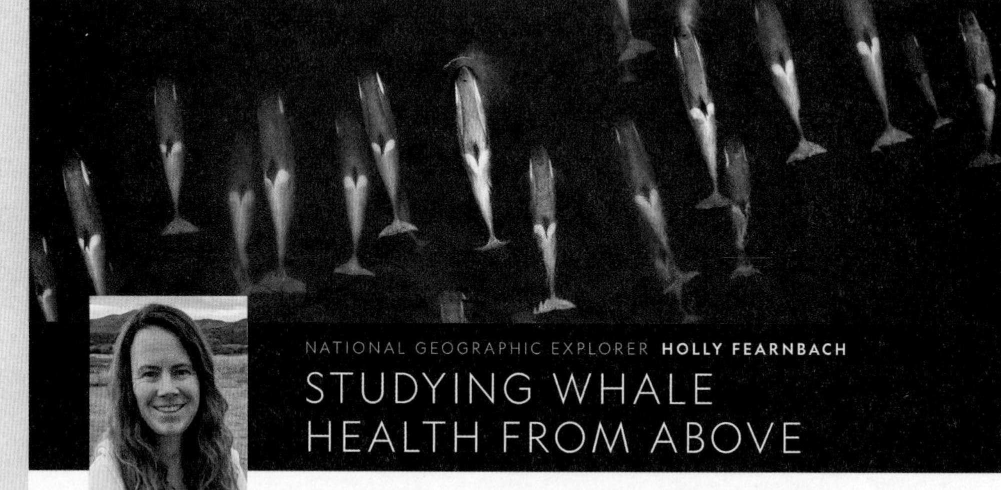

Dr. Holly Fearnbach is a leader in the field of marine mammal health.

VIDEO 10-4 Go online to view drone footage of Southern Resident killer whales captured by Fearnbach and a team of researchers.

Zoologist Dr. Holly Fearnbach's research focuses on assessing the health of whale populations around the world. "I've always been drawn to the ocean and fascinated by marine wildlife," Fearnbach says. Immediately after high school, Fearnbach decided to pursue a career in marine biology—and she has never looked back. Today, Fearnbach is the marine mammal research director of a Pacific Northwest nonprofit called SR^3 (Sealife Response + Rehab + Research).

Fearnbach and her colleagues pioneered the use of drones to collect aerial images of free-ranging whales. They study the whales' health using photogrammetry, which is the science of making measurements from photographs. They fly the drone at an altitude of more than 30 meters above the whales and position their research boat several hundred meters away, enabling the team to collect aerial images without disturbing the whales.

Fearnbach spends most of her year assessing the health of Southern Resident killer whales (SRKWs). This endangered population of only 73 individuals has a core summer habitat in the Salish Sea between Puget Sound, Washington and Southern Vancouver Island, British Columbia. Three key threats have been identified for SRKWs: physical and acoustic disturbance from vessels, contaminants in their food and water, and declines in the availability of Chinook salmon, their primary prey.

Fearnbach and her team analyze the aerial images to assess the health of every whale in the population. They identify each whale using distinct pigmentation patterns on the white saddle patches on the whale's back and sides. Then they collect a series of biometric measurements that are used to estimate body length, monitor growth, and assess body condition. These measurements give a picture of the overall health of both the individual whale and the population as a whole. The datasets are also provided to management groups to help inform decisions aimed at population recovery.

"My greatest motivation is the hope that our research efforts will continue informing conservation and management measures to protect marine ecosystems around the world," Fearnbach says.

THINKING CRITICALLY

Identify Patterns *Understanding the health of whales helps researchers understand the health of the ecosystem in which they live. Why are biometric measurements useful in assessing whale population health?*

DIFFERENTIATED INSTRUCTION | English Language Learners

Distinguishing /wh/, /w/, and /h/ Read the title of the article, and repeat the words *whale* and *health*. Have students focus on the initial sounds: /wh/ and /h/. Repeat slowly, distinguishing the two sounds. Point out the spellings for each. Then repeat the process with *whale* and *water*, distinguishing /wh/ and /w/. Guide students to notice the sounds as they listen to the text and watch the video.

Beginning Read word pairs from the text and have students identify the word with the target sound: /wh/ (*whale, marine*), /w/ (*research, wildlife*), or /h/ (*photograph, hundred*).

Intermediate Read paragraph 2 and have students raise a hand when they hear the /wh/, /w/, or /h/ sound. Pause for students to identify it. Then have students use one of the words in a sentence.

Advanced Read paragraph 3, and have students raise a hand when they hear the /wh/ sound (*whale, white, whole*). Then have them explain Fearnbach's research, using each word as they do so.

TYING IT ALL TOGETHER
THE INCREDIBLE NONSTOP FLIGHT OF THE BAR-TAILED GODWIT

HOW DO ANIMALS RESPOND TO CHANGES IN THE ENVIRONMENT?

In this chapter, you learned how an animal's body systems interact to help it survive and respond to changes in its environment. You also learned about the benefits of social interactions and group behavior. The Case Study described migratory behavior in the bar-tailed godwit, a type of shorebird. On average, these birds fly 29,000 km (18,000 miles) annually. While the length of the bar-tailed godwit's annual migration is impressive, the Arctic tern (**Figure 10-23**) flies nearly 90,000 km (56,000 miles) annually. However, unlike the bar-tailed godwit, the Arctic tern makes several stops during its journey from the North Pole to the South Pole.

Much about bird migration remains unknown, including how birds are able to navigate long distances. It is thought that birds use a combination of tools to navigate. These tools include celestial clues, such as the position of the sun, moon, and stars, to help them find their way. Research also indicates a protein called Cry4 in the eyes of birds allows them to "see" Earth's magnetic fields. This information can be used as a map and compass. Birds orientate themselves based on the strength of the magnetic field and its inclination, that is, the angle at which the magnetic field is tilted with respect to the surface of Earth.

In this activity, you will conduct research to identify the body system interactions involved in bird migration.

Obtain Information

1. Research the internal and external cues that trigger migratory behaviors in birds. Use your research to identify how these cues affect bird behavior.

Develop a Model

2. Choose a migratory bird species. Identify specific body systems and structures that support its migratory behavior. Use your research to describe how these body systems interact together.
3. Develop a model that illustrates these interactions.

Construct an Explanation

4. Make a presentation that uses digital media to share your findings, including your model, with your class.

FIGURE 10-23
Arctic terns have one of the longest migratory paths of all animals on Earth. Every year, these birds navigate from the Arctic Circle to Antarctica and back again.

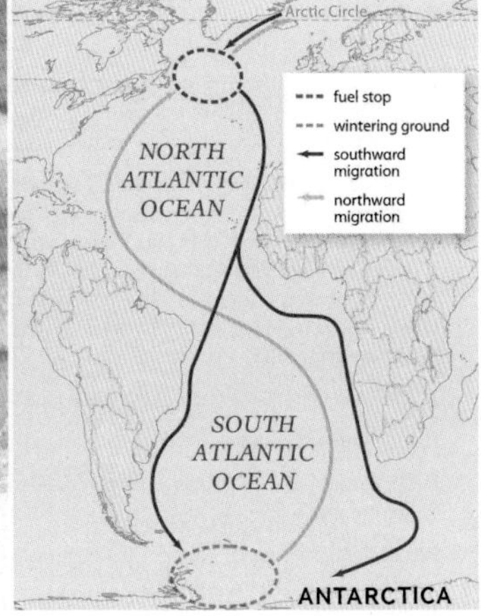

SCIENCE AND ENGINEERING PRACTICES
Developing and Using Models

Migratory Systems To help students prepare for this activity, have them select a migratory bird. Many ducks, geese, and crane species migrate, but students should not limit themselves to these examples. Once students have chosen which bird to research, help them plan the steps that they will follow to gather information. They will likely need to know the migratory routes and the distance that the birds travel, as well as where they nest to raise their young. Students should also research the body systems and structures that their birds have that support their migratory practices. Once students have developed an understanding of their bird species, have them list the body systems and structures they plan to model. Suggest students build a prototype first, but students can develop their models using the materials of their choice. Ensure students remember to incorporate their models into their digital presentations.

TYING IT ALL TOGETHER

ELABORATE

The Case Study discusses how the body systems of the bar-tailed godwit and its behaviors allow it to make its annual migration. In the Tying It All Together activity, students will conduct research to explain how the body system interactions support migration.

This activity is structured so that students may begin once they complete Section 10.3. **Step 1** has students researching the internal and external cues that trigger migration. **Step 2** requires that students identify the body systems and structures of a migratory bird and explain how those structures support migration. In **Step 3**, students build a model to illustrate the interactions. **Step 4** directs students to make a presentation using digital media to share their findings with the class.

Alternatively, this can be a culminating project at the end of the chapter. The research done for this activity reflects processes that students may use to collect evidence in the Unit 3 activity.

Go online to access the Student Self-Reflection and Teacher Scoring rubrics for this activity.

EVALUATE

REVIEW KEY CONCEPTS

1. *epithelial; nervous; connective; muscle;* **DOK 1**
2. *brain; sensory receptors;sensory receptors; muscles and gland; muscles and glands* **DOK 2**
3. *C, D* **DOK 1**
4. *D* **DOK 2**
5. *circulatory; skeletal; urinary; reproductive; digestive; respiratory; lymphatic; endocrine; muscular* **DOK 2**
6. *C* **DOK 2**
7. *A* **DOK 1**
8. *Sample answer: Food enters through the mouth and is broken down by chewing and saliva. It passes through the esophagus to the stomach, then enters the small intestine where most absorption of nutrients takes place. Digestive juices from the pancreas, liver, and gall bladder enter the small intestine and help break down and absorb nutrients. Nutrients move into the cells of the intestine and pass into the bloodstream via the circulatory system. The remains of the food enter the large intestine, where water is reabsorbed before food waste is eliminated.* **DOK 1**
9. *Sample answer: Advantages: warning others of threats; makes it easier for individuals to raise young and protect breeding animals. Disadvantages: competition for food, water, and other resources as well as increased risk of disease and parasitism.* **DOK 2**

CHAPTER 10 **REVIEW**

REVIEW KEY CONCEPTS

1. **Classify** Place each description of a tissue function into the appropriate category: nervous, connective, epithelial, or muscle.
   - lines, covers, and protects
   - sends and receives signals
   - provides structure and elasticity
   - contracts and relaxes to move body parts
2. **Classify** Label each role in maintaining homeostasis with the responsible body system. Choose from: sensory receptors; brain; and muscles and glands.
   - send(s) signals to effectors
   - monitor(s) body for changes
   - send(s) signals to control center
   - receive(s) signals from control center
   - take(s) action to restore homeostasis
3. **Identify** Which are examples of negative feedback mechanisms? Select all correct answers.
   A. A change in the external environment has an undesirable effect on the body.
   B. A body system operates to help the body maintain a stable internal environment.
   C. An increase in blood pressure triggers processes that decrease the blood pressure.
   D. An increase in body temperature triggers processes that decrease body temperature.
   E. A torn blood vessel attracts platelets that begin clotting around the tear; the platelets release chemicals that attract even more platelets.
4. **Explain** Wild pigs live in groups called sounders. Sounders are typically territorial, meaning they will not share their territory with pigs from other sounders. However, studies suggest that sounders are more willing to share territory with other sounders in areas where food is highly abundant. What is the most likely explanation for this?
   A. Well-fed animals have no need to live in groups.
   B. Competition is higher when there is abundant food.
   C. Abundance of food makes a habitat less desirable for the group.
   D. Benefits of territorial behavior are lower when resources are highly abundant
5. **Classify** Label each function with the animal body system that is primarily responsible for the function.
   - transports materials around the body
   - contains tissue where new blood cells are produced
   - helps maintain fluid and pH balance and excretes wastes
   - produces cells used for reproduction; may protect developing embryo
   - breaks down food so that nutrients can be absorbed; eliminates leftover wastes
   - brings in oxygen to be absorbed by the blood and exchanged with carbon dioxide
   - important in the immune response, returns fluid to the blood, and includes the spleen
   - produces hormones to cause the body to respond to stimuli and to control internal organs
   - moves bones; moves food through digestive tract and blood through blood vessels; helps generate heat
6. **Explain** Which statement reflects a typical cost of group behavior for animals that travel in herds?
   A. Individuals can alert other members of the herd when a predator is detected.
   B. When the herd flees, a predator may become confused and have trouble selecting an individual.
   C. Competition between group members for mates, food, and water can make it more difficult for some individuals to survive.
   D. All members receive the same benefits of living in the group, preventing stronger individuals from passing on their genes at a higher rate.
7. **Identify** Which of the following are chemical signals that convey information between individuals of the same species?
   A. pheromones
   B. neurotransmitters
   C. hormones
   D. releasers
8. **Explain** Describe how the circulatory and digestive systems interact so that nutrients from food can be transported through the body.
9. **Infer** Zebras live in large herds that roam the African savanna. Discuss two advantages and two disadvantages to zebras of living in a group.

308 CHAPTER 10 ANIMAL SYSTEMS

CRITICAL THINKING

10. *Sample answer: The protective behavior of the sterile workers benefits the hive by preventing the entire colony, including the queen, from being destroyed by the bear. The sterile workers cannot reproduce, so their genes are only carried into the next generation if their relatives (including the queen, potential future queens, and males) survive and continue to reproduce. If the queen behaved the same way, she would die upon stinging the bear. Without a queen, the colony would be unlikely to survive for very long even if it survived the temporary threat of the bear.* **DOK 3**
11. *Sample answer: The ability of others to replicate an experiment is an important part of the scientific process.* **DOK 3**
12. *Sample answer: The nervous and endocrine systems work together to detect stimuli and send signals that cause the body to respond appropriately. For example, the nervous system sends impulses through neurons to cause responses such as muscle contraction. This allows an animal to respond rapidly to stimuli. The endocrine system produces hormones. Hormones travel throughout the bloodstream and affect cells that have appropriate receptors to which the hormones can bind.* **DOK 3**
13. *Sample answer: Darker colors absorb more light energy than light colors, so the body temperature of the lizard is likely to increase as the lizard's skin darkens. A possible negative feedback mechanism would involve a decrease in the secretion the melanocyte-stimulating hormone once a certain body temperature is reached. This*

CRITICAL THINKING

10. **Explain** Queen honeybees lay both fertilized and unfertilized eggs. The fertilized eggs become females and the unfertilized eggs become males. Except for the occasional female that becomes a queen, adult female honeybees are sterile workers. If an animal such as a bear threatens the hive, worker bees will sting the bear and die soon after. How does this altruistic behavior benefit the group? If the queen behaved the same way, would the same benefit occur? Explain why or why not.

11. **NOS** **Scientific Investigations** A group of students is conducting an investigation to provide evidence that feedback mechanisms maintain homeostasis during exercise. Why is it important that the group's experimental design is clear and easy to follow?

12. **Synthesize** Identify the two major animal body systems that work together to control the body's responses to stimuli. Explain how these systems work together and complement each other.

13. **Predict** On a cool day, a particular lizard can regulate its body temperature by increasing secretion of a hormone that stimulates melanocytes. Melanocytes are cells that contain melanin, a pigment that darkens the lizard's skin. Predict how this color change will affect the lizard's body temperature. Describe a possible negative feedback mechanism that could prevent the lizard's body temperature from changing too much.

MATH AND ENGLISH LANGUAGE ARTS CONNECTIONS

Read the Connections box titled "Can You Hear Me Now?" in Section 10.4. Use evidence from the passage to answer questions 1–3.

1. **Cite Textual Evidence** Do you think whale songs are instinctive or learned? Explain your answer and argue whether it is advantageous or disadvantageous when whales encounter loud noise from human activities.
2. **Cite Textual Evidence** What body system is responsible for producing the hormone that serves as evidence that whales experience stress from the noise of human activities? Explain how these hormones function in the body.
3. **Reason Abstractly and Quantitatively** Use the information in the article to construct a graph modeling whale hormone levels between June 2000 and June 2002. How do you think this curve would compare to other periods of decreased or increased shipping levels, such as during a pandemic or an economic boom? How could you test your hypothesis?
4. **Write Explanatory Text** Male tungara frogs adjust their calls in response to the presence or absence of bats. Explain why they do this and describe a situation in which these frogs would be highly likely to give conspicuous calls at a high frequency.

REVISIT RAINFOREST CONNECTIONS

Gather Evidence In this chapter, you have seen how animal systems interact to help them perform specific functions essential for life. In Section 10.2, you learned that most animals can be organized into cells, tissues, organs, and organ systems. In Section 10.3, you learned that these organ systems work together to perform the functions animals need to maintain homeostasis. In Section 10.4, you learned that group behaviors often enhance an individual's and species' likelihood to survive and reproduce.

1. Choose an animal that inhabits a tropical rainforest and lives in a social group. Describe one behavior this species exhibits and explain how it aids in the species' overall survival.
2. What other questions do you have about the ways in which animals in the rainforest communicate? What else do you need to know to construct an explanation about the systems that allow animals to survive, grow, reproduce, and respond to their rainforest habitat?

REVISIT THE ANCHORING PHENOMENON

1. *Sample answer: Howler monkeys are a primate species that live in large social groups. When predators or rival monkeys are near, howler monkeys give loud whooping calls that alert others in their social group of potential danger.* **DOK 3**
2. *Sample answer: Rainforest seem like noisy places; how do animals hear each other over the din? How do rainforest animals use coloration as a way to communicate?* **DOK 3**

would reduce the amount of melanin and cause the lizard's skin to lighten, absorbing less heat from its surroundings and preventing the lizard from overheating. **DOK 4**

MATH AND ELA CONNECTIONS

1. *Sample answer: The passage suggests that whales can adjust their calls in response to a change in the environment. This suggests that whale calls are learned behaviors that are flexible and complex, not simple instinctive behaviors. This is advantageous to the whales when they need to adjust to human noise.* **DOK 3**
2. *Sample answer: Hormones are produced by the endocrine system. Hormones regulate and control body functions.* **DOK 3**
3. *Sample answer: Answers will vary, but the curve should begin at a specific level on the vertical axis and run relatively horizontally with a major dip in September 2001, followed by an increase as shipping levels returned to normal. Students should reason that other decreases in shipping would lead to similar decreases in whale cortisol levels, whereas increases in shipping levels might lead to increases in whale cortisol levels. These hypotheses would need to be tested empirically by collecting and analyzing whale feces and shipping data regularly over long periods of time and looking for patterns and correlations.* **DOK 3**
4. *Sample answer: To avoid attracting bats, the frogs reduce the complexity and frequency of their calls when bats are present. This increases their chance of survival but decreases their chance of attracting females, which prefer complex calls. When bats are absent, male frogs are more likely to give conspicuous calls at a high frequency to increase their ability to attract a female.* **DOK 2**

Class Discussion

Encourage students to practice speaking, listening, and collaborative skills when discussing these questions as a class or in smaller groups as they review the topics in Unit 3.

1. How do fungi communicate with other organisms in their ecosystems? What can genome sizes tell us about the complexity of the different types of organism classes?
2. Consider the different ways that plants respond to changes in their environment. How do active and passive mechanisms allow for the plant to change? How are these responses different from those of animals? How are they similar?
3. Consider the prairie dogs that call out warnings to their family groups. Why would this altruistic behavior develop if it increases the danger to the caller? How do other animals respond to changes in the environment?

Go Online

Three Performance Tasks for Unit 3 are available on MindTap to assess students' mastery of the following NGSS Performance Expectations:

- Task 1: HS-LS1-2
- Task 2: HS-LS1-2 and LS1-3
- Task 3: HS-LS2-8

Go Online

The Virtual Investigation, *Communication in the Rainforest*, includes observational evidence that students may find useful for supporting their claims about how rainforest species communicate.

UNIT 3 **SUMMARY**

INTERACTIONS IN LIVING SYSTEMS

CHAPTER 8
DIVERSITY OF LIVING SYSTEMS

How are the characteristics of an organism related to its place in an ecosystem?

- Phylogenetic analyses have allowed scientists to understand relatedness between species on the basis of genetic sequences.
- Genetic evidence has expanded the number of known bacteria and archaea species.
- Protists are incredibly diverse. Many protists are more closely related to plants and animals than to each other.
- Fungi participate in a wide range of ecological roles, ranging from symbioses that encourage plant growth and communication to parasitic relationships that cause other organisms to spread their spores.
- Viruses have DNA or RNA genomes surrounded by a protein capsid. Viruses require host cell processes to replicate their genomes and make proteins to assemble more viruses.

CHAPTER 9
PLANT SYSTEMS

How do plants respond to changes in the environment?

- Plant organization includes cells, simple and complex tissues, organs, and organ systems. Plants are categorized as vascular or nonvascular based on whether they have transport systems.
- Plants alternate between sporophyte and gametophyte generations.
- Passive and active mechanisms drive the movement of fluids through vascular tissue. Osmosis and active transport drive the movement of sugars through phloem cells.
- Plant hormones control growth, development, and homeostasis. Environmental cues trigger movement and other responses.

CHAPTER 10
ANIMAL SYSTEMS

How do animals respond to changes in the environment?

- All animals are multicellular eukaryotes. Most animals have specialized cells, tissues, and organ systems to carry out specialized functions.
- Negative and positive feedback loops help to maintain homeostasis.
- Behavior allows animals to respond to changes in their environment. All behaviors have costs and benefits. Social behaviors such as living in a group help to increase an individual's and species' chances to survive and reproduce.

Go Online **VIRTUAL INVESTIGATION**

Communication in the Rainforest
How do rainforest species communicate?
Explore a Puerto Rican rainforest and discover the many species that live there.

DIFFERENTIATED INSTRUCTION | English Language Learners

Sharing Information Have students work in mixed groups to share information about the content taught in the lessons and give examples for each bullet. Encourage group members to build on one another's ideas.

Beginning Provide sentence frames to help students share information: *An example of _______ is _______. In _______, we learned that _______.*

Intermediate Provide sentence frames to help students add relevant information: *Another example is _______. We also learned that _______. In a similar way, _______.*

Advanced Provide sentence frames to help students build on one another's ideas: *When you said _______, it reminded me of _______. You mentioned _______, which is similar to _______. Another thing we learned about _______ is _______.*

HOW DO RAINFOREST SPECIES COMMUNICATE?

Argue From Evidence In this unit, you learned about the various categories into which scientists group living things based on similarities between the structures and systems their bodies employ for survival. An organism's survival in an ecosystem involves relationships, characteristics, and behaviors that communicate information about threats and resource availability. Organisms interact with their own species, but they also communicate with organisms of other species using sensory signals such as visual cues, sound, and smell, as well as by exchanging chemicals.

The El Yunque National Forest is the only tropical rainforest in the U.S. National Forest System. Its location on the small island of Puerto Rico exposes this forest to the recurring threat of hurricanes that can disrupt its ecosystems. In 2017, Hurricane Maria hit the island as a Category 4 storm with wind speeds as high as 250 kilometers per hour (155 mph) and a 1.5-meter (5-ft.) storm surge. Some scientists predict that recovery in the hardest hit areas of the rainforest will take a century.

National Geographic Explorer Dr. Nicole Colón Carrión studies how symbiotic fungi can support the restoration of the native forest. While fungi are not always visible, they send signals into the environment and communicate with other organisms by releasing chemical compounds. Some fungi form networks between the roots of different trees.

Endophytic fungi colonize plant tissues and produce substances that stimulate the plant's growth, improve its resistance to drought, and protect it from predators and pathogens. In turn, the plants provide the fungi with nutrients they need to grow and survive.

C Claim Make a claim about how interactions between rainforest organisms of different species transmit resources or information that help each species survive.

E Evidence Use the evidence you gathered throughout the unit to support your claim.

R Reasoning To help illustrate your reasoning, develop a model that illustrates how the organisms' systems enable them to interact.

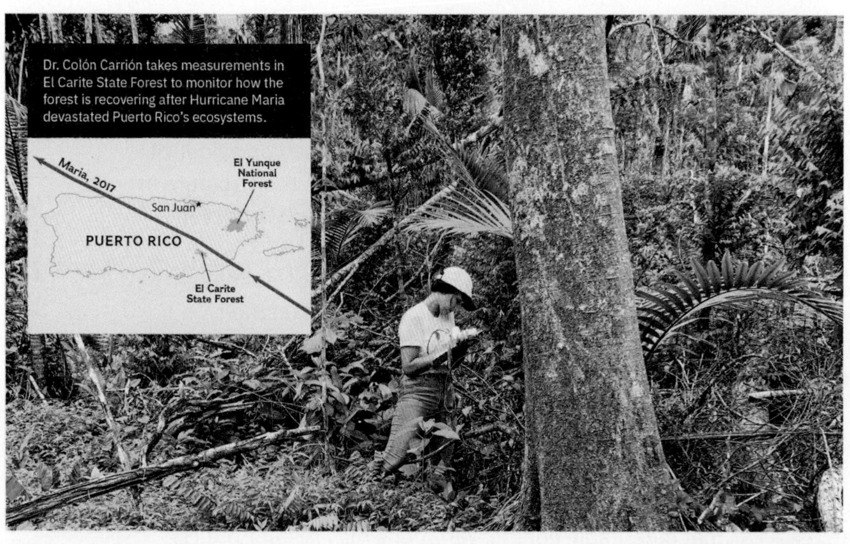

Dr. Colón Carrión takes measurements in El Carite State Forest to monitor how the forest is recovering after Hurricane Maria devastated Puerto Rico's ecosystems.

DIFFERENTIATE INSTRUCTION | English Language Learners

Varying Sentences As students write their claim and support it with evidence, encourage them to vary sentence lengths and use connecting words to combine ideas. Explain that this can make their writing clearer and more interesting for the reader.

Beginning Have students review their writing and look for phrases to combine. Remind them that they can connect related ideas or additional information with *and*. They can use *but* to connect different or contrasting ideas.

Intermediate Have students combine shorter sentences into compound sentences. Remind them to add a comma before the conjunction (*and, but, or,* and so on). Point out that students should keep some shorter sentences for variety.

Advanced Have students edit their writing so they use a mix of shorter and longer sentences. Suggest that their complex sentences use connecting words such as *although, because,* or *while*. The second sentence in the third paragraph (beginning *While fungi...*) is an example.

REVISIT THE ANCHORING PHENOMENON

C Claim
E Evidence
R Reasoning

This Unit Activity asks students to formulate a claim to answer the question posed at the beginning of the unit, cite evidence they have obtained throughout the unit, and use the concepts they have learned to explain how this evidence supports their claim.

C Formulating a Claim Students should write their answer to the question "How do rainforest species communicate?" in the form of a single sentence. Examples of claims that students might make using terms and concepts from the unit chapters include: "Mycorrhizal fungi communicate by releasing chemicals." or "Animals communicate using sight, touch, sound, and chemical signals."

E Citing Evidence Have students review their answers to the Rainforest Communication questions in Sections 8.3, 9.2, and 10.4 and in each Chapter Review as they determine what evidence to use to support their claim. If you wish to have students look for evidence from sources other than this text, they can find useful information on National Geographic, as well as the National Forest Foundation.

R Explaining Reasoning In addition to their explanatory writing, models that show organism interactions, such as food webs and communication diagrams, are convenient structures for students to build their explanations around.

Go online to access Student Self Reflection and Teacher Scoring rubrics for this activity.

Unit Anchoring Phenomenon: How can we slow the spread of a virus?

Use the Driving Question to help frame the Anchoring Phenomenon as an investigatable subject and motivate student learning. Leverage the virus connection prompts wihtin each chapter to connect concepts back to the units' Driving Question, supporting students in gathering evidence and asking their own research questions so they are equipped to complete the Unit Activity.

NATIONAL GEOGRAPHIC

Meet the Explorer

Daniel Streicker is an ecologist who studies infectious diseases using vampire bats. Watch Unit Video 2, Explorers at Work: Daniel Streicker to engage student interest in contagious disease research and the Anchoring Phenomenon.

Virtual Investigation

Fighting a Viral Pandemic Students learn about SARS-CoV-2, the cause of the COVID-19 pandemic. They gather evidence and make decisions to help protect people in the fictional nation of Capsidia.

NGSS Progression

Middle School

- **MS-LS3-1** Develop and use a model to describe why structural changes to genes (mutations) located on chromosomes may affect proteins and may result in harmful, beneficial, or neutral effects to the structure and function of the organism.
- **MS-LS3-2** Develop and use a model to describe why asexual reproduction results in offspring with identical genetic information and sexual reproduction results in offspring with genetic variation.
- **MS-LS4-4** Construct an explanation based on evidence that describes how genetic variations of traits in a population increase some individuals' probability of surviving and reproducing in a specific environment.
- **MS-LS4-5** Gather and synthesize information about technologies that have changed the way humans influence the inheritance of desired traits in organisms.

High School

- **HS-LS1-1** Construct an explanation based on evidence for how the structure of DNA determines the structure of proteins, which carry out the essential functions of life through systems of specialized cells.
- **HS-LS2-7** Design, evaluate, and refine a solution for reducing the impacts of human activities on the environment and biodiversity.
- **HS-LS3-1** Ask questions to clarify relationships about the role of DNA and chromosomes in coding the instructions for characteristic traits passed from parents to offspring.
- **HS-LS3-2** Make and defend a claim based on evidence that inheritable genetic variations may result from (1) new genetic combinations through meiosis, (2) viable errors occurring during replication, and/or (3) mutations caused by environmental factors.
- **HS-LS3-3** Apply concepts of statistics and probability to explain the variation and distribution of expressed traits in a population.
- **HS-ETS1-3** Evaluate a solution to a complex real-world problem based on prioritized criteria and trade-offs that account for a range of constraints, including cost, safety, reliability, and aesthetics as well as possible social, cultural, and environmental impacts.

Claim, Evidence, Reasoning Students can make a claim about the unit phenomenon, gather evidence, and revisit their claim periodically to evaluate how well the evidence supports it. The Driving Question presented in the Case Study of each chapter can get students invested in chapter topics and in working toward answering the unit's Driving Question, "How can we slow the spread of a virus?" In the Unit Activity, students can practice scientific reasoning and argumentation to show how the evidence supports their claim.

Follow the Anchoring Phenomenon How can we slow the spread of a virus?

<table>
<tr><th>Gather evidence with . . .</th><th>Chapter 11</th><th>Chapter 12</th><th>Chapter 13</th><th>Unit Activity</th></tr>
<tr><td>CASE STUDY</td><td>How is genetic information transformed and regulated?</td><td>How does genetic information change and get passed from generation to generation?</td><td>How can genetic information be used in medical and other scientific applications?</td><td rowspan="8">Revisit the unit's Anchoring Phenomenon of the cellular mechanisms that enable a virus to express genes and reproduce.

Claim, Evidence, Reasoning Students use the evidence they gathered throughout the unit to support their claim with reasoning.
Go online to access Student Self Reflection and Teacher Scoring rubrics for this activity.
English Language Learners Describing and Explaining</td></tr>
<tr><td>MINILAB</td><td>How can I demonstrate DNA replication, transcription, or translation?</td><td>How can we model the probability of inheriting a trait?</td><td>How does the percentage of people vaccinated impact the number of people infected with a disease?</td></tr>
<tr><td>LOOKING AT THE DATA</td><td>Students will analyze and interpret data to learn about mRNA half-life and organism complexity.</td><td>Students will analyze and use data to predict offspring blood type and identify family members that could donate blood.</td><td>Students will analyze the effectiveness of clinical trials to determine the likelihood of job opportunities in the research and development of gene therapies.</td></tr>
<tr><td>TYING IT ALL TOGETHER
REVISIT THE CASE STUDY</td><td>Students will create a model to describe how a genetic coding is expressed in animal coat color.</td><td>Students will research a dog breed and develop a model to predict the genotypes of the offspring from heterozygous parents.</td><td>Students will design solutions to a conservation problem using genetic technologies.</td></tr>
<tr><td>Chapter Review: Revisit Viral Spread Connections</td><td>Identify genes that need to be encoded within the SARS-CoV-2 genome to make new virus particles.</td><td>Propose how mutations affect a virus's ability to infect and the impact of that on science, technology, and society.</td><td>Consider how genetic technologies are used to help fight diseases.</td></tr>
<tr><td>Virtual Investigation</td><td colspan="3">In the wake of a pandemic, students take on the role of investigators developing a response to help protect people in a fictional nation.</td></tr>
<tr><td>Chapter Investigation A</td><td>How can we model the building blocks of life?</td><td>How do alleles determine the phenotype of an organism?</td><td>How can you use the properties of DNA to figure out who committed the "crime" of taking a bite out of a cookie?</td></tr>
<tr><td>Chapter Investigation B</td><td>How can genes be expressed and regulated in organisms?</td><td>How can we map genes using traits that are inherited together?</td><td>Can we replicate the process scientists engineered to insert foreign DNA into bacteria?</td></tr>
</table>

Planning Your Investigations

Each chapter features a Minilab and two full Investigations that offer hands-on opportunities for students to engage in Science and Engineering Practices.

Advance Preparation For Chapter 11, have a variety of materials available for students to choose from for their demonstrations. For Chapter 13, you will need to collect materials and containers for student models. Advance preparation notes for all labs are included in the Teacher's version. Advance lab preparation notes for all labs are included in the Teacher's version.

Chapter 11	Title	Time	Standards
Minilab	*Modeling DNA Replication, transcription, And translation*	30 minutes	SEP Asking Questions and Defining Problems CCC Scale, Proportion, and Quantity
Investigation A **Guided Inquiry**	*Investigating the Building Blocks of Life*	50 minutes	HS-LS3-1
Investigation B **Open Inquiry**	*Regulation of Gene Expression*	100 minutes	HS-LS3-1
Chapter 12	**Title**	**Time**	**Standards**
Minilab	*Modeling Inheritance*	30 minutes	SEP Asking Questions and Defining Problems
Investigation A **Guided Inquiry**	*Design an Organism*	50 minutes	HS-LS3-3
Investigation B **Open Inquiry**	*Mapping Fruit Fly Genes Through Linkage*	90 minutes	HS-LS3-3
Chapter 13	**Title**	**Time**	**Standards**
Minilab	*Herd Immunity*	30 minutes	SEP Constructing Explanations and Designing Solutions CCC Influence of Science, Engineering, and Technology on Society and the Natural World
Investigation A **Guided Inquiry**	*DNA Evidence*	100 minutes	HS-ETS1-3
Investigation B **Engineering**	*Fluorescent Genes*	100 minutes	HS-ETS1-3

Assessment Planning

UNIT 4 PERFORMANCE TASKS

Four performance-based assessments target the NGSS Performance Expectations with three-dimensional activities to measure student mastery. Rubrics are available online for Performance Tasks.

Performance Tasks

	Title	Overview	PEs Addressed	Use After
1	*How does a single-gene trait disappear and reappear in a subsequent generation?*	Students investigate a genetic disorder about which they conduct research and analyze data about inheritance patterns.	HS-LS3-1	Chapter 11
2	*What caused the unusual skin discoloration in the people living in a rural area?*	Students make claims supported by evidence about about a condition and then predict how the condition can be avoided in the future.	HS-LS3-2	Chapter 12
3	*What are the risks and benefits of genetically engineered food?*	Students collect and analyze data about GMO foods and argue for and against their use.	HS-LS3-3	Chapter 13
4	*How will we curb the spread of mosquito-borne disease?*	Students research a mosquito-borne disease and evaluate methods for controlling it.	HT-ETS1-2	Chapter 13

Additional Resources

Use a search engine to find these resources on the internet.

VIDEOS

"Study finds trauma effects may linger in body chemistry of next generation" PBS Newhour

A 6-minute video describes new research on survivors of the Holocaust that shows how catastrophic events can alter our body chemistry, and how these changes can transmit to the next generation.

Search: "newshour trauma inherited"

"3 ways to prepare society for the next pandemic" TED Talk

Infectious disease epidemiologist Jennifer B. Nuzzo speaks for 14 minutes about how lessons from the Great Baltimore Fire of 1904 should inform our pandemic responses.

Search: "nuzzo ted pandemic"

ARTICLES

"Despite odds, fish species that bypasses sexual reproduction is thriving" Washington University School of Medicine

An article that looks at the Amazon molly, an all-female fish species that reproduces asexually.

Search: "asexual fish Amazon"

"How Genetic Disorders Are Inherited" PBS Learning Media

An infographic elaborates on the heritability of genetic disorders.

Search: "tdc02 doc inherited"

ENGAGE

Unit Storyline

In this unit, students will answer the question, "How can we slow the spread of a virus?" Each chapter provides part of the answer to this Driving Question.

Chapter 11 *How is genetic information transformed and regulated?*

Students learn about the processes of DNA replication and mRNA transcription. Students then learn how the different structures of RNA cooperate in the formation of proteins.

Chapter 12 *How does genetic information change and get passed from generation to generation?*

Students learn about the different mechanisms at the cellular level that produce genetic variation. Students then learn about genetic and chromosomal mutations and how mutations can affect trait expression, before moving on to the difference between Mendelian inheritance, incomplete dominance, and codominance.

Chapter 13 *How can genetic information be used in medical and other scientific applications?*

Students learn about how genetic engineering has been used to modify plants and animals. The focus then shifts to how vaccines function and the process required for their development.

Unit Activity Students revisit the Anchoring Phenomenon to make a claim about the Driving Question and apply reasoning to the evidence they have gathered throughout the unit.

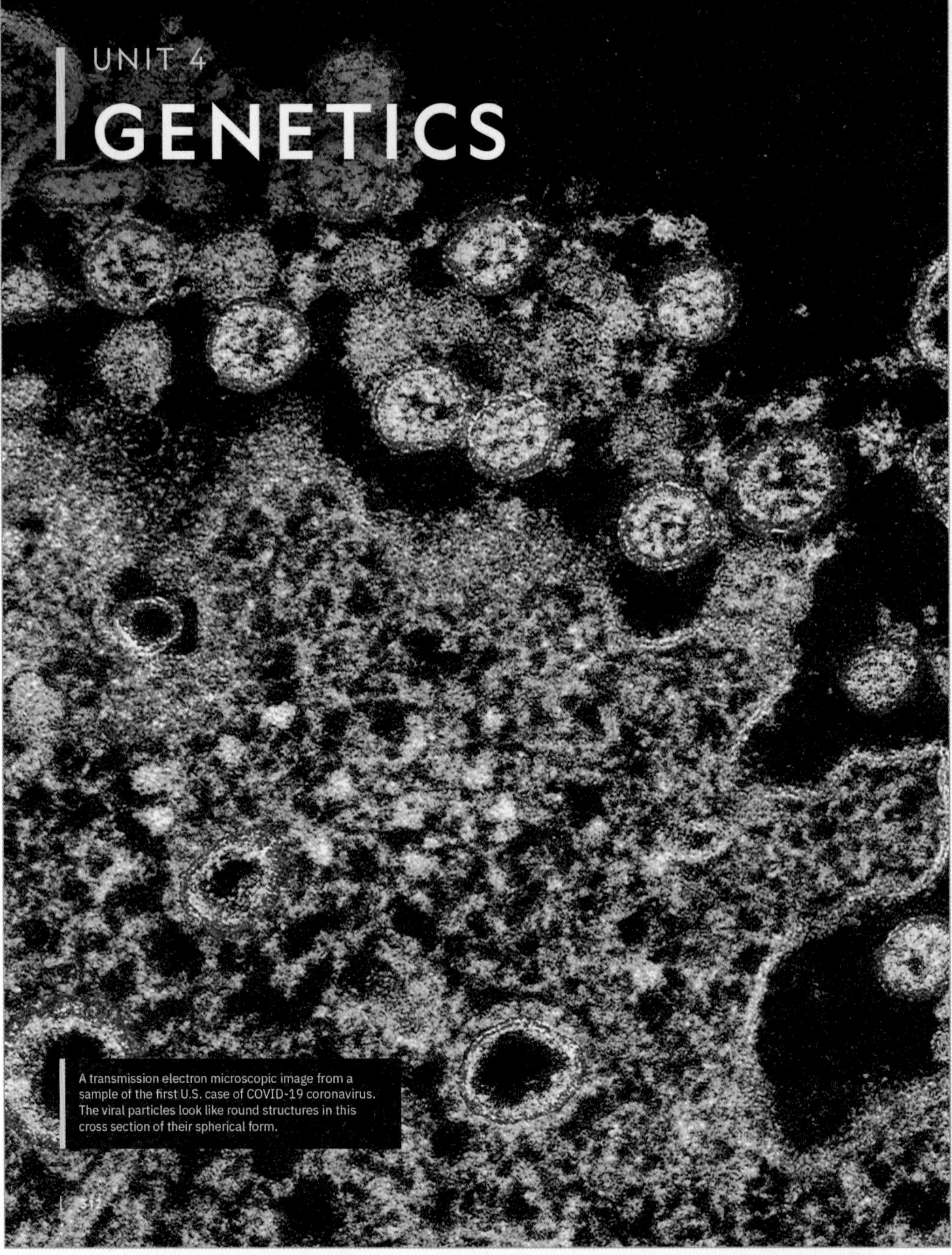

A transmission electron microscopic image from a sample of the first U.S. case of COVID-19 coronavirus. The viral particles look like round structures in this cross section of their spherical form.

SCIENCE BACKGROUND

COVID-19 Origins COVID-19 is a disease caused by the SARS-CoV-2 virus. The exact origin of the SARS-CoV-2 virus is unknown. What is known is that the first reporting of the virus occurred in November of 2019. By December of 2019, the first large cluster outbreak was reported in Wuhan, China. Due to the rapid rate that the virus spread and its easy airborne transmission from human to human, it quickly became a worldwide pandemic.

COVID-19 Replication As the proteins within a COVID-19 viral cell combine with the hydrogen bonding of the different amino acids, it causes them to fold into specific shapes. These spikes are the perfect size and shape to attach to the angiotensin-converting enzyme 2 receptor located on the outer surface of cells in the lungs, arteries, heart, kidney, and intestines. Once the COVID cell is attached to the host cell, the genetic material is transferred into the cell and takes over the replication system.

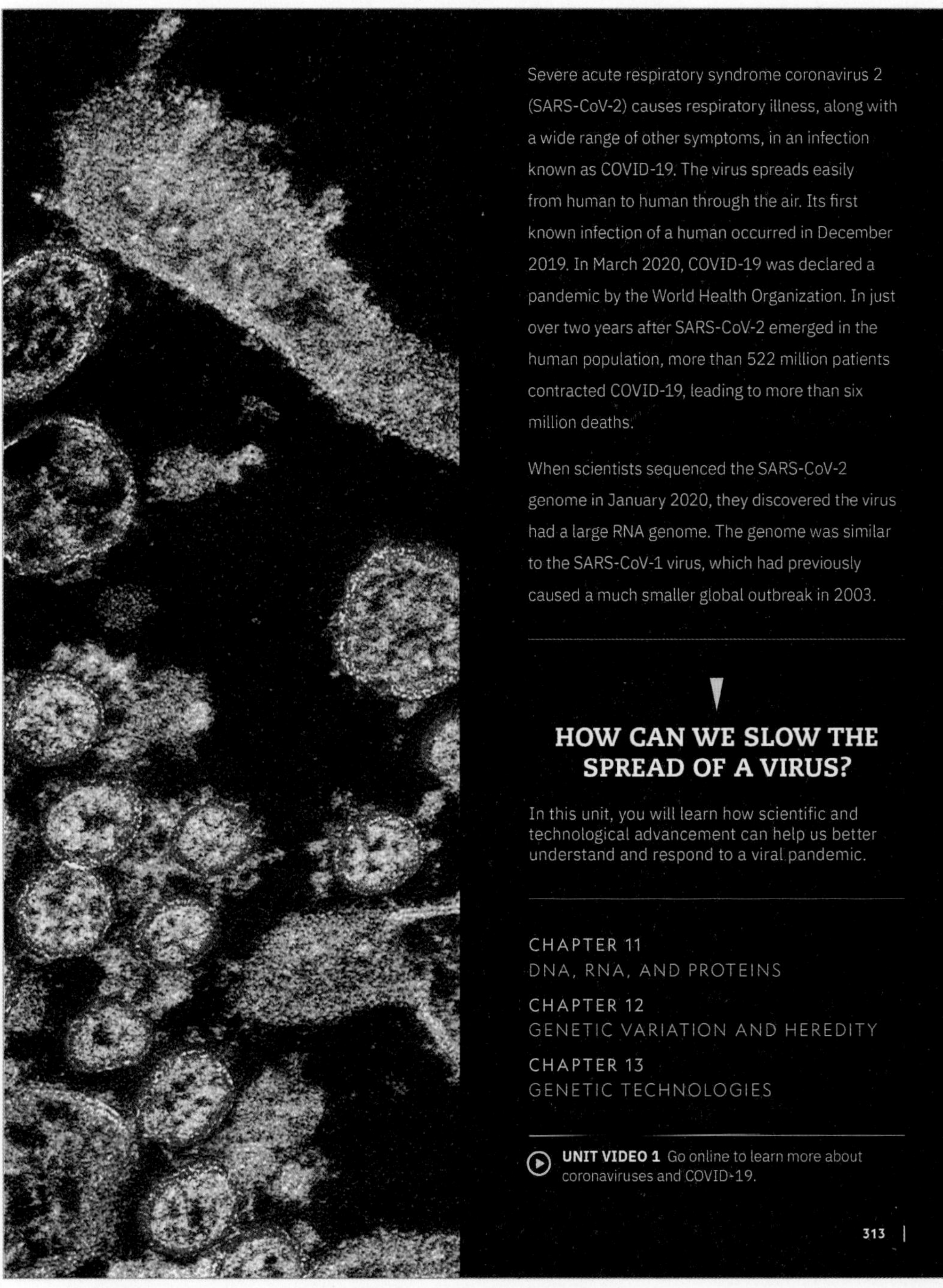

Severe acute respiratory syndrome coronavirus 2 (SARS-CoV-2) causes respiratory illness, along with a wide range of other symptoms, in an infection known as COVID-19. The virus spreads easily from human to human through the air. Its first known infection of a human occurred in December 2019. In March 2020, COVID-19 was declared a pandemic by the World Health Organization. In just over two years after SARS-CoV-2 emerged in the human population, more than 522 million patients contracted COVID-19, leading to more than six million deaths.

When scientists sequenced the SARS-CoV-2 genome in January 2020, they discovered the virus had a large RNA genome. The genome was similar to the SARS-CoV-1 virus, which had previously caused a much smaller global outbreak in 2003.

HOW CAN WE SLOW THE SPREAD OF A VIRUS?

In this unit, you will learn how scientific and technological advancement can help us better understand and respond to a viral pandemic.

UNIT VIDEO 1 Go online to learn more about coronaviruses and COVID-19.

313

Introduce the Anchoring Phenomenon

Viruses are miscroscopic structures that bind to host cells. Each virus consists of either an RNA or DNA genome, which is surrounded by a protective, virus-coded protein coat. Viruses are grouped by several characteristics, such as their size and shape, chemical composition, and structure of the genome, as well as their mode of replication. Examining the structure and function of reproduction systems, including viruses, enables students to apply what they learn throughout the unit to a complex situation.

Driving Question Have students make a list of things they think they would need to know about viruses to answer the Driving Question. The photo and caption may prompt them to ask, "How does a viral cell reproduce?" To analyze the systems that allow organisms to reproduce, students will need to focus eukaryotic cells and the processes used for cell reproduction. What they learn in this unit about reproduction and how genetic material is transferred will inform their understanding of how to slow the spread of a virus.

About the Photo The technology used to produce this magnified view is a electron microscope. The images are produced by scanning the sample with a focused beam of electrons. As the electrons interact with the atoms from the sample, signals with information about the topography and composition of the sample are produced, which can then be converted to an image.

Video

Time: 1:27

Use **Unit Video 1** to show students a timeline of key events of the beginning of the Covid-19 pandemic and an overview of what coronaviruses are.

Go Online VIRTUAL INVESTIGATION

Fighting a Viral Pandemic

Time about 50 minutes

Objectives Students will learn about SARS-CoV-2, the cause of the COVID-19 pandemic. They will gather evidence and make decisions to help protect people in the fictional nation of Capsidia.

Explore and Learn Students enter the situation room and laboratory to obtain evidence that will help them better understand the COVID-19 pandemic and make decisions.

Collect Data In the situation room, students hear from experts to better understand the spread of the virus and analyze available options for vaccinating the population. In the laboratory, they learn about how the virus infects cells and about techniques used by scientists to track mutations and new variants.

Analyze and Report Students answer questions about the investigation, incorporating qualitative and quantitative data in a report to support their analyses.

EXPLORER
DANIEL STREICKER

ENGAGE

About the Explorer

National Geographic Explorer Daniel Streicker is an ecologist who studies infectious diseases. He has primarily focused on diseases originating in vampire bats from the Peruvian mountains. Streicker researches how an effective rabies vaccine might work in vampire bats and hopes that the information collected from vampire bats can help improve the understanding of how organisms host and spread pathogens to other species. You can learn more about Daniel Streicker on the National Geographic Society's website: www.nationalgeographic.org.

On the Map

Blood Drive Students should understand that many infectious diseases have animal origins. The Peruvian Mountains is home to over 122 species of bats. The mountains extend across parts of Peru, Bolivia, Venezuela, Colombia, and Ecuador to the southern parts of Argentina and Chile. There are over 21,000 mountains located in Peru. The most prominent mountain range, the Andes, runs across the center of the country from north to south.

THINKING CRITICALLY

Research *Look for logic in students' suggestions. They may include ideas such as repeated testing of the virility of viruses in captive bat populations, comparing viruses in bats from areas with and without disease outbreaks, and in-depth comparisons of the aspects of bats and humans affected by the viruses.*

NATIONAL GEOGRAPHIC EXPLORER **DANIEL STREICKER**

BLOOD DRIVE

VACCINATING VAMPIRE BATS IN THE PERUVIAN MOUNTAINS

It's the stuff legends are made of—a mammal that comes out at night to feed only after the moon sets. It flies flawlessly through the complete darkness, guided by sound and smell. Near a farm, it lands silently and skulks along the ground on long legs. When the bat finds its preferred prey, such as a cow, it hops or flies onto the cow's back, holding on tightly as the cow switches its tail and jerks its head around trying to dislodge the intruder, but to no avail. The bat's infrared sensors tell it where warm blood flows most closely to the surface of the skin. The cow, no longer bothered, goes back to sleep and the vampire bat prepares to feed.

Infectious disease ecologist Dr. Daniel Streicker marvels at how perfectly adapted the vampire bat is to its lifestyle of feeding only on blood. The mammal does not suck blood through its teeth as popularized by vampire folklore. Instead, its triangular teeth are tiny and as sharp as needles. Prey animals barely feel a pinch when a bat cuts a small divot in the skin. As the bat laps up blood, its tongue secretes an anticoagulant that prevents the blood from clotting. When it finishes feeding, the bat simply flies away.

Although Streicker admires the vampire bat's adaptations, he's more interested in studying them as a reservoir for the rabies virus. In biology, a reservoir is an organism that hosts a pathogen and spreads it to other species. For all mammals, the virus that causes rabies (*Rabies lyssavirus*) is deadly. Rabies has existed for thousands of years with no known cure once symptoms set in. Streicker researches vampire bats in Peru, where most colonies live in caves, mines, buildings, bridges, and wells. Infected vampire bats feeding on livestock create a perfect route for the virus to spread.

Streicker cites alarming data about the many thousands of livestock that die from rabies each year and the disturbing trend of the disease spreading to humans. In response, the Peruvian government and those of many other countries, from Mexico to Argentina, have killed vampire bats to reduce the spread. Their method involves a poison-laced petroleum jelly smeared on the back of captured bats. Treated bats return to the colony, where they are groomed by bats who ingest the poison and die. Streicker notes that little evidence exists that this method helps to control the spread of rabies. He proposes another solution—petroleum jelly mixed with an oral rabies vaccine instead of poison.

Streicker and his team have collected data that show where the disease might appear next. They found that rabies was moving in a steplike manner down one mountain valley and then down another. The patterns of old and new cases showed them how they could predict when rabies would reach an area so they could warn farmers. The farmers could then make decisions about getting the vaccine for their livestock and themselves.

Interest in the study of bats that harbor viruses resurfaced with the emergence of COVID-19, a virus thought to have crossed over from bats to humans. Streicker notes that many bats can live with viruses that are lethal in humans, such as SARS and Middle East respiratory syndrome (MERS). But, does bat tolerance of viruses result in genetic changes in viruses that cause them to be more dangerous to humans? That's something Streicker says needs more study.

THINKING CRITICALLY

Research *Suggest ways that Streicker and other scientists might further study the question of whether viruses become more dangerous to humans after living in bat populations.*

314

DIFFERENTIATED INSTRUCTION | English Language Learners

Evaluating Information Briefly explain vampires in legends (creatures that bite people with their fangs and feed on their blood). Have students evaluate the information the text provides to show that vampire bats are different from these legends.

Beginning Ask: What do vampire bats feed on? What do they use? How are they dangerous to humans? What helps prevent this danger?

Intermediate Ask: How are vampire bats different from vampire legends in the way they feed? How are they different in their danger to humans? How do you think legends cause people to feel about vampire bats?

Advanced Ask: How are vampire bats different from vampires as described in legends? Do you think the legends affect how people treat bats? How?

Top: Although Dr. Daniel Streicker leads a lab in Glasgow, Scotland, he spends much of his time in the Amazon studying bats. **Bottom left:** Streicker has proposed that healthy virus-carrying bats could lead to the development of new treatments for humans. **Bottom right:** Vampire bats live in Central and South America, feeding mostly on the blood of livestock.

UNIT VIDEO 2 Go online to watch our interview with Streicker and learn more about his career and research.

315

Explorers at Work

Time: 6:40

Unit Video 2 interviews Daniel Strieker, who describes the large percentage of human pathogens that originated in other species. He focuses on factors that promote or impede the passage of pathogens between species. He outlines his research methodology in the field where his team collects blood and saliva and the lab where they conduct genetic sequencing to look for rabies antibodies and signs of other pathogens to predict the risk to humans. He describes his work as both ecologically interesting and of great medical importance.

Connect to Careers

Epidemiologists Scientists who study infectious outbreaks are referred to as epidemiologists. An epidemiologist investigates the disease by identifying the cause of disease along with people that may be at risk of the disease. They also seek ways to minimize or even stop the spread of the disease. A master's degree in public health or a related field is required to become an epidemiologist. Governments, universities, hospitals, and health departments all employ epidemiologists.

DIFFERENTIATED INSTRUCTION | English Language Learners

Understanding Implicit Ideas After checking comprehension of ideas stated in the video, guide students to make inferences based on the information they heard. Replay parts of the video as needed.

Beginning Have students complete sentence frames: Bats spread (rabies). People use (poison) to stop the spread. Dr. Streicker uses (vaccines) instead. He wants to (protect) people from the spread of rabies without (killing) the bats.

Intermediate Have students explain how Dr. Streicker's approach is different from government methods. Then, have them discuss why he might use this method.

Advanced Have students explain why Dr. Streicker uses vaccines instead of poison to stop the spread of rabies. Then, have them discuss why it is important to research viruses in bat populations.

Three-Dimensional Learning

The practices, core ideas, and crosscutting concepts presented in this chapter's text, investigations, and resources provide support to address the following Performance Expectations: **HS-LS1-1** and **HS-LS3-1.**

Science and Engineering Practices	Disciplinary Core Ideas	Crosscutting Concepts
Asking Questions and Defining Problems (HS-LS3-1)	**LS1.A:** Structure and Function (HS-LS1-1, HS-LS3-1) **LS3.A:** Inheritance of Traits (HS-LS3-1)	Cause and Effect (HS-LS3-1) Structure and Function (HS-LS1-1)

Contents		Instructional Support for All Learners	Digital Resources
ENGAGE			
316–317	CHAPTER OVERVIEW **CASE STUDY** How is genetic information transferred and regulated?	**Social-Emotional Learning** CCC Structure and Function **Historical Connection** Batesian Mimicry **English Language Learners** Use Context Clues for Listening	
EXPLORE/EXPLAIN			
318–325 DCI LS1.A LS3.A	**11.1** GENETIC INFORMATION • Discuss how experimentation led to the discovery of the structure and function of DNA. • Relate the structures of DNA, mRNA, and proteins to their functions in an organism. • Analyze a model of gene expression to describe the transfer and translation of genetic information.	**Vocabulary Strategy** Roots and Affixes SEP Constructing Explanations and Designing Solutions CONNECTIONS Social-Emotional Learning **Connect to Mathematics** Model with Mathematics **Visual Support** Structure of DNA **Differentiated Instruction** Leveled Support **Crosscurricular Connections** Chemistry **Connect to English Language Arts** Write Explanatory Text **English Language Learners** Spelling Patterns **In Your Community** Prion Diseases **On Assignment** American Alligators SEP Planning and Carrying Out Investigations	
324	EXPLORER DR. CARTER CLINTON	**Connect to Careers** Genetic Anthropologist **Social-Emotional Learning** Social Awareness SEP Analyzing and Interpreting Data	*Video 11-1*

Contents		Instructional Support for All Learners	Digital Resources
EXPLORE/EXPLAIN			
326–333 DCI LS1.A LS3.A	**11.2 REPLICATION, TRANSCRIPTION, AND TRANSLATION** • Relate the structure of DNA to its replication. • Describe the process and outcomes of transcription. • Explain the process of translation using codons.	**Vocabulary Strategy** Semantic Map **SEP** Developing and Using Models **Visual Support** HeLa Cells, Models, Polysomes, Shape of tRNA **Address Misconceptions** Cellular DNA **English Language Learners** Vocabulary Strategies, Roots and Affixes **CCC** Systems and System Models **Connect to English Language Arts** Write Explanatory Text, Synthesize Information **CER** Revisit the Anchoring Phenomenon **Connect to Mathematics** Reason Quantitatively **Differentiated Instruction** Leveled Support **In Your Community** Ribosome Factories	*Video 11-2* **SIMULATION** Gene Expression *Video 11-3*
334	**MINILAB** MODELING DNA REPLICATION, TRANSCRIPTION, AND TRANSLATION		
335–341 DCI LS1.A LS3.A	**11.3 REGULATING GENE EXPRESSION** • Compare gene expression in prokaryotes and eukaryotes. • Explain the function of introns and exons. • Describe environmental factors that can affect gene expression.	**Vocabulary Strategy** Word Wall, Word Wall Followup **English Language Learners** Asking and Answering Questions, Expressing Opinions and Ideas **Visual Support** Operon Regulation, Eukaryote Gene Regulation **CASE STUDY** Ask Questions **SEP** Developing and Using Models **Differentiated Instruction** Leveled Support **Address Misconceptions** One Gene, Many Proteins **Connect to English Language Arts** Write Explanatory Text **In Your Community** RNAi and Pest Control, Epigenetics and 9/11	**Investigation B** Regulation of Gene Expression (100 minutes or 2 days) **Investigation A** Investigating the Building Blocks of Life (50 minutes)
342	**LOOKING AT THE DATA** DECAY OF mRNA MOLECULES		
ELABORATE			
343	**TYING IT ALL TOGETHER** How is Genetic Information Transferred and Regulated?	**In Your Community** Camouflage **SEP** Developing and Using Models	
Online	**Investigation A** Investigating the Building Blocks of Life **Investigation B** Regulation of Gene Expression	Guided Inquiry (50 minutes) Open Inquiry (100 minutes or 2 days)	MINDTAP Access all your online assessment and laboratory support on MindTap, including: sample answers, lab guides, rubrics, and worksheets. PDFs are available from the Course Companion Site.
EVALUATE			
344–345	**Chapter 11 Review**		
Online	**Chapter 11 Assessment** **Performance Task 1** *How Does a Single-Gene Trait Disappear and Reappear in a Subsequent Generation?* (HS-LS3-1)		

CHAPTER 11

ENGAGE

This chapter builds on prior knowledge of DNA in Chapter 5 that covered how genetic information is expressed. Here, students begin with the early history of experimentation that led scientists to our present-day knowledge of the structure and function of DNA. Students are then introduced to the processes of DNA replication and mRNA transcription, followed by how the different structures of RNA cooperate in the formation of proteins.

About the Photo *Heliconius* butterflies are known for their bright color patterns that serve not only to warn predators of their toxicity but also to attract mates. However, what makes them attractive to scientists is that over 400 different wing patterns are found among 46 species, providing an exciting opportunity to explore how phenotypic diversity is generated within the context of adaptive radiation.

Social-Emotional Learning

As students progress through the chapter, invite them to look for contexts and opportunities to practice some of the five social and emotional competencies. For example, **self-awareness skills**, such as a growth mindset, help students confront complex ideas. Having strong **relationship skills** helps students to proactively seek or offer support when a new concept is challenging. Encourage the development of **responsible decision-making**, such as showing curiosity and open-mindedness, as well as having students challenge themselves to use information and facts to make reasoned judgments.

CHAPTER 11

DNA, RNA, AND PROTEINS

The patterns on the wings of this *Heliconius erato notabilis* butterfly are the result of the different colors of individual scale cells.

The bright colors that make the patterns on this butterfly are determined by gene expression. The butterfly's genes ultimately code for proteins that affect the appearance of colors on its wings. The expression of those genes is a complex process that involves many factors.

Chapter 11 supports the NGSS Performance Expectations **HS-LS1-1** and **HS-LS3-1.**

CROSSCUTTING CONCEPTS | Structure and Function

DNA Understanding the relationship between structure and function will be key to students' understanding of genetics and the replication, transcription, and translation of genetic information. Students must understand how the structure of DNA is critical to genetic variation and the passing of traits from one generation to another.

The double-helix structure of DNA constructed with nucleotide units containing a nitrogen base, a five-carbon sugar, and a phosphate group is the foundation for the mechanisms of genetics. The arrangement of nucleotide building blocks in a gene specifies the order of amino acids in the protein it encodes. This chapter specifically covers the structure of DNA and the central role of nucleotides in the replication, transcription, and translation of DNA. Show students a model or diagram of DNA and discuss initial observations and predictions of how the structure might relate to its function.

CASE STUDY

EXPRESSING THE GENETIC CODE

HOW IS GENETIC INFORMATION TRANSFERRED AND REGULATED?

A butterfly's wing is covered with thousands of tiny scales that give the butterfly its characteristic colors and patterns. Each scale is produced by a single epithelial cell whose genes dictate the structure of the scale and the pigments produced. The scales' structures give some species their iridescent blue or purple colors, and pigments color the scales with shades and mixtures of red, yellow, and black. Scientists have long studied how these genes are expressed and what causes the similarities and differences in wing decorations between individual butterflies and between butterfly species.

Experiments performed in the early 1900s showed that temperature and light can influence gene expression during wing development. For example, when caterpillars of the genus *Vanessa* were placed under red, green, or blue light, differences in wing color emerged as the caterpillars matured into butterflies. Those that had been exposed to red light had brightly colored wings. However, exposure to green light resulted in dusky wings, and exposure to blue or no light resulted in pale wings.

Over a century has passed, yet questions remain about how and why butterfly wings vary. Scientists are now able to investigate them at much smaller scales due to major advancements in our knowledge of how gene expression works. In recent experiments with *Heliconius* butterflies, a genus that signals their toxicity to predators with dramatic multicolored wing patterns, researchers have identified four genes that control the color and pattern variations in all *Heliconius* species. Further investigation into how the same set of genes are expressed in different butterfly species may open up our understanding of how these species evolved.

Ask Questions *In this chapter, you will learn how the processes of gene expression transfer genetic information. As you read, generate questions about factors that affect these processes.*

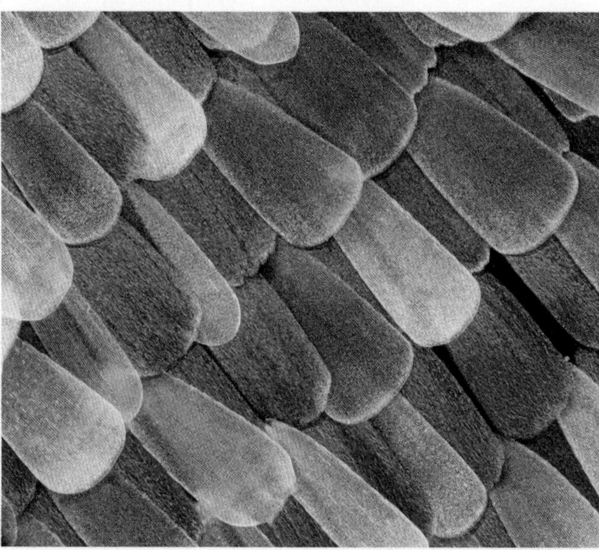

FIGURE 11-1 This light micrograph image shows the color and iridescence of scales on the wing of a butterfly.

DIFFERENTIATED INSTRUCTION | English Language Learners

Use Context Clues for Listening Read aloud from the Case Study and have students listen for context clues to help them understand a particular word. Reread as needed so students can identify context clues.

Beginning Read the third sentence of the first paragraph. Have students identify context clues for *pigment* (*color, shades, mixtures*). Tell students that iridescent means "shiny."

Intermediate Read the second paragraph and have students listen for context clues that help them understand *exposed* and *exposure* ("placed under").

Advanced Read the first sentence and have students identify context clues for *scales*. Then, read the second sentence of the third paragraph and have students explain how context clues help them understand the different meaning of *scales* used in this context.

CASE STUDY

ENGAGE

Gene Expression The influence of temperature and light on gene expression in butterfly wings is a good opportunity to assess students' prior knowledge of gene expression and the transfer of genetic information. Ask students how they think environmental factors such as temperature and light can affect gene expression and how the diversity of wing patterns benefits the species overall.

Over the past decade of study, research has shown that central to wing pattern variation is nearby non-coding regions that control expression during wing development, not the limited number of genes. In the Tying It All Together activity, students will develop and use a model to show how other environmental factors affect gene expression in an animal's coloration.

Ask Questions Students should revisit the Case Study as they read the chapter to make connections with the content. See a specific suggestion in Section 11.3.

Historical Connection

Batesian mimicry Beginning in 1848, *Heliconius* butterflies were studied extensively in the Amazon Basin by Henry Walter Bates, who is best known for his discovery of mimicry. His observations of the *Heliconius* wing patterns and the rapid changes noted as he traveled across the area led to the identification of what is now known as Batesian mimicry. In this type of mimicry, harmless organisms take on similar color patterns to toxic or unpalatable organisms to ward off predators.

11.1

GENETIC INFORMATION

LS1.A, LS3.A

EXPLORE/EXPLAIN

This section begins with the discovery of DNA. Students are then introduced to the structure of DNA and the complementary nature of base-pairings, as well as the differences between DNA and RNA that result in different functions. Building on this, the transfer of information from DNA to mRNA and the translation of that information to a protein product are discussed.

Objectives

- Discuss how experimentation led to the discovery of the structure and function of DNA.
- Relate the structures of DNA, mRNA, and proteins to their functions in an organism.
- Analyze a model of gene expression to describe the transfer and translation of genetic information.

Pretest Use **Questions 1** and **2** to identify gaps in background knowledge or misconceptions.

Vocabulary Strategy

Roots and Affixes Model how to use word parts to determine meaning. For example, the prefix *trans-* is used in *transcription* and *translation*. Display both words and underline the prefix that means "across, over, or beyond." Then underline the suffix *–tion*, or "state of being." Point out that the root *script* means "write." Guide students to use the word parts to define the word *transcription* as "the process of copying or writing again in another place." Suggest students use a similar thought process for other terms.

Infer *Students might say that chemists would have worked on the chemical structure of the molecule. Biologists might have used observations to determine that information is inherited and combined by parents. Scientists could have used modeling, as well as microscopy and other visualization techniques to determine structures of the DNA molecules.*

11.1 GENETIC INFORMATION

KEY TERMS

gene
gene expression
replication
RNA
transcription
translation

Many sports competitions include multiple referees who watch the athletes from different angles and work together to make sure the game is played fairly. In a similar way, science questions are typically solved with input from multiple scientists from different research perspectives.

Infer *Consider what you already know about DNA. What different areas of science and techniques do you think might have been involved in understanding the structure and function of DNA molecules?*

DNA Discovery and Function

The substance that is now known as DNA was first described in 1869 by a Swiss biochemist named Friedrich Miescher. From experimentation, he knew it was not protein and that it was rich in nitrogen and phosphorus, but its function was unknown. Sixty years later, British bacteriologist Frederick Griffith unexpectedly found a clue to the function of DNA. Griffith was studying pneumonia-causing bacteria in the hope of making a vaccine. He isolated two strains (types) of the bacteria, one harmless (R), the other lethal (S). Griffith used R and S cells in a series of experiments testing their ability to cause pneumonia in mice (**Figure 11-2**). He discovered that heat destroyed the ability of lethal S bacteria to cause pneumonia, but it did not destroy their hereditary material. That hereditary material could be transferred from dead S cells to live R cells. The transformation was permanent and was passed onto offspring: even after several generations, descendants of transformed R cells retained the ability to kill mice.

FIGURE 11-2 Frederick Griffith discovered the hereditary information passed between two strains of bacteria (R and S).

Experiment 1	**Experiment 2**	**Experiment 3**	**Experiment 4**
When R cells were injected into mice, the bacteria multiplied, but the mice remained healthy.	When S cells were injected into mice, the bacteria multiplied, and the mice died.	When heat-killed S cells were injected into mice, the mice remained healthy.	When heat-killed S cells and live R cells were both injected into mice, the mice became fatally ill and contained living S cells.
Conclusion R cells were harmless.	**Conclusion** S cells caused mice to develop fatal pneumonia.	**Conclusion** Heat-killed S cells were harmless.	**Conclusion** The hereditary information from the heat-killed S cells was transferred to the R cells.

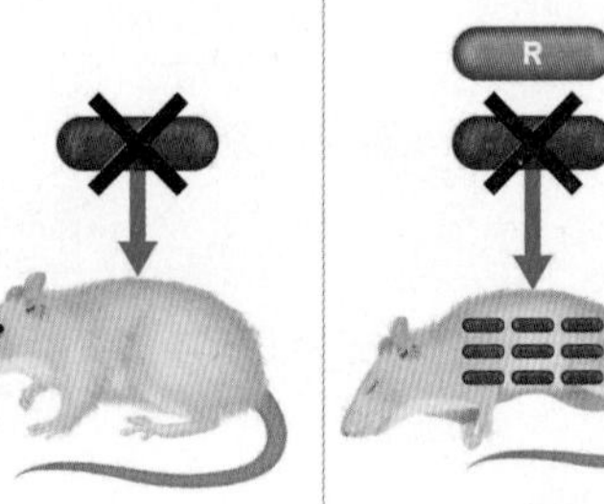

SCIENCE AND ENGINEERING PRACTICES
Constructing Explanations and Designing Solutions

Griffith's Experiment Have students analyze **Figure 11-2** and consider how this model of Griffith's experiment can be used to construct an explanation of the existence of hereditary material. Briefly discuss with students what makes a good explanation and the importance of being evidence-based. Remind students that Griffith did not start out looking for a transformative factor. He was looking for better ways to fight pneumonia when, as often happens in science, his experimental results led to new questions.

After talking through the steps in Griffith's experiment, have students work together in small groups to develop an explanation of the results that support Griffith's conclusions that hereditary material can be transferred in what he called the *transformative process*.

CONNECTIONS

WHAT IS IN A GENOME? It is well understood that genomes include the genes that code for important functions and the traits that make organisms unique. For example, the human genome has been estimated to include 20,000–30,000 genes, most coding for multiple proteins. (Learn how individual genes can code for more than one protein in Section 11.3.) Even with that tremendous number of genes, those coding sequences represent only about one percent of the three billion base pairs in the human genome.

What is the rest of the DNA in the genome doing? While most of the noncoding DNA does not have any understood function and has long been considered "junk DNA," scientists continue to work and improve our understanding of the origins and roles for those sequences. Here are some of the things scientists are learning about the rest of the human genome:

- **Small RNAs** Many parts of the genome code for small RNA molecules that are important for cell function and gene regulation. Some of these RNAs are introduced in Section 11.3.
- **Gene Expression Regulation** Specific DNA sequences in the genome regulate gene expression. A variety of sequences, including promoters, enhancers, silencers, and insulators have been identified to interact with other transcription factors described in Section 11.3.
- **Transposons** These sequences of DNA, which are hundreds or thousands of nucleotides long, can duplicate themselves and spontaneously "jump" between or within chromosomes.
- **Pseudogenes** These are copies of genes that are elsewhere in the genome and tend to be mutated so that they no longer produce a functional protein.
- **Endogenous Retroviruses** Scientists have estimated that 8 percent of the human genome originated with retroviruses. Retrovirus DNA is incorporated into the host genome as a part of the retroviral life cycle.
- **Telomeres** Sequences of DNA at the end of chromosomes are called telomeres. Each time a chromosome is copied, telomeres get shorter, which seems to be related to a shortening lifespan of a cell.

Canadian-American bacteriologist Oswald Avery and his colleagues subsequently sought to determine what substance "transformed" the R cells. They extracted lipid, protein, and nucleic acids from S cells and then used a process of elimination to determine which substance transformed bacteria. Treating the extract with enzymes that destroy lipids and proteins did not destroy the substance that transformed the cells. The researchers realized that the substance they were seeking must be a nucleic acid. That is, it must be either DNA or RNA. DNA-degrading enzymes destroyed the ability to transform cells, but RNA-degrading enzymes did not. Therefore, DNA must have been the substance that transformed the cells.

The result surprised the researchers, who, along with most other scientists, thought proteins passed genetic information from parent to offspring. They reasoned that individual characteristics are diverse, and proteins are the most diverse of all biological molecules. DNA was widely assumed to be too simple because it has only four nucleotide components. In 1944, after years of supporting results, scientists finally published the conclusion that DNA is the molecule that is inherited, passing genetic information from parent to offspring. Subsequent experiments by other scientists confirmed that the body cells of any individual of a species contain precisely the same amount of DNA.

CCC **Cause and Effect** *Describe at least one line of experimental evidence that confirmed DNA as the molecule that is inherited between generations of cells.*

Science Background

The Human Genome Project Beginning in 1990, scientists from around the world collaborated on an enormous project to identify and map the entire nucleotide sequence in human DNA. A human cell has 46 chromosomes, with 6 billion nucleotide pairs. However, only 3 billion needed to be sequenced to construct an accurate map of the human genome since chromosome pairs are similar. Still, this project took 13 years to complete, ending in 2003.

Already, the Human Genome Project has identified genes associated with genetic disorders, as well as genes that are precursors to diseases such as cancer and heart conditions. This has led to improved early detection of diseases, custom therapeutics, and gene therapies.

CCC **Cause and Effect** *Avery and colleagues used destructive enzymes to systematically eliminate every possible heritable molecule and determined that destroying the DNA was the only thing that stopped information from being transferred from parent to offspring cells.*

CONNECTIONS | What is in a Genome?

Social-Emotional Learning As students review the *CONNECTIONS* feature, remind them that genetics is about each of them as individuals and what makes them unique. As students grapple with complex language and processes, have them practice *self-awareness* of their belief in how much they can learn. Tell students that they can control their mindset by believing in their ability to learn and achieve goals.

Relate the idea of embracing challenges to the experimentation and journey of discovery taken by scientists. Discuss with the class what kind of mindset and skills it would take for scientists, such as Griffith and Avery, to pursue the challenges of discovering unknown factors of heredity.

Talk about the enormous task of learning about and mapping the human genome and how it took a collaborative effort over a period of 13 years. Explain that, even for scientists, struggle is part of the learning process, and the goal is to improve understanding step by step.

Connect to Mathematics

Model with Mathematics Rosalind Franklin played a critical role in the discovery of the structure of DNA. In the early 1950s, Franklin focused X-rays on thin strands of DNA and recorded the patterns formed when the X-rays were diffracted by the repeating structure of the DNA crystal. Patterns had to be collected for each orientation of the crystal and mathematically analyzed. Because modern computers had not been invented yet, it took Franklin a year to complete the complex calculations by hand to model the double-helical structure of DNA.

Visual Support

Structure of DNA Direct students' attention to **Figure 11-3** and ask what patterns they can identify. They should observe that A is always paired with T, and C is always paired with G. Ask them what they can infer from this consistent pattern; e.g., there are equal amounts of adenine and thymine and equal amounts of cytosine and guanine (Chargaff's rules). They should also observe the sugar phosphate backbone and the hydrogen bonds that hold the two strands together. Ask students why there is a difference in the number of bonds in each pairing. (*The A and T pairs form two hydrogen bonds, while the G and C pairs form three.*)

Structure and Information in DNA, RNA, and Proteins

DNA DNA is the molecule that makes up the *genomes* of all living organisms and many viruses. DNA is a polymer of four types of nucleotides connected by a phosphate backbone (**Figure 11-3**). Each has a deoxyribose sugar, three phosphate groups, and a nitrogen-containing base: adenine (A), guanine (G), thymine (T), or cytosine (C). Nucleotide structure was understood in the early 1900s, but it took 50 years to puzzle out how they were arranged into the familiar double helix structure of DNA. British chemist Rosalind Franklin generated the breakthrough data, demonstrating the double helix and calculating the diameter and other characteristics of the molecule. These data ultimately led to American biologist James Watson and British biophysicist Francis Crick's first accurate model of DNA structure in 1953 (**Figure 11-3a**). Rosalind Franklin was not given credit for her contribution until after her death.

Notice how the two strands of DNA in **Figure 11-3b** match up. They are *complementary*. The base of each nucleotide on one strand pairs, through hydrogen bonding, with a partner base on the other. This base-pairing pattern (A to T, G to C) is the same in all molecules of DNA. How can just two kinds of base pairings give rise to the incredible diversity of traits we see among living things? Even though DNA is composed of only four nucleotides, the order in which the nucleotide bases occur along a strand—the DNA sequence—varies tremendously among species. DNA molecules can be hundreds of millions of nucleotides long, so their sequence can encode a massive amount of information. DNA sequence variation is the basis of traits that define species and distinguish individuals within a species.

FIGURE 11-3
A replica of the physical model generated by Watson and Crick (a) demonstrates how double-stranded DNA (b) twists into a double helix structure.

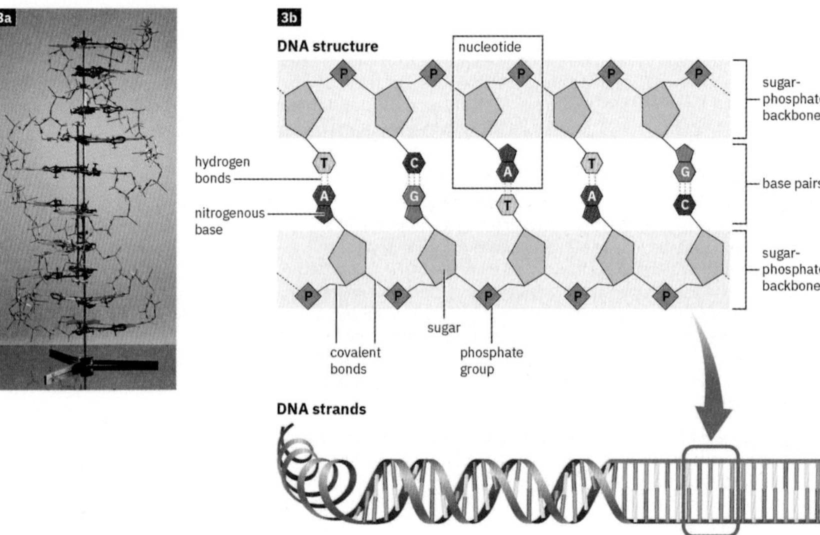

320 CHAPTER 11 DNA, RNA, AND PROTEINS

DIFFERENTIATED INSTRUCTION | Leveled Support

Struggling Students Provide students with materials that they can manipulate to create a kinesthetic-tactile experience as they build a model of DNA. Working in pairs or groups, instruct students to label each part using the terms *nucleotide*, *nitrogenous base* (labeled as *adenine*, *guanine*, *thymine*, and *cytosine*), *hydrogen bond*, and *deoxyribose sugar*. Encourage students to practice pronouncing each word on the labels as they are attached.

Advanced Learners For students with a good understanding of DNA structure, ask them how DNA, with only four nucleotides, can carry information that results in a tremendous number and variety of traits. Discuss other examples of codes that can transmit information, such as the Morse code systems of dots and dashes, computer language using zeros and ones, or even the alphabet that can be combined to make innumerable words. Challenge students to create a message using a code and exchange it with others to decipher.

RNA The nature of information represented by the sequence of nucleotides in a DNA molecule occurs in hundreds or thousands of units called **genes**. The DNA sequence of a gene encodes (contains instructions for building) an RNA or protein product. Converting the information encoded by a gene into a product starts with RNA synthesis, which is called **transcription**. During transcription, enzymes use a strand of DNA as a template to synthesize a complementary strand of **RNA**.

Most of the RNA inside cells occurs in single strands that are similar in structure to single strands of DNA. For example, both are chains of four kinds of nucleotides. Like a DNA nucleotide, an RNA nucleotide consists of three phosphate groups, a sugar, and one of four bases. However, there are some differences (**Table 11-1**). The sugar in an RNA nucleotide is a ribose (ribonucleotide), and in a DNA nucleotide it is a deoxyribose (deoxyribonucleotide). Three bases—adenine, cytosine, and guanine—occur in both RNA and DNA nucleotides, but the fourth base differs between the two molecules. In RNA, the fourth base is uracil (U), which is only slightly different from the fourth base in DNA (thymine, T).

TABLE 11-1. Comparing DNA and RNA

Characteristic	DNA	RNA
form	double-stranded double helix	most are single stranded
monomers	deoxyribonucleotides	ribonucleotides
sugar	deoxyribose	ribose
bases	adenine (A), guanine (G), cytosine (C), thymine (T)	adenine (A), guanine (G), cytosine (C), uracil (U)
base pairing	A-T, G-C	A-U, G-C
variations	none	rRNA, tRNA, mRNA, and more
function	stores genetic information	protein synthesis, carries information, other roles

Small differences in structure between DNA and RNA give rise to very different functions. While DNA's critical function is to store a cell's genetic information, a cell makes several kinds of RNA, each with a different function. Three types of RNA have roles in protein synthesis, which will be detailed in the next section. Ribosomal RNA (rRNA) is the main component of ribosomes, which are the organelles that combine amino acids into polypeptide chains. Transfer RNA (tRNA) delivers the amino acids to ribosomes, one by one, in the order specified by a messenger RNA (mRNA). mRNA was named for its function as the "messenger" between DNA (stored in the nucleus) and the proteins encoded by it (synthesized in the cytoplasm).

CCC Structure and Function *RNA and DNA have very slight structural differences that generally make RNA strands less stable and more likely to participate in chemical reactions than DNA strands. How is this reflected in the typical functions of each molecule?*

Crosscurricular Connections

Chemistry In 1948, Erwin Chargaff published the results of his experiments that became known as Chargaff's Rules. Taking small samples of DNA from different organisms, Chargaff broke each sample of DNA down into its four units and separated the units using paper chromatography. Each unit has a different attraction to the paper and a different solubility in solvent that runs up the paper. These differences result in each unit moving to a different location on the paper. After separation, Chargaff recovered each unit and measured how much UV radiation the material absorbed, which correlates with the amount of the unit. His data showed the relationship of guanine and cytosine units and adenine and thymine units. He also found that the composition of DNA varies from one species to another.

Connect to English Language Arts

Write Explanatory Text Review **Table 11-1** with students and discuss any questions they may have. Explain that tables are a visual way to convey a lot of information; however, sometimes details are left out. Using the table, have students translate the information into an explanatory paragraph that compares DNA and RNA with at least three additional supporting details.

CCC Structure and Function
Students might say that RNA is involved in more temporary roles or those that probably involve chemical reactions, such as protein synthesis. DNA's primary function is to store information in a stable form for the organism to duplicate and use as needed for gene expression.

DIFFERENTIATED INSTRUCTION | English Language Learners

Spelling Patterns As students write their explanatory paragraph, have them check their spelling carefully. Encourage students to spell syllable by syllable and to notice similar word parts. Before students begin, have them review spelling for new vocabulary in pairs.

Beginning Have pairs underline the similar ending in *adenine, guanine, cytosine*, and *thymine*. Pronounce the ending *-ine* for them to repeat.

Intermediate Have pairs divide the words *deoxyribonucleotides* and *deoxyribose* into syllables and underline the syllables that are similar. Then, have each pronounce and spell the words from memory for their partner to check.

Advanced Have pairs build related words by adding and taking away word parts, for example, *nucleotide, ribonucleotide, deoxyribonucleotide, deoxyribose*, and *ribose*. Have students use each word in a sentence.

Science Background

Quaternary Structures Point out in **Figure 11-4** the primary, secondary, tertiary, and quaternary protein structures. Note the protein made up of two polypeptide chains. Many proteins have only three levels of structures; however, some are made up of multiple amino acid chains and are called quaternary structures. Hemoglobin functions to carry blood from the lungs to the rest of the body due to its quaternary structure that consists of two pairs of peptides held together by hydrogen bonds. Each peptide holds an iron-containing organic molecule called a *heme* that can bind to oxygen. The structure of the quaternary protein allows hemoglobin to bind tightly enough to oxygen to carry it throughout the body but loosely enough to let it go in the body's tissues.

Protein As described in Chapter 5, proteins are composed of chains of amino acids, known as polypeptides (**Figure 11-4**). Each amino acid in a protein chain is encoded by a three-nucleotide sequence on an mRNA. mRNA guides the assembly of the amino acid chain in a process of protein synthesis called **translation**. Proteins form complex structures and combine with other molecules to perform most of the work in cells.

FIGURE 11-4 At its simplest primary structure, a protein is a chain of peptides. Those peptides interact to form complex secondary and tertiary structures that are required to carry out reactions in cells.

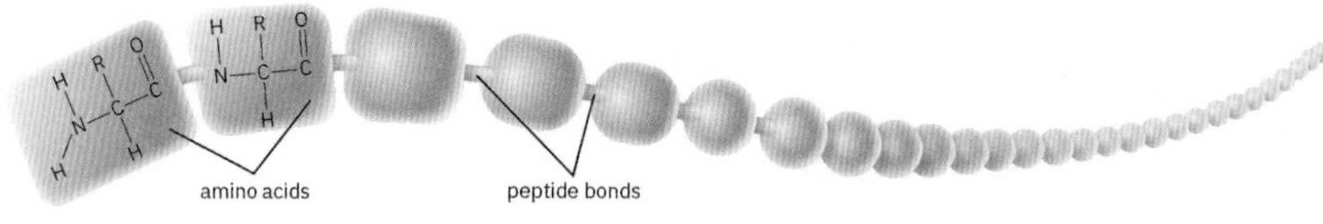

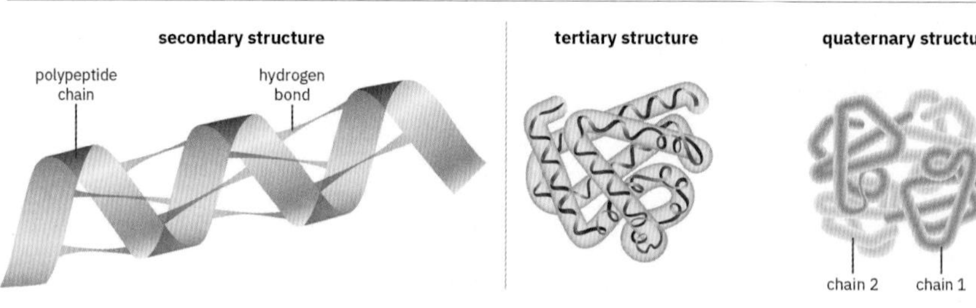

Information Transfer

Most genetic information in living organisms is decoded and transferred in the process of gene expression. In **gene expression**, information is transferred as the sequences of the DNA genome are transcribed into an mRNA intermediary. The mRNA is then translated to a protein product that performs functions in the cell (**Figure 11-5**).

However, the transfer of information in the cell is much more complicated than just the one path of gene expression. In this chapter, you will learn how RNAs serve many additional important functions in cells. You will also learn how DNA and protein structures serve complex functions beyond just being a sequence of informational nucleotides or amino acids.

FIGURE 11-5 In a typical model of gene expression, the information from DNA is transferred to RNA in transcription. That information in the RNA is then transferred into a protein through translation.

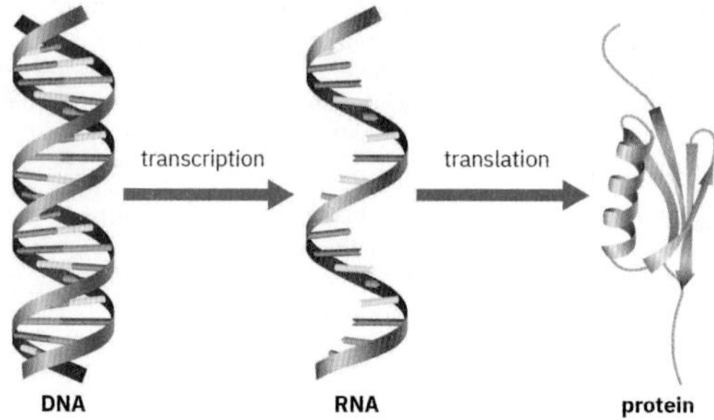

Figure 11-6 shows the ways we know genetic information can be copied and transferred. In normal cell processes, information is transferred by DNA **replication** (the copying of DNA), transcription of DNA to RNA, and translation of RNA to protein.

Additional paths for information transfer shown in the figure either have only been generated in special laboratory experiments outside living cells, or are used by infectious agents. For example, viruses exploit a variety of mechanisms and encode their own proteins that perform unique functions most living cells are not capable of, such as reverse transcription and RNA replication. *Prions* are infectious forms of normal cellular proteins that can multiply by inducing normal proteins to misfold and become infectious.

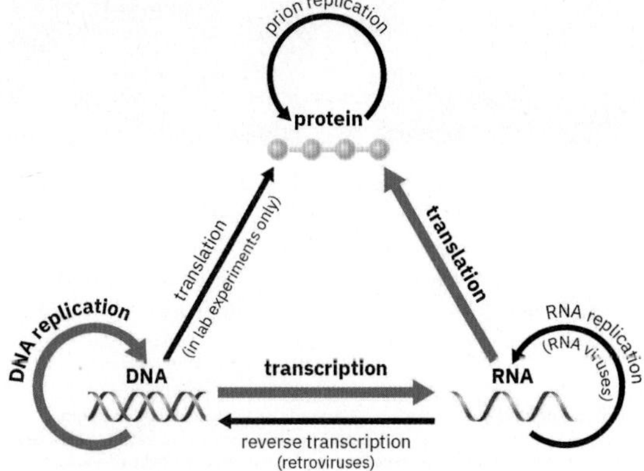

FIGURE 11-6 DNA replication, RNA transcription, and protein translation are emphasized with blue arrows to indicate they are the most important transfers of information in biological systems.

SEP Use a Model *How does the model in **Figure 11-6** demonstrate the transfer of information from DNA to protein in living cells? Notice there are no arrows pointing out from protein. What do you think that might mean?*

11.1 REVIEW

1. **Relate** A small segment of one DNA strand is made up of sugar, phosphate, adenine, and cytosine. Which components make up the matching segment in the second strand? Select all correct answers.

   A. adenine C. guanine E. sugar

   B. cytosine D. phosphate F. thymine

2. **Identify** Which processes are involved in normal gene expression? Select all correct answers.

   A. DNA replication D. transcription

   B. prion replication E. translation

   C. RNA replication

3. **Distinguish** Identify the type of RNA that fulfills each role.
   - assemble amino acids into polypeptide chains
   - deliver amino acids to ribosomes in a particular order
   - guide sequence of polypeptides during protein synthesis

In Your Community

Prion Diseases Proteins have complex structures that affect their function. As prions can induce otherwise healthy proteins in the brain to fold abnormally, the function of the protein is then affected. This results in a group of serious neurodegenerative diseases classified as prion diseases. Relatively rare, these diseases cause progressive changes in memory, behavior, and movement most often resulting in death. Prion diseases affect humans and other animals and can be transmitted through contaminated food, as well as through inherited mutations. Examples include Creutzfeldt-Jakob Disease and fatal familial insomnia in humans, chronic wasting disease (CWD) in cervids (the deer family), and bovine spongiform encephalopathy (BSE) in cattle. A cow with BSE may act nervous or violent, which is why BSE is often called "mad cow disease." Prion diseases are always fatal, and there are no treatments or vaccines.

SEP Use a Model *The arrow from DNA to RNA represents the process of transcription, and the arrow from RNA to protein represents translation, which make up gene expression. There are no arrows from protein to DNA or RNA, so information must not be transferable in that direction.*

11.1 REVIEW

1. *C, D, E, F* **DOK 2**
2. *D, E* **DOK 1**
3. *rRNA: assemble amino acids into polypeptide chains; tRNA: deliver amino acids to ribosomes in a particular order; mRNA: guide sequence of polypeptide chains during protein synthesis* **DOK 2**

EXPLORER
DR. CARTER CLINTON

About the Explorer

National Geographic Explorer Dr. Carter Clinton is a postdoctoral scholar at Penn State University, specializing in genetic anthropology with interests in African and African American genomes. You can learn more about Carter Clinton and his research on the National Geographic Society's website: www.nationalgeographic.org.

Connect to Careers

Genetic Anthropologist Genetic anthropology provides an opportunity to cross an interest in science with an interest in history. As a genetic anthropologist, careers can focus on forensics that might include jobs such as analyzing DNA left at a crime scene.

Social-Emotional Learning

Social Awareness Students will know that some enslaved African Americans sought freedom by traveling to the North. However, few will know that in colonial times about 20 percent of New Yorkers were enslaved Africans who were forced to construct much of the city. As this topic arises, allow small groups to discuss and ask questions. Encourage students to *show concern for others' feelings* and to *demonstrate empathy and compassion* for themselves and for others.

Video

Time: 2:47

In **Video 11-1**, Dr. Clinton explains what drives his work and what stories his analyses provide. Ask students what interests them about his research and other ways it might be applied.

THINKING CRITICALLY

Ask Questions *Look for logical responses that reference the longevity of DNA samples. Students might describe usefulness in unraveling mysteries such as what happened to the Anasazi, answering questions about the relatedness of various species, or locations of tombs of ancient Egyptian rulers.*

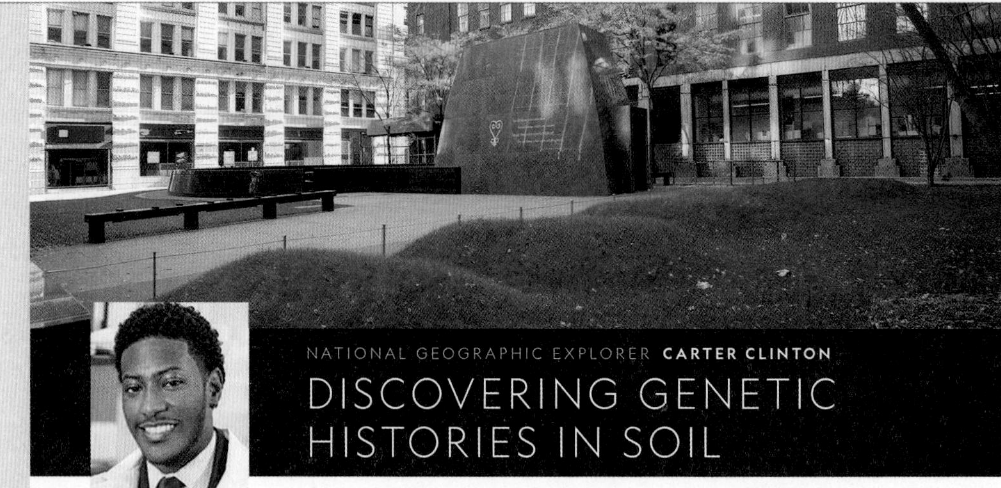

NATIONAL GEOGRAPHIC EXPLORER CARTER CLINTON
DISCOVERING GENETIC HISTORIES IN SOIL

Dr. Carter Clinton works to identify and analyze DNA from historic samples, to better understand the past and improve present disparities in genetic data.

VIDEO 11-1 Go online to learn more about Clinton's research and motivations.

It is dusk on a November evening in 1847. The travelers move swiftly but quietly through the brambles, avoiding the shrubs' prickly thorns as best they can. It's taken them a few weeks to reach this Maryland forest from the South Carolina plantation where they had been enslaved. They stop at the base of a huge tree, whose hollowed trunk provides shelter and traps the heat of a small fire so that they might rest before continuing northward.

Fast forward to a sunny day in November, 170 years later. Biological anthropologist Dr. Carter Clinton pauses before the very same huge tree that is hollowed out at the base. Clinton has studied the movements of people on the Underground Railroad and knows a location like this would have provided needed cover. He thinks that many people may have stopped here for the night, so it may be a good spot to collect DNA they left behind.

Back in the lab, Clinton extracts the DNA from the soil and then performs next generation sequencing to determine the genomic sequences isolated from these samples. From these data, he can tell where an individual was from as well as hair color, face shape, age, sex, and other details. Computer algorithms also tell him about how the people lived, and possibly how they died.

Clinton is continuing similar work in what he describes as the most out-of-the-ordinary project he's ever worked on: sampling DNA from the New York African Burial Ground in Lower Manhattan. First discovered in 1991, the New York City burial grounds were located about 8 meters below the surface and stretched over an area equivalent to five football fields. An estimated 15,000 individuals are buried there, more than half of them children. Researchers excavated 419 individuals, carefully cataloging each sample of soil gathered. Over six months, Clinton's team sifted and analyzed the samples, obtaining small particles on which to perform DNA analysis. Eventually, they were successful in detecting genomic DNA in burial soil samples. From those DNA analyses, Clinton was able to identify human-associated bacteria, including disease-causing bacteria that might have been a cause of death. He is currently working to identify human DNA in the samples. He hopes to use his research to address health disparities in African Americans and fill in data gaps for African-descended people in genomic databases. To do so, he'll keep digging for answers.

THINKING CRITICALLY

Ask Questions *Pose a question or problem that might be answerable using techniques similar to Clinton's research on detecting and analyzing DNA in soil.*

SCIENCE AND ENGINEERING PRACTICES
Analyzing and Interpreting Data

African American Burial Ground When the remains of both enslaved and free African Americans were interred, they were either wrapped in a cloth shroud or placed in a wood coffin. During the nearly 200-year period when the site was undiscovered, the shrouds and wooden coffins disintegrated, leaving the skeletal remains exposed to soil that settled inside different parts of the skeletons.

Ask students to propose different kinds of data that could be gathered. Then, explain that Dr. Clinton performed soil chemistry, bacterial DNA, and geospatial analysis. From this data, Dr. Clinton discovered what each individual farmed and ate, what they died from, and other features related to lifestyle.

From these discoveries, Dr. Clinton learned about the conditions of African Americans in New York during the era of slavery and some of the ways they contributed to the culture and history of the United States.

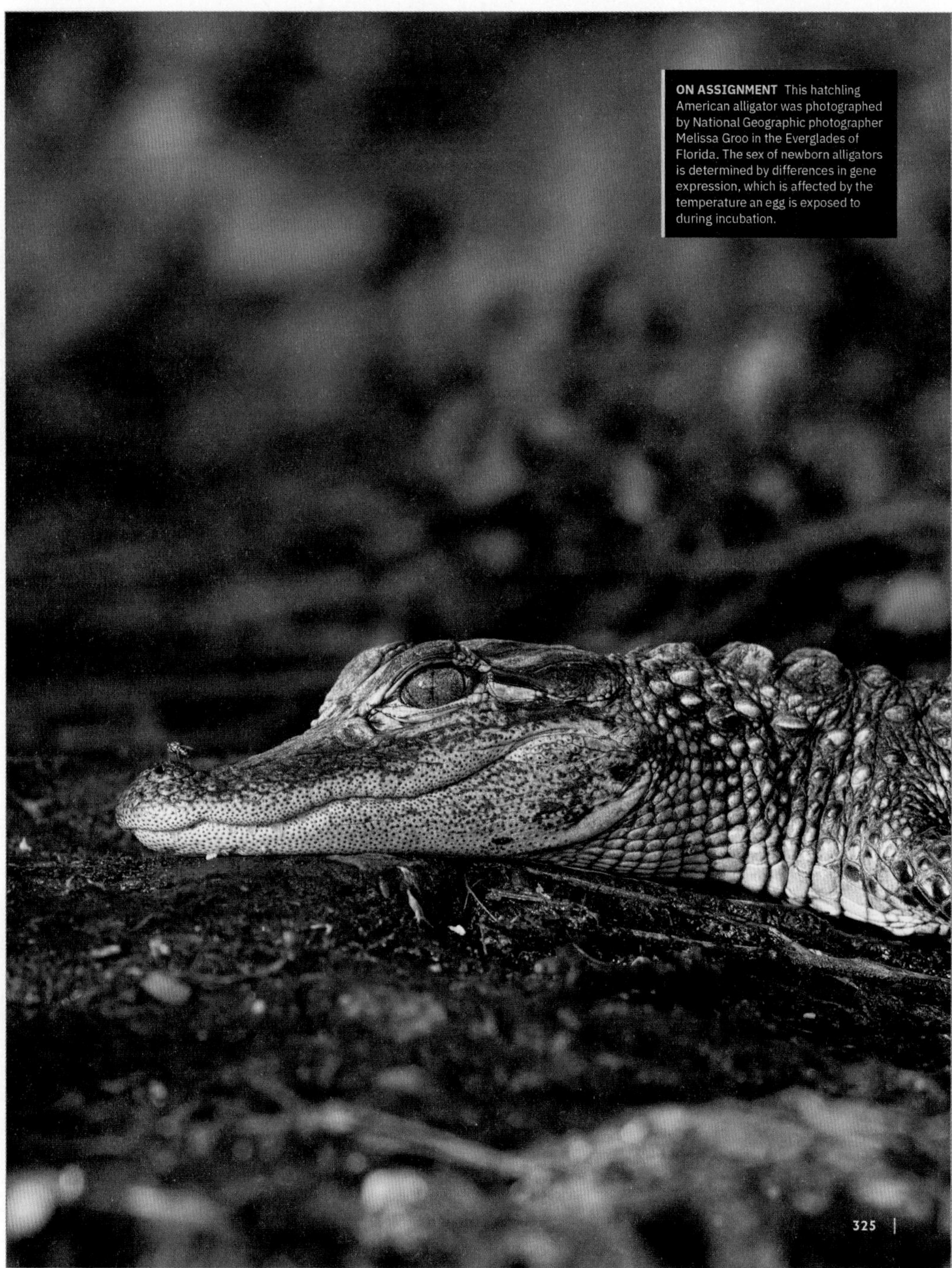

ON ASSIGNMENT This hatchling American alligator was photographed by National Geographic photographer Melissa Groo in the Everglades of Florida. The sex of newborn alligators is determined by differences in gene expression, which is affected by the temperature an egg is exposed to during incubation.

ON ASSIGNMENT

American Alligators Melissa Groo is a wildlife photographer, writer, and conservationist. She is passionate about wildlife and their habitats and uses her talents to educate people about conservation. The American alligator photographed here is native to the southeast United States, primarily Florida and Louisiana. The young hatch from eggs that are guarded by the mother during the incubation period. During this period, the nest may be covered with soil or sand to maintain a constant temperature. If the nest temperature rises above 34°C (93°F), most of the offspring will be male. If the nest temperature falls below 30°C (86°F), most of the offspring will be female.

SCIENCE AND ENGINEERING PRACTICES
Planning and Carrying Out Investigations

Role of the Environment in Reproduction For fifty years, scientists have known that the gender of the offspring of turtles, alligators, and crocodiles are determined by temperature during the incubation period of the eggs. Ask students to propose different experiments that a scientist might do to better understand this phenomenon and discuss their ideas. Students might suggest measuring the gender ratios of offspring at different temperatures, determining the point in development after which the gender will no longer change in response to a change in temperature, and studying the animals' behavior and survival to determine possible reasons why different gender ratios are advantageous at different temperatures.

11.2

REPLICATION, TRANSCRIPTION, AND TRANSLATION

LS1.A, LS3.A

EXPLORE/EXPLAIN

This section walks students through the processes of DNA replication, transcription, and translation. It begins with an introduction to the process of copying DNA that ensures the exact genetic information is passed to the next generation. Students move on to transcription, learning how information encoded in a DNA segment can be used to make an RNA molecule. The section ends with translation of information by ribosomes to synthesize proteins.

Objectives

- Relate the structure of DNA to its replication.
- Describe the process and outcomes of transcription.
- Explain the process of translation using codons.

Pretest Use **Question 6** to identify gaps in background knowledge or misconceptions.

Vocabulary Strategy

Semantic Map Have students create a semantic map for the vocabulary in this section. Begin by choosing a key term that is central to the topic. Have students write that word in the center of a page or diagram it on the board. As they read through each section, ask them to add each new word to the map.

Explain *Students might say that replication reflects copy and paste functions on a computer. Transcription is similar to having someone write down someone's words as they are talking. Translation is like converting one language to another, or using sign language to communicate spoken words into hand gestures.*

11.2

KEY TERMS

anticodon
codon
DNA polymerase
primer
RNA polymerase

REPLICATION, TRANSCRIPTION, AND TRANSLATION

The information in DNA is copied as cells replicate, slightly changed in transcription from DNA to RNA, and entirely recoded in the process of translation. The name for each process reflects a critical aspect of how the process occurs.

Explain *Generate an analogy to explain the general processes of replication, transcription, and translation.*

DNA Replication

As discussed in Sections 7.1 and 7.2, when a cell undergoes mitosis to reproduce, it divides (**Figure 11-7**). The two descendant cells must inherit a complete and accurate copy of their parent's genome, or they will not function like the parent cell. In preparation for division, the cell copies its chromosomes so that it contains two complete sets: one for each of its future offspring. The copying process, which is called *replication*, ensures the continuity of genetic information from one generation to the next.

FIGURE 11-7 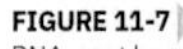
DNA must be replicated for cells to divide and pass on genetic information to offspring.

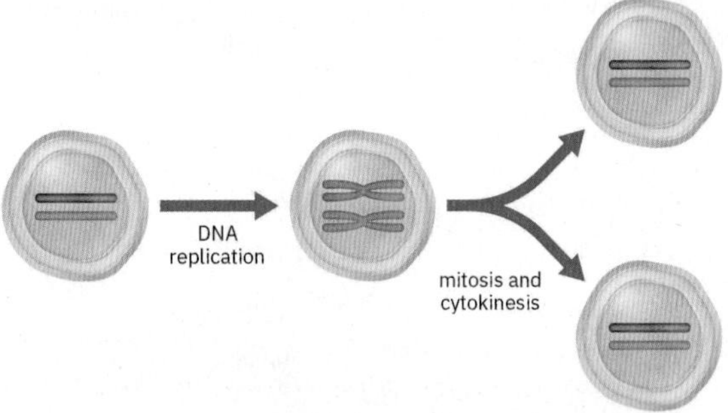

Separation of the DNA Strand Before DNA replication, a chromosome consists of one molecule of DNA—one double helix consisting of two complementary DNA strands (**Figure 11-8**). The replication process is triggered when an initiator complex assembles and begins to unwind the double helix. An enzyme called helicase breaks the hydrogen bonds that hold the two DNA strands together, forming a replication bubble as shown in **Figure 11-9**.

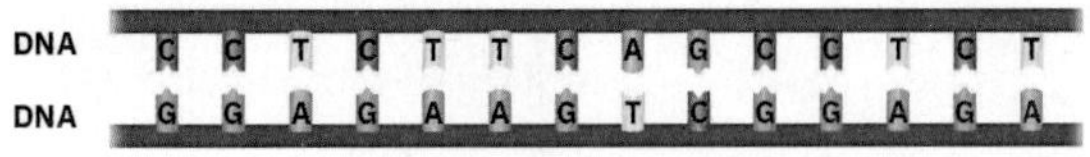

FIGURE 11-8
In DNA, C (cytosine) pairs with G (guanine) and A (adenine) pairs with T (thymine).

SCIENCE AND ENGINEERING PRACTICES
Developing and Using Models

Modeling DNA Engage students in a discussion about templates as guides for the creation of one or more things in the same way. Discuss examples of different types of templates.

Have students work in small groups to create their own models of a DNA molecule using available materials to represent the sugars, phosphates, and different nitrogen bases. Material used for nitrogen bases should be available in different colors for ease of use (e.g., clay in different colors). Once their models are completed, students can use their models to actively reproduce each process in this section. If possible, they might create a stop motion film of each process using their model.

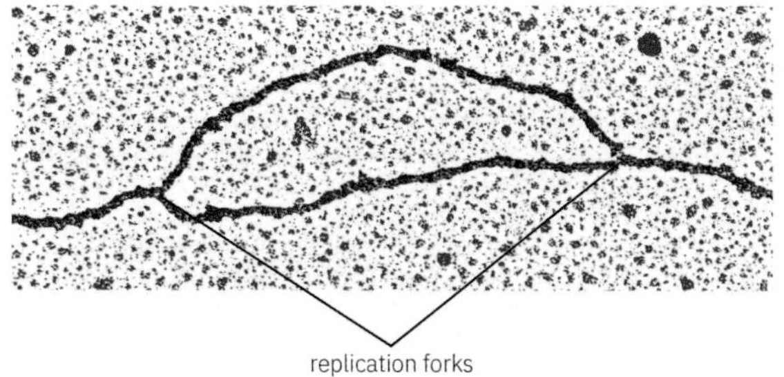

FIGURE 11-9 This colored transmission electron microscope image shows DNA from a HeLa cell undergoing DNA replication. Replication forks, where the DNA unwinds to allow new strands to be formed, are indicated.

Assembling a New DNA Strand **Figure 11-10** shows many of the important factors involved in DNA replication. When the helicase separates the strands for replication, an enzyme attaches to each DNA strand to build a short sequence of nucleotides on which replication starts. This short sequence of nucleotides is called a **primer**. Multiple primers attach in different places along the original, or template, DNA strands.

DNA polymerases assemble around the primers and attach free-floating nucleotides to the new nucleotide chain. At each position, the nucleotide added has a nitrogen base complementary to the exposed base on the template strand: A (adenine) to T (thymine) and C (cytosine) to G (guanine). Hydrogen bonds connect each pair of complementary bases as DNA replication proceeds, ultimately producing two new double-stranded DNA molecules from one. The DNA sequence of each new strand is complementary to its template strand.

The unwinding of the DNA strand happens in both directions at the same time. Likewise, replication happens at both forks (both ends of the replication bubble) at the same time. On the leading strand, DNA replication begins at the primer and proceeds in one direction, building a continuous strand. On the lagging strand, replication occurs in segments in the opposite direction. On the lagging strand, primers attach at multiple locations, so multiple molecules of DNA polymerases can add nucleotides to each primer at the same time. The enzyme *DNA ligase* seals any gaps on the lagging strand, so the new DNA strands are continuous.

VIDEO

VIDEO 11-2 Go online to see how molecular data has been used to model DNA replication.

CCC Structure and Function *Identify the leading and lagging strand on the diagram in* ***Figure 11-10****. How do you know which strand is which?*

FIGURE 11-10 At a replication fork, the helicase enzyme unwinds DNA. New DNA is generated as the strands of the original double helix are replicated.

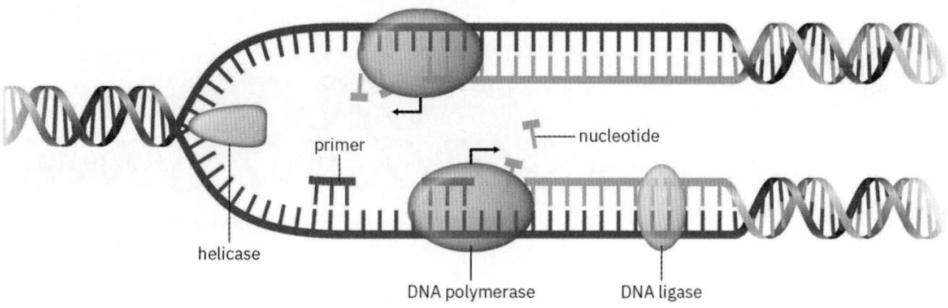

DIFFERENTIATED INSTRUCTION | English Language Learners

Vocabulary Strategies Use physical models to help students understand new vocabulary. Introduce the concept of DNA replication using a zipper. As you unzip it, tell students to refer to **Figure 11-10**. Give students zippers to work with.

Beginning Ask students to relate parts of the zipper to what they see in **Figure 11-10** (*the teeth = complementary base pairs, slider handle = helicase; opening of zipper = unwinding of DNA and formation of two replication forks*).

Intermediate Using **Figure 11-10** and the zipper, have students identify the differences between the leading and lagging strands. Tell them to make a drawing that includes helicase, DNA polymerase, DNA ligase, and primer.

Advanced Have students use **Figure 11-10**, as well as their zipper, to compare and contrast the leading and lagging strands using a T-chart. Tell students to use their chart to write a summary of the similarities and differences.

Visual Support

HeLa Cells Remind students of HeLa cells from Chapter 7, which gave context to their understanding of the cell cycle and its regulation. The significance of Henrietta Lacks's story is that her cells were characterized by uncontrolled cell division. Point out that **Figure 11-9** shows how the DNA from a HeLa cell replicates by unwinding and allowing new strands to be formed. Tell students that DNA in human cells has about 3.2 billion base pairs. Ask students if the DNA that is replicated today is the same as the DNA first isolated from Henrietta Lacks's tissue and give reasoning for their answers. Then discuss how mistakes in DNA replication can occur and potential outcomes of those mistakes.

Address Misconceptions

Cellular DNA Ask students if a skin cell and a nerve cell have the same or different DNA. Students may think that cells that look different and have different functions within the same organisms have different DNA. This is a good time to reiterate to students that every somatic (body) cell has the same DNA due to the process of replication. Each time a cell divides, the two daughter cells will have the same genetic information. Other factors determine which genes are expressed for specialized cells.

Video

Time: 1:05

Use **Video 11-2** to present a three-dimensional action model showing the mechanisms of DNA replication, which can help students visualize the steps they read in the text. Once you have run the video through completely, you may choose to play it again, stopping the video after each step and asking students to describe the events depicted.

CCC Structure and Function *The leading strand is the top strand because it is a continuous piece. The lagging strand is the bottom strand on the figure because it is made up of multiple short pieces of RNA.*

Visual Support

Models Point out to students that while models are very helpful in visualizing processes that cannot be directly observed, they are not perfect. The relative and actual sizes of components are not always accurate in drawings. For example, **Figure 11-12**, as well as others, are not to scale. The promoter and gene sequences in the transcription diagram are very short. Explain to students that these sequences are actually much longer than the number of nucleotides shown here. During elongation, the RNA polymerase spans about 35 base pairs.

Transcription

DNA stores the information for multiple forms of RNA (mRNA, rRNA, and tRNA). Transcription is the process by which information encoded in a gene, or DNA segment, is used to synthesize an RNA molecule. Similar to DNA replication, during transcription, a strand of DNA functions as the template for synthesis of a new RNA molecule. For RNA transcription, slightly different base-pairing rules are followed, compared with DNA replication, because uracil is substituted for thymine (**Figure 11-11**). Cytosine (C) pairs with guanine (G), and adenine (A) pairs with uracil (U). A nucleotide can be added to the end of a growing RNA molecule only if the base pairs with the corresponding nucleotide of the template DNA. Thus, the new RNA is complementary in sequence to the DNA template.

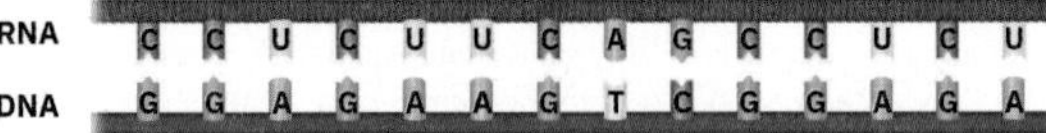

FIGURE 11-11 In RNA, C (cytosine) pairs with G (guanine), and A (adenine) pairs with U (uracil).

In contrast to DNA replication, only a defined region of the DNA strand is used as a template for transcription, not the entire strand. So, only the particular segment of DNA is transcribed when it is needed. Also, transcription produces a single strand of RNA, not a double strand as with DNA replication.

Similar in function to DNA polymerases, enzymes called **RNA polymerases** are used in transcription. A regulatory site, called a *promoter,* is a short DNA sequence that serves as a binding site for RNA polymerase.

Transcription begins when proteins that recognize a promoter help RNA polymerase bind to it (**Figure 11-12**). After binding, the polymerase starts moving along the DNA, unwinding the double helix to reveal the template DNA sequence. The enzyme adds free RNA nucleotides into a chain, in the order determined by the DNA sequence.

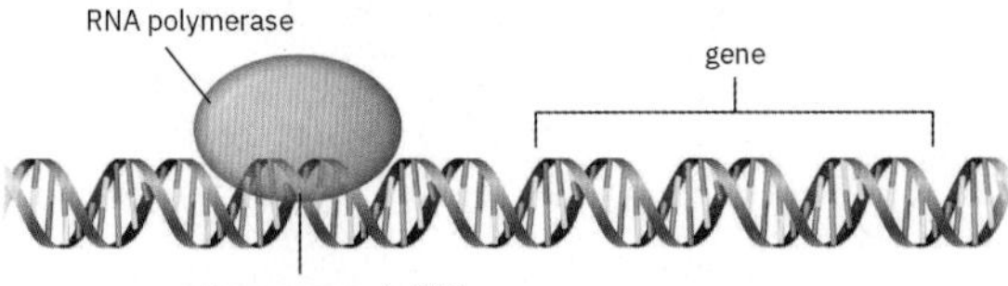

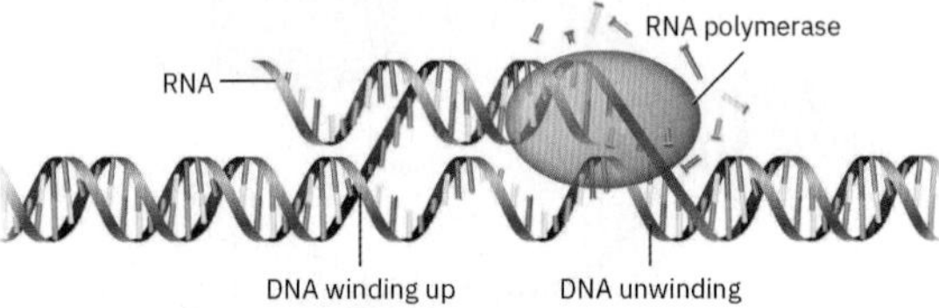

FIGURE 11-12 RNA polymerase binds a promoter sequence in the DNA. The DNA is unwound, and the polymerase adds nucleotides to form a new RNA strand as it moves along the gene sequence.

328 CHAPTER 11 DNA, RNA, AND PROTEINS

CROSSCUTTING CONCEPTS | Systems and System Models

Central Dogma The central dogma of molecular biology explains the flow of genetic information from DNA to RNA to protein. Proposed by Francis Crick in 1958, the central dogma states that DNA contains the information needed to make all of the body's proteins and the cell utilizes RNA as the messenger so that DNA can be protected in the nucleus. An exact RNA copy of needed DNA segments is transcribed and then carried out of the nucleus to ribosomes where it is translated into a protein in the process of gene expression. Discuss with students how the central dogma models transcription and translation. Encourage students to apply the central dogma to explain why transcription produces a single strand of RNA while replication produces a double strand of DNA.

When the RNA polymerase reaches the end of the gene region, it disconnects from the DNA template and the newly synthesized RNA strand. The DNA template re-forms a double helix, and the RNA product is processed according to its function (usually mRNA, rRNA, or tRNA). Messenger RNA (mRNA) is the molecule that codes for a protein to be produced. Unlike the model in **Figure 11-12**, many RNA molecules can be made at once from a single gene, as shown in **Figure 11-13**.

SEP Construct an Explanation *Describe the similarities and differences between replication and transcription.*

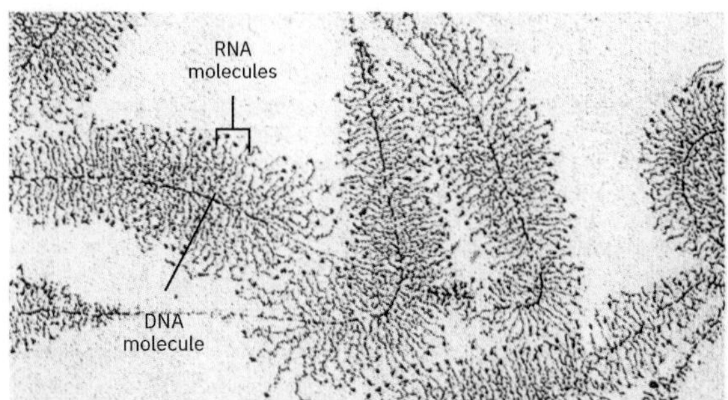

FIGURE 11-13 This electron micrograph shows RNA molecules being transcribed in huge numbers, with individual genes being transcribed multiple times simultaneously.

Translation

The information encoded by a gene is ultimately translated by ribosomes, using mRNA to direct synthesis of a polypeptide, or protein. This process is called translation. Translation occurs in the cytoplasm of all cells where there are many free amino acids, tRNAs, and ribosomal subunits available to participate in the process. Often, several ribosomes attach to a single mRNA to translate new protein molecules at one time (**Figure 11-14**).

mRNA Coding Messenger RNA (mRNA), transfer RNA (tRNA), and ribosomal RNA (rRNA) interact to translate DNA's information into a protein. mRNA is essentially a temporary, abridged copy of a gene. Its job is to carry the gene's protein-building information to the other two types of RNA during translation. The protein-building information carried by an mRNA is a sequence of genetic "words" that occur one after another along its length. Like the words of a sentence, a series of these genetic words can form a meaningful parcel of information—in this case, a sequence of amino acids that constitutes the primary structure of a protein.

VIRAL SPREAD

Gather Evidence *SARS-CoV-2 is a virus with a single-stranded RNA genome that resembles a messenger RNA to a host cell. The genome encodes genes for the structural proteins to assemble new viruses as well as some additional enzymes. What molecules and processes does it need from the host cell to make new viruses? What additional enzymes do you think it needs to provide to replicate itself?*

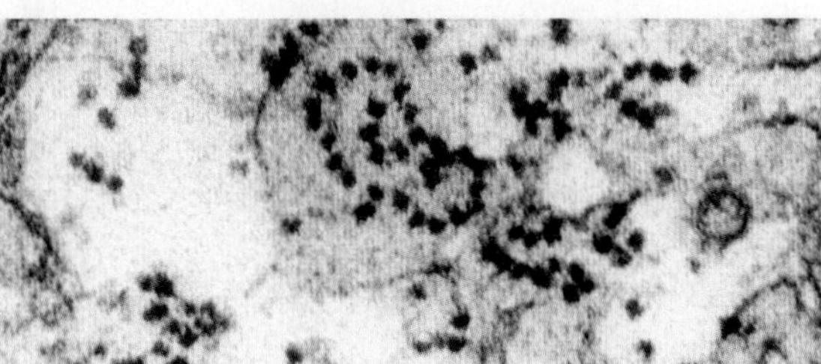

FIGURE 11-14 This electron micrograph shows multiple ribosomes translating single mRNA molecules at one time. These are known as polysomes.

Connect to English Language Arts

Write Explanatory Text To benefit all students in understanding the roles of mRNA, tRNA, and rRNA, ask them to create a three-column chart that compares the three forms, including the structure and function of each. Using the chart, have each student write 1–2 paragraphs summarizing the information. Students can add further information to their charts or summaries as they learn more throughout the lesson.

Visual Support

Polysomes A single mRNA transcript can be transcribed by several ribosomes simultaneously. This creates a complex called a *polysome*. Ask students to identify the polysomes and the mRNA in **Figure 11-14** and to explain how they made their identification. Ask students how polysomes can be an advantage to the cell. (*They can quickly produce large amounts of proteins without expending the energy to create new mRNA.)*

SEP Construct an Explanation
Students might say that both processes involve unwinding the DNA strand to make new complementary molecules and that both involve polymerase enzymes. Transcription makes an RNA molecule instead of DNA and only copies a targeted segment of the DNA strand.

REVISIT THE ANCHORING PHENOMENON

Gather Evidence *Students might say that because the virus genome looks like a messenger RNA, it can have its proteins translated in the cytoplasm of the host cell. However, cells do not replicate RNA typically, so the virus probably needs to bring its own enzymes to duplicate its genome.*

As the COVID-19 pandemic brought to light, it is important for people to understand the molecules and processes of viral infection. Coronaviruses contain a positive, single-stranded RNA genome packaged inside a protein shell, or capsid, that contains three proteins: the membrane protein, the envelope protein, and the spike protein.

During the pandemic, the spike protein came to be identified as a key characteristic of the virus, as this structure allows the COVID-19 virus to bind to the host receptor and penetrate the host cell.

Connect to Mathematics

Reason Quantitatively Have students calculate the largest number of codons, or three-letter sequences, that can be made using the four letters representing the four bases. ($4^3 = 4 \times 4 \times 4 = 64$). Discuss with students how this calculation could give scientists a clue to the size of codons. (*A sequence of four letters would result in 256 combinations, too many to code for 20 amino acids; a two-letter sequence would result in only 16 amino acids.*)

Science Background

Redundant Codons Point out to students that all amino acids, except tryptophan and methionine, are coded for by more than one codon. Rather than being a mistake, this actually offers an advantage. The redundant codons are usually different at the third base. The advantage is that, if there is a mutation, or error, during translation, there is a good chance that the altered codon will still code for the same amino acid and still function as it should. Discuss with students how this affects the function of the protein originally coded for. (*It reduces the chance that the mutation will change the function of the protein.*)

Each genetic "word" in an mRNA is three bases long, and each is a code—a **codon**—for a particular amino acid. The sequence of bases in a triplet determines which amino acid the codon specifies (**Figure 11-15**). For instance, the codon UUU specifies the amino acid phenylalanine, and UUA specifies leucine. With four possible bases (G, A, U, or C) in each of the three positions of a codon, there are a total of 64 mRNA codons. Collectively, the 64 codons constitute the genetic code. These codons specify a total of 20 naturally occurring amino acids, so some amino acids are specified by more than one codon. For instance, the amino acid tyrosine (tyr) is specified by two codons: UAU and UAC.

Other codons signal the beginning and end of a protein-coding sequence. The first AUG in an mRNA almost always serves as the signal to start translation. AUG is also the codon for methionine, so methionine is almost always the first amino acid in new polypeptides. The codons UAA, UAG, and UGA do not specify an amino acid. These are signals that stop translation, so they are called stop codons. A stop codon marks the end of the protein-coding sequence in an mRNA.

FIGURE 11-15
This chart describes which three-letter RNA codons match each of the 20 naturally occurring amino acids, as well as the stop codons that end translation.

first base of codon	second base: U		second base: C		second base: A		second base: G		third base of codon
U	UUU	phenylalanine (phe)	UCU	serine (ser)	UAU	tyrosine (tyr)	UGU	cysteine (cys)	U
	UUC		UCC		UAC		UGC		C
	UUA	leucine (leu)	UCA		UAA	STOP	UGA	STOP	A
	UUG		UCG		UAG		UGG	tryptophan (trp)	G
C	CUU	leucine (leu)	CCU	proline (pro)	CAU	histidine (his)	CGU	arginine (arg)	U
	CUC		CCC		CAC		CGC		C
	CUA		CCA		CAA	glutamine (gln)	CGA		A
	CUG		CCG		CAG		CGG		G
A	AUU	isoleucine (ile)	ACU	threonine (thr)	AAU	asparagine (asn)	AGU	serine (ser)	U
	AUC		ACC		AAC		AGC		C
	AUA		ACA		AAA	lysine (lys)	AGA	arginine (arg)	A
	AUG	methionine (met)	ACG		AAG		AGG		G
G	GUU	valine (val)	GCU	alanine (ala)	GAU	aspartic acid (asp)	GGU	glycine (gly)	U
	GUC		GCC		GAC		GGC		C
	GUA		GCA		GAA	glutamic acid (glu)	GGA		A
	GUG		GCG		GAG		GGG		G

DIFFERENTIATED INSTRUCTION | Leveled Support

Struggling Students Working in pairs, give students several different hypothetical DNA sequences, such as AUGACCGAACUUUUAAUCUAA. Have students determine the complementary mRNA code. For further practice, student pairs can create sequences using **Figure 11-15** and exchange them with their partners to encode.

Advanced Learners Working in pairs, give students several different hypothetical DNA sequences, with at least one missing the start or stop codon. Have students determine the complementary mRNA, tRNA, and amino acid sequences, as well as identify the sequence with the missing start or stop codon.

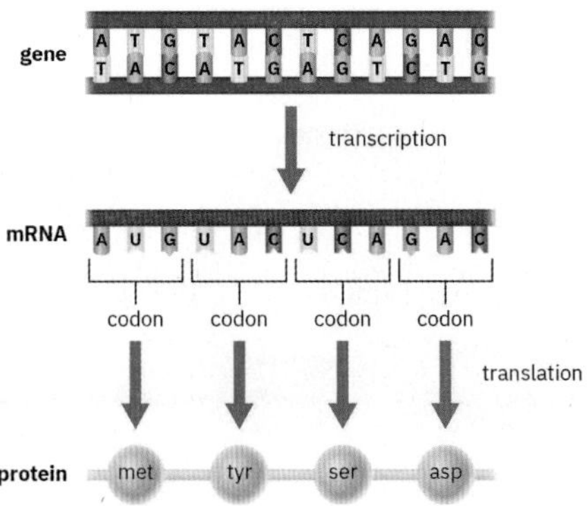

FIGURE 11-16 Use this model to trace the information in the DNA sequence as it is transcribed into an mRNA and then translated into an amino acid sequence of a protein.

SEP Use a Model *The DNA strand that is used to make an mRNA molecule is often referred to as the template strand. The other strand that is similar to the mRNA sequence is often called the complementary strand. In **Figure 11-16**, which DNA strand is the template strand and which is the complementary strand?*

tRNA and rRNA Codons occur one after the next in an mRNA. When the mRNA is translated, the order of its codons determines the order of amino acids in the resulting polypeptide. Thus, the DNA sequence of a gene is transcribed into the nucleotide sequence of an mRNA, which is in turn translated into an amino acid sequence (**Figure 11-16**).

Go Online SIMULATION

Gene Expression Go online to step through the processes of transcription and translation.

Ribosomes interact with transfer RNAs (tRNAs) to translate the sequence of codons in an mRNA into a polypeptide. Each tRNA has two attachment sites. The first is an **anticodon**, which is a triplet of nucleotides that base-pairs with an mRNA codon. The other attachment site binds to an amino acid—the one specified by the codon. Transfer RNAs with different anticodons carry different amino acids. During translation, these tRNAs deliver amino acids to a ribosome. As the amino acids are delivered, the ribosome joins them into a new polypeptide.

A ribosome has two subunits, one large and one small (**Figure 11-17**). Each subunit consists of rRNA and associated structural proteins. During translation, a large and small ribosomal subunit come together to form a complete ribosome on an mRNA. Ribosomal RNA is one example of RNA with enzymatic activity. The rRNA components of a ribosome (not the protein components) catalyze the formation of peptide bonds that connect amino acids during protein synthesis.

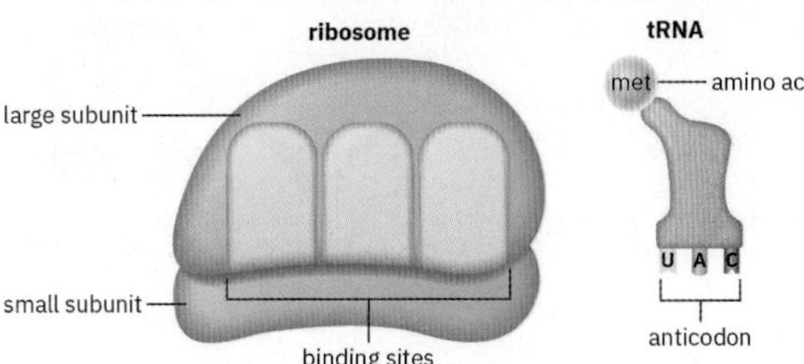

FIGURE 11-17 The ribosome is a complex molecular machine made out of many proteins and RNAs. tRNAs match corresponding sequences of the mRNA sequence at binding sites within the ribosome.

Visual Support

Shape of tRNA Transfer RNA is shown in **Figure 11-17**. It is a single strand of RNA that ranges in length from 60 to 95 nucleotides. This strand loops back on itself to create a structure that supports its function. This structure has two sites of attachment. One site, the anticodon, binds to the codon on the mRNA molecule. The arm of the structure that is not looped attaches to a specific amino acid.

SEP Use a Model *Students might say that the template strand is the bottom strand starting with TAC, and the complementary strand is the top strand starting with ATG.*

Go Online **SIMULATION**

Gene Expression After students have gone through the simulation, have them talk with a partner to compare the simulations with the text figures and descriptions of transcription and translation. Have each pair explain the benefits and drawbacks they noted between these model types to the class in terms of understanding these processes.

DIFFERENTIATED INSTRUCTION | English Language Learners

Roots and Affixes Have students use prefixes to understand the words *transcription, translation, anticodon,* and *subunit*. Explain that the prefix *trans-* means "across," the prefix *anti-* means "opposite or against," and the prefix *sub-* means "under."

Beginning Have students use the prefixes to define each word. Allow them to use a dictionary or glossary, but have them put the definitions in their own words.

Intermediate Have students use the prefixes and context clues in the text to help them define all four words and use each in a sentence.

Advanced Have students write an explanation of the meaning of each word based on the prefix and what they have learned from the text.

In Your Community

Ribosome Factories In many communities, factories exist where products are put together using assembly lines. In these systems, products move along an assembly line where specific parts are added in the correct places. The product keeps moving along the assembly line until all of the correct parts are in place. Have students write an explanatory paragraph that describes how ribosomes are similar to factory assembly lines that may be present in their community.

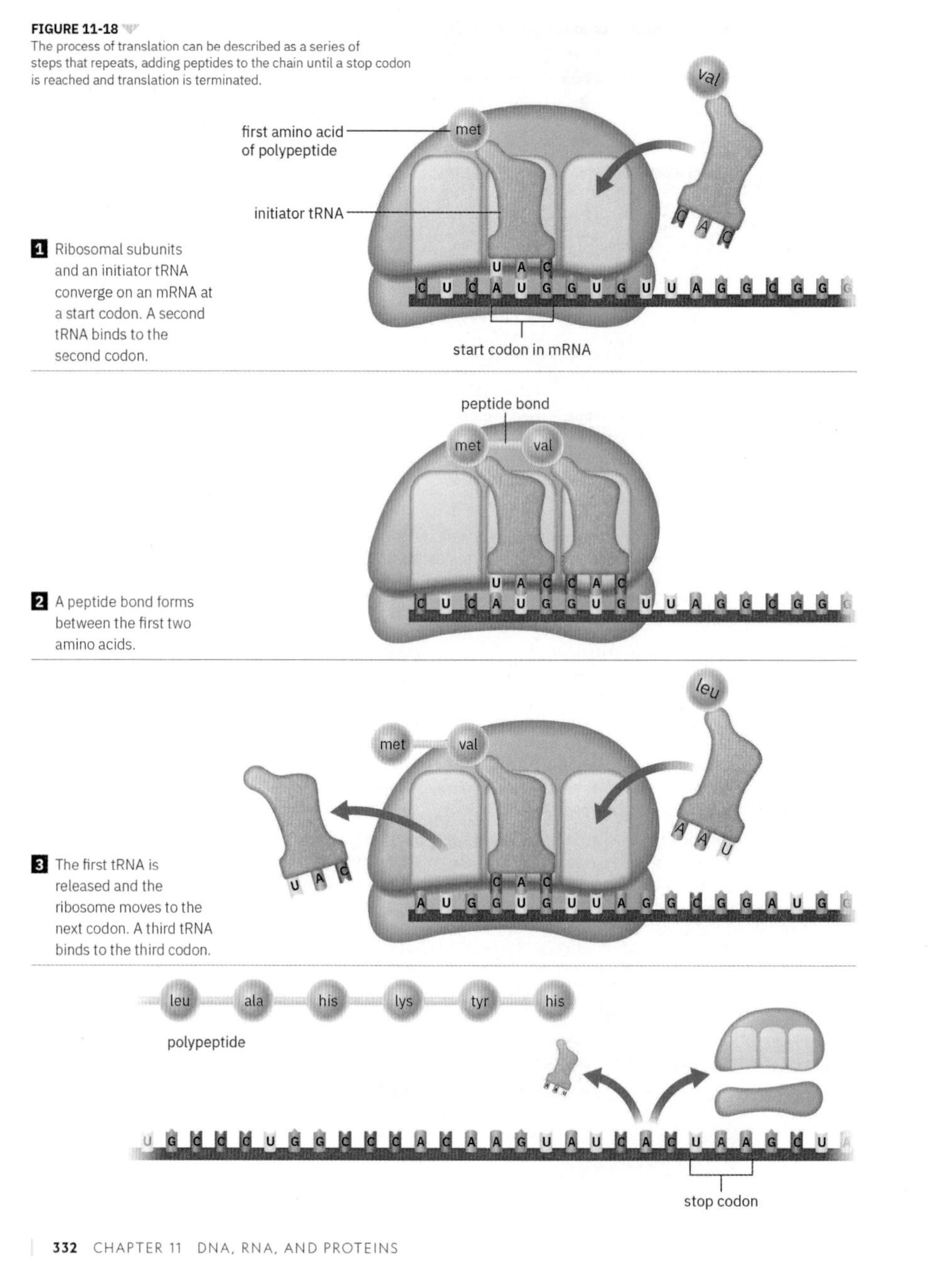

first amino acid of polypeptide
met
val
initiator tRNA
1 Ribosomal subunits and an initiator tRNA converge on an mRNA at a start codon. A second tRNA binds to the second codon.
start codon in mRNA
peptide bond
2 A peptide bond forms between the first two amino acids.
leu
3 The first tRNA is released and the ribosome moves to the next codon. A third tRNA binds to the third codon.
leu ala his lys tyr his
polypeptide
stop codon
332 CHAPTER 11 DNA, RNA, AND PROTEINS

FIGURE 11-18 The process of translation can be described as a series of steps that repeats, adding peptides to the chain until a stop codon is reached and translation is terminated.

Steps of Translation In a eukaryotic cell, ribosome biosynthesis occurs in the nucleus, and then subunits are transported through nuclear pores into the cytoplasm where translation occurs. In a prokaryotic cell, translation may begin on an mRNA even as it is still being transcribed.

- Translation begins when a small ribosomal subunit binds to an mRNA, and the anticodon of an *initiator tRNA* base pairs with the mRNA's first AUG codon. Then, a large ribosomal subunit joins the small subunit.
- The complex of molecules is now ready to carry out protein synthesis (**Figure 11-18**). The ribosome moves along the mRNA and assembles a polypeptide. Most initiator tRNAs carry methionine, so the first amino acid of most new polypeptide chains is methionine. Another tRNA joins the complex when its anticodon pairs with the second codon in the mRNA, bringing the second amino acid. The ribosome then catalyzes formation of a peptide bond between the first and second amino acids.
- As the ribosome moves to the next codon, it releases the first tRNA. Another tRNA brings the third amino acid to the complex as its anticodon base-pairs with the third codon of the mRNA. A peptide bond forms between the second and third amino acids.
- The second tRNA is released, and the ribosome moves to the next codon. Another tRNA brings the fourth amino acid to the complex as its anticodon base-pairs with the fourth codon of the mRNA, adding another peptide to the chain.
- The amino acid chain continues to elongate. Translation terminates when the ribosome reaches a stop codon in the mRNA. The mRNA and the polypeptide detach from the ribosome, and the ribosomal subunits separate from each other. Translation is now complete.

CCC **Structure and Function** *Ribosomes add as many as 20 amino acids per second. While the main functions of the ribosome are all performed by RNA, the process would proceed much more slowly without the protein components of the ribosome. How do you think those structures might contribute to the efficiency of the ribosome?*

VIDEO

VIDEO 11-3 Go online to view a simulation of translation that is based on molecular modeling data.

11.2 REVIEW

1. **Analyze** For each segment of template DNA strand, determine the corresponding sequence of nucleotides that the DNA polymerase enzyme will synthesize during DNA replication.
   - TGA
   - CCT
   - CTC
   - GAC
2. **Analyze** For each segment of mRNA, determine the original sequence of nucleotides in the DNA strand before transcription.
   - ACU
   - GAA
   - GAG
   - CUG
3. **Contrast** Label each statement as being true for replication, transcription, or both.
   - Polymerase enzymes carry out this process.
   - This process produces one strand of nucleic acids.
   - Bases involved include the four nucleic acids A, C, G, T.
   - At least one part of a strand of nucleic acids serves as a template.
4. **Evaluate** Use **Figure 11-15** to convert the codons into amino acids.
   - ACU
   - GGA
   - GAG
   - CUG

11.2 REVIEW

1. *(lt) ACT; (lb) GGA; (rt) GAG; (rb) CTG* **DOK 2**
2. *(lt) TGA; (lb) CTT; (rt) CTC; (rb) GAC* **DOK 2**
3. *Replication: Bases involved include the four nucleic acids A, C, G, T; transcription: This process produces one strand of nucleic acids; both: Polymerase enzymes carry out this process. At least one part of a strand of nucleic acids serves as a template.* **DOK 2**
4. *(lt) threonine; (lb) glycine; (rt) glutamic acid; (rb) leucine* **DOK 3**

Connect to English Language Arts

Synthesize Information Share several flowchart templates with students. Tell them to study **Figure 11-18** and read the steps in translation. Working in groups, have students use a template of their choice to summarize the process of translation in a flowchart. Explain that sometimes an arrow going back to a previous box is used for repeating processes. Remind them that they should include the start and stop codons and include different types of RNA.

Video

Time: 0:32

Use **Video 11-3** to present an animation that shows the synthesis of proteins in cells. You can use this animation to help students better understand the roles of mRNA and tRNA and the processes by which proteins are constructed by ribosomes. As they watch the video, pause it and have students explain what they are observing.

Science Background

Antibiotics and Translation Antibiotics are commonly used to treat infectious diseases. One of the advantages of antibiotics is that they have specificity, i.e., they are able to eradicate the bacteria without harming the human. Some antibiotics work by inhibiting translation. They may block the synthesis of the bacterial cell wall, which is critically important to the bacteria but not to humans. Antibiotics also work to inhibit all bacterial protein synthesis. Because of the difference in ribosomes between prokaryotes and eukaryotes, some antibiotics bind only to bacterial ribosomal proteins. When bacteria are unable to make proteins, the bacteria die and the infections they cause are stopped.

CCC **Structure and Function** *Students might say that the proteins probably help position all the molecules in the most efficient location to take in tRNAs and mRNA and link amino acids together. Any molecules that are not in the right place might slow down the reading of the mRNA or the moving of tRNAs through the ribosome.*

MINILAB
MODELING DNA REPLICATION, TRANSCRIPTION, AND TRANSLATION

Students will work with a model of DNA and the related components, along with their digital devices, to demonstrate DNA replication, transcription, or translation. They will demonstrate how DNA strands have a very specific order and how that order is precisely replicated, transcribed, or translated. They will demonstrate a small segment of DNA replication, translation, or transcription for a process that occurs on a vastly larger scale in all organisms.

Time: 30 minutes

Materials and Advance Preparation

- Have a variety of materials available for students to choose from for their demonstrations. A collection of craft supplies, as well as various types of building block toys, would be useful for students to be able to quickly establish a plan for modeling their process.

Procedure

Before **Step 1**, decide how you will determine which groups will demonstrate which process. This can include assigning specific processes to students or letting them decide on their own.

In **Step 2**, remind students that the information they need can be found in their text.

In **Steps 3** and **4**, remind students that they have limited time for this lab and that it is expected that their model might not represent some of the details.

Results and Analysis

1. *Sample answer for DNA replication: Demonstrating DNA replication required a DNA strand, RNA bases, DNA bases, helicase, primase, DNA polymerase, and RNA polymerase.* **DOK 1**
2. *Sample answer for DNA replication: The steps of DNA replication are: 1) DNA unwinds; in the presence of helicase, DNA "unzips" and creates a replication fork with a leading strand and a lagging strand. 2) In the presence of primase, RNA bases are added that are complementary to the DNA leading strand, creating an RNA primer. 3) DNA polymerase binds to the primer; DNA bases are added, from the 5' end to the 3' end. On the leading strand, the DNA bases are added continuously. On the lagging strand, which runs in the opposite direction, 3' to 5', they are added in a series of small chunks.* **DOK 2**
3. *Sample answer for DNA replication: Our model represented the DNA strand and showed how the DNA strand was "unzipped" by helicase. We demonstrated how DNA replication works with DNA polymerase on the leading strand but did not include information about the slightly different process required for the lagging strand because of the bases needing to be added from the 5' end to the 3' end.* **DOK 3**
4. *Sample answer for DNA replication: DNA replication is not dependent on the other two processes. However, during transcription, a complementary RNA strand is copied from the DNA strand. During translation, the RNA strand is "translated" to determine the specific amino acids that make up a particular protein.* **DOK 3**

MINILAB

MODELING DNA REPLICATION, TRANSCRIPTION, AND TRANSLATION

SEP Asking Questions and Defining Problems How can replication, transcription, or translation be demonstrated?

Modeling the process of DNA replication, transcription, and translation are good ways to gain a better understanding of these processes as well as to explain them to others. In this activity, you will work with a model of DNA and related components and determine a way to demonstrate DNA replication, transcription, or translation using your model.

Suggested Materials

- camera, tablet, computer
- craft paper, poster paper
- modeling clay, craft sticks, building blocks
- pipe cleaners, bendable wire
- scissors, glue, transparent tape

Procedure

1. Determine which one of these processes you will demonstrate: DNA replication, transcription, or translation.
2. Think about the steps involved in your process, and create a list of the components that are necessary for your demonstration. Not all details of the process need to be represented. Make a note of any steps you simplified or left out because of limited time and materials.
3. Working from your list, decide on the method and items you will use to demonstrate your process. Ideas include, but are not limited to, acting out the process with props, recording a video, or creating a stop-action animation.
4. Create your demonstration of the process you picked, either DNA replication, translation, or transcription. Include narration, if possible, or captions that explain the process.
5. Once you are done, clean up and return unused materials to the materials station.

Results and Analysis

1. **Identify** For the process you have chosen, what components are necessary in order to demonstrate your chosen process?
2. **Sequence** Describe the sequence of events represented by your demonstration.
3. **Evaluate** Consider the strengths and limitations of the model you presented. Which components and events of the process were you able to demonstrate most clearly? What elements or events did you choose to simplify or omit?
4. **Construct an Explanation** How does the process you demonstrated work with the other two processes to accomplish gene expression?

334 CHAPTER 11 DNA, RNA, AND PROTEINS

REGULATING GENE EXPRESSION

Pilots have many ways to adjust an airplane's speed in the air. They can adjust the engine's thrust, or they can change altitude and direction to fly with or against prevailing winds. An airplane's wings can change shape to increase or reduce drag through the air. Similarly, cells have many different mechanisms to speed up, slow down, or otherwise modify gene expression.

Infer *Considering what you know about gene expression, how do you think a cell is likely to increase or decrease the expression of particular genes?*

11.3

KEY TERMS

- epigenetic
- exon
- intron
- operon
- RNA interference
- transcription factor

Gene Expression Regulation in Prokaryotes

Although most prokaryotes do not have the need to express different genes for the purposes of differentiation, they do have a life cycle and the need to rapidly respond to changes in the environment. Prokaryotes respond to environmental changes by adjusting gene expression, mainly by changing the amount of transcription. For example, when a preferred nutrient becomes available, a bacterium begins transcribing genes whose products allow the cell to use the nutrient. When the nutrient is no longer available, transcription of those genes stops. Thus, the cell does not waste energy and resources producing gene products that are not needed.

In bacteria, genes that are used together often occur in a group, one after the next on the chromosome. One *promoter* sequence precedes the group, so all of the genes are transcribed together, producing a single RNA strand. Their transcription can be controlled in a single step. Typically, this control involves a *repressor* protein that can bind to an *operator* sequence in the DNA. A group of genes together with a promoter and one or more operators are collectively called an **operon**. Operons were discovered in bacteria, but they also occur in archaea and eukaryotes, including individual cells within multicellular organisms.

The *lac* Operon *Escherichia coli* (*E. coli*) bacteria live in the gut of mammals, consuming nutrients that travel past as food moves through the intestines. Their carbohydrate of choice is glucose, but they can make use of other sugars such as lactose, a disaccharide in milk.

The *lac* operon includes three genes that encode proteins involved in lactose metabolism and a promoter flanked by two operators (**Figure 11-19**). Bacteria make these three proteins only when lactose is present.

***lac* operon in the *E. coli* chromosome**

operator | promoter | operator | gene 1 | gene 2 | gene 3

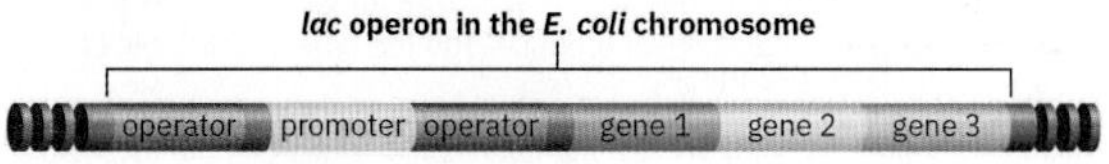

FIGURE 11-19 This *lac* operon model includes three genes under the control of one promoter and two operators.

CHAPTER INVESTIGATION

Regulation of Gene Expression
How can genes be expressed and regulated in organisms?
Go online to explore this chapter's hands-on investigation to study how gene expression can be regulated in the bacteria *Serratia marcescens*.

DIFFERENTIATED INSTRUCTION | English Language Learners

Asking and Answering Questions Have students work in pairs to ask and answer questions about new key vocabulary and concepts.

Beginning Have students individually write the words *operator*, *repressor*, and *promoter* on index cards and write phrases on the back to explain their functions. Then have partners ask each other: *What does (an operator) do*? They should use their phrases to answer.

Intermediate Have students take turns asking questions about the operator, repressor, and promoter. They should ask questions that begin with *What*? *How*? *Why*? Encourage students to answer in complete sentences.

Advanced Have students ask and answer questions about the regulation process of the *lac* genes, including the function of the operator, repressor, and promoter. Encourage students to build on one another's ideas by adding information to each other's answers.

11.3

REGULATING GENE EXPRESSION

LS1.A, LS3.A

ENGAGE

Students have learned that different proteins are made using information stored in genes. However, at each point in time a cell only needs certain proteins. This section explains how cells regulate gene expression to control when each protein is made.

Objectives

- Compare gene expression in prokaryotes and eukaryotes.
- Explain the function of introns and exons.
- Describe environmental factors that can affect gene expression.

Pretest Use **Questions 4** and **5** to identify gaps in background knowledge or misconceptions.

Vocabulary Strategy

Word Wall Create a word wall with the new vocabulary terms presented in this lesson. Post all of the words together on the first day or add new words as they come up. Use the word wall along with a question of the day that students answer using one or more terms.

Infer *Students might say that cells must be able to increase or decrease transcription or translation rates to change the amount of mRNA or protein produced for each gene.*

CHAPTER INVESTIGATION B

Guided Inquiry *Regulation of Gene Expression*

Time: 100 minutes or 2 days

In this investigation, students will observe temperature-regulated gene expression of prodigiosin in *S. marcescens*.

Go online to access detailed teacher notes, answers, rubrics, and lab worksheets.

Visual Support

Operon Regulation Walk students through the steps in **Figure 11-20**, comparing the diagram when lactose is absent and when it is present. Ask students how the repressor acts like a padlock and what happens when the "padlock" is removed.

Science Background

Discovering the *Lac* Operon François Jacob and Jacques Monod first proposed the idea of the *lac* operon. From 1957 to 1959, they conducted experiments with Arthur Pardee at the Pasteur Institute in Paris, France. These experiments looked at how *E. coli* controlled the processes of enzyme production. For their work, Jacob and Monod were awarded the Nobel Prize in Physiology or Medicine in 1965.

Although this work provided evidence to support the idea that a molecule mediated the production of proteins from DNA, it did not explain how the genetic information stored by DNA was used by the ribosome to synthesize proteins. Messenger RNA had yet to be discovered. Ask students: *What process could help scientists make this discovery?*

SEP Use a Model *Students might say that disrupting the operator regions might stop the repressor from binding, which would allow transcription to occur regardless of the presence of lactose. If the promoter region was broken and the polymerase couldn't bind, the operon could not be transcribed.*

CASE STUDY
EXPRESSING THE GENETIC CODE

Ask Questions Students should revisit the Case Study as they read the chapter to make connections with the content. Prompt students to think about why environmental changes can effect gene expression. Why is it important that the transfer of genetic information be regulated in organisms? Discuss answers as a class. Invite students to ask for clarification as needed and allow other students to explain their answers to the questions.

20a Lactose absent
In the absence of lactose, a repressor binds to the two operators. Binding prevents RNA polymerase from attaching to the promoter, so transcription of the operon genes does not occur.

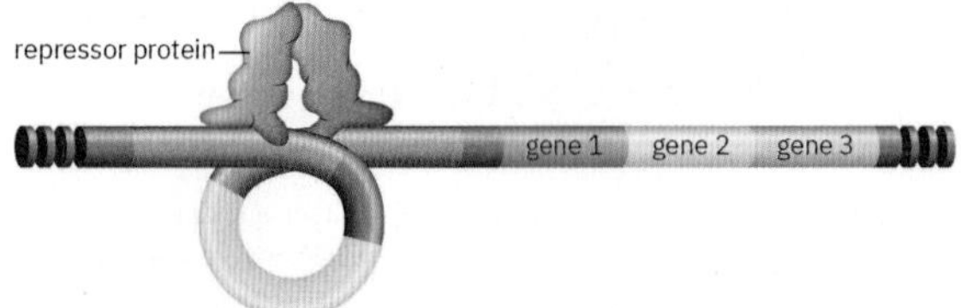

20b Lactose present
When lactose is present, some is converted to a form that binds to the repressor such that it releases the operators.

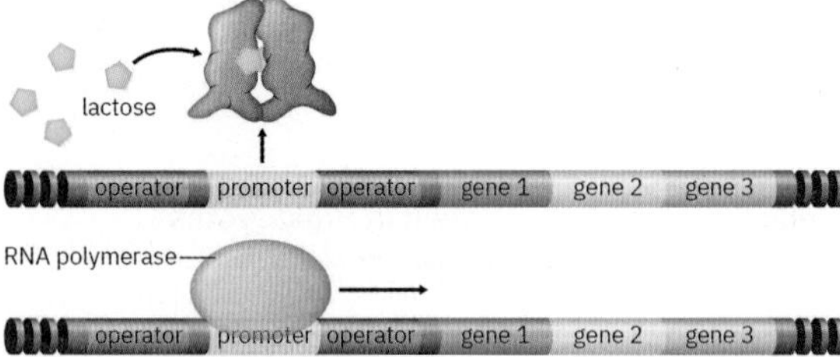

RNA polymerase can now attach to the promoter and transcribe the operon genes.

RNA polymerase
operator
promoter
operator
gene 1
gene 2
gene 3

FIGURE 11-20
This model of the *lac* operon demonstrates how a repressor protein works to regulate expression of genes, depending on the presence or absence of lactose molecules.

When lactose is not present, a repressor binds to the two operators and twists the region of DNA with the promoter into a loop (**Figure 11-20a**). RNA polymerase cannot access the twisted-up promoter, so the *lac* operon's genes cannot be transcribed. When lactose is present (**Figure 11-20b**), some of it is converted to another sugar that binds to the repressor and changes its shape. The altered repressor releases the operators and the looped DNA unwinds. The promoter is now accessible to RNA polymerase, and transcription of lactose-metabolizing genes begins.

The presence of glucose also inhibits transcription of the *lac* operon. Because glucose metabolism is a shorter pathway than lactose metabolism, it requires less energy and fewer resources. Therefore, when both lactose and glucose are present, the cells will use up all of the available glucose before switching to lactose metabolism.

SEP Use a Model *Using the model of the lac operon, explain what you think might happen to its function if one of the operator regions was disrupted. What if the promoter region was broken in some way?*

CASE STUDY
EXPRESSING THE GENETIC CODE

Ask Questions *Generate questions about how the mechanisms of gene regulation could help organisms express the particular genes they need, and how different factors in the environment might affect that regulation.*

Gene Expression Regulation in Eukaryotes

Responding to internal and external change is critical to life. Cellular responses often involve adjustments to the timing and rate of gene expression. Consider how a typical differentiated human body cell uses only about 10 percent of its genes. Cells differentiate during development as they begin to express different genes. The set of genes a cell expresses determines the molecules it will produce, and the type of cell it will be.

All steps of gene expression are regulated. Many of these regulation points are indicated in **Figure 11-21**. Each cell is a very complex system. Everything from the position of a gene in the nucleus to the stability of its final protein products can influence the level of expression for each gene.

SCIENCE AND ENGINEERING PRACTICES
Developing and Using Models

The *Lac* Operon Provide students with materials such as different colors of clay and bendable straws, or you may allow them to use a computer program, to create a model of the *lac* operon. Working in small groups, have students create and use their models to explain the regulation of the *lac* operon to the class. When students have completed their models, have them make a bullet point summary that explains the process.

Remind students that lactose acts as a switch that turns the operon on and off. They may choose to include such analogies in their designs if they serve a purpose to make the concept easier to understand.

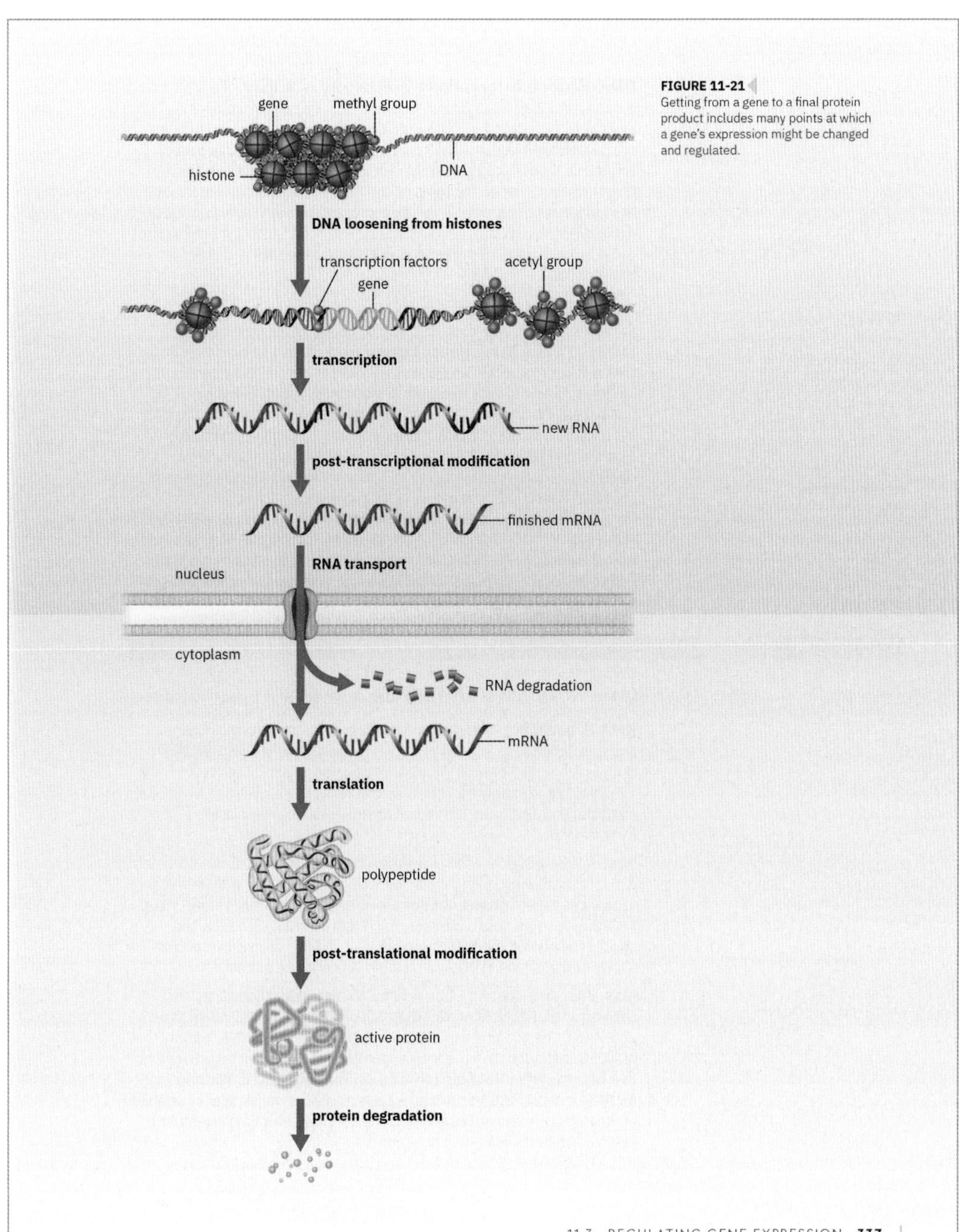

FIGURE 11-21 Getting from a gene to a final protein product includes many points at which a gene's expression might be changed and regulated.

Science Background

Barr Bodies On rare occasions, an entire chromosome may remain condensed, tightly coiled around histones and, therefore, inaccessible to RNA polymerase. An example of this is found in one of the X chromosomes of female mammals. Although female mammals have two homologous X chromosomes, only one is available for transcription in any given cell. The other X chromosome remains tightly condensed, showing up as a dark spot in the nucleus, known as a Barr body. This a mechanism of whole chromosome-gene regulation that balances expression between XX and XY cells in mammals.

Visual Support

Eukaryote Gene Regulation Walk through **Figure 11-21** with students, breaking it down into steps. The concepts in this diagram will be covered in detail in the text that follows. You may go over this visual with students to give them an overview of the processes involved, and it will work well as an anchor to guide them through learning more detailed information. Encourage students to focus on the structures and changes that occur. As students read the text that follows, have them return to **Figure 11-21** and discuss with a partner how the relevant information is depicted in the diagram.

Science Background

Junk DNA Junk DNA refers to noncoding regions of DNA that have no identified instructions or function. When RNA polymerase moves along a gene, the entire gene is transcribed. The transcribed RNA contains introns and must be edited through a splicing process to exit the nucleus as mRNA. This means that the introns, or "junk" sequences, must be spliced out and the exons pasted back together. Discuss with students how this compares with the job of a film editor who must cut scenes from a movie.

Vocabulary Strategy

Word Wall Followup Students are introduced to a number of new vocabulary words in this section that are related and tell a "story." If you are using a word wall, you may have them add to that or they can create a diagram in their notebooks. Have them draw the diagram in **Figure 11-22**, adding the words *activator* and *repressor* to the sites on the diagram to where they may bond using a word bubble and line to connect them. Tell students to write a short description of the actions of activators and repressors.

DNA-Histone Interactions In eukaryotic cells, much of a DNA strand is tightly wrapped around *histones*. A histone is a type of protein that associates with DNA and helps organize and condense it into a smaller space. Only DNA regions that have been unwound from histones are accessible to RNA polymerase. Modifications to histone proteins make them loosen or tighten their grip on the DNA wrapped around them, and this influences transcription. For example, adding acetyl groups ($-COCH_3$) to histones loosens DNA so transcription can occur. Adding methyl groups ($-CH_3$) to histones (methylating them) tightens the DNA and prevents transcription.

Transcription Factors A cell can change the gene products it produces (and the rate of production) by changing the transcription factors it makes. **Transcription factors** are proteins that bind directly to DNA and affect whether and how fast a gene is transcribed. *Repressors* are transcription factors that slow or stop transcription, either by preventing RNA polymerase from accessing the promoter or by blocking the enzyme's progress. Some repressors work by binding directly to a *promoter* (**Figure 11-22**). Others bind to a DNA sequence called a *silencer* to reduce transcription. *Activators* are transcription factors that speed up transcription by helping RNA polymerase access a promoter. Many activators work by binding to the promoter. Others bind to DNA sequences called *enhancers*, which, like silencers, may be very far away from the gene they affect. A silencer or enhancer operating on a distant gene can inappropriately affect a nearby gene. Insulators prevent this from occurring. An *insulator* is a region of DNA that, upon binding a transcription factor, can block the effect of a silencer or enhancer on a neighboring gene.

FIGURE 11-22
Repressors slow or stop transcription by binding to a promoter or silencer. Activators encourage transcription of nearby genes by binding to an enhancer. Insulators prevent activators or repressors from inappropriately affecting transcription of a neighboring gene.

gene | promoter | insulator | silencer | enhancer | promoter | gene

RNA Processing Transcription in eukaryotes occurs in the nucleus, and translation occurs in the cytoplasm. A newly transcribed mRNA can pass through pores of a nuclear envelope only after it has been modified appropriately. Mechanisms that delay these post-transcriptional modifications also delay delivery of the mRNA to the cytoplasm for translation.

Most eukaryotic genes (and many prokaryotic ones) consist of both exons and introns. **Exons** contain the actual coding information, with intervening sequences called **introns**. Introns are removed from a newly transcribed RNA through a process called splicing (**Figure 11-23**). Exons can be spliced together in different combinations, and this *alternative splicing* allows a single gene to encode multiple versions of a protein.

After alternative splicing, additional mRNA processing adds a guanine cap and a tail made out of a long stretch of adenine bases, known as a poly-A tail. The guanine cap is bound to the end of the mRNA in a way that prevents degradation of the messenger RNA, and it interacts with factors that help export mRNA from the nucleus. The poly-A tail is also important for the export of mRNA from the nucleus as well as translation and stability of the mRNA. Typically, the poly-A tail will shorten over time, eventually ending with the degradation of the mRNA.

DIFFERENTIATED INSTRUCTION | Leveled Support

Struggling Students Have students work in groups to practice editing a sequence by cutting out the introns, or noncoding sequences, and pasting together the exons to decode the message. Tell them to first cut out the blue letters and then group the remaining letters by threes starting on the left.

THECATRAMPDQANANDAQZDMOPTETHERAT

(THECATRANANDATETHERAT → THE CAT RAN AND ATE THE RAT)

Advanced Learners Have students construct a model of the *lac* operon, including the promoter, repressor, and operator using available materials or creating a digital model. Then have students use the model to create a digital presentation that explains how the expression of the *lac* operon is controlled.

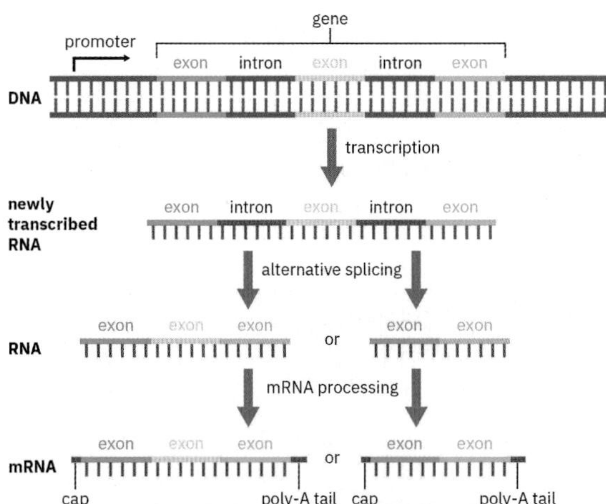

FIGURE 11-23 During post-transcriptional modifications, introns are removed, and exons are spliced together. Messenger RNAs are also given a tail made from a series of adenine bases and a cap made from a modified guanine.

SEP Argue From Evidence *In the human genome, there are an average of 8.8 exons and 7.8 introns per gene. Many genes code for several different proteins. If the gene in* ***Figure 11-23*** *codes for two different proteins, imagine how many different combinations could be generated by a gene with 9 exons and 8 introns. Why do you think this flexibility might be useful? Think about how cells need to react to changes and how different cells serve different functions.*

mRNA Stability and Translation Homeostasis in a cell depends on tight regulation of the proteins being produced in a cell at a given time, which in turn depends on the mRNA molecules available for translation. The longer an mRNA persists, the more protein molecules can be made from it. Some mRNAs last for several hours in cytoplasm, whereas others are degraded within a minute or two. How long a particular mRNA persists depends mainly on the proteins attached to it. Some binding proteins prevent translation until an mRNA has been delivered to its destination. Other proteins produced in response to environmental signals can attach to and stabilize (or destabilize) specific mRNAs.

Protein Stability and Modifications Some proteins last a long time in a cell, and others are broken down almost immediately after synthesis. The amount of a particular protein in a cell depends on the rate of synthesis and the rate of breaking down. Regulatory proteins such as transcription factors typically have a rapid turnover. A cell can shift its activities quickly by activating molecules that break down these proteins. Many newly synthesized polypeptide chains must be modified before they become functional. There are a variety of post-translational modifications that inhibit, activate, or stabilize many molecules, including enzymes that participate in transcription and translation.

CHAPTER INVESTIGATION

Investigating the Building Blocks of Life

How can we model the building blocks of life?

Go online to explore this chapter's hands-on investigation to design your own model of gene expression and its regulation.

CHAPTER INVESTIGATION A

Guided Inquiry *Investigating the Building Blocks of Life*

Time: 50 minutes

Students will use building blocks to model the process of protein synthesis from a given DNA sequence.

Go online to access detailed teacher notes, answers, rubrics, and lab worksheets.

Address Misconceptions

One Gene, Many Proteins At one time scientists thought that each gene coded for a single protein or for a single subunit of a protein. Today scientists now know there are mechanisms that modify the structure of a protein following translation and result in changes to its function or regulate its behavior. There are many types of post-translational modifications—two common modifications include phosphorylation and ubiquitylation. Phosphorylation adds a phosphate group to certain amino acids (serine, threonine, or tyrosine) in a target protein, which alters the structure/function of the protein. Because this modification is reversible, it can act as a switch to turn the function of the protein on or off. Phosphorylation is involved in regulating processes such as the cell cycle. Ubiquitylation, or the addition of one or more ubiquitin groups, plays an important role in protein degradation.

SEP Argue From Evidence *Students might say that if the cell only has to transcribe one gene to generate many different products, that could help the cell react quickly to changes in the environment. Also, one gene could be expressed differently in different cell types that have different needs.*

Connect to English Language Arts

Write Explanatory Text Walk through **Figure 11-24** with students and discuss each step in the process. Explain the meaning of any abbreviations that may not be clear; e.g., RISC stands for RNA Induced Silencing Complex, which can target expression of almost any gene. After all questions have been addressed, have students use the diagram to write a paragraph explaining the process of mRNA interference as a regulatory process.

In Your Community

RNAi and Pest Control Throughout history insects and other pests have destroyed entire crops, leaving people without food. Pesticides are now used to control pests, but pests can eventually develop resistance and pesticides can have deleterious effects on people and the environment. Scientists think RNA interference, or RNAi, can be used to control pests. RNAi can be triggered to selectively inhibit the expression of specific genes in a target organism, essentially shutting down genes that are essential for the organism's growth, development, and reproduction.

Take Action While RNAi might reduce the use of pesticides in the future, how can people protect themselves from pesticide exposure in the present? Discuss strategies with students. They might suggest checking with school officials and local agencies to ensure that notifications of pesticide application are posted in public areas to keep people out. Tell students that in some states, laws require notification of residents before pesticide applications. Is this true in your state? Tell students that they can check with local agencies and news outlets to learn about community-wide pesticide applications, such as mosquito spraying.

FIGURE 11-24 Short pieces of noncoding RNA, often called microRNAs, serve important regulatory functions through processes such as this RNA interference pathway.

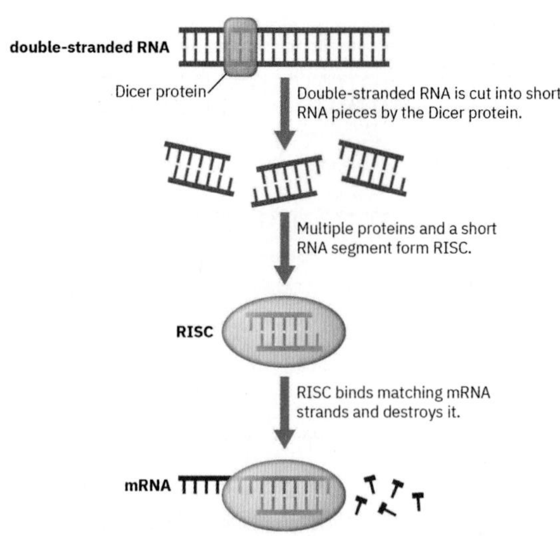
double-stranded RNA
Dicer protein
Double-stranded RNA is cut into short RNA pieces by the Dicer protein.
Multiple proteins and a short RNA segment form RISC.
RISC
RISC binds matching mRNA strands and destroys it.
mRNA

RNA Interference Small regulatory RNAs can also impact mRNA stability and translation of specific transcripts. A microRNA is a short non-coding RNA molecule that is complementary in nucleotide sequence to part of the target mRNA. MicroRNAs can be generated by proteins such as Dicer, and these microRNAs combine with other proteins to form a complex such as RISC, as shown in **Figure 11-24**. Base pairing between the microRNA and mRNA transcript forms a small double-stranded region, triggering a cascade of events known as **RNA interference**. Depending on how the sequences match, the enzyme complexes that recognize the double-stranded RNA may either prevent translation or facilitate mRNA degradation. RNA interference is used by both prokaryotes and eukaryotes as a mechanism of gene regulation, and it is used by scientists to target genes of interest in research studies.

FIGURE 11-25 In the DNA of differentiated cells, a methyl group (the red CH_3 in the diagram) is most often attached to a cytosine nucleotide that is next to a guanine nucleotide. When the cytosine on one strand is methylated, enzymes tend to also methylate the other strand.

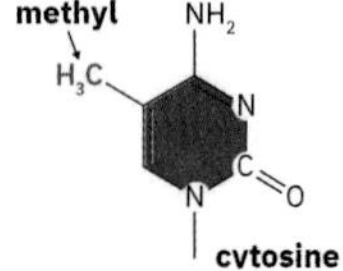
methyl
NH_2
H_3C
N
C
O
N
cytosine

Epigenetics and Environmental Effects on Gene Expression

Recall that adding methyl groups to histones silences DNA transcription. Direct methylation of DNA nucleotides also suppresses transcription. In eukaryotes, methyl groups are usually added to a cytosine that is followed by a guanine (**Figure 11-25**). When the cytosine on one DNA strand is methylated, enzymes methylate the cytosine on the other strand. Once a particular nucleotide has become methylated in a cell's DNA, it will usually stay methylated in the DNA of the cell's descendants.

DNA methylation is necessary for cell differentiation, so it begins very early in embryonic development. Genes actively expressed in cells of a developing embryo become silenced as their promoters get methylated, and this silencing is the basis of selective gene expression that drives cell differentiation.

DIFFERENTIATED INSTRUCTION | English Language Learners

Expressing Opinions and Ideas Read aloud the information in "In Your Community" and have small groups discuss their opinions about pest control and insecticides based on this information.

Beginning Have students complete these sentence frames: *Insecticides are helpful because they kill ______. They are harmful for ______ and the ______. RNAi is a better way to control pests because it ______ genes that insects need to ______.*

Intermediate Have students discuss their answers to these questions: *How is pest control helpful? What harmful effects does it have? How does RNAi change pest control? Is this helpful or harmful? Why?*

Advanced Have students discuss both the pros and cons of insecticides and how RNAi can address the problems with insecticides.

During development, each cell's DNA continues to acquire methylations. Which sites are methylated in a genome varies by the individual. This is because methylation is influenced by environmental factors experienced during an individual's lifetime. Some of these factors increase methylation of particular genes and others reduce methylation. Patterns of DNA methylation in an individual's cells change after exposure to environmental contaminants. For example, cigarette smoke substantially alters methylation of certain promoters in a pattern that also occurs in cancer cells. Some other examples of factors that can affect methylation patterns include

- alcohol consumption
- BPA consumption
- cancer
- cigarette smoke
- heavy metal exposure
- lead exposure
- low nutritional intake
- obesity
- phthalate consumption
- radiation exposure
- stress
- sun exposure
- temperature

These and other factors that influence DNA methylation patterns can have multigenerational effects. When an organism reproduces, it passes its DNA to offspring. Methylation of parental DNA is normally "reset" in gametes (eggs and sperm) and after fertilization, but this reprogramming does not remove all of the parental methyl groups. Methylations acquired during an individual's lifetime can be passed to future offspring. Inheritable, stable changes in gene expression that occur through chromosomal changes rather than DNA sequence changes are **epigenetic**. DNA methylations are epigenetic, as are histone modifications, some proteins that modify histones, and some noncoding RNAs that influence gene expression.

SEP Construct an Explanation *Some methylation changes to DNA are reversible, including some effects of cigarette smoke, which causes a reduction in methylation in certain genes. Why do you think those changes might readily be reversed if an individual stops smoking?*

11.3 REVIEW

1. **Classify** Identify the types of proteins that enable or inhibit transcription.
   - activators
   - promoters
   - repressors
   - enhancers
   - silencers
2. **Infer** What is an advantage and a disadvantage of a long-living mRNA molecule?
3. **Explain** How is the regulation of gene expression different in eukaryotes versus prokaryotes?
4. **Apply** Which of the following statements apply to the *lac* operon?
   - **A.** RNA polymerase converts lactose to glucose.
   - **B.** Glucose binds the repressor to allow transcription.
   - **C.** The repressor twists the promoter region into a loop.
   - **D.** The genes encode three proteins needed for lactose metabolism.

11.3 REGULATING GENE EXPRESSION **341**

In Your Community

Epigenetics and 9/11 On September 11, 2011, two airplanes hijacked by terrorists slammed into the World Trade Center buildings in New York City. Among the people near the World Trade Center at the time of the event were 100 women who were over six months pregnant. Their experience caused them to develop post-traumatic stress disorder (PTSD). People exposed to extremely traumatic events resulting in PTSD have abnormally low levels of the stress hormone cortisol and the low level persists. The pregnant women who experienced 9/11 still had abnormally low levels of cortisol a year later. Moreover, although the women's babies were not yet born on 9/11, they also had low levels of cortisol. Discuss with students how this supports the idea that patterns of DNA methylation passed to offspring constitute a form of genetic memory.

Take Action Discuss the implications of epigenetics for communities that experience stressful events. Ask students: *What kinds of events occur in our community that could cause PTSD? What actions can be taken by our community to reduce the occurrence of these stressful events?*

SEP Construct an Explanation
Students might say that if the environmental factor, such as cigarette smoke, is reducing the amount of methylation, removing the smoke should allow those genes to be methylated to the appropriate levels.

11.3 REVIEW

1. *enable transcription: activators, promoters, enhancers; inhibit transcription: repressors, silencers* **DOK 2**
2. *More protein molecules can be made from a long-living mRNA molecule, but if an mRNA molecule lasts too long, too many proteins of a particular type are made that the cell might not need.* **DOK 2**
3. *Gene expression in eukaryotes involves some additional complications, such as DNA-histone interactions and regulation of RNA transport out of the nucleus. In prokaryotes with no nucleus, much of the regulation is at the transcription level.* **DOK 3**
4. *C, D* **DOK 2**

LOOKING AT THE DATA
DECAY OF mRNA MOLECULES

In this activity, students will be using mathematical skills to read and interpret graphs showing the frequency of mRNA molecule half-lives. Students may be familiar with the concept of half-life from prior studies of fossils. Similar to the way in which scientists have studied the half-life of mRNA molecules in different species, the half-life of a radioactive element is the length of time required for half the radioactive atoms in a sample to decay. As in mRNA half-life among different species, each radioactive element has a different half-life.

SAMPLE ANSWERS

1. *E.coli: most common 4 min; minimum 2 min; maximum 14 min; S. cerevisiae: most common 20 min; minimum 5 min; maximum 60 min; H. sapiens: most common 7.5 hours; minimum 2.5 hours; maximum 30 hours* **DOK 2**
2. *Sample answer: The more complex the organism, the longer the mRNA half-life.* **DOK 2**
3. *Sample answer:* E. coli *have a relatively short life cycle and must respond to changes in their environment. Rapid adjustments to gene expression allow changes to happen quickly. A mature* H. sapiens *cell is typically no longer dividing and is not typically subjected to rapid environmental changes.* **DOK 3**
4. *Sample answer: part a: Claim: mRNAs with longer half-lives have more protein produced from them. Part b: Evidence needed: The mRNA half-life does not directly tell us anything about the rate of translation for that mRNA, so measuring the protein levels for the products of the mRNAs would be useful evidence.* **DOK 3**

LOOKING AT THE DATA

DECAY OF mRNA MOLECULES

SEP Mathematical Thinking Gene expression occurs at different rates in the cells of different species.

One of the ways gene expression is controlled is through the timely destruction of mRNA molecules that are no longer needed. After a relatively short period of time, mRNA molecules will degrade into component parts that can be incorporated into new RNA molecules. The lifetime of mRNA is related to the time available for the translation of genetic information into proteins.

Scientists have studied mRNA degradation in the cells of many organisms to better understand the regulation of genetic processes. **Figure 1** and **Figure 2** show the results of three studies that measured the lifetime of mRNA molecules in a single-celled bacterium, a single-celled eukaryote, and a multicellular eukaryote. The half-life of mRNA molecules in all three species showed normal distributions, typically with single peaks and measurable ranges. The half-life is the time it takes for the amount of mRNA present to decrease by half.

FIGURE 1 This graph shows the frequency of mRNA molecule half-life in a prokaryote, the bacterium *Escherichia coli.*

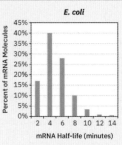

E. coli

Percent of mRNA Molecules: 0%, 5%, 10%, 15%, 20%, 25%, 30%, 35%, 40%, 45%

mRNA Half-life (minutes): 2, 4, 6, 8, 10, 12, 14

FIGURE 2 These two graphs show the frequency of mRNA molecule half-life in the eukaryotes yeast (*Saccharomyces cerevisiae*) and humans (*Homo sapiens*).

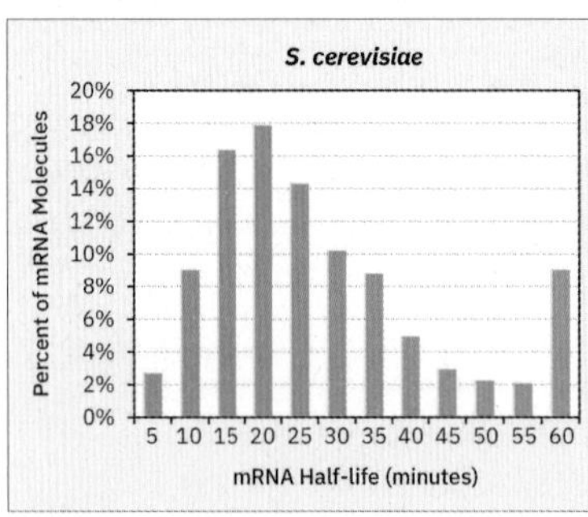

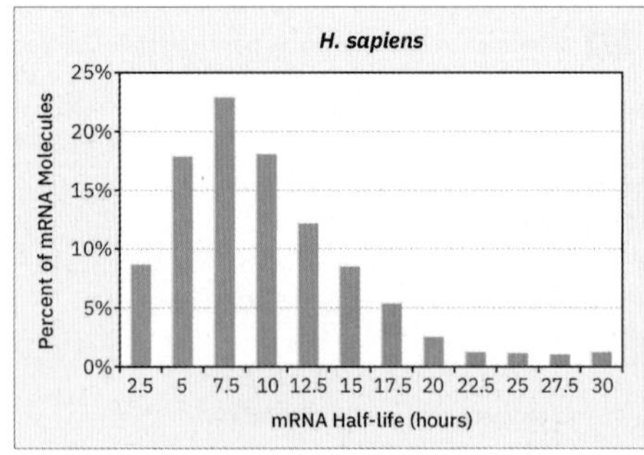

1. **Analyze Data** Determine the most common, minimum, and maximum half-life for *E. coli*, *S. cerevisiae*, and *H. sapiens*.
2. **Identify Patterns** Use **Figures 1** and **2** to describe a relationship between the mRNA half-life and organism complexity.
3. **Predict** Why might it be advantageous for an *E. coli* cell to degrade most of its mRNAs so quickly, compared with a mature *H. sapiens* cell?
4. **Argue From Evidence** State a claim about the relationship between an mRNA's half-life and the amount of protein produced. What additional evidence would you need to support your claim?

TYING IT ALL TOGETHER
EXPRESSING THE GENETIC CODE

HOW IS GENETIC INFORMATION TRANSFERRED AND REGULATED?

In this chapter, you learned how genes are expressed by translating the information contained in the nucleotide sequence of DNA into instructions for building proteins that perform specific functions. The butterflies described in the Case Study are examples of organisms in which variations in the expression of genes that control their pigmentation can occur due to environmental factors.

The changing of seasons commonly affects the expression of color genes in animals that live in cold climates. Although many arctic animal species stay white all year, some, including weasels, arctic foxes, and snowshoe hares, have brown fur in summer that turns white in winter. Ptarmigans, a genus of chicken-like birds native to the northern tundra, molt from brown or gray to completely white plumage as the weather turns cold.

By analyzing snowshoe hare skin samples collected during molting stages, scientists have identified more than 600 genes that are expressed differently over time. Determining which of these genes dictate changes in the hare's coat color requires advanced genetic techniques, but the fundamental process of gene expression is the same for all of them.

In this activity, you will develop and use a model to construct an explanation of how an animal's fur color changes seasonally. Work individually or in a group to complete these steps.

Ask Questions

1. Review the questions you generated throughout the chapter about how genetic information is transmitted. Identify any questions whose answers help you understand how an animal's appearance is determined by its DNA. If needed, refine some of your existing questions or generate new questions to structure your explanation.

Develop and Use a Model

2. Make a model that describes how a gene that codes for an animal's color is expressed.

Construct an Explanation

3. Use your model of gene expression to explain how the animal's color changes with the seasons.

FIGURE 11-26 The fur of the least weasel (*Mustela nivalis*) changes color in winter, as its habitat turns white with snow and ice.

SCIENCE AND ENGINEERING PRACTICES
Developing and Using Models

Genes and Seasonal Life Events To help students think through a possible genetic model for seasonal fur color change, ask them to list other seasonal life events. Students may identify reproduction, hibernation, and migration. Ask them what external inputs could control the timing of these annual events. Then, ask them to identify the processes that cells use to regulate gene expression. Review histones, transcription factors, post-transcriptional modifications, RNA interference, mRNA stability and the role of binding proteins, post-translational modifications, and protein degradation. Ask students to analyze the change in fur color and to hypothesize which processes they think would regulate seasonal gene expression of fur color. Tell them they can make a 3-dimensional model out of available materials or draw a 1-dimensional model.

TYING IT ALL TOGETHER

ELABORATE

The Case Study at the beginning of the chapter focused on patterns in butterfly wings. The genes for a single epithelial cell dictate the structure of the pigments that are produced. Experiments show that temperature and light can influence gene expression in butterfly wings. Throughout this chapter, students have learned about processes that can influence gene expression. In the Tying It All Together activity, students examine how environmental factors affect the expression of genes in other organisms.

This activity is structured so that students may complete the steps as they study the chapter content. **Step 1** correlates to questions generated throughout the chapter. Section 11.3 will be of particular help as students construct models of how a gene that codes for an animal's color is expressed in **Step 2**. Students will then need to synthesize ideas from all sections to complete **Step 3**.

Alternatively, this can be a culminating project at the end of the chapter. The gathering of evidence here to support a claim or argument is a process that students may find useful in the Unit 4 Activity.

Go online to access the Student Self-Reflection and Teacher Scoring rubrics for this activity

In Your Community

Camouflage The ability to blend in to the environment and not be seen can be the difference between life or death for certain animals. Animals living in environments with extreme seasonal differences are well known for related seasonal color changes. However, some animals such as chameleons living in more tropical environments can quickly change their colors to blend in. Camouflage even happens underwater. Certain crustaceans change color to match the type of algae or plant they are hanging out on.

Encourage students to research local examples of animals with seasonal color differences or other examples that interest them.

CHAPTER 11 **REVIEW**

EVALUATE

REVIEW KEY CONCEPTS

1. *C* **DOK 1**
2. *RNA: uses the base uracil; DNA: uses the base thymine; typically double stranded; Both: carries genetic information; uses the base guanine; has a sugar phosphate backbone* **DOK 2**
3. *polymerase: elongates the new DNA molecule; primase: builds a short starting nucleotide sequence; helicase: unwinds the DNA molecule; ligase: seals up gaps in the DNA backbone* **DOK 1**
4. *A* **DOK 1**
5. *D* **DOK 2**
6. *Sample answer: The DNA molecule is wrapped around histones, proteins that help organize and condense the DNA molecule. In order for the DNA to be transcribed, it has to be unwound from the histones. To prevent genes from being expressed, histones can be modified to more tightly hold the DNA, which will prevent unwinding and transcription of a region of the DNA.* **DOK 2**
7. *Sample answer: Some traits can be controlled and passed on through gene regulation rather than a change in the genetic code. External changes to the DNA molecule, other than the nucleotide sequence, can determine which genes are expressed or suppressed. For example, if an individual experiences a lack of nutrition, this can alter methylation of certain genes. That methylation can sometimes be passed on to offspring, meaning those genes will also be expressed differently in offspring even though the genetic code did not change.* **DOK 1**
8. *Complementary DNA: CCT GAT GGC; mRNA: CCU GAU GGC; peptide: proline, aspartic acid, glycine* **DOK 2**
9. *B* **DOK 2**
10. *A, C, D, E* **DOK 2**
11. *D* **DOK 2**
12. *A* **DOK 2**
13. *B, E* **DOK 3**

CHAPTER 11 **REVIEW**

REVIEW KEY CONCEPTS

1. **Identify** In DNA, which nucleotide is complementary with adenine?
   - **A.** cytosine (C)
   - **B.** guanine (G)
   - **C.** thymine (T)
   - **D.** uracil (U)
2. **Classify** Indicate whether each characteristic describes RNA, DNA, or both.
   - uses the base uracil
   - uses the base guanine
   - uses the base thymine
   - is typically double stranded
   - has a sugar phosphate backbone
   - carries genetic information
3. **Label** Match the descriptions of the process in DNA replication to the enzyme that performs it: polymerase, helicase, primase, or ligase.
   - unwinds the DNA molecule
   - elongates the new DNA molecule
   - seals up gaps in the DNA backbone
   - builds a short starting nucleotide sequence
4. **Calculate** Using **Figure 11-15**, determine which amino acid is coded for by CAU.
   - **A.** histidine
   - **B.** methionine
   - **C.** serine
   - **D.** threonine
5. **Explain** Which describes the role of tRNA?
   - **A.** holds the ribosome together
   - **B.** brings the amino acid sequence from the DNA
   - **C.** builds the peptide bond between the amino acids
   - **D.** delivers the appropriate amino acid based on the codon
6. **Summarize** What role do histones play in gene regulation?
7. **Describe** How can genetic traits be affected epigenetically? Give a specific example to support your answer.
8. **Identify Patterns** If a DNA sequence is GGA CTA CCG, what would the order of the bases be on (a) the complementary DNA strand and (b) the mRNA built from the sequence? (c) What would be the amino acids in the polypeptide?
9. **Predict** Which statement best describes the status of the *lac* operon when a large amount of glucose is present?
   - **A.** Genes of the *lac* operon are transcribed to metabolize glucose.
   - **B.** The operators of the *lac* operon are bound to inhibit lactose metabolism.
   - **C.** The repressor protein binds glucose, so the DNA in the *lac* operon is not accessible for transcription.
   - **D.** Polymerase is attached to the *lac* operon promoter region, so lactose can be metabolized as well.
10. **Identify** What methods can a eukaryotic cell use to regulate gene expression? Select all correct answers.
    - **A.** Adjust transcription factors.
    - **B.** Destroy sections of DNA that are not needed.
    - **C.** Break down mRNA molecules at different rates.
    - **D.** Break down protein products at different rates.
    - **E.** Prevent or promote the unwinding of DNA sections.
11. **Sequence** Two ribosomal subunits and a tRNA with methionine are positioned at the start codon of an mRNA. What is the next step in translation?
    - **A.** A peptide bond is formed.
    - **B.** The mRNA molecule unwinds.
    - **C.** The tRNA releases the methionine.
    - **D.** Another tRNA binds to the next codon.
12. **Compare** In which of these ways is DNA replication similar to transcription?
    - **A.** It occurs inside the nucleus.
    - **B.** It copies the entire strand of DNA.
    - **C.** Another DNA molecule is produced.
    - **D.** Uracil is incorporated into the molecule.
13. **Explain** What role do hydrogen bonds play in DNA replication? Select all correct answers.
    - **A.** They make it easier to separate the strands.
    - **B.** They enable the correct base pairings to occur.
    - **C.** They signal the cell to know when to replicate itself.
    - **D.** They assist the helicase molecule in unwinding the DNA.
    - **E.** They hold the two strands of the DNA molecule together.

CRITICAL THINKING

14. *Sample answer: Just like only a few letters are needed to write a large number of different words, the four bases in DNA can be combined and sequenced differently to create different kinds of proteins. Because the bases are grouped into threes, there are 64 possible codons that the DNA could code for. These 64 codons can be put into different orders and can be strung together in varying lengths, so this allows for a very diverse group of proteins to be produced.* **DOK 2**
15. *Sample answer: Miescher took the first step in identifying the existence of DNA. He described the general characteristics of the molecule, but he did not know its function. Griffith's research helped establish that genetic information could be passed between cells even if the cell with the information had been killed. He hypothesized that this information was contained in molecules within the cell, but he did not know which molecules. Avery and colleagues were able to determine that the molecule that served this function was the DNA that Miescher had discovered much earlier.* **DOK 2**

CRITICAL THINKING

14. **Explain** If DNA has only four different bases, how can it be used to produce so many different proteins?

15. NOS **Scientific Knowledge** Describe how the discoveries of Miescher, Griffith, and Avery built on each other to help advance our understanding of DNA.

16. **Explain** What happens when a ribosome reaches a stop codon?

17. **Contrast** Compare activators and repressors in gene regulation.

MATH AND ENGLISH LANGUAGE ARTS CONNECTIONS

1. **Synthesize Information** Describe what the words *transcription* and *translation* mean in other contexts outside of science, and relate those meanings to what they mean in protein synthesis.

2. **Write a Function** Write an equation to describe the relationship between the number of nucleotides (n) in a DNA sequence and the number of amino acids (a) in the resulting polypeptide.

3. **Infer from Statistics** If a codon is randomly generated, what amino acids are most likely to be coded for? Which are least likely? Refer to **Figure 11-15**.

4. **Write Explanatory Text** Francis Crick said, "Rather than believe that Watson and Crick made the DNA structure, I would rather stress that the structure made Watson and Crick." Using what you know about the history of the discovery of DNA and what DNA does, how do you interpret this quote?

5. **Reason Abstractly** In geometry, complementary angles fit together to form an angle that measures exactly 90 degrees. Use this mathematical definition and **Figure 11-3** to help explain why base pairs in nucleotides are called complementary.

REVISIT VIRAL SPREAD

Gather Evidence In this chapter, you learned about how cells transfer information from their genomes to the RNA and protein products that are important for life. You have also learned here and in Section 8.4 that viruses have their own genomes. Viral genomes vary widely and require different help from their host cells to be expressed, but the overall goal remains the same: to express genes and replicate. Look at the structure of SARS-CoV-2, which was first presented in Section 8.4.

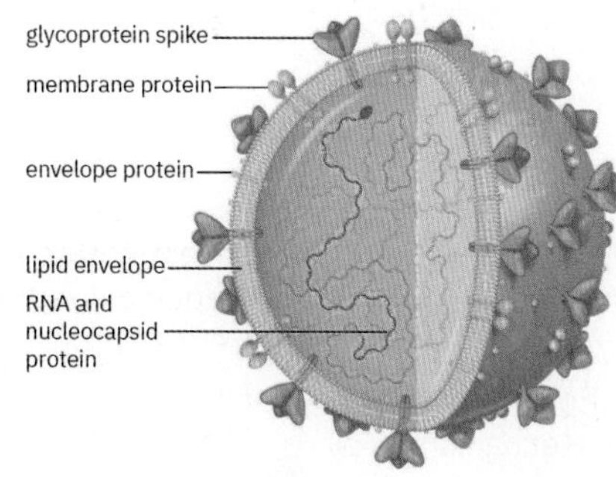

1. Observe the structures shown in the provided model of the SAR-CoV-2 viral particle. What are some of the genes that need to be encoded within the SARS-CoV-2 genome to make new virus particles? Which genes do you think code for proteins that are important for infecting a host cell?

2. What other questions do you have about how SARS-CoV-2 infects and replicates within host cells?

REVISIT THE ANCHORING PHENOMENON

1. *Sample answer: The viral genome would need to encode genes for spike proteins, envelope proteins, membrane proteins, and capsid proteins. As an RNA virus, it would also need to code for a specialized enzyme for viral RNA replication since animal cells do not need to replicate RNA. The genes that code for RNA and for glycoprotein spikes are important for infecting a host cell.* **DOK 3**

2. *Sample answer: Does the viral RNA only act as mRNA or does a virus include other functional RNAs, too? How does a cell respond when new viral particles are generated? Which of the host's resources does the virus rely on to make new virus particles?* **DOK 3**

16. *The stop codon does not add an amino acid to the peptide. Translation stops, the ribosomal subunits separate, and the completed peptide is released.* **DOK 2**

17. *Sample answer: Activators and repressors are both transcription factors that work by binding to the DNA and affecting the rate of transcription. Activators speed up transcription, and repressors slow down transcription.* **DOK 2**

MATH AND ELA CONNECTIONS

1. *Sample answer: Transcription can refer to a written record of what was said. This is similar to transcription in the cell, where information is copied from DNA to RNA. Translation can refer to converting information from one language to another. In the cell, translation occurs when the information in mRNA is used to construct a protein from amino acids. This is similar to taking words and translating them into a different language.* **DOK 3**

2. *Sample answer:* $a = n/3$ *or* $n = 3a$ **DOK 3**

3. *Sample answer: Leucine, serine, and arginine are most likely, with a 6/64 chance of being coded for. Methionine and tryptophan are least likely, with 1/64 chance each.* **DOK 3**

4. *Sample answer: The discovery of the structure of DNA was a monumental step forward in our scientific understanding. The discovery "made" Watson and Crick in that it established their careers as scientists. Alternatively, DNA is the instructions for all life and therefore all life, including Watson and Crick, are made by DNA.* **DOK 3**

5. *Sample answer: Complementary base pairs fit together in a precise way, like puzzle pieces, to form the DNA molecule. The C–G base pair fits together to make a size and shape that is the same as the A–T base pair. When paired together, complementary bases add up to the exact size needed to join the sugar molecules that form the backbone of the DNA molecule, similar to the way that complementary angles always add up to exactly 90 degrees.* **DOK 3**

Three-Dimensional Learning

The practices, core ideas, and crosscutting concepts presented in this chapter's text, investigations, and resources provide support to address the following Performance Expectations: **HS-LS3-2** and **HS-LS3-3.**

Science and Engineering Practices	Disciplinary Core Ideas	Crosscutting Concepts
Asking Questions and Defining Problems Analyzing and Interpreting Data (HS-LS3-3) Constructing Explanations and Designing Solutions Engaging in Argument from Evidence (HS-LS3-2) Scientific Investigations Use a Variety of Methods	**LS3.A:** Inheritance of Traits **LS3.B:** Variation of Traits (HS-LS3-2, HS-LS3-3)	Patterns Cause and Effect (HS-LS3-2) Scale, Proportion, and Quantity (HS-LS3-3) Science is a Human Endeavor (HS-LS3-3)

Contents		Instructional Support for All Learners	Digital Resources
ENGAGE			
346–347	**CHAPTER OVERVIEW** **CASE STUDY** How does genetic information change and get passed from generation to generation?	**Social-Emotional Learning** CCC Cause and Effect **On the Map** Evolution of Dogs **English Language Learners** Silent Reading	*Video 12-1*
EXPLORE/EXPLAIN			
348–356 DCI LS3.A LS3.B	**12.1 MEIOSIS** • Describe the relationship between the genes, alleles, and traits. • Compare the advantages and disadvantages of asexual and sexual reproduction. • Explain how sexual reproduction results in greater genetic diversity among offspring.	**Vocabulary Strategy** Frayer Squares **English Language Learners** Taking Notes on a Text **Visual Support** Karyotypes, Relate Diagrams, Phases of Meiosis, Mechanisms of Genetic Variation CCC Structure and Function **Connect to English Language Arts** Compare and Contrast, Write an Argument CCC Cause and Effect **Address Misconceptions** Asexual Advantages, Mechanisms of Genetic Variation **BIOTECHNOLOGY FOCUS** In Your Community **Connect to Mathematics** Reason Quantitatively CCC Cause and Effect	**Investigation B** Mapping Fruit Fly Genes Through Linkage (90 minutes) *Video 12-2*
357	**EXPLORER** KRISTEN RUEGG	**Connect to Careers** Ornithologist **On the Map** Bird Migration **English Language Learners** Formal and Informal Language	

Contents		Instructional Support for All Learners	Digital Resources
EXPLORE/EXPLAIN			
358–363 DCI LS3.B	**12.2 MUTATIONS** • Explain how mutations can affect trait expression. • Describe the characteristics of genetic and chromosomal mutations. • Relate genetic mutations to greater genetic diversity within a population.	**CER** Revisit the Anchoring Phenomenon **Vocabulary Strategy** Sentence Frames **Social-Emotional Learning** Social Awareness **Visual Support** Genetic Mutations **English Language Learners** Use Connecting Words, Use Academic Language **Crosscurricular Connections** Earth Science **On Assignment** Black Panther	**Virtual Investigation** Fighting a Viral Pandemic
364–369 DCI LS3.B	**12.3 MENDELIAN INHERITANCE** • Differentiate between genotype and phenotype. • Evaluate possible genotypes from a genetic cross. • Outline how a recessive allele is expressed.	**Vocabulary Strategy** Word Web **SEP** Analyzing and Interpreting Data **Visual Support** Mendel's Observations, Dihybrid Cross **Differentiated Instruction** Leveled Support **Connect to English Language Arts** Key Ideas and Details **Address Misconceptions** Genetic Crosses **Connect to Mathematics** Analyzing Data	*Video 12-3* **SIMULATION** Punnett Square
370	**LOOKING AT THE DATA** BLOOD TYPE COMPATIBILITY		
371–375 DCI LS3.B	**12.4 OTHER PATTERNS OF INHERITANCE** • Compare the difference between Mendelian inheritance, incomplete dominance, and codominance. • Relate how environmental factors affect phenotype. • Describe how sex-linked traits are expressed in male and female offspring.	**Vocabulary Strategy** Use Prior Knowledge **CCC** Patterns **Connect to English Language Arts** Present Claims and Findings **Differentiated Instruction** Leveled Support **Connect to Mathematics** Define Quantities for Modeling **CASE STUDY** Ask Questions **English Language Learners** Use Reasoning **Visual Support** Genetic Cross **SEP** Analyzing and Interpreting Data	**Investigation A** Design an Organism (50 minutes)
376	**MINILAB** MODELING INHERITANCE		
ELABORATE			
377	**TYING IT ALL TOGETHER** How does genetic information change and get passed from generation to generation?	**Connect to English Language Arts** Develop and Strengthen Writing **English Language Learners** Citing Sources	Self-Reflection and Scoring Rubrics
Online	**Investigation A** Design an Organism **Investigation B** Mapping Fruit Fly Genes through Linkage	Guided Inquiry (50 minutes) Guided Inquiry (90 minutes)	MINDTAP Access all your online assessment and laboratory support on MindTap, including: sample answers, lab guides, rubrics, and worksheets. PDFs are available from the Course Companion Site.
EVALUATE			
378–379	**Chapter 12 Review**		
Online	**Chapter 12 Assessment** **Performance Task 2** *What caused the unusual skin discoloration in the people living in a rural area?* (HS-LS3-2)		

CHAPTER 12

ENGAGE

In middle school, students will have learned about basic structural changes to genes and how structural change may be harmful, helpful, or have no affect on the organism. Students will have been exposed to genetic crosses. They will also have an understanding of asexual and sexual reproduction and how asexual reproduction results in offspring with identical genetic information and sexual reproduction results in offspring with genetic variation from their parents. This chapter expands on that prior knowledge by exploring heredity and genetic variation.

Students will learn about the different mechanisms at the cellular level that produce genetic variation and explore the advantages of genetic diversity in sexual reproduction. They will examine how genetic and chromosomal mutations can affect trait expression, and they will learn about different inheritance patterns.

About the Photo This photo shows examples of genetic variation in dog breeds. The features displayed in the photo show examples of Mendelian inheritance, incomplete dominance, and codominance within an individual species. To garner interest for this chapter's content, use the photo to encourage students to activate prior knowledge related to inheritance and genetic variability.

Social-Emotional Learning

As students progress through the chapter, invite them to look for contexts and opportunities to practice some of the five social and emotional competencies. For example, **social awareness skills** can help them find ways to recognize strengths in others as the class learns about and discusses genetic diversity. As they learn about mutations (Section 12.2), remind them to practice **self-awareness** and **relationship skills** by providing an environment where students can openly discuss inheritance involving genetic mutations.

CHAPTER 12

GENETIC VARIATION AND HEREDITY

The American Kennel Club currently registers around 200 dog breeds. This number does not take into account the immense variety of mixed breeds that exist.

Despite the great variety among dog breeds, they share 99.9 percent of their DNA. A dog's genome contains three billion base pairs. Though 0.1 percent might seem like a tiny difference, it actually adds up to three million base pairs. Within those base pairs lie the genetic variations that make each breed–and each individual dog–unique from all others.

Chapter 12 supports the NGSS Performance Expectations **HS-LS3-2** and **HS-LS3-3**.

346 CHAPTER 12 GENETIC VARIATION AND HEREDITY

CROSSCUTTING CONCEPTS | Cause and Effect

Relationship of Genotype and Phenotype In this chapter, students explore the concept of cause and effect as they learn about sexual reproduction, inheritance, and genetic variation on a molecular level. Scientists analyze patterns that can provide invaluable clues about cause-and-effect relationships. Gaining a deeper understanding of the mechanisms of inheritance can enable predictions and, in the case of genetic mutations, can be used to find treatments and even cures. Chapter 12 explores the how and why of inheritance and trait expression. You can reinforce this crosscutting concept in this lesson by helping students make the connection between the molecular level of inheritance, such as genotype, and observable traits (phenotype).

CASE STUDY
FOR THE LOVE OF DOGS

HOW DOES GENETIC INFORMATION CHANGE AND GET PASSED FROM GENERATION TO GENERATION?

Large, small, long hair, short hair, perky ears, floppy ears—the list of dog trait variations goes on and on. However, despite their differences, all dogs belong to the same species, *Canis lupus familiaris*. Research indicates that all dogs descended from a gray wolf (*Canis lupus*) ancestor. Although how and why remain unclear, most scientists agree that dogs were likely domesticated sometime between 14,000 and 40,000 years ago.

Today, there are more than 340 recognized dog breeds around the world. Most of these breeds were developed within the past 150 years. Through selective breeding by humans, these dogs exhibit desired traits in both their appearance and behavior. For example, while some dogs are bred to exhibit certain physical traits, such as short legs or wiry hair, others are bred for certain skills, such as herding, retrieving, or guarding. These traits are embedded in a dog's genetic code. When dogs with the same traits are bred together, their resulting offspring will often–but not always–exhibit the same characteristics as their parents.

But if all domesticated dogs evolved from a wolf ancestor, what accounts for such a wide variety of traits seen today? One of the key characteristics of domesticated dogs is tameness. In the late 1950s, Dmitry Belyaev, a Russian geneticist, hypothesized that by initially selecting the tamest wolves to live with them, early humans were unknowingly choosing a single trait that could be genetically linked to a wide range of traits, including changes in body shape, behavior, and physical functioning. To test his hypothesis, Belyaev chose to work with wild foxes, a species related to dogs. He found that within only eight generations of domestication, the tamed fox population exhibited changes in fur color and developed floppy ears and curved tails, traits common in many dog species. Belyaev's team also discovered that an increase in tameness was associated with changes in two developmental milestones: the tamer fox kits opened their eyes several days earlier than those born in the wild, and their fear response appeared almost a month later than typically seen in wild foxes. The scientists hypothesize that these changes, if also exhibited by the first domesticated dogs, would have made them more open to interacting with humans without fear. Belyaev's population of tame foxes continues to be studied today. Recent discoveries include the finding that foxes bred for tameness and those bred for aggressiveness show a significant difference in gene expression related to neurologic processing.

Ask Questions *As you read this chapter, generate questions about the mechanisms that lead to genetic variation.*

FIGURE 12-1 The silver fox is a species that has been bred for tameness. Foxes and modern dogs belong to the canine group that also includes wolves, coyotes, and jackals.

VIDEO 12-1 Go online to watch a brief history of humans' relationship with dogs.

DIFFERENTIATED INSTRUCTION | English Language Learners

Silent Reading Have students read the Case Study silently. Have them annotate the text by underlining main ideas, circling words they are not sure of, and writing a question mark by sentences they have trouble understanding.

Beginning Have pairs use a dictionary to look up words they are not sure of. Help students rephrase sentences they had trouble understanding or break them into smaller parts.

Intermediate Have pairs use context clues to look up words they are not sure of. Have them try to rephrase the sentences they marked in their own words. Then have partners discuss these sentences.

Advanced Have pairs take turns explaining main ideas in the text. Have them share sentences they had trouble understanding and help each other clarify.

CASE STUDY

ENGAGE

The situation described in the Case Study can be used to assess students' prior knowledge of genetic diversity. Ask students to identify a trait that is representative of a specific dog breed. Encourage students to discuss how this could happen.

Revisit the Case Study after each section to have students identify how sexual reproduction can result in greater genetic diversity. In the Tying It All Together activity, students will research how breeders use genetics to produce dog breeds with specific traits.

Ask Questions Students should revisit the Case Study as they read the chapter to make connections with the content. See a specific suggestion in Section 12.4.

Video

Time: 2:08

Use **Video 12-1** to discuss selective breeding in the evolution of dogs from wolves, including desired traits and humans' active role in breeding dogs.

On the Map

Evolution of Dogs An international team of scientists used genome sequencing to reveal how our furry companions evolved from wild wolf to domesticated dog. Led by Ya-Ping Zhang and Peter Savolainen, the team examined the genome sequences of 58 wolves and dogs and discovered that the evolution from wolf to dog occurred in two distinct phases. The initial phases began approximately 33,000 years ago with origins in China. The second phase took place 18,000 years later. Migration of dogs out of Southeast Asia occurred after the second evolutionary phase.

Human Connection Some studies show many similarities between gene sequences in dogs and humans, such as sequences related to behavior and diet. This is thought likely because, living in such close proximity, both were subjected to similar selective pressures over time. Without similar studies of other domestic species, however, it is unclear whether these similarities are unique.

12.1

MEIOSIS

LS3.A, LS3.B

EXPLORE/EXPLAIN

This section provides a review of traits, genes, and alleles, compares asexual and sexual reproduction, introduces the process of meiosis, and reviews the different mechanisms at the cellular level that produce genetic variation. It also covers the advantages of genetic diversity in sexual reproduction.

Objectives

- Describe the relationship between genes, alleles, and traits.
- Compare the advantages and disadvantages of asexual and sexual reproduction.
- Explain how sexual reproduction results in greater genetic diversity among offspring.

Pretest Use **Questions 1, 2, 3, 5,** and **7** to identify gaps in background knowledge or misconceptions.

Vocabulary Strategy

Frayer Squares As students work through the key terms in this section, have them create a Frayer square for each one. Students should draw a square and divide it into four boxes, with the word in the center. In the boxes they should write the definition of the word, an example of the word, a non-example, and additional relevant information, such as images or characteristics. As an example, make a Frayer square for trait as a class.

Make a Claim *Students might say that the parental DNA sequences result in similarities in the traits of offspring.*

12.1

KEY TERMS

allele
asexual reproduction
gamete
meiosis
sexual reproduction
trait

MEIOSIS

Take a look around you. What living things do you see? When you are outside, you might see an even greater variety of organisms—plants, humans, and other animals. What makes every organism unique is its genetic code, or genome. Information encoded in DNA is the basis of visible traits that define species and distinguish individuals. Humans, trees, insects, and all organisms have a genome that is unique.

Make a Claim *Make and defend a claim about the basis of similarities and differences in characteristics of a particular species.*

Traits, Genes, and Alleles

Members of a species have the same **traits**, or inherited characteristics, because they have the same genes. A *gene* is a section of DNA that codes for a certain protein or RNA molecule. Genes are the most basic unit of heredity. A continuous piece of DNA is made up of many genes and is called a *chromosome* (**Figure 12-2**). Genes are the units of information that are passed from one generation to the next. Every gene has a locus, or specific position on a chromosome. All the genes of an organism are called its *genome*. A genome is a complete set of genetic instructions that include all the information needed to build the individual organism and allow it to develop and grow.

FIGURE 12-2 A strand of DNA is made up of many genes. In eukaryotes, DNA is organized into linear chromosomes that are duplicated.

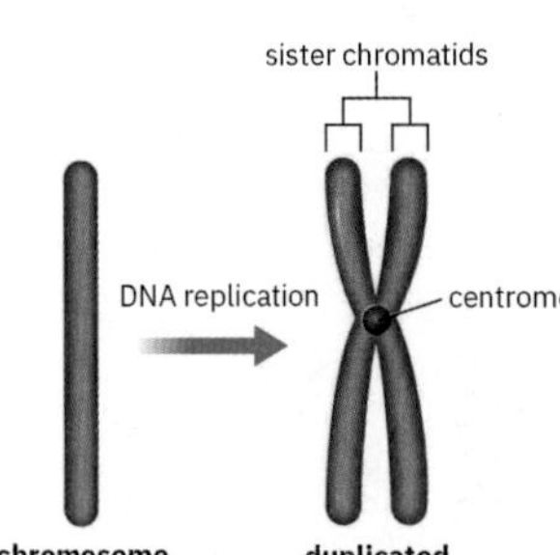

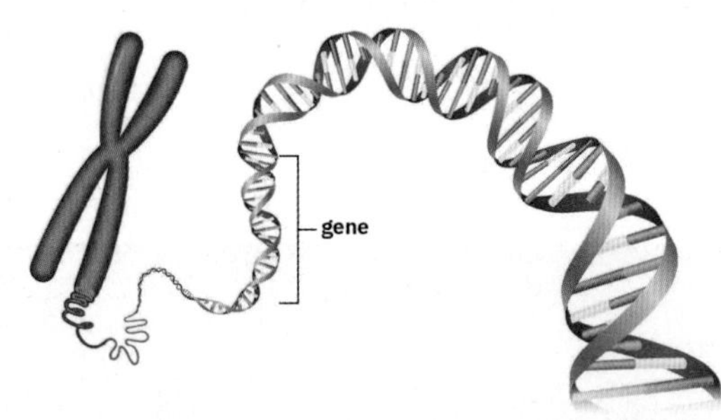

In eukaryotes, a DNA molecule is organized into a chromosome contained in the nucleus of the cell. During cell division, a chromosome replicates. A replicated chromosome is made up of two identical DNA molecules, called sister chromatids, that are joined at the centromere.

In prokaryotes, DNA is contained in the cytoplasm of the cell. Most prokaryotes have a single, circular chromosome. The cytoplasm also contains plasmids, which are small circles of DNA that carry a few genes. Plasmid genes are not required for the survival and reproduction of the organism. However, a plasmid can provide advantages to the cell, such as resistance to antibiotics.

DIFFERENTIATED INSTRUCTION | English Language Learners

Taking Notes on a Text Encourage students to create a T-chart to organize what they learn about meiosis. Have them label the columns "Traits," "Genes," and "Alleles." Then, as they read, have students write down facts that they learn about each. For example, in the "Traits" column, they might write about characteristics that are inherited or passed down through DNA. In the "Alleles" column, they might include information about how alleles are different forms of the same gene. As students move through the chapter and activities, have them refer to their chart to add additional details.

Beginner Have students write words and short phrases. Allow them to take notes in their native language as needed.

Intermediate Have students take notes in phrases and then key ideas in complete sentences on a separate sheet of paper.

Advanced Have students use their notes to explain each section of the text.

Your somatic, or body, cells contain pairs of homologous chromosomes. There are 23 homologous pairs of chromosomes in humans, shown in **Figure 12-3**. A karyotype is an image of an individual cell's chromosomes. To make a karyotype, cells taken from the individual are stained so the chromosomes can be distinguished under a microscope. The chromosomes are then digitally rearranged according to size, shape, and length.

Of the 23 pairs of chromosomes, 22 pairs are autosomes, or chromosomes that contain genes for characteristics that are not directly related to the sex of the individual. The last pair are called sex chromosomes because they directly control the development of sexual characteristics. The sex chromosomes are referred to as X and Y, or the XY system, and determine the biological sex of an individual. Typically, organisms with two X chromosomes (XX) are female, and organisms with XY chromosomes are male. The XY system occurs in humans, most mammals, and some reptiles and plants.

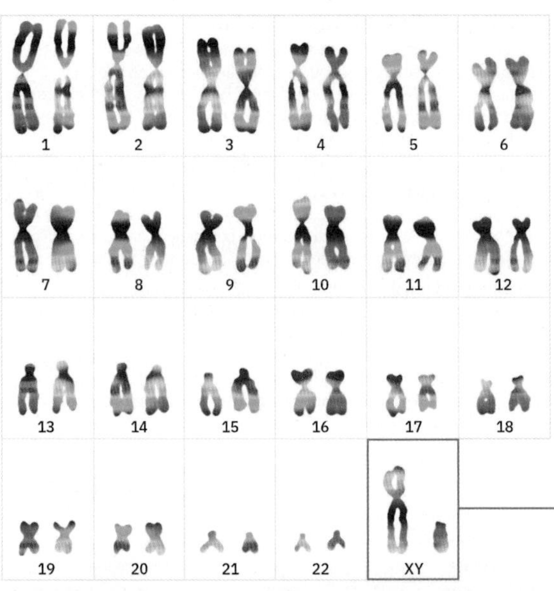

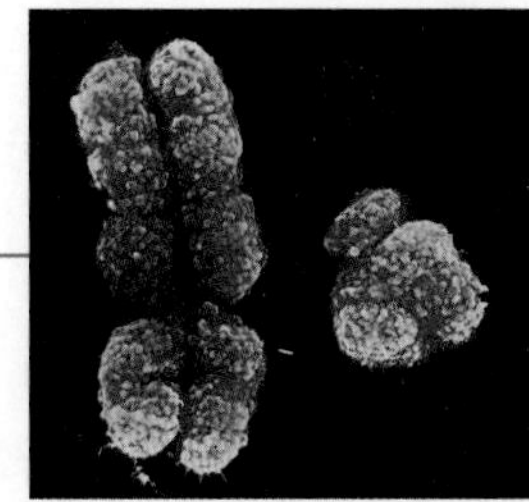

FIGURE 12-3 Humans have 23 homologous pairs of chromosomes. A karyotype shows the complete set of 23 chromosomes in the cells of humans. As shown in the inset scanning electron micrograph, the X chromosome is about three times the size of the Y chromosome.

The two chromosomes of each homologous pair have the same genes, but their DNA sequence is not identical. This is because an individual inherits chromosomes from each parent who differ genetically. Thus, the DNA sequence of any of an individual's genes may vary from a corresponding gene on a homologous chromosome.

Visual Support

Karyotypes Explain to students that a karyotype, as shown in **Figure 12-3**, is a tool used by medical specialists to identify abnormal numbers of chromosomes or abnormalities in their structure. Structural abnormalities include deletion, duplication, inversion, substitution, or translocation, all of which can alter a portion of the chromosome.

The karyotype colors result from the use of special stains to reveal the banding pattern. Each band is numbered to help identify specific parts of a chromosome. Have students compare the XY chromosomes of the karyotype in **Figure 12-3** with the structure shown in the micrograph. Point out that karyotypes are prepared from cells that are in the metaphase portion of mitosis when DNA is most condensed. Refer students to **Figure 7-11** as a refresher on the phases of mitosis.

Science Background

Chromosome Maps As time allows, display chromosome maps published by the National Institutes of Health, which can be found through an Internet search with keywords such as "NIH chromosome map." Although developed several years ago, they give information about the relative sizes of chromosomes by listing the number of genes and base pairs that were known at the time. Scrolling through the various maps, students will learn the location of genes for recognizable diseases such as various cancers, asthma, and diabetes. The maps may also spark interest in learning more about other kinds of genetic diseases.

CROSSCUTTING CONCEPTS | Structure and Function

DNA Replication After viewing **Figure 12-2** and **Figure 12-3**, help students reaffirm their understanding of the replication process by allowing them to make their own duplicated chromosome. Provide yarn and three pieces of pipe cleaner to each student. Explain that the yarn represents the DNA in each chromosome. To reinforce that DNA is made up of many genes, students should color or wrap tape around sections of the yarn. Then, have them wrap the first pipe cleaner in their yarn (coded DNA). Refer students again to **Figure 12-3** and have small groups discuss how to recreate an exact copy of the DNA located on the chromosome they created. Students should understand they will need to unwrap their chromosome and use the yarn as reference to create a second, identical piece of yarn. Then, they would repeat the process to create a duplicate chromosome.

Connect to English Language Arts

Compare and Contrast This section connects asexual and sexual reproduction to the concept of genetic variation. Have students write an informative summary of the information presented for each type of reproduction. Encourage students to use a Venn diagram to collect information about similarities and differences.

Genes occur in pairs on homologous chromosomes. Different forms of the same gene are called **alleles** (**Figure 12-4**). Traits are inherited characteristics that derive from the genes that are inherited. However, almost every shared trait varies among individuals of a species. Alleles of the shared genes are the basis of this variation.

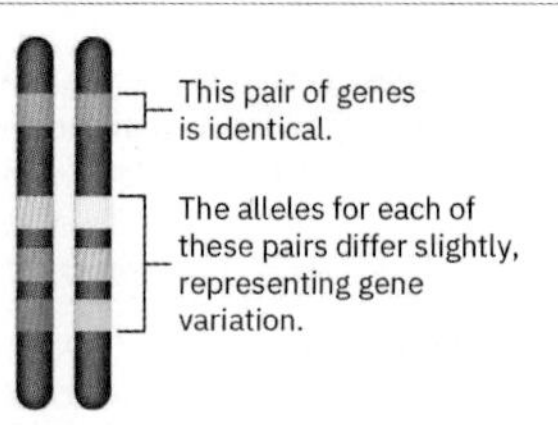
This pair of genes is identical.
The alleles for each of these pairs differ slightly, representing gene variation.

FIGURE 12-4 Corresponding colored patches indicate corresponding DNA sequences in a homologous chromosome pair.

Asexual and Sexual Reproduction

Chapter 7 introduced mitosis and cytoplasmic division as part of the processes of asexual reproduction. Not all species can use this reproductive mode. Many eukaryotes reproduce sexually, either exclusively or most of the time. **Sexual reproduction** is the process in which offspring arise from two parents and inherit genes from both.

All organisms transmit genetic information to their offspring either through asexual reproduction or sexual reproduction. Sexual reproduction mixes genetic information from two parents. Such variation is almost always advantageous, because genetically diverse populations are more resilient to environmental change than populations with low genetic diversity.

Asexual reproduction produces genetically identical individuals, so populations of asexually reproducing organisms typically have low genetic diversity. This is because an asexually reproducing individual passes all of its genes to every one of its offspring. All the offspring are basically clones because they have the same alleles as their one parent. For comparison, only about half of a sexually reproducing organism's genes are passed to each offspring (**Table 12-1**).

TABLE 12-1. Comparison of Asexual and Sexual Reproduction

Attribute	Asexual reproduction	Sexual reproduction
type of cell division	mitosis	meiosis and mitosis
number of parents	1	2
parental cell(s)	diploid cell	haploid gametes
inherited genes	100% from 1 parent	50% from each parent
offspring genes	genetically identical	not genetically identical
advantages	does not require a partner	high genetic diversity
disadvantages	no genetic diversity	typically requires a partner
organisms	prokaryotes, such as bacteria and archaea, and eukaryotes, such as many plants, fungi, and some animals	eukaryotes, including most animals and some plants and fungi

CROSSCUTTING CONCEPTS | Cause and Effect

Sexual Reproduction Have students create a flowchart graphic organizer using arrows to show the relationship between genes, alleles, and traits. Refer students to **Figure 12-4**. Encourage them to find the key terms in both the figure and in the text. Then, challenge students to synthesize the information into their diagram. Student diagrams should resemble the following:

DNA → genes → alleles → traits

Allow for open discussion with varying diagrams.

Consistency is a good evolutionary strategy in a favorable, unchanging environment. Alleles that help an organism survive and reproduce in the environment do the same for its descendants. However, most environments are constantly changing, and change is not always favorable. Individuals that are identical are equally vulnerable to challenges.

In a changing environment, genetic diversity in sexual reproducers is an advantage. Sexual reproduction randomly mixes up the genetic information of two parents who have different alleles. The offspring of sexually reproducing organisms inherit different combinations of parental alleles, so they differ from one another and from their parents. Some of the offspring may be perfectly suited to a new environmental challenge. Thus, as a group, they have a greater chance of surviving environmental change than clones.

SEP Analyze Data *Table 12-1 identifies one advantage of asexual reproduction as not requiring a partner. Identify at least one other advantage of asexual reproduction.*

CHAPTER INVESTIGATION

Mapping Fruit Fly Genes Through Linkage

How can we map genes using traits that are inherited together?

Go online to explore this chapter's hands-on investigation to map some traits on fruit fly chromosomes.

BIOTECHNOLOGY FOCUS

FOLLOW THE MAP A karyotype is a collection of chromosomes in the cells of an individual. A genome map identifies all the genes and the distance between genes on a chromosome. Scientists map genomes to compare the DNA of closely related species to establish and analyze their genetic differences. The differences help scientists to catalogue the biodiversity of organisms on the planet. This process is called comparative genomics.

Comparative genomics includes comparing single genes and gene fragments to study their function and help establish relationships among species. Scientists are building databases of genome sequences to share knowledge and support the progress of this scientific research. For example, researchers study where genetic variants associated with specific diseases or characteristics are located on the chromosomes. Genes or gene sequences with a known location on a chromosome are called genetic markers. Physical maps identify the physical distance between the known gene sequences. The distance is expressed as the number of base pairs between the sequences. This information helps scientists to identify and isolate genes.

As scientists learn more about a particular genome, its map becomes more accurate and detailed. A genome map is not a final product but a work in progress. Genetic mapping is a tool scientists use to identify genes and to understand their function.

There are two basic ways of mapping a genome: genetic mapping and physical mapping.

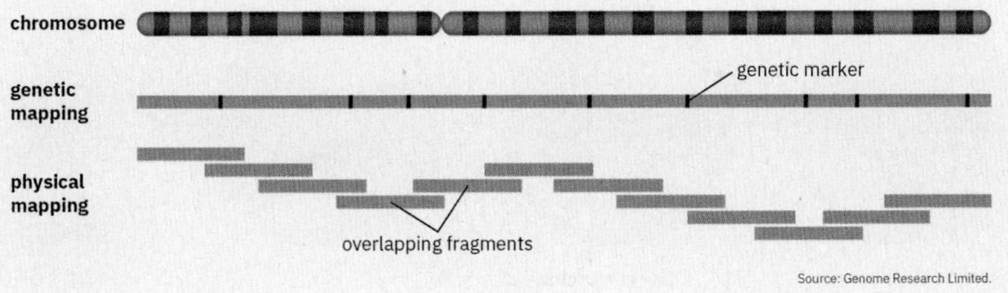

Source: Genome Research Limited.

BIOTECHNOLOGY FOCUS | Follow the Map

In Your Community Comparative genomics enables scientists to compare sequences of the human genome with genomes of other organisms. This allows researchers to identify any areas of similarity and difference found within the genomes being compared, which can provide evidence for evolutionary changes and relationships. It is also helpful for gaining a better understanding of the structure and function of human genes, which can then be used to fight human diseases. Ask students how different their DNA is from that of a human relative, another random human, or another kind of animal? Use student responses to lead into the significance of the information discussed in the Biotechnology Focus. After reading, encourage students to relate this information to themselves and people in their communities by having them reflect on how it could change modern medicine if doctors determine treatment needs based on individual genetic information.

CHAPTER INVESTIGATION B

Guided Inquiry *Mapping Fruit Fly Genes Through Linkage*

Time: 90 minutes or 2 days

Students are given test crosses between two different recessive traits for fruit flies (*D. melanogaster*) and the phenotypes of the offspring of the test cross. Using these data, students form part of the chromosome map of the fruit flies.

Go online to access detailed teacher notes, answers, rubrics, and lab worksheets.

Address Misconceptions

Asexual Advantages Students may not consider asexual reproduction to be advantageous due to the greater genetic diversity produced from sexual reproduction. Help them to understand that asexual reproduction can also be beneficial. Refer students to the information in **Table 12-1**. Allow them time to brainstorm situations when it would be helpful for an organism to reproduce asexually. Have students validate each item on their brainstormed list by providing an example as evidence for each item. Encourage students to participate in a class debate using their brainstormed ideas and examples. Challenge them to provide evidence, in the form of examples, for their point of view.

SEP Analyze Data *Sample answer: Besides not requiring a partner, asexual reproduction is advantageous because it is faster, as seen in bacterial growth for example.*

Connect to Mathematics

Reason Quantitatively The number of haploid chromosomes is tied to the number of diploid chromosomes. Direct students to **Figure 12-5** and have them look for the algebraic expression associated with diploid (2*n*) and haploid (*n*) cells. Remind students that meiosis halves the number of chromosomes, so it makes sense that the expression for haploid cells is *half* of diploid cells. To reinforce this concept, have students work in pairs to complete the table. If necessary, fill in the first row (*78, 39*).

Organism	Chromo-somes	Diploid Cells (2*n*)	Haploid Cells (*n*)
Dog	78		
Fruit Fly	8		
Spinach	12		
Kangaroo	16		
Giraffe	30		
Great White Shark	92		
Carrots	18		
Strawberry	56		

Visual Support

Relate Diagrams Ensure students understand that **Figure 12-5** is a summary of **Figure 12-6** and **Figure 12-7**. Point out that the top portion of **Figure 12-5** relates to the beginning of prophase I, while the middle portion relates to the end of telophase I and the beginning of prophase II. The bottom portion of the diagram relates to the end of telophase II.

FIGURE 12-5 During meiosis, a diploid (2*n*) cell divides into four haploid (*n*) cells.

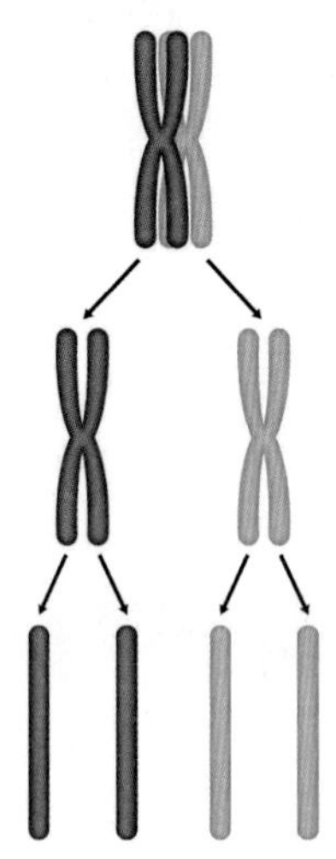

The Phases of Meiosis

Genetic variation occurs in many ways, including the recombination of genes during two processes: meiosis and fertilization by sexual reproduction. **Meiosis** is a nuclear division process. During meiosis, diploid cells become haploid sex cells, called **gametes**. Meiosis halves the number of chromosomes to maintain the proper number of chromosomes through successive generations (**Figure 12-5**).

Before meiosis begins, DNA is copied during the S phase of the cell cycle. During meiosis I, homologous chromosomes separate, producing two haploid cells with duplicated chromosomes. Meiosis I can be described in distinct phases, each of which is a series of gradual changes (**Figure 12-6**).

- **Prophase I** The chromosomes condense, and homologous chromosomes pair up. The nuclear envelope breaks up. Microtubules extend from opposite poles in all directions to form a spindle. The microtubules attach chromosomes of the homologous pair to opposite spindle poles.
- **Metaphase I** The microtubules push and pull the chromosomes until they are aligned midway between the spindle poles.
- **Anaphase I** Microtubules of the spindle move the homologous chromosomes of each pair away from one another and toward opposite spindle poles.
- **Telophase I** Two sets of chromosomes reach the spindle poles, and a new nuclear envelope forms around each set as the DNA loosens up. The two new nuclei are haploid (*n*); each contains one complete set of chromosomes. The sister chromatids stay together. The cytoplasm divides.

In some cells, meiosis II occurs immediately after meiosis I. In other cells, a period of protein synthesis—but no DNA replication—intervenes between the divisions.

FIGURE 12-6 During meiosis I, a diploid nucleus is divided into two haploid nuclei.

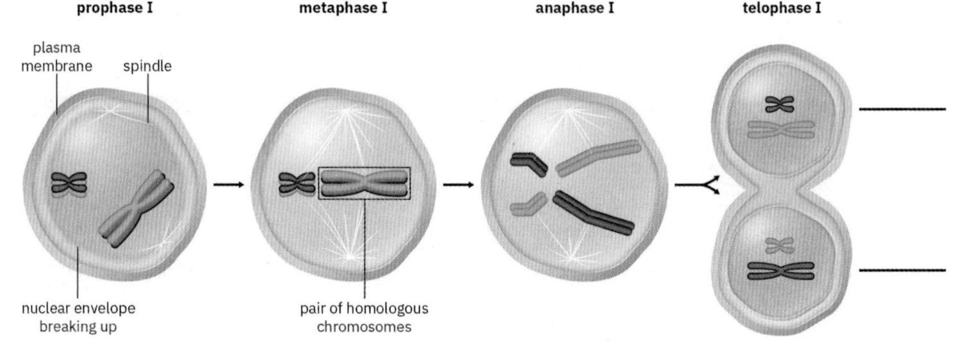

352 CHAPTER 12 GENETIC VARIATION AND HEREDITY

Meiosis II (**Figure 12-7**) proceeds simultaneously in both nuclei that formed in meiosis I.

- **Prophase II** The chromosomes condense, and the nuclear envelope breaks up. A new spindle forms. By the end of prophase II, microtubules attach each chromatid to one spindle pole, and its sister chromatid to the opposite spindle pole.
- **Metaphase II** The microtubules push and pull the chromosomes, aligning them in the middle of the cell.
- **Anaphase II** The spindle microtubules move the sister chromatids apart and toward opposite spindle poles.
- **Telophase II** The chromosomes reach the spindle poles. New nuclear envelopes form around the four clusters of chromosomes as the DNA loosens up. Each of the four nuclei that form are haploid (n), with one set of unduplicated chromosomes. The cytoplasm divides.

Comparing Mitosis and Meiosis Meiosis is an important process that produces new genetic combinations in each of the four daughter cells. Meiosis also ensures that offspring produced by sexual reproduction contain the correct number of chromosomes. If the chromosome number changes, so does the individual's genetic instruction.

The first part of meiosis is similar to mitosis. A cell duplicates its DNA before either nuclear division process begins. However, meiosis sorts the chromosomes into new nuclei not once but twice, so it results in the formation of four haploid nuclei.

CCC **Cause and Effect** *Why is it important that human gametes have half a set of DNA instead of a full set of DNA? Use evidence and scientific reasoning to support your claim.*

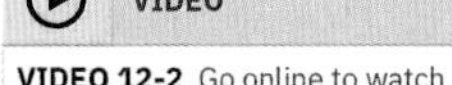

VIDEO 12-2 Go online to watch meiosis in action.

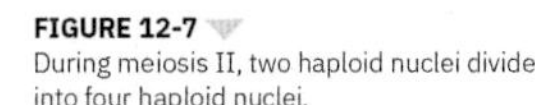

FIGURE 12-7 During meiosis II, two haploid nuclei divide into four haploid nuclei.

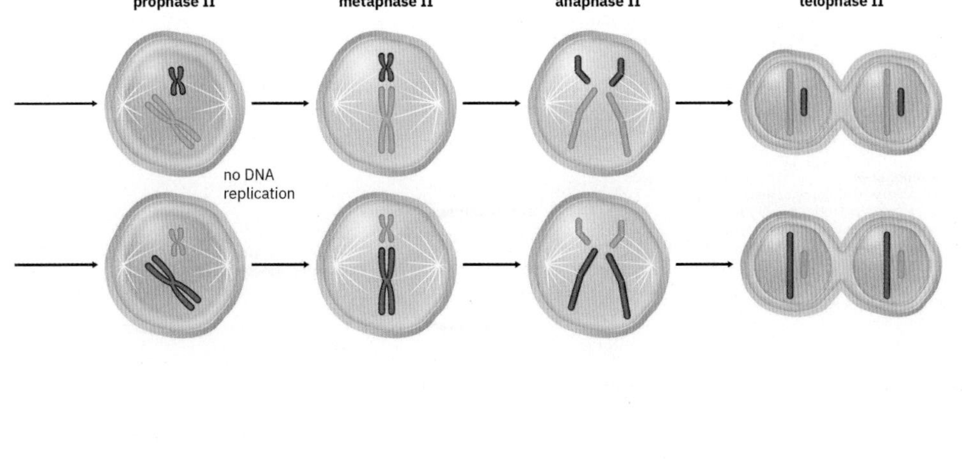

Visual Support

Phases of Meiosis Students may look at **Figure 12-6** and **Figure 12-7** and become overwhelmed with the different phases of meiosis being modeled. To help them make sense of the information in the figures, have them create an interactive model of the process. Provide students with index cards. Assign each student one meiosis phase. Have them reference both figures as they record any relevant information from their phase on the index card. Each card should include phase title, description of what happens during phases, and a model. Then, have students go around the class looking for the person that represents the phase before and after their phase. When complete, groups should have an ordered chain of the eight phases of meiosis.

Video

Time: 1:08

Use **Video 12-2** to observe meiosis in crane fly cells. Initiate a discussion among students about the model of meiosis in the text and what they observed in the video. Ask: *What are the challenges of modeling a process?* Have students explain the benefits of the model compared to the video observation.

CCC **Cause and Effect** *Gametes must have half a set of DNA to ensure that offspring have the correct number of chromosomes. Gametes join during fertilization, so if they had more than half a set of DNA, the resulting offspring would have more than 46 chromosomes.*

CROSSCUTTING CONCEPTS | Cause and Effect

Sexual Reproduction Remind students that humans reproduce sexually, which means they half of their DNA from a sperm cell and half from an egg cell. To help students use evidence to support their claims, first have them review **Figure 12-5** and **Figure 12-6** looking for key words or models that support their claim. For example, **Figure 12-6** shows a diploid nucleus being divided into two haploid nuclei. Encourage students to cite these examples as evidence in their claim.

Visual Support

Mechanisms of Genetic Variation Help students process the different mechanisms that result in genetic variation by organizing relevant information from the section into a four-column T-chart. Have students label the column headings: Crossing Over, Fertilization, Segregation, and Independent Assortment. Students should include summarized information from the text, as well as a model of each process. Prior to reading, have students review **Figure 12-8** through **Figure 12-11**. Have students look for any similarities from the models. Then, as they read, have them pay attention to the text for any reference to their observed similarities from the models. Encourage students to use information from their chart as evidence for any claims they make during class discussions.

Mechanisms of Variation in Meiosis and Fertilization

In addition to the random mating that occurs in many species, there are a number of mechanisms at the cellular level that produce genetic variation.

Crossing Over Early in prophase I, all chromosomes in the cell condense. When they do, each is drawn close to its homologous partner, so that the chromatids align along their length. This tight, parallel orientation favors crossing over, a process in which a chromosome and its homologous partner exchange corresponding pieces of DNA (**Figure 12-8**). Homologous chromosomes may swap any segment of DNA along their length, although crossovers tend to occur more frequently in certain regions of a chromosome. Swapping segments of DNA shuffles alleles between homologous chromosomes. It breaks up the combinations of alleles that occurred on the parental chromosomes and makes new ones on the chromosomes that end up in gametes.

Crossing over introduces novel combinations of alleles among offspring. It is a required process—meiosis will not finish unless it happens. The rate of crossing over varies among species and among chromosomes.

FIGURE 12-8 During prophase I of meiosis, crossing over occurs.

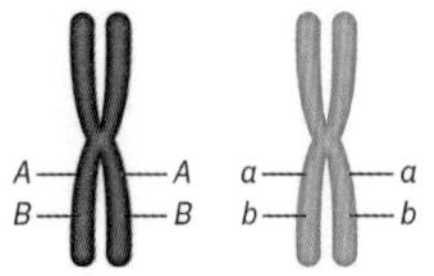

In this example, one gene has alleles *A* and *a*; the other gene has alleles *B* and *b*.

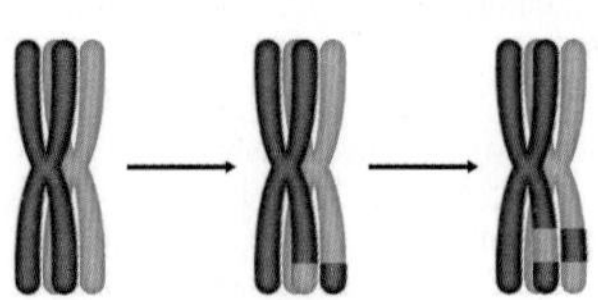

Close contact between homologous chromosomes promotes crossing over, in which nonsister chromatids exchange corresponding pieces.

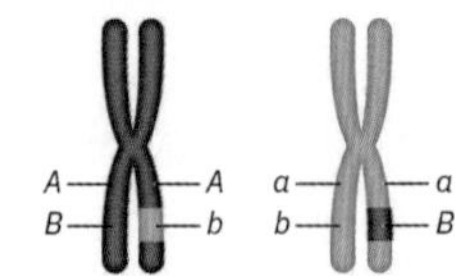

After crossing over, paternal and maternal alleles have become mixed up on homologous chromosomes.

Fertilization Sexual reproduction requires that diploid organisms produce haploid gametes through meiosis. Fertilization is the union of two haploid cells, one from each parent (**Figure 12-9**). When sexual reproduction results in fertilization, gametes unite to produce new, unique individual organisms. A male gamete fuses with a female gamete at fertilization to produce a diploid cell called a zygote. A zygote is the first cell of a new individual. Sexual reproduction results in increased genetic diversity within a species because of the random assortment of combined alleles from both parents.

FIGURE 12-9 At fertilization, two gametes meet to form a zygote. Meiosis halves the chromosome number, and fertilization restores it. If meiosis did not precede fertilization, the chromosome number would double with every generation.

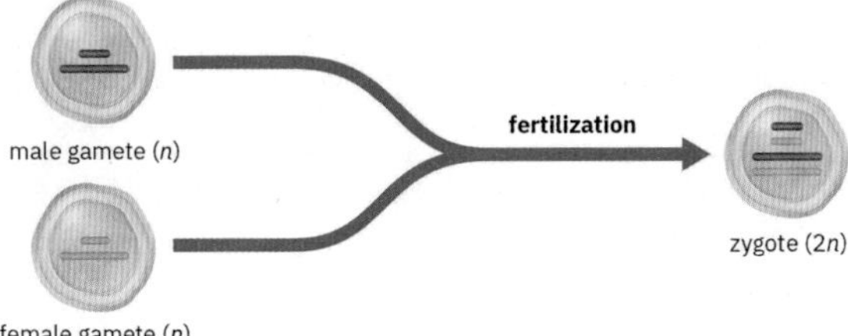

354 CHAPTER 12 GENETIC VARIATION AND HEREDITY

Segregation The new nuclei that form in meiosis I receive a complete set of chromosomes, but whether a new nucleus ends up with the maternal or paternal version of a chromosome is random. The chance that the maternal or paternal version of any chromosome will end up in a particular nucleus is 50 percent because of the way the spindle segregates homologous chromosomes.

In prophase I, after crossing over has occurred, microtubules fasten homologous chromosomes of each pair to opposite spindle poles. There is no pattern to the attachment of the maternal or paternal chromosome to a particular pole. Microtubules extending from a spindle pole attach to the centromere of the first chromosome they contact, regardless of whether it is maternal or paternal (**Figure 12-10**).

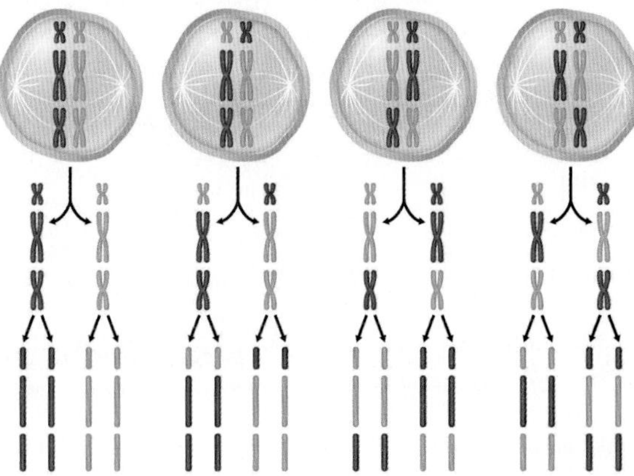

FIGURE 12-10 Segregation of two chromosome pairs, one from each parent, occurs during meiosis. For simplicity, assume no crossing over, so all sister chromatids are identical.

The homologous chromosomes of two pairs can be divided between two spindle poles in four possible ways. There are eight possible combinations of maternal and paternal chromosomes in the two nuclei that form at telophase I. There are eight possible combinations of maternal and paternal chromosomes in the four nuclei that form at telophase II.

CCC Cause and Effect *Explain how chromosome segregation during meiosis is one mechanism for why offspring are not exact replicas of their parents.*

Independent Assortment During meiosis, two alleles at one gene locus tend to be sorted into gametes independently of alleles at other loci. During metaphase I, the homologous chromosomes arrange themselves randomly in the center of the cell. This determines the genes each gamete will receive because only one of the two homologous chromosomes is moved into each newly forming cell.

Address Misconceptions

Mechanisms of Genetic Variation Students may think that each mechanism being discussed works independently from the others, such that genetic variation occurs through only one of these processes. Ensure that students understand that each mechanism occurs during a specific stage of meiosis. Have them create a cause-and-effect chart to complete as they read about each mechanism. For example, the first sentence in the Crossing Over section tells us that crossing over occurs during prophase I. This information tells us the effect (what happens): crossing over occurs during prophase I. As students read, they should read with the intent of understanding the cause (why it happens). For students that struggle to find this information, have them think about what effect this would have on the cell if the step did not occur.

CCC Cause and Effect *Sample answer: During segregation, the distribution of maternal and paternal chromosomes in a cell is random, giving rise to eight possible combinations of chromosomes in the four nuclei that form at telophase II.*

CROSSCUTTING CONCEPTS | Cause and Effect

Segregation Remind students that chromosome segregation is completely random with equal chances of receiving maternal or paternal DNA. Because segregation is random, the offspring are not exact replicas of their parents. Refer students to **Figure 12-10**. Have them identify evidence from the figure that explains how chromosome segregation results in genetic diversity. For example, the text below the figure states, "There are eight possible combinations of maternal and paternal chromosomes..." This statement supports the claim that chromosome segregation results in genetic diversity. Guide students to evaluate the information before and after this statement to find evidence that could possibly explain the "why."

Connect to English Language Arts

Write an Argument As students observed from **Figure 12-8** through **Figure 12-10**, there are multiple mechanisms on the cellular level that contribute to genetic diversity in sexual reproduction. Students also learned that these mechanisms work in partnership with each other, not independently from each other. Lead a discussion with students that highlights the role of crossing over, fertilization, segregation, and independent assortment (**Figure 12-11**) in genetic diversity. As you discuss with the class, place relevant information within a multi-circle Venn diagram on the board. Then, guide students to write an argument using the information from the center of the diagram as evidence for how sexual reproduction tends to give rise to greater genetic diversity among offspring.

SEP Argue From Evidence
Independent assortment of chromosomes produces a mixture of chromosomes from both parents in each gamete. Crossing over increases genetic variation by producing new gene combinations. Fertilization of any egg with any sperm produces even more combinations of chromosomes.

In meiosis I, either chromosome of a pair may get attached to either spindle pole. With two pairs of homologous chromosomes, there are two ways in which the maternal and paternal chromosomes can become attached to the two spindle poles, as shown in **Figure 12-11**. Two nuclei from each scenario show four possible combinations of alleles in the nuclei that form after meiosis I. Thus, when sister chromatids separate during meiosis II, the gametes that result have one of the four possible combinations of alleles.

FIGURE 12-11 Independent assortment during meiosis results in new combinations of genes. Here, we can track alleles on different chromosomes.

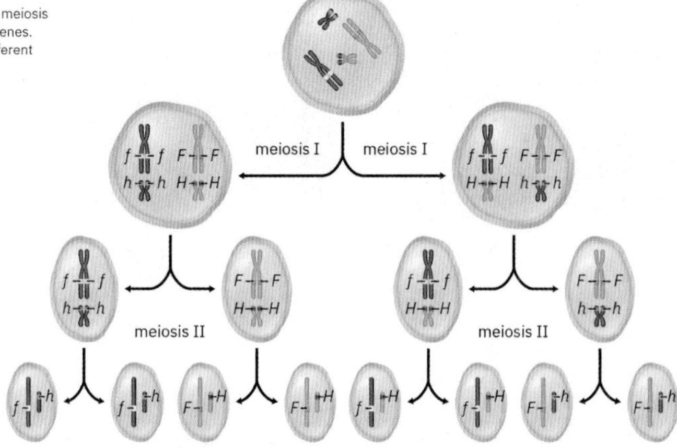

SEP Argue From Evidence *Describe two pieces of evidence to support the claim that sexual reproduction tends to give rise to greater genetic diversity among offspring.*

12.1 REVIEW

1. **Classify** Label each description as relevant to mitosis, meiosis, or both.
   - A cell's DNA is duplicated.
   - Cell division results in four daughter cells.
   - Offspring exhibit more genetic variation.
   - This process begins with diploid cells.
   - The products of the steps are gametes for sexual reproduction.
   - A single cell divides in a series of distinct phases.
2. **Explain** How do crossing over, segregation, and independent assortment provide evidence for increased genetic variation?
3. **Identify** Which statement represents an advantage of sexual reproduction over asexual reproduction?
   A. Reproduction occurs without a partner.
   B. Offspring are genetically identical to the parents.
   C. Genetic diversity increases from one generation to the next.
   D. Offspring have a decreased chance of surviving environmental change.
4. **Analyze** What is the difference between the genetic material on two sister chromatids and the genetic material on homologous chromosomes?

12.1 REVIEW

1. *meiosis: Cell division results in four daughter cells. Offspring exhibit more genetic variation. The products of the steps are gametes for sexual reproduction. both: A cell's DNA is duplicated. This process begins with diploid cells. A single cell divides in a series of distinct phases.* **DOK 2**
2. *Sample answer: Crossing over introduces new combinations of alleles among offspring. During segregation, the distribution of maternal and paternal chromosomes is random. In independent assortment, new combinations of genes are formed. All of these mechanisms result in genetic variation.* **DOK 1**
3. *C* **DOK 3**
4. *Sample answer: The genetic material on two sister chromatids is identical because each sister chromatid is half of a duplicated chromosome. On homologous chromosomes it is similar because they are two separate chromosomes, one from each parent. The genes on homologous chromosomes code for the same structures and functions, but some of the instructions may vary, resulting in variations in traits.* **DOK 2**

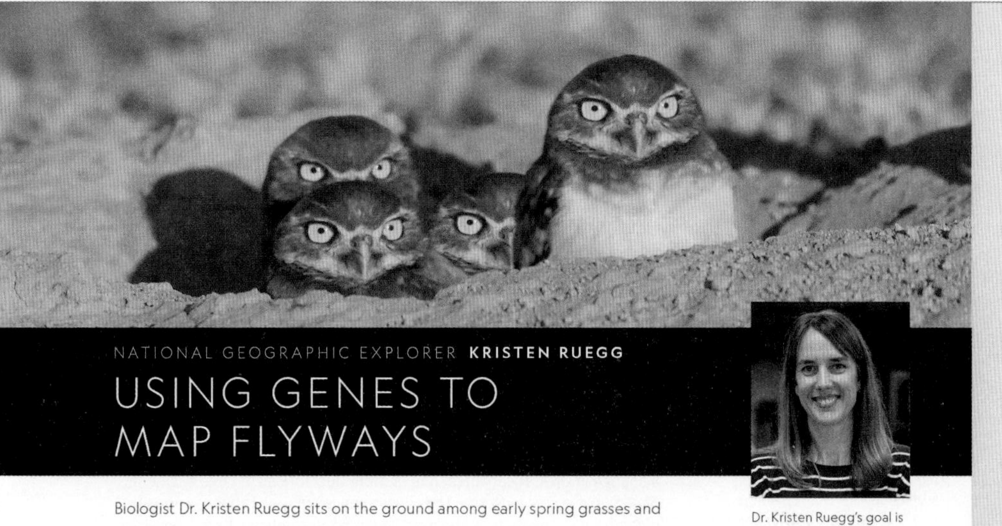

NATIONAL GEOGRAPHIC EXPLORER **KRISTEN RUEGG**

USING GENES TO MAP FLYWAYS

Dr. Kristen Ruegg's goal is to collect DNA from 100 species of migratory birds for the Bird Genoscape Project.

Biologist Dr. Kristen Ruegg sits on the ground among early spring grasses and pulls an American robin from a soft, cloth bag. She holds it in her left hand, belly up, with its head between her index and middle fingers. She then gently but firmly grasps it above the legs with her other hand and turns it upright. The bird looks up at her and chirps with a sharp *"peek-peek-peek."* Ruegg looks it in the eyes and replies warmly, "Yes. I see you."

The American robin is one of the species in the Bird Genoscape Project, an effort Ruegg directs. The project is much like a human ancestry database—a collection of genetic profiles to which an individual's genes can be compared. Ruegg will go on to pluck two of the bird's tail feathers, which she says is easier than pulling a hair from your chin. Then, she opens her hand and the bird flies off. Birds lose and replace feathers regularly, so these will soon grow back.

Ruegg slips the feathers in clean envelopes, records the location and date, and places the envelopes in a cooler. She's extremely careful to not touch the tiny bit of flesh at the quill's tip where the DNA she wants for the Genoscape Project will be extracted. The goal of the project is to map the routes of 100 genetically distinct populations of migratory birds by 2025. The project's urgency is underscored by data that shows about 30 percent of North American birds have vanished over the last 50 years.

Ruegg thinks that mapping the migration routes of different populations based on their distinct genetic makeup holds keys to developing conservation strategies in Earth's changing climate. In actuality, many birds identified as "North American birds" spend only about three months in North America during their breeding season. The rest of the year they winter in warmer semitropical climates. An individual bird also tends to breed and winter in the same locations. A robin that breeds in eastern Canada likely winters in Florida, and one that breeds in Alaska likely winters in northern Mexico. Genetic markers from the DNA of an individual bird's tail feathers can tell where it is from and its migration route. Knowing the migration route helps to determine where environmental challenges exist and predict how bird ranges may change as the climate warms.

THINKING CRITICALLY

Identify Patterns *What are some reasons that populations of the same species might develop distinct genetic profiles?*

DIFFERENTIATED INSTRUCTION | English Language Learners

Formal and Informal Language Explain that formal language is used in situations such as classroom presentations and academic writing, while informal language is used with people you know well. Point out that contractions are often used in informal language and give some examples (such as *she's*, a contraction of *she is*).

Beginning Have students choose a sentence from the article and retell it as if they were talking to a friend. Encourage them to use at least one contraction.

Intermediate Have students summarize the article as if they were telling the information to a friend. They can either speak or write it. Check that students use informal language, including contractions.

Advanced Have students write two summaries of the article: one in formal language and one in informal language. Have them discuss the audience that would be appropriate for each. Check that they have used contractions correctly.

EXPLORER
KRISTEN RUEGG

About the Explorer

National Geographic Explorer Dr. Kristen Ruegg studies the DNA from migratory birds for the Bird Genoscape Project. Her research involves collecting DNA samples from birds to create a collection of genetic profiles that can be used to compare hundreds of migratory birds. She hopes that by mapping the migration routes of different birds, she may be able to develop conservation strategies. You can learn more about Kristen Ruegg and her research on the National Geographic Society's website: www.nationalgeographic.org.

Connect to Careers

Ornithologist Scientists who study birds in their natural habitats are referred to as ornithologists. Ornithologists study birds in their natural habitats as well as laboratory settings, including their physical and behavioral attributes, flight characteristics, and migration patterns. Understanding migration patterns is especially helpful when working in endangered species conservation. Have students discuss how an ornithologist might find information about migration patterns beneficial when working to preserve endangered species.

On the Map

Bird Migration Do an Internet search with keywords "migration flyways U.S." to show students the four predominant flyway routes traveled by migratory birds in North America. Encourage students to identify the flyway route closest to their location. Then, have them research a species that travels this route and discuss the advantages this specific route affords the species.

THINKING CRITICALLY

Identify Patterns *Sample answer: Populations of the same species can develop distinct genetic profiles when they are separated. For example, migration to different environments can result in adaptations that differ between the populations.*

12.2

MUTATIONS

LS3.B

EXPLORE/EXPLAIN

This section provides a review of causes of genetic mutations, introduces the concept of genetic and chromosomal mutations, and explores the relationship between mutations and increased genetic diversity.

Objectives

- Explain how mutations can affect trait expression.
- Describe the characteristics of genetic and chromosomal mutations.
- Relate genetic mutations to greater genetic diversity within a population.

Pretest Use **Question 6** to identify gaps in background knowledge or misconceptions.

Predict *Students might say that a genetic mutation could change the code of the mRNA. This could alter the protein that is produced, which will affect the trait of the individual.*

CCC Cause and Effect *A mutation changes the sequence of nucleotides in DNA. This may or may not cause a change in the codons. If the mutation results in a new codon, this may result in a different amino acid. If the amino acid sequence is changed, it could result in the protein folding differently. A protein that is folded differently could result in a new or altered trait.*

Go Online VIRTUAL INVESTIGATION

Fighting a Viral Pandemic In the situation room and the laboratory, students will gather evidence and make decisions to help protect the people of Capsidia, a fictional nation.

12.2

KEY TERMS

mutagen
mutation

MUTATIONS

Genetic variation occurs in many ways. So far you have learned about recombination of genes during meiosis and by fertilization in sexual reproduction. Genetic variation also occurs via mutations. These happen randomly and cause a range of effects, from unnoticeable changes to structural and functional defects or genetic disease.

Predict *How could a change in the way a protein is made affect the trait of an individual?*

VIRAL SPREAD

Gather Evidence *How might mutations in the SARS-CoV-2 surface spike protein affect the ability of the virus to infect people?*

Causes of Mutations

Spontaneous errors during DNA replication are often repaired by enzymes. When proofreading and repair mechanisms fail, an error results in a permanent change in the DNA sequence of a chromosome. A **mutation** is a change in DNA. It can affect one or more nucleotides or large segments of genes on a chromosome. The consequences of these changes appear in the proteins made from the modified code instructions. Repair enzymes cannot fix a mutation after the altered DNA strand has been replicated. Thus, a mutation is passed on to the next generation.

In addition to spontaneous events, physical and chemical environmental agents, called **mutagens**, can increase the frequency of mutation in organisms. For example, ultraviolet (UV) radiation from sunlight exposure can cause long-term damage to DNA. Infectious agents, such as some viruses and bacteria, and chemical agents found in cigarettes and some preservatives in processed foods, have also been linked to mutations.

CCC Cause and Effect *Explain how mutations can affect protein synthesis, and, in turn, the expression of a trait.*

Gene Mutations

Some mutations occur from changes in the sequence of nucleotides of a single gene, typically during DNA replication. The result is a newly replicated DNA strand that is not exactly complementary to its parent strand. A nucleotide may get deleted during DNA replication, or an extra one may be inserted. Most of these replication errors occur simply because DNA polymerases work very fast. Mistakes are inevitable, and some DNA polymerases make a lot of them.

Frameshift Mutations When one or more nucleotides in a DNA sequence are either inserted or deleted, the reading frame is altered. As a result, the codon base sequences are altered, and the synthesized protein does not match the protein originally intended (**Figure 12-12**).

Luckily, most DNA polymerases proofread their work. They can correct a mismatch by reversing the synthesis reaction to remove the mispaired nucleotide and then resume synthesis in the forward direction. Replication errors also occur after DNA is broken or otherwise damaged because DNA polymerases do not copy damaged DNA very well. In most cases, enzymes and other proteins can repair damaged DNA before replication begins.

Go Online VIRTUAL INVESTIGATION

Fighting a Viral Pandemic
How can we slow the spread of a virus?
Enter the situation room and make decisions to protect your community from a deadly virus that is mutating.

REVISIT THE ANCHORING PHENOMENON

Gather Evidence *The surface proteins of a virus interact with host cells when they infect them. Any mutations that cause differences in the amino acid sequence may cause a virus to be able to more easily or less easily infect human cells. If, for example, the mutation allows the virus to enter cells more easily, the mutated virus might more easily infect people.*

Vaccine development can take between 10 to 15 years to be ready for public administration. In sharp contrast, the COVID-19 vaccine took less than two years to develop, complete clinical trials, and go through regulation approval, which led many people to fear the vaccine had been rushed and not properly vetted. Technologies that allowed scientists from around the world to virtually share and discuss experimental data played a major role in shortening the vaccine development time. As communication technologies advance, it is likely that scientists will continue to be able to respond rapidly to future outbreaks.

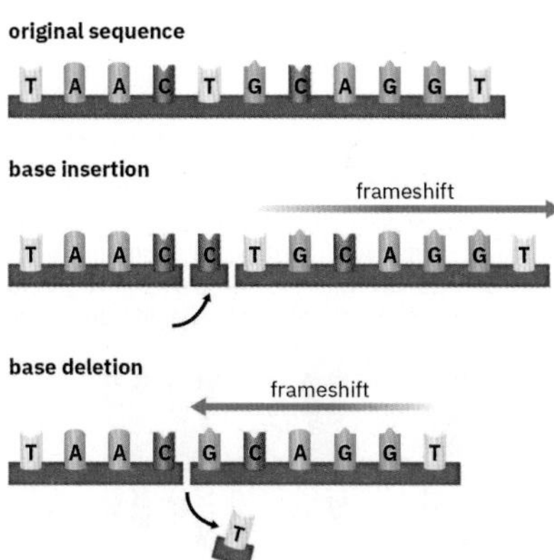

FIGURE 12-12 A frameshift mutation can have drastic consequences because it garbles the genetic message in the same way that grouping a series of letters garbles the meaning of a sentence.

Point Mutations Point mutations are gene mutations caused by the substitution of one nucleotide for another (**Figure 12-13**). These nucleotide, or base, substitutions generally have three possible effects on the amino acid sequence, which may alter the structure and function of the resulting protein.

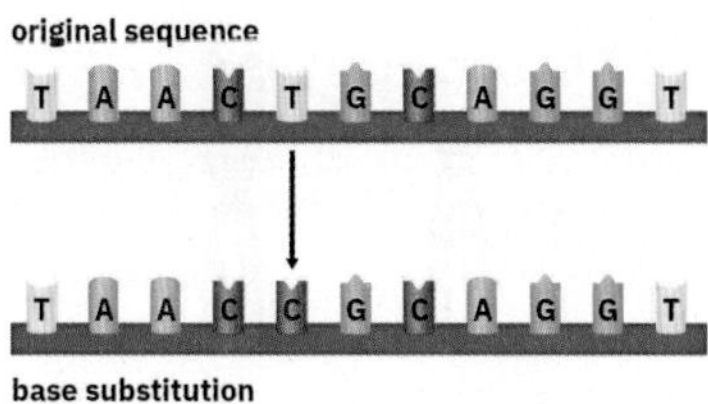

FIGURE 12-13 Point mutations occur when one base is substituted for another in a sequence.

- **Silent Mutation** The base substitution in the altered codon corresponds to the same amino acid. Although a silent mutation does not change the amino acid produced or present a visible change, it can still alter a protein's function.
- **Missense Mutation** The base substitution in the altered codon corresponds to a different amino acid. Sickle cell disease is caused by a missense mutation.
- **Nonsense Mutation** The base substitution in the altered codon corresponds to a stop signal. As a result, the protein is shortened and likely nonfunctional. Duchenne muscular dystrophy, cystic fibrosis, and hemophilia are different diseases caused by nonsense mutations.

DIFFERENTIATED INSTRUCTION | English Language Learners

Using Connecting Words Support students in connecting their ideas with transition words and phrases as they discuss causes and effects of mutations in pairs or small groups.

Beginning Have students state the cause of a nonsense mutation in a simple sentence. Then, have them state an effect (sickle cell disease). Then, have them use the word *because* to link the effect and the cause in one sentence.

Intermediate Have students complete these sentence frames: *A nonsense mutation happens when _______. This can cause _______ such as _______.*

Advanced Have students explain the causes and effects of different kinds of point mutations. Tell them to connect ideas with these words or phrases: *because, as a result, cause, effect, affect.* Point out that *effect* is a noun and *affect* is a verb.

Vocabulary Strategy

Sentence Frames Use sentence frames to support students at different academic language proficiency levels to learn to talk about science. As students work through this section, have them work in pairs to use sentence frames, like the ones that follow, to discuss the topics found in the section.

A change in DNA is a _______. (*mutation*) Physical and chemical agents that can increase the frequency of mutation in organisms are _______. (*mutagens*)

Social-Emotional Learning

Social Awareness As students learn about genetic mutations, invite them to look for contexts and opportunities to practice social awareness. Help them find ways to demonstrate empathy and compassion as the class learns about and discusses genetic and chromosomal mutations. Many well-studied mutations present in the form of genetic diseases and disorders. Students may not be aware that a classmate has firsthand experience with conditions related to genetic or chromosomal mutations. In addition, students should be reminded to show concern for the feelings of others if students decide to share their personal experiences with genetic diseases and disorders.

Visual Support

Genetic Mutations Students may look at **Figure 12-12** and **Figure 12-13** and not understand the difference between the types of genetic mutations being modeled. To help students make sense of the information in these figures, have them analyze each figure to identify where the error is occurring and what type of error is occurring—insertion, deletion, or substitution. With **Figure 12-12**, ensure students understand that the second and third lines of the image represent two possible ways that the error could occur. Some students may incorrectly interpret this diagram to mean that the first coding error that occurs results in the second coding error.

12.2 MUTATIONS 359

Science Background

Gene Therapy Gene therapy is the process of altering or replacing defective genes inside the body with healthy ones to cure disease or improve the body's ability to fight disease. Scientists are currently researching how to replace genes that cause specific medical problems such as cystic fibrosis, adding genes to help the body to fight off diseases such as cancer, or turning off genes that cause problems such as diabetes.

Students have learned that variations to a gene can disrupt how proteins are made, which can result in health problems. Students might be surprised to learn that scientists use viruses, typically thought of as having a negative influence on the body, to deliver these artificial genes. Refer students to **Figure 8-27** to remind them how viruses reproduce by infecting the host cells, resulting in changes to how the infected cell functions. Encourage students to use the model of viral reproduction to explain how scientists can use this same process to alter defective genes.

Chromosomal Mutations

Chromosomal changes occur in either segments of chromosomes or whole chromosomes. These types of mutations result in changes to the amount of genetic material or the structure of the chromosome.

Gene Translocation In gene translocation, a chromosome segment moves to a nonhomologous chromosome. Often translocations are reciprocal in that the two nonhomologous chromosomes both exchange segments (**Figure 12-14**). Translocation is different from crossing over, in which homologous chromosomes swap segments during prophase I.

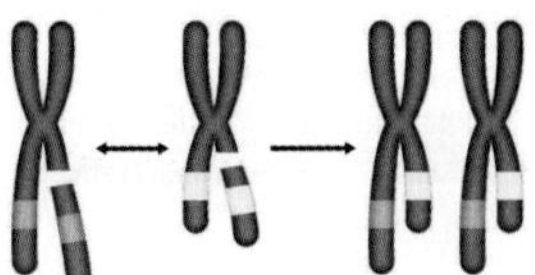

FIGURE 12-14 A segment of one chromosome moves to a nonhomologous chromosome during meiosis, resulting in gene translocation. Often segments are swapped, resulting in reciprocal translocation.

Gene Duplication During meiosis, crossing over results in homologous chromosomes exchanging DNA segments. Sometimes the chromosomes do not align with each other and the exchange of DNA segments is uneven, resulting in one chromosome with two copies of a gene or genes. This process is called gene duplication (**Figure 12-15**). The chromosome that lost the segment has undergone gene deletion.

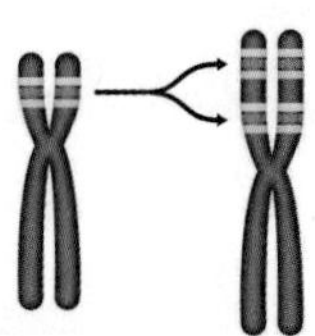

FIGURE 12-15 A segment of a chromosome is repeated when homologous chromosomes do not align properly during meiosis, resulting in gene duplication.

Gene Inversion Sometimes genes on chromosomes switch places. A chromosome segment breaks off and reattaches in reverse order (**Figure 12-16**). This type of structural change, called a gene inversion, may not affect a carrier's health if the break point does not occur in a gene or control region. The individual's cells still contain their full complement of genetic material. However, fertility may be compromised because a chromosome with an inversion does not pair properly with its homologous partner during meiosis. Crossovers between the mispaired chromosomes produce other chromosome abnormalities that reduce the viability of offspring.

FIGURE 12-16 A segment of a chromosome is flipped in the opposite orientation during meiosis, resulting in gene inversion.

360 CHAPTER 12 GENETIC VARIATION AND HEREDITY

DIFFERENTIATED INSTRUCTION | English Language Learners

Using Academic Language If students model, illustrate, and use new academic language, they will better understand and remember it. Review *translocation*, *duplication*, and *inversion*. Explain that the *translocation* means to change location. Show a piece of paper being moved from one location in the room to another. The noun *duplication* means to make a copy. Draw a basic image, such as a square or triangle. Then, *duplicate* the image on a second piece of paper. Finally, explain that the *inversion* means to reverse or flip locations. Show an image being flipped upside down.

Beginning Working in pairs, have students create a model of each term.

Intermediate Working in pairs, have students use the information provided about each new word and write a definition using their own words.

Advanced Working in pairs, have students write a short paragraph using the words *translocation*, *duplication*, and *inversion*.

Genetic Variation

Mutations make new alleles, which results in small differences in DNA sequences that make every individual unique. Once new alleles arise, meiosis and sexual reproduction combine different alleles in new ways to increase genetic variation. Mutations account for the variation we see in human hair color, skin color, height, shape, behavior, and susceptibility to disease. The same is true for individuals in other species.

Fruit flies are a model organism for the study of genetic variation. Some common mutations seen in fruit flies include wing structure and eye color variations. Mutant wings may be short or backward. Fruit flies are known for their red eyes, although mutations cause certain individuals to have golden or white eyes (**Figure 12-17**).

Often mutation has a negative connotation. In contrast to variations that cause disease, there are many more examples of mutation variations that are neither beneficial nor harmful, just different.

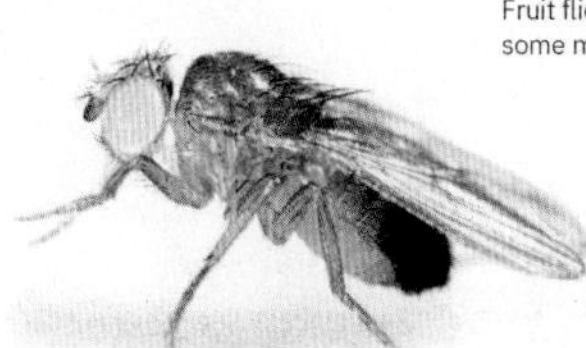

FIGURE 12-17 Fruit flies typically have red eyes, though some mutant individuals have white eyes.

12.2 REVIEW

1. **Argue From Evidence** Use evidence to support the claim that genetic mutations can increase the genetic variation in populations of organisms.
2. **Explain** If a gene mutation changes the structure of a protein, how could it affect that organism's traits?
3. **Identify** Which statement is true of mutations?
   - A. They only result from genetic factors.
   - B. They always change the DNA sequence.
   - C. They only result from environmental factors.
   - D. They are always detrimental to the organism.
4. **Compare** Which type of mutation has the greater impact on the resulting proteins?
   - A. point mutations because they always change the amino acid sequence
   - B. point mutations because they can sometimes change the resulting protein
   - C. frameshift mutations because they substitute one nucleotide for another
   - D. frameshift mutations because they often shift the entire sequence following the mutation

Crosscurricular Connections

Earth Science Students who have previously studied earth science and the fossil record may have prior knowledge of how species evolve because of genetic mutations. Remind students that genetic mutations occur on a molecular level, but the trait expression occurs on the physical level. Lead a class discussion using examples from the fossil record and have students debate what types of mutations within a population of organisms would be reflected in the fossil record. Possible examples include horses, turtles, and humans.

Fossil Evidence In Section 14.2 students will go into detail about fossils and the fossil record as a line of evidence that supports evolution of organisms.

12.2 REVIEW

1. *Mutations are changes in the genetic sequence of an organism's DNA. They can result in differences in observable traits. Mutations that are heritable can be passed from parent to offspring. Heritable mutations introduce genetic variation in future generations and can change a population over time.* **DOK 3**
2. *Sample answer: Gene mutations are changes in the nucleotide sequence of genes that can result in a different protein being synthesized. A change in protein structure could affect the traits the organism expresses.* **DOK 2**
3. *B* **DOK 2**
4. *D* **DOK 2**

ON ASSIGNMENT

Black Panther National Geographic photographer Shannon Wild is an award-winning wildlife photographer who spends entire days stalking elusive animals. Melanistic leopards, a rarely seen color variant of spotted Indian leopards, still have spots although they only can be seen in bright sunshine.

A typical day for Wild begins early, collecting her gear and heading to the park before sunrise in her "office," an open-bed truck with a bench seat and her gear. Video gear is heavy, and the tripod and batteries needed to keep it running all day are too. She keeps her camera in standby mode, as there will be no time for preparation once she sees what she is hoping to. Even the warning calls of the Langur monkey may not give enough time. Wild has been tracking the black panther long enough that she can distinguish the monkeys' warning call for tigers, who are more of a nuisance, from warnings for panthers, which can follow the monkeys into the trees. The day she shot this photo, she saw the panther before the monkeys did. She admits that most days are not as successful as this one, especially given her target, but she films a wide range of animals and always has work to do later, downloading her data cards and charging the batteries before an exhausted night's sleep.

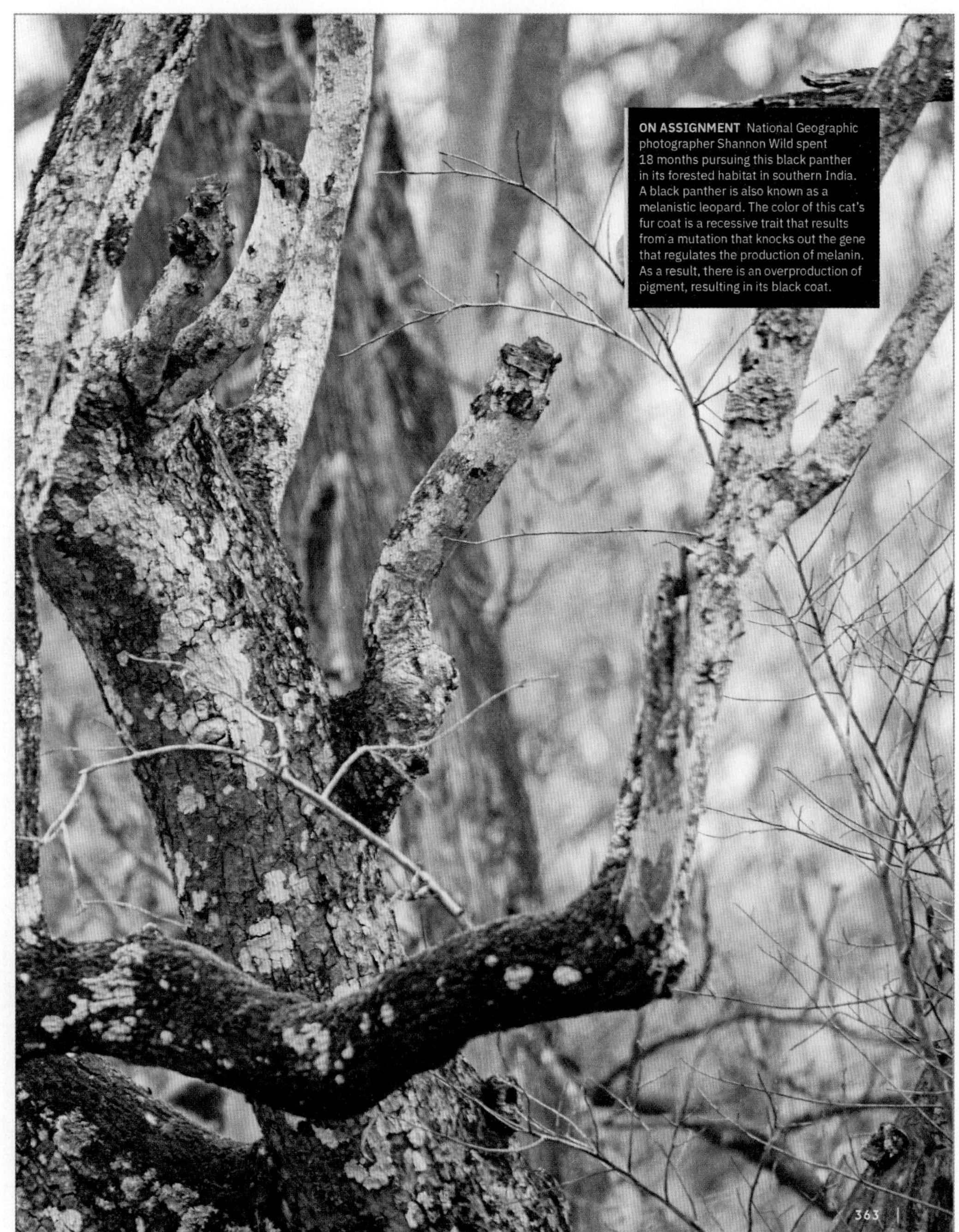

ON ASSIGNMENT National Geographic photographer Shannon Wild spent 18 months pursuing this black panther in its forested habitat in southern India. A black panther is also known as a melanistic leopard. The color of this cat's fur coat is a recessive trait that results from a mutation that knocks out the gene that regulates the production of melanin. As a result, there is an overproduction of pigment, resulting in its black coat.

Science Background

Melanin Production Melanin is the key pigment responsible for the diverse pigmentation found in animal and human skin, hair, and eyes. Melanin pigment is produced by melanocytes and is then transmitted to nearby epidermal cells that produce keratin. Here, it acts as an internal sunscreen to protect animals from ultraviolet (UV) damage.

12.3

MENDELIAN INHERITANCE

LS3.B

EXPLORE/EXPLAIN

This section provides a review of Gregor Mendel's experiments, introduces the concept of genotype and phenotype, and explores the relationship between genetic crosses and possible allele combinations.

Objectives

- Differentiate between genotype and phenotype.
- Evaluate possible genotypes from a genetic cross.
- Outline how a recessive allele is expressed.

Pretest Use **Question 4** to identify gaps in background knowledge or misconceptions.

Vocabulary Strategy

Word Web As students work through this section, suggest they begin a word web with the term *alleles* at the top or center. Then, encourage them to create branches for *genotype, homozygous, heterozygous, phenotype, dominant,* and *recessive*. Encourage students to look for relationships by grouping related terms.

Predict *Sample answer: A list of traits that I can observe in an dog, for example, are hair color, eye color, and type of hair (curly or straight).*

Video

Time: 1:51

Use **Video 12-3** for an overview of genetics and heredity, including Mendel's discoveries. Have students pair up and explain to their partner the relationship between genetics and heredity, or inheritance of traits.

12.3 MENDELIAN INHERITANCE

KEY TERMS

dominant
genotype
heterozygous
homozygous
phenotype
recessive

In 1865, Austrian monk Gregor Mendel was carefully breeding thousands of pea plants. By keeping detailed records of how traits passed from one generation to the next, Mendel was collecting evidence of how inheritance works. At the time, no one knew that hereditary information (DNA) is divided into discrete units, called genes, an insight that is critical to understanding how traits are inherited.

Predict *Make a list of traits you could observe in a plant or animal to show how traits are passed down from one generation to another.*

VIDEO

VIDEO 12-3 Go online to learn more about the study of genetics.

Mendel's Experiments

Mendel cultivated garden pea plants. This species is naturally self-fertilizing, which means each plant's flowers produce male and female gametes that form a zygote that can develop into a viable embryo (Section 9.3). To study inheritance, Mendel had to carry out controlled mating between individuals with specific traits (**Figure 12-18**). Mendel kept individual pea plants from self-fertilizing by removing the pollen-bearing parts (anthers) from their flowers. He then cross-fertilized the plants by brushing the egg-bearing parts (carpels) of their flowers with pollen from other plants. He collected and planted seeds from the cross-fertilized individuals and recorded the traits of the resulting pea plant offspring.

FIGURE 12-18 Mendel cultivated and bred garden pea plants in his experimental procedure.

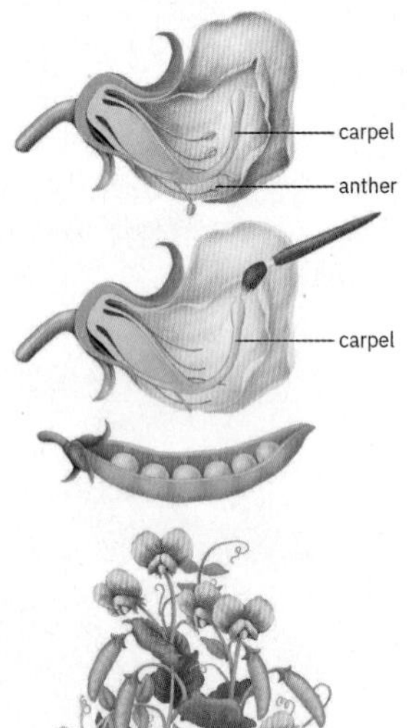

In this example, the white flower's pollen-producing anthers are cut off to prevent it from self-fertilizing. Then, pollen from the anther of the purple flower is brushed onto the egg-producing carpel of the white flower.
carpel
anther
carpel
In this example, every plant that arises from the cross has purple flowers. Predictable patterns such as this are evidence of how inheritance works.

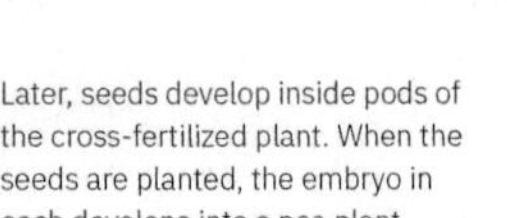

Later, seeds develop inside pods of the cross-fertilized plant. When the seeds are planted, the embryo in each develops into a pea plant.

SCIENCE AND ENGINEERING PRACTICES
Analyzing and Interpreting Data

Mendel's Observations Show students photos of a field of pea plants or individual ones so they get a sense of scale of Mendel's study. Then, refer students to **Table 12-2**. Explain to them that the information in the table summarizes claims made by Gregor Mendel after years of collecting, analyzing, and interpreting data from around 30,000 plants. Have students think about what the data that Mendel used to draw these conclusions would have looked like, based on the description of his experiments. Once students have read about genetic crosses, expand the discussion to have students describe the testcross results that would have supported Mendel's claims in **Table 12-2**. Students should understand that in a cross of two heterozygous plants, about 75% of Mendel's offspring counts would have the phenotype listed in the Dominant form column of the table.

Many of Mendel's experiments, which are called crosses, started with plants of pure breed for particular traits, such as white flowers or purple flowers. Purebred for a trait generally means that all offspring have the same form of the trait as the parent(s), generation after generation. For example, all offspring of pea plants that are purebred for white flowers also have white flowers.

Mendel cross-fertilized purebred pea plants for different forms of a trait and discovered that the traits of the offspring often appear in predictable patterns. Mendel's meticulous work tracking pea plant traits led him to correctly conclude that hereditary information passes from one generation to the next in discrete units.

Mendel's Observations

Mendel discovered hereditary units, which we now call genes, almost a century before the discovery of DNA. Today, we know that individuals of a species share certain traits because their chromosomes carry the same genes.

Diploid cells have pairs of homologous chromosomes, so they have two copies of each gene. The two copies of any gene may be identical, or they may vary as alleles. **Genotype** is the particular set of alleles an individual carries. If two alleles at a specific locus on the gene are the same, they are **homozygous**. If two alleles at a specific locus on the gene are different, they are **heterozygous**. Genotype is the basis of **phenotype**, which refers to the individual's observable traits. Mendel studied seven traits in garden pea plants (**Table 12-2**).

TABLE 12-2. Mendel's Seven Pea Plant Traits

Trait	Dominant form	Recessive form
seed shape	round	wrinkled
seed color	yellow	green
pod texture	smooth	wrinkled
pod color	green	yellow

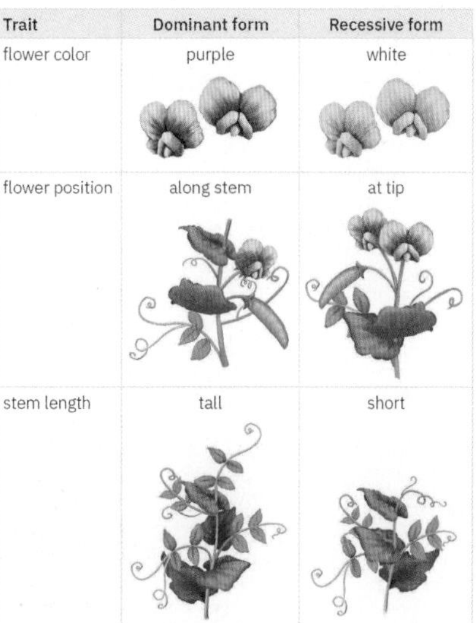

Trait	Dominant form	Recessive form
flower color	purple	white
flower position	along stem	at tip
stem length	tall	short

Visual Support

Mendel's Observations Students may look at **Table 12-2** and not recognize the trait being compared. Prior to students reading the information, review all column headings from **Table 12-2** with the class. Pay particular attention to the trait column. Focus on the characteristics that may be unfamiliar to students, such as flower position or pod texture.

DIFFERENTIATED INSTRUCTION | Leveled Support

Struggling Students Some students may not understand the subtle differences in the traits shown in **Table 12-2**. To help these students, bring in items that can be used to represent each difference represented in the table. For example, with pod shape, bring in one item that is smooth to the touch and one that is wrinkled. Then, have students examine and touch each item to feel what the different pod shapes might have felt or looked like to Gregor Mendel. Guide them in a discussion about how their observations could be used to sort and organize the items. Relate this to how Mendel sorted and organized his pea plants into the seven traits shown in **Table 12-2**.

Advanced Learners Students may be familiar with Mendel and his pea plant experiments. Allow these students to discuss the vocabulary terms *genotype*, *phenotype*, *homozygous*, and *heterozygous*. Then have them apply these terms to the examples displayed in **Table 12-2**.

Go Online **SIMULATION**

Punnett Square Student can review demonstrations of monohybrid and dihybrid crosses along with a blank tool that they can use to generate their own crosses and see probabilities of each outcome.

Connect to English Language Arts

Key Ideas and Details This section introduces the concept of dominant and recessive traits through genetic crosses. Have students write an informative summary of the information presented for each type of genetic cross. Encourage students to use a graphic organizer as they read to ensure they are collecting similar key information for each type of cross. Graphic organizer categories could include cross type, definition, uses, and a model for each cross.

Go Online SIMULATION

Punnett Square Go online to generate your own test crosses.

Modeling Genetic Crosses

The phenotype of a heterozygous individual for a particular trait depends on its two different alleles. In many cases, the effect of one allele influences the effect of the other, and the outcome is reflected in the individual's phenotype. A **dominant** allele is the allele that is expressed when two different alleles or two dominant alleles are present. A **recessive** allele is the allele that is expressed only when two copies are present. A dominant allele is often represented by a capital letter, in this case *F*. A recessive allele is represented by a lowercase letter, such as *f* (**Figure 12-19**).

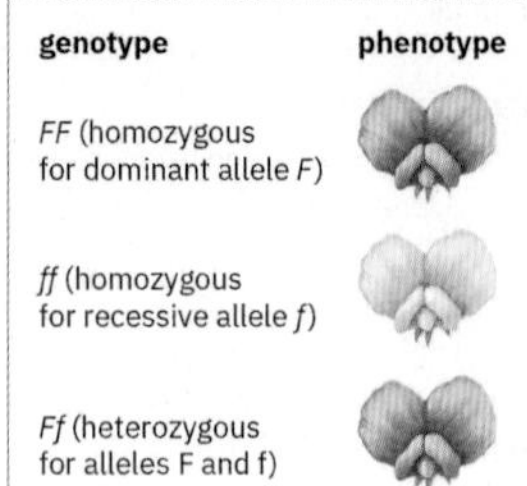

FIGURE 12-19 Genotype gives rise to phenotype. In this example, the dominant allele *F* specifies purple flowers and the recessive allele *f*, specifies white flowers.

A grid called a Punnett square is helpful for predicting the outcomes of crosses. This grid was developed by British geneticist R. C. Punnett in the early 1900s to track the combination of alleles possible from parental gametes to their offspring. In sexually reproducing organisms, fertilization occurs when a male and female gamete fuse. The gametes are haploid, but the resulting zygote is diploid. A Punnett square is a useful model to show this fusion and the probable outcomes for a specific gamete combination.

female gametes

male gametes

	F	*f*
F		
f		*ff*

	F	*f*
F	*FF*	*Ff*
f	*Ff*	*ff*

FIGURE 12-20 To make a Punnett square, parental gametes are placed on the top and left sides of a grid. Each square is filled with the combination of alleles that would result if the gametes in the corresponding row and column met up.

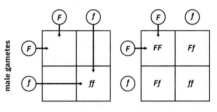

Test Crosses We can determine whether an individual is homozygous (*FF* or *ff*) or heterozygous (*Ff*) for a trait using test crosses (**Figure 12-20**). For example, by crossing a known homozygous recessive (*ff*) individual with an individual having a dominant trait (but unknown genotype), the pattern of traits among the offspring of the cross can reveal whether the tested individual is heterozygous or homozygous. If all offspring of the test cross have the dominant trait, then the parent with the unknown genotype is homozygous (*FF*) for the dominant allele. If some of the offspring have the recessive trait, then the parent is heterozygous (*Ff*). Dominance relationships between alleles determine the phenotypic outcome of a monohybrid cross.

DIFFERENTIATED INSTRUCTION | Leveled Support

Struggling Students Students may struggle to understand phenotypes resulting from the genetic cross. Review with them the parent alleles. For the parents, have students circle the dominant alleles and underline the recessive alleles. Allow students to discuss and determine the phenotype for each parent. Encourage them to label the phenotype next to each parent label. Then, have them apply the same treatment to the offspring allele combinations.

Advanced Learners Students who understand the concept of modeling genotype and phenotype from genetic crosses may be interested in creating additional crosses to show additional generations. Challenge students to transfer this information to a pedigree chart.

366 CHAPTER 12 GENETIC VARIATION AND HEREDITY

Monohybrid Cross To make a monohybrid cross, we would start with individuals that are purebred for distinct variations of a single trait. In pea plants, flower color (*F*, purple; *f*, white) is one example of a trait with two distinct variations. A cross between such individuals (*FF* × *ff*) yields hybrid offspring (*Ff*), all with the same set of alleles governing the trait. A cross between two of these F_1 (first generation) hybrids is the monohybrid cross (*Ff* × *Ff*). The frequency at which the two traits appear in the F_2 (second generation) offspring offers information about a dominance relationship between the two alleles (**Figure 12-21**).

The 3:1 pattern is evidence that purple and white flower color are specified by alleles with a clear dominance relationship: purple is dominant and white is recessive. In this example, there will be roughly three purple-flowered plants for every white-flowered one. This example illustrates a pattern so predictable that it can be used as evidence of a dominance relationship between alleles. The phenotype ratios in the F_2 offspring of Mendel's monohybrid crosses were all close to 3:1.

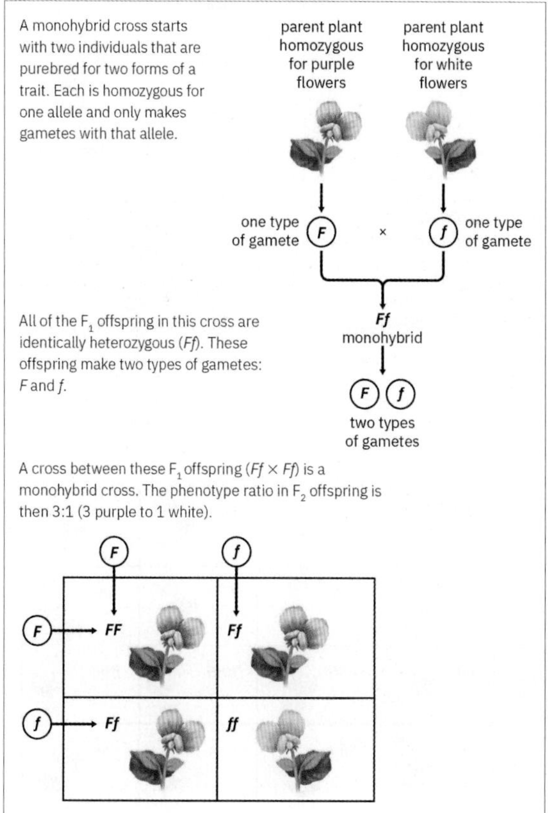

A monohybrid cross starts with two individuals that are purebred for two forms of a trait. Each is homozygous for one allele and only makes gametes with that allele.
parent plant homozygous for purple flowers
parent plant homozygous for white flowers
one type of gamete
F
×
f
one type of gamete
All of the F1 offspring in this cross are identically heterozygous (Ff). These offspring make two types of gametes: F and f.
Ff
monohybrid
F f
two types of gametes
A cross between these F1 offspring (Ff × Ff) is a monohybrid cross. The phenotype ratio in F2 offspring is then 3:1 (3 purple to 1 white).
F
f
F
FF
Ff
f
Ff
ff

FIGURE 12-21 A cross between F_1 offspring (*Ff* × *Ff*) is a monohybrid cross. The phenotype ratio in F_2 offspring is 3:1 (3 purple to 1 white) in this example.

Address Misconceptions

Genetic Crosses Students often think that possible allele combinations from a genetic cross are absolute, meaning that if the genetic cross shows a one in four chance of producing a white flower, then every fourth flower will be white. Refer students to the genetic cross shown in **Figure 12-21**. Explain to students that the one in four chance refers to the *individual* outcome. So, in a cross with alleles *Pp* × *Pp*, there is a one in four chance that the *individual* offspring will have white flowers, not that for every four offspring exactly three will produce purple flowers and one will produce white flowers.

Visual Support

Dihybrid Cross Students may struggle to make sense of the four gametes shown in **Figure 12-22**, as the initial genetic cross that resulted in the possible gametes is not shown. For students that struggle with this connection, walk through the pairing of the alleles for each trait. Remind students there are two traits being paired, color represented by *F* or *f* and stem length represented by *H* or *h*.

	fh	*fh*
FH	*FHfh*	*FHfh*
FH	*FHfh*	*FHfh*

Dihybrid Cross An individual heterozygous for two alleles at two loci is called a dihybrid. A cross between two such individuals is a dihybrid cross. As with a monohybrid cross, the frequency of traits appearing among the offspring of a dihybrid cross depends on the dominance relationships between the alleles.

To make a dihybrid cross, we would start with individuals that are purebred for two different traits. In pea plants, flower color (*F*, purple; *f*, white) and plant height (*H*, tall; *h*, short) are examples of two traits each with two distinct variations. Making a dihybrid cross using these genes begins with one parent purebred for purple flowers and tall stems (*FFHH*) and one parent purebred for white flowers and short stems (*ffhh*) (**Figure 12-22**).

FIGURE 12-22 A dihybrid cross starts with two individuals that are purebred for two different traits.

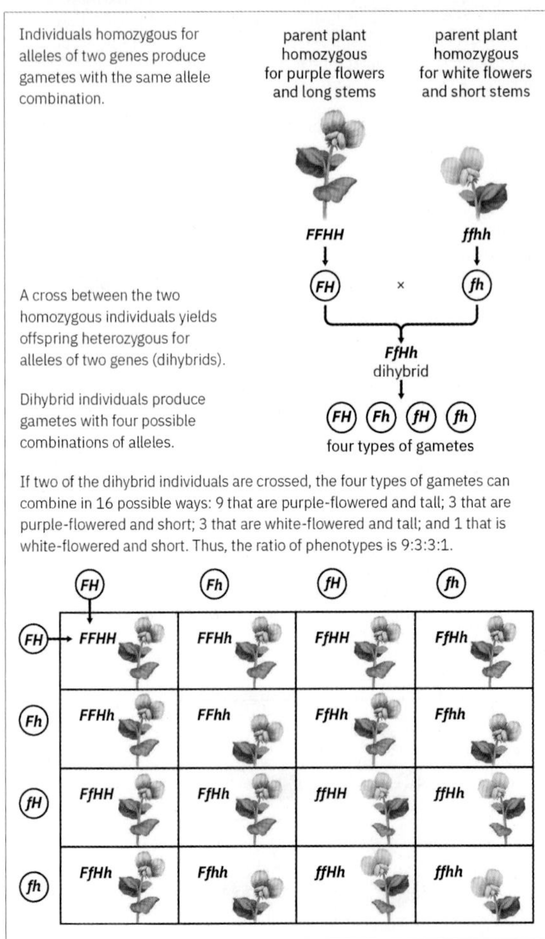

Individuals homozygous for alleles of two genes produce gametes with the same allele combination.
parent plant homozygous for purple flowers and long stems
parent plant homozygous for white flowers and short stems
FFHH
ffhh
FH
×
fh
A cross between the two homozygous individuals yields offspring heterozygous for alleles of two genes (dihybrids).
FfHh
dihybrid
Dihybrid individuals produce gametes with four possible combinations of alleles.
FH
Fh
fH
fh
four types of gametes
If two of the dihybrid individuals are crossed, the four types of gametes can combine in 16 possible ways: 9 that are purple-flowered and tall; 3 that are purple-flowered and short; 3 that are white-flowered and tall; and 1 that is white-flowered and short. Thus, the ratio of phenotypes is 9:3:3:1.
FH
Fh
fH
fh
FH
FFHH
FFHh
FfHH
FfHh
Fh
FFHh
FFhh
FfHh
Ffhh
fH
FfHH
FfHh
ffHH
ffHh
fh
FfHh
Ffhh
ffHh
ffhh

SCIENCE AND ENGINEERING PRACTICES
Analyzing and Interpreting Data

Understanding Crosses As students observed from the genetic crosses shown in **Figure 12-20** through **Figure 12-22**, multiple mechanisms on the cellular level contribute to genetic diversity and trait expression. Have small groups analyze **Figure 12-22**. Then, have them identify the cellular mechanism (allele combination) and the resulting trait expression. Encourage students to utilize the terms *genotype*, *phenotype*, *homozygous*, and *heterozygous* in their group discussions. Guide students to use information from their discussions to identify mechanisms that influence the genome of an organism.

Calculating Genetic Probabilities

The frequency at which traits associated with alleles appear among the offspring depends on whether one of the alleles is dominant over the other. Scientists use probability, a branch of mathematics, to calculate the chances for a particular outcome. In genetics, scientists use probability to determine the genotype of offspring from a specific cross.

$$\text{Probability} = \frac{\text{the number of ways a specific outcome can occur}}{\text{the number of total possible outcomes}}$$

For example, the probability of human offspring having the *XX* or the *XY* sex chromosome is typically 50 percent or ½. The mother can donate only X chromosomes, so the father determines the sex of the offspring with the Y chromosome. The father, being *XY*, could donate either an X chromosome, in which case the offspring would be *XX*, or a Y chromosome, in which case the offspring would be *XY* (**Figure 12-23**).

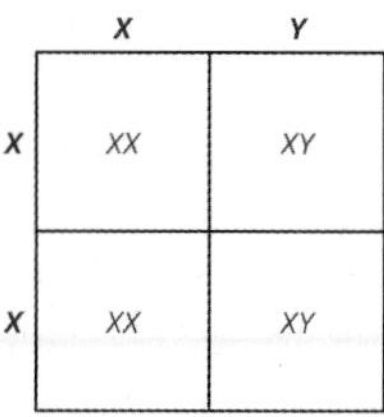

FIGURE 12-23 The probability that a human offspring will be either *XX* or *XY* is typically 50 percent.

12.3 REVIEW

1. **Distinguish** Describe the differences between genotype and phenotype.
2. **Calculate** What is the genotypic ratio of homozygous dominant to heterozygous to homozygous recessive offspring in a cross between parents with the genotypes *Ff* × *Ff*?
3. **Identify** List the genotypes of possible gametes from a heterozygous organism *WwRr*.
4. **Relate** Why is a recessive allele only expressed when the organism is homozygous for that allele?

Connect to Mathematics

Analyzing Data To help students identify the relationship between the information in **Figure 12-23** and the probability equation, have students label the input and output values on both the cross and the probability equation. Have students use the probability equation to calculate genotypic and phenotypic probabilities for the cross in **Figure 12-22.**

12.3 REVIEW

1. *Sample answer: The genotype of an organism is the particular set of alleles an individual carries. Alleles at a specific locus are heterozygous if there is one dominant allele and one recessive allele. Alleles at a specific locus are homozygous if both alleles are either dominant or recessive. Genotype is the basis of phenotype, or an organism's observable traits.* **DOK 2**
2. *1:2:1* **DOK 2**
3. *WR, Wr, wR, wr* **DOK 2**
4. *Sample answer: Recessive alleles are masked by the presence of the dominant allele. So, a heterozygous organism will show the dominant phenotype for that allele, not the recessive phenotype. Only a homozygous recessive organism will show the recessive phenotype for that allele.* **DOK 2**

12.3 MENDELIAN INHERITANCE 369

LOOKING AT THE DATA
BLOOD TYPE COMPATIBILITY

Connect this activity to the science and engineering practices of analyzing and interpreting data and using mathematical thinking. Students analyze patterns in a pedigree chart to identify blood types and possible allele combinations for parents, given the offspring phenotype. Students also apply their pedigree chart analysis to calculate blood type probability within a specific generation.

In Your Community

Blood Donation Encourage all students to find out more about their own blood type, which blood types are compatible with their own, which blood types they might donate to, and the relative rarity of certain blood types.

Individuals as young as 16 can donate blood with a parent or guardian's permission, and at age 17 without as long as they meet certain height and weight requirements, which are available on the Internet at the American Red Cross. Interested students of age can research local facilities at which they can donate blood.

SAMPLE ANSWERS

1. *The possible blood types for this match would be A, B, AB, or O. The pedigree demonstrates two of the four possible blood types.* **DOK 2**
2. *father: AO, mother: AO* **DOK 2**
3. *Sample answer: This offspring has blood type A. Relatives who can safely donate blood to this offspring must be phenotype O or A. This includes individuals 2, 5, 6, 10, and 11.* **DOK 3**

LOOKING AT THE DATA

BLOOD TYPE COMPATIBILITY

SEP **Mathematical Thinking** Blood donors and recipients must have compatible blood types, even if they are close relatives.

Blood type is a heritable trait. The three alleles are I^A, I^B, and i. The antigens (I or i) expressed in the red blood cells contribute to the determination of an individual's blood type. **Table 1** shows how allele combinations result in an offspring's blood type.

TABLE 1. Blood Types and Alleles

Blood type	Allele combinations	
	Homozygous	Heterozygous
A	AA	AO
AB	n/a	AB
B	BB	BO
O	OO	n/a

A donor-recipient chart that shows blood type compatibility is shown in **Figure 1**. Each square links a donor on the horizontal axis and a recipient on the vertical axis. Y means a blood transfer is medically safe while N means it is not. For example, a person with blood type AB can safely receive a blood transfusion from a family member with blood type O, A, B, or AB. However, a recipient with blood type B can only safely receive blood from a donor of blood type O or B.

Figure 2 shows the blood type pedigree chart for a fictitious family. Squares represent males, and circles represent females. The letters inside each shape represent the family member's blood type. In the first generation, a male (1) with blood type AB reproduces with a female (2) with blood type O. They produce three offspring in the second generation: one female of blood type B (4) and two males with blood type A (5, 6).

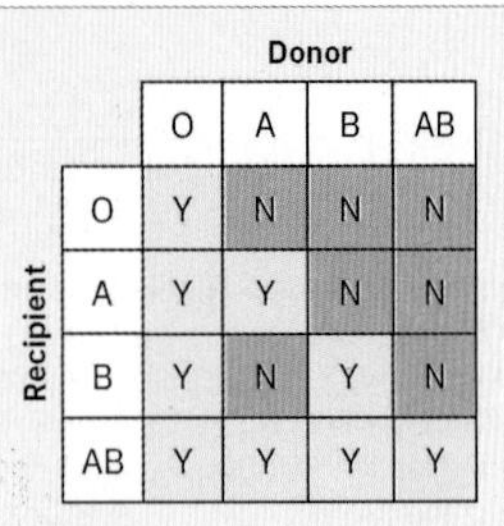

FIGURE 1 A donor-recipient chart shows blood type compatibility between two individuals.

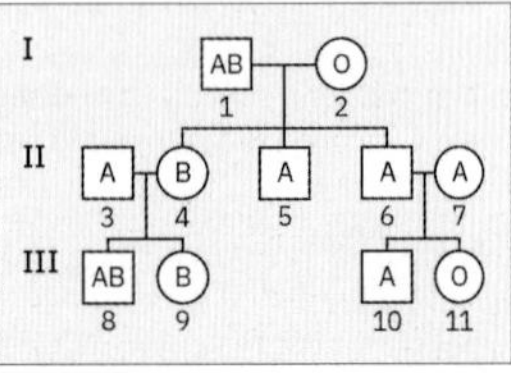

FIGURE 2 A blood type pedigree chart shows the blood types of family members.

1. **Analyze Data** Determine the probability of blood types in generation III for the offspring of the pairing of heterozygous male 3 with heterozygous female 4. Does the pedigree match the possibilities?
2. **Evaluate** Use the phenotypes of male 10 and female 11 in the pedigree to determine the genotype of their father and mother.
3. **Predict** A homozygous male with blood type A (10) and homozygous female (11) produce a female offspring with iron-deficiency anemia. Determine the offspring's blood type and identify all the family members who could donate blood to them.

OTHER PATTERNS OF INHERITANCE

12.4

KEY TERMS

codominance
incomplete dominance

Most human traits do not follow the rules of Mendelian inheritance with predictable outcomes. Long past the early days of genetic studies, we now know there is much more to genetic variation than the straightforward patterns of dominant and recessive relationships.

Predict *How might patterns of inheritance and environmental factors influence a trait, such as height in humans?*

Incomplete Dominance

In an inheritance pattern called **incomplete dominance**, one allele is not fully dominant over the other, so the heterozygous phenotype is an intermediate blend of the two homozygous phenotypes. One example is the gene that affects flower color in snapdragon plants (**Figure 12-24**). One allele of the gene (*R*) encodes an enzyme that makes a red pigment. The enzyme encoded by an allele with a mutation (*r*) cannot make any pigment. Plants homozygous for the *R* allele (*RR*) make a lot of red pigment, so they have red flowers. Plants homozygous for the *r* allele (*rr*) make no pigment, so their flowers are white. Heterozygous plants (*Rr*) make only enough red pigment to color their flowers pink.

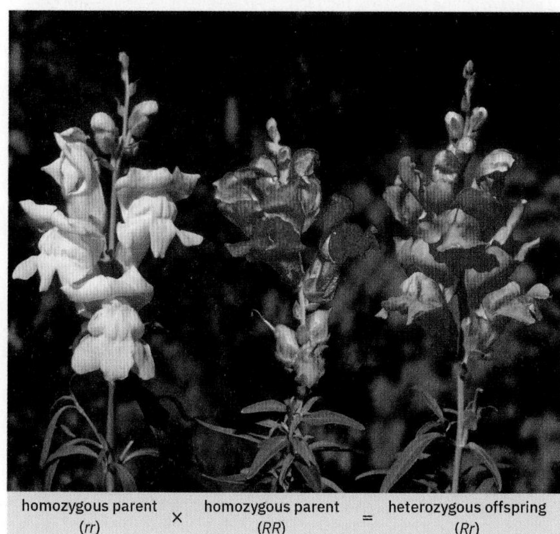

FIGURE 12-24 Snapdragons are a plant that demonstrate incomplete dominance. When a red-flowered snapdragon plant is crossed with a white-flowered one, the resulting offspring have pink flowers.

CROSS CUTTING CONCEPTS | Patterns

Inheritance Patterns Use **Figure 12-24** as a springboard for discussion of how Mendel might have been fortunate to observe pea plants instead of snapdragons. Mendelian inheritance patterns control the expression of many traits; however, students will learn in this section how other patterns of inheritance come into play.

Have groups work together to organize information about Mendelian inheritance, incomplete dominance, codominance, multiple alleles, and polygenic traits into a graphic organizer, such as a Venn diagram or T-chart. Then, have students relate each type of inheritance to Mendelian inheritance. Encourage students to focus on how each inheritance type is both similar and different from Mendelian inheritance.

12.4

OTHER PATTERNS OF INHERITANCE

LS3.B

EXPLORE/EXPLAIN

This section provides a review of dominance, introduces the concepts of incomplete dominance and codominance, and explores the role multiple alleles and sex-linked traits have on trait expression. The role of environmental factors on gene expression is also reviewed.

Objectives

- Compare the difference between Mendelian inheritance, incomplete dominance, and codominance.
- Relate how environmental factors affect phenotype.
- Describe how sex-linked traits are expressed in male and female offspring.

Pretest Use **Question 8** to identify gaps in background knowledge or misconceptions.

Vocabulary Strategy

Using Prior Knowledge Ask students what they already know about the term *dominance*. Use their prior knowledge of the term to complete a class KWL chart. Then, have students use this information to make sense of the upcoming terms *incomplete dominance* and *codominance*. Add these terms to the KWL chart. Create a class definition for the terms and post the chart in the classroom. As students work through this section, refer them back to the KWL chart and the class definitions to help apply the meaning of the terms in the context of their learning.

Predict *Students might say that inherited genes and eating more would influence height.*

Connect to English Language Arts

Present Claims and Findings Incomplete dominance and codominance are more common than students realize. Have students research examples of incomplete and codominance. Possible examples of codominance include flowers with two-colored petals or spotted fur coat for cats or dogs. Examples of incomplete dominance include pink-colored flowers or animals with wavy fur.

Once students have completed their research, allow them to create a presentation with images for each type of inheritance pattern. Encourage students to present their images in small groups while the students take turns identifying the examples as incomplete dominance or codominance. Enourage respectful discussion if group members disagree on an example type. Specifically, group members should ask probing questions about the type of dominance to the student presenting, allowing them time to reflect on the example image. The student can then choose to defend or modify their selection using evidence to support their image choice.

Science Background

Blood Identification Scientist Karl Landsteiner is responsible for creating the identification system of the human blood type groups. In 1901, he performed testing and identified three blood types, A, B, and C, which later evolved into blood type O. His testing involved mixing red cells and serum from each of his staff. What he discovered was the serum of some people clumped the red cells of others. This important observation provided an explanation for why some blood transfusions were successful while others were not. Refer students to **Table 12-3** in Looking at the Data. Have small groups identify which combinations clumped.

CCC Patterns *With incomplete dominance, neither of the alleles is completely dominant over another. The heterozygotes show the blending of traits and show a mix of both phenotypes.*

Codominance

In an inheritance pattern called **codominance**, two versions of an allele, one from each parent, are equally expressed in heterozygous individuals. Neither allele is dominant or recessive. Codominance means that neither allele can mask the expression of the other allele. The gene that affects hair color in shorthorn cattle is an example (**Figure 12-25**).

FIGURE 12-25 Shorthorn cattle are animals that demonstrate codominance. When a white-haired cow is crossed with a red-haired cow, the resulting offspring will have speckled white and red hair.

white coat color homozygous parent (*WW*) ×

red coat color homozygous parent (*RR*) =

roan coat color heterozygous offspring (*WR*)

Multiple Alleles

Human blood type is controlled by multiple alleles (**Table 12-3**), two of which are codominant (*AB*). Alleles *A* and *B* are codominant when paired. The *O* allele is recessive when paired with either the *A* or *B* allele. The three alleles are called I^A, I^B and *i*. Both I^A and I^B result in a protein, called an antigen, on the surface of red blood cells. Allele *i* is recessive and does not result in an antigen.

TABLE 12-3. Combinations of Alleles (Genotype) for Human Blood Type (Phenotype)

Blood type (Phenotype)	Genotype	Can donate blood to:	Can receive blood from:
O	*ii* (*OO*)	A, B, AB, and O (universal donor)	O
AB	I^AI^B	AB	A, B, AB, and O (universal receiver)
A	I^AI^A or I^Ai (I^AO)	AB, A	O, A
B	I^BI^B or I^Bi (I^BO)	AB, B	O, B

If your genotype is *AA* or *AO*, your blood type is A. If your genotype is *BB* or *BO*, it is type B. If your genotype is *OO*, then your blood type is O. The antigens (*I* or *i*) expressed in the red blood cells also contribute to the determination of blood type.

CCC Patterns *Explain in your own words how incomplete dominance and codominance differ from what is called Mendelian inheritance, which involves dominant and recessive traits.*

DIFFERENTIATED INSTRUCTION | Leveled Support

Struggling Students Student may struggle with the concept of polygenic traits. They have previously learned that traits are influenced by one specific gene. To introduce the concept of multiple genes influencing a trait, have students reflect on the height of people who are genetically related. Then, ask them to think about why all members of that group are not the exact same height. Explain that height is influenced by three genes and six total alleles and that the resulting phenotype comes from the cumulative influence of the six codominant alleles. Relate this example to the coat color example shown in **Figure 12-26.**

Advanced Learners Students who understand the concept of polygenic traits may be interested in exploring how the majority of human traits are controlled by multiple genes. Have students brainstorm a list of as many human traits as they can think of. Then have them sort the list into Mendelian or polygenic traits.

Polygenic Traits

In a common inheritance pattern called polygenic inheritance, one trait is affected by multiple genes. Your height is an example of a polygenic trait, in which multiple genes contribute to the overall phenotype observed. The height genes you inherit from your parents accumulate, and the final height you are likely to reach is due in part to the cumulative effect of these genes. Scientists have discovered more than 600 genes that affect height. These complex traits show a continuous range of phenotypes from very short to very tall, making a bell-shaped curve when graphed.

In a phenomenon called *epistasis*, the expression of one gene is affected by the expression of one or more independently inherited genes. One example is genes related to pigments, called melanins, that produce different fur color in Labrador retriever dogs. The brown and black fur comes from a dark brown form of melanin and the yellow fur comes from a reddish form of melanin. Alleles of two of these genes determine whether the individual has black, brown, or yellow fur (**Figure 12-26**). The product of one gene (*TYRP1*) helps make the brown melanin. A dominant allele (*B*) of this gene results in a higher production of brown melanin than the recessive allele (*b*). A different gene (*MC1R*) affects which type of melanin is produced. A dominant allele (*E*) of this gene triggers production of the brown melanin. Its recessive partner (*e*) carries a mutation that results in production of the reddish form. Dogs homozygous for the *e* allele are yellow because they only make the reddish melanin.

CASE STUDY
FOR THE LOVE OF DOGS

Ask Questions *Generate questions about the inheritance or expression of traits in dogs.*

FIGURE 12-26
In a phenomenon called epistasis, interactions among products of two genes affect fur color in Labrador retrievers.

	EB	*Eb*	*eB*	*eb*
EB	*EEBB*	*EEBb*	*EeBB*	*EeBb*
Eb	*EEBb*	*EEbb*	*EeBb*	*Eebb*
eB	*EeBB*	*EeBb*	*eeBB*	*eeBb*
eb	*EeBb*	*Eebb*	*eeBb*	*eebb*

Sex-Linked Traits

Recall that humans have 23 pairs of chromosomes: 22 pairs of autosomes, or nonsex chromosomes, and one pair of sex chromosomes. These sex chromosomes—X and Y—contain different genes, which make a unique pattern of inheritance. Many of the genes seen on the X chromosome do not have corresponding genes on the Y chromosome, simply because the Y chromosome is much smaller.

The genes located on an X or Y chromosome are referred to as sex-linked genes. Males have only one copy of the Y chromosome, so any recessive gene on a Y chromosome will be expressed. Any recessive gene on an X chromosome also will be expressed in males because there is no second X chromosome to mask the recessive allele's expression.

Connect to Mathematics

Define Quantities for Modeling As students observe **Figure 12-26**, remind them that the likelihood of a trait being expressed is calculated by dividing the specific outcome by the total number of possible outcomes. Have students calculate the probability of black, brown, and yellow coats from the data in **Figure 12-26**. Have them order the fur colors from least likely to most likely to occur. Then, have them convert the ordered list from fur color to allele combinations that result in each specific fur color. Have students use their ordered allele combination list to explain why certain phenotypes are more/less likely to occur.

CASE STUDY
GENETICS TO THE RESCUE!

Ask Questions To help students make connections with the content, have them identify a list of desirable traits in dogs that breeders have intentionally bred. For example, dogs that do not shed would be a desirable trait for prospective owners with animal allergies, which has resulted in the popularity of poodle crosses, or "doodles." Students should use this list to develop their questions.

DIFFERENTIATED INSTRUCTION | English Language Learners

Using Reasoning Have students use deductive reasoning and *if...then* statements to help them understand and explain epistasis using the visuals in **Figure 12-26.**

Beginning Have students point to combinations of alleles in the visual and describe the outcome. Provide this sentence frame: *If a dog has ________ and ________, then it will have ________ fur.*

Intermediate Have students explain each color of fur using an *if...then* statement. Provide this frame: *If the dog has ________, then its fur will be ________.*

Advanced Have students explain the differences in fur color using *if...then* statements.

12.4 OTHER PATTERNS OF INHERITANCE 373

CHAPTER INVESTIGATION A

Guided Inquiry *Design an Organism*

Time: 50 minutes

Students use probability to explain the variation and distribution of traits in a population of organisms they design. They design and construct an organism based on a genotype they develop. They identify three physical traits, determine the possible alleles for those traits, and decide what variation of the trait the alleles represent and the patterns of dominance. They then predict possible outcomes of various genetic combinations using a dihybrid cross.

Go online to access detailed teacher notes, answers, rubrics, and lab worksheets.

Visual Support

Genetic Cross Students may struggle to make sense of the information in **Figure 12-27**. Have them draw the cross in their notebooks. Then, instruct them to use two different colored highlighters to color code the cross. Remind students that *homozygous* homozygous means the alleles are the same. Have them select and highlight, in one color, the parent alleles located on the outside of the cross that are homozygous. Students should then add the label "dominant for red-green color blindness." Next, have students select and highlight, in two different colors, the parent alleles that are *heterozygous*. Students should then add the label "normal vision." Have students continue the color coding from the parent alleles through to the offspring. Students should then discuss which color(s) could result in offspring with red-green color blindness.

SEP Analyze and Interpret Data
The couple will not have color-blind children.

CHAPTER INVESTIGATION

Design an Organism
How do alleles determine the phenotype of an organism?
Go online to explore this chapter's hands-on investigation about variation in alleles and designing your own organism.

Females have double the number of genes located on an X chromosome, but they do not need double the number of their associated proteins. A process known as X inactivation solves this dilemma. Only one X chromosome is active, while the other is silenced or has very few active genes. X inactivation results in more balanced gene expression between males and females.

Red-green color blindness is an example of a trait caused by a sex-linked gene that occurs more often in males. The dominant allele that produces normal vision is represented by the C superscript (X^C). The recessive allele that is responsible for red-green color blindness is represented by the c superscript (X^c), as shown in **Figure 12-27**.

FIGURE 12-27 A cross between a female homozygous dominant for red-green color blindness and a male with normal vision.

	X^C	Y
X^c	X^cX^C	X^cY
X^c	X^cX^C	X^cY

SEP Analyze and Interpret Data *Using the Punnett square in **Figure 12-27**, determine the probabilities that a couple will have a son or daughter with color blindness.*

Environmental Factors

The environment affects the expression of many genes, which in turn affects phenotype, including physical and behavioral traits. Today, we understand that genetics (nature) and environmental factors (nurture) both play a substantial role in the behavioral traits of humans and other organisms.

In plants, a flexible phenotype gives immobile individuals an ability to thrive in diverse habitats. For example, genetically identical yarrow plants can grow to different heights at different altitudes. More challenging temperature, soil, and water conditions are typically encountered at higher altitudes. Differences in altitude are also correlated with changes in the reproductive mode of yarrow. Plants at higher altitude tend to reproduce asexually, and those at lower altitude tend to reproduce sexually.

Environmental cues trigger adjustments in gene expression that change an individual's form and function to suit its current environment. For example, a water flea that swims to the bottom of a pond can survive the low oxygen conditions by turning on expression of genes involved in the production of hemoglobin. Hemoglobin is a protein that carries oxygen, and it enhances the individual's ability to absorb oxygen from the water. Other environmental factors also affect water flea phenotype. For example, the presence of insect predators causes water fleas to form a protective pointy helmet and lengthened tail spine. Individual water fleas also switch between asexual and sexual modes of reproduction depending on the season.

SCIENCE AND ENGINEERING PRACTICES
Analyzing and Interpreting Data

Sex Linked Traits Refer students to **Figure 12-27**. Explain to them that color blindness is expressed through the X^c allele, which is recessive. A recessive genotype can only occur in females who inherit an X^c allele from both parents. Males only have one X allele, so any male that inherits the X^c allele would express color blindness. Since two different situations are presented, have students examine the female only offspring in **Figure 12-27** first for the expression of color blindness. Then, have them examine the figure again for male expression of color blindness.

Environmental cues trigger adjustments in gene expression that change an individual's form. In many mammals, seasonal changes in temperature and length of day affect the production of melanin and other pigments that provide skin and fur with their color. These species have different color phases in different seasons (**Figure 12-28**). Hormonal signals triggered by seasonal changes cause fur to be shed, and new fur grows back with different types and amounts of pigments deposited in it. The resulting change in phenotype provides the animals with seasonally appropriate camouflage from predators.

FIGURE 12-28
Environmental cues, such as temperature changes, can trigger changes in melanin and other pigments that can alter fur color between white (a) and brown (b) in hares.

12.4 REVIEW

1. **Calculate** If two pink-flowered snapdragons are crossed, what is the phenotypic ratio of the offspring? Support your ratio with the percentages for each phenotype.
2. **Distinguish** Why is a sex-linked trait on an X chromosome expressed differently in male and female offspring?
3. **Identify** Sort the examples of genetic and environmental factors that affect phenotype.
   - carcinogens
   - parental DNA
   - genotype
   - temperature
4. **Explain** Two species of hare occupy an area that experiences four seasons. Hare A has white fur in the winter and brown fur in the spring. Hare B has brown fur all year round. Which of these types of hare has a more beneficial phenotype? Explain your answer.

Environmental Factors Students learn that environmental factors, such as the habitat a species lives in, can influence the expression of many genes. **Section 4.3** discusses environmental factors that limit growth within a population.

Science Background

Epigenetics Epigenetics focuses on discovering how behaviors, along with environmental factors, affect gene expression through reversible, chemical modifications to the DNA (without altering the nucleotide sequence itself). This impacts how a DNA sequence is read by molecular machinery in the cell. Lifestyle factors, such as diet, physical activity, and work habits have been linked to changing epigenetic patterns. Allow small groups to discuss how epigenetics is both similar to and different from a genetic mutation.

12.4 REVIEW

1. *1:2:1; 25%, 50%, 25%* **DOK 2**
2. *Sample answer: The recessive trait on a sex-linked allele will always be expressed in a male because they only have one X chromosome. However a female has two X chromosomes, so a recessive allele may not be expressed.* **DOK 2**
3. *genetic: parental DNA, genotype; environmental: carcinogens, temperature* **DOK 2**
4. *Sample answer: The hare with the beneficial phenotype is the hare that changes color because its white color during the winter and its brown color during the spring would help it blend into its environment and therefore better hide from predators.* **DOK 2**

MINILAB
MODELING INHERITANCE

In this activity, students will model a cross for two heterozygous parents. The students will flip coins to explain the variation of expressed traits. They will hold a tally of their results and apply concepts of statistics and probability to determine genotype and phenotype of offspring.

Time: 30 minutes

Materials and Advance Preparation

- If you think students will have trouble flipping the coins and catching them, have them flip the coin inside a box, such as a shoebox.

Procedure

In **Step 1**, make sure students understand the purpose of the coin flip: to calculate the probability of two independent events occurring at the same time—that is, the probability that the offspring will get the *R* allele or the *r* allele from each parent.

In **Steps 2, 3,** and **4**, students will complete the data table.

In **Step 5**, students will do the coin flip a total of 25 times.

MINILAB

MODELING INHERITANCE

SEP Analyzing and Interpreting Data How can we model the probability of inheriting a trait?

Most organisms are diploid, meaning they carry two sets of chromosomes. In diploid organisms, an individual has two alleles for most genes. Each set of alleles includes an allele from the mother and an allele from the father. In this activity, you will use probability calculations to explain the variation and distribution of traits in a cross between two heterozygous individuals.

Fruit flies are model organisms for genetic study because they reproduce quickly and are easy to handle, among other reasons.

Materials
- coins, 2
- black marker

Procedure

The inherited gene you will model is red or white eye color in fruit flies. Red eye color (*R*) is dominant, and white eye color is recessive (*r*). For a cross between heterozygous parents, you will calculate the probability that the offspring will get either the red eye color allele or the white eye color allele from each parent.

1. Each coin represents the alleles from each parent. In this activity, the parents are heterozygous, so heads equals red eyes and tails equals white eyes.
2. Enter the possible genotypes for a heterozygous cross *Rr* × *Rr* in the data table.

Genotype	Phenotype	Offspring

3. Enter the corresponding phenotypes for the genotypes in the data table.
4. Flip both coins together. Enter a tick mark in the offspring column in the data table based on the outcome of the coin flip.
5. Repeat Step 4 an additional 24 times. In the end, you will have data for 25 offspring.

Results and Analysis

To calculate the probability of an event, use the formula

$$\text{Probability} = \frac{\text{number of ways a specific event can occur}}{\text{number of total possible outcomes}}$$

To calculate the probability of two independent events occurring together, multiply the probability of the individual events. So, the probability of flipping heads is 1/2. Therefore, the probability of flipping two heads together is 1/2 × 1/2 = 1/4.

1. **Calculate** What is the probability of each genotype?
2. **Calculate** What is the probability of each phenotype?
3. **Analyze** How do your calculations in Question 1 compare with the data you collected?
4. **Evaluate** Describe and evaluate the model used in this lab.

Results and Analysis

1. **Calculate** *Sample answer for genotypes:*

   RR frequency $= \frac{5}{25} = 20\%$

   Rr frequency $= \frac{14}{25} = 56\%$

   rr frequency $= \frac{6}{25} = 24\%$
   DOK 2

2. **Calculate** *Sample answer for phenotypes:*

   red eye frequency $= \frac{5+14}{25} = 76\%$

   white eye frequency $= \frac{6}{25} = 24\%$
   DOK 2

3. **Analyze** *Sample answer: We expected the phenotypes to be 75 percent red eyes and 25 percent white eyes. Our data almost exactly reflects that. For genotypes, we expected 25 percent of the population to have RR and 50 percent of the population to have Rr. The RR genotype ended up being slightly less, and the Rr genotype ended up being slightly more. We think this is because 25 flips of the coin is not a very high number. If we were to flip the coin 100 or 200 times, the percentages would be closer to what we expect.* **DOK 3**

4. **Evaluate** *Sample answer: We flipped coins to simulate which allele each parent donates to the offspring. The model is useful to understand the basic probability of allele inheritance. The model is also limited because it does not account for environmental variables, for example.* **DOK 3**

TYING IT ALL TOGETHER
FOR THE LOVE OF DOGS

HOW DOES GENETIC INFORMATION CHANGE AND GET PASSED FROM GENERATION TO GENERATION?

In this chapter, you learned how traits are passed from parents to offspring through the transfer of genes. You also learned how mutations can affect gene expression. The Case Study described how scientists learned that animal domestication through careful breeding can result in offspring with desired characteristics, such as certain physical characteristics or the ability to herd, retrieve, or hunt.

Today, several consumer brands offer genetic testing for dog owners interested in learning more about the ancestry of their pets. These genetic tests can not only help to pinpoint the breeds that make up a mixed-breed dog (**Figure 12-29**), but they can also provide insight into potential health issues. These companies only analyze a fraction of the billions of base pairs in a dog's genome, focusing on certain short sequences of base pairs that are known to be associated with a particular dog breed. However, there can be a significant amount of genetic variation even within a single purebred dog breed. For example, a Corgi bred for herding has different genes from one bred to be a household pet. The accuracy of breed identification continues to advance as more data are collected, analyzed, and added to a company's repository of genetic information.

In this activity, you will develop models and use evidence to explain how the genetic information passed from parents to their offspring determines which traits they exhibit. Work individually or in a group to complete these steps.

Gather Evidence

1. Choose a dog breed, and research the traits it was bred to have. Which of these traits are Mendelian?

Develop and Use a Model

2. Choose one of the Mendelian traits, such as coat color, and develop a model to predict the genotypes of the offspring from parents that are heterozygous for that trait.

Argue From Evidence

3. Agouti is a type of fur pattern that shows more than one color on each strand of fur. Research the four alleles that control this fur pattern. What can you conclude about the way this trait is inherited?

FIGURE 12-29
Leroy is a mixed-breed dog that was adopted from a small dog rescue in New Jersey. Leroy's adoption papers listed him as a Cairn terrier mix. Genetic test results showed otherwise.

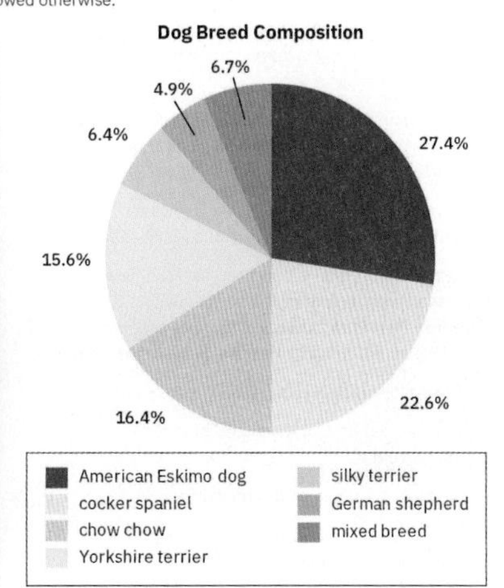

DIFFERENTIATED INSTRUCTION | English Language Learners

Citing Sources As students research the traits of a dog breed, have them use language for citing research sources. Have students individually research traits and share their facts with their group. Have groups take notes. Then, have groups decide which traits are Mendelian and develop their model.

Beginning Provide this sentence frame: *(Source) says that (information).* Have students list traits in words or phrases.

Intermediate Provide these sentence frames: *According to _______, the traits of _______ are _______. I read in ______ that _______.*

Advanced Have students cite their sources as they discuss traits of their dog breed. Encourage them to also cite sources in the text (for example, by page or section of the text) as they decide which traits are Mendelian and develop their model.

TYING IT ALL TOGETHER

ELABORATE

The Case Study discusses how animals can be bred for the gene expression of specific traits. In the Tying It All Together activity, students research how breeders use genetics to produce dog breeds with specific traits.

This activity is structured so that students may complete the steps as a culminating project at the end of the chapter. **Step 1** has students gather evidence on a specific dog breed and the traits it was bred for. In **Step 2**, students develop a model for one of the Mendelian traits to predict the genotypes of the offspring from parents. In **Step 3**, students communicate information from their models by making a digital presentation. The analysis practiced here reflects processes that students may use to support a claim in the Unit 4 activity.

Go online to access the Student Self-Reflection and Teacher Scoring rubrics for this activity.

Connect to English Language Arts

Develop and Strengthen Writing Have students use their model to write an informative summary of the possible genetic combinations for the trait they selected. Encourage students to cite evidence from their model as they identify and explain each potential allele combination resulting from heterozygous parents. Written explanations should include parent allele combinations, possible offspring allele combinations, the probability of each offspring genotype, and the resulting phenotype.

CHAPTER 12 **REVIEW**

EVALUATE

REVIEW KEY CONCEPTS

1. *A, B, D* **DOK 1**
2. *Sexual: Genetic variation is introduced; Meiosis plays a role in the reproductive process; Offspring from two parents inherit genes from both. Asexual: Mitosis is the reproductive process; All parental genes are passed to offspring; Genetically identical offspring are produced.* **DOK 2**
3. *D* **DOK 2**
4. *point mutation: A replication error results in the insertion of an adenosine in place of a cytosine in a DNA strand; gene duplication: A misalignment of homologous chromosomes results in DNA segments of unequal size because one chromosome has two copies of a gene; gene translocation: A faulty crossing over event results in a gene for one chromosome being swapped for a gene from another chromosome.* **DOK 2**
5. *B* **DOK 3**
6. *0.50* **DOK 2**
7. *FfHh and FFHH; ffHh and ffHH* **DOK 2**
8. *C* **DOK 1**
9. *C, D* **DOK 2**

CRITICAL THINKING

10. *Sample answer: Both increase genetic diversity among offspring of sexually reproducing organisms, whereas crossing over involves physical exchange of DNA between chromosomes during meiosis and segregation involves separation of maternal and paternal chromosomes during meiosis so that each gamete receives only one or the other inherited chromosome, but not both.* **DOK 2**
11. *Sample answer: Boveri and Sutton would not have been able to identify chromosomes as the material that carries genetic information, slowing the progress of understanding the mechanism of heredity. Their work made it possible for scientists to ask different questions about the chemical nature and makeup of chromosomes.* **DOK 3**

CHAPTER 12 **REVIEW**

REVIEW KEY CONCEPTS

1. **Describe** Which statements are true about alleles in a eukaryotic organism? Select all correct answers.
   - **A.** They can be dominant or recessive.
   - **B.** They are alternative forms of a gene.
   - **C.** They are both inherited from the female parent.
   - **D.** They contribute to the phenotype of an organism.
   - **E.** They are located on the same chromosome.
2. **Contrast** Match each statement with either sexual reproduction or asexual reproduction.
   - Genetic variation is introduced.
   - Mitosis is the reproductive process.
   - All parental genes are passed to offspring.
   - Genetically identical offspring are produced.
   - Meiosis plays a role in the reproductive process.
   - Offspring from two parents inherit genes from both.
3. **Summarize** Which of the following is the outcome of meiosis for a parent cell?
   - **A.** two diploid gametes
   - **B.** two haploid gametes
   - **C.** four diploid gametes
   - **D.** four haploid gametes
4. **Classify** Match each statement to the type of mutation it describes: gene duplication, point mutation, gene translocation.
   - A faulty crossing over event results in a gene for one chromosome being swapped for a gene from another chromosome.
   - A misalignment of homologous chromosomes results in DNA segments of unequal size because one chromosome has two copies of a gene.
   - A replication error results in the insertion of an adenine in place of a cytosine in a DNA strand.
5. **Analyze** A plant that is purebred for blue flowers is crossed with a plant of the same species that is purebred for yellow flowers. Offspring produced from the cross all have yellow flowers. Explain this outcome.
   - **A.** The yellow and blue traits are codominant.
   - **B.** Yellow is a dominant trait, and blue is a recessive trait.
   - **C.** Blue is a dominant trait, and yellow is a recessive trait.
   - **D.** The yellow and blue traits demonstrate incomplete dominance.
6. **Predict** Sickle cell anemia is a genetic disease caused by an autosomal recessive mutation in the gene for hemoglobin. This means an individual must have two copies of the sickle cell mutation to be afflicted with the disease. Individuals who inherit only one copy of the sickle cell mutation are carriers of the disease but are not afflicted with any of the symptoms. If a woman who has sickle cell anemia has a child with a man who is a carrier, what is the probability that the child will have sickle cell anemia?
7. **Relate** In pea plants, the purple flower trait (*F*) is dominant, and the white flower trait (*f*) is recessive. Also, the tall stem trait (*H*) is dominant, and the short stem trait (*h*) is recessive. Which pairs of pea plants have the same phenotype? Select all correct answers.
   - **A.** *FFhh* and *ffHH*
   - **B.** *FfHh* and *FFHH*
   - **C.** *ffhh* and *Ffhh*
   - **D.** *FFHH* and *FFhh*
   - **E.** *ffHh* and *ffHH*
8. **Identify** Keratin is a protein found in the hair of mammals. Dogs that are homozygous for a particular allele for keratin have very curly hair. Dogs that are homozygous for a different allele for keratin have straight hair. Dogs that are heterozygous for these two alleles have wavy hair. What type of inheritance pattern does this example represent?
   - **A.** codominance
   - **B.** epistasis
   - **C.** incomplete dominance
   - **D.** polygenic inheritance
9. **Describe** Which statements about sex-linked traits are true? Select all correct answers.
   - **A.** Only females express sex-linked trait.
   - **B.** Only males express sex-linked traits.
   - **C.** Sex-linked traits are determined by genes found on X and Y chromosomes.
   - **D.** An individual can be a carrier of a sex-linked trait without expressing the trait.
   - **E.** Sex-linked traits follow the rules of Mendelian inheritance in a predictable pattern.

378 CHAPTER 12 GENETIC VARIATION AND HEREDITY

12. *Sample answer: Both types involve changes to the DNA sequence. In either mutation type, a beneficial, harmful, or no phenotypic change may occur. Gene mutations occur from changes in the sequence of nucleotides of a single gene, typically during DNA replication, resulting in a newly replicated DNA strand that is not exactly complementary to its parent strand. Chromosomal mutations occur in either segments of chromosomes or whole chromosomes, resulting in changes to the amount of genetic material or the structure of the chromosome.* **DOK 2**
13. *Sample answer: Alleles can be dominant or recessive for a trait. Each offspring receives one allele from each parent. Thus, offspring can be homozygous or heterozygous for that trait. A dominant genotype, whether heterozygous or homozygous, will have the same phenotype. An example is flower color in pea plants. Purple flower color (P) is dominant; white flower color (p) is recessive. Plants with the homozygous PP genotype and the heterozygous Pp genotype will both have purple flowers.* **DOK 2**

CRITICAL THINKING

10. **Compare** Describe one similarity and one difference between crossing over and segregation.

11. **NOS Scientific Investigation** Gregor Mendel conducted heredity experiments on pea plants. Thirty-five years later, Theodor Boveri and Walter Sutton used microscopes to study the behavior of chromosomes during gamete formation and fertilization. They concluded that chromosomes were likely responsible for carrying genetic information from parents to offspring. How do you think scientific understanding of heredity would have been affected if microscope technology had not been developed?

12. **Compare** How are gene mutations similar to and different from chromosomal mutations?

13. **Explain** Why is it possible for two organisms to have the same phenotype for a particular trait but different genotypes? Provide an example as part of your response.

14. **Identify Cause and Effect** Butterfly eggs from the same population were placed in different environments. The larvae hatched, grew, and underwent metamorphosis. The adult butterflies that emerged had different markings on their wings. How can this difference be explained?

MATH AND ENGLISH LANGUAGE ARTS CONNECTIONS

1. **Reason Abstractly and Quantitatively** How many different combinations are there with two pairs of homologous chromosomes? Explain your reasoning.

Read the following passage. Use evidence from the passage to answer the question.

The azoospermia factor (AZF) regions of the human Y chromosome contain genes that code for proteins critical to sperm development. Gene deletions sometimes occur in the AZF regions during meiosis. This leads to the production of sperm carrying these mutations. While these sperm are capable of producing healthy offspring, the mutations are carried to the next generation.

2. **Cite Textual Evidence** Explain how a man could be infertile from this type of mutation. Cite information from the text to support your answer.

3. **Write Explanatory Text** Mendel began his experiments with purebred parent pea plants. Purebred means that a purple-flowering pea plant only produces offspring that have purple flowers when allowed to self-fertilize. Why was this critical in allowing Mendel to conclude that traits were passed from one generation to the next in discrete units?

Read the following passage. Use evidence from the passage to justify your answer to the question.

A human female inherits one X chromosome from her father and the other from her mother. When the embryo reaches about 1000 cells, a process called X inactivation occurs. Each cell renders one of the two X chromosomes inactive. There is no preference for which X chromosome is inactivated; each is inactivated roughly 50 percent of the time. From that point on, all of the descendant cells continue to inactivate the same X chromosome.

4. **Write an Argument** How can you explain why a female who inherits the gene for color blindness from her father does not express the color-blind trait?

5. **Reason Abstractly and Quantitatively** A man has blood type A and is heterozygous for the I^A and i alleles. He fathers a child with a woman with blood type *B* who is heterozygous for the I^B and i alleles. What are the possible blood types that the offspring could have? Calculate the probability of each blood type. Probability is calculated by dividing the number of ways a specific outcome can occur by the total number of possible outcomes.

REVISIT VIRAL SPREAD

Gather Evidence In this chapter, you learned about genetic variation that is introduced through processes related to meiosis and mutations. You also learned in Section 12.2 that a DNA mutation can alter the structure and function of the resulting protein. Similar to those mutations, errors in replicating the RNA genome of SARS-CoV-2 can introduce significant changes in the proteins that make up the virus particle.

1. The spike protein of SARS-CoV-2 enables the virus to infect human cells and is the target of most COVID-19 vaccines. The D614G mutation changes a single amino acid in the spike protein, making the spike proteins more stable. How do you think a mutation such as D614G can impact science, technology, and society?

2. What other questions do you have about how mutations can affect pandemics such as COVID-19?

REVISIT THE ANCHORING PHENOMENON

1. *Sample answer: Understanding the structure of a virus and how it functions helps scientists gain insight into such things as vaccine technologies. Understanding how mutations such as D614G work can help researchers modify vaccines. Having vaccines that work can be used to fight some human infections, such as COVID-19.* **DOK 3**

2. *Sample answer: Do mutations always make a disease more severe? How do mutations alter the effectiveness of vaccines? Do important mutations happen on other proteins beyond spike proteins?* **DOK 3**

14. *Sample answer: Environmental conditions can affect gene expression, which can influence phenotype. Environmental cues trigger adjustments in gene expression that change an individual's form and function to suit its current environment.* **DOK 2**

MATH AND ELA CONNECTIONS

1. *Sample answer: A diploid cell with two homologous pairs of chromosomes has two equally possible arrangements for the parental chromosomes during metaphase I. These two arrangements lead to gametes with four equally possible combinations of chromosomes.* **DOK 3**

2. *Sample answer: During gamete formation, a gene deletion occurred in the AZF region of one cell. If this mutated sperm fertilized the egg, the male offspring produced (the son) would inherit the deletion, causing him to be infertile.* **DOK 3**

3. *Sample answer: It was important for Mendel to know that the traits he was testing for were homozygous in the parental pea plants so that he could observe any changes in the offspring resulting from his testcrosses. Mendel controlled the pollination and the testcrosses to observe the results of specific crosses. This made it possible for him to draw conclusions about how these traits were inherited as discrete factors that passed from the parental generation to the offspring.* **DOK 3**

4. *Sample answer: The gene for color blindness is located on the X chromosome. The female inherited one X chromosome from each parent. Due to X inactivation, the maternal X is silenced in half of a female's cells. The paternal X chromosome is silenced in the other half of her cells. So, the gene for color blindness would be expressed only in cells in which the paternal X chromosome is active. The other half of the cells will not have the gene for color blindness, so they will produce the normal protein product of the gene. The female will not be color blind but may have reduced color vision.* **DOK 3**

5. *Type A 25%; Type B 25%; Type AB 25%, Type O 25%* **DOK 3**

GENETIC TECHNOLOGIES

Three-Dimensional Learning

The practices, core ideas, and crosscutting concepts presented in this chapter's text, investigations, and resources provide support to address the following Performance Expectations: **HS-LS2-7** and **HS-ETS1-3**.

Science and Engineering Practices	Disciplinary Core Ideas	Crosscutting Concepts
Asking Questions and Defining Problems Constructing Explanations and Designing Solutions (HS-LS2-7, HS-ETS1-3)	**LS1.A:** Structure and Function **LS2.C:** Ecosystem Dynamics, Functioning, and Resilience (HS-LS2-7) **ETS1.A:** Defining and Delimiting Engineering Problems **ETS1.B:** Developing Possible Solutions (HS-LS2-7, HS-ETS1-3)	Patterns Structure and Function Science Addresses Questions About the Natural and Material World Influence of Science, Engineering, and Technology on Society and the Natural World (HS-ETS1-3)

Contents		Instructional Support for All Learners	Digital Resources
ENGAGE			
380–381	CHAPTER OVERVIEW **CASE STUDY** How can genetic information be used in medical and other scientific applications?	**Social-Emotional Learning** CCC Influence of Science, Engineering, and Technology on Society and the Natural World **On the Map** Great Plains **English Language Learners** Monitoring and Self-Correcting	
EXPLORE/EXPLAIN			
382–391 DCI LS1.A	**13.1** TOOLS IN GENETIC TECHNOLOGY • Describe some of the tools scientists use to work with DNA. • Relate genomics and the tools used for genome analysis.	**Vocabulary Strategy** Knowledge Rating Scale **English Language Learners** Using Visuals **Visual Support** Restriction Enzymes SEP Developing and Using Models **Connect to Mathematics** Reason Quantitatively CER Revisit the Anchoring Phenomena **Visual Support** Replicating DNA **Differentiated Instruction** Students with Disabilities **Crosscurricular Connections** Physical Science **Differentiated Instruction** Leveled Support **Connect to English Language Arts** Evaluate a Science or Technical Text **English Language Learners** Analyzing a Text CCC Influence of Science, Engineering, and Technology on Society and the Natural World **Address Misconceptions** Identical Twins **On Assignment** Click Beetle Larvae **Address Misconceptions** Fluorescence **Crosscurricular Connections** Chemistry CCC Structure and Function	**Investigation A** DNA Evidence (100 minutes) **SIMULATION** PCR and Gel Electrophoresis

Contents		Instructional Support for All Learners	Digital Resources
EXPLORE/EXPLAIN			
392–402 DCI ETS1.A ETS1.B	**13.2 GENETIC ENGINEERING** • Identify how genetic engineering is used to modify different kinds of organisms. • Describe how somatic cells are used to develop new organisms. • Explain how gene therapy might be used to correct a genetic disorder. • Evaluate if CRISPR technology is a a more effective process for gene therapy.	**English Language Learners** Sounds and Language Patterns **Vocabulary Strategy** Understanding in Context **Connect to Mathematics** Reason Quantitatively CCC Influence of Science, Engineering, and Technology on Society and the Natural World **In Your Community** Endocrine Disruptors CONNECTIONS In Your Community **Address Misconceptions** Cloning SEP Constructing Explanations and Designing Solutions **Connect to English Language Arts** Write an Argument **Differentiated Instruction** Economically Disadvantaged Students **In Your Community** Gene Therapy **English Language Learners** Read Silently and Take Notes **Address Misconceptions** Targeted Gene Treatments **Differentiated Instruction** Leveled Support **Visual Support** CRISPR Base Editing SEP Engaging in Argument from Evidence **CASE STUDY** Ask Questions	**Investigation B** Fluorescent Genes (100 minutes) *Video 13-1* *Video 13-2*
403	**EXPLORER** PARDIS SABETI	**Connect to Careers** Medical Parasitologist	

Contents	Instructional Support for All Learners	Digital Resources
EXPLORE/EXPLAIN		
404–405 **LOOKING AT THE DATA** GENETIC THERAPY CLINICAL TRIALS		
406–415 DCI ETS1.A ETS1.B **13.3 VACCINE DEVELOPMENT** • Summarize how the immune system protects individuals from infection. • Describe how vaccines are used to protect the population from infectious diseases. • Compare the differences between the types of vaccine technologies that have been developed and implemented. • Evaluate the other measures that can be used to support vaccines.	**Vocabulary Strategy** Semantic Map SEP Developing and Using Models **Connect to Mathematics** Reason Abstractly **English Language Learners** Narrating Past Events **In Your Community** Public Health Departments **Connect to Mathematics** Evaluate Reports Based on Data SEP Constructing Explanations and Designing Solutions **English Language Learners** Prior Knowledge and Experience **Connect to English Language Arts** Research CCC Influence of Science, Engineering, and Technology on Society and the Natural World **Differentiated Instruction** Leveled Support **Connect to Mathematics** Use Units to Understand Problems **Address Misconceptions** COVID-19 Vaccine CER Revisit the Anchoring Phenomena **Visual Support** Vaccines **In Your Community** Pathogen Reservoirs CCC Cause and Effect	 *Video 13-3*

Contents		Instructional Support for All Learners	Digital Resources
EXPLORE/EXPLAIN			
416	**MINILAB** HERD IMMUNITY		
ELABORATE			
417	**TYING IT ALL TOGETHER** How can genetic information be used in medical and other scientific applications?	**SEP** Constructing Explanations and Designing Solutions	Self-Reflection and Scoring Rubrics
Online	**Investigation A** DNA Evidence **Investigation B** Fluorescent Genes	Guided Inquiry (100 minutes) Engineering Investigation (100 minutes)	MINDTAP Access all your online assessment and laboratory support on MindTap, including: sample answers, lab guides, rubrics, and worksheets. PDFs are available from the Course Companion Site.
EVALUATE			
418–419	**Chapter 13 Review**		
Online	**Chapter 13 Assessment** **Performance Task 3** *What are the risks and benefits of genetically engineered food?* (HS-LS3-4) **Performance Task 4** *How will we curb the spread of mosquito-borne disease?* (HS-ETS1-1, HS-ETS1-2)		

CHAPTER 13

ENGAGE

In middle school, students learned about how humans have used technology to influence the traits of plants and animals through selective breeding. Students also learned about the science and engineering practices used to define, develop, and optimize design solutions. This chapter expands on that prior knowledge by studying specific methods scientists use to work with DNA.

Students will dive deeper into how genetic engineering has been used to modify plants and animals and will learn about how vaccines function and the process required for their development.

About the Photo This photo shows Elizabeth Ann, a cloned black-footed ferret. Because so few black-footed ferrets existed at the start of the breeding program, the population was severely inbred, meaning it had limited genetic variation. Low genetic variation often leads to an increased probability of inherited diseases, as well as limited immunity to infectious disease. Scientists hope that Elizabeth Ann will introduce genetic variation into the captive breeding population, which can then effect change in the wild. To prepare students for this chapter's content, ensure that they understand that emerging technologies allow scientists to find new solutions to problems in society and the natural world.

Social-Emotional Learning

As students progress through the chapter, invite them to look for contexts and opportunities to practice some of the five social and emotional competencies. For example, as students work through the content in the chapter, support them in their **responsible decision-making**. Help them understand how scientists find solutions to help solve problems.

In addition, as students learn about vaccine development (Section 13.3 Minilab), remind them to practice **self-awareness** and **relationship skills**. Support students in communicating effectively and examining prejudices and biases.

CHAPTER 13

GENETIC TECHNOLOGIES

Elizabeth Ann is a clone of an endangered species, the black-footed ferret, which is native to North America.

13.1 TOOLS IN GENETIC TECHNOLOGY
13.2 GENETIC ENGINEERING
13.3 VACCINE DEVELOPMENT

Over the past few decades, with increasing success, scientists have been able to study the effects of changing a gene by manipulating DNA. The application of genetic editing technology to problems involving conservation and human health is a major part of the field of genetic engineering.

Chapter 13 supports the NGSS Performance Expectation **HS-ETS1-1** and **HS-ETS1-3**.

380 CHAPTER 13 GENETIC TECHNOLOGIES

CROSSCUTTING CONCEPTS | Influence of Science, Engineering, and Technology on Society and the Natural World

Scientific Advances In this chapter, students explore how scientific advances in the fields of genetic engineering and vaccine research are being used by scientists to solve problems in society and the natural world. Students will learn about tools scientists use to work with DNA and how their discoveries are being applied. Students will also explore the development and distribution of vaccines. Some fields of biology, especially biomedical engineering, are focused on discovering new ways to alter genes for various uses. Chapters 6 and 7 in Unit 2 introduce the concept of DNA in the cell and how it is replicated during cellular division. Chapter 11 in this unit examines the discoveries that allowed scientists to determine how DNA codes for traits. This crosscutting concept is further reinforced in Chapters 15 and 16 of Unit 5 where students learn how scientific discoveries from the past have allowed scientists to learn more about phenomena in the natural world.

CASE STUDY
GENETICS TO THE RESCUE!

HOW CAN GENETIC INFORMATION BE USED IN MEDICAL AND OTHER SCIENTIFIC APPLICATIONS?

An organism that is genetically identical to its parent, typically as a result of asexual reproduction, is a clone. Cloning is a widespread agricultural strategy for ensuring that crops bred for specific desirable traits continue to have those traits. For example, bananas in grocery stores are often clones of the Cavendish variety.

Animal cloning is being tested as a solution to global problems of species conservation. When most of the individuals in an isolated population are directly related, the genetic diversity needed to adapt to changing conditions is limited. One way to introduce more genetic diversity is to transplant individuals from elsewhere into the endangered population to breed. However, in cases where existing populations are so small that no such animals are available, conservation scientists explore cloning genetically distinct individuals from the cells of previous members of the species as an alternative.

A black-footed ferret was the first endangered animal to be cloned in the United States. The North American native species was thought to have become extinct in the 1970s due to the eradication of prairie dogs, their main prey, by hunters and ranchers. However, a small colony of 18 ferrets found in Wyoming in 1981 were used to start a captive breeding program. Only seven of them bred. Thus, all black-footed ferrets in the wild population of several hundred are closely related. This ferret population is additionally threatened by sylvatic plague, an often fatal flea-borne bacterial disease.

In 2020, scientists attempted to again rescue the species. They made a clone using the frozen cells of a female black-footed ferret named Willa that died in 1988. A surrogate domestic ferret, implanted with an embryo made from eggs containing genetic material extracted from these cells, gave birth to Elizabeth Ann, Willa's clone. The cloned ferret's DNA, while identical to its parent's DNA, differs from that of all other living black-footed ferrets. Breeding Elizabeth Ann and reintroducing her descendants into the wild is anticipated to widen the gene pool and improve the species' odds of survival.

Ask Questions *In this chapter, you will learn about fundamental tools and techniques for manipulating genetic material. As you read, generate questions that would help determine when genetic technology should be used to solve medical or scientific problems.*

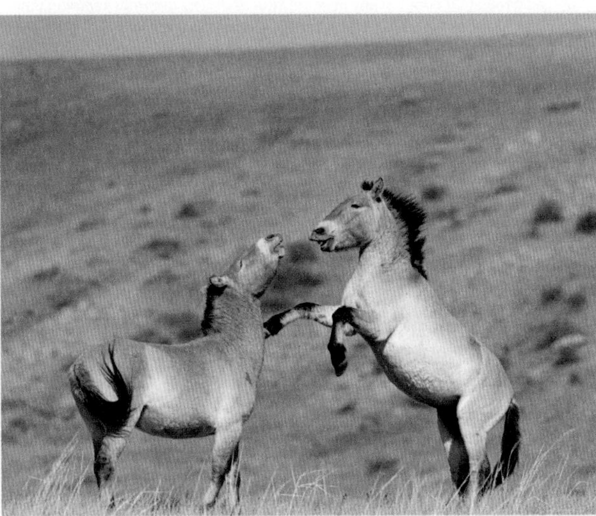

FIGURE 13-1 Przewalski's horse is a critically endangered species of wild horse native to Mongolia. It was successfully cloned in 2020.

CASE STUDY **381**

DIFFERENTIATED INSTRUCTION | English Language Learners

Monitoring and Self-Correcting Have students read the case study aloud in pairs or small groups. Have them monitor their language production and pause to self-correct if they make a mistake. Then, have them reread the phrase or sentence and continue.

Beginning Have small groups each read a paragraph, one sentence at a time. Encourage students to scan the sentence before they start and look for punctuation to help them break the sentence into its parts.

Intermediate Have students take turns reading a few sentences at a time. If they make a mistake, encourage them to read to the end of the sentence and then go back and reread it correctly before they continue on.

Advanced Have students read a paragraph at a time. Tell students to try to read at a natural pace, not too fast or too slow.

CASE STUDY

ENGAGE

The situation described in the Case Study can be used to assess students' prior knowledge of the importance of genetic diversity in a population. Students should have learned about asexual reproduction but may be unaware that humans have developed technologies that allow them to artificially clone organisms. Have students explain how introducing Elizabeth Ann to the population will increase genetic variation within the population. Students should understand that she will introduce new genes, increasing genetic diversity within the species.

In the Tying it All Together activity, students will develop a solution that uses genetic engineering to conserve a hypothetical animal population.

Ask Questions Students should revisit the Case Study as they read the chapter to make connections with the content. See a specific suggestion in Section 13.2.

On the Map

Great Plains Use a map of the United States to point out the Great Plains region of the United States. Share with students that the historical range of the black-footed ferret spanned from Canada to Texas and Arizona. It also included parts of Utah and Virginia. Have students discuss the different problems facing the black-footed ferret and possible solutions to those problems. Students may mention that there is little genetic variation in the population and that the population is also threatened by disease.

Human Connection While farmers do not have issues with the black-footed ferret, they do have problems with their primary source of food: prairie dogs. Prairie dogs are considered an agricultural pest by many because their burrows strip the grass and other plant life around an area. They can quickly destroy a crop or grass field, so farmers and ranchers often advocate eradicating prairie dog populations. However, prairie dogs are a keystone species and their eradication leads to several different changes in the ecosystem.

13.1

TOOLS IN GENETIC TECHNOLOGY

ETS1.B

EXPLORE/EXPLAIN

This section provides an overview of the tools available that allow genetic researchers to work with DNA and introduces genomics and the methods used for genome analysis.

Objectives

- Describe some of the tools scientists use to work with DNA.
- Relate genomics to the tools used for genome analysis.

Pretest Use **Question 3** to identify gaps in background knowledge or misconceptions.

Analyze *Sample Answer: Both types of tools are probably designed to work for very specific tasks. Genetic tools may be very small and we might not be able to see them at work, compared with a spatula or wrench that you can hold in your hand.*

Vocabulary Strategy

Knowledge Rating Scale Have students make a table of the key terms with four columns labeled 1, 2, 3, and 4. Before students begin reading, they should rate their knowledge of each term:

1 = I have never heard the word.
2 = I have heard the word but do not know its meaning.
3 = I am familiar with the word.
4 = I know the word, its meaning, and could explain it to others.

Upon completing the reading, students should revisit the scale and rate the terms again. Have students explain each word to a partner.

13.1

KEY TERMS

bioinformatics
DNA cloning
DNA profiling
gel electrophoresis
genomics
microarray
polymerase chain reaction (PCR)
recombinant DNA
restriction enzyme

TOOLS IN GENETIC TECHNOLOGY

In a pastry chef's kitchen or an automobile mechanic's shop, there are a wide variety of very specialized tools, each designed to perform a particular task. A pastry chef uses one type of spatula to scrape batter in a bowl and a different type to apply frosting to a cake. An automobile mechanic needs the right type of wrench to remove an oil filter. Similarly, scientists have spent decades developing and honing a set of tools for use in genetic technology.

Analyze *What are some likely similarities and differences between kitchen or workshop tools and the tools used in a genetics lab?*

Working With DNA

The basic tools that unlock most genetic technologies start with the ability to manipulate DNA. Scientists have repurposed enzymes borrowed from nature as tools to cut, copy, paste, edit, sequence, and thus better understand DNA.

Restriction Enzymes In the 1950s, excitement over the discovery of DNA's structure gave way to challenges as scientists attempted to determine the order of nucleotides in a molecule of DNA. Identifying a single base among thousands or millions of others turned out to be a major technical challenge. Research in a seemingly unrelated field yielded a solution when scientists discovered how some bacteria resist infection by bacteriophage viruses. The defense mechanism of the bacteria involves enzymes that chop up any injected viral DNA. The enzymes restrict viral growth. Therefore, they were named **restriction enzymes**. A restriction enzyme cuts DNA wherever a specific nucleotide sequence occurs (**Figure 13-2**). For example, the enzyme *Eco*RI (named after *Escherichia coli*, the bacterium from which it was isolated) cuts DNA at the nucleotide sequence G-A-A-T-T-C. Other restriction enzymes cut at different sequences. The discovery of restriction enzymes made it possible to cut chromosomal DNA into manageable chunks.

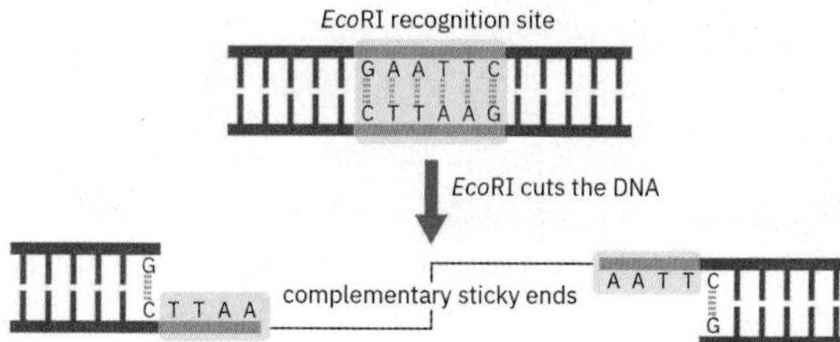

FIGURE 13-2 The restriction enzyme *Eco*RI recognizes a specific base sequence in DNA and cuts it. *Eco*RI leaves single-stranded tails that are known as "sticky ends."

DIFFERENTIATED INSTRUCTION | English Language Learners

Using Visuals As students read the text in this section, have pairs use the visuals to help them retell what they have learned. Remind students to read DNA sequences from left to right. Explain that visuals should also generally be viewed from left to right and top to bottom.

Beginning Encourage students to follow the letters and arrows with a finger as they read any sequences or labels.

Intermediate Have pairs use single sentences or short phrases to describe what is happening in each visual. Remind them to follow the order of the arrows.

Advanced Have pairs use the graphics to summarize the text in the corresponding section. Encourage students to point to relevant parts of the graphic.

Recombinant DNA Many restriction enzymes, including *Eco*RI, leave single-stranded tails on DNA fragments. The tails are called "sticky ends" because two DNA fragments stick together when their matching tails base-pair. Scientists realized that because the chemical structure of DNA is the same in all organisms, complementary tails will base-pair regardless of the source of DNA. Another enzyme, *DNA ligase*, can be used to seal the gaps between hybridized DNA fragments, forming continuous DNA strands (**Figure 13-3**). Using appropriate restriction enzymes and DNA ligase, DNA can be cut and pasted to combine fragments from different sources. The result, a hybrid molecule that consists of genetic material from two or more organisms, is called **recombinant DNA**. This ability to use enzymes to manipulate DNA fragments was a major advance in genetic technology.

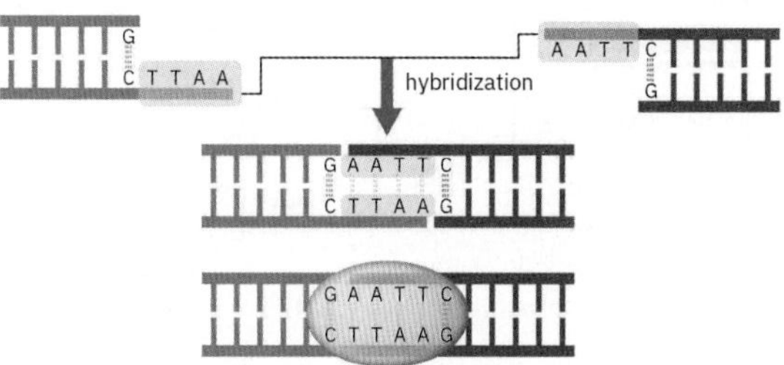

FIGURE 13-3 When DNA fragments from two sources are cut with *Eco*RI and mixed together, their matching sticky ends base-pair. DNA ligase seals the gaps between hybridized DNA fragments. The resulting hybrid molecules are called recombinant DNA.

DNA Cloning Making recombinant DNA is the first step in DNA cloning. **DNA cloning** is a set of laboratory methods that uses living cells to mass-produce specific DNA fragments. A fragment of DNA is cloned by inserting it into a *vector*, which in this context is a DNA molecule used to shuttle foreign DNA fragments into host cells. As shown in **Figure 13-4**, plasmids can be used as cloning vectors by introducing them into a bacterial host. When the bacterial cell divides, the recombinant plasmid is replicated and distributed to descendant cells in the same way as the bacterium's own genetic material. Recombinant vectors are introduced into cells that can be easily grown in the laboratory to yield huge populations of genetically identical descendants. Each of these clones contains a copy of the recombinant vector. The vector can be extracted in quantity, and the cloned DNA fragment can then be manipulated using restriction enzymes and other tools.

FIGURE 13-4 DNA fragments are inserted into plasmid vectors with the use of restriction enzymes and DNA ligase.

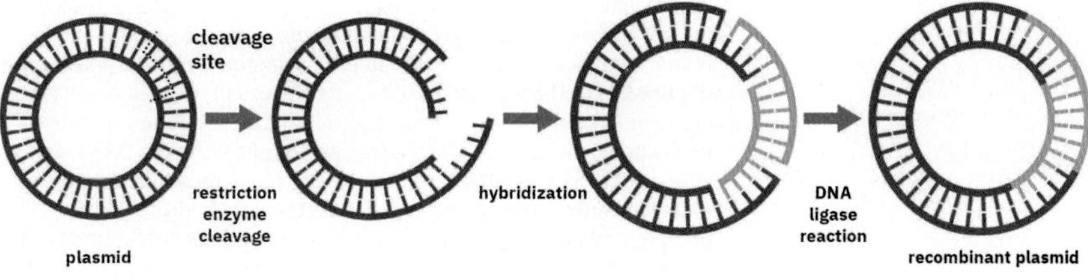

SEP Construct an Explanation *Plasmid DNA vectors often include a gene for antibiotic resistance. Why do you think that might be useful?*

SCIENCE AND ENGINEERING PRACTICES
Developing and Using Models

Genetic Research Models Point out to students that scientists often use multiple models to show different parts of a process. Guide them to use the multiple models shown in **Figures 13-2** to **13-4** to understand the basic tools necessary to conduct genetic research. While the structures shown in the first two figures are not identical to the structures shown in the third figure, have students point out similarities between them. Students may identify color, the ladder structure, or even the shape of the cleavage site as features that tie the three structures together. Have student groups make their own version of **Figure 13-4** that includes the restriction enzyme nucleotide codes for *Eco*RI. Students should align the cleavage site of *Eco*RI to where the DNA is cut.

Visual Support

Restriction Enzymes Bacteria use restriction enzymes to cut up plasmids that are introduced by bacteriophages, but they protect their own DNA with the use of methyltransferases. However, the unique cutting ability of restriction enzymes allows scientists to use them for gene recombination and cloning. Students may not realize how **Figures 13-2, 13-3,** and **13-4** connect. After they read the three sections associated with the figures, have them identify where the process shown in **Figure 13-2** occurs in **Figure 13-4**. Students should then repeat this with **Figure 13-3** and **Figure 13-4**. Share with students that various classes of restriction enzymes exist, but only one class, the one that *Eco*RI belongs to, binds and cuts at specific sites. Have students consider what might happen if a restriction enzyme that cut at random was used for the processes shown in the figures. Ask them to discuss why a random cut site insertion would be useful to scientists.

Plasmids Students learn that plasmids can be used as a vector to introduce DNA into a bacterial host. In **Section 6.1**, students are first introduced to plasmid structure and function in prokaryotic cells.

SEP Construct an Explanation *Being resistant to antibiotics increases the likelihood of survival of a bacterium that has the plasmid DNA in its genome. Scientists can identify which bacterial cells have successfully incorporated the plasmid DNA by exposing the bacteria to the antibiotic, so the cells that do not have plasmid DNA die and the ones that do survive.*

Connect to Mathematics

Reason Quantitatively Students should be familiar with exponential growth but may not connect that with how 30 PCR cycles can lead to over 1 billion copies of the targeted DNA. Direct students to the pattern shown at the bottom of **Figure 13-5**. In the example, Round 1 shows the number of copies after one round of PCR. Round 2 shows the number of copies after 2 rounds of PCR. Guide students to recognize that what they are seeing is exponential growth. Have them identify the equation that can be used to calculate the number of copies of DNA after each round. Then, have them calculate the actual theoretical yield of copies after 30 rounds of PCR.
(2^x; $2^{30} = 1,073,741,824$)

DNA Replication Students learn that PCR allows scientists to produce large quantities of DNA through artificial replication using heat and *Taq* polymerase. **Section 11.2** discusses how gene expression drives cellular differentiation that results in tissue development.

Science Background

PCR Techniques The rapid replication of DNA achieved using PCR makes it useful for a variety of different fields, such as forensics, diagnostics, and the development of gene therapy. In forensics, it can be useful in developing evidence related to a crime, and in diagnostics, it can be used to determine whether a specific DNA sequence is present. To support students in constructing an explanation as to why PCR is useful in a variety of techniques, guide them in an exercise where they identify different ways that larger amounts of DNA are more useful than small amounts. Students should be able to identify that it is difficult to obtain results when sample sizes are too small but that PCR allows for samples to be increased to testable levels.

SEP Develop a Solution *PCR copies a sample of genetic material many times, so if there is a pathogen present in very small amounts, it would be easier to detect in a much larger sample.*

Polymerase Chain Reaction Scientists often need to generate large quantities of a DNA sequence they want to study. Making recombinant DNA to grow in host bacteria is not necessarily the quickest or most precise method. **Polymerase chain reaction** (PCR) is a technique used to mass-produce, or amplify, copies of a particular section of linear DNA. One PCR cycle can double the number of copies of the targeted section of DNA. Thirty PCR cycles may amplify that number a billionfold (**Figure 13-5**).

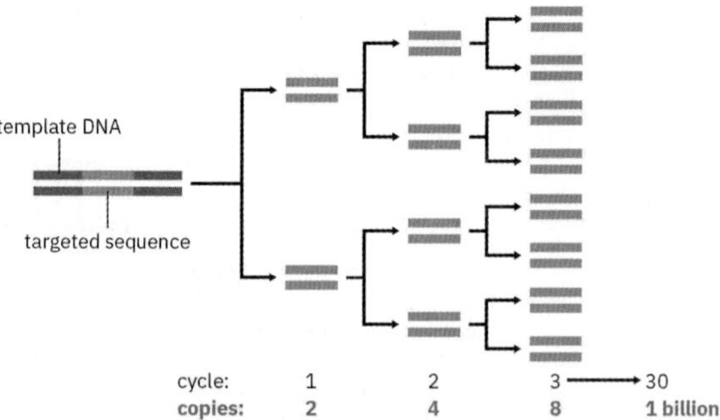

FIGURE 13-5 In one cycle of PCR, each strand of DNA that contains the targeted section may be copied to form a new strand.

VIRAL SPREAD

Gather Evidence *RT-PCR is a PCR technique that relies on reverse transcriptase (RT) to make a DNA copy of an RNA strand, which can then be amplified. How could this technology be used for diagnosing cases of SARS-CoV-2 infection?*

The starting material for PCR is any sample of DNA with at least one molecule of the desired DNA sequence. The source of the DNA might be a mixture with multiple DNA samples, a hair left at a crime scene, or a mummy—essentially, any substance that has DNA in it.

PCR is DNA replication in a test tube (Section 11.2). Like replication in a cell, DNA strands are separated, primers bind the DNA, and a DNA polymerase enzyme builds complementary strands. However, rather than a helicase enzyme, the PCR technique uses high temperatures to separate DNA strands. The temperatures required to do this would damage the DNA polymerases of most organisms, so PCR reactions make use of *Taq* polymerase, which comes from the bacterial species *Thermus aquaticus*. These bacteria live in hot springs and hydrothermal vents, so their polymerase tolerates high temperatures. Like other DNA polymerases, *Taq* polymerase recognizes primers that have matched and hybridized with the DNA template as places to start DNA synthesis.

PCR requires two DNA primers designed to base-pair at opposite ends of the sequence of the sample DNA that scientists want to amplify. These primers are mixed with the DNA sample, nucleotides, and *Taq* polymerase (**Figure 13-6**). The reaction mixture is then exposed to repeated cycles of high and low temperatures. A few seconds at high temperature disrupts the hydrogen bonds that hold the two strands of the sample DNA together, so every molecule of DNA unwinds into single strands. As the temperature of the mixture is lowered, the single DNA strands hybridize with the primers, and the *Taq* polymerase builds complementary DNA strands. The newly synthesized DNA is a copy of the targeted section. When the mixture is reheated, all of the DNA strands separate to begin another cycle.

SEP Develop a Solution *How might PCR be used to diagnose whether a patient is infected with a particular pathogen?*

REVISIT THE ANCHORING PHENOMENON

Gather Evidence *Students might say that RT-PCR technique could be used used for diagnosing cases of SARS-CoV-2 infection*

RT-PCR is used for the diagnosis of several different diseases. For example, it can be used to diagnose influenza infections and differentiate between A and B strains. In 2022, a company in India launched a combined RT-PCR test kit that can detect and discern between seven diseases, including malaria, chikungunya, dengue, Zika, leishmaniasis, leptospirosis, and salmonellosis infections.

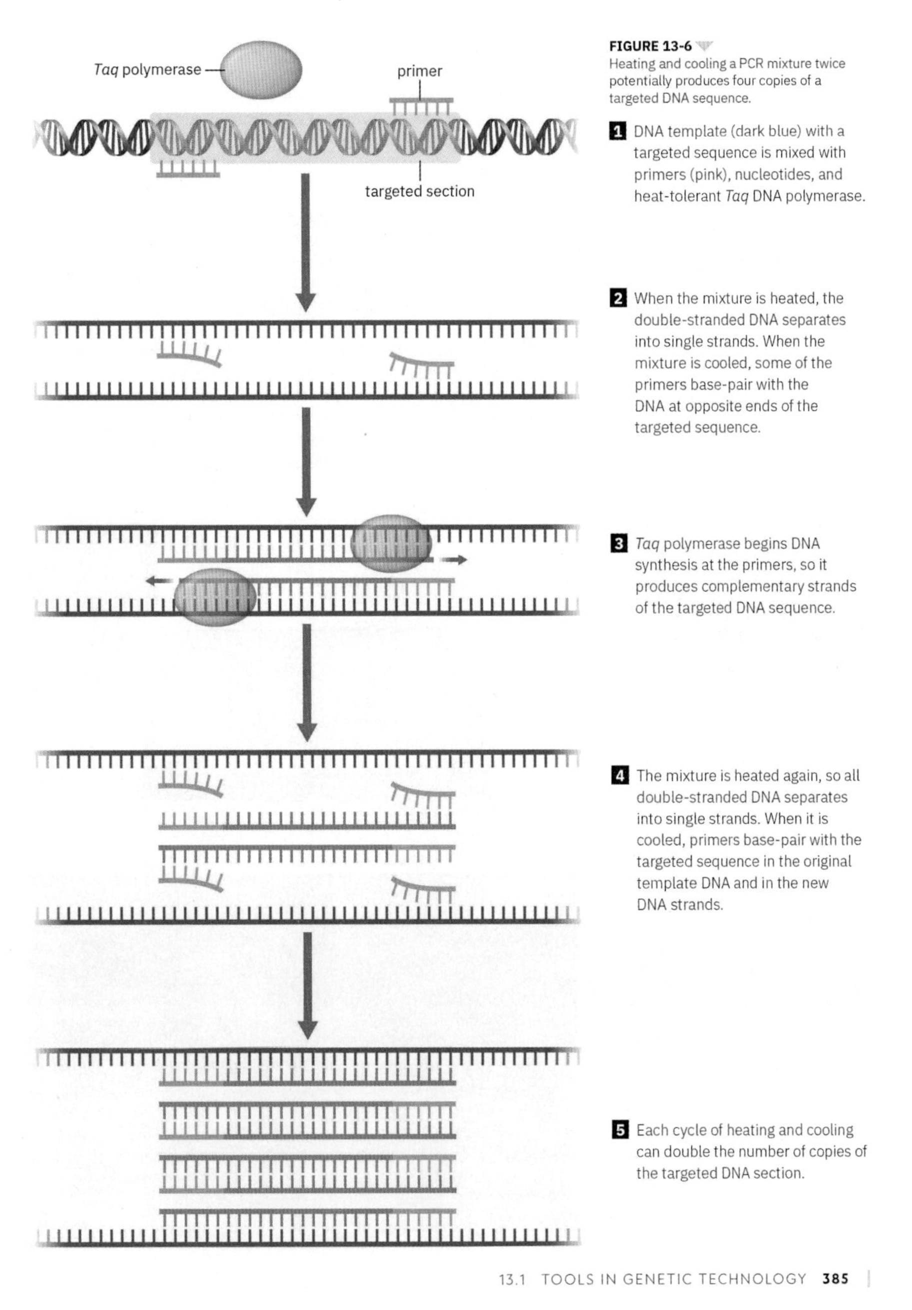

FIGURE 13-6 Heating and cooling a PCR mixture twice potentially produces four copies of a targeted DNA sequence.

1 DNA template (dark blue) with a targeted sequence is mixed with primers (pink), nucleotides, and heat-tolerant *Taq* DNA polymerase.

2 When the mixture is heated, the double-stranded DNA separates into single strands. When the mixture is cooled, some of the primers base-pair with the DNA at opposite ends of the targeted sequence.

3 *Taq* polymerase begins DNA synthesis at the primers, so it produces complementary strands of the targeted DNA sequence.

4 The mixture is heated again, so all double-stranded DNA separates into single strands. When it is cooled, primers base-pair with the targeted sequence in the original template DNA and in the new DNA strands.

5 Each cycle of heating and cooling can double the number of copies of the targeted DNA section.

Visual Support

Replicating DNA Tell students that DNA replication using PCR is a pattern that is repeated over and over until scientists generate enough DNA for their needs. The process consists of three steps that are repeated: denaturation, annealing, and extension. Step 1 in **Figure 13-6** represents the materials in the PCR mixture before the process begins. Inform students that the model artwork is not shown to scale, as primers and DNA sequences used during PCR are generally much longer than those that are shown in the model. Remind students that models are rarely exact representations of a system or process but focus instead on the processes to help others understand how they occur. Have students discuss how the process would repeat after Step 5.

DIFFERENTIATED INSTRUCTION | Students with Disabilities

Visual Impairments Students who have visual impairments may struggle to understand concepts related to PCR without the visual support provided in **Figure 13-5** and **Figure 13-6.** Students who have partial visual impairments can be supported by providing large printouts of the models throughout the lesson. You can also support students with visual impairments by allowing them to make sensory connections using other modes to present the content. Students can be supported with audio recordings of detailed descriptions of each model. You may also choose to provide students with tactile versions of the models created with printouts that have been outlined with glue.

13.1 TOOLS IN GENETIC TECHNOLOGY 385

Go Online SIMULATION

PCR and Gel Electrophoresis
Students can move through the steps of PCR and gel electrophoresis to assess the success of their reaction products. Encourage them to discuss their results and factors that came into play.

Crosscurricular Connections

Physical Science Pure water does not conduct a charge, which is why the solution that is used to fill the gel electrophoresis apparatus must contain electrolytes. Electrolytes produce ions, which conduct electricity through the water. The gel electrophoresis apparatus produces a constant electric field between the oppositely charged ends. This field is uniform and, in turn, generates an electric force on the charged phosphate groups of the DNA backbone. When making the gel, scientists can decide how small they would like the pores to be by using different ratios of materials. Smaller pores have a greater influence on how quickly molecules will migrate across the gel because the movement of larger molecules is impeded. DNA is loaded into the negatively charged end of the apparatus, and the field will move the particles toward the positively charged end. **Figure 13-7** shows a small-scale gel electrophoresis apparatus. High-throughput capillary electrophoresis is an advanced version of this technology that enables scientists to measure extracted DNA more precisely.

Science Background

DNA Sequencing Next-generation sequencing enables rapid sequencing of genomes using microarrays. However, earlier methods used a process called *Sanger sequencing*, or the chain termination method, which works similarly to PCR. However, in addition to the ingredients used in PCR, Sanger sequencing includes dyed, chain-terminating versions of each nucleotide. When the chain-terminating version of the nucleotide binds to the template DNA, replication stops. Using a process called *capillary gel electrophoresis*, a computer is used to detect the fragment lengths and attached dyes. This allows the DNA sequence of the sample to be determined.

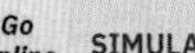

Go Online SIMULATION

PCR and Gel Electrophoresis Go online to use PCR and gel electrophoresis to amplify DNA and check your results.

Gel Electrophoresis In a sample of DNA that has been cut into fragments by enzymes or that contains specific segments that scientists want to study, it is useful to separate the DNA fragments by length. A technique called **gel electrophoresis** uses an electric field and a semisolid, porous gel to accomplish this. Each gel has wells that hold samples of DNA. The wells position the samples so they can move through the gel material during the procedure.

Figure 13-7 shows the apparatus for gel electrophoresis. A power source generates the electric field by charging electrodes positioned at opposite ends of the gel slab positively at one end and negatively at the other end. The phosphate groups along the DNA backbone give DNA molecules an overall negative charge. When a charged particle is placed in an electric field, it is attracted toward the opposite charge. Thus, negatively charged DNA fragments placed in wells near the negative electrode will all be pulled through the gel toward the positive electrode. However, DNA fragments of different lengths travel through the gel at different rates. Smaller fragments move farther toward the positive electrode than larger fragments, which stay closer to the negative electrode. **Figure 13-8** shows a pattern resulting from a gel electrophoresis experiment. Each band represents a DNA fragment of a particular size from a sample.

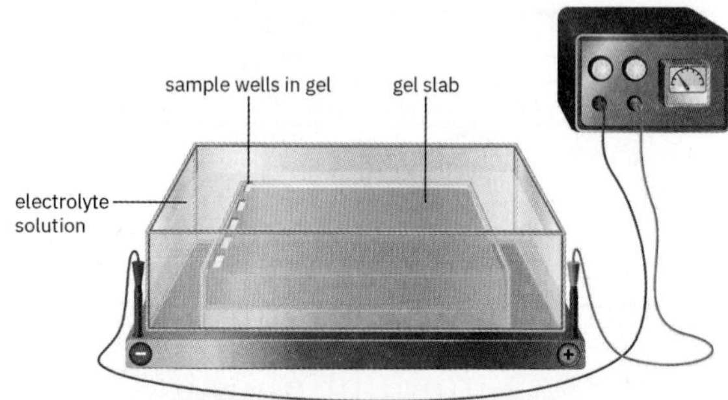

FIGURE 13-7 Gel electrophoresis uses an electric field to separate fragments of DNA based on size. In the gel, smaller fragments will move farther than larger ones. Once they have been separated, specific fragments can be cut out of the gel to use in further studies. For the model shown, samples would be loaded into the wells shown on the left side of the gel, near the negatively charged black electrode. Samples would be pulled through the gel toward the positively charged red electrode.

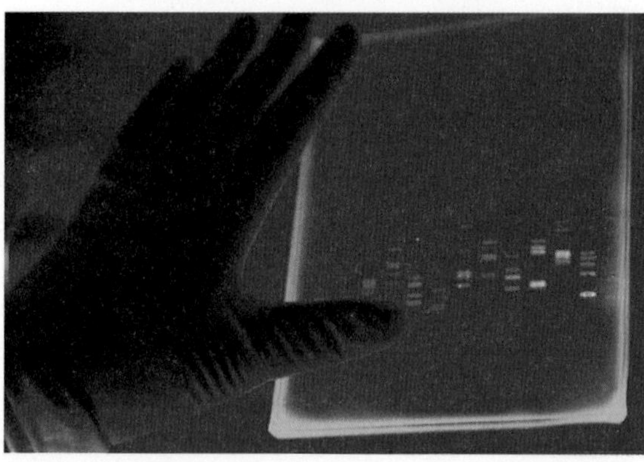

FIGURE 13-8 The patterns made by separated DNA fragments in the gel give scientists information about the sequences present in the sample. A solution of DNA molecules of various known lengths, called a DNA ladder, is typically placed in one of the wells next to experimental DNA samples. The pattern made by an unknown sample can be compared to the known solution to determine fragment sizes. In this image, the wells in the gel are at the bottom of the photo, so the smallest DNA fragments are toward the top of the gel.

DIFFERENTIATED INSTRUCTION | Leveled Support

Struggling Students Remind students that DNA travels through the gel from the negatively charged end where it is loaded to the positively charged end. Students may wonder why smaller fragments move more quickly. Share with students that the gel acts like a sieve, allowing smaller fragments to move more rapidly. Have students predict what would happen if the gel was allowed to run for a long time. (*The smaller fragments would move through the entire gel and into the electrolyte solution.*)

Advanced Learners Ask students to consider why pure water would not work as the solution in gel electrophoresis. Have students identify the elements that exist in pure water (oxygen and hydrogen) and discuss whether it can hold a charge. Have students consider why smaller fragments move through the gel faster than large DNA fragments. Share with students that scientists can make the gels have different consistencies. Have students predict whether a denser gel would allow DNA to travel faster or slower. (*slower*)

Genomes and Genome Analysis

The manipulation and amplification of DNA led to other technologies. Most important were techniques to determine DNA sequences. Studying the DNA of many organisms and sequencing their genomes has generated huge quantities of data, leading to new fields of statistical analyses and a better understanding of evolutionary relationships among organisms.

Genomics With continuing improvements in DNA sequencing technology, scientists began to sequence entire genomes of living organisms, starting with the bacterium *Haemophilus influenzae* in 1995. Three years later, the first genome to be sequenced for a multicellular organism was that of the roundworm *Caenorhabditis elegans*. Meanwhile, the Human Genome Project was launched in 1990 and completed in 2003. Sequencing the human genome paved the way for many other projects, including efforts to determine gene function. Because all organisms are descended from shared ancestors, all genomes are related to some extent. We see evidence of such genetic relationships by aligning sequences. Some regions of DNA are extremely similar across many species (**Figure 13-9**). Similarities among genomes allow us to discover functions of human genes by studying their counterparts in other species.

FIGURE 13-9 This is a region of the gene that codes for a DNA polymerase. Nucleotides that are different from the human sequence are highlighted.

```
758 GATAATCCTGTTTTGAACAAAAGGTCAAATTGCTGAATAGAAA-GTCTTGATTAACTAAAAGATGTACAAAGTGGAATTA 836 Human
752 GATAATCCTGTTTTGAACAAAAGGTCAAATTGCTGAATAGAAA-GTCTTGATTAACTAAAAGATGTACAAAGTGGAATTA 830 Mouse
751 GATAATCCTGTTTTGAACAAAAGGTCAAATTGCTGAATAGAAA-GTCTTGATTAACTAAAAGATGTACAAAGTGGAATTA 829 Rat
754 GATAATCCTGTTTTGAACAAAAGGTCAAATTGCTGAATAGAAA-GTCTTGATTAACTAAAAGATGTACAAAGTGGAATTA 832 Dog
782 GATAATCCTGTTTTGAACAAAAGGTCAAATTGCTGAATAGAAA-GTCTTGATTAACTAAAAGATGTACAAAGTGGAATTA 860 Chicken
758 GATAATCCTGTTTTGAACAAAAGGTCAAATTGCTGAATAGAAA-GTCTTGATTAAGTAAAAGATGTACAAAGTGGAATTA 836 Frog
823 GATAATCCTGTTTTGAACAAAAGGTCAGATTGCTGAATAGAAAAGGCTTGATTAAAGCAGAGATGTACAAAGTGGACGCA 902 Zebrafish
763 GATAATCCTGTTTTGAACAAAAGGTCAAATTGTTGAATAGAGACGCTTTGATAAAGCGGAGGAGGTACAAAGTGGGACC- 841 Pufferfish
```

The study of whole-genome structure and function is called **genomics**. This field has yielded many insights into gene function and evolutionary relationships among organisms. For example, comparing primate genomes revealed a change in chromosome structure that occurred during the evolution of our species. Comparing genomes also showed us that structural changes in chromosomes are not entirely random. Some specific alterations are much more common than others. This is because chromosomes that break tend to do so in particular locations. Human, mouse, rat, cow, pig, dog, cat, and horse chromosomes have undergone several translocations at these breakage "hot spots" during evolution. In humans, chromosome abnormalities that contribute to the progression of cancer also occur at the very same breakage hot spots.

Studying the products of a genome is even more difficult than deciphering the genome itself because the genome is constant but gene expression is not. The amounts and types of mRNA and proteins in a cell can differ dramatically by cell type and over time—an outcome of gene expression regulation (Section 11.3). Scientists study the *transcriptome* (mRNAs) and the *proteome* (proteins) to observe all mRNAs and proteins expressed at a particular moment. Scientists may choose to study the full set of mRNA or proteins expressed by an organism, or they may choose to just observe the expression in a particular type of cell within an organism. Understanding patterns and shifts in mRNA and protein production gives scientists additional ways to view cellular function.

Connect to English Language Arts

Evaluate a Science or Technical Text Genomics is a continuously progressing field. Where the first human genome took years to sequence, new machines can sequence the genome in about five hours. This allows scientists to study and compare the DNA sequences of a variety of organisms more rapidly.

Have students read the first paragraph in the genomics section and examine the evidence shown in **Figure 13-9**. Students should discuss and evaluate the conclusions in the text and use the evidence to corroborate or challenge the conclusions. Students should recognize that within the region shown, the first five organisms' nucleotide sequences are exactly the same and these organisms include four mammals and one bird. Have students discuss the differences they see between the sequences for the frog, zebrafish, and pufferfish and what conclusions they might draw.

Gene Expression Regulation
Students learn that proteomics is the study of proteins encoded by the genome and the context under which certain proteins are formed and expressed. **Chapter 11** introduced students to how DNA is replicated, transcribed, and translated into proteins and the various ways that genetic information is regulated.

DIFFERENTIATED INSTRUCTION | English Language Learners

Analyzing a Text Have students use a graphic organizer to help them think through the conclusions in the text and the supporting evidence in both the text and the visual. Have them draw a box for the conclusions across the top and then make notes in two columns below it: one for evidence from the text and one for evidence from the visual. Then, have them compare notes with a partner and add details as needed.

Beginning Have students take notes in short words and phrases or copy sentences from the text. They could also take notes in their first language if they wish.

Intermediate Have students take notes in phrases or complete sentences.

Advanced Have students take notes in complete sentences that explain the ideas in the text and the visual.

13.1 TOOLS IN GENETIC TECHNOLOGY 387

Science Background

Microarray Technology The various kinds of microarrays use different types of probes, but the text focuses on DNA microarrays. SNP chips are a kind of DNA microarray that is used to detect genetic polymorphisms that may exist in a population. These are sensitive enough that they can recognize when a genetic sequence has a single point substitution and can test hundreds of thousands, or even millions, of sequences per chip. Assays typically use multiple baseline samples from individuals that are normalized to make a reference set and the sample of the targeted individuals to be tested.

Traditional microarrays allowed for the use of only two fluorescent probes, one for the sample and one for the control. Advanced techniques allow for the use of multiple colors, permitting more than one test sample to be run at the same time.

The major applications of microarrays include gene expression profiling, genotyping, and DNA sequencing. Gene expression profiling determines the level of gene expression that exists within the sample. Genotyping determines the genetic variants that an individual has, while sequencing can determine the actual sequence of the DNA.

Bioinformatics Genomics and proteomics require computational tools for analyzing and interpreting biological data. The science of developing these tools is called **bioinformatics**. This field has profoundly shifted the focus of biology research. Instead of trying to understand a complex biological system by studying its parts, scientists can now study the system as a whole. Until recently, for example, the role of genetics in medicine was limited to the diagnosis of single-gene disorders. Bioinformatics is bringing us closer to unraveling conditions that have complex inheritance patterns, such as cancer, diabetes, and cardiovascular disease.

Microarrays One important tool for performing a large number of genetic tests at once is the **microarray**, which can test for the presence of thousands of DNA sequences in a sample at once. **Figure 13-10** demonstrates a simple model of the principle that underlies microarrays and similar genetic technologies. In a basic microarray, known sequences of DNA are fixed to wells or spots on a plate. A sample of DNA is introduced to many wells at once. When DNA sequences in the sample match a microarray spot and hybridize, a fluorescent probe is activated. Changes in the fluorescent light in each cell can be measured to identify matches. Microarray-related technology can be used in many applications. For example, it can be used to quickly test whole genomes for important mutations, or to detect what and how many mRNA molecules are present in a cell at a given time.

DNA Profiling About 99 percent of your DNA is exactly the same as everyone else's DNA. The shared part is what makes you human; the differences make you a unique member of the species. Identifying an individual by his or her DNA is called **DNA profiling**.

If you compared your DNA with another human's, regardless of race or ethnicity, an estimated 2.97 billion nucleotides of the two sequences would be identical. The remaining 30 million nonidentical nucleotides are sprinkled throughout your chromosomes. Conserved regions of DNA are those sequences that tend to be the same between individuals. Conserved sequences tend to vary at single nucleotides in particular locations.

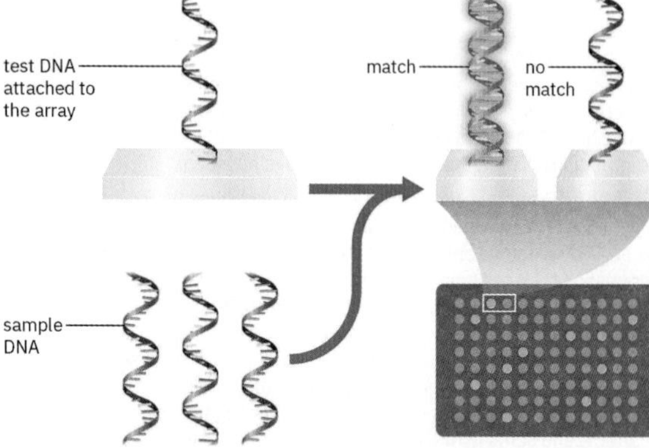

FIGURE 13-10 Microarrays can detect a large number of specific DNA sequences in a sample at one time. Known test DNA sequences are attached to each well or spot on the array. When a sequence from the sample matches one of the test sequences, a fluorescent probe can be detected. Microarrays can use multiple types of fluorescent probes to generate different colors depending on the conditions of the test.

CROSSCUTTING CONCEPTS | Influence of Science, Engineering, and Technology on Society and the Natural World

Medical Testing Modern medicine is dependent on the technological advances that exist in the field. Genomics is a rapidly expanding field that influences society. Emphasize to students that the techniques mentioned have a variety of uses in the medical field, but also in anthropological studies and criminology. To exemplify this, describe a number of well-known uses of genome analysis. Examples may include testing for BRCA alleles, fetal genomic testing, genetic disorder carrier testing, and pharmacogenetic testing.

Have students discuss how the advances in medical testing have influenced society and have them come up with additional applications of the technologies discussed that are not in the lesson. Describe how advances in computers and computation allows for the faster analysis of samples than in the past. Have students describe ways that these scientific advances can affect our understanding of the natural world.

A base pair substitution that occurs in a measurable percentage of a population is called a *single-nucleotide polymorphism*, or SNP (pronounced SNIP). Alleles of most genes differ by single nucleotides, and differences in alleles are the basis of the variation in human traits. Those differences are so unique that they can be used to identify you.

Another method of DNA profiling involves analysis of short tandem repeats in an individual's chromosomes. These repeated sequences of nucleotides usually occur at predictable locations, but the number of times a sequence is repeated differs among individuals. For example, one person's DNA may have 15 repeats of the nucleotides T-T-T-T-C at a certain location. Another person's DNA may have only four repeats. DNA polymerases may repeat or skip over these regions during DNA replication, so the number of repeats can increase or decrease from one generation to the next.

Unless two people are identical twins, the chance that they have identical short tandem repeats at even three locations is one in a quintillion (10^{18}), which is far more than the number of people who have ever lived. Therefore, an individual's array of short tandem repeats is unique.

Geneticists compare short tandem repeats on Y chromosomes to determine relationships among male relatives and to identify ethnic heritage. They also track mutations that accumulate in populations over time by comparing DNA profiles of living humans with those of people who lived long ago. These studies allow scientists to study population dispersals that happened in the ancient past.

SEP Develop a Solution *Construct a problem or a question related to genealogy or forensic science that could be solved with the help of DNA evidence. How could you use the genetic tools described in this section to solve the problem or answer the question? How would you begin if you had only a tiny sample of DNA available?*

CHAPTER INVESTIGATION

DNA Evidence

How can you use the properties of DNA to figure out who committed the "crime" of taking a bite out of a cookie?

Go online to explore this chapter's hands-on investigation about the use of DNA evidence in forensics.

13.1 REVIEW

1. **Sequence** Order the steps of DNA cloning.
   - Identify a segment of DNA to clone.
   - The segment is inserted into the plasmid.
   - Bacterium replicates the recombinant DNA.
   - Plasmid is taken up by a single-cell bacterium.
   - Restriction enzymes cut the DNA plasmid and segment.
   - DNA ligases seal the gaps between the segment and plasmid.
2. **Identify** Which of these qualities of DNA enables gel electrophoresis to work?
   - A. double stranded
   - B. negatively charged
   - C. coils up into chromosomes
   - D. contains only four types of nucleotide bases
3. **Justify** Which statement describes an advantage of PCR over DNA cloning for producing copies of a DNA segment?
   - A. Hydrogen bonds that hold two strands of DNA together can be disrupted at high temperature.
   - B. Polymerase from bacteria that thrive in hot springs and thermal vents is essential for the process to work.
   - C. Genetic material can be exponentially mass produced very quickly without requiring living host cells to carry out replication.
   - D. DNA polymerases of most organisms would be damaged at the high temperature required to separate double strands of DNA.
4. **Explain** Why can most individuals be identified using short tandem repeats in their chromosomes?

CHAPTER INVESTIGATION A

Guided Inquiry *DNA Evidence*

Time: 100 minutes or 2 days plus advance preparation and incubation time

Students perform gel electrophoresis to analyze DNA samples from a mock "crime scene"—a cookie with a bite out of it.

Go online to access detailed teacher notes, answers, rubrics, and lab worksheets.

Address Misconceptions

Identical Twins A common misconception is that because identical twins have exactly the same DNA, they have the same gene expression. Students may also believe that it is impossible to tell the difference between identical twins using genetic testing.

To guide students in their understanding, explain that at the point when the embryo splits into two, the two embryos have identical genetic information. However, environmental factors can affect gene expression, as well as epigenetic changes. Somatic cell mutations during development can also cause differences between the twins. Have students consider why one identical twin may be shorter or taller than the other, even though they have the same genes. Share with students that new testing methods allow for scientists to be able to identify differences between twins, as exemplified by the studies of astronaut Scott Kelly, who spent a year on the ISS away from his twin Mark.

SEP Develop a Solution *Sample Answer: If I was doing genealogy research on the Internet and found someone who I thought was my cousin, DNA tests could be used to determine whether we are related. We could use PCR to amplify a sequence of DNA for each of us, plus other samples for comparison. Then, we could use DNA profiling to compare our SNP or short tandem repeats for that sequencing to see whether we have similar patterns.*

13.1 REVIEW

1. *Identify a segment of DNA to clone.; Restriction enzymes cut the DNA plasmid and segment.; The segment is inserted into the plasmid.; DNA ligases seal the gaps between the segment and plasmid.; Plasmid is taken up by a single-cell bacterium; Bacterium replicates the recombinant DNA.* **DOK 2**
2. *B* **DOK 2**
3. *C* **DOK 2**
4. *The probability that two people have identical short tandem repeats is extremely low unless they are identical twins with the same genetic material.* **DOK 2**

ON ASSIGNMENT

Click Beetle Larvae National Geographic photographer Ary Bassous is a Brazilian surgeon and award-winning photographer. His interest in photography began in the 1970s, but he began to specialize in scientific photography after graduating from medical school. This photo was taken on an overnight stay during the beginning of the rainy season in Emas National Park, located in Central Brazil. Click beetle larvae, seen in the mound, are similar to fireflies in their bioluminescence. However, unlike fireflies, their luminescence does not have a flashing pattern. Click beetles stay constantly illuminated to lure prey, unless threatened.

Address Misconceptions

Fluorescence A common misconception is that bioluminescence is the same as fluorescence. Fluorescence requires an external source of light and occurs when energy from that source is absorbed by the subject then reflected back. Fluorescence is not seen in complete darkness. However, bioluminescence, a subset of chemiluminescence, occurs when chemical reactions in an animal allow for light production.

ON ASSIGNMENT Braving jaguars and lightning during humid nights in Emas National Park, Brazil, National Geographic photographer Ary Bassous captured the glow of bioluminescent click beetle larvae (*Pyrearinus termitilluminans*) that live near the outer surfaces of termite mounds. For several weeks each year, the beetles light up to lure their prey of flying termites and other insects. Genes that code for bioluminescent compounds in nature have been used in genetic engineering experiments.

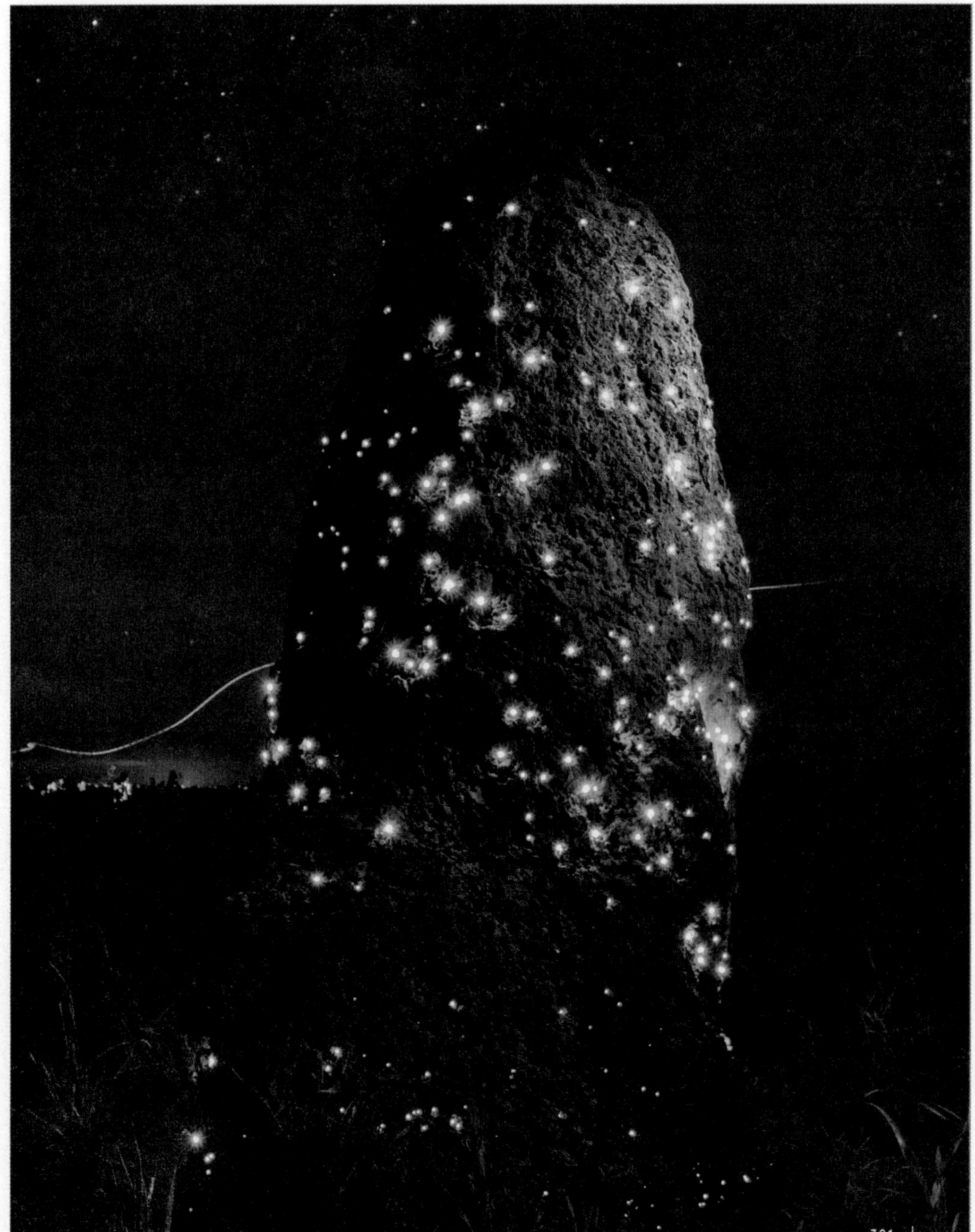

Crosscurricular Connections

Chemistry Students who have studied chemistry should understand that chemical reactions can result in a release of energy. In bioluminescent organisms, this released energy is in the form of light. Light energy as produced by the click beetle larvae is the result of an organic compound called *luciferin* interacting with oxygen that is catalyzed by an enzyme called *luciferase*. Luciferin can be any molecule that emits energy as a flash of light when it reacts with oxygen. Assess students' prior knowledge of biological chemical reactions by having students identify whether the reaction is endothermic or exothermic. Students should identify that it is an exothermic reaction, as there is a release of energy.

CROSSCUTTING CONCEPTS | Structure and Function

Luciferin Bioluminescence is an interesting phenomenon because it can be used to warn or evade predators, find a mate, or lure prey. *P. termitilluminans,* seen in the image, is just one species of hundreds of bioluminescent organisms, but it is rare because it is one of only a few bioluminescent terrestrial animals. It has a gene that codes for the production of luciferin, the organic compound that emits light when oxidized. Other organisms that produce luciferin also are bioluminesce; however, not all organisms that are bioluminescent produce luciferin. Some of these organisms consume other organisms or have symbiotic relationships with bioluminescent organisms that provide them with their glow. Have students discuss the processes that allow for the organisms to glow. Students should understand that the genes that code for luciferin and luciferase are expressed by specific cells and that the expression is not continuous.

13.2

GENETIC ENGINEERING

ETS1.B

EXPLORE/EXPLAIN

This section introduces genetically modified organisms, discusses cloning, explores possible uses of gene therapy and genome editing, and explains CRISPR technology.

Objectives

- Identify how genetic engineering is used to modify organisms.
- Describe how somatic cells are used to develop new organisms.
- Explain how gene therapy might be used to correct a genetic disorder.
- Evaluate if CRISPR is a more effective gene therapy process.

Pretest Use **Questions 1, 2, 4,** and **5** to identify gaps in background knowledge or misconceptions.

Analyze *Sample Answer: Any phenotype coded by a gene could be changed by editing that gene. If you wanted to edit a gene to have a particular flower color, there would not be many ethical issues, but if you wanted to edit the genes of a human baby to have a particular eye color, there could be lots of concerns. Which genes are okay to change? Given the complexity of human gene expression, will changing one gene cause any other problems for the child?*

Science Background

Model Organisms *S. cerevisiae* and *E. coli* are model organisms that have been extensively studied to use in human gene studies. When a disease-linked gene is found, scientists can study how the gene functions in the models and search for possible tools or treatments that can be used in humans. Most models are easy to breed, mature quickly, and can produce many descendants quickly. Ask students to identify why these features are important in a model organism. Students should say that diseases may be rare in people. However, in a model organism, the genes can be quickly replicated and in large volumes, allowing scientists to study them.

13.2

KEY TERMS

clone
CRISPR
gene drive
gene therapy
genetic engineering
genetically modified organism
somatic cell nuclear transfer

GENETIC ENGINEERING

Using a variety of techniques, scientists can modify an organism's genome in order to find the genetic basis of a trait or determine what traits arise from a specific change in a DNA sequence. Altering a gene may inhibit or inactivate a protein's function, or it may cause the protein to be more active or to perform a new function. Investigating these effects opens many new doors in our scientific understanding of the living world and unlocks numerous possible benefits to humanity. However, there are also many limitations and ethical considerations related to genetic engineering.

Analyze *What is one genetic trait or problem you have learned about in genetics that you think could probably be altered by changing an organism's genome? Can you think of any possible ethical questions that could arise from making the change?*

Genetically Modified Organisms

Intentional breeding of selected organisms can alter genomes over time, as was the case with Mendel's pea plants or the domestication of dogs. However, that works only if individuals with the desired traits will interbreed. The tools that allow scientists to manipulate DNA take gene-swapping to an entirely different level. **Genetic engineering** is a process by which an individual organism's genome is deliberately modified with the intent of changing the organism's phenotype. An individual with a genome that has been modified in this way is a **genetically modified organism**, or GMO. The processes that generate GMOs often change an organism's genome in a way that can be passed onto offspring.

Microorganisms Most genetic engineering involves the yeast *Saccharomyces cerevisiae* and various species of bacteria, often *Escherichia coli*. Both types of cells have the metabolic machinery to make complex organic molecules and are easy to cultivate in the lab. Both organisms also make excellent model organisms for research projects because of their well-understood genomes.

Genetically engineered microorganisms are common sources of medically important proteins. People with diabetes were among the first beneficiaries of such genetically modified organisms. Insulin for treatment of diabetes was once extracted from animals, but animal insulin provoked an allergic reaction in some people. Human insulin, which does not provoke allergic reactions, has been produced by *E. coli* engineered to express the human insulin gene since 1982. GMOs that have a gene from a different species inserted into their DNA, such as these bacteria with human genes, are called *transgenic* organisms.

Microorganisms are also engineered to make proteins used in foods. For example, we use enzymes produced by modified microorganisms to improve the taste and clarity of fruit juice, to slow bread staling, and to modify fats. Cheese was traditionally made with an enzyme (chymosin) extracted from calf stomachs. Today, many cheeses are made with calf chymosin produced by transgenic yeast.

392 CHAPTER 13 GENETIC TECHNOLOGIES

DIFFERENTIATED INSTRUCTION | English Language Learners

Sounds and Language Patterns Point out that the silent *e* at the end gives the *e* the long-vowel sound, which is the difference in pronunciation between *gene* and *genome* and *genetic* or *genetically*. Guide students to identify both sounds with the word part *gen-* to recognize that these words are related.

Beginning Read the following words and have students identify words that sound like gene: *genome, genetic engineering, gene therapy, genetically modified organism,* and *gene drive.*

Intermediate Have pairs group the words by sound (gene, genome, gene drive, gene therapy and genetics, genetic engineering, genetically modified organism). Then, have them define each term, keeping in mind that each is related to the word *gene*.

Advanced Have pairs take turns explaining how two of the terms from this list are related: *gene, genome, gene drive, gene therapy, genetic engineering,* and *genetically modified organism*. Remind them to be careful to pronounce the words correctly.

Plants As crop production expands to keep pace with human population growth, many farmers have begun to rely on genetically modified crop plants. Genes are often introduced into plant cells by way of *Agrobacterium tumefaciens*, a bacterium that infects plants. *A. tumefaciens* carries a plasmid with genes that cause tumors to form on a host plant, so the plasmid is called the tumor-inducing (Ti) plasmid. Tumor-inducing genes are replaced with foreign or modified genes, and then the recombinant plasmid is used as a vector to deliver the genes into plant cells. Whole plants can be grown from cells that integrate a recombinant plasmid into their chromosomes (**Figure 13-11** and **Figure 13-12**).

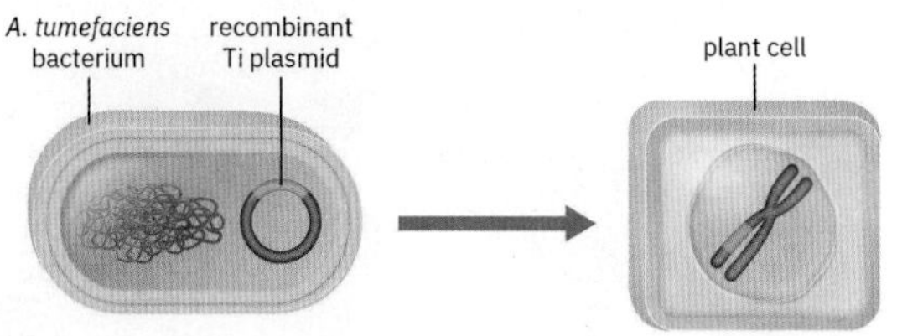

FIGURE 13-11 When *A. tumefaciens* containing a recombinant Ti plasmid infects a plant cell, the Ti plasmid is able to insert DNA into the plant cell's genome.

Some genetically modified crops carry genes that impart resistance to plant diseases or pests. Farmers that follow organic growing methods often spray their crops with spores of *Bt* (*Bacillus thuringiensis*), a species of bacterium that makes a protein toxic to some insect larvae. Alternatively, farmers can grow crops that have been genetically modified to produce the toxic *Bt* protein. These crops include soybeans, corn, cotton, and even eggplants. Cells of the engineered plants produce the *Bt* protein, and larvae die shortly after eating their first and only GMO meal.

Worldwide, hundreds of millions of acres are planted in GMO crops. The majority of these crops are corn, sorghum, cotton, soybeans, canola, and alfalfa that have been genetically engineered to be resistant to the herbicide glyphosate. Rather than tilling the soil or spraying a variety of targeted herbicides to control weeds, farmers can spray their fields with glyphosate, which kills the weeds but not the GMO crops. Genes that confer glyphosate resistance have appeared in weeds and other wild plants, as well as in unmodified crops—which means that recombinant DNA can (and does) escape into the environment.

SEP Construct an Explanation *How do you think the genes for glyphosate resistance could become incorporated into wild plants?*

FIGURE 13-12 These rice plants have been modified with DNA introduced by Ti plasmids. The plants are first grown in nutrient agar. When the rice plants have grown large enough in the laboratory, the modified seedlings are transplanted into containers and eventually into controlled plots in a paddy field.

CROSSCUTTING CONCEPTS | Influence of Science, Technology, and Engineering on Society and the Natural World

Plant Vaccines *A. tumefaciens*, modeled in **Figure 13-11**, is used to introduce vectors into plants. Tobacco plants easily express foreign proteins introduced by *A. tumefaciens* in a process known as *agroinfiltration*. This process can produce vaccines for some animals and is of interest in the mass production of vaccines against human infectious diseases. *A. tumefaciens* can be used to transfer glyphosate tolerance to crop plants. Have students consider how the glyphosate is spreading into the unmodified plants. They should connect back to plant reproduction and understand that the modified gene is present in the gametes of the GMO plants and can be introduced to unmodified plants during pollination. Then, ask them how the gene might be found in weeds and why that is a potential problem. They might identify that the gene could be introduced to the weeds through other bacteria, but if the gene is present in the weeds, the herbicide will not kill them.

Vocabulary Strategy

Understanding in Context Students may struggle differentiating between the different key terms that include the phrase or prefix for "gene." Suggest that they make sure to take their notes in context. Instead of just writing a definition and attempting to memorize it, they should use the surrounding phrases, ideas, statements, and other words to build their understanding for each one. This will support them in building a science vocabulary and making sense using reasoning and deductive thinking as they learn about the topic.

Connect to Mathematics

Reason Quantitatively Students may not understand the size of an acre, nor how large hundreds of millions of acres would be. To help students understand the amount of land that is planted with GMO crops, have them consider the size of a football field. Share with students that an acre is 43,560 square feet and that a football field is about a third larger than an acre, at about 57,600 square feet. Explain that the state of Texas is approximately 170,000,000 acres. Have students calculate the number of square feet in Texas and the approximate number of football fields that would fit in the state of Texas. Have them then infer how many football fields represent the "hundreds of millions of acres" referred to in the text. *(Approximately 7.405 trillion square feet in Texas, more than 128,559 football fields)*

SEP Construct an Explanation
Sample Answer: Wild plants that happen to survive being sprayed by glyphosate could already have a gene or mutation that gives them resistance, and when they reproduce, that gene would be passed on to their offspring, making more of the population glyphosate-resistant.

Science Background

Traditional Selective Breeding Humans have modified gene variants in plants and animals for millennia. However, prior to the development of genetic engineering, the modification of organisms was limited to the crossbreeding of plants and animals that had desired characteristics. For example, a rancher might choose to breed larger cattle to produce more meat to increase profit. These organisms might pass down genes that promoted increased growth to their offspring. Over time, larger animals would become more common in the herd. However, inbreeding could lead to an increase of genetic conditions within the herd. Also, since the organisms are closely related, diseases could easily spread through the population. As students read this section, have them relate how human modifications affect organisms, whether through traditional practices or using modern technology.

In Your Community

Endocrine Disrupters Endocrine disrupters interfere with the endocrine system because they mimic the body's hormones. These chemicals are found in many products and easily enter the body through air, food, water, and skin contact. Some of these chemicals include phytoestrogens, per- and polyfluoroalkyl substances (PFAS), Bisphenol A (BPA), and dioxins, which are manufacturing biproducts found in some products and have been released into the environment. It is believed that at least 95 percent of the U.S. population has been exposed to PFAS, a chemical used in water repellents and industrial applications. Endocrine disrupters are believed to lead to problems in human reproduction, certain cancers, problems with neurological development, thyroid issues, and other health problems.

Take Action Have students identify products with endocrine disrupters that they may be exposed to. Then, find safe replacements that they may consider.

CONNECTIONS

GOLDEN RICE While many GMO crops are engineered to confer herbicide or pesticide resistance, some other genetic modification efforts aim to solve problems associated with nutrition. Deficiency in vitamin A can cause blindness in children, often followed by death. In many parts of the world, vegetables that are rich in vitamin A are not readily available. Meanwhile, rice is a staple crop that is easy to obtain in many areas.

Since the 1990s, scientists have attempted to engineer a variety of rice that will address vitamin A deficiency. The original golden rice included two genes that are critical for the synthesis of beta-carotene, which is a precursor of vitamin A. One gene comes from a daffodil plant, and the other comes from the soil bacterium *Erwinia uredovora*. With later generations of golden rice, scientists have continued to hone the genes used to increase the synthesis and storage of beta-carotene.

Vitamin A deficiency is one of the most common nutrient deficiencies affecting children around the world, with an estimated thirty percent of children under five years old experiencing a deficiency. Despite many efforts to implement vitamin A supplement programs, this deficiency is still a tremendous global problem, requiring multiple approaches to solve. The World Health Organization estimates there are 500,000 cases of childhood blindness every year, and that half of those children die within a year of going blind.

Although golden rice has been reviewed and deemed safe by several countries, including Australia, Canada, and the United States, it has not been produced widely. In 2021, the Philippines was the first country to approve golden rice production on a commercial scale.

Beta-carotene gives golden rice its characteristic color.

Animals Genetically modified animals can be produced by a variety of methods. Two methods that have been used often in research either inactivate genes or add new genes. For example, "knock-out" mice are generated by inserting DNA into specific areas in the genome to inactivate targeted genes. Other genetically modified animals are made by injecting recombinant DNA into a fertilized egg and implanting the resulting embryo with its new gene into a surrogate to complete its development. More recently, scientists have utilized CRISPR techniques to modify animal genes. This method is described in detail later in this section.

Genetically modified mice are invaluable in research. Scientists have discovered the functions of many human genes by inactivating the same genes in experiments using knock-out mice. Genetically modified mice are also used as models of human diseases. For example, researchers inactivated the molecules involved in the control of glucose metabolism, one by one, in mice. Studying the effects of these modifications in mice resulted in much of our understanding of how diabetes works in humans.

Genetically engineered animals other than mice are also useful in research, and some make molecules that have medical applications. For example, various transgenic goats produce proteins used to treat cystic fibrosis, heart attacks, and blood-clotting disorders. Rabbits make human interleukin-2, a protein that triggers immune cells to divide and is used as a treatment for some cancers.

394 CHAPTER 13 GENETIC TECHNOLOGIES

CONNECTIONS | Golden Rice

In Your Community Explain to students that vitamin A is a fat-soluble vitamin that supports vision and cell growth, division, and differentiation, among a variety of other functions, in the body. But, it can be toxic in high levels. The body converts beta-carotene, which is a carotenoid, or pigment, from food sources into vitamin A. The body will only convert beta-carotene to vitamin A as needed, which makes food-derived beta-carotene the safest source of vitamin A. Have students discuss the importance of vitamin A in small children and pregnant women. Students should infer that a vitamin A deficiency could lead to problems with cell growth and division, impeding the growth of children and fetuses. Have students consider why rice was likely chosen as the carrier for the gene. They may identify that it is one of the largest food crops in the world and that it can last longer in storage than fresh fruits and vegetables.

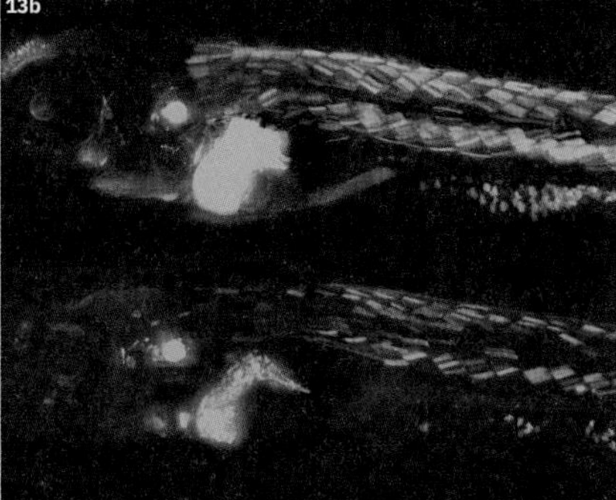

Some animals are genetically engineered to carry mutations associated with human disorders—multiple sclerosis, cystic fibrosis, diabetes, cancer, or Huntington disease, for example. Researchers produce engineered animals to study the disorders and potential treatments without experimenting on humans. However, the modified animals often suffer the same symptoms of the conditions as humans do.

Other genetic engineering experiments have attempted to tackle environmental issues. For example, endocrine disruptors are a group of chemicals that can interfere with endocrine, or hormone-related, systems in the body and cause negative health effects. These chemicals can enter the environment and water supplies from pesticides, plastics, and other wastes. As a method for detecting these pollutants, a gene from another organism has been inserted into the genome of a zebrafish (**Figure 13-13a**). This gene encodes a green fluorescent protein, thus engineering the zebrafish to glow green in the presence of endocrine disrupters (**Figure 13-13b**).

Scientists also engineer food animals. For example, genetic engineering has produced trout with extra muscle, pigs with environmentally friendly low-phosphate feces, chickens that do not transmit bird flu, and cows that do not get mad cow disease. In 2015, the FDA approved the first GMO animal for use as human food: a transgenic Atlantic salmon. These fish have been engineered to carry a gene promoter from ocean pout (a type of fish) that governs a growth hormone gene from a Chinook salmon. The engineered fish grow to full size about twice as fast as their unmodified counterparts. The engineered salmon are all female, are sterile, and have been approved only for specific aquaculture uses.

Many people do not think livestock should be genetically engineered. Others see it as an extension of thousands of years of acceptable animal husbandry practices. The techniques have changed, but the intent has not: humans continue to have an interest in improving our livestock. Either way, tinkering with the genes of animals raises a host of ethical dilemmas.

STSE Science, Technology, and Society *What ethical questions do you think scientists should consider when generating genetically modified organisms? Do the ethical considerations change depending on the goals of the technological application?*

FIGURE 13-13 
Zebrafish (a) are an important model organism for genetic studies. Genetically modified zebrafish (b) glow with green fluorescent protein.

CHAPTER INVESTIGATION

Fluorescent Genes
Can we replicate the process scientists engineered to insert foreign DNA into bacteria?
Go online to explore this chapter's hands-on investigation about genetic engineering.

VIDEO

VIDEO 13-1 Go online to learn about some of the many organisms that use fluorescent compounds in nature.

CHAPTER INVESTIGATION B

Engineering Design *Fluorescent Genes*

Time: 100 minutes or 2 days

Students perform a genetic engineering experiment by using bacterial transformation to introduce fluorescent genes into *Escherichia coli* (*E. coli*).

Go online to access detailed teacher notes, answers, rubrics, and lab worksheets.

STSE Science, Technology, and Society *Sample Answer: Scientists could ask whether the genetic modification adversely affects the organism's quality of life and whether the modification could somehow spread into the environment, such as by plant pollination.*

Video

Time: 2:28

Video 13-1 shows biofluorescent sharks that were discovered while studying biofluorescent coral. Pause the video at the 2-minute mark and ask students to explain why they need to use light to record the biofluorescent sharks. Students should understand that biofluorescent sharks have structures that allow them to absorb and emit certain wavelengths, but they do not produce light like the click beetle larvae do.

Address Misconceptions

Cloning A common misconception about cloning is that it produces an organism that is an exact copy of the original organism. However, while the genetics of a cloned organism may be identical to the donor animal, the two actual organisms are unlikely to have the exact same traits. To support students in rectifying this misconception, have them discuss possible reasons as to why the organisms would not be exactly the same. They should be able to connect with prior knowledge that gene expression can be affected by environmental factors, including the hormones and other chemicals that may be present in the surrogate during fetal development.

Science Background

Epigenetics The study of epigenetics focuses on how gene expression is controlled, and known processes can affect gene expression. Methylation attaches methyl groups to DNA to recruit gene repressors or prevent transcription factors from binding to DNA, thereby decreasing or preventing gene expression. DNA that is methylated typically has decreased gene expression, while demethylation increases gene expression. Acetylation attaches acetyl groups to DNA, which prevents it from being tightly wrapped around histones and increases gene expression. In **Chapter 11**, students learned about epigenetics and its relationship to cell differentiation. Remind students that DNA methylation is necessary for cell differentiation and helps control which genes are expressed in the process. Have students consider how epigenetic factors may be involved in cloning failures. Students should connect that these may occur when the somatic cell reprogramming fails to remove epigenetic controls that are specific to the cell type.

Cloning Organisms

All cells descended from the same fertilized egg inherit the same DNA. That DNA is like a master blueprint for directing the development of an individual's body. Consider how identical twins have identical DNA, so their bodies develop the same way. Identical twins are the product of a natural process called *embryo splitting*. The first few divisions of a fertilized egg form a tiny ball of cells that sometimes splits spontaneously. If both halves of the ball develop independently, identical twins result. Animal breeders have long exploited this phenomenon with a technique called artificial embryo splitting. A tiny ball of cells grown from a fertilized egg is teased apart into two halves that develop as separate embryos. The embryos are implanted in surrogate mothers, who give birth to identical twins.

Twins produced by embryo splitting are genetically identical to one another, but they are not identical to either parent. This is because in humans and other animals, twins get their DNA from two parents that typically differ in their DNA sequence.

Somatic Cell Nuclear Transfer "Cloning" means making an identical copy of something, and it can refer to deliberate interventions intended to produce an exact genetic copy of an organism. Animal breeders who want an exact copy of a specific individual use **somatic cell nuclear transfer** (SCNT), a laboratory procedure in which an unfertilized egg's nucleus is replaced with the nucleus of a donor's somatic cell (**Figure 13-14**). A somatic cell is a body cell, as opposed to a reproductive cell. If all goes well, the transplanted DNA directs the development of an embryo, which is then implanted into a surrogate mother. The animal born to the surrogate is a **clone**—a genetic copy—of the donor. Clones produced by SCNT have the same features as their adult donors, and there are other benefits. SCNT can yield many more offspring in a given time frame than traditional breeding, and offspring can be produced from a donor animal that is castrated or even dead.

FIGURE 13-14 Somatic cell nuclear transfer replaces the nucleus of an egg with a donor nucleus.

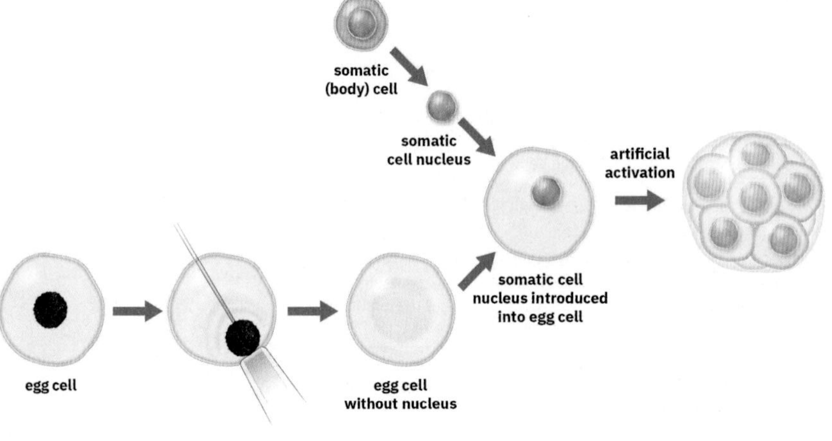

SCIENCE AND ENGINEERING PRACTICES
Constructing Explanations and Designing Solutions

Cloning Failures Improperly reprogrammed somatic cells leading to abnormal gene expression are often a cause of cloning failures. Students may wonder why somatic cells are used when they appear to pose such serious problems in cloned organisms. They may also wonder if using DNA from gametes may produce healthier clones. Direct students to **Figure 13-14** and ask them to think about whether the egg cell, shown in the first part of the image, is haploid or diploid. Then, have them answer the question about the cell after the egg nucleus is removed and the somatic cell has been inserted. Have students explain why another egg nucleus could not just be inserted instead. Students should explain that inserting another egg nucleus after the first one was removed would only produce a haploid cell, which could not develop into an organism. Students should also explain that adding a second egg nucleus to the first one would not lead to a cloned organism, even if it worked.

FIGURE 13-15 Dolly was the first mammal to be cloned from an adult cell.

Reproductive cloning refers to SCNT and any other technology that produces clones of an animal from a single cell. SCNT has been used since 1997, when a lamb named Dolly (**Figure 13-15**) was cloned from a mammary cell of an adult sheep. Dolly looked and acted like a normal sheep, but she died early, most likely because she was a clone.

The outcome of SCNT is sometimes unpredictable. Depending on the species, few implanted embryos may survive until birth. For many of the initial attempts at SCNT, most of the clones that did survive had serious health problems such as enlarged organs and obesity. Cloned mice developed lung and liver problems, and almost all died prematurely. Cloned pigs tended to limp and have heart problems; some developed without important parts.

SCNT works because the DNA in a somatic cell contains all the information necessary to build the individual all over again. However, a somatic cell does not automatically start dividing to form an embryo. This is because during early development, an embryo's cells start using different subsets of their DNA. As they do, the cells become different in form and function through the process of differentiation. In animals, this process is usually a one-way path, which means that once a cell has become specialized, all of its descendant cells will be specialized the same way. By the time a liver cell, muscle cell, or other differentiated cell forms, many genes in its DNA are no longer expressed. To clone an adult, the DNA of a somatic cell must be reprogrammed. The parts that trigger embryonic development are turned back on so the cells have the same potential for gene expression as a normal developing embryo. Scientists still have a lot to learn, but SCNT techniques have improved to the extent that health problems are much less common in animals cloned today.

SEP Construct an Explanation *Consider what you know about DNA replication and other processes in the cell. What are some events that might occur after somatic cell nuclear transfer that could result in an unexpected outcome?*

Connect to English Language Arts

Write an Argument After students read about cloning, have them discuss issues that can arise from the use of somatic cells in cloning to produce organisms such as Dolly, seen in **Figure 13-15**. Students should write an argument and support their argument using evidence from the text and prior knowledge to explain whether or not they would clone a pet.

Differentiation and DNA Students learn here that using cells that have differentiated can lead to problems in cloned animals. **Chapter 7** introduces the concept of differentiation and how environmental factors affect what genes may be expressed, while **Chapter 11** discusses how differentiated cells express a subset of genes that are specific for the cell type.

SEP Construct an Explanation
Sample Answer: After the nucleus is planted in the new cell, the cell has to replicate its DNA and divide successfully. Possibly mutations could be introduced during DNA replication, and chromosomal mutations could occur during mitosis if chromosomes are not divided properly.

DIFFERENTIATED INSTRUCTION | Economically Disadvantaged Students

Access to Resources Students who are economically disadvantaged may have difficulty complete such an assignment as the argument due to gaps in their background knowledge and lack of access to informational resources, such as the Internet or time to spend in the school library. If the assignment is unable to be completed during class time, provide an alternative study time within the school day during which students can access computers and informational resources and have discussion time with classmates.

In Your Community

Gene Therapy While gene therapy is beginning to make its way into certain fields of medicine, it is still a relatively rare form of treatment. Most gene therapy treatments focus on cancer, with many using virus-associated vectors to introduce new genes. However, candidates may have immunity to these vectors, reducing the success rate of the treatment. Encourage students to find out if there are any local medical centers that use gene therapy to treat patients and find out the kinds of conditions they treat.

Gene Therapy and Genome Editing

We know of more than 15,000 serious genetic disorders. Genetic disorders cause a large number of infant deaths each year, along with a significant number of hospitalizations for genetic disorders that are survivable. Drugs and other treatments can minimize the symptoms of some genetic disorders, but modifying the genome to correct the mutation would be the only real cure. **Gene therapy** is the transfer of DNA into an individual's body cells to correct a genetic disorder or treat a disease.

Genetic Disorders and Abnormalities Many genetic disorders are caused by single-gene mutations (**Figure 13-16**). Studying human inheritance patterns has given scientists many insights into how genetic disorders arise and progress, and how to treat them. Surgery, prescription drugs, hormone replacement therapy, and dietary controls can minimize and in some cases eliminate the symptoms of a genetic disorder. Some disorders can be detected early enough to start countermeasures before symptoms develop. Most hospitals in the United States now screen newborns for certain mutations. For example, one mutation causes phenylketonuria, or PKU, in one of every 10,000–15,000 newborns. The mutations affect an enzyme that converts one amino acid (phenylalanine) to another (tyrosine). Without this enzyme, the body becomes deficient in tyrosine, and phenylalanine accumulates to high levels. The imbalance inhibits protein synthesis in the brain, which can ultimately result in permanent intellectual disability. Restricting all intake of phenylalanine can slow the progression of PKU, so routine early screening has resulted in fewer individuals experiencing the symptoms of the disorder.

FIGURE 13-16 Autosomal abnormalities and disorders are caused by genetic problems that occur on the numbered chromosomes in the genome, as opposed to an X or Y sex chromosome.

Inheritance Pattern: autosomal dominant

Disorder/Abnormality	Primary Symptoms
achondroplasia	dwarfism
aniridia	eye defects
Huntington disease	nervous system degeneration
Marfan syndrome	abnormal/missing connective tissue
progeria	drastic premature aging

Inheritance Pattern: autosomal recessive

Disorder/Abnormality	Primary Symptoms
albinism	absence of pigmentation
cystic fibrosis	difficulty breathing, lung infections
Ellis-van Creveld syndrome	dwarfism, heart defects, polydactyly
phenylketonuria (PKU)	mental impairment
sickle cell anemia	anemia, swelling, frequent infection
Tay-Sachs disease	mental and physical deterioration, early death

DIFFERENTIATED INSTRUCTION | English Language Learners

Read Silently and Take Notes Have students read these two pages silently and take notes on each paragraph in their own words. If students get stuck on a word or concept, have them raise a hand and provide assistance by clarifying or rewording the text. Once they have completed their reading, they should share their notes with a partner and add to their notes or make changes as needed.

Beginning Encourage students to write notes in their native language and supplement them with symbols.

Intermediate These students may use a graphic organizer or table to take notes on the content.

Advanced Have students write short sentences and paraphrase their notes as they work through the paragraphs. Students can translate their notes into a brief paragraph about how gene therapies can target genetic disorders and abnormalities.

Gene Therapy Gene therapy is being tested as a treatment for a wide variety of diseases and disorders, but not all diseases and disorders are readily treatable through these methods. Some disorders arise because a needed protein is not produced or is malfunctioning. Gene therapy can provide a healthy, working gene to replace that lost function. Other gene therapy techniques work by silencing mutated genes that are causing a disorder. Potential targets for gene therapy treatments include

- AIDS
- Alzheimer disease
- cancer (several types)
- cystic fibrosis
- heart failure
- hemophilia A
- inherited eye, ear, immune disorders
- muscular dystrophy
- Parkinson disease
- sickle cell anemia

The biggest challenge in gene therapy is developing a method that efficiently delivers functioning copies of genes to the affected tissue so the proteins can be produced in the cells where they are needed. For example, to treat leukemia, a potentially fatal cancer of bone marrow cells, scientists have used a viral vector to insert a gene into immune cells extracted from the leukemia patient. When the modified cells are reintroduced into the patient's body, the inserted gene directs the destruction of cancer cells.

Despite the successes, the outcomes of manipulating a gene for use in a living individual can be unpredictable. For example, X-linked severe combined immunodeficiency (SCID-X1) is a severe genetic disorder that stems from a mutated allele of the *IL2RG* gene on the X chromosome. The gene normally encodes a receptor important for signaling the immune system. Without treatment, people affected by this disorder can survive only in germ-free isolation tents because they cannot fight infections. Researchers used a genetically engineered virus to insert unmutated copies of *IL2RG* into cells taken from the bone marrow of 20 boys with SCID-X1. Each child's modified cells were infused back into his bone marrow. Within months of treatment, 18 of the boys were able to leave their isolation tents. Gene therapy had repaired their immune systems. However, five of the boys later developed leukemia and died. Developers of the gene therapy could not have known that the gene targeted for repair would cause cancer. The viral vector had inserted the gene into the chromosome at a site near a proto-oncogene (Section 7.1). The insertion triggered transcription of the gene, resulting in the onset of leukemia.

SEP Evaluate a Solution *Age-related macular degeneration (AMD) is a leading cause of blindness. Many genes are related to AMD, but scientists have identified just a few specific genes as potential gene therapy targets. One trial in particular is attempting to block an immune system gene that generates a protein that attacks cells of the retina. Do you think AMD is a disorder that is likely to be solved with gene therapy? What are some of the challenges? How do you think the treatment would be delivered?*

Address Misconceptions

Targeted Gene Treatments While targeted gene therapies have produced promising results, misconceptions surround their effects. Students may believe that if a gene is modified, the patient will be cured. Help students understand some conditions might be controlled, yet if the patient has children, they may inherit the condition. Encourage students to consider the differences that exist between germ cells versus the genes targeted in somatic cells. Have students discuss how the genes can still be passed on even if the parent is treated.

Proto-Oncogenes While gene therapy appears promising, sometimes unintended outcomes occur. Students learn about genes involved in cell growth that can mutate into oncogenes, called *proto-oncogenes*, in **Section 7.1.**

SEP Evaluate a Solution *Sample Answer: AMD could maybe be partially solved by gene therapy. You could deliver the modified genes directly to the site where the problem is occurring (the eye). A challenge could be that there are many genes contributing to the condition, so even if you can solve one genetic problem, others may continue to cause problems.*

Science Background

CRISPER-Cas9 In addition to inserting DNA, as shown in **Figure 13-17**, CRISPR-Cas9 can also be used to remove mutations or enhance genetic traits in gametes and embryos. Beta thalassemia is a condition that causes a type of anemia and other conditions during infancy. It can be caused by a single point mutation in one of two genes. An unsuccessful attempt to cure a beta thalassemia mutation in human embryos led to an international summit in 2015. Have students consider why there would be a concern about editing genes in embryos. Students should understand that scientists do not yet understand all of the risks associated with germline editing, and it could lead to birth defects or other issues that would be inheritable by any future generations.

RNA Interference Pathway CRISPR technology allows for the use of modified prokaryote RNA to recognize sequences in DNA that can be targeted. Students learn about RNA interference and how microRNA that is complementary to mRNA can bind to it, causing RNA interference by both prokaryotes and eukaryotes in **Section 11.3**.

Retroviruses CRISPR technology allows for the recognition and removal of HIV DNA that has been integrated into a patient's DNA. Students learn about how HIV replication uses reverse transcriptase to produce viral DNA that is inserted into a host's genome in **Section 8.4**.

CRISPR Technology

Gene therapy with viral vectors has some risks. As was the case for the SCID-X1 treatments, a recombinant viral vector inserts randomly into the genome. This can have unpredictable consequences. By taking advantage of a system that prokaryotes use to defend themselves against viral infection, scientists have developed a powerful and precise method of editing DNA called **CRISPR** (pronounced CRISPER) that is safer. CRISPR is named for a DNA sequence that, when transcribed, helps a prokaryote combat viral infection through an RNA interference pathway that recognizes and destroys viral DNA (Section 11.3).

Editing Gene Sequences The CRISPR technique uses a combination of molecules, or a complex, that combines an RNA sequence, a *nuclease* enzyme (Cas9), and a DNA sequence. The RNA guides the complex to a specific sequence of DNA in the chromosome, where the nuclease enzyme cuts both strands of DNA (**Figure 13-17**). Then, the new DNA sequence is used to repair the break. The guide RNA and new DNA sequences can be designed to precisely target and change essentially any part of a genome in a living cell. Researchers deliver the guide RNA, the nuclease enzyme, and the new DNA into cells in the form of genes carried by a plasmid vector. A cell that expresses the edited genes will have both of its chromosomes edited, and the change will be passed to all of the cell's descendants.

CRISPR is more accurate and efficient than viral vectors or any other method of altering genomes. It is fast, easy, inexpensive, and works in almost any cell. The first generation of CRISPR techniques sometimes cut DNA in unintended locations. However, a newer generation of CRISPR base-editing techniques (**Figure 13-18**) improves accuracy and uses different enzymes to perform in-place editing of individual nucleotides without introducing unexpected breaks in the DNA.

FIGURE 13-17 The first generation of CRISPR technology cut target DNA strands to introduce new DNA sequences. The guide RNA leads the complex to the target sequence. The nuclease enzyme cleaves the DNA, and a new DNA sequence is inserted at the cleavage site.

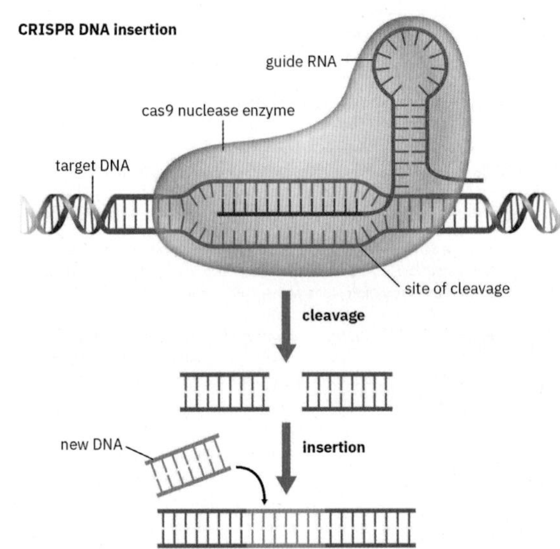

400 CHAPTER 13 GENETIC TECHNOLOGIES

DIFFERENTIATED INSTRUCTION | Leveled Support

Struggling Students To aid students in understanding how the CRISPR DNA insertion process shown in **Figure 13-17** works, have them develop their own graphic organizer to show how the system works. Help them write out a description of each step shown in the figure. Students can also make a graphic organizer showing the DNA cloning process introduced at the beginning of the lesson and compare and contrast the processes.

Advanced Learners Students can review the content that supports **Figure 13-17** and consider how the process is similar to and different from DNA cloning as introduced in **Figure 13-14**. Encourage students who are curious about the acronym CRISPR to briefly search the topic. They should find that it stands for clustered regularly interspaced short palindromic repeats. Share with students that Cas stands for CRISPR-associated and that often, proteins are named for their structures or functions.

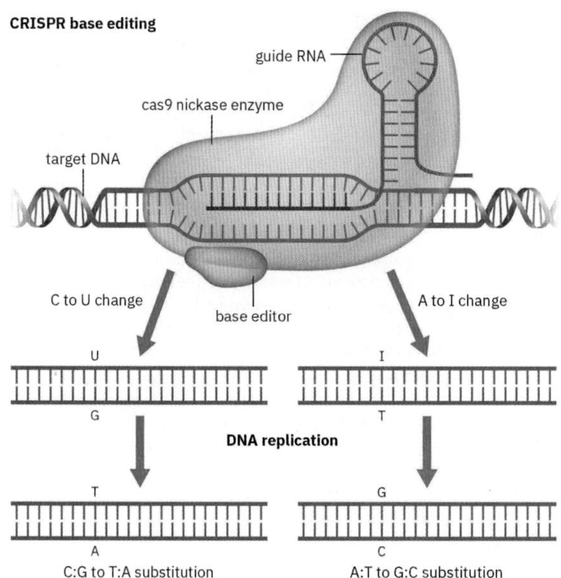

FIGURE 13-18 Newer CRISPR base-editing techniques can swap individual nucleotides. Unlike the cas9 nuclease that cuts both strands of DNA in earlier generations of CRISPR, the cas9 nickase enzyme cuts only one strand of the DNA at a targeted location. A base editor molecule can then change particular nucleotides on that side of the double strand. DNA replication then replaces the nucleotide on the opposite strand, completing the base editing process. The "I" in this diagram represents inositol, which serves as an intermediary in the conversion of A:T to G:C.

Applications of CRISPR These CRISPR methods pioneered by scientists such as American biochemist Jennifer Doudna (**Figure 13-19**) are useful in a variety of applications. For example, crop plants can be enhanced without leaving traces of the vector DNA. HIV DNA that has been integrated into an infected patient's cells (Section 8.4) can be edited out to prevent relapse after remission. Removing mutations from somatic cells can permanently cure many genetic disorders. The technique has been used successfully to repair mutations that cause muscular dystrophy, achondroplasia, and cancer in cells taken from human patients.

VIDEO

VIDEO 13-2 Go online to see Dr. Jennifer Doudna discuss the CRISPR method.

FIGURE 13-19 In 2020, Jennifer Doudna (pictured) and Emmanuelle Charpentier were awarded the Nobel Prize in Chemistry for their discovery of the CRISPR method of genome editing.

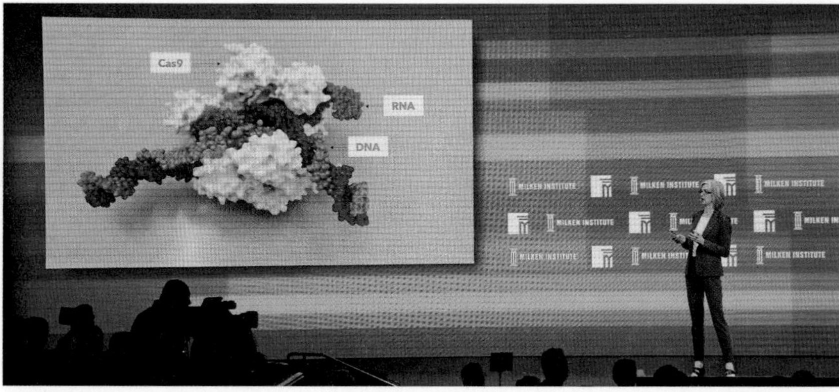

Visual Support

CRISPR Base Editing There is potential for base editing technologies in the treatment of blood disorders, especially those that are caused by single nucleotide mutations. Students may struggle to understand the process that changes the bases. Have students work in pairs to examine **Figure 13-17** and **Figure 13-18**. While the enzymes have different names in the two figures, guide students to understand that they have very similar functions, with the primary difference being that the nickase enzyme does not cause a complete cut, while the nuclease enzyme does. Then, have the pairs of students identify the difference between the complexes shown in the two images. They should identify that there is a base editor on the complex in the second figure. Have students make their own flowchart showing how a C:G pair becomes an A:T. Then, have them explain why there is a U in the example. Students should compare the function of uracil to the inositol that is used in the second conversion example.

Video

Time: 2:20

Video 13-2 features biochemist Dr. Jennifer Doudina who discusses research into and application of the CRISPR technology for manipulating DNA. After students watch the video, have them replay to match segments of the video to the steps shown in **Figure 13-17** and **Figure 13-18**.

SCIENCE AND ENGINEERING PRACTICES
Engaging in Argument from Evidence

Compare Using Evidence Students should recognize examples of DNA cloning, CRISPR DNA insertion, and CRISPR base editing. Ask students to compare and contrast the three techniques. Project a triple Venn diagram and have students identify characteristics that each process may share with each other and characteristics that are unique to each one. Project **Figures 13-14**, **13-17**, and **13-18** if students need visual support. Once students have properly sorted each characteristic, have them write a one-paragraph argument about which gene editing technique shows the most promise for human use. They should support their argument with evidence from the text and the class-collected observations.

Students can also explain which technique could be the most problematic. They should explain reasoning.

CASE STUDY
GENETICS TO THE RESCUE!

Ask Questions To help students make connections with the content, have them identify a list of problems that could occur with the use of CRISPR, as well as the benefits that the application of CRISPR provides. Students should use this list to develop their questions.

SEP Evaluate a Solution *Sample Answer: Eliminating mosquitoes from an environment could save millions of lives, but there are some tradeoffs. Mosquitoes probably help pollinate plants and are a food source for other organisms in an ecosystem.*

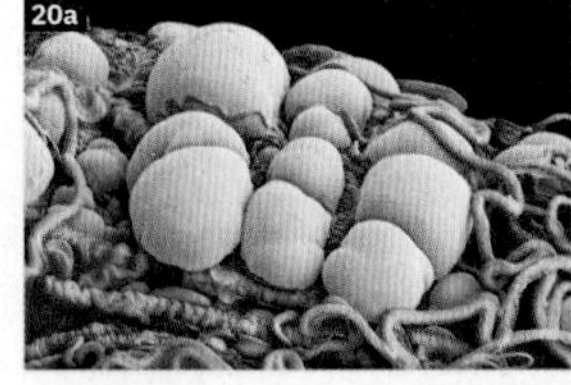

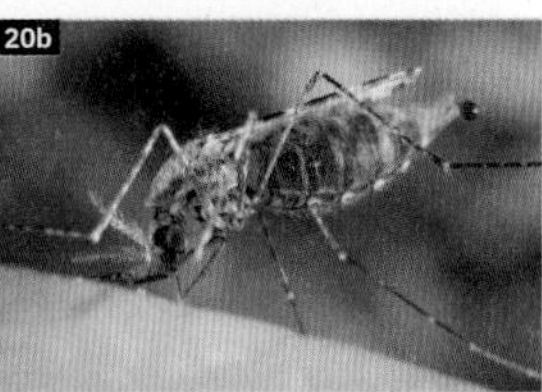

FIGURE 13-20 *Plasmodium* parasites (a) reproduce in the gut of the *Anopheles* mosquito (b) and are transferred to humans in the mosquito's saliva.

CASE STUDY
GENETICS TO THE RESCUE!

Ask Questions *Generate questions that the scientific and medical communities should consider when determining whether a CRISPR method should be approved to treat a human disease or disorder.*

Gene Drives CRISPR was the first successful method of making **gene drives**, which are genes that can "drive" themselves into other chromosomes and spread through a population. The eradication of malaria is one possible application of this technology. This disease infects more than two million people each year, about one-fifth of whom die. Most are children under the age of five.

Already, mosquitoes have been engineered to carry genes that boost their immune systems to prevent *Plasmodium*, the malaria parasite, from growing in their guts (**Figure 13-20b**). With a typical gene therapy, the engineered genes are diluted out of a population because only some of the mosquito's offspring inherit them. However, the outcome is different in mosquitoes engineered with CRISPR and CRISPR-based gene drives. When one of the engineered mosquitoes mates with an unmodified mosquito, 100 percent of their offspring inherit the engineered genes. This is because the CRISPR system that delivered the modification is part of the engineered mosquitoes' chromosomes, and it also edits homologous chromosomes in zygotes. If the engineered mosquitoes were to be released in the wild, the worldwide population of mosquitoes would probably be modified within a few years, eliminating malaria.

SEP Evaluate a Solution Aedes aegypti *mosquitoes carry several viruses that cause deadly disease in humans. Scientists have produced mosquitoes with genes that prevent female offspring from surviving to adulthood. How might eliminating adult female mosquitoes affect an ecosystem, and what are some tradeoffs that would be made if we were to use genetically engineered* A. aegypti *to prevent human disease?*

13.2 REVIEW

1. **Identify** Which are common goals of genetically modifying plants? Select all correct answers.
   A. increasing crop productivity
   B. developing medicinal compounds
   C. growing crops resistant to herbicides
   D. spreading genes to animals that eat the plants
   E. cultivating plants that are resistant to insects
2. **Describe** What are some goals of scientists who clone animals?
3. **Explain** What characteristics make a genetic disorder a good candidate for gene therapy?
4. **Sequence** List the steps in the correct order for the process of using CRISPR to edit a gene.
   - The cas9 enzyme cuts the DNA.
   - A target DNA sequence to modify is identified.
   - A plasmid delivers DNA, RNA, and enzymes to a cell.
   - A new DNA sequence is inserted into the place where the DNA was cut.
   - The RNA guides the complex to the target DNA sequence and hybridizes.

402 CHAPTER 13 GENETIC TECHNOLOGIES

13.2 REVIEW

1. *A, B, C, E* **DOK 2**
2. *Sample answers may include duplicating championship animals, saving endangered species, bringing back extinct species, and others.* **DOK 2**
3. *Sample answer: Disorders that are caused by a single gene are good targets for gene therapy. It is also helpful for the tissues affected by the disorder to be easily accessible for administration of treatment. If introduction of a functional version of a gene via gene therapy can alleviate the problem, the disorder may be a good candidate.* **DOK 2**
4. *A target DNA sequence to modify is identified; a plasmid delivers DNA, RNA, and enzymes to a cell; the enzyme cuts the DNA; RNA guides the enzyme to the target DNA sequence and hybridizes; a DNA template sequence repairs the cell's cut DNA.* **DOK 2**

NATIONAL GEOGRAPHIC EXPLORER **PARDIS SABETI**

CURING MALARIA WITH MATH

Computational biologist Dr. Pardis Sabeti uses math to mine the human genome to solve problems in human health.

An individual suffering from malaria sweats with a high fever and shakes with chills. The *Anopheles* mosquito feasts. With its bite, the mosquito takes in a small amount of blood containing parasites that mix with its saliva. For its next meal, the mosquito lands on a young child and injects the parasite into the child's bloodstream. A few weeks later, the child falls ill with the same symptoms as the other infected individual, but the outcome is much worse. The young child has no immunity to the infection and will likely die.

Computational biologist Dr. Pardis Sabeti has long been on the trail of infections such as malaria. She develops computer algorithms that analyze changes in DNA related to the spread of infectious microbes. Because the malaria parasite's genetic makeup is extremely diverse, it can become drug resistant by adapting quickly to new environmental conditions that result from drug treatment. By studying microbial genetic diversity, Sabeti can predict where a disease outbreak might occur next and inform more effective drug strategies.

A person's genetic makeup may play a role in how sick they become if infected. Sabeti wanted to test whether gene mutations that had occurred recently in the human population might provide an evolutionary advantage, such as resistance to the malaria parasite. Differentiating between recent mutations and those that appeared much longer in the past would require analyzing DNA sequences in the surrounding regions. In 2001, Sabeti developed a breakthrough algorithm that enabled this analysis.

After identifying many mutations that indicated relatively recent natural selection, Sabeti had to determine which ones were relevant. She and her colleagues developed another algorithm that allowed them to narrow down a region of 10,000 mutations to only 10 or so that were biologically meaningful to test.

Sabeti has continued to extend her algorithms to study other diseases, including Ebola, Lassa fever, and COVID-19. Pinpointing mutations in survivors and infection-resistant individuals helps identify genes that are important in the human response to infection. "So much of the physical world has been explored," Sabeti says. "But the deluge of data I get to investigate really lets me chart new territory. Most of my work may happen at a computer, but it's still a new and very exciting frontier."

THINKING CRITICALLY

Explain *Why is computational analysis crucial for genomic analysis?*

EXPLORER
PARDIS SABETI

About the Explorer

National Geographic Explorer Pardis Sabeti uses computer algorithms to predict disease outbreaks. She analyzes recently identified mutations in microbial DNA to determine which to target with drug and prevention strategies. She is the co-founder of an educational platform called *Operation Outbreak*, which focuses on outbreak prevention and preparedness. You can learn more about Pardis Sabeti and her research on the National Geographic Society's website: www.nationalgeographic.org.

Science Background

Malaria About 50 percent of the world's population is at risk of contracting malaria, caused by the microscopic *Plasmodium* parasite. Of the different species of *Plasmodium* that can infect humans, *P. falcicparum*, found in Africa, and *P. vivax*, found throughout the rest of the world, pose the highest risk to humans. Children under the age of five on the African continent are disproportionately at risk and account for about 80 percent of malaria deaths.

Connect to Careers

Medical Parasitologist Medical parasitologists focus on how parasites cause disease in humans. They use various fields of study in their fight against parasites, including immunology, pathology, and epidemiology. Medical parasitologists may work for local, state, and national government agencies, as well as international organizations. They may also be hired by private and nonprofit organizations, as well as universities.

THINKING CRITICALLY

Explain *Sample Answer: DNA molecules contain a tremendous number of nucleotide bases and genes. Across many samples, the data pools are incredibly large. Computational analysis helps identify meaningful genes and sequences for detailed study.*

LOOKING AT THE DATA
GENETIC THERAPY CLINICAL TRIALS

Connect this activity to the Science and Engineering Practices of using mathematics and computational thinking and analyzing and interpreting data. Students analyze two graphs: the increasing numbers of clinical trials for gene therapy and the costs of clinical trials by phase and disease types. Students use mathematical principles to analyze the patterns and organize the data and are expected to use the evidence they gather to make their own claim about the future of clinical trials.

Connect to Mathematics

Evaluate Reports Based on Data Have students analyze the graphs shown in **Figures 1** and **2** and consider how the Phase III trial numbers compare. Students should glean that in **Figure 1**, the Phase III trials make up the smallest number of trials in each year, but in **Figure 2**, they are the most expensive trials to conduct outside of Phase IV trials. Have students discuss reasons why this might be the case. Students should recognize that preclinical trials are conducted in animals, cells, and tissues but that clinical trials are first conducted on small numbers of people. As the trial sizes in humans become larger, they cost more money to conduct.

LOOKING AT THE DATA

GENETIC THERAPY CLINICAL TRIALS

SEP Mathematical Thinking Despite setbacks and increasing costs, development of genetic medical therapies is rising.

In 2021, a phase III trial for a gene therapy treatment of Huntington disease (a degenerative brain disease) was ended because no therapeutic effect was demonstrated, when comparing treated patients with a placebo group. The community of individuals affected by the disease was devastated because the treatment had produced promising initial results. Many treatments that enter clinical trials do not make it as far as phase III. Clinical trials are costly, but they perform the important function of determining what treatments are effective and safe for use in most patients.

There are several phases in most clinical trials. The preclinical phase includes laboratory research on animals, tissue, or cell samples. Institutional Review Boards vet the trial plans to ensure safety protocols are followed. Phase I begins with six to ten participants with special attention on treatment side effects. Phase II and phase III involve more participants, from 25–50 to 100–200, and are intended to evaluate safety, confirm benefits, and prove treatment is effective. Following these trial phases, the FDA conducts a thorough review prior to approving the treatment for public use. Some clinical trials continue into phase IV, in which they study long-term effects on all participants.

FIGURE 1 The number of clinical trials for genetic therapies by year over the last decade has shown a steady increase.

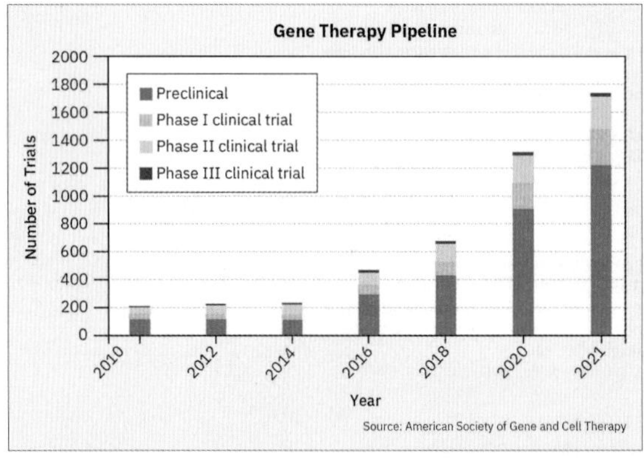

SCIENCE AND ENGINEERING PRACTICES
Analyzing and Interpreting Data

Analyzing Graphs Have students read the first paragraph on the page. Draw their attention to the phrase "Many treatments that enter clinical trials do not make it as far as Phase III." Have students make an assessment as to why this is so and lead a discussion about the gene therapy techniques they learned about in this section. Students should understand that most of the techniques discussed in the lesson are heavily tested in non-human models before testing ever begins on humans. Only successful laboratory trials move to the next phase. After students complete the reading, guide them to **Figure 1** and ask them to analyze the graph and summarize the data represented. Have students use evidence from the graph to support a claim about the percentage of all trials that are Phase III trials. Students should recognize that there are a minuscule number of Phase III trials when compared to the preclinical trials.

Figure 1 shows the number of gene therapy clinical trials at different phases over the last decade. **Figure 2** shows the average cost breakdown of the phases in all clinical trials, including gene therapies and other treatment methods, for a selection of common therapeutic areas.

FIGURE 2
The cost of conducting a clinical trial ranges from tens of millions to hundreds of millions of dollars and depends on the phase and therapeutic area.

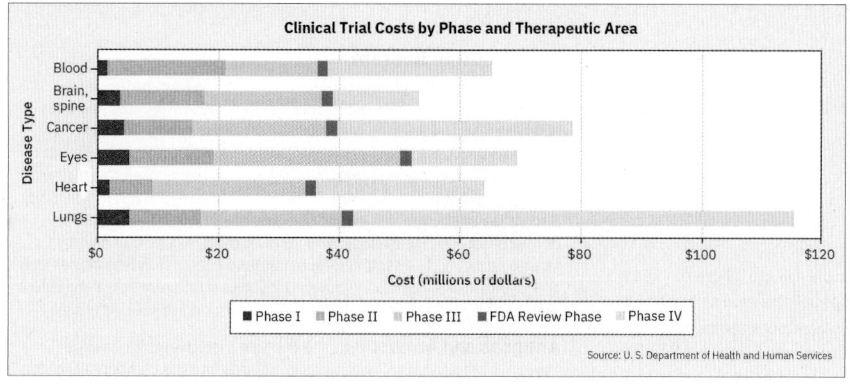

1. **Identify Patterns** Approximately how many times more preclinical trials were there in 2021 than in 2010? Compare this to how clinical trials in at least two other phases have changed over the decade.
2. **Organize Data** For each therapeutic area, rank the phases of clinical trials by average cost. Which phases tend to be the most and least expensive?
3. **Compare** Rank the therapeutic areas by total cost.
4. **Argue From Evidence** Use the data to support or refute the claim that in the future, there will be many job opportunities in the research and development of gene therapies.
5. **Nature of Science** Why do you think so many phases are required before the FDA approves treatment for the public? Why do you think companies are willing to invest all this money before being able to sell the treatment?

SAMPLE ANSWERS

1. *Sample answer: In 2021, there were about 12 times as many preclinical trials than in 2010. The number of clinical trials in the preclinical phase has increased dramatically. In comparison, the number of Phase III clinical trials has remained about the same. The number of Phase I clinical trials has grown but not as much as preclinical.* **DOK 2**
2. *Sample answer: The phase that tends to be most expensive is the Phase IV, where long-term effects of treatments are studied among all participants, and the least expensive is FDA Review. This phase requires review of results and does not involve human participants.* **DOK 2**
3. *lungs; cancer; eyes; blood; heart; brain, spine* **DOK 2**
4. *Sample answer: With a large amount of money being invested in gene therapy development, there should be an increase in opportunities to work in this field.* **DOK 3**
5. *Sample answer: The treatment must be tested in laboratory systems before introduction to humans. Then, safety and efficacy for human recipients must be determined before the treatment is administered to a larger number of people. Companies invest in these trials because they will make a large return on their investment if they can develop a successful therapy.* **DOK 3**

13.3

VACCINE DEVELOPMENT

ETS1.A, ETS1.B

EXPLORE/EXPLAIN

This section provides an overview of the immune system, explains the importance of vaccines to public health, discusses the different kinds of vaccine technologies, and explores how vaccine efficacy can be supported using other measures.

Objectives

- Summarize how the immune system protects individuals from infection.
- Describe how vaccines are used to protect the population from infectious diseases.
- Compare the differences between the types of vaccine technologies that have been developed and implemented.
- Evaluate the other measures that can be used to support vaccines.

Identify *Sample answers: I use an umbrella or raincoat to keep myself dry in a storm, and I wear a seatbelt when riding in a car. Everyone follows the same traffic laws to reduce the risk of traffic accidents. My community pool donated resources to help people who need food or clothing.*

CCC Structure and Function *Sample Answer: Each antibody has 2 binding sites that can bind antigens. The shape and charge pattern that form on each binding site makes them bind only specific antigens.*

Vocabulary Strategy

Semantic Map As students begin the section, suggest that they make a semantic map to visualize the relationships between the key terms. Have students use the glossary to define the key terms in a preliminary map, perhaps using "immunity" in the center and the other terms arrayed around it. Then, allow them time to make the connections between the terms as they learn about them. This will support their understanding of relationships that exist between the immune system and vaccines.

13.3

KEY TERMS

antibody
antigen
immunity
pathogen
reservoir
vaccine

VACCINE DEVELOPMENT

The genetic technologies that underlie modern vaccines are complex. An idea that is simpler to understand is the human need to protect ourselves from discomfort and illness and to work together to help others in our communities. Vaccines require both thoroughly tested genetic technology and community cooperation to be most effective.

Identify *Name at least one way you use a tool or technology to make yourself safer or more comfortable. Then describe at least one way you have observed people working together to help a community.*

The Immune System

The human immune system protects an individual from **pathogens**—bacteria, viruses, or other organisms that can cause infection. The body's ability to resist infection is generally known as **immunity**. An immune response starts with the detection of **antigens**, which are molecules or particles that the immune system recognizes as being foreign to the host. The first level of defense for the immune system is called *innate immunity*. It includes the body's immediate responses, such as increasing body temperature and releasing general immune system cells that destroy foreign particles. The second level of immune response is less immediate and is known as *adaptive immunity*. It involves a different set of immune system cells and antibodies that target and respond to specific antigens.

Antigens and Antibodies The immune system employs a wide variety of specialized cell types and structures, but the components of the immune system most relevant to understanding vaccines are antibodies. An **antibody** is a Y-shaped antigen receptor protein made by specialized immune cells to help the body fight infections. Many antibodies circulate in blood, where they bind to antigens and activate pathways that destroy associated pathogens. Antibody binding can also prevent pathogens and toxins from attaching to body cells. An antibody molecule consists of four polypeptides (**Figure 13-21**). When polypeptide chains fold up together as an antibody, two antigen-binding sites form at the tips of the Y. These sites bind only to specific antigens that match their shape and charge patterns.

FIGURE 13-21 An antibody molecule includes four polypeptide chains joined in a Y shape. Two antigen-binding sites form at the tips of the Y.

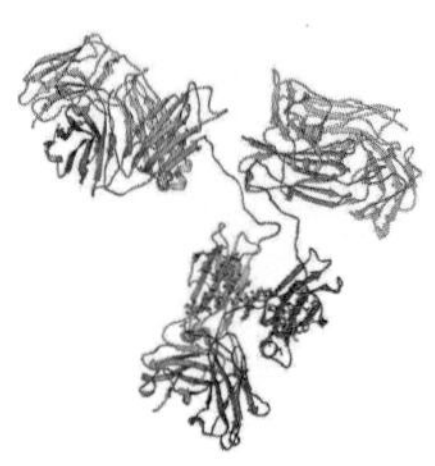

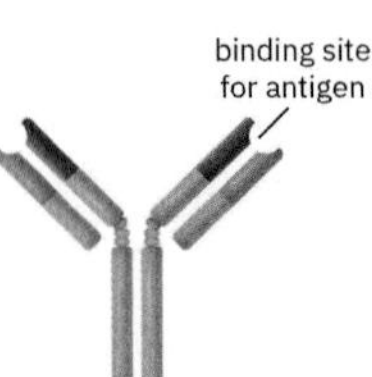

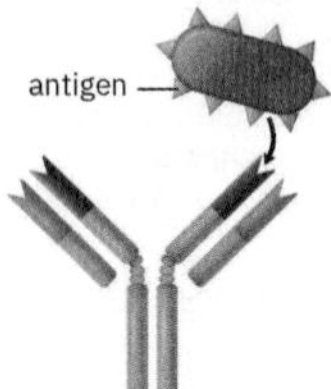

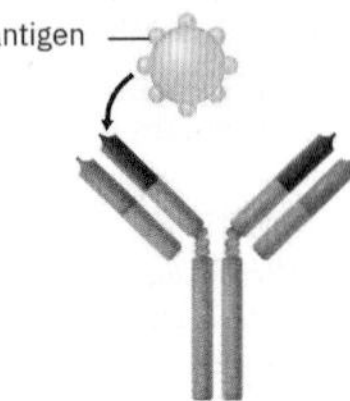

CCC Structure and Function *Look at the structures in **Figure 13-21**. How do the shape and properties of the polypeptides that make up an antibody help it recognize and bind a specific antigen?*

406 CHAPTER 13 GENETIC TECHNOLOGIES

SCIENCE AND ENGINEERING PRACTICES
Developing and Using Models

Antibodies and Antigens Remind students that models are not exact replicas of an object but instead symbolize relationships or interactions. Adaptive immunity depends on specific antigens and antibodies. Use **Figure 13-21** to guide students in a conversation about how antibodies only bind to antigens that fit in their variable region. Have students discuss why specific binding sites would be important when fighting infections that an individual may have been exposed to in the past. Students should develop their own model to explain what would happen if an individual were exposed to a novel antigen. Student models should show the first exposure to the antigen, mention the response by the innate immune system, and explain how the immune response varies between the primary and secondary exposures to the antigen.

When the immune system encounters a new antigen, its immune cells form a lasting antibody memory that allows the system to respond more readily the next time it detects the same antigen. This means that more immune cells and antibodies are available to quickly respond the second time the antigen invades (**Figure 13-22**).

FIGURE 13-22
Adaptive immunity generates a fast response after a second exposure to an antigen.

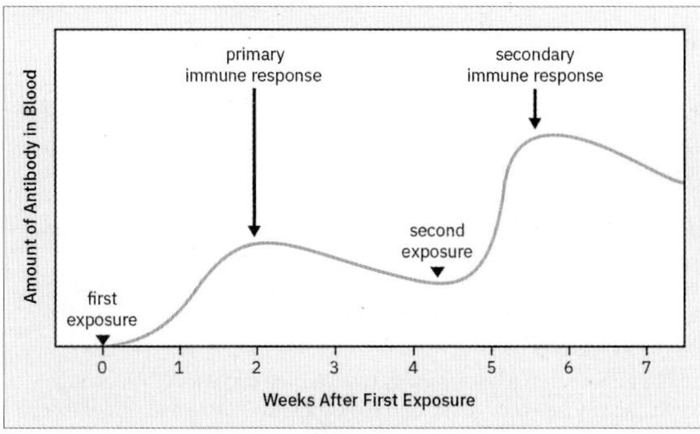

Vaccines and Public Health

Immunization refers to procedures designed to induce immunity. In active immunization, a preparation that contains an antigen—a **vaccine**—is injected or given orally. This immunization elicits a primary immune response, just as an infection would. A second immunization, or booster, elicits a secondary immune response for enhanced protection. Vaccines use a variety of technologies to modify and manipulate pathogen genomes to stimulate this immune response.

Discovery of Vaccines The first vaccine was developed as a result of desperate attempts to survive epidemics of smallpox, a severe disease that kills up to one-third of the people who contract it. Before 1880, no one knew what caused infectious diseases or how to protect anyone from getting them, but there were clues. For instance, survivors of smallpox only rarely contracted the disease a second time. At the time, the idea of acquiring immunity to smallpox was extremely appealing. People risked their lives by poking bits of smallpox scabs or pus from smallpox sores into their own skin. Some people survived these crude practices, but many others did not. This practice of deliberately infecting someone in hopes of contracting a mild infection to gain immunity is known as *variolation*, and it dates back to 15th-century China.

By 1774, it was known that women who milked cows usually did not get smallpox after they had contracted cowpox, a mild disease that affects humans as well as cattle. An English farmer collected pus from a cowpox sore on a cow's udder and poked it into the arms of his pregnant wife and two small children. All members of his family survived the smallpox epidemic, but their neighbors pelted them with rocks, convinced the family would turn into cows.

Connect to Mathematics

Reason Abstractly Encourage students to analyze the graph shown in **Figure 13-22** and examine the slope of the increase between the first exposure and the primary immune response and the second exposure and the secondary immune response. Have students identify how the slopes are different. They should be able to point out that the amount of antibody in blood increased more rapidly and to a higher level after the secondary response. Students should use this evidence to confirm that an adaptive immune response allows for a more rapid response to antigens upon later exposure.

DIFFERENTIATED INSTRUCTION | English Language Learners

Narrating Past Events As students read "Discovery of Vaccines," have them write a short retelling of each historical example. Remind students to use the past tense for events that took place in the past. Check that students are using correct past-tense verbs.

Beginning Provide sentence frames with students to fill in past-tense verbs for each example. Provide a word bank as needed. Depending on students' need for support, use either past-tense forms or base verbs in the word bank.

Intermediate Have students write a sentence or two to summarize each example. Make sure students narrate what happened using past-tense verbs.

Advanced Have students paraphrase each example, using about the same level of detail as in the text but using their own words. Have students share their writing with a partner to check past-tense verbs.

In Your Community

Public Health Departments The origins of public health departments began with sanitation. By the early 1800s, it was known that sanitation was an important factor in public health, with regions that had large populations and poor sanitation experiencing larger outbreaks of various diseases, such as smallpox, cholera, and typhoid. It was estimated that 1 in 10 people died of smallpox in London. This led to the development of sanitation departments that over many decades led to health departments. Now, federal, state, and local health departments are integral for the management of public health.

Take Action Have students identify the health department that services your region and learn about the programs and services they provide.

NOS Scientific Investigation *Sample answer: Edward Jenner introduced a potentially dangerous virus to a child. Modern medical research ethics require consent and should begin with adult patients who can make responsible decisions. Now, nearly all medical research involving humans requires approval and is regulated tightly.*

Connect to Mathematics

Evaluate Reports Based on Data Epidemiologists use the number of cases of a disease to determine whether an outbreak is occurring. Have students examine the graph in **Figure 13-23** and identify the two years that had obvious outbreaks. Have students suggest possible reasons that the years following an outbreak had fewer cases. Students may infer that after an outbreak occurred, it was likely that there was an increase in immunity and vaccinations, leading to a lower incidence of disease in subsequent years. Point out to students that 2020 also coincided with the COVID-19 pandemic, which likely decreased the number of cases because of social-distancing measures.

Twenty years later, British physician Edward Jenner introduced liquid from a cowpox sore into the arm of a healthy boy. Six weeks later, Jenner inoculated the boy with liquid from a smallpox sore. The boy did not get smallpox. Jenner's experiment and additional confirming studies showed directly that the cowpox virus elicits immunity to smallpox. Jenner named his procedure "vaccination," after the Latin word for cowpox (*vaccinia*). A smallpox vaccine that was developed later eradicated the disease, and the last known case of naturally occurring smallpox was in 1977.

NOS Scientific Investigation *How do you think ethical standards for medical research have changed since the time of Edward Jenner?*

Public Health Today, we know that the use of cowpox for immunization was effective against smallpox because the two diseases are caused by closely related viruses. Antibodies produced during an infection with one of the viruses recognize antigens of both. Building on this knowledge, vaccines for many other infectious diseases have since been developed. Vaccines greatly reduce rates of infection, the degree of associated suffering, and death as a result of infectious disease. Many nations where vaccines are available have very successful child vaccination programs that have nearly eliminated devastating diseases, such as measles.

Public confidence is a necessary part of vaccine success. When enough individuals refuse vaccination, outbreaks of preventable and sometimes fatal diseases occur. Measles had nearly been eradicated in the United States by the year 2010, with only 63 reported cases, but cases spiked to 1282 in the year 2019 (**Figure 13-23**). A significant number of parents are choosing not to vaccinate their children, most citing fears about vaccine safety. Safety concerns reported about one vaccine are sometimes applied to all vaccines, even those with millions of uses and proven safety. One concern that childhood vaccines cause an increased risk of autism has no scientific basis. The 1998 scientific paper that first proposed this risk was later shown to contain falsified data, and the lead author of the paper lost his medical license as a result of this fraud. The scientist was attempting to discredit one particular vaccine to sell more of a vaccine he was financially invested in. Rigorous studies of thousands of individuals have repeatedly failed to show a link between vaccinations and autism.

FIGURE 13-23 Measles outbreaks have occurred more often in recent years.

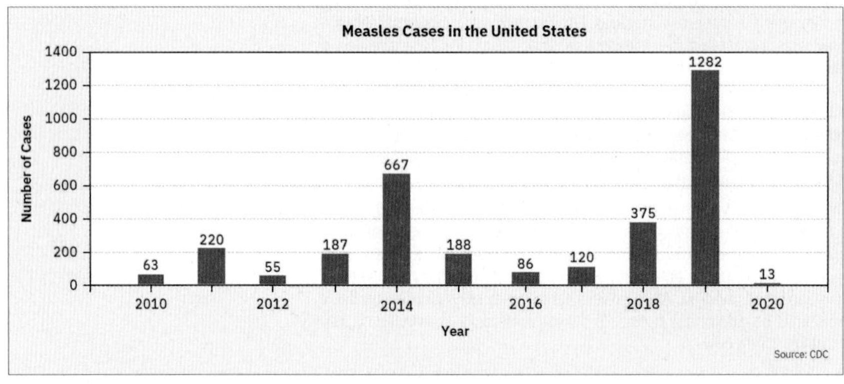

408 CHAPTER 13 GENETIC TECHNOLOGIES

SCIENCE AND ENGINEERING PRACTICES
Constructing Explanations and Designing Solutions

Vaccine Boosters Successful vaccination is dependent on the adaptive immune system. Typically, a vaccinated person is protected against the disease. However, occasionally a person may still become infected after vaccination. This can be due to a number of factors, including prolonged close contact with someone who is infected or a decreased immune response. Share with students that one reason smallpox was eradicated was because it is considered a relatively stable virus. As they complete the section, help students construct an explanation as to why some vaccinations last a lifetime while other vaccinations need regular boosters. Students should also consider how vaccinations that require boosters increase the global challenges of vaccine delivery. Students should be able to connect that viruses that reproduce rapidly and mutate at high rates are more likely to need vaccine boosters. These viruses require more involved distribution processes and record keeping than single vaccinations.

Global Challenges of Vaccine Delivery Pathogens can be easily transmitted as people travel around the globe. However, most of the technology to develop and produce large quantities of vaccines exists primarily in countries with well-funded and established medical and science infrastructure. Other parts of the world do not have these resources or the money to purchase the vaccines. Solving a global problem such as a pandemic disease requires a robust global response that addresses this inequality. The support of international organizations is critical for the delivery of vaccines to the places that need them most (**Figure 13-24**).

FIGURE 13-24 Workers in Burundi unload donations of SARS-CoV-2 vaccines in October 2021.

Key regulatory agencies involved in vaccine distribution and policy in the United States and worldwide are the Centers for Disease Control and Prevention (CDC) and the World Heath Organization (WHO), respectively.

Vaccines are delivered in many ways, and the ideal method varies depending on the constraints of the situation. Many vaccines are delivered as part of childhood vaccination programs. Others, such as the seasonal influenza and COVID-19 vaccines, must be administered as needed. However, not every nation or area has robust vaccination programs or the infrastructure necessary to distribute vaccines.

Administering and distributing vaccines poses many challenges. Most vaccines are administered via injection into the muscle, which requires a trained nurse. Many vaccines have storage requirements, such as stable cold temperatures. A variety of approaches have been studied in attempts to address difficulties in storage, transportation, and administration of vaccines. Among these approaches are vaccines that are stable at a variety of temperatures, as well as vaccines that can be administered orally or inhaled. Other attempts have involved developing edible vaccines in genetically modified fruits or vegetables, as well as making a live bacterial vector that will spread through a population, expressing vaccine proteins but not causing illness.

STSE Science, Technology, and Society *Why is it important to ensure that all countries around the globe have access to sufficient quantities of vaccines, as well as resources for delivering them to the population?*

Science Background

COVID-19 Vaccine Challenges One of the challenges that scientists and medical professionals faced with the distribution of one of the early COVID-19 vaccines was that it had to be kept below -70 °C to remain viable. This required specialized ultra-cold freezers for both transport and storage. However, many rural and international medical centers did not have access to these devices. This delayed the distribution of the vaccines in certain regions. Once students complete the reading on this page, write *criteria* and *constraints* on the board. Have students identify the criteria and constraints that have to be considered when determining how vaccine programs distribute vaccinations.

STSE Science, Technology, and Society *Sample answer: From a scientific perspective, as long as a disease is present in one population, the pathogen could potentially mutate and spread again. From an ethical perspective, nations that have resources to protect their own population should be helping deliver vaccines and other support to people who cannot afford to develop, purchase, or distribute their own.*

DIFFERENTIATED INSTRUCTION | English Language Learners

Prior Knowledge and Experience Have students discuss what they know about vaccines and health organizations in their native country (or another country they are familiar with) to help them understand global challenges of vaccine delivery.

Beginning Ask: *In (country), are there vaccine programs for children? Who gives vaccines? Would it be difficult to transport vaccines or store them at cold temperatures?*

Intermediate Ask: *How are vaccines distributed in (country)? What vaccines are given to most children? What issues might (country) face when storing or transporting vaccines?*

Advanced Have small groups discuss more than one country and compare vaccination distribution and possible issues.

Connect to English Language Arts

Research Vaccine technologies used today are far more advanced than those first used to inoculate people against smallpox. Have students select two other diseases identified in **Table 13-2** and the descriptions of the vaccine types on the page to explain how these vaccines are different from early examples of inoculation. Students should briefly research the illnesses and vaccine types and synthesize the information they find from multiple sources. If students struggle to find sources, suggest they begin their research using www.cdc.gov and www.nih.gov. If time allows, have students share what they have learned with the class.

CCC Structure and Function *Sample answer: A viral pathogen contains all of the proteins and structures that a virus needs to replicate and spread. In a vaccine, the virus is either inactivated so it cannot replicate, or only a part of the virus structure that is important to immunity is delivered.*

Vaccine Technologies

Since the smallpox vaccine was developed, vaccine technology and delivery has advanced immensely. Public health efforts and vaccine requirements have greatly reduced and nearly eliminated many diseases (**Table 13-1**). Smallpox is the only disease that has been eradicated—or eliminated—worldwide.

Vaccine Types Many different types of vaccines can be developed, depending on the nature of the pathogen and the body's immune response to infection. The following are types of vaccines that have been successfully developed and used:

- **Inactivated Vaccines** These vaccines involve delivering whole virus particles that have been inactivated so they cannot reproduce.
- **Live-attenuated Vaccines** These vaccines are made from whole live viruses that have been modified to not cause disease.
- **Messenger RNA (mRNA) Vaccines** These vaccines directly deliver mRNA (encoding a viral protein) in a lipid shell. The mRNA induces the body to produce an antigen that will elicit an immune response against the pathogen.
- **Subunit Vaccines** These vaccines directly introduce the antigen or nonreplicating parts of a pathogen, which will elicit an immune response against the pathogen.
- **Toxoid Vaccines** These vaccines elicit an immune response against a toxic component made during an infection instead of the infecting pathogen.
- **Viral-vector Vaccines** These vaccines introduce a different, safe virus to deliver genetic material that will then induce the body to make antigens that will elicit an immune response to the pathogen virus.

CCC Structure and Function *What is the difference between a viral pathogen and a vaccine that uses a virus or part of a virus to elicit an immune response?*

TABLE 13-1. Examples of Key Vaccines

Disease	Pathogen	Common vaccine type(s)	Outcomes
COVID-19	coronavirus	mRNA or viral-vector	varies by formulation, typically >90% effective at preventing infection, and a high rate of protection against severe illness
HPV	papillomavirus	subunit	results (2021) show 87% reduction in cervical cancer in the United Kingdom
measles	morbillivirus	live-attenuated	97% effective at protecting against infection (CDC)
polio	poliovirus	inactivated	eliminated in United States, 99.9% decrease in cases worldwide
seasonal influenza	influenza virus	inactivated, live-attenuated, subunit	CDC reports 40%–60% seasonal reduction in illness
smallpox	variola virus	live-attenuated	eradicated in 1977; samples exist only in laboratories
tetanus	*Clostridium tetani* (bacteria)	toxoid	WHO reported a 96% decrease in childhood deaths from 1988 to 2017
whooping cough	*Bordetella pertussis* (bacteria)	subunit	80%–90% effective at protecting against infection (CDC)

CROSSCUTTING CONCEPTS | Influence of Science, Engineering, and Technology on Society and the Natural World

Vaccine Advancements Scientific advances in vaccine research and gene manipulation have allowed scientists to develop new ways to prevent existing and emerging diseases. Early inoculations against smallpox used scab material containing live virus from infected individuals. As the science progressed, new technologies were used to solve the problem of infectious diseases. Even now, scientists continuously analyze and modify medical findings to increase safety and improve efficacy. Scientists must also consider any unforeseen consequences that may arise from new technologies and how they may affect society. Have students discuss how medical advances may affect some segments of society differently from others.

Influenza Vaccines There are four types of influenza viruses (A, B, C, and D). Types A and B are responsible for seasonal flu epidemics. Influenza A viruses are also responsible for historic pandemics as well as avian and swine influenza infections. Subtypes of influenza are determined by two different proteins on the surface of the virus, hemagglutinin and neuraminidase (**Figure 13-25**), indicated by H and N. Influenza has a genome with eight RNA segments, and different viruses often swap genome segments to generate new, unique virus subtypes that can consistently evade the human immune system. Each year, there are approximately one billion cases of influenza worldwide, resulting in 290,000–650,000 related deaths.

VIDEO

VIDEO 13-3 Go online to learn about influenza subtypes and how the variety of subtypes affects vaccine development.

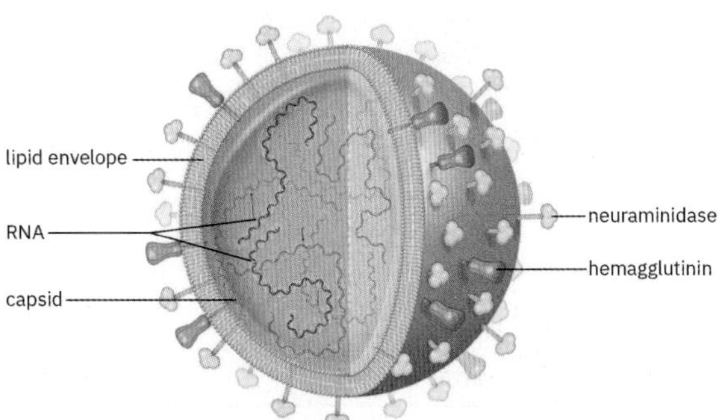

FIGURE 13-25 Influenza subtypes are referred to based on the types of hemagglutinin (H) and neuraminidase (N) proteins present on the surface of the virus.

Producing seasonal influenza vaccines is a year-round process (**Figure 13-26**). Scientists observe the dominant subtypes in the southern hemisphere's winter to anticipate the upcoming influenza season in the northern hemisphere, and vice versa. Vaccines are produced twice a year, for the southern and northern hemisphere influenza seasons. Influenza vaccines are produced through a variety of methods. For most vaccines, viruses are grown in eggs or mammalian cell lines, and they are then inactivated before purification for vaccine use. Live-attenuated influenza vaccines are also grown in eggs. Not all viruses or influenza subtypes are able to be grown using these methods.

FIGURE 13-26 Outside of tropical regions, influenza tends to peak every winter season. The data show positive specimens for the northern hemisphere. The 2009–2010 H1N1 swine flu pandemic and the 2020–2021 winter during the COVID-19 pandemic are both represented in this graph's timeframe.

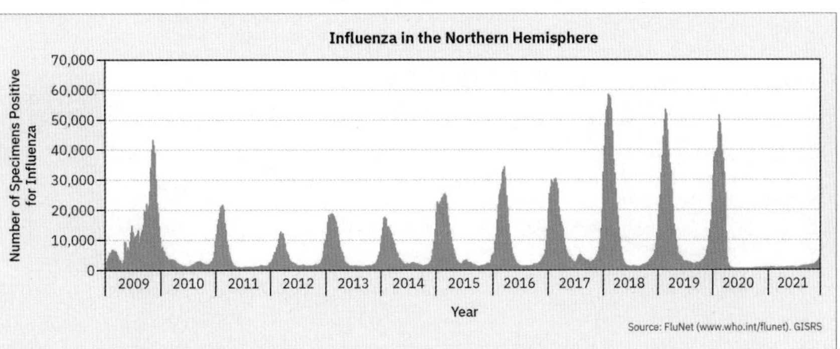

DIFFERENTIATED INSTRUCTION | Leveled Support

Struggling Students Help students understand why producing the seasonal flu vaccine is a year-round process by reminding them that the northern and southern hemispheres experience opposite seasons. By observing the most common strains that exist in the southern hemisphere during its winter season, scientists can use this to predict the most common strains that will be transmitted across the northern hemisphere when winter rolls around.

Advanced Learners Ask students to examine **Figure 13-26** and look for a pattern in the timing of peak infection annually between 2010 and 2020. Students should identify that the peaks generally occurred right after the beginning of the year. Then, have students examine the peak that is associated with the 2009–2010 flu season and identify how it was different. Have students consider why infections might have peaked earlier.

Science Background

Influenza Typically, if a flu variant is similar to one that a person has previously been exposed to, the individual is less likely to get sick. Influenza type A is known to cause disease in humans but is mainly found in wild birds. The swine flu pandemic in 2009–2010, also known as the A(H1N1) swine flu, was a Type A influenza virus but consisted of a unique combination of avian, swine, and human flus. Most young adults had not previously been exposed to a similar virus, and the existing flu vaccines did not protect against it, leading to more severe illness in this population. Most older individuals showed antibodies against the virus, indicating that they had encountered a similar variant in the past. Have students discuss why influenza variants occur.

Virus Mutations Students learn about the types of influenza viruses and subtypes that can recombine to evade the immune system. In **Section 8.4**, students are first introduced to how viral reassortment and mutations can lead to novel influenza strains.

Video

Time: 2:53

Video 13-3 shows influenza subtypes and how the variety of subtypes affects vaccine development. Guide a discussion to compare the rapidly changing influenza virus and need for yearly flu shots to students' initial experiences with the Covid-19 vaccines and need for boosters when new variants were identified. Elicit from students their ideas about how the Covid-19 vaccine may be handled differently a few years after the outbreak than it was at the beginning.

CCC Patterns *Sample answer: In places with cold winters, people spend more time inside together, and maybe the colder and dryer air affects transmission. In the winter where people wore masks and stayed farther apart to reduce COVID transmission, influenza cases dropped. Wearing masks and staying farther away from each other, particularly when sick, could help reduce transmission of respiratory viruses.*

Connect to Mathematics

Use Units to Understand Problems Students may underestimate the number of cases shown in **Figure 13-27** upon first glance. Remind students that when evaluating data, they need to check the units for the scale in a graph. Guide students to the label on the *y*-axis and make sure that they understand that the graph shows the number of cases per 100,000 people, not just the number of cases. Check student understanding of the scale by asking about how many cases there were on July 3 in the unvaccinated population if there were a total of 400,000 people (*about 400*).

Address Misconceptions

COVID-19 Vaccine There are several misconceptions about the COVID-19 vaccines that have led to vaccine hesitancy. One such misconception is that the vaccine can alter DNA because it contains mRNA or viral vectors. Students may think that because genetic material is injected, the genetic material will be integrated into their DNA. To guide students in their understanding, explain that the two different types of vaccines function in different ways, but neither integrates the genetic material into their genomes. The mRNA vaccine delivers genetic code for cells to produce the necessary proteins that induce immunity, but mRNA is not able to enter the nucleus. It is transcribed in the cytoplasm before being broken down by enzymes. The DNA carried by the viral vector vaccine does enter the nucleus but does not have all the components needed to make a new virus and does not integrate into the DNA.

VIRAL SPREAD

Gather Evidence *Vaccines for SARS-CoV-2 were generated very quickly, compared with traditional methods of vaccine development that lead to inactivated, attenuated, and subunit vaccines. How do you think scientists' increased understanding of mRNA contributed to the rapid development of both mRNA and adenovirus-vector vaccines? How might this technology help us respond quickly to new variants or future epidemics and pandemics?*

Another type of influenza vaccine uses recombinant DNA technology. Large amounts of hemagglutinin protein are produced to make a subunit vaccine. Influenza vaccines typically cover multiple H and N subtypes, in hopes of more effectively protecting us against a variety of possible infections. Vaccines are key in combating seasonal influenza and providing better outcomes for most patients. They work best in combination with other precautions that can be taken to reduce the spread of illness when an individual is sick.

CCC Patterns *Why do you think influenza and other respiratory viruses tend to be transmitted more often in the winter season of each hemisphere? What measures could be taken to reduce transmission?*

COVID-19 Vaccines COVID-19 has caused millions of deaths, along with significant long-term health concerns for some patients who survived infection. Daily case data for the United States show the trends over the first two years of the pandemic (**Figure 13-27**). The COVID-19 pandemic demanded a more rapid response than traditional vaccine development allows. Vaccines have demonstrated effectiveness in reducing transmission and negative outcomes (such as hospitalization) for vaccinated individuals (**Figure 13-28**).

FIGURE 13-27
Spikes in the number of daily COVID cases in the United States were associated with the Delta (mid 2021) and Omicron (early 2022) variants of SARS-CoV-2.

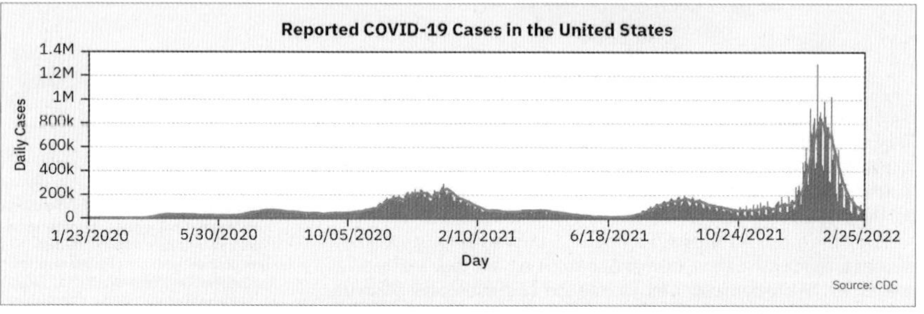

FIGURE 13-28
While the administration of vaccines in the United States did not stop all viral transmission for vaccinated individuals, particularly in the case of new variants, rates of hospitalization and death are significantly lower among the vaccinated population.

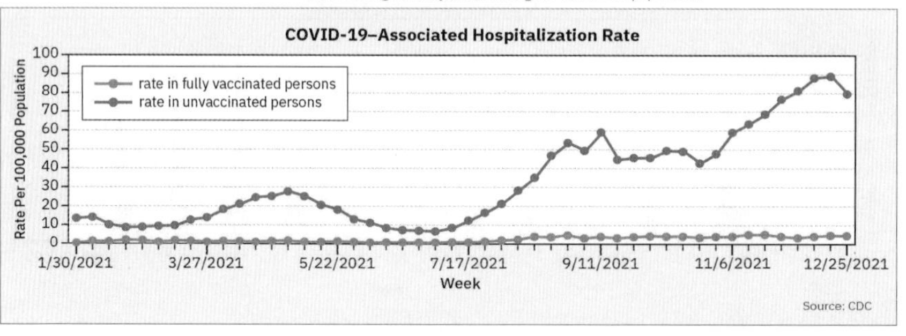

REVISIT THE ANCHORING PHENOMENON

Gather Evidence *Sample answer: Scientists' understanding of mRNA allowed them to generate RNA sequences that could be delivered to patients and generate the proteins needed to start an immune response. The ability to quickly generate new or modified sequences for new variants or other types of viruses should make it even easier to optimize and quickly manufacture future vaccines.*

Smallpox is the first and only disease to be fully eradicated through vaccination, but it took nearly two centuries. It took about fifteen years to develop an influenza vaccine. As scientists continue to learn more about how flu viruses change, the vaccine also has to change.

Vaccine research involving mRNA started in the 1970s. By the 1990s, they were being tested in mice, and the first clinical trials for the technology began in the 2010s. However, scientists rapidly built on this preexisting technology to develop the COVID-19 vaccine within two years.

SARS-CoV-2 is the causative agent of COVID-19. The virus has a spike protein on its surface that is recognized as an antigen by the immune system. Adenovirus-vector and mRNA vaccine delivery methods were quickly developed and tested in an effort to slow down the virus's rapid transmission. Both methods rely on delivering genetic material into a host cell (**Figure 13-29**). The viral-vector versions of the COVID-19 vaccine deliver genetic material with an adenovirus that interacts with host cells to deliver DNA encoding the antigen. mRNA vaccines involve synthesizing an RNA molecule that encodes the viral spike antigen. The mRNA is delivered in lipid nanoparticles that can be absorbed by host cells, which then express the antigen and stimulate the production of antibodies to protect against the virus. The speed gained by delivering genetic material, rather than having to grow virus particles and purify their proteins, gives mRNA delivery a significant advantage over other vaccine preparation methods.

Although the mRNA and adenovirus-vector vaccines for COVID-19 were developed quickly, both use established technologies. Viral-vector technology has been successfully used in gene therapy, and mRNA vaccines and lipid nanoparticles have both been under study since 1990. It is likely that the success of the COVID-19 vaccine will lead to more scientific advancements and new vaccines that use mRNA. In addition to these newer genetic vaccine technologies, manufacturers have also developed more traditional inactivated and subunit vaccines for SARS-CoV-2.

FIGURE 13-29 COVID-19 vaccine mRNA encoding the viral spike protein antigen can be delivered using adenovirus vectors (a) or lipid nanoparticles (b).

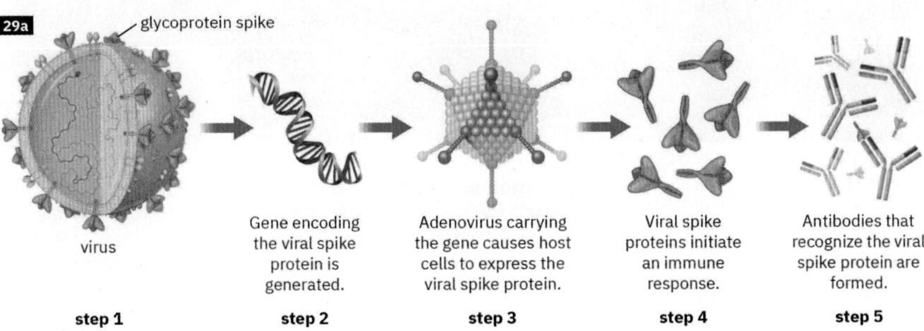

29a
glycoprotein spike
virus
Gene encoding the viral spike protein is generated.
Adenovirus carrying the gene causes host cells to express the viral spike protein.
Viral spike proteins initiate an immune response.
Antibodies that recognize the viral spike protein are formed.
step 1
step 2
step 3
step 4
step 5

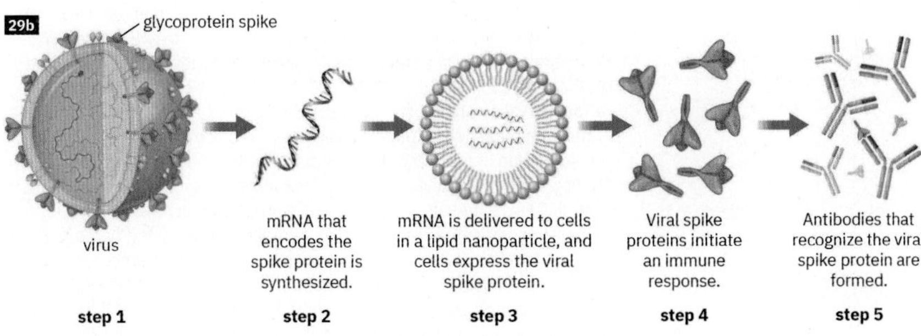

29b
glycoprotein spike
virus
mRNA that encodes the spike protein is synthesized.
mRNA is delivered to cells in a lipid nanoparticle, and cells express the viral spike protein.
Viral spike proteins initiate an immune response.
Antibodies that recognize the viral spike protein are formed.
step 1
step 2
step 3
step 4
step 5

Visual Support

Vaccines Pair students and have them examine **Figure 13-29**. Have them point out the steps that are different between **Figure 13-29a** and **13-29b**. Students may need additional information for step 3 of both processes for complete understanding. For **Figure 13-29a**, guide students to understand that recombinant DNA encoding the spike protein is packaged in the adenovirus vector (step 2). The adenovirus vector infects cells and delivers the DNA, which enters the nucleus to be transcribed into mRNA that will be translated in the cytoplasm of the cell to make viral spike proteins (step 3). Have students discuss how the lipid nanoparticle method differs. In this case, mRNA for the spike protein is made directly and packaged in lipid nanoparticles (step 2). Once injected, these particles will fuse to target cell membranes, releasing the mRNA into the cell where it is directly translated to produce viral spike proteins (step 3).

13.3 VACCINE DEVELOPMENT **413**

In Your Community

Pathogen Reservoirs There are a number of common animal reservoir species that can transmit diseases between each other and humans. Anthrax is found in sheep but can infect humans. Trichinosis is caused by parasitic roundworms that are often harbored by pigs but can also infect humans and other animals. The environment can also be a reservoir for disease-causing agents. For example, *Legionella*, the bacterium that causes Legionnaires' disease, occurs naturally in freshwater ecosystems such as lakes and streams. Environmental reservoirs can be especially problematic, as they can be difficult to pinpoint and cannot be cleared by vaccination.

Take Action Have students identify diseases that exist in their community, either in reservoir species or in environmental reservoirs. Students should take steps to find out how to protect themselves from these diseases.

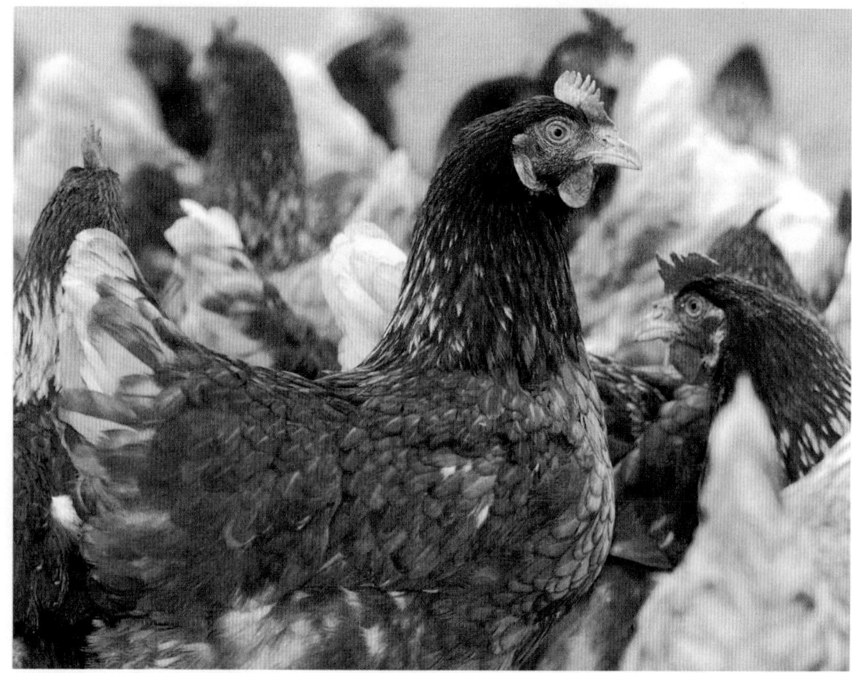

FIGURE 13-30 Chickens are often vaccinated to protect the flock from disease and to reduce transmission of certain pathogens to humans.

Vaccination of Reservoir Species Most human pathogens can also infect other species. Organisms that can be infected and transmit pathogens to humans are known as **reservoirs**. In some cases, vaccinating these reservoir species may be a method that can protect humans from infection. For example, animals raised as food sources may transmit pathogens to humans, and vaccinations of the animals have successfully been shown to protect humans who encounter them (**Figure 13-30**).

The challenge is different with wild populations of animals, however. For example, coronaviruses infect a variety of species. It is most likely that SARS-CoV-2 has a natural reservoir in bats, and the virus has also been shown to infect other species, such as white-tailed deer. For some pathogens, such as the virus that causes rabies, scientists are making efforts to vaccinate the natural reservoir (bats) to limit the spread of the virus to humans. While effective vaccines already exist for some pathogens in nonhuman hosts, it is difficult to vaccinate an entire bat population. To be effective, vaccines for wild reservoir species will have to distribute themselves. Scientists are working on several approaches. One way of generating transmissible vaccines is to make an attenuated pathogen that will elicit an immune response but not cause disease. Another approach involves putting part of the target pathogen genome into a safe vector virus that can spread through the population. Some successful examples of immunizations of pigs have been reported to limit transmission of an infection caused by *Salmonella enterica* from pigs to humans.

414 CHAPTER 13 GENETIC TECHNOLOGIES

CROSSCUTTING CONCEPTS | Cause and Effect

Viruses and Growing Populations Cause-and-effect relationships are especially important to consider in disease prevention and reducing the spread of disease. As the world population grows, more space for food production is needed to feed the growing population. However, as humans encroach on new areas and live in closer quarters with each other and other species, scientists believe that there will be an increase in the number of pathogens that emerge. Ask students to think of ways that diseases spread between animals and people and how the impact of such diseases can be minimized. Students should consider reservoirs as origins of pathogens, infectious dose, the mode of transmission, and herd immunity in their explanation of how to decrease the risk of a novel pathogen causing an outbreak.

Supporting Vaccines With Other Measures

Slowing down the spread of disease requires multiple parallel efforts that involve everyone from individuals, to small communities, to international agencies and governments. A significant challenge with any public health crisis is related to communication of scientific ideas to the broader public. As an example of effective communication, **Figure 13-31** demonstrates the Swiss cheese model for mitigating the transmission of an illness that is spread through airborne droplets. Just one approach, such as vaccines, is often not enough to protect against illness. Depending on the type of disease and how it is transmitted, there are many factors to consider.

- **Infectious Dose** The amount of virus or other pathogen necessary to cause infection varies depending on the pathogen. Generally, the less pathogen an individual is exposed to, the less likely the individual is to be infected or experience severe illness.
- **Mode of Transmission** Solutions for reducing the spread of pathogens must reflect the mode of transmission. Different pathogens can be transmitted through blood, food, water sources, or droplets in the air.
- **Herd Immunity** For some pathogens, a level of immunity can be reached such that the infections become very rare, or even eradicated for some populations. Herd immunity is important because it enables a community to protect vulnerable individuals who cannot get vaccinated and most need protection from severe disease.

STSE Science, Technology, and Society *What can you do to prevent transmission of diseases and protect other people in your community?*

FIGURE 13-31 This model illustrates how each method of prevention against transmission of an airborne virus has holes, but together, they provide a robust defense against disease transmission. Notice how these prevention methods involve a variety of both shared and personal responsibilities.

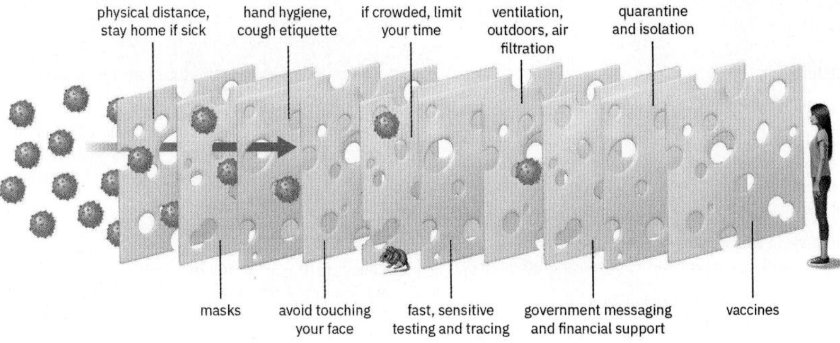

13.3 REVIEW

1. **Distinguish** What is the relationship between vaccines and antibodies?
2. **Explain** Why is public participation and confidence necessary for the success of a vaccine?
3. **Identify** Which of the following statements accurately describes the COVID-19 vaccines?
   A. They deliver whole virus to host cells.
   B. They deliver genetic material to host cells.
   C. They can be transmitted between organisms.
   D. They deliver inactivated parts of a virus to the host's cells.

Science Background

Mode of Transmission Pathogens infect individuals through a variety of methods. The table below provides examples of pathogens and their various modes of transmission.

Illness/Pathogen	Mode of transmission
influenza	droplets
Ebola	contact with fluids
E. coli	contaminated food or water, direct contact
malaria	mosquito bites
chickenpox	direct contact, droplets
cholera	contaminated food or water, contaminated undercooked shellfish

Have students select one of the illnesses or pathogens in the table and consider which steps in the Swiss cheese model would be effective at minimizing infection with the disease. If students identify that a step would not be helpful, they should make sure to support their claim with evidence.

STSE Science, Technology, and Society *Sample Answer. I can help educate people about how disease spreads and share ways severity can be reduced or transmission can be prevented. I can advocate for people who need help and resources. If I am sick, I can wear masks and minimize my exposure to other people.*

13.3 REVIEW

1. *Sample answer: Vaccines work by "teaching" the host's antibodies to recognize a specific type of antigen, either by introducing a part of the antigen or genetic material that directs cells to produce the antigen so that the antibodies are prepared to respond to it.* **DOK 2**
2. *Sample Answer: For contagious infectious diseases, if a virus cannot survive in a vaccinated host, it will not be transmitted to a new host, so the amount of transmission is reduced by having a greater number of people vaccinated. Even if the disease is not contagious, vaccination can also reduce the severity of the disease in an infected host.* **DOK 3**
3. *B* **DOK 2**

MINILAB
HERD IMMUNITY

Students work with models of populations with different vaccination rates and explore how this impacts the spread of a virus and the development of herd immunity and strong protection from infectious diseases. They will also consider how well their model compares to real-world vaccination rates and the spread of diseases.

Time: 30 minutes

Advance Preparation

- Select objects for the models that students can easily grab a handful of and that both colors of objects are the same size and shape. Be sure students cannot see into the containers, and the containers should be big enough for students to fit their hand but small enough to easily grab a handful of objects.

Procedure

Before **Steps 1 and 2**, you can save time by having the tubs and labels prepared in advance.

For **Steps 4**, **5**, and **6**, it could save time if you already have the objects separated and counted.

In **Step 11**, each student is only collecting data once for each vaccination group. Then, they are calculating the average infection rate their group or class obtained for each vaccination group.

Results and Analysis

1. **Calculate** *Sample answer: The average infection rate for the 30 percent vaccinated was 65 percent. The average infection rate for the 50 percent vaccinated was 60 percent. The average infection rate for the 95 percent vaccinated was 4 percent.* **DOK 2**
2. **Explain** *Sample answer: It would be realistic to have a tub with 0 percent vaccinated because this would be comparable to a situation in which a new virus is introduced into a population and no vaccine is available for that virus. It would not be realistic to have a tub with 100 percent vaccination because there are almost always situations in which a certain percentage of the population is not able to receive a vaccine, such as newborns, the elderly, or those with immune systems that are weak because of other illnesses.* **DOK 2**

MINILAB

HERD IMMUNITY

SEP Constructing Explanations and Designing Solutions How does the percentage of people vaccinated impact the number of people infected with a disease?

Vaccines provide a "preview" of a specific pathogen that allows the immune system to prepare itself to recognize the pathogen when it sees it again. In this minilab, you will work with models of populations that have different percentages of vaccinations and explore how this affects the spread of a virus.

Materials
- opaque tubs (3)
- opaque plastic bags or lids (3)
- paper (3 pieces)
- 300 small objects of 2 different colors (suggestions include colored cotton balls, toy building blocks, small beads or beans, etc.)
- masking tape or labels
- marker
- scissors

Procedure

1. Set the three opaque tubs side by side.
2. Label one tub "30% vaccinated," one "50% vaccinated," and one "95% vaccinated." Each tub represents a population with a different percentage of vaccinated people.
3. Determine which color object represents vaccinated people and which represents unvaccinated people.
4. Place 30 "vaccinated" objects and 70 "unvaccinated" objects in the tub labeled "30% vaccinated."
5. Place 50 "vaccinated" objects and 50 "unvaccinated" objects in the tub labeled "50% vaccinated."
6. Place 95 "vaccinated" objects and 5 "unvaccinated" objects in the tub labeled "95% vaccinated."
7. Cover each tub with a lid or plastic bag and then cut a small opening in each lid so that a hand can reach into the tub.
8. Each student in a group should take a turn being a virus. The "virus" reaches into each tub, one at a time, and "infects" as many people as possible by grabbing a handful of the objects and pulling them out of the tub.
9. After the objects are drawn out of the tub, count the total number of objects you pulled out, and then count the number of unvaccinated individuals that became infected. Create a data table and record these data in your biology notebook.
10. Return the objects to the tub from which they were taken.
11. Repeat Steps 8 through 10 until everyone has had the opportunity to withdraw objects from all three tubs. Shake or stir up the objects in the container between students.

Results and Analysis

1. **Calculate** Based on the data you collected, what is the average infection rate of objects (people) for each of the tubs?
2. **Explain** Would it be realistic to have a tub with 0 percent vaccinated objects? Would it be realistic to have a tub with 100 percent vaccinated objects? Why?
3. **Identify** How do infected people spread disease in their community and when they travel outside their community?
4. **Compare** Imagine a real-life scenario where an infection is spreading in your community. What factors are represented well in your model, and what factors are not?

3. **Identify** *Sample answer: Infected people spread the disease the same way within their community as they do outside their community. Depending on the disease, infected people can spread to others through direct contact or indirect contact. When an infected person leaves their community, they have a greater chance of infecting more new people.* **DOK 2**
4. **Compare** *Sample answer: Just as in the real world, the model shows how nonvaccinated people can become easily infected. It also shows how having the majority of a population vaccinated can keep infection rates really low. In our models, however, we only had one "virus" move into a population at a time. What the models do not show is that each time a person is infected, they become a carrier of and transmitter of the virus. So, instead of just one virus trying to infect the population, with each infection, there are more and more viruses circulating in the population—so, in our model's terms, more hands reaching into the population at the same time trying to infect people.* **DOK 3**

TYING IT ALL TOGETHER ENGINEERING
GENETICS TO THE RESCUE!

HOW CAN GENETIC INFORMATION BE USED IN MEDICAL AND OTHER SCIENTIFIC APPLICATIONS?

Somatic cell nuclear transfer (SCNT) was used to produce Elizabeth Ann, the black-footed ferret clone described in this chapter's Case Study. This chapter also introduced other techniques, including PCR, recombinant DNA, and CRISPR. These are commonly used in studying genetics and can also be applied to tackle a variety of problems. When considering genetic engineering as part of a solution, there are often ethical concerns associated with changing the genes of living and future organisms. There are also trade-offs to be made between the effectiveness of the solution and the potential impact of unpredictable outcomes.

For instance, genetic disorders in animals are usually observed in domesticated species that have been bred to have specific traits. These disorders occur more frequently in certain breeds as a consequence of the increased likelihood of offspring inheriting a recessive trait from both parents. Genetic engineering can rapidly produce livestock animals with traits that make them profitable in the marketplace but may render them susceptible to genetic disorders. This poses ethical dilemmas regarding the role of humans in preserving the value and quality of animal life.

In this activity, you will develop genetic engineering solutions for conserving a hypothetical wild animal population. Work individually or in a group to complete these steps.

Define the Problem

1. Consider a hypothetical wild animal species whose global population is endangered by a deadly pathogen. Suppose that it is possible, at a reasonable cost, to genetically engineer an individual of the species to resist infection by the pathogen. List some criteria and constraints that should be met by a solution designed to conserve this species. Include time, safety, social, ethical, and environmental factors in your list.

Design Solutions

2. Design at least two different solutions to the conservation problem, using genetic technologies you learned about in this chapter.
3. Construct and use models to explain how each solution would work.

Evaluate Solutions

4. Evaluate each solution in terms of how well it satisfies the criteria and constraints you defined. Decide which is the best solution according to your evaluation.

FIGURE 13-32 Naked foal syndrome is caused by a recessive trait that renders offspring mostly hairless and causes other health issues. Typically offspring with this disorder die within weeks or months of birth. This Akhal-Teke colt shown survived longer than most affected horses.

TYING IT ALL TOGETHER

ELABORATE

The Case Study discusses how SCNT was used to produce a clone of an endangered ferret and how genetic engineering techniques are being used by humans for a variety of purposes. In the Tying It All Together activity, students will develop a genetic engineering solution to conserve a hypothetical wild animal population.

This activity is structured so that students may begin once they complete Section 13.2. **Step 1** has students consider the problem of a hypothetical animal species at risk of extinction due to a deadly pathogen. **Step 2** corresponds to the content in Sections 13.1 and 13.2 and expects students to design two solutions to the conservation problem. **Step 3** asks students to use models to explain how the solution would work. Students will then evaluate each solution as to how well they address the criteria and constraints in **Step 4**.

Alternatively, this can be a culminating project at the end of the chapter. The engineering practices defined here reflect the reasoning processes that students may use to find a solution for the Unit 4 Activity.

Go online to access the Student Self-Reflection and Teacher Scoring rubrics for this activity.

SCIENCE AND ENGINEERING PRACTICES
Constructing Explanations and Designing Solutions

Considering Solutions Help students prepare for this activity by having them select a hypothetical animal species and the pathogen that would infect the species. Some students may find it helpful to describe the symptoms of the illness that the animal would develop. Students should take the time to make a T-chart of the criteria and constraints of their solution, and to consider how the different solutions may have ethical, societal, and environmental ramifications. Once students have defined their problem fully, they may begin with designing their solutions. The exercise in describing the symptoms of the pathogen may help students decide the type of treatment they would need. Students may pull from any of the genetic engineering techniques that were discussed in the lesson. Once students have designed their solutions, they should construct a model of each that they can use to describe their solutions to a partner. Students choose to evaluate their solutions on their own, or to evaluate each other's solutions.

CHAPTER 13 **REVIEW**

EVALUATE

REVIEW KEY CONCEPTS

1. *A* **DOK 1**

2. *A, B, E* **DOK 2**

3. *template: It is double-stranded; It separates when the reaction is heated; primer: It is single-stranded; polymerase: It is an enzyme; It adds new nucleotides to the end of DNA molecules.* **DOK 1**

4. *B, D, E* **DOK 2**

5. *D* **DOK 1**

6. *C, D* **DOK 1**

7. *Sample answer: Cloning results in two organisms that are genetically identical. Genetic engineering results in one organism whose genome is different from what it was before the organism was genetically engineered.* **DOK 2**

8. *C, D, E* **DOK 1**

9. *Sample answer: Vaccination produces an immune response similar to the response produced by a natural infection, but usually without also producing the symptoms of the infection. Vaccines achieve this by introducing dead or weakened pathogens or antigens into the body, or by providing instructions for body cells to produce antigens that will provoke an immune response. Once the body learns to recognize the antigens associated with a pathogen, it can produce antibodies to defend itself against natural infections. Meanwhile, the antigens introduced by the vaccine are unable to replicate, so they do not pose a risk of serious infection.* **DOK 1**

CHAPTER 13 **REVIEW**

REVIEW KEY CONCEPTS

1. **Identify** A molecular biologist uses restriction enzymes to cut a section of genomic DNA, inserts the fragment into the plasmid of a bacterium, and then allows the bacterium to reproduce. The offspring of the bacterium all contain the recombinant plasmid. Which of these procedures best describes the scientist's method?
   - **A.** DNA cloning
   - **B.** DNA profiling
   - **C.** Gel electrophoresis
   - **D.** Polymerase chain reaction (PCR)
2. **Compare** Two samples of DNA migrate the same distance on an electrophoresis gel. What must be true about the DNA molecules in the two samples? Select all correct answers.
   - **A.** They have the same length.
   - **B.** They have the same charge.
   - **C.** They have the same sequence.
   - **D.** They have the same number of cytosine bases.
   - **E.** They have the same number of nucleotides.
3. **Describe** Decide whether each of these statements best describes the template, the primer, or the polymerase in a PCR reaction.
   - It is an enzyme.
   - It is single stranded.
   - It is double stranded.
   - It separates when the reaction is heated.
   - It adds new nucleotides to the end of DNA molecules.
4. **Contrast** Which of these statements correctly describes how the genomes of two unrelated humans would most likely differ? Select all correct answers.
   - **A.** The chemical composition of their genomes would differ.
   - **B.** They would have different numbers of short tandem repeats.
   - **C.** The genomic content inherited from ancestral species would differ.
   - **D.** The sequence of the nucleotide bases in their genomes would be different.
   - **E.** The single-nucleotide polymorphisms (SNPs) in their genomes would differ in length.
5. **Define** Which statement best describes how clones are produced through somatic cell nuclear transfer (SCNT)?
   - **A.** A somatic cell's nucleus is removed, and then the nucleus of a fertilized egg is transferred to the somatic cell.
   - **B.** A fertilized egg's nucleus is removed, and then the nucleus of a somatic cell is transferred to the fertilized egg.
   - **C.** An unfertilized egg's nucleus is removed, and then the nucleus of a fertilized egg is transferred to the unfertilized egg.
   - **D.** An unfertilized egg's nucleus is removed, and then the nucleus of a somatic cell is transferred to the unfertilized egg.
6. **Identify** Which of these conditions would most likely be treatable with gene therapy? Select all correct answers.
   - **A.** muscle loss due to advanced age
   - **B.** congestion caused by a respiratory virus
   - **C.** mental impairment due to phenylketonuria
   - **D.** difficulty breathing caused by cystic fibrosis
   - **E.** an infection of the stomach lining by bacteria
7. **Compare** How is cloning an organism different from genetically engineering an organism?
8. **Identify** Which of these vaccine technologies involve the delivery of whole viruses to a patient? Select all correct answers.
   - **A.** toxoid vaccines
   - **B.** subunit vaccines
   - **C.** inactivated vaccines
   - **D.** viral-vector vaccines
   - **E.** live-attenuated vaccines
   - **F.** messenger RNA vaccines
9. **Describe** Explain how vaccines are able to induce immunity without causing severe illness in the patient.

418 CHAPTER 13 GENETIC TECHNOLOGIES

CRITICAL THINKING

10. *2* **DOK 2**

11. *Sample answer: More variation occurs in noncoding regions of genomic DNA because these regions are not transcribed and translated. Mutations in these regions are less likely to cause a change in phenotype, and therefore less likely to cause a detrimental effect on the organism.* **DOK 3**

12. *Sample answer: Although clones are genetically identical to their parents, environmental factors and random differences in gene expression also play a role in phenotypes. Environmental factors can cause some genes to be expressed or repressed, resulting in physical or behavioral differences.* **DOK 3**

13. *Answers may vary. Sample answer: One possible advantage is that cows are large animals that produce a lot of milk, so manufacturers might be able to produce larger amounts of insulin more cheaply or efficiently. Another possible advantage is that making the protein in an organism that is more similar to humans (another eukaryote) might make a better insulin.* **DOK 3**

14. *Sample answer: Answers may vary. Students may argue that scientists should collect data on measles incidence and measles vaccination rates in communities around the world. If their hypothesis is correct, they should find a correlation between the location of outbreaks and communities with low rates of vaccination.* **DOK 3**

CRITICAL THINKING

10. **Reason Quantitatively** AluI is a restriction enzyme that recognizes the sequence AGCT and makes a single cut between the guanine and cytosine of both DNA strands. A researcher adds AluI to a sample of DNA with the following sequence:

```
TCAGCTTAACAGCTGG
AGTCGAATTGTCGACC
```

After the reaction is complete, she separates the products by gel electrophoresis. How many bands will appear on her gel?

11. **Explain** Single-nucleotide polymorphism mutations are more likely to occur in noncoding DNA rather than in coding DNA. Why do you think stable unique polymorphisms are more likely to be in noncoding regions rather than in the middle of gene sequences?

12. **Explain** In somatic cell nuclear transfer (SCNT) experiments, scientists have found that the clones do not always look or behave like the parent from which they are cloned. Why might this be the case?

13. **Evaluate** People with diabetes are unable to produce enough insulin, a hormone that regulates blood sugar levels. To combat the disease, diabetics are often prescribed insulin that is produced by bacteria that have been genetically engineered to produce human insulin. Scientists have also investigated alternative methods for producing insulin, such as using recombinant DNA technology to engineer cows to secrete human insulin in their milk. This insulin can then be isolated from the milk and purified for use as treatment. What is one possible advantage of manufacturing recombinant insulin in cow's milk as opposed to using bacteria?

14. **Analyze** Measles is a highly contagious disease that is caused by a virus. In the United States, most people are vaccinated against measles as children, and cases are extremely rare. However, in recent years, outbreaks of measles have become more common in the United States. Some scientists hypothesize that these increases are due to the reluctance of some people to vaccinate their children. What data should scientists collect to test this hypothesis? What would the data show if the hypothesis is correct?

MATH AND ENGLISH LANGUAGE ARTS CONNECTIONS

1. **Model With Mathematics** A scientist is using PCR to amplify a segment of DNA. The initial sample contains five copies of a segment of DNA that can hybridize with the pair of primers added to the reaction. After six cycles, how many copies will be produced by the reaction?

2. **NOS Science, Technology, and Society** In 2018, a Chinese scientist used CRISPR technology to produce the world's first known genetically engineered humans. In the experiment, a gene conferring resistance to HIV was edited into the genomes of human embryos. This experiment was extremely controversial and instigated great debate among medical ethicists. What is one question that medical ethicists should answer before recommending this procedure for wider practice?

3. **Use Sources in Multiple Formats** Examine the base sequences in **Figure 13-9**. Suppose a laboratory wants to determine whether a sample of DNA came from a frog or a rat. Based on the information in the chapter, would it be possible to make this determination by using PCR to amplify this location followed by gel electrophoresis of the products? Explain your answer.

4. **Synthesize Information** During the first year of the COVID-19 pandemic in 2020, governments around the world imposed restrictions on travel and policies to encourage masking and social distancing in public. These policies were intended to reduce community spread of the COVID-19 virus. During this time, public health experts observed a much lower-than-usual incidence of seasonal influenza, which is caused by a different virus than COVID-19. What conclusions can be drawn from these observations?

REVISIT VIRAL SPREAD

Gather Evidence In this chapter, you learned about how genetic technologies are employed in research and to help solve problems facing society, including viral pandemics.

1. How many genetic technologies can you identify that may have played an important role in understanding and developing a response to the COVID-19 pandemic?
2. What additional questions do you have about genetic technologies and their role in public health?

MATH AND ELA CONNECTIONS

1. *320* **DOK 2**
2. *Sample answer: Answers may vary. Ethicists should consider whether a procedure that will affect an embryo's whole life, as well as the lives of every person descended from the embryo, should be performed without the consent of the embryo. They should also consider whether the potential benefits of such procedures outweigh the risks, some of which may be unforeseen.* **DOK 3**
3. *Sample answer: No. Although there is a single substitution difference between frogs and rats at this locus, the two segments of DNA are the same size. The PCR product of sequences from both animals would be the same size and would produce bands that migrate the same distance on an electrophoresis gel.* **DOK 3**
4. *Sample answer: Since the abnormally low incidence of influenza correlated with the same period of time as the public safety measures, it is reasonable to conclude that these measures were effective in decreasing the rate of influenza transmission, and probably other viral diseases that spread through close contact between hosts.* **DOK 3**

REVISIT THE ANCHORING PHENOMENON

1. *Answers will vary. Genomic sequencing could be used to identify the virus and mutations in the genome. PCR could be used to isolate genes for study and to detect and track infections. Bioinformatics would be useful to track the spread of the virus and coordinate public health responses. Scientists also had to use these genetic technologies to develop an mRNA vaccine against SARS-CoV-2.* **DOK 2**
2. *Sample answer: How can CRISPR be used to develop tests and therapies? How can new vaccine technology lead to vaccines for diseases like HIV or cancer? How could gene editing or cloning be used to develop new treatments or cures for various disease? How could bioinformatics improve public health outcomes?* **DOK 3**

UNIT 4 SUMMARY

Class Discussion

Encourage students to practice speaking, listening, and collaborative skills when discussing these questions as a class or in smaller groups as they review the topics in Unit 4.

1. What role do transcription and translation play in DNA replication? How can DNA be used to make different proteins? How can a cell regulate gene expression?
2. Why does sexual reproduction result in greater genetic diversity? Describe the difference between genotype and phenotype. What are sex linked traits? How can genetic mutations have a positive influence on gene expression?
3. How can gene therapy be used to treat inherited diseases? Describe the relationship between vaccines and cellular reproduction.

Go Online

Four Performance Tasks for Unit 4 are available on MindTap to assess students' mastery of the following NGSS Performance Expectations:

- Task 1: HS–LS3–1
- Task 2: HS–LS3–2
- Task 3: HS–LS3–3
- Task 4: HT–ETS1–2

Go Online

The Virtual Investigation, **Fighting a Viral Pandemic**, includes critical thinking evidence that students may find useful for supporting their claims about how to slow the spread of a virus.

UNIT 4 **SUMMARY**

GENETICS

CHAPTER 11
DNA, RNA, AND PROTEINS

How is genetic information transferred and regulated?

- A series of independent experiments proved that DNA is the hereditary molecule in cells.
- Replication, transcription, and translation are mechanisms by which genetic information is moved and changed in living organisms.
- Transcription converts DNA to mRNA. Ribosomes translate RNA into proteins.
- Regulation of gene expression occurs at many points in the cell and can adjust to environmental and developmental conditions.

CHAPTER 12
GENETIC VARIATION AND HEREDITY

How does genetic information change and get passed from generation to generation?

- Sexually reproducing organisms distribute chromosomes to offspring through meiosis, which introduces genetic variation.
- Nucleotide and chromosomal mutations may be silent or may result in advantageous or disadvantageous changes.
- Gregor Mendel described basic rules of inheritance.
- The proportion of genes and traits passed from one generation to the next can often be modeled with Punnett squares.
- Complex inheritance patterns and gene expression include codominant, incomplete dominant, and sex-linked traits.
- Environmental factors can affect the expression of many genes.

CHAPTER 13
GENETIC TECHNOLOGIES

How can genetic information be used in medical and other scientific applications?

- Scientists use restriction enzymes and polymerases to cut, paste, and duplicate DNA through DNA cloning and PCR.
- Genome analyses help scientists study relatedness of organisms, understand how different genes work, and identify individuals.
- Scientists use genetic tools to modify or clone animals and plants.
- Gene therapy techniques, such as CRISPR, can replace gene function or repair gene sequences.
- Vaccine technologies take advantage of immune system functions to protect against disease.

Go Online VIRTUAL INVESTIGATION

Fighting a Viral Pandemic
How can we slow the spread of a virus?
Enter the situation room and make decisions to protect your community from a deadly virus.

DIFFERENTIATED INSTRUCTION | English Language Learners

Main Ideas and Details Have students work in small groups to identify the main idea of each chapter and the most important details that support it.

Beginning Have individual students use the wording in the chapter question to help them write a main idea, for example: *Genetic information is transferred by ______ and regulated by ______.* Then, have them share their central ideas with group members.

Intermediate Have group members work together to write a main idea for each chapter and list at least one example for each bullet point.

Advanced Have students write a main idea for each chapter along with two or three important details that support it. Then, have them discuss how the details support the main idea.

UNIT 4 **ACTIVITY**

HOW CAN WE SLOW THE SPREAD OF A VIRUS?

Argue From Evidence In this unit, you learned how genetic information functions in organisms and how it is used in technological applications. You have learned how many of these mechanisms apply to COVID-19 and the SARS-CoV-2 virus that causes it.

The human immunodeficiency virus (HIV) is an RNA retrovirus (Section 8.4). HIV is transmitted through contact with bodily fluids. It attacks immune cells of the human host, making the infected individual susceptible to other infections. This condition is known as acquired immunodeficiency syndrome (AIDS). AIDS was first recognized as a global epidemic in the 1980s.

In 2020, approximately 37.7 million people worldwide were living with HIV. About 73 percent of those individuals had access to medical treatments. Antiretroviral therapies allow people with HIV to live longer and healthier lives, with a significantly reduced risk of transmission. Both new HIV infections and AIDS-related deaths per year have decreased significantly since the early 2000s, but there are still many challenges worldwide.

Numerous efforts by organizations, such as the United Nations and the World Health Organization, have been important in curbing the AIDS epidemic through education and the distribution of treatments. Education is a critical aspect of the international response to HIV and AIDS. It leads to reduced transmission and increased adherence to treatment regimens. The more scientists and the public understand a virus, the easier it is to stop its spread. Efforts to develop an effective vaccine have been ongoing since the virus was isolated and characterized in 1984, with an mRNA vaccine in development.

HIV and SARS-CoV-2 are viruses with different modes of replication and transmission, but many of the same factors apply. Both viruses spread easily and mutate in a way that can potentially dodge vaccines. Slowing the spread of any virus relies on global cooperation as well as individuals caring for their community and working together to take precautions.

Claim Make a claim about the role of science and technology in combating SARS-CoV-2 and the COVID-19 pandemic.

Evidence Use the evidence you gathered throughout the unit to support your claim.

Reasoning Use your understanding of genetic mechanisms, mutations, and genetic and antiviral technology to address the goals and challenges associated with stopping a pandemic.

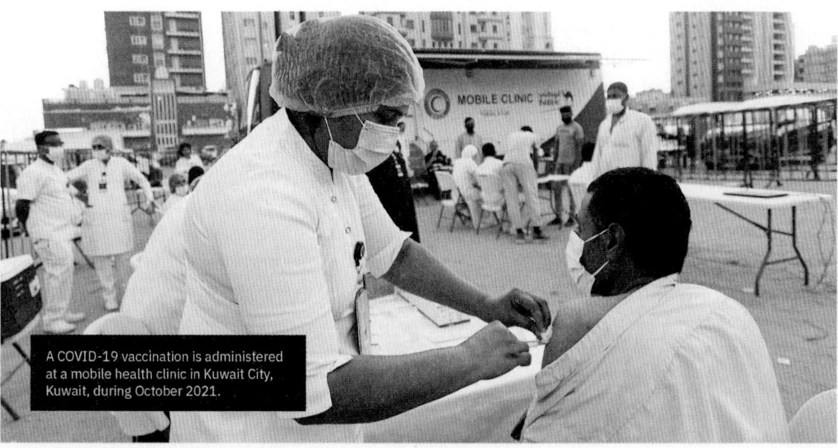

A COVID-19 vaccination is administered at a mobile health clinic in Kuwait City, Kuwait, during October 2021.

UNIT 4 **ACTIVITY**

REVISIT THE ANCHORING PHENOMENON

C Claim

E Evidence

R Reasoning

This Unit Activity asks students to formulate a claim to answer the question posed at the beginning of the unit, cite evidence they have obtained throughout the unit, and use the concepts they have learned to explain how this evidence supports their claim.

C Formulating a Claim Students should write their answer to the question "How can we slow the spread of a virus?" in the form of a single sentence. Examples of claims that students might make using terms and concepts from the unit chapters include: "Viral cells reproduce through asexual reproduction" or "Gene therapy can be used to treat genes with mutations."

E Citing Evidence Have students review their answers to the Viral Spread questions in Sections 11.2 and 13.3 and in each Chapter Review as they determine what evidence to use to support their claim. If you wish to have students look for evidence from sources other than this text, they can find useful information on National Geographic, as well as the National Forest Foundation.

R Explaining Reasoning In addition to their explanatory writing, models that show inheritance patterns, such as genetic crosses, are convenient structures for students to build their explanations around.

Go online to access the Student Self-Reflection and Teacher Scoring rubrics for this activity.

DIFFERENTIATED INSTRUCTION | English Language Learners

Describing and Explaining As students write their claim and evidence, encourage them to include specific descriptions and explanations to help readers understand their ideas. As they use new vocabulary, have them think about questions like these: *What does it do? How does it work? What is it for?*

Beginning Have students think about terms or concepts that relate to the topic and define each one by describing or explaining it. Have them use a dictionary or glossary for support.

Intermediate Have students brainstorm details and choose three to five to include. Have them write notes in phrases and then write a final draft with complete sentences.

Advanced Have students use transitions such as *through, in order to, as a way of* and so on to help them explain or describe ways to slow the spread of a virus.

Unit Anchoring Phenomenon: Hummingbird Evolution

Use the Driving Question to help frame the Anchoring Phenomenon as an investigable subject and motivate student learning. Leverage the hummingbird prompts within each chapter to connect concepts back to the unit's Driving Question, supporting students in gathering evidence and asking their own research questions so that they are equipped to complete the Unit Activity.

NATIONAL GEOGRAPHIC

Meet the Explorer

Anusha Shankar is an ecologist who studies how hummingbirds use torpor as an energy strategy as opposed to regenerative sleep. Watch **Unit Video 2**, Explorers at Work to engage student interest in tropical ecosystem interactions and the Anchoring Phenomenon.

Virtual Investigation

Hummingbirds on the Move Students go into the field learn about a wide variety of hummingbirds. They gather evidence to describe how different species of hummingbirds are uniquely adapted to their environments.

NGSS Progression

Middle School

- **MS-LS2-5** Evaluate competing design solutions for maintaining biodiversity and ecosystem services.*
- **MS-LS4-1** Analyze and interpret data for patterns in the fossil record that document the existence, diversity, extinction, and change of life forms throughout the history of life on Earth under the assumption that natural laws operate today as in the past.
- **MS-LS4-2** Apply scientific ideas to construct an explanation for the anatomical similarities and differences among modern organisms and between modern and fossil organisms to infer evolutionary relationships.
- **MS-LS4-3** Analyze displays of pictorial data to compare patterns of similarities in the embryological development across multiple species to identify relationships not evident in the fully formed anatomy.
- **MS-LS4-4** Construct an explanation based on evidence that describes how genetic variations of traits in a population increase some individuals' probability of surviving and reproducing in a specific environment.
- **MS-LS4-5** Gather and synthesize information about the technologies that have changed the way humans influence the inheritance of desired traits in organisms.
- **MS-LS4-6** Use mathematical representations to support explanations of how natural selection may lead to increases and decreases of specific traits in populations over time.

High School

- **HS-LS2-7** Design, evaluate, and refine a solution for reducing the impacts of human activities on the environment and biodiversity.
- **HS-LS3-1** Ask questions to clarify relationships about the role of DNA and chromosomes in coding the instructions for characteristic traits passed from parents to offspring.
- **HS-LS3-3** Apply concepts of statistics and probability to explain the variation and distribution of expressed traits in a population.
- **HS-LS4-1** Communicate scientific information that common ancestry and biological evolution are supported by multiple lines of empirical evidence.
- **HS-LS4-2** Construct an explanation based on evidence that the process of evolution primarily results from four factors: (1) the potential for a species to increase in number, (2) the heritable genetic variation of individuals in a species due to mutation and sexual reproduction, (3) competition for limited resources, and (4) the proliferation of those organisms that are better able to survive and reproduce in the environment.
- **HS-LS4-3** Apply concepts of statistics and probability to support explanations that organisms with an advantageous heritable trait tend to increase in proportion to organisms lacking this trait.
- **HS-LS4-4** Construct an explanation based on evidence for how natural selection leads to adaptation of populations.
- **HS-LS4-5** Evaluate the evidence supporting claims that changes in environmental conditions may result in: (1) increases in the number of individuals of some species, (2) the emergence of new species over time, and (3) the extinction of other species.
- **HS-LS4-6** Create or revise a simulation to test a solution to mitigate adverse impacts of human activity on biodiversity.*
- **HS-ETS1-2** Design a solution to a complex real-world problem by breaking it down into smaller, more manageable problems that can be solved through engineering.
- **HS-ETS1-3** Evaluate a solution to a complex real-world problem based on prioritized criteria and trade-offs that account for a range of constraints, including cost, safety, reliability, and aesthetics, as well as possible social, cultural, and environmental impacts.
- **HS-ETS1-4** Use a computer simulation to model the impact of proposed solutions to a complex real-world problem with numerous criteria and constraints on interactions within and between systems relevant to the problem.

Claim, Evidence, Reasoning Students can make a claim about the unit phenomenon, gather evidence, and revisit their claim periodically to evaluate how well the evidence supports it. The Driving Question presented in the Case Study of each chapter can get students invested in chapter topics and in working toward answering the unit's Driving Question, "How did hummingbirds become adapted to their environments?" In the Unit Activity, students can practice scientific reasoning and argumentation to show how the evidence supports their claim.

Follow the Anchoring Phenomenon How did hummingbirds become adapted to their environments?

Gather evidence with . . .	Chapter 14	Chapter 15	Chapter 16	Unit Activity
CASE STUDY	How do we know how species have evolved?	How can changes in the environment generate diverse populations of organisms?	Why do some new species emerge while others disappear?	Revisit the unit's Anchoring Phenomenon of how natural selection and changing environmental conditions affect populations of species **Claim, Evidence, Reasoning** Students use the evidence they gathered throughout the unit to support their claim with reasoning. **Go online** to access the Student Self-Reflection and Teacher Scoring rubrics for this activity. **English Language Learners** Claims and Evidence
MINILAB	How can you form an evolutionary tree of related species based on fossils?	How does fur color impact a prey population?	How does a human-caused environmental change affect biodiversity?	
LOOKING AT THE DATA	Students will analyze and interpret data to learn about the forensic radiometric dating.	Students will use mathematical thinking to learn about how allele frequencies change over time.	Students will analyze the ecological footprints of different countries to construct an argument for reducing a country's ecological footprint.	
TYING IT ALL TOGETHER REVISIT THE CASE STUDY	Students will research paleoproteomics and explain how scientists are using the field to reconstruct evolutionary histories of different organisms.	Students will research the reproduction rate of bedbugs and apply their findings to explain how a mutation could spread through the population.	Students will develop a model and construct an explanation about how an elephant population can change with regard to tusklessness.	
Chapter Review: Revisit Rainforest Connections	Reflect on evidence that supports the evolution of hummingbirds over time.	Reflect on how hummingbird feather colors benefit female hummingbirds	Reflect on how changing environmental conditions affect populations of hummingbirds.	
Virtual Investigation	Students take on the role of an ornithologist, going into the field to gather evidence about hummingbird adaptations.			
Chapter Investigation A	How can you use genetic information to determine how closely related organisms are?	Simulate genetic drift through the bottleneck effect and the founder effect to demonstrate how the gene pool in a population can change.	What effect does environmental change have on speciation?	
Chapter Investigation B	How can you use biological evidence to identify ancient organisms?	Why do we need different types of antibiotics?	How can wildlife crossings or corridors help a species access all parts of its habitat?	

Planning Your Investigations

Each chapter features a Minilab and two full Investigations that offer hands-on opportunities for students to engage in Science and Engineering Practices.

Advance Preparation For Chapter 16 Investigation A and the Minilab, you will need to purchase beads or use hole punch-outs from colored paper. Advanced preparation notes for all labs are included in the Teacher's version.

Chapter 14	Title	Time	Standards
Minilab	*Organizing Fossil Evidence*	30 minutes	SEP Obtaining, Evaluating, and Communicating Information CCC Patterns HS-LS4-1
Investigation A **Guided Inquiry**	*What Lived Here?*	50 minutes	HS-LS4-1
Investigation B **Guided Inquiry**	*Comparing Genetic Information Among Organisms*	50 minutes	HS-LS4-1
Chapter 15	**Title**	**Time**	**Standards**
Minilab	*Hawks and Mice*	30 minutes	SEP Constructing Explanations and Designing Solutions SEP Analyzing and Interpreting Data CCC Patterns
Investigation A **Guided Inquiry**	*Genetic Drift*	50 minutes	HS-LS4-3
Investigation B **Open Inquiry**	*Mapping Fruit Fly Genes Through Linkage*	50 minutes	HS-LS4-3
Chapter 16	**Title**	**Time**	**Standards**
Minilab	*Modeling Human-Caused Changes in the Environment*	30 minutes	CCC Stability and Change
Investigation A **Guided Inquiry**	*Modeling Speciation*	80 minutes	HS-LS4-5
Investigation B **Open Inquiry**	*Wildlife Crossings and Corridors*	100 minutes over 2 days	HS-ETS1-2 HS-ETS1-3 HS-LS2-7 HS-LS4-6

Assessment Planning

UNIT 5 PERFORMANCE TASKS

Five performance-based assessments target the NGSS Performance Expectations with three-dimensional activities to measure student mastery. Rubrics are available online for Performance Tasks.

Performance Tasks

	Title	Overview	PEs Addressed	Use After
1	*How can we determine evolutionary relationships?*	Students research a list of mammals and determine the evidence for their evolutionary relationships.	HS-LS4-1	Chapter 14
2	*How does bacterial evolution affect public health globally?*	Students collect and analyze data about AMR bacteria and research methods to mitigate their spread.	HS-LS4-2, HS-LS4-3	Chapter 15
3	*How is climate change altering species evolution?*	Students obtain and evaluate information on evolutionary changes taking place in a species over time.	HS-LS4-4, HS-LS4-3	Chapter 16
4	*How do human-induced changes in the environment affect different species?*	Students conduct research on human-based changes to the environment as well as changes in population numbers of two or more species coexisting in the same area.	HS-LS4-5	Chapter 16
5	*What kind of artificial reef is most effective at preserving and restoring biodiversity?*	Students define the problem of threats to coral reef ecosystems and evaluate the effectiveness of various artificial reef solutions.	HS-LS4-6, HS-ETS1-4	Chapter 16

Additional Resources

Use a search engine to find these resources on the Internet.

VIDEOS

"Why bacteria out-evolve us with antibiotic resistance," TEDx Talk

This twelve-minute explanatory video shows how fast bacterial evolution occurs, and how it surprised the scientific community.

Search: "TEDx bacteria out-evolve"

"Four billion years of evolution in six minutes" TED Talk

This six-minute video dispels misconceptions of evolution, and encourages watchers to understand that humans are not the goal of evolution.

Search: "TED evolution in six minutes"

ARTICLES

"Humans have 'stressed out' Earth far longer, and more dramatically, than realized" National Geographic

This article discusses how a study of ancient pollen showed that human activities have impacted Earth's ecosystems for millennia.

Search: "humans stressed out earth National Geographic"

Unit Storyline

In this unit, students will answer the question, “How did hummingbirds become adapted to their environments?” Each chapter provides part of the answer to this Driving Question.

Chapter 14 *How do we know how species have evolved?*

Students learn about common ancestry and biological evolution. Students learn that genetic variation is connected to evolution and consider what types of evidence supports evolution. Students learn about fossil and geological evidence and then about developmental, anatomical, and genetic evidence.

Chapter 15 *Why do populations evolve over time?*

Students study the four principles of the theory of evolution and explain using an example of how natural selection explains evolution. Students analyze how different mechanisms of evolution change the distribution and frequency of alleles and associated traits in populations.

Chapter 16 *Why do new species evolve while other species disappear?*

Students learn how organisms survive in changing environments. The lesson introduces the mechanisms that lead to speciation and then moves on to the topic of extinction. Students learn about the impact of humans on the environment and consider how these impacts can be reduced.

Unit Activity Students revisit the Anchoring Phenomenon to make a claim about the Driving Question and apply reasoning to the evidence they have gathered throughout the unit.

UNIT 5

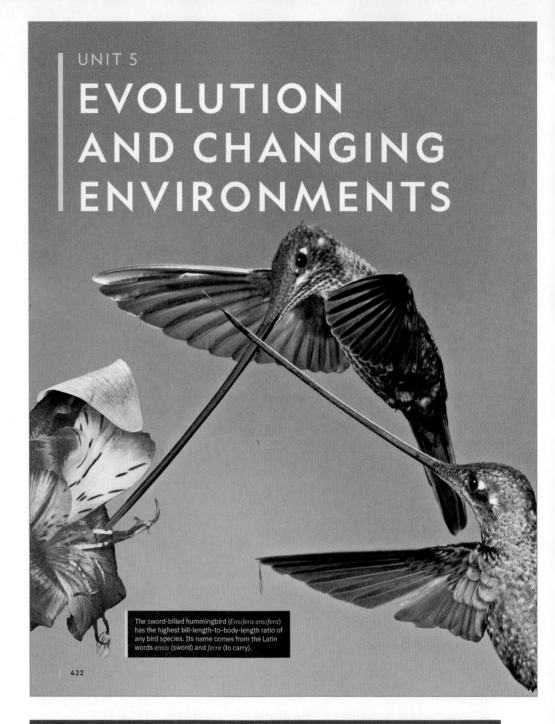

EVOLUTION AND CHANGING ENVIRONMENTS

The sword-billed hummingbird (*Ensifera ensifera*) has the highest bill-length-to-body-length ratio of any bird species. Its name comes from the Latin words *ensis* (sword) and *ferre* (to carry).

422

SCIENCE BACKGROUND

Taxonomy of the Sword-Billed Hummingbird

phylum: Chordata, class: Aves, order: Apodiformes, family: Trochilidae, genus: *Ensifera*

Sword-billed Hummingbirds The common name “hummingbird” refers to the hum that is produced by the rapid beating of their wings. These birds are often named because of their characteristics or the locations where they are found. The sword-billed hummingbird is the only bird whose beak is longer than its body, not including the tail. They are the largest species of hummingbird, averaging 14 cm in length, not including the bill. Their exceptionally long bills hold an even longer tongue, which is used to feed from nectar of brightly colored flowers. These hummingbirds feed on flowers with long corollas because their beaks can reach the sweet nectar deep in the flower that most birds cannot reach. These birds also feed on some small spiders and insects. During mating season, females can eat up to 2,000 insects a day.

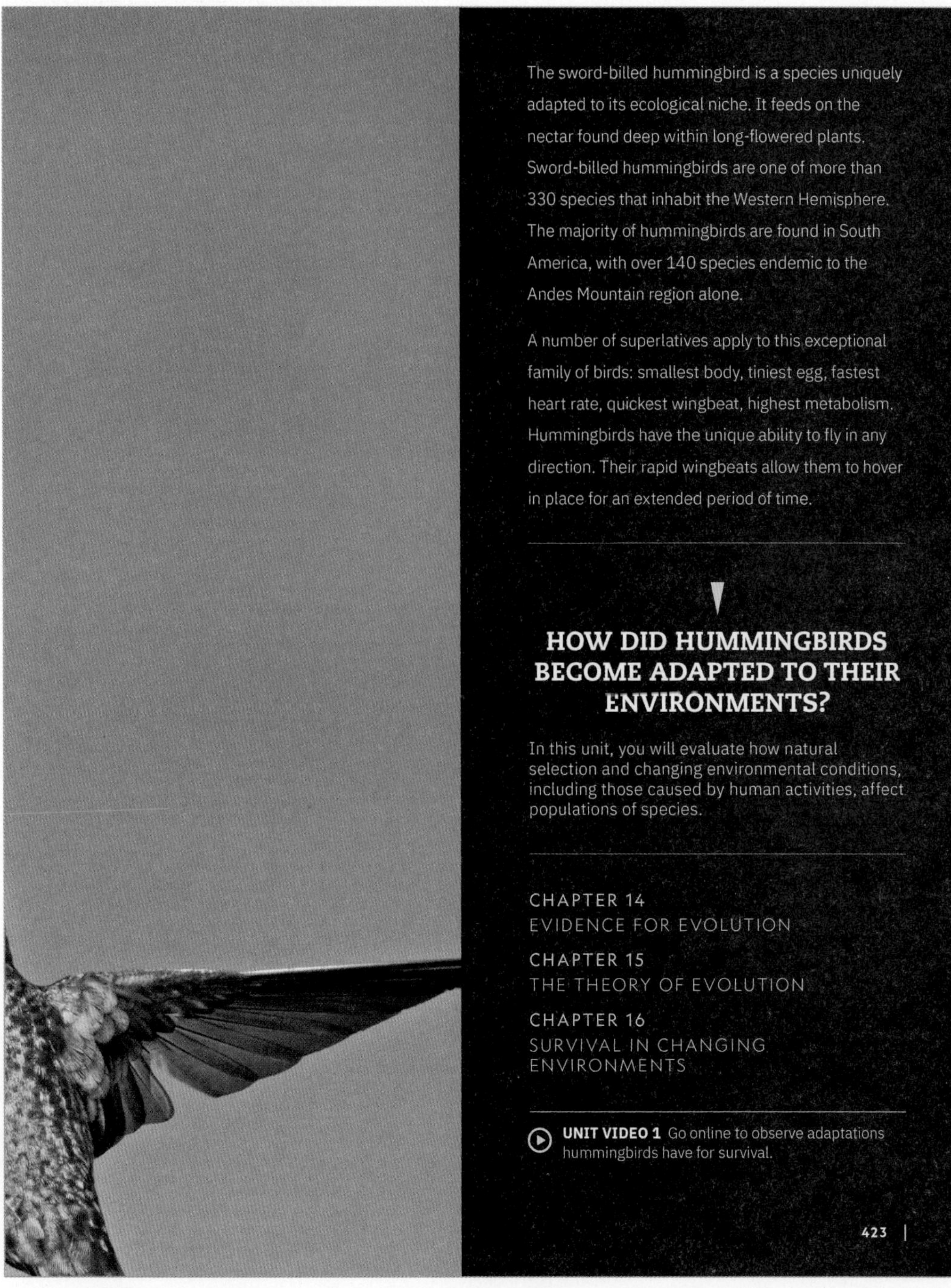

The sword-billed hummingbird is a species uniquely adapted to its ecological niche. It feeds on the nectar found deep within long-flowered plants. Sword-billed hummingbirds are one of more than 330 species that inhabit the Western Hemisphere. The majority of hummingbirds are found in South America, with over 140 species endemic to the Andes Mountain region alone.

A number of superlatives apply to this exceptional family of birds: smallest body, tiniest egg, fastest heart rate, quickest wingbeat, highest metabolism. Hummingbirds have the unique ability to fly in any direction. Their rapid wingbeats allow them to hover in place for an extended period of time.

HOW DID HUMMINGBIRDS BECOME ADAPTED TO THEIR ENVIRONMENTS?

In this unit, you will evaluate how natural selection and changing environmental conditions, including those caused by human activities, affect populations of species.

UNIT VIDEO 1 Go online to observe adaptations hummingbirds have for survival.

423

Go Online VIRTUAL INVESTIGATION

Hummingbirds on the Move

Time about 50 minutes

Objectives Students will learn about a wide variety of hummingbirds. They will gather evidence to describe how different species of hummingbirds are uniquely adapted to their environments.

Explore and Learn Students experience a variety of habitats where hummingbirds live in North and South America, learning about some of the unique circumstances scientists work in.

Collect Data Students will learn about how hummingbirds evolved to live all over much of South and North America They will explore how hummingbirds interact with their habitats near the Southwest Research Station in Arizona, and then students will go into the lab to learn about how scientists analyze hummingbird energy usage.

Analyze and Report Students answer questions about the investigation, incorporating qualitative and quantitative data in a report to support their analyses.

Introduce the Anchoring Phenomenon

Changing habitats exist in every ecosystem, and populations need to change along with them. These changes naturally occur over long periods of time, but due to human interactions, the world is changing faster than ever. Examining how populations change along with their environments in the Anchoring Phenomenon encourages students to apply what they learn throughout the units to a type of problem that is occurring globally.

Driving Question Have students make a list of things they think they need to know about hummingbirds to answer the Driving Question. The photo and caption may prompt them to ask, "How did hummingbirds develop such long beaks?" To analyze how populations evolve with their environments, students should focus on how variation in alleles and associate traits, natural selection, and changing environments affect species. What students learn in this unit about evolution and surviving in changing environments will inform their understanding of organisms in their environments.

About the Photo Sword-billed hummingbirds are solitary birds that only come together to mate. Males establish territories where they chase off competitors who eat their food sources and they defend their territories from other males and large insects.

Video

Time: 2:30

Unit Video 1 from the BBC explains how sword-bills are used to exploit a food source that other hummingbirds cannot, and why there is a drawback to the long beak length.

EXPLORER
ANUSHA SHANKAR

ENGAGE

About the Explorer

National Geographic Explorer Anusha Shankar is an ornithologist. She studies how hummingbirds manage their energetic needs given that these birds have some of the highest metabolic rates of all vertebrates. She also has an interest in how changing environments are changing plant distribution and how these changes could affect hummingbird survival. You can learn more about Anusha Shankar on the National Geographic Society's website: www.nationalgeographic.org.

On the Map

Andes Mountains Point out the Andean regions on the west coast of South America on a map and tell students that the Andes is the longest continental mountain range in the world. It is made up of a succession of parallel mountain ranges that creates a wide range of different habitats. Some animals can live at any altitude in the region and can be found throughout the mountain range, while other species are specialized and can only live at certain altitudes. Many different species of hummingbirds are found throughout the different habitats in the Andes Mountains.

THINKING CRITICALLY

Infer *Sample answer: Torpor aids in survival by allowing an animal to conserve energy for short periods of time.*

NATIONAL GEOGRAPHIC EXPLORER **ANUSHA SHANKAR**

THE HUMMINGBIRD'S SECRET TO SURVIVAL

EXAMINING EXTREME ENERGY CONSERVATION

As they flit and fly through the air, hummingbirds seem to defy the laws of gravity. They hover in midair, floating above the delicate flowers that provide them with nectar. According to ecologist Dr. Anusha Shankar, hovering is one of the most energetically expensive activities in the animal world. Because hovering uses up so much energy, hummingbirds have an enormous appetite. In order to survive, many species must consume two to three times their body weight in nectar and small insects every day.

MAXIMIZING ENERGY CONSERVATION One of Shankar's primary research goals is to understand how hummingbirds balance their daily energy needs. As part of her dissertation research, Shankar studied hummingbird energy budgets. She focused particularly on their use of torpor, an evolutionary strategy some animals use to conserve energy. Hummingbirds lapse into torpor at night or when their energy stores are very low. If sitting on a branch, the bird appears unresponsive and hunched up. Its beak is in the air, its eyes are closed, and it is barely breathing. Unlike hibernation that can last for days or weeks, daily torpor is short term. It typically lasts for only a few hours.

Shankar studies how different species of hummingbirds use torpor as an energy-saving strategy in different climates. Torpor saves the bird energy by drastically lowering its metabolism and body temperature. A hummingbird's normal body temperature is around 40 °C (104 °F). When it enters torpor, its body temperature lowers to that of its surroundings or less. The bird's body temperature could be as low as 3–10 °C (37–50 °F). Research indicates that a hummingbird conserves 65 to 92 percent of its energy every hour it is in torpor. The longer it is in torpor, the more energy it conserves.

Unlike during sleep, the body doesn't repair and regenerate itself during torpor. Shankar wondered exactly how much time hummingbirds spent in energy-saving torpor versus regenerative sleep. More recent research indicates that hummingbirds are able to regulate their body temperature at an intermediate state between deep torpor and sleep. This state is called shallow torpor. She hopes that further research in this area will help scientists to understand the physiological functions of sleep compared to torpor in both birds and other animals.

SURVIVAL IN A CHANGING CLIMATE In addition to her research focused on torpor, Shankar is interested in understanding how changing environmental conditions could affect hummingbird populations. As nectar feeders, hummingbirds play a primary role as pollinators for many plant species. Shankar's collaborators are combining plant distribution maps, climate data, and research on hummingbird behavior to model how a warming planet could affect hummingbird survival. Shankar hopes to expand this research to study the health of other bird species.

THINKING CRITICALLY

Infer *What is the evolutionary benefit of an energy-conserving mechanism such as torpor?*

424

CROSSCURRICULAR CONNECTIONS

Environmental Science The realization that habitat loss is placing many species at risk of extinction has led planners to work with scientists, conservationists, and land owners in certain parts of the world to save threatened and endangered species in developed and rural areas.

Depending on how cities are planned and built, regions within the same city can have a 20-degree difference in temperature. Areas with higher temperatures are less hospitable to wildlife. Wildlife habitats are often fragmented in cities, which breaks up populations and puts them at risk. Planners can reduce materials such as concrete and asphalt and increase green space in cities to lower the temperature and connect wildlife habitats.

Habitat fragmentation is also being addressed in rural areas. At the Atlantic Forest of the Northeast Project in Brazil, forest restorations are underway to connect fragmented habitats to save species. The restoration of forests in some regions has led to the rapid return of many insect species that had previously disappeared.

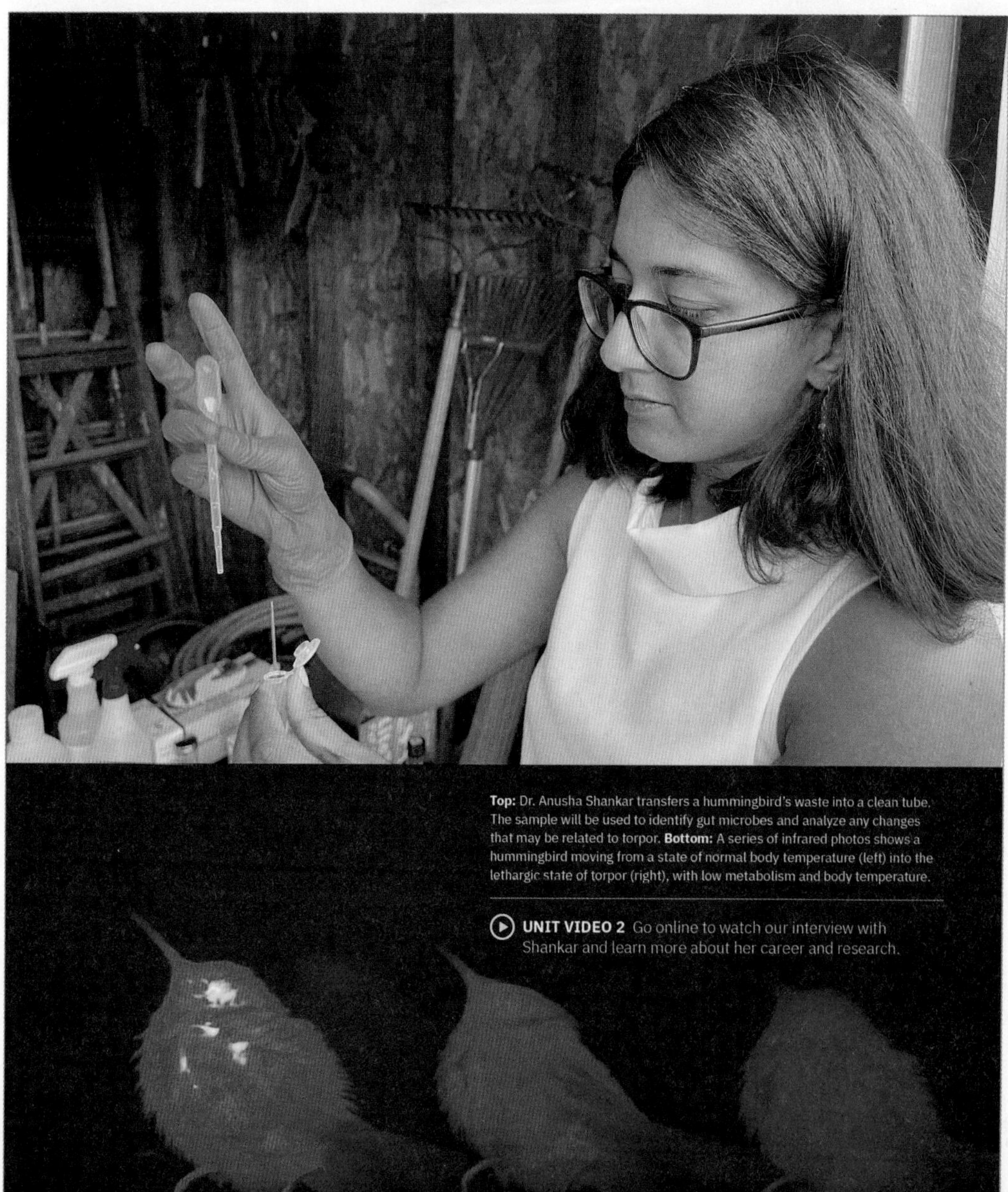

Top: Dr. Anusha Shankar transfers a hummingbird's waste into a clean tube. The sample will be used to identify gut microbes and analyze any changes that may be related to torpor. **Bottom:** A series of infrared photos shows a hummingbird moving from a state of normal body temperature (left) into the lethargic state of torpor (right), with low metabolism and body temperature.

UNIT VIDEO 2 Go online to watch our interview with Shankar and learn more about her career and research.

Explorers at Work

Time: 8:51

Unit Video 2 interviews Anusha Shankar, who introduces students to an animal's energy budget and how that energy is spent. She expands on this idea in relation to her research focus on hummingbirds and their use of torpor.

Connect to Careers

Evolutionary Ornithologist Ornithology is the study of birds. Evolutionary ornithologists study how avian species have changed over time, using phylogenetics, evolutionary evidence of feathers from fossils, and a variety of other methods. They may use traditional field biology techniques, genetic testing, and technology such as drones and field cameras to collect data. Evolutionary ornithologists typically need a bachelor or master's degree in biology. Evolutionary ornithologists who lead research projects or work in specialized positions typically have a master's degree or higher degree. Universities, museums, and wildlife and conservation organizations hire evolutionary ornithologists to study birds in the field and in laboratory settings.

Wildlife Rehabilitator As environments continue to change, animals are more affected by human activities. Wildlife rehabilitators care for injured and ill animals and assist residents with animal conflicts. They work for nonprofit or governmental agencies to promote conservation of species and educate the public about wildlife. Wildlife rehabilitators are required to have knowledge about ecology, biology, and medical care, but a college degree is not always required.

DIFFERENTIATED INSTRUCTION | English Language Learners

Demonstrate Listening Comprehension Have students watch the video in pairs or small groups and then collaborate to answer questions about it. Provide students with questions in advance so they can listen for the answers and take notes using words, phrases, and sketches. Encourage them to rewatch parts of the video as needed.

Beginning Allow students to take notes in their native language as they watch the video. Supply questions that can be answered in a word or phrase or with yes or no.

Intermediate Have students rewatch and add to their notes. Supply questions that can be answered with a sentence, such as *How? Why? What happened?*

Advanced Supply questions that students can answer in a few sentences, adding details to their explanations from their notes.

EVIDENCE FOR EVOLUTION

Three-Dimensional Learning

The practices, core ideas, and crosscutting concepts presented in this chapter's text, investigations, and resources provide support to address the following Performance Expectations: **HS-LS3-1** and **HS-LS4-1**.

Science and Engineering Practices	Disciplinary Core Ideas	Crosscutting Concepts
Obtaining, Evaluating, and Communicating Information (HS-LS4-1) Scientific Knowledge is Based on Empirical Evidence Science Models, Laws, Mechanisms, and Theories Explain Natural Phenomena (HS-LS4-1)	**LS2.C:** Ecosystem Dynamics, Functioning, and Resilience **LS4.A:** Evidence of Common Ancestry and Diversity (HS-LS4-1)	Patterns (HS-LS1-4) Scientific Knowledge Assumes an Order and Consistency in Natural Systems (HS-LS4-1)

Contents		Instructional Support for All Learners	Digital Resources
ENGAGE			
426–427	**CHAPTER OVERVIEW** **CASE STUDY** How do we know how species have evolved?	**Social-Emotional Learning** CCC Patterns **On The Map** Near the Equator **English Language Learners** Drawing Conclusions	
EXPLORE/EXPLAIN			
428–434 DCI LS4.A	**14.1 LINES OF EVIDENCE** • Explain how evolution occurs at large and small scales. • Differentiate the five lines of evidence that support evolution. • Apply cladistics to explain evolutionary relationships.	**Vocabulary Strategy** Roots and Affixes **Differentiated Instruction** Leveled Support **Connect to Mathematics** Reason Quantitatively **In Your Community** Using Evidence **English Language Learners** Using Reasoning **On the Map** Wadi Al-Hitan CER Revisit the Anchoring Phenomenon **Visual Support** Changes in Characteristics Over Time **English Language Learners** Using Visuals for Listening **English Language Learners** Vocabulary Strategies	
434	**MINILAB** ORGANIZING FOSSIL EVIDENCE		
435–445 DCI LS4.A PS1.C ESS2.A	**14.2 FOSSIL AND GEOLOGICAL EVIDENCE** • Sequence steps in the process of fossilization. • Explain different methods used to date fossils. • Differentiate three different types of evidence used to determine evolutionary relationships and the history of life on Earth.	**Vocabulary Strategy** Word Wall **English Language Learners** Modeling Vocabulary **Connect to English Language Arts** Write about Sequence **In Your Community** Finding Fossils **CASE STUDY** Ask Questions **Differentiated Instruction** Leveled Support **Connect to Mathematics** Reason Quantitatively	*Video 14-1*

Contents		Instructional Support for All Learners	Digital Resources
EXPLORE/EXPLAIN			
435–445	**14.2 FOSSIL AND GEOLOGICAL EVIDENCE** (continued)	**Differentiated Instruction** Students with Disabilities **On the Map** Where in the World? **English Language Learners** Monitoring and Self-Correcting **Connect to Mathematics** Reason Quantitatively **Crosscurricular Connections** Physical Science **Differentiated Instruction** Leveled Support **CONNECTIONS** Social-Emotional Learning **Connect to English Language Arts** Write Explanatory Text **Connect to Mathematics** Model with Mathematics **English Language Learners** Possessives **On the Map** Fossil Hunting in the Badlands **On Assignment** The Badlands	*Video 14-1*
443	**LOOKING AT THE DATA** FORENSIC RADIOMETRIC DATA		
446–453 DCI LS4.A	**14.3 DEVELOPMENTAL, ANATOMICAL, AND GENETIC EVIDENCE** • Explain how developmental evidence shows evolutionary relationships. • Identify anatomical evidence for evolution and distinguish among homologous, analogous, and vestigial structures. • Summarize how advancements in genetic evidence build knowledge about evolution.	CCC Structure and Function **Vocabulary Strategy** Roots and Affixes **Connect to English Language Arts** Draw Evidence from Texts SEP Constructing Explanations and Designing Solutions **Address Misconceptions** Divergence and Convergence CCC Structure and Function **English Language Learners** Roots and Affixes CCC Patterns	**Investigation A** What Lived Here? (50 minutes) **INTERACTIVE FIGURE** Evolutionary Evidence of Anatomy **Investigation B** Comparing Genetic Information Among Organisms (50 minutes) *Video 14-2*
452	**EXPLORER** DR. BRENDA LARISON	**Connect to Careers** Evolutionary Biologist	
ELABORATE			
453	**TYING IT ALL TOGETHER** How do we know how species have evolved?	**Differentiated Instruction** Economically Disadvantaged Students	Self-Reflection and Scoring Rubrics
Online	**Investigation A** What Lived Here? **Investigation B** Comparing Genetic Information Among Organisms.	Guided Inquiry (50 minutes) Guided Inquiry (50 minutes)	MINDTAP Access all your online assessment and laboratory support on MindTap, including: sample answers, lab guides, rubrics, and worksheets. PDFs are available from the Course Companion Site.
EVALUATE			
454–455	**Chapter 14 Review**		
Online	**Chapter 14 Assessment** **Performance Task 1** *How can we determine evolutionary relationships?* (HS-LS3-1)		

CHAPTER 14

ENGAGE

In the previous unit, students learned how gene expression and variation explains change over time within a population. Chapter 14 begins by connecting genetic variation with evolution and the scales at which evolution occurs, and then presents evidence for evolution. Students will explore how fossils form and how to interpret fossils and associated geological evidence. Students will also explore other evidence for evolution from genetics, molecular biology, and embryology and anatomy.

About the Photo This photo shows a three-toed sloth in a tree, where it spends around 90 percent of its time hanging upside down. That amount of time hanging upside down could cause breathing problems for other mammals but not sloths. Their internal organs are connected to their ribs and hips by fibrous membranes, preventing them from pressing down on their lungs. Students can find out more about the unusual characteristics that evolved in sloths from an Internet search using the keyword "slothopedia."

Social-Emotional Learning

Some students may come to class with previously developed ideas about evolution and bring strong opinions and emotions about the topic. There may be several social-emotional competencies that can be used depending on the robustness of the conversation. Practicing **self-awareness** skills can help students identify their personal identities and their emotions. Applying **relationship skills** will help students communicate effectively and resolve conflicts constructively.

Remind students that a theory is a time-tested explanation for observations in the natural world that are supported by evidence, some of which they will be learning in this chapter. That idea contrasts with the vernacular use of the word *theory* as a "guess."

CHAPTER 14

EVIDENCE FOR EVOLUTION

Three-toed sloths are found throughout the tropical forests of Central America and South America.

A sloth is a slow-moving mammal that spends most of its lifetime living within the branches of its tropical treetop habitat. Algae that live within its fur help to camouflage the sloth from predators. While sloths may in some ways look similar to monkeys, they are actually more closely related to anteaters and armadillos.

Chapter 14 supports the NGSS Performance Expectation **HS-LS4-1.**

CROSSCUTTING CONCEPTS | Patterns

Species Evolution Tell students to look for three main patterns of species evolution as they go through the chapter. These are divergent, convergent, and parallel evolution. Divergent evolution occurs when populations in a species evolve to be increasingly different, particularly as they adapt to new habitats. Convergent evolution occurs when different species with different lineages evolve analogous features. Parallel evolution occurs when two or more distinct species evolve in similar ways.

CASE STUDY
SOLVING THE MYSTERY OF SLOTH EVOLUTION

HOW DO WE KNOW HOW SPECIES HAVE EVOLVED?

If sloths physically resemble monkeys, why have scientists concluded that they are more closely related to armadillos? Looking beyond surface-level resemblances, scientists have found evidence in the anatomical and molecular makeup of these animals, as well as in the history revealed by fossils. Modern and extinct sloths, anteaters, and armadillos are members of a superorder called Xenarthra. The name *Xenarthra* derives from Greek words meaning "strange joints" and refers to extra joints found in the lower spines of all members of this group. All Xenarthrans have fewer teeth compared to other mammals their size. Sloths and armadillos have simple teeth that lack enamel and anteaters have no teeth at all. In addition, genetic evidence shows that all Xenarthrans have the same three consecutive amino acid deletions in a particular eye protein.

Xenarthrans are one of the oldest groups of mammals. Fossil evidence indicates they originated in South America 55 million to 65 million years ago. They diversified into numerous species while the continent was geographically isolated. The different species spread to Central and North America eight million years ago.

Ancient Xenarthrans flourished until about 10,000 years ago when the majority of species became extinct during the Ice Age, likely due to overhunting by humans.

Unlike their modern tree-dwelling equivalents, the ancient ancestors of sloths were giant ground dwellers. The first giant ground sloth fossil (**Figure 14-1**) was discovered in Argentina by Manuel Torres in 1788. This elephant-sized species was named *Megatherium americanum*. In 1797, a naturalist presented a paper to the American Philosophical Society about the discovery of a large claw and limb bone found in a cave in West Virginia. The naturalist initially identified the fossils as belonging to an extinct Ice Age cat species. However, upon recognizing similarities between the fossils and *M. americanum*, they were recategorized as belonging to a species of giant ground sloth. The ox-sized species was named *Megalonyx jeffersonii* in honor of the naturalist, Thomas Jefferson, then vice president of the United States. Fossils of *M. jeffersonii* have been recovered from 150 sites across the United States as well as from sites in northwestern Canada and western Mexico.

Ask Questions *As you read, generate questions about the lines of evidence scientists use to determine the evolutionary history of a species.*

FIGURE 14-1 *Megatherium americanum* means "great beast from America." This vegetarian mammal could stand and walk on its hind legs. With a standing height of 3.5 m (12 ft.), it is the largest bipedal mammal of all time.

DIFFERENTIATED INSTRUCTION | English Language Learners

Drawing Conclusions Have pairs read the first paragraph and discuss this question: *Why have scientists concluded that sloths are related to armadillos?* Have them read the second and third paragraphs and answer this question: *How are sloths from long ago different from sloths today?*

Beginning For the first question, have students underline phrases in the text that give evidence scientists used to draw this conclusion. For the second question, have them make a two-column chart with phrases that tell similarities and differences.

Intermediate Have students answer each question in complete sentences. Have them reread several sentences from the text that helped them answer.

Advanced Have pairs explain their answers to each question using details from the text. Encourage them to cite the text, for example: *According to the Case Study, ___. The text says that ___, so ___.*

CASE STUDY

ENGAGE

This Case Study focuses on the evolution of sloths and the evidence that supports these changes. The Case Study also shows how the study of evolution has changed over time. Early scientists, such as Linnaeus and Darwin, used organisms' characteristics and fossils to study how evolution occurs. Scientists now use molecular biology and genetics to study evolutionary relationships. Without modern evolutionary evidence, scientists might still think that sloths are closely related to monkeys. In the Tying It All Together activity, students will describe how technology is used to uncover evolutionary relationships among organisms.

Ask Questions Students should revisit the Case Study as they read the chapter to make connections with the content. See a specific suggestion in Section 14.2.

On the Map

Near the Equator Sloths are found in the tropical rainforests of Central America and northern South America. The Central American rainforest is located near the equator in countries such as Costa Rica, Nicaragua, and Panama, meaning the climate is usually wet and warm. One particular type of sloth, the pygmy three-toed sloth, is found in only one location on a small island off the coast of Panama.

Human Connection With microscopic grooves on their fur, sloths provide ideal growing conditions for a variety of organisms, such as algae, fungi, bacteria, and even cockroaches. In their effort to find new therapeutics for diseases like cancer and bacterial infections, scientists are turning to the fungal communities that sloths harbor. Scientists have reported that the fungi found in sloth fur have bioactivity against agents that cause Chagas disease, malaria, and other pathogenic bacteria.

14.1

LINES OF EVIDENCE

LS4.A

EXPLORE/EXPLAIN

This section begins by helping students make the connection between what they have learned about genetic variation and evolution. Students compare macroevolution and microevolution in a tawny owl population to examine the empirical evidence that supports evolution and learn how that evidence was pieced together. Students wrap up with a look at evolutionary relationships and the application of cladistics. As a class, discuss how changes in whale structure over time support evolution.

Objectives

- Explain how evolution occurs at large and small scales.
- Differentiate the five lines of evidence that support evolution.
- Apply cladistics to explain evolutionary relationships.

Pretest Use **Questions 1, 2,** and **4** to identify gaps in background knowledge or misconceptions.

Vocabulary Strategy

Roots and Affixes Have students break down the words *microevolution* and *macroevolution*, explaining that the prefix *micro-* means small and *macro-* means large. Tell students to use both words in a sentence.

Ask Questions *Sample answer: How are biologists able to determine the age of fossils? How do biologists determine which species are related to one another? What evidence do scientists use to build a species' evolutionary history?*

14.1

KEY TERMS

evolution
macroevolution
microevolution

LINES OF EVIDENCE

You may already know that whales are mammals, but you may not know that biologists have concluded that whales are descended from an ancient ancestor that lived on land. What types of evidence do you think scientists might have found that led them to this conclusion? What types of evidence and thought processes do scientists use to determine the evolutionary relationships among different species?

Ask Questions *What questions do you have about how biologists study the history of living things?*

Connecting Genetic Variation With Evolution

Evolution is the means by which all of Earth's species have descended from a single common ancestor. From the time the first living cell appeared on this planet billions of years ago to the incredible variety of both simple and complex organisms we know today, the process of evolution has been at work.

Genetic variation provides the raw material of evolutionary change. The main sources of variation are the mutations that occur within genomes as well as the new genetic combinations made during meiosis when crossover occurs. These processes result in a stockpile of unique alleles within a population's gene pool, any of which can suddenly be favored if environmental conditions change.

Consider also the effects of sexual reproduction on evolution. Though sexual reproduction produces no new alleles, the sheer number of genes that are shuffled during the formation of a zygote produces an enormous range of different combinations. If conditions change, any one of the genotypes produced could suddenly become advantageous.

Evolution Occurs at Different Scales

Evolutionary change occurs in populations at both large and small scales. Whether evolution occurs over a long period of time (macroevolution) or in a relatively short period of time (microevolution), the underlying mechanisms remain the same.

Macroevolution refers to major evolutionary events that occur in groups of species over geological time, or millions of years. Macroevolution can be thought of as the accumulation of microevolutionary change over an incredibly long period of time. These events are large phenotypic changes, such as the appearance of wings with feathers that evolved in birds. The phenotypic changes are so great that the organisms (and their descendants) that possess them are assigned to a whole new taxonomic category, such as the changes that caused transition from fishes to amphibians, or amphibians to reptiles. Macroevolution involves the origin of major taxonomic categories more general than the species level.

In contrast to macroevolution, **microevolution** is change in allele frequency within a single species or population. Microevolution accounts for the differences among the various populations of a species and, in some cases, can account for the origin of new species.

DIFFERENTIATED INSTRUCTION | Leveled Support

Struggling Students To help students understand the concept of evolution, connect it to an everyday life experience. Even in students' lifetimes, they have likely seen things change over time. For example, most students are aware of changes in cell phones. Have students work in pairs and choose an item in which they can trace back changes through the years. If need be, they can research the changes in phone features. Have students create either a print or digital time line. Tell them to explain if the changes were to the whole phone or just parts of the phone.

Advanced Learners Students may choose an everyday item that has changed during their lifetime. Working in pairs, have them research the item of their choice, possibly including how the item changed prior to their lifetime. Have students create either a print or digital time line showing these changes, as well as explaining the causes and effects of the change.

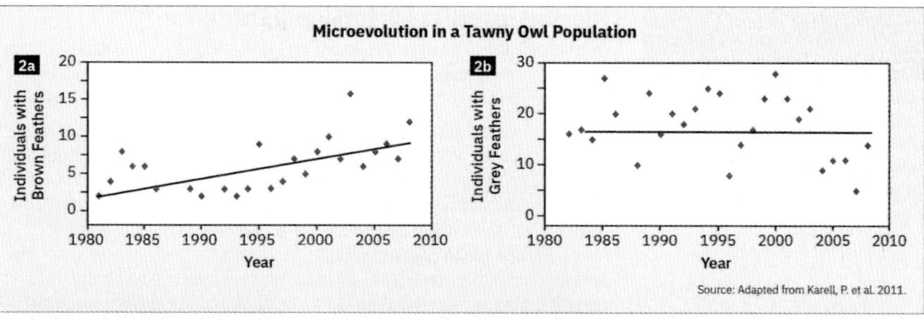

FIGURE 14-2 Brown feathers (a) are becoming more common in populations of tawny owls in Finland, as grey feathers (b) become less advantageous in a warming climate.

Feather color in a population of tawny owls in Finland provides one example of microevolution in action (**Figure 14-2**). This heritable trait has two variations: dominant brown feathers and recessive grey feathers. In the past, brown feathers had been selected against, as grey feathers helped the owls to better camouflage themselves in Finland's snowy conditions. However, as the climate warms, milder winters are producing less snow in Finland. As a result, brown-feathered owls are becoming more common in the population.

Empirical Evidence

Scientific knowledge is based on empirical evidence, which is information acquired by observation or experimentation. Because most evolutionary change occurs over thousands to millions of years, much of the evidence for evolution is based on observations, such as comparisons between ancient and modern organisms. Evolutionary biologists are a bit like detectives, using clues to piece together history that no human witnessed. Five lines of evidence that support evolution include

- **Fossils** The fossil record connects long-dead organisms with their modern-day counterparts.
- **Developmental Patterns** Some animal embryos share similarities during early development.
- **Anatomy** Evidence for actual modification of species is found by comparing the anatomy of various organisms.
- **Molecular Biology and Genetics** A shared genetic code supports the idea that life-forms descended from a common ancestor.
- **Biogeography** The global distribution of plants and animals helps identify when and where species originated and evolved.

Connect to Mathematics

Reason Quantitatively Help students develop their skills in gathering information from graphs or scatter plots. Explain that **Figure 14-2** is similar to a line graph in that it represents quantitative data points. However, since the lines are drawn through the data points and are not directly connected, they show trends in the data. Review the elements of the scatter plot in **Figure 14-2**. Choose one of the points on the graph and write the actual coordinates, asking students what the numbers represent. Ask them to explain the relationship between the *x*- and *y*-values. Have students work in pairs to identify what trend they see and what information they can gather. Have them compare two points between the graphs and explain how this can be used as evidence for microevolution.

In Your Community

Using Evidence Ask students what they think the word *evidence* means and how they have seen it used in everyday life. In a crime scene drama, evidence is collected to determine what happened and who may have committed the crime. It may include things such as fingerprints, DNA samples, or other physical signs. Because this evidence is directly observed and gathered through human senses or through analysis, it is classified as empirical evidence. Give students an example of when evidence might be needed in everyday life and allow them to work through the example to decide what kind of evidence would be needed to prove what happened. For example, you could give them the following scenario: the lettuce in a garden is disappearing leaf by leaf. What evidence could be collected to determine what happened?

DIFFERENTIATED INSTRUCTION | English Language Learners

Using Reasoning Support students as they use inductive reasoning to solve the scenario posed in "In Your Community." Point out that scientists use a similar process to draw conclusions about evolution based on evidence.

Beginning Have students draw evidence they might find, such as lettuce with a bite out of it, animal fur, or a hole in a fence around the garden. Supply vocabulary as needed. Ask: *Based on this evidence, what made the lettuce disappear?*

Intermediate Have students describe two or three pieces of evidence and then state a conclusion in a complete sentence. Provide a sentence frame: *Based on the evidence, the lettuce disappeared because* ___.

Advanced Have students role-play detectives explaining the evidence they found and how they drew their conclusion.

On the Map

Wadi Al-Hitan Wadi Al-Hitan translates to Whale Valley, but this valley is not found beneath the sea. Wadi Al-Hitan is located in Ibsheway, Egypt, part of the Western Desert. It provides one of the richest bodies of evidence showing the emergence of whales as ocean dwellers after a previous life on land. This site is considered the most important collection in the world with over 400 fossil skeletons of whales and other vertebrates, including fossils that show the youngest archaeocetes in the last stages of losing their hind limbs. Other evidence in the same area has allowed scientists to also determine the surrounding environment and ecology of the region at the time the whales existed.

SEP Communicate Information *Sample answer: Modern cetacean skeletons show remnants of pelvis and hind limbs common to land animals. A transitional fossil was found that showed a link between cetaceans and aquatic animals. Genetic analysis showed that cetaceans are more closely related to hooved animals that walked on land than to other groups of animals.*

HUMMINGBIRD EVOLUTION

Gather Evidence *Multiple lines of evolution support the evolution of hummingbirds. Identify three types of evidence that scientists have used to determine the evolutionary history of hummingbirds.*

Piecing Together Evolutionary Evidence

Our knowledge of evolution is based on making logical inferences from the best available evidence. This knowledge has been greatly enhanced by our understanding of genetics and recent advancements in biotechnology.

Whales provide one example of how scientists have pieced together fossil, anatomical, and molecular evidence to fully understand their evolutionary history. Whales and their relatives such as dolphins and porpoises make up the group called cetaceans. Modern cetaceans have skeletons with remnants of a pelvis which would have at one time connected to hind limbs (**Figure 14-3a**), so evolutionary biologists had long inferred that the ancestors of this group walked on land. However, no one had found fossils demonstrating a clear link between cetaceans and ancient land animals, so the rest of the story remained hypothetical.

Intact fossil skeletons of extinct whalelike *Dorudon atrox* had been discovered, but these animals were clearly cetaceans with little resemblance to land animals. *Dorudon* lived about 37 million years ago and like modern whales, it was a tail swimmer and fully aquatic. Its tiny hind limbs could not have supported the animal's huge body out of water (**Figure 14-3b**).

In the early 1990s, DNA sequence comparisons suggested that modern cetaceans are more closely related to artiodactyls than to other groups. Artiodactyls are hooved mammals with an even number of toes (two or four) on each foot. Modern representatives of the lineage include camels, hippopotamuses, pigs, antelopes, and sheep.

The DNA findings were controversial at the time. Some scientists thought the ancestors of cetaceans were Mesonychids, a lineage of carnivorous mammals. The ancestors of artiodactyls were a lineage of herbivores that looked a bit like tiny deer with long tails. Either way, a seemingly unimaginable number of skeletal and physiological changes would have been required for small, land-based animals to evolve into whales with gigantic bodies uniquely adapted to swimming in the deep ocean.

Then, in 2000, paleontologist Philip Gingerich and his colleagues unearthed complete fossilized skeletons of two cetaceans from a 47-million-year-old rock formation in the Sulaiman Mountains in central Pakistan. Later named *Artiocetus clavis* and *Rodhocetus balochistanensis* (**Figure 14-3c**), both animals had whalelike skulls and robust hind limbs. The bodies of both animals were built for swimming with their feet—not their tails as whales do—and their ankle bones had clearly distinctive features of artiodactyls. This evidence settled the debate about cetacean ancestry. *Rodhocetus* and *Artiocetus* were not direct ancestors of modern whales, but their telltale ankle bones mean they were long-lost relatives. Both ancient cetaceans were offshoots of the artiodactyl-to-modern-whale lineage as it transitioned from life on land to life in water.

The oldest ancestor of living whales is considered to be *Pakicetus attocki* (**Figure 14-3d**), which lived around 50 million years ago. Fossils of this strange mammal with a whalelike head, webbed feet, and a wolf-sized body were discovered in 1983. The mammal was later found to have an ear bone similar to whales and ankle bones similar to artiodactyls, clearly linking it to modern cetaceans.

SEP Communicate Information *Summarize the evidence that modern whales evolved from animals that lived on land.*

430 CHAPTER 14 EVIDENCE FOR EVOLUTION

REVISIT THE ANCHORING PHENOMENON

Gather Evidence *Sample answer: Evidence that supports the evolution of hummingbirds includes fossil, anatomical, and genetic evidence.*

Following the end of the Cretaceous period, dinosaurs and many plants became extinct, opening the door for other plants and animals to diversify. This is about the time that angiosperms evolved, as well as many new species of insects. The coevolution of flowering plants and insects made it possible for both to survive. Unlike other plants, most angiosperms depend on insects, birds, and other animals for pollination and the pollinators require food that is supplied by pollen and nectar from the plant. The hummingbird is an example of coevolution in that it has several adaptations—such as a long, slender beak, the ability to hover, and the ability to see certain colors well—that allowed it to reach the nectar that is in long, slender flowers.

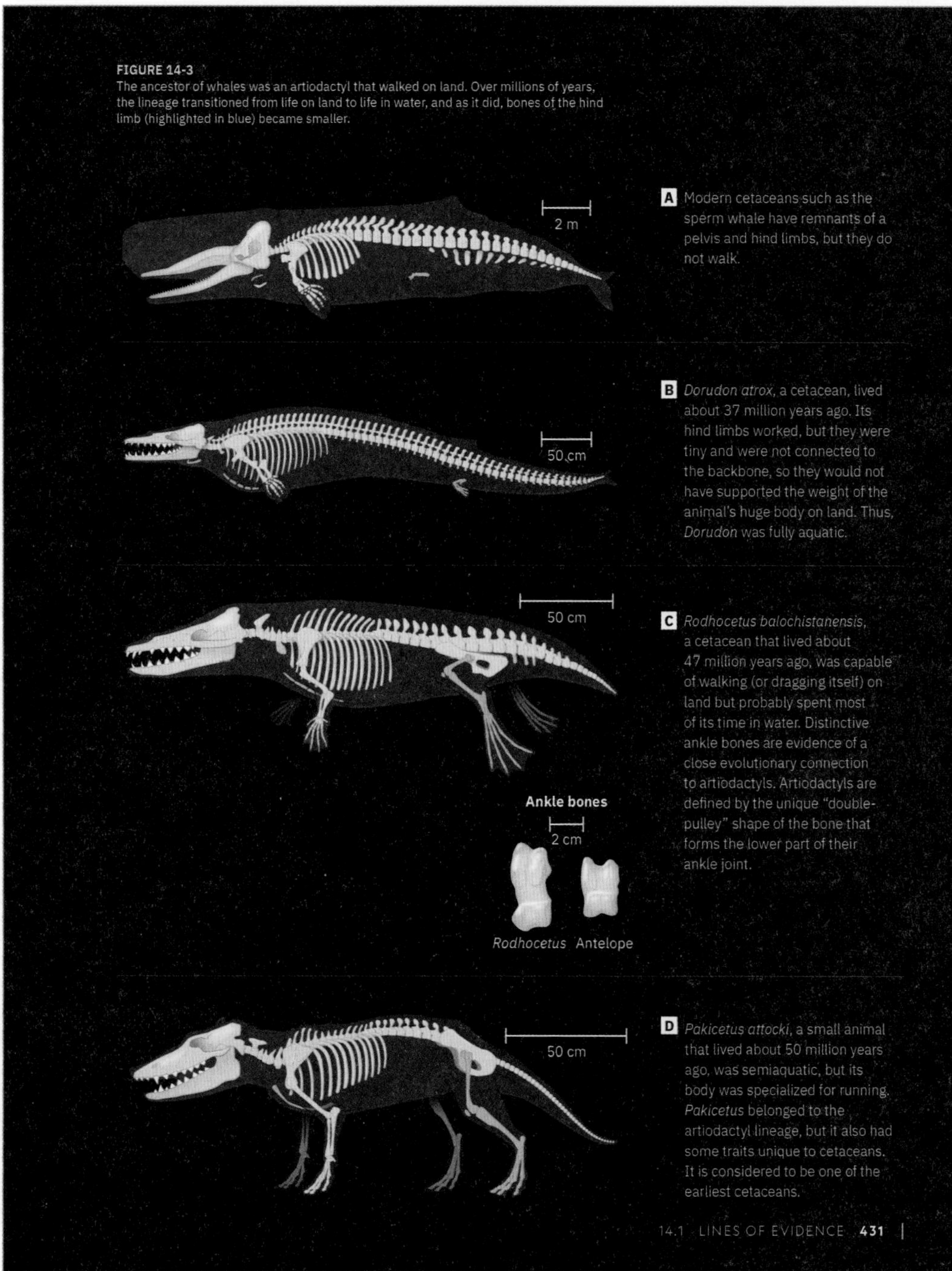

FIGURE 14-3
The ancestor of whales was an artiodactyl that walked on land. Over millions of years, the lineage transitioned from life on land to life in water, and as it did, bones of the hind limb (highlighted in blue) became smaller.
2 m
A Modern cetaceans such as the sperm whale have remnants of a pelvis and hind limbs, but they do not walk.
50 cm
B Dorudon atrox, a cetacean, lived about 37 million years ago. Its hind limbs worked, but they were tiny and were not connected to the backbone, so they would not have supported the weight of the animal's huge body on land. Thus, Dorudon was fully aquatic.
50 cm
C Rodhocetus balochistanensis, a cetacean that lived about 47 million years ago, was capable of walking (or dragging itself) on land but probably spent most of its time in water. Distinctive ankle bones are evidence of a close evolutionary connection to artiodactyls. Artiodactyls are defined by the unique "double-pulley" shape of the bone that forms the lower part of their ankle joint.
Ankle bones
2 cm
Rodhocetus
Antelope
50 cm
D Pakicetus attocki, a small animal that lived about 50 million years ago, was semiaquatic, but its body was specialized for running. Pakicetus belonged to the artiodactyl lineage, but it also had some traits unique to cetaceans. It is considered to be one of the earliest cetaceans.
14.1 LINES OF EVIDENCE 431

Visual Support

Changes in Characteristics Over Time Before reading the descriptions provided, have students observe the four diagrams in **Figure 14-3**. As a class, discuss the observable characteristics and changes to those characteristics from one diagram to the next. If guidance is needed, direct their attention to head size and changes in limb structure. Make a list of student observations (*the first whales had long skulls, large teeth, hind legs*) and discuss how each observation can serve as evidence for whale evolution.

Science Background

Dinosaurs to Birds Explore more about macroevolution by asking students what they think the relationship between dinosaurs and birds might be. Some will already be familiar with *Archaeopteryx*, the first fossil of a birdlike animal. With teeth, toes, and claws, this fossil could have been classified as a dinosaur except for the fact that it had feathers. Modern birds descended from dinosaurs that included the vicious *T. Rex*. Each feature of today's bird—small stature (compared to dinosaurs), feathers, and wings—evolved over time.

DIFFERENTIATED INSTRUCTION | English Language Learners

Using Visuals for Listening Read the text in "Piecing Together Evolutionary Evidence" aloud as students look **Figure 14-3**. As your read each section, have students point to the corresponding part of the visual. Use the visuals to check students' comprehension of the text.

Beginning Ask and have students point to answer: *Which fossil shows a modern whale? In the second fossil, what body part shows this animal lived in the water? In the third fossil, what body parts does this animal use for swimming? In the fourth fossil, what two body parts are similar to whales?*

Intermediate Ask: *What does the first fossil show? In the second fossil, how do scientists know this animal lived in the water? In the third fossil, how did this animal swim? What did scientists conclude based on the fourth fossil?*

Advanced Have students explain what each fossil helped scientists learn about the evolution of whales.

Science Background

Connect to DNA Structure Students should recall that DNA is the genetic material of all living things. It contains a complete set of instructions for making the proteins that determine the characteristics of an organism. The most accurate way to determine evolutionary relationships among species is to compare their DNA or the proteins they are responsible for making. This field of study is called comparative genomics. A comparison of two species' DNA begins with a description of general features of their genomes, such as size, number of chromosomes, and number of genes. Next, homologous genes are identified. Homologous genes are genes inherited in two species from a common ancestor. The homologous genes can then be sequenced to determine the specific differences in DNA between the species.

Connecting Evolutionary Relationships

Classifying life's tremendous diversity into a series of taxonomic ranks (Chapter 1) is a useful endeavor, in the same way that it is useful to organize a contact list in alphabetical order: The result is convenient. Traditional (Linnaean) classification schemes have ranked species into successively higher groups based on shared traits—birds have feathers, cacti have spines, and so on. However, these rankings do not necessarily reflect evolutionary relationships.

Phylogeny Today's biologists work from the premise that every living thing is related if you just look back far enough in time. Grouping species according to evolutionary relationships is a way to fill in the details of this bigger picture of evolution. Thus, reconstructing *phylogeny*, which is the evolutionary history of a species or a group of species, is a priority. Phylogeny is a kind of genealogy that follows evolutionary relationships through time.

Humans were not around to witness the evolution of most species, but as you will see in this chapter, there is plenty of evidence to help us understand ancient events. Consider how each species bears traces of its own unique evolutionary history in its characters. A *character* is a quantifiable, heritable trait such as the number of segments in a backbone, the nucleotide sequence of ribosomal RNA, or the presence of wings. Evolutionary biologists group (rather than rank) species based on shared characters. They focus on what makes the species share the traits in the first place: a common ancestor. Common ancestry is determined by a *derived trait*—a character that is present in a group under consideration, but not in any of the group's ancestors.

A grouping whose members share one or more defining derived traits is called a *clade*. This grouping consists of an ancestor (in which a derived trait evolved) together with all of its descendants. It is the relative newness of a derived trait that defines a clade. Humans and bacteria use some of the same proteins to repair DNA, for example, but humans and bacteria are not close relatives. As another example, consider how alligators look a lot more like lizards than like birds (**Figure 14-4**). While the similarity in appearance does indicate shared ancestry, it is a more distant relationship than alligators have with birds. A unique set of traits that include a gizzard and a four-chambered heart evolved in the lineage that gave rise to alligators and birds, but not in the lineage that gave rise to lizards.

FIGURE 14-4 Lizards (a) look quite similar in appearance to alligators (b). However, alligators share more derived traits with, and are more closely related to, birds (c).

DIFFERENTIATED INSTRUCTION | English Language Learners

Vocabulary Strategies Teach students the expression "family tree" and have them draw one. Model your own family tree before students begin. Remind students that a cladogram is an evolutionary tree. Guide students to compare and contrast their family trees with **Figure 14-5**.

Beginning Have students draw a family tree and label family members. Supply vocabulary as needed. Point out similarities and differences to the cladogram in **Figure 14-5**.

Intermediate Have students draw a family tree and point out close and distant relationships. Have them do the same with the cladogram in **Figure 14-5**.

Advanced Have students draw a family tree and explain how it is similar to and different from the cladograms in **Figure 14-5**.

Cladistics All species are interconnected in the big picture of evolution. An evolutionary biologist's job is to figure out where the connections are. Making hypotheses about evolutionary relationships among clades is called *cladistics*. A *cladogram* is an evolutionary tree that visually summarizes a hypothesis about how a group of clades are related (**Figure 14-5**). Each line is a lineage, which may branch into two lineages at a node. The *node* represents a common ancestor of two lineages, and the two lineages that emerge from a node are called *sister groups*.

Evolutionary history does not change because of events in the present. A species' ancestry remains the same no matter how it evolves. However, as with traditional taxonomic rankings, we can make mistakes when we group organisms based on incomplete information. Thus, a clade or cladogram may change when new discoveries are made. As with all hypotheses, the more data in support of an evolutionary grouping, the less likely it is to require revision.

FIGURE 14-5 Evolutionary connections among clades are represented as lines on a cladogram. Sister groups emerge from a node, which represents a common ancestor (a). A cladogram can be viewed as "sets within sets" of derived traits. Each set, which includes an ancestor together with all of its descendants, is a clade (b).

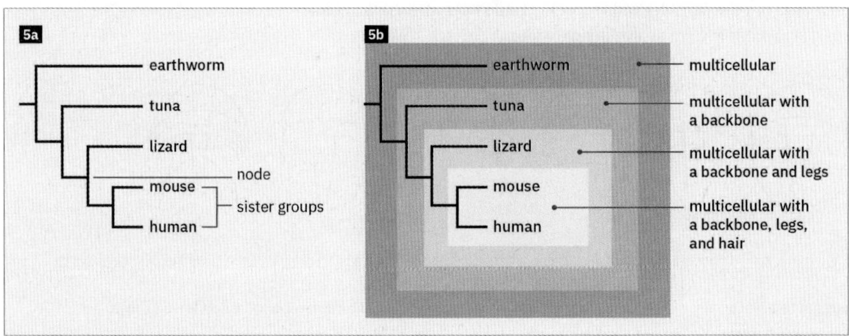

14.1 REVIEW

1. **Define** What is empirical evidence?
   - A. information that is based on opinions
   - B. information that is common knowledge
   - C. information that is found in popular literature
   - D. information that is gathered by observation or experimentation
2. **Identify** Which lines of evidence support evolution? Select all correct answers.
   - A. common anatomical structures
   - B. similarities in physical appearance
   - C. comparable developmental stages
   - D. matching genetic sequence patterns
   - E. sequence of fossils in geologic record
3. **Compare** What is the difference between microevolution and macroevolution?
4. **Apply** Which scenario is an example of microevolution?
   - A. A tadpole develops into a frog.
   - B. A tree loses its leaves in the fall.
   - C. A population of bacteria is resistant to antibiotics.
   - D. A brown dog has a puppy that is white with brown spots.

Science Background

Cladograms and Phylogenetic Trees Sometimes cladograms and phylogenetic trees are used interchangeably; however, there are differences between the two. Discuss these two tools with students, as they are likely to encounter and use both. After reading about cladistics and reviewing **Figure 14-5**, ask students for their understanding of a cladogram (*an evolutionary tree showing shared derived characters that suggest how groups of clades are related*). Show students an example of a phylogenetic tree, which also describes evolutionary relationships, and point out that these diagrams are more complex and are used to represent the evolutionary relationships among different groups of organisms. See Appendix E for additional support.

14.1 REVIEW

1. *D* **DOK 2**
2. *A, C, D, E* **DOK 2**
3. *Microevolution occurs on a small scale, such as within the gene pool of a single population. Macroevolution occurs on a large scale across different species and over millions of years.* **DOK 2**
4. *C* **DOK 2**

MINILAB
ORGANIZING FOSSIL EVIDENCE

Students observe the patterns in trilobite fossil sketches and, based on their observations, they analyze and evaluate evidence of common ancestry and biological evolution. They then use this evidence to develop an evolutionary time line.

Materials and Advance Preparation

- Students can use materials or applications approved by you to develop an evolutionary time line. You can copy the images from the minilab to make a digital evolutionary tree, or students can redraw the fossil images to present their digital evolutionary tree. You can also have students draw sketches of the trilobites on a poster or in their notebooks.
- The sketches provided in the activity do not belong to actual trilobites. They are imaginary examples drawn based on real fossils. Images of actual trilobite fossils are readily available on the Internet. You may find these images to be useful as part of a classroom discussion about fossil evidence and changes in trilobite appearance over time.

Procedure

In **Step 2**, trilobite fossil 3 is the simplest species. It is the only example without a central vertical section.

For **Step 2** and **Step 3**, a sample arrangement of fossils is provided online in MindTap.

Results and Analysis

1. **Analyze** *Sample answer: Fossil 3 has the most basic structure, so I think it is the oldest. Fossil 5 has the same shape as fossil 3 but has the central region, so I think fossil 5 evolved from fossil 3. From here, our tree branches into two. The first branch has fossils 1, 4, and 6 in order. These have the same general wide head and gradually increasing number of spikes. The second branch has fossil 2. This has a narrow body.* **DOK 3**
2. **Predict** *Sample answer: Fossil 7 is the most recent species. I added a spiked tail to have a species that could have evolved from fossil 7. This organism's environment may have more predators than its ancestors encountered, so the spike tail is a beneficial defense against predators.* **DOK 3**
3. **Predict** *Sample answer: Fossil 3 is the oldest species. I removed a segment from its center and made it a bit shorter. Conditions may have changed in the ancestor's environment that led to a long, more segmented body in fossil 3 to give it more mobility so it is easier to move around and find food.* **DOK 3**

MINILAB

ORGANIZING FOSSIL EVIDENCE

SEP Construct Explanations How can you form an evolutionary tree of related species based on fossils?

Trilobites are a group of extinct sea animals that existed over a period of millions of years. They were arthropods that had an exoskeleton, a segmented body, and segmented legs. Trilobites fossilize well, so scientists have been able to classify and study tens of thousands of trilobite species.

In this activity, you will study images of fossils that are related to trilobites. You will consider which species could have given rise to other species in order to form an evolutionary tree.

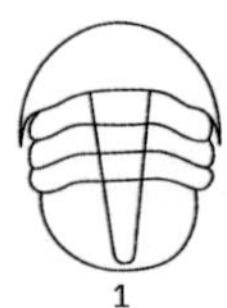
1

2

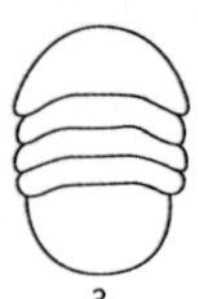
3

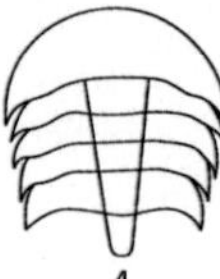
4

5

6

Suggested Materials

- computer/tablet
- poster board or paper
- scissors
- ruler
- glue
- craft materials

Procedure

1. Study the fossil images. These fossils are related species of trilobites.
2. Based on the physical forms you see, make an evolutionary tree that orders the species from oldest to most recent. Redraw the fossil images as necessary. As you make your tree, consider that your tree may have multiple branches. Some species tend to evolve into more specialized and complex shapes if it benefits their chance for reproduction and survival. For this activity, use the most basic trilobite as your oldest species.
3. Add the appropriate labels to your evolutionary tree using the following list. Some of these labels will be used more than once, and some may not be needed at all. Note: *mya* means "million years ago."
   - Cambrian period trilobite—521 mya
   - Ordovician period trilobite—460 mya
   - Devonian period trilobite—400 mya
   - Carboniferous period trilobite—300 mya
   - Permian period trilobite—255 mya

Results and Analysis

1. **Analyze** Describe the reasoning for how you organized the fossils to represent an evolutionary pathway. What linked each species to the next?
2. **Predict** Study the trilobite that is the most recent in your evolutionary pathway. Draw a sketch of a trilobite that might have evolved from that species. Describe how your species is different from its ancestor and what change in conditions may have led to those differences.
3. **Predict** Study the trilobite that is the oldest in your evolutionary pathway. Draw a sketch that could be the ancestor of this species. Describe how your ancestor species is different from its descendants and what change in conditions may have led to those differences.

FOSSIL AND GEOLOGICAL EVIDENCE

Trilobites were marine animals that first appeared at the beginning of the Cambrian period, about 542 million years ago. These prolific arthropods persisted for 300 million years, more than double the length of time dinosaurs roamed our planet. Fossilized remains of trilobites have been found on every continent.

Infer *What do you think the discovery of trilobite fossils in the Midwest region of the United States tells us about how that region's environmental conditions have changed over geologic time?*

14.2

KEY TERMS

fossil
gradualism
half-life
index fossil
radiometric dating
trace fossil
uniformitarianism

Fossil Formation

Fossils have long been recognized as rock-hard evidence of earlier forms of life. Not only do they give us clues about what certain organisms looked like, but they also give us a glimpse of what environmental conditions were like millions of years ago. **Fossils** are mineralized bones, teeth, shells, seeds, spores, or other long-lasting parts of an organism that lived in the past. **Trace fossils** such as footprints and other impressions, nests, burrows, trails, eggshells, or feces are evidence of activities. For example, the spacing between fossilized footprints provides evidence of how an ancient animal moved. Fossilized feces, called coprolites, provide evidence of an extinct animal's diet. **Figure 14-6** shows a variety of fossil types.

 VIDEO

VIDEO 14-1 Go online to learn more about fossils and how they are formed.

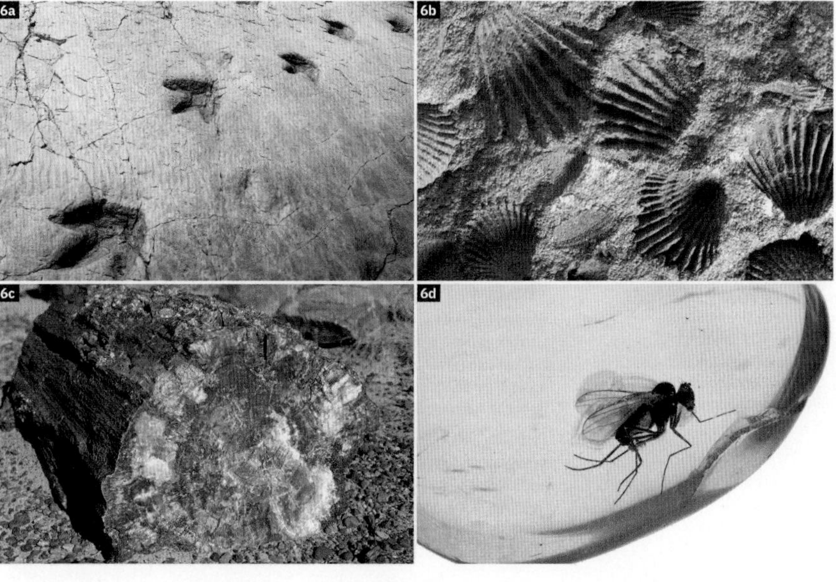

FIGURE 14-6 Examples of fossils include fossilized dinosaur tracks (a), molds or casts of fossil shells (b), petrified wood (c), and insects preserved in amber (d).

DIFFERENTIATED LEARNING | English Language Learners

Modeling Vocabulary Use models to help students understand the terms *impression* and *cast*. Have students press a small object into modeling clay to make an impression. Explain that when an impression hardens, it forms a *mold*. Have students fill the impression with quick-drying plaster of Paris, let it dry, and peel off the clay to reveal a *cast*.

Beginning Provide sentence frames for students to explain the models: *I press an object in to make an (impression). When it hardens, it is a (mold). Minerals fill the (impression). When they harden, they make a (cast).*

Intermediate Have students describe their models in complete sentences using the words *impression*, *mold*, and *cast*.

Advanced Have students explain how the meanings of *impression*, *mold*, and *cast* are different, using their models as examples.

14.2

FOSSIL AND GEOLOGICAL EVIDENCE

LS4.A, PS1.C, ESS2.A

EXPLORE/EXPLAIN

This section presents information on fossils and describes the process of fossilization step by step. Students will explore how the tools of index fossils and radiometric dating are used to interpret the fossil record. They will also explore the chronology of the history of life on Earth as revealed by the fossil record.

Objectives

- Sequence steps in the process of fossilization.
- Explain different methods used to date fossils.
- Differentiate three different types of evidence used to determine evolutionary relationships and the history of life on Earth.

Pretest Use **Question 3** to identify gaps in background knowledge or misconceptions.

Vocabulary Strategy

Word Wall Have students begin a word wall by writing "evolution" in the center. Explain that they will add new words as they learn them in the context of the lesson, connecting them to the word in the center. Tell them that they may also connect words with arrows if they see a relationship between them. Students can add definitions, images, or other relevant material to the word wall if it helps them remember the word and its meaning.

Video

Time: 4:08

Use **Video 14-1** to learn about two major fossil types, how fossils form, and the stories they can tell about Earth's history.

Infer *Students might say that the region was likely covered by ocean at some point during its geologic history.*

Connect to English Language Arts

Write About Sequence Review **Figure 14-7** with students. Have students ask and answer questions about each step. Then, ask students to retell in order how organisms become fossilized in an explanatory text. Explain that since they are describing steps in a process, they should use sequence words, such as *first, next, then,* or *finally*.

In Your Community

Finding Fossils After discussing fossils and trace fossils, encourage students to explore their surroundings and look for fossils in their communities. Remind students that a fossil can be any evidence of a living thing. Show them a leaf imprint, or a picture of one, and talk about why it is considered a fossil. Have students look for images online using the keyword *fossil* or *local fossils,* or visit a local museum. There are also some virtual tours available though national museums, such as the Smithsonian National Museum of Natural History and the American Museum of Natural History. Students may work in groups to create a display of what they find, with a written explanation of what it is and where it was found.

CASE STUDY
SOLVING THE MYSTERY OF SLOTH EVOLUTION

Ask Questions To help students revisit the Case Study and make connections with the content, have them use the diagram on this page to generate questions about the lines of evidence scientists use to determine the evolutionary history of species.

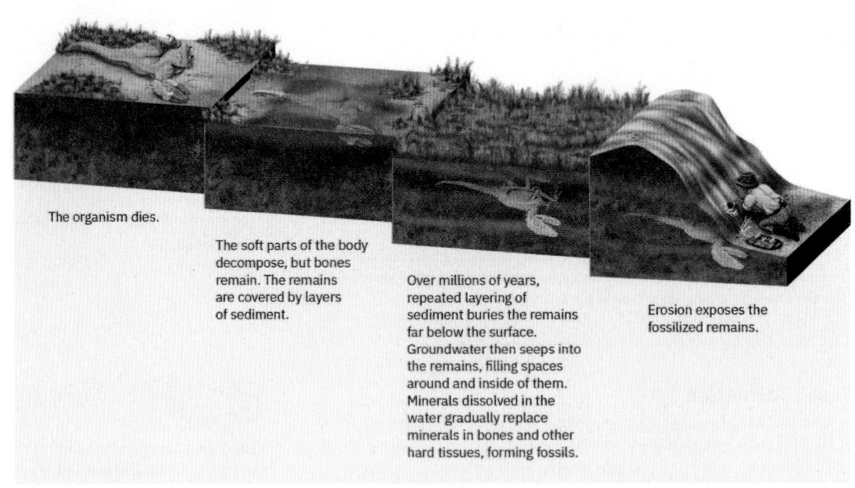

The organism dies.
The soft parts of the body decompose, but bones remain. The remains are covered by layers of sediment.
Over millions of years, repeated layering of sediment buries the remains far below the surface. Groundwater then seeps into the remains, filling spaces around and inside of them. Minerals dissolved in the water gradually replace minerals in bones and other hard tissues, forming fossils.
Erosion exposes the fossilized remains.

FIGURE 14-7 Very few organisms become fossils. A particular set of conditions is required for the fossilization process to occur.

The process of fossilization is shown in **Figure 14-7**. This process typically follows these steps:

- Fossilization begins when an organism or its traces become covered by sediments, mud, or ash.
- Groundwater then seeps into the remains, filling spaces around and inside of them.
- Minerals dissolved in the water gradually replace minerals in bones and other hard tissues.
- Mineral particles that settle out of the groundwater and crystallize inside cavities and impressions form detailed imprints of internal and external structures.
- Sediments that slowly accumulate on top of the site exert increasing pressure. After a very long time, extreme pressure transforms the mineralized remains into rock.
- Over time, the fossilized remains may become exposed as a result of the geologic processes of uplift and erosion.

CASE STUDY
SOLVING THE MYSTERY OF SLOTH EVOLUTION

Ask Questions *Generate questions about the evidence the fossil record provides in support of the evolution of sloth species.*

The Fossil Record

We have fossils for more than 250,000 known species. Considering the current range of biodiversity, there must have been many millions more, but we will never know all of these species. Why not? The odds are against finding evidence of an extinct species. Few individuals of any species become fossilized.

Typically, when an organism dies, its remains are destroyed quickly by scavengers. Organic materials decompose in the presence of moisture and oxygen. Thus, remains that escape scavenging endure only if they dry out, freeze, or become encased in a material such as sap, tar, or mud. Most remains that do become fossilized are crushed or scattered by erosion and other geologic processes.

436 CHAPTER 14 EVIDENCE FOR EVOLUTION

DIFFERENTIATED INSTRUCTION | Leveled Support

Struggling Students Working in pairs, have students use different materials to create a video storyboard or graphic art of how an organism becomes a fossil after it dies. Each frame of their art should include a description and indicate the correct sequence.

Advanced Learners Working in pairs, have students use different materials to create a video storyboard or graphic art of how a different type of organism becomes a fossil after it dies. For example, they may choose a marine organism or an organism trapped in a material such as amber or tar. Have students write an audio script that describes the process to go with their storyboard or art. Then, record it for a presentation.

For us to know about an extinct species that existed long ago, we have to find a fossil of it. At least one specimen had to be buried before it decomposed or something ate it. The burial site had to escape destructive geologic events and end up in a place we can find today. Most ancient species had no hard parts to fossilize, so we do not find much evidence of them. For example, there are many fossils of bony fishes and mollusks with hard shells, but few fossils of the jellyfish and soft worms that were probably much more common. Also, think about relative numbers of organisms. Fungal spores and pollen grains are typically released by the billions. By contrast, the earliest humans lived in small bands and few of their offspring survived. The odds of finding even one fossilized human bone are much smaller than the odds of finding a fossilized fungal spore.

Most fossils are found in layers of sedimentary rock that form as rivers wash silt, sand, volcanic ash, and other materials from land into the sea. Mineral particles in the materials settle on the seafloor in horizontal layers that vary in thickness and composition. After many millions of years, the layers of sediments become buried and compacted into layered sedimentary rock. Geologic processes can tilt the rock and lift it far above sea level, where the layers may become exposed by the erosive forces of water and wind (**Figure 14-8**).

Biologists study sedimentary rock formations to understand the historical context of ancient life. Rock layers provide a kind of clock or timing device for scientists to date fossils. In a multilayered formation, the more recent layers lie on top of the older layers. Knowing the rate of sediment deposition allows scientists to provide relative dates for given fossils. If, for example, you know it took 20 million years for each rock layer to form, then a dinosaur fossil found four layers down was alive approximately 80 million years ago.

CCC Patterns *What characteristics do you think are shared by species that are more likely to be represented in the fossil record?*

FIGURE 14-8
It takes thousands, even millions, of years for sedimentary rock layers to form out of sand, soil, and other materials. The youngest rock layer that makes up the Grand Canyon is 270 million years old. The oldest layers date back 1.8 billion years.

DIFFERENTIATED INSTRUCTION | Students with Disabilities

Visual Impairments After instruction on the fossil record, have students use clay and plastic figures to model layering of fossil-bearing sedimentary rocks.

Explain to the class that they should decide beforehand which figures will be included in each layer. If needed, guide students with a visual impairment to select what figures they will include in each layer of their model. Help with arranging the figures in the order of the layer they will appear in, from bottom layer to top layer.

Each layer should be made with a different color of clay to make them easy to differentiate. Students with a visual impairment can use clay with different textures. Prepare batches of clay with large stones, small stones, cornmeal, dried peas, dried beans of different sizes, or other materials that would be easy to distinguish by touch.

All students should make their model by placing one or more plastic figures in each layer. Ask them to then identify the pattern of oldest to youngest.

Science Background

Homo sapiens Scientists have long used the discovery of *Homo sapiens* fossils at two separate sites in Ethiopia, dating back 195,000 years ago and 160,000 years ago, to support the claim that our species arose in that area close to the time of the older fossil. However, some discoveries raised questions. In 1961, workers discovered a fossil of an ancient human skull at a mining site in Jebel Irhoud, Morocco, in northern Africa. Later, a team of scientists from the Max Plank Institute for Evolutionary Anthropology in Germany visited the site and found more human fossils and tools. Using many different techniques, the team dated the rock layer where the fossils were found to 350,000 to 280,000 years ago. The dating of the fossils and tools in Jebel Irhoud raises questions about the age and origin of *Homo sapiens*. Discuss with students how the discovery of human fossils and tools may differ from the discovery of fossils of other organisms.

Connect to Mathematics

Reason Quantatively Layers of fossil-containing sedimentary rock can provide information about the relative ages of Earth's organisms. Relative age can determine if one fossil is younger or older than another because the lowest layer in a rock formation was the first to be laid down, making it the oldest—rather like the firstborn child in a family is the oldest. Each layer above a layer is younger than the lower layer. This is called the *law of superposition*. A fossil's absolute age can be estimated from the amount of sediment deposited above the fossil. Have students solve the problem: If a fossil is found five layers down and the top layer took 1 million years to form, layer 2 took 5 million years, layer 3 took 10 million years, and layers 4 and 5 each took 15 million years, what is the estimated age of the fossil? (*46 million years*)

SEP Patterns *Students might say that the species that existed for a hundred million years is more likely to be a part of the fossil record because there is a greater likelihood that an individual of the species met the requirements for becoming a fossil.*

Science Background

Trilobites Trilobites are referred to as index fossils. They first appeared in the fossil record about 521 million years ago during the Early Cambrian period and lasted for almost 300 million years, until the end of the Permian period. Trilobites were marine arthropods covered by an exoskeleton that protected their soft parts. Most trilobites lived in shallow water and thrived as a species. Their fossils are found worldwide, on every continent where Paleozoic outcroppings exist. Although they appeared quickly and existed for a relatively short period of time, they thrived and today are one of the most readily identified fossils on Earth. After sharing this information with students, have them identify which statements show that trilobites meet the criteria for being classified as an index fossil.

On the Map

Where in the World? Although trilobites lived just about everywhere, there are a few locations that are notable for trilobite fossils. For example, the Burgess Shale in the Canadian Rockies are known for late Middle Cambrian fossils, with some trilobite fossils that have soft parts preserved. Morocco is well-known for Devonian trilobites. In North America, well-known sites are found in Utah, Oklahoma, Ohio, California, and New York. Have students research these localities and create a map showing the widespread existence of these index fossils.

Index Fossils

Scientists use a number of different sources of evidence to connect rocks of the same age. **Index fossils** are useful pieces of evidence because they can be used to precisely determine sedimentary rock ages. The most useful index fossils share the following characteristics:

- The fossils are abundant in rock layers.
- The fossils are found all over Earth.
- The organisms existed as a species or genus for a relatively short period of geologic time.
- The fossils are easily identified in the field.

Marine organisms are some of the best index fossils because they spread rapidly and widely throughout their watery environments. Species that existed only for a short period of geologic time enable scientists to more precisely date the rock layers in which they are found.

Fossils of *Neptunea tabulata* (**Figure 14-9a**) date to the Quaternary period. Ammonites, trilobites, and brachiopods were particularly widespread over geological time. Today, many species from these groups are used as index fossils to determine the relative age of rock layers. The ammonite genus *Dactylioceras* (**Figure 14-9b**) is dated to the Jurassic period, the trilobite genus *Paradoxides* (**Figure 14-9c**) is dated to the Cambrian period, and the brachiopod genus *Mucrospirifer* (**Figure 14-9d**) is dated to the Devonian period.

FIGURE 14-9
Examples of index fossils include *Neptunea tabulata* (a) ammonites (b), trilobites (c), and brachiopods (d).

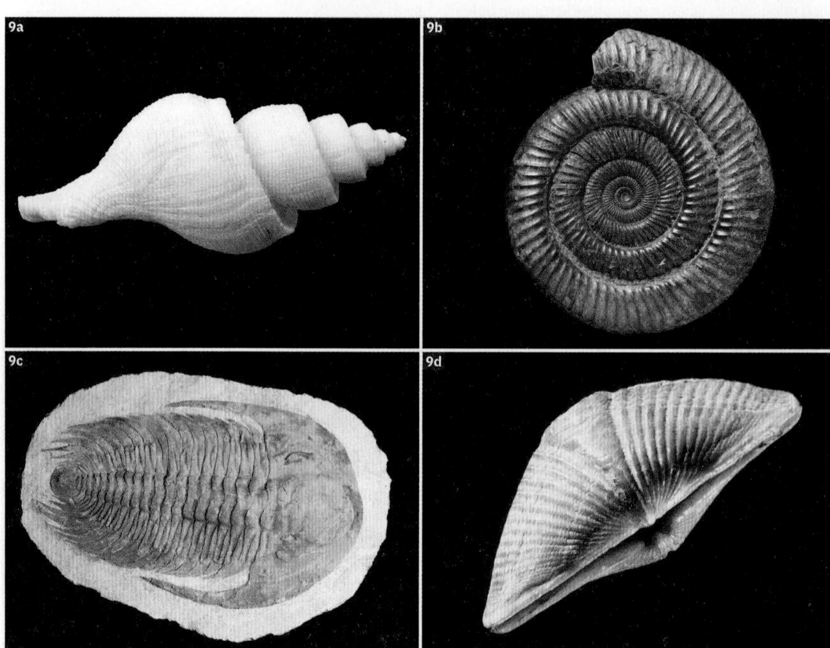

DIFFERENTIATED INSTRUCTION | English Language Learners

Monitoring and Self-Correcting Model reading aloud from the text, pausing to sound out an unfamiliar word and check understanding. Then have students do the same as they read the text aloud in pairs or small groups.

Beginning After modeling, have small groups read parts of the text aloud. Encourage students to sound out new or difficult words. After they read, discuss the meanings of any words they struggled with.

Intermediate Working in pairs or small groups, have students take turns reading parts of the text aloud. Encourage them to help each other sound out difficult words and discuss the meanings of new vocabulary. They could also use resources such as a dictionary or online translator

Advanced Working in pairs, have students read aloud a paragraph or two at a time until they have read the complete text. Have students sound out new words and explain their meanings.

Radiometric Dating

Researchers use the predictability of radioactive decay to reveal the age of a fossil or other ancient object. This method, called **radiometric dating**, reveals the age of a material by measuring its radioisotope content (**Figure 14-10**). Each type of radioisotope decays at a constant rate into predictable products called daughter elements. The time it takes for half of the atoms in a sample of a radioisotope to decay is called its **half-life**.

Almost all carbon on Earth is in the form of ^{12}C. A tiny but constant amount of ^{14}C forms in the upper atmosphere and becomes incorporated into molecules of carbon dioxide. The ratio of ^{14}C to ^{12}C in atmospheric CO_2 is relatively stable. Because carbon dioxide is life's main source of carbon, the same ratio is maintained in the body of a living organism. After the organism dies, no more carbon is incorporated into its body. The ratio of ^{14}C to ^{12}C in the organism's remains declines over time as the ^{14}C decays. The half-life of ^{14}C is known (5730 years) so the ratio of ^{14}C to ^{12}C in the organism's remains can be used to calculate how long ago it died. Carbon isotope dating can be used to find the age of a biological material less than 60,000 years old. In older material, all of the ^{14}C has decayed. The age of older fossils can be estimated by dating volcanic rocks in lava flows above and below the fossil-containing layer of sedimentary rock.

The original source of most rock on Earth is magma, a hot, molten material deep under Earth's surface. When magma cools, it hardens and becomes rock. As this occurs, the atoms in the magma join and crystallize as different kinds of minerals, each with a characteristic structure and composition. Zircon is a mineral that consists mainly of zirconium silicate molecules. Some of the molecules in a newly formed zircon crystal have uranium atoms substituted for zirconium atoms. Uranium is a radioactive element with a half-life of 4.5 billion years. Its daughter element, thorium, is another radioactive element. Thorium decays at a predictable rate into another radioactive element, and so on until the atom becomes lead, a stable element. Over time, uranium atoms disappear from a zircon crystal, and lead atoms accumulate in it. The ratio of uranium atoms to lead atoms in a zircon crystal can be used to calculate its age. Using this technique, we know that the oldest known terrestrial rock, a tiny zircon crystal from the Jack Hills in Western Australia, is 4.404 billion years old.

SEP Interpret Data *According to **Figure 14-10**, what percentage of any radioisotope remains after two of its half-lives have passed?*

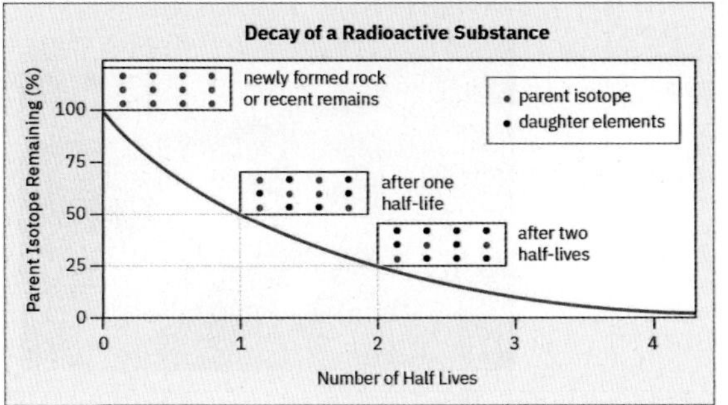

FIGURE 14-10 This generalized graph shows how the ratio of parent isotope and daughter elements changes as each half-life has passed.

Connect to Mathematics

Thorium-230 Thorium-230 is a decay product of the radioisotope uranium-234. Uranium-234 (U-234) is taken into living coral, but thorium-230 (Th-230) is not. The decay of U-234 results in a build-up of Th-230 in samples of older coral and provides an accurate measure of time due to the known decay rate of U-234. The ratio of U-234/Th-230 allows scientists to date the original formation of the coral. The half-life of Th-230 is 75,000 years and can be used to date objects up to about 500,000 years. Ask students to calculate the age of a coral fossil in which 75 percent of the Th-230 has decayed.
(75 percent is two half-lives, or 75 × 2 = 150,000 years old.)

Crosscurricular Connections

Physical Science Some students may need a refresher on basic chemistry concepts. Remind students that the atomic number of an element is the number of protons in the nucleus. The number of protons does not vary for the atoms of a specific element; however, the number of neutrons can vary. Atoms of the same element that differ in the number of neutrons are called *isotopes*.

The mass number of an isotope is the total number of protons and neutrons in the nucleus. For example, the mass number of the most common carbon isotope is 12. This isotope has six protons and six neutrons and is written as carbon-12.

SEP Interpret Data *25 percent*

DIFFERENTIATED INSTRUCTION | Leveled Support

Struggling Students Provide students with a set of about 72 chips that are two different colors. Explain to students that they will use the data represented in **Figure 14-10** to make a model of half-life using different colored chips to represent the parent isotope and daughter elements. They can begin by drawing a grid similar to that in **Figure 14-10** and then arrange their chips on the grid. Have students ask and answer questions about the grid with partners.

Advanced Learners Provide students with a set of about 72 chips that are two different colors and a stopwatch or timer. Tell them that they will use a hypothetical half-life of one minute. One student can act as a recorder and mark the start time. After one minute, students should turn 32 chips over. Ask students to discuss what this represents (*these chips have decayed and are stable*). After another minute, students should turn over 16 more chips. This should continue for a total of six minutes or six half-lives.

Science Background

Changing the Dating Carbon dating is about to be recalibrated. This method is based on data collected from studies of tree rings, lake and ocean sediments, and corals and stalagmites. The way carbon dating works remains the same—measuring the amount of natural radioactive carbon in a plant or animal after it dies. However, this dating method assumes a constant amount of carbon-14 in the environment, and that has changed. The addition of tons of excess carbon dioxide from the burning of fossil fuels has been shown to affect this method of dating. The recalibration will likely have an effect on the estimated ages of many fossil discoveries, including the oldest modern human fossils.

Reconstructing Earth's Climate History

Scientists can also use other data to support the fossil record and radiometric findings. Proxy data, such as ice cores and tree rings, provide an indirect view of past conditions on Earth.

- **Ice Cores** Over hundreds and thousands of years, ice on mountaintops and at the Poles forms from the accumulation of snowfall. Scientists can drill deep into the ice to collect ice cores. These ice cores are made up of distinct layers. Each layer contains air bubbles, oxygen isotopes, and dust that scientists can analyze to determine past volcanic activity, atmospheric composition, temperature, precipitation, and wind patterns.
- **Tree Rings** Long-lived tree species such as redwoods and bristlecone pines add wood over centuries, typically one ring per year. If you know the year in which the tree was cut, you can determine when a particular ring formed by counting the rings backward from the outer edge. Tree-ring cores can also be taken from living trees. Tree rings can be used to estimate average annual rainfall; to date archaeological ruins; and to gather evidence of wildfires, floods, landslides, and glacier movements. Thickness, isotope content, and other features of a ring offer clues about the environmental conditions that prevailed during the year it formed. For example, during periods of drought, the rings are thinner compared to periods of heavy precipitation, when the rings are thicker.

CONNECTIONS

REFLECTIONS OF A DISTANT PAST The foundation for scientific understanding of the history of life is this premise: Natural phenomena that occurred in the ancient past can be explained by the same physical, chemical, and biological processes that operate today. Consider what caused the dinosaurs to become extinct at the end of the Cretaceous period. Most scientists accept the hypothesis that dinosaurs died in the aftermath of a catastrophic meteorite impact. No humans were around 66 million years ago to witness the impact, so how could anyone know what happened? The event is marked by an unusual, worldwide formation of sedimentary rock called the Cretaceous–Paleogene (or K–Pg) boundary sequence. There are plenty of dinosaur fossils below this formation. Above it, in layers of rock that were deposited more recently, there are no dinosaur fossils, anywhere.

The K–Pg boundary sequence consists of an unusual clay that is rich in iridium, an element rare on Earth's surface but common in meteorites. The clay also contains shocked quartz and small glass spheres called tektites. Both form when quartz or sand (respectively) undergoes a sudden, violent application of extreme pressure. The only processes on Earth known to produce these minerals are nuclear explosions and meteorite impacts.

Geologists hypothesized that the K–Pg boundary layer must have originated with extraterrestrial material. The geologists began looking for evidence of a meteorite that hit Earth 66 million years ago—one big enough to cover the entire planet with its debris. In 1991, they found it: an impact crater roughly the size of Ireland off the coast of the Yucatán Peninsula. To make a crater this big, a meteorite 10 km (6.2 miles) wide would have slammed into Earth with the force of 100 trillion tons of dynamite—enough to cause an ecological disaster of sufficient scale to wipe out almost all life on Earth.

The K–Pg boundary sequence (dark band) is a global sedimentary rock formation that formed 66 million years ago.

CONNECTIONS | Reflections of a Distant Past

Social-Emotional Learning Changes in life forms over Earth's history is one that invites different opinions and hypotheses. For example, a second scientific hypothesis suggests that dinosaurs and other organisms below the K-Pg boundary died off gradually due to environmental change already occurring by the time the asteroid hit Earth. However, proponents of the two different hypotheses still find common ground in the idea of global climate change as a cause for mass extinction.

This is a good opportunity for students to practice multiple social-emotional skills, beginning with *self-awareness* and the ability to understand individual thoughts and ideas and how they influence beliefs. At the same time, students can practice *relationship skills* as they listen respectfully to the opinion of others and resolve conflicts constructively.

It is also an opportunity to teach how to differentiate between fact and opinion and the role of evidence in doing so.

The Geologic Time Scale

Similar sequences of sedimentary rock layers occur around the world. The geologic time scale is a chronology of Earth's history that correlates these sequences with time (**Figure 14-11**). Transitions between layers of rock mark boundaries between time intervals. Each layer's composition offers clues about conditions that prevailed on Earth when the layer was deposited. For example, fossils are a record of life during the period of time that sediments in a layer accumulated.

Similar to how a year is divided into smaller increments including months, weeks, and days, so too is the geologic time scale split into smaller intervals. Units of geologic time include the following:

- Eons, which extend over 500 million years or more, are the longest interval of geologic time.
- Eras extend over several hundred million years.
- Periods extend over ten to several hundred million years. Periods are associated with specific layers of rock.
- Epochs last tens of millions of years.
- Ages extend over several million years.

Era	Period	Epoch	Millions of years ago	Major events
Cenozoic	Quaternary	Holocene	* 0.01	humans
		Pleistocene	1.65	new mountains
	Tertiary	Pliocene	5.3	early humans
		Miocene	23.7	apes
		Oligocene	36.6	monkeys
		Eocene	57.8	grasslands
		Paleocene	66.4	early primates
Mesozoic	Cretaceous		* 144	flowering plants, insects, and reptiles diversify
	Jurassic		208	first birds
	Triassic		245	first mammals
Paleozoic	Permian		* 286	first conifer trees
	Carboniferous		320	first reptiles and forests
			360	age of amphibians
	Devonian		408	first insects and amphibians
	Silurian		* 438	first jawed fishes
	Ordovician		* 505	first land plants
	Cambrian		545	first fishes
Precambrian			900	first invertebrate animals
			3500	photosynthesis in bacteria
			4600	origin of Earth

* mass extinctions

FIGURE 14-11 This diagram shows the history of life on Earth. Note that the great expansion of life-forms occurred during the Cambrian period more than 500 million years ago.

DIFFERENTIATED INSTRUCTION | English Language Learners

Possessives Support students in using possessives correctly as they write their explanatory text. Remind students that most singular nouns form the possessive with *'s*, while most plural nouns form the possessive with *s'*. Go over examples that are likely to trip up students, for example: *Holmes's geologic time scale* and *Hutton and Lyell's work*.

Beginning Write *scientist* and *scientists* and have students form the possessive by adding *'s* or *s'*. Repeat for other nouns from the time line in "Connect to English Language Arts" instructional note. Ask students to use at least three of these in their writing.

Intermediate Underline mistakes in students' use of possessives, and have them correct their own work. Alternatively, tell students how many mistakes they made and have them find and correct their mistakes.

Advanced Have partners check each others' work and underline mistakes in possessives. Then have them correct their own mistakes.

Connect to English Language Arts

Write Explanatory Text The first geologic time scale to use absolute dates was published in 1913 by the British geologist Arthur Holmes. However, like many discoveries in science, it was built on the work of earlier scientists. The geologic time scale is an important tool for scientists to build a time line of Earth's history. The time scale was not invented by one person. Instead, it grew out of necessity as a way to organize the continuous growth in scientists' body of knowledge about the age of Earth. In 1669, Danish scientist Nicolaus Steno noted that sedimentary rocks were laid down in layers with the younger layers on top of older layers. This understanding was later added to with the works of British geologists James Hutton in 1795 and Charles Lyell in the early 1800s.

Have students discuss and then write a short essay on how the development of the geologic time scale represents the nature of science as scientists build on the ideas of others.

Connect to Mathematics

Model with Mathematics Discuss with students that the geologic time scale is similar to a calendar or clock in that it divides time into units, such as eons, eras, periods, epochs, and ages. Work with students to develop a model of the geologic time scale using a clock to represent the entire history of Earth. Each hour on the clock represents 350 million years. Using midnight as the formation of Earth, have them predict the time on the clock that would represent the appearance of the first photosynthetic life forms (*about 6:00 a.m.*). Ask questions such as what time on the clock would represent when the first animals with backbones would appear (*about 9:15 p.m.*). Continue with examples such as the appearance of reptiles at about 10:30 p.m. and mammals commonly found at about 11:40 p.m. Ask students when they think early humans would appear *(about 40 seconds before midnight)*.

Science Background

Darwin's Library The HMS *Beagle*, the ship that carried Charles Darwin on his extraordinary journey that would change scientific thinking, housed a 400-volume library in the cabin that Darwin stayed in. Among the volumes were those chosen by Darwin to carry on his journey, including John Milton's *Paradise Lost*, and Charles Lyell's *Principles of Geology*, among others. Throughout his five-year journey, Darwin's detailed observations provided evidence in support of Lyell's ideas that Earth's surface was shaped gradually over time through observable Earth processes that were continuously operating.

Early Explorers Students learn that different theories developed and changed over time as more explorations of Earth gathered new evidence. **Chapter 15** provides more detail about Charles Darwin's discoveries and ideas.

NOS Scientific Knowledge *Students may say that uniformitarianism provided a time frame (millions of years) for evolution to occur.*

The Present Is the Key to the Past

During the 18th and 19th centuries, prior to the advent of radiometric dating, it was commonly thought that Earth was only several thousand years old and the species and landforms on Earth were the same as they ever had been. The discovery of fossils and other geological evidence made many scientists of the time question this viewpoint.

In the late 1700s, Scottish geologist James Hutton reasoned that the changes he observed in landforms were the result of changes that occurred over a long period of time. This became known as the principle of **gradualism**. Unlike the theory of *catastrophism*, which states that major catastrophic events such as floods and volcanoes shaped Earth's surface, Hutton argued that many of Earth's features, such as deep canyons, resulted from slow processes.

British geologist Charles Lyell was also a proponent of gradualism. In the 1830s, he expanded on Hutton's work by proposing what is now known as the principle of **uniformitarianism**. This principle states that geologic processes that shape Earth's surface, such as erosion, are the same today as they were in the past. Lyell published his ideas in *The Principles of Geology*, a three-volume book. This book would prove to be quite influential on other scientists of the time, including Charles Darwin, which is described in detail in Chapter 15.

NOS Scientific Knowledge *Why do you think Lyell's principle of uniformitarianism proved to be so influential on scientists' understanding of evolution?*

14.2 REVIEW

1. **Compare** Use the descriptions to determine whether each organism would be ideal or not ideal as an index fossil.
   - Organism is abundant in rock layers.
   - Organism was limited to a single area.
   - Organism was widespread geographically.
   - Organism lived for hundreds of millions of years.
   - Organism is similar in appearance to other fossils.
2. **Calculate** Based on its carbon-14 to carbon-12 ratio, a fossil is determined to be 11,500 years old. Use the half-life of carbon-14 (5730 years) to determine the percentage of original carbon-14 that remains in the fossil.
3. **Contrast** Explain the differences between the geologic principles of uniformitarianism and catastrophism.
4. **Sequence** A rock formation that contains four layers with fossils has been greatly changed by geologic forces. Assuming the fossils are near the oldest possible age limit, use **Figure 14-11** to sort the layers into chronological order based on the fossils they contain.

Layer M – bird and amphibian fossils	Layer P – mammal and fish fossils
Layer O – fish fossils only	Layer T – primate and flowering plant fossils

5. **Identify** Which are examples of trace fossils? Select all correct answers.
   - **A.** fossilized claw
   - **B.** fossilized burrow
   - **C.** dinosaur footprints
   - **D.** fossilized leaf
   - **E.** insects trapped in amber

14.2 REVIEW

1. *not ideal: organism lived for hundreds of millions of years, organism is similar in appearance to other fossils, organism was limited to a single area; ideal: organism was widespread geographically, organism is abundant in rock layers* **DOK 2**
2. *25%; 25 percent* **DOK 3**
3. *Catastrophism states that Earth's features were formed primarily by a few disastrous, violent events such as volcanic eruptions and meteor impacts. Uniformitarianism is a form of gradualism that says many of Earth's features are due to the slow geologic processes we see occurring today.* **DOK 2**
4. *T; M; P; O* **DOK 2**
5. *B, C* **DOK 2**

LOOKING AT THE DATA

FORENSIC RADIOMETRIC DATING

SEP **Analyze and Interpret Data** Traditionally, carbon dating has been considered an archaeological tool, not a forensic one. It is used to date tree, plant, and animal fossils that are between 500 and 50,000 years old. However, since the mid-1950s, forensic scientists have been able to use the same technique to date skeletal and tissue remains.

The concentration of ^{14}C relative to ^{12}C atoms in the atmosphere has remained relatively constant for thousands of years. In 1955, the radiometric carbon ratio began to rise sharply just 10 years after the first atmospheric nuclear tests by the United States. Nuclear testing creates neutrons that can interact with ^{14}N to produce the unstable ^{14}C isotope. **Figure 1** shows the measured atmospheric $^{14}C/^{12}C$ relative to the natural level up to the turn of the century. **Figure 2** shows the number of nuclear weapon tests by all countries since 1945.

FIGURE 1
The relative atmospheric radiometric carbon ratio $^{14}C/^{12}C$ has changed since the 1950s.

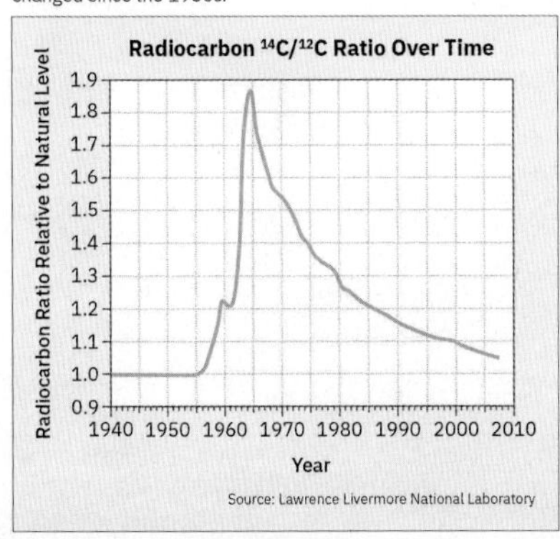

FIGURE 2
Nuclear tests by all countries were conducted above ground between 1945 and 1962.

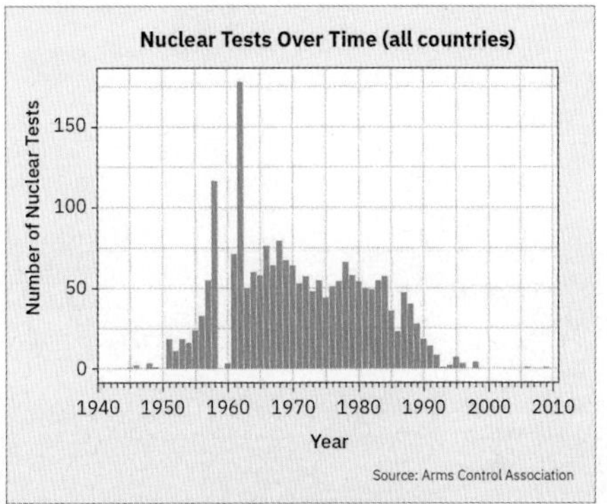

1. **Compare** In which years did the atmospheric $^{14}C/^{12}C$ and number of above-ground nuclear tests peak?
2. **Relate** Describe the relationship between atmospheric $^{14}C/^{12}C$ and the number of above-ground nuclear tests conducted during the period.
3. **Explain** The half-life for spontaneous decay of ^{14}C is 5730 years. How does radiometric dating work and why does this nuclear lifetime make dating bones from organisms that died very recently a challenge?
4. **Infer** Scientists argue that as excess atmospheric ^{14}C continues to be incorporated into the biosphere and hydrosphere, future radiocarbon dating will underestimate the age of fossil remains of organisms alive today. Why might this be the case?

LOOKING AT THE DATA

FORENSIC RADIOMETRIC DATA

Connect this activity to the science and engineering practice of analyzing and interpreting data. Students analyze a line graph showing the ratio of radiocarbon isotopes over time and a bar graph showing the number of nuclear tests over the same time period. Students use their interpretation of the data to determine the date of birth of individuals from hypothetical human remains.

SAMPLE ANSWERS

1. *1963; 1962* **DOK 2**
2. *Sample answer: The first nuclear test was in 1945 and atmospheric radiocarbon ratio began to rise in 1955. Almost immediately after the peak of nuclear testing in 1962, the atmospheric radiocarbon ratio also peaked. Then as the number of nuclear tests decreased and testing moved to below the ground, the radiocarbon ratio began to decrease. Since 2000, nuclear testing has been almost non-existent, and the radiocarbon ratio has been declining closer to the natural level that existed before 1955.* **DOK 2**
3. *Sample answer: When an organism dies, its ratio of $^{14}C/^{12}C$ begins to decrease as the ^{14}C spontaneously decays. It takes 5730 years for half of the existing ^{14}C isotopes to decay into ^{14}N. An organism that died recently will have only had enough time for tiny amounts of ^{14}C to decay and this may not be easily measured.* **DOK 3**
4. *Sample answer: Since the half-life of ^{14}C is 5730 years, it will take more than 5000 years for the excess radiocarbon from nuclear testing to decay into ^{14}N. The fossils of organisms alive today will begin with a greater proportion of ^{14}C than organisms alive during the pre-nuclear era. This means the radiocarbon dating will underestimate the age of fossils.* **DOK 3**

On the Map

Fossil Hunting in the Badlands The Badlands have long been known for the wealth of fossils found there. Mammalian fossils include the brontotheres, an order of mammals that includes horses, rhinoceroses, and tapirs. The most common fossils are those of oreodonts. These even-toed hooved animals are related to today's bison and bighorn sheep and are distantly related to the camel.

Fossils of oreodont teeth show that they were flat and low; this shape is useful for chewing and grinding plants. Ask students what they can infer about the environment of the Badlands at the time these animals existed.

Human Geography The area known as the Badlands was occupied for 11,000 years by Native Americans and are a big part of the history of the land. In addition to fossils, other artifacts are found that tell the story of how these people lived. Evidence of campfires along the stream banks can be found, along with tools for hunting and butchering game. Have students research the history of the Lakota people in the area and how they used the land.

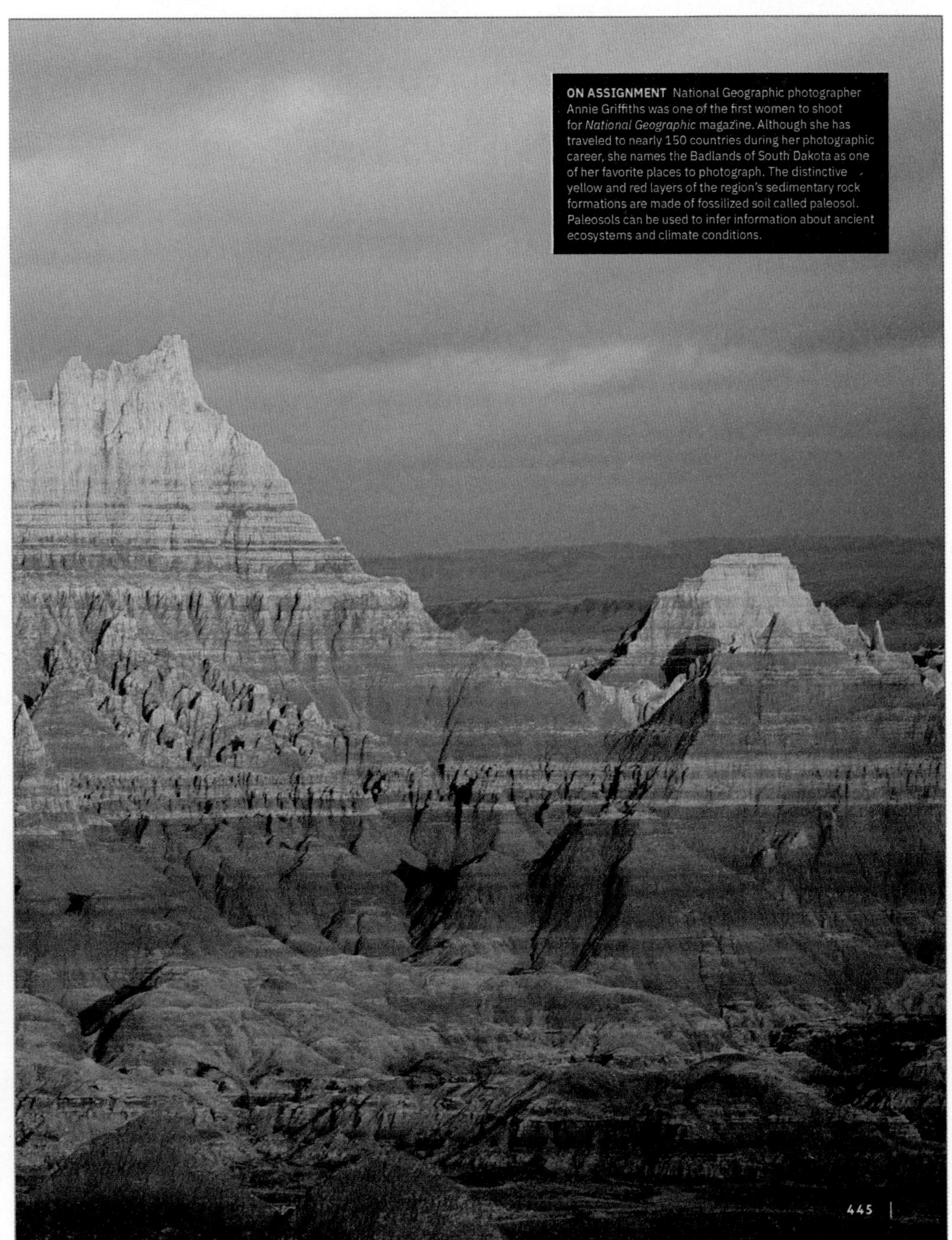

ON ASSIGNMENT

The Badlands Annie Griffiths is a contributing photographer for *National Geographic,* as well as other magazines. She also uses her photography skills to fundraise for and aid organizations and programs that empower women and girls around the world. As part of this commitment, she is the founder and Executive Director of Ripple Effect Images, a collective of photographers who document stories of women in the developing world.

This image, taken in the Badlands of South Dakota, shows the stratified layers of sedimentary rock that contain one of the world's wealthiest sources of fossils. The fossils found here tell the incredible story of this region, which was once home for ancient horses and rhinos.

You can learn more about Annie Griffiths at the National Geographic Society's website: https://www.anniegriffiths.com.

14.3

DEVELOPMENTAL, ANATOMICAL, AND GENETIC EVIDENCE

LS4.A

ENGAGE

Evidence from fossils and the stories told within layers of sedimentary rock reveal Earth's history. This section explains that further evidence for evolution, and evolutionary connections between species, is found through embryological patterns of development, comparisons of body structures and functions, and genomic comparisons. These lines of evidence add to our body of knowledge and provide more information about the evolutionary history of species.

Objectives

- Explain how developmental evidence shows evolutionary relationships.
- Identify anatomical evidence for evolution and distinguish among homologous, analogous, and vestigial structures.
- Summarize how advancements in genetic evidence build knowledge about evolution.

Pretest Use **Questions 6** and **7** to identify gaps in background knowledge or misconceptions.

Infer *Sample answer: Scientists could use genetic sequencing techniques to analyze similarities and differences among the genomes of different species to determine how related they might be.*

CHAPTER INVESTIGATION A

Open Inquiry *What Lived Here?*

Time: 50 minutes

Students analyze genetic, anatomical, radiometric, and fossil evidence to model some of the work that a paleontologist might conduct to identify an unknown organism.

Go online to access detailed teacher notes, answers, rubrics, and lab worksheets.

14.3

DEVELOPMENTAL, ANATOMICAL, AND GENETIC EVIDENCE

KEY TERMS

analogous structure
homologous structure
vestigial structure

Fossil evidence helped scientists begin to put together a picture of the history of life on Earth. Biological evidence such as developmental patterns and anatomical similarities helps scientists visualize the evolutionary connections between different species. Advancements in molecular and genetic sequencing techniques have allowed scientists to get an even clearer picture of the evolutionary relationships among organisms.

Infer *What genetic evidence might scientists use to analyze evolutionary relationships among species?*

CHAPTER INVESTIGATION

What Lived Here?
How can you use biological evidence to identify ancient organisms?

Go online to explore this chapter's hands-on investigation about the biological lines of evidence that support evolution.

Developmental Evidence

In general, the more closely related animals are, the more similar their development is. For example, all vertebrates go through a stage during which a developing embryo has four limb buds, a tail, pharyngeal arches, and divisions of the body that give rise to the backbone and associated skin and muscle (**Figure 14-12**).

While these structures look similar to one another during early stages of development, over time they diverge in structure and function. For example, pharyngeal arches, which form as a series of pouches on the side of a developing embryo's head, differentiate into gills in fish and ear and throat structures in mammals. Also, like all vertebrate embryos, humans have a tail early in development. The tail is typically absorbed by about eight weeks, forming the coccyx, or tailbone. Vertebrates have similar patterns of embryonic development because the same master regulator genes direct the process. Embryonic development in animals is coordinated by layer after layer of master gene expression.

FIGURE 14-12
The embryos of a mouse (a), alligator (b), and chicken (c) all share similarities during early development.

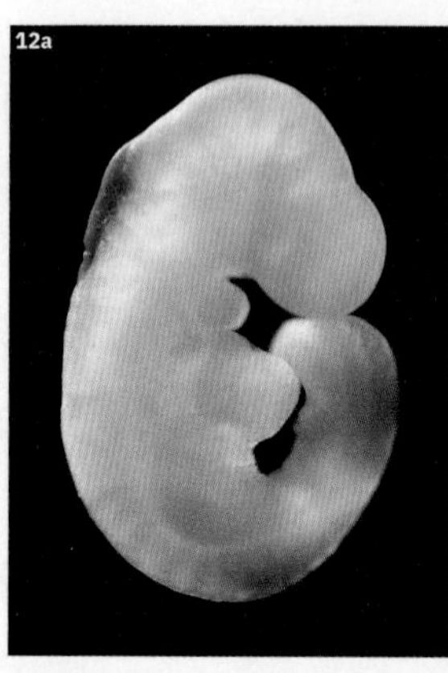

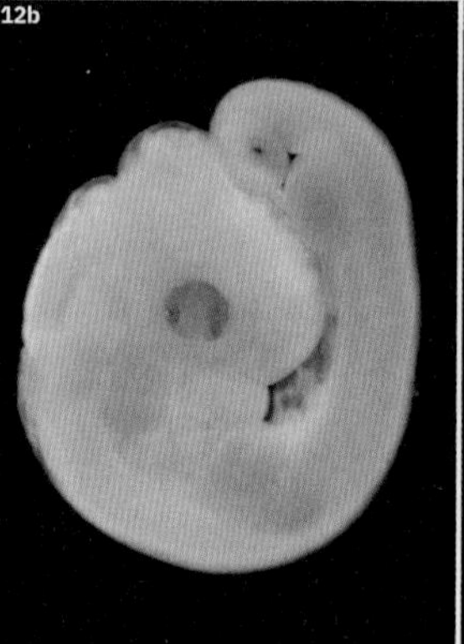

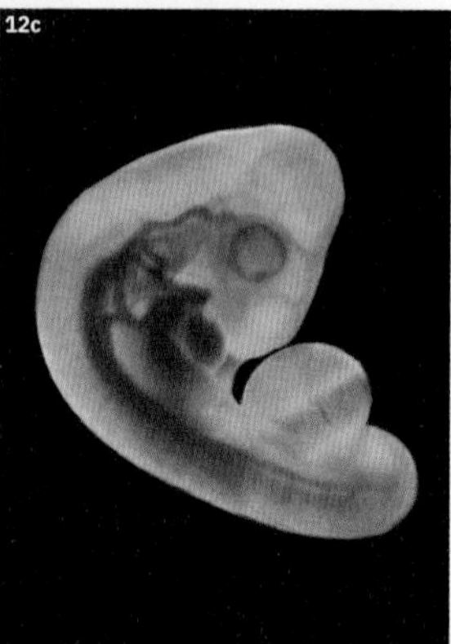

446 CHAPTER 14 EVIDENCE FOR EVOLUTION

CROSSCUTTING CONCEPTS | Structure and Function

Homologous and Analogous Structures As students learn about biological evidence that shows evolutionary relationships, they need to be aware of the concept of structure and function.

Have students examine **Figure 14-12** and identify the similarities in structure. Discuss the importance of the similarities (*it shows that similar genes are at work and indicates that these organisms share a common ancestor*). Then, have them examine **Figure 14-13** and compare the homologous structures in the limbs of various organisms. The limbs are remarkably similar in structure but are adapted for performing different tasks that enable each organism to survive in its environment. Finally, have students examine **Figure 14-14** and compare the analogous structures, which are features of organisms that carry out similar functions but with different structures indicating different origins.

Students can also share other homologous and analogous structures that they may be familiar with.

The failure of any single master regulator gene to function properly can result in a drastically altered body plan, an outcome that often has severe consequences on the organism's health. Because a mutation in a master gene can unravel development completely, these genes tend to be highly conserved. This means the genes have remained relatively unchanged for a long period of time. Even among lineages that diverged a very long time ago, many master regulator genes retain similar sequences and functions.

SEP Construct an Explanation *How do similar structures during development provide evidence of a common ancestor among all vertebrates?*

Anatomical Evidence

Similarities in developmental patterns provide one source of biological evidence for common ancestry among organisms. Comparisons of body structures and their functions are another strong source of evidence. These sources are used to support scientists' understanding of evolutionary relationships among organisms.

Homologous Structures Comparing the anatomy of various organisms can provide evidence for descent from a common ancestor, or in other words, a shared evolutionary lineage. **Homologous structures** are body parts that appear similar in separate lineages because they evolved in a common ancestor. These structures may be used for different functions in different groups, but the same genes direct their development.

For example, the forelimbs of a bird, a human, and a sea lion, though quite different in appearance, have a strikingly similar arrangement of internal structures (**Figure 14-13**). Bird wings, human arms, and sea lion flippers are used in entirely different ways for different functions—flying, lifting, swimming, to name just a few. Given these different functions, there is no mechanical reason for them to be similar. However, they mirror each other in the structure and positioning of internal elements such as bones, nerves, blood vessels, and muscles. The similarities in these forelimb structures suggest that these organisms all share a common ancestor. In other words, this forelimb structural pattern was so useful that it was retained as new species evolved over time.

Go Online INTERACTIVE FIGURE

FIGURE 14-13 Go online to learn more about the evolutionary evidence left behind in the anatomy of different organisms.

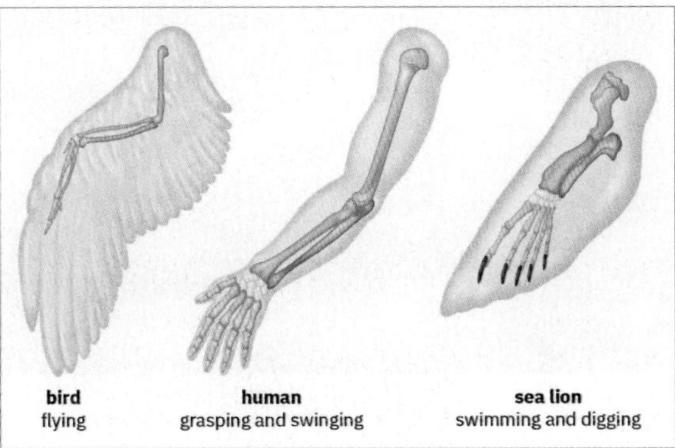

FIGURE 14-13 Bird, human, and sea lion limbs all function in different ways. However, their underlying structures are very similar in appearance.

SCIENCE AND ENGINEERING PRACTICES
Constructing Explanations and Designing Solutions

Models of Homologous Structures To help students prepare for the interactive figure, have them look up homologous structures in the forelimbs of an eagle, bat, crocodile, horse, and human. Using their research for guidance, students should make models of each forearm using different colors for the humerus, radius and ulna, wrist bones, and digits. Have them use these models to construct an explanation about how each organism's structure relates to its function. Provide guiding questions, such as: *How are the differences in the forelimbs of the horse and the human related to their function? How did the forelimbs of the bat and eagle develop differently to support flight?*

SEP Construct an Explanation
Sample answer: Similarities among vertebrate embryos that develop into adults with very different body forms indicate they share a common ancestor.

Vocabulary Strategy

Roots and Affixes Display the words *analogous* and *homologous*. Point out that the suffix *-ous* forms an adjective. Give students an example of a word analogy, for example: Dog is to puppy as cat is to kitten. Guide students to conclude that an analogy shows how things are similar. Have students write what they think analogous structures are based on the word parts. Then, point out the prefix *homo-* and tell students that it means "the same." Have students write what they think homologous structures are. Have them revise their definitions as they go through the lesson.

Go Online INTERACTIVE FIGURE

Evolutionary Evidence of Anatomy
Students can use the interactive version of **Figure 14-13** to view descriptions of different parts of the anatomy and demonstrate homologous, analogous, and vestigial structures.

Connect to English Language Arts

Draw Evidence From Texts The forearms of the bird, sea lion, and human shown in **Figure 14-13** are just one example of homologous structures. Challenge students to find more examples. They may work in groups to brainstorm ideas and then research homologous structures online. When they are finished, have each group prepare at least one diagram similar to **Figure 14-13**, using different colors to compare structures, and then present it to the class. Possible responses include: tailbone in humans and monkeys; vertebrae in all mammals; and pelvises in humans, dogs, and cats. Allow time for groups to practice their presentations to be sure they are clear, concise, and logical.

Address Misconceptions

Divergence and Convergence To help students understand how homologous and analogous structures are evidence of the processes of divergent evolution and convergent evolution, draw a model on the board. Begin by drawing two arrows that begin at the same point and then diverge. Point out that to *diverge* means to move apart. Then, draw two arrows that begin at two different points and converge at one point. Point out that to *converge* means to come together. Ask students what each model represents and what types of structures are formed. (*Divergence results in homologous structures; convergence results in analogous structures.*)

SEP Construct an Explanation
Sample answer: The vestigial structures do not entirely disappear from a population because they do not impede an organism's ability to survive or reproduce.

Analogous Structures Body parts that have a similar function but did not arise from a common ancestor are called **analogous structures**. The wings of insects, for example, perform the same function as the wings of birds (**Figure 14-14**). However, bird wings are supported by bone structure, while insect wings have no underlying skeletal support. Analogous organs are of evolutionary interest because they show that organisms with different ancestries can adapt in similar ways to similar environmental demands. This process is called *convergent evolution*.

FIGURE 14-14 Though the functions of the insect wings and bird wings are the same, their underlying structures are very different. This is an example of convergent evolution.

Vestigial Structures Body parts that no longer have a function are called **vestigial structures**. Humans, for example, have vestigial tail bones and muscles for moving the ears. Blind cavefish (**Figure 14-15**) live in total darkness where eyesight is not necessary for survival. In these unusual fish, the fish embryos initially have eyes during development. The eyes later disintegrate and become covered by skin and connective tissue.

FIGURE 14-15 The nonfunctional eyes of a blind cavefish are an example of vestigial structures.

SEP Construct an Explanation *Why do you think vestigial structures do not entirely disappear from a population?*

CROSSCUTTING CONCEPTS | Structure and Function

Vestigial Organs Engage students in a discussion about human vestigial features. Ask students if anyone has had their appendix or wisdom teeth removed. Guide students in a discussion as to why the appendix or wisdom teeth can be removed, typically with no harmful effects. Have students make a two-column chart listing human vestigial features and what they think their use once was. (*Both the appendix and wisdom teeth are related to chewing and digesting coarse plants and tough uncooked food.*) Another human vestigial feature is the coccyx.

Genetic Evidence

Over time, mutations change a genome's DNA sequence. Most of these mutations are neutral. In other words, they have no effect on an individual's survival or reproduction. The more recently two lineages diverged, the less time there has been for unique mutations to accumulate in the DNA of each one. This is why the genomes of closely related species tend to be more similar than those of distantly related species, a general rule that can be used to estimate relative times of divergence. Thus, similarities in the nucleotide sequence of a shared gene (or in the amino acid sequence of a shared protein) can be used to study evolutionary relationships. Biochemical comparisons like these may be combined with structural comparisons, in order to provide data for hypotheses about shared ancestry.

Pseudogenes Recall from Section 11.1 that pseudogenes are copies of genes that exist elsewhere in the genome but no longer produce a functional protein. Pseudogenes can sometimes reveal information about ancient genes and the rate of gene duplication over long periods of time.

mRNA Extremely strong evidence in support of common ancestors for almost all life-forms comes from the genetic code. For example, the codon UUU in mRNA, which codes for the amino acid phenylalanine, occurs in organisms as diverse as shrimp, humans, bacteria, and tulips. This kind of molecular evidence strongly suggests that all life descended from common ancestors, one-celled organisms, that originally developed the genetic code. Once initiated, the code was found to be so useful that it was never abandoned over billions of years.

Mitochondrial DNA As shown in **Figure 14-16**, the mitochondria in most animals are inherited intact from only one of the parents (most often the mother, in humans). Mitochondrial chromosomes are circular pieces of DNA that do not undergo crossing over, so any changes in their DNA sequence occur only by mutation. Thus, in most cases, differences in mitochondrial DNA (mtDNA) sequences among maternally related people are due to new mutations. mtDNA has been a particularly useful tool in tracing the origin, evolution, and migration of human populations.

CHAPTER INVESTIGATION

Comparing Genetic Information Among Organisms

How can you use genetic information to determine how closely related organisms are?

Go online to explore this chapter's hands-on investigation about the genetic evidence that supports evolution.

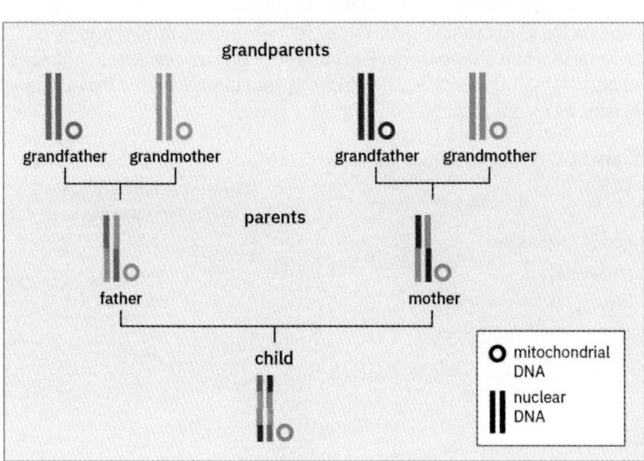

FIGURE 14-16 Mitochondrial DNA is typically passed from one generation to the next only through the maternal line. In contrast, each generation inherits a unique combination of nuclear DNA from both their parents. Because of the reduced complexity, mitochondrial DNA is much easier to trace through generations than nuclear DNA is.

CHAPTER INVESTIGATION B

Guided Inquiry *Comparing Genetic Information Among Organisms*

Time: 50 minutes

Students compare partial nucleotide sequences for the alpha-hemoglobin protein in humans and other animals. They also compare partial cytochrome b amino acid sequences to make their own observations about genetic evidence.

Go online to access detailed teacher notes, answers, rubrics, and lab worksheets.

Science Background

Last Universal Common Ancestor

Based on the study of genomes and the universality of the genetic code and amino acid chirality, scientists hypothesize that all known organisms trace back to a last universal common ancestor (LUCA). Scientists who research LUCA have identified ancient genes using phylogenetic trees. Initially, scientists used the three-domain framework to try to find genes that were present in ancient archaea, bacteria, and eukaryotes to guide their hypothesis of what LUCA might have been like. However, new discoveries in phylogenetics and new metagenomic data have led to questions about the accuracy of the three-domain framework. Because eukaryotes are chimeras that possess archaeal ribosomes in the cytosol and bacterial ribosomes in mitochondria, some scientists think there should only be two domains.

DIFFERENTIATED INSTRUCTION | English Language Learners

Roots and Affixes Review the words *multicellular* and *pluripotent* using the word parts *multi-* ("more than one") and *plur-* ("many, several"). Have students use the meaning of *pseudo-* ("false, unreal") to help them explain the meaning of *pseudogenes*.

Beginning Write each word in two parts (*multi/cellular, pluri/potent, pseudo/genes*) in random order. Explain the meanings of *multi-*, *pluri-*, and *pseudo-*. Then, have students match. Finally, have students write each word and its definition in their notebooks.

Intermediate Have students complete sentence frames to use the word part to define the word, for example: The word part *multi-* means ___, so *multicellular* means ___.

Advanced Have students use the word parts to explain the meaning of each word in as much detail as they can.

Science Background

Muscle Proteins in Yeast It seems surprising, but researchers have found a yeast gene that codes for the protein myosin. These proteins are best known for being found in the muscle cells of multicellular organisms, including humans, and play a role in muscle contraction and, therefore, movement. However, yeasts do not have muscles nor are they known to be motile. As it turns out, certain cellular components within yeasts do move and it is due to the presence of myosin. The original form of the protein allowed parts of cells to move about. As life diversified, changes to the myosin genes over time allowed them to adapt to forms that allow the muscles of the body to contract and produce movement.

SEP Analyze Data *The song sparrow is most closely related to honeycreepers.*

```
...CRDVQFGWLIRNLHANGASFFFICIYLHIGRGIYYGSYLNK--ETWNIGVILLLTLMATAFVGYVLPWGQMSFWG...honeycreepers (10)
...CRDVQFGWLIRNLHANGASFFFICIYLHIGRGIYYGSYLNK--ETWNVGIILLLALMATAFVGYVLPWGQMSFWG...song sparrow
...CRDVQFGWLIRNIHANGASFFFICIYLHIGRGLYYGSYLYK--ETWNVGVILLLTLMATAFVGYVLPWGQMSFWG...Gough Island finch
...CRDVNYGWLIRYMHANGASMFFICLFLHVGRGMYYGSYTFT--ETWNIGIVLLFAVMATAFMGYVLPWGQMSFWG...deer mouse
...CRDVHYGWIIRYMHANGASMFFICLFMHVGRGLYYGSYLLS--ETWNIGIILLFTVMATAFMGYVLPWGQMSFWG...Asiatic black bear
...CRDVNYGWLIRNLHANGASFFFICIYLHIGRGLYYGSYLYK--ETWNIGVVLLLLVMGTAFVGYVLPWGQMSFWG...bogue (a fish)
...TRDVNYGWIIRYLHANGASMFFICLFLHIGRGLYYGSFLYS--ETWNIGIILLLATMATAFMGYVLPWGQMSFWG...human
...MRDVEGGWLLRYMHANGASMFLIVVYLHIFRGLYHASYSSPREFVWCLGVVIFLLMIVTAFIGYVLPWGQMSFWG...thale cress (a plant)
...ETDVMNGWMVRSIHANGASWFFIMLYSHIFRGLWVSSFTQP--LVWLSGVIILFLSMATAFLGYVLPWGQMSFWG...baboon louse
...MRDVHNGYILRYLHANGASFFFMVMFMHMAKGLYYGSYRSPRVTLWNVGVIIFTLTIATAFLGYCCVYGQMSHWG...baker's yeast
```

FIGURE 14-17 Partial amino acid sequences of cytochrome *b* from 19 species are aligned in this diagram. The honeycreeper sequence is identical in 10 species of honeycreeper. Amino acids that differ in the other species are shown in red. Dashes indicate insertion or deletion mutations.

Proteins Another form of evidence is found in the proteins of various organisms. For example, even organisms that are only remotely related, such as humans, baker's yeast, and a type of bird called the honeycreeper, all have the protein cytochrome *b*. This protein functions in all these species' mitochondria (**Figure 14-17**). In fact, because this protein developed billions of years ago, mutations have built up during that enormous time period to make different versions of cytochrome *b*.

SEP Analyze Data *Based on the amino acid sequences shown in* ***Figure 14-17****, which species is the most closely related to honeycreepers?*

Scientists use these differences to map evolutionary pathways. The scientists assume that the longer it has been since two organisms diverged, or took separate evolutionary pathways, the greater the differences in their cytochrome *b* molecules. Thus, two species that have very different cytochrome *b* molecules can be identified as having diverged in the far distant past. On the other hand, species with similar cytochrome *b* molecules are characterized as being more closely related and therefore sharing a more recent common ancestor.

DNA Sequencing With DNA sequencing, scientists have also discovered that the more similar two genomes are with respect to nucleotide sequences, the more closely related the organisms are that share those sequences. As **Table 14-1** shows, humans and chimpanzees are thought to be closely related and share a recent common ancestor because their genomes are so similar. Humans and lemurs, on the other hand, are not as closely related. Researchers use data from DNA sequencing to map which organisms are more closely related and to determine when different species diverged. Molecular evidence continues to provide many missing puzzle pieces to help us understand the evolution of species on our planet.

TABLE 14-1. Differences in Nucleotide Sequences Between Humans and Other Species

Species pair	Percentage difference in nucleotide sequences
human-chimpanzee	2.5%
human-gibbon	5.1%
human-Old World monkey	9.0%
human-New World monkey	15.8%
human-lemur	42.0%

CROSSCUTTING CONCEPTS | Patterns

As shown in **Figure 14-17**, scientists currently determine how closely organisms are related by comparing the patterns of the sequences of a small number of genes that organisms have in common. Ask students to identify some possible limitations of this method. Explain that not all organisms have genes in common that have been identified, and sometimes there are significant differences in the patterns of similarities among different genes.

Because of these limitations, scientists are working on better methods to determine how organisms are related by comparing whole genomes. The feature frequency profile (FFP) method is similar to the method used to compare texts. In FFP, the genome sequence is expressed with a two-letter alphabet that corresponds to the purine nucleic acids (A and G) and pyrimidine nucleic acids (T and C). The frequency of the possible sequences of lengths of 18 base pairs are then calculated and compared. This method has helped scientists classify some bacteria and viruses that had previously been unclassified.

Body Development and Homeotic Genes

Homeotic genes are master regulator genes that control the development of anatomical structures in animals, plants, and fungi. *Hox* genes are a subset of homeotic genes in some animals that direct the head-to-tail formation of body segments and structures during embryonic development (**Figure 14-18**). Insects have 10 *Hox* genes, and vertebrates have multiple versions of all of them. One insect *Hox* gene, called *antennapedia*, triggers leg formation in insect embryos. Thus, it normally determines the identity of the thorax (the part with legs). Humans and other vertebrates have a version of *antennapedia* called *HoxC6*. Expression of *HoxC6* in a vertebrate embryo causes ribs to develop on a vertebra, so it normally determines the identity of the back. Without *HoxC6* expression, vertebrae of the neck and tail develop, but ribs do not.

SEP Construct an Explanation *How could mutations in* Hox *genes lead to the diversification of animal species?*

 VIDEO

VIDEO 14-2 Go online to learn how genetic evidence helped scientists understand snake evolution.

FIGURE 14-18 The genes that direct the development of structures in a fruit fly are variations of the same genes that determine a human's head-to-tail development, but they are expressed in different patterns.

14.3 REVIEW

1. **Compare** Label the statements as consistent with analogous structures or homologous structures.
   - The organisms with a similar structure share a common ancestor.
   - The structures develop in the same way in the compared organisms.
   - A body part has the same function but different structures in the compared organisms.
   - A body part has the same structures but different functions in the compared organisms.
2. **Explain** How do similarities in DNA or protein sequences provide evidence that two species share a common ancestor?
3. **Identify** Which types of data are most effective as evidence for shared ancestry?
   A. vestigial structures
   B. amino acid sequences
   C. DNA from mitochondria
   D. analagous structures
   E. embryonic development patterns
4. **Explain** Why might a mutation in a *Hox* gene result in a major difference in an organism's adult form?

Science Background

Homeotic Mutations When a homeotic mutation occurs, it can mix up body parts. This results in fruit flies with legs growing on the head or antennae on the mouth. These unusual body features are what led scientists to discover homeotic genes. Named homeobox genes, or *Hox* genes, scientists found that a defect in a single gene could lead to these strange transformations. The take-away from this work is that every cell in an organism carries all of the genetic information necessary to build every part of the entire organism.

Video

Time: 2:55

Use **Video 14-2** to explain to students the mutations that occurred in snakes, pythons in particular, that caused pythons to lose their legs through the course of evolution whereas four-legged lizards maintain their leg development genetics.

14.3 REVIEW

1. *analogous: A body part has the same function but different structures in the compared organisms. homologous: The organisms with a similar structure share a common ancestor. The structures develop in the same way in the compared organisms. A body part has the same structures but different functions in the compared organisms.* **DOK 2**
2. *Species that have a high percentage of shared genes or amino acid sequences are more likely to be closely related than species that do not have shared genes or amino acid sequences.* **DOK 2**
3. *B, C, E* **DOK 2**
4. *Sample answer: Hox genes direct the head-to-tail formation of body segments and structures during embryonic development in some animals. If there were a mutation in one of these genes, a structure might not develop at all or could develop in a different location or on a different body segment, resulting in a different body form as an adult, if the mutation is not lethal.* **DOK 3**

EXPLORER
DR. BRENDA LARISON

About the Explorer

National Geographic Explorer Dr. Brenda Larison is an Assistant Adjunct Professor and Assistant Researcher at the University of California, Los Angeles in the Center for Tropical Research and Department of Ecology and Evolutionary Biology. Her interest and focus on research includes both evolution and conservation, which leads to her studies on zebra striping. You can learn more about Dr. Larison and her research on the National Geographic Society's website: www.nationalgeographic.org.

Science Background

Zebra Stripes Scientists such as Dr. Larison are interested in why zebras have stripes. Some ideas are that stripes help repel biting flies, help with thermoregulation, and support social behaviors. In one study, scientists disguised domestic horses in zebra-striped coats and studied them along with zebras and regular horses. The scientists observed that far fewer horseflies landed on zebras and zebra-disguised horses than landed on regular horses.

Connect to Careers

Evolutionary Biologist One of the many areas of biological sciences is evolutionary biology. Careers in evolutionary biology explore the processes and patterns of biological evolution, especially in relation to the diversity of organisms and how they change over time. Some careers require a four-year bachelor's degree. Other careers, such as Dr. Larison's field of evolutionary biology, require a master's or doctorate degree in addition to a bachelor's degree.

THINKING CRITICALLY

Justify *Reviewing the research and testing of questions that have been answered can bring to light new or more information. Testing using methods other than those used originally can confirm conclusions or result in conflicting data.*

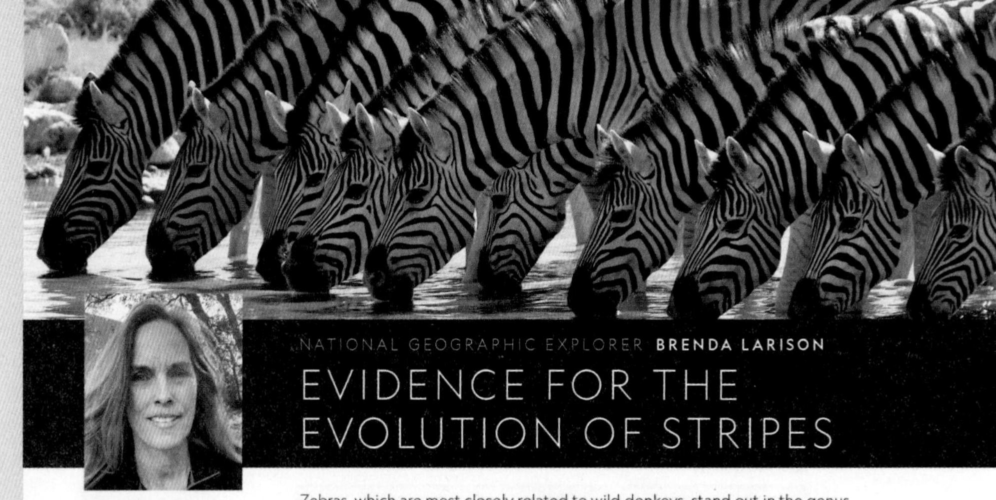

NATIONAL GEOGRAPHIC EXPLORER **BRENDA LARISON**

EVIDENCE FOR THE EVOLUTION OF STRIPES

Dr. Brenda Larison's research, which focuses on populations of plains zebras in sub-Saharan Africa, addresses questions related to both evolution and conservation.

Zebras, which are most closely related to wild donkeys, stand out in the genus *Equus* mainly because of their distinctive coats. What environmental changes spurred the survival of ancestors with stripes of any kind? What genetic changes resulted in such a bold trait with several variations? Biologist Dr. Brenda Larison studies plains zebras to answer those questions.

Larison has been revisiting some of the more popular explanations for the zebras' striped patterns that scientists have tested over time. Among these hypotheses are that stripes support social behaviors, aid in thermoregulation, help in the evasion of predators, or in the avoidance of biting flies that carry fatal diseases. As part of her research, Larison and her team devised a test that pitted some of the hypotheses against one another.

"Providing evidence in support of one hypothesis does nothing to disprove other hypotheses unless the hypotheses are mutually exclusive," Larison says. "The various hypotheses for why zebras have stripes are not mutually exclusive. It can be true that stripes deter biting flies, and that they also aid in other ways, such as predator avoidance or thermoregulation. The only way to eliminate the other hypotheses is to disprove them and no one has successfully done that as yet."

Larison's team collected and analyzed data about the relationship between stripe patterns and the environment from 16 sites across the plains zebras' range. The team looked at stripe numbers, thickness, length, and boldness of color to develop a stripe index. While their data provided support for the thermoregulation hypothesis, the reason for this correlation remains unclear. Many factors may be involved, but it is possible that stronger stripe patterns are beneficial in warmer tropical areas where flies carry more disease.

Because individual striping patterns are as unique as fingerprints, Larison can collect tissue samples and connect an individual's stripe phenotype to the genetic variation related to that stripe pattern. This information might help shed light on the advantages striping provides. Researchers are trying to determine whether genes related to striping are selected differently for zebras in different environments. This genetic evidence may at last provide a definitive answer to why striping evolved in zebras.

THINKING CRITICALLY

Justify *Why might scientists revisit problems even if others have tested and drawn logical conclusions?*

TYING IT ALL TOGETHER
SOLVING THE MYSTERY OF SLOTH EVOLUTION

HOW DO WE KNOW HOW SPECIES HAVE EVOLVED?

In this chapter, you learned about the lines of evidence that support evolution and how scientists use this evidence to build the evolutionary history of a species. The Case Study described some of the evolutionary history of sloths and their relatives.

Modern-day sloths include six species that are divided into two main groupings: *Choleopus* and *Bradypus*. Members of each genus are distinguished by the number of toes on their front feet: *Bradypus* have three toes and *Choleopus* have two toes. Initially it was thought that these two groups branched from the same common ancestor. However, while both their appearance and behavior are quite similar, it turns out the two groups are only distantly related (**Figure 14-19**).

In 2019, the results of two independent studies about the evolution of sloths were published. One relied on ancient DNA data while the other used ancient collagen (protein) sequence information. Both studies confirmed that three-toed sloths were in fact not a sister group to two-toed sloths. Instead, they are more closely related to a group of extinct ground sloths that includes *Megatherium* and *Megalonyx*.

Two-toed sloths, on the other hand, are more closely related to a group of extinct ground sloths called *Mylodontidae*. The evolution of three-toed and two-toed sloths therefore is an example of convergent evolution, in which the two lines, though similar in appearance, evolved independently of one another.

In this activity, you will research how advancements in technology have aided our understanding of evolutionary relationships among organisms. Work individually or in a group to complete these steps.

Gather Evidence

1. Research the field of paleoproteomics. Compare the reliability of ancient protein evidence and ancient DNA evidence, taking into consideration the environments in which different fossils are found.

Scientific Knowledge

2. Find out how scientists used proteomic data to change our understanding of the evolutionary relationships among sloth species. What other species' evolutionary histories have been revised due to advancements in molecular techniques?

Communicate Information

3. Develop a multimedia presentation to present your findings to your class.

FIGURE 14-19
Comparisons of DNA and protein sequences among extinct and living sloth species changed scientists' understanding of how sloths were related. 19a shows the old scheme and 19b shows the new scheme.

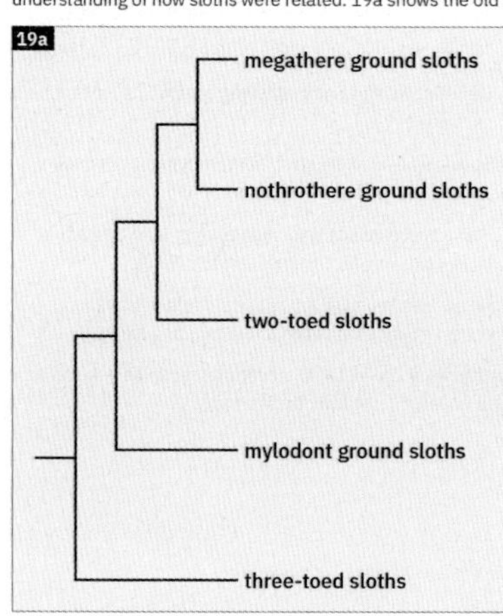

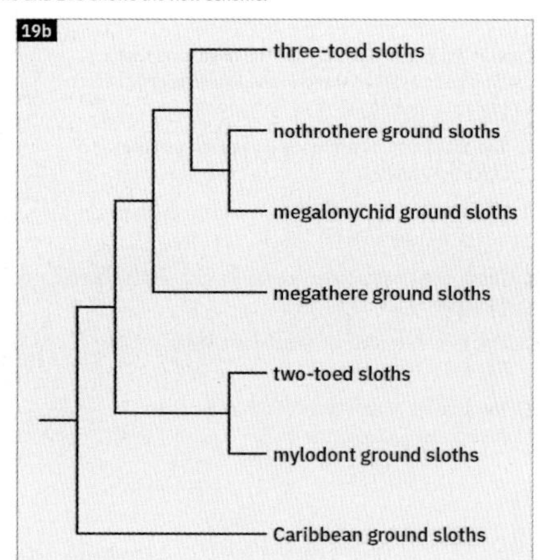

DIFFERENTIATED INSTRUCTION | Economically Disadvantaged Students

Sharing Supplies and Skills Make sure all students have equal access to resources, such as books and the Internet for their research. Toward this goal, set up some time for student groups to work together with shared supplies in the classroom or in the school's media center or library.

Additionally, encourage students working within groups to be aware of their teammates strengths and talents. Those with talents in illustration, graphic design, or computer design may take the lead with the media. Students with great research skills can take the lead in gathering information.

TYING IT ALL TOGETHER

ELABORATE

The Case Study focused on the evolution of sloths and posed the question of how we know how species have evolved. To get students started, the Case Study included some introductory information from several different lines of evidence. In the Tying It All Together activity, students will extend their knowledge as they research the field of paleoproteomics and apply it here.

This activity is structured so that students may complete the steps as they study the chapter content. **Step 1** correlates with the chapter opener, as well as content in Sections 14.1 through 14.3. Proteomics, being the large-scale study of proteins encoded and expressed by whole genome(s), aligns particularly with the content in Section 14.3, which will help students complete **Step 2**. Students will then need to synthesize ideas from all sections and their research to complete **Step 3**.

Alternatively, this can be a culminating project at the end of the chapter. The gathering of evidence here to support a claim or argument is a process that students may find useful in the Unit 5 Activity.

Go online to access the Student Self-Reflection and Teacher Scoring rubrics for this activity.

Science Background

Proteomics The first studies in proteomics began in 1975. This discipline is characterized as the large-scale study of proteins. Scientists study the extensive and diverse properties of proteins in order to gather detailed information about their structure and function and their performance in biological systems in healthy and diseased organisms.

CHAPTER 14 REVIEW

EVALUATE

REVIEW KEY CONCEPTS

1. *microevolution: Brown owls become more common as climate change diminishes snow cover; A strain of bacteria with antibiotic resistance becomes more common after treatment with antibiotics. Macroevolution: Whales descended from an ancestor that lived on land; Over millions of years, numerous species of dinosaurs lived on Earth.* **DOK 2**
2. *C* **DOK 1**
3. *fossils: bone; tooth. trace fossils: coprolite; footprint; nest* **DOK 2**
4. *A, C, E* **DOK 2**
5. *analogous: the fin of a fish and the flipper of a whale; the leg of a caterpillar and the leg of a dog; homologous: the wing of a bat and the flipper of a whale; the leg of a kangaroo and the leg of a chicken* **DOK 2**
6. *D* **DOK 2**
7. *D* **DOK 2**
8. *A, D* **DOK 1**
9. *Sample answer: Vestigial structures may not fully disappear if there is no evolutionary advantage to losing them. For example, the eyes of a blind cavefish have no function, but they do not prevent the fish from moving, developing, or reproducing. A cavefish with a mutation that prevents its eyes from developing is no more likely to survive and reproduce.* **DOK 2**

CHAPTER 14 REVIEW

REVIEW KEY CONCEPTS

1. **Classify** Determine whether each statement represents an example of microevolution or macroevolution.
   - Whales descended from an ancestor that lived on land.
   - Over millions of years, numerous species of dinosaurs lived on Earth.
   - Brown owls become more common as climate change diminishes snow cover.
   - A strain of bacterium with antibiotic resistance becomes more common after treatment with antibiotics.
2. **Define** Which line of evidence for evolution involves observing the formation and growth of an embryo?
   A. anatomy
   B. biogeography
   C. developmental patterns
   D. genetics
3. **Classify** Sort the examples into fossils and trace fossils.
   - bone
   - coprolite
   - footprint
   - nest
   - tooth
4. **Explain** What factors increase the likelihood that a species will be represented in the fossil record? Select all correct answers.
   A. The species had hard body parts such as bones, seeds, or shells.
   B. The species lived in an area with many predators and scavengers.
   C. The species had a large population and existed over a long period of time.
   D. The species lived in an area where there are very few sedimentary rocks.
   E. The species lived in an area protected from destructive geologic events.
5. **Classify** Indicate whether the following pairs of structures are homologous or analogous.
   - the fin of a fish and the flipper of a whale
   - the leg of a caterpillar and the leg of a dog
   - the wing of a bat and the flipper of a whale
   - the leg of a kangaroo and the leg of a chicken
6. **Contrast** The embryos of fish and humans both have pharyngeal arches. In fish, they develop into gills. What happens to the pharyngeal arches as a human embryo develops?
   A. They develop into lungs.
   B. They disappear as the embryo develops.
   C. They have no function; they are vestigial.
   D. They develop into ear and throat structures.
7. **Explain** Why is carbon-14 dating an ineffective way to determine the age of a rock?
   A. The daughter isotope of carbon-14 decay is not detectable.
   B. Carbon-14 does not decay, so it cannot be used for radioactive dating.
   C. Carbon is present only in living things, so it cannot be measured in rocks.
   D. Most rocks are too old to be measured using the half-life of carbon-14 atoms.
8. **Explain** Which statements are supported by the evidence for evolution? Select all correct answers.
   A. All life on Earth descends from a common ancestor.
   B. Genetic variation among living organisms cannot be explained.
   C. Species never truly go extinct; they simply evolve and adapt, taking new forms.
   D. Populations of species change over long periods of time as they adapt to their environment.
   E. All organisms alive today are closely related to a common ancestor from a few million years ago.
9. **Justify** Why do vestigial structures, such as eyes in a blind cavefish, not fully disappear?

454 CHAPTER 14 EVIDENCE FOR EVOLUTION

CRITICAL THINKING

10. *Sample answer: Sexual reproduction can produce new combinations of genetic information. Although the alleles for traits existed before, they may have never existed in the same individual before. This allows for selection to occur for combinations of traits that are advantageous.* **DOK 2**
11. *Sample answer: The rock layers at the bottom are the oldest and would contain the oldest fossils. The uppermost rock layers formed more recently when they were deposited on top of the lower layers. Thus, the youngest fossils would be found in the uppermost rock layers.* **DOK 2**
12. *Sample answer: The principle of gradualism states that the processes that shape the face of Earth are due to small changes that accumulate over a long time. The presence of the K-Pg boundary challenges gradualism because it provides evidence of a meteor impact that caused a sudden, drastic change that quickly altered life on Earth. However, many of the processes that shape Earth are gradual. Catastrophic events like the meteor strike are rare exceptions.* **DOK 2**
13. *Sample answer: Master genes direct the process of embryonic development and are highly conserved across closely related species because any mutations to these genes are usually lethal. Thus, the sequences of master genes among different birds were probably very similar. Genes for feather color are less likely to be conserved because mutations in genes for feather color might often have an advantageous effect, which is why different birds have developed diversity in feather color.* **DOK 3**

CRITICAL THINKING

10. **Analyze** How does sexual reproduction contribute to evolution?
11. **Predict** As a river erodes a canyon, it exposes layers of rock that are different in color and composition. Which rock layers would you expect to contain the oldest fossils? The youngest? Justify your prediction.
12. **Relate** Does the presence of the K-Pg boundary support or challenge the principle of gradualism? Explain your reasoning.
13. **Contrast** A scientist compared the DNA sequence of a master gene for development among many different bird species. Then she compared the DNA sequence of a gene for feather color. How would these two comparisons differ? Explain.
14. **Explain** Why are amino acid sequences used to determine evolutionary relationships between species?
15. **NOS Scientific Knowledge** How do multiple lines of evidence provide support for biological evolution? Use examples in your explanation.

MATH AND ENGLISH LANGUAGE ARTS CONNECTIONS

1. **Reason Quantitatively** Carbon-14 dating is used to determine that a fossil is about 28,650 years old. Given that the half-life of C-14 is 5730 years, what percentage of the original C-14 is left in the sample?
2. **Reason Quantitatively** A fossil is determined to have 12.5% of the original carbon-14. The half-life of carbon–14 is 5730 years. How old is the fossil in years?
3. **Write an Argument** A scientist is comparing the cytochrome *b* gene sequences of three species. Species X and Y have nearly identical sequences, with only a few differences, whereas the sequence of species Z is quite different. What does this suggest about the evolutionary relationships between the three organisms? Explain your reasoning.
4. **Write Informative Text** Define "convergent evolution" using an example.

► REVISIT HUMMINGBIRD EVOLUTION

Gather Evidence In this chapter, you have seen multiple lines of evidence that support common ancestry and biological evolution. In Section 14.2, you learned about fossil and geological evidence. In Section 14.3, you learned about the developmental, anatomical, and genetic evidence.

1. Which lines of evidence best support the evolution of hummingbirds over time?
2. What other questions do you have about the evidence that supports common ancestry and the evolution of hummingbirds?

► REVISIT THE ANCHORING PHENOMENON

1. *Sample answer: Evidence that supports the evolution of hummingbirds includes fossil, anatomical, and genetic evidence.* **DOK 3**
2. *Sample answer: What environmental conditions led to the evolution of so many different species of hummingbirds? Where has fossil evidence of hummingbirds been found?* **DOK 3**

14. *Sample answer: Certain homologous proteins are found in many organisms, even those that are very distantly related to one another. Homologous proteins have very similar amino acid sequences because they serve the same function; any major changes in the sequence would compromise the protein's function. However, genetic mutations that do not affect protein function can build up in a species over time. Comparing slight differences in the amino acid sequences of homologous proteins can help scientists estimate how long ago two organisms diverged from a common ancestor.* **DOK 3**
15. *Sample answer: Evidence that supports biological evolution comes from the fossil record, biogeography, DNA, developmental patterns, and anatomical comparisons. Because this evidence comes from different scientific disciplines and spans across different time and geographical scales, it provides overwhelming support for evolution.* **DOK 2**

MATH AND ELA CONNECTIONS

1. *3.13%* **DOK 2**
2. *17,190* **DOK 2**
3. *Sample answer: These results suggest that species X and Y are closely related. They diverged from a common ancestor not long ago, so there has not been much time for genetic mutations to accumulate in the genes responsible for producing cytochrome b. In contrast, species X and Y are only distantly related to species Z. Their common ancestor dates to a much more distant past, so more mutations have accumulated in the gene sequences for cytochrome b.* **DOK 2**
4. *Sample answer: Convergent evolution occurs when two separate groups of organisms develop similar methods for adapting to the same environmental demands. For example, the wings of insects and the wings of birds are both adaptations that enable flight. However, there are fundamental differences in the structure of insect and bird wings because they do not have the same evolutionary origin.* **DOK 2**

THEORY OF EVOLUTION

Three-Dimensional Learning

The practices, core ideas, and crosscutting concepts presented in this chapter's text, investigations, and resources provide support to address the following Performance Expectations: **HS-LS3-3, HS-LS4-2, HS-LS4-3,** and **HS-LS4-4.**

Science and Engineering Practices	Disciplinary Core Ideas	Crosscutting Concepts
Asking Questions and Defining Problems Analyzing and Interpreting Data (HS-LS3-3, HS-LS4-3) Using Mathematics and Computational Thinking Constructing Explanations and Designing Solutions (HS-LS4-2, HS-LS4-4) Scientific Investigations Use a Variety of Methods Scientific Knowledge is Based on Empirical Evidence Scientific Knowledge is Open to Revision in Light of New Evidence	**LS3.B:** Variation of Traits (HS-LS3-3) **LS4.A:** Evidence of Common Ancestry and Diversity **LS4.B:** Natural Selection (HS-LS4-2, HS-LS4-3) **LS4.C:** Adaptation (HS-LS4-2, HS-LS4-3, HS-LS4-4)	Patterns (HS-LS4-3) Cause and Effect (HS-LS4-2, HS-LS4-4) Scientific Knowledge Assumes an Order and Consistency in Natural Systems (HS-LS4-4)

Contents		Instructional Support for All Learners	Digital Resources
ENGAGE			
456–457	**CHAPTER OVERVIEW** **CASE STUDY** Why do populations evolve over time?	**Social-Emotional Learning** CCC Cause and Effect **Historical Connection** How Old Are Bedbugs? **English Language Learners** Shared Reading	*Video 15-1*
EXPLORE/EXPLAIN			
458–466 DCI LS4.B LS4.C	**15.1 DEVELOPING THE THEORY OF EVOLUTION** • Compare the differences and similarities between artificial and natural selection. • Communicate evidence that supports the theory of evolution. • Evaluate the evidence for natural selection as a mechanism for evolution.	**Vocabulary Strategy** Word Wall **Differentiated Instruction** Leveled Support **On the Map** Environmental factors **English Language Learners** High-Frequency Words, Narrating with Details **Visual Support** Plant Ancestry, Beak Function SEP Constructing Explanations and Designing Solutions, Engaging in Argument from Evidence **On the Map** Darwin's Journey **Connect to English Language Arts** Synthesize Information From Text and Graphics, Analyze an Argument **CASE STUDY** Ask Questions CCC Patterns **BIOTECHNOLOGY FOCUS** Take Action	*Video 15-2* **Virtual Investigation** Hummingbirds on the Move **Investigation B** Mapping Fruit Fly Genes Through Linkage (add time)

Contents		Instructional Support for All Learners	Digital Resources
EXPLORE/EXPLAIN			
466	**MINILAB** HAWKS AND MICE		
467–475 DCI LS3.B LS4.A LS4.B LS4.C	**15.2 EVOLUTION IN POPULATIONS** • Describe how populations evolve due to environmental factors. • Construct an explanation as to how the process of natural selection leads to adaptations. • Relate change in distribution of traits in a population to type of natural selection.	**Vocabulary Strategy** Frayer Squares **English Language Learners** Vocabulary Strategies **Address Misconceptions** Sample Size SEP Analyzing and Interpreting Data **Connect to Mathematics** Reason Quantitatively **Differentiated Instruction** Students with Disabilities **Visual Support** Directional Selection CER Revisit the Anchoring Phenomenon **Visual Support** Disruptive Selection CCC Cause and Effect **On Assignment** Sea Snails	**INTERACTIVE FIGURE** Selection Patterns
474	**LOOKING AT THE DATA** TRACKING EVOLUTION		
476–481 DCI LS3.B LS4.A	**15.3 OTHER MECHANISMS OF EVOLUTION** • Describe evolution mechanisms that can influence allele frequency. • Analyze how mechanisms of evolution affect genetic diversity in both small and large populations.	**Vocabulary Strategy** Semantic Map **Differentiated Instruction** Leveled Support **English Language Learners** Sayings and Expressions **Visual Support** Founder Effect CCC Cause and Effect **Connect to English Language Arts** Key Ideas and Details	*Video 15-3* **Investigation A** Genetic Drift (50 minutes)
480	**EXPLORER** Christopher C. Austin	**Connect to Careers** Molecular Geneticist	
ELABORATE			
481	**TYING IT ALL TOGETHER** Why do populations evolve over time?	SEP Constructing Explanations and Designing Solutions	Self-Reflection and Scoring Rubrics
Online	**Investigation A** Genetic Drift **Investigation B** Mapping Fruit Fly Genes Through Linkage	Guided Inquiry (50 minutes) Open Inquiry (50 minutes)	MINDTAP Access all your online assessment and laboratory support on MindTap, including: sample answers, lab guides, rubrics, and worksheets. PDFs are available from the Course Companion Site
EVALUATE			
482–483	**Chapter 15 Review**		
Online	**Chapter 15 Assessment** **Performance Task 2** *How does bacterial evolution affect public health globally?* (HS-LS4-2, HS-LS4-3)		

CHAPTER 15

ENGAGE

In middle school, students have learned how availability of resources and changes to abiotic and biotic factors can influence populations. They should have an understanding that genetic variation within a population allows some individuals to thrive. Students will also have learned how natural selection affects specific traits within a population over time. This chapter expands on that prior knowledge by exploring the different mechanisms of evolution.

Students will learn about the theory of evolution by natural selection and how it impacts our overall understanding of evolution. Students will explore the relationship between natural selection and genetic variation and analyze the distribution of traits within a population. They will examine how the frequency of alleles in populations is changed by natural selection.

About the Photo This photo shows a magnified view of a bedbug, *Cimex lectularius*. Fossil records found at archaeological sites tell scientists that bedbugs have been around for over 115 million years. It is possible that they could have used dinosaurs or ancient birds as hosts. To garner interest for this chapter's content, ensure that students understand that fossil records allow scientists to make observations about biological evolution for organisms both small and large.

Social-Emotional Learning

As students progress through the chapter, invite them to look for contexts and opportunities to practice some of the five social and emotional competencies. For example, as students work through the content in the chapter, support them in their **responsible decision-making**. Help them understand how scientists make observations to obtain evidence.

In addition, as students learn about tracking evolution, (Looking At The Data), remind them to practice **self-awareness** and **relationship skills**. Support students in communicating effectively and examining prejudices and biases.

CHAPTER 15

THE THEORY OF EVOLUTION

Bedbugs are not known to transmit or spread disease but they can cause itchiness, secondary skin infection, and anxiety to anyone dealing with an infestation.

Bedbugs are found all around the world. Scientists are studying various bedbug species to better understand the evolutionary relationships among them. Bedbugs have existed since at least the time of dinosaurs, and each species tends to be specialized to a single host type.

Chapter 15 supports the NGSS Performance Expectations **HS-LS3-3**, **HS-LS4-2**, **HS-LS4-3**, and **HS-LS4-4**.

456 CHAPTER 15 THE THEORY OF EVOLUTION

CROSSCUTTING CONCEPTS | Cause and Effect

Natural Selection and Genetic Diversity In this chapter, students explore the concept of cause and effect as they learn about different ways in which evolution can occur, the process of natural selection, and how environmental factors can influence how a population changes over time.

Scientists studying evolution observed cause-and-effect relationships that provided clues about the mechanisms of evolution. A deeper understanding of these mechanisms provides a framework that explains changes in genetic diversity within a population and why natural selection is not random.

You can reinforce this crosscutting concept in this lesson by helping students make the connection between the types of natural selection and changes in genetic diversity within a population.

CASE STUDY
DON'T LET THE BEDBUGS BITE!

WHY DO POPULATIONS EVOLVE OVER TIME?

Insects represent about 80 percent of animal species on Earth. They provide the best example of biodiversity on the planet with at least 900,000 documented species. The historical success of insect evolution, which has persisted over enormous spans of geologic time to develop such diversity, is remarkable. Two factors that contribute to their survival success include their ability to adapt to changing environments and their high reproductive capacity. In other words, most insects can reproduce quickly and have a huge number of offspring.

Like all organisms, insects are just trying to survive. Take, for example, the bedbug (*Cimex lectularius*). These pesky prowlers are attracted to warmth, feed on blood, and can be hard to find given their small size and their habit of staying hidden (**Figure 15-1**). They are a public health pest because they cause a variety of health issues such as minor to severe allergic reactions to their bites, secondary infections of the skin from the bite reaction, and even mental health impacts for those living in an infested home. The United States Environmental Protection Agency includes bedbugs on their official List of Pests of Significant Public Health Importance.

Bedbugs were nearly eradicated in the United States in the 1950s through the use of a pesticide called DDT, which was banned in the 1970s. A new type of pesticide called pyrethroids works by attacking an insect's nervous system, but over a very short time bedbugs evolved resistance to them. Bedbugs (as well as all insects and humans) have tiny pores in cell membranes that trigger nerve impulses. Pesticides bind to the pores, locking them in an open position, which leads to paralysis.

Biologists have discovered that only a couple of specific mutations in one pore protein gene are enough to make a bedbug 250 times more resistant to pyrethroids. These mutations may change the pore so that the insecticide can no longer bind to it effectively and may change the way the pore responds when the insecticide binds. Such mutations arise randomly and are favored when the environment in which a population lives changes, such as when a bed is sprayed with pesticide.

If only a couple bedbugs in a population carry mutations that cause resistance, they will be better able to survive and reproduce and will pass the mutations on to their offspring. As this process continues through several generations, the population may evolve such that every individual carries the resistance mutations, surviving the changing pesticides. However, if the population does not carry any of the advantageous resistance mutations, the pyrethroid treatment will wipe out the bedbug population.

Ask Questions *In this chapter, you will learn how the theory of evolution was developed based on evidence in population evolution over time. Generate questions and gather evidence to support the claim that natural selection leads to adaptations in populations.*

FIGURE 15-1 Bedbugs resemble many other types of bugs. Finding bugs in your bed might be a clue that you have a bedbug infestation. Before you panic, there are many internet resources to help identify region-specific bedbug characteristics.

VIDEO 15-1 Go online to learn more about bedbugs.

DIFFERENTIATED INSTRUCTION | English Language Learners

Shared Reading Model reading the first paragraph of the text. Pause to sound out unfamiliar words (such as *capacity*) and to think aloud to look for context clues. Ask questions to check comprehension: *How do we know insect evolution is successful? Why is it successful?* Then, have students do shared reading in pairs.

Beginning Have students reread the first paragraph, taking turns reading a sentence at a time. Have them underline details that help them answer the two questions you modeled.

Intermediate Have students do a shared reading of paragraphs 2 and 3. As needed, model reading a sentence for students to repeat. Have students answer these questions: *Why are bedbugs pests? How have people tried to get rid of them?*

Advanced Have students do a shared reading of paragraphs 4 and 5. Have them explain how bedbugs' evolution helps them survive.

CASE STUDY

ENGAGE

The Case Study can be used to assess students' prior knowledge of mutations and that they can be beneficial, harmful, or neutral. Students may know about negative mutations that cause disease and illness. In the Case Study, students learn about a positive mutation, pesticide resistance, that allows bedbugs to survive and reproduce. Revisit the Case Study after each section to have students identify the relationships among natural selection, environmental factors, inheritance, and organism adaptations.

In the Tying It All Together activity, students will research the reproduction rate of bedbugs to construct an explanation for how resistance to a specific pesticide can spread genetically throughout a population.

Ask Questions Students should revisit the Case Study as they read the chapter to make connections with the content. See a specific suggestion in Section 15.1.

Video

Time: 2:10 minutes

Use **Video 15-1** to show a brief overview of bedbugs. Share with students that bedbugs are insects that feed on a host (a parasite). Have students discuss the mechanisms that allow a bedbug to feed on its host without waking the host.

Historical Connection

How Old Are Bedbugs? The origin of bedbugs was at first linked to bats and humans. Then, fossil data showed that bedbugs are much older—they walked next to or hitched a ride with dinosaurs during the Cretaceous period. Ask students "How could bedbugs have survived the extinction event that killed the dinosaurs?" Discuss bedbugs' potential to adapt to new environments.

15.1

DEVELOPING THE THEORY OF EVOLUTION

LS4.B, LS4.C

EXPLORE/EXPLAIN

This section provides a review of early ideas about evolution and introduces evidence in support of the theory of evoution. It also reviews the concept of artificial selection and discusses the theory of natural selection in terms of fitness and adaptation.

Objectives

- Compare the differences and similarities between artificial selection and natural selection.
- Communicate the evidence that supports the theory of evolution.
- Evaluate the evidence for natural selection as a mechanism for evolution.

Pretest Use **Questions 1, 2,** and **3** to identify gaps in knowledge or misconceptions.

Vocabulary Strategy

Word Wall Create a word wall with the new vocabulary terms students will need to know. Post terms as they are introduced in the section. Refer back to the word wall at the end of the section to encourage students to use the terms to summarize their learning. Students can also make their own personal word wall in their notebooks to use when studying.

Predict *Students might say that many types of evidence, such as comparing physical features, fossil evidence, and geographical information allow us to reconstruct evolutionary events.*

15.1

KEY TERMS

adaptation
biogeography
fitness
natural selection

DEVELOPING THE THEORY OF EVOLUTION

When you are developing an idea about how a process works, you might start with a basic question you want to answer. You might predict a certain outcome, behavior, or response to a stimulus or a change to a situation. The main role of a scientist is to ask questions to develop an understanding about a phenomenon. Gathering evidence is part of understanding and explaining a phenomenon. For example, hundreds of years ago naturalists and scientists were asking questions about and documenting observations of relationships among organisms. They developed many different theories based on their evidence to explain why and how organisms change over time.

Predict *How do we know that organisms evolve over time to adapt to a changing environment?*

Early Ideas About Evolution

Biologically, evolution can be described as genetic changes in a population over many generations. Recall that a population is a group of the same species that lives in the same geographical area at the same time.

British naturalist Charles Darwin is widely recognized as the individual who proposed the theory of evolution. He defined evolution as "descent with modification." However, related ideas were developing long before his time (**Figure 15-2**).

FIGURE 15-2 Many proposals over the course of several hundred years have led to our current understanding of evolution.

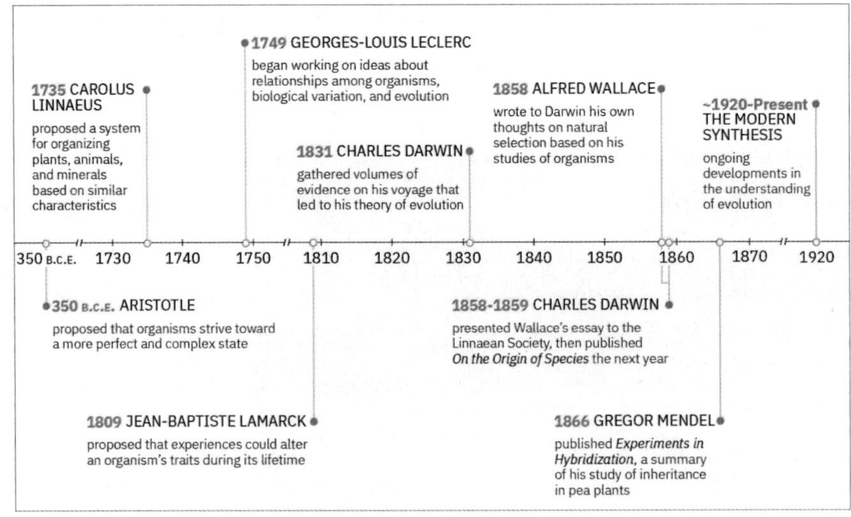

1735 CAROLUS LINNAEUS
proposed a system for organizing plants, animals, and minerals based on similar characteristics
1749 GEORGES-LOUIS LECLERC
began working on ideas about relationships among organisms, biological variation, and evolution
1831 CHARLES DARWIN
gathered volumes of evidence on his voyage that led to his theory of evolution
1858 ALFRED WALLACE
wrote to Darwin his own thoughts on natural selection based on his studies of organisms
~1920-Present THE MODERN SYNTHESIS
ongoing developments in the understanding of evolution
350 B.C.E. 1730 1740 1750 1810 1820 1830 1840 1850 1860 1870 1920
350 B.C.E. ARISTOTLE
proposed that organisms strive toward a more perfect and complex state
1858-1859 CHARLES DARWIN
presented Wallace's essay to the Linnaean Society, then published On the Origin of Species the next year
1809 JEAN-BAPTISTE LAMARCK
proposed that experiences could alter an organism's traits during its lifetime
1866 GREGOR MENDEL
published Experiments in Hybridization, a summary of his study of inheritance in pea plants

DIFFERENTIATED INSTRUCTION | Leveled Support

Struggling Students Students may struggle to understand some of the hypotheses that led to the current understanding of how evolution occurs. Refer students to **Figure 15-2**. Explain that time lines can be a helpful tool for understanding chronological events. Encourage students to make connections in the time line. Have students review information for the first two dates on the time line. Then, have them identify new information presented in the second date. Ask students to determine if the new information from the second date confirmed or contradicted the evolution proposal presented in the first date. Continue this process as students read the additional dates and information displayed on the time line.

Advanced Learners Students who understand the concept of evolution may be interested in creating their own evolutionary time line. Encourage students to recreate the time line in **Figure 15-2** and identify any patterns or cause-and-effect relationships in the original time line.

In the 1900s, people had different thoughts about heredity. For example, some thought that hereditary material must be a type of fluid that was blended at fertilization and passed on to offspring. Today we know that DNA stores hereditary information and that genes are the units passed on from one generation to the next.

Darwin did not know how heredity worked, but his detailed observations led him and others to recognize and question changes in organisms over time. Meanwhile, the little-known Austrian monk, Gregor Mendel, was developing what would become the fundamental laws of inheritance.

NOS Scientific Knowledge *How does the abbreviated time line in **Figure 15-2** support the idea that theories can be revised over time as new evidence is discovered?*

Asking Questions In the 1800s, European naturalists traveled the globe exploring tens of thousands of plants and animals. Their observations and records showed patterns in where species live and similarities in body plans. These explorers were pioneers in **biogeography**, the study of patterns in the geographic distribution of species and communities. Some of the patterns raised questions that could not be answered within the framework of how they thought species came to be.

- How could a flightless bird species have similar features to species living on the other side of impassable mountain ranges or across vast expanses of open ocean (**Figure 15-3**)?

FIGURE 15-3 The emu, rhea, and ostrich are native to three different continents but share a common ancestor.

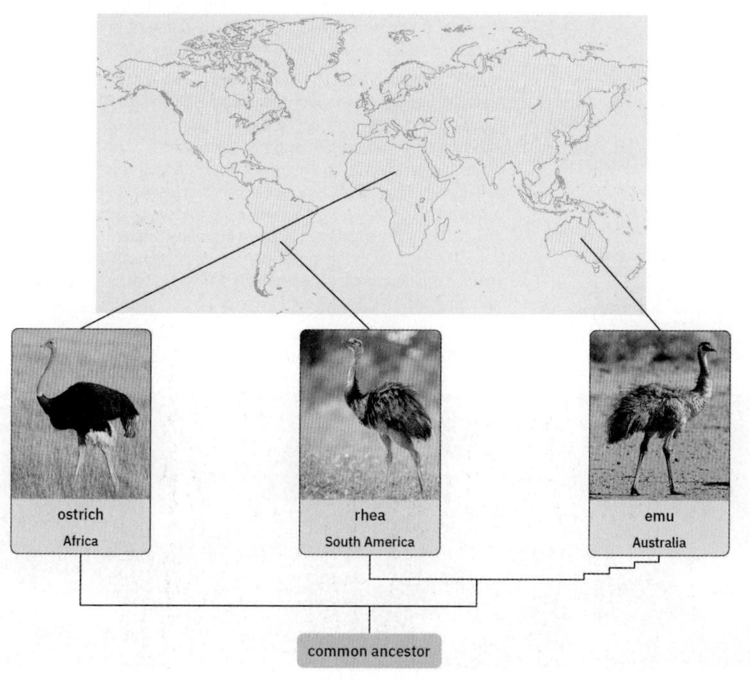

On the Map

Environmental Factors Environmental factors play a large role in an organism's ability to survive and reproduce. Use a world map to point out the locations of the birds with common ancestry discussed in **Figure 15-3**. Challenge students to think about the environments these birds are found in. Are the environments similar? Different? Lead a class discussion about the environmental conditions for each location. Have students think about the temperature, population density, light availability, and water sources for each location. Then, challenge them to determine how these conditions could relate to development of similar features for the flightless bird species found in the three different regions of the world.

NOS Scientific Knowledge *Students might say that the time line shows how scientists were organizing and recording their observations and thoughts about relationships between organisms. They were making observations about living things before Darwin's time. For example, Alfred Wallace came to a similar conclusion as Darwin about natural selection though they did not work together. Gregor Mendel discovered mechanisms of inheritance apart from any work that was happening in evolution research.*

DIFFERENTIATED INSTRUCTION | English Language Learners

High-Frequency Words As students read this part of the text, have them pause to describe and compare environmental conditions and species in the text and pictures. Before students begin, brainstorm adjectives that students can use to describe temperature, climate, shape, body parts, and so on.

Beginning Provide students with a word bank of adjectives for each section of the text. Have them draw each species or location and list the adjectives that describe it. Then have them point to similarities and differences.

Intermediate Have pairs work together to list similarities and differences of species discussed in each section. Supply vocabulary as needed.

Advanced Have pairs or small groups create their own word bank of descriptive words and features as they read. Then have them use their word bank to discuss similarities and differences they learned about.

Visual Support

Plant Ancestry Students may look at the plants in **Figure 15-4** and assume they are closely related based on their similarities in appearance and the kind of location where they are found. Explain that observations of anatomical features and the environment are not the only evidence used to determine ancestry. Lead a class discussion related to modern technologies that can be used to prove or disprove ancestry. Encourage students to think about how developments in technology provide additional evidence of potential ancestry and that this additional evidence must be synthesized with previous observations to provide an accurate assessment of how evolution occurs.

Genetic Evidence Students learn how different lines of evidence are applied to support common ancestry. Refer students back to **Chapter 14** as needed to review genetic evidence for evolution.

Subgroups of these birds are related species, but they are native to very distant geographic locations. They are unlike most other birds in several unusual anatomical features, including long, muscular legs and an inability to fly. They also share strikingly similar DNA sequences. All these data provide evidence to support the evolutionary relationship of these birds.

- How could organisms be very similar in some features but very different in others (**Figure 15-4**)?

FIGURE 15-4
The saguaro cactus (*Carnegiea gigantea*) is native to the Sonoran Desert of Arizona (a) and the African milk barrel plant (*Euphorbia horrida*) is native to the Great Karoo desert of South Africa (b) share similarities in appearance though are unrelated species.

The saguaro cactus and the African milk barrel plant live in hot deserts where water is seasonally scarce. Both have rows of sharp spines that deter herbivores, and both store water in their thick, fleshy stems. However, their reproductive parts are very different, so these plants are not as closely related as their outward appearance might otherwise suggest.

In 1831, when he was 22, Charles Darwin joined a five-year survey expedition to South America on the ship HMS *Beagle*. During his voyage, Darwin found many unusual fossils and saw diverse species living in environments that ranged from the sandy shores of remote islands to plains high in the Andes. Among the thousands of specimens Darwin collected during his voyage were *Glyptodon* fossils. These mammals are extinct, but they have many unusual traits in common with modern armadillos. Armadillos also live only in places where *Glyptodons* once lived. If *Glyptodons* were ancient relatives of armadillos, then perhaps traits of their common ancestor had changed in the line of descent that led to armadillos.

- How could extinct *Glyptodons* have so many unusual traits in common with modern armadillos (**Figure 15-5**)?

FIGURE 15-5
The *Glyptodon* (a) and armadillo (b) are ancient relatives. Though separated in time by millions of years, these animals share a limited geographic distribution and unusual traits, including a shell and helmet of keratin-covered bony plates similar to crocodile and lizard skin.

SCIENCE AND ENGINEERING PRACTICES
Constructing Explanations and Designing Solutions

Uniformitarianism Evidence is required to support a valid explanation or viewpoint. Prior to having students gather evidence related to Darwin's theory of how new species emerge, ensure that they understand the theory of uniformitarianism. Remind them that uniformitarianism states that Earth's geologic processes occurred in the same manner and with approximately the same intensity over time. This consistent, slow change resulted in the gradual changes we see in our current landscape. Refer students to **Figure 15-7**. Have them make observations of both images. Then, challenge them to look for evidence in support of Darwin's theory. For example, the fossils found in each layer of sediment show distinct similarities. Because the species were found in different layers, they would have been deposited in different time periods.

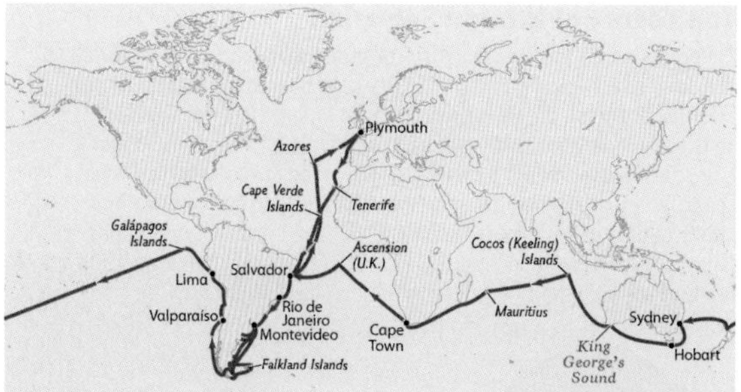

FIGURE 15-6 Darwin's voyage aboard the HMS *Beagle* was a five-year trip that went around the world.

Looking at the Evidence During Darwin's journey on the HMS *Beagle*, he read *Principles of Geology*, a new and popular book by Charles Lyell. Lyell, a British geologist, was a proponent of what became known as the theory of uniformitarianism, the idea that gradual, everyday geologic processes such as erosion could have sculpted Earth's current landscape over great spans of time. The theory challenged the prevailing belief that Earth was 6000 years old. By Lyell's calculations, it must have taken millions of years to sculpt Earth's surface. Darwin's exposure to Lyell's ideas gave him insights into the history of the regions he encountered on his journey (**Figure 15-6**).

The first stop for the HMS *Beagle* journey was the Cape Verde Islands. It was there that Darwin observed sediments enclosed by lava flows and raised above the sea level, but with fossils similar to the shells in the sea nearby (**Figure 15-7**). These layers held fossils of simple marine organisms that were clearly related to one another, but their form changed slightly from one layer to the next. With these observations and Lyell's principles in mind, Darwin became convinced that the surface of Earth changes slowly and gradually over time.

SEP Construct an Explanation *Darwin considered the geologic theory of uniformitarianism as support for his theory of how new species emerge. What do you think led him to this conclusion?*

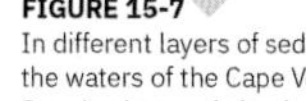

FIGURE 15-7 In different layers of sediment high above the waters of the Cape Verde Islands (a), Darwin observed simple marine organism fossils (b).

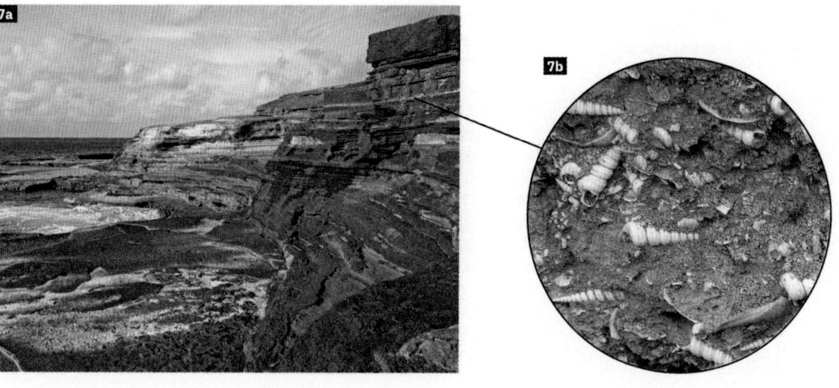

On the Map

Darwin's Journey Point out the route Darwin traveled along the coasts of South America, Africa, and Australia. Share with students that much of the route of Darwin's voyage focused on areas that were near the equator. Have students discuss the different climates for each of the areas visited during the five-year voyage and if Darwin's observation might vary based on the season in an area during the time of his visit. Students may mention that since the route was near the equator, the climates would be very similar, making it helpful for comparing species in varying locations.

SEP Construct an Explanation
Students might say Darwin realized that, like geologic processes, the process of evolution requires vast amounts of time. Geologic evidence also showed that Earth is extremely old, which for Darwin suggested that there had also been enough time for millions of species to emerge and either evolve into new species or go extinct.

DIFFERENTIATED INSTRUCTION | English Language Learners

Narrating with Details Have students narrate a fictional account of Darwin's journey of discovery based on facts using information from the section *Looking at the Evidence* and the paragraphs before and after. Have them take notes on each part of the journey to help them tell the story in detail. Remind them to tell the events in order.

Beginning Prompt students to use visuals to help them retell Darwin's journey. Point to **Figure 15-6** and ask: *Where did he go first?* Point to the images at the bottom of the page and ask: *What did he see there?*

Intermediate Have students write a sentence about each part of the journey and note down at least two details about what Darwin observed. Have them use their notes to narrate Darwin's journey.

Advanced Have students explain each part of the journey using details from the text that help listeners picture where Darwin went and what he saw.

Visual Support

Beak Function Refer students to **Figure 15-8**. Elicit which bird in the array is thought to be the ancestral species. Then, have students observe the beak shape of each finch and allow time for pairs to discuss the attributes of the particular food each eats. Prompt them to discuss how each bird's beak shape might be more suited to that particular kind of food than a different one in the array. Finally, have pairs suggest which finch might be better able to adapt to a new food supply if it suddenly found itself in a different locale with only that food supply.

Video

Time: 2:10

Use **Video 15-2** to show students various examples of Darwin's finches in their habitats and, in particular, see how the woodpecker finch uses tools to help it gather food.

SEP Argue From Evidence *Students might say that natural selection is not random because individuals with traits that are better adapted for their environment have a better chance of surviving and reproducing than do individuals without those traits. Natural selection drives a population to select for advantageous traits based on the environment. The environment controls the direction of natural selection. When the environment changes, different traits may become advantageous. The population responds with advantageous trait selection to survive and reproduce.*

The Theory of Natural Selection

Reflecting on his journey, Darwin started thinking about how individuals of a species often vary a bit in the details of shared traits such as size, coloration, and so on. He saw such variation among finch species on isolated islands of the Galápagos archipelago. This island chain is separated from South America by hundreds of kilometers of open ocean, so most species living on them had little to no opportunity for interbreeding with mainland populations. The birds on the Galápagos Islands resembled finch species in South America, but unique traits suited many of them to their particular island habitats (**Figure 15-8**).

FIGURE 15-8 On his voyage to the Galápagos, Darwin noticed differences in finch beaks that were related to function in terms of food eaten.

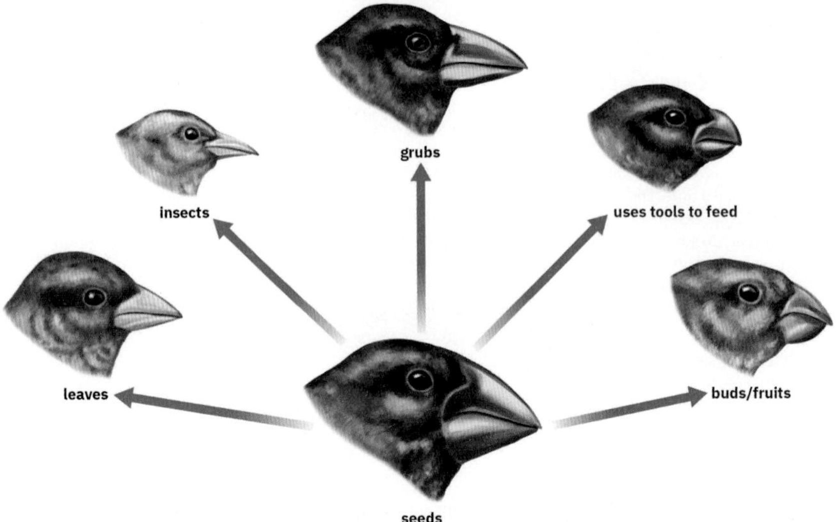

VIDEO

VIDEO 15-2 Go online to learn more about Galápagos finches.

SEP Argue From Evidence *Darwin thought that the development of these different types of finches was not a random process. Use the evidence from the food sources and beak types in* ***Figure 15-8*** *to support this claim.*

Darwin was familiar with *artificial selection*, which is intentionally breeding individuals with desired or advantageous traits over generations. In artificial selection, humans act as the selective agent. Darwin reasoned that natural environments could similarly "select" certain traits. Naturalists and scientists before him thought that organisms changed as a result of need or experience. The key to Darwin's insight was that individuals did not change but that individuals were selected because of favorable characteristics that matched changes in their environment. If a trait favors survival and reproduction, then individuals with that trait would mate and have a higher probability of passing on the favorable trait to their offspring than would individuals that do not carry that trait.

SCIENCE AND ENGINEERING PRACTICES
Engaging in Argument from Evidence

Natural Selection Scientists cannot create a theory for a phenomenon without evidence that shows the theory is valid. The theory of natural selection states that organisms that are better adapted to their environment are more likely to survive and reproduce. Over time, these adaptive traits become more common in the population. From this theory, an observer can predict that when new traits arise in a population the process is not random. Students must gather evidence using their observations of **Figure 15-8** to support this claim. Have groups discuss each beak shape. Encourage students to discuss how the shape of the beak would influence the finch's ability to secure food and ultimately reproduce. Encourage students to use their prior knowledge about genetic diversity and how traits are passed down to offspring in future generations as a piece of evidence in support of the claim that natural selection is not random.

Principles of Natural Selection Darwin realized that individuals with adaptive traits would tend to leave more offspring than their less-fit rivals, a process he named **natural selection.** He understood that natural selection causes a population to change—in other words, it drives evolution. **Table 15-1** summarizes this reasoning.

If an individual has an adaptive trait that makes it better suited to an environment, then it is better able to survive. If an individual is better able to survive, then it has a better chance of living long enough to produce offspring. If individuals with an adaptive trait produce more offspring than other individuals, then the frequency of the adaptive trait in the population will increase over successive generations—the population will evolve.

TABLE 15-1. Summary of Darwin's Main Principles of Natural Selection

competition in populations	• Natural populations have an inherent capacity to increase in size over time. • As a population expands, resources that are used by its members (such as food and living space) eventually become limited. • When resources are limited, members of a population compete for them.
genetic variation	• Individuals of a species share certain traits. • Individuals of a natural population vary in the details of those shared traits. • Shared traits have a heritable basis in genes. Slightly different forms of those genes (alleles) give rise to variation in shared traits.
overproduction	• A certain form of a shared trait may make individuals with that trait better able to survive. • Individuals of a population that are better able to survive tend to leave more offspring.
adaptation	• An allele associated with an adaptive trait tends to become more common in a population over time.

CASE STUDY
DON'T LET THE BEDBUGS BITE!

Ask Questions *Generate questions about the role of natural selection in the evolution of pesticide resistance in bedbug populations.*

Darwin wrote out his hypothesis of evolution by natural selection in the late 1830s but was so conflicted about its implications that he collected evidence for 20 years without sharing it. Meanwhile, British naturalist Alfred Wallace, who was studying wildlife in the Amazon Basin and the Malay Archipelago, wrote to Darwin and Lyell about patterns in the geographic distribution of species. When he was ready to publish his own ideas about evolution, he sent them to Darwin for advice. To Darwin's shock, Wallace had come up with the same hypothesis: that evolution occurs by natural selection.

In 1858, the hypothesis of evolution by natural selection was presented at a scientific meeting. Darwin and Wallace were credited as authors, but neither attended the meeting. The next year, Darwin published *On the Origin of Species by Means of Natural Selection*, which presented detailed evidence in support of the hypothesis, which eventually became the theory of natural selection.

CCC **Patterns** *Why might a particular beak shape or size become more common in a finch species over time? Use the four principles of natural selection from* ***Table 15-1*** *to support your answer.*

CROSSCUTTING CONCEPTS | Patterns

Food Availability Remind students that food availability is essential to the survival of a species. Have students work in groups to discuss how a specific trait, such as beak shape or size, relates to the principles of natural selection. Assign each group one principle from **Table 15-1**. Have students synthesize information from **Table 15-1** and **Figure 15-8** to explain why the trait might have become more common in a finch species over time. For example, from the adaptation principle in **Table 15-1**, a student could argue that as availability of seeds decreases in a specific environment, the allele associated with the adaptive beak trait for a food source that is abundant in the environment will become more common in the finch population over time. Observations of the beak traits and environments should show patterns that relate food source and beak type. The patterns can be used to identify cause-and-effect relationships.

Connect to English Language Arts

Synthesize Information from Text and Graphics This section discusses the main principles of natural selection. Have students read the first paragraph in the natural selection section and examine the information shown in **Table 15-1**. Students should discuss and evaluate the conclusions in the text and use evidence in **Figure 15-8** to corroborate or challenge the conclusions. Students should recognize that natural selection is not random, as those with adaptive traits are more likely to reproduce, which results in more offspring with the adaptive traits. Have students discuss the adaptive traits from **Figure 15-8** and how each trait was better suited for the finch's environment and therefore increased the survival of finches who had that trait.

CASE STUDY
DON'T LET THE BEDBUGS BITE!

Ask Questions *To help students make connections with the content, have them relate what they know about bedbugs to each principle summarized in* **Table 15-1**. *Students might make a bulleted list for each principle that outlines how bedbugs display that principle. They should use this outline to develop their questions.*

CCC **Patterns** *Individuals in a population with a beak type that is more suitable for eating the available food will have a competitive advantage. These individuals would be better suited to the environment and would survive to reproduce offspring with the advantageous trait. Over time, population would have more individuals with the shared advantageous trait.*

Connect to English Language Arts

Analyze an Argument Have students write an essay that explains how modern developments helped to confirm the theories of Darwin, Wallace, and Lyell. Explain that modern evidence might be the discovery of a new fossil. It also might be the result of new or improved tools and technologies, such as DNA sequencing. Have students state a clear claim at the beginning of the text and support it with evidence. Instruct them to write a conclusion that restates the claim and sums up their most important points.

Go Online VIRTUAL INVESTIGATION

Hummingbirds on the Move After learning about hummingbird evoution in South and North America, students will explore how hummingbirds interact with their habitats near the Southwest Research Station in Arizona. Encourage students to make connections between their observations and the idea of fitness.

BIOTECHNOLOGY FOCUS

PUT DOWN THOSE PHONES! Technology has helped us to advance in many different ways. Some technology has reduced human's needs to rely on some genetic adaptations. For example, humans can generally live in most any climate by using various types of shelter and clothing. Humans can also travel great distances thanks to many types of transportation that help us to move about. Our ancestors did not have the variety of shelter, clothing, and transportation we have today.

Over time, different technologies have changed some of the work and behavior patterns of our lives. We rely on machines more and more to do physical and mental work that humans used to do. For example, a calculator can complete complex mathematics within seconds. Smartphones allow us to access libraries of information with just a word search. As a result, there is less of a burden to retain as much information.

As we adapt to life with more sophisticated technologies, certain physical and mental activities we have done may no longer be essential. Over time, the brain could favor analyzing skills and rely less on its capability to retain information. The changes in our behavioral patterns that are caused by these technological advances could prove to be the primary driving forces behind the next stage of evolution for our species. It is worth noting, however, that we are still limited in part by our biology and genetics despite greater flexibility that some technological advances permit.

Technology can be a powerful source of information and distraction.

Go Online VIRTUAL INVESTIGATION

Hummingbirds on the Move
How did hummingbirds become adapted to their environments?
Grab your binoculars and head into the field to gather evidence about hummingbird adaptations.

Fitness and Adaptation In any natural population, some individuals have variations of shared traits that make them better suited to their environment than other individuals. In nature, the environment generates the selective pressure that determines if a trait is passed on or not. The ability of populations to undergo adaptive evolution depends on the presence of genetic variation for advantageous traits. An **adaptation** is a heritable trait that improves an organism's chances of surviving and producing offspring in a particular environment. **Fitness** is a measure of an individual's ability relative to other members of its population to survive and produce offspring.

A population tends to grow until it exhausts resources in its environment. As that happens, competition for those resources intensifies among the population's members. Individuals with forms of shared, heritable traits that make them more competitive for the resources tend to produce more offspring. Thus, adaptations impart greater fitness and become more common in a population over generations. This process in which environmental pressures result in the differential survival and reproduction of individuals of a population is natural selection. In turn, natural selection is one of the mechanisms that drives evolution, or the change in a population over time. Natural selection drives adaptive evolution.

BIOTECHNOLOGY FOCUS | Put Down Those Phones!

Take Action The digital revolution began in the 1970s. This time marked the transition from mechanical technology to digital technology, such as computers and wireless devices. Digital technology enables progress in many fields. But the progress technology brings comes at a cost. It can decrease social interactions and creativity and cause health issues. For example, prolonged screen exposure can result in complications with vision, such as eye fatigue, blurred vision, or double vision. Challenge students to reduce the amount of time spent on technology for one week. Have them record daily observations related to their vision and note if a reduced amount of screen time has any effect on their vision. Then, lead a class discussion about how human vision might change or adapt over time as humans continue to use digital devices.

The Modern Synthesis

Charles Darwin launched evolutionary biology with the help of Alfred Wallace, Charles Lyell, and many other scientists before them. Darwin's work considered evidence from a wide range of sciences including geology, paleontology, and genetics (inheritance) to show that living things are related to one another by common descent. Despite having no idea what a gene was or how heredity worked, Darwin successfully introduced natural selection as a mechanism of evolution.

Darwin supported his hypothesis of natural selection on a foundation of evidence he and others gathered over time from observations and experiments. Recurring evidence confirmed that each generation of a species was full of variations. Some variations helped organisms survive and reproduce, and those were passed on to their offspring. Heredity was later understood to be the transmission of genes from generation to generation.

The modern synthesis is a unified explanation for the theory of evolution that combines natural selection with modern genetics. The theory of evolution is a fundamental framework in which we scientifically explain phenomena of living things. Scientists today use the modern synthesis as a tool for asking questions about biology and the history of life.

SEP Argue From Evidence *Explain how modern evidence confirms the theory of evolution.*

CHAPTER INVESTIGATION

Evolution of Antibiotic Resistance in Bacteria
Why do we need different types of antibiotics?
Go online to explore this chapter's hands-on investigation about the evolution of antibiotic resistance in bacteria.

15.1 REVIEW

1. **Relate** Define the term *biogeography* in your own words and explain how the science initiated the study of evolution.
2. **Compare** Describe the key difference between artificial selection and natural selection.
3. **Apply** Classify each factor as essential or nonessential for natural selection to occur.
   - adaptations
   - environmental change
   - geographic isolation
   - inheritance of traits
   - migration
4. **Identify** What are the main principles of natural selection? Select all correct answers.
   A. Some organisms are resistant to environmental changes.
   B. Many species' traits are based on genes that were inherited from parents.
   C. Organisms, even those within the same species, compete for resources.
   D. Certain traits make organisms better suited for reproduction and survival.
   E. Adaptations to the conditions in the environment increase an organism's fitness.

CHAPTER INVESTIGATION B

Guided Inquiry *Evolution of Antibiotic Resistance in Bacteria*

Time: 115 minutes over 3 days

Students will carry out an investigation over the course of three days to observe bacterial growth in the presence and absence of ampicillin to make a claim as to why a single antibiotic is not effective against all bacteria.

Go online to access detailed teacher notes, answers, rubrics, and lab worksheets.

Science Background

Modern Synthesis The Modern Synthesis was developed by a number of renowned biologists in the 1930s and 1940s. The Modern Synthesis proposed changes in how evolution and evolutionary principles were regarded. It offered a new definition of evolution focusing on changes in allele frequencies within populations, which emphasized the genetic basis of evolution.

SEP Argue From Evidence *Today, scientists understand more about DNA and the genetic code than ever before. DNA comparisons can provide evidence for relationships between species. Biogeographers today travel to even the most remote places to improve global distribution maps of organisms that reflect evolution and geological change.*

15.1 REVIEW

1. *Sample answer: Biogeography is the study of species patterns based on geographic location. When biogeographers could not explain observed patterns between geographically isolated species, they began asking new questions about how separated species could have so many similarities.* **DOK 2**
2. *Sample answer: In artificial selection, humans act as the selective agent, breeding for specific desired traits. In natural selection, the environment is the selective agent. Organisms that are better suited to environmental factors survive and reproduce.* **DOK 2**
3. *essential: adaptations; inheritance of traits; non-essential: environmental change, geographic isolation, migration* **DOK 2**
4. *B, C, D, E* **DOK 1**

MINILAB
HAWKS AND MICE

In this minilab, students will simulate a predator-and-prey situation in which the prey have different colored fur that makes it easier or more difficult for a hawk to see them.

They will consider how variation among organisms in a population leads to differences in performance among individuals. They will also apply the results of their simulation to possible future generations of mice in their simulation. Additionally, they will consider how patterns they observe in their simulation could be scaled to explain natural phenomena.

Materials and Advance Preparation

- *The black and white speckled squares should be as close as possible to newsprint so that they will blend in more than the white or black squares do. If black and white newsprint is difficult to find, search for newsprint on the Internet to print out and use.*
- *The two containers for this activity should be ones in which students can easily place and remove the squares of paper. Avoid glass containers, as this activity is timed and students may be moving quickly. The containers used to hold all of the paper squares should not be clear.*

Procedure

In **Step 1**, explain that genetics is not as simple as what is being simulated here. This is a challenge of simulations and models. Before **Steps 2** and **3**, you can save time by having the supplies divided up and counted out for the number of groups that you anticipate. For **Step 5**, emphasize that students are not allowed to run in the lab. Though they have a limited time, obtaining the mice is not a race. In **Step 12**, explain that the 90 squares represent a second generation of mice. Instead of having an even number of white, black, and speckled mice, there is now a population based on the mice that remained after those from the first generation were captured by the hawk.

MINILAB

HAWKS AND MICE

SEP **Analyze Data** How does fur color impact a prey population?

Any population of organisms has variance within it. Organisms that have favorable variations are better adapted to their environment and will survive and reproduce in greater numbers. In this activity, you will simulate the effect of predation by a hawk on a population of field mice.

Materials

- newspaper, sheet (2)
- paper square, white, 3 x 3 cm (50)
- paper square, black, 3 x 3 cm (50)
- paper square, black-and-white speckled, 3 x 3 cm (90)
- container, clear, small
- container, dark, small
- timer

Procedure

1. Spread the sheets of newspaper open at one end of a table.
2. Mix and randomly spread out 30 of each color of paper squares (mice) on the newspaper (field).
3. Place the clear container, which represents the hawk's nest, at the other end of the table or on a different table. It should be far enough away that several steps need to be taken to get from the newspaper to the container.
4. Select one person to be the hawk (have them stand by the nest), another to operate the timer, and another to count the mice caught by the hawk.
5. The hawk has one minute to pick up as many mice as possible. The hawk may only pick up one mouse at a time and must place it in the nest before returning to pick up another mouse.
6. When the time is up, record the number of mice of each type left in the field (newspaper) in your biology notebook.
7. Switch roles within the group, return all the squares to the field, and repeat Steps 5 and 6. Do this as many times as necessary for each person within the group to play the role of the hawk.
8. Remove all the squares from the newspaper.
9. Using the data collected from each hawk, determine the average number of mice for each color remaining in the field.
10. Using the averages, place an equivalent number of each color of squares in a dark container.
11. Draw out 10 squares at random from your extra squares (those not in the containers or in the field) and place them on the newspaper.
12. Repeat Step 11 until you have 90 squares on the newspaper.
13. Repeat Steps 5 through 7 as many times as necessary for each person in your group to play the role of the hawk.
14. Using the data collected, determine the average number of mice for each color remaining in the field.
15. Clean up and return materials to their stations.

Results and Analysis

1. **Calculate** Based on the data you collected, what is the average number of each color of mice left after the hawk has captured the mice? Do the results surprise you? Explain why or why not.
2. **Explain** Were there any similarities or differences between the data collected from the first population of mice and the second population of mice?
3. **Draw Conclusions** How might the mouse population change over time from natural selection?

466 CHAPTER 15 THE THEORY OF EVOLUTION

Results and Analysis

1. **Calculate** *Sample answer: The average number of white squares left was 10, the average number of black squares left was 15, and the average number of speckled squares left was 20. I thought that the white and black squares selected would be about equal to each other, so the actual numbers were different than what I thought they would be. I did, however, expect that there would be more speckled squares left than either black or white, and I was correct about that.* **DOK 2**
2. **Explain** *Sample answer: In both the first and the second populations of mice, more of the white and black mice were captured than the speckled mice.* **DOK 2**
3. **Draw Conclusions** *Sample answer: If more of the white and the black mice are captured by the hawks than the speckled mice, there will be fewer white and black mice left in the population to reproduce. That means that those left are more likely to reproduce with a mouse of a different color or a speckled mouse, producing more speckled mice. Consequently, over time, the mouse population is likely to shift from a balance of the three different colors to being predominantly speckled.* **DOK 3**

EVOLUTION IN POPULATIONS

You may have noticed patterns in your everyday environment. For example, there are weather patterns associated with the four seasons, daily patterns involving the moon and tides, and patterns tied to animal behaviors, including your own!

Analyze *What environmental changes suggest that patterns can be explained by evolution?*

15.2

KEY TERMS

- allele frequency
- directional selection
- disruptive selection
- gene pool
- stabilizing selection

Genetic Variation

Evolution occurs in populations, not individuals. Individuals do not evolve during their lifetimes. Populations evolve over generations as heritable differences become either common or rare in a population. There are five known ways that populations can evolve. You learned about natural selection in the previous section. Natural selection acts on genetic variations that introduce different traits in an individual organism. If the trait is advantageous and helps the individual survive and reproduce, the trait is more likely to be passed to the next generation.

Gene Pool of a Population A population's **gene pool** is the total genetic material of all the individuals that make up that population at a given time (**Figure 15-9**). The gene pool includes all possible alleles, or variations of individual genes, within the population. Recall that an allele is a variation of a gene for a trait (Section 12.1). Genetic variation occurs when genes are recombined during meiosis and sexual reproduction. Recombination introduces genetic variation at the cellular level. Sexual reproduction results in increased genetic diversity within a species because of the random assortment of combined alleles from both parents.

FIGURE 15-9 Variation in shared traits among individuals is mainly an outcome of variations in alleles that influence those traits (a). Snail species, such as *Cepaea nemoralis* (b), provide an example of shell color variations.

9a

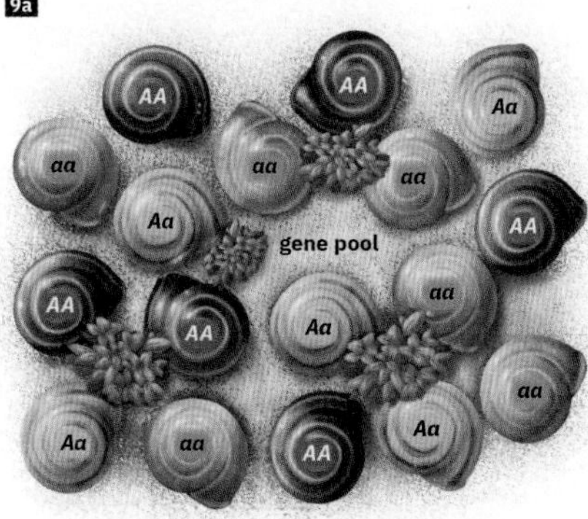

9b

15.2

EVOLUTION IN POPULATIONS

LS3.B, LS4.A, LS4.B, LS4.C

EXPLORE/EXPLAIN

This section provides a review of genetic variability and mutations, introduces the role of environmental factors on genetic diversity, and explores the different types of natural selection—directional, disruptive, and stabilizing selection.

Objectives

- Describe how populations evolve due to environmental factors.
- Construct an explanation as to how the process of natural selection leads to adaptations.
- Relate change in distribution of traits in a population to type of natural selection.

Pretest Use **Questions 4** and **6** to identify gaps in background knowledge or misconceptions.

Vocabulary Strategy

Frayer Squares As students work through the key terms in this section, have them create a Frayer square for each one. Students should draw a square and divide it into four boxes, with the word in the center. In the boxes they should write the definition of the word, an example of the word, a non-example, and additional relevant information, such as images or characteristics. As an example, make a Frayer square for *gene pool* as a class.

Analyze *Students might say that evolutionary processes can shape patterns of genetic variation and adaptations within and between species.*

DIFFERENTIATED INSTRUCTION | English Language Learners

Vocabulary Strategies Students need to regularly review new vocabulary in order to internalize and fully understand it. After students read this part of the text, review the terms *directional selection, disruptive selection*, and *stabilizing selection*.

Beginning Have students write and illustrate *directional selection* in their notebooks. Model as needed. For example, students could draw an arrow to show "direction." They could write *phenotype* above it and draw Earth under it to help them recall that directional selection relates to a specific phenotype being selected over time due to environmental changes.

Intermediate Have students draw a normal bell curve and one with a disruption. Model as needed. Have pairs use the pictures to restate the definition of *disruptive selection*.

Advanced Have students explain why *stabilizing selection* keeps a population *stable*, or without change.

Address Misconceptions

Sample Size Students may not be familiar enough with statistical sampling to understand that a sampling of three individuals, as shown in **Figure 15-10**, may not apply to a larger sample. Remind students that the greater the sample size, the more valid the data. Explain to students that if the allele combinations shown in **Figure 15-10** were applied to a larger sample of 100 deer mice, approximately 66 of the mice would present with allele combination *BB* and 33 with *bb*. However, these numbers could vary significantly because a small sample might not accurately represent the distribution of alleles in the whole population.

SEP Analyze Data *Students should calculate an allele frequency of the B gene as 0.5 or 50 percent. Students should notice that the allele frequency for B is the same as the allele frequency for b. However, mice with the B gene, whether homozygous or heterozygous, better match the surrounding environment and are less likely than the mice with the homozygous b gene to be caught by a predator.*

Mutations Genetic recombination by sexual reproduction is a source of genetic variation. However, mutations in gametes are the ultimate source of genetic variation in an organism's genome. Changes in the environment favor mutations that provide even a slight advantage, such that the frequency of that mutation tends to increase in a population over time. Thus, mutations are another way that populations can evolve because they are a source of new alleles in a gene pool.

Allele Frequency Members of a population in the same area breed with one another more often than they breed with members of other populations, so their gene pool is basically isolated. Different combinations of alleles in a gene pool can be formed when organisms mate and have offspring. For example, deer mice chromosomes have different alleles for fur color. A change within a deer mouse population gene pool would be a new variation of fur color within the population (**Figure 15-10**).

FIGURE 15-10 Differences in allele combinations cause differences in fur color in mice.

The frequency of a particular allele for fur color depends on the environment in which the population lives. **Allele frequency** is the proportion of one allele relative to all the alleles for that trait in a population's gene pool. Allele frequency is expressed as a percentage or a fraction. For example, if half of the chromosomes carried by members of a population have a particular allele, then the frequency of that allele in the population is 50 percent, or 0.5.

$$\text{allele frequency} = \frac{\text{number of a particular allele}}{\text{total number of alleles}}$$

Allele frequency can also be expressed as a percentage by multiplying the frequency by 100. The frequencies of all the different alleles in a population should equal 1.0, or 100 percent.

SEP Analyze Data *Calculate the allele frequency of the B gene in* ***Figure 15-10*** *expressed as a percentage. Explain the frequency result in terms of the population of mice in that environment.*

SCIENCE AND ENGINEERING PRACTICES
Analyzing and Interpreting Data

Data Displays Displaying information in varying forms can be beneficial for identifying and understanding relationships between different variables in a data set. As students use the equation to find the value for allele frequency, the results will be displayed in a decimal format. To answer the question, students will need to convert a decimal to a percentage. Remind them to multiply the allele frequency value by 100 to find the percent value. Have students compare both sets of data. Ask if one set of information was easier for them to analyze.

Types of Natural Selection

Natural selection affects the frequency of alleles that influence a trait by increasing, decreasing, or maintaining particular variations of the trait in a population. The genetic variation on which natural selection acts may occur randomly, but natural selection itself is not random at all. The survival and reproductive success of an individual is directly related to the ways its inherited traits function in the context of its local environment.

Natural selection is a major driver of evolution, but it is not the only one. Allele frequencies in a population can also change for other reasons. Individuals moving into and out of populations bring their alleles with them. Environmental events and other factors that reduce the number of individuals in a population can also affect allele frequencies.

Normal Distribution A normal distribution of traits across a population looks like a bell curve (**Figure 15-11**). The shape of the curve shows that the traits of the majority of the individuals in a population are close to the mean, and there are fewer individuals have extreme traits.

FIGURE 15-11
Normal distribution is a bell curve, showing a distribution that occurs naturally in many situations.

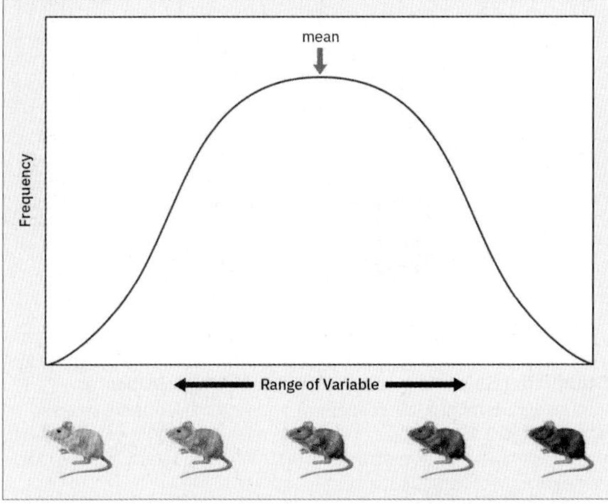

In populations, natural selection favors advantageous phenotypes that increase reproductive success. This results in favorable genes becoming more widespread in a population over time. *Microevolution* is the observable change in allele frequency in a population over time. It occurs constantly on a small scale within a single population. Microevolution occurs because natural selection acts on an organism's phenotype, which is often largely a product of genotype. When a phenotype produced by certain alleles helps individuals in a population survive and reproduce more successfully than other individuals, natural selection can increase the frequency of the helpful alleles from one generation to the next. Natural selection can shift the phenotype distribution in a population in three basic ways.

15.2 EVOLUTION IN POPULATIONS 469

Connect to Mathematics

Reason Quantitatively Determining trends in data is necessary for evaluating data collected from an experiment or study. As students observe **Figure 15-11**, remind them that a bell curve shows the normal distribution of a data set with the highest point representing the mean, median, or mode. The width of the curve reveals information related to the standard deviation from the mean. Explain that bell curves can also be left skewed or right skewed.

DIFFERENTIATED INSTRUCTION | Students with Disabilities

Auditory Impairments Students with auditory impairments benefit from visual supports to help encourage mastery of concepts. When developing an understanding of the types of natural selection, suggest they begin a word web with the term *natural selection* at the top or center. Then, encourage them to create branches for the terms d*irectional selection, stabilizing selection,* and *disruptive selection.* Encourage students to look for relationships by adding and grouping related terms, examples, and images.

Visual Support

Directional Selection Refer students to **Figure 15-12**, the graphical model for directional selection. Have them examine the images of the finches and describe how they are different. Likely responses are finch color and beak length. Help students determine the trait modeled by reviewing the text and the labels on the horizontal and vertical axes and on the curves. Help students see that without obtaining this information first, they may think the range of variability on the horizontal axis is for change in finch color, when it is actually change in beak length. Once students identify the trait modeled in the graph, ask students to explain the meaning of the two curves and the direction in which the trait changes. Make sure students understand the two curves represent the distribution of traits in a population at two points in time that are separated by many years. Then, lead a class discussion on how varying beak size relates to food availability.

Go Online INTERACTIVE FIGURE

Selection Patterns Students can use the interactive versions of **Figure 15-12**, **Figure 15-13**, and **Figure 15-14** to manipulate the graphs to see how distributions change over time.

Go Online INTERACTIVE FIGURE

Selection Patterns Go online to observe how the prevalence of traits can change over time in a population.

HUMMINGBIRD EVOLUTION

Gather Evidence *The sword-billed hummingbird is a species uniquely adapted to its ecological niche. Identify one advantage and one disadvantage of extreme adaptations in organisms such as the sword-billed hummingbird.*

Directional Selection In **directional selection**, some phenotypes are favored over others as the environment changes over time. A classic case of directional selection occurred in Galápagos finches (**Figure 15-12**). As available food sources changed over time, so did the beak lengths of finches in the population. When there were fewer insects and the birds relied on seeds as food sources, finches with larger beaks that could break seeds survived. When more insects were available for food, finches with smaller, longer beaks were better able to catch insects and survive.

FIGURE 15-12
With directional selection, forms of a trait at one end of a range of variation are adaptive to environmental changes.

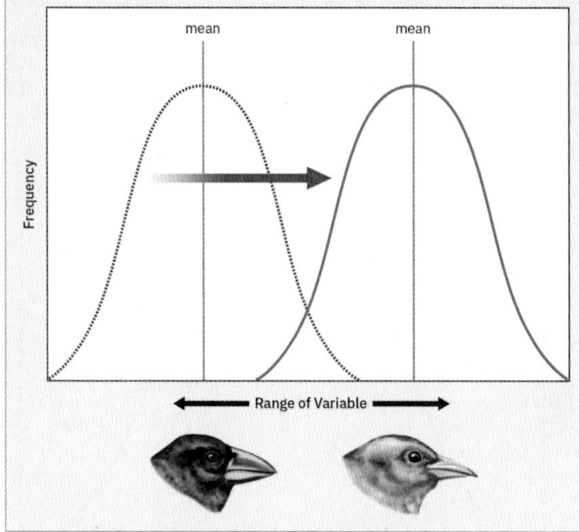

Stabilizing Selection In the case of **stabilizing selection**, an intermediate form of a trait is adaptive, and extreme forms are selected against (**Figure 15-13**). Most populations are at least fairly well adapted to their environment, so scientists think that stabilizing selection is the most common form of natural selection.

For example, environmental pressures maintain an intermediate body mass in populations of sociable weavers, which are birds that build large communal nests in the African savanna. Studies indicate that optimal body mass in sociable weavers is a trade-off between the risks of starvation and predation. Birds that have less fat are more likely to starve than plump birds. However, birds that have more fat spend more time eating, which in this species means foraging in open areas where they are easily accessible to predators such as snakes and birds of prey. Birds with more fat are also more attractive to predators and not as agile when escaping. Thus, predators are agents of selection that eliminate the individual birds with the most fat. Birds of intermediate weight have the selective advantage, and as a result they make up the bulk of sociable weaver populations.

REVISIT THE ANCHORING PHENOMENON

Gather Evidence *Students might say that one advantage of an extreme adaptation is exclusive access to a food resource. A disadvantage would be that if the food source was severely reduced or eliminated, the organism would be challenged to find an alternative food resource or may not be able to compete successfully for other food sources.*

Hummingbirds are nectarivores, which means they primarily feed on the nectar found within flowers. In the rainforest habitats where many species live, these flowers are first come, first serve. The sword-billed hummingbird has a beak that can reach up to 12 cm in length. This incredibly long beak allows it to feed from the flowers of *Passiflora mixta*, which has a floral tube with a depth up to 15 cm. The sword-billed hummingbirds are the only hummingbird species that can feed from this flower. In turn, *Passiflora mixta* relies on sword-billed hummingbirds for pollination.

FIGURE 15-13
With stabilizing selection, extreme forms of a trait are eliminated, and an intermediate form is maintained.

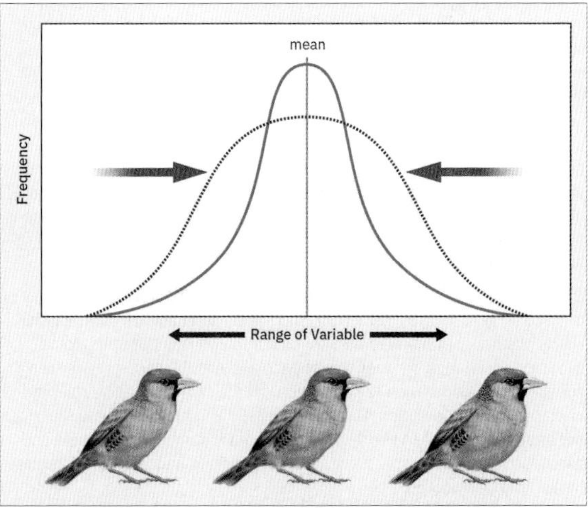

Disruptive Selection The opposite of stabilizing selection, **disruptive selection** favors the extremes rather than the center of the population (**Figure 15-14**). For example, beak sizes of black-bellied seedcracker birds in central Africa are either 12 millimeters wide or wider than 15 millimeters. Birds with a bill size between 12 and 15 millimeters are uncommon. Large-billed and small-billed seedcrackers inhabit the same geographic range, and they breed randomly with respect to bill size.

FIGURE 15-14
Disruptive selection eliminates midrange forms of a trait and maintains extreme forms at both ends of the range of variation.

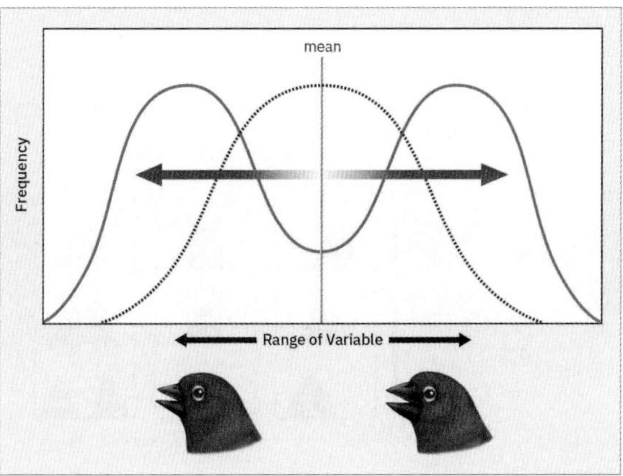

15.2 EVOLUTION IN POPULATIONS 471

Visual Support

Disruptive Selection The distribution of traits displayed in **Figure 15-14** may be difficult for students to interpret. To help with interpretation, have students label the left peak of the blue curve "adapted for soft seeds" and the right peak of the blue curve "adapted for hard seeds." Ask students to analyze the distribution of these two traits in the populations shown in the blue and red curves. Ask students to describe the current environment. Both small seeds and hard seeds are available year round, but during the dry season, both seed types are scarce. Birds with small beaks and birds with large beaks are more efficient in opening their preferred seed type and outcompete birds with medium-sized beaks. Guide students to understand that due to selection in this environment, the frequency of alleles for small and large beaks have increased in the population. The distribution curve is very unlike a normal distribution curve and hence is "disruptive."

CROSSCUTTING CONCEPTS | Cause and Effect

Genetic Diversity Students explore the concept of cause and effect as they learn about environmental factors that can influence genetic diversity. Emphasize to students that changes in an environment that influence food availability, ability to reproduce, and/or the ability to be camouflaged have the greatest impact on survival of the species. Focus student attention as they read about the different types of natural selection on how changes in the environment can influence the genotype and phenotype of a species. As students examine **Figure 15-12** and **Figure 15-13**, have them review how the phenotypes changed over time and if the variation in traits increased or decreased. Have students explain how the environmental change and the mechanism of natural selection caused the changes in the distribution and variation shown in the graphs.

Science Background

Natural Disasters Forest fires and other natural disasters can also influence how well plant and animal species survive in an environment. Changes in the environment can favor one or more adaptations in which certain individuals might be better suited for survival. It is estimated that the Australian bushfires of 2020 affected 28 million acres of habitat and harmed over 3 billion animals. Have students discuss which individuals might be best suited to survive a natural disaster. (*those with certain genetic variations that enable them to survive in the changed environment; those that are most able to successfully compete for resources*)

The existence of both forms of individuals arises from environmental factors that affect feeding performance. The birds feed mainly on the seeds of two types of sedge, a grasslike plant. One sedge produces hard seeds and the other produces soft seeds. Small-billed birds are better at opening the soft seeds, but large-billed birds are better at cracking the hard ones. Both hard and soft sedge seeds are abundant during the semiannual wet seasons. At these times, all seedcrackers feed on both seed types. During the region's dry seasons, the seeds become scarce. As competition for food increases, each bird focuses on eating the seeds that it opens most efficiently. Small-billed birds feed mainly on soft seeds, and large-billed birds feed mainly on hard seeds. Birds with intermediate-sized bills cannot open either type of seed as efficiently as the other birds, so they are less likely to survive the dry seasons.

Genetic Equilibrium

Evolutionary changes occur in a population due to natural selection, mutations, genetic drift, gene flow, and sexual selection. In contrast, the Hardy-Weinberg principle states that if a population is not undergoing evolutionary change, frequencies of each allele in the gene pool remain constant from generation to generation. This is called genetic equilibrium, or stability, in a population. It is an unrealistic and idealized state that does not happen in nature. A Hardy-Weinberg equilibrium, in which allele and genotype frequencies stay constant from generation to generation requires the following conditions:

- **Random Mating** Individuals within the population must mate with one another at random and not select their mates on the basis of genotype.
- **No Mutations** There can be no new alleles generated by mutations, nor duplication or deletion of genes.
- **Large Population Size** A large population size prevents random fluctuations that are more likely to occur in a small population.
- **No Migration or Immigration** No individuals enter or leave the population, a situation that could change allele frequency.
- **No Natural Selection** If natural selection were to occur, certain genotypes would be favored over others, and the allele frequencies would change.

15.2 REVIEW

1. **Explain** Why is it that populations, not individuals of a species, evolve?
2. **Justify** Support the claim that natural selection is not random.
3. **Identify** The sets of graphs show how the normal distribution of traits in three populations (L, M, O) changes over time. Identify the type of natural selection underway in each population.

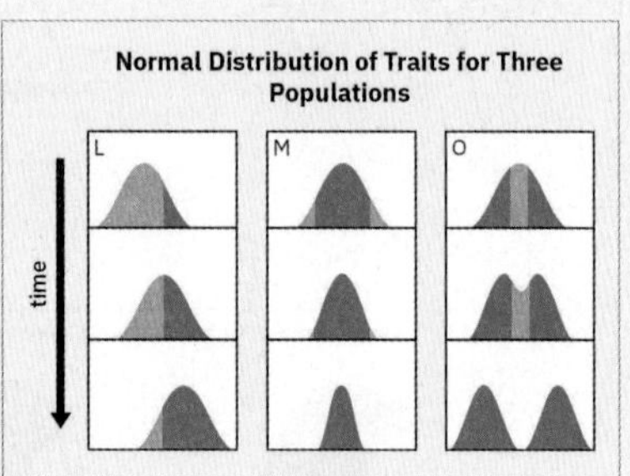

15.2 REVIEW

1. *Sample answer: Natural selection does not cause individuals to change their genetic traits. Natural selection does allow individuals with certain advantageous traits to survive and reproduce offspring with those traits. Individuals without the advantageous traits are not as likely to survive and reproduce. Over time, the population will contain more individuals with the advantageous traits than individuals without these traits.* **DOK 2**
2. *Sample answer: Natural selection acts on traits that benefit individuals that can survive and reproduce, which is directly related to the biotic and abiotic factors in the environment in which the population lives.* **DOK 2**
3. *L: directional selection; M: stabilizing selection; O: disruptive selection* **DOK 2**

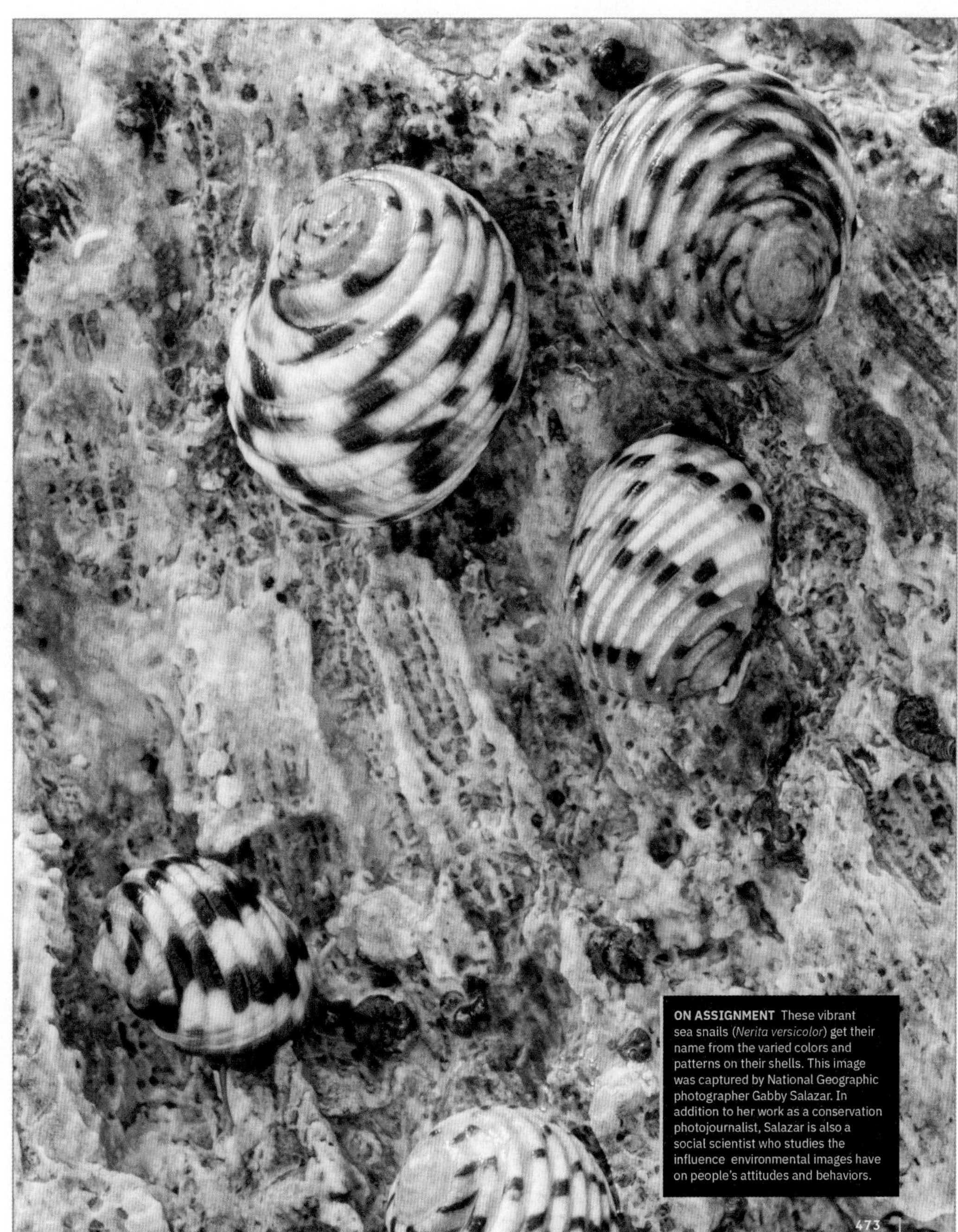

ON ASSIGNMENT These vibrant sea snails (*Nerita versicolor*) get their name from the varied colors and patterns on their shells. This image was captured by National Geographic photographer Gabby Salazar. In addition to her work as a conservation photojournalist, Salazar is also a social scientist who studies the influence environmental images have on people's attitudes and behaviors.

ON ASSIGNMENT

Sea Snails National Geographic photographer Gabby Salazar is a social scientist and a conservation photojournalist. She has traveled throughout North America and to over 40 countries to create images of nature and wildlife. Sea snails are similar to slugs in their taxonomic class with the primary difference between slugs and sea snails being a visible external shell. The external shell of the sea snail comes in a variety of shapes and sizes, with the largest living sea snail shell measuring 91 cm and the smallest measuring less than 1 mm. The shells of snails are complex and develop at different rates. The rate at which a sea snail shell grows is affected by different factors. These include temperature and depth of the water, food availability, and isotopic oxygen levels.

Find out more about Gabby Salazar's photography at her personal website or nationalgeographic.org.

LOOKING AT THE DATA
TRACKING EVOLUTION

Connect this activity to the Science and Engineering Practice of using mathematics and computational thinking. Students make calculations for the genotypic frequencies for coat color in different generations of a population of deer mice. Students draw conclusions about the evolution of coat color in the population based on their calculations and make inferences about the causes of change in coat color.

SAMPLE ANSWERS

1. *dark brown 1976/2080 = 95%; light brown. 104/2080 = 5%; dark brown 6115/8736 = 70%; light brown 2621/8736 = 30%; dark brown 27824/56784 = 49%; light brown 28960/56784 = 51%* **DOK 2**
2. *O: 5%; T: 30%; W: 51%* **DOK 2**
3. *p^2=60%; 2pq = 35%; p2=20%; 2pq = 50%; p2=8%; 2pq = 41%* **DOK 2**
4. *Sample answer: Generation O has the same genotype frequencies as Generation M. Generation T shows differences in the genotype percentages from the original population, so it is the earliest generation that shows signs of evolution toward more light-brown deer mice.* **DOK 3**
5. *Sample answer: Both the homozygous dominant genotype and heterozygous dominant genotype have the same phenotype, so it is not possible to distinguish between them.* **DOK 3**
6. *Sample answer: Light-brown homozygous recessive deer mice blend in well with the light-colored soil in Nebraska, so by natural selection, they are more likely to evade predators than their homozygous or heterozygous dark-brown relatives.* **DOK 3**

LOOKING AT THE DATA

TRACKING EVOLUTION

SEP Mathematical Thinking The population of deer mice in areas of Nebraska where glaciers once deposited light-colored soil tend to have a large proportion of individuals with homozygous recessive coat color. Coat color is based on a single gene where dark brown is the dominant allele (**Figure 1**). Allele frequencies can be predicted mathematically using a Punnett square and the Hardy-Weinberg equation. The Hardy-Weinberg principle states that genotype frequencies in any population remain constant over time in the absence of evolutionary influences. Mathematically, this equilibrium is represented with two equations

$$p^2 + 2pq + q^2 = 1$$

$$p + q = 1$$

where p^2 represents the frequency of the homozygous dominant genotype, q^2 represents the frequency of the homozygous recessive genotype, and $2pq$ represents the heterozygous genotype. The corresponding allele frequencies are then p and q. The frequency of all possible genotypes in a population must equal 1. If allele frequency can be found using the equation $p + q = 1$, then $p^2 + 2pq + q^2 = 1$.

FIGURE 1 *Peromyscus maniculatus* have two main coat colors. Mice with light brown coats (a) are found in Nebraska Sand Hills and mice with dark brown ones are found elsewhere (b).

Figure 2 shows a graphical model of this principle for a single-gene trait based on dominant *A* and recessive *a* alleles. The area of the square represents all the possible genotypes in the population and must be equal to 1.

A summary of the variables in this Hardy-Weinberg model is as follows:

- p = the frequency of dominant allele *A*
- q = the frequency of recessive allele *a*
- p^2 = the frequency of homozygous dominant genotype *AA*
- q^2 = the frequency of homozygous recessive genotype *aa*
- $2pq$ = the frequency of heterozygous genotype *Aa*

For example, a population of 100 deer mice (generation M) has five deer mice with light brown coat color. The frequency q^2 will be

$$q^2 = 5/100 = 0.05$$

Therefore, five percent of the population has the genotype *aa*.

The allele frequency $q = \sqrt{0.05} = 0.22$. Then, using the second equation

$$p + q = 1$$
$$p = 1 - q$$
$$p = 1 - .22 = 0.78$$
$$p^2 = (0.78)^2 = 0.60$$

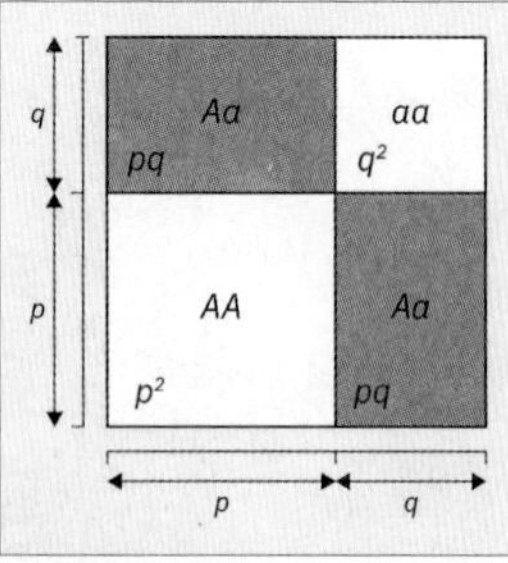

FIGURE 2 The Hardy-Weinberg equation using a Punnett square model shows allele distribution.

DIFFERENTIATED INSTRUCTION | English Language Learners

Classroom Language As needed, support students in understanding how to read the equations and mathematical symbols shown in *Looking at the Data* in words. Review words for symbols, such as *squared* and *square root*. Have students listen as you verbally describe expressions using standard classroom language, for example, the first equation for **Figure 2**: *Q squared equals five over one hundred, which equals zero point zero five.*

Beginning Have students say the equations for **Figure 1** and read the text that describes what each variable represents.

Intermediate Have students say the equations for **Figure 2** and explain in complete sentences what each variable represents.

Advanced Have students explain the model and example in **Figure 2**, including the equations used.

Therefore, 60 percent of the population has the homozygous genotype *AA*.

The predicted allele frequencies are p = 0.78 and q = 0.22. To calculate the predicted genotypic frequencies from the predicted allele frequencies, we use

$$2pq = 2(0.78)(0.22) = 0.35$$

Therefore, 35 percent of the population has the heterozygous genotype.

Generation M is made up of 60 homozygous dark brown deer mice, 35 heterozygous dark brown deer mice, and 5 homozygous light brown deer mice. If the population is not evolving, the genetic variation remains in equilibrium and these proportions will remain constant over time.

Table 1 shows the number of deer mice of different coat colors across three additional generations. The generations are not consecutive but cover the gene pool over dozens of years.

TABLE 1. A Deer Mouse Population Over Time

Population	Number of dark brown mice	Number of light brown mice
generation O	1976	104
generation T	6115	2621
generation W	27,824	28,960

1. **Calculate** Use the population data to determine the percentage of dark brown coat color and light brown coat color in deer mice of each generation. Round your answer to the nearest percent.
2. **Analyze Data** Calculate the genotype frequency (q^2) for homozygous recessive deer mice in each generation. Round your answer to the nearest percent.
3. **Use a Model** Calculate the expected percentage of homozygous dark brown (p^2) and heterozygous dark brown ($2pq$) deer mice expected in each generation using the Hardy-Weinberg equations.
4. **Evaluate** Which is the earliest generation that shows evidence for evolution from generation M? Explain how you know.
5. **Summarize** Why does looking at the observable traits of a population over time not give a complete picture of evolution occurring in a population?
6. **Infer** What are the possible reasons for a change in the coat color of deer mice populations over time?

Connect to Mathematics

Reason Quantitatively Have students analyze the data shown in **Table 1** and consider how the populations compare over time. Students should notice that in **Table 1**, as the time increases, the population growth of the dark- and light-brown fur color of the deer mice also increases. Draw student attention to Generation T and Generation W. Have students discuss reasons why there might be such a significant jump in population from Generation T to Generation W. Students should recognize that over time, the changes to the environment may affect traits displayed.

15.3

OTHER MECHANISMS OF EVOLUTION

LS3.B, LS4.A

EXPLORE/EXPLAIN

This section introduces the evolution mechanisms of gene flow, genetic drift, and sexual selection and explores how lack of genetic diversity can affect different population sizes.

Objectives

- Describe evolution mechanisms that can influence allele frequency.
- Analyze how mechanisms of evolution affect genetic diversity in both small and large populations.

Pretest Use **Question 5** to identify gaps in background knowledge or misconceptions.

Vocabulary Strategy

Semantic Map Have students create a semantic map for the vocabulary in this section. Begin by choosing a key term that is central to the topic. Have students write that word in the center of a page. As they read through each section, encourage students to add new words to the map. The words can be vocabulary terms or other key terms relevant to the content in the section. The words can be connected to the central term and to each other to show any relationships among them.

Explain *Student answers will vary. Students should use an example supported by new evidence to explain how their understanding changed.*

15.3

KEY TERMS

bottleneck effect
founder effect
gene flow
genetic drift
sexual selection

OTHER MECHANISMS OF EVOLUTION

When you spend time figuring out how something works, one thing that you do (whether or not you realize it!) is gain an understanding of the mechanisms involved. This occurs whenever you learn how to play a video game, use a computer application, or create a new art piece. Mechanisms, such as rules, variables, materials, and environmental conditions, affect how things work.

Scientists also study mechanisms as part of their understanding of natural processes and developing theories. For example, measuring the selection of certain traits in populations that allow species to adapt and thrive over time has helped researchers to more fully understand the processes of evolution.

Explain *Identify one concept or phenomenon in science where your understanding changed after you came across new evidence.*

Gene Flow

Natural selection and genetic variation (mainly mutations) are two mechanisms of evolution. In this section, you will learn about the remaining three mechanisms: gene flow, genetic drift, and sexual selection. All of these mechanisms can cause a population to evolve over time. All of the mechanisms of evolution may act to some extent in any natural population. Changes in a population's allele frequency across generations may result from several evolutionary mechanisms acting at the same time. For example, a mutation may produce a new allele, which is then favored (or disfavored) by natural selection.

Individuals tend to mate or breed most frequently with other members of their own population. However, not all populations of a species are completely isolated from one another, and nearby populations may occasionally interbreed. Also, individuals sometimes leave one population and join another. **Gene flow** is the movement of alleles between populations. Whether individuals from nearby populations of the same species interbreed or move from one population to the other, gene flow occurs, and it can change or stabilize allele frequencies.

Genetic Drift Allele frequencies also change within a population by random chance events. For example, consider a population of 10 individuals in which an allele for a particular trait exists with a frequency of 20 percent. That means two of the 10 individuals in the population are likely to carry that allele. Because of the very small sampling size, there is a reasonable chance that those two individuals could be eliminated purely by accident, thus losing the allele permanently.

Now consider a second population of 10,000 individuals with the same allele that has the same frequency of 20 percent. That means 2000 individuals in the population are likely to carry that allele. The chances that all 2000 individuals that carry this allele would be eliminated is much less likely than in the first population.

DIFFERENTIATED INSTRUCTION | Leveled Support

Struggling Students To help students in understanding how the bottleneck effect and founder effect can influence a population, have them review **Figure 15-16** and **Figure 15-17**. Then, have students add a real-world scenario to an image to help provide context for the process being modeled. Students could add a drought to **Figure 15-16** to better understand how a population can decrease. Students could add a natural disaster, such as a forest fire, to **Figure 15-17** to better understand how a population could be separated into two or more smaller groups.

Advanced Learners Have students review the information related to **Figure 15-17** and consider how the founder effect is similar to and different from the bottleneck effect described in **Figure 15-16**. Encourage students to consider how each effect would influence genetic diversity in the populations that continue.

From this situation, we see that there is a greater likelihood that the allele will be lost from the smaller population than from the larger population. This phenomenon is called **genetic drift**, or the change in allele frequency due to chance alone. A population can "drift" toward or away from certain alleles purely by chance, regardless of whether they provide an advantage.

CCC Cause and Effect *Why does the loss of genetic diversity have a greater effect in a small population compared to a large population?*

Bottleneck Effect A reduction in population size so severe that it reduces genetic diversity is called the **bottleneck effect**. For example, northern elephant seals underwent a bottleneck during the late 1890s, when overhunting left only about 20 individuals of this species alive. The bottleneck effect is an extreme example of genetic drift because with so few individuals, the frequency of alleles is either greatly reduced or some alleles are eliminated (**Figure 15-15**).

Hunting restrictions have allowed the population of the northern elephant seal to recover (**Figure 15-16**), but genetic diversity among its members has been greatly reduced. The bottleneck and subsequent genetic drift eliminated many alleles that had been present in the population.

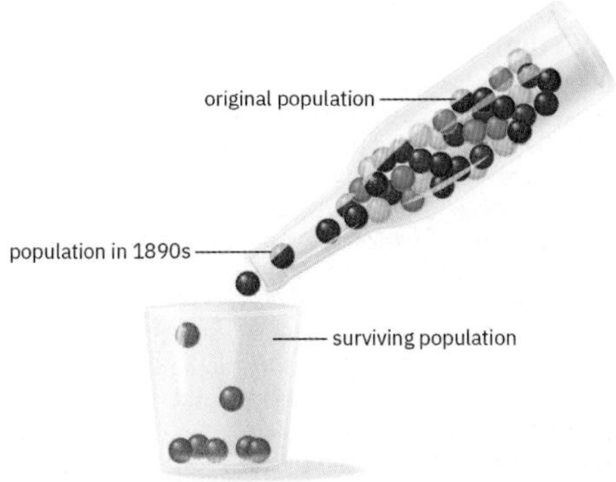

FIGURE 15-15 Beads in a bottle is one way to model the bottleneck effect. Random chance determines which individual beads fall out of the bottle when it is tipped.

FIGURE 15-16 The northern elephant seal suffered from the bottleneck effect when most were killed for their blubber, which was used in lamp oil.

CCC Cause and Effect *Students might say that the loss of genetic diversity is possible in all populations but is greater. However, in a population with fewer individuals, any loss of individuals is greater and the gene pool is smaller.*

Science Background

Bottleneck Effect In addition to human activities, other natural events such as famines, flooding, fire, and droughts can also contribute to a reduction in population size. Encourage students to think about how both natural and human factors could increase the bottleneck effect and potentially lead to extinction. Encourage students to look at both the pros and cons of the bottleneck effect. For example, it could promote greater genetic fitness. But it could also lower genetic diversity, which then decreases a species' ability to adapt and survive when the environment changes.

DIFFERENTIATED INSTRUCTION | English Language Learners

Sayings and Expressions Idioms often have a different meaning than their literal meaning, so they may be difficult for language learners to understand. Identify and explain idioms as students come across them in the text, for example, *bottleneck* in the section *Bottleneck Effect.*

Beginning Have students point to their own *neck*. Then have them point to the *neck* of the bottle in **Figure 15-15**. Point out that not many balls can get through the bottle because the neck is thin. Relate this to the reduction in population in a bottleneck effect.

Intermediate Have students draw a line between the words *bottle* and *neck* and use the diagram to explain what it is. Ask: *What happens when the neck of the bottle is small? In a bottleneck effect, what happens to genetic diversity when the population is reduced?*

Advanced Have students use **Figure 15-15** to explain what a *bottleneck* is and how it relates to the bottleneck effect in science.

Visual Support

Founder Effect Help students process the graphic model for the founder effect in **Figure 15-17** by reviewing the allele types contained within the original, founding, and descendent populations. Have students analyze the frequency of the allele represented by the yellow dots in the different populations. The frequency of these alleles is similar among the populations. Then, have students analyze the frequency of the alleles represented by the green, blue, red, and purple dots. The frequency of these alleles is different among the populations. Have students compare the green allele in descendant population A and the purple allele in descendant population B to the original population. Have students use their frequency analysis to describe in general terms how the founder effect changes the genetic diversity of descendant populations compared to the original populations. Then, have students consider how the size of the founding population affects the extent of the change in genetic diversity.

Video

Time: 2:16

Use **Video 15-3** to exemplify lekking behavior in greater sage grouses, which occurs in other bird species as well, including prairie chickens, musk ducks, hermit hummingbirds, and peacocks. A few mammals also lek, most of which are hooved, such as fallow deer. Some species of insects, frogs, and fish also carry out this behavior.

Explain to students that the word *lek* is a Swedish word, likely from the Swedish *lekställe*, or "mating ground." The behavior occurs in species in which the female requires no resources from the male other than good genes. It also reduces risk to the female as she expends less energy searching for a mate and has less chance of predation or parasite transmission.

Founder Effect The **founder effect** occurs when a few individuals break off from an established population and form a new colony that is fully separated from its original population (**Figure 15-17**). In this case, the new population can drift in unexpected ways. For example, if the small group happens to possess very unusual alleles, those alleles will have a high frequency in the new colony, much greater than that in the original population.

FIGURE 15-17 Genetic variation of a new founding population compared to the original population will vary. The different colored beads represent different alleles in the populations.

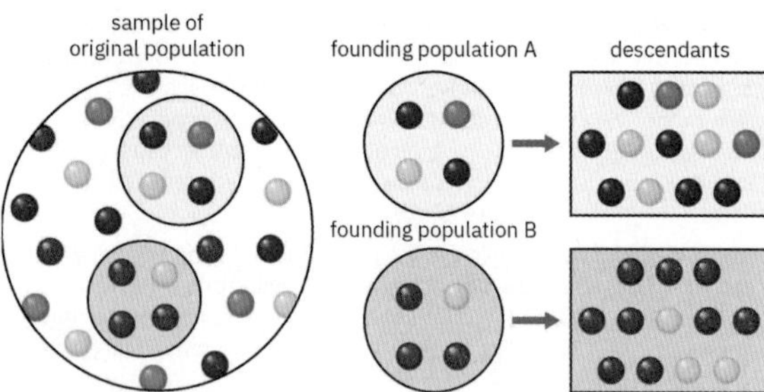

Sexual Selection

Apart from the genetic mechanisms (mutations, gene flow, genetic drift) that introduce variations in a population, random mating introduces variation as well. Not all evolution is driven by selection for traits that enhance survival. Competition for mates is another selective pressure that can shape form and behavior. In some species, individuals of one sex (often males) are more colorful, larger, or more aggressive than individuals of the other sex. These traits can take energy and time away from activities that enhance survival, and some actually compromise survival.

Sexual selection is a mechanism of evolution in which the evolutionary winners out-reproduce others of a population because they are better at securing mates. With this mode of natural selection, the most adaptive forms of a trait are those that help individuals defeat rivals for mates or are most attractive to the opposite sex.

For example, the females of some species cluster in defensible groups when they are sexually ready, and males compete for sole access to the groups. This is called lekking, and is a classic model for studies of sexual selection because males compete with one another to attract a female, and the female chooses her mate (**Figure 15-18**).

FIGURE 15-18 A female sage grouse (center) judges the courtship displays of three male sage grouse as they call and dance around her.

 VIDEO

VIDEO 15-3 Go online to learn more about the behavior called lekking.

478 CHAPTER 15 THE THEORY OF EVOLUTION

CROSSCUTTING CONCEPTS | Cause and Effect

Sexual Selection Students explore the concept of cause and effect as they learn about how sexual selection can influence genetic diversity. Emphasize to students that characteristics that influence an individual's ability to attract a mate and reproduce significantly influence the survival of the species. Focus student attention as they read about the quest of the males on attracting a mate and how time spent in that quest potentially comes at a cost. Have students create a T-chart of pros and cons for males as they seek to attract a mate. Then, have them repeat the process for females. As students list out the pros and cons from both sides, encourage them to think about how each characteristic could influence adaptations or possibly contribute to extinction of the species.

Males or females that are choosy about mates act as selective agents on their own species. The females of some species shop for a mate among males that display species-specific cues, such as a highly specialized appearance or courtship behavior. The cues often include flashy body parts or movements—traits that attract predators and in some cases are a physical hindrance. However, to a female member of the species, such displays may imply health and vigor, two traits that are likely to improve her chances of bearing healthy, vigorous offspring. Selected males pass alleles for their attractive traits to the next generation of males, and females pass alleles that influence mate preference to the next generation of females. Highly exaggerated traits, such as a male peacock's tail, can be an evolutionary outcome (**Figure 15-19**).

CCC Cause and Effect *Why is the trade-off of an individual expending energy to find a mate worthwhile for a population?*

CHAPTER INVESTIGATION

Genetic Drift

How can genetic drift affect a population?

Go online to explore this chapter's hands-on investigation about the ways in which population's gene pool can change.

FIGURE 15-19 A male peacock displays his elaborate tail to attract a female mate.

15.3 REVIEW

1. **Classify** Sort the processes into those that can affect the allele frequencies and those that do not.
   - asexual reproduction
   - gene flow
   - genetic drift
   - meiosis
   - mitosis
   - mutation
   - natural selection
   - sexual selection
2. **Explain** How does the bottleneck effect lead to evolution?
3. **Identify** Which mechanism of evolution can keep two distinct populations genetically alike?
   A. gene flow
   B. genetic drift
   C. mutations
   D. natural selection
4. **Explain** Why does sexual selection often result in the evolution of elaborate traits, such as a male peacock's tail?

CHAPTER INVESTIGATION A

Open Inquiry *Genetic Drift*

Time: 50 minutes

Students apply statistics and probability to demonstrate and draw conclusions about how the gene pool in a population can change. They will design procedures to simulate genetic drift to gather evidence and calculate allele frequencies for various populations.

Go online to access detailed teacher notes, answers, rubrics, and lab worksheets.

Connect to English Language Arts

Key Ideas and Details This section introduces the concept of sexual selection and how selective mating can also produce variations in a population. Have students write an informative summary describing the advantages and challenges associated with selective mating. Encourage students to use a graphic organizer as they read to ensure they are collecting similar information for each item.

CCC Cause and Effect *Students might say that expending energy is worthwhile because the mating payoff is more likely to result in reproduction, resulting in an increase in the population.*

15.3 REVIEW

1. *affect allele frequencies: gene flow, genetic drift, meiosis, mutation, natural selection, sexual selection; do not affect allele frequencies: asexual reproduction, mitosis* **DOK 2**
2. *Sample answer: A random event or environmental disturbance dramatically reduces the size of a population. The surviving individuals reproduce, and the population increases with a smaller gene pool.* **DOK 2**
3. *A* **DOK 2**
4. *Sample answer: An elaborate trait such as a male peacock's large tail is a sign to potential mates that the displaying male is in good health, which could lead to the production of healthy, vigorous offspring.* **DOK 2**

EXPLORER
CHRISTOPHER C. AUSTIN

About the Explorer

National Geographic Explorer Christopher Austin uses research to collect data and answer genetic evolution questions. He is analyzing the evolution and environments of green-blooded skink species to try to determine the advantage that is conferred by this rare kind of blood. His initial conclusion is that green blood protects the skinks from certain kinds of parasites. You can learn more about Christopher Austin and his research on the National Geographic Society's website: www.nationalgeographic.org.

Science Background

Malaria Treatment Malaria is a disease caused by parasites and is currently treatable with prescription drugs that kill the parasites. Parasites have a complex life cycle, and there are five different species of parasites that infect humans. In order to correctly treat malaria, doctors must identify the species of parasite that is present in your blood. Then, they choose the medication that will be most successful in killing that specific species. Chloroquine phosphate and artemisinin-based combination therapies are the most common treatments for malaria.

Connect to Careers

Molecular Geneticist Molecular geneticists investigate the structure and function of genes and how differences in the expression of DNA molecules leads to genetic variation among organisms and contributes to human disease. Geneticists may work for local, state, and national government agencies, as well as international organizations. They may also work in private and nonprofit organizations, as well as universities.

THINKING CRITICALLY

Infer *All of the other skink species have red blood, so the ancestor would have had red blood and a mutation must have occurred.*

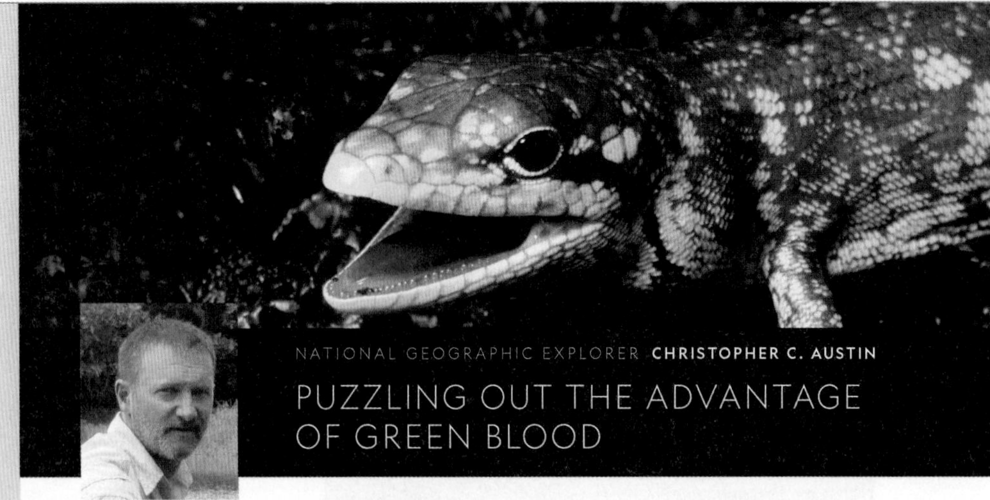

NATIONAL GEOGRAPHIC EXPLORER **CHRISTOPHER C. AUSTIN**

PUZZLING OUT THE ADVANTAGE OF GREEN BLOOD

Dr. Christopher Austin's research focuses on the use of molecular genetic data sets to answer evolutionary questions.

More than 100 species of skinks live on New Guinea, an island off the coast of Australia. Their smooth-scaled, slender appearance and behavioral habits are not especially different from those of other skinks. But biologist Dr. Christopher Austin took a special interest in certain ones in 1990. What made them so unique? Their bright, lime-green blood.

Until recently, the few described green-blooded skink species on the island have been grouped together in the genus *Prasinohaema*. Yet, Austin's research shows that the species are not closely related. They live in very different habitats and some lay eggs while others give birth to live young. He thinks the trait arose in all members of this group from a common red-blooded ancestor. But what advantage did it confer that caused its selection?

While not common, green blood in animals is not unique to these skink species. It occurs in some species of insects, frogs, and fish as well. However, it doesn't occur in any other reptiles or in any birds or mammals. The green color is caused by the green bile pigment called biliverdin that overwhelms the red color of the red blood cells present. This and other liver pigments are toxic in even small amounts in all other reptiles, birds, and mammals. The green bile present in the skinks is 15 to 20 times higher than the level that would kill a human. The high concentration of biliverdin in the blood colors not only the blood green but also the muscles, bones, and tongues of the skinks.

Austin's team examined 51 species including six with green blood. They could determine which species were those that had evolved on New Guinea and which descended from Australian lineages. By comparing segments of DNA, they reconstructed the evolutionary history of those relationships that included the green-blooded skinks. Surprisingly, based on the branching indicated by the DNA analysis, the trait arose independently four different times in species with a strict New Guinea heritage.

To solve the puzzle of green blood's evolutionary advantage, Austin's team is currently investigating how it might protect the skinks against blood parasites such as malaria. A similar human bile pigment, bilirubin, is toxic to human malaria parasites.

THINKING CRITICALLY

Infer *Even though the green-blooded skink species are not closely related, why do you think Austin states they all descended from a red-blooded ancestor?*

TYING IT ALL TOGETHER
DON'T LET THE BEDBUGS BITE!

WHY DO POPULATIONS EVOLVE OVER TIME?

In this chapter, you learned how natural selection and other mechanisms drive adaptation in populations over time in response to changes in the environment. The bedbugs described in the Case Study are an example of a species that adapts relatively quickly to its changing environment (**Figure 15-20**).

Natural selection occurs when there is genetic variation in the population. Scientists recognize that bedbug populations have been primed with the right sort of genetic variation by their evolutionary history of extensive exposure to DDT. Like pyrethroids, DDT kills insects by acting on certain pores in their nerve cells, causing paralysis. Many of the same mutations that protect bedbugs against DDT also protect it from pyrethroids. Though such mutations were probably extremely rare at the outset, the widespread use of DDT in the 1950s and 1960s drove such mutations to be more common in bedbugs through the process of natural selection. While DDT is rarely used today because of its harmful environmental effects, these mutations are still present in modern bedbug populations. Because of the action of natural selection in the past (favoring resistance to DDT), many bedbug populations today evolve resistance to pyrethroids rapidly.

In this activity, you will use evidence you gather to explain how a population can evolve over time.

Ask Questions

1. Review the evidence you gathered and the questions you generated throughout the chapter about the mechanisms of evolution. Relate these questions to what you have learned about bedbugs.

Analyze Data

2. Research data on the reproduction rate of bedbugs.

Construct an Explanation

3. Explain how a mutation causing resistance to pyrethroids would spread through a population of bedbugs that are being treated with the pesticide. Make sure to include the concepts of genetic variation, natural selection, and inheritance in your explanation.

FIGURE 15-20
The total development process of bedbugs takes about 37 days at optimal temperatures above 22 °C (72 °F) and with readily available blood meals.

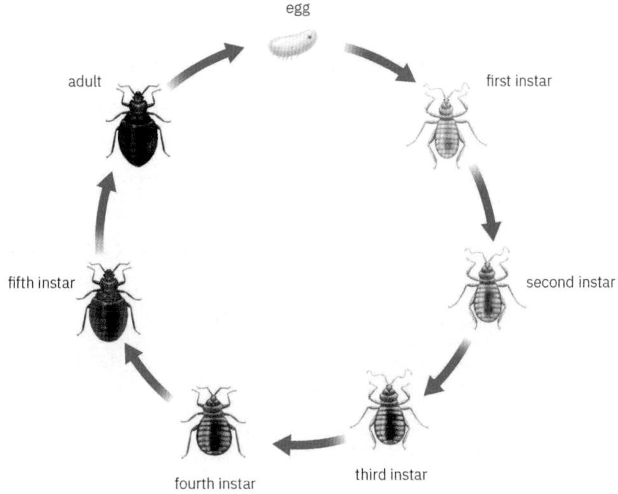

TYING IT ALL TOGETHER

ELABORATE

The Case Study discusses how evolutionary history of bedbugs has contributed to their resistance to the pesticides DDT and pyrethroids. In the Tying It All Together activity, students will gather evidence to explain how the population of bedbugs evolved over time. This activity is structured so that students may begin once they complete Section 15.3. **Step 1** has students gather evidence used to answer questions about mechanisms of evolution and relate them to bedbug evolution. **Step 2** requires students to research information related to reproduction rates of bedbugs. **Step 3** asks students to apply their research information to what they have learned about genetic variation, natural selection, and inheritance to explain how a mutation causing resistance to pyrethroids would spread through a population of bedbugs that is being treated with the pesticide.

Go online to access the Student Self-Reflection and Teacher Scoring rubrics for this activity.

Science Background

Human Impact of Bedbugs While bedbugs do not typically transmit diseases, medical issues can still arise from these little blood suckers. For example, rashes, irritation, and secondary skin infections, such as impetigo, ecthyma, and lymphangitis, can develop at or near the bite site.

SCIENCE AND ENGINEERING PRACTICES
Constructing Explanations and Designing Solutions

Pesticide Resistance To help students prepare for researching and collecting evidence to explain why a mutation for pesticide resistance has spread throughout the population, have them first define which kinds of information they need. Students must understand the mechanisms of evolution and determine which mechanisms apply to changes in the traits of the bedbug. Students should make a T-chart of the mechanisms of evolution and consider how the mechanisms are related and work together. Students must also research the reproduction rate of bedbugs. Once they have gathered evidence from the chapter and researched the reproduction process of bedbugs, have them synthesize the information. Encourage the use of cause-and-effect graphic organizers to help in this process. Students may also benefit from creating a model of the process. Recommend a pedigree chart to show inheritance over generations.

CHAPTER 15 REVIEW

EVALUATE

REVIEW KEY CONCEPTS

1. *C, D, E* **DOK 1**
2. *B* **DOK 2**
3. *Sample answer: Fitness describes an individual organism's ability to survive and reproduce in its environment. Adaptations are traits that improve an organism's fitness.* **DOK 2**
4. *A* **DOK 1**
5. *D* **DOK 1**
6. *C* **DOK 1**
7. *B* **DOK 1**
8. *Sample answer: This is an example of the bottleneck effect. The distribution of traits in this population was altered by an event that randomly shrunk the gene pool. Alleles for short needles became more prevalent in the population not because short needles were an advantageous adaptation, but because a relatively small number of trees survived the forest fire and short-needled trees happened to be well represented among the survivors. When these short-needled survivors reproduced, they passed the allele on to a high percentage of their offspring, which repopulated the area.* **DOK 2**
9. *A* **DOK 1**

CHAPTER 15 **REVIEW**

REVIEW KEY CONCEPTS

1. **Identify** Which of these ideas helped Darwin formulate his theory of evolution? Select all correct answers.
   - **A.** Individuals develop different traits during their lifetime.
   - **B.** DNA is a molecule that passes inherited traits from parents to offspring.
   - **C.** Fossils of ancient organisms share features with some modern animals.
   - **D.** Features of Earth's surface gradually change over very long periods of time.
   - **E.** Similar species living in different habitats have unique traits suited for where they live.
2. **Analyze** Which is an example of a change that has resulted from natural selection?
   - **A.** a racehorse that runs faster after training on a practice track for a year
   - **B.** a population of wildflowers that produces more nectar than previous generations
   - **C.** a flock of seagulls that searches for food scraps in human garbage left on the beach
   - **D.** a new variety of apple produced by crossing a sweet variety with a drought-resistant variety
3. **Contrast** In evolutionary biology, how is fitness related to adaptation? How are they different?
4. **Define** What does overproduction mean in the context of natural selection?
   - **A.** More individuals are produced than can be supported by the environment, so competition occurs.
   - **B.** More genes than necessary are produced in an organism's genome, so some genes are not expressed.
   - **C.** More mutations are produced than can be supported by a population, so some mutations will be less fit.
   - **D.** More resources than necessary are produced by the environment, so a population will grow more fit over time.
5. **Define** What is a gene pool?
   - **A.** all of an individual's genetic material
   - **B.** the total genetic material in an ecosystem at a given time
   - **C.** the total genetic material of all species on Earth at a given time
   - **D.** the total genetic material of all individuals in a population at a given time
6. **Classify** Peppered moths are commonly found in the British Isles. Most are either black in color, which provides camouflage on dark surfaces, or white in color, which provides camouflage on light surfaces. Gray moths also exist but are extremely rare. What could explain this?
   - **A.** artificial selection
   - **B.** directional selection
   - **C.** disruptive selection
   - **D.** stabilizing selection
7. **Identify** During a heavy rainstorm, several rabbits are carried away by rushing floodwaters and deposited several miles downriver from their original habitat. A handful of survivors establishes a new population there. Why might the new population have less genetic diversity than the original population?
   - **A.** bottleneck effect
   - **B.** founder effect
   - **C.** genetic drift
   - **D.** sexual selection
8. **Classify** A wildfire burns through part of a forest, killing most of the Eastern white pine trees (*Pinus strobus*) in the population. Fifty years later, the area has a large number of Eastern white pine trees, most less than fifty years old. Compared to other areas of the forest, a very large percentage of these trees has an allele that causes them to grow short needles. What evolutionary effect does this demonstrate? Justify your answer.
9. **Define** Which description best fits gene flow?
   - **A.** the transfer of alleles between two populations
   - **B.** the transfer of alleles from parents to offspring
   - **C.** the change in allele frequency when a population is greatly reduced in size
   - **D.** the change in allele frequency when a small population loses an allele randomly

CRITICAL THINKING

10. *Sample answer: Deuteranopia likely does not significantly reduce the fitness of affected individuals. If individuals with the gene for colorblindness are not disadvantaged in terms of surviving and reproducing, this trait will persist in the population at some level.* **DOK 3**
11. *Sample answer: It is probably not very often applied to natural populations. In general, populations do not satisfy the assumptions of Hardy-Weinberg equilibrium. Many species selectively choose mates, violating random mating. Genetic mutations do occur in most if not all organisms. Although many species have large populations, Hardy-Weinberg equilibrium best models an infinite population, which is impossible. It is natural for individuals to immigrate and emigrate from a population, which would violate another principle of Hardy-Weinberg Equilibrium. Finally, natural selection is always taking place because there is competition in the environment. In short, natural conditions virtually guarantee that there will always be some fluctuation in a population's allele frequency.* **DOK 3**
12. *Sample answer: Sexual reproduction increases genetic variation in progeny by recombination of genes on chromosomes and independent assortment of those chromosomes. This genetic variation drives the appearance of new traits in populations, increasing the likelihood that favorable adaptations will arise. Sexual reproduction increases the likelihood that natural selection will drive evolution in populations.* **DOK 2**
13. *Sample answer: Populations evolve over many generations, making it difficult for any one scientist to observe evolution in progress. Since the process of evolution works today in the same way it worked in the past, the fossil record allows scientists to examine evolution over millions of years simultaneously.* **DOK 2**

CRITICAL THINKING

10. **Analyze** Deuteranopia is a form of color blindness in humans that affects a small percentage of males. It can result from a single mutation on the X chromosome. Individuals who inherit this condition have trouble distinguishing red and green colors. Explain why natural selection may not have yet eliminated this trait from the human gene pool.

11. **Explain** The Hardy-Weinberg equilibrium model can be used to predict the frequency of alleles in a gene pool. Is this model applicable to most natural populations? Explain why or why not.

12. **Explain** The first organisms on Earth reproduced asexually. It was not until about two billion years ago that sexual reproduction began to evolve in prokaryotes. Today, nearly all eukaryotes reproduce sexually. How does sexual reproduction increase the reproductive fitness of organisms?

13. **NOS Scientific Knowledge** In Darwin's research, he examined many fossils to explain how organisms evolve into new species. Why is it necessary for evolutionary biologists to study the fossil record to learn about how species evolve?

MATH AND ENGLISH LANGUAGE ARTS CONNECTIONS

Darwin, in On the Origin of Species, *writes:*

The laws governing inheritance are for the most part unknown; no one can say why the same peculiarity in different individuals of the same species, or in different species, is sometimes inherited and sometimes not so; why the child often reverts in certain characteristics to its grandfather or grandmother or more remote ancestor; why a peculiarity is often transmitted from one sex to both sexes, or to one sex alone, more commonly but not exclusively to the like sex.

1. **Cite Textual Evidence** What clues suggest that Darwin did not have an understanding of modern genetics when he wrote this passage?

2. **Write an Argument** Human activity has resulted in the introduction of many species into new environments. In some cases, this results in invasive species that quickly colonize the new ecosystem and outcompete native species. Using what you know about evolutionary fitness, explain what factors would make a particular species likely to become invasive.

3. **Reason Abstractly** Suppose a herd of goats colonizes an area and starts eating the foliage. Before the goats arrived, shrub A and shrub B had approximately the same height distribution. However, because the goats cannot reach very high, shrubs with an allele that makes them tall prove to be more reproductively fit over time. In addition, the goats have a slight preference for shrub A over shrub B because its leaves taste better. Identify the type of selective pressure exerted by the goats on the shrubs and compare how the two populations of shrubs would likely evolve over a long period of time.

4. **Use Mathematics** A biologist surveys the alleles of a gene that affects fur color in a population. She finds four alleles at this locus in total. In her survey, she finds 74 copies of allele *A*, 115 copies of allele *B*, 65 copies of allele *C*, and 146 copies of allele *D*. What is the allele frequency of allele A? Express your answer as a percentage rounded to the nearest tenth of a percent.

REVISIT HUMMINGBIRD EVOLUTION

Gather Evidence In this chapter, you learned about changes in populations over time. In Section 15.1 you learned about natural selection as a mechanism for adaptive evolution. In Section 15.2 you learned about mutations as a mechanism of genetic variation. In Section 15.3 you learned about other mechanisms of evolution including genetic drift, gene flow, and sexual selection.

1. Hummingbirds have adapted to survive in very specific habitats. Describe how flowers may have also adapted over time to suit hummingbirds. How would those adaptations benefit the flowers?

2. What other questions do you have about adaptations in hummingbirds?

REVISIT THE ANCHORING PHENOMENON

1. *Sample answer: As hummingbird beaks change over time, flower shapes also change to match the beak shapes of specific hummingbird species. The hummingbirds gain exclusive access to the nectar of plants specifically suited to their beak shape, and the flowers benefit by having the hummingbirds transfer pollen from one flower to another, helping the flowers reproduce.* **DOK 3**

2. *Sample answer: How are a hummingbird's wings different from the wings of other birds? Given its energy requirements, are a hummingbird's cells different from those of other organisms? What special adaptations do migrating hummingbirds have compared to hummingbirds that do not migrate?* **DOK 3**

MATH AND ENGLISH LANGUAGE ARTS CONNECTIONS

1. *Sample answer: The passage states that many principles about inherited traits are unknown or cannot be explained. However, today we know that traits are coded by genes that are inherited from an individual's parents. Each individual inherits only half of the genes from any parent, which explains why some traits are sometimes inherited and sometimes not. Some genes are recessive, which explains why a child can revert to characteristics from his or her grandparents. Still other genes are sex-linked because they are located on the X or Y chromosome, which explains why some traits are more common in one sex but not the other. Darwin must not have known about these principles when he wrote the passage, which makes sense because they had not been discovered yet.* **DOK 3**

2. *Sample answer: To become invasive, a nonnative species would need to have adaptations that are well suited to the new environment, and perhaps better suited than adaptations possessed by native species. Having such adaptations makes the species more fit to survive, thrive, and reproduce than native species.* **DOK 3**

3. *Sample answer: The goats exert directional selective pressure on both shrubs because one characteristic (tallness) is favored over others. The mean heights of both Shrub A and Shrub B would increase over time. However, because the goats prefer the taste of Shrub A, they will exert a stronger selective pressure on Shrub A and its distribution would likely shift more rapidly toward the tall trait.* **DOK 3**

4. *18.5%* **DOK 2**

SURVIVAL IN CHANGING ENVIRONMENTS

Three-Dimensional Learning

The practices, core ideas, and crosscutting concepts presented in this chapter's text, investigations, and resources provide support to address the following Performance Expectations: **HS-LS2-7, HS-LS4-5, HS-LS4-6, HS-ETS1-2, HS-ETS1-3,** and **HS-ETS1-4.**

Science and Engineering Practices	Disciplinary Core Ideas	Crosscutting Concepts
Developing and Using Models Analyzing and Interpreting Data Constructing Explanations and Designing Solutions (HS-LS2-7, HS-ETS1-2, HS-ETS1-3) Engaging in Argument from Evidence (HS-LS4-5) Obtaining, Evaluating, and Communicating Information	**LS2.A:** Interdependent Relationships in Ecosystems **LS2.C:** Ecosystem Dynamics, Functioning, and Resilience (HS-LS2-7) **LS4.A:** Evidence of Common Ancestry and Diversity **LS4.B:** Natural Selection **LS4.C:** Adaptations (HS-LS4-5, HS-LS4-6) **LS4.D:** Biodiversity and Humans (HS-LS2-7, HS-LS4-6) **ETS1.A:** Defining and Delimiting Engineering Problems **ETS1.B:** Developing Possible Solutions (HS-LS2-7, HS-LS4-6, HS-ETS1-3, HS-ETS1-4) **ETS1.C:** Optimizing the Design Solution (HS-ETS1-2)	Cause and Effect (HS-LS4-5, HS-LS4-6, HS-ETS1-3) Scale, Proportion, and Quantity Systems and System Models (HS-ETS1-4) Science Addresses Questions About the Natural and Material World Influence of Engineering, Technology, and Science on Society and the Natural World

Contents		Instructional Support for All Learners	Digital Resources
ENGAGE			
484–485	CHAPTER OVERVIEW **CASE STUDY** Why do new species emerge while other species disappear?	**Social-Emotional Learning** CCC Cause and Effect **On the Map** Gorongosa National Park **English Language Learners** Making Inferences Using Text and Graphics	

Contents		Instructional Support for All Learners	Digital Resources
EXPLORE/EXPLAIN			
486–491 DCI LS4.C LS4.D	**16.1 SPECIATION** • Describe the mechanisms of speciation. • Explain how adaptive radiation can change a lineage. • Differentiate the possible patterns in evolution that can exist between species.	**English Language Learners** Requesting Assistance **Vocabulary Strategy** Graphic Organizers **On the Map** Panama Canal **Differentiated Instruction** Leveled Support **On the Map** Eurasian Rosefinch **Visual Support** Bird Bills CCC Cause and Effect **Connect to English Language Arts** Develop and Strengthen Writing **Address Misconceptions** Evolutionary Perfection CER Revisit the Anchoring Phenomenon **Visual Support** Graduated vs. Punctuated Equilibrium	*Video 11-1* **Investigation A** Modeling Speciation (80 minutes)
492–499 DCI LS2.C LS4.C LS4.D	**16.2 EXTINCTION** • Identify the different kinds of extinction that lead to the elimination of species. • Evaluate the protections provided to endangered and threatened species. • Differentiate the causes of species decline.	**Vocabulary Strategy** Knowledge Rating Scale SEP Using Mathematics and Computational thinking **Address Misconceptions** Mass Extinctions **On Assignment** Sudan the White Rhinoceros **Address Misconceptions** Species vs. Subspecies **Connect to English Language Arts** Write a Response **English Language Learners** Combining Ideas **CASE STUDY** Tuskless Elephants **In Your Community** Habitat Loss **English Language Learners** Environmental Print **Differentiated Instruction** Students with Disabilities	*Video 16-1*

Contents		Instructional Support for All Learners	Digital Resources
EXPLORE/EXPLAIN			
499	**EXPLORER** ÇAĞAN ŞEKERCIOĞLU	**Connect to Careers** Ornithologist **English Language Learners** Listening for Main Idea and Details	
500–509 DCI LS2.C LS4.D ESS2.D ESS3.A	**16.3 HUMAN IMPACT ON THE ENVIRONMENT** • Explain how human population growth impacts resource use. • Describe the effects of water pollution and air pollution caused by human activities. • Analyze how human activities lead to land degradation and habitat loss.	**Vocabulary Strategy** Using Prior Knowledge **Connect to Mathematics** Use Units to Understand Problems **Differentiated Instruction** Leveled Support **Visual Support** Pollution Sources **English Language Learners** Describing and Explaining with Details **Connect to Mathematics** Reason Quantitatively CCC Cause and Effect **Connect to English Language Arts** Research **In Your Community** Local Climate Change **English Language Learners** Research **Visual Support** Air Pollutants SEP Scientific Knowledge is Open to Revision in Light of New Evidence **Address Misconceptions** Decreasing Biodiversity **English Language Learners** Predicting SEP Engaging in Argument from Evidence **On the Map** Global Desertification	*Video 16-2*

Contents		Instructional Support for All Learners	Digital Resources
EXPLORE/EXPLAIN			
509	**LOOKING AT THE DATA** ECOLOGICAL FOOTPRINTS		
510–517 **DCI** LS4.D ESS3.A ESS3.C ESS3.D ETS1.A ETS1.C	**16.4 REDUCING HUMAN IMPACT ON THE ENVIRONMENT** • Describe how sustainability can help with the management of limited resources. • Analyze how air and water pollution can be decreased. • Evaluate design solutions to manage ecosystem degradation and habitat loss. • Explain why biodiversity should be maintained.	**Vocabulary Strategy** Semantic Map **Differentiated Instruction** Economically Disadvantaged Students **In Your Community** Solar Panels **Connect to Mathematics** Reason Abstractly **Connect to English Language Arts** Assess Reasoning and Evidence **CCC** Cause and Effect **English Language Learners** Using Context Clues **CONNECTIONS** Using Technology **Crosscurricular Connections** Earth Science	**Investigation B** Wildlife Crossings and Corridors (100 minutes) *Video 16-3* **SIMULATION** Saving the Reefs
516	**MINILAB** MODELING HUMAN-CAUSED CHANGES IN THE ENVIRONMENT		
ELABORATE			
517	**TYING IT ALL TOGETHER** Why do new species emerge while other species disappear?	**SEP** Constructing Explanations and Designing Solutions	Self-Reflection and Scoring Rubrics
Online	**Investigation A** Modeling Speciation **Investigation B** Wildlife Crossings and Corridors	Guided Inquiry (50 Min) Engineering Design (100 Min)	MINDTAP Access all your online assessment and laboratory support on MindTap, including: sample answers, lab guides, rubrics, and worksheets. PDFs are available from the Course Companion Site
EVALUATE			
518–519	**Chapter 16 Review**		
Online	**Chapter 16 Assessment** **Performance Task 3** *How is climate change altering species evolution?* (HS-LS4-3, HS-LS4-4) **Performance Task 4** *How do human-induced changes in the environment affect different species?* (HS-LS4-5) **Performance Task 5** *What kind of artificial reef is most effective at preserving and restoring biodiversity? (HS-LS4-6)*		

CHAPTER 16

ENGAGE

In middle school, students learned about natural selection and how it leads to the increase and decrease of different traits in a population. Students have also learned how natural selection over many generations leads to the evolution of a population. This chapter expands on that prior knowledge by explaining how changing environments and human actions affect selective pressures on organisms.

Students will be introduced to speciation and how populations change when they are reproductively isolated. Students will also examine patterns in extinction.

About the Photo This image shows a group of female tuskless African elephants. Tusklessness is caused by an X-linked mutated gene and is lethal in males. Female tuskless elephants with one copy of the gene survive. To prepare students for this chapter, ask them to use scientific terminology to describe how a trait can be selected for even when detrimental to some individuals in a population.

Social-Emotional Learning

As students progress through the chapter, invite them to look for contexts and opportunities to practice some of the five social and emotional competencies. For example, as students work through content in the chapter, support them in their **relationship skills**. Help them practice effective communication and to seek and offer support and help when needed.

In addition, as students learn about human impacts on the environment, ("Human Impact on the Environment" and "Reducing Human Impact on the Environment"), remind them to practice **self-management** and **responsible decision-making**. Support students in setting personal goals and identifying solutions for personal and social problems.

CHAPTER 16

SURVIVAL IN CHANGING ENVIRONMENTS

This herd of savanna elephants, led by a tuskless matriarch, lives in Addo Elephant National Park in Eastern Cape, South Africa.

16.1 SPECIATION
16.2 EXTINCTION
16.3 HUMAN IMPACT ON THE ENVIRONMENT
16.4 REDUCING HUMAN IMPACT ON THE ENVIRONMENT

Although all African elephants were once grouped together as a single species, DNA analysis has shown that forest elephants (*Loxodonta cyclotis*) and savanna elephants (*Loxodonta africana*) are two different species that diverged millions of years ago. Forest elephants are smaller in size and live in smaller family groups than elephants that live in the grassy, open savanna.

Chapter 16 supports the NGSS Performance Expectations **HS-LS4-5**, **HS-LS4-6**, **HS-LS2-7**, and **HS-ETS1-4**.

CROSSCUTTING CONCEPTS | Cause and Effect

Cause and Effect in Ecosystems This chapter focuses on the cause-and-effect relationships in ecosystems and how natural selection results in changes in a population. It examines the natural cause-and-effect relationships that occur within an ecosystem, as well as cause-and-effect relationships that are a direct result of human interactions with an ecosystem. Some fields of biology, such as ecology, conservation biology, evolutionary biology, and environmental biology, study how these cause-and-effect relationships affect organisms at a variety of system levels. Chapter 10 of Unit 3 reviews the cause-and-effect relationships that exist within an individual. Chapters 11 and 12 take a deeper dive into the cause-and-effect relationships that exist on the genetic and phenotypic levels of an individual. You can further reinforce this crosscutting concept in this lesson at a higher level by having students organize information about how environmental factors can affect the genetic information that exists in a population.

CASE STUDY
TRACKING TUSKLESS ELEPHANTS

WHY DO NEW SPECIES EMERGE WHILE OTHER SPECIES DISAPPEAR?

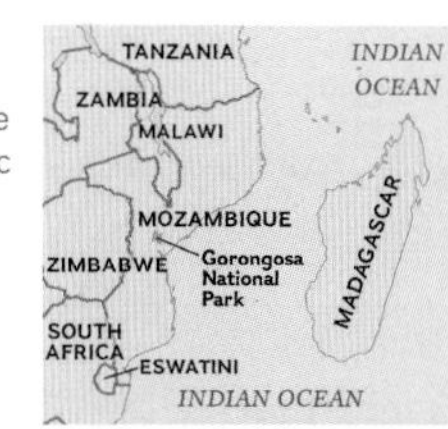

Along with a trunk and large floppy ears, an elephant's tusks are one of its most characteristic features. Elephants use their tusks for a variety of essential tasks, including digging for water, removing edible bark from trees, and self defense.

The ivory in elephant tusks has long been prized for its appearance and durability, and it is easily carved and polished into practical and ornamental objects. In some traditions, ivory powder is believed to have healing properties. Poachers seeking to profit from selling ivory illegally hunt and kill elephants to remove their tusks. Older, larger elephants are preferred for their longer, heavier tusks. Despite long-standing international regulations banning the trade of ivory, poaching continues to endanger many elephant populations.

A 2014–2015 aerial census of savanna elephants in 18 African countries found a 30 percent decline in the region's elephant population. In Mozambique alone, the population declined by 53 percent in just five years.

Poaching not only decreases the size of a population. It can also influence the animals' traits. Although both male and female African elephants typically have tusks, between 2 and 4 percent of female elephants are naturally tuskless. In Mozambique's Gorongosa National Park, scientists have observed a change in this trait over several generations. Using historical film footage, they estimated that prior to Mozambique's civil war, which began in 1977, 18 percent of the park's female elephants had no tusks. Over 15 years of war, the conflicting sides turned to poaching to finance their efforts. Ultimately, they killed 90 percent of the park's elephants. Counts taken in 2018 determined that about half of the female elephants that survived the war were tuskless, and 33 percent of the survivors' daughters were also tuskless (**Figure 16-1**). These findings suggested that not having tusks increased the likelihood that an elephant would survive the war. Genetic analysis indicates that one of the genes for tusklessness is linked to the X chromosome.

Preliminary data indicate that elephants without tusks eat different types of plants than tusked elephants. Because elephant herbivory shapes the landscape of the savanna, these differences may have further implications for how African elephant populations evolve in a changing environment.

Ask Questions *In this chapter, you will learn how new species emerge and why species become extinct. As you read, generate questions about how human activity can influence a population's genetic makeup.*

FIGURE 16-1
The poaching of tusked elephants in Mozambique during the country's civil war led to an increase in the number of tuskless females in the population following the war's end.

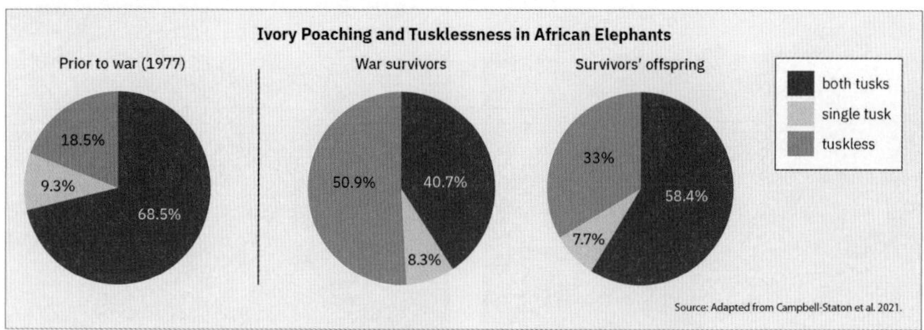

DIFFERENTIATED INSTRUCTION | English Language Learners

Making Inferences Using Text and Graphics Have students read the Case Study in pairs or small groups of mixed language proficiency. Then, have them use the text and circle graphs to make an inference to answer this question: *Are tusked elephant populations changing because of human activities?*

Beginning Have students make an inference by answering yes or no and support it with details from the graphs and the text. Provide sentence frames for describing the graphs, for example: *Before the war, _______ percent of elephants had _______.* Guide students to underline details in the text that support their inference.

Intermediate Have students support their inference with two or three details paraphrased from the text and at least two pieces of information from the graphs.

Advanced Have students explain their inference in detail and support it with details from the text as well as information from the graphs.

CASE STUDY

ENGAGE

The situation described in the Case Study can be used to assess students' prior knowledge of how human interactions affect traits in a population. Students have learned about artificial selection, where humans intentionally influence the characteristics of organisms through selective breeding. Students may be unaware of the consequences in populations caused by other human actions. Have students consider why elephants without tusks would have different diets than elephants with tusks.

The changing environments found in the elephants' habitats provide another context for students to consider as they read the chapter. In the Tying It All Together, students will use a model to develop an explanation of how tusklessness can lead to speciation or extinction.

Ask Questions Students should revisit the Case Study as they read the chapter to make connections with the content. See a specific suggestion in Section 16.2.

On the Map

Gorongosa National Park Use the map of Southeast Africa to point out Mozambique and Gorongosa National Park, located at the southern end of the African Rift Valley. The large 4000-square-kilometer preserve provides a variety of environments that support many organisms in forest and savannah ecosystems. Lead a discussion about how preserving land can support species conservation.

Human Connection A long period of war in Mozambique began in 1962 and lasted through 1992. In 1975, the nation gained independence and experienced a brief peace before falling into a civil war in 1977. This war lasted an additional 15 years. During the wars, the high cost of ivory led to the slaughter of tusked elephants to raise funds for the conflict. In addition to poaching, African elephants are also in danger of extinction due to increasingly fragmented habitats and land degradation due to human activities.

16.1

SPECIATION

LS4.C, LS4.D

EXPLORE/EXPLAIN

This section provides an introduction to the mechanisms of speciation, a discussion on adaptive radiation, and an overview of patterns in evolution that can lead to changes in a population.

Objectives

- Describe the mechanisms of speciation.
- Explain how adaptive radiation can change a lineage.
- Differentiate the possible patterns in evolution that can exist between species.

Pretest Use **Questions 2, 3, 5, 6,** and **8** to identify gaps in background knowledge or misconceptions.

Identify *Students might say that species may differentiate if they are unable to successfully reproduce together or if resource shortages cause a population to separate.*

CHAPTER INVESTIGATION A

Guided Inquiry *Modeling Speciation*

Time: 50 minutes

Students will follow a step-by-step procedure to investigate how changes in environmental conditions may result in an increase in the number of individuals in a species, cause the emergence of new species, and cause the extinction of species.

Go online to access detailed teacher notes, answers, rubrics, and lab worksheets.

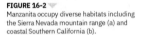

16.1

KEY TERMS

adaptive radiation
coevolution
reproductive isolation
speciation

SPECIATION

Earth is inhabited by millions of species that evolved over long stretches of time. All of the species alive today have not always existed. In fact, the vast majority of species that have ever existed on Earth are now extinct. Even species such as *Homo sapiens* (modern humans) are evolved versions of ancestor species. Evolutionary processes lead to adaptations within species and sometimes result in new species. The driving force behind evolutionary change is survival to reproduce.

Identify *What factors could lead to a species diversifying into a new species?*

CHAPTER INVESTIGATION

Modeling Speciation
What effect does environmental change have on speciation?
Go online to explore this chapter's hands-on investigation about speciation.

Mechanisms of Speciation

When two populations of a species do not interbreed, the number of genetic differences between them increases because mutation, natural selection, and genetic drift occur independently in each population. Over time, the populations may become so different that we classify them as different species. The emergence of a new species is an evolutionary process called **speciation**.

Reproductive Isolation Every time speciation happens, it occurs in a unique way. This means that each species is a product of its own unique evolutionary history. However, there are recurring patterns. **Reproductive isolation** is the end of gene flow between populations and occurs when members of those populations can no longer sexually reproduce. Several mechanisms of reproductive isolation prevent successful interbreeding, and, thus, reinforce differences between diverging populations.

- **Ecological Isolation** Adaptation to different environmental conditions may prevent closely related species from interbreeding. For example, there are over 90 species of manzanita (**Figure 16-2**). One species that lives on dry, rocky hillsides is better adapted for conserving water. Another species, which requires more water, lives in coastal regions where water stress is not as intense. The physical separation makes interbreeding unlikely.

FIGURE 16-2
Manzanita occupy diverse habitats including the Sierra Nevada mountain range (a) and coastal Southern California (b).

DIFFERENTIATED INSTRUCTION | English Language Learners

Requesting Assistance Encourage students to ask for help to understand vocabulary or concepts as they read each section of text.

Beginning Have students mark words they do not understand and ask: *What does ______ mean?* Allow students to discuss the text in their native language if needed. Encourage them to ask for the corresponding English words: *How do you say ______ in English?*

Intermediate Have students mark words they do not understand and ask: *What does ______ mean?* After answering, prompt students to restate the definition in their own words or point to an example in the images on the page.

Advanced Have students work in pairs to ask and answer questions with *How?* and *Why?* For example, they could ask: *How are ______ and ______ different? How does ______ work? Why does ______ cause species to ______?* Have them use details in the text to answer.

- **Temporal Isolation** Some closely related species cannot interbreed because the timing of their reproductive behaviors differ. One example is the periodical cicada. Three cicada species reproduce every 17 years. Each has a sibling species with nearly identical form and behavior that emerges on a 13-year cycle instead of a 17-year cycle. Sibling species have the potential to interbreed, but they can only get together once every 221 years.
- **Behavioral Isolation** Differences in behavior can prevent mating between related animal species. For example, many animal species engage in courtship displays before mating. In a typical pattern, the female recognizes the sounds and movements of a male of her species as the start of mating, but females of different species do not.
- **Mechanical Isolation** The size or shape of an individual's reproductive parts can prevent it from mating with members of related species. For example, closely related species of sage plants grow in the same areas, but their flowers are specialized for different animal pollinators, so cross-pollination rarely occurs.

Allopatric Speciation Genetic changes that lead to a new species can begin with ecological or physical separation between populations. With allopatric speciation, a physical barrier arises and separates two populations, ending gene flow between them. Whether a geographic barrier can block that gene flow depends on how the species moves, such as by swimming, walking, or flying. It also depends on how it reproduces, such as internal fertilization or pollen dispersal.

A geographic barrier can arise over millions of years, such as when mountains form, or it can happen much quicker. The Great Wall of China (**Figure 16-3**) is an example of a barrier that arose relatively abruptly. DNA sequence comparisons show that trees, shrubs, and herbs on either side of the wall are diverging genetically.

FIGURE 16-3
The longest stretch of the Great Wall that still exists today cuts across 8850 km (5500 miles) of land. It was built during the Ming dynasty (1368–1644).

Vocabulary Strategy

Graphic Organizers Encourage students to make a graphic organizer for each section of text to show how the words are related. For example, they can make a chart with each type of reproductive isolation, its definition, and an example. As another option, students could make a chart listing each term and then its causes and effects.

Science Background

The Great Wall of China A paper published in 2003 discussed a study of a plant species that was separated by the Great Wall of China about 600 years ago. Researchers compared the subpopulations on either side of the wall with a control group and found that the subpopulations had a much higher level of variance than expected. This study and others have changed the once-prevalent idea that the process of speciation occurs over a long time. Share with students that scientists have found that wind-pollinated plant species showed less variance than animal-pollinated plant species and ask them to discuss why this might be the case.

On the Map

Panama Canal The Panama Canal was completed in 1914. It is a 65-kilometer (40-mile) artificial waterway that connects the Caribbean Sea and the Pacific Ocean to shorten nautical journeys between the eastern and western coasts of the United States. Prior to its construction, ships sailing between the eastern and western coasts of the Americas would travel around the southern tip of South America. Have students discuss how the construction of the canal affected society in the United States and elsewhere. Then, have students identify possible ways the separation of land by the canal may have affected different species. Guide a discussion about the impacts of poplations being separated by the canal and the new opportunity for species once separated to then come together via the canal.

SEP Communicate Information *Sample answer: A change in environmental conditions may result in some members of a population not migrating to their typical breeding grounds.*

FIGURE 16-4 The Isthmus of Panama forms a geographic barrier that prevents species of snapping shrimp from intermingling. Over time, this reproductive isolation has led to the development of separate species in the Pacific and Atlantic Oceans.

Natural geographic barriers usually arise much more slowly. For example, it took millions of years of tectonic plate movements to bring the two continents of North and South America close enough to collide. The land bridge where the two continents now connect is called the Isthmus of Panama. When this isthmus formed about three million years ago, it cut off the flow of water and gene flow among populations of aquatic organisms. It separated one large ocean into what are now the Pacific and Atlantic Oceans (**Figure 16-4**).

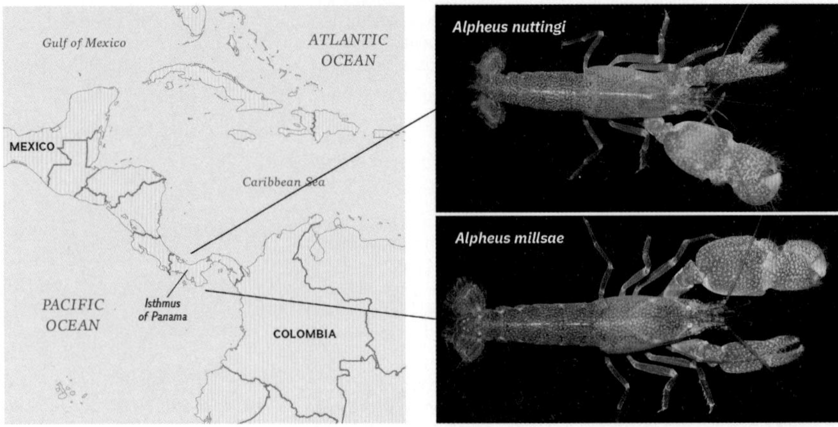

This geographic isolation led to the speciation of snapping shrimp. The species on either side of the Isthmus of Panama appear to be very similar to one another. However, without the ability to cross the geographic barrier, the species have become genetically distinct. For example, *Alpheus nuttingi* are found in the western Atlantic Ocean while *Alpheus millsae* inhabit the eastern Pacific Ocean. Instead of mating when they are brought together, male and female shrimp snap their claws at one another aggressively. Because they are unable to successfully mate together, the snapping shrimp are considered to be separate species.

Sympatric Speciation In this type of speciation, a new species arises within the same range as its ancestral population. Sympatric speciation is common only in plants. During meiosis, the chromosomes of some plants fail to separate and incorrectly produce diploid (2*n*) gametes. This condition, called polyploidy, refers to the possession of more than the proper complement of chromosomes. Polyploidy is a major factor in plant evolution. Polyploid gametes can develop into mature plants, even though they have double the number of chromosomes that a gamete should have. These individuals can self-fertilize and reproduce with themselves and other polyploid individuals. But they cannot sexually reproduce with any plants from the original parent generation because the two organisms now have different numbers of chromosomes.

SEP Communicate Information *Describe a possible scenario in which two populations of the same species might become reproductively isolated from one another.*

DIFFERENTIATED INSTRUCTION | Leveled Support

Struggling Students Clarify for students that allopatric speciation occurs in populations that have been separated, but sympatric speciation occurs within a species in the same area due to reproductive isolation. To help students understand the differences between the two, guide them to use a graphic organizer where they compare and contrast the requirements and results of the two types of speciation. Help students understand that sympatric speciation can occur through different processes and that polyploidy is the most common type of sympatric speciation.

Advanced Learners While polyploid cells are rare in animals, they are found in a few species of amphibians and fish. Have students consider how a failure to divide into haploid cells would affect zygotes. Guide students to realize that it could affect the relative volume of DNA in comparison to the rest of the cell. Have students discuss how the increased size and surface-to-volume ratio of polyploid cells might affect homeostasis and feedback mechanisms.

Adaptive Radiation

With **adaptive radiation**, one lineage rapidly diversifies into several new species. Adaptive radiation can occur after a population colonizes a new environment that has a variety of different habitats and niches and few competitors. Speciation occurs as adaptations to the different habitats evolve. Hawaiian honeycreepers arose by adaptive radiation (**Figure 16-5**).

FIGURE 16-5
The bills of these bird species are adapted to feed on insects, seeds, fruits, sap, burrowing insects, and nectar in floral cups. All Hawaiian honeycreepers descended from a Eurasian rosefinch that probably resembled modern rosefinches.

SEP **Argue From Evidence** *How do the Hawaiian honeycreepers shown in **Figure 16-5** provide evidence for adaptive radiation?*

A geologic or climatic event that eliminates some species from a habitat can spur adaptive radiation. Species that survive the event then have access to newly opened niches and resources from which they had previously been excluded. This is the way mammals were able to undergo an adaptive radiation after the dinosaurs were eliminated 66 million years ago. Adaptive radiation may also occur after a key innovation evolves. A key innovation is an adaptive trait that allows an organism to exploit a habitat more efficiently or in a new way. The evolution of lungs offers an example, because lungs were a key innovation that opened the way for an adaptive radiation of vertebrates that could live on land.

Patterns in Evolution

Evolution happens over a long time period. Environmental changes influence evolutionary processes, and organisms can sometimes adapt to such changes to survive and reproduce. Sometimes environmental changes, such as a physical barrier, result in the emergence of a new species. In other rare cases, two or more species competing for resources in shared habitats develop an evolutionary connection that leads them to independent adaptions in response to each other.

On the Map

Eurasian Rosefinch The Eurasian rosefinch, also known as the common rosefinch, has a distribution that ranges across Europe and Asia. During their breeding season, they are generally found in the northern regions of Europe and Asia, but they migrate south to regions in Southeast Asia, India, and areas of North America for the winter. Have students consider how the rosefinch ancestor ended up in the Hawaiian archipelago.

Visual Support

Bird Bills There are thought to have been at least 56 different species of Hawaiian honeycreepers. Now, all but 18 species are extinct due to the arrival of humans. The bottom area of **Figure 16-5** shows five different honeycreepers that have very distinct bills. Students should compare each of the bills of the Hawaiian honeycreepers to the Eurasian rosefinch shown at the top of the figure. Tell students that the rosefinch feeds on seeds, berries, nectar, shoots, fruits, insects, and larvae. Have students identify how the different bill shapes would have enabled the different species to exploit the different food sources available in Hawaii.

SEP **Argue From Evidence** *Sample answer: The different species of honeycreepers have different bill shapes, which are related to the types of foods they eat.*

CROSSCUTTING CONCEPTS | Cause and Effect

Natural Selection Understanding the cause-and-effect relationships in natural selection is essential in understanding how new species can arise from a common ancestor. Have students consider how a natural disaster can lead to both sympatric speciation and adaptive radiation through different mechanisms. Have students discuss in pairs or small groups the causes of these types of diversification and describe the effects they can have on individual species. Emphasize to students that evolution is dependent on natural selection as well as the other interactions that exist within ecosystems. Physical barriers that arise over time, the migration of organisms, and natural disasters, such as wildfires and volcanic eruptions that rapidly change the landscape, are events that change the dynamics within an environment and the organisms that live there. Students may also discuss how a natural disaster that is common in your area might lead to sympatric speciation.

Connect to English Language Arts

Develop and Strengthen Writing Have students choose one of the patterns of evolution in this section and write a paragraph explaining what it is and describing an example of it. Once students have written their paragraphs, have partners trade paragraphs and review each other's writing. Students should suggest ways their partner could clarify or reorganize ideas, as well as mark proofreading corrections to spelling and grammar. Then, have students each revise their own work and make a final copy of their paragraph.

SEP Argue From Evidence *Coevolution is only an example of mutualism if both species benefit from the relationship.*

Address Misconceptions

Evolutionary Perfection A common misconception around speciation and natural selection is that there is an evolutionary goal that will lead to a perfect species, forgetting that as the species changes, so does the environment. The concepts of coevolution and the evolutionary arms race can be used to rectify this misconception. Have students consider the snails and the crabs described in the text. As the snails develop stronger shells and the crabs develop sharper teeth and stronger claws over time, eventually the two populations may hit a point where the defensive and offensive strategies negatively affect individuals' ability to survive and reproduce.

FIGURE 16-6 The Darwin's star orchid has a 30-cm-long nectar tube called a spur (a). It can be accessed only by the especially long proboscis of a Wallace's sphinx moth (b).

HUMMINGBIRD EVOLUTION

Gather Evidence *Explain the role of coevolution in the diversification of hummingbird species.*

Coevolution In **coevolution**, two species that have close ecological interactions evolve in response to changes in each other. One species acts as an agent of selection on the other, and each adapts to changes in the other. Over evolutionary time, the two species may become so interdependent that they can no longer survive without the other.

One example of coevolution is the development in some insects of an especially long proboscis, a long narrow mouthpart used to sip nectar from a flower (**Figure 16-6**). The plant has likewise developed an adaptation called a floral nectar spur, which holds nectar deep within the flower.

The development of these complementary structures is beneficial to both organisms. The plant benefits by pollinator specialization, maintaining a narrow range of pollinators that can access the nectar deep within the flower. Animals must either be small enough to enter the spur or have a sufficiently long and narrow mouthpart to sip nectar from the spur. Charles Darwin observed this phenomenon in Madagascar orchids during his voyage. He correctly predicted the existence of a moth with a sufficiently long proboscis to pollinate the flower.

SEP Argue From Evidence *When is coevolution an example of a mutualistic symbiotic relationship?*

Evolutionary Arms Race In some cases, coevolution occurs between two species that compete with one another, particularly those that are in predator-prey relationships. For example, the thick shells of modern snails were an evolutionary response to predation by crabs. The crabs, in turn, have evolved to have sharper teeth and more powerful claws to crack the shells of the snails' thicker shells.

Punctuated Equilibrium How fast does evolution occur? The traditional model is one of gradualism, or incremental change, in which change takes place over millions rather than thousands of years. Gradualism presumes that favorable mutations take a long time to develop. Thus, the sequence in which, for example, giraffes' necks were to elongate might take dozens of mutations that occur only after millions of years.

REVISIT THE ANCHORING PHENOMENON

Gather Evidence *Sample answer: Many hummingbirds have special adaptations, such as beak length or shape, that allow them to feed from only certain plants. These hummingbird species have coevolved with the plants and in turn aid plants in reproduction through the transfer of pollen.*

Scientists do not have much information on ancestral species of hummingbirds because of their small size and fragile bones. The oldest fossils are found in Europe and are about 35 million years old. The sparse fossil record suggests hummingbirds reached South America 22 million years ago and then diversified and moved north. Much like the case with the star orchid and the Wallace's sphinx moth, the birds with bills best adapted to the flowers they fed on would have been most successful, and the flowers that were better pollinated by these birds would have evolved with the birds.

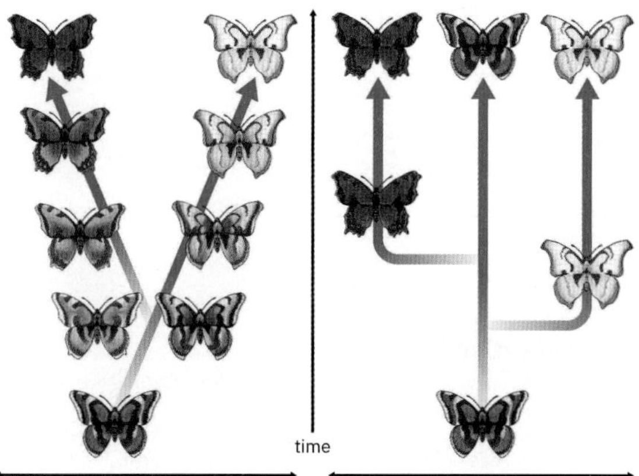

FIGURE 16-7 Gradualism assumes a large number of intermediate forms in the development of a species. With punctuated equilibrium, the change occurs rapidly with only a minimal number of intermediate forms.

The *punctuated equilibrium* model of evolution proposes that change is in fact quite rapid, occurring over a short period of time (a few thousand rather than millions of years) as ecological niches open up. Once these spurts of change occur, they are followed by much longer periods of stasis, or no change.

In fact, there is evidence that both models (**Figure 16-7**) might be correct for different time periods in the past. In some cases, noticeable change occurs only after millions of years. In others, the changes might have taken place in a much shorter period of time followed by a period of no change.

16.1 REVIEW

1. **Classify** Sort the examples into the type of reproductive isolation each represents.
   - Left-coiling snails cannot mate with right-coiling snails.
   - A male *Maratus nemo* spider conducts a ritual dance before mating and being killed by his mate.
   - *Rana boylii* mate during spring and summer while the co-existing *Rana draytonii* mate from winter into spring.
   - The common blackbird lives and breeds in forests while the closely related ring ouzel prefer to breed in open land.
2. **Explain** How can natural selection lead to speciation?
3. **Identify** Which factors are required for allopatric speciation in which a population becomes separated? Select all correct answers.
   A. migration
   B. natural selection
   C. a physical barrier
   D. genetic mutations
   E. development of adaptations
4. **Distinguish** How are allopatric speciation and adaptive radiation similar and different?

Visual Support

Graduated vs. Punctuated Equilibrium Students may wonder how both gradualism and punctuated equilibrium can work together. Guide students in a discussion about how the speciation events they have learned about, such as allopatric speciation, can lead to rapid change in the variation that exists within a population. Share that one method that could lead to a mixture of gradualism and punctuated equilibrium is a temporary period of isolation for a part of the population that is eventually reversed. Have students examine **Figure 16-7**. When a population is large, new mutations and traits would spread through the population slowly, so gradualism or even stasis would occur. However, if a portion of the population was isolated, the new mutations and traits would spread through the isolated population more rapidly, leading to a quicker rate of evolution. If the isolation ended before full differentiation occurred, new mutations would flow between the two previously separated populations and new traits could become more prevalent.

16.1 REVIEW

1. *behavioral isolation: A male Maratus nemo spider conducts a ritual dance before mating and being killed by his mate. ecological isolation: The common blackbird lives and breeds in forests, while the closely related ring ouzel prefers to breed in open land. mechanical isolation: Left-coiling snails cannot mate with right-coiling snails. temporal isolation: Rana boylii mate during spring and summer, while the co-existing Rana draytonii mate from winter into spring.* **DOK 2**
2. *Sample answer: Natural selection is a process that increases the frequency of an advantageous trait in a population over time. This can result in organisms that are more likely to survive and reproduce and may give rise to a new and distinctly different species.* **DOK 2**
3. *A, C, E* **DOK 1**
4. *Sample answer: Both mechanisms lead to new species, but allopatric speciation requires physical separation, while adaptive radiation can occur within an environment that has a variety of habitats and conditions.* **DOK 2**

16.2

EXTINCTION

LS2.C, LS4.C, LS4.D

EXPLORE/EXPLAIN

This section introduces the types of extinction, discusses endangered and threatened species, and explains different reasons that lead to the decline of species.

Objectives

- Identify the different kinds of extinction that lead to the elimination of species.
- Evaluate the protections provided to endangered and threatened species.
- Differentiate the causes of species decline.

Pretest Use **Question 4** to identify gaps in background knowledge or misconceptions.

Explain *Students might say that if a species goes extinct, it could result in the collapse of a food web if there are not any species in the same trophic level that could replace it.*

Vocabulary Strategy

Knowledge Rating Scale Have students make a table of the key terms with four columns labeled 1, 2, 3, and 4. Before students begin reading, they should rate their knowledge of each term:

1 = I have never heard the word.
2 = I have heard the word, but do not know its meaning.
3 = I am familiar with the word.
4 = I know the word, its meaning, and could explain it to others.

Upon completing the reading, students should revisit the scale and rerate the terms. At the end, have students discuss the meaning of each word with a partner to be sure they understand it.

SEP Construct an Explanation *The extinction of some species opens up niches for other species to occupy.*

16.2

KEY TERMS

background extinction
endangered species
extinction
habitat fragmentation
mass extinction
non-native species
threatened species

EXTINCTION

The passenger pigeon population once numbered in the billions across the United States and Canada. However, by the early 1900s, the bird had disappeared from the wild as a result of unrestricted hunting. The last known passenger pigeon died in captivity at the Cincinnati Zoo in 1914.

Explain *What could be an ecological consequence of a species going extinct?*

Types of Extinction

Extinction occurs when a species is eliminated from Earth. It is a critically important concept for evolution. While extinction marks the end of the existence of a species, it also presents opportunity for existing species to fill niches that become vacant. Extinction stimulates diversification into new species.

A low, steady level of extinction has continuously taken place throughout Earth's history. This type of extinction is called **background extinction**. Due to random events such as earthquakes, floods, or the introduction of new predators or competitors, some species simply do not survive under new environmental conditions and become extinct.

By current estimates, more than 99 percent of all species that ever lived are now extinct. In addition to continuing extinctions of individual species, the fossil record indicates that there have been more than five **mass extinctions** during which entire lineages were wiped out. These include five catastrophic events in which the majority of species on Earth went extinct (**Figure 16-8**). In most cases, these mass extinctions probably were caused by a massive environmental catastrophe. In other cases, there might be biological causes of extinctions, such as the introduction of a destructive new species. Humans, in fact, are a species that already has disrupted Earth's ecosystems and has been responsible for many thousands of extinctions. In the past, each period of mass extinction has been followed by a period of increasing diversity, in which thousands of new species have developed to fill newly emptied niches.

SEP Construct an Explanation *Why is mass extinction correlated with increased diversity?*

FIGURE 16-8 Geologic fossil evidence indicates that five major extinctions (indicated by the vertical bars) have occurred on Earth. They caused great loss but also greater biodiversity.

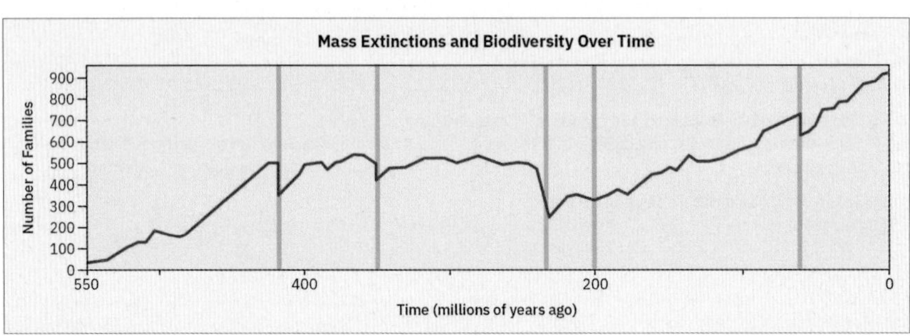

SCIENCE AND ENGINEERING PRACTICES
Using Mathematics and Computational Thinking

Mass Extinctions On the board, write out the following mass extinction events:

- Ordovician-Silurian extinction/444 million years ago: 85% of species lost
- Late Devonian extinction/383 million years ago: 75% of species lost
- Permian-Triassic extinction/252 million years ago: 96% of species lost
- Triassic-Jurassic extinction/201 million years ago: 80% of species lost
- Cretaceous-Paleogene extinction/66 million years ago: 76% of species lost

Have students analyze **Figure 16-8** with a partner and compare the data on the graph to the data on the board. Have students calculate the percent decrease after each extinction event using the data on the graph and ask why the numbers don't align. Students should recognize that the units in the graph are the numbers of families that declined, while the list identified the numbers of species that declined.

Scientists estimate that the current extinction rate is about 1000 times that of the typical background extinction rate. Unlike most previous mass extinctions, which resulted from a massive catastrophe such as an asteroid impact, humans are driving the current rise in extinctions. Our actions will determine the extent of the losses.

A species is considered extinct if repeated, extensive surveys of its known range fail to turn up signs of any individuals. A species is "extinct in the wild" if the only known members of the species are in captivity. Some species are difficult to find, so occasionally a population of species previously thought to be extinct or extinct in the wild is rediscovered. However, this does not happen very often.

Endangered and Threatened Species

An **endangered species** is one in which its population has dwindled to the point where it is in imminent danger of extinction. Species whose populations are extremely low and are of concern for possible extinction in the future are called **threatened species**.

The International Union for Conservation of Nature (IUCN) manages a Red List of Threatened Species. This global list was first established in 1964. Listed species are sorted into three categories: critically endangered, endangered, and vulnerable. Several criteria are assessed to determine whether a species should be added to the list, such as the current species population and whether the species has a restricted geographic range.

In many ways, Earth's living species are its most valuable resources. Consider, for example, the more than 120 known medically effective substances derived from plants. Examples include morphine and other pain relievers from poppy plants, colchicine from the meadow saffron that treats gout and some types of cancer (**Figure 16-9**), and quinine from a tropical tree bark that treats malaria.

Scientists estimate that as many as one-third of all plant families might be lost by the end of the 21st century, in most cases because of human activity. With perhaps millions of plant species still undiscovered, who knows how many more critically important medicines and foods will be lost forever from the world because of extinction. Additionally, countless animals are also likely to disappear.

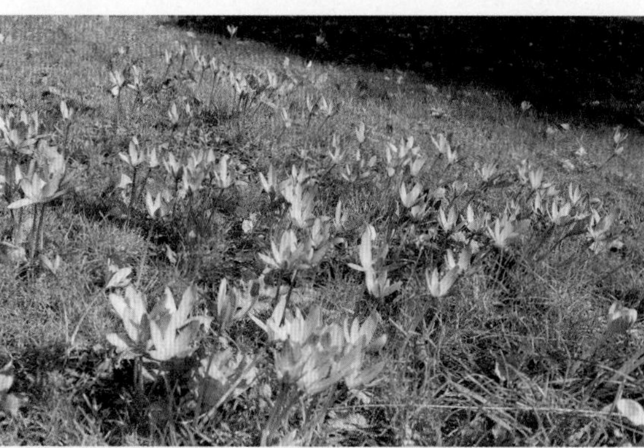

FIGURE 16-9 Colchicine, a compound derived from flowers of the meadow saffron, has potent anti-cancer properties.

Address Misconceptions

Mass Extinctions A common misconception is that mass extinctions are caused by sudden and catastrophic events, such as an asteroid striking the planet. To help students rectify this misconception, it may be useful to discuss how the interacting spheres of Earth can amplify events that lead to lasting changes that decrease biodiversity over time. Walk students through how a large volcanic eruption can lead to a mass extinction event. While many organisms in nearby ecosystems will die from the released gases and magma, other organisms in more distant ecosystems will die when gases and ashes spread out and change atmospheric conditions.

Science Background

Passenger Pigeons It is thought that at one time, passenger pigeons were the most populous species of bird in the world. Scientists estimate that the population peaked between 3 and 5 billion individuals and then rapidly declined until the species went extinct in 1914. Prior to the arrival of Europeans in North America, scientists think the genetic diversity declined within the population because beneficial mutations were so successful, which led to the decline or disappearance of other variants. While the species may have been more vulnerable to change due to lack of genetic diversity, it is human actions that ultimately led to their extinction. Humans hunted passenger pigeons for food and to protect their crops, and the invention of the train and telegraphs made it easier for hunters to track, trap, and shoot flocks. Human actions also destroyed the nesting grounds of passenger pigeons. Conservation efforts were initiated in 1900, but it was too late to save the species. Have students consider how species decline has been caused by human actions other than hunting.

ON ASSIGNMENT

Sudan the White Rhinoceros

National Geographic photographer Ami Vitale is known for her ability to immerse herself in the stories that she covers. The goal for this project was to shift away from the stories of war and conflict and to focus on the bond between humans and the natural world. She first met Sudan in 2009 in the Czech Republic where he lived in a zoo but was being prepared for transport back to Kenya. After 10 years, she was with Sudan and his caretakers when his final moments came as the last male northern white rhino.

You can learn more about Sudan and Vitale at the National Geographic society website: www.nationalgeographic.org.

Science Background

Northern White Rhino The last two northern white rhinos, both female, are in the Ol Pejeta Conservancy in Kenya and are kept under constant guard to protect them from poachers. However, the subspecies is functionally extinct. In the example of the northern white rhinos, make sure that students understand that it is a subspecies and not the species that is functionally extinct. Lead students in a discussion on how a species or subspecies can be considered extinct even when a few individuals remain. Students should understand that since there are no males left, natural reproduction is no longer an option. Scientists are attempting to use assisted reproductive technologies to save the subspecies, but this effort raises ethical questions. One fear is that the existence of technological methods to save species will be used to justify reduced efforts to address underlying causes of extinction.

ON ASSIGNMENT National Geographic photographer Ami Vitale began her career documenting conflict zones. An assignment in 2009 to photograph the transport and release of one of the world's last living male northern white rhinos named Sudan changed her focus from war zones to wildlife and environmental stories. In this photograph, taken in 2019, a wildlife ranger comforts Sudan moments before his death.
494

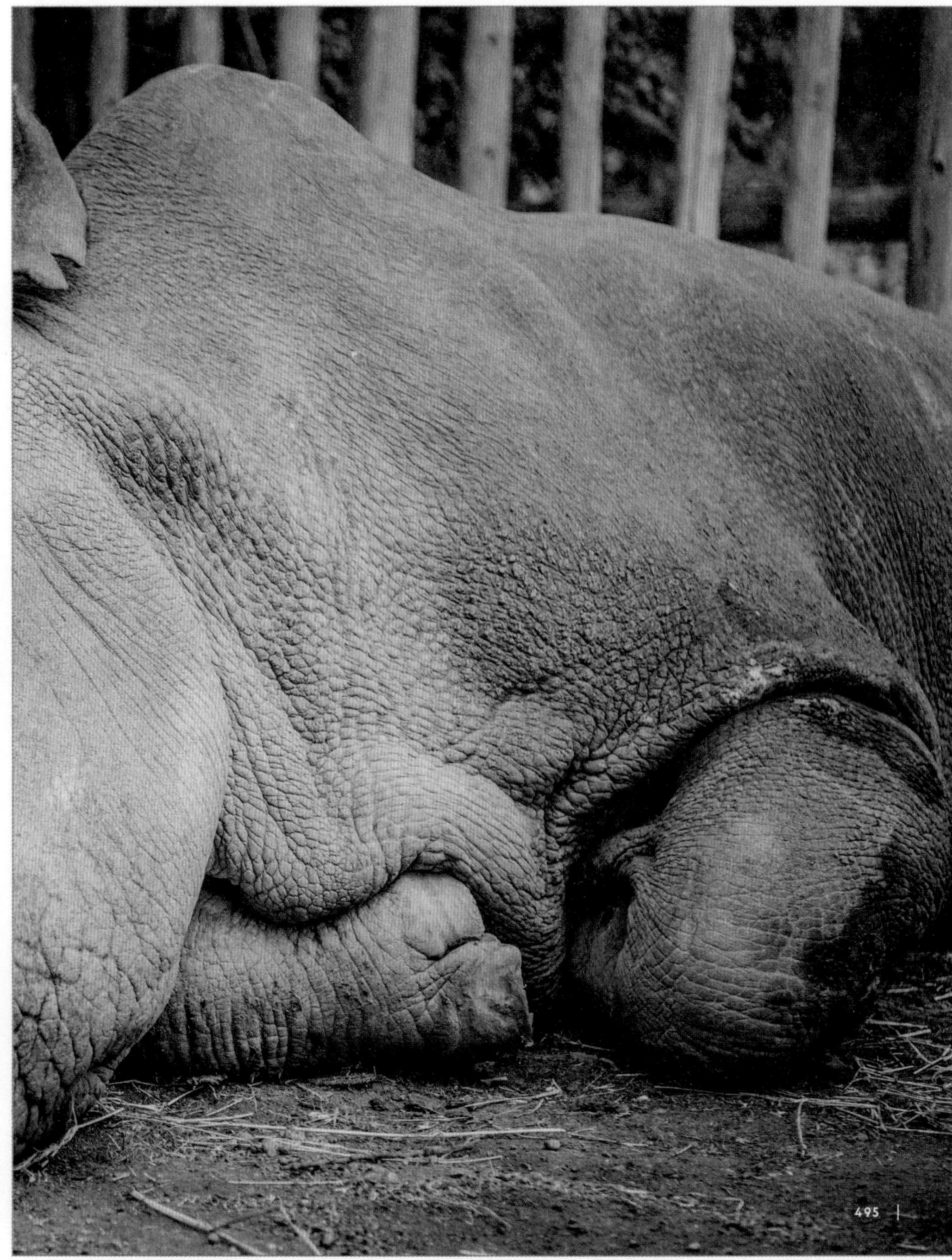

Address Misconceptions

Species vs. Subspecies As students know, members of different species may be able to interbreed, such as horses and donkeys, yet their offspring (mules and hinnys) are not fertile. Mules and hinnys are nearly always sterile due to the mismatch of their chromosomes that prevents pairing during meiosis.

Students may have a less clear understanding of what subspecies are. Different subspecies can interbreed and produce fertile offspring. Typically, they do not have the chance to, however, because of geographic isolation, as in the case of the northern and southern white rhinos. Often, phenotypic characteristics differ due to the effects of this isolation. The northern whites are smaller with a straight back, flat skull, and shorter front horn. The larger southern whites have a concave back, concave skull, more body hair, and a longer front horn. Both subspecies have the same scientific name: *Ceratotherim simum*.

Students may liken the idea of subspecies to breeds of dogs. While similar, the term *breed* is applied to results of selective breeding by humans to produce organisms with predictable characteristics that differ from their wild counterparts.

Connect to English Language Arts

Write a Response Have students write a personal response to the photo that describes the situation, gives some background about it, and also shares how they feel about it. Encourage students to use descriptive words and opinion words to effectively share their feelings about the photo and about the extinction of the northern white rhinos.

DIFFERENTIATED INSTRUCTION | English Language Learners

Combining Ideas Support students in combining phrases, clauses, and sentences as they write their response to the photo. Provide sentence frames to support students in learning and using language patterns.

Beginning Provide sentence frames such as the following: *In the photo, the rhinoceros _______, and the caretaker _______. The photo makes me feel _______ because _______.*

Intermediate Then, guide them to review their work for similar ideas that can be combined with *and*, contrasting ideas that can be combined with *or*, or causes and effects that can be linked with *because*.

Advanced Have students write their response and then exchange with a partner to look for ideas that could be combined. Provide connecting words as needed.

CASE STUDY
TUSKLESS ELEPHANTS

Ask Questions To help students revisit the Case Study and make connections with the content, have them identify which causes of species decline are actively affecting elephant populations. Students should use their list to develop their questions.

In Your Community

Habitat Loss Since 1970, scientists estimate that 60 percent of mammals, birds, reptiles, and fish have been wiped out worldwide due to human impact. Similarly, over time, the continental United States has lost nearly 60 percent of its natural habitats, leading to the extinction of at least 500 species across the country. Provide students with historical maps and current topographic maps of your region and have students identify regions that have experienced habitat loss. Allow students to use technology to find if there are any local plants or animals that are threatened and any protections that may be in place to preserve them.

CCC Scale, Proportion, and Quantity *Sample answer: Nonpoint source runoff from an agricultural area could end up in a reservoir or other water source used as a source of drinking water for a community.*

CASE STUDY
TRACKING TUSKLESS ELEPHANTS

Ask Questions *Generate questions about how the poaching of elephants for their tusks may have influenced the increase in tuskless elephants over time.*

Causes of Species Decline

In the United States, the listing and management of threatened and endangered species is managed by the U.S. Fish and Wildlife Service (USFWS). Keep in mind that not all rare species are threatened or endangered. Some species have always been uncommon.

Habitat Loss and Fragmentation Humans harm species indirectly by altering or destroying species' habitats. Because most species require a specific type of habitat, any degradation, fragmentation, or destruction of that habitat reduces population numbers.

An *endemic species* occurs only in the area in which it evolved. Endemic species are more likely to go extinct as a result of habitat degradation than species that have spread to many areas. For example, Pyne's ground plum (**Figure 16-10a**) is a flowering plant that lives only in a cedar forest near the rapidly growing city of Murfreesboro in middle Tennessee. The plant is threatened by conversion of its habitat to homes and industrial use. Texas blind salamanders (**Figure 16-10b**) are among the species endemic to Edwards Aquifer, a series of water-filled, underground limestone formations located in the Texas Hill Country. Excessive withdrawals of water from the aquifer, in combination with water pollution, threaten the salamander and other species in the aquifer.

Habitat fragmentation forms islands of suitable habitat scattered across what was once a continuous range. Buildings, roads, and fences can fragment a habitat. Thus, a large population can be broken into smaller populations in which inbreeding is more likely to occur. Inbreeding can have deleterious genetic effects. Fragmentation also prevents individuals from accessing essential resources. For example, roads, fences, and houses prevent spotted turtles endemic to the eastern United States from moving among wetlands to feed and hibernate. The wetlands still exist, but turtles cannot safely move between them.

CCC Scale, Proportion, and Quantity *Why is it important to maintain all parts of a species' habitat?*

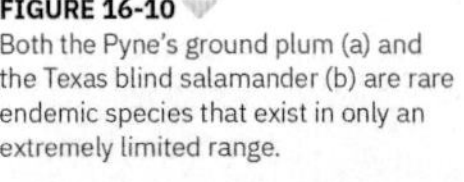

FIGURE 16-10 Both the Pyne's ground plum (a) and the Texas blind salamander (b) are rare endemic species that exist in only an extremely limited range.

DIFFERENTIATED INSTRUCTION | English Language Learners

Environmental Print Use the information on Habitat Loss and Fragmentation and Overharvesting as a starting point for a discussion of warning signs and rules students might see in everyday life, for example: "No fishing." "Do not pick the flowers." "Do not feed the wildlife." "Wild animals don't make good pets!" Or in camping, "Take only pictures, leave only footprints."

Beginning Have students draw these signs, including an illustration that shows they understand the meaning.

Intermediate Have students discuss where they might see these signs and suggest others they have seen. Ask: *How do the signs help species?*

Advanced Have students explain why these signs are important. Provide sentence frames for conditional sentences as needed: *If people ______, then the environment might ______.*

FIGURE 16-11 
The critically endangered white abalone is one of seven abalone species that live off the coast of California.

VIDEO

VIDEO 16-1 Go online to learn more about the overharvesting of white abalone.

Overharvesting In some cases, people directly reduce a species' number. Consider that when European settlers first arrived in North America, they found more than 30 million bison roaming the Western plains. By the late 1800s, that population had been reduced to less than 1000 individuals. It was only through the maintenance of captive populations, the development of nature reserves, and protective measures that restricted hunting that bison were able to rebound. Today, the bison population is around 500,000, although many live on private property. Only about 20,000 bison roam without restriction on tribal, state, and federal lands.

We continue to overharvest many species. One such species is the white abalone (**Figure 16-11**), a type of mollusk native to kelp forests off the coast of California. Heavy harvesting of this species during the 1970s reduced the population to about one percent of its original size. As a result, the white abalone fishery was closed in 1996. In 2001, it became the first invertebrate to be listed as endangered by the USFWS.

Although some white abalone remain in the wild, their population density remains too low for effective reproduction. The species' only hope for survival is a program of captive breeding, which began in 2001. After major setbacks, including the loss of 95 percent of the initial population of captive-bred juveniles due to bacterial infection and other factors, the first population of captive-bred juvenile white abalone were released back into the wild in 2019.

Species are overharvested not only as food, but also for use in traditional medicine, for the pet trade, and for ornamentation. Some orchids prized by private collectors have become nearly extinct in the wild. Rhinoceroses are killed by poachers to obtain their horns. The majority of rhinoceros horns are used to make traditional Chinese medicines. Possessing a whole rhinoceros horn is also seen by some people to be a status symbol.

Video

Time: 3:40 seconds

Use **Video 16-1** to explain why the white abalone has continued to decline after harvesting stopped. Pause the video at 0:56 and ask students to interpret the graphs with a partner. Students should understand that the two graphs show the decline at two different times and across different populations.

Science Background

Ocean Harvests Overharvesting is exceptionally problematic in the world's oceans given past and current fishing practices that include longline fishing and the use of certain kinds of nets. These practices have brought species of fish that are sold for food, such as sardines and Atlantic cod, to the brink of extinction. However, these practices also lead to the decline of marine animals that are caught and killed but not sold for food. Examples of this bycatch are dolphins, sea lions, whales, and sea turtles. Describe a few harmful fishing practices to students and facilitate a discussion around how overfishing affects ocean food chains. Students should identify how collapsing populations of targeted fish will affect other populations, as well as discuss how bycatch affects the larger food web and ecosystem.

DIFFERENTIATED INSTRUCTION | Students with Disabilities

Auditory Impairments Students with auditory impairments may struggle to follow videos shared with the entire class. To support these students when presenting videos, make sure to present the video with closed captioning on. You may also choose to allow them to observe the videos on a mobile device or tablet so they can pause the video if they need additional time to read the captioning.

Place students with auditory impairments in mixed groups to work together to analyze the graphs shown during the video that discuss abalone population sizes. Students should summarize the actions that scientists are using to conserve the species. Encourage students to use the communication means that most effectively support all students in the group.

Science Background

Kudzu Known as "the vine that ate the South," kudzu is a species of plant that is native to regions of Asia. While originally imported as an ornamental shade plant, in the early 1900s, the government paid farmers to utilize the plant for erosion control and nitrogen fixation. The species then rapidly spread because it easily outcompeted native plant species. Kudzu now covers wide swaths of forest and fields primarily in the South and Midwest, decimating other plant life and disrupting habitats. If kudzu is not common in your area, use an Internet keyword search such as "kudzu overgrowth" to show students the incredible blanket it forms over native species and have them discuss why kudzu is an invasive species.

Non-Native Species A community can change dramatically after the introduction of a non-native species. As described in Chapter 3, a **non-native species** is a species that evolved in one community and then dispersed from its home and became established elsewhere. The rate of non-native species introductions has accelerated along with the pace of global travel. Florida is home to more than 500 non-native species. In Hawaii, it is estimated that more than 45 percent of plant species are non-native. Some of these species were brought in as food crops, as ornamentals for gardens, or as a source of textiles. Other species arrived as stowaways along with cargo from distant regions.

Invasive Species Many non-native species have little impact on their adoptive community, but some of them are invasive. An invasive species is a non-native species that harms members of its new community. In some cases, the arrival of a non-native species can lead to the extinction of endemic species. Invasive species often have a far greater impact in their new community than they did in the region from which they originated. For example, invasive feral cats in Australia have had a devastating impact on endemic populations of ground-dwelling birds and small mammals. In fact, it is estimated that feral cats consume two billion reptiles, birds, frogs, and mammals each year.

When a species leaves its community of origin behind, it also leaves behind the competitors, predators, and parasites that coevolved with it and that helped keep its population in check. If the invasive species is a parasite, predator, or herbivore, it also leaves behind hosts or prey that had coevolved with it and had defenses against it. Its new hosts are often defenseless against it. As a result, an invasive species often reaches a higher population density in its new home than it achieved in its old one.

16.2 REVIEW

1. **Describe** Use the data provided in **Figure 16-8** to support the claim that life has thrived and diversified on Earth, despite extinction of species.
2. **Compare** What is the difference between background extinction and mass extinction?
3. **Explain** Why are native species more at risk of extinction than non-native species that have spread to many areas?
4. **Sequence** Rank the scenarios from least to most destructive for organisms living in a habitat. Use 1 to designate the least destructive scenario.
   - repaving existing walking and biking paths
   - developing a network of natural walking paths
   - building a narrow one-way road with a slow speed limit
   - designating the land as a protected conservation area
   - constructing a road with high walls to protect wildlife from crossing

16.2 REVIEW

1. *The number of species on Earth has been increasing over time. From 438 Mya to 243 Mya, the number of species remained relatively the same but since then, the number of species has been on a consistent increasing slope.* **DOK 2**
2. *245 Mya ; 438 Mya; 66.5 Mya; 360 Mya; 208 Mya* **DOK 2**
3. *Sample answer: Since native species occur only in the areas where they evolved, they are more sensitive to environmental changes and habitat degradation than species that have adapted to live in multiple locations. This puts native species more at risk of extinction because they are less likely to be able to adapt to drastic or sudden changes.* **DOK 2**
4. *designating the land as a protected conservation area; developing a network of natural walking paths; repaving existing walking and biking paths; building a narrow one-way road with a slow speed limit; constructing a road with high walls to protect wildlife from crossing* **DOK 2**

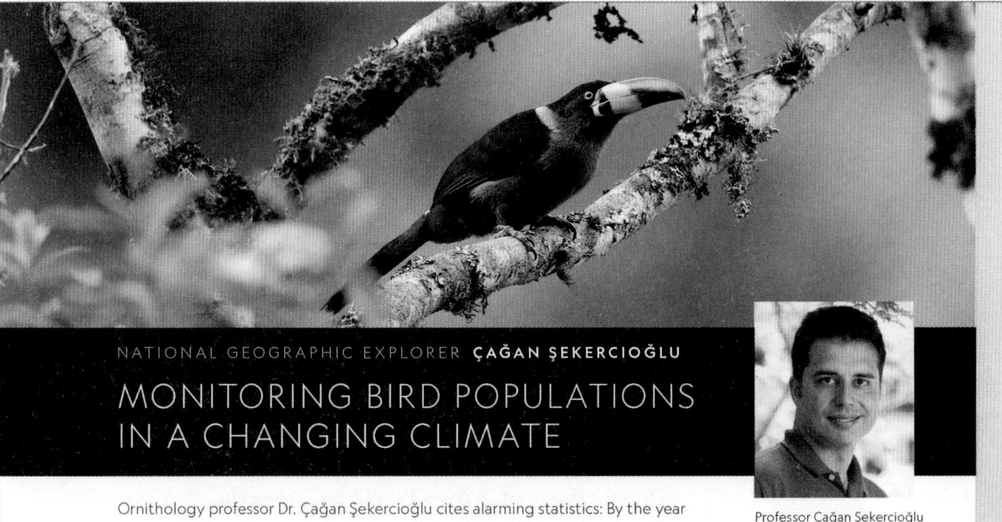

NATIONAL GEOGRAPHIC EXPLORER **ÇAĞAN ŞEKERCIOĞLU**

MONITORING BIRD POPULATIONS IN A CHANGING CLIMATE

Professor Çağan Şekercioğlu is not only a biologist; he is also an accomplished nature photographer. He thinks people need to feel the personal, emotional connection photography can provide to support facts and figures.

Ornithology professor Dr. Çağan Şekercioğlu cites alarming statistics: By the year 2100, as many as 25 percent of the world's bird species could go extinct. This adds up to around 2800 unique bird species. Şekercioğlu cites two main reasons for the potential extinctions—climate change and habitat loss. And he notes that it's not a question of whether these species will become extinct, but how quickly.

In 2000, Şekercioğlu developed a database of all the world's bird species, and he has been updating and analyzing it constantly since then. He collects extensive data about the ecology, conservation, distribution, and biogeography of species from field data and literature. The database allows him to analyze specific factors related to population changes. From that, he tries to answer questions about which species are more likely to go extinct based on climate change's impact on habitat availability. He warns that as the environment warms over 2 °C, the number of bird extinctions increases significantly.

Şekercioğlu notes that most of the world's bird species live in the tropics. However, very little data from those regions exists, so he says that their risk of extinction is greatly underestimated. He's watching tropical mountain birds in Costa Rica very closely. As the climate warms, the birds move higher and higher up the mountain in search of habitat and into smaller and smaller ranges. He monitors birds in farming areas as well to determine the effects of forest loss and land use. His study shows that populations of insect-eating birds are shrinking fast. He notes that the birds are so specialized in the habitats in which they feed that individuals will not cross a small field to get to a larger forested area. He says that when birds are lost, so are the ecosystem services they provide. As a result, the ecosystem deteriorates.

Today, despite all the environmental setbacks occurring around the world, Şekercioğlu strongly believes that it's the general public's lack of environmental awareness that is the biggest problem. Far too many people are unable to realize the grave nature of the issues facing birds and biodiversity in general. He is committed to improving this situation.

THINKING CRITICALLY

Analyze *How does Şekercioğlu use data analysis to draw his conclusions about the potential for bird species extinctions?*

16.2 EXTINCTION **499**

DIFFERENTIATED INSTRUCTION | English Language Learners

Listening for Main Idea and Details Listening for the general meaning and main points can help focus students' listening and improve their understanding of the text. As they listen a second time, students can then listen for details that support the main ideas.

Beginning Read the first paragraph of "Monitoring Bird Populations in a Changing Climate" aloud. Have students complete a sentence frame for the main idea: *By 2100, up to ______ percent of ______ could ______*. Read the paragraph aloud again and have students listen for two reasons why this might happen.

Intermediate Read the second paragraph aloud. Have students state the main idea or most important point. Then, read the paragraph again and have students list one or two details.

Advanced Read the third paragraph aloud and have students explain the main idea and details that support it. Reread the paragraph and have students add any details they missed the first time.

EXPLORER
ÇAĞAN ŞEKERCIOĞLU

About the Explorer

National Geographic Explorer Dr. Çağan Şekercioğlu (ch-HAN shay-KER-jew-loo) developed a database to analyze traits and ecological factors that affect the survival of bird species. He uses practices such as bird banding and radio tracking to gather data that shows how birds act in their changing habitats. You can learn more about Çağan Şekercioğlu and his research on the National Geographic Society's website: www.nationalgeographic.org.

Science Background

Bird Banding Bird banding has been used for centuries to track and identify birds. Modern banding involves using numbered plastic or aluminum tags that are applied to the legs of birds caught in specialized nets or traps. If the bird is caught again in the future, the presence of the band can be used to determine data about the bird's life, such as age and where it has traveled since it was first caught.

Connect to Careers

Ornithologist Scientists who study birds are ornithologists. Ornithologists study bird habitats, physiology, and behavior. To support bird conservation efforts, ornithologists study migration routes and reproduction needs and monitor certain bird populations using tools such as tracking devices and drones. They may work for local, state, or national agencies, nonprofit conservation organizations, environmental education agencies, and universities.

THINKING CRITICALLY

Analyze *Sample answer: The database includes huge amounts of data about nearly every species and relates the data to certain factors. By continuing to collect data, he can sort according to a given factor to make relationships.*

16.3

HUMAN IMPACT ON THE ENVIRONMENT

LS2.C, LS4.D, ESS2.D, ESS3.A

EXPLORE/EXPLAIN

This section explains how human population growth affects resource use, discusses the effects of water and air pollution, and provides an overview of how human activities to obtain resources can lead to land degradation and habitat loss.

Objectives

- Explain how human population growth impacts resource use.
- Describe the effects of water pollution and air pollution caused by human activities.
- Analyze how human activities lead to land degradation and habitat loss.

Identify *Students might say that riding in a vehicle to school adds carbon dioxide to the atmosphere; using water to brush teeth or shower impacts the amount of available fresh water; using a smartphone utilizes energy to keep it charged.*

Vocabulary Strategy

Using Prior Knowledge Students will have been introduced to how human activities affect the environment in their elementary and middle school science classes. They may also be aware of air or water pollution or other human impacts on the environment in their community. As students are introduced to the key terms in this section, have them make connections with their prior knowledge and identify other examples of the terms with which they are familiar.

16.3

KEY TERMS

deforestation
desertification
ecological footprint
pollutant

HUMAN IMPACT ON THE ENVIRONMENT

Humans rely on natural resources, such as food, energy, and building materials sourced from the ecosystems in which we live. Like other organisms, we need basic things like food and shelter. However, as humans have advanced over time, so have our needs and expectations. In modern times, we have the ability to prepare a variety of foods, construct elaborate buildings, and travel long distances in relatively short time using multiple modes of transportation. Every decision we make can have a lasting impact on Earth's ecosystems and the species that live within them.

Identify *List some decisions you made today that have an impact on the ecosystem in which you live.*

Human Population Growth and Resource Use

For most of its history, the human population grew very slowly (**Figure 16-12**). The growth rate began to pick up about 10,000 years ago, and it soared during the past two centuries. Three trends promoted the large increases. First, humans expanded into new habitats. Second, they developed technologies that increased the carrying capacity of their habitats. Third, they sidestepped limiting factors that typically limit population growth.

What is Earth's carrying capacity for humans? There is no simple answer to this question. For one thing, we cannot predict what new technologies may arise or the effects they will have. For another, different types of societies require different amounts of natural resources to sustain them. *Natural resources* are natural materials used by humans, such as water, forests, animals, and minerals.

FIGURE 16-12 The graph shows that where it once took hundreds or thousands of years to add one billion people to the population, it now takes a little over a decade.

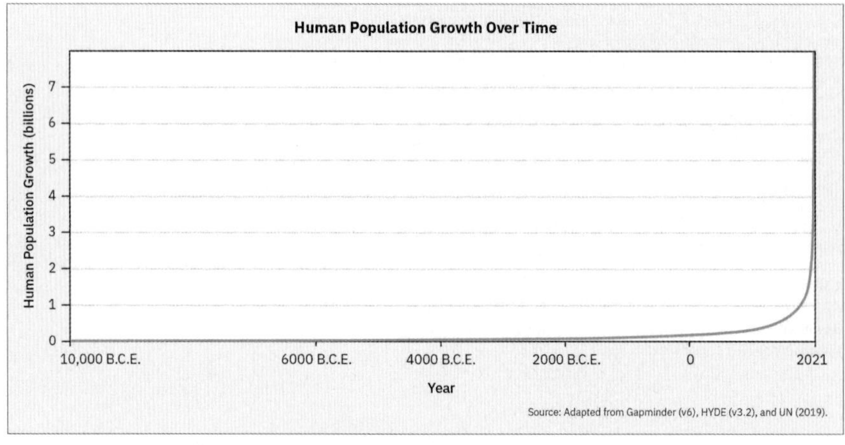

500 CHAPTER 16 SURVIVAL IN CHANGING ENVIRONMENTS

On a per capita (per individual) basis, people in highly developed countries use far more natural resources than those in less developed countries, and they also generate more waste and pollution.

Renewable and Nonrenewable Resources Natural resources can be categorized as either renewable or nonrenewable.

- *Renewable resources* are natural resources that can be replenished at the same rate at which they are used. Some examples are wind energy, solar energy, hydroelectric energy, and geothermal energy.
- *Nonrenewable resources* are natural resources that cannot be replenished at the same rate at which they are used. Some examples are fossil fuels and minerals.

Figure 16-13 compares the global consumption of energy, including both renewable and nonrenewable sources of energy. As shown in the graph, the consumption of nonrenewable energy sources, such as oil, coal, and gas, has risen dramatically since the 1950s. Renewable energy sources, such as wind and solar energy, have become significantly more popular within the past few decades.

The local availability of natural resources influences the type of energy generated and used in a particular region. For example, coal has historically been a significant source of energy in the Appalachian coal region and Wyoming, where coal deposits are abundant. Areas that receive large amounts of sunlight are excellent candidates for solar power generation. Locations that experience high or constant winds are ideal sites for wind farms.

SEP **Analyze Data** *Explain the trend in energy consumption shown in **Figure 16-13** in terms of human population growth. Include the terms* renewable energy *and* nonrenewable energy *in your explanation.*

FIGURE 16-13
This graph shows how global energy consumption has changed over time for 10 different energy sources.

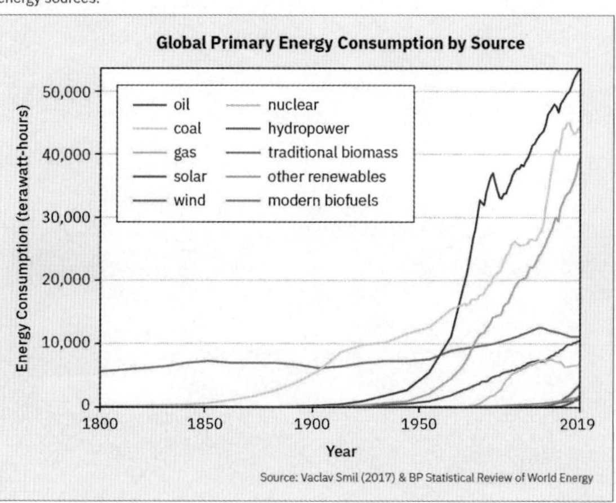

Connect to Mathematics

Use Units to Understand Problems Have students analyze the trendlines shown in **Figure 16-13** and consider how global energy resources have changed over time. Have students identify sources of traditional biomass, such as wood fuels and animal waste, which was the primary source for energy prior to the increase in oil, coal, and gas consumption. Lead a discussion around the time period when coal began being more widely used than traditional biomass and have them consider why oil consumption rapidly increased to outpace the consumption of coal.

SEP **Analyze Data** *Sample answer: As human population increases, so does energy consumption. Some renewable energy resources, such as biomass, are not a new trend. Still, renewable energy resources are not consumed as much as nonrenewable energy resources.*

DIFFERENTIATED INSTRUCTION | Leveled Support

Struggling Students Students might have difficulties understanding how renewable energy resources could possibly catch up with the amount of nonrenewable energy resources used. Guide students to understand that rapid increases in oil, coal, and gas use occurred because technological advances spurred by monetary profits increased the ability to obtain and use these resources. Point out that applying the same mindset can result in rapid increases in renewable energy resource use as well.

Advanced Learners Ask students to examine **Figure 16-13** and look for a pattern in the data to extrapolate how long it might take for wind power to catch up if it increased at the same rate as oil did between 1900 and 1970. Have students consider with a partner if they think that increasing the availability of renewable energy resources will lead to a decrease in the use of nonrenewable energy resources or just to an increase in total energy resources.

Visual Support

Pollution Sources Point and nonpoint source pollution are the two classes identified by the United States Environmental Protection Agency (EPA). Have students examine the commercial buildings on the left and the residential buildings on the right of **Figure 16-14**. Have them identify how wastes from these regions enter the river. Students should understand that wastes can get washed off of roads and into the ground before entering the water table. They may also point out that septic systems that are not maintained can leak into the water table and make their way to the river and other bodies of surface water.

SEP Scale, Proportion, and Quantity *Sample answer: Nonpoint source runoff from an agricultural area could end up in a reservoir or other water source used as a source of drinking water for a community.*

Water Pollution

Natural or synthetic substances released into the soil, air, or water in greater than natural amounts are called **pollutants**. The presence of a pollutant disrupts the biological processes of organisms that evolved in its absence or that are adapted to lower levels of the pollutant.

As shown in **Figure 16-14**, water pollution originates from many sources.

- *Point sources* are easily identifiable sources of water pollution. For example, a factory or wastewater treatment plant that discharges pollutants into water is a point source. These pollutants are easiest to control because the location of the source is known, and action can be taken to contain it.
- *Nonpoint sources* come from the widespread release of pollutants. For example, leaked motor oil that pollutes waterways originates from vehicles on many roads and driveways. Because of their widespread nature, it is much more difficult to eliminate nonpoint sources of water pollution than it is to eliminate point sources of water pollution.

SEP Scale, Proportion, and Quantity *Explain how a point or nonpoint source of water pollution could affect an ecosystem far away from where you live.*

FIGURE 16-14 Point sources of water pollution are easily identified. In contrast, it is difficult to pinpoint from where nonpoint sources of water pollution originate.

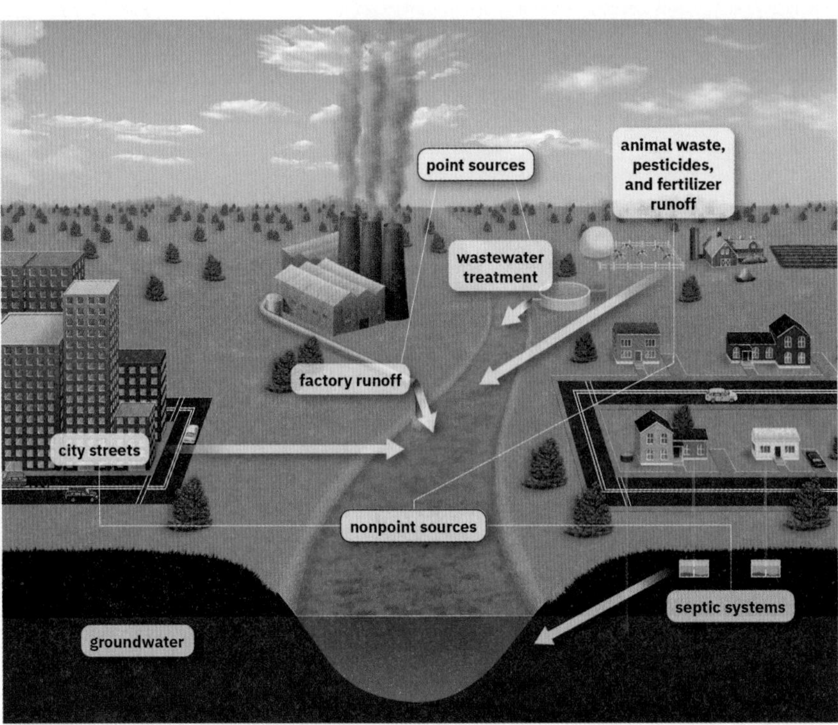

DIFFERENTIATED INSTRUCTION | English Language Learners

Describing and Explaining with Details Have students verbally describe the water pollution sources in **Figure 16-14** and explain their causes and effects. Encourage students to include as much detail as they can, using the visual for support.

Beginning Have students describe the sources in their native language. Then, help them translate words and phrases into English. Then have them identify each source as point or nonpoint.

Intermediate Have partners take turns selecting point and nonpoint sources in the figure. They should describe the source and explain how it affects water.

Advanced Have students work together to explain the causes and effects of water pollution using the point and nonpoint sources in the figure.

Solid Waste Seven billion people use and discard a lot of material. Where does all the waste go? Historically, much of it was buried in the ground or dumped out at sea. We now know that these practices have negative impacts on both biodiversity and human health.

Piling up or burying garbage on land can contaminate groundwater, such as when lead from discarded batteries seeps into aquifers. Modern landfills in developed countries are lined with clay or plastic to prevent toxins in the trash from reaching the groundwater. However, many landfills in less-developed countries have no such barrier. Toxins leached from trash enter the groundwater and can turn up in waterways and in wells that supply drinking water.

Discarding solid waste in oceans threatens marine life. In the United States, it is no longer legal to dump solid waste into the sea. Nevertheless, trash constantly enters coastal waters. Scientists estimate that the world's oceans currently contain 269,000 tons of floating plastic. The Great Pacific Garbage Patch is part of the five offshore plastic accumulation zones in the world's oceans and is located halfway between Hawaii and California. It covers an approximate surface area of 1.6 million square kilometers. This area is twice the size of Texas and three times the size of France. Much of the solid waste that ends up in the ocean is made up of discarded fishing gear (**Figure 16-15**). It is estimated that 640,000 tons of this type of waste is dumped or lost in the ocean every year.

 VIDEO

VIDEO 16-2 Go online to learn how National Geographic Explorer Dr. Heather Koldewey implemented a solution for discarded fishing nets.

FIGURE 16-15 Lost or discarded fishing gear, which includes rope, nets, traps, crab pots, and fishing line, is a major source of ocean pollution.

Video

Time: 3:36 seconds

Use **Video 16-2** to explain how humans are working to collect plastic nets that are damaging ecosystems in the Philippines. Once students have completed watching the video, ask them to estimate how much single-use plastic they use weekly and identify ways they can cut down on its use.

Connect to Mathematics

Reason Quantitatively Share with students that about 60 percent of plastic produced has a lower density than seawater. Have students calculate how many tons of garbage had to be introduced into the ocean for 269,000 tons to be floating at the surface. *(approximately 450,000 tons; 269,000 is about 60 percent of the total 100 percent, which would be 40 percent more [450 x .6 = 270])*

CROSSCUTTING CONCEPTS | Cause and Effect

Point and Nonpoint Pollution Once students have completed reading this spread, have them consider how the disposal of solid wastes described in the text create point and nonpoint pollution sources. Emphasize to students that human actions can have wide-ranging effects on the environment. To exemplify this concept, draw a T-chart on the board and lead students through an activity where they identify a pollutant found in solid waste and describe the effects it has on the environment. Students may identify pollutants from the text or prior knowledge. Guide them through a discussion where they identify changes that could be made to minimize the effects of the pollutants on the environment.

Connect to English Language Arts

Research Once students have completed the reading on this spread, have them research a species that is thriving due to climate change and a species that is declining. Have them write three to four paragraphs where they synthesize their findings from multiple resources and explain why climate change affects the two species differently. You may choose to encourage students to select one plant and one animal species or have students select different kinds of species within the same kingdom. Remind students to cite the sources where they found their information.

In Your Community

Local Climate Change Some regions are experiencing greater effects from climate change than others. For example, many South Central states have an increase in the frequency of severe weather events and many areas situated on the ocean have an increase in coastal erosion.

Take Action Encourage students to track how the weather patterns in your region have changed over the past 100 years. Provide students with materials to make a poster showing how the changes in the weather are affecting the local environment.

Air Pollution

Fossil fuels such as petroleum, natural gas, and coal supply most of the energy used by developed countries. Burning these nonrenewable fuels contributes to climate change and acid rain. Natural gas that leaks from wells and pipes contributes to global warming. Pollutants that enter the atmosphere can harm organisms when they fall to Earth as dust or in precipitation. Common air pollutants include sulfur dioxides that are released by coal-burning power plants and nitrogen oxides that are released by combustion of gasoline and oil.

Global Climate Change Ongoing climate change affects ecosystems worldwide. As you learned in Chapter 2, average temperatures are increasing, with the more pronounced rise at temperate and polar latitudes. This rising temperature is elevating sea level by melting sea ice and glaciers, adding meltwater to the sea. In the past century, sea level has risen about 20 cm. As a result, some coastal wetlands and low-lying islands are disappearing under water.

Temperature changes are important cues for many species. Abnormally warm spring temperatures are causing deciduous trees to produce leaves earlier, and spring-blooming plants to flower earlier. Animal migrations and breeding seasons are also shifting. The variety of species in communities is changing as warmer temperatures allow some species to expand their range to higher latitudes or elevations that were once too cold for them. Not all species can move or disperse quickly to a new location, and warmer temperatures are expected to drive some of these species to extinction.

Temperature affects evaporation, winds, and currents, so many weather patterns are expected to change as the world continues to warm. For example, warmer temperatures are correlated with extremes in rainfall patterns: periods of drought interrupted by unusually heavy rains. In addition, warmer seas tend to increase the intensity of, and thus the damage incurred by, hurricanes (**Figure 16-16**).

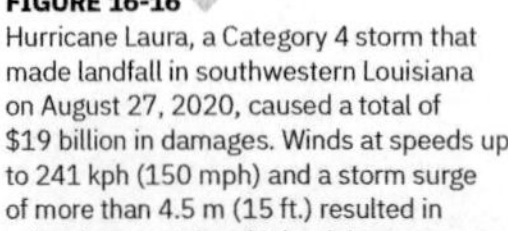

FIGURE 16-16 Hurricane Laura, a Category 4 storm that made landfall in southwestern Louisiana on August 27, 2020, caused a total of $19 billion in damages. Winds at speeds up to 241 kph (150 mph) and a storm surge of more than 4.5 m (15 ft.) resulted in extensive coastal and inland damage.

DIFFERENTIATED INSTRUCTION | English Language Learners

Research As students research species for the Connect to English Language Arts activity, support them in selecting search terms and finding research sources that use both academic and accessible language. Model how to list or cite sources.

Beginning Have students use dictionaries or online translators to help them select search terms related to their research topic. Have students list their sources.

Intermediate Have students write a list of questions to guide their research. Provide sources that use accessible language where possible. Have students list their sources.

Advanced Have students take notes as they research, paraphrasing or directly quoting sources and citing the source where they found each piece of information.

Some species actually might thrive in a changed climate. For example, the growing season in Canada might increase because of climate change, resulting in more productive farming and more species diversity. Southern regions, on the other hand, might be devastated by heat, drought, and dramatic and unpredictable changes in precipitation patterns.

Ozone Depletion Ozone (O_3) is a form of oxygen that is a pollutant in the lower atmosphere where we live, but it blocks ultraviolet radiation from the sun in the stratosphere. Ultraviolet radiation is linked to several health problems in humans, including cataracts, skin cancer, eye problems, and a weakened immune system.

Scientists determined that excessive use of chlorofluorocarbons (CFCs) had significantly destroyed the stratospheric ozone. CFCs were widely used as aerosol-can propellants, air conditioner/refrigerator coolants, and industrial solvents. As the ozone layer became thinner, more ultraviolet radiation was reaching Earth's surface. Researchers reported declining populations of microscopic algae and fish around Antarctica that were related to the increased radiation. The good news is that by the 1990s, most nations agreed to ban and phase out CFCs and other harmful compounds by the end of the century. Since that time, the ozone layer has gradually recovered, but scientists estimate that it might take 50 years to determine whether the problem is completely solved.

Near-Ground Ozone Pollution While ozone does not naturally occur in the lower atmosphere, it can form when compounds released by burning or evaporating fossil fuels are exposed to sunlight (**Figure 16-17**). Warm temperature speeds this reaction, so ground-level ozone varies daily (it is higher in the daytime) and seasonally (it is higher during the summer). The ozone that forms in the lower atmosphere never reaches the ozone layer where it would be useful. Instead, ozone that forms in the lower atmosphere acts as a harmful pollutant. It irritates the eyes and respiratory tracts of animals and interferes with plant growth.

FIGURE 16-17
Ozone in the lower atmosphere forms from the interaction between air pollutants and sunlight. These air pollutants originate from a variety of sources.

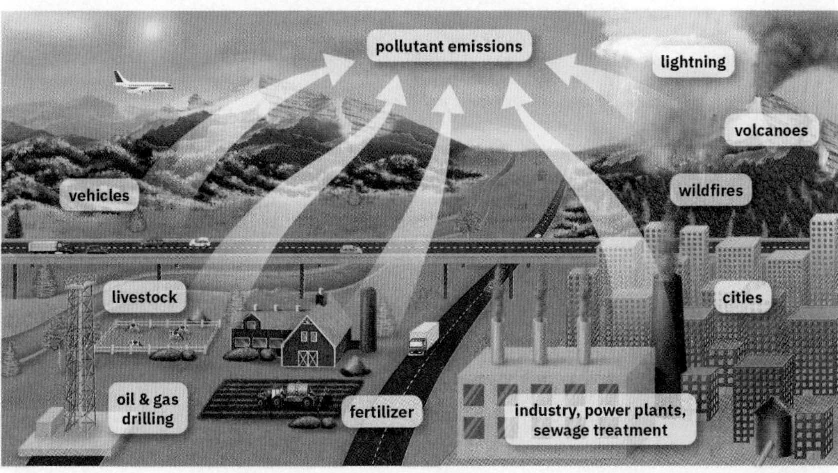

Visual Support

Air Pollutants Students may already have prior knowledge of some of the sources of air pollution shown in **Figure 16-17**. However, they may wonder how some of the other sources cause air pollutants. Guide the class in an activity where they make a chart to list the air pollutant sources shown in the figure and explain what pollutants are produced by each source. Students should understand that fossil fuel combustion in vehicles and power plants emit pollutants such as carbon dioxide, sulfur dioxides, and nitrogen oxides. Students may be surprised to learn that livestock produce air pollutants that are responsible for 14.5 percent of global greenhouse gas emissions. Students may also be surprised to learn that lightning causes a series of chemical reactions that produce ground-level ozone, another air pollutant. Encourage them to work in groups to use technology to find which air pollutants are produced by the other sources.

SCIENCE AND ENGINEERING PRACTICES
Scientific Knowledge is Open to Revision in Light of New Evidence

Ozone Layer In the 1980s, a large area of depleted ozone called the "ozone hole" was discovered over Antarctica. The depletion was determined to be caused by chlorofluorocarbons, but prior to this discovery, the scientific community proposed a number of different possible causes such as meteorological conditions and solar cycle phenomena. Remind students that scientific knowledge is open to revision based on new evidence or reinterpretation of previous evidence. Have students discuss how scientific argumentation is dependent on how strongly evidence supports a proposed explanation. Have students identify a few scientific topics that are currently in flux because the evidence is still under investigation.

Science Background

Mining Pollution In addition to land degradation, mining extraction and processing operations pollute air and waterways. The EPA estimates that mining in the western United States has polluted more than 40 percent of the watersheds in the west and thousands of dollars per day are spent treating damage caused by each contaminated mine facility. Place students in groups and have them select a mineral that is mined in your region. Students should identify how the mining operations affect the environment, how the mined resource is used, and the waste products that are produced by the extraction and processing of the mineral.

Address Misconceptions

Decreasing Biodiversity A common misconception is that climate change is the primary driver of decreasing biodiversity worldwide. However, many scientists now think that the overexploitation of resources that results in habitat loss and destruction is a greater risk to biodiversity. To remedy this misconception, guide students in a discussion about how resource overexploitation affects the environment. You may choose to draw a graphic organizer, like a mind map, on the board and have students identify how resource exploitation affects organisms in a specific location. Guide students to understand that identifying the causes of decreasing biodiversity in specific areas requires investigation of climate change and habitat loss and degradation.

CCC Cause and Effect *Sample answer: Many types of surface mining completely destroy ecosystems. Waste materials from mining can pollute nearby waterways.*

Land Degradation and Habitat Loss

Many human activities related to obtaining resources threaten species by degrading land or destroying habitats. Obtaining resources for the global population requires large-scale operations that greatly impact ecosystems and biodiversity.

Mining Minerals are mined worldwide, and global trade makes it difficult to know the source of the raw materials in products you buy. Also, resource extraction in developing countries is often carried out under regulations that are less strict or less stringently enforced than those in the United States. As a result, the environmental impact of mining is even greater in these countries. There are four main methods of mining.

- In-situ mining involves dissolving a mineral resource underground and then processing it at the surface.
- Placer mining involves sifting minerals from sediments in streambed deposits.
- Surface mining includes open-pit mining (**Figure 16-18**), strip mining, and mountaintop removal. In this mining method, mineral deposits are removed near Earth's surface.
- Underground mining involves extracting mineral resources from deep within Earth.

Surface mining strips an area of vegetation and soil, forming an ecological dead zone. Mining typically puts particulate matter into the atmosphere, makes mountains of rocky waste, and contaminates nearby waterways.

SEP Cause and Effect *Identify two ways that mining affects the ecosystem in which it is located.*

FIGURE 16-18
The Cowal gold mine is a large open-pit mine in New South Wales, Australia.

DIFFERENTIATED INSTRUCTION | English Language Learners

Predicting Have students use the headings and photos in the Land Degradation and Habitat Loss text to predict what they will learn from the text. Then, have them read to confirm their predictions.

Beginning Have students look at each photo to understand the meanings of the terms *mining* and *deforestation*. Prompt them to make a prediction: *What happens to species' habitats in each place?*

Intermediate Explain that the word *degradation* comes from the verb *degrade*, meaning "to make worse." Have students use the headings and pictures to predict how mining and deforestation degrade the land.

Advanced Have students predict causes and effects of mining and deforestation. Have them list their predictions and correct or add to them as they read.

Deforestation The clearing of forest areas is called **deforestation**. It causes enormous damage to both local and global environments. Forests cover 30 percent of Earth's land surface. Trees perform an important service in the storage of carbon. Tropical rainforests in particular play an important role in the removal of carbon dioxide from the atmosphere. Nevertheless, half of the world's rainforests are now gone, and an area of forest equal to about 20 football fields is being cut down on Earth every minute. At this rate, the world's rainforests will be eliminated in 100 years, if left unchecked. The prime causes of deforestation are agriculture, logging, and ranching with subsistence agriculture (**Figure 16-19**).

The Amazon is the world's largest rainforest and has sustained damage from subsistence farming, logging, ranching, and other destructive practices. Other areas, especially in Southeast Asia, are facing similar losses from destructive practices, such as burning and fragmenting forests and replacing them with vast plantations. The destruction of rainforests seriously damages the environment in many ways:

- Forests hold enormous amounts of water in the roots and trunks of trees. Areas that have been cleared suffer from floods that wash away or erode soil and cause great destruction.
- Trees release a large amount of water into the atmosphere through transpiration, which evaporates and returns as precipitation. Once cleared, an area becomes prone to drought as it has no way to retain water, and rainfall declines.
- Once a tropical forest has been logged, the resulting nutrient losses and drier, hotter conditions can make it impossible for tree seeds to germinate or for seedlings to survive. Thus, deforestation can be difficult to reverse.

FIGURE 16-19 This aerial view of a portion of the Amazon rainforest in Brazil shows a patch of forest surrounded by farmland deforested for use as cattle pasture.

Science Background

Amazon Rainforest Students will have been introduced to the Amazon Rainforest in previous science classes and may know that it is the largest tropical rainforest on the planet. One in ten species on Earth calls it home, and it covers an area of about 6,000,000 square kilometers. It covers about 40 percent of Brazil's land area and contains the largest river basin in the world. In spite of initiatives to save rainforests, significant loss of the Amazon Rainforest is occurring. Tens of thousands of fires were set between 2019 and 2021 to clear large swaths of the rainforest and use the land to produce agricultural products. Many of these products, such as soybeans, palm oil, and beef, are sold globally. Additional forest continues to be cleared because rainforest soil is poor and its ability to support agriculture is quickly depleted. It is thought that if the current rate of deforestation continues, about 27 percent of the Amazon will be deforested by 2030. This is troubling because the Amazon is a carbon sink—it stores billions of tons of carbon dioxide from the atmosphere annually, buffering temperatures and regulating climate. Evidence suggests that deforestation has a major impact on global climate.

SCIENCE AND ENGINEERING PRACTICES
Engaging in Argument from Evidence

Resource Control and Use One argument made by leaders of nations where rainforests are being deforested is that protected lands are resources that they should have the opportunity to use and develop. Most countries where rainforests exist are considered developing nations, and some people think that each country should determine how they use their resources. Leaders in many developed nations are concerned about the larger ramifications that will result from rainforest loss, but high consumption in these countries contributes to the problem. Have students consider the products that they use that may originate from cleared land in the Amazon, such as palm oils or soybean oils found in snacks. Lead a discussion with the class where they evaluate the arguments made by the developing nations and how their choices are connected to the decisions made in these developing nations. Allow students time to consider and evaluate the arguments from both sides.

On the Map

Global Desertificaton Use an Internet keyword search such as "desertification map USDA" to find their global desertification vulnerability map to download and display for students. Discuss areas of the world that are most prone to this phenomenon. Students may be surprised to see the degree of vulnerability of the western half of the continental United States.

Encourage students to discuss modifications humans can make in these regions to reverse desertification. Guide them to understand that the replanting of native plants can often encourage native animals to return to these areas. Students may also do their own Internet research on desertification solutions at National Geographic or other authoritative sources.

FIGURE 16-20 Large sand dunes form from the deposition of silt and sand along the riverbanks of the Brahmaputra River Basin in the Shannan Region of Tibet.

Desertification Although deserts naturally expand and contract over geological time as climate conditions vary, human activities sometimes result in the rapid conversion of a grassland or woodland to desert. This process is called **desertification**. As human populations increase, greater numbers of people are forced to farm in areas that are not suitable for agriculture. In other places, people allow livestock to overgraze in grasslands. In both cases, the result can be habitat degradation. Drought encourages desertification, which results in more drought (**Figure 16-20**). Plants cannot thrive in a region where the topsoil has blown away.

16.3 REVIEW

1. **Explain** What factors have allowed the human population to grow exponentially?
2. **Classify** Categorize each resource as a renewable or nonrenewable resource.
   - forests
   - minerals
   - fossil fuels
   - Earth's internal heat
   - naturally flowing water
3. **Argue** How does global climate change affect biodiversity?
4. **Predict** Which suggested actions would likely decrease land degradation and habitat loss? Select all correct answers.
   - **A.** increasing water use
   - **B.** planting cover crops to prevent erosion
   - **C.** implementating rotational grazing to prevent overgrazing
   - **D.** implementing climate change initiatives that reduce fossil fuel use
   - **E.** reusing or recycling materials that are made from mined resources

16.3 REVIEW

1. *Sample answer: Advancements in a variety of technologies such as medicine, modes of transportation, and crop production have allowed the human population to grow exponentially.* **DOK 1**
2. *renewable: Earth's internal heat, naturally flowing water, forests; nonrenewable: minerals, fossil fuels* **DOK 2**
3. *Sample answer: Changes in climate affect species' migration patterns and ability to find the resources they need to survive. Climate change also has a long-term impact on weather patterns, including temperature, precipitation, and wind patterns.* **DOK 3**
4. *B, C, D, E* **DOK 2**

LOOKING AT THE DATA

ECOLOGICAL FOOTPRINTS

SEP **Analyze and Interpret Data** Ecological footprint analysis is one widely used method of measuring and comparing resource use. An **ecological footprint** is the amount of Earth's surface required to support a particular level of resource production, consumption, and waste processing. It includes the amount of area required to grow crops, graze animals, produce forest products, catch fish, construct and maintain buildings, and take up any carbon emitted by burning fossil fuels.

In 2021, the per capita global footprint for the human population was 2.7 hectares, or about 6.7 acres. That is the area of just over five football fields per person. Ecological footprint analysis by the Global Footprint Network, a nonprofit research organization, suggests that the human population is currently living well beyond its ecological means, and is in the process of racking up a deficit that will take its toll on future generations. **Table 1** compares the ecological footprints of nine countries. **Table 2** compares their human populations.

TABLE 1. Comparing Ecological Footprints

Country	Hectares per capita (1 hectare = 2.47105 acres)
Brazil	3.11
Canada	8.17
China	3.38
France	5.14
Great Britain	7.93
India	1.16
Japan	5.02
Mexico	2.89
United States	8.22

Source: Global Footprint Network

TABLE 2. Comparing Populations

Country	Population (in thousands)
Brazil	212,559.41
Canada	38,005.24
China	1,410,929.36
France	67,391.58
Great Britain	67,215.29
India	1,380,004.39
Japan	125,836.02
Mexico	128,932.75
United States	329,484.12

Source: World Bank

1. **Organize Data** Sort the data in **Table 1** by ecological footprint size. Which country has the largest ecological footprint? Which has the smallest?
2. **Analyze** Sort the data in **Table 2** by population size. How does each country's population size compare to the size of its ecological footprint?
3. **Construct an Argument** Use the information provided in the passage and your analysis of the data to construct an argument for reducing a country's ecological footprint. In which two or three countries should maximum efforts to reduce their populations' ecological footprint be focused? Explain why.
4. **Predict** Which suggested changes in behavior would likely increase a country's ecological footprint? Select all correct answers.
   - A. using waterways in addition to roadways
   - B. eating less meat and more plant-based foods
   - C. ridesharing and using more public transportation
   - D. replacing paper and wood products with plastic products
   - E. transporting waste to the deep ocean rather than burying it

LOOKING AT THE DATA
ECOLOGICAL FOOTPRINTS

Connect this activity to the Science and Engineering Practices of analyzing and interpreting data and engaging in argument from evidence. Students analyze the ecological footprint per capita in a number of countries and determine which countries have the highest and lowest per capita resource use and the countries that have the largest and smallest population. Students then construct an argument using this data as evidence to determine which countries need to decrease their ecological footprint per capita.

SAMPLE ANSWERS

1. *The country with the largest ecological footprint is the United States. The country with the smallest ecological footprint is India.* **DOK 2**
2. *Sample answer: There is not always a correlation between population size and ecological footprint size. For example, India has the second highest population but the lowest ecological footprint. In contrast, Canada has the second highest ecological footprint but the lowest relative population.* **DOK 2**
3. *Answers will vary. Sample answer: Efforts should be made to reduce the ecological footprint in countries that have a relatively low population size compared to the size of their ecological footprint, such as Canada and Great Britain, as this means they are likely using more resources than warranted and at an unsustainable rate.* **DOK 1**
4. *Use waterways in addition to roadways; Replace paper and wood products with plastic products; Transport waste to the deep ocean rather than burying it.* **DOK 1**

16.4

REDUCING HUMAN IMPACT ON THE ENVIRONMENT

LS4.D, ESS3.A, ESS3.C, ESS3.D, ETS1.A, ETS1.C

EXPLORE/EXPLAIN

This section provides an overview of how achieving goals of sustainability and biodiversity requires careful management of limited resources and decreases in pollution, ecosystem degradation, and habitat loss.

Objectives

- Describe how sustainability can help with the management of limited resources.
- Analyze how air and water pollution can be decreased.
- Evaluate design solutions to manage ecosystem degradation and habitat loss.
- Explain why biodiversity should be maintained.

Pretest Use **Question 7** to identify gaps in background knowledge or misconceptions.

Identify *Sample answers: turning the lights off when rooms are not in use, installing a schoolyard vegetable garden, recycling and/or reusing materials*

Vocabulary Strategy

Semantic Map The topics covered in this section all depend on how human impacts on the environment can be reduced through *sustainability*. Have students build a semantic map with the word *sustainability* at the center. As they work through the other key terms in the section, they can add them to the map with an explanation on how they tie back to the central term.

16.4

KEY TERMS

ecological restoration
ecosystem service
sustainability
wildlife corridor

REDUCING HUMAN IMPACT ON THE ENVIRONMENT

An understanding that all life on Earth draws upon the same limited resources and that the health of the environment affects human well-being has given rise to a call for more sustainable practices. The first step is to recognize the environmental impact of every action we take.

Identify *What is one practice that could be implemented at your school to reduce the human impact on the environment?*

Managing Limited Resources

Living sustainably begins with recognizing the environmental impact of every action we take. **Sustainability** means using resources in a way that meets the needs of the current human population without harming the environment so that future generations will also be able to meet their own needs (**Figure 16-21**). People in industrial nations use large quantities of resources that must be extracted, produced, and transported before they are consumed and eliminated as waste. All these actions use energy, impact the environment, and have negative effects on biodiversity. However, we rely on many of these resources, and we can establish ways to minimize their use as part of living sustainably and preserving biodiversity. Sustainability begins with being aware of the problems, learning about the causes, and designing solutions.

FIGURE 16-21 Many fisheries, including the salmon fishery in Alaska, have strict enforcement of regulations to prevent the overharvesting of fish populations.

510 CHAPTER 16 SURVIVAL IN CHANGING ENVIRONMENTS

DIFFERENTIATED INSTRUCTION | Economically Disadvantaged Students

REVIEW Sustainable Practices
Students who come from economically disadvantaged backgrounds often live in food deserts, meaning they have limited access to fresh foods and vegetables. These students may eat low-cost, prepared foods with extensive packaging. Ask students if they have ever observed the phrasing "Made with 10 percent post-consumer recycled products" and what kinds of products they have seen the phrasing on. Challenge students to explain what the phrasing means and how using products that are made from recycled products reduce human impacts on the environment. Guide students to understand that they can further reduce their impact on the environment by recycling packaging.
Consider researching if community gardens exist in your area and share this information with students. Ask how these gardens are beneficial to the environment. Have students identify other ways that they could reduce their impacts on the environment.

Managing Air and Water Pollution

Minimizing energy use is also part of living sustainably. Fossil fuels such as petroleum, natural gas, and coal supply most of the energy used by developed countries. Burning these nonrenewable fuels contributes to global warming and acid rain. Extracting and transporting fossil fuels add environmental costs. For example, oil that escapes from ocean-drilling operations harms aquatic species. Natural gas that leaks from wells and pipes contributes to global warming. Even nuclear energy production, which produces no carbon dioxide, has occasional accidents that allow dangerous radioactive material to escape into the environment.

Alternative Energy Reducing greenhouse gas emissions will be a challenge. However, international efforts are underway to increase the efficiency of processes that require fossil fuel, to shift to alternative energy sources such as solar and wind power, and to develop innovative ways to capture and store carbon dioxide before it leaves a power plant.

Solar technologies convert sunlight into electrical energy either through photovoltaic panels or through mirrors that concentrate solar radiation. This electrical energy can be used immediately or stored for later use. Solar energy systems must be integrated into homes, businesses, and existing electrical grids, typically as an additional or backup energy source. More individual residents and businesses are opting to install solar panels on rooftops across the United States. Utilities, too, are building large solar power plants to provide energy to all customers connected to the grid.

Solar energy is a clean, renewable, and essentially inexhaustible source of energy. However, there are downsides. Industrial-scale solar farms can pose a threat to birds that sometimes mistake the array of solar panels or mirrors at such facilities for a lake (**Figure 16-22**).

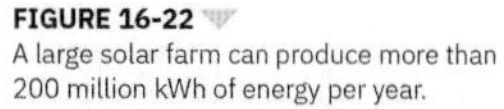

FIGURE 16-22
A large solar farm can produce more than 200 million kWh of energy per year.

In Your Community

Solar Panels States have different laws, regulations, and initiatives to increase the production of solar energy. Some states are implementing laws that require all new homes be outfitted with solar panels. States may also regulate how solar energy can be captured and used by residents, with some states stipulating that excess electricity generated from a home's solar panels can only be sold to a utility provider.

Take Action Have students find out the laws, regulations, and initiatives around solar power in your state. You may choose to have students also look up the laws, regulations, and initiatives in surrounding states and compare and contrast how consumers can obtain and use solar panels in the different states.

Connect to Mathematics

Reason Abstractly To support students in their understanding of the size of the Horse Hollow Wind Energy Center, shown in **Figure 16-23**, relate the size of an acre to a familiar space. Share with students that a football field is about 1.3 acres and have them calculate how many football fields would fit in the Hollow Wind Energy Center. *(about 35,544 football fields)*

Connect to English Language Arts

Assess Reasoning and Evidence Conservationists call for a variety of different methods to reduce human impacts on environments. As students complete the section on Reducing Human Impact on the Environment, have them analyze the reasoning and evidence provided for each sustainability measure to explain how they would work together to support biodiversity in a region. Have them also consider how some of these initiatives do not actually accomplish what they promise. Oftentimes, consumers think that everything they place in a recycling bin is recycled, but the outcomes vary depending on the material and the recycling process. Have students find out the percentage of plastic, metal, glass, and paper that gets recycled. Have them create a presentation to share their analysis with the class. They should include visual aids, such as graphs and charts, to support their ideas.

SEP Design Solutions *Sample answer: I can decrease the amount of nonrenewable resources I use; I can increase the amount of material I recycle; I can participate in activities that aid in ecosystem restoration.*

FIGURE 16-23 The Horse Hollow Wind Energy Center spans 47,000 acres of land in central Texas.

Wind energy is the process by which the wind is used to generate electricity or mechanical power. Mechanical power can be used for such things as grinding grain or pumping water, or a generator can convert the mechanical power into electricity. Wind turns the blades of a turbine around a rotor, which spins a generator, which generates electricity.

Wind is another clean, renewable, inexhaustible source of energy. However, wind farms require miles of open land (**Figure 16-23**), are quite large and loud, and can be dangerous to flying animals.

Waste to Energy Energy recovery from waste is the conversion of nonrecyclable waste materials into usable heat, electricity, or fuel. This process is often called waste to energy. Confined and controlled burning of solid waste destined for landfills can recover energy from the waste-burning process. This generates a renewable energy source and reduces carbon emissions by offsetting the need for energy from fossil sources and reduces methane generation from landfills.

Reduce, Reuse, Recycle Nonrenewable mineral resources are used in electronic devices such as phones, computers, and televisions. Reducing consumption by fixing existing products is a sustainable resource use, as is recycling. Reuse and recycling reduce the need for extracting nonrenewable resources, and they also help keep material out of landfills. In 2020, the United States recycled 69 million tons of material that would otherwise have ended up in landfills.

Finding alternative building materials such as engineered timber and metal composites are becoming more common in new construction projects. Composite materials such as fiber-reinforced plastic and alternative metal alloys are gaining popularity in commercial construction as well. Composites can be more durable than steel, and repairing damaged composite components is often less costly and requires less heavy machinery.

SEP Design Solutions *What are two ways you can reduce your impact on the ecosystem in which you live?*

CROSSCUTTING CONCEPTS | Cause and Effect

Human Effects All human action in the outdoors can affect organisms and the environment. For example, hikers are encouraged to "leave no trace"; however, even when the most care is taken, plants or other organisms are stepped on or animals avoid areas where people have passed through. Ask students to think of ways that their minor actions outdoors can affect other organisms. They should consider the effects of unnatural lighting and noise as well as movements. Have them discuss ways they can minimize their impacts, such as collecting wastes and disposing of them appropriately, using focused night lighting, minimizing loud noises, or determining which items they can reduce, reuse, and recycle.

Take Action If a recycling or lunchroom composting program is available at your school, have students consider how they might increase its visibility or drive participation within the school.

Managing Ecosystem Degradation and Habitat Loss

Land and habitat preservation can seem beyond an individual's ability, but it is not. There are many things that individuals can do, from reducing, reusing, and recycling their own materials to planting trees and butterfly gardens. On a larger scale, individuals can form groups that are actively involved at the local, national, and even global levels to help support environmental causes. It is important to protect ecosystems because they provide a number of important services. These **ecosystem services** can be divided into four categories:

- **Provisioning Services** These are the material benefits we get from ecosystems, such as food, fuel, water, and wood.
- **Supporting Services** These are the services that are necessary for all other ecosystem services, such as providing living spaces for species.
- **Regulating Services** These services include ecosystem processes such as water filtration by wetlands and crop pollination by insects.
- **Cultural Services** These are the intrinsic benefits we receive from ecosystems, such as natural spaces for recreational activities, as well as their aesthetic and inspirational aspects.

Reversing Land Degradation There are many rehabilitation strategies designed to address ecosystem degradation:

- development of large-scale composting programs to replenish organic soil to restore overgrazed land
- reforestation by planting trees to restore forests and habitats
- building terraces on steep land for crop cultivation to prevent soil erosion from rainfall runoff and increase water retention on the slopes of hills and mountains (**Figure 16-24**)

FIGURE 16-24
Building terraces allows farmers to cultivate rice on the steep hillsides in Mu Cang Chai, Vietnam.

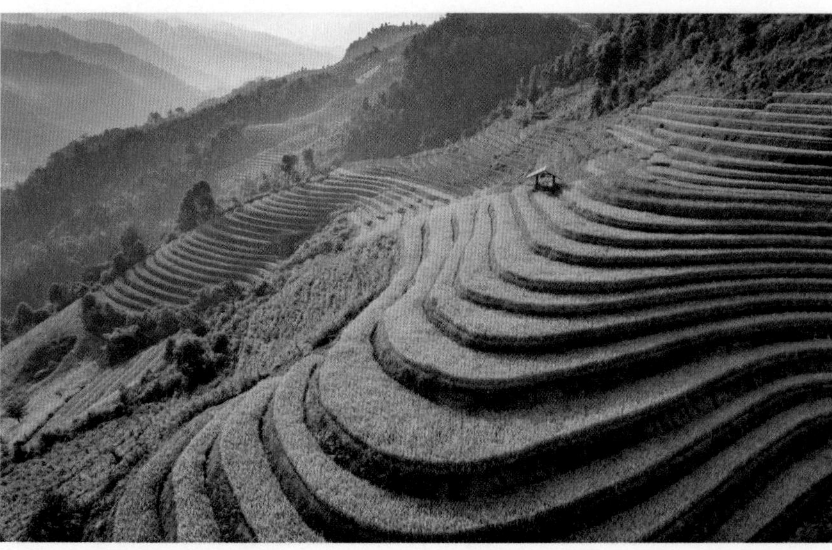

Science Background

Farming Terraces The practice of terrace farming is thought to have been first used by the Incas in South America. Today, it is often associated with rice cultivation in the Philippines, Vietnam, and Japan, although it is also used for a variety of other crops. There are two types of terrace farming. Storage terraces are primarily built to collect and store water until it flows into the ground. This is the type that is typically seen in rice paddies. Gradient terraces are designed to slow down runoff. Generally, the slopes of terraces are seeded with grasses to maintain their shape, and the terraces follow the contour of the land.

DIFFERENTIATED INSTRUCTION | English Language Learners

Using Context Clues Remind students that if they come across unfamiliar words, they can look for context clues, or words and phrases nearby that help them understand the word. On this page, model looking for context clues for provisioning (food, fuel, water, wood) and concluding that provisioning means "supplying needs." Have students work in pairs or small groups to find context clues for other unfamiliar words.

Beginning Have students flag context clues that help them understand *supporting*, *regulating*, and *cultural*. Then, have them check their work in a dictionary.

Intermediate Have students use context clues to write definitions for *supporting* and *regulating*. Then, have them use each word in a sentence that shows the meaning.

Advanced Have students use context clues to figure out unfamiliar words in the section "Reversing Land Degradation," such as *rehabilitation*, *composting*, *runoff*, and *retention*.

CHAPTER INVESTIGATION B

Engineering Design *Wildlife Crossings and Corridors*

Time: 100 minutes or 2 days

Students will design a wildlife crossing or corridor to reduce the adverse impacts of human activity on a population, and they will then evaluate and refine their solution before producing a model.

Go online to access detailed teacher notes, answers, rubrics, and lab worksheets.

Video

Time: 3:38 seconds

Use **Video 16-3** to show a variety of animals crossing a highway using a wildlife corridor near Parleys Summit in Utah. Have students compare how the small rodent crosses between rocks at time 0:32 to how the larger animals cross during the day. Students should consider how the design of the wildlife corridor was intentional to increase usage of the pathway by a variety of species.

SEP Design Solutions *Sample answer: Producing paper products from recycled materials could help reduce the demand for trees to be cut down by the paper industry.*

CHAPTER INVESTIGATION

Wildlife Crossings and Corridors
How can wildlife crossings or corridors help a species access all parts of its habitat?
Go online to explore this chapter's hands-on investigation about wildlife corridor design.

Restoring Habitats In the United States, most of the protected land resources are undeveloped lands with various forms of protection status. These include national conservation areas, national parks, national forests, national wildlife refuges, and wildernesses. Worldwide, many biodiverse regions have been protected in ways that benefit local people. Ecosystems that are damaged by overuse need more than conservation measures to sustain biodiversity.

Ecological restoration is the process of renewing a natural ecosystem that has been degraded or destroyed. For example, artificial reefs are being used to help restore coral reefs damaged by human activity. Consider also the ecological restoration occurring in Louisiana's coastal wetlands. These ecosystems are being destroyed by upstream dams that hold back sediments that should replace sediment the marshes lose to the sea. Channels cut through the marshes for oil exploration and extraction have encouraged erosion, and the rising sea level threatens to drown existing plants. Restoration efforts now underway in Sabine National Wildlife Refuge have the goal of reversing some of those losses (**Figure 16-25a**).

VIDEO

VIDEO 16-3 Go online to watch a wildlife corridor in use by several different species of animals.

As more roads and buildings continue to create barriers across natural lands, more habitats will become fragmented. This fragmentation forces animals to migrate to unfamiliar territory or prevents them from accessing their migratory routes entirely. In some places, engineers and ecologists have developed **wildlife corridors** designed to help local migratory species to navigate between human-constructed barriers such as roads and housing developments. These corridors, such as the one shown in **Figure 16-25b**, help to divert animals from crossing highways, busy roads, and other areas where their traditional migratory patterns intersect.

SEP Design Solutions *How might engineers have broken down the larger problem of habitat fragmentation into smaller problems to work on?*

FIGURE 16-25 Restoring marsh (a) in areas that had become open water increases animal abundance and overall species richness of the area. Wildlife corridors (b) have been designed to mitigate the effects of habitat fragmentation and allow animals to pass safely from area to area.

514 CHAPTER 16 SURVIVAL IN CHANGING ENVIRONMENTS

CONNECTIONS | Citizen Science

Using Technology The Seek app was developed by iNaturalist to allow anyone to contribute to science by collecting and sharing data with repositories, such as the Global Biodiversity Information Facility in Copenhagen. This facility provides biodiversity data to researchers and people around the world so they can assess biodiversity in different regions. Using geotagged smartphone images of plants and animals, researchers collect data showing how climate change is affecting organisms, where invasive species are spreading, and the identity of organisms that can transmit disease in humans. This information can be used to set priorities for conservation and health initiatives in different regions.

Take Action Plan a BioBlitz! Use the planning sheets found at https://www.nationalgeographic.org/projects/bioblitz to create an event to collect observational data using the iNaturalist or Seek apps, or on paper.

CONNECTIONS

CITIZEN SCIENCE Documenting Earth's existing biodiversity and monitoring the effects of climate change and other human-caused factors are enormous undertakings. Scientists need to know the locations and sizes of a species' population in order to monitor changes in its range and numbers. Given the vast number of existing species, gathering such information is a task too big to be accomplished by scientists alone.

Fortunately, modern technology makes it easy to crowdsource data—to gather and organize information collected independently by a large number of individuals. Anyone with a camera can photograph organisms they encounter and then upload information about their sighting to an online database. Researchers then access this information.

Consider the iNaturalist project, an online social network for sharing information about biodiversity. The goal of iNaturalist is to foster a public appreciation of biodiversity, while also gathering scientifically valuable biodiversity data. The iNaturalist app allows users to upload photos of organisms they observe, along with the time and location of the observation. Users do not need to know exactly what they saw; in fact, they are encouraged to post unknown organisms.

Members of the iNaturalist community, which includes individuals with specialized expertise, collaborate in identifying the species in the uploaded photos. The resulting data are made available to anyone who is interested.

Since its launch in 2008, iNaturalist has collected more than 86 million observations documenting sightings of over 345,000 different species from locations throughout the world. To learn more about iNaturalist, visit iNaturalist.org or download the iNaturalist app.

Citizen scientists provide valuable data to scientists through their observations of the organisms around them.

Maintaining Biodiversity

Why should we protect biodiversity? From a selfish standpoint, doing so is an investment in our future. Healthy ecosystems are essential to the survival of our species. Other organisms produce the oxygen we breathe and the food we eat. They remove excess carbon dioxide from the air and decompose and detoxify wastes. Plants take up rain and hold soil in place, preventing erosion and reducing the risk of flooding. Wild species produce medically valuable compounds that we are still discovering. Wild relatives of crop plants are reservoirs of genetic diversity that plant breeders draw on to protect and improve crops.

There are ethical reasons to preserve biodiversity too. All living species are the result of an ongoing evolutionary process that stretches back billions of years. Each species has a unique combination of traits, and extinction removes that collection of traits from the world forever. The more diverse a biological system is, the greater is its capacity to recover from damage. This applies at all levels of life.

Go Online SIMULATION

Saving the Reefs Go online to evaluate solutions for the restoration of coral reef habitats.

16.4 REVIEW

1. **Identify** Describe two alternative energy resources.
2. **Synthesize** How might an increase in green spaces contribute to sustainability?
3. **Evaluate** What are some design considerations that would need to be evaluated when deciding on alternative materials or rehabilitation strategies?

Crosscurricular Connections

Earth Science Students may recall from their Earth science classes that the biosphere, where plants and animals live, is closely connected to the atmosphere, the hydrosphere, and the geosphere, which they may recall from Chapter 2. Have students consider how a disruption in any of these systems can affect the biodiversity in a region. Have them identify the organisms that remove carbon dioxide from the atmosphere and how the atmosphere would be affected if these organisms were no longer able to perform this service. Have students also consider how deforestation affects the geosphere. Have them work with a partner or small group to identify other processes that are affected when biodiversity is not maintained.

Go Online SIMULATION

Saving the Reefs Students work within the simulation to develop a decision matrix about approaches to coral reef restoration.

16.4 REVIEW

1. *Two sources of alternative energy are solar power and wind power. Wind and sunlight are both examples of renewable resources.* **DOK 1**
2. *Sample answer: More green spaces, especially in urban and suburban areas, helps maintain natural habitats for more biodiverse populations of organisms. Humans benefit from green spaces as places to enjoy nature. Air quality improves with more green space, which benefits human health and the environment.* **DOK 2**
3. *Sample answer: Alternative materials may be more costly to produce, and they may have other downsides such as using different natural resources. Rehabilitation strategies should take into account new potential impacts to the environment.* **DOK 2**

MINILAB
MODELING HUMAN-CAUSED CHANGES IN THE ENVIRONMENT

In this activity, students will model the effect of human activities on biodiversity by analyzing the effect of dam construction on a population of beetles.

Time: 30 minutes

Materials and Advance Preparation

The beads in the experiment can be replaced by other small-colored objects, such as paper hole punch-outs. Students can also put letters on the map to represent the beetle populations.

Procedure

In **Step 2**, ask the students which color beetle would be more suited to live in each location. Then, have the students put more of that color beetle there. Students should have at least one bead of each color in both lower and higher elevations on both sides of the river.

In **Step 4**, make sure students remove the beads that are in flooded areas. They should not push them to dry areas.

MINILAB

MODELING HUMAN-CAUSED CHANGES IN THE ENVIRONMENT

SEP Develop and Use a Model How does a human-caused environmental change affect biodiversity?

Environmental change, including change due to human activity, can cause changes in populations, which affect a region's biodiversity. Human-caused habitat destruction can also disrupt ecosystems and threaten the survival of some species. In this activity, you will model how construction of a dam affects a species of beetles living in an area.

The construction of a dam drastically changes the environment around it.

Materials

- white poster board or sheet of paper
- color markers
- green beads (10)
- brown beads (10)
- yellow beads (10)
- blue construction paper
- scissors

Procedure

In this investigation, you will model the effects of a dam on a population of beetles living in a valley. The beetles will be represented by green, brown, and yellow beads.

1. On the white poster board, use the markers to draw an image of a river with banks on either side. One side of the river has a more desert-like habitat, while the other side of the river has a more forested habitat.
2. Distribute the different-colored beads onto the image. Each bead represents a group of similarly colored beetles. Do not put any beetles on the river. Place beads of all colors on both sides of the river. You do not need to use a specific number of each colored bead or even all the beads. If you think a bead of specific color would have an easier time surviving in a location, put more beads of that color there.
3. Count and record the number of beads on each side of the river. Keep the beads on the image, and do not move them.

A dam built downriver causes an artificial lake to form around the riverbed, causing areas to become flooded.

4. To model this lake, use the scissors to cut out the lake from the blue construction paper. You can make the lake whatever shape you want it to be.
5. Overlay the lake on the drawing of the river. To do this, start at one of the sides. As you place the lake on the river, remove the beads underneath it. This represents the beetles that died as a result of the formation of the lake.
6. Count and record the remaining beads on each side of the river.

Results and Analysis

1. **Represent Data** Make a data table that shows the number of each colored beetle on each side of the river before and after the dam was built.
2. **Analyze** How did the dam affect the beetle population? Consider the number and type of beetles affected.
3. **Analyze** How did the dam affect the biodiversity of the beetle population? Explain your reasoning.
4. **Predict** What changes might occur within the beetle populations on the drier side of the lake compared to those on the forested side of the lake after 5, 25, and 50 years?
5. **Predict** What might happen to the beetle populations if the river and the lake dried up?

516 CHAPTER 16 SURVIVAL IN CHANGING ENVIRONMENTS

Results and Analysis

1. **Represent Data** *Sample data table* **DOK 1**

Beetle color	*North side before the dam is built*	*North side after the dam is built*	*South side before the dam is built*	*South side after the dam is built*
green	6	4	4	1
brown	7	5	3	3
yellow	3	2	7	4
total	16	11	14	8

2. **Analyze** *Sample answer: The dam and destruction of beetle habitat caused the population to decrease 11 beetles (30 to 19). Beetle color was also affected. More green and yellow beetles were lost.* **DOK 3**
3. **Analyze** *Sample answer: Before the dam, the population consisted of 33 percent of each color of beetle. The dam destroyed more of the green and yellow beetles and fewer of the brown beetles. In the resulting population, the genes of the green and yellow beetles became scarcer. This reduced the biodiversity of the population.* **DOK 2**
4. **Predict** *Sample answer: Green beetles were more suited to living in lower elevations. Yellow beetles lived near the border of lower and higher elevations, and brown beetles lived in the higher elevations. Much of the lower-elevation areas are gone, and the remaining parts are fragmented. The model shows what would happen right after the flooding but does not consider the fragmented low-elevation areas. We predict the population of green beetles will continue to decrease 5 years in. In 25 years, we expect the green beetles may disappear from the population. In 50 years, we expect both green and yellow beetles may disappear from the population.* **DOK 3**
5. **Predict** *Sample answer: The removal of the dam would increase the habitat of the beetle in the short term. Without a river, however, we expect the populations of green and yellow beetles would continue to dwindle. In the medium term, the lack of water and plants would also affect brown beetles. So if the river and the lake dried up, we predict the overall population would actually become smaller and it would be mostly brown beetles.* **DOK 3**

TYING IT ALL TOGETHER
TRACKING TUSKLESS ELEPHANTS

WHY DO NEW SPECIES EMERGE WHILE OTHER SPECIES DISAPPEAR?

In this chapter, you learned how human activity can lead to changes in the biodiversity of an ecosystem. The Case Study described how ivory poaching has resulted in an increased proportion of tuskless female elephants in Mozambique's Gorongosa National Park. Savanna elephants first reproduce at about 12 years old and give birth every three to four years. The offspring of tuskless mothers are twice as likely to be female than male, and not a single tuskless male elephant has been seen in the park for over five decades. These clues directed scientists looking for the genetic origins of tusklessness to investigate X-linked genes involved in tooth and gum formation that may be lethal if inherited by male offspring.

In this activity, you will apply genetic and evolutionary principles to make predictions regarding an elephant population. Work individually or in a group to complete these steps.

Develop a Model

1. Suppose all the males in an elephant population have tusks and half of the females are tuskless. If the gene for tusklessness is linked to the X chromosome, recessive, and lethal to male offspring, how frequently would you expect the trait of tusklessness to occur in the population after three generations?

Construct an Explanation

2. Under what conditions could the presence of the gene for tusklessness in an elephant population lead to speciation?
3. Under what conditions could the presence of the gene for tusklessness in an elephant population lead to extinction?
4. African forest elephants, such as the one shown in **Figure 16-26**, take almost twice as long to reach reproductive age, with five to six years between offspring. The graph shows projected forest elephant population growth modeled using the natural birth and death rates of forest elephants in Dzanga, Central African Republic. How would the effects of poaching on forest elephant populations compare to savanna elephant populations? Explain your reasoning.

FIGURE 16-26
African forest elephants can be identified by their smaller size and downward-pointing tusks. The graph compares the projected growth rates of savanna and forest elephant populations.

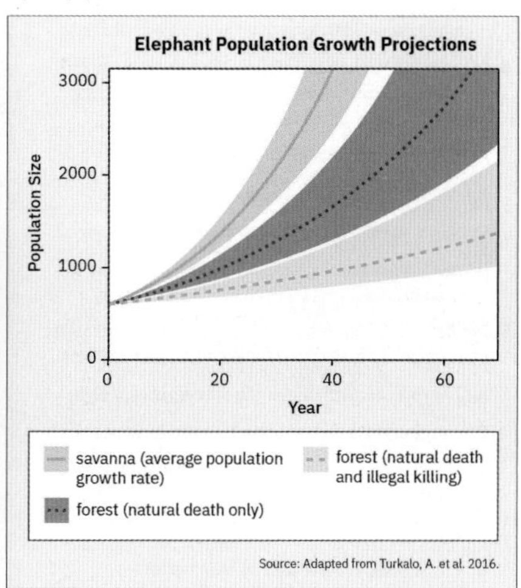

TYING IT ALL TOGETHER

ELABORATE

The Case Study examines how ivory poaching is resulting in an increase in female African elephants being born without tusks. In the Tying It All Together activity, students will develop a model to simulate how frequently tusklessness will occur in the population and construct an explanation for how the presence of the trait could lead to different outcomes.

This activity is structured so that students may begin as they work through the lesson. **Step 1** has students developing a model based on certain characteristics about the tusklessness trait. **Step 2** corresponds to the content in Section 16.1 and asks students to explain how the trait can lead to speciation. **Step 3** corresponds to the content in section 16.2 and asks students to explain how the trait could lead to extinction. **Step 4** has students consider the graph that projects elephant population growth models and compare the populations of forest and savanna elephants. Alternatively, the activity can be a culminating project at the end of the chapter. The modeling and analysis practiced here reflects a reasoning process that students may use to support a claim in the Unit 5 Activity.

Go online to access the Student Self-Reflection and Teacher Scoring rubrics for this activity.

SCIENCE AND ENGINEERING PRACTICES
Constructing Explanations and Designing Solutions

How Will the Population Change? To help students prepare for this activity, provide them with an overview of Punnett squares and sex-linked traits, first introduced in **Chapter 12**. Some students may find it helpful to review the minilab from that chapter where they modeled the probability of inheriting a trait. Students should use evidence from each section and outside resources to apply the scientific ideas and evidence introduced through the lesson to provide an explanation of how the tusklessness trait will change the population. They should link evidence from the graph to the reasoning provided throughout the lesson to make a claim that they can support about the effects of poaching and human impacts on forest and savanna elephant populations. Once students have completed constructing their arguments, have them work with a partner to ensure that their arguments are supported with evidence.

CHAPTER 16 **REVIEW**

EVALUATE

REVIEW KEY CONCEPTS

1. *A, C, D* **DOK 1**
2. *Sample answer: In allopatric speciation, a physical barrier arises and divides a population, ending gene flow between them and resulting in speciation over time. Sympatric speciation, which is common only in plants, does not involve a physical barrier, but occurs when some plants in a population produce diploid gametes during meiosis. This results in polyploid offspring that can self-fertilize and breed with other polyploid plants, but not with normal plants. The end of gene flow between polyploid and diploid plants results in speciation.* **DOK 2**
3. *B* **DOK 1**
4. *C* **DOK 2**
5. *B, C, E* **DOK 2**
6. *Sample answer: With the loss of so many species during the mass extinction event, surviving organisms face fewer predators and fewer competitors for resources. This provides these organisms a chance to move into new habitats, where they undergo adaptive radiation as they fill ecological niches vacated by species lost to extinction.* **DOK 3**
7. *A, C, E* **DOK 2**
8. *A, B, D* **DOK 2**
9. *Sustainable living: switching from nonrenewable to renewable energy sources, using irrigation systems that deliver water directly to plant roots, reducing food waste, recycling electronic components. Unsustainable living: leaving lights on when not in use, switching from biodegradable packaging to plastic packaging, spraying crops with large amounts of pesticides, converting natural areas into strip mines.* **DOK 2**
10. *A, D, E* **DOK 2**

CRITICAL THINKING

11. *Sample answer: Yes, this is reasonable because predator animals rely on prey for survival. If new species of prey animals emerge, they will provide new feeding opportunities and therefore new niches for predator species to fill. This results in the emergence of new predator species adapted to the increased diversity of prey species.* **DOK 3**
12. *Sample answer: Native populations will likely experience an increase in extinction rate because of competition from or predation by nonnative species. Native species often lack adaptations to protect themselves because they have not coevolved with those species.* **DOK 3**
13. *Sample answer: Industrialization has increased habitat destruction, which has put many species at risk of extinction. It is important for people to understand the connection between industrialization and habitat destruction because personal choices can reduce the negative effects of industrialization. For example, people can reduce their consumption of manufactured goods and support recycling efforts, thereby reducing the incentive for industries to seek new resources in natural areas. People can also support public policies that regulate industries, preventing them from using excessive amounts of natural resources and regulating pollutants and land degradation that can destroy natural habitats.* **DOK 3**

CHAPTER 16 **REVIEW**

REVIEW KEY CONCEPTS

1. **Identify** Which situations can lead to speciation? Select all correct answers.
   - **A.** A physical barrier separates a population into two new populations.
   - **B.** Interbreeding takes place between individuals from two different populations.
   - **C.** Some individuals breed at a different time than others in the same population.
   - **D.** Increases in flooding change soil conditions in part of the habitat of a plant population.
   - **E.** Females from one population respond to courtship displays of males from a different population.
2. **Contrast** How is allopatric speciation different from sympatric speciation?
3. **Describe** Which phrase would best describe the fossil record of a line of related organisms that followed a punctuated equilibrium model of evolution?
   - **A.** small, incremental changes at a constant rate over long periods of time
   - **B.** long periods of little change interspersed with short periods of large changes
   - **C.** short periods of little change interspersed with long periods of small, incremental changes
   - **D.** short periods of small, incremental changes interspersed with long periods of large changes
4. **Identify** Which human impact on biodiversity most directly prevents wildlife from accessing resources needed for their survival?
   - **A.** capture of animals for the pet trade
   - **B.** introduction of nonnative species to new areas
   - **C.** degradation and fragmentation of natural areas
   - **D.** overharvesting of species through hunting and fishing
5. **Predict** Which factors would increase the vulnerability of a species to extinction? Select all correct answers.
   - **A.** The reproduction rate of the species is very high.
   - **B.** The species is preyed upon by an invasive species.
   - **C.** The species is endemic to a small geographic area.
   - **D.** The species can easily migrate to new geographic areas.
   - **E.** The species is harvested by humans for food, medicine, or other uses.
6. **Explain** Why does the number of species on Earth increase after a mass extinction event ends?
7. **Identify** What factors contribute to a country's per capita ecological footprint? Select all correct answers.
   - **A.** carbon emissions
   - **B.** form of government
   - **C.** natural resource usage
   - **D.** total land area
   - **E.** waste produced
8. **Cause and Effect** Which outcomes would be expected if humans significantly reduced greenhouse gas emissions? Select all correct answers.
   - **A.** Acid rain would diminish.
   - **B.** The rate of global warming would decrease.
   - **C.** Fewer renewable energy sources would be used.
   - **D.** Changes in rainfall and storm patterns would be observed.
   - **E.** The ecological footprints of developed countries would increase.
9. **Classify** Identify whether each action contributes to sustainable or unsustainable living.
   - reducing food waste
   - recycling electronic components
   - leaving lights on when not in use
   - converting natural areas into strip mines
   - spraying crops with large amounts of pesticides
   - switching from nonrenewable to renewable energy sources
   - switching from biodegradable to plastic packaging
   - using irrigation systems that deliver water directly to plant roots
10. **Identify** Which strategies can help reverse land degradation and deforestation? Select all correct answers.
    - **A.** using terraced land for growing crops
    - **B.** encouraging vehicle off-roading activities
    - **C.** opening public lands to cattle and sheep grazing
    - **D.** restricting commercial development using zoning laws
    - **E.** finding alternatives to wood in manufacturing and construction

518 CHAPTER 16 SURVIVAL IN CHANGING ENVIRONMENTS

CRITICAL THINKING

11. **Justify** A biologist hypothesizes that adaptive radiation of predator animals will follow an increase in prey diversification. Does this seem to be a reasonable hypothesis? Explain.

12. **Predict** Climate change is expected to increase the rate at which species migrate into new ecosystems. Explain what effect this will have on the extinction rate of native populations in those ecosystems.

13. **Relate** What is the relationship between industrialization and habitat destruction? How important is it for people to understand this relationship? Explain your reasoning.

14. **NOS Science Knowledge** What social and economic benefits do humans enjoy as a result of preserving biodiversity?

MATH AND ENGLISH LANGUAGE ARTS CONNECTIONS

Read the following passage. Use evidence from the passage to answer the question.

As land plants evolved from aquatic ancestors, they benefitted from several key innovations and subsequent adaptive radiations. The most recent of these innovations was the evolution of flowers. Flowers are reproductive structures that provide attractions for pollinators as well as protection for fertilized embryos. Data show the relative distribution of species belonging to the flowering plants compared to those belonging to nonflowering plant groups.

Plant type	Terrestrial plant groups	Relative distribution (%)
nonflowering plants	bryophytes and others	10.3
	gymnosperms	0.3
flowering plants	angiosperms	89.4

1. **Evaluate a Science or Technical Text** Evaluate whether the data support the claim in the text that flowers were a key innovation during the evolution of land plants. Explain your reasoning.

2. **Argue From Evidence** Using facts provided in the table, state and support a claim about the importance of sustaining biodiversity in ecosystems.

Facts about Freshwater Ecosystems	
contain 0.01% of Earth's water	support 33% of vertebrates and 10% of all animal species
occupy < 1% of Earth's surface	trap and break down pollutants, toxins, and heavy metals

Read the following passage. Use evidence from the passage to answer the question.

Before Europeans colonized North America, the Florida panther (*Puma concolor coryi*) was abundant throughout the southeastern United States. Currently, scientists estimate that only 120 to 230 adult panthers remain in the wild, with their current range restricted to southern Florida. In the 36 years from 1981 to 2017, the leading cause of death for Florida panthers was vehicular collisions. Before 2000, the rate of panther roadkill was four or fewer per year. Since 2000, the rate has increased substantially, with a high of 34 in 2016.

3. **Cite Textual Evidence** Propose a solution to aid in the restoration efforts for the Florida panther. Cite information from the text to support your answer.

REVISIT HUMMINGBIRD EVOLUTION

Gather Evidence In this chapter, you have seen how changing environmental conditions lead to an increase in the individuals of some species, the emergence of new species, and the extinction of other species. In Section 16.1, you learned about the mechanisms of speciation. In Section 16.2, you learned about factors that lead to extinction. In Sections 16.3 and 16.4, you learned about the impact humans have on the environment and ways that this impact can be reduced.

1. How are changing environmental conditions affecting populations of hummingbirds?
2. What other questions do you have about how hummingbirds became adapted to their environment? What else do you need to know to understand hummingbird speciation?

REVISIT THE ANCHORING PHENOMENON

1. *Sample answer: Changing environmental conditions may alter the migration patterns of some species of hummingbirds. Changing environmental conditions could also affect the availability of resources hummingbirds need to survive in their habitats and could result in their moving to different habitats or becoming extinct.* **DOK 3**
2. *Sample answer: What environmental factors most affect the speciation of hummingbirds? How have changes in the environment affected hummingbird speciation? How are hummingbird species changing today?* **DOK 3**

14. *Sample answer: Most people place a high value on the preservation of natural landscapes and ecosystems. Many people enjoy recreational activities such as hiking, camping, and other outdoor activities that bring them into contact with species in their natural environment. Such activities also provide economic benefits to the tourism and travel industries associated with biologically diverse ecosystems. These industries provide jobs and sustain local economies. Preserving biodiversity ensures that the recreational and inspirational value of natural areas remains available to people in the future.* **DOK 3**

MATH AND ELA CONNECTIONS

1. *Sample answer: The data support the claim. Key innovations lead to adaptive radiation and proliferation of new species to fill new niches in the environment. Flowering plant species outnumber nonflowering species by almost nine to one (89.4 percent of the total compared to 10.6 percent). This suggests that the development of flowers gave some plants an evolutionary advantage, allowing greaters proliferation.* **DOK 2**
2. *Sample answer: It is very important to sustain biodiversity so that ecosystems continue to function properly and support life on Earth. For example, freshwater ecosystems contain only a tiny percentage of Earth's water and take up less than one percent of Earth's surface, yet freshwater ecosystems play a large role in supporting life on Earth. 10% of all animals and about one-third of all vertebrates rely on freshwater ecosystems for survival.* **DOK 3**
3. *Sample answer: Wildlife corridors could be very useful in helping to restore Florida panther populations. Human activity has been the main cause in the near extinction of this species. According to the data, the rate of panther death due to vehicle collisions is increasing. This says that panther habitats are fragmented by human development, requiring them to cross roads often. Building wildlife corridors would provide safe passage and help reestablish a healthy population.* **DOK 2**

UNIT 5 SUMMARY

Class Discussion

Encourage students to practice speaking, listening, and collaborative skills when discussing these questions as a class or in smaller groups as they review the topics in Unit 5.

1. How are multiple lines of evidence used to determine biological evolution? What can fossils show about the change in organisms over time? How can homologous and analogous structures help determine evolutionary relationships? What do genetic sequences reveal about evolutionary relationships?
2. Consider a population of organisms that is split into two groups due to a natural disaster. The new populations have an unequal distribution of alleles. How do you expect the new populations will change over time? What is this phenomenon and how does it differ from other mechanisms of evolution?
3. Consider that a sewage system lining fails. What would be the ramifications to the local environment? How would this affect local organisms and humans? Explain possible solutions humans could put into place that would reduce the impact to the environment.

Go Online

Five Performance Tasks for Unit 5 are available on MindTap to assess students' mastery of the following NGSS Performance Expectations:

- Task 1: HS-LS4-1
- Task 2: HS-LS4-2 and HS-LS4-3
- Task 3: HS-LS4-4 and HS-LS4-3
- Task 4: HS-LS4-5
- Task 5: HS-LS4-6 and HS-ETS1-4

Go Online

The Virtual Investigation, **Hummingbirds on the Move**, includes observational evidence that students may find useful for supporting their claims about how hummingbirds have adapted to their environments.

UNIT 5 **SUMMARY**

EVOLUTION AND CHANGING ENVIRONMENTS

CHAPTER 14
EVIDENCE FOR EVOLUTION

How do we know how species have evolved?

- Common ancestry and biological evolution are supported by multiple lines of evidence including the fossil record, developmental patterns, anatomical structures, and DNA sequences.
- Fossils are typically found in sedimentary rock. The most recent fossils are found above layers that contain older fossils.
- Homologous structures are derived from a common ancestor. Analogous structures evolved independently in different lineages.
- Evolutionary relationships can be determined by comparing amino acid or nucleotide sequences from different organisms.

CHAPTER 15
THE THEORY OF EVOLUTION

Why do populations evolve over time?

- Natural selection is one process that drives evolution. The four principles of natural selection are competition, heritable genetic variation, overproduction, and adaptation.
- Natural selection acts on genetic variations. An advantageous trait helps individuals survive and reproduce.
- Directional selection, stabilizing selection, and disruptive selection are three ways that natural selection can change the distribution of a trait in a population.
- Other mechanisms that can lead to the evolution of populations include gene flow, genetic drift, and sexual selection.

CHAPTER 16
SURVIVAL IN CHANGING ENVIRONMENTS

Why do new species emerge while other species disappear?

- Changes in environmental conditions may result in population increases, speciation, or extinction.
- Engineering solutions, such as optimizing renewable resources, can be used to reduce the negative effects of human activities on the environment and biodiversity.

Go Online **VIRTUAL INVESTIGATION**

Hummingbirds on the Move
How did hummingbirds become adapted to their environments?
Go into the field and gather evidence about hummingbird adaptations.

520 UNIT SUMMARY

DIFFERENTIATED INSTRUCTION | English Language Learners

Share Information in Groups Have students work in mixed groups to summarize the content taught in the lessons and give examples from the text for each bullet on the page. Student questions and answers will vary based on their proficiency levels. Examples include:

Beginning Provide groups with sentence frames to develop their questions and answers, for example: *Chapter ______ is about ______. In the section ______, we learned that ______.*

Intermediate Have students ask questions with *What? How?* and *Why?* Encourage them to answer in complete sentences.

Advanced Have students ask and answer questions and then ask follow-up questions to build on one another's ideas

HOW DID HUMMINGBIRDS BECOME ADAPTED TO THEIR ENVIRONMENTS?

Argue From Evidence In this unit, you learned that evolution occurs in populations of species over time as they adapt to changes in their environment. The diversity of hummingbirds is one example of how certain species are able to rapidly expand into open ecological niches.

Using DNA data, scientists constructed a family tree for hummingbirds and their closest relatives. They determined that the branch leading to hummingbirds arose about 40 million to 50 million years ago. That is when they split from their sister group, the swifts and treeswifts. Scientists think that this likely occurred in Europe or Asia. Hummingbird fossils that date back 30 million to 45 million years ago have been found in Germany, France, and Poland.

If hummingbirds first evolved in Eurasia, how did they get to South America? Scientists hypothesize that ancestral hummingbirds likely migrated across the Bering Land Bridge, which once connected Asia to Alaska. They traveled from there down to what is now South America. Over the course of 22 million years—a relatively short period of evolutionary time—a common ancestor gave rise to the 338 species of hummingbirds that exist today. Hummingbirds returned to North America around 12 million years ago, and others began to inhabit the Caribbean around five million years ago. Research indicates that new species of hummingbirds continue to arise, though at a much slower pace than they have in the past. Scientists estimate that over the next several million years, the total number of hummingbird species could double as they maximize the number of ecological niches available to them.

 **Claim** Make a claim about how climate change may affect hummingbird speciation in the future.

 **Evidence** Use the evidence you gathered throughout the unit to support your claim.

 **Reasoning** To help illustrate your claim, develop a model that illustrates how changes in a hummingbird's environment could affect its ability to survive and reproduce.

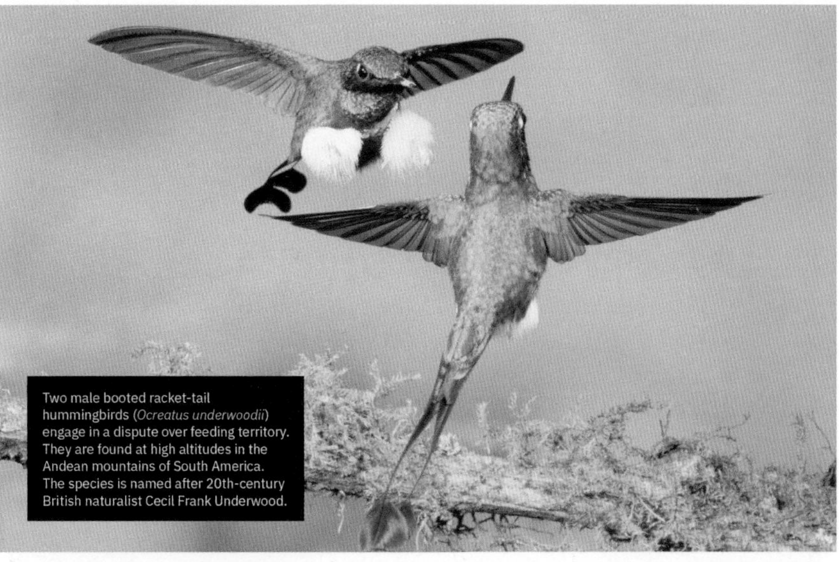

Two male booted racket-tail hummingbirds (*Ocreatus underwoodii*) engage in a dispute over feeding territory. They are found at high altitudes in the Andean mountains of South America. The species is named after 20th-century British naturalist Cecil Frank Underwood.

REVISIT THE ANCHORING PHENOMENON

This Unit Activity asks students to formulate a claim to answer the question posed at the beginning of the unit, cite evidence they have obtained throughout the unit, and use the concepts they have learned to explain how this evidence supports their claim.

C Formulating a Claim Students should write their answer to the question "How did hummingbirds become adapted to their environments?" in the form of a single sentence. Examples of claims that students might make using terms and concepts from the unit chapters include: "Hummingbirds changed as they maximized the ecological niches available to them" or "Hummingbirds coevolved with the flowers they fed on."

E Citing Evidence Have students review their answers to the Hummingbird Evolution questions in Sections 14.1, 15.2, and 16.1 and in each Chapter Review as they determine what evidence to use to support their claim. If you wish to have students look for evidence from sources other than this text, they can find useful information on National Geographic, as well as the National Forest Foundation.

R Explaining Reasoning In addition to their explanatory writing, models that show how hummingbirds could be affected by the changing environment are convenient structures for students to build their explanations around.

Go online to access the Student Self-Reflection and Teacher Scoring rubrics for this activity.

DIFFERENTIATED INSTRUCTION | English Language Learners

Claims and Evidence Have students use a graphic organizer to structure their paragraph, with a box for students to write a sentence for the claim, three boxes for evidence, and a box for explaining their reasoning.

Beginning Provide sentence frames or sample claims for students to choose from. Have them work in pairs or small groups to look for evidence. Have students sketch a model.

Intermediate Have students write their claim and evidence and then exchange their writing with a classmate. Have partners check that the evidence relates to the claim. Finally, have students briefly describe their model.

Advanced After students write their claim and evidence and develop their model, have them work in pairs to explain how their model relates to the claim and evidence they wrote. Allow pairs time to revise their work based on their discussions. Remind them to use transition words.

APPENDIX CONTENTS

LABORATORY SAFETY AND PROCEDURES

Working Safely in the Laboratory

Working in the lab will help you master both the content of biology and the processes of science. But the laboratory can be a dangerous place if safety rules are not followed. You must be responsible for your safety and the safety of others.

General Safety Rules Study these symbols and the warnings or guidelines they represent in the lab. Look for them in each lab to indicate a safety risk to you and your fellow students where you should use extra caution.

1. Thoroughly read each lab several times before you begin. Follow instructions exactly. Ask your teacher if you have questions.
2. Only perform authorized activities. Do not work in the lab or handle equipment without supervision.
3. Use safety equipment specified in each lab.
   - Wear safety goggles when working with chemicals, burners, or any substance that might get into your eyes.
   - Wear the appropriate gloves when handling any toxic or hazardous agents.
   - Wear a lab apron or lab coat when working with chemicals or heated substances.
   - When using lab equipment and chemicals, keep your hands away from your body, mouth, eyes, and face.
4. Never eat or drink in the lab. Do not inhale chemicals or taste them.
5. Use caution with liquids. If a spill occurs, immediately ask your teacher for the proper cleanup procedure. Never pour chemicals or other substances into a sink or trash can.
6. Tie back chin-length or longer hair.
7. Remove or tie back any loose clothing or jewelry that can hang down and come in contact with chemicals or a flame. Avoid wearing sandals or other open-toed shoes.
8. Know the locations of safety showers and eyewashes and how to use them.
9. Wash your hands with soap and water after performing an investigation.

Injury	Response
cuts and abrasions	Stop the bleeding by applying direct pressure. Call your teacher immediately.
fainting	Leave the person lying down. Loosen any tight clothing. Call your teacher immediately. Keep people away.
solid foreign matter in the eye	Close or cover the eye. Call your teacher immediately.
poisoning	Note the suspected substance. Call your teacher immediately.
spills on skin	Flush with large amounts of water or use the safety shower. Call your teacher immediately.
chemical splashes in the eye	Use the eyewash to flush with plenty of water. Call your teacher immediately.

APPENDIX A
Laboratory Safety and Procedures

Science Background

Lab Safety *"As professionals, teachers of science have a duty or standard of care to ensure the safety of students, teachers, and staff."* This statement from the National Science Teaching Association is important to keep in mind as you introduce students to the guidelines for laboratory (lab) safety. You can find it with an Internet keyword search such as "NSTA lab safety."

Giving students the opportunity to experience science through investigations and labs is essential to learning not only content but also the processes of science, but more importantly it is essential that the safety of all persons involved in lab procedures is the priority.

Prior to conducting any investigations or labs, have students demonstrate understanding of safety procedures by successfully answering the practice questions provided. If there are long periods of time between lab experiences, it may be useful to revisit the safety guideless prior to allowing students access to the investigation and the lab materials, especially if fire and/or newly introduced chemicals are going to be used.

During investigations and labs, it is essential that you supervise all students/groups; this can be easily accomplished by circulating and interacting with students/groups during the lab.

DIFFERENTIATED INSTRUCTION | English Language Learners

Environmental Print Environmental print includes signs, symbols, and other print students see in everyday life. Discuss the meanings of the safety symbols and labels. Use realia or sketches to show related safety equipment.

Beginning Have students draw the symbols and write the labels on separate index cards and match the symbols to the labels. Encourage students to explain the symbols in their first language. Then restate their explanations in English.

Intermediate Have students explain in a sentence or two what the symbol and label mean. Supply vocabulary for safety equipment or lab equipment such, as *goggles, apron, gloves, knife, beaker,* and *test tube,* as needed.

Advanced Have students work in small groups. Have them take turns pointing to a symbol, reading the label, and then describing a situation in which they might need the safety rule.

APPENDIX A (continued)

Heating and Fire Safety

 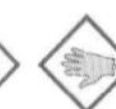  

1. Maintain a clean work area and keep all materials away from flames. Store backpacks, coats, and other personal items away from the immediate lab bench area.
2. Use only heat-resistant glassware for heating. Prior to using glassware, check for chips or cracks and notify your teacher of any damaged glassware. Never heat glassware that is not thoroughly dry. Use a wire screen to protect glassware from any flame.

3. Never reach across a flame.
4. Ask your teacher how to use a Bunsen burner. Never leave a heat source unattended.
5. Point the opening of a test tube or bottle that is being heated away from you and others.
6. Never heat a liquid in a closed container. The increase in pressure may cause the container to break, injuring you or others.
7. Hot glass looks like cold glass. Before touching a container that has been heated, hold the back of your hand close to the container. If you can feel heat, the container is too hot to handle. Use a clamp, tongs, heat-resistant gloves, or a hot pad to move the container.
8. If a fire breaks out or your clothing catches fire, smother it with a fire blanket or roll on the ground. NEVER RUN.

Working With Chemicals and Tools

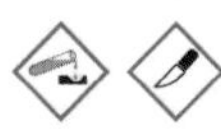    

1. Use materials only from properly labeled containers. Read chemical labels carefully.
2. Discard extra chemicals as your teacher directs. Notify your teacher if chemicals are splashed or spilled.
3. When diluting acids, always add acid to water. Never pour water into acid.
4. Do not discard matches or other solid materials in the sink.
5. Use a lubricant when inserting glass tubing or thermometers through stoppers. Protect your hand with a paper towel or a folded cloth.
6. Be careful when handling sharp objects such as pins and dissecting probes.

DIFFERENTIATED INSTRUCTION | Students with Disabilities

Accessibility All students should have access to investigations and labs. This may require adjustments to the lab setting. For example, students with special needs can be partnered with another student with the understanding that they are to contribute, although it may be in a modified capacity. If lab benches are too high for student accessibility, bring in a lower table so that students can still participate.

Handling Living Organisms

1. Treat all living things with respect. Your teacher will instruct you on how to handle each organism that is brought into the classroom. Do not touch any organism without permission. Many plants are poisonous or have thorns, and even tame animals may bite or scratch if alarmed.
2. Animals should be handled only when necessary. If an animal is excited or frightened, pregnant, feeding, or with its young, special handling is required.
3. Treat all microorganisms as if they were harmful. Use antiseptic procedure as directed by your teacher when working with microbes. Dispose of microbes as your teacher directs.
4. Clean your hands thoroughly after handling any organism or object that has been in contact with an organism.

5. Wear gloves when handling any animal. Report animal bites, scratches, or stings to your teacher at once.

Cleaning Up the Laboratory

1. After working in the lab, clean your work area, and return all equipment to its proper place.
2. Turn off all faucets and burners. Check that you have turned off the gas jet as well as the burner.
3. Dispose of chemicals and other materials as directed by your teacher.
4. Wash your hands thoroughly before leaving the lab.

Practice *Describe how you and your classmates would stay safe in each of the following situations.*

1. An investigation displays these safety symbols at the beginning:

2. Your team is to design your own plan to test various cleaning materials in a forensics investigation.
3. Your team is about to carry out an investigation into the effect of heat on a few chemical reactions.
4. One of your teammates accidently cuts a finger with a knife while cutting carrots for an investigation.
5. It is the period before lunch, and you are testing the breakdown rates of digestive enzymes on various food materials, including mints, crackers, and small cakes.

PRACTICE

1. *Sample answer: We should wear our eye protection, handle glassware safely, and take special precautions with the chemicals we are using.*
2. *Sample answer: We should carefully read the labels of the chemicals we are testing for cautions. We should also wear eye protection and aprons. When finished, we should find out from our teacher how to dispose of the chemicals and wash our hands as well as the area where we did our testing.*
3. *Sample answer: Before we set up the Bunsen burner, we will clear away all loose materials from the area of the burner; this includes tying back long hair and removing or tying back any loose clothing or jewelry. We will use only heat-resistant glass and be careful to not touch the glass with our bare hands. After we are finished, we will make sure the burner and gas line are turned off.*
4. *Sample answer: We grab a paper towel to apply pressure to the cut and immediately get our teacher.*
5. *Sample answer: Although I may be hungry right before lunch, we do not eat or drink in the lab. Food used in an investigation should not be eaten even if it seems safe.*

Visual Support

Lab Equipment The skills given on these pages seem obvious; however, keep in mind that this may be the first time that students have been exposed to investigations or labs that require the use of this equipment. Display the images and go through the steps to ensure that students are familiar with the equipment prior to using. This can be accomplished through a practical exam where students demonstrate understanding by showing the proper use of equipment, not just through the answering of written questions. The proper use of equipment increases safety in the lab.

Skills for Using Laboratory Equipment

The laboratory equipment you will encounter in biology investigations may require a little practice to help you make clear observations and collect useful data.

Using a Compound Microscope Review the parts of a typical microscope used in biology investigations and their functions for the best viewing experience and care of the microscope.

Clean the lenses with lens paper only. Other types of paper may scratch the lens. Be careful not to touch the lenses with your fingers, which can leave smudges.

When finished viewing, lower the stage, click the low power objective into position, and remove the slide. Switch off any light source and cover the microscope.

1. **Eyepiece** Contains a magnifying lens you look through. Keep both eyes open as you look through the eyepiece to avoid eyestrain.
2. **Arm** Supports the body tube. Place the arm toward you during viewing. To move the microscope, grasp the microscope here and place another hand under the base.
3. **Stage** Supports the slide. Place the area of the slide to be viewed over the hole and secure the slide with the stage clips.
4. **Coarse Adjustment Knob** Focuses the image under low power. Use this adjustment knob first to raise the low-power or shortest objective about 2 cm above the stage before placing the object on the stage. Then use this knob to bring the object into focus.
5. **Fine Adjustment Knob** Sharpens the image under high and low magnification. Only use this adjustment when the high-power or longest objective is in place. Using the coarse adjustment knob can easily crash the high-power objective into the slide, breaking the slide and damaging the objective.
6. **Base** Supports the microscope. Place one hand under the base while the other is on the arm when moving the microscope.
7. **Body Tube** Connects the eyepiece to the revolving nosepiece.
8. **Revolving Nosepiece** Holds and turns the objectives into viewing position. Gently turn the revolving nosepiece in either direction to click an objective into place.
9. **Objectives** Allow varying powers of magnification with lenses. Choose the low-power or shortest objective first. After using the coarse and then fine adjustment to focus on the slide, turn the revolving nosepiece to change to medium or high power.
10. **Stage Clips** Holds a slide in place. Gently push each end of the slide under the clips.
11. **Diaphragm** Regulates the amount of light passing up toward the eyepiece. Use the lever or dial to open or close the diaphragm.
12. **Light Source** Produces light or reflects light up toward the eyepiece. Adjust the source to direct as much light as possible through the lenses.

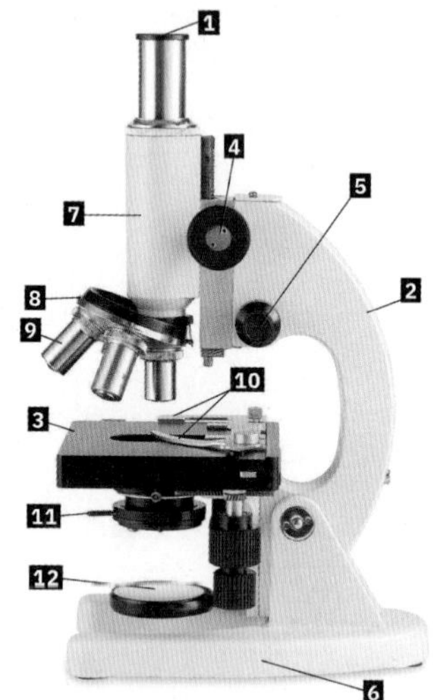

Stereomicroscopes These microscopes are designed to view larger specimens at lower magnification (usually 6× to 50×) and are common in biology laboratories. They have many of the same features as compound microscopes. The main difference is that they have binocular eyepieces for viewing with both eyes at the same time for greater depth perception.

Making a Wet Mount Slide

1. Obtain a clean microscope slide and a coverslip.
2. Using tweezers, place the specimen in the middle of the slide. The specimen must be thin enough for light to pass through it.
3. Place a drop of water on the specimen.
4. Hold a clean coverslip by its edges and place it at one edge of the drop of water. Slowly lower the coverslip until it lies flat on top of the specimen and the water. Lowering the coverslip from one side to the other minimizes or eliminates air bubbles under the coverslip. If air bubbles are present, gently tap the surface of the coverslip with a pencil eraser. You will not be able to see a specimen through an air bubble.
5. If you have too much water on the slide, touch the edge of a paper towel to the edge of the coverslip. This will draw off extra water, force air out of the mount, and further flatten the wet mount.

Staining a Wet Mount Specimen A stain may be added to a wet mount slide to make the specimen on the slide more visible. Follow these steps to stain a wet mount.

1. Collect a drop of stain with a dropper or pipette.
2. Add one drop of stain at the edge of the coverslip.

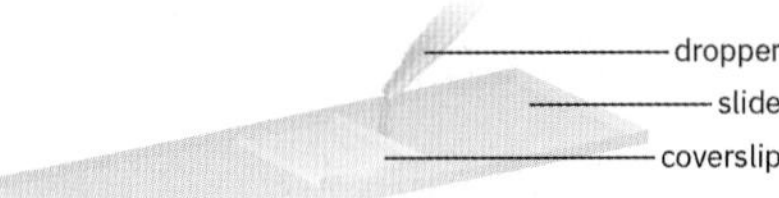

3. Hold a small piece of paper towel or lens paper with forceps. Touch the paper to the edge of the coverslip opposite the stain. The paper causes the stain to be drawn under the coverslip, staining the specimen.

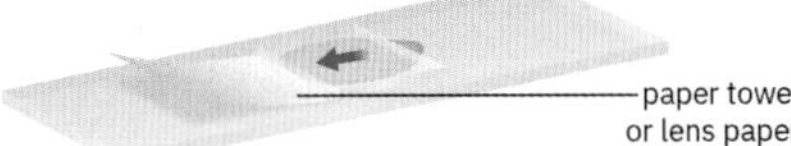

Using a Graduated Cylinder A graduated cylinder is used to measure the volume of a liquid. A graduated cylinder is a cylindrical container marked with lines from bottom to top. Most have lines indicating milliliters (mL), although the volume of graduated cylinders ranges from 10 mL to 4 liters (L). To use a graduated cylinder properly, follow these steps.

1. Place the cylinder on a flat, level surface.
2. Move your head so that your eye is level with the surface of the liquid.
3. Read the mark closest to the liquid level. In glass and some plastic cylinders, the surface of the liquid forms a curve called a meniscus. Read the mark closest to the bottom of the meniscus to determine the volume in a glass cylinder.

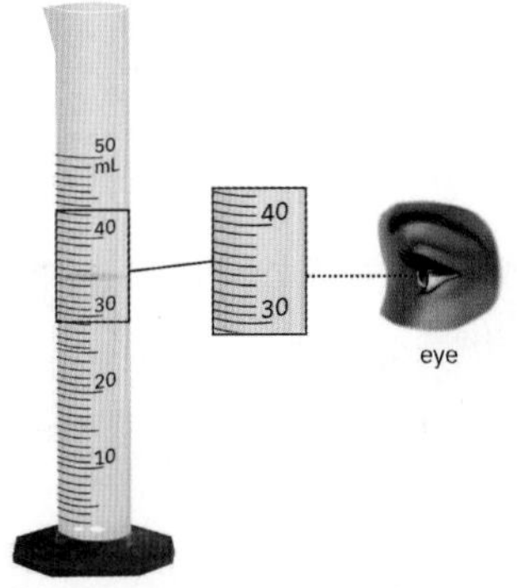

APPENDIX A: LABORATORY SAFETY AND PROCEDURES 527

APPENDIX A (continued)

Using a Triple Beam Balance A triple beam balance is used to find the mass of objects in grams (g). On one side of the balance is the pan where the object to be massed is placed. The other side of the balance is made up of a set of beams ending in a pointer that points toward a scale. Each beam has a mass, or rider, that slides along the beam. When the pointer is aligned with 0 on the scale, the masses on the beams equal the mass of the object on the pan. Follow these steps to properly mass an object.

1. Set the balance to zero by sliding all the riders back to the zero point.
2. Check that the pointer on the right swings an equal distance above and below the zero point on the scale. If the swing is unequal, turn the adjustment knob until the swing is equal.
3. Place the object to be massed on the pan. Slide the rider with the largest mass along its beam until the pointer drops below zero. Then move that rider back a notch. Repeat the process on the second beam.
4. The 1-gram slider goes from 0 to 10 g. Marks between each whole number show portions of a gram. Each mark represents 0.1 g. Push the 1-g slider until the pointer points directly at the zero on the scale.
5. Add the masses indicated on the three beams to find the mass of the object. To get the most precise measurement, count the increments past the nearest whole number on the 1-g slider. If the mass is pointing between two marks, add a value of 0.05 to your measurement.

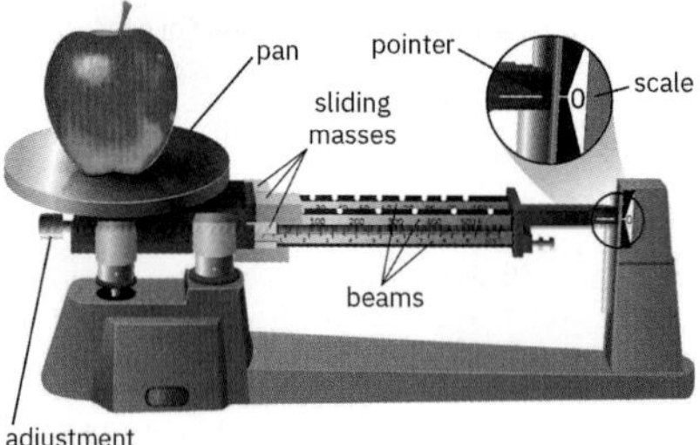

Massing Chemicals Never pour chemicals directly on the pan of any scale. Instead, follow these steps:

1. To find the mass of an empty beaker, place it on the scale and wait until the mass is displayed.
2. Press the tare button. The display will reset to zero.
3. Pour the dry or liquid chemical into the beaker. The display will show the mass of the chemical without including the mass of the beaker.
   - If you remove the beaker to add the chemical, the display will show the mass of the beaker as a negative number. When you place the beaker containing the chemical back onto the scale, the display will show only the mass of the chemical in the beaker.

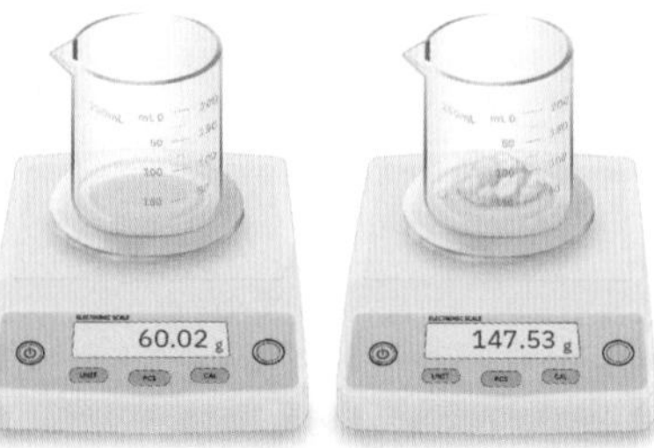

If you are using a balance, calculate the mass of the chemical by subtracting the mass of the empty beaker from the combined mass of the beaker and chemical.

Practice *Check your understanding of how to use laboratory equipment.*

1. Why should you use a compound microscope's coarse adjustment knob only with the low-power objective?
2. Why would you place the coverslip by its edges onto the water drop instead of dropping it down from the top?
3. Why is it important to read the measurement of fluid in a graduated cylinder directly from the side?
4. How would a measurement be affected if you did not zero out the riders and the pointer of a triple beam balance before taking a measurement?
5. Why should you mass dry chemicals by putting them in a beaker first instead of directly on the scale?

528 APPENDIX A

PRACTICE

1. *Sample answer: The coarse adjustment knob moves the nosepiece farther faster, and if used with the high-power objective, it could cause the objective to smash into the slide, damaging both the slide and the objective.*
2. *Sample answer: As the cover slip touches the water, air bubbles move out the other side. If we drop the coverslip straight down, air bubbles can be trapped in the middle.*
3. *Sample answer: Reading it at any other angle can result in a false reading or not being able to observe the meniscus.*
4. *Sample answer: The measurement will be greater or less than the actual measurement.*
5. *Sample answer: It adds a step, but the chemical could damage the scale if placed on it directly. Chemicals may also fall off the scale and be lost after measuring.*

DATA ANALYSIS GUIDE

Experimental Design and Data Collection

Scientists investigate natural phenomena by conducting experiments. Often, the scientist has a question about the phenomenon. The question could relate to the phenomenon's cause or effect or what happens during the event. Pondering the phenomenon, the scientist develops a *hypothesis,* or prediction about the answer to the question. Then, the scientist outlines an experimental plan or series of actions that will determine if the prediction is correct.

Plans allow scientists to collect data in organized ways that can be analyzed, such as in **Figure 1**. In a scientific experiment, the *independent variable* is a factor that differs among the experimental groups. For example, a scientist might apply different amounts of nitrogen per hectare (2.47 acres) to groups of crop plants to investigate how this nutrient affects plant height. In this experiment, the amount of nitrogen the plants receive is the independent variable. The height of the plants is the *dependent variable*; it is the observed or measured result of the experiment. The scientist's hypothesis predicts how the independent variable will affect the dependent variable.

Figure 1 A scientist measures the growth of plants in both the experimental group and the control group as evidence for the conclusion drawn from the data.

Another part of an experiment is a *control group*. In **Table 1**, the group of plants that received 0.0 kg/ha of nitrogen is the control group. Including a control group allows a scientist to compare the experimental groups, in which the independent variable was altered, to the control group, in which the independent variable was not altered.

Table 1. Amount of Nitrogen and Plant Growth

Nitrogen applied (kg/ha)	Plant height (cm)
0.0	27.4
37.5	37.2
75.0	41.9
112.5	42.3
150.0	43.5
187.5	44.1

Factors other than the independent variable should ideally be kept constant across all groups in an experiment. For example, the plant groups used in the nitrogen study should all receive the same amount of rainfall, sunlight, and nutrients. This type of consistency, however, does not always occur. So, it is crucial for scientists to identify possible sources of error in an experiment. Sources of error include mistakes made by humans, issues with equipment used to make measurements, and natural variations that cannot be avoided.

One way to account for unwanted variation in scientific experiments is to repeat the experiment many times. In the nitrogen study, the scientist would probably study many large areas of plants instead of only a few small areas. Using a large sample size decreases the likelihood that the results of the experiment happened by chance.

In the process of conducting experiments, scientists record observations and measurements called data. Scientists use tools such as data tables and computer programs to organize data. This organization makes it possible for scientists to analyze data from an experiment and draw conclusions about its results. **Table 1** shows how data from the nitrogen experiment mentioned earlier might be organized. The changes in the independent variable and dependent variables are clearly indicated, and the units of measurement are included.

APPENDIX B
Data Analysis Guide

Science Background

SEP Connection Use this guide to reinforce the importance of applying Science and Engineering Practices (SEPs) to the content being addressed. For students to see the connection between SEPs and investigations and laboratory (lab) experiences, conduct an activity where students match the SEP to the section in the guide. For example, paragraph two discusses the importance of planning investigations, which corresponds with the SEP "planning and carrying out investigations." Encourage students to identify how a lab or critical thinking question connects to an SEP. When students know and understand the "why" of what they are doing, they are more likely to find and make connections between content and process.

APPENDIX B (continued)

The data collected during an experiment may be qualitative or quantitative. The data collected in the nitrogen experiment is *quantitative data* because it is expressed as numbers. Data that is not easily expressed as numbers, such as the data in **Table 2**, is called qualitative data.

Table 2. Colors of Common Substances

Substance	Color
ammonia	off-white
salt water	clear
water	clear
milk	white
lemon juice	yellow

Practice *Consider how a scientist would carry out an investigation of the effects of carbonic acid on a species of algae.*
The scientist wants to know how increasing levels of carbonic acid in ocean water affect the number of algae present in the water. Write a plan for this investigation that includes an independent variable, a dependent variable, a control group, and one or more experimental groups. Draw one or more data tables that the scientist could use to organize observations or measurements collected in the investigation.

Units of Measurement

Quantitative data collected during an experiment should always contain units of measurement. Including units allows people to understand the meaning of the data and possibly repeat the experiment. Scientists use a system of units called the International System (or SI from the French *Système International*) to ensure that scientists around the world use consistent units of measurement. SI, which is based on the metric system of units, was developed to replace multiple measurement systems with a single, standardized system.

Table 3 shows the seven basic units of the SI system. The base unit for mass is kilogram. The prefix *kilo-* means 1000, or 10^3. So, a kilogram is equivalent to 1000 grams. When using a graduated cylinder to measure volume, you might record your measurements in milliliters (mL). The prefix *milli-* means 0.001, or 10^{-3}. So, 1 milliliter is 1/1000th of a liter. **Table 4** shows other prefixes used with SI units.

Table 3. Base Units of the SI System

Measurement	Base unit
length	meter (m)
mass	kilogram (kg)
time	second (s)
electric current	ampere (A)
temperature	kelvin (K)
light intensity	candela (cd)
amount of substance	mole (mol)

Table 4. Prefixes Used with SI Base Units

Prefix	Symbol	Meaning	Order of magnitude
giga-	G	1,000,000,000	10^9
mega-	M	1,000,000	10^6
kilo-	k	1000	10^3
hecto-	h	100	10^2
deka-	da	10	10^1
	base unit	1	10^0
deci-	d	0.1	10^{-1}
centi-	c	0.01	10^{-2}
milli-	m	0.001	10^{-3}
micro-	μ	0.000001	10^{-6}
nano-	n	0.000000001	10^{-9}

Practice *Convert each measurement below by writing the equivalent measurement for the unit that is shown.*

1. 100 km = ________ Gm
2. 61976.569 cm = ________ nm
3. 89 mm = ________ hm
4. 30856775814913673 m = ________ Mm
5. 16.09344 hm = ________ cm
6. 149.5981 Gm = ________ km
7. 5.5 km = ________ mm
8. 2.5426924 cm = ________ dam
9. 2.2633485173216473 mm = ________ μm
10. 5 nm = ________ km

PRACTICE

Look for logic in the flow of the procedure and methods that will enable comparisons about the levels of carbonic acid and number of algae. Data tables should reflect quantitative data.

PRACTICE

a. *0.001 Gm*
b. *619765690000 nm*
c. *0.00089 hm*
d. *30856775814.913673 Mm*
e. *160934.4 cm*
f. *149598100 km*
g. *5500000 mm*
h. *0.0025426924 dam*
i. *2263.3485173216473 μm*
j. *0.000000000005 km*

Scientific Notation

The numbers we must work with in scientific measurements are often very large or very small; thus it is convenient to express them using powers of 10. For example, the number 1,300,000 can be expressed as 1.3×10^6, which means multiply 1.3 by 10 six times, or

$$1.3 \times 10^6 = 1.3 \times \underbrace{10 \times 10 \times 10 \times 10 \times 10 \times 10}_{10^6 = 1 \text{ million}}$$

A number written in scientific notation always has the form:

a number (between 1 and 10) times
the appropriate power of 10

To represent a large number such as 20,541 in scientific notation, we must move the decimal point in such a way as to achieve a number between 1 and 10 and then multiply the result by a power of 10 to compensate for moving the decimal point. In this case, we must move the decimal point four places to the left.

2 0 5 4 1
4 3 2 1

to give a number between 1 and 10.

2.0541

To compensate for moving the decimal point four places to the left, we must multiply by 10^4. Thus

$$20{,}541 = 2.0541 \times 10^4$$

As another example, the number 1985 can be expressed as 1.985×10^3. To end up with the number 1.985, which is between 1 and 10, we had to move the decimal point three places to the left. To compensate for that, we must multiply by 10^3. Some other examples are given below.

Number	*Exponential notation*
5.6	5.6×10^0 or 5.6×1
39	3.9×10^1
943	9.43×10^2
1126	1.126×10^3

So far, we have considered numbers greater than 1. How do we represent a number such as 0.0034 in exponential notation? First, to achieve a number between 1 and 10, we start with 0.0034 and move the decimal point three places to the right.

0.0034
1 2 3

This yields 3.4. Then, to compensate for moving the decimal point to the right, we must multiply by a power of 10 with a negative exponent, in this case, 10^{-3}. Thus

$$0.0034 = 3.4 \times 10^{-3}$$

In a similar way, the number 0.00000014 can be written as 1.4×10^{-7}, because going from 0.00000014 to 1.4 requires moving the decimal point seven places to the right.

In dealing with exponents, you must first learn to enter them into your calculator. First the number is keyed in and then the exponent. There is a special key that must be pressed just before the exponent is entered. This key is often labeled EE or exp. For example, the number 1.56×10^6 is entered as follows:

Press	*Display*	
1.56	1.56	
EE or exp	1.56	00
6	1.56	06

To enter a number with a negative exponent, use the change-of-sign key, usually marked +/−, after entering the exponent number. For example, the number 7.54×10^{-3} is entered as follows:

Press	*Display*	
7.54	7.54	
EE or exp	7.54	00
3	7.54	03
+/−	7.54	−03

Once a number with an exponent is entered into your calculator, the mathematical operations are performed exactly the same as with a "regular" number. For example, the numbers 1.0×10^3 and 1.0×10^2 are multiplied as follows:

Press	*Display*	
1.0	1.0	
EE or exp	1.0	00
3	1.0	03
×	1	03
1.0	1.0	
EE or exp	1.0	00
2	1.0	02
=	1	05

APPENDIX B: DATA ANALYSIS GUIDE **531**

APPENDIX B (continued)

Practice *Complete the table by converting each measurement to either scientific notation or the equivalent number.*

Number	Number in scientific notation
0.000045	
1,750,539	
	3.92174×10^5
	1.8×10^{-4}
	8.43×10^{-8}

Accuracy and Precision

Accuracy refers to how close a measurement is to the accepted value. For example, a tool used to check the accuracy of triple beam balances might have a mass of exactly 50.0 g. The closer a measurement is to this accepted mass, the greater the accuracy of the measurement. Precision refers to how consistent a set of measurements is. The first target in **Figure 2** represents both accuracy and precision. The darts on this target are on the bullseye (accurate) and close together (precise).

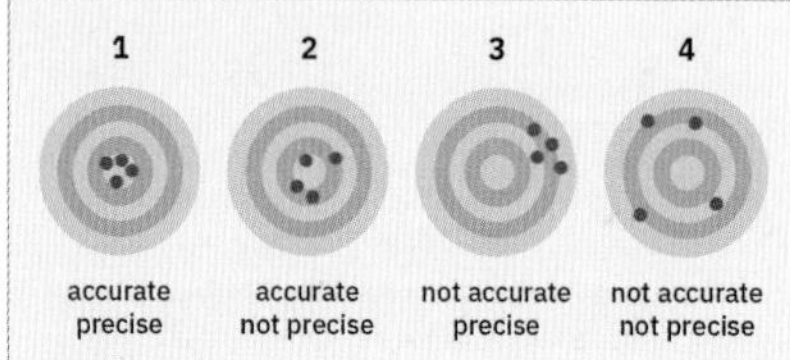

Figure 2 The targets represent varying degrees of accuracy and precision.

In a similar way, **Table 5** shows data collected by group 1, which are both accurate and precise. Group 1's measurements are very close to the accepted mass of 50.0 g, and they are very similar to one another. The next target represents less precision than the first. Similarly, group 2's data are less precise than group 1's data. The third target, like group 3's data, reflects precision but not accuracy. All the measurements are similar, but they are far from the accepted value of 50.0 g. The fourth target and the data gathered by group 4 show neither accuracy nor precision; the data are neither correct nor consistent.

Table 5. Measurements of Mass

Group 1: Mass (g)	Group 2: Mass (g)	Group 3: Mass (g)	Group 4: Mass (g)
50.1	48.6	75.0	12.1
50.0	49.5	74.9	80.4
49.9	50.3	75.1	73.2
49.8	51.1	75.0	24.6
50.2	47.0	75.2	19.7

One way to determine the accuracy of a measurement is to calculate percent error, which can be found using the following formula:

$$\%\ \text{Error} = \frac{|\text{experimental value} - \text{accepted value}|}{\text{accepted value}} \times 100\%$$

For example, if the accepted value was 50.0 g and a measurement gathered during an experiment (the experimental value) was 49.3 g, the percent error would be calculated as follows:

$$\%\ \text{Error} = \frac{|49.3\ \text{g} - 50.0\ \text{g}|}{50.0\ \text{g}} \times 100\%$$

$\%\ \text{Error} = 1.4\%$

Thus, the percent error is 1.4 percent, which is small, indicating an accurate measurement.

Practice *Four groups conducted multiple trials to collect data of an object that has an accepted value of 80.00 g.*

Determine which group's data were both accurate and precise, which group's were accurate but not precise, which group's were precise but not accurate, and which group's were neither accurate nor precise. Explain your reasoning.

Group 1: Mass (g)	Group 2: Mass (g)	Group 3: Mass (g)	Group 4: Mass (g)
19.36	120.00	80.16	77.76
128.64	119.84	80.00	79.20
117.12	120.16	79.84	80.48
39.36	120.00	79.68	81.76
31.52	120.32	80.32	75.20

PRACTICE

Row 1: 4.5×10^{-5}

Row 2: 1.750539×10^6

Row 3: 392,174

Row 4: 0.00018

Row 5: 0.0000000843

PRACTICE

Group 1: incorrect and inconsistent, Group 2: precise, Group 3: accurate and precise, Group 4: accurate. The accepted value of the object is 80 g, meaning that accurate data will include measurements that are as close as mathematically possible to 80 g; a precise measurement refers to how close measurements are to each other, regardless of accepted value; accurate and precise measurements would be mathematically as close as possible to the accepted value AND the measurements would be close to each other. Incorrect and inconsistent data would be neither close to the accepted value nor close to other measurements within the data set.

Connect to Mathematics

Computational Thinking The mathematical background of students varies. This section can be used as reinforcement or to teach the basic mathematical concepts that are necessary for scientific understanding. The use of mathematical concepts is also a great place to make cross-curricular connections. Consider planning a project with the math department that allows students to apply the skills they are learning in math with the skills and content that they are learning in science.

Graph Construction and Analysis

Once data from an experiment have been collected and organized, the data are analyzed to determine what conclusions can be made. This analysis may involve mathematical calculations, graphing, or statistical analysis.

Most graphs follow certain guidelines. For one, a graph has both an *x*-axis and a *y*-axis. The number values on both axes may be positive or negative. The title of a graph should describe the data displayed on the graph. To determine the scale for each axis, it is useful to identify the maximum and minimum values in the data set. For example, the minimum value in the first column of **Table 6** is 15 seconds, and the maximum value is 90 seconds. Thus, as shown in **Figure 3**, the scale for the *x*-axis extends from 0 to 100 and is divided by increments of 20. The scale for the *y*-axis is determined in the same way.

Table 6. Exercise and Heart Rate

Time exercising (s)	Heart rate (bps)
30	74
15	71
60	86
45	82
75	91
90	99

Figure 3 The range of measurements represented by the collected data helps determine the scale of each axis.

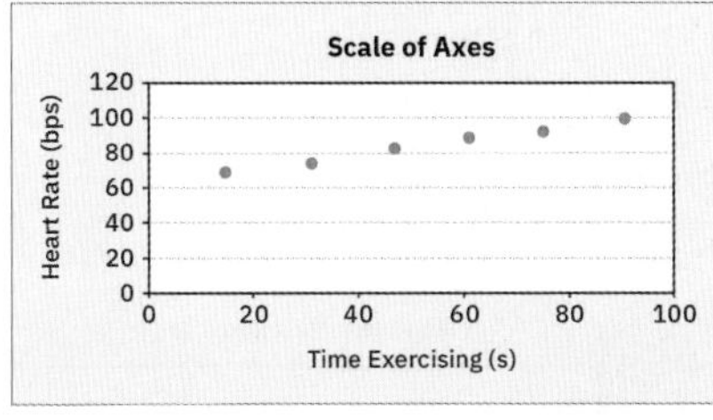

A graph may have multiple *y*-axes or multiple *x*-axes. **Figure 4** shows a graph with two *y*-axes. When interpreting this type of graph, pay careful attention to which set of points corresponds to which axis. On this graph, the purple line corresponds to the scale shown on the right *y*-axis. The orange line corresponds to the scale shown on the left *y*-axis.

Figure 4 Each axis on a graph shows different scales.

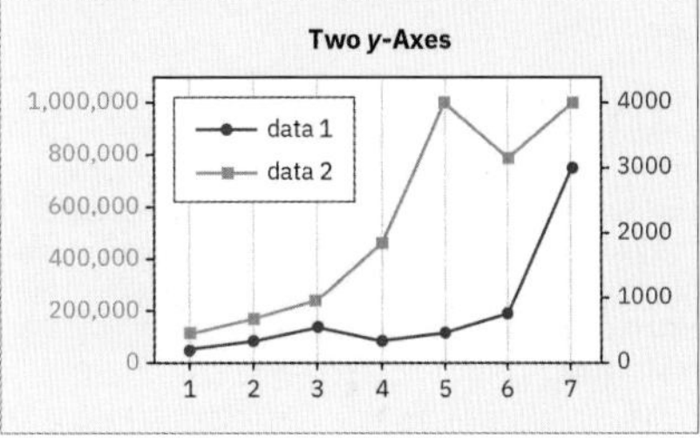

When points are plotted on a graph, a mathematical relationship between the points may be present. An equation may be used to show the relationship among the points. **Figure 5** shows a graph where the points form a straight line. The equation for a straight line (a linear equation) can be represented in the general form

$$y = mx + b$$

where *y* is the dependent variable, *x* is the independent variable, *m* is the slope, and *b* is the intercept with the *y*-axis.

To illustrate the characteristics of a linear equation, $y = 3x + 4$ is plotted in **Figure 5**. For this equation, $m = 3$ and $b = 4$. Note that the *y*-intercept occurs when $x = 0$. In this case the *y*-intercept is 4, as can be seen from the equation ($b = 4$).

APPENDIX B (continued)

The slope of a straight line is defined as the ratio of the rate of change in *y* to that in *x*:

$$m = \Delta y/\Delta x$$

For the equation $y = 3x + 4$, *y* changes three times as fast as *x* (because *x* has a coefficient of 3). So the slope in this case is 3. This can be verified from the graph. For the triangle shown in **Figure 5**,

$$\Delta y = 50 - 14 = 36 \text{ and } \Delta x = 15 - 3 = 12$$

So

$$\text{slope} = \Delta y/\Delta x = 36/12 = 3$$

This example illustrates a general method for obtaining the slope of a line from the graph of that line. Simply draw a triangle with one side parallel to the *y*-axis and the other side parallel to the x-axis, as shown in **Figure 5**. Then determine the lengths of the sides to get Δy and Δx, respectively, and compute the ratio $\Delta y/\Delta x$.

Figure 5 Direct relationships occur when both variables increase or decrease at the same time.

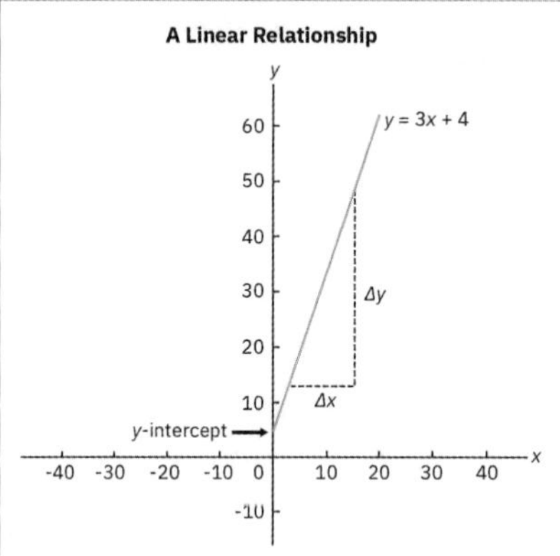

The relationship shown in **Figure 5** is also a direct relationship because, as one variable increases, the other variable increases. A direct relationship is also present when one variable decreases and the other decreases.

In an inverse relationship, as one variable increases, the other decreases, or vice versa. **Figure 6** shows a line that represents an inverse relationship. Sometimes the points on a graph do not form a neat line. But it may be appropriate to find a line of best fit. **Figure 6** shows a scatter plot with a line of best fit. When finding a line of best fit, it is also possible to calculate an R^2 value. This value tells you how well the line of best fit works as a predictor of *y* (if given *x*).

Figure 6 A line of best fit is similar to an average of the data.

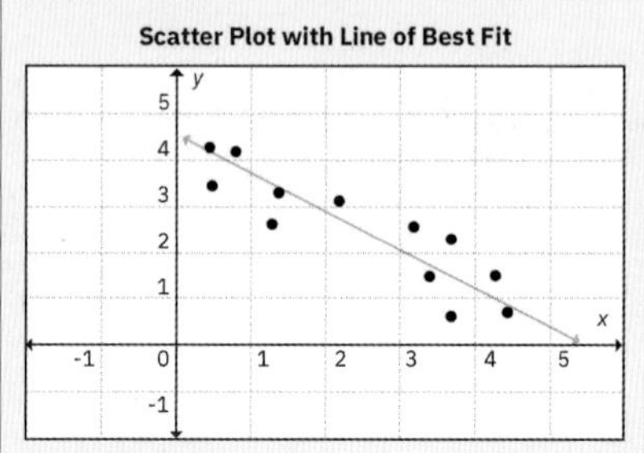

Sometimes, there is a nonlinear relationship between the points on a graph. **Figure 7** shows such a graph. An equation may still be used to represent the relationship between the points, but the equation is different from the equation used to represent linear relationships.

Figure 7 A line of best fit is similar to an average of the data.

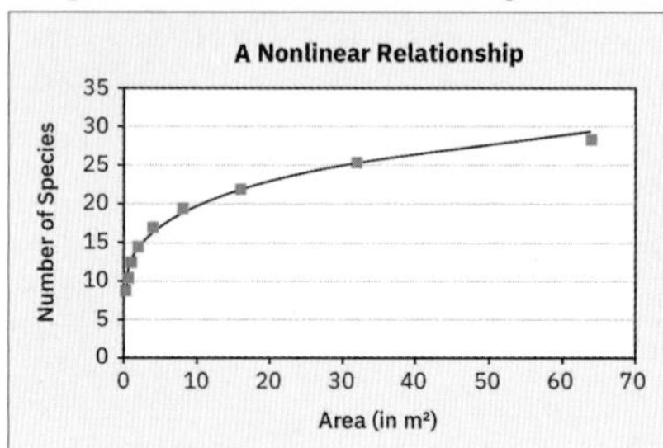

534 APPENDIX B

Another type of graph used to represent data is a bar graph. This type of graph is best for data that can be placed in separate groups or categories. **Figure 8** shows a bar graph. The lines extending from the ends of the bars are error bars. These shows the calculated error associated with each data point. The error may represent a standard or range shown by the data or a degree of uncertainty.

Error may also be represented as a number plus or minus some amount. For example, the range of values for the 46.6 ± 5.0 can be calculated as follows:

maximum: $46.6 + 5.0 = 51.6$
minimum: $46.6 - 5.0 = 41.6$

Figure 8 Possible errors related to data collection may be shown by the extensions on the bars.

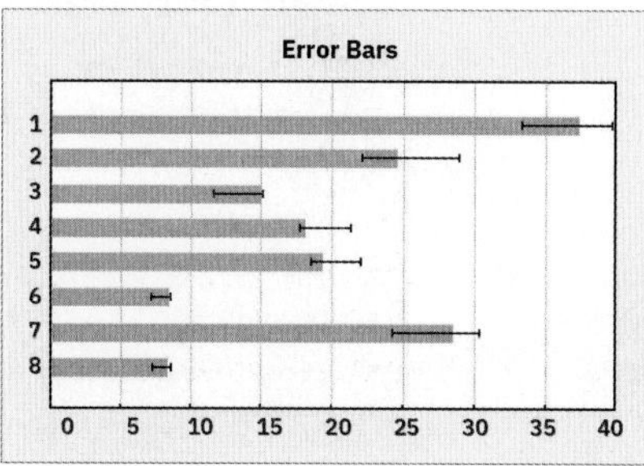

In some cases, the scales on the *x*-axis and/or *y*-axis may be in the form of a logarithmic scale. This means the values on the scale represent powers of 10. **Figure 9** shows a graph that uses a logarithmic scale on the *y*-axis. This type of graph is best for representing very large or very small numbers or numbers that vary by large amounts. Using a logarithmic scale can also make it easier to identify trends on a graph. In this case, the graph shows how, at first, the population size grows slowly and then grows very quickly over regular time intervals.

Figure 9 Very small and very large numbers are often shown with a logarithmic scale.

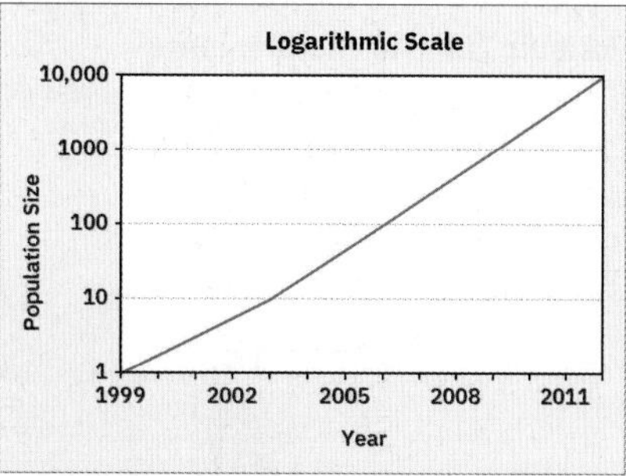

Practice *Describe the two graphs. Explain what each shows and what conclusions you might draw.*

1

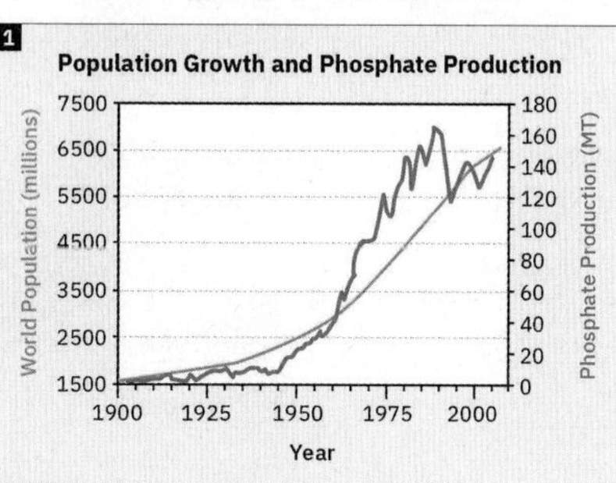

2

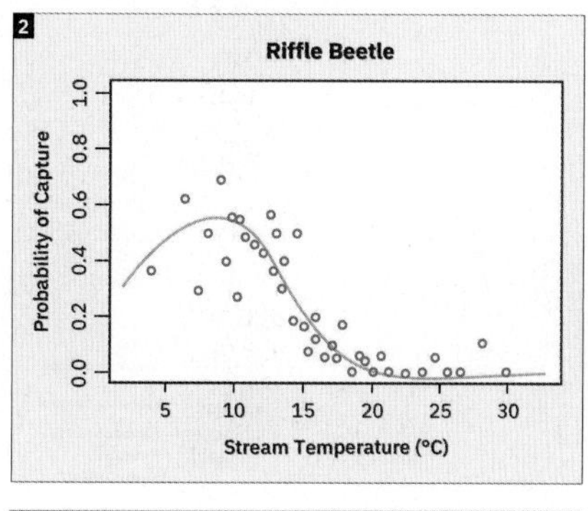

PRACTICE

Graph 1: Sample answer: This is a graph with two y axes that compare the increase in the amount of fertilizer being produced with the growth of the world population. One might conclude that more fertilizer used for crops enabled greater population growth.

Graph 2: Sample answer: This graph shows a scatter plot of a nonlinear relationship between stream temperatures and the likelihood of capturing a riffle beetle. The data show that they are most likely to be captured when stream temperature is between 5 and 15 degrees.

APPENDIX C
The Periodic Table and Biochemistry

Visual Support

Periodic Table Allow students access to the periodic table either by using individual versions for each student or by displaying a large classroom copy. A very basic understanding of elements and atomic structure is necessary, but an understanding beyond the elements necessary for life is outside the assessment boundaries for Biology.

THE PERIODIC TABLE AND BIOCHEMISTRY

Periodic Table of the Elements

The periodic table displays all chemical elements arranged in order of increasing atomic number (the number of protons in the nucleus). Atomic number increases across a row, or period. The elements in each column, or group, share some properties. More than 90 of the elements occur naturally on Earth. Scientists have made the other elements by smashing atoms together at very high speeds, and they continue to make more.

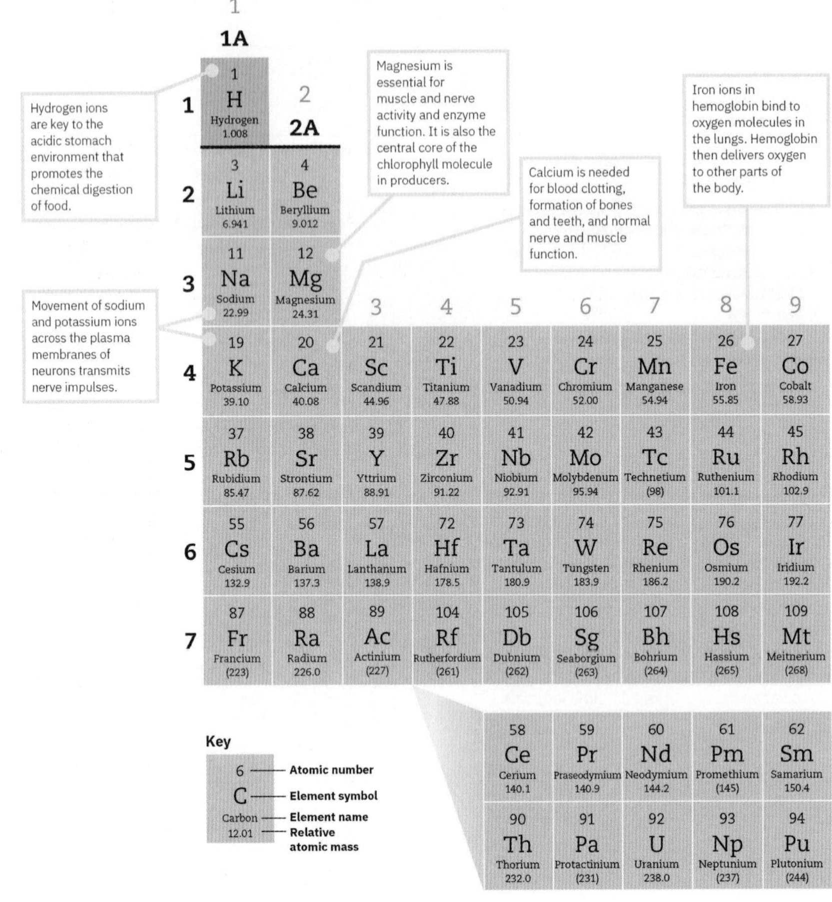

In this periodic table, metals are shaded blue, metalloids green, and nonmetals orange. Hydrogen, shaded brown, does not belong to any group, and its properties are not similar to those of the group 1A elements below it.

Some sets of elements have their own names. Group 1A is known as the alkali metals; group 2A, the alkaline earth metals; group 7A, the halogens; and group 8A, the noble gases. Elements 57 through 71 are the lanthanides (also called rare earth elements), and elements 89 through 103 are the actinides. Most of these elements typically are extracted from the sixth and seventh rows of the table for display purposes.

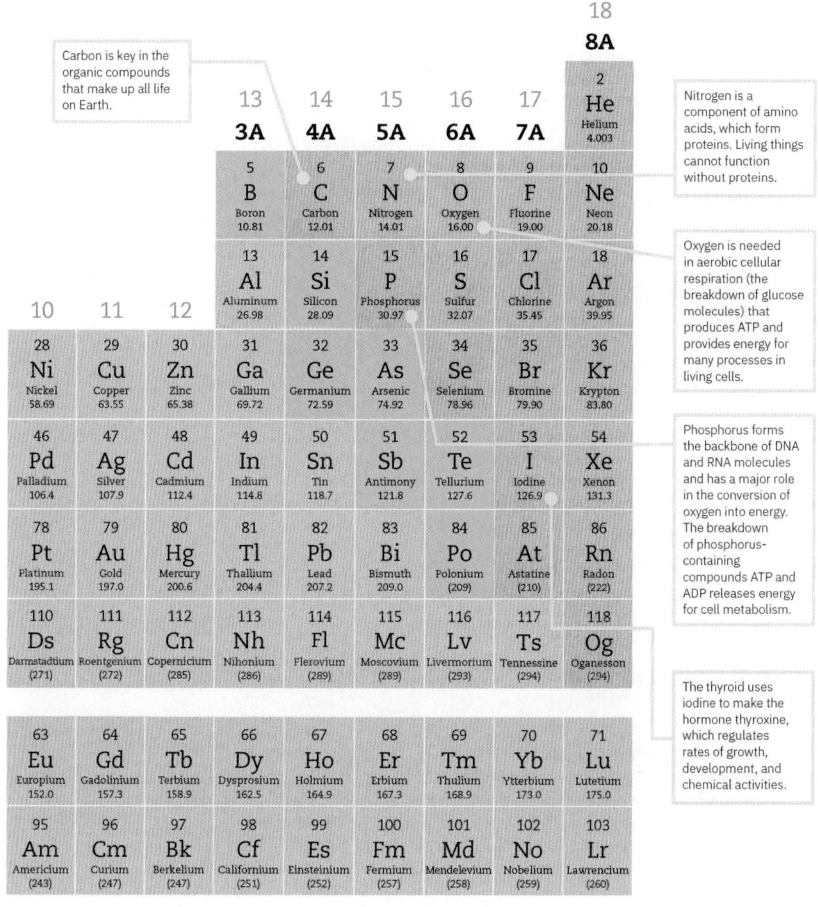

APPENDIX C: THE PERIODIC TABLE AND BIOCHEMISTRY **537**

Visual Support

Atomic Structure These pages should be used in conjunction with Chapter 5. Consider using this appendix as an introduction to what students will learn in Chapter 5 or as a reference tool if your curriculum does not include the Chapter 5 content.

Show students the connection between atomic structure and the structure of monomers and polymers. For example, ask the students to determine the chemical compound of glucose by having them count the number of carbon, hydrogen, and oxygen atoms present in **Figure 1**. This will allow students to discover that glucose is $C_6H_{12}O_6$. Have students look at each monomer and relate their structure to the chemical make-up of the associated polymer. Ask students to identify polymers by elements. For example, ask students to identify mystery polymers by listing the elements present.

APPENDIX C (continued)

Organic Molecules

All organisms take up carbon compounds from the environment and use them to build the molecules of life. Biological molecules are made up of carbon, hydrogen, nitrogen, and a few other elements. Carbon is the key element. Organic molecules have monomers, or single units of sugars, fatty acids, amino acids, and nucleotides, as building blocks. These monomers bond to form polymers, such as DNA, cellulose, and proteins. Biological molecules can be quite large polymers whose structure and function arise from the order, orientation, and interaction of its subunits.

See Chapter 5: Molecules in Living Systems *for more information about organic molecules.*

Carbohydrates Carbohydrates are organic compounds made of carbon, hydrogen, and oxygen. The monomer in **Figure 1**, glucose, is one of two building blocks of most common carbohydrates, such as amylose. These compounds are involved in cellular metabolism, releasing energy that organisms need to survive. Complex carbohydrates can have chains of hundreds to thousands of monomers.

Figure 1 Glucose forms the basis of complex carbohydrates.

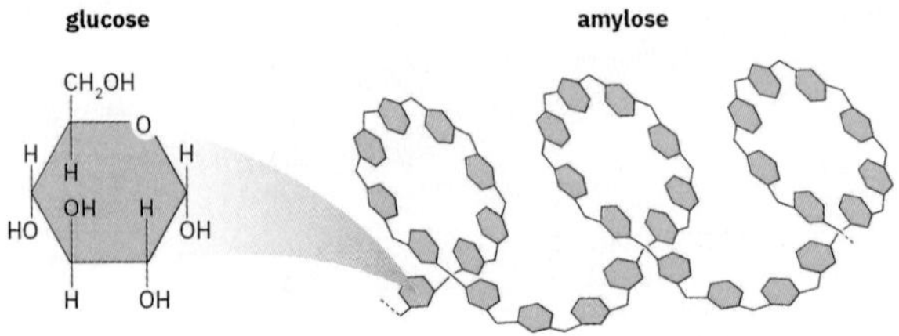

Lipids Lipids are comprised of fats, oils, and waxes that are relatively insoluble, or difficult to dissolve, in water. Composed mainly of carbon, hydrogen, and oxygen atoms, lipids function primarily to store energy. Typically, lipids are made up of one or more hydrocarbon chains such as that shown in **Figure 2** and are attached to a single functional group that includes oxygen.

Figure 2 The fatty acid stearic acid displays the general structure of a lipid.

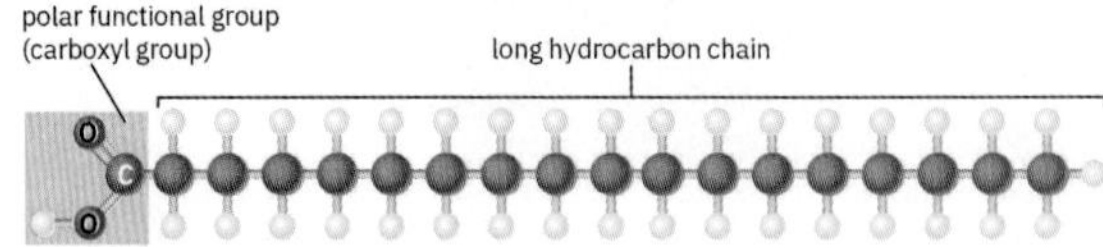

Amino Acids and Proteins Amino acids are small organic compounds with the general structure shown in **Figure 3a**. In most amino acids, all three groups are covalently bonded to the same carbon atom. Cells make the thousands of different proteins they need from only 20 amino acid monomers. Each type of amino acid has a different side-chain R group that determines its bonding pattern in secondary and tertiary protein structures. R groups fall into four categories based on their overall polarity or distribution of charge.

Proteins generally contain 50 to 1000 amino acids linked together by covalent bonds called peptide bonds. The order of the bonded amino acids dictates a given protein's unique primary structure, as shown in **Figure 3b**. Proteins perform many cellular functions, such as the structural components of cells and their neighbors in multicellular organisms. Proteins also serve as enzymes that speed up countless chemical reactions and as the many regulatory proteins that control signals between and within cells.

Figure 3 Differences in the side-chain R-group differentiate the 20 amino acids (a). Proteins with fewer than 100 amino acids are considered small proteins (b).

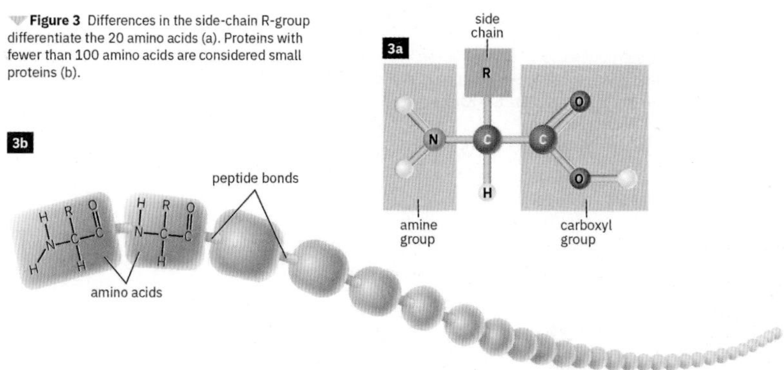

Nucleotides and Nucleic Acids Nucleotides can act alone to carry energy or act as monomers linked together to form the nucleic acids DNA and RNA, as shown in **Figure 4**. DNA and RNA code for protein synthesis and transmission of hereditary characteristics, and they control cell activities in all living things. Sugars and phosphate groups make the backbone of each strand. A DNA molecule forms when two such strands coil in the shape of a double helix. Information coded in DNA molecules is recorded in the sequence of the four nucleotide bases—adenine (A), thymine (T), guanine (G), and cytosine (C).

Figure 4 Millions of linked nucleotides form a strand of DNA or linear chromosome.

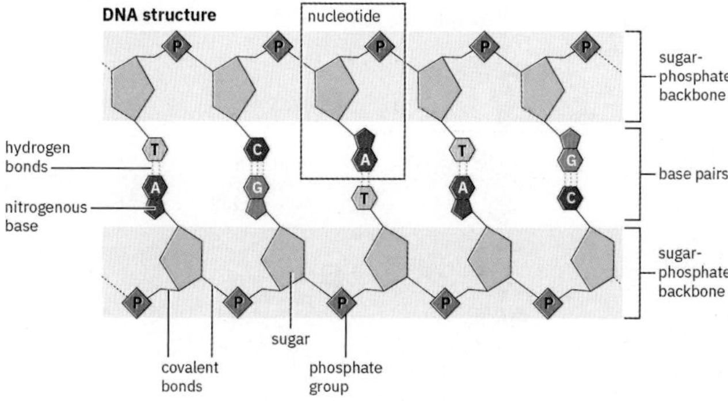

APPENDIX C: THE PERIODIC TABLE AND BIOCHEMISTRY 539

APPENDIX D
Cellular Processes

Science Background

The content in this appendix goes deeper into the processes summarized in Chapter 6. If your curriclum includes this level of detail, be sure to take into consideration the background knowledge of your students. This appendix can be used for remediation or for extension based on students' level of understanding and the depth of presentation you prefer.

Help students to see the connection between these complex cycles to the way that their bodies produce energy. Start with a generalized understanding that food is first broken down into its biological molecules, one of which is glucose, which undergoes glycolysis to give the body pyruvate. Clarification of this first step can help students to see how the food they eat is related to these processes.

Vocabulary Strategy

Roots and Suffixes Break the word glycolysis into its root components: glyco = sugar and lysis = breaking.

CELLULAR PROCESSES

Cellular Respiration

Section 6.3 provides an overview of the process of cellular respiration and describes the three major stages—glycolysis, the Krebs cycle, and the electron transport chain. The models in this section outline key steps in each of the stages of cellular respiration.

Glycolysis Glycolysis takes place in the cytoplasm of cells and makes two very important contributions. First, glycolysis produces some energy in the form of ATP as well as electron carrier molecules (NADH) that are needed in later stages of cellular respiration. Second, glycolysis produces pyruvate, which helps start the Krebs cycle. The Krebs cycle, in turn, produces molecules that are needed in the next stage of metabolism to produce even more ATP. Cells cannot use glucose directly to produce more energy, so pyruvate is an important intermediate molecule.

As shown in **Figure 1**, the first step of glycolysis uses energy. Using some energy in this step allows the cell to gain more energy later. Each glucose molecule gets split into two, so each subsequent reaction will happen twice for each glucose molecule.

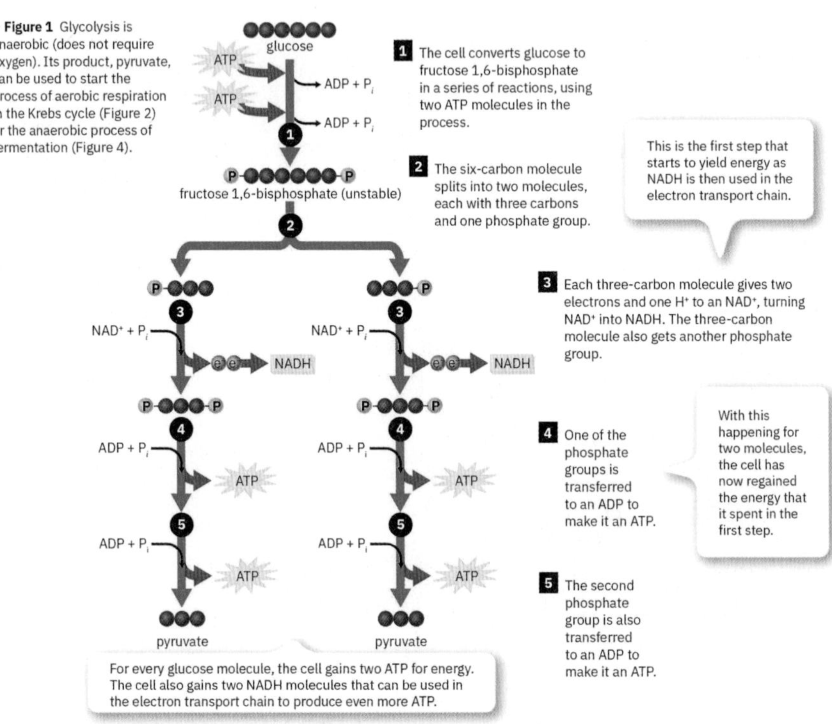

glucose
ATP
ATP
ADP + P$_i$
ADP + P$_i$
1 The cell converts glucose to fructose 1,6-bisphosphate in a series of reactions, using two ATP molecules in the process.
fructose 1,6-bisphosphate (unstable)
2 The six-carbon molecule splits into two molecules, each with three carbons and one phosphate group.
This is the first step that starts to yield energy as NADH is then used in the electron transport chain.
NAD$^+$ + P$_i$
NADH
3 Each three-carbon molecule gives two electrons and one H$^+$ to an NAD$^+$, turning NAD$^+$ into NADH. The three-carbon molecule also gets another phosphate group.
ADP + P$_i$
ATP
4 One of the phosphate groups is transferred to an ADP to make it an ATP.
With this happening for two molecules, the cell has now regained the energy that it spent in the first step.
5 The second phosphate group is also transferred to an ADP to make it an ATP.
pyruvate
pyruvate
For every glucose molecule, the cell gains two ATP for energy. The cell also gains two NADH molecules that can be used in the electron transport chain to produce even more ATP.

Figure 1 Glycolysis is anaerobic (does not require oxygen). Its product, pyruvate, can be used to start the process of aerobic respiration in the Krebs cycle (Figure 2) or the anaerobic process of fermentation (Figure 4).

SCIENCE AND ENGINEERING PRACTICES
Developing and Using Models

Make Models Allow students to physically manipulate a model of the processes for better understanding.

Either give students or have them make a model of a glucose molecule by creating a drawing of the molecule. The drawing can be very simple (6 circles connected) or can be more complex depending on the course level. Make sure that the model is large enough for the students to manipulate. Use this model to walk students through the processes in order.

For advanced students, you may want to have them include as much detail as possible. In partners, sitting at a table or by pushing two desks together, have students create and manipulate models of all molecules involved.

Make connections to the site within a cell where these processes take place. Allow students to act out or physically move to different areas of the room while they manipulate their models.

Acetyl-CoA Formation The two molecules of pyruvate move into a mitochondrion. Follow steps 1 and 2 in **Figure 2** to see how Coenzyme A helps form acetyl-CoA, the starting molecule for the Krebs cycle.

Krebs Cycle Steps 3 through 8 in **Figure 2** detail the Krebs cycle, which starts when a four-carbon oxaloacetate molecule combines with the two-carbon acetyl-CoA to form a six-carbon molecule of citric acid. So, the cycle is sometimes called the citric acid cycle. The steps of the Krebs cycle harvest H^+ ions and their companion electrons from the participating molecules while removing carbon atoms that are no longer useful.

In one turn of the Krebs cycle, two carbon atoms enter in the form of acetyl-CoA and two molecules of carbon dioxide are released. This restores the four-carbon oxaloacetate molecule, so it is ready to begin another turn of the cycle. Four electron carrier molecules are produced: three NADH and one $FADH_2$. One molecule of ATP is produced.

Other important products result from the Krebs cycle, namely the electron carrier molecules NADH and $FADH_2$. These molecules are needed by the cell in the next stage of cellular respiration, the electron transpot chain, where they help the cell produce many more ATP molecules. Additional products formed during the Krebs cycle are also used in the biosynthesis of amino acids and fatty acids.

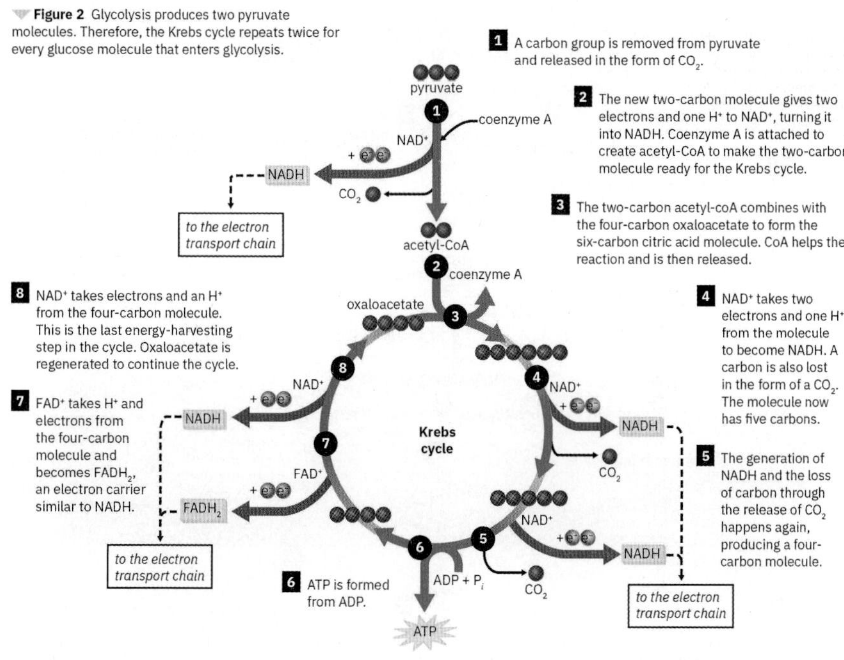

1 A carbon group is removed from pyruvate and released in the form of CO_2.
2 The new two-carbon molecule gives two electrons and one H^+ to NAD^+, turning it into NADH. Coenzyme A is attached to create acetyl-CoA to make the two-carbon molecule ready for the Krebs cycle.
3 The two-carbon acetyl-coA combines with the four-carbon oxaloacetate to form the six-carbon citric acid molecule. CoA helps the reaction and is then released.
4 NAD^+ takes two electrons and one H^+ from the molecule to become NADH. A carbon is also lost in the form of a CO_2. The molecule now has five carbons.
5 The generation of NADH and the loss of carbon through the release of CO_2 happens again, producing a four-carbon molecule.
6 ATP is formed from ADP.
7 FAD^+ takes H^+ and electrons from the four-carbon molecule and becomes $FADH_2$, an electron carrier similar to NADH.
8 NAD^+ takes electrons and an H^+ from the four-carbon molecule. This is the last energy-harvesting step in the cycle. Oxaloacetate is regenerated to continue the cycle.
pyruvate
coenzyme A
NAD^+
NADH
CO_2
to the electron transport chain
acetyl-CoA
coenzyme A
oxaloacetate
Krebs cycle
NAD^+
NADH
CO_2
FAD^+
$FADH_2$
ADP + P_i
ATP

Figure 2 Glycolysis produces two pyruvate molecules. Therefore, the Krebs cycle repeats twice for every glucose molecule that enters glycolysis.

APPENDIX D: CELLULAR PROCESSES **541**

APPENDIX D (continued)

Electron Transport Chain The electron transport chain shown in **Figure 3** uses the electron carriers NADH and $FADH_2$ that were produced in the Krebs cycle to make ATP for the cell. For each starting molecule of glucose, two ATP were produced during glycolysis and two ATP were produced during two turns of the Krebs cycle. In the electron transport chain, 26 to 28 ATP are produced through a process that moves hydrogen ions (H^+) across the inner membrane of the mitochondrion. This process of moving a chemical component to set up a concentration gradient that can power the production of ATP is called chemiosmosis. Here, it is the movment of ions across a membrane.

The set of proteins that make up the electron transport chain, as well as the chemiosmosis protein ATP synthase, are embedded in the inner membrane of the mitochondria. The membrane separates the mitochondrial matrix from the intermembrane space, preventing molecules, ions (like H^+), and electrons from moving across freely. The inner membrane acts as a dam, holding more H^+ on one side than the other. The electron transport chain and chemiosmosis are like a hydropower plant, harnessing the flow of the electrons and H^+ (from the more concentrated side to the less concentrated side) for energy to power the production of ATP.

The electron transport chain takes the electrons from NADH and $FADH_2$ to supply energy for setting up the H^+ concentration gradient. In doing so, NAD^+ and FAD are released so that they can be used again in glycolysis and the Krebs cycle.

In the last step, oxygen acts as the final electron acceptor. Because the electron transport chain requires oxygen to function, cellular respiration is often called "aerobic respiration."

Figure 3 The electron transport chain produces the most ATP and can be thought of as the "payoff" stage of cellular respiration.

1 NADH and $FADH_2$ donate electrons that move from the mitochondrial matrix to the inner membrane's electron transport chain. The newly formed NAD^+ and FAD can be used again where needed.

4 Hydrogen ions move from where they are highly concentrated in the intermembrane space back out into the mitochondrial matrix. The hydrogen ions can only pass through special channels created by the protein ATP synthase. As the H^+ moves through the ATP synthase, it spins like a rotor. The energy of the motion adds a phosphate group to ADP to produce ATP.

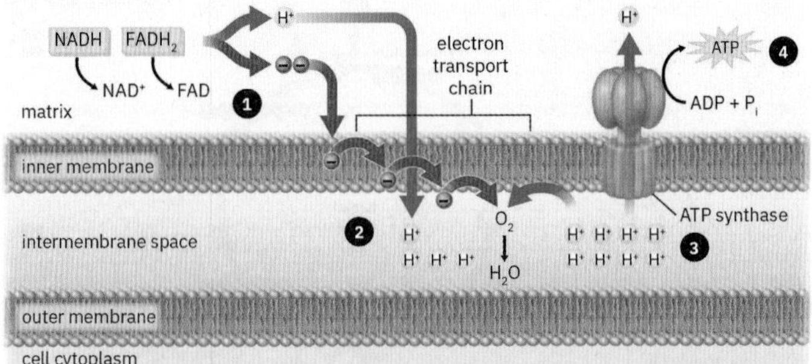

2 As electrons flow through the proteins in the electron transport chain towards the intermembrane space, the proteins use the energy to pump H^+ from the mitrochondrial matrix to the intermembrane space.

3 In the last step of the chain, the electrons combine with two H^+ and half of an oxygen molecule to create H_2O. Creating H_2O keeps the concentration of electrons in the intermembrane space low, maintaining the concentration gradient that drives electrons to move through the electron transport chain.

542 APPENDIX D

Fermentation

If the cell has enough oxygen, the NADH made from glycolysis is constantly being turned back into NAD^+ as it is used in the electron transport chain. However, if no oxygen is present, the electron transport chain cannot function, so the NADH cannot drop off its electrons and H^+ there. That means NAD^+ is not being regenerated for the cell to use in glycolysis, so glycolysis cannot occur.

Figure 4 shows fermentation, a way to turn NADH back into NAD^+ by moving the electrons and H^+ from NADH onto the pyruvate molecules formed from glycolysis. This process does not produce ATP on its own, and it turns the pyruvate into molecules that are hard for the cell to use. Fermentation products can also be toxic in large quantities. However, fermentation can keep glycolysis going so that at least some ATP is being generated.

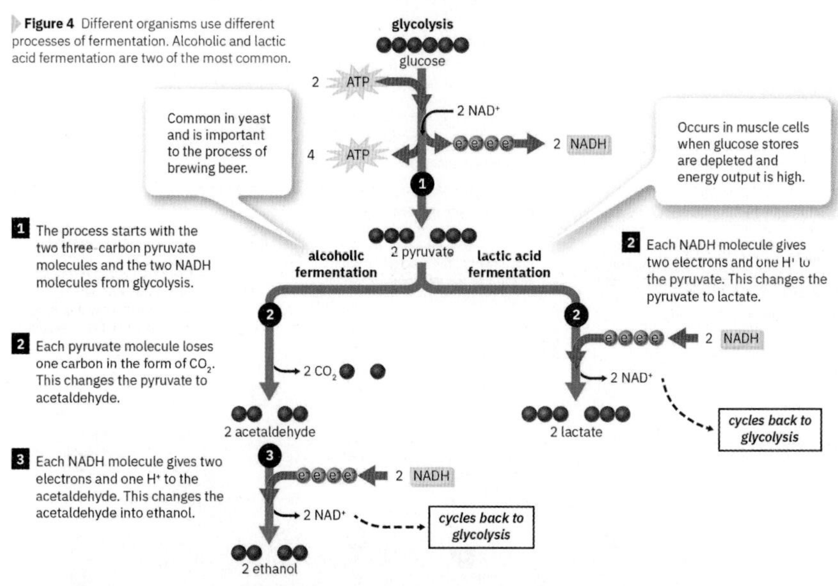

glycolysis
glucose
2 ATP
2 NAD+
4 ATP
2 NADH
Common in yeast and is important to the process of brewing beer.
Occurs in muscle cells when glucose stores are depleted and energy output is high.
1 The process starts with the two three-carbon pyruvate molecules and the two NADH molecules from glycolysis.
alcoholic fermentation
2 pyruvate
lactic acid fermentation
2 Each NADH molecule gives two electrons and one H+ to the pyruvate. This changes the pyruvate to lactate.
2 Each pyruvate molecule loses one carbon in the form of CO2. This changes the pyruvate to acetaldehyde.
2 CO2
2 NADH
2 NAD+
2 acetaldehyde
2 lactate
cycles back to glycolysis
3 Each NADH molecule gives two electrons and one H+ to the acetaldehyde. This changes the acetaldehyde into ethanol.
2 NADH
2 NAD+
cycles back to glycolysis
2 ethanol

Figure 4 Different organisms use different processes of fermentation. Alcoholic and lactic acid fermentation are two of the most common.

Practice *Refer to the processes in cellular respiration as you answer these questions.*

1. How many total ATP can be produced from one glucose molecule? Trace the processes from glycolysis through the Krebs cycle and end with the electron transport chain. Hint: assume that the electrons from one NADH can power the production of 2.5 ATP from chemiosmosis and that one $FADH_2$ can power the production of 1.5 ATP.
2. How is fermentation an "alternate pathway" for cellular respiration. Why is it helpful to the cell even though no ATP is produced through the fermentation process?
3. Heterotrophic organisms such as humans must take in food molecules for energy. Autotrophs such as plants can make their own food through photosynthesis. Does this mean that autotrophs do not need to carry out cellular respiration? Explain.

PRACTICE

1. *2 ATP*
   Glycolysis produces 4 ATP molecules from one molecule of glucose, but it needs 2 ATP molecules to accomplish this.
   Krebs Cycle: 2 ATP
   Krebs cycle: Two turns for one glucose molecule gives 2 ATP molecules.
   Electron Transport Chain: 28 ATP
   2 NADH from glycolysis, 2 NADH from the formation of acetyl-CoA and 6 NADH from the Krebs cycle equals 10 NADH that could produce about 25 ATP. Two FADH2 from the Krebs cycle could produce about 3 ATP. That gives a total of about 28 ATP from the electron transport chain.
2. *When oxygen is present, cellular respiration follows a path from glycolysis to the Krebs cycle and finally through the electron transport chain. When oxygen is not present, the alternate pathway from glycolysis to fermentation occurs because it resupplies the NAD+ that glycolysis needs. In other words, fermentation does not produce ATP, but it does produce the NAD+ that glycolysis needs to produce ATP.*
3. *Although autotrophs can make their own food, they still need to extract the energy from that food to run cellular processes. So, autotrophs do perform cellular respiration. Even though plants release oxygen to the environment through photosynthesis, they do need some oxygen to perform cellular respiration just as we do.*

Address Misconceptions

Photosynthesis Students understand that photosynthesis occurs in plants, but they may not understand that there are two processes, one of which requires light energy and one that does not require light energy. Emphasize that the first cycle of photosynthesis produces the energy necessary for the second process.

A common misconception is that photosynthesis is the only process necessary in plants. It is important to connect the concept that plants produce their own food (glucose) through the process of photosynthesis but that plant cells also contain mitochondria and therefore use all of the processes of cellular respiration.

APPENDIX D (continued)

Photosynthesis

Section 6.3 provides an overview of the process of photosynthesis and summarizes it as a series of light-dependent and light-independent reactions. The models in this section detail key steps in those two series.

Light-Dependent Reactions The light-dependent reactions shown in **Figure 5** can be thought of as similar to the electron transport chain in cellular respiration. Both use energy from moving electrons to gradually create a gradient, or area that is strongly positive with a high concentration of H^+, across a membrane. At the end, that gradient is used to turn ADP into ATP. The main difference is that instead of using electrons and energy from breaking down molecules of glucose (stored in molecules that carry energy and electrons, like NADH and $FADH_2$), light-dependent reactions take electrons from water and use energy from light to power the process.

Another difference is that in cellular respiration, the electron transport chain and ATP synthesis are at the end of the entire respiration process. But in light-dependent reactions, the electron transport chain and ATP synthesis occur in the first part of photosynthesis.

Figure 5 Chloroplasts contain stacks of disk-shaped thylakoids surrounded by a fluid called stroma. The thylakoid membrane holds chlorophylls and protein complexes used in the light-dependent reactions.

1. An electron is taken from H_2O by photosystem II. This turns water into two H^+ (which are released into the thylakoid compartment) and half of an O_2, which combines with an oxygen atom from another H_2O molecule and is released as O_2 byproduct.
2. Photosystem II uses energy from sunlight to give the electron more energy. This is known as excitation of the electron.
3. The excited electron is then moved from photosystem II to an electron transport chain in the thylakoid membrane.
4. As the electron moves through the molecules in the electron transport chain, the molecules use the movement of the electron to pump H^+ through the membrane, from the outer stroma into the inner thylakoid compartment.
5. Once the electron gets to the end of the transport chain, it is transferred to photosystem I, which uses light energy to excite it again.
6. The excited electron moves from photosystem I to a second electron transport chain in the thylakoid membrane.
7. The electron is moved to an $NADP^+$ on the stroma side of the membrane. They combine with an H^+ to form NADPH. NADPH and $NADP^+$ also hold and let go of electrons and H^+.
8. Energy released as the electron moves through this second chain is used to pump H^+ into the thylakoid compartment. This causes a concentration gradient, with a higher concentration of H^+ on the thylakoid compartment side than on the stroma side.
9. H^+ flow through the ATP synthase molecule, from the high-concentration thylakoid compartment side to the low-concentration stroma side. The energy from the flow of H^+ is used by the ATP synthase molecule to turn ADP into ATP.

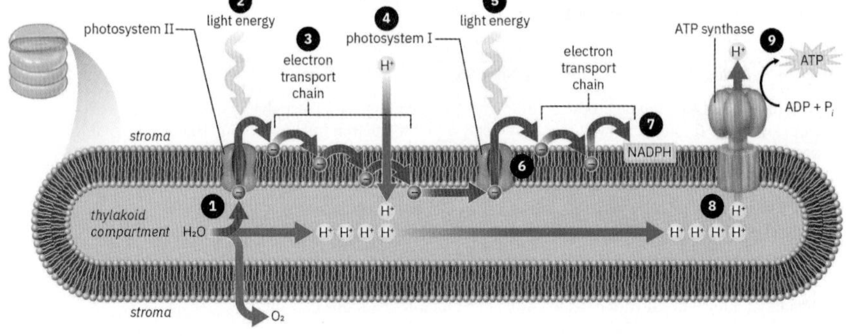

544 APPENDIX D

CROSSCUTTING CONCEPTS | System and System Models

Systems and System Models There are multiple systems all interacting through all of these processes. Help students to see that the processes are a system working within the system of the cell, which is working within the system of the organism. Ultimately, the system of interdependency between plants and animals is illustrated through these processes. Plants produce glucose that animals eat and then return to the system through digestion.

Light-Independent Reactions **Figure 6** shows the light-independent reactions, often called the Calvin cycle after one of its pioneering researchers. Energy from the ATP produced in the light-dependent reactions of photosynthesis is used to "fix" carbon by creating carbon molecules out of CO_2, in what can be thought of almost as a "reverse Krebs cycle."

The Calvin cycle is unusual in that multiples of each molecule are needed to complete a full cycle. In the first half of the cycle, three five-carbon molecules are used to create six three-carbon molecules. However, only one of those three-carbon molecules is released to be used to form glucose. The other five three-carbon molecules are combined to create three five-carbon molecules that are needed to re-start the process. Eighteen carbon atoms are involved in the balanced version of the cycle.

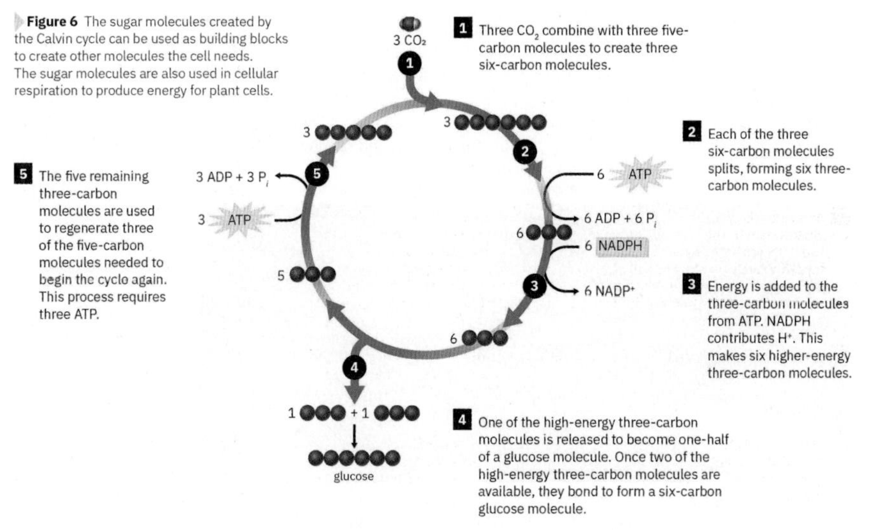

3 CO₂
1 Three CO₂ combine with three five-carbon molecules to create three six-carbon molecules.
2 Each of the three six-carbon molecules splits, forming six three-carbon molecules.
6 ATP
6 ADP + 6 Pᵢ
6 NADPH
6 NADP⁺
3 Energy is added to the three-carbon molecules from ATP. NADPH contributes H⁺. This makes six higher-energy three-carbon molecules.
4 One of the high-energy three-carbon molecules is released to become one-half of a glucose molecule. Once two of the high-energy three-carbon molecules are available, they bond to form a six-carbon glucose molecule.
glucose
5 The five remaining three-carbon molecules are used to regenerate three of the five-carbon molecules needed to begin the cycle again. This process requires three ATP.
3 ADP + 3 Pᵢ
3 ATP

Figure 6 The sugar molecules created by the Calvin cycle can be used as building blocks to create other molecules the cell needs. The sugar molecules are also used in cellular respiration to produce energy for plant cells.

Practice *Refer to the processes involved in photosynthesis as you answer these questions.*

1. The light-dependent reactions convert light energy to chemical energy. Where does the light energy come from? What form of chemical energy is produced? How is this useful to the cell?

2. Plants take in carbon dioxide and water from the environment to make sugar building blocks, and they release oxygen back into the environment, as shown in the equation.

$$6CO_2 + 6H_2O \rightarrow C_6H_{12}O_6 + 6O_2$$

Identify the place in the light-dependent or light-independent reaction where each input and output of the process is important.

PRACTICE

1. *The light-dependent reactions use energy from sunlight (or maybe from grow lights indoors) to produce chemical energy in the form of NADPH and ATP. Both of these forms of energy can be used in the light-independent reactions to make sugar molecules.*

2. *Carbon dioxide enters the light-independent reaction cycle to combine with the starting 5-carbon molecule. Water supplies H^+ and electrons in the light-dependent reactions. Glucose is the product of the light-independent reactions. Oxygen is released from the light-dependent reactions after the hydrogen is removed for use in the electron transport chain.*

APPENDIX E
Classification and Relatedness of Species

Science Background

Organization Classification is a topic that is changing due to advances in biotechnology. Have students research some of the ways that biotechnology is impacting the classification of organisms. Students could start by looking at the way human DNA technology is changing the conversation of human relatedness and then move on to how similar technology is used to determine the relatedness of seemingly unrelated organisms.

CLASSIFICATION AND RELATEDNESS OF SPECIES

Taxonomic Systems Over Time

For thousands of years, people sought to bring order to the vast array of life on Earth. Of course, at that time, classification extended only to those organisms visible to the naked eye. Of those organisms, most people focused on classifying plants according to their medicinal properties. Aristotle is considered the first to classify animals as well.

In the 1500s, optical lenses enabled a closer look at the structure of organisms, and taxonomic efforts progressed. In the mid-1700s, Carolus Linnaeus built on the work of earlier taxonomists to develop the nomenclature, or naming system, that is still used today with some modifications. Linnaeus grouped species into genera as Aristotle did. However, based on similarities, he then grouped genera into orders, orders into classes, and classes into kingdoms. Today, the most widely accepted taxonomic levels are reflected in **Table 1**, with domain being the most general and species the most specific. As significant differences have been discovered, organisms have been reclassified and new levels of classification have been inserted, such as groups, superorders, and subphyla.

Figure 1 The species name *Balaenoptera musculus* refers to the blue whale and no other living thing.

Table 1. Classification of the Blue Whale, *Balaenoptera musculus*

Level	Classification	Description
domain	Eukarya	complex cell structures enclosed within membranes
kingdom	Animalia	animals
phylum	Chordata	chordates (possessing a nerve cord)
subphylum	Vertebrata	vertebrates (with backbone and spinal column)
class	Mammalia	mammals
order	Cetacea	whales and dolphins
suborder	Mysticeti	baleen whales (possessing baleen plates instead of teeth)
family	Balaenopteridae	rorquals (possessing pleated throat grooves)
genus	*Balaenoptera*	finback whales
species	*Balaenoptera musculus*	blue whale

Current advancements in technology enable taxonomists to use a variety of genetic tools to determine relatedness among species. Findings based on similarities in DNA and protein sequences, ribosomal RNA, and mitochondrial DNA enable a more detailed picture of relatedness. Therefore, taxonomic classification is an ever-changing science. The snapshot shown in **Figure 2** is one way scientists currently classify the relatedness of life on Earth. New discoveries and different ways of looking at the data could change scientists' understanding of relatedness in the future.

Figure 2 The three domains of life have a common ancestor.

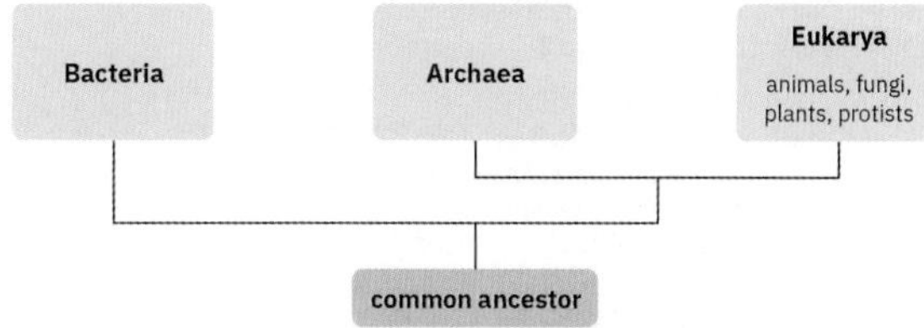

Bacteria

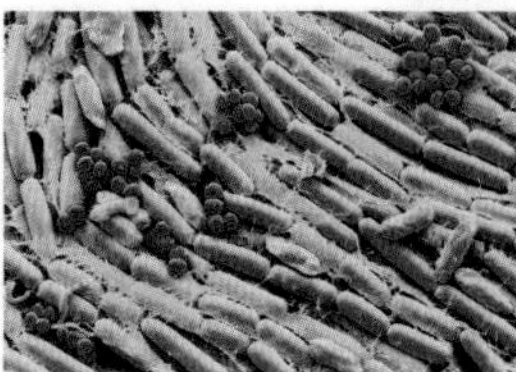

Archaea

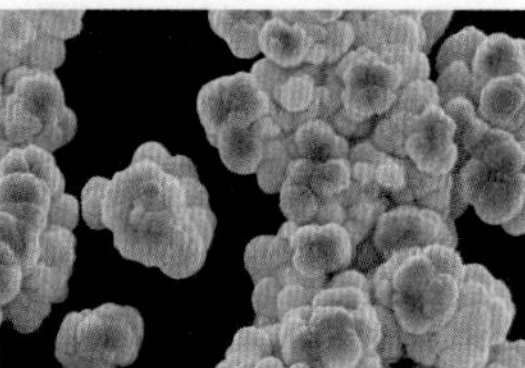

Eukarya

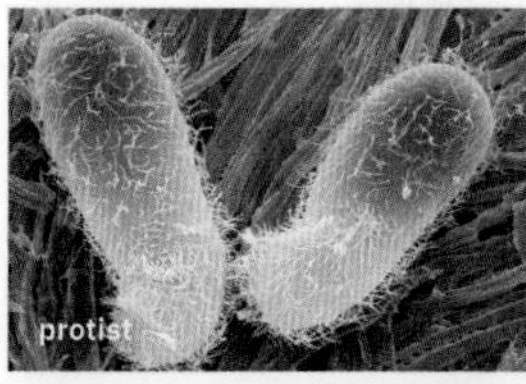

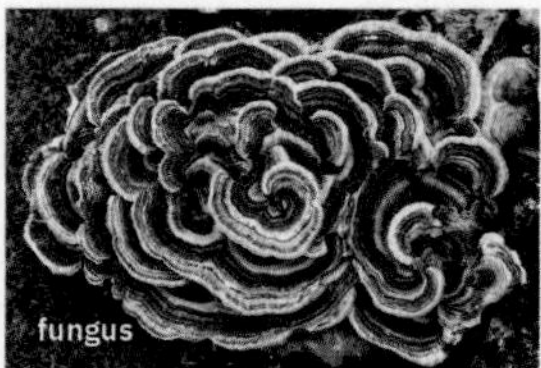

APPENDIX E: CLASSIFICATION AND RELATEDNESS OF SPECIES 547

Visual Support

Compare and Contrast Have students work in partners to compare **Figure 3** with **Figure 5**. Ask them to find similarities and differences in structure and in what each is showing. Have students determine when the use of one may be beneficial over the use of the other or to explain how the two diagrams support each other. The idea is for students to see the importance of visual representation.

APPENDIX E (continued)

Phylogenetics and Relatedness

With advancements in biotechnology, scientists began to look at taxonomic relationships more as evolutionary pathways that provided evidence of common ancestry and how lineages diverged. Research into molecular and cellular structure also led to the conclusion that species grouped into the domain Eukarya were more closely related to species in domain Archaea than to those in domain Bacteria.

The representation shown in **Figure 3** was developed in 2009 using ribosomal RNA (rRNA) sequences because these sequences occur in all cells and organelles. They are also the most conserved large sequences in nature. Although much has been learned since 2009, scientists still agree on the evolutionary relationships shown in **Figure 3** and described in the boxes below.

Domain Bacteria All bacteria are prokaryotes and unicellular, but they are extremely diverse. They vary in shape, with some being rodlike and others being spherical. Still others are curved, forming shapes from commas to spirals. They differ in their modes of nutrition or how they obtain energy as well. They may be phototrophic, chemotrophic, autotrophic, or heterotrophic. They also vary greatly in growth requirements related to temperature, pH, oxygen, and osmotic pressure. The cell wall varies somewhat in bacteria, but in nearly all species it is made of a large molecule called peptidoglycan. This layer's thickness determines whether species are Gram-positive or Gram-negative. Bacteria are extremely important ecologically, both to the environment and to the individual organisms with which they form symbiotic relationships.

Domain Archaea All archaea are prokaryotic and unicellular, but they are also extremely diverse. They can be found in the most extreme habitats on Earth, such as hydrothermal vents, extremely acidic or salty soils, hot springs, and anerobic environments. Most archaea, however, are simply spread throughout the environment. Much of what scientists know about archaea comes from isolating and analyzing nucleic acids from the environment rather than from cultures grown in labs, which is extremely difficult to do. Such studies show that the organization of DNA around histone proteins connects archaea more closely to eukaryotes than bacteria. Another distinguishing factor is that while the cell walls of archaea vary in structure, none contain peptidoglycan. Also, the RNA involved in transcription and the process of protein synthesis is much more like that of eukaryotes than bacteria.

Domain Eukarya Eukaryotes can be unicellular or multicellular, but cells of all organisms in this diverse group have a membrane-bound nucleus. While eukaryotes are broadly divided into four kingdoms—Protista, Fungi, Plantae, and Animalia—their evolutionary relationships are not as neatly defined. Protista is no longer considered a valid grouping given its lack of a unifying characteristic. In fact, any given protist may be more closely related to fungi, plants, or animals than to other protists. Therefore, this text uses the system of classifying eukaryotes into supergroups, a level between domain and kingdom. **Figure 4** below shows one way scientists look at eukaryote supergroups.

Figure 4 Using supergroups better depicts the evolutionary relationships of eukaryotes than earlier kingdom classification systems.

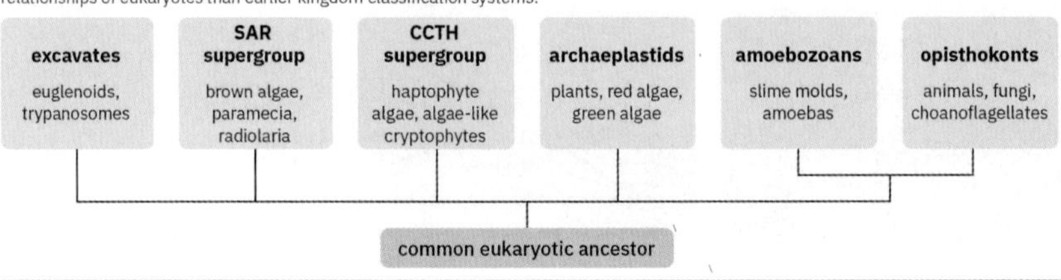

548 APPENDIX E

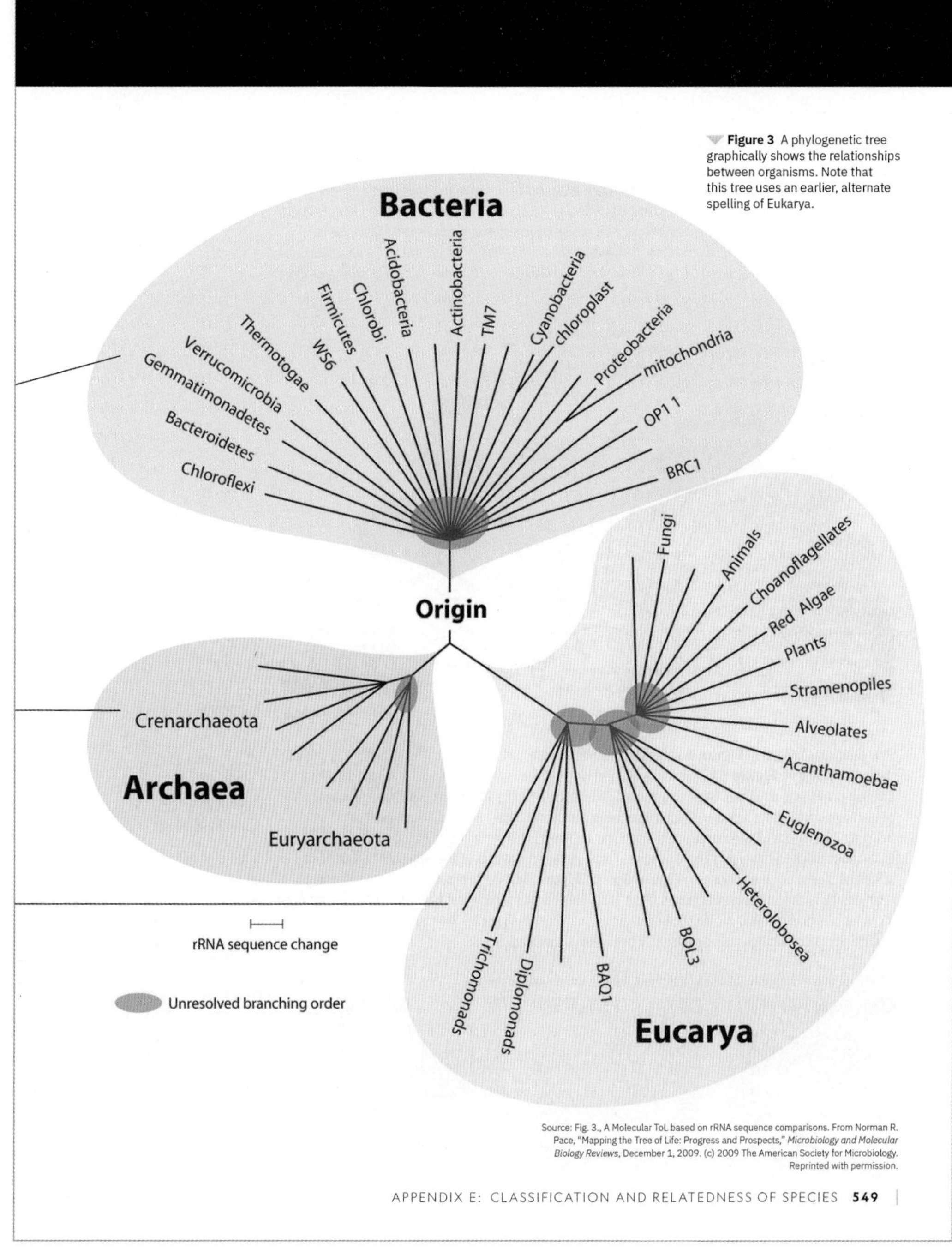

Figure 3 A phylogenetic tree graphically shows the relationships between organisms. Note that this tree uses an earlier, alternate spelling of Eukarya.

Source: Fig. 3., A Molecular ToL based on rRNA sequence comparisons. From Norman R. Pace, "Mapping the Tree of Life: Progress and Prospects," *Microbiology and Molecular Biology Reviews*, December 1, 2009. (c) 2009 The American Society for Microbiology. Reprinted with permission.

APPENDIX E: CLASSIFICATION AND RELATEDNESS OF SPECIES 549

APPENDIX E (continued)

Depicting Relatedness

Given that the vast array of life on Earth arose from a single common ancestor, the challenge becomes how to depict that relatedness and how to change the depiction given ongoing discoveries. The ability to computer-analyze huge molecular datasets feeds this challenge. For example, recent work has resulted in an even newer idea that the six supergroups of eukaryotes should be divided further into eleven or as many as fourteen groups, as shown in **Figure 5**. Time will tell how other scientists will react and whether they reach consensus.

Figure 5 Follow the pathways to determine which supergroups are more closely related. For example, this model shows that scientists think Ancyromonadida are more closely related to Amoebozoa than to Telonemia.

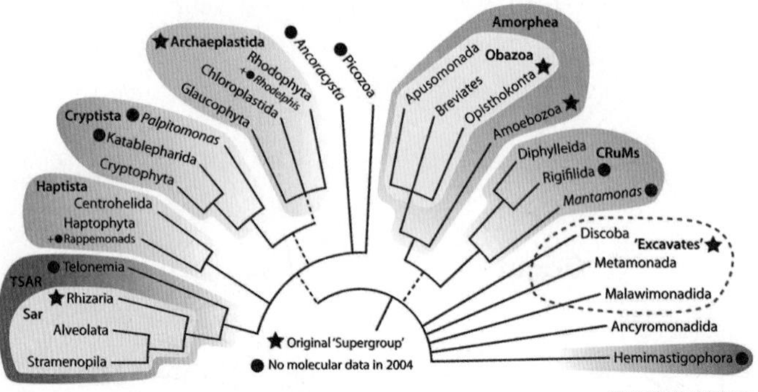

The diagram shown in **Figure 5** refers to a small section of a circular tree diagram, such as the one shown in **Figure 6a**. A circular evolutionary tree is often used to show relatedness among hundreds or thousands of species. The center of the diagram denotes the root, or earliest common ancestor. Each branching point indicates the common ancestor from which others in the connected sections evolved. The innermost point of the circle is farthest back in time and the branches flow outward toward more recent times. Other styles of trees, such as those in **Figure 6b** and **Figure 6c**, are typically used for showing relationships among fewer species. A self-contained branching section of a tree represents a clade, a grouping that includes a common ancestor and all the descendants. In any style of tree, the time scale can vary greatly between branches.

Figure 6 A circular tree (a), a diagonal tree (b), and a square-corner tree (c).

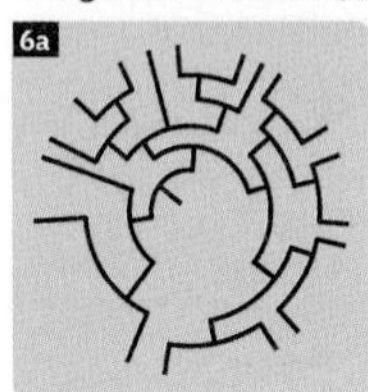

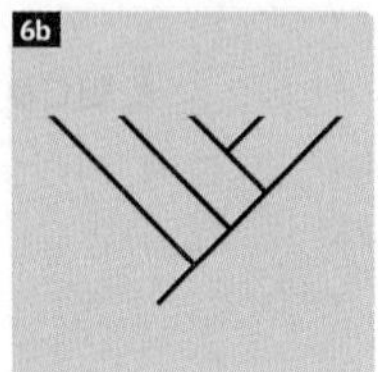

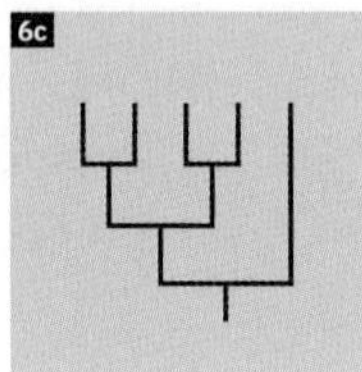

550 APPENDIX E

One challenge of depicting relatedness is that we have data for relatively few species. Of the estimated 8.7 million plant and animal species on Earth, only about 1.2 million have been identified and described. Overall, scientists have sequenced the genomes of about 3500 complex life forms. Researchers at the University of Texas used data from approximately 3000 species to create the tree of life in **Figure 7**, which includes all three domains and illustrates the complexity of showing relatedness of all living things. Most of the species in this diagram represent plants, animals, fungi, and protists. Yet, along with the millions of complex species yet to be described, a recent study estimates that between 0.8 and 1.6 million prokaryotic species exist, of which approximately 200,000 have been sequenced. Only a few of these are included in the dataset for this tree. Use **Figure 8** to find out more about a few of the species represented here.

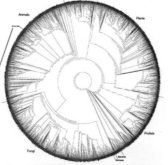

Figure 7 Use a strong magnifier to identify other species cataloged here that have a relatively close relationship to *Homo sapiens*.

Source: From Elizabeth Pennisi, "Modernizing the Tree of Life," Science, 300(5626), June 13, 2003, pp. 1692-1697. Figure: Roundabout. Credit: David M. Hillis, Derrick Zwicki, and Robin Gutell, University of Texas, Austin. Reprinted with permission from AAAS and David M. Hillis.

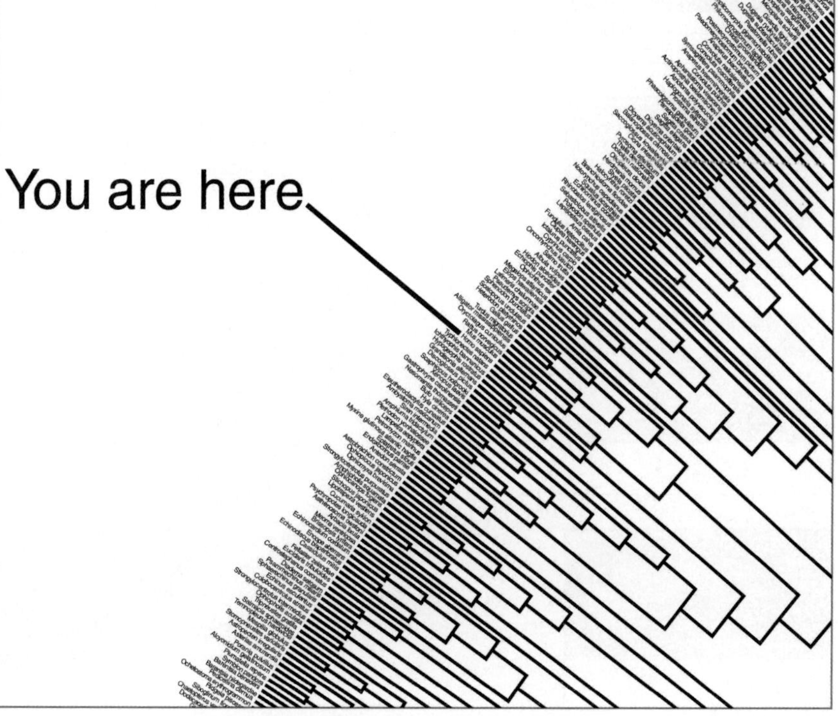

CROSSCUTTING CONCEPTS | Patterns

Conceptual Understanding Guide students to understand the general concept over the specific content. Depictions of organism relatedness are a way for humans to put order to the world. The visual representations, although based on data, are meant to give an overall image of how organisms are interrelated. Have students focus on the types of visual representation as opposed to the organismal traits that define the visual. Have them answer questions such as: "*Why is it important to understand how organisms are related*?" or "*How would you represent relatedness*?"

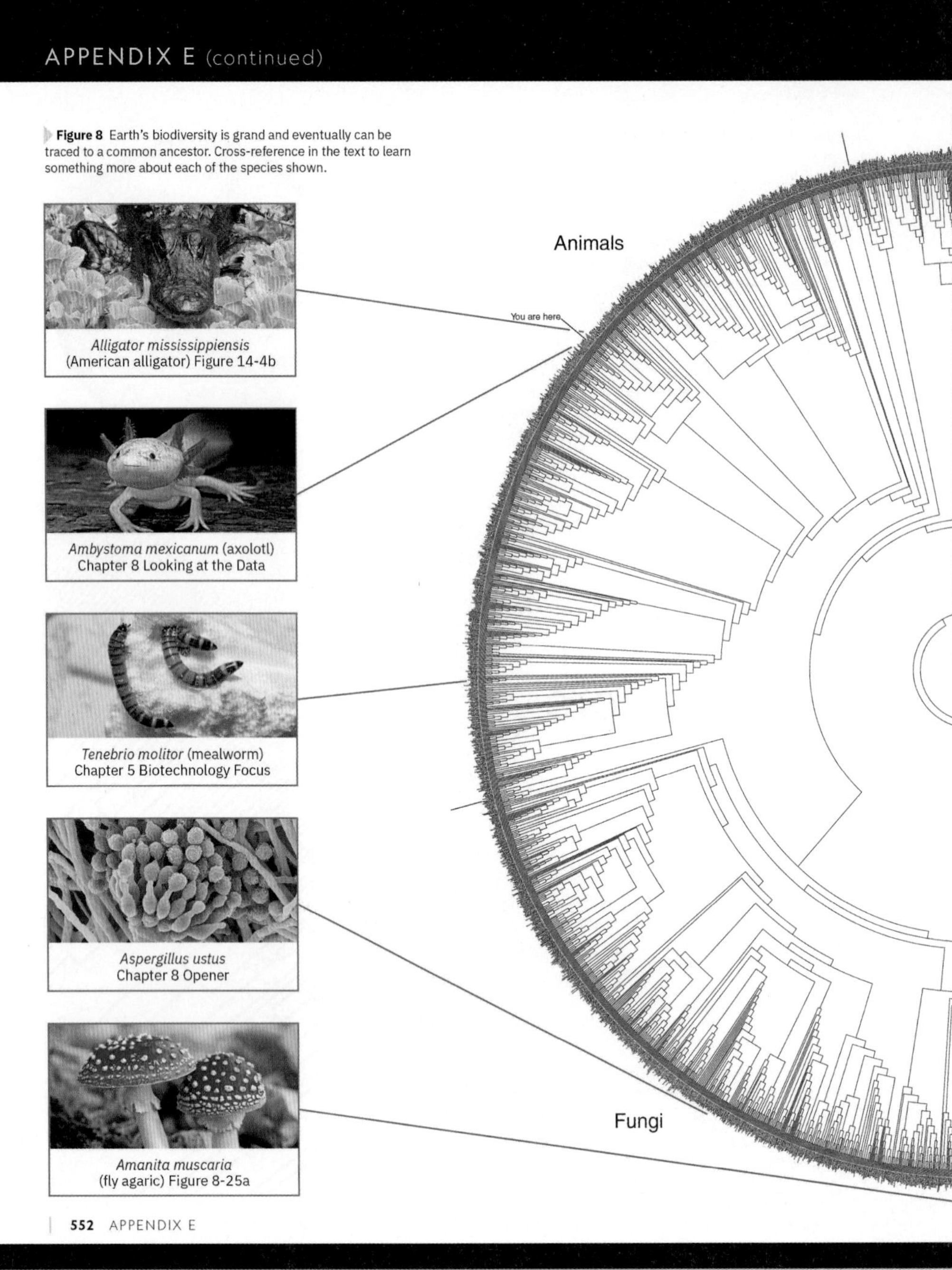

Figure 8 Earth's biodiversity is grand and eventually can be traced to a common ancestor. Cross-reference in the text to learn something more about each of the species shown.

Alligator mississippiensis (American alligator) Figure 14-4b

Ambystoma mexicanum (axolotl) Chapter 8 Looking at the Data

Tenebrio molitor (mealworm) Chapter 5 Biotechnology Focus

Aspergillus ustus Chapter 8 Opener

Amanita muscaria (fly agaric) Figure 8-25a

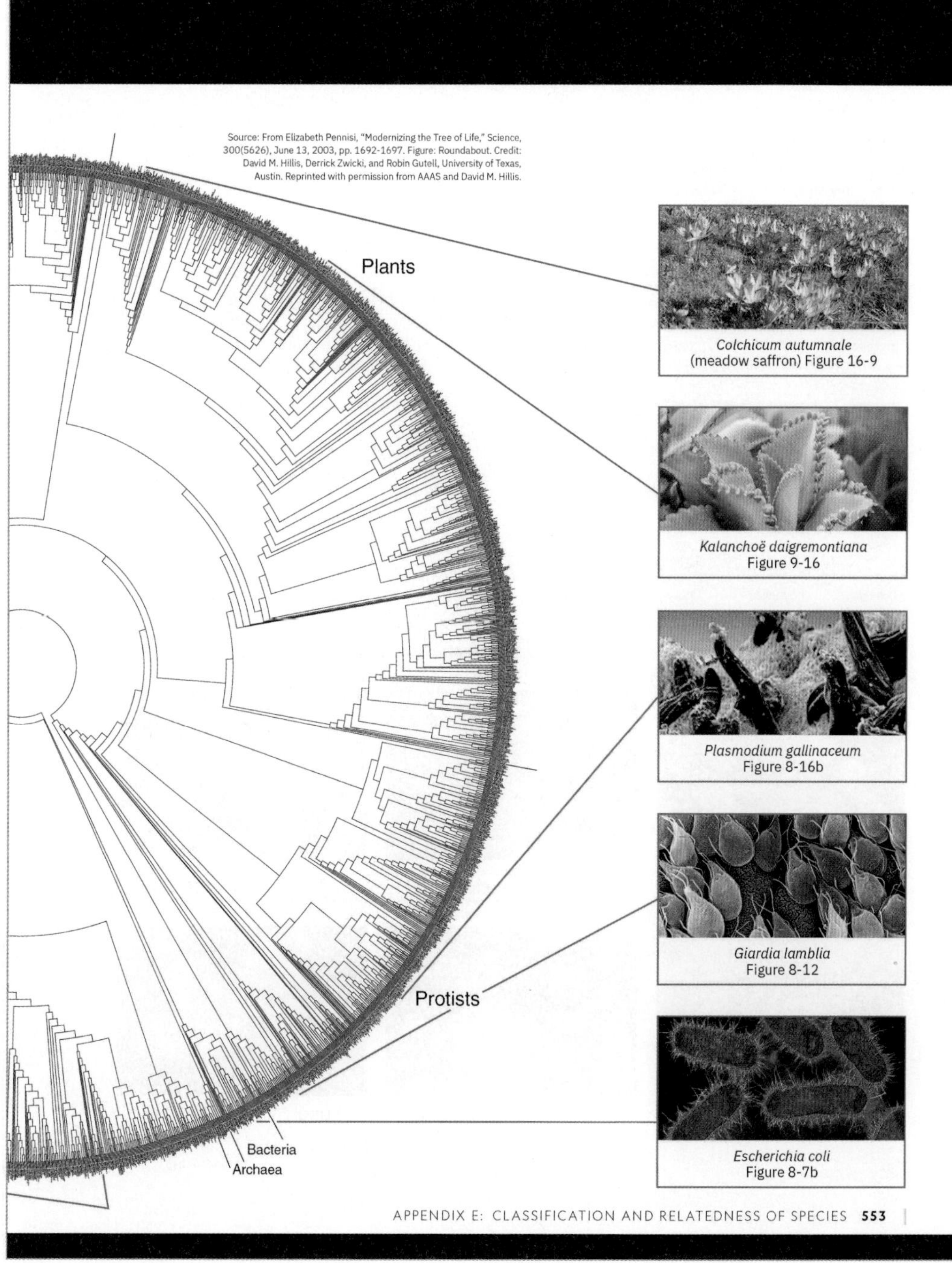

Source: From Elizabeth Pennisi, "Modernizing the Tree of Life," Science, 300(5626), June 13, 2003, pp. 1692-1697. Figure: Roundabout. Credit: David M. Hillis, Derrick Zwicki, and Robin Gutell, University of Texas, Austin. Reprinted with permission from AAAS and David M. Hillis.
Plants
Protists
Bacteria
Archaea
APPENDIX E: CLASSIFICATION AND RELATEDNESS OF SPECIES 553

Colchicum autumnale (meadow saffron) Figure 16-9

Kalanchoë daigremontiana Figure 9-16

Plasmodium gallinaceum Figure 8-16b

Giardia lamblia Figure 8-12

Escherichia coli Figure 8-7b

APPENDIX E (continued)

Diversity of Life

It's hard to imagine the incredible variation among Earth's life forms, much less show it in a snapshot. Use **Table 2** to compare the basic characteristics of the domains and kingdoms. Then, get a glimpse of this vast array in the images and cross-reference to other parts of the text to find out more.

Table 2. Comparison of domain and kingdom characteristics.

Domain	Kingdom	Characteristics				
		Cell type	Cell structure	Body type	Nutrition	Examples
Bacteria	Eubacteria	prokaryotic	cell wall of peptidoglycan	unicellular	autotrophic and heterotrophic	*Micrococcus luteus, Dietzia kunjamensis, Treponema pallidum*
Archaea	Archaebacteria	prokaryotic	cell wall with no peptidoglycan	unicellular	autotrophic and heterotrophic	Halophilic archaea, *Sulfolobus* archaea
Eukarya	Protista	eukaryotic	mixed	unicellular and multicellular	autotrophic and heterotrophic	diatoms, foraminifera, slime molds
Eukarya	Plantae	eukaryotic	cell wall of cellulose	unicellular and multicellular	autotrophic	liverworts, spruce, trilliums
Eukarya	Fungi	eukaryotic	cell wall of chitin	unicellular and multicellular	heterotrophic	bread mold, violet webcap, bleeding Hydnellum
Eukarya	Animalia	eukaryotic	no cell wall	multicellular	heterotrophic	spotted jelly, garden snail, puffins

Bacteria

See Chapter 8: *Find out more about bacteria in Section 8.1.*

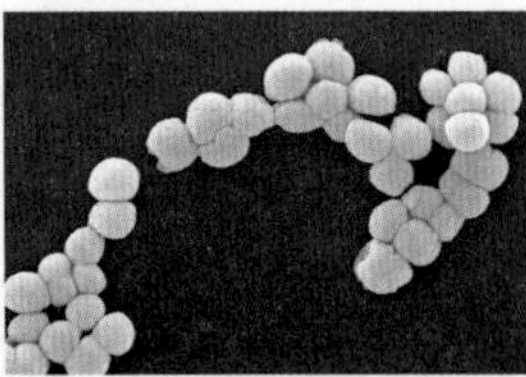

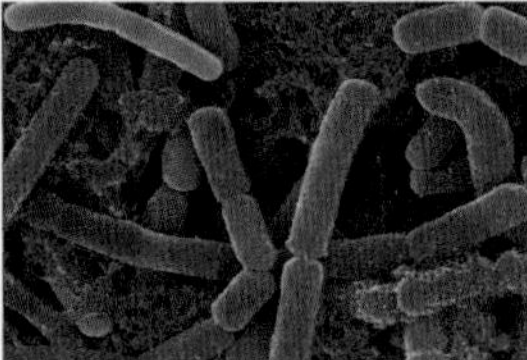

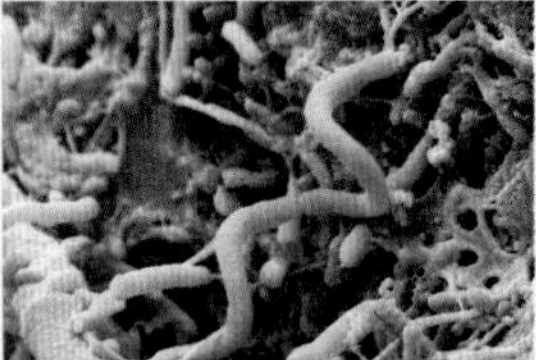

Archaea

See Chapter 8: *Find out more about archaea in Section 8.1.*

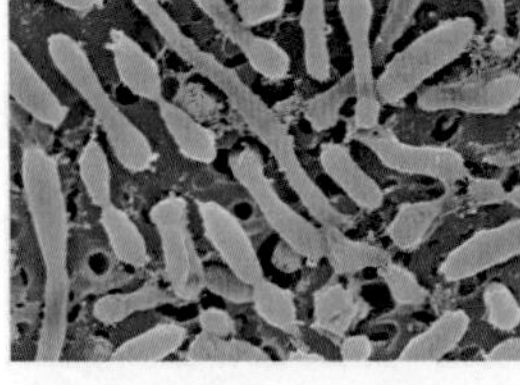

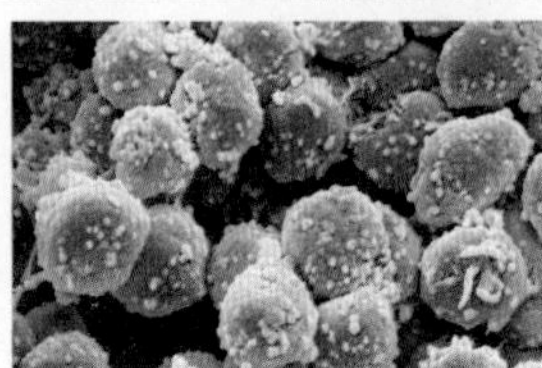

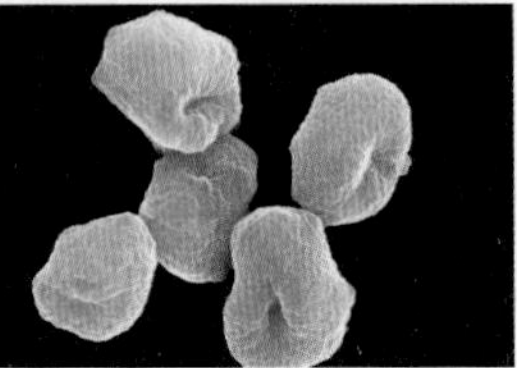

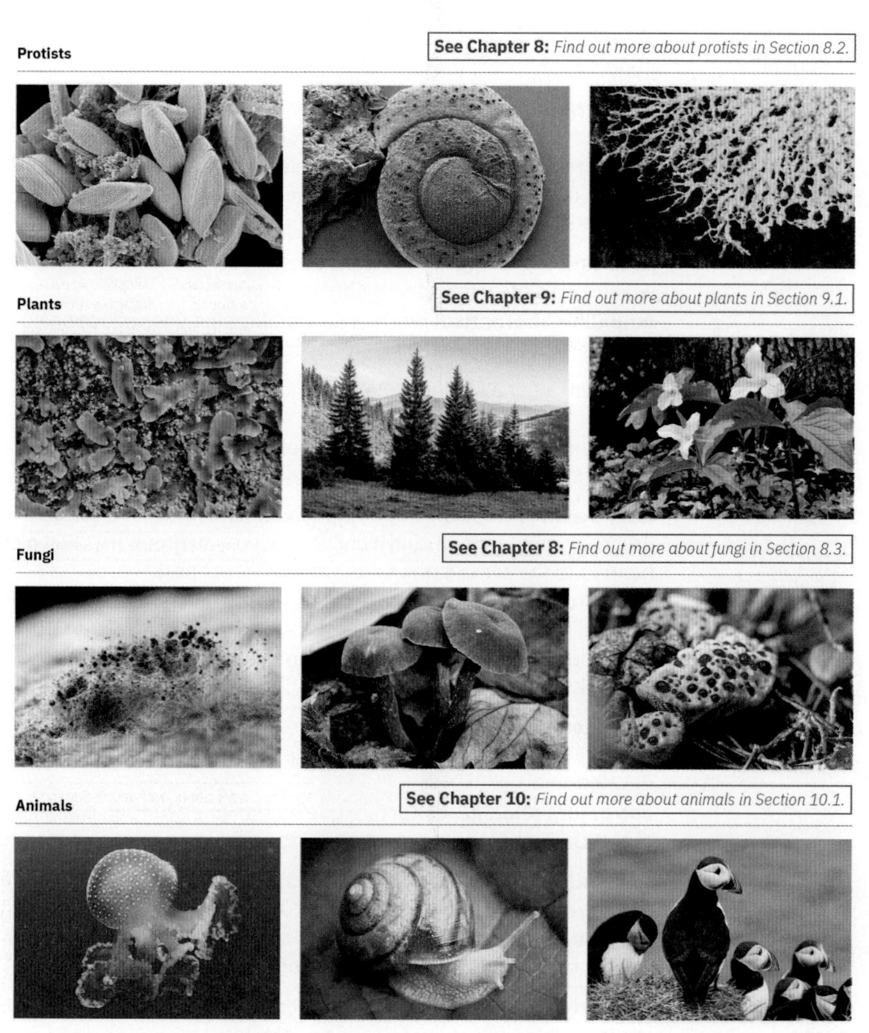
Protists
See Chapter 8: Find out more about protists in Section 8.2.
Plants
See Chapter 9: Find out more about plants in Section 9.1.
Fungi
See Chapter 8: Find out more about fungi in Section 8.3.
Animals
See Chapter 10: Find out more about animals in Section 10.1.
APPENDIX E: CLASSIFICATION AND RELATEDNESS OF SPECIES 555

GLOSSARY

A

abiotic factor a nonliving factor in an ecosystem, such as temperature, wind, soil, and sunlight
factor abiótico factor inerte de un ecosistema, como la temperatura, el viento, el suelo y la luz solar

acid a substance that releases hydrogen ions in water
ácido sustancia que libera iones de hidrógeno en el agua

activation energy minimum amount of energy required to start a reaction
energía de activación cantidad mínima de energía que se necesita para provocar una reacción

activator a transcription factor that increases the rate of transcription when it binds to a promoter or enhancer
activador factor de la transcripción que aumenta la tasa de transcripción al unirse a un promotor o potenciador

active site an area of an enzyme that accepts a substrate and catalyzes its breakdown or its reaction with another substrate
sitio activo área de una enzima que acepta un sustrato y cataliza su descomposición o su reacción con otro sustrato

active transport energy-requiring mechanism in which a transport protein pumps a solute across a cell membrane against the solute's concentration gradient
transporte activo mecanismo que requiere energía, donde una proteína de transporte bombea un soluto a través de una membrana celular en dirección opuesta al gradiente de concentración del soluto

adaptation a heritable trait that improves an organism's chances of surviving and producing offspring
adaptación rasgo hereditario que mejora las posibilidades de que un organismo sobreviva y produzca descendencia

adaptive immunity a set of immune defenses that can be tailored to specific pathogens as an organism encounters them during its lifetime; characterized by self/nonself recognition, specificity, diversity, and memory
inmunidad adaptativa conjunto de defensas inmunitarias que se adaptan a patógenos específicos a medida que los organismos los encuentran durante su vida; se caracterizan por el reconocimiento de lo propio y lo ajeno, la especificidad, la diversidad y la memoria

adaptive radiation a pattern of macroevolution in which a lineage undergoes a burst of genetic divergences that gives rise to many species
radiación adaptiva patrón macroevolutivo en el que un linaje experimenta un aumento explosivo de divergencias genéticas que da origen a muchas especies

adenosine triphosphate (ATP) a nucleotide that consists of an adenine base, a ribose sugar, and three phosphate groups; functions as a subunit of RNA and as a coenzyme in many reactions; important energy carrier
adenosín trifosfato (ATP) nucleótido compuesto de una base de adenina, un azúcar ribosa y tres grupos de fosfato; funciona como subunidad del ARN y como coenzima en muchas reacciones; es un importante portador de energía

adhesion an attraction between molecules of different substances
adhesión atracción entre las moléculas de sustancias diferentes

aerobic cellular respiration a type of cellular respiration that requires oxygen
respiración aeróbica celular tipo de respiración celular que requiere oxígeno

allele one form of a gene with slightly different DNA sequence; may encode different versions of the gene's product
alelo forma de un gen con una secuencia de ADN ligeramente distinta; puede codificar distintas versiones del producto del gen

allele frequency the proportion of one allele relative to all the alleles for that trait in a population's chromosomes
frecuencia alélica en los cromosomas de una población, la proporción de un alelo en relación con todos los alelos de una característica dada

alternative splicing a post-translational RNA modification process in which exons are joined in different combinations
empalme alternativo proceso de modificación postraducción en el cual los exones se unen en distintas combinaciones

altruism a behavior that benefits others at the expense of the individual performing it
altruismo comportamiento que beneficia a otros a expensas del individuo que lo lleva a cabo

amino acid a small organic compound that consists of a carboxyl group, an amine group, and a characteristic side group (R), all typically bonded to the same carbon atom; monomer of proteins
aminoácido pequeño compuesto orgánico que consta de un grupo carboxílico, un grupo amino y un grupo característico adjunto (R), los cuales están unidos al mismo átomo de carbono; monómero de proteínas

analogous structure a body structure that is similar in function or appearance to another in a different lineage but not similar in origin or development
estructura análoga estructura que es similar en función o aspecto a otra de un linaje distinto, pero que no es similar en origen ni desarrollo

anaphase a stage of mitosis during which sister chromatids separate and move toward opposite spindle poles
anafase etapa de la mitosis durante la cual las cromátidas hermanas se separan y se mueven hacia polos opuestos del huso

antibody a y-shaped antigen receptor protein made by B cells; immunoglobulin; e.g., IgA, IgG, IgE, IgM, IgD
anticuerpo proteína receptora de antígenos que tiene forma de Y y está hecha de células B; inmunoglobulina; p. ej., IgA, IgG, IgE, IgM, IgD

anticodon a set of three nucleotides that base-pairs with an mRNA codon in a tRNA
anticodón conjunto de tres nucleótidos que se empareja en la base con un codón de ARNm en un ARNt

antigen a molecule or particle that the immune system recognizes as nonself; its presence in the body triggers an immune response
antígeno molécula o partícula que el sistema inmune reconoce como ajena; su presencia en el cuerpo desencadena una respuesta inmune

apex consumer a consumer at the top trophic level that eats other consumers but is not typically consumed by others
consumidor ápice consumidor que está en el nivel trófico más alto y que se come a otros consumidores, pero generalmente sin que los otros se lo coman

apoptosis a predictable, controlled process of cell self-destruction; programmed cell death
apoptosis proceso predecible y controlado de autodestrucción celular; muerte celular programada

artificial selection the selection by humans of traits that are desirable in plants or animals and breeding only those individuals that possess the desired traits
selección artificial selección por parte de los seres humanos de rasgos deseables en plantas o animales mediante el cruce exclusivo de aquellos individuos que poseen los rasgos deseados

asexual reproduction a reproductive mode of eukaryotes by which offspring arise from a single parent only
reproducción asexual modo reproductivo de los eucariotas en el que la descendencia surge de un solo progenitor

atom the fundamental unit of which elements are composed
átomo unidad fundamental de la cual se componen los elementos

ATP see adenosine triphosphate
ATP ver adenosín trifosfato

autotroph an organism that makes its own food from abiotic sources, such as sunlight or inorganic materials; also known as a producer
autótrofo organismo que produce su propio alimento a partir de fuentes abióticas, como luz solar o materiales inorgánicos; conocido también como *productor*

B

background extinction the continuous, low-level extinction of species that has been evident throughout much of the history of life
extinción de fondo la extinción continua y de bajo nivel de especies que ha sido evidente a lo largo de una gran parte de la historia de la vida

bacteriophage a type of virus that infects bacteria
bacteriofago tipo de virus que infecta bacterias

base a substance that accepts hydrogen ions in water
base sustancia que acepta iones de hidrógeno en el agua

binary fission a method of asexual reproduction in which one prokaryotic cell divides into two identical descendant cells
fisión binaria método de reproducción asexual en el cual una célula procariota se divide en dos células descendientes idénticas

biodiversity the variety and variability of life on Earth; a region's biodiversity is measured at three levels: the genetic diversity within species, species diversity, and ecosystem diversity
biodiversidad variedad y variabilidad de la vida en la Tierra; la biodiversidad de una región se mide en tres niveles: la diversidad genética dentro de las especies, la diversidad de especies y la diversidad de ecosistemas

biogeochemical cycle a process by which matter cycles from the living world to the nonliving physical environment and back again; examples of biogeochemical cycles include the carbon cycle, the nitrogen cycle, and the phosphorus cycle
ciclo biogeoquímico proceso a través del cual la materia transita entre el mundo vivo, el entorno físico no vivo y de regreso al mundo vivo; los ciclos del carbono, del nitrógeno y del fósforo son ciclos biogeoquímicos

biogeography the study of patterns in the geographic distribution of species and communities
biogeografía el estudio de patrones en la distribución geográfica de especies y comunidades

bioinformatics the use of computational tools for analyzing and interpreting biological data
bioinformática uso de herramientas computacionales para analizar e interpretar datos biológicos

biological engineer an engineer that uses principles of biology to design solutions to problems
ingeniero biológico ingeniero que aplica los principios de la biología para diseñar soluciones a problemas

biology the scientific study of life
biología el estudio científico de la vida

biomass the total dry mass of all the living organisms within a given area at a specific time
biomasa el total de masa seca de todos los organismos vivos dentro de un área dada en un momento específico

biomass pyramid a model depicting the total dry mass of all the living organisms at each trophic level within a given area at a specific time
pirámide de biomasa modelo que muestra el total de masa seca de todos los organismos vivos de cada nivel trófico dentro de un área dada en un momento específico

biome a region characterized by its climate and dominant vegetation
bioma región que se caracteriza por su clima y vegetación dominante

biosphere all of Earth's organisms and the regions where organisms live
biosfera todos los organismos de la Tierra y las regiones donde los organismos viven

biotic factor a living thing in an ecosystem
factor biótico un ser vivo en un ecosistema

biotic potential a measure of how much a population could grow under ideal conditions; measured by its per capita growth rate as determined only by the reproductive characteristics of the species
potencial biótico medida de cuánto podría crecer una población bajo condiciones ideales; se mide por su crecimiento per cápita, que se determina únicamente por las características reproductivas de la especie

bottleneck effect a reduction in population size so severe that it reduces genetic diversity; a type of genetic drift
efecto de cuello de botella reducción tan drástica del tamaño de la población que reduce la diversidad genética; tipo de deriva genética

bulb a short plant stem covered with overlapping scales that stores reserve energy during unfavorable growth conditions; the lower portion of a bulb gives rise to new roots
bulbo tallo corto de una planta cubierto con escamas superpuestas y que almacena energía de reserva durante condiciones de crecimiento desfavorables; en la parte baja de un bulbo se generan raíces nuevas

C

capsid the protein shell containing the genetic material of a virus
cápside envoltura de proteína que contiene el material genético de un virus

carbohydrate an organic compound that consists of carbon, hydrogen, and oxygen
carbohidrato compuesto orgánico de carbono, hidrógeno y oxígeno

carpel the reproductive organ of a flower that produces the female gametophyte; consists of an ovary, stigma, and usually a style
carpelo órgano reproductivo de una flor que produce el gametofito femenino; está compuesto por un ovario, un estigma y un estilo

carrying capacity the maximum number of individuals that a population's environment can support indefinitely
capacidad de carga número máximo de individuos que el medio ambiente de una población puede soportar de manera indefinida

catalyst a substance that regulates the speed at which a chemical reaction occurs without affecting the end point of the reaction and without being used up as a result of the reaction
catalizador sustancia que regula la rapidez de una reacción química sin afectar el final de la reacción y sin ser consumido como resultado de la reacción

catastrophism the principle stating that infrequent catastrophic geologic events alter the course of Earth history, in contrast to the principle of gradualism
catastrofismo principio según el cual los sucesos geológicos catastróficos e infrecuentes alteran el curso de la historia de la Tierra, en contraste con el principio de gradualismo

cell differentiation a process by which cells become specialized during development; occurs as different cells in an embryo begin to use different subsets of their DNA
diferenciación celular proceso por el cual las células se especializan durante el desarrollo; ocurre cuando las células de un embrión empiezan a usar diferentes subconjuntos de su ADN

cell membrane a lipid bilayer that encloses the cytoplasm; separates a cell from its external environment
membrana celular una bicapa lipídica que encierra el citoplasma; separa a la célula de su ambiente externo

cellular respiration a process that harvests energy from an organic molecule to store it in the usable form of ATP
respiración celular proceso que cosecha la energía de una molécula orgánica para almacenarla como ATP que se puede usar

cellulose a complex carbohydrate that consists of parallel chains of glucose monomers
celulosa carbohidrato complejo compuesto de cadenas paralelas de monómeros de glucosa

centromere a constricted region where sister chromatids attach to each other
centrómero región restringida donde se unen las cromátidas hermanas

character a quantifiable, heritable trait such as the number of segments in a backbone or the nucleotide sequence of ribosomal RNA
carácter rasgo cuantificable y hereditario, como el número de segmentos que hay en la columna vertebral o la secuencia de nucleótidos de ARN ribosomal

charge a property of matter that results in electrical forces; like charges repel, and opposite charges attract
carga propiedad de la materia que resulta en fuerzas eléctricas; las cargas iguales se repelen y las cargas opuestas se atraen

chemical bond a force that holds atoms in a molecule or substance together
enlace químico fuerza que une los átomos de una molécula o sustancia

chemical reaction a process in which substances change into different substances through the breaking and forming of chemical bonds
reacción química proceso en el que las sustancias se convierten en otras sustancias al romperse y formarse enlaces químicos

chemoautotroph an organism that uses carbon dioxide as its carbon source and obtains energy by oxidizing inorganic molecules
quimioautótrofo organismo que usa dióxido de carbono como fuente de carbono y oxida moléculas inorgánicas para obtener energía

chemoheterotroph an organism that obtains energy and carbon by breaking down organic compounds
quimioheterótrofo organismo que descompone compuestos orgánicos para obtener energía y carbono

chemosynthesis a process by which sugars are made by organisms that use chemicals as an energy source rather than sunlight
quimiosíntesis proceso en el que organismos producen azúcares utilizando productos químicos como fuente de energía en lugar de luz solar

chitin a nitrogen-containing polysaccharide that composes fungal cell walls and arthropod exoskeletons
quitina polisacárido que contiene nitrógeno y compone las paredes celulares de los hongos y los exoesqueletos de los artrópodos

chloroplast an organelle of photosynthesis in plants and photosynthetic protists; light-dependent reactions occur at its inner thylakoid membrane; light-independent reactions, in the stroma
cloroplasto orgánulo de la fotosíntesis en plantas y protistas fotosintetizadores; en su membrana tilacoide interior se dan reacciones que dependen de la luz; las reacciones que son independientes de la luz ocurren en el estroma

choanoflagellate the protist group most closely related to animals
coanoflagelados el grupo protista más cercano a los animales

chordate an animal with an embryo that has a notochord, dorsal nerve cord, pharyngeal gill slits, and a tail that extends beyond the anus
cordado animal cuyo embrión tiene notocorda, tubo neural hueco en la posición dorsal, hendiduras branquiales y cola que se extiende más allá del ano

chromosome a structure that consists of DNA together with associated proteins; carries part or all of a cell's genetic information
cromosoma estructura que consta de ADN y proteínas asociadas; contiene una parte o toda la información genética de una célula

chytrid a fungus with flagellated spores
quítrido hongo con esporas flageladas

clade a group of taxa whose members share one or more defining derived traits
clado grupo de taxones cuyos miembros comparten uno o más rasgos derivados

cladistics a system of classification that arranges organisms by shared characteristics and inferred evolutionary relationships
cladística sistema de clasificación que ordena los organismos por características compartidas y relaciones evolutivas inferidas

cladogram an evolutionary tree diagram that visually summarizes a hypothesis about how a group of clades are related
cladograma diagrama evolutivo ramificado que resume visualmente una hipótesis sobre las relaciones entre un grupo de clados

clone a genetic copy of an organism
clon copia genética de un organismo

club fungus a fungus that produces spores on club-shaped structures during sexual reproduction
basidiomiceto hongo que produce esporas con estructuras que tienen forma de botella durante la reproducción sexual

clumped distribution a population distribution pattern in which individuals are closer to one another than would be predicted by chance alone
distribución agrupada patrón de distribución de la población en el que los individuos están más cerca de lo que se podría predecir por azar únicamente

codominance an effect in which the full and separate phenotypic effects of two alleles are apparent in heterozygous individuals
codominancia efecto en el que los efectos fenotípicos totales y separados de dos alelos son aparentes en individuos heterocigotos

codon a nucleotide base triplet that codes for an amino acid or stop signal during translation in an mRNA
codón triplete de bases de nucleótidos que, durante la traducción de un ARNm, codifica un aminoácido o una señal de parada

coevolution the joint evolution of two closely interacting species; a macroevolutionary pattern in which each species is a selective agent for traits of the other
coevolución evolución conjunta de dos especies que interactúan de manera cercana; patrón macroevolutivo en el que cada especie es un agente selectivo de los rasgos de la otra

cohesion an attraction between molecules of the same substance
cohesión atracción entre moléculas de la misma sustancia

cohesion-tension theory an explanation of how transpiration creates a tension that pulls a cohesive column of water upward through xylem
teoría de la cohesión-tensión explicación de cómo la transpiración crea una tensión que jala una columna cohesiva de agua hacia arriba a través de la xilema

colonial organism an organism composed of many similar cells, each capable of living and reproducing on its own
organismo colonial organismo compuesto por muchas células similares, cada una de las cuales es capaz de vivir y reproducirse por su cuenta

community the collection of populations of all species in a defined area
comunidad conjunto de poblaciones de una especie en un área definida

competition a ecological relationship in which two organisms try to obtain the same resource
competencia relación ecológica en la cual dos organismos tratan de obtener el mismo recurso

complementary a strand in which the base of each nucleotide pairs, through hydrogen bonding with a partner base on another strand of DNA
complementaria hebra en la cual la base de cada nucleótido se empareja, por medio de un enlace de hidrógeno, con una base asociada de otra hebra de ADN

constraint a factor that limits the possible solutions that could be developed to solve a problem
restricción factor que limita las soluciones posibles que se pueden desarrollar para resolver un problema

consumer an organism that gets its energy and nutrients from eating other organisms
consumidor organismo que come otros organismos para obtener energía y nutrientes

control center the body system that gathers information from sensory receptors and sends information to the body's effectors; the main control center in vertebrates is the nervous system
centro de control sistema del cuerpo que reúne información de los receptores sensoriales y envía información a los efectores del cuerpo; el centro de control principal de los vertebrados es el sistema nervioso

convergent evolution the independent evolution of similar structures that carry on similar functions in two or more organisms of widely different, unrelated ancestry
evolución convergente evolución independiente de estructuras similares que cumplen con funciones similares en dos o más organismos de ascendencia notoriamente distinta y no relacionada

covalent bond a type of chemical bond in which two atoms share electrons
enlace covalente tipo de enlace químico en el cual dos átomos comparten electrones

CRISPR a method of editing DNA using an RNA sequence and nuclease enzyme
CRISPR método de edición de ADN en el que se usa una secuencia de ARN y enzima nucleasa

criterion a standard that an engineering solution must meet to be considered successful
criterio estándar que una solución de ingeniería debe cumplir para ser considerada exitosa

cuticle a waxy covering over the epidermis of the aboveground portion of land plants that reduces water loss from plant surfaces
cutícula cubierta cerosa sobre la epidermis de la parte aérea de las plantas terrestres que reduce la pérdida de agua en la superficie de las plantas

cytokinesis a process in which a eukaryotic cell divides in two after mitosis or meiosis; cytoplasmic division
citocinesis proceso en el que una célula eucariota se divide en dos después de la mitosis o meiosis; división citoplásmica

cytoplasm a semifluid substance enclosed by a cell's plasma membrane
citoplasma substancia semilíquida contenida por la membrana plasmática de una célula

cytoskeleton a network of microtubules, microfilaments, and intermediate filaments that support, organize, and move eukaryotic cells and their internal structures
citoesqueleto red de microtúbulos, microfilamentos y filamentos intermedios que apoyan, organizan y mueven a las células eucariotas y sus estructuras internas

D

decomposer an organism that breaks down decaying organic matter into simpler compounds that other organisms can use
descomponedor organismo que descompone materia orgánica deteriorada, reduciéndola a compuestos más simples que les sirven a otros organismos

deforestation the removal of forest without adequate replanting
deforestación tala de bosques sin reforestación adecuada

denature to modify the molecular structure, and therefore the structure, of a protein
desnaturalizar modificar la estructura molecular y, por tanto, la estructura de una proteína

density-dependent factor a factor that slows the growth of a population by causing an increase in death rate or a decrease in birth rate as the size of the population increases
factor dependiente de la densidad factor que reduce el crecimiento de la población al causar un aumento en la tasa de mortalidad o una reducción en la tasa de natalidad a medida que aumenta el tamaño de la población

density-independent factor a factor for which the likelihood of occurrence and the magnitude of its effect is not influenced by the density of the affected population
factor independiente de la densidad factor en el que la densidad de la población afectada no influye en la probabilidad de que algo suceda ni en la magnitud de su efecto

deoxyribonucleic acid (DNA) the nucleic acid that carries hereditary information in its sequence; double helix structure consists of two chains (strands) of deoxyribonucleotides (adenine, guanine, thymine, cytosine)
ácido desoxirribonucleico (ADN) el ácido nucleico que contiene la información hereditaria en su secuencia; estructura de doble hélice que consiste en dos cadenas (hebras) de desoxirribonucleótidos (adenina, guanina, timina, citosina)

derived trait a character that is present in a clade under consideration, but not in any of the clade's ancestors
rasgo derivado rasgo que está presente en un clado bajo consideración, pero que no está presente en los antepasados del clado

desertification the conversion of grassland or woodland to desert
desertificación la conversión de un pastizal o bosque en desierto

detritus dead particulate organic matter
detrito materia orgánica particulada muerta

diffusion the spontaneous spreading of molecules or atoms through a fluid or gas
difusión la propagación espontánea de moléculas o átomos por medio de un fluido o gas

directional selection a mode of natural selection in which forms of a trait at one end of a range of variation are adaptive
selección direccional modo de selección natural en el cual las formas de un rasgo en un extremo de un rango de variación son adaptativas

disruptive selection a mode of natural selection in which forms of a trait at both ends of a range of variation are adaptive, and intermediate forms are selected against
selección disruptiva modo de selección natural en el que las formas de un rasgo a ambos extremos de un rango de variación son adaptativas y se selecciona contra las formas intermedias

DNA see deoxyribonucleic acid
ADN ver ácido desoxirribonucleico

DNA cloning a set of laboratory methods that uses living cells to mass produce specific DNA fragments
clonación de ADN conjunto de métodos de laboratorio en el que se usan células vivas para producir en serie fragmentos específicos de ADN

DNA ligase an enzyme that seals strand breaks or gaps in double-stranded DNA
ADN ligasa enzima que sella rupturas o brechas en las hebras del ADN de doble cadena

DNA polymerase an enzyme that carries out DNA replication; it uses a DNA template to assemble a complementary strand of DNA
ADN polimerasa enzima que lleva a cabo la replicación del ADN; usa un cebador de ADN para ensamblar una hebra complementaria de ADN

DNA profiling identifying an individual by analyzing the unique parts of his or her DNA
perfil de ADN o huella genética identificación de un individuo mediante el análisis de las partes únicas de su ADN

dominant an allele that is always expressed when it is present
dominante alelo que siempre se expresa cuando está presente

E

ecological disturbance an event of intense environmental stress that occurs over a relatively short time period and causes a large change in the affected ecosystem
perturbación ecológica suceso de estrés ambiental intenso que ocurre en un período relativamente corto y causa un cambio grande en el ecosistema afectado

ecological footprint the area of Earth's surface required to sustainably support a particular level of development and consumption
huella ecológica el área de la superficie de la Tierra que se necesita para mantener un nivel determinado de desarrollo y consumo de manera sostenible

ecological niche an organism's role within its habitat; it includes how an organism utilizes materials in its environment as well as how it interacts with other organisms
nicho ecológico el papel de un organismo dentro de su hábitat; incluye la manera en que un organismo usa los materiales de su medio ambiente y su manera de interactuar con otros organismos

ecological pyramid a model that shows the distribution of energy and matter at each trophic level in a food chain or food web
pirámide ecológica modelo que muestra la distribución de energía y materia en cada nivel trófico de una cadena o red alimenticia

ecological restoration the process of assisting in the recovery of an area in an effort to restore an ecosystem that has been damaged or destroyed
restauración ecológica el proceso de apoyar la recuperación de un área en un esfuerzo por restaurar un ecosistema que ha sido dañado o destruido

ecological succession the sequence of changes in a plant community over time
sucesión ecológica secuencia de cambios en una comunidad de plantas a lo largo del tiempo

ecosystem a biological community interacting with its environment
ecosistema comunidad biológica que interactúa con su medio ambiente

ecosystem service a benefit people derive from ecosystems
servicio de ecosistema beneficio que la gente obtiene de los ecosistemas

effector a muscle or gland that acts in response to instructions from the nervous system
efector músculo o glándula que actúa en respuesta a instrucciones del sistema nervioso

electron a negatively charged subatomic particle that moves around the nucleus of an atom
electrón partícula subatómica con carga negativa que se mueve alrededor del núcleo de un átomo

electron transport chain the final step of aerobic respiration which produces most of the ATP; occurs at the inner mitochondrial membrane in eukaryotes, and the infoldings of the inner cell membrane in prokaryotes
cadena de transporte de electrones paso final de la respiración aeróbica que produce la mayoría del ATP; ocurre en la membrana mitocondrial interna en eucariotas y en los pliegues internos de la membrana celular de los procariotas

element a pure substance that consists only of atoms with the same characteristic number of protons
elemento sustancia pura que consta solo de átomos que tienen el mismo número característico de protones

embryo splitting a natural process in which the developing cells of an early embryo split spontaneously; both halves develop independently into identical twins
división de embriones proceso natural en el cual las células en desarrollo de un embrión temprano se dividen de manera espontánea; ambas partes se desarrollan de manera independiente como gemelos idénticos

endangered species a species that faces extinction in all or a part of its range
especie en peligro especie que está en riesgo de extinción en su totalidad o en parte de su zona de distribución

endemic species a species that occurs only in the area in which it evolved
especie endémica especie que ocurre solo en el área donde evolucionó

endergonic a reaction that requires a net input of free energy
reacción endergónica reacción que requiere una entrada neta de energía libre

endocytosis the process by which a cell takes in a small amount of extracellular fluid (and its contents) by the ballooning inward of the plasma membrane
endocitosis proceso por el cual una célula introduce una pequeña cantidad de fluido extracelular (y su contenido) a través del englobamiento de la membrana plasmática hacia su interior

endophyte a fungal organism that lives within a plant host; many are symbiotes
endófito organismo fúngico que vive dentro de una planta hospedadora; muchos son simbiontes

endoplasmic reticulum a membrane-enclosed organelle that is a system of sacs and tubes extending from the nuclear envelope; smooth ER makes phospholipids, stores calcium, and has additional functions in some cells; ribosomes on the surface of rough ER make proteins
retículo endoplásmico orgánulo contenido dentro de una membrana que es un sistema de sacos y tubos que salen de la envoltura nuclear; el RE liso produce fosfolípidos, almacena calcio y tiene funciones adicionales en algunas células; los ribosomas en la superficie del RE rugoso producen proteínas

endospore resistant resting stage of some soil bacteria
endospora etapa de reposo resistente en ciertas bacterias del suelo

endosymbiont an organism that lives inside another organism and does not harm it
endosimbionte organismo que vive dentro de otro organismo sin causarle daño

endosymbiont theory a theory that mitochondria and chloroplasts evolved from bacteria that entered and lived inside a host cell
teoría de la endosimbiosis teoría según la cual las mitocondrias y los cloroplastos evolucionaron de una bacteria que ingresó y vivió dentro de una célula hospedadora

energy pyramid a model that shows the energy that flows through each trophic level in a given ecosystem
pirámide de energía modelo que muestra la energía que fluye por cada nivel trófico de un ecosistema dado

engineering the application of scientific and mathematical ideas to design solutions to problems
ingeniería la aplicación de ideas científicas y matemáticas para diseñar soluciones a problemas

enzyme a biological catalyst such as a protein that speeds up a reaction without being changed by it
enzima catalizador biológico, como una proteína, que acelera una reacción sin cambiarla

enzyme-substrate complex an enzyme with substrates bound at its active sites
complejo enzima-sustrato enzima con sustratos unidos a sus sitios activos

epigenetic a heritable and stable change in gene expression that occurs through chromosomal changes rather DNA sequence changes
epigenético cambio estable y hereditario en expresión génica que ocurre a través de cambios cromosómicos y no a través de cambios en la secuencia de ADN

equilibrium a state in which forward and reverse reactions proceed at equal rates
equilibrio estado en el que las reacciones directas e inversas suceden a velocidades iguales

eukaryote any single-celled or multicellular organism whose cells contain a distinct, membrane-bound nucleus; a protist, fungus, plant, or animal
eucariota cualquier organismo unicelular o multicelular cuyas células contienen un núcleo diferenciado cubierto por una membrana; protista, hongo, planta o animal

eusocial an animal species that lives in a multigenerational group in which many sterile workers cooperate in all tasks essential to the group's welfare; only a few members of the group produce offspring
eusocial especie animal que vive en un grupo multigeneracional en el que muchos trabajadores estériles cooperan en todas las labores esenciales para el bienestar del grupo; solo unos cuantos miembros del grupo producen crías

evolution the cumulative genetic change in a population of organisms from generation to generation; leads to differences among populations and explains the origin of all of the organisms that exist today or have ever existed
evolución cambio genético acumulativo en una población de organismos de una generación a la siguiente; conduce a diferencias entre poblaciones y explica el origen de todos los organismos que existen o que han existido

exergonic a reaction that ends with a net release of free energy
reacción exergónica reacción que termina con una liberación de energía libre neta

exocytosis a process by which a cell expels a vesicle's contents to extracellular fluid
exocitosis proceso por el cual una célula expulsa los contenidos de una vesícula hacia el fluido extracelular

exon a nucleotide sequence that remains in an RNA after post-transcriptional modification
exón secuencia nucleotídica que permanece en el ARN después de una modificación postranscripcional

exponential growth a population growth pattern in which a population grows at a constant rate of increase over a period of time; exponential growth produces a characteristic J-shaped curve
crecimiento exponencial patrón de crecimiento de población en el que una población crece a un ritmo que va en aumento constante durante un período de tiempo; el crecimiento exponencial produce una curva característica en forma de J

extinction the disappearance of a species; occurs when the last individual of a species dies
extinción desaparición de una especie; ocurre cuando muere el último individuo de una especie

F

facilitated diffusion a passive transport mechanism in which a solute follows its concentration gradient across a membrane by moving through a transport protein
difusión facilitada mecanismo de transporte pasivo en el que un soluto sigue su gradiente de concentración a través de una membrana al moverse por medio de una proteína de transporte

fatty acid an organic molecule with a polar carboxyl functional group "head" and a long nonpolar hydrocarbon "tail"
ácido graso molécula orgánica con una "cabeza" carboxílica polar y una "cola" hidrocarbonada no polar

feedback the process by which the outputs of a system influence the system's behavior
retroalimentación proceso mediante el cual las salidas de un sistema influyen en el comportamiento del sistema

feedback mechanism a change within a system that intensifies the conditions that caused its occurrence (positive feedback) or causes a response that reverses the change (negative feedback)
mecanismo de retroalimentación cambio dentro de un sistema que intensifica las condiciones que causaron su ocurrencia (retroalimentación positiva) o causa una respuesta que revierte el cambio (retroalimentación negativa)

fermentation the anaerobic glucose-breakdown pathway that produces ATP without the use of electron transfer chains
fermentación trayectoria de descomposición anaeróbica de la glucosa que produce ATP sin el uso de cadenas de transferencia de electrones

fitness a measure of an individual's ability to survive and produce offspring relative to other members of its population
aptitud física medida de la capacidad de un individuo para sobrevivir y producir descendencia en comparación con el resto de los miembros de su población

flower a specialized reproductive shoot of a flowering plant (angiosperm)
flor brote reproductivo especializado de una planta con flores (angiosperma)

fluid mosaic model the model of a cell membrane as a two-dimensional fluid of mixed composition
modelo de mosaico fluido modelo de una membrana celular como un fluido bidimensional de composición mixta

food web a complex system of feeding relationships in an ecosystem
red alimentaria sistema complejo de relaciones de alimentación en un ecosistema

fossil the physical evidence of an organism that lived in the ancient past
fósil evidencia física de un organismo que vivió en el pasado antiguo

fossil fuel the combustible deposits in the Earth's crust; fossil fuels are composed of the remnants of prehistoric organisms that existed millions of years ago; examples include oil, natural gas, and coal
combustible fósil depósitos combustibles en la corteza de la Tierra; los combustibles fósiles están compuestos de los restos de organismos prehistóricos que existieron hace millones de años; entre ellos están el petróleo, el gas natural y el carbón

founder effect a reduction in genetic variation that results from a small population colonizing a new area; a type of genetic drift
efecto fundador reducción en la variación genética que resulta cuando una población pequeña coloniza un espacio nuevo; tipo de deriva genética

functional group a group of atoms that, when added to a hydrocarbon, gives the compound a particular property
grupo funcional grupo de átomos que, al ser agregados a un hidrocarburo, le dan al compuesto una propiedad particular

fungus (plural *fungi*) a single-celled or multicellular eukaryotic consumer that digests food outside the body, then absorbs the resulting breakdown products; has chitin-containing cell walls
hongo consumidor eucariota unicelular o multicelular que digiere alimentos fuera del cuerpo para luego absorber los productos descompuestos resultantes; sus paredes celulares contienen quitina

G

gamete a mature, haploid reproductive cell; e.g., a sperm or an egg
gameto célula reproductiva haploide madura; p. ej., un espermatozoide o un óvulo

gametophyte generation the stage in the life cycle of a plant during which the plant's cells undergo mitosis to produce gametes
generación de gametofitos etapa en el ciclo de vida de una planta durante la cual sus células producen gametos a través de la mitosis

gel electrophoresis a technique for separating DNA fragments according to length using an electric field generated in a semi-solid gel
electroforesis en gel técnica de separación de fragmentos de ADN de acuerdo con su longitud a través de un campo eléctrico generado en un gel semisólido

gene a chromosomal DNA sequence that encodes an RNA or protein product
gen secuencia de ADN cromosómico que codifica un producto de ARN o proteína

gene drive a gene capable of inserting itself into other chromosomes and spreading through a population
impulso genético gen capaz de insertarse en otros cromosomas y distribuirse en una población

gene expression the multistep process by which the information in a gene guides assembly of an RNA or protein product; includes transcription and translation
expresión génica proceso de varias etapas en el que la información de un gen guía el ensamblaje de un producto de ARN o proteína; incluye la transcripción y traducción

gene flow the movement of alleles between populations
flujo genético movimiento de alelos entre poblaciones

gene pool all the alleles of all the genes in a population; a pool of genetic resources
acervo génico todos los alelos de todos los genes de una población; acervo de recursos genéticos

gene therapy a treatment method for a genetic defect or disorder by transferring a normal or modified gene into the affected individual
terapia génica método de tratamiento para un defecto o trastorno genético por medio de la transferencia de un gen normal o modificado al individuo afectado

genetic drift a change in allele frequency due to chance alone
deriva genética cambio en la frecuencia alélica que se da solamente por azar

genetic engineering a process by which an individual's genome is altered deliberately, in order to change its phenotype
ingeniería genética proceso por el cual el genoma de un individuo es alterado deliberadamente con el fin de cambiar su fenotipo

genetically modified organism (GMO) an organism deliberately modified by genetic engineering
organismo modificado genéticamente (OMG) organismo deliberadamente modificado a través de la ingeniería genética

genome an organism's complete set of genetic material
genoma conjunto completo de material genético de un organismo

genomics the study of whole-genome structure and function
genómica el estudio de la estructura y función del genoma completo

genotype the particular set of alleles that is carried by an individual's chromosomes
genotipo el conjunto particular de alelos que llevan los cromosomas de un individuo

genus (plural *genera*) a group of species that share a unique set of traits; also the first part of a species name
género grupo de especies que comparten un conjunto único de rasgos; también es la primera parte del nombre de una especie

glomeromycete a fungus that partners with plant roots; its hyphae grow inside cell walls of root cells
glomeromycetes hongos que se asocian con las raíces de las plantas; sus hifas crecen dentro de las paredes celulares de las células de las raíces

glucose a simple sugar with the chemical formula $C_6H_{12}O_6$; often serves as a carbohydrate monomer with a ring structure
glucosa azúcar simple con la fórmula química $C_6H_{12}O_6$; puede servir como monómero de carbohidrato con estructura de anillo

glycogen a complex carbohydrate that consists of highly branched chains of glucose monomers
glucógeno carbohidrato complejo que consiste en cadenas de monómeros de glucosa muy ramificadas

glycolysis a set of reactions that convert glucose to two pyruvate for a net yield of two ATP and two NADH
glicólisis conjunto de reacciones que convierten la glucosa en dos piruvatos para un rendimiento neto de dos ATP y dos NADH

Golgi body a membrane-enclosed organelle that modifies proteins and lipids, then sorts the finished products into vesicles; shaped like a stack of pancakes
aparato de Golgi orgánulo contenido en una membrana, que modifica proteínas y lípidos y luego ordena los productos terminados en vesículas; tiene la forma de una pila de discos

gradualism in biology, the idea that evolutionary change of a species is because of a slow, steady accumulation of changes over time; in geology, the principle that geologic changes result from slow processes rather than sudden events
gradualismo en biología, la idea de que el cambio evolutivo en una especie se debe a la acumulación lenta y estable de cambios en el tiempo; en la geología, el principio según el cual los cambios geológicos son el resultado de procesos lentos en lugar de sucesos repentinos

gravitropism a directional response to gravity
gravitropismo respuesta direccional a la gravedad

greenhouse effect the natural warming of the lower atmosphere by greenhouse gases
efecto invernadero calentamiento natural de la atmósfera inferior por gases invernadero

greenhouse gas a gas, such as carbon dioxide, methane, or water vapor, that warms the lower atmosphere
gas invernadero gases como el dióxido de carbono, el metano o el vapor de agua, que calientan la atmósfera inferior

H

habitat the natural environment or place where an organism, population, or species lives
hábitat medio ambiente natural o lugar donde vive un organismo, población o especie

habitat fragmentation the break up of a continuous habitat into smaller pockets by the intrusion of human alterations to the land (e.g., buildings, roads)
fragmentación del hábitat rompimiento de un hábitat continuo en secciones más pequeñas como consecuencia de alteraciones humanas a la tierra (p. ej., edificios, caminos)

half-life the characteristic time it takes for half of a quantity of a radioisotope to decay
vida media tiempo típico que tarda media cantidad de radioisótopo en descomponerse

halophile an organism adapted to life in a highly salty environment
halófilo organismo adaptado a la vida en un medio ambiente muy salado

heliotropism a directional change in position in response to the changing angle of the sun throughout the day
heliotropismo cambio de dirección en respuesta al ángulo cambiante de la luz del sol durante el día

heterotroph an organism that gets its energy and nutrients from eating other organisms; also known as a consumer
heterótrofo organismo que obtiene su energía y nutrientes al comerse a otros organismos; también conocido como consumidor

heterozygous possessing two different alleles of the same gene
heterocigoto que posee dos alelos del mismo gen

histone a type of protein that associates with DNA and structurally organizes eukaryotic chromosomes
histona tipo de proteína que se asocia con el ADN y organiza estructuralmente los cromosomas eucariotas

homeostasis a process in which organisms keep their internal conditions within tolerable ranges by sensing and responding appropriately to change
homeostasis proceso en el que organismos mantienen sus condiciones internas dentro de rangos tolerables al sentir cambios y responder adecuadamente a ellos

homologous chromosome in a nucleus, one of a pair of chromosomes that have the same length, shape, and genes
cromosoma homólogo en un núcleo, cada miembro de un par de cromosomas que tienen la misma longitud, forma y genes

homologous structure a body structure similar to another in a different lineage because the structure evolved in a common ancestor
estructura homóloga estructura corporal que es similar a una de otro linaje debido a que la estructura evolucionó en un ancestro común

homozygous possessing a pair of identical alleles
homocigoto que posee un par de alelos idénticos

hormone one of many chemical messengers in multicellular organisms that usually travels in body fluids to target cells, where it combines with receptors and affects some aspect of metabolism, growth, or reproduction
hormona uno de muchos mensajeros químicos en organismos multicelulares que normalmente viajan en fluidos corporales hacia células objetivo, donde se combinan con receptores y afectan algún aspecto del metabolismo, el crecimiento o la reproducción

hydrocarbon a substance made entirely of carbon and hydrogen
hidrocarburo sustancia compuesta completamente de carbono e hidrógeno

hydrogen bond a force of attraction between a hydrogen atom in one molecule and a highly attractive atom in another molecule
enlace de hidrógeno fuerza de atracción entre un átomo de hidrógeno y un átomo altamente atractivo de otra molécula

hydrotropism a directional growth of plant roots in response to water
hidrotropismo crecimiento direccional de las raíces de una planta en respuesta al agua

hyperaccumulator a plant that can take up and store high concentrations of metals from soil or water
planta hiperacumuladora planta que puede absorber y almacenar altas concentraciones de metales del suelo o agua

hypertonic a solution with a solute concentration greater than that of the solution with which it is compared
hipertónica solución con una concentración de soluto mayor que la solución con la cual se le compara

hypha (plural *hyphae*) one filament in a fungal mycelium
hifa filamento en un micelio fúngico

hypotonic a solution with a solute concentration less than that of the solution with which it is compared
hipotónica solución con una concentración de soluto menor que la solución con la cual se le compara

I

immunity the body's ability to resist and fight infections
inmunidad la capacidad del cuerpo para resistir y combatir infecciones

incomplete dominance an inheritance pattern in which one allele is not fully dominant over another, so the heterozygous phenotype is an intermediate blend between the two homozygous phenotypes
dominancia incompleta patrón de herencia en el que un alelo no es completamente dominante sobre otro, por tanto, el fenotipo heterocigoto es una combinación intermedia entre los dos fenotipos homocigotos

index fossil a fossil that dates the layers where it is found because it came from an organism that is abundantly preserved in rocks, was widespread geographically, and existed as a species or genus for only a relatively short time
fósil índice fósil que data las capas del sitio donde fue encontrado porque viene de un organismo preservado en abundancia en las rocas, que estuvo geográficamente esparcido y existió como especie o género por un tiempo relativamente corto

inflammation a part of innate immunity: a local response to tissue damage or infection; characterized by redness, warmth, swelling, and pain
inflamación parte de la inmunidad innata: respuesta local a daños a tejidos o infección; se caracteriza por enrojecimiento, calor, hinchazón y dolor

inheritance the transmission of DNA to offspring
herencia transmisión de ADN a la descendencia

initiator tRNA a type of tRNA that delivers amino acids to a ribosome during translation
ARNt iniciador tipo de ARNt que distribuye aminoácidos a un ribosoma durante la traducción

innate immunity a set of immediate, general responses (complement activation, phagocytosis, inflammation, fever) that defend the body from infection in multicellular organisms
inmunidad innata conjunto de respuestas inmediatas generales (activación de complemento, fagocitosis, inflamación, fiebre) que defiende al cuerpo de la infección en organismos multicelulares

instinctive behavior an innate response to a simple stimulus
comportamiento instintivo respuesta innata ante un estímulo simple

insulator a region of DNA that, upon binding a transcription factor, can block the effect of a silencer or enhancer on a neighboring gene
aislador región del ADN que, al unirse a un factor de transcripción, puede bloquear el efecto de un silenciador o potenciador en un gen vecino

intermediate disturbance hypothesis a hypothesis that states that species richness is greatest when physical and biological disturbances are moderate in their intensity or frequency
hipótesis de la perturbación intermedia hipótesis según la cual la riqueza de especies es mayor cuando las perturbaciones físicas y biológicas son de intensidad y frecuencia moderada

interphase the period in the life cycle of a cell in which there is no visible mitotic division; period between mitotic divisions in which cell performs normal growth and metabolic functions
interfase período en el ciclo de vida de una célula en el que no hay división mitótica visible; período entre divisiones mitóticas durante el cual la célula desempeña funciones normales de metabolismo y crecimiento

interphase in a eukaryotic cell cycle, the interval during which a cell grows, roughly doubles the number of its cytoplasmic components, and replicates its DNA in preparation for division
interfase en una célula eucariota, el intervalo durante el cual una célula crece, duplica aproximadamente el número de sus componentes citoplásmicos y replica su ADN para prepararse para la división

intron a nucleotide sequence that intervenes between exons and is removed during post transcriptional modification
intrón secuencia de nucleótidos que interviene entre exones y es retirada durante la modificación postranscripcional

invasive species a non-native species that outcompetes and threatens the existence of native organisms in an ecosystem
especie invasiva especie no nativa, más competitiva y que amenaza la existencia de organismos nativos en un ecosistema

invertebrate an animal that does not have a backbone
invertebrado animal que no tiene columna vertebral

ion an atom that has a different number of electrons than protons
ion átomo que tiene un número de electrones diferente del de protones

ionic bond a type of chemical bond formed by the force of attraction between oppositely charged ions
enlace iónico tipo de enlace químico formado por la fuerza de atracción entre iones de carga opuesta

isotonic a solution with identical concentrations of solute and solvent molecules
isotónico solución con concentraciones idénticas de moléculas de soluto y solvente

isotope a variation of an element with the same number of protons (or the same atomic number) but a different number of neutrons (and thus a different atomic mass) than another variation of the same element
isótopo variación de un elemento que tiene el mismo número de protones (o el mismo número atómico) pero un número diferente de neutrones (y, por tanto, una masa atómica diferente) que otra variación del mismo elemento

K

keystone species a species that has a disproportionately large effect on community structure relative to its abundance
especie clave especie que, en comparación con su abundancia, tiene un efecto desproporcionadamente grande sobre la estructura de una comunidad

Krebs cycle a cyclic pathway that harvests energy from acetyl–CoA; part of aerobic respiration; also called citric acid cycle
ciclo de Krebs trayectoria cíclica que cosecha energía del acetil-CoA; parte de la respiración aeróbica; también conocido como ciclo del ácido cítrico

L

learned behavior a behavior that is modified by experience
comportamiento aprendido comportamiento modificado por la experiencia

lichen a composite organism consisting of a fungus and green algae or cyanobacteria
liquen organismo compuesto que consta de un hongo y un alga o cianobacteria

life history a pattern that describes how an individual organism uses resources and energy for growth, survival, and reproduction over the course of its lifetime
historia de vida patrón que describe la manera en que un organismo individual usa recursos y energía para crecer, sobrevivir y reproducirse a lo largo de su vida

light-dependent reactions the first stage of photosynthesis that requires the presence of light; light energy is converted to chemical energy in the form of ATP and NADPH
reacciones dependientes de la luz primera etapa de la fotosíntesis que requiere la presencia de luz; la energía luminosa se convierte en energía química en forma de ATP y NADPH

light-independent reactions the second stage of photosynthesis that does not directly require light; use ATP and NADPH to build sugar molecules such as glucose from carbon dioxide and water.
reacciones no dependientes de la luz segunda etapa de la fotosíntesis que no requiere de luz directa; usa ATP y NADPH para construir moléculas de azúcar; p. ej., glucosa a partir de dióxido de carbono y agua

lignin a material that strengthens cell walls of vascular plants
lignina material que fortalece las paredes celulares de las plantas vasculares

lipid a fatty, oily, or waxy organic compound that is partly or completely nonpolar
lípido compuesto orgánico graso, aceitoso o ceroso que es parcial o completamente no polar

logistic growth a population growth pattern in which a population grows slowly at first, followed by a period of exponential growth before leveling off at a stable size; logistic growth produces a characteristic S-shaped curve
crecimiento logístico patrón de crecimiento de la población en el que la población crece lentamente al principio, seguido por un período de crecimiento exponencial antes de nivelarse al alcanzar un tamaño estable; el crecimiento logístico produce una curva típica en forma de S

lysogenic pathway the replication pathway of a bacteriophage in which viral DNA becomes integrated into the host's chromosome and is passed to the host's descendants
ciclo lisogénico trayectoria de replicación de un bacteriófago en la cual el ADN se integra al cromosoma del anfitrión y es transmitido a los descendientes del anfitrión

lysosome an enzyme-filled vesicle that breaks down cellular wastes and debris
lisosoma vesícula llena de enzima que descompone desperdicios celulares y desechos

lytic pathway the replication pathway of a bacteriophage in which a virus immediately replicates in its host and kills it
ciclo lítico trayectoria de replicación de un bacteriófago en la cual un virus se replica dentro de su anfitrión y lo mata

M

macroevolution evolutionary patterns and trends on a larger scale than microevolution; e.g., adaptive radiation, coevolution
macroevolución patrones evolutivos y tendencias a una escala mayor que la de la microevolución; por ejemplo, la radiación adaptativa, la coevolución

mark-recapture sampling a method in which a sample group of members of a population are captured and marked; the proportion of marked organisms in a second group, captured at a later time, is used to estimate the size of the population
muestreo de marca y recaptura método en el que un grupo muestra de los miembros de una población es capturado y marcado; la proporción de organismos marcados en un segundo grupo, capturado en un momento posterior, se usa para estimar el tamaño de la población

mass extinction a sudden, catastrophic event during which a significant percentage of all life-forms on Earth become extinct
extinción masiva suceso repentino y catastrófico en el que se extingue un porcentaje significativo de todas las formas de vida en la Tierra

meiosis division of the cell nucleus that produces haploid cells; produces gametes in animals and spores in plants
meiosis división del núcleo celular que produce células haploides; produce gametos en animales y esporas en plantas

meristem a zone of undifferentiated parenchyma cells, the divisions of which give rise to new tissues during plant growth
meristemo zona de células de parénquima indiferenciadas, cuyas divisiones dan lugar a nuevos tejidos durante el crecimiento de la planta

metabolism all of the enzyme-mediated reactions in a cell
metabolismo todas las reacciones mediadas por enzimas de una célula

metaphase a stage of mitosis at which all chromosomes are aligned midway between spindle poles
metafase etapa de la mitosis en la cual todos los cromosomas se alinean en la línea media del huso

methanogen a prokaryotic organism that produces methane gas (CH_4) as a metabolic by-product
metanógeno organismo procariota que produce gas metano (CH_4) como producto secundario de su metabolismo

microarray a laboratory tool used to detect the expression of thousands of DNA sequences at the same time
microarreglo herramienta de laboratorio que detecta la expresión de miles de secuencias de ADN al mismo tiempo

microevolution changes in allele frequencies that occur within a population over successive generations
microevolución cambios en la frecuencia alélica que ocurren dentro de una población a lo largo de generaciones sucesivas

mitochondrion (plural *mitochondria*) a double-membraned organelle that produces ATP by aerobic respiration in eukaryotes
mitocondria orgánulo de doble membrana que produce ATP a través de la respiración celular en los eucariotas

mitosis a division of the cell nucleus resulting in the distribution of a complete set of chromosomes to each daughter cell; occurs in stages (prophase, metaphase, anaphase, and telophase); basis of body growth, tissue repair, and (in some organisms) asexual reproduction
mitosis división del núcleo celular que resulta en la distribución de un juego completo de cromosomas a cada célula hija; ocurre en etapas (profase, metafase, anafase y telofase); es la base del crecimiento del cuerpo, reparación de tejidos y (en algunos organismos) de la reproducción asexual

mitotic spindle a temporary structure consisting of microtubules that moves chromosomes during mitosis
huso mitótico estructura temporal que consiste en microtúbulos que mueve los cromosomas durante la mitosis

model a tool for representing an idea or an explanation, or an analagous system that can be used to test a hypothesis
modelo herramienta para representar una idea o explicación, o un sistema análogo que sirve para probar una hipótesis

molecule a collection of atoms bonded together that behave as a unit
molécula colección de átomos enlazados que se comportan como unidad

monomer a molecule that is used as a subunit of a larger molecule
monómero molécula que se usa como subunidad de una molécula más grande

monosaccharide a simple carbohydrate molecule that is often a monomer for a complex carbohydrate; also referred to as a simple sugar
monosacárido molécula de carbohidrato simple que a menudo es un monómero para un carbohidrato complejo; también se conoce como azúcar simple

mutagen any physical or chemical agent that is capable of producing mutations
mutágeno cualquier agente físico o químico que es capaz de producir mutaciones

mutation a permanent change in the DNA sequence of a chromosome
mutación cambio permanente en la secuencia de ADN de un cromosoma

mycelium mass of threadlike filaments (hyphae) that make up the body of a multicellular fungus
micelio masa de filamentos filiformes (hifas) que forman el cuerpo de un hongo multicelular

mycorrhiza (plural *mycorrhizae*) a mutually beneficial partnership between a fungus and a plant root
micorriza asociación mutuamente benéfica entre un hongo y la raíz de una planta

N

natural resource a natural material used by humans, such as water, forests, animals, and minerals
recurso natural material natural usado por humanos, como el agua, los bosques, los animales y los minerales

natural selection a major mechanism of evolution: the differential survival and reproduction of individuals of a population based on differences in shared, heritable traits; outcome of environmental pressures
selección natural uno de los mecanismos principales de la evolución: la supervivencia y reproducción diferenciales entre individuos de una población según las distinciones de los rasgos hereditarios que comparten; es el resultado de presiones ambientales

near-uniform distribution a population distribution pattern in which individuals are more evenly spaced than would be expected by chance
distribución casi uniforme patrón de distribución de una población en el cual los individuos están espaciados de manera más pareja de lo que se esperaría por azar

negative feedback mechanism a change that causes a response that reverses the change; important mechanism of homeostasis
mecanismo de retroalimentación negativa cambio que genera una respuesta que revierte el cambio; mecanismo importante de la homeostasis

neutral an atom with exactly the same number of electrons as protons; it carries no charge
neutral átomo que tiene exactamente el mismo número de electrones que de protones; no lleva carga

neutron a neutral subatomic particle that occurs in the nucleus of an atom
neutrón partícula subatómica que ocurre en el núcleo de un átomo

node in a phylogenetic tree, represents a common ancestor
nodo en un árbol filogenético, representa un ancestro común

non-native species a species that evolved in one community and later became established in a different one
especie no nativa especie que evolucionó en una comunidad y más adelante se estableció en otra distinta

nonenveloped virus a virus particle in which its outermost layer is the protein capsid
virus sin envoltura partícula de virus en la cual la capa exterior es la proteína cápside

nonpoint source a source of water pollution that is widespread in the environment
contaminación difusa fuente de contaminación de agua que está extendida en el medio ambiente

nonpolar a chemical bond or molecule in which charge is distributed evenly
no polar enlace químico o molécula cuya carga está distribuida de manera pareja

nonrenewable resource a natural resource that cannot be replenished at the same rate at which it is used
recurso no renovable recurso natural que no puede ser reabastecido a la misma tasa que se usa

nonvascular plant a plant that does not have xylem and phloem; a bryophyte such as a moss
planta no vascular planta que no tiene xilema ni floema; una briófita, como un musgo

nucleic acid polymer of nucleotides; DNA or RNA
ácido nucleico polímero de nucleótidos; ADN o ARN

nucleotide a molecule that consists of a monosaccharide ring bonded to a nitrogen-containing base and one, two, or three phosphate groups
nucleótido molécula que consta de un anillo monosacárido unido a una base que contiene nitrógeno y uno, dos o tres grupos fosfato

nucleus of an atom, central core area occupied by protons and neutrons; of a eukaryotic cell, organelle with a double membrane that holds the cell's DNA
núcleo de un átomo, el área central ocupada por protones y neutrones; de una célula eucariota, un orgánulo de doble membrana que contiene el ADN de la célula

O

oncogene a gene that helps transform a normal cell into a tumor cell
oncogén gen que ayuda a transformar una célula normal en una célula tumoral

operator a DNA binding site for a repressor; part of an operon
operador el sitio donde un represor se une al ADN; parte de un operón

operon a group of genes together with a promoter–operator DNA sequence that controls their transcription
operón grupo de genes unidos a una secuencia promotora-operadora que controla su transcripción

organ a structure that consists of tissues engaged in a collective task in multicellular organisms
órgano estructura compuesta de tejidos que cumplen una función colectiva en los organismos multicelulares

organ system a set of interacting organs that carry out a particular body function in multicellular organisms
sistema de órganos conjunto de órganos que interactúan para llevar a cabo una función corporal específica en los organismos multicelulares

organelle a structure that carries out a specialized metabolic function inside a cell
orgánulo estructura que desempeña una función metabólica especializada dentro de una célula

organic compound a chemical compound that consists mainly of carbon and hydrogen
compuesto orgánico compuesto químico que está formado principalmente por carbono e hidrógeno

osmosis the diffusion of water across a selectively permeable membrane; occurs in response to a difference in solute concentration between the fluids on either side of the membrane
ósmosis difusión del agua a través de una membrana selectivamente permeable; ocurre en respuesta a una diferencia en concentración del soluto los entre fluidos que hay a cada lado de una membrana

ovary in flowering plants, the enlarged base of a carpel, inside which one or more ovules form and eggs are fertilized; in animals, the egg-producing gonad
ovario en las plantas con flores, la base agrandada del carpelo, dentro del cual se forman uno o más óvulos y son fertilizados; en los animales, la gónada que produce óvulos

ovule a seed plant reproductive structure in which the female gametophyte forms and develops; matures into a seed after fertilization
óvulo estructura reproductiva de las plantas espermatofitas donde se desarrolla y forma el gametofito femenino; madura en forma de semilla después de la fertilización

P

parenchyma a simple plant tissue composed of living, thin-walled cells; has various specialized functions such as photosynthesis
parénquima tejido simple en las plantas; está compuesto de células vivas con paredes delgadas; tiene varias funciones especializadas, como la fotosíntesis

passive transport a membrane-crossing mechanism that requires no energy input
transporte pasivo mecanismo de movimiento a través de membranas que no requiere energía

pathogen disease-causing agent (virus or cellular organism
patógeno agente que causa enfermedad (virus u organismo celular)

pellicle an outer layer of plasma membrane and elastic proteins that protects and gives shape to many unwalled, single-celled protists
película capa exterior formada por membrana plasmática y proteínas elásticas que protege y da forma a muchos protistas unicelulares que no cuentan con paredes

per capita growth rate the rate of change in population size, expressed in terms of the increase in population per existing individual; may be calculated by subtracting the population's per capita death rate from its per capita birth rate
tasa de crecimiento per cápita tasa de cambio en el tamaño de la población, expresada en términos del aumento en la población en relación con los individuos existentes; se calcula restando la tasa de mortalidad per cápita de la población de su tasa de natalidad per cápita

peristalsis wavelike smooth muscle contractions that propel food through the digestive tract
peristalsis contracciones musculares como ondas que impulsan el alimento por el tracto digestivo

peroxisome an enzyme-filled vesicle that breaks down fatty acids, amino acids, and toxic substances
peroxisoma vesícula llena de enzimas que descompone ácidos grasos, aminoácidos y sustancias tóxicas

petal a modified, often brightly colored, leaf of a flower
pétalo hoja modificada y a veces colorida de una flor

pH a measure of the number of hydrogen ions in a water-based fluid
pH medida del número de iones de hidrógeno que tiene un fluido basado en agua

phenotype an individual's observable traits
fenotipo rasgos observables de un individuo

pheromone a chemical that serves as a communication signal between members of an animal species
feromona químico que sirve como señal de comunicación entre miembros de una especie animal

phloem a complex vascular tissue of plants; its living sieve elements compose sieve tubes that distribute sugars through the plant body; each sieve element has an associated companion cell that provides it with metabolic support
floema tejido vascular complejo que tienen las plantas; sus elementos cribosos vivientes componen tubos cribosos que distribuyen azúcares a lo largo del cuerpo de la planta; cada elemento criboso tiene una célula acompañante que le da apoyo metabólico

phosphate group a functional group that consists of a phosphorus atom bonded to four oxygen atoms
grupo fosfato grupo funcional que consta de un átomo de fósforo unido a cuatro átomos de oxígeno

photoautotroph an organism that obtains carbon from carbon dioxide and energy from light
fotoautótrofo organismo que obtiene carbono del dióxido de carbono y energía de la luz

photoheterotroph an organism that obtains carbon from organic compounds and energy from light
fotoheterótrofo organismo que obtiene carbono de compuestos orgánicos y energía de la luz

photoperiodism a biological response to seasonal changes in the relative lengths of day and night
fotoperiodismo respuesta biológica a cambios estacionales en la duración relativa del día y la noche

photosynthesis the biological process that captures light energy and transforms it into the chemical energy of organic molecules (such as glucose), which are made from carbon dioxide and water; used by plants, algae, and several kinds of bacteria
fotosíntesis proceso biológico que captura energía luminosa y la transforma en la energía química de moléculas orgánicas (como la glucosa), que se componen de dióxido de carbono y agua; este proceso lo siguen las plantas, las algas y varios tipos de bacterias

phototropism a directional response to light
fototropismo respuesta direccional a la luz

phylogenetic analysis a group of methods to estimate relatedness between organisms by comparing genetic sequences
análisis filogenético grupo de métodos para estimar el nivel de relación entre organismos mediante la comparación de sus secuencias genéticas

phylogeny an evolutionary history of a species or group of species
filogenia historia evolutiva de una especie o grupo de especies

phytoplankton a community of tiny drifting or swimming photosynthetic organisms
fitoplancton comunidad de diminutos organismos fotosintéticos que flotan o nadan

phytoremediation the use of plants to extract chemical contaminants from soil or water
fitorremediación uso de plantas para extraer contaminantes químicos del suelo o el agua

pioneer species a species whose traits allow it to colonize new or newly vacated habitats
especie pionera especie cuyos rasgos le permiten colonizar hábitats nuevos o recientemente desocupados

plot sampling a sampling method in which the number of members of a population occupying a sample area or volume of their range is counted directly in order to estimate the size of the population
muestreo de parcelas método de muestreo en el cual el número de integrantes de una población que ocupa un área de muestra o volumen de muestra de su rango se cuenta directamente para estimar el tamaño de la población

pluripotent the capability of stem cells to develop into any of the cell types present in an organism
pluripotente capacidad de las células madre de desarrollarse como cualquier tipo de célula presente en un organismo

point source a single, easily identifiable source of water pollution
fuente puntual fuente única de contaminación del agua que es fácil de identificar

polar a chemical bond or molecule in which charge is distributed unevenly
polar enlace químico o molécula cuya carga está distribuida de manera dispareja

pollen grain a walled, immature male gametophyte of a seed plant
grano de polen en las plantas que producen semillas, un gametofito masculino que tiene pared externa

pollination the arrival of pollen on a stigma or other pollen-receiving reproductive part of a seed plant
polinización en las plantas que producen semillas, la llegada del polen a un estigma u otra parte reproductiva receptora de polen

pollutant a harmful substance released into the environment by human activities
contaminante sustancia dañina liberada al ambiente por actividades humanas

polymer a molecule that consists of repeated monomers
polímero molécula que está compuesta de monómeros repetidos

polymerase chain reaction (PCR) laboratory method that rapidly generates many copies of a specific section of DNA
reacción en cadena de la polimerasa (PCR, en inglés) método de laboratorio que genera muchas copias de un segmento específico de ADN de manera rápida

polysaccharide a molecule formed by three or more simple sugar molecules (monosaccharides)
polisacárido molécula formada por tres o más moléculas de azúcar simple (monosacáridos)

population a group of individual organisms of the same species that interbreed with one another
población grupo de organismos individuales de la misma especie que se reproducen entre sí

population density the average number of individuals in a population per unit area or volume
densidad de población número promedio de individuos que tiene una población por unidad de área o volumen

population distribution the location of individuals in a population relative to one another
distribución de la población la ubicación de individuos en una población en relación con los otros individuos

population size the total number of individuals in a population
tamaño de población número total de individuos que tiene una población

positive feedback mechanism a response that intensifies the conditions that caused its occurrence
mecanismo de retroalimentación positiva respuesta que intensifica las condiciones que causaron que sucediera

predation an ecological relationship in which one species captures, kills, and eats another
depredación relación ecológica en la cual una especie captura, mata y se come a otra

pressure flow theory an explanation of how a difference in turgor between sieve elements in source and sink regions pushes sugar-rich fluid through a sieve tube
hipótesis de flujo de presión explicación de cómo una diferencia de turgencia entre elementos cribosos en regiones de fuente y sumidero impulsa fluidos ricos en azúcar a través de un tubo criboso

primer a short, single strand of DNA or RNA that base-pairs with a specific DNA sequence and serves as an attachment point for DNA polymerase
partidor hebra de ADN o ARN simple y corta que se aparea con una secuencia específica de ADN y sirve como punto de unión para la ADN polimerasa

prion an infectious protein
prion proteína infecciosa

producer an organism that makes its own food from abiotic sources, such as sunlight or inorganic materials
productor organismo que produce su propio alimento a partir de fuentes abióticas como luz solar o materiales inorgánicos

product a molecule that is produced by a reaction
producto molécula que se produce por una reacción

prokaryote an informal term for a single-celled organism without a nucleus; a bacterium or archaeon
procariota término informal para referirse a un organismo unicelular que no tiene núcleo; bacteria o arquea

promoter a sequence that is a site where RNA polymerase binds and begins transcription in DNA
promotor secuencia que es un sitio donde el ARN polimerasa se une y comienza la transcripción en el ADN

prophase stage of mitosis during which chromosomes condense and become attached to a newly forming spindle
profase etapa de la mitosis en la que los cromosomas se condensan y se unen a un huso recién formado

protein a large, complex organic compound composed of amino acid subunits; proteins are the principal structural components of cells
proteína compuesto orgánico complejo grande que se compone de subunidades de aminoácidos; las proteínas son el principal componente estructural de las células

proteome the study of all the proteins encoded by a genome within a specific context
proteoma el estudio de todas las proteínas codificadas por un genoma dentro de un contexto específico

protist a common term for a eukaryote that is not a plant, animal, or fungus
protista término común para un eucariota que no es una planta, animal u hongo

proto-oncogene a gene that, by mutation, can become an oncogene
protooncogén gen que, por mutación, se convierte en oncogén

proton a positively charged subatomic particle that occurs in the nucleus of an atom
protón partícula subatómica con carga positiva que se encuentra en el núcleo de un átomo

pseudopod a temporary protrusion that helps some eukaryotic cells move and engulf prey
pseudópodo protuberancia temporal con la que ciertas células eucariotas se mueven y envuelven a sus presas

punctuated equilibrium the concept that evolution has periods of inactivity (i.e., periods of little or no change within a species) followed by very active phases, so that major adaptations or clusters of adaptations appear suddenly in the fossil record
equilibrio puntuado concepto que establece que la evolución tiene períodos de inactividad (es decir, períodos de poco o ningún cambio dentro de una especie) seguidos por fases muy activas, de tal manera que las grandes adaptaciones o cúmulos de adaptaciones aparecen repentinamente en el registro fósil

pyramid of numbers a model that compares the relative number of individual organisms at each trophic level
pirámide de números modelo en que se compara el número relativo de organismos individuales a cada nivel trófico

pyruvate a three-carbon product of glycolysis
piruvato producto de la glicólisis que contiene tres átomos de carbono

R

radioisotope an isotope with an unstable nucleus
radioisótopo isótopo con un núcleo inestable

radiometric dating a method of estimating the age of a rock or fossil by measuring the content and proportions of a radioisotope and its daughter elements
datación radiométrica método para estimar la edad de un fósil o roca a través de la medición del contenido y proporciones de sus elementos hijos

random distribution a population distribution pattern that occurs when occurs when resources are distributed uniformly through the environment, and proximity to others neither benefits nor harms individuals
distribución aleatoria patrón de distribución que ocurre cuando los recursos están uniformemente distribuidos en el medio ambiente y su proximidad con otros ni beneficia ni daña a los individuos

range the geographical area in which a population of organisms lives
rango área geográfica en la cual vive una población de organismos

reactant a molecule that enters a reaction and is changed by participating in it
reactante molécula que entra en una reacción y cambia al participar en ella

recessive an allele that is not expressed in the heterozygous state
recesivo alelo que no se expresa en el estado heterocigoto

recombinant DNA a DNA molecule that consists of genetic material from two or more organisms
ADN recombinante molécula de ADN que consiste en material genético de dos o más organismos

renewable resource a natural resource that can be replenished at the same rate at which it is used
recurso renovable recurso natural que puede ser reabastecido al mismo ritmo en que se usa

replication the process by which a cell duplicates all of its DNA before it divides
replicación proceso a través del cual una célula duplica todo su ADN antes de dividirse

repressor a transcription factor that slows or stops transcription
represor factor de transcripción que detiene o reduce la velocidad de la transcripción

reproductive cloning a laboratory procedure such as SCNT that produces animal clones from a single somatic cell
clonación reproductiva procedimiento de laboratorio, como SCNT, que produce clones animales a partir de una sola célula somática

reproductive isolation the end of gene flow between populations; part of speciation
aislamiento reproductivo el fin del flujo genético entre poblaciones; parte de la especiación

reservoir an organism that can be infected with pathogens transmissible to humans
reservorio organismo que puede ser infectado con patógenos transmisibles a los humanos

resilience in ecology, the ability of an ecological community to regain its original structure and function following a disturbance
resiliencia en la ecología, la capacidad de una comunidad ecológica para recuperar su estructura y función originales después de una perturbación

resistance in ecology, the ability of an ecological community to remain unchanged when affected by a disturbance
resistencia en la ecología, la capacidad de una comunidad ecológica de permanecer inalterada al ser afectada por una perturbación

restriction enzyme an enzyme that cuts DNA wherever a specific nucleotide sequence occurs
enzima de restricción enzima que corta el ADN en cualquier lugar donde ocurre una secuencia nucleótida específica

retrovirus an RNA virus that uses the enzyme reverse transcriptase to produce viral DNA in a host cell
retrovirus virus ARN que usa la enzima transcriptasa inversa para producir ADN viral en una célula anfitriona

rhizome a fleshy plant stem that grows horizontally along or under the ground
rizoma la parte carnosa del tallo de una planta que crece horizontalmente a lo largo o debajo del suelo

ribonucleic acid (RNA) a single-stranded nucleic acid molecule that functions mainly in protein synthesis from DNA templates
ácido ribonucleico (ARN) molécula de ácido nucleico de cadena sencilla que funciona principalmente en la síntesis de proteínas a partir de moldes de ADN

RNA see ribonucleic acid
ARN ver ácido ribonucleico

RNA interference a mechanism in which RNA molecules interfere with gene expression by preventing translation of specific messenger RNAs
ARN interferente mecanismo donde las moléculas de ARN interfieren con la expresión génica al prevenir la traducción de ARNs mensajeros específicos

RNA polymerase an enzyme that carries out transcription
ARN polimerasa enzima que lleva a cabo la transcripción

root system the organ system of a plant that absorbs water and dissolved minerals; roots generally grow below ground
sistema de raíces sistema de órganos de una planta que absorbe agua y minerales disueltos; las raíces generalmente crecen bajo tierra

S

sac fungus a fungus that forms spores in a sac-shaped structure during sexual reproduction; sac fungi are the most diverse fungal group
ascomicetos hongos que durante la reproducción sexual forman esporas en una estructura en forma de bolsa; los ascomicetos son los más diversos del grupo fúngico

seed an embryo sporophyte of a seed plant packaged with nutritive tissue inside a protective coat
semilla embrión esporófito de una semilla con tejido nutritivo dentro de una capa protectora

selective gene expression the expression of differing subsets of an organism's genes by different cells
expresión génica selectiva expresión de diferentes subconjuntos de genes de un organismo por células diferentes

selective permeability a characteristic of a cell membrane in which some substances are able to cross it more easily than others
permeabilidad selectiva característica de una membrana celular en la cual unas sustancias son más capaces de atravesarla que otras

sensory receptor a cell or cell component that responds to a specific stimulus, such as temperature or light
receptor sensorial célula o componente celular que responde a un estímulo específico, como la temperatura o luz

sepal one of a set of green, leaflike portions of a plant that serve as the outer covering of unopened flowers
sépalo parte de un conjunto de porciones verdes que parecen hojas de una planta; funcionan como cobertura de flores que no se han abierto

sexual reproduction reproductive mode by which offspring arise from two parents and inherit genes from both
reproducción sexual modo reproductivo en el que la descendencia procede de dos progenitores y hereda genes de ambos

sexual selection a mode of natural selection in which some individuals out reproduce others of a population because they are better at securing mates
selección sexual modo de selección natural donde unos individuos se reproducen más que otros en la población porque son mejores para conseguir parejas

shoot system the organ system of a plant consisting of stems, leaves, and reproductive organs; generally the aboveground parts of a plant
sistema de brotes sistema de órganos de una planta que está compuesto de tallos, hojas y órganos reproductivos; generalmente, son las partes de la planta que están arriba de la tierra

SI system the internationally standardized system of measurements
sistema internacional sistema de medidas estandarizado internacionalmente

sieve elements the living cells that make up sugar conducting sieve tubes of phloem; each sieve tube consists of a stack of sieve elements that meet end-to-end at sieve plates
elementos cribosos células vivas que componen los tubos cribosos de floema que conducen el azúcar; cada tubo criboso consta de una pila de elementos cribosos que se unen de extremo a extremo en las placas cribosas

sieve tube the sugar-conducting tube of phloem; consists of stacked sieve elements
tubo criboso tubo del floema que conduce el azúcar; está compuesto de elementos cribosos amontonados

silencer a DNA sequence where a repressor may bind to prevent RNA polymerase from transcribing a gene
silenciador secuencia de ADN a la que un represor se puede unir para evitar que el ARN polimerasa transcriba un gen

single nucleotide polymorphism (SNP) a one-nucleotide DNA sequence variation carried by a measurable percentage of a population
polimorfismo de nucleótido único (SNP, en inglés) variación de secuencia de ADN de un nucleótido que lleva un porcentaje medible de una población

sister chromatid one of two identical DNA molecules of a duplicated eukaryotic chromosome, attached at the centromere
cromátida hermana cada una de las dos moléculas idénticas de ADN de un cromosoma eucariota duplicado, unida por el centrómero

sister group the two lineages that emerge from a node on a cladogram
grupo hermano cada uno de los dos linajes que surgen de un nodo en un cladograma

solute a dissolved substance
soluto sustancia diluida

solution a uniform mixture of solute completely dissolved in solvent
solución mezcla uniforme de un soluto completamente disuelto en un solvente

solvent a liquid in which other substances dissolve
solvente líquido en el que otras sustancias se disuelven

somatic cell nuclear transfer (SCNT) a reproductive cloning method in which the DNA of a donor's body cell is transferred into an enucleated egg; produces clones of the donor
transferencia nuclear de células somáticas (SCNT, en inglés) método de clonación reproductiva en el cual el ADN de la célula del cuerpo donador es transferido a un óvulo desnucleado; produce clones del donador

sori (singular *sorus*) clusters of spore-producing capsules on a fern leaf
soros cúmulos de cápsulas productoras de esporas en la hoja de un helecho

speciation the emergence of a new species
especiación surgimiento de nuevas especies

species a group of organisms with similar structural and functional characteristics that in nature breed only with one another and have a close common ancestry; a group of organisms with a common gene pool
especie grupo de organismos con características estructurales y funcionales similares que en la naturaleza se aparean únicamente entre sí y tienen ascendencia común cercana; grupo de organismos con un acervo genético común

species evenness the relative abundance of each species in a community
igualdad de especies abundancia relativa de cada especie en una comunidad

species richness the number of different species in a community
riqueza de especies número de especies diferentes en una comunidad

spore a reproductive cell that gives rise to individual offspring in plants, algae, fungi, and certain protozoa
espora célula reproductiva que produce descendientes individuales en plantas, algas, hongos y ciertos protozoarios

sporophyte generation the stage in the life cycle of a plant during which the plant's cells undergo meiosis to produce spores
generación esporófita etapa en el ciclo de vida de una planta durante la cual las células producen esporas a través de meiosis

stabilizing selection a mode of natural selection in which an intermediate form of a trait is adaptive, and extreme forms are selected against
selección estabilizadora modo de selección natural donde una forma intermedia de un gen es adaptativa y se selecciona en contra de formas extremas

stamen the reproductive organ of a flower that consists of a pollen-producing anther and, in most species, a filament
estambre órgano reproductivo de una flor, que consta de una antera productora de polen y, en la mayoría de las especies, un filamento

starch a complex carbohydrate that does not dissolve readily in water; main form of stored sugars in plants
almidón carbohidrato complejo que no se disuelve fácilmente en agua; forma principal de azúcares almacenados en plantas

stem cell a cell that can divide to produce more stem cells or differentiate into a specialized cell type
célula madre célula que puede dividirse para producir más células madre o diferenciarse en un tipo de célula especializada

stigma the upper part of the carpel of a flower; adapted to receive pollen
estigma parte superior del carpelo de una flor; adaptado para recibir polen

stolon a stem that branches from a plant's main stem and grows horizontally on, or just under, the ground
estolón tallo que brota del tallo principal de la planta y crece horizontalmente sobre la tierra o justo debajo de esta

stomata (singular *stoma*) tiny gaps between pairs of guard cells in a plant's epidermis; allow internal tissues to exchange gases with the air; can be closed to prevent water loss
estomas pequeños huecos entre pares de células de guarda en la epidermis de una planta; permiten que los tejidos internos intercambien gases con el aire; se pueden cerrar para evitar la pérdida de agua

style the tube that connects the ovary to the stigma of a flower
estilo tubo que conecta el ovario con el estigma de una flor

substrate a molecule that an enzyme acts upon and converts to a product; reactant in an enzyme-catalyzed reaction
sustrato molécula sobre la cual una enzima actúa y la convierte en producto; reactante en una reacción catalizada por enzimas

survivorship curve a plot that shows how many individuals born during the same interval remain alive over time
curva de supervivencia gráfica que muestra cuántos individuos que nacen en el mismo intervalo permanecen vivos a lo largo del tiempo

sustainability a way of using resources to meet human needs without deep impacts to the environments that produce those resources
sostenibilidad forma de usar recursos para satisfacer necesidades humanas sin impacto profundo en los medioambientes que producen esos recursos

symbiosis a close ecological relationship in which one species lives in or on another species in a commensal, mutualistic, or parasitic relationship
simbiosis relación ecológica cercana donde una especie vive dentro de otra especie o sobre ella, en una relación comensal, mutualista o parasitaria

system a combination of interacting parts that form a complex whole
sistema combinación de partes que interactúan y forman una entidad compleja

T

taxonomy the science of naming, describing, and classifying organisms
taxonomía ciencia de nombrar, describir y clasificar organismos

telophase a stage of mitosis during which chromosomes arrive at opposite spindle poles, decondense, and become enclosed by a new nuclear envelope
telofase etapa de la mitosis durante la cual los cromosomas llegan a polos opuestos del huso, se descondensan y se encapsulan en una nueva envoltura nuclear

tension in transpiration, the continuous negative pressure created by evaporation of water from the surface of a leaf or stem
tensión en la transpiración, la presión negativa continua creada por la evaporación del agua de la superficie de una hoja o tallo

thermophile an organism adapted to life at a very high temperature
termófilo organismo adaptado a la vida a temperaturas muy altas

thigmotropism plant movement or growth in a direction influenced by contact
tigmotropismo movimiento o crecimiento de una planta en una dirección por influencia del contacto

threatened species a species whose population is low enough for it to be at risk of becoming extinct, but not low enough that it is in imminent danger of extinction
especie amenazada especie cuya población es suficientemente baja para que esté en riesgo de extinción, pero no lo suficientemente baja para enfrentar peligro inminente de extinción

thylakoid a flat membranous sac occuring in stacks inside the chloroplast; where light energy is converted into ATP and NADPH used in carbohydrate synthesis
tilacoide saco plano y membranoso que se presenta en pilas dentro del cloroplasto; sitio donde la energía luminosa se convierte en el ATP y NADPH que se usan en la síntesis de carbohidratos

trace fossil the physical evidence from the ancient past of an organism's activities
icnofósiles evidencia física del pasado antiguo de las actividades de un organismo

tracheid a component of xylem; a tapered cell that dies when mature; pitted walls that remain interconnect with other tracheids to form water-conducting tubes
traqueida componente del xilema; célula estrecha que muere cuando madura; paredes picadas que permanecen interconectadas con otras traqueidas para formar tubos conductores de agua

trade-off a compromise in the solution to a problem where one feature, quality, or factor is favored at the cost of another
compensación compromiso en la solución a un problema donde una característica, cualidad o factor es favorecido a expensas de otro

trait a characteristic inherited by an offspring from its parent
rasgo característica heredada por un descendiente de parte de un progenitor

transcription the process in which enzymes use DNA as a template to assemble RNA; part of gene expression
transcripción proceso en el que las enzimas usan ADN como modelo para ensamblar ARN; parte de la expresión génica

transcription factor a regulatory protein that influences transcription by binding directly to DNA; e.g., an activator or repressor
factor de transcripción proteína reguladora que influye sobre la transcripción al unirse directamente al ADN; por ejemplo, un activador o represor

transcriptome the study of all the messenger RNA molecules in a genome
transcriptoma estudio de todas las moléculas de ARN mensajero en un genoma

transgenic a genetically modified organism that carries a gene from a different species
transgénico organismo modificado genéticamente que lleva un gen de una especie diferente

translation the process in which a polypeptide chain is assembled from amino acids in the order specified by an mRNA
traducción proceso en el que una cadena de polipéptidos es ensamblada a partir de aminoácidos en el orden especificado por un ARNm

transpiration the evaporation of water from aboveground plant parts
transpiración evaporación del agua en plantas que viven encima del suelo

trophic level the group of organisms in an ecosystem which occupy the same level in a food chain
nivel trófico grupo de organismos de un ecosistema que ocupan el mismo lugar en una cadena alimenticia

tropism a directional movement toward or away from a stimulus that triggers it
tropismo movimiento direccional hacia o en contra de un estímulo que desencadena ese movimiento

tuber a fleshy storage structure made from the stem of a plant; may grow buds on its surface that potentially develop into new plants
tubérculo estructura carnosa de almacenamiento hecha del tallo de una planta; muchos tienen brotes en su superficie que potencialmente se desarrollan para convertirse en plantas nuevas

U

uniformitarianism a geologic principle that states geologic processes operate consistently throughout time
uniformismo principio geológico que establece que los procesos geológicos operan de manera consistente a lo largo del tiempo

V

vaccine a preparation introduced into the body in order to elicit immunity to a specific antigen
vacuna preparación introducida al cuerpo para despertar inmunidad a un antígeno específico

vacuole a large, fluid-filled vesicle that isolates or breaks down waste, debris, toxins, or food
vacuola vesícula grande, llena de fluido, que aísla o descompone desechos, toxinas o alimentos

variolation a practice of deliberately infecting a person with a pathogen in the hope they develop a mild infection and acquire natural immunity
variolización práctica de infectar a una ersona con un patógeno de manera deliberada con la esperanza de desarrollar una infección ligera y obtener inmunidad natural

vascular plant a plant that has xylem and phloem
planta vascular planta que tiene xilema y floema

vascular tissue the plant tissue that carries water and nutrients through the plant body
tejido vascular el tejido de la planta que transporta agua y nutrientes por todo el cuerpo de la planta

vector in biotechnology, a molecule that can carry foreign DNA into host cells; in epidemiology, an organism that transmits an infectious pathogen to another organism
vector en la biotecnología, una molécula que puede transportar ADN foráneo a una célula anfitriona; en la epidemiología, un organismo que transmite un patógeno infeccioso a otro organismo

vertebrate an animal with a backbone
vertebrado animal con columna vertebral

vesicle a saclike, membrane enclosed organelle; different kinds store, transport, or break down their contents
vesícula orgánulo contenido en una membrana con forma de saco; distintos tipos almacenan, transportan o descomponen sus contenidos

vessel a water-conducting tube consisting of a stack of xylem cells that meet end to end at perforation plates
vaso tubo conductor de agua que consiste en una pila de células del xilema que se unen de extremo a extremo en las placas de perforación

vestigial structure an evolutionary remnant of a formerly functional structure
estructura vestigial resto evolutivo de una estructura que antes era funcional

viral envelope an outer membrane of a virus particle derived from the host cell in which it formed
envoltura vírica membrana externa de una partícula de virus derivada de la célula anfitriona donde se formó

viral particle a virus that is not inside a host cell
partícula viral virus que no está dentro de una célula anfitriona

W

wildlife corridor an area of habitat that connects wildlife populations separated by human activities or structures
corredor ecológico área de hábitat que conecta poblaciones de fauna separadas por actividades o estructuras humanas

xylem a complex vascular tissue of plants; cell walls of its dead tracheids and vessel elements form tubes that distribute water and mineral ions through the plant body

xilema tejido vascular complejo en las plantas; las paredes celulares de traqueidas muertas y elementos de los vasos forman tubos que distribuyen agua e iones minerales por el cuerpo de la planta

Z

zooplankton a community of tiny drifting or swimming heterotrophic organisms

zooplancton comunidad de diminutos organismos heterotróficos que flotan o nadan

zygote the first cell of a new individual

cigoto la primera célula de un individuo

zygote fungus a fungus that forms a thick-walled zygospore during sexual reproduction

zygomycota hongo que forma una zigospora con paredes gruesas durante la reproducción sexual

INDEX

Page numbers in **bold** indicate key terms. An *f* refers to a figure; *t* refers to a table.

A

E

H

Q

R

S

T

U

V

ACKNOWLEDGEMENTS

National Geographic Learning gratefully acknowledges the contributions of the following National Geographic Explorers and affiliates to our program:

Katie Amato, Biological anthropologist
Diva Amon, Deep-sea biologist
Christopher C. Austin, Biologist
Carter Clinton, Biological anthropologist
Nicole Colón Carrión, Biologist and ecologist
Camila Espejo, Veterinarian
Holly Fearnbach, Zoologist
Andrian Gajigan, Biochemist
Pablo Garcia Borboroglu, Marine biologist and penguin conservationist
Stanley Gehrt, Urban wildlife researcher
Shane Gero, Marine biologist
David Gruber, Marine biologist
Federico Kacoliris, Biologist
Brenda Larison, Biologist
Heather Joan Lynch, Quantitative ecologist
Emily Otali, Primatologist
Robert A. Raguso, Biologist
Kristen Ruegg, Biologist
Pardis Sabeti, Computational biologist
Çağan Şekercioğlu, Ornithology professor
Anusha Shankar, Ecologist
Daniel Streicker, Infectious disease ecologist
Varun Swamy, Ecologist
Gerard Talavera, Evolutionary biologist
Rosa Vásquez Espinoza, Chemical biologist and Amazon conservationist
Branwen Williams, Oceanographer and climate scientist
Kimberly Williams-Guillén, Conservationist
Robert Wood, Roboticist
Rae Wynn-Grant, Carnivore ecologist

National Geographic Learning gratefully acknowledges the contributions of the following National Geographic Photographers to our program:

Ary Bassous
Annie Griffiths
Melissa Groo
Charlie Hamilton James
Frans Lanting
David Liittschwager
Ian Nichols
Thomas Peschak
Jim Richardson
Gabby Salazar
Joel Sartore
Babak Tafreshi
Stefano Unterthiner
Edson Vandeira
Ami Vitale
Shannon Wild

PHOTOGRAPHIC CREDITS

Front Cover Krishan Lad/EyeEm/Getty Images.

Back Cover Yuhko Okada/EyeEm/Getty Images.

IV (tl) © Mark Thiessen; (bl) Courtesy Dr. Catherine L. Quinlan, Ed.D.; **VI** (tl) © Jon Sanders/National Geographic Image Collection; (tc1) © Novus Select/Biographic/National Geographic Image Collection; (tc2) Courtesy of Christopher Austin; (tr) © Carter Clinton/National Geographic Image Collection; (cl1) Courtesy of Varun Swamy; (cl2) © Nicole Colón Carrión; (cl3) © Global Penguin Society; (c1) © Dr. Manuel Ruiz-Aravena; (c2) Courtesy of Holly Fearnbach; (c3) Courtesy of Stanley Gehrt; (c4) © Mark Thiessen/National Geographic Image Collection; (cr1) Courtesy of Andrian Gajigan; (cr2) © Rebecca Drobis/National Geographic Image Collection; (bl) Courtesy of Federico Kacoliris; (bc) © Mark Thiessen/National Geographic Image Collection; (br1) © Tsalani Lassiter; (br2) © Brenda Larison; **VII** (tl) Courtesy of Heather Lynch; (tc) Courtesy of Emily Otali; (tr1) © Yash Sondhi; (tr2) © Neil Losin; (cl) © Chiun-Kai Shih/National Geographic Image Collection; (c1) © Stephanie King/University of Michigan; (c2) Courtesy of Anusha Shankar; (c3) © Rebecca Hale/National Geographic Image Collection; (c4) © Àngels Codina; (cr1) Courtesy of Daniel Streicker; (cr2) Courtesy of Branwen Williams; (bc1) Courtesy of Kim Williams-Guillén; (bc2) © Drobis,Rebecca/National Geographic Image Collection; **VIII** John Hyde/Design Pics/Getty Images; **IX** Apostolos Giontzis/Alamy Stock Photo; **X** © Klaus Nigge/National Geographic Image Collection; **XI** Dr Paul Andrews, University Of Dundee/Science Source; **XII** Dennis Kunkel Microscopy/Science Source; **XIII** Buiten-Beeld/Alamy Stock Photo; **XIV** James Schwabel/Alamy Stock Photo; **XV** © Revive & Restore; **XVI** Parkol/Shutterstock.com; **XVII** Eye of Science/Science Source; **XVIII-XIX** (spread) Krishan Lad/EyeEm/Getty Images; **2** Franco Banfi/Minden Pictures; **5** (tl) © Joel Sartore/National Geographic Image Collection; (tr) Ignacio Palacios/Stone/Getty Images; (cl) Kwangshin Kim/Science Source; (cr) Steve Gschmeissner/Science Source; (bl) © Robert Harding Picture Library/National Geographic Image Collection; (br) Biosphoto/Alamy Stock Photo; **7** Michael Durham/Minden Pictures; **8** Jtas/Shutterstock.com; **9** © Gerard Talavera; **10** (cl) © Peter J. Bryant; (cr) ImageBroker/Alamy Stock Photo; (bl) Michael Durham/Minden Pictures; (br) Philip Bird LRPS CPAGB/Shutterstock.com; **13** © The Royal Society Publishing; **14** © The Regents of the University of Michigan; **17** (tl) Peter Pankiw/Shutterstock.com; (tr) Peter Darcy/500px Prime/Getty Images; (cl) © Richard Seeley/National Geographic Image Collection; (cr) Blickwinkel/Alamy Stock Photo; **19** (tl) Nano Creative/Science Source; (tr) Dennis Kunkel Microscopy/Science Source; (cl) Sojibul/Shutterstock.com; (cr) Natalya Chernyavskaya/Shutterstock.com; **20** (tc) Reuters/Alamy Stock Photo; (bl) © Sergio Pitamitz/National Geographic Image Collection; (br) Sportpoint/Alamy Stock Photo; **22-23** (spread) © Thomas Peschak/National Geographic Image Collection; **24** (t) Mauritius Images GmbH/Alamy Stock Photo; (tl) Courtesy of Kim Williams-Guillén; **25** (bl) © David Gruber/Project CETI; (br) © Project CETI; **28-29** (spread) © Monterey Bay Aquarium 2021; **31** (tc) NOAA Office of Exploration and Research; (bl) © Solvin Zankl; (br) Office of Ocean Exploration and Research, Exploration of the Gulf of Mexico 2014/NOAA; **32** John Hyde/Design Pics/Getty Images; **33** Bryan and Cherry Alexander/Science Source; **36** Craig Tuttle/Corbis/Getty Images; **37** Leonid Ikan/Shutterstock.com; **38** Stephan Morris/Shutterstock.com; **39** (bl) Oxford Scientific/Photodisc/Getty Images; (bc) Roger Tidman/Corbis Documentary/Getty Images; (br) Chameleonseye/iStock/Getty Images; **42**(tl) © Tsalani Lassiter; (tc) Don Grall/Photodisc/Getty Images; **43** Menno Schaefer/Shutterstock.com; **45** (c) Nataliya Hora/Shutterstock.com; (br) GMVozd/E+/Getty Images; **47** © Joel Sartore/National Geographic Image Collection; **48** CampPhoto/iStock/Getty Images; **55** European Space Agency; **56** (tc) Iunewind/Shutterstock.com; (tr) Patrick J. Endres/Corbis Documentary/Getty Images; **57** Earleliason/E+/Getty Images; **60** Mike Lewelling/NPS Photo/Nps.gov; **61** (bl) Jim Peaco/NASA's Scientific Visualization Studio/National Park/NASA; (br) Jim Peaco/NASA's Scientific Visualization Studio/National Park/NASA; **62** Nicola Ferrari/Alamy Stock Photo; **63** (tc) Gabriel Barathieu/Minden Pictures; (bc) Simon Colmer/Alamy Stock Photo; **64** Manoj Shah/Stone/Getty Images; **65** (tc) © Hernán Povedano; (tr) Courtesy of Federico Kacoliris; **67** Cyril Ruoso/Minden Pictures; **69** © Frans Lanting/National Geographic Image Collection; **70** Mathess/iStock/Getty Images; **71** Imagebroker/Alamy Stock Photo; **72** Ammit/Alamy Stock Photo; **73** Cavan Images/Alamy Stock Photo; **75** John Sullivan/Alamy Stock Photo; **76** James Osmond/Alamy Stock Photo; **77** Don Johnston_ON/Alamy Stock Photo; **78** E. R. Degginger/Science Source; **79** Peter Haigh/Alamy Stock Photo; **82** Apostolos Giontzis/Alamy Stock Photo; **83** Jaymi Heimbuch/Minden Pictures; **84** Karine Aigner/Nature Picture Library; **85** Sarah09/Alamy Stock Photo; **86** (bl) © Frans Lanting /National Geographic Image Collection; (bc) Wildscotphotos/Alamy Stock Photo; (br) Swissdrone/Shutterstock.com; **88** George D. Lepp/Corbis Documentary/Getty Images; **90** Images of Africa Photobank/Alamy Stock Photo; **91** Zoonar GmbH/Alamy Stock Photo; **92** (bl) Thomas Kline/Design Pics/Getty Images; (br) Boris Diakovsky/Alamy Stock Photo; **94** (tl) Courtesy of Heather Lynch; (tc) Courtesy of Heather Lynch; **97** (tl) Cultura Creative RF/Alamy Stock Photo; (bl) Wirestock, Inc./Alamy Stock Photo; **99** (tc) Bestber/Shutterstock.com; (tr) Wesdotphotography/Alamy Stock Photo; **100-101** (spread)© Edson Vandeira/National Geographic Image Collection; **103** Nature Picture Library/Alamy Stock Photo; **107** © NOAA ONMS/Ocean Exploration Trust; **108-109** (spread) © M. Oeggerli/Pathology, University Hospital Basel and School of Life Sciences, FHNW.; **111** (tc) Courtesy of Katie R. Amato; (bl) Juniors Bildarchiv GmbH/Alamy Stock Photo; (br) Courtesy of Katie R. Amato; **112** © Klaus Nigge/National Geographic Image Collection; **113** Tom & Pat Leeson/Science Source; **121** © Jennifer Hayes/National Geographic Image Collection; **122** Phil Degginger/Alamy Stock Photo; **123** Hans Heinz/Shutterstock.com; **126** © Jasper Doest/National Geographic Image Collection; **135** © Jim Richardson/National Geographic Image Collection; **136** (t) © Lvcas Fiat/Projeto Mantis; (tl) © Stephanie King/University of Michigan; **142** Paul Gunning/Science Source; **143** Jonathan Plant/Alamy Stock Photo; **145** (bl) Cyril Ruoso/JH Editorial/Minden Pictures; (br) Ritvars/Shutterstock.com; **148** Vojce/iStock/Getty Images; **149** Courtesy of Argonne National Laboratory; **151** Steve Gschmeissner/Science Photo Library/Getty Images; **153** Biophoto Associates/Science Source; **155** Jennifer Waters/Science Source; **156** (bl) Don W. Fawcett/Science Source; (br) Biophoto Associates/Science Source; **157** (cl) Keith R. Porter/Science Source; (bl) Dr. Kari Lounatmaa/Science Source; **159** Eye of Science/Science Source; **160-161** (spread) © Charlie Hamilton James/National Geographic Image Collection; **164** (bl) Dennis Kunkel Microscopy/Science Source; (bc) Dennis Kunkel Microscopy/Science Source; (br) Dennis Kunkel Microscopy/Science Source; **165** Hermann Schillers, Prof. Dr. H.Oberleithner, University Hospital Of Muenster/Science Source; **169** Izabella Joubert/EyeEm/Getty Images; **173** inset Steve Gschmeissner/Science Source; (tl) Richard T. Nowitz/Science Source; (tr) Nicolas Asfouri/AFP/Getty Images; inset Andrew Syred/Science Source; (bl) NOAA; **176** (t) © Gil S. Jacinto; (tl) Courtesy of Andrian Gajigan; **177** (bl) Jenson/Shutterstock.com; (br) © University of Bordeaux, CNRS, Centre de Recherche Paul Pascal/T. Beneyton; **180** Dr Paul Andrews, University Of Dundee/Science Source; **181** (bl) Pictorial Press Ltd/Alamy Stock Photo; (br) BSIP/Universal Images Group/Getty Images; **185** From "Expression of the epidermal growth factor receptor (EGFR) and the phosphorylated EGFR in invasive breast carcinomas." http://

breast-cancer research.com/content/10/3/R49; **186** Robert Henno/Alamy Stock Photo; **188** (bc) Don. W. Fawcett/Science Source; (br) Andrew Syred/Science Source; **190** (tl) Ed Reschke/Photolibrary/Getty Images; (tr1) Michael Abbey/Science Source; (tr2) Michael Abbey/Science Source; (cl1) Ed Reschke/Stone/Getty Images; (cl2) (cr1) (cr2) (br) Michael Abbey/Science Source; **192** Dr Torsten Wittmann/Science Source; **193** (cl) Nature Picture Library/Alamy Stock Photo; (cr) Christopher Biggs/Shutterstock.com; (bl) GFC Collection/Alamy Stock Photo; (br) Ondrej Vavra/Shutterstock.com; **194** (cl) Roger J. Bick & Brian J. Poindexter/UT-Houston Medical School/Science Source; (c) SPL/Science Source; (cr) SPL/Science Source; (bl) Steve Gschmeissner/Science Source; (br) Daniel Schroen, Cell Applications Inc/Science Source; **196** (cl) © Triarch/Visuals Unlimited, Inc.; (cr) © Biodisc/Visuals Unlimited, Inc.; (bl) Dr. Keith Wheeler/Science Source; (br) Emilio Ereza/Alamy Stock Photo; **197** © David Liittschwager/National Geographic Image Collection; **198** (t) © Dr. Manuel Ruiz-Aravena; (tl) © Dr. Manuel Ruiz-Aravena; **199** Dr. Gopal Murti/Science Source; **203** Dennis Kunkel Microscopy/Science Source; **204-205** (spread) Brian Overcast/Alamy Stock Photo; **207** (t) Courtesy of Varun Swamy; (bl) Pete Oxford/Minden Pictures; (br) Courtesy of Varun Swamy; **208** Dennis Kunkel Microscopy/Science Source; **209** NASA; **213** (cl) Frank Fox/Science Source; (cr) Charles V. Angelo/Science Source; **214** (tl) James Cavallini/Science Source; (tr) CNRI/Science Source; (bl) © Heide Schulz-Vogt/Leibniz Institute for Baltic Sea Research; (br) Eye of Science/Science Source; **216** (cl) Stem Jem/Science Source; (cr) Juniors Bildarchiv GmbH/Alamy Stock Photo; **217** Ralph White/Corbis Documentary/Getty Images; **220** Ami Images/Science Source; **221** (cl) Eye of Science/Science Source; (bl) Biophoto Associates/Science Source; **222** © Brian Skerry/National Geographic Image Collection; **223** (tl) M. I. Walker/Science Source; (tr) BSIP SA/Alamy Stock Photo; (bl) Frank Fox/Science Source; **224** (cl) London School of Hygiene & Tropical Medicine/Science Source; (cr) Matthijs Wetterauw/Alamy Stock Photo; **225** (tl) M. I. Walker/Science Source; (tr) Udeyismail/Shutterstock.com; (cl) Eye of Science/Science Source; **226** Science History Images/Alamy Stock Photo; **229** (bl) Eye of Science/Science Source; (br) Professor Sir David Read FRS; **230** (bl) Prisma/Dukas Presseagentur GmbH/Alamy Stock Photo; (br) Ed Reschke/Photodisc/Getty Images; **231** Reza Saputra/iStock/Getty Images; **232** Michel Viard/Jacana/Science Source; **233** (tl) Arie v.d. Wolde/Shutterstock.com; (cr) Galina Bondarenko/Shutterstock.com; (bl) © Dennis, David M./Animals Animals; (br) JerHetrick/Shutterstock.com; **235** (cl) Paul Starosta/Stone/Getty Images; (cr) Akira Kaede/Photographer's Choice RF/Getty Images; (bl) Andrew Syred/Science Source; (br) © Etienne Littlefair/Minden Pictures; **238** Biozentrum, University of Basel/Science Source; **239** Petit Format/Science Source; **241** (tl) Cellou Binani/AFP/Getty Images; (tr) CAMR/A. Barry Dowsett/Science Source; (bl) James Cavallini/Science Source; **243** © Gabriele Galimberti/National Geographic Image Collection; **244** (t) © Ronan Donovan; (tl) Courtesy of Emily Otali; **245** NASA; **248** United States Forest Service; **249** Vasilis Ververidis/Alamy Stock Photo; **251** Scott Camazine/Science Source; **252** (bl) Nichole Casebolt/Shutterstock.com; (bc1) ShadeDesign/Shutterstock.com; (bc2) Memitina/iStock/Getty Images; (br) Gloomy Photography/Shutterstock.com; **253** Dr. Keith Wheeler/Science Source; **254** AntiMartina/E+/Getty Images; **255** (cl) Wild Horizon/Universal Images Group/Getty Images; (c) Amit kg/Shutterstock.com; (cr) Kamomeen/Shutterstock.com; **256** Nick Garbutt/Steve Bloom Images/Alamy Stock Photo; **257** Dr. Keith Wheeler/Science Source; **259** Power and Syred/Science Source; **262** © Babak Tafreshi/National Geographic Image Collection; **263** Little Dinosaur/Moment/Getty Images; **264** (tl) Jeff Holcombe/Shutterstock.com; (tc) Ed Reschke/Stone/Getty Images; (tr) Tami Ruble/Alamy Stock Photo; **265** Rieger Bertrand/Hemis/Alamy Stock Photo; **267** Andrew E Gardner/Shutterstock.com; **268** Aimpol Buranet/Shutterstock.com; **269** (cl) Blickwinkel/Alamy Stock Photo; (cr) MichaelGrantPlants/Alamy Stock Photo; (bl) FotoDuets/Shutterstock.com; (br) Inga Gedrovicha/Shutterstock.com; **270** (t) © Robert Raguso; (tl) © Yash Sondhi; (bl) Martin Shields/Alamy Stock Photo; (br) Franz Krenn/Okapia/Science Source; **273** Ornitolog82/iStock/Getty Images; **274** DieterMeyrl/E+/Getty Images; **275** Eric Mischke/EyeEm/Getty Images; **276** Panther Media GmbH/Alamy Stock Photo; **277** © ANP/Hollandse Hoogte/Ko Hage; **280** Buiten-Beeld/Alamy Stock Photo; **281** Daniel Ruthrauff/USGS; **282** (cl) Dennis Sabo/Shutterstock.com; (cr) Ozgur Kerem Bulur/Shutterstock.com; (bl) Kurit Afshen/Shutterstock.com; (br) MarieHolding/iStock/Getty Images; **283** (tl) Ifish/E+/Getty Images; (tc) Selim Kohen/500px/Getty Images; (tr) Pulse/The Image Bank/Getty Images; **285** HeritagePics/Alamy Stock Photo; **286** Lightspruch/iStock/Getty Images; **289** Eye of Science/Science Source; **290** Eye of Science/Science Source; **291** James Oliver/Digital Vision/Getty Images; **293** In Green/Shutterstock.com; **295** All Canada Photos/Alamy Stock Photo; **296** Blickwinkel/Alamy Stock Photo; **297** Steve Bloom Images/Alamy Stock photo; **298** Blickwinkel/Alamy Stock Photo; **299** (cl) Scott Camazine/Science Source; (br) Sylvain Cordier/Stockbyte/Getty Images; **300** Tony Wu/Minden Pictures; **301** All Canada Photos/Alamy Stock Photo; **302** Jim Brandenburg/Minden Pictures; **303** Stan Tekiela Author/Naturalist/Wildlife Photographer/Getty Images; **304** Paul Souders/Stone/Getty Images; **305** © Stefano Unterthiner/National Geographic Image Collection; **306** (t) © John Durban/Holly Fearnbach; (tl) Courtesy of Holly Fearnbach; **307** Winfried Wisniewski/The Image Bank/Getty Images; **311** © Félix Colón Ruíz; **312-313** (spread) CDC/Science Source; **315** (t) © Daniel Streicker/National Geographic Image Collection; (bl) © Daniel Streicker/National Geographic Image Collection; (br) Courtesy of Daniel Streicker; **316** James Schwabel/Alamy Stock Photo; **317** Frank Fox/Science Source; **320** Science & Society Picture Library/SSPL/Getty Images; **324** (t) National Park Service; (tl) © Carter Clinton/National Geographic Image Collection; **325** © Melissa Groo/National Geographic Image Collection; **327** Professor Oscar Miller/Science Source; **329** (cl) Don W. Fawcett/Science Source; (bl) Biology Pics/Science Source; **343** (bl) Hans Verburg/Alamy Stock Photo ; (br) Risto0/istock/Getty Images; **346** Dora Zett/Shutterstock.com; **347** Ressormat/Shutterstock.com; **349** (bl) Omikron/Science Source; (br) Biophoto Associates/Science Source; **357** (t) RichardSeeley/Istock/Getty Images; (tr) © Neil Losin; **361** Hermann Eisenbeiss/Science Source; **368-369** (spread) © Shannon Wild; **371** John Kaprielian/Science Source; **372** (tl) Richard P Long/Alamy Stock Photo; (tc) Harvey Wood/Alamy Stock Photo; (tr) Wayne Hutchinson/Alamy Stock Photo; **373** Grossemy Vanessa/Alamy Stock Photo; **375** (tl) Jim Cumming/Shutterstock.com; (tr) Colleen Gara/Moment/Getty Images; **377** © Susan Bronson; **380** © Revive & Restore; **381** Kristel Richard/NaturePL/Science Source; **386** Eurelios/Science Source; **390-393** (spread) © Ary Bassous; **393** (bl) Pascal Goetgheluck/Science Source; (br) KorradolYamsattham/Istock/Getty Images; **394** Reuters/Alamy Stock Photo; **395** (tl) © University of Exeter; (tr) Slowmotiongli/Istock/Getty Images; **397** The Asahi Shimbun/Getty Images; **401** Frederic J. Brown/AFP/Getty Images; **402** (tl) IanDagnall Computing/Alamy Stock Photo; (tr) Dennis Kunkel Microscopy/Science Source; **403** (t) © Pardis Sabeti; (tr) © Chiun-Kai Shih/National Geographic Image Collection; **406** Science Photo Library/Alamy Stock Photo; **409** Tchandrou Nitanga/AFP/Getty Images; **414** Freder/E+/Getty Images; **417** © Bauer et al./The Genetics Society of America; **421** Xinhua News Agency/Getty Images; **422-433** (spread) © Nelson, Alan G./Animals Animals; **425** (t) Courtesy of Anusha Shankar; (b) Courtesy of Anusha Shankar; **426** Parkol/Shutterstock.com; **427** © Trustees of the Natural History Museum, London; **429** (cl) Chris Packham/Minden Pictures; (cr) Sduben/iStock/Getty Images; **432** (bl) Stephiii/iStock/Getty Images; (bc) Chris Mattison/Minden Pictures; (br) Ondrej Prosicky/Alamy Stock Photo; **435** (cl) Age Fotostock Spain, S.L./Alamy Stock Photo; (cr) Science History Images/Alamy Stock Photo; (bl) Francois Gohier/Science Source; (br) Nature Picture Library/Alamy Stock Photo; **437** © Pete Mcbride/National Geographic Image Collection; **438** (cl) Historic Collection/Alamy Stock Photo; (cr) The Natural History Museum/Alamy Stock Photo; (bl) Sinclair Stammers/Science Source; (br) Ted Kinsman/Science Source; **440** Francois Gohier/Science Source; **444-445** (spread) © Annie Griffiths/National Geographic Image Collection; **446** (bl) Sinclair Stammers/Science Source; (bc) © Catherine May; (br) Oxford Scientific/The Image Bank/Getty Images; **448** (cl) Joao Ponces/Alamy Stock Photo; (cr) MriyaWildlife/iStock/Getty Images; (bc) Breck P. Kent/Animals Animals; **452** (t) Iumen-digital/Shutterstock.com ; (tl) © Brenda Larison; **456** (tc) Eye of Science/Science Source; **457** (bl) Dmitry Bezrukov/iStock/Getty Images; **459** (bl) Its About Light/Design Pics/Getty Images; (bc) Mike Powles/Stone/Getty Images; (br) Tier Und Naturfotografie J und C Sohns/Photographer's Choice RF/Getty Images; **460** (tl) B.A.E. Inc./Alamy Stock Photo; (tr) Jose A. Bernat Bacete/Moment/Getty Images; (bl) Millard H. Sharp/Science Source; (br) © George Grall/National Geographic Image Collection; **461** (bl) Arterra/Universal Images Group/Getty Images; (br) © Eduardo Mayoral Alfaro; **464** Robert Alexander/Archive Photos/Getty Images; **467** Elena Shavlovska/Shutterstock.com; **473** © Gabby Salazar/National Geographic Image Collection; **474** (tr) Rick & Nora Bowers/Alamy Stock Photo; (cr) Karel Bock/iStock/Getty Images; **477** Juergen & Christine Sohns/Minden Pictures; **478** All Canada Photos/Alamy Stock Photo; **479** Imageplotter/Alamy Stock Photo; **480** (t) © Christopher Austin; (tl) Courtesy of Christopher Austin; **484** Peter Chadwick/Gallo Images ROOTS RF collection/Getty Images; **486** (bl) Cavan Images/Getty Images; (br) David Madison/Photodisc/Getty Images; **487** Aaron Geddes Photography/Moment/Getty Images; **488** (tr) © Arthur Anker; (cr) © Arthur Anker; **489** (tl) Johannes Melchers/Minden Pictures; (cl) NPS Photo/Alamy Stock Photo; (cl) Robby Kohley/USFWS; (c) Researchgate/U.S. Geological Survey; (cr) Michael Walther/Alamy Stock Photo; (cr) Photo Resource Hawaii/Alamy Stock Photo; **490** (tl) Carmen Hauser/Shutterstock.com; (tr) The Natural History Museum/Alamy Stock Photo; **493** AYImages/iStock/Getty Images; **494-495** (spread) © Ami Vitale/National Geographic Image Collection; **496** (bl) Byron Jorjorian/Alamy Stock Photo; (br) Dante Fenolio/Science Source; **497** NOAA; **499** (t) © Çağan Şekercioğlu; (tr) © Rebecca Hale/National Geographic Image Collection; **503** Magnus Larsson/iStock/Getty Images; **504** Bloomberg Creative/Getty Images; **506** Alf Manciagli/Shutterstock.com; **507** Sue Cunningham Photographic/Alamy Stock Photo; **508** Imaginechina Limited/Alamy Stock Photo; **510** Alaska Stock/Alamy Stock Photo; **511** Henglein and Steets/Cultura/Getty Images; **512** © Joel Santore/National Geographic Image Collection; **513** Jung Getty/Moment/Getty Images; **514** (cr) Ken Hurst/Alamy Stock Photo; (bc) Steve_Gadomski/iStock/Getty Images; **515** Erik Reis/Alamy Stock Photo; **516** Ahei/iStock/Getty Images; **517** Jwngshar Narzary/EyeEm/Getty Images; **521** (bc) Hal Beral/Corbis/Getty Images; **524** (cl) Adrian Sherratt/Alamy Stock Photo; (br) Imran Khan's Photography/Shutterstock.com; **525** (bl) Lacey Barber/Shutterstock.com; **526** (br) Winston Link/Shutterstock.com; (bl) Jon Rehg/Shutterstock.com; **546** Eco2drew/iStock/Getty Images Plus/Getty Images; **547** (tl) Steve Gschmeissner/Science Photo Library/Getty Images; (tr) Power and Syred/Science Source; (cl) Andrew Syred/Science Source; (cr) Roberto Sorin/Shutterstock.com; (bl) Kimberly Boyles/Shutterstock.com; (br) Christian Puntorno/Shutterstock.com; **549** © Norman R. Pace/American Society For Microbiology; **550** © Burki, F. Roger, A. J., Brown, M. W., & Simpson, A. G. B. (2020). "Review: The New Tree of Eukaroytes." Trends in Ecology & Evolution, 35(1), p. 48; **551** (bc) © David M. Hillis, Derrick Zwickl, and Robin Gutell, University of Texas/The American Association for the Advancement of Science; **552** (tl1) Chris Mattison/Minden Pictures; (tl2) Paul Starosta/Stone/Getty Images; (cl1) BSIP/Universal Images Group/Getty Images; (cl2) Jonathan Plant/Alamy Stock Photo; (cl3) Dennis Kunkel Microscopy/Science Source; (c) Dennis Kunkel Microscopy/Science Source; (cr) Eye of Science/Science Source; (bl1) Dennis Kunkel Microscopy/Science Source; (bl2) Arie v.d. Wolde/Shutterstock.com; (bc) Steve Gschmeissner/Science Source; (br) Eye of Science/Science Source; **553** (tl) Micro Discovery/Corbis Documentary/Getty Images; (tc) Steve Gschmeissner/Science Source; (tr1) Ed Reschke/Stone/Getty Images; (tr2) AYImages/iStock/Getty Images; (tr3) Aimpol Buranet/Shutterstock.com; (cl1) Fatahdijava/Shutterstock.com; (cl2) Greg Dimijian/Science Source; (c1) Mike Pellinni/Shutterstock.com; (c2) Drew Rawcliffe/Shutterstock.com; (cr1) Malachi Jacobs/Shutterstock.com; (cr2) Julija Kumpinovica/Shutterstock.com; (cr3) BSIP SA/Alamy Stock Photo; (cr4) Ami Images/Science Source; (bl) Silvergull/Shutterstock.com; (bc) Poirot55/Shutterstock.com; (br1) CatalinT/Shutterstock.com; (br2) Eye of Science/Science Source.